Graduate Texts in Physics

Graduate Texts in Physics publishes core learning/teaching material for graduate- and advanced-level undergraduate courses on topics of current and emerging fields within physics, both pure and applied. These textbooks serve students at the MS- or PhD-level and their instructors as comprehensive sources of principles, definitions, derivations, experiments and applications (as relevant) for their mastery and teaching, respectively. International in scope and relevance, the textbooks correspond to course syllabi sufficiently to serve as required reading. Their didactic style, comprehensiveness and coverage of fundamental material also make them suitable as introductions or references for scientists entering, or requiring timely knowledge of, a research field.

More information about this series at http://www.springer.com/series/8431

Peter Mulser

Hot Matter from High-Power Lasers

Fundamentals and Phenomena

Springer

Peter Mulser
University of Technology Darmstadt
Institute of Applied Physics
Darmstadt, Germany

ISSN 1868-4513 ISSN 1868-4521 (electronic)
Graduate Texts in Physics
ISBN 978-3-662-61179-1 ISBN 978-3-662-61181-4 (eBook)
https://doi.org/10.1007/978-3-662-61181-4

This Springer imprint is published by the registered company Springer-Verlag GmbH, DE part of Springer Nature.
The registered company address is: Heidelberger Platz 3, 14197 Berlin, Germany

Preface

Zusammenhänge müssen nicht wirklich bestehen,
aber ohne sie würde alles zerfallen

Robert Menasse in *Die Hauptstadt*

To Whom is the Book Addressed?

The book deals with what happens when a high-power laser of intensities from 10^{10} to 10^{22} Wcm^{-2} interacts with matter from the density of foams of some 10^{20} cm^{-3} up to precompressed solids of several 10^{25} particles per cm^3. High-power lasers have opened a new era of atomic and nuclear physics, of solid state and high pressure research, and of new particle acceleration schemes and intense radiation sources. Radiation pressure in the laboratory competes with pressures in collapsing cosmic objects and exceeds pressures in the interior of main sequence stars. Homogeneous black body radiation temperatures of 300 eV have been generated. Matter has been heated, to start from the hot solid of 10 eV, up to the hottest plasma of 10 MeV. For the first time, such extremes bring astrophysics to the laboratory.

In Chap. 1, the reader is acquainted with the basic aspects of matter composed of a positive fluid of ions and a negative fluid of free electrons held together by their electric charges and interpenetrating each other to form the new state of a locally neutral plasma. Its dynamics is widely governed by collective effects, a property the newcomer in the field has to become familiar with. This is the first obstacle to be faced. Our brains look everywhere for structure and shape and find instead to a large extent continuous flows and transitory forms. In detail, field ionization and collisional heating, plasma oscillations, radiation pressure effects, plasma profile steepening, electron thermal conduction, and first steps in superintense laser interaction with solid samples and microstructured targets are the subject of this introductory chapter to the field of hot matter.

The question arises on how to model laser-target dynamics. One way is to study the motion of the single positive and negative particles in representative fields: Gyromotions and drifts in static electric and magnetic fields, quiver motion of the electrons in the high frequency electromagnetic laser field and the resulting ponderomotive force on the single electron. The plasma response to the latter is at the origin of a whole variety of laser plasma phenomena. It confers the laser generated plasma a characteristic imprint. The intense fields necessitate also a fully relativistic treatment. The single particle description is the more successful the closer to reality the fields are modelled. This is the content of Chap. 2.

The approach complementary to the single particle motion is the fluid dynamic description of plasma and hot matter in its self-generated fields. Here, the two fluid model and its merging into a single fluid, where appropriate, show their power. Once the sources in the form of charge and current densities are known, they yield results that are macroscopically correct. Modelling of the sources represents a major effort. The formulation of exact and approximate fluid conservation laws, extended also to the relativistic domain, is the subject of Chap. 3.

Under the irradiation by intense fluxes of energy matter is excited to extreme behaviour all but in equilibrium. Once, however, after a relaxation phase, it has turned into the new equilibrium the powerful instrument of phenomenological and statistical thermodynamics applies. Since its use is permanently accompanied by the question of its applicability great emphasis is concentrated on the governing principles and subtleties of thermostatistics in this Chap. 4.

One of the most basic and far reaching concepts adaptable to structureless charged matter is the concept of waves. Here, in the Chap. 5 the first, and still the best field theory in the form of Maxwell's equations is at our disposal. The reader is introduced to the basic types of waves and their properties in the homogeneous and inhomogeneous plasma, as they are of electromagnetic transverse type and of high and low frequency longitudinal electrostatic and hydromagnetic kind.

The accelerated and streaming plasma is subject to hydrodynamic instabilities of Rayleigh-Taylor and Kelvin-Helmholtz type. The radiation pressure associated with the high intensity laser beam leads to parametric wave-wave coupling, realized as stimulated Brillouin and Raman scattering, and to back action of the deformed plasma on the modulation and self- focusing of the laser beam itself. The unstable behaviour of plasma under acceleration and radiation pressure makes it necessary to dedicate a whole chapter to their description, Chap. 6.

At low laser intensities transport phenomena, to mention first absorption of the laser beam and thermal conduction, are collision-dominated and are understood as local phenomena. The so-called Coulomb logarithm plays a dominant role and is, therefore, highlighted. With increasing laser power all transport phenomena become noncollisional and collective. As particular examples, computer simulations of interaction with matter at relativistic intensities are presented. Nonlocal transport phenomena are the realm of numerical modelling. The strong laser coupling with matter gives origin to new kinds of transport phenomena under extremes. Collisional transport and simulations are presented in Chap. 7.

The high-power laser is a unique source of secondary coherent and incoherent radiation, of high electromagnetic harmonics, of black body radiation, of X-ray line and bremsstrahlung radiation, and of gammas. Photons are the most noble kind of matter. They do not interact with themselves but they do so strongly with charged matter. They have got energy and momentum and make widely use of these properties to heat matter to extreme temperatures and to impress pronounced structures onto matter. They generate superthermal electrons which, in turn, pass part of their energy to the acceleration of ions in the Megavolt regime and another part into a wide spectrum of secondary radiation. This is Chap. 8.

The last part, Chap. 9, is dedicated to high-power laser applications: Terahertz radiation, X-ray lasing, short wavelength radiation from harmonics, novel schemes of ion and wake field electron acceleration, and laser induced inertial fusion. High-power laser interaction with matter leads to new states of matter and opens a wide new field in physics of radiation, plasma production, and hot matter.

The present work is intended as a textbook for students, a help for the researcher at the desk, and an aid for the experimentalist and the engineer in the laboratory. The specialist may find the glossary of basic formulas and numerical constants helpful. To offer something useful to all readers the exposure starts from simple models and intuitive pictures, proceeds gradually to more elaborated schemes and ends, possibly, with specific applications. Ending the presentation with the state of the art in present research is one concern of the author. The prerequisites of knowledge are at Bachelors level: Basic knowledge of classical mechanics, Maxwell's equations, phenomenological thermodynamics, and thermostatistics, basic quantum mechanics. All concepts lying beyond are developed in the book when and where they are needed. Use of physical intuition, possibly concise presentation combined with adequate formal rigor has been a guide for the writer. Formulas may come and pass, pictures persist.

How to Read the Book?

Nobody has enough time to read an entire book. It is the reader's free choice how many pages he is able to persevere and when to stop reading and assimilating. The newcomer may have much interest in the fundamental outline of a subject and perhaps an interest in solving some of the exercises in section of Problems and in answering some questions in the section Self-assessment. In general both are not difficult, except a few of them. The glossary is the index of the most important and most frequently used definitions and formulas in analytic work and computer simulations, and in estimates accompanying the experiment. Their origin and limits of applicability are easily found by their numbers which are the same as in the text. Both, the student and the advanced researcher may find them useful. Finally, going through the assessment may stimulate the student and the professor; the latter as an aid for preparing his own questions, certainly more original and deeper, in the students' examinations.

The best book is that which is fun to read. For such a purpose the author presented a modest collection of arguments generally not found in a textbook of plasma physics, like a simplified derivation of Landau damping (almost all authors follow Landau's procedure on shifting the integration contour), a discussion of the Coulomb logarithm, a criterion on the validity of the classical Maxwell equations, on the quantization of the radiation field, on the moon as a Schrödinger problem (why does the moon not fall onto the earth like the excited electron decays in the atom? The moon does fall!), on the importance, usefulness, and ubiquity of adiabatic invariants in physics, in particular in thermodynamics (entropy is an adiabatic invariant), Feynman's optical Bloch model (just for its beauty), etc.. The student (and anybody else?) may wonder why there is a factor of 2 in the denominator of the magnetic pressure, and in the Alfvén velocity there it is missing; Newton had the analogous problem with the sound speed. The moon, looked at as a hydrogen quantum system, may find itself in a Rydberg state of what order of magnitude, and this principal quantum number would change by how much after the impact of 1 microgram mass at 1 km speed? Can a free electron gain energy over an entire cycle of oscillation from a monochromatic electric wave? What is the physics behind uphill acceleration? Which energy is negative in a negative energy wave? How to modify Maxwell's equations to make them compatible with Newtonian mechanics? How does the relativistic Doppler effect read in the homogeneous medium? Why is the Lagrangian of a mass point the difference of kinetic and potential energy and not their sum or some other function of them? A collection of problems and questions may serve as a test but they can never replace a scientific discussion. Plasma physics is interdisciplinary. Cross connections to the physics as a whole are strong incentives.

Citations

There exists an avalanche of publications to the subject of the book, excellent ideas, excellent quality. The author's aim has been to cite some first papers (not systematically done) and some very new papers. Sometimes the criterion was not to cite papers which are referenced all the time by most of the authors, so no further need. After all the author is aware of the fact that too many references may disturb the flux of reading a textbook; he tries to limit them thereby being unjust in the sense that many most excellent papers remain unreferenced.

Nomenclature

The use of mathematical symbols throughout the text is standard. The only exception is made with the scalar product of vectors by omitting the dot between the vector symbols, e.g. $\mathbf{ab}$ stands for $\mathbf{a} \cdot \mathbf{b}$ and $(\mathbf{a}\nabla)\mathbf{b}$ is used for the derivative $\mathbf{a} \cdot \nabla\mathbf{b}$ of vector $\mathbf{b}$ along $\mathbf{a}$.

Acknowledgements

The continuous assistance of Dr. Markus Rosenstihl concerning all problems with the computer is gratefully acknowledged in the first place. The author is further indebted for multiple help in computer problems to Ibrahim El Idrissi and to Christian Kolb. For numerous scientific discussions, multiple thanks go to Prof. Gernot Alber at the Technical University of Darmstadt, to Prof. Dieter Bauer at the University of Rostock, and to Dr. Klaus Eidmann from the Max Planck Institute for Quantum Optics in Munich.

The author is particularly indebted to Ute Heuser from the Springer Verlag in Heidelberg for her valuable advise with respect to structuring and editing, for her continuous encouragement and for her numerous suggestions.

Writing a book is a major enterprise. As such it contains a personal aspect. The author is very much indebted to Charlotte Tiedt. She has been all the time a helpful and encouraging friend and has followed the progress of the book with great patience.

Darmstadt, Germany | Peter Mulser

Contents

1 Hot Matter from High-Power Lasers 1
1.1 Laser and Ion Beam Generated Hot Matter 1
1.1.1 High Power Lasers 2
1.1.2 Wavelengths of Common High Power Lasers 2
1.1.3 The Bird of the Laser Plasma 3
1.1.4 Multiphoton and Field Ionization 5
1.2 Basic Properties of the Laser Plasma 11
1.2.1 Collisional Absorption and Plasma Heating 11
1.2.2 Thermalization 19
1.2.3 Ideal Plasma 20
1.3 The Dynamics of the Laser Plasma 22
1.3.1 Basic Elements of Plasma Dynamics 23
1.3.2 Fully Developed Plasma Dynamics 40
1.4 Superintense Laser-Matter Interaction 47
1.4.1 Collisionless Absorption 48
1.4.2 Microstructured Targets 56
1.5 Summary 58
1.6 Problems 60
1.7 Self-assessment 62
1.8 Glossary 65
1.9 Further Readings 68
References 69

2 Single Particle Motion 73
2.1 Non-relativistic Regime 74
2.1.1 Electron in the Electromagnetic Wave 74
2.1.2 Lagrangian and Hamiltonian Description of Motion 82
2.1.3 Charged Particle Motion in Crossed Static Fields 97
2.1.4 Slow Motions and Adiabatic Invariants 101
2.1.5 Poincaré–Cartan Invariant and the Adiabatic Theorem 109

2.1.6 The Ponderomotive Force 117
2.1.7 Particle Trapping 136
2.1.8 Binary Collisions 137
2.2 Relativistic Regime 142
2.2.1 Essential Relativity 142
2.2.2 Scalars, Contravariant, and Covariant Quantities 162
2.3 Summary 164
2.4 Problems 166
2.5 Self-assessment 171
2.6 Glossary 173
2.7 Further Readings 177
References 177

3 Laser Induced Fluid Dynamics 179
3.1 Conservation Laws 180
3.1.1 Particle and Mass Conservation 180
3.1.2 Navier–Stokes Equation 182
3.1.3 Energy Conservation 187
3.1.4 Two-Fluid Model of the Fully Ionized Plasma 189
3.1.5 Standard Form of the Conservation Equations 195
3.1.6 Collective Ponderomotive Force Density 196
3.1.7 The Lagrangian Picture of the Fluid 200
3.1.8 Kinetic Foundation of Diluted Fluids 203
3.2 Relativistic Fluid Dynamics 212
3.2.1 Ideal Fluid Dynamics 212
3.2.2 Moment Equations 215
3.3 Similarity Solutions 219
3.3.1 Dimensional Analysis 220
3.3.2 Riemann Invariants 232
3.3.3 The Plane Shock Wave 236
3.3.4 From Ablation to Radiation Pressure Under Heat Flow 244
3.4 Summary 260
3.5 Problems 262
3.6 Self-assessment 266
3.7 Glossary 269
3.8 Further Readings 273
References 273

4 Hot Matter in Thermal Equilibrium 275
4.1 Phenomenological Approach to Entropy 277
4.1.1 The Fundamental Laws of Thermodynamics 277
4.1.2 Properties and Applications of Entropy 282
4.1.3 Thermodynamic Potentials 289

4.2 LTE: The Local Thermodynamic Equilibrium 290
4.2.1 Evolution to Thermal Equilibrium 290
4.3 Essentials of Thermostatistics: Classical Systems 296
4.3.1 The Fundamental Principle of Equilibrium Thermodynamics 296
4.4 Essentials of Thermostatistics: Quantum Systems 312
4.4.1 The Density Matrix 312
4.4.2 Ideal Systems 316
4.5 From Warm Dense Matter to Hot Dense Plasma 331
4.5.1 The Equation of State of Dense Matter 332
4.6 Summary 343
4.7 Problems 347
4.8 Self-assessment 350
4.9 Glossary 353
4.10 Further Readings 359
References 360

5 Waves in the Ideal Plasma 361
5.1 The Maxwell Equations 362
5.2 Wave Equations 364
5.2.1 The Poynting Theorem 366
5.2.2 Maxwell's Stress Tensor T 366
5.3 Covariant Electrodynamics 367
5.3.1 The Electromagnetic Field Tensor 367
5.3.2 Lorentz Scalars and Lorentz Invariant Operators 368
5.3.3 Gauge Transformations 369
5.4 Eigenmodes of the Uniform Plasma 371
5.4.1 The Unmagnetized Fully Ionized Plasma 371
5.4.2 The Magnetized Fully Ionized Plasma 399
5.5 Waves in the Inhomogeneous Plasma 406
5.5.1 From the Transverse Wave to the Classical Photon 408
5.5.2 Wave Amplitudes 415
5.5.3 High Frequency Energy Fluxes 423
5.5.4 Collisional Absorption in Special Density Profiles 429
5.5.5 Summary 431
5.6 Problems 436
5.7 Self-assessment 438
5.8 Glossary 440
5.9 Further Readings 443
References 443

6 Unstable Fluids and Plasmas . . . 445
6.1 Fluid Dynamic Instabilities and Unstable Waves . . . 445
6.1.1 Basic Unstable Phenomena . . . 445
6.1.2 Summary: Plasma Modes, Energy Densities, and Fluxes . . . 455
6.2 Mode Conversion: Resonance Absorption . . . 457
6.2.1 Inhomogeneous Stokes Equation . . . 459
6.2.2 Linear Resonance Absorption . . . 463
6.2.3 Comparison with Experiments . . . 471
6.3 Nonlinear Resonance Absorption . . . 473
6.3.1 Wave Breaking . . . 473
6.3.2 Hot Electron Generation . . . 480
6.3.3 Kinetic Theory of Wave Breaking . . . 485
6.4 Resonant Three Wave Interactions . . . 490
6.4.1 Overview and Physical Picture . . . 491
6.4.2 The Doppler Effect . . . 493
6.4.3 Growth Rates . . . 506
6.4.4 Parametric Amplification of Pulses . . . 519
6.4.5 Quasi-particle Conservation and Manley-Rowe Relations . . . 521
6.4.6 Light Scattering at Relativistic Intensities . . . 522
6.4.7 Self Focusing and Filamentation . . . 527
6.4.8 Modulational Instability . . . 533
6.5 Summary . . . 536
6.6 Problems . . . 539
6.7 Self-assessment . . . 541
6.8 Glossary . . . 543
6.9 Further Readings . . . 546
References . . . 546

7 Transport in Plasma . . . 551
7.1 Collision Frequencies . . . 552
7.1.1 Screening . . . 552
7.1.2 Reduction of Simultaneous Interactions to a Sequence of Collisions . . . 553
7.1.3 Jackson's Model of Coulomb Interaction . . . 553
7.1.4 The Oscillator Model of Uniform Drift . . . 554
7.1.5 Debye Shielding . . . 560
7.2 Collisional Absorption in the Thermal Plasma . . . 561
7.2.1 The Ballistic Model of Collisional Absorption . . . 562
7.2.2 The Dielectric Model of Collisional Absorption . . . 566
7.2.3 Ion Beam Stopping . . . 575
7.2.4 Collision Frequency in the Classical Plasma . . . 578
7.2.5 Supplements . . . 582

7.3 Collisionless Absorption from Overdense Plasma Surfaces 588
7.3.1 Overview and Purpose 588
7.3.2 Anharmonic Resonance 590
7.3.3 1D PIC Simulations of Relativistic Laser-Overdense Matter Interaction 599
7.3.4 Fast Electrons and Energy Partition 604
7.4 On Scaling Laws of the "Hot Electrons" 608
7.5 Pressure-Viscosity Tensor and Friction in Plasma 611
7.5.1 Coefficients of Viscosity 611
7.5.2 Friction 613
7.6 Particle Diffusion and Thermal Conduction 614
7.6.1 Thermal Conduction 616
7.6.2 κ_e from Boltzmann Equation 617
7.7 *Nonideal Plasma: The BBGKY Hierarchy 619
7.7.1 The Liouville Equation and Its Reduced Moments 619
7.8 Summary 624
7.9 Problems 625
7.10 Self-assessment 626
7.11 Glossary 627
7.12 Further Readings 630
References 631

8 Radiation from Hot Matter 633
8.1 The Radiating Plasma 634
8.1.1 The Quantized Maxwell Field 634
8.1.2 The Optical Bloch Model 640
8.1.3 Coherent Effects 644
8.1.4 Spontaneous Radiation from Single Particles 648
8.1.5 Bremsstrahlung from the Thermal Plasma 653
8.2 Radiation Transport 655
8.2.1 The Transport Equation 656
8.2.2 Thermal Radiation from a Plane Layer 660
8.2.3 Diffusion Model of Radiation Transport 665
8.3 Radiation Reaction 667
8.4 Summary 669
8.5 Problems 671
8.6 Self-assessment 672
8.7 Glossary 673
8.8 Further Readings 675
References 675

9 Applications of High Power Lasers . . . 677
9.1 The Nonlinear Response of the Plasma to the Laser . . . 677
9.2 Generation of Radiation . . . 678
9.2.1 Terahertz Radiation . . . 678
9.2.2 X Ray Lasing . . . 681
9.2.3 High Harmonic Generation . . . 683
9.3 Controlled Nuclear Fusion . . . 691
9.3.1 Plan and Requirements . . . 691
9.3.2 Compressional Pellet Heating . . . 693
9.3.3 Fast Ignition . . . 695
9.4 Ion Acceleration by TNSA . . . 696
9.4.1 Dynamic Model of Acceleration . . . 697
9.4.2 Static Models of Acceleration . . . 700
9.5 Radiation Pressure Acceleration (RPA) . . . 704
9.6 Wake Field Acceleration . . . 707
9.6.1 The Nonlinear Wake . . . 708
9.6.2 Energy Gain from the Electron Plasma Wave . . . 709
9.6.3 Nonlinear Bubble and Monoenergetic Beams . . . 709
9.7 Thomson Scattering as a Plasma Diagnostic Tool . . . 713
9.7.1 Basics . . . 713
9.8 Digression On: Classical or Quantum Treatment? . . . 714
9.8.1 A Strong Statement . . . 715
9.8.2 High E and High T Criteria . . . 717
9.8.3 The Interplay of Quantum Theory and Classical Physics . . . 719
9.8.4 Conclusion . . . 719
9.9 Problems . . . 720
9.10 Self-assessment . . . 721
9.11 Glossary . . . 722
9.12 Further Readings . . . 724
References . . . 725

Index . . . 729

Chapter 1
Hot Matter from High-Power Lasers

1.1 Laser and Ion Beam Generated Hot Matter

There are as many types of hot matter and plasmas as technical methods to produce them: Discharges, radiation induced plasmas, pressure generated plasmas, plasmas from dynamic processes, and from particle beams. The widest class of plasmas, concerning their spatial extension as well as their variety, long living, and stable confinement, is found in the cosmos [1]. A special class of plasmas on earth is represented by conducting solids. There the high Fermi pressure of the electrons is neutralized by the Coulomb attraction of the ions, in contrast to the large scale cosmic and laboratory confinement by magnetic fields, gravitation, or the inertia of matter.

Plasmas can be produced by all kinds of intense energy sources. Here, the laser plays a special role. Photons do not interact with each other and can therefore be focused to arbitrary high energy density. As they interact with charged matter, preferentially with the light electrons, high power lasers are capable of producing extremely hot plasmas. At equal energy photons exhibit the highest momentum per particle of all matter. As the resulting radiation pressure in the laboratory may exceed the gas pressure in the center of the sun it is not surprising that the laser induces a whole variety of stable and unstable nonlinear structures in the plasma and generates fast electrons, and accelerates electron bunches up to several GeV on the length of 1 cm only. Fast electron jets in turn give rise to collimated intense radiation sources. High power laser beams are made of low energy photons. In concomitance severe limits are imposed to them in penetrating dense matter. For this reason laser generated plasmas are by far less dense in the mean than solids but hotter than any other plasma on earth.

Dense, compressed matter and nonideal plasmas can be generated with intense beams of heavy ions. A particular advantage is their well defined spatial range due to their stiffness and the Bragg peak; it makes them a powerful instrument for hot matter production and technical and medical applications. Among the latter cancer therapy is on the top of the list.

The alternative to the high power laser is the free electron laser (XFEL) with its energetic photons up to 25 keV, intensities by some $10^{19}\,\mathrm{Wcm}^{-2}$, and extremely

P. Mulser, *Hot Matter from High-Power Lasers*, Graduate Texts in Physics,
https://doi.org/10.1007/978-3-662-61181-4_1

high brilliance, however modest energy per X ray bunch of 25 – 100 fs duration. The XFEL is the ideal instrument for producing *warm dense matter* (WDM) up to a few eV and to generate new kinds of plasma states by extremely fast electronic transitions (e. g. *nonthermal melting*) in the few tens femtosecond domain. The free electron laser is a unique instrument for atomic and nuclear physics research; the high power long wavelength laser shows its prominence as a generator of all kinds of radiation from Terahertz to hardest gammas.

The particle accelerator has been the most successful scientific tool of the past century. The laser is the most successful scientific tool of the present century.

1.1.1 High Power Lasers

The physics of high power laser interaction is considered in the intensity range I from 10^{10} to some $10^{22}\,\mathrm{Wcm^{-2}}$. Matter exposed to ion beams is considered in the beam energy range from 1 to 100 MeV per nucleon for a variety of charge states.

The characteristic pulse length at moderate laser intensities up to the order of $10^{16}\,\mathrm{Wcm^{-2}}$ is from 1 ns ($=10^{-9}$ s) to several 10 ns. Picosecond lasers extend up to $10^{20}\,\mathrm{Wcm^{-2}}$ and play in the sub ns time domain down to 1 ps ($=10^{-12}$ s). Experiments with laser pulses in the relativistic intensity range $I \simeq 10^{18} - 10^{22}\,\mathrm{Wcm^{-2}}$ are performed from 5 fs ($=10^{-15}$ s, shortest pulse in the near infarred) to hundreds of fs. Laser pulses of attosecond ($=10^{-18}$ s) length need a broad bandwidth and are therefore composed of high harmonics from fs Ti:Sa and Nd laser pulses. The high intensities at all pulse lengths are reached by beam focusing from 1 to 100 wavelengths in diameter.

1.1.2 Wavelengths of Common High Power Lasers

Laser typ	λ	ω	$\hbar\omega$
CO_2 laser	10600 nm	$1.78\times10^{14}\,\mathrm{s^{-1}}$	0.12 eV
Iodine laser	1315 nm	$1.46\times10^{15}\,\mathrm{s^{-1}}$	0.96 eV
Neodymium (Nd) laser	1060 nm	$1.78\times10^{15}\,\mathrm{s^{-1}}$	1.17 eV
Titanium-Saphir (Ti:Sa) laser	800 nm	$2.36\times10^{15}\,\mathrm{s^{-1}}$	1.55 eV
3rd harmonic (Ti:Sa) laser	260 nm	$7.17\times10^{15}\,\mathrm{s^{-1}}$	4.65 eV
Krypton-Fluorid (KrF) laser	248 nm	$7.59\times10^{15}\,\mathrm{s^{-1}}$	5.0 eV
	Free Electron Laser (FEL)		
FLASH (DESY Hamburg)	4.2–45 nm	4.5×10^{17}–$1.1\times10^{16}\,\mathrm{s^{-1}}$	304–28.4 eV
XFEL (DESY Hamburg)	0.05 nm	$3.8\times10^{19}\,\mathrm{s^{-1}}$	23 keV

1.1.3 *The Bird of the Laser Plasma*

1.1.3.1 Modelling of the Laser Field

Before the availability of intense laser beams the dynamics of matter in the radiation field, in particular spectroscopy and optics, was adequately described by taking the fields of the unperturbed matter as the leading quantities and the radiation field as a small disturbance. With the dynamics of matter in the intense fields the situation has been reversed, the laser provides for the main field and the atomic fields are the perturbing quantity. Except rare situations the radiation field $\mathbf{E}(\mathbf{x}, t)$ can be modelled as a plane monochromatic wave of amplitude $\hat{\mathbf{E}}$, frequency ω, and wave vector $\mathbf{k}$,

$$\mathbf{E}(\mathbf{x}, t) = \hat{\mathbf{E}}e^{i(\mathbf{kx}-\omega t)}. \tag{1.1}$$

It is a classical, unquantized field obeying Maxwell's equations. The intensity I is the Poynting vector $\mathbf{S}$ averaged over one oscillation,

$$\mathbf{I} = \varepsilon_0 c^2 \overline{\mathbf{E} \times \mathbf{B}} = \frac{1}{2}\varepsilon_0 c \mathbf{k}^0 \hat{\mathbf{E}}\hat{\mathbf{E}}^*; \quad \mathbf{k}^0 = \mathbf{k}/|\mathbf{k}|; \quad \varepsilon_0 = 8.85 \times 10^{-12}\,\mathrm{IU}. \tag{1.2}$$

The field amplitude is given numerically by

$$\hat{E}\,[\mathrm{Vcm}^{-1}] = 27.5 \times (I\,[\mathrm{Wcm}^{-2}])^{\frac{1}{2}}. \tag{1.3}$$

For comparison, in the hydrogen atom the electron on its first Bohr orbit "sees" the field $E = 4.5 \times 10^9\,\mathrm{Vcm}^{-1}$, corresponding to the laser intensity $I = 3 \times 10^{16}\,\mathrm{Wcm}^{-2}$. The amplitude $\hat{E} = 4 \times 10^{12}\,\mathrm{Vcm}^{-1}$ from the actual $I = 2 \times 10^{22}\,\mathrm{Wcm}^{-2}$ is the highest macroscopic field on earth. A third important quantity is the mean oscillation energy W of the free electron in the linearly polarized laser field,

$$W = \frac{e^2}{4m_e\omega^2}\mathbf{E}\mathbf{E}^* \sim I\lambda^2. \tag{1.4}$$

W is 1.0 keV at $10^{16}\,\mathrm{Wcm}^{-2}$ and 96 keV at $10^{18}\,\mathrm{Wcm}^{-2}$, both Nd. In circular polarization it is twice these values. For the relativistic expressions of W see Chap. 2.

1.1.3.2 Laser Induced Breakdown in Matter

When a laser of $10^{10} - 10^{11}\,\mathrm{Wcm}^{-2}$ is focused in air a brilliant flash of bluish-white light appears at the lens focus, accompanied by a distinctive cracking noise: A *gas breakdown* has occurred and a hot plasma, 100 eV (= 10^6 K) has formed in sub-ns time. This phenomenon has been reported first by Maker et al. in 1963 [2] and subsequently by numerous investigators in all details, see [3] and the references

Fig. 1.1 Particle-in-cell simulation of plasma production from solid hydrogen by laser. Parallel laser beam impinges from right, intensity $I = 10^{15}\,\mathrm{Wcm}^{-2}$. Dark bow is the compressed cold matter in the shock travelling into the target. Dots: rarefied plasma flowing against the laser beam. Shock width is broadened by numerical diffusion in the artificial viscosity

therein. The plasma spark shows a marked threshold behaviour of the incident laser intensity. When the laser intensity is reduced to its threshold value, gas breakdown becomes a sporadic event, the threshold intensity for initiating the breakdown can vary up to a factor of 2. At intensities well above threshold spark ignition occurs easily and in a reproducible manner. The stochastic behaviour at threshold induces to assume that it is intimately connected with the stochastic presence of a first few free electrons in the focus volume of the gas. Systematic investigations with ns Nd lasers at fundamental and second harmonic wavelengths $\lambda = 1064\,\mathrm{nm}$ and $\lambda = 532\,\mathrm{nm}$, respectively, in *pure* gases of pressures between 150 and 3000 Torr show typical thresholds between $I_{\mathrm{thr}} = 10^{12}$ and $10^{14}\,\mathrm{Wcm}^{-2}$. The dependence on pressure p decreases as $I_{\mathrm{thr}} \sim p^{-n}$, with $n = 0.78$ at $2\omega_{\mathrm{Nd}}$ and $n = 0.69$ at the fundamental ω_{Nd} for hydrogen, $n = 0.65$ for air, and is much weaker, $n \approx 0.4$ in other gases [4]. As a general experience dielectric matter in all phases, gas, liquid, solid, transforms rapidly into plasma as soon as a threshold intensity is exceeded, irrespective of the photon energy. At equal intensities plasma formation with a CO_2 laser may sometimes happen to be faster than with the Nd laser of ten times shorter wavelength, just contrary to what is known from the linear photo effect.

As a typical example the transformation of a solid hydrogen sample into *laser plasma* under the action of a Nd laser beam is illustrated by Fig. 1.1 when focused to an intensity of $I = 10^{15}\,\mathrm{Wcm}^{-2}$ in a spot of 50 μm radius size. At the very beginning the flat target is transparent to the laser light. After breakdown has occurred somewhere beneath the target surface violent ionization of the hydrogen by electron impact sets in. The free electron density soon reaches values exceeding a *critical density* just there and blocks the beam from propagating further. If initially this critical zone has a wider extension it is forced to reduce to a layer of a fraction of a wavelength thickness by the free electron density increasing up to the density of the bound electrons in the solid. From this instant on an equilibrium establishes between further absorption and heating to higher pressures, and, concurrently, attenuating the pressure increase by expansion forward against the laser beam and backward by compressing the cold solid to form a *shock wave* propagating into the solid. With ongoing time a crater forms in the target by plasma ablation from the shock and rarefying into the vacuum, and cold matter receding in the compressed shock. Shock front and plasma ablation zone remain attached to each other by lateral expulsion of plasma and pushing aside accumulated cold target matter, marked by the dark bow shock in the Figure.

1.1.4 Multiphoton and Field Ionization

Mysterious First Free Electron

The question about the origin of the first few unbound electrons leads into rich and fascinating physics immediately once trying to answer the problem decently. The electron heating mechanism by electron-ion collisions (*collisional heating*, *inverse bremsstrahlung*) and subsequent thermal ionization can work only if a few free electrons in the region of high laser intensity are present. Typical ionization energies E_I of atoms and molecules range from about 4 to 25 eV (Cs 3.9, H 13.6, He 24.6 eV), and hence none of the long wavelength high-power lasers of the Table above are capable of directly photo-ionizing them, except Cs by the KrF or the 3rd harmonic Nd laser. In a perfectly neutral environment at threshold intensity the only possible mechanism is *multiphoton ionization*. It consists of the "simultaneous" absorption of $n \geq E_I/\hbar\omega$ photons to satisfy the atomic process

$$A + n\hbar\omega \rightarrow A^+ + e^-. \tag{1.5}$$

As long as both, photon energy $\hbar\omega$ and ionization energy E_I are large compared to the energy W of the electron oscillating in the laser field, multiphoton ionization can be treated quantum mechanically by the perturbation technique, see, e.g. [5]. The n-photon process starts with the nth order Dirac perturbation theory. The single bound electron oscillates over many cycles in the laser field of frequency ω before becoming free, the product $\omega\tau_I$, τ_I the ionization time, is much larger than unity. A typical measure of τ_I is expressed by the Rabi frequency $\omega_R = \mu\hat{E}/\hbar$, μ dipole moment. So at $I_{\mathrm{Nd}} = 10^{12}\,\mathrm{Wcm}^{-2}$ for $\omega_R \simeq \omega_{\mathrm{Nd}}/40$ and $\omega_{\mathrm{Nd}}\tau_I \simeq 250$ results. With increasing laser intensity *above threshold ionization* (ATI), level shift due to the dynamical Stark effect and higher level excitation become relevant [5, 6]. In concomitance, higher order diagrams can no longer be disregarded a priori and the standard perturbation method may no longer be adequate. Various approaches have therefore been used at moderately high intensities ($I \lesssim 10^{16}$ Wcm^{-2}) [7]. Furthermore, as was explicitly stressed in [8], the adequacy of a perturbation analysis strongly depends on the rise time of the laser pulse also. Ionization in stronger laser fields may end in considerable complexity. Meanwhile a rich specialized literature exists on the subject.

The essence of multiphoton ionization can be made clear in classical terms. Under the influence of the laser field the electron cloud in the atom is modulated at the laser frequency ω. Subsequently impinging coherent photons impose themselves a modulation at frequency ω on the already existing ω modulation. Thus, a 2ω modulation arises on which a new, 3ω modulation is impressed by subsequent coherent photons, etc. The n times modulated electron cloud of the atom or molecule emits photons of $n\hbar\omega$ which in turn can be auto-absorbed by the electron to get free in an Auger like process. If now an $r\omega$ modulation resonates with an atomic level the atom can accumulate, in linear terms, an arbitrarily high amount of energy to facilitate remarkably the multiphoton transition. Thereby a high density of resonant levels may strongly cooperate. Beam coherence is essential. Chaotic light creates, on the average, $\sqrt{n}\hbar\omega$ photons only out of the same photon density.

The ionization cross sections depend sensitively on the individual matrix elements between virtual states, and may change by orders of magnitude when the laser frequency or a multiple of it approaches a transition frequency $\omega_{ij} = (E_i - E_j)/\hbar$ of two energy levels [5, 9]. Fortunately, as the photon number n needed for ionization increases, the ω-dependence greatly decreases and approaches a nonresonant behavior; the ionization probability P_n assumes the structure

$$P_n \simeq \sigma_n I^n \tag{1.6}$$

in a wide range of intensities. It can be shown by several independent arguments that the nth root of the generalized cross section σ_n for multiphoton ionization, with the contributions from higher order diagrams included, $\sigma_n^{1/n}$ is almost a constant [8], and therefore ln P_n plotted as a function of ln I is a straight line, as was confirmed by numerous experiments [10–13]. Deviations from this behaviour only occur at resonances, close to saturation (when $P_n \to 1$), or when *nonsequential ionization* [14] is important, see Fig. 1.2. It is apparent that the measured He^{2+} yield is many orders of magnitude higher than expected from a sequential "single active electron" ionization process, sketched by the solid graphs in the plot. At around 10^{15} Wcm^{-2} the He^{2+} curve changes slope in the nonsequential "knie" to merge with the single active electron graph. It happens when the production of the previous charge state

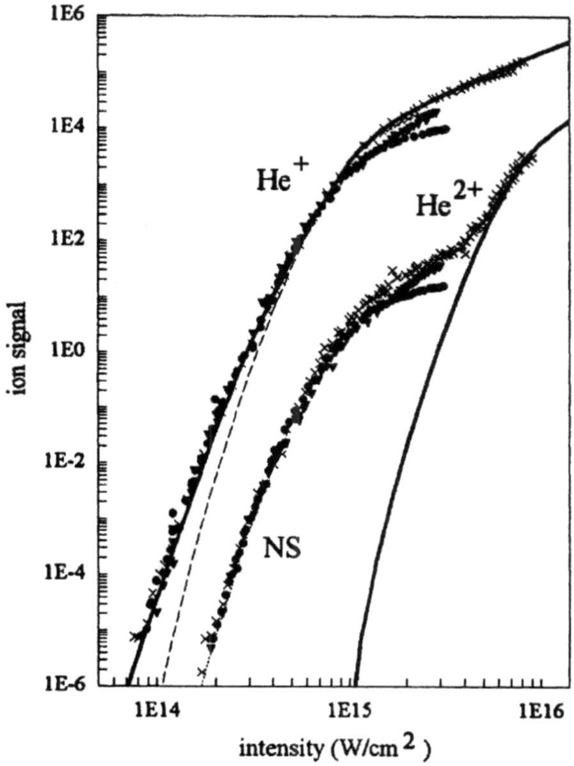

Fig. 1.2 Experimental ion yields for He^+ and He^{2+} after the interaction with a 160 fs 780 nm laser pulse. The solid lines are the theoretically expected yields when a sequential, single active electron ionization scenario is assumed. It is seen that below 10^{15} Wcm^{-2} the measured He^{2+} yield is many orders of magnitude greater before it merges with the theoretical prediction. The deviation from the sequential rate (solid curve) is the so-called nonsequential ionization (NSDI) "knee". From [14]

He^+ enters into its saturation stage, indicating a strong correlation of the two electrons in the ionization dynamics. The observed increase in ionization beyond saturation in Fig. 1.2 stems from the increase of the spark in time [15].

The calculated and measured thresholds for appreciable multiphoton ionization lie all above 10^{12} Wcm^{-2} or an order of magnitude higher and there is no doubt that in very pure atomic gases (and probably very pure liquid or solid dielectrics with extremely clean surfaces) these are the thresholds for plasma formation by focused laser beams [16]. On the other hand, it is known that normally breakdown occurs at much lower intensities, sometimes as low as 10^9 Wcm^{-2} [17]. From this discrepancy the question arises where the "first" electron comes from. Although in general this is an unsolved question, many reasons for the presence of a few free electrons before the arrival of the laser pulse can be given: Ionization by UV light from outside or from the flash lamps of the laser, aerosols or dust particles carrying very weakly bound electrons, negative surface charges on solids. Densely spaced energy levels in molecules may facilitate multiphoton ionization, or two-step ionization-dissociation processes which, for instance in Cs, require much lower laser intensity. Hence, no general answer to the question is to be expected, nor would such one be very convincing. Rather the search for further individual well-defined effects

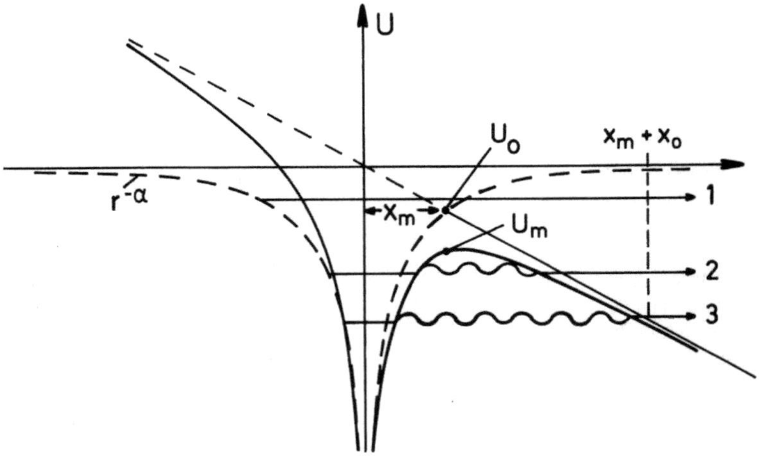

Fig. 1.3 Deformation of the atomic potential $U \sim r^{-\alpha}$ by the time-dependent laser field. The electron from the energy level 1 is free, electrons 2 and 3 exhibit finite tunnelling probabilities. x_m is the distance of the maximum of U from the nucleus; $U_0 = U_m/2$. The tunnelling probability decreases rapidly with increasing difference $eU_m - \epsilon$

is indicated which may lead to breakdown threshold lowering in the actual case under consideration [18–22].

1.1.4.1 Field Ionization

Under the action of a strong field the first bound electron is removed from the nucleus in a fraction of an oscillation period. This is revealed already by a simple classical estimate. A bound electron is typically confined within a distance $d = 0.12$ nm in nearly all solid targets (d = lattice constant) or isolated atoms. In moving across a lattice constant in the laser field $E = \hat{E}\cos\omega t$ it gains the energy $\Delta\epsilon = eEd$. Its maximum value is $\Delta\epsilon_{max} = e\hat{E}d$, i.e., $\Delta\epsilon_{max} = 33$ and 330 eV at 10^{16} and $10^{18}\,\text{Wcm}^{-2}$, respectively. In any case this is larger than the ionization energy E_I of an outer electron. The time dependence of the laser field can be suppressed since, in the ground state the internal frequency $\omega_e \simeq E_I/\hbar$ is generally much higher than the laser frequency. In addition, the laser field imparts an additional high velocity to the electron so that the time t in E may be treated merely as a parameter even for excited states. For illustration the minimum crossing time τ when starting from zero velocity is $\tau = (2m_e d/e\hat{E})^{1/2}$. At $I = 10^{16}\,\text{Wcm}^{-2}$ $\tau = 2 \times 10^{-16}$ s; at $I = 10^{18}\,\text{Wcm}^{-2}$ it is 10 times shorter. The cycle times for Nd and KrF are 3.5×10^{-15} and 8×10^{-16} s.

In a first approach to the quantum picture field ionization may be treated in the Coulomb field that is deformed by the laser, see Fig. 1.3. The atomic potential is assumed to follow the power law $U(r) \sim r^{-\alpha}$, $0 < \alpha < 2$. With the laser field superposed in x direction it assumes along $r = x$

$$eU = -ZKr^{-\alpha} - eEx; \quad K = \frac{e^2}{4\pi\varepsilon_0} \tag{1.7}$$

with Z the effective ion charge. Maximum potential U_m, radial distance x_m, and U_0 are

$$eU_\mathrm{m} = -ZKx_\mathrm{m}^{-\alpha} - eEx_\mathrm{m}, \quad x_\mathrm{m} = \left(\frac{\alpha ZK}{eE}\right)^{1/(\alpha+1)}, \quad U_0 = U_\mathrm{m}/2. \tag{1.8}$$

For the Coulomb potential U holds $\alpha = 1$ and $x_\mathrm{m} = x_C = (ZK/eE)^{\frac{1}{2}}$, $U_\mathrm{m} = U_C = -2Ex_\mathrm{m}$. The variations of x_m and U_m with α are best seen in the following representation of (1.8)

$$x_\mathrm{m}^\alpha = (\alpha x_C^2)^{1/(1+\alpha)}, \quad U_\mathrm{m} = \left(\alpha^{-\alpha/(1+\alpha)} + \alpha^{1/(1+\alpha)}\right)\left(\frac{ZK}{e}\right)^{1/(1+\alpha)} E^{\alpha/(1+\alpha)}.$$

Close to U_m, $U(x)$ is well approximated by a parabola with its vertex at x_m. A classical electron can escape from the atom if its energy level lies above $U_\mathrm{m} = 2U_0$. The resulting ionization rate turns out to be too low. The reason is that the electron captured in the atom is a quantum mechanical entity endowed with the capability of tunnelling through the potential wall and to be slowed down above its top. This behaviour can be accounted for by the tunnelling probability T, i.e. the transmission factor, through the parabolic potential barrier of our model. For energies $\epsilon \gtreqless U_\mathrm{m}$ it is determined by [23],

$$T = \frac{1}{1+\mathrm{e}^{-\zeta}}, \quad \zeta = 2\pi\left(\frac{m_e}{k}\right)^{1/2}\frac{\epsilon - eU_\mathrm{m}}{\hbar}, \quad k = -e\left(\frac{\partial^2 U}{\partial x^2}\right)_{x_\mathrm{m}} = -\alpha(\alpha+1)\frac{eU_0}{x_\mathrm{m}^2}. \tag{1.9}$$

More precisely, the expression of T is valid for $|\epsilon/eU_\mathrm{m} - 1| \lesssim 1/10$. For $\epsilon = eU_\mathrm{m}$ the transmission factor is $T = 1/2$, against $T_\mathrm{classical} = 1$. Thus, there is enough time for an electron with bound energy $\epsilon \geq eU_\mathrm{m}(E = \hat{E})$ to be ionized. In Fig. 1.4 $U_\mathrm{m}(E = \hat{E})$ as a function of the effective ion charge Z for the laser intensities from $I = 10^{16}$ to $10^{24}\,\mathrm{Wcm}^{-2}$ and the ionization potentials of the isolated atoms C, Al, Cu, Ag, and Au as functions of the real charge state Z are reported. Under the assumption that the potentials have approximately hydrogen-like structure, i.e., $Z_\mathrm{eff} \simeq Z$, the various ionization degrees by multiphoton absorption are determined from the intersection points in the figure. For example, the laser intensity $I = 10^{18}\,\mathrm{Wcm}^{-2}$ is capable of producing the minimum ionization stages C^{4+}, Al^{9+}, Cu^{15+}, Ag^{18+}, Au^{22+}. At 100 times higher intensity the result is C^{6+}, Al^{11+}, Cu^{26+}, Ag^{37+}, Au^{51+}.

The tunnelling time τ_I for an electron in an energy state $\epsilon \simeq eU_\mathrm{m}$ is given by

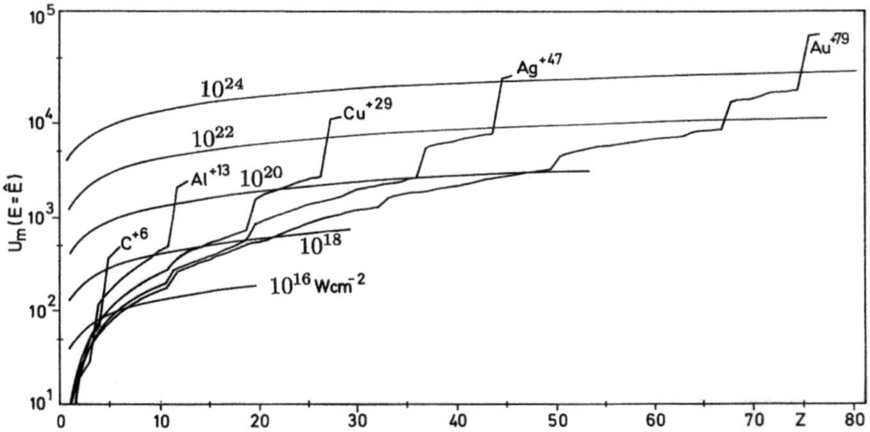

Fig. 1.4 Field ionization of C, Al, Cu, Ag, and Au in the potential assumed Coulomb-like. the maxima U_{m} of the ionic potentials are determined for the laser intensities $I = 10^{16} - 10^{24}\,\mathrm{Wcm}^{-2}$ as functions of Z when the laser field reaches its maximum at $E = \hat{E}$. The degree of ionization of the isolated atoms results from the intersection points of the corresponding graphs

$$\tau_I = \frac{1}{T}\left(\frac{m_e}{2|\epsilon|}\right)^{1/2} \times \min(d, 2x_{\mathrm{m}}), \tag{1.10}$$

where d is the interionic distance. Under the condition of $\tau_I \ll \pi/\omega$, τ_I can be regarded as the mean lifetime of the bound energy state ϵ. The potential maxima U_{m} as well as their positions x_{m} vary like $Z^{1/2}$ for fixed laser intensity.

The influence of tunnelling is studied by calculating T and τ_I for various energies ϵ in the neighbourhood of eU_{m}. Evaluation of T from (1.9) shows that only very close to U_{m} the criterion $\tau_I \ll \pi/\omega$ is well fulfilled for energies $\epsilon < eU_{\mathrm{m}}$. At $I_{\mathrm{Nd}} = 10^{16}\,\mathrm{Wcm}^{-2}$ an energy lower by 5% only than eU_{m} leads still to short enough times for static tunnelling at $Z \geqslant 5$, no longer however for KrF at the same intensity. Tunnelling at $\epsilon < eU_{\mathrm{m}}$ becomes more significant with increasing intensity at low Z-values and decreases with Z increasing. In general, beyond laser intensities $I \simeq 10^{17}\,\mathrm{Wcm}^{-2}$ the electron behaves almost classical, tunnelling does not play a major role.

A rough criterion for discerning which of the two processes prevails, multiphoton or field ionization, may be found in the Keldish parameter γ_K [24]

$$\gamma_K = \left(\frac{E_I}{2W}\right)^{1/2}. \tag{1.11}$$

It can be interpreted as the ratio of the tunnelling time and the laser period, $\gamma_K = \omega\tau_I$. It indicates whether the tunnelling process is fast on the inneratomic time scale or the laser field reverses sign before tunnelling is completed. Hence, $\gamma_K > 1 \Rightarrow$ multiphoton ionization, $\gamma_K < 1 \Rightarrow$ tunnelling ionization.

Initial Kinetic Temperature

For several applications (e.g., spectroscopy, X-ray laser development) plasmas of very high ionization degree at low electron temperatures are of great interest for violent recombination, and eventual population inversion of metastable levels are to be expected. For an estimate the determination of the electron temperature just immediately after field ionization has occurred is of interest. An effective kinetic temperature T_0 is found in the following way.

In the static classical picture an electron is free as soon as the laser field amplitude $\hat{E}$, the so called *critical field* is such that $|eU_{\rm m}|$ equals the ionization energy E_I. The electron may arrive at $x_{\rm m}$ with a kinetic energy lower than its ground state kinetic energy $E_{\rm kin} = \alpha\epsilon/(\alpha - 2)$ and subsequently gain additional kinetic energy downhill up to the point where it becomes a free particle. During this ionization process the electron undergoes a Stark shift from the laser field at the nucleus and an adiabatic lowering of its undisturbed energy state ϵ by the fraction $(r_Z/x_{\rm m})^{2/3}$, r_Z Bohr radius (for the adiabatic invariance of $r\epsilon^{1/2}$ the reader may see Chap. 2). Both effects and the kinetic energy of the freed electron have been calculated in terms of a classical Hamiltonian [25]. As a result, no electron escapes below a threshold of $E_{\rm kin} = |\epsilon|/3$, however an upper bound for it does not exist. Accompanying Monte Carlo calculations with random initial conditions have provided additional confirmation. As a rule, under the action of a strong laser field both classical as well as numerous quantum calculations in the Coulomb field led to the average ejection energy [26]

$$\langle E_{\rm kin}\rangle = \frac{1}{2}E_I = k_B T_0. \tag{1.12}$$

Separate analysis has to show how close T_0 is to a true equilibrium electron temperature T_e. The reader may be surprised for the low average ejection energy. Though it has a simple explanation. The maximum potential depression $U_{\rm m}$ exhibits the longest opening time and the shortest velocity of the quasi-free electron. Note, the free electron assumes oscillation velocity zero when the laser field reaches its extrema.

1.2 Basic Properties of the Laser Plasma

1.2.1 Collisional Absorption and Plasma Heating

Multiphoton and field ionization lead to a rapidly growing electron population in matter. At a given *threshold* the two processes go over into ionization by electron-atom and electron-molecule collisions of the swift thermal electrons and induce an avalanche like increase of their concentration. High electron concentration of particle density n_e in the laser focus leads to self-trapping of the electron cloud by the ions: The mutual collective Coulomb attraction becomes such as to inhibit the escape of an appreciable fraction of electrons out of the laser beam focus. The criterion for self-trapping is determined by the *Debye length* λ_D,

$$\lambda_D = \left(\frac{\varepsilon_0 k_B T_e}{n_e e^2}\right)^{1/2} \tag{1.13}$$

to be smaller than the minimum diameter d of the ion cloud. Once $\lambda_D/d \ll 1$ is fulfilled the ion cloud results *quasineutral*, $|n_e - Zn_i| \ll n_e$, and its properties are no longer determined by its shape. The cloud has formed a *thermal plasma*, characterized by its intrinsic properties of electron and ion densities n_e, n_i, and electron and ion temperatures T_e, T_i. Numerically the Debye length is given by

$$\lambda_D[\mathrm{cm}] = 6.9\left(\frac{\mathrm{T_e[K]}}{\mathrm{n_e[cm^{-3}]}}\right)^{1/2} = 743\left(\frac{\mathrm{T_e[eV]}}{\mathrm{n_e[cm^{-3}]}}\right)^{1/2}. \tag{1.14}$$

The Debye length tells over what distance an equilibrium is established between the volatility of the thermal electrons and their restoring force by the attraction of the quasistatic ions. To determine this equilibrium length imagine a stripe of homogeneous plasma of width λ_D. In equilibrium an electron which escapes across spends the work $-eE\lambda_D$. On the other hand, on the average this is just the mean thermal energy $m_e v_{\mathrm{th}}^2/2 = k_B T_e/2$ the electron is able to deliver transversally. The equilibrium electric field E follows from Maxwell's equation $\partial_x E = -e(n_e - Zn_i)/\varepsilon_0$. Under the reasonable estimate of

$$\partial_x E = \frac{E}{\lambda_D}, \quad -\frac{e}{\varepsilon_0}(n_e - Zn_i) = -\frac{e}{\varepsilon_0}\frac{n_e}{2} \;\Rightarrow\; E = -\frac{e}{\varepsilon_0}\frac{n_e}{2}\lambda_D.$$

Substitution of this expression for E in the energy relation above yields λ_D,

$$-eE\lambda_D = \frac{1}{2}k_B T_e, \quad E = -\frac{e}{\varepsilon_0}\frac{n_e}{2}\lambda_D \;\Rightarrow\; \lambda_D^2 = \frac{\varepsilon_0 k_B T_e}{n_e e^2}.$$

Uncorrelated thermal motion necessarily induces fluctuations in n_e, however they extend not much farther than a Debye length. Thus, on macroscopic dimensions *quasineutrality* implies in the absence of external forces that $n_e + Zn_i$ can be identified with $2n_e$ or $2Zn_i$, whereas the difference $n_e - Zn_i$ to be set to zero is only allowed if the electric field from the charge imbalance can be neglected.

A numerical example may illustrate why the plasma is quasineutral over distances much larger than λ_D. Assume $n_e = 10^{20}\mathrm{cm}^{-3}$ and $Zn_i - n_e = 10^{-6}n_e$. The charge imbalance creates a voltage over the distance $d = 0.1\mathrm{cm}$ of

$$\Delta V = \frac{e}{2\varepsilon_0}(Zn_i - n_e)\mathrm{d}^2 = 9 \times 10^5\,\mathrm{V}.$$

In a thermal plasma an electron temperature of 1 MeV is needed to produce such a potential difference.

The plasma composed of high Z ions will be partially ionized only. Its properties, like electrical and thermal conductivity and viscosity, will depend on the interaction of electrons and ions with ions of various degree of ionization and with neutrals. However, in moderately high Z atoms the encounters between electrons and ions may determine the plasma properties already at its *partial ionization*. In such a case the so called model of the *fully ionized plasma* applies (although it may be ionized to a few percent only). If not stated differently, here one species of ions of average charge Z is assumed for simplicity. Once the conditions of a fully ionized plasma are fulfilled further heating of the plasma by the laser is determined by the electrical conductivity of the electrons. The irreversible process of heating means *absorption* of radiation and transfer of the amount of energy eventually to internal plasma energy. In the particle picture it is the annihilation of a photon and creation of a plasmon or the energy transfer of the photon directly to an electron. The latter case is the important process of *collisional absorption*. It is the inverse of what happens to an electron when it is slowed down under the emission of a photon by bremsstrahlung. Therefore collisional absorption is named synonymously *inverse bremsstrahlung*.

1.2.1.1 Collisional Absorption and the Drude Model

Collisional interaction of a particle with one or more other particles is a short range event. It occurs during a time interval which is short compared with the change of trajectory by exterior forces. The prototype of a collision is that of hard spheres. The hard sphere model is used here because it reveals all essential aspects of a collision, like irreversibility and energy transfer. To this aim let us consider Fig. 1.5. The oscillating laser field has the direction indicated by the double arrow. In the elastic collision of an electron of relative velocity $\mathbf{v}$ with an ion only its component normal to the surface of the hard sphere is affected. At the instant t_0 of the collision it turns into its negative value. In a thermal plasma with the ions at rest electron-ion collisions do not affect their distribution function $f(\mathbf{v}_e)$ because of equal likelihood for a collision of $\mathbf{v}_e$ and of $-\mathbf{v}_e$ in the absence of a drift. In presence of the monochromatic laser field $\mathbf{E}(\mathbf{x}, t) = \hat{\mathbf{E}} \cos(\mathbf{kx} - \omega t)$ the single electron experiences a drift velocity,

$$\mathbf{v}(\mathbf{x}, t) = \mathbf{v}_e + \hat{\mathbf{v}} \sin(\mathbf{kx} - \omega t); \quad \hat{\mathbf{v}} = \frac{e}{m\omega}\hat{\mathbf{E}}.$$

Again, for symmetry reasons only the component $\mathbf{v}(\mathbf{x}, t)_\parallel = \mathbf{v}_{e\parallel} + \hat{\mathbf{v}} \sin(\mathbf{kx} - \omega t)$ parallel to the laser field is of relevance; the component perpendicular to it because of no drift is not affected by the collision. If the collision happens at the instant $t = t_0$, at an arbitrarily small instant $t = t_0 + \varepsilon$ later the following *irreversible* transition has happened,

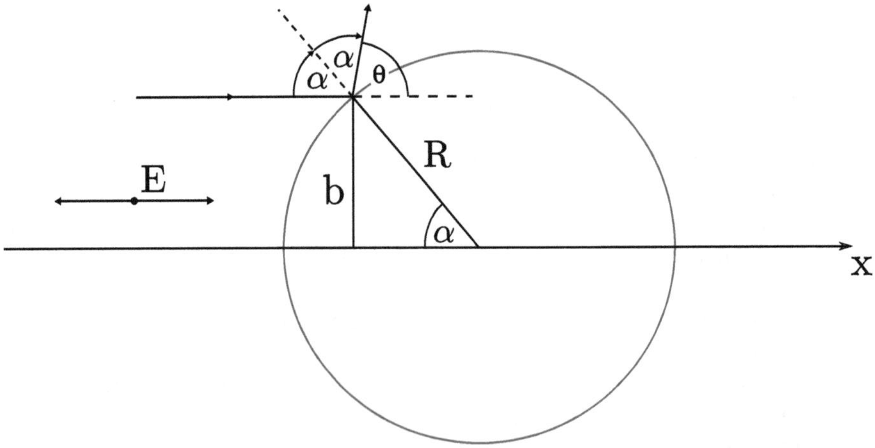

Fig. 1.5 Elastic collision of a point charge of incident velocity component $\mathbf{v}_{\text{inc}} = \mathbf{v}_{e\|} + \hat{\mathbf{v}} \sin(\mathbf{kx} - \omega t_0)$ with a hard sphere of radius $R = r_e + r_i$. Impact parameter $b = R \sin \alpha$, scattering angle $\vartheta = \pi - 2\alpha$. Horizontal double arrow indicates direction of the laser **E** field. Irreversible velocity change occurs for the projection of $\mathbf{v}_{\text{inc}}$ into direction of R (dashed)

$$[\mathbf{v}_{e\|} + \hat{\mathbf{v}} \sin(\mathbf{kx} - \omega t_0)] \cos \alpha \;\Rightarrow\; -[\mathbf{v}_{e\|} + \hat{\mathbf{v}} \sin(\mathbf{kx} - \omega t_0)] \cos \alpha. \tag{1.15}$$

It expresses the fact (i) that in the elastic collision only the component normal to the surface is reflected, and (ii) that the reflection occurs *instantaneously*. The resulting energy gain is to be calculated in the reference system $S'[\mathbf{v} = -\hat{\mathbf{v}} \sin(\mathbf{kx} - \omega t_0) \cos \alpha]$ at time $t = t_0 - \varepsilon$ in which the oscillatory velocity before the collision is zero (see exercise). Thus, (1.15) reads

$$\mathbf{v}_{e\|} \;\Rightarrow\; -[\mathbf{v}_{e\|} + 2\hat{\mathbf{v}} \sin \varphi(t_0)] \cos \alpha; \quad \varphi(t_0) = \mathbf{kx} - \omega t_0.$$

In the collision the electron gains the net energy

$$\frac{1}{2} m_e \left\{ [\mathbf{v}_{e\|} + 2\hat{\mathbf{v}} \sin \varphi(t_0)]^2 \cos^2 \alpha - \mathbf{v}_{e\|}^2 \right\} = 2 m_e \hat{\mathbf{v}}^2 \sin^2 \varphi(t_0) \cos^2 \alpha. \tag{1.16}$$

The collision happens at all instants of one laser period with equal probability. Hence, averaging in t yields the energy gain $m_e \hat{\mathbf{v}}^2 \cos^2 \alpha$ per collision. The *differential cross section* σ_Ω of colliding hard spheres of radii r_e, r_i is $\sigma_\Omega = (r_e + r_i)^2/4$ and does not depend on the scattering angle ϑ. The *total cross section* is $\sigma_t = 4\pi\sigma_\Omega = R^2\pi$; $R = r_e + r_i$. Note, $\cos \alpha = -\sin \vartheta/2$. Hence, averaging in ϑ yields the energy gain $\mathcal{E}_e$ of twice the mean oscillation energy W each hard sphere collision,

$$\mathcal{E}_e = \frac{1}{2} m_e \hat{\mathbf{v}}^2 = 2W. \tag{1.17}$$

Folding with a *Maxwellian* electron distribution

$$f_0 = \left(\frac{\beta}{\pi}\right)^{3/2} \mathrm{e}^{-\beta v_0^2}, \quad \beta = \frac{m_e}{2k_B T_e} = \frac{1}{2v_{\mathrm{th}}^2} \tag{1.18}$$

yields $\langle v_0 \rangle = 4\pi \int v_0^3 f_0 \, dv_0 = (8/\pi)^{1/2} v_{\mathrm{th}}$, and the average collision frequency

$$\nu = (8\pi)^{1/2} n_i R^2 v_{\mathrm{th}}. \tag{1.19}$$

Using this, the average energy an electron acquires per unit time is

$$\frac{\mathrm{d}E_e}{\mathrm{d}t} = \frac{\mathrm{d}}{\mathrm{d}t}\left(\frac{3}{2}k_B T_e\right) = 2W\nu = n_i \sigma_t \left(\frac{8}{\pi}\frac{k_B T_e}{m_e}\right)^{1/2} \frac{e^2}{2m_e \varepsilon_0 c \omega^2} I = \alpha_e I, \tag{1.20}$$

with the *absorption coefficient* α_e per single electron

$$\alpha_e = \left(\frac{2k_B}{\pi}\right)^{1/2} \frac{n_i e^2}{m_e^{3/2} \varepsilon_0 c \, \omega^2} \sigma_t T_e^{1/2}. \tag{1.21}$$

In passing from W to I in (1.20) it has been tacitly assumed that the refractive index of the partially ionized plasma is not far from unity.

Electron-ion collisions are governed by the soft Coulomb interaction where the interaction time is finite. As will be shown in the chapter on transport only those electrons contribute to the collision frequency and absorption for which the interaction time is shorter than $T/3$, $T = 2\pi/\omega$ laser period. Thus, efficient Coulomb collisions can be considered as *quasi-instantaneous*. The expression of the collision frequency (1.19) in Coulomb interactions will depend on the effective sphere radius R which reduces with the thermal electron velocity; R shrinks proportionally to T_e^{-1}. An additional underlying assumption hidden in (1.19) is $|\hat{\mathbf{v}}| \ll v_{\mathrm{th}}$. Only then ν is an intrinsic property of the plasma.

Plasma breakdown. As the mean electron energy increases, the excitation and ionization cross sections σ_{ex}, σ_I also grow. An avalanche process starts and *plasma breakdown* occurs when the inequality

$$\frac{dE_e}{dt} = \alpha_e I - \nu_I E_I - \frac{d}{dt} E_{\mathrm{loss}} \geq 0$$

is fulfilled for a significant electron density n_e formed already. Here, $\nu_I = n_i \langle \sigma_I v_0 \rangle$ is the mean ionization frequency, $n_i = n_e/Z$, E_I is the ionization energy, and E_{loss} comprises all energy losses due to hot electron diffusion (thermal conduction), ion and atom excitation, radiation losses, recombination, and energy loss by expansion of the electron gas. For an approximate breakdown criterion the losses can be disregarded in not too small foci (diameter $\geq 50 - 100\,\lambda$) and ν_I may be determined from the Lotz formula [30, 44]

$$\sigma_I = 2.8\pi a_0^2 \left(\frac{E_H}{E_I}\right)^2 \frac{\ln(u+1)}{u+1}, \quad u = \frac{E_e - E_I}{E_I},$$

$$\langle \sigma_I v_0 \rangle \, [\mathrm{cm}^3\,\mathrm{s}^{-1}] = 6 \times 10^{-8} \left(\frac{E_H}{E_I}\right)^{3/2} \beta_I^{1/2} \mathrm{e}^{-\beta_I} \mathrm{Ei}(-\beta_I), \;\; \beta_I = \frac{E_I}{k_B T_e}. \qquad (1.22)$$

$E_H = 13.6\,\mathrm{eV}$ (hydrogen), a_0 is the Bohr radius, E_I the ionization energy, and $\mathrm{Ei} = -\int_{\beta_I}^{\infty} \mathrm{e}^{-x^2} dx$ is the exponential integral. Breakdown occurs when $I = I_0$,

$$I_0 \simeq \frac{n_i \langle \sigma_I v_0 \rangle}{\alpha_e} E_I \qquad (1.23)$$

is satisfied. Multiphoton or field ionization does not play a role in the breakdown process at low laser intensities $I \leqslant I_0$.

The Drude model of electrical conductivity. Guided by the success of the kinetic theory of gases Drude has proposed the equation of motion for the *average* electron velocity $\mathbf{u}$

$$\frac{\mathrm{d}\mathbf{u}}{\mathrm{d}t} + \nu \mathbf{u} = \frac{e}{m_e}(\mathbf{E} + \mathbf{u} \times \mathbf{B}); \quad \mathbf{u} = \int \mathbf{v} f_e(\mathbf{v}) \mathrm{d}\mathbf{v} \qquad (1.24)$$

to explain the conductivity of metals with the ions at rest [45]. The term $\nu\mathbf{u}$ is the *dynamical friction*. It is *irreversible* as seen from substituting t by $-t$ in the equation. In the absence of an external electric field the drift $\mathbf{u}$ vanishes because the thermal electron distribution is isotropic. As a consequence of this symmetry $\mathbf{u}$ is parallel to the comoving laser field $\mathbf{E}' = \mathbf{E} + \mathbf{u} \times \mathbf{B}$. The coefficient ν plays the key role in collisional plasma heating by laser. It has the dimension of $[t^{-1}]$. If the electron motion in the hard sphere model is cast into the form (1.24) ν identifies with (1.19). Equation (1.24) can be considered as the defining relation of the collision frequency in the specific context. In the form presented here it is Galilei invariant. $\mathbf{E}' = \mathbf{E} + \mathbf{u} \times \mathbf{B}$ is the electric field comoving with the average velocity $\mathbf{u}$. It holds $|\mathbf{u} \times \mathbf{B}| \leq (|\mathbf{u}|/c)|\mathbf{E}|$. For $|\mathbf{u}|/c \ll 1$ the Lorentz term $\mathbf{u} \times \mathbf{B}$ can be neglected.

The Drude model has proven enormously successful in solid state and plasma physics, in the theory of conducting fluids, and in optics of conductors and insulators. The connection of ν with the electron current density $\mathbf{j}$, the electric conductivity σ, the refractive index η, and the absorption coefficient α in *Beer's law* $I(x) = I_0 \exp(-\int \alpha dx)$ is given by

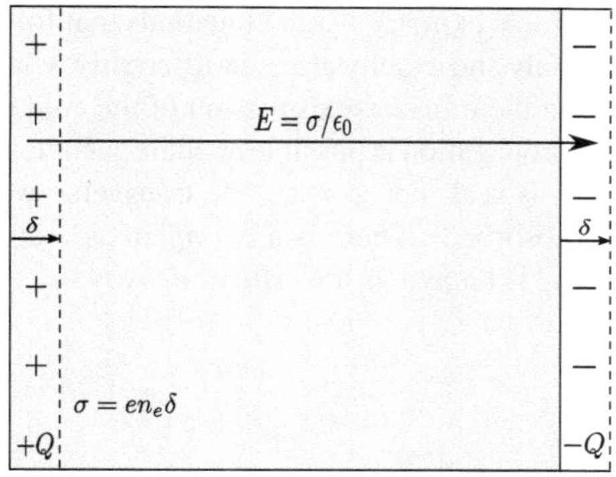

Fig. 1.6 Plasma frequency: By the shift δ of the electron block an electric field of strength $E = en_e\delta/\varepsilon_0$ is created. It forces the electrons to oscillate at the plasma frequency ω_p against the ions

$$\mathbf{j} = -en_e\mathbf{u} = \sigma(\mathbf{E} + \mathbf{u} \times \mathbf{B}); \quad \eta^2 = 1 + i\frac{\sigma}{\varepsilon_0\omega} = 1 - \frac{\omega_p^2}{\omega^2}\frac{1}{1 + i\nu/\omega}; \quad \alpha = 2k_0\Im\eta. \tag{1.25}$$

The quantity ω_p is the *plasma frequency*, i.e., the resonance frequency at which the cold plasma ($T_e = 0$) oscillates,

$$\omega_p^2 = \frac{n_e e^2}{m_e\varepsilon_0}; \quad \omega_p^2\,[\mathrm{s}^{-2}] = 3.2 \times 10^9 n_e\,[\mathrm{cm}^{-3}]; \quad \omega_p\,[\mathrm{s}^{-1}] = 5.64 \times 10^4 (n_e\,[\mathrm{cm}^{-3}])^{1/2}. \tag{1.26}$$

For the derivation of these expressions, and limits of validity of (1.25) see Chap. 5.

The plasma frequency is one of the fundamental plasma parameters. Therefore here already a simple derivation is given. Consider a cold homogeneous plasma slab of thickness d and infinite extension transversally, see Fig. 1.6. By the small longitudinal excursion $\delta \ll d$ of the electrons the plasma slab is polarized, it transforms into a capacitor of surface charge density $\sigma = en_e\delta$. The electric field of the capacitor is simply $E = \sigma/\varepsilon_0$. The single electron moves according to the law of a harmonic oscillator in δ

$$m_e\ddot{\delta} = -eE \;\Rightarrow\; \ddot{\delta} = -\frac{n_e e^2}{m_e\varepsilon_0}\delta \;\Rightarrow\; \ddot{\delta} + \omega_p^2\delta = 0$$

with eigenfrequency ω_p given by (1.26).

Transverse plasma frequency. The previous derivation of ω_p holds for a displacement δ in longitudinal direction. In (1.25) ω_p refers to transverse excitation by a laser. Where is the restoring force in this case, or is ω_p here purely accidental? First, there is a difference between a longitudinal and a transverse electron plasma oscillation. If the longitudinal displacement is assumed as

$\delta = \hat{\delta}\exp(ikx - i\omega t)$ one finds that for all k values the cold plasma oscillates only and exactly at ω_p, indifferently what the wave number k is (see Problems). In the transverse excitation of the cold plasma (1.25), existence of waves and propagation is possible at all frequencies $\omega > \omega_p$ for which the refractive index η is real. For $\omega < \omega_p$ the transverse wave is damped exponentially (but not absorbed). There is a *cut off* at $\omega = \omega_p$. The corresponding electron density n_e is known as the *critical density* n_c,

$$n_c = \frac{\varepsilon_0 m_e}{e^2}\omega^2 = 1.75 \times 10^{21}\left(\frac{\omega}{\omega_{\mathrm{Ti:Sa}}}\right)^2 \,[\mathrm{cm}^{-3}]. \tag{1.27}$$

Beyond n_c matter is opaque to the laser beam; no direct heating is possible. In the longitudinal oscillation the restoring force is caused by space charges, in the transverse excitation it originates from the interplay of the electric with the magnetic field. In the longitudinal electric field ω_p is the resonance frequency of the cold plasma, in the transverse E field ω_p is not a resonance frequency.

It is instructive to apply an alternating electric field $E(t) = \hat{E}\exp(-i\omega t)$ from outside on the plasma oscillator into direction of δ,

$$\ddot{\delta} + \omega_p^2\delta = -\frac{e}{m_e}\hat{E}\exp(-i\omega t) \;\Rightarrow\; \delta(t) = \frac{e\hat{E}\exp(-i\omega t)}{\omega^2 - \omega_p^2}. \tag{1.28}$$

The plasma slab exhibits violent growth in amplitude at resonance $\omega = \omega_p$. Out of resonance the oscillation reaches a steady state.

The Drude ansatz permits an alternative proof of the absorption of energy per laser cycle. Multiplication of (1.24) with $m_e\mathbf{u}$ and integration over one laser cycle T yields

$$\frac{1}{2}m_e\frac{\mathrm{d}\mathbf{u}^2}{\mathrm{d}t} + \nu m_e\mathbf{u}^2 = -e\mathbf{E}\mathbf{u} \quad\Rightarrow\quad \frac{1}{2}m_e\Delta\mathbf{u}^2 + \nu m_e\mathbf{u}^2 T = -e\mathbf{E}\mathbf{u}T.$$

Under steady state conditions $m_e\Delta\mathbf{u}^2$ is zero and $m_e\mathbf{u}^2 = 2W = -e\mathbf{E}\mathbf{u}$ is the work supplied per unit time by the laser field in agreement with the result from the hard sphere model. Electron-electron collisions do not contribute to plasma heating. In the reference system oscillating with $\mathbf{u} = \hat{\mathbf{v}}\sin(\mathbf{kx} - \omega t)$ the quiver motion disappears from the electron-electron encounters and leaves the electron-electron collision frequency ν_{ee} unchainged. There is only the indirect effect of the electron distribution function $f_e(\mathbf{v}_0)$ in that a high electron-electron collision frequency ν_{ee} accelerates their thermalization. The formation of a *supergaussian* electron distribution $f_e(\mathbf{v}_0) \sim \exp[-(\mathbf{v}_0/\kappa_e)^\alpha]$, $\alpha > 2$, is the best known consequence of a too low ν_{ee} [46], see Chap. 7.

1.2.2 Thermalization

The hard sphere model is excellent for introducing the basic concepts of the kinetic theory and for revealing the underlying physics. In order to obtain realistic numbers for the single quantities, like the total collision cross section σ_t and the collision frequency ν_{ei}, the Coulomb potential must be used. Without proof here we give some useful formulas and refer the reader for details to Chaps. 2 and 7. With the scattering angle ϑ and the collision parameter b in the reference system of the mass center (see Fig. 1.7) the *differential Coulomb cross section* is

$$\sigma_\Omega = \sigma(\vartheta) = \frac{b_\perp^2}{4\sin^4(\vartheta/2)}, \quad \tan\frac{\vartheta}{2} = \frac{b_\perp}{b}, \quad b_\perp = \frac{Ze^2}{8\pi\varepsilon_0 E_r} = 0.7\frac{Z}{E_r[\mathrm{eV}]}\,\mathrm{nm}. \tag{1.29}$$

E_r is the electron energy in the center of mass system, $b_\perp$ is the collision parameter for perpendicular deflection. The total collision cross section and the electron-ion collision frequency in presence of a weak external field ($|\hat{\mathbf{v}}| < v_{\mathrm{th}}$; more exact: $\ll$) result as

$$\sigma_t = \pi b_\perp^2 \ln\Lambda, \quad \nu_{ei} = \frac{4}{3}(2\pi)^{1/2}\left(\frac{Ze^2}{4\pi\varepsilon_0 m_e}\right)^2\left(\frac{m_e}{k_B T_e}\right)^{3/2} n_i \ln\Lambda$$

$$\nu_{ei}\,[\mathrm{s}^{-1}] = 3.6\times\frac{Zn_e[\mathrm{cm}^{-3}]}{(T_e[\mathrm{K}])^{3/2}}\ln\Lambda. \tag{1.30}$$

In its simplest form, and generally used version, the coefficient $\alpha = 2k_0\Im\eta$ of *collisional absorption* is

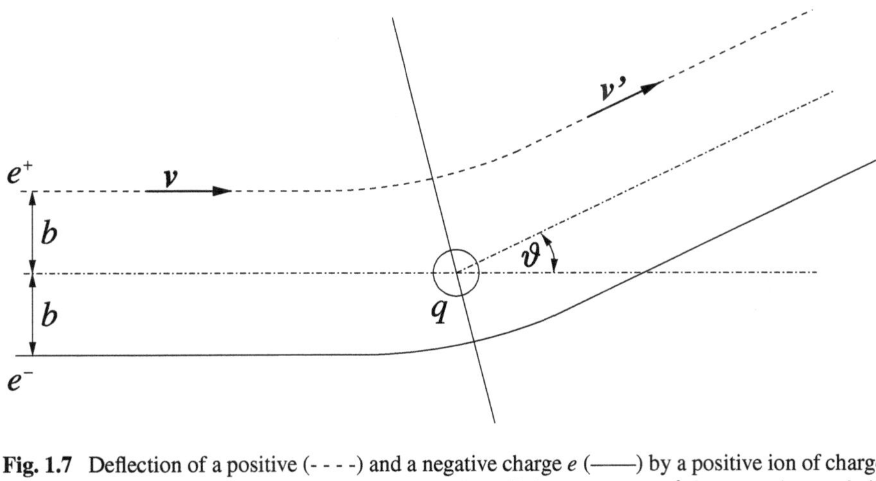

Fig. 1.7 Deflection of a positive (- - - -) and a negative charge e (——) by a positive ion of charge $q = Ze$. The relative velocity is $\mathbf{v} = \mathbf{v}_e - \mathbf{v}_i$; b is the collision parameter, ϑ the scattering angle in the center of mass system; $|\mathbf{v}'| = |\mathbf{v}|$

$$\alpha = n_e\alpha_e = \frac{\omega_p^2}{\omega^2 c}\nu_{ei} = \frac{n_e}{n_c}\frac{\nu_{ei}}{c}, \quad \eta = 1, \ \Im\eta \ll 1$$

$$\alpha[\mathrm{cm}]^{-1} = 1.2 \times 10^{-10}\,\frac{\mathrm{n_e}}{\mathrm{n_c}} \times \frac{\mathrm{Zn_e[cm^{-3}]}}{(\mathrm{T_e[K]})^{3/2}}\ln\Lambda. \tag{1.31}$$

The *Coulomb logarithm* ln Λ in laser plasmas ranges from 2 to approximately 5, in magnetic fusion plasmas it lies between 10 and 20. The electron-electron collision frequency ν_{ee} is close to ν_{ei}. Since ν_{ee} and ν_{ei} are of central importance for understanding collisional laser plasma interaction (absorption) the reader should know one of the following formulas by heart,

$$\nu_{ei} \simeq (10-20) \times \frac{Z^2 n_i[\mathrm{cm}^{-3}]}{(T_e[\mathrm{K}])^{3/2}}\,\mathrm{s}^{-1} = (1-2)\times 10^{-5}\frac{Z^2 n_i[\mathrm{cm}^{-3}]}{(T_e[\mathrm{eV}])^{3/2}}\,\mathrm{s}^{-1}. \tag{1.32}$$

It follows from (1.24) that $\tau_{ei} = 1/\nu_{ei}$ is the average time for single electron collisions to add up to 90^0 deflection. The electron-electron collision time $\tau_{ee} = 1/\nu_{ee}$ is a measure for the formation of the electron temperature T_e in the sense of thermalization of the electron fluid in the time $\tau \gg \tau_{ee}$, and no existence of T_e in the opposite case $\tau \ll \tau_{ee}$. If $\tau \simeq \tau_{ee}$ holds, only a more accurate estimation can give the correct answer on the degree of thermalization. The ions thermalize in a characteristic time $\tau_{ii} \simeq \tau_{ee}(m_i/m_e)^{1/2}/Z^3$ because, for protons, the collision cross section σ_{pp} is the same as σ_{ee} for electrons but their mean velocities are reduced by the square root of the mass ratio at equal energies of the colliders and $\tau_{pp} = \frac{1}{\nu_{pp}} = \frac{1}{n_p\langle\sigma_{pp}v_p\rangle}$. The ion temperature T_i becomes equal to T_e in about $\tau_{eq} \simeq \tau_{ee}m_i/(Z^2 m_e)$ owing to the energy transfer ratio m_e/m_i in a collision in comparison to the energy exchange in one electron-electron collision. Hence, the three times for equilibration stay to each other in the ratios

$$\tau_{ee} : \tau_{ii} : \tau_{eq} = 1 : \left(\frac{m_i}{m_e}\right)^{1/2}\frac{1}{Z^3} : \frac{m_i}{m_e}\frac{1}{Z^2}. \tag{1.33}$$

Isotropization of the electron distribution function is increasing by electron-ion collisions, in particular if the ions are highly charged. An additional contribution to increase of ν_{ei} may originate from inelastic collisions of not fully stripped ions.

1.2.3 *Ideal Plasma*

1.2.3.1 Ideality

The regions of a plasma generated by high power laser which are directly accessible by the laser beam, the so called *underdense region*, and the fraction heated by electron thermal conduction, constituting the *overdense region*, are generally hot enough to represent an *ideal plasma*. In analogy to the neutral ideal gas the plasma is considered *ideal* if the potential energy between the particles can be ignored and the internal energy of a plasma element is the sum of the kinetic energies only of its constituents. Customarily this property is characterized by the *ideality parameter* Γ or, equivalently, by the inverse number of particles in the *Debye sphere* $g = 1/N_D$,

$$\Gamma = \frac{\langle E_{\text{pot}} \rangle}{\langle E_{\text{kin}} \rangle} = \frac{e^2/4\pi\varepsilon_0\lambda_D}{(3/2)k_B T_e}; \quad g = \frac{1}{N_D} = \frac{3}{4\pi n_e \lambda_D^3} = \frac{9}{2}\frac{\langle E_{\text{pot}} \rangle}{\langle E_{\text{kin}} \rangle} \simeq 4\Gamma. \tag{1.34}$$

The ideal plasma is characterized by $\Gamma < 0.1$, or $g < 0.4$ corresponding to 2 electrons in the Debye sphere. To be precise it refers to the *classical ideal plasma*. The electrons in a good conductor, like Al, Cu, Ag, at room temperature may also be treated in a first approach as a an ideal plasma. However in this case it is the *ideal Fermi gas* that applies. It is dominated by the *Fermi energy* E_F of the free electrons,

$$E_F = (3\pi^2)^{2/3}\frac{\hbar^2}{2m_e}n_e^{2/3} = 3.65 \times 10^{-15}\left(n_e\,[\text{cm}^{-3}]\right)^{2/3}\ \text{eV}. \tag{1.35}$$

Typical values range between 5 and 10 eV. When compressed, say by a factor of 10^3, E_F increases with the 2/3 power of the electron density, in our example by a factor of 100. The ideal Fermi gas belongs to the *degenerate plasmas*. Only if the temperature $k_B T \gg E_F$ it turns over into a classical fluid. In the degenerate ideal plasma screening is determined by the number density of the quantized free electron states. For $k_B T_e \ll E_F$, the Debye length is given by

$$\lambda_D = (3\pi^2)^{1/3}\left(\frac{\varepsilon_0}{3m_e}\right)^{1/2}\frac{\hbar}{e}\,n_e^{-1/6} = 3.7 \times 10^{-5}\left(n_e\,[\text{cm}^{-3}]\right)^{-1/6}\ \text{cm} \tag{1.36}$$

with $k_B T_e$ replaced by E_F and $T_i = 0$. In the laser produced plasma the electron density is highest in the overdense thermal conduction zone, however rarely exceeding solid state density. In the underdense region n_e is by two orders of magnitude lower, E_F does hardly reach 1.5 eV. As we shall see this tells that from high power lasers in the intensity interval from $10^{10} - 10^{22}$ Wcm^{-2} $k_B T_e \gg E_F$ the laser plasma is classical. Note the ion Fermi energy is by the factor m_e/m_i smaller. Even at room temperature the ions are not degenerate.

1.2.3.2 The Laser Plasma as an Ideal Classical Fluid

The outline above suggests to describe the laser plasma by a classical fluid, with possible quantum corrections for the compressed low temperature regions. The fluid is capable of local space charges, high electrical currents, and electronic thermal conduction. So, one arrives quite naturally at the model of *two interpenetrating fluids*, one for the electrons and one for the ions, both tied together by electric and magnetic fields. The fields are determined by the Maxwell equations. This *two fluid model* of the laser plasma has turned out extremely successful. The next question is how to describe the internal energies and pressures of the fluid components. Let us make the simplest choice of two ideal gases, with ε_e and ε_i for the *inner energy densities* per unit volume and p_e and p_i for the *pressures*,

$$\varepsilon_e = \frac{3}{2}p_e, \quad p_e = n_e k_B T_e, \quad \varepsilon_i = \frac{3}{2}p_i, \quad p_i = n_i k_B T_i, \quad n_e = Zn_i. \tag{1.37}$$

The correctness of the choice can be controlled by the corresponding expressions with the next order corrections included [47],

$$\varepsilon_e = \frac{3}{2}n_e k_B T_e \left(1 - \frac{g}{3\pi}\right), \quad p_e = n_e k_B T_e \left(1 - \frac{g}{6\pi}\right). \tag{1.38}$$

For example, at $n_e = 10^{20}$ cm^{-3} and $T_e = 100$ eV, $T_i = 0$, one obtains $\lambda_D = 7.4$ nm, $N_D = 172$, $g = 0.006$, $E_F = 0.08$ eV. In this case the relative corrections to ε_e and p_e are less than 10^{-3}. In order to see how reliable the expressions (1.37) are against quantum corrections the reduced *deBroglie wavelength* λ_B and the reduced *thermal deBroglie wavelength* λ_{th},

$$\lambda_B = \frac{\hbar}{m_e v} = \frac{0.185}{(E[\text{eV}])^{1/2}}\,[\text{nm}], \quad \lambda_{\text{th}} = \frac{\hbar}{m_e v_{\text{th}}} = \frac{0.34}{(k_B T_e[\text{eV}])^{1/2}}\,[\text{nm}], \tag{1.39}$$

$$v_{\text{th}}[\text{cms}^{-1}] = \left(\frac{k_B T_e}{m_e}\right)^{1/2} = 4.2 \times 10^7 (T_e[\text{eV}])^{1/2} \tag{1.40}$$

have to be compared; they represent the adequate scale. For classical systems the interelectron distance $\text{d} = n_e^{-1/3}$ must be an order of magnitude larger than λ_B in order to resolve the single electron as an individual entity.

1.3 The Dynamics of the Laser Plasma

When a dielectric target of solid density is exposed to an intense laser beam, absorption of the radiation may occur by multiphoton and field ionization, but also by collisions of some adsorbed or loosely bound electrons present in the sample. Subsequently, in the breakdown phase the laser radiation may either be entirely absorbed according to Beer's law over a length L without ever reaching a critical electron density anywhere. This may happen at modest intensities and is further supported by the onset of rarefaction of the heated plasma due to expansion. Or, with increasing probability of the laser pulse rising in intensity, the rapidly growing electron density becomes overcritical and prevents radiation from penetrating further. Even more dramatic, by collisional ionization the electron avalanche may increase faster than it can expand. The consequence is a fast reduction of the absorption length L which, depending on the laser intensity, may shrink until it is reduced to the skin length of radiation of a fraction of the wavelength. The light reaching the critical density is totally reflected there and partially absorbed by collisions on its way back. The critical position is unstable towards resonant coupling to longitudinal electron plasma oscillations. This phenomenon is well known under the name of *resonance absorption*. It leads to increased laser beam-matter interaction.

Metallic targets and semiconductors are overdense since the beginning, radiative interaction is limited to the skin layer at the surface. Continuous heating of the surface layer facilitates rarefaction by violent target expansion and, in concomitance, receding of the critical density front. With long enough pulses the target is heated in depth in a way that resembles much a deflagration wave. Density dependent absorption, plasma pressure increase and plasma expansion compete with each other. Which process prevails depends crucially on temporal laser pulse rise in time, but also on maximum beam intensity. Particularly interesting from the physical point of view is the beam-target interaction at relativistic intensities beyond $10^{17} - 10^{18}\,\mathrm{Wcm}^{-2}$ in the fs time regime. In a fast rising beam of good prepulse-main pulse *contrast* the first target layer is fully ionized instantaneously but expansion sets in typically after 100 fs only. On the basis of standard optics total beam reflection should occur because collisional absorption is negligibly small and linear resonance absorption does not take place owing to $n_e \gg n_c$ everywhere. The experiment, however tells that well above 70% of the incident light may be absorbed [48, 49]. At present the *collisionless absorption* mechanism is basically understood in its main features, however a distinct classification of the single processes and their intensity dependent onset and dominance is not available yet.

1.3.1 Basic Elements of Plasma Dynamics

In this place a few concepts and facts on the dynamics of the laser plasma must be known. As a rule the (kinetic) temperature of the electrons and their pressure are considerably higher than that of the ions. Frequently, the ion temperature can be assumed equal to zero without sacrificing physical insight. Expansion makes the laser plasma very inhomogeneous in electron density and, to a minor extent, in electron temperature. The electrons are quick and have the tendency to escape from the plasma cloud. However they do not separate by more than a local Debye length; *quasineutrality* does not allow more. Here we are faced with the first principle of plasma dynamics. In contrast, neutral gas mixtures demix during expansion.

1.3.1.1 Thermoelectric Field

Beyond the distance of a Debye length λ_D the plasma keeps quasineutral, i.e., $Zn_i \simeq n_e$, because of the strong coupling of the positive and the negative charges. As a consequence the outflow velocities of both components must be the same, $\mathbf{u}_i = \mathbf{u}_e$. It is intuitively clear that a strong back holding electrostatic field $\mathbf{E}$ must build up to inhibit separation. We determine it here because it is an essential requirement of quasineutrality and we will meet it time and again throughout all chapters.

Consider a plasma layer of thickness dx. The net force $f dx$ on it is the pressure difference of the electron pressure acting onto both sides, thus

$$f dx = p_e(x) - p_e(x + dx) = p_e(x) - [p_e(x) + \partial_x p_e dx] = -\partial_x p_e dx \;\Rightarrow\; f = -\frac{\partial p_e}{\partial x}.$$

Watch the direction the pressure acts. f is a force *density*, i.e., a force per unit length and unit area. The other force to keep the balance is the electrostatic field E, times the number of charges: $f' dx = -eEn_e dx$. Owing to the inertia of the ions which is four orders higher than that of the electrons the forces $f,\ f'$ must balance,

$$f + f' = -eEn_e - \frac{\partial p_e}{\partial x} = 0 \;\Rightarrow\; E = -\frac{1}{en_e}\frac{\partial p_e}{\partial x}; \quad \mathbf{E} = -\frac{\nabla p_e}{en_e}. \tag{1.41}$$

The last expression on the RHS is the vectorial version of the *thermoelectric field* E. Quasineutrality is valid for extensions L exceeding the Debye length. In that case n_e can be expressed by Zn_i owing to $n_e \simeq Zn_i$. In the isothermal plasma the thermoelectric field is proportional to the electron density variation, the Debye layer included.

1.3.1.2 Rarefaction Wave

Imagine a homogeneous sphere of radius R heated suddenly up to a constant temperature $T_e = T_i = T$ at which the ideal gas pressure law holds. How long will the sphere remain confined at original density by its own inertia? To determine the *inertial confinement time* τ we need the combination of two relevant physical quantities which, combined in a suitable way, yield the dimension of τ. The confinement time will certainly depend on R. Hence we need another quantity of the dimension of a velocity (and only one in order to get a unique result!). Such an intrinsic quantity is the sound velocity c_s. For motivation of this choice one may argue further that disintegration of the sphere will depend on the pressure and the density, more precisely, on the ratio pressure to ion mass because in the ideal plasma twice the pressure divided by twice the ion mass must give the same result. The reader may also convince himself that there is no other independent quantity governing the dynamics of expansion of the sphere into the vacuum. How expansion works is sketched in Fig. 1.8 by the distributions of density and temperature, each representing a spherical *rarefaction wave* at a fixed time. The rarefaction of density and the temperature cooling start at $t = t_0$ from the surface of the sphere. The rarefaction edge propagates with constant sound speed c_s towards the center. The sound speed is

$$c_s = \left(\alpha \frac{p}{\rho}\right)^{1/2}. \tag{1.42}$$

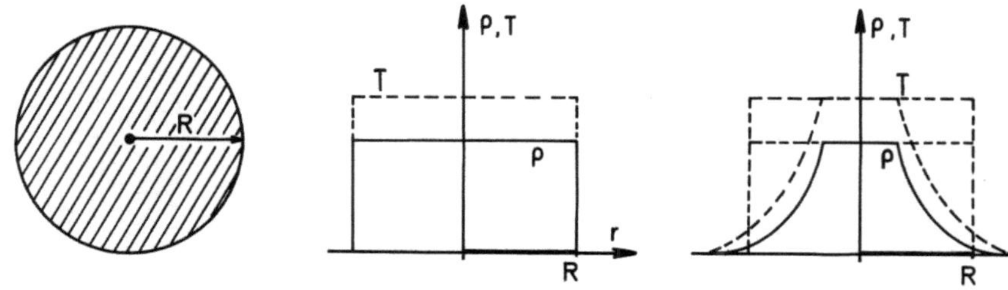

Fig. 1.8 Rarefaction wave. If the pressure at the border of the sphere of radius R is released at $t_0 = 0$ a rarefaction wave of density ρ and temperature T starts from there and moves with sound velocity c_s towards the center. The time τ from (1.43) is a suitable measure for the time during which the high density gas remains confined by its own inertia

α is a factor of order unity; it is determined from electron heat conduction. Between $R = 0$ and the onset of rarefaction matter is at rest and density ρ and temperature T are held at their initial values. From there on towards increasing r values the flow velocity u gradually increases towards its maximum value $u = u_F$ at the fluid-vacuum interface. Density and temperature at this point are zero. As a consequence u_F remains constant owing to missing driving pressure. In the ideal monoatomic gas at any infinitesimal time ϵ after t_0 the front velocity u_F has reached its final value of $3c_s$, as will be derived in Chap. 3. When the rarefaction edge has reached the center of the sphere the density starts decreasing there very rapidly and the density and temperature profiles flatten. Therefore it is reasonable to measure the confinement time as the time which elapses until the density disturbance has reached the center of the sphere,

$$\tau = \frac{R}{c_s}. \tag{1.43}$$

One may ask how τ depends on geometry, for instance on plane or cylindrical shape of the hot plasma sample. With respect to the rarefaction edge there is the argument that as long as the local curvature radius is larger than its thickness the fluid flow is locally onedimensional and the advancement of the edge is determined by the local sound speed and is independent of geometry.

1.3.1.3 Plane Shock Wave

The momentum which can be transmitted by a linear sound wave of intensity I to the medium is given by the *wave pressure* p_w,

$$p_w = \frac{I}{c_s}(1 + R), \tag{1.44}$$

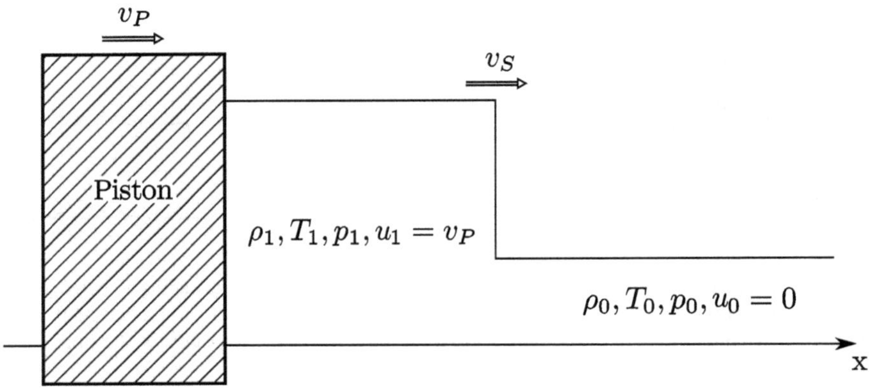

Fig. 1.9 Shock wave driven by a piston moving at speed $v_P = u_1 =$ const propagates with shock velocity v_S into an undisturbed gas at rest of density ρ_0, temperature T_0, and pressure p_0

where R is the reflection coefficient and c_s is the sound speed. A hot plasma expanding from the surface of dense matter, e.g. of a solid, acts like a piston of very high strength. The effect is the generation of a compression wave the intensity of which is far beyond the limit of a linear sound wave. A compression wave into a quasineutral fluid steepens up until a discontinuity or a steep, narrow transition region has formed. For example, when an ideal gas of density ρ_0, temperature T_0, and pressure p_0 at rest is compressed by a piston moving at constant speed v_P a compression wave with spatially constant values ρ_1, T_1, p_1, $u_1 = v_P$ develops which is separated from the undisturbed gas by a transition layer, a few mean free path lengths thick. Such a structure is called a *shock wave*, see Fig. 1.9. It propagates with the *shock velocity* v_S. In the ideal plasma v_S is supersonic. It guarantees the longitudinal stability of its step like structure: Any small rarefaction originating at the shock front can propagate only at sound velocity c_s and is therefore immediately overtaken by the shock front. Only after slowing down to sound velocity a rarefaction wave can start from the discontinuity and lead finally to its decay.

In Fig. 1.9 the piston is assumed to proceed at constant velocity v_P from left to right. The shock front divides matter into the undisturbed region characterized by the variables ρ_0, T_0, p_0, $u_0 = 0$ in front of it and into the compressed region determined by the analoguous quantities ρ_1, T_1, p_1, u_1 behind it. The piston pushes with pressure p_P. According to Newton's third law $p_P = p_1$. From $v_P =$ const follows $v_P = u_1$. The piston pressure p_P equals the change of momentum per unit time at the shock front which is the mass $\rho_0 v_S$ brought from rest to the velocity u_1 plus the undisturbed pressure p_0. Hence $p_P - p_0 = \rho_0 v_S u_1$. For simplicity and convenience we introduce the *shock compression* ratio $\kappa = \rho_1/\rho_0$. Then,

$$p_P - p_0 = \rho_0 v_S u_1 = \rho_0 v_S^2 \left(1 - \frac{1}{\kappa}\right), \quad u_1 = v_S \left(1 - \frac{1}{\kappa}\right), \quad \kappa = \frac{\rho_1}{\rho_0}. \tag{1.45}$$

This expresses just Newton's second law. In addition to mass and momentum conservation a third relation for the energy must be fulfilled. By eliminating the velocities from the set of the three conservation equations, for the internal energies ε per unit mass a relation between shocked region 1 and undisturbed region 0 in Fig. 1.9 results,

$$\epsilon_1 - \epsilon_0 = \frac{1}{2}(p_0 + p_1)\left(\frac{1}{\rho_0} - \frac{1}{\rho_1}\right). \tag{1.46}$$

This is the celebrated *Hugoniot relation*. It relates the involved thermodynamic variables of the undisturbed material in front of the shock wave to those of the shocked matter of index 1. As only state variables are involved it is of general validity. In the *strong shock* the undisturbed pressure and internal energy are set to zero, $p_0 = 0,\ \epsilon_0 = 0$.

The laser-heated down streaming plasma acts as a powerful semi-transparent piston driving a strong shock into the solid target. Plasma pressures in the multi Megabar range are easily generated by *laser ablation*. The shock "resisting" to such pressures is speeded up to several 10^7 cms^{-1}. The deformation of the flat target surface into the shape of a crater by the powerful ns laser pulse is the most evident proof of such high pressure generation, see Fig. 1.10. There is an important regime of ns laser-dense target interaction which is determined by the critical density: The laser beam is absorbed in a restricted zone in front of the critical point and a quasi steady plasma flow builds up with a flat maximum of the temperature there. This implies that the flow fulfils the *Chapman–Jouguet condition* of a deflagration process, i.e., at the maximum of T the plasma crosses the velocity of sound. Somewhere in the dense target the flux is subsonic, whereas far out in the rarefied plasma it is supersonic. Thus, in a not too divergent flow transition through the sonic point can only occur at the maximum of T. Here we can assume that the plasma flow in the interaction zone is one dimensional (1D) and plane. As a rule, such an assumption is correct in an absorption interval of depths d that does not exceed approximately the radius of the laser focal spot. The plasma sound velocity (1.42) is taken isothermal for the electrons and adiabatic for the ions, the average degree of ionization is assumed Z-fold, $n_e = Zn_i$, hence

$$c_s^2 = \frac{p_e + \gamma p_i}{\rho} = \frac{k_B}{m_i}(ZT_e + 5T_i/3); \qquad \gamma = \frac{f+2}{f}. \tag{1.47}$$

Thereby γ is the *adiabatic coefficient* and f the number of degrees of freedom. The *ablation pressure* of the outflowing plasma in the plane flow region is P_a,

$$P_a = p + \rho u^2 = \text{const.} \tag{1.48}$$

This is the simple formula of the thrust on the plasma-driven rocket in the reference system of the rocket (dense target): in the combustion chamber P_a is all pressure p and u is zero; subsequently, along the diluted exhaust the pressure gradually transforms into momentum flow ρu^2 and the local pressure p reduces to zero. In principle P_a

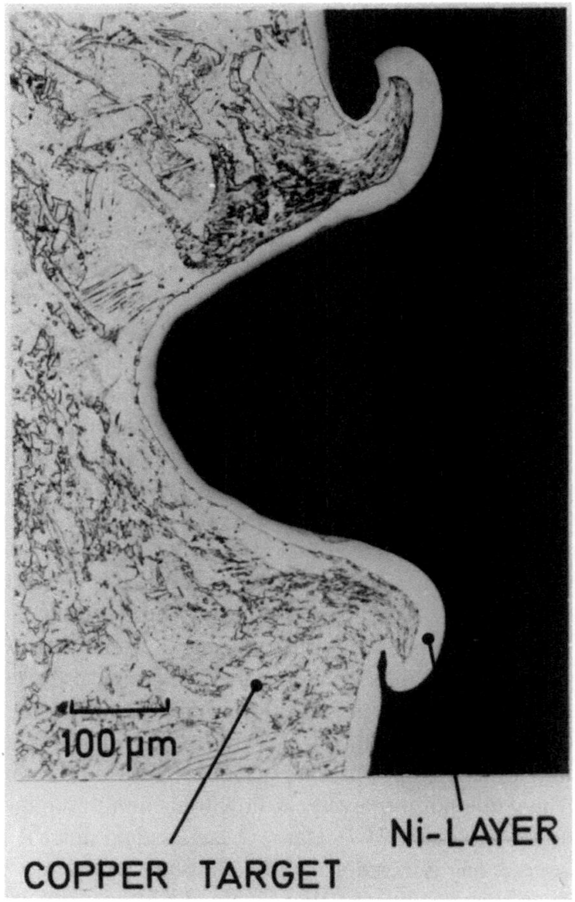

Fig. 1.10 Crater produced in copper by a focused Nd laser beam of $E = 10$ J, $\tau = 10$ ns [50]. Only a small fraction of the crater volume is transformed into hot plasma; most of it is the work of plasma pressure deformation. Ni-Layer is for protection only

can be evaluated at any position in the plane flow region. At the critical point density ρ_c and flow velocity u_c are known best; therefore this position is privileged. Thus,

$$P_a = p_c + \rho_c c_s^2 = n_c k_B(T_e + T_i/Z) + n_c k_B(T_e + 5T_i/3Z) \simeq 2p_c \simeq 2\rho_c c_s^2. \tag{1.49}$$

The critical pressure can be assigned the effective temperature $T_{\text{eff}} = T_e + T_i/Z$ in order to obtain compact expressions. With the absorbed intensity $I_a = (1 - R)I$ the energy conservation reads

$$I_a = \rho_c c_s \left(\varepsilon_e + \varepsilon_i + \frac{1}{2}c_s^2 + \frac{p_c}{\rho_c} \right) = 3n_c k_B T_{\text{eff}} c_s \simeq \frac{3n_c}{m_i^{1/2}}(k_B T_{\text{eff}})^{3/2} = 3Z\rho_c c_s^3. \tag{1.50}$$

The term p_c/ρ_c must be added to the internal and the kinetic energy density in the bracket; $c_s p_c$ represents the work of the plasma ablated per unit time against the ejected plasma in front of it. A rigorous derivation of the general conservation equations will be given in Chap. 3. The ionization energy can be neglected in general. In the case of heavy elements however, this should be checked by an appropriate estimate. Combination of the two conservation (3.215) and (1.50) leads to the steady state ablation pressure expression as a function of the absorbed laser intensity I_a,

$$P_a = 2\rho_c c_s^2, \;\; I_a = 3Z\rho_c c_s^3 \;\Rightarrow\; P_a = 0.96\rho_c^{1/3}\left(\frac{I_a}{Z}\right)^{2/3} \sim \left(\frac{I_a}{\lambda Z}\right)^{2/3}. \quad (1.51)$$

The dependence on ρ_c or the laser wavelength λ is weak. A more sophisticated expression taking into account heat conduction and laser radiation pressure will be presented in Sect. 3.3.4.

The knowledge of ablation pressure is important for calculating shock strength in dense matter and investigating equations of state [51], foil acceleration [52], pellet compression for inertial confinement fusion (ICF) [53–55], and hole boring in fast ignition [56]. Fast ablation of matter by powerful lasers or particle beams, e.g. light or heavy ions, is one of the methods to create pressures in the Megabar range [57]. Above a certain energy supply in solids and liquids a phase transition occurs to the gaseous state; at even higher energies direct transformation of matter into the plasma state is achieved. This latter process occurs at laser intensities beyond the *breakdown threshold* under skipping the intermediate liquid (when starting from solid) and gaseous phases. Currently the highest pressures in the laboratory are obtained by laser ablation. Already at comparatively low laser pulse energies pressures up to 50 Mbars were measured in shock waves launched in solid foils by direct irradiation [52, 58]. By using ultrashort laser pulses their energy content can be significantly reduced, e.g. a 120 fs pulse of only 30 mJ was able to produce a 3 Mbar shock [59]. Higher pressures were reported in impact experiments using laser-accelerated foils [31, 60]. In numerical simulations of laser-accelerated colliding foils the calculated pressure maxima reached up to 2 Gbar [32], and from experiments the authors concluded that shock pressures over 400 Mbar are achievable [33]. Subsequently 750 Mbar planar shocks in gold target foils impacted by X-ray driven gold flyer foils were measured by Cauble et al. [34]. Much higher pressures are achievable in converging shocks, as for example 11 Gbar in a laser pellet compression experiment [35]. Intense shock generation represents an extremely significant tool to study equations of state of hot dense matter in radiation physics and astrophysics. For such purposes the shock front has to be planar. The best quality in this respect up to now was obtained by matter ablation induced by thermal radiation from laser heated *hohlraums* [36, 37]. As the laser beam smoothing technique has gradually improved, shock generation by direct laser drive has gained increasing attention and is now being used with growing success [38–40]. In this way the unavoidable losses of pressure in indirect drive are bypassed. For such reasons and for inertial confinement fusion (ICF) research (pellet compression and spark ignition [41–61]) ablative pressure generation by laser has repeatedly attracted the interest of numerous theoreticians for many years.

The gas dynamic aspect of ablation pressure was studied in great detail in 1D models and with two-dimensional corrections, taking also anomalous heat conduction and fast electron production into account [41]. There are two situations accessible to an analytic treatment: sudden impact heating before hydrodynamic motion sets in, and the opposite case, i.e. after a steady state has developed. If the flow of the ablated material is quasi-stationary the ablation pressure is easily estimated if the *Mach number* $M = u/c_s$, u flow velocity, is known at one fixed density value. For this reason in a number of papers the problem of the critical Mach number, i.e., the Mach number at special points, was investigated in plane and spherical flows [57, 62–68]. Several arguments were used, in plane and spherical geometry, to show that the Jouget point, that is $M = 1$, is located at the critical point or in the overdense region [65]. The analysis of spherical flow is rendered more difficult by the appearance of a characteristic radius as a new parameter. Gitomer et al. [64] solved the problem for high laser flux densities by the guidance of numerical calculations and obtained simple expressions for the ablation pressure. They also showed that in spherical geometry the mass flow at the critical point can be either subsonic or supersonic depending on the limitation of heat flux $\mathbf{q}_e$.

No universally accepted ablation pressure model exists so far. There are contradictions in details between the various models themselves and between the models and the experiments. In some experiments agreement was found more or less with a whole class of simple theoretical pictures but not with those which are believed to be particularly accurate [69]. Quite in general it can be stated that the 2/3 power dependence on laser absorbed intensity (1.51) has never really been confirmed by experiments. For not well understood reasons exponents from 0.56 to 0.62 appear more likely. One of the most obvious deviations might come from lateral electronic heat flow. To the reader wishing to go more into detail at moderate laser intensities [41, 52, 67, 70] are particularly recommended. There exists a considerable amount of published material on ablation pressure, mainly from the last century and concentrated on low Z materials. Nevertheless the degree of apparent contradictions and confusion seems to be very high.

A remarkable recent experimental study on high-Z droplets accompanied by analytical and thorough numerical 2D simulations under inclusion of radiation transport has been undertaken by Kurilovich et al. [71]. The authors find scaling laws on energy on the droplet $E_d^{0.583}$ and the radiation corrected energy scaling $E_{dr}^{0.724}$. None of the published previous models seem to be able to explain their scaling laws. The deviations are attributed essentially to the unsteady evolution of ablation and to the radiative losses which significantly modify the power of the pressure scaling. Deviations in ablation pressure due to radiation losses and unsteady hydrodynamic behaviour have been measured in aluminum in the intensity range $5 \times 10^{12} - 5 \times 10^{13}$ Wcm^{-2}[72]. Measurements of pressure in diamond with the 3rd harmonic Nd up to 10 Mbar with intensities up to $I = 7 \times 10^{13}$ Wcm^{-2}have led to the ablation pressure scaling $I^{0.71}$ [73].

1.3.1.4 Ponderomotive Profile Steepening

The force p_L per unit area a laser pulse of intensity I exerts onto a surface at normal incidence from vacuum is

$$p_L = (1 + R)\frac{I}{c}. \tag{1.52}$$

R is the reflection coefficient, p_L is the so called *radiation pressure*. In the plasma I is a continuous intensity distribution $I(\mathbf{x}, t)$ given by the modulus of (1.2). The *ponderomotive force* $\mathbf{f}_p$ the intensity I exerts onto the single electron is the negative gradient of the cycle averaged electron oscillation energy,

$$\mathbf{f}_p = -\nabla W. \tag{1.53}$$

The relationship tells that W is the *ponderomotive potential*. It follows from the fact that a free electron can neither absorb nor emit a photon; its oscillation energy difference, when shifted from a position of high intensity to a position of low intensity, can only go into kinetic energy of the electron. The force on the unit plasma volume results as

$$\boldsymbol{\pi} = n_e \mathbf{f}_p = -\varepsilon_0 \frac{\omega_p^2}{4\omega^2} \nabla |\mathbf{E}|^2. \tag{1.54}$$

For a detailed derivation see Chap. 2. The ablation pressure P_a must be generalized to include also the radiation pressure p_π at an arbitrary position $(\mathbf{x})$. Thereby we can limit ourselves to 1D plane geometry. Like the plasma pressure $p(x)$ the radiation pressure $p_\pi(x)$ is the force density $\boldsymbol{\pi}$ integrated from a position where it is zero in the overdense plasma downstream to an arbitrary position x in the 1D interval, $p_\pi(x) = \int \pi dx$. It is solved with the aid of the stationary wave equation,

$$\partial_{xx}\hat{E} + k_0^2\eta^2\hat{E} = \partial_{xx}\hat{E} + \frac{\omega^2}{c^2}\left(1 - \frac{n_e}{n_c}\right)\hat{E} = 0.$$

Multiplication with $\hat{E}' = \partial_x \hat{E}$ yields the desired result

$$p_\pi(x) = -\int \frac{Zn}{4n_c}\partial_x(\varepsilon_0\hat{E}^2)dx = -\frac{\varepsilon_0}{4}\hat{E}^2 - \frac{\varepsilon_0 c^2}{4\omega^2}\hat{E}'^2, \tag{1.55}$$

$$\partial_x P_a = \partial_x(\rho u^2 + p - p_\pi) = 0 \quad \Leftrightarrow \quad P_a = \rho u^2 + p - p_\pi = \text{const.} \tag{1.56}$$

For further consideration a look at Fig. 1.11 is in order. Under the action of plasma heating and expansion alone the density is flat as indicated by the dashed line and the plasma scale length at the critical point $L_p = n_c/|\partial_x n_c|$ is of the order of the focal radius $d/2$. Instead, $\pi(x)$ oscillates in space with the partially standing wave and decays at the critical point with the much smaller scale length $\lambda = \lambda_0\eta \sim \lambda_0,\ \lambda_0 \ll L_p$. Therefore steepening by the gradient of p can be ignored. Again, the temperature distribution can be assumed sufficiently flat with its maximum in between x_c and x_m of the maximum $\hat{E}^2(x_\mathrm{m})$. Hence, The Jouguet point with $M = 1$ lies at x_m. Under steady state conditions $\rho u = \mathrm{const}$ and

$$\frac{1}{L_c} = \left|\frac{1}{\rho}\frac{\partial \rho}{\partial x}\right|_c = \left|\frac{1}{u}\frac{\partial u}{\partial x}\right|_c \Rightarrow \rho u \frac{\partial u}{\partial x} = \frac{\rho u^2}{L_c} = \left|\frac{\partial p_\pi}{\partial x}\right|_c = \frac{p_\pi \eta}{\lambda_0}\Big|_c. \tag{1.57}$$

The term containing $\hat{E}'^2$ in (1.55) can be disregarded. With the laser intensity I holds $\varepsilon_0 \hat{E}^2 \simeq 2I(1+R)/c$ and with (1.55) in (1.57) follows

$$L_c = \frac{\rho_c c_s^2}{p_\pi \eta}\lambda_0 = 2\frac{n_c}{Z}k_B T_\mathrm{eff}\frac{c\lambda_0}{(1+R)I\eta}. \tag{1.58}$$

More correctly, the overall reflection coefficient R should be replaced by its value $R^{1/2}$ at x_c.

As will be discussed in Chap. 3 at the standard intensities of ns laser pulses the radiation pressure p_π contributes to P_a at most by 10% and is therefore often neglected in the plasma dynamics. This is incorrect because its presence leads to considerable density profile steepening, see the bold curve in Fig. 1.11, with consequences for collisional absorption and the onset of plasma instabilities. Even more surprising at first glance is the lowering of P_a due to accounting for p_π. Depression of P_a has overcome at incident laser intensity $I > 2.5 p_c c$. Beyond this (approximate) limit the radiation pressure p_π dominates the plasma pressure at the critical point p_c in (1.56). Lowering of P_a has a natural explanation. Steady state flow requires $\rho_c M_c = \rho(x_\mathrm{m}),\ M_c < 1$. In the absence of radiation pressure $M_c = 1$; hence, $(p + \rho u^2)_c > (p - p_\pi + \rho u^2)_c$. An extreme example of profile steepening is presented in Fig. 1.17 from a particle in cell simulation of irradiance $I\lambda_0^2 = 5 \times 10^{21}\,\mathrm{Wcm^{-2}\mu m^2}$. At low laser intensities it may happen that absorption is nearly entirely collisional and localized in the underdense plasma, i.e. $R = 0$. In this situation $\hat{E}^2$ is flat and no ponderomotive profile steepening develops. The ablation pressure scales differently, to be shown in a further place on plasma dynamics. The first experimental proofs on profile steepening were reported by [74–76]. For a detailed study of profile structures and steepening see [77].

1.3.1.5 Electron Heat Wave

A major impact on plasma energy balance and dynamics is provided by electronic *heat conduction*. It is defined as energy transport without mass flow. In the case an electron temperature can be indicated locally, $T_e = T_e(\mathbf{x})$, the thermal energy flow density $\mathbf{q}_e$ can be expressed by means of the local temperature gradient as follows

$$\mathbf{q}_e = -\kappa_e \nabla T_e. \tag{1.59}$$

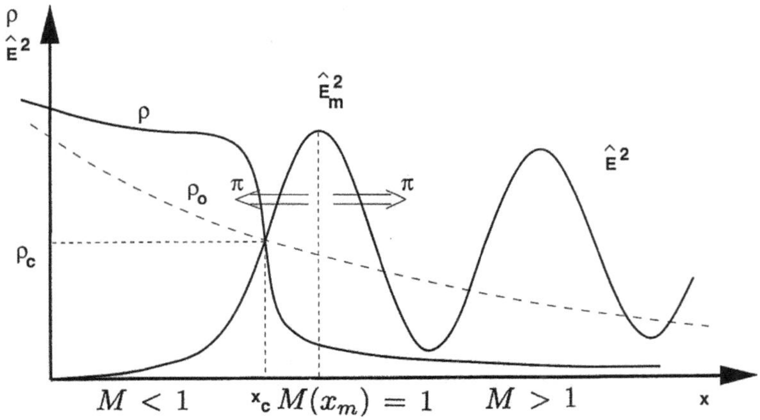

Fig. 1.11 Ponderomotive density profile steepening in the critical region (bold graph); $\hat{E}^2$ partially standing laser wave; x_c critical point; M Mach number; ρ_0 density distribution under condition $\pi = 0$ (dashed). Transition from the subsonic to supersonic flow occurs, as a rule, between x_c and the position x_m of the first maximum of $\hat{E}^2$, in the figure at $x = x_\mathrm{m}$

This is the *Fourier ansatz* of heat conduction, κ_e is the *heat conduction coefficient*. Its justification in terms of the mean free path λ_e is as follows. The energy flux transported through a plane at position x orthogonal to grad T_e per unit time is $q_+ = -(n_e/2)\langle[(3/2)k_B(T_e + \lambda_e\partial_x T_e)]v_e\rangle$ down the density gradient and $q_- = +(n_e/2)\langle[(3/2)k_B(T_e - \lambda_e\partial_x T_e)]v_e\rangle$ the temperature gradient up. The net flow is

$$q_e = q_+ - q_- = -\frac{n_e}{2}\langle\lambda_e v_e\rangle \times 3k_B\partial_x T_e = -\frac{n_e}{2}\langle\tau_e v_e^2\rangle \times 3k_B\partial_x T_e; \qquad \tau_e = \frac{\lambda_e}{v_e},$$

τ_e is the individual time between two collisions, the inverse $\nu_e = 1/\tau_e$ is the individual collision frequency. Mean free path λ_e and velocity v_e are to be taken into direction of the gradient, hence $\langle v_e^2\rangle = k_B T_e/2m_e$. After averaging $\mathbf{q}_e$ becomes

$$\mathbf{q}_e = -\frac{p_e}{\nu_{ee} + \nu_{ei}}\frac{k_B}{m_e}\nabla T_e = -\kappa_e\nabla T_e. \tag{1.60}$$

It becomes clear from this formula that electron heat conduction is dominated by the fast electrons in the thermal distribution since $\kappa_e \sim T_e^{5/2}$. On the other hand, fast electrons exhibit a much longer mean free path than slower ones; consequently, averaging of single quantities as v_th, λ_e, etc., instead of their products underestimates κ_e. Its true value for $Z = 1$ is larger by a factor of 3–4 [78],

$$\mathbf{q}_e = -\kappa_e\nabla T_e; \quad \kappa_e = \eta_e\frac{p_e}{\nu_{ei}}\frac{k_B}{m_e} = \kappa_0 T_e^{5/2}, \tag{1.61}$$

$$\kappa_0 = 1.8 \times 10^{-5}\,[\mathrm{g\,cm\,s^{-3}K^{-7/2}}], \qquad \eta_e = 3.16.$$

As Z increases η_e increases to $\eta_e = 13$. The coefficient κ_0 depends on the charge state as $\kappa_0 \sim Z^{-1}$ but is almost independent of density [dependence is only through the Coulomb logarithm in (1.61)]. It is essential for (1.61) to be valid that $\lambda_e \ll L = T_e/|\nabla T_e|$ holds everywhere. Another limit on the heat flux $\mathbf{q}_e$ is imposed by the condition that it can never exceed the energy flux into the half space $\{\mathbf{u}_+\}$, i.e. $\mathbf{q}_{\max} = n_e m_e \int \mathbf{u}^2 \mathbf{u} f_e(\mathbf{x}, \mathbf{u}, t)\, d\mathbf{u}_+/2$, f_e electron distribution function. For a Maxwellian distribution f_0 this yields

$$q_{\max} = \left(\frac{2}{\pi}\right)^{1/2} k_B T_e v_{\mathrm{th}} \simeq k_B T_e v_{\mathrm{th}}; \qquad \left(\frac{2}{\pi}\right)^{1/2} = 0.8. \tag{1.62}$$

The early plasma generation and heating by high contrast laser beams is dominated by electron heat diffusion before rarefaction sets in. It reflects the fact that at the beginning the heat front penetrates with the average thermal electron velocity whereas plasma motion sets in with the sound speed of the ions. Their ratio is characterized by $(m_i/m_e)^{1/2} \geqslant 43$. The time for heating the ions by the electrons is even much longer, see (1.33).

In a first step we solve the idealized problem of heating by the instantaneous release of the energy Q_0 per unit area at $t = 0$ on a plane of a solid, liquid, or gas from which it diffuses into the half space interior. With the constant specific heat c_v per unit volume the energy balance reduces to a nonlinear heat diffusion equation as follows

$$c_v \frac{\partial T_e}{\partial t} = \frac{\partial}{\partial x} q_e = \kappa_0 \frac{\partial}{\partial x}\left(T_e^{5/2} \frac{\partial T_e}{\partial x}\right) \Rightarrow \frac{\partial T}{\partial t} = a \frac{\partial}{\partial x}\left(T^{\alpha} \frac{\partial T}{\partial x}\right); \quad a = \frac{\kappa_0}{c_v}. \tag{1.63}$$

In the diffusion equation the exponent 5/2 is replaced by α because with suitable values for it the equation is also capable of describing radiation fronts [79, 80]. For that reason the index e is dropped in T. The equation can also be written as

$$\frac{\partial T}{\partial t} = \frac{a}{\alpha + 1} \frac{\partial^2}{\partial x^2} T^{\alpha+1}. \tag{1.64}$$

It is easy to show that for $\alpha > 0$ the heat front coordinate $x_T(t)$ is finite. For $\alpha = 0$ (constant thermal conductivity) $x_T = \infty$ for all $t > 0$.

Three Examples of Heat Waves*

Plane heat wave under sudden supply of energy. There is the search for a solution $T(x, t)$ with the initial temperature $T(x, t = 0) = 0$ in the half space $x \geq 0$. For fixed α, T can only depend on $q_0 = Q_0/c_v$, a, x, and t. Their dimensions are

$$[T] = \mathrm{K\,(Kelvin)}, \quad [x] = \mathrm{m}, \quad [t] = \mathrm{s}, \quad [q_0] = \mathrm{K\,m}, \quad [a] = \mathrm{K^{-\alpha} m^2 s^{-1}}. \tag{1.65}$$

From the property that the solution must be invariant with respect to scale transformations it follows that the variables involved must be expressible as *dimensionless quantities*. The assertion is the content of the *Buckingham theorem* on similarity, to be treated in Chap. 3. A dimensionless variable is

$$\pi = \frac{x}{(aq_0^\alpha t)^{1/(\alpha+2)}}$$

and it is the only independent one. The heat front $x_T(t)$ and the maximum temperature T_0 (at $x = 0$) are uniquely determined from (1.65),

$$x_T(t) = A(aq_0^\alpha t)^{1/(\alpha+2)}, \quad T_0(t) = B\left(\frac{q_0^2}{at}\right)^{1/(\alpha+2)}.$$

With $BT(x,t) = T_0 f(\pi)$ (1.63) reduces to the ordinary differential equation

$$(\alpha+2)\frac{d}{d\pi}\left(f^\alpha\frac{df}{d\pi}\right) + \pi\frac{df}{d\pi} + f = 0. \tag{1.66}$$

The solution is

$$T(x,t) = T_0\left(1 - \frac{x^2}{x_T^2}\right)^{\frac{1}{\alpha}}, \quad x_T = A(aq_0^\alpha t)^{1/(\alpha+2)}, \quad \int_0^{x_T} T dx = q_0,$$

$$A = \left[2\frac{\alpha+2}{\alpha}\left(\frac{2\Gamma(\frac{3}{2}+\frac{1}{\alpha})}{\Gamma(\frac{1}{2})\Gamma(\frac{1}{\alpha}+1)}\right)^\alpha\right]^{\frac{1}{\alpha+2}}; \quad \alpha = 5/2 \Rightarrow A = 1.480,$$

$$B = A^{2/\alpha}\left(\frac{\alpha}{2(\alpha+2)}\right)^{1/\alpha},$$

$$T_0(t) = \frac{q_0}{x_T(t)}\frac{2\Gamma(\frac{3}{2}+\frac{1}{\alpha})}{\Gamma(\frac{1}{2})\Gamma(\frac{1}{\alpha}+1)}, \quad \langle T\rangle = \frac{q_0}{x_T(t)}. \tag{1.67}$$

For electron heat conduction with $\alpha = 5/2$ the quantities read

$$T = T_0\left(1 - \frac{x^2}{x_T^2}\right)^{2/5}, \; x_T = A\, q_0^{5/9}\left(\frac{\kappa_0}{c_v Z}t\right)^{2/9}, \; \frac{T_0}{\langle T\rangle} = 1.223. \tag{1.68}$$

As α increases the temperature assumes a rectangle-like distribution. At the beginning the heat diffusion speed tends to infinity and soon after it slows down drastically (Fig. 1.12).

Steady state heat front. Another relevant situation is given when the heat applied to the surface is delivered continuously with the power law

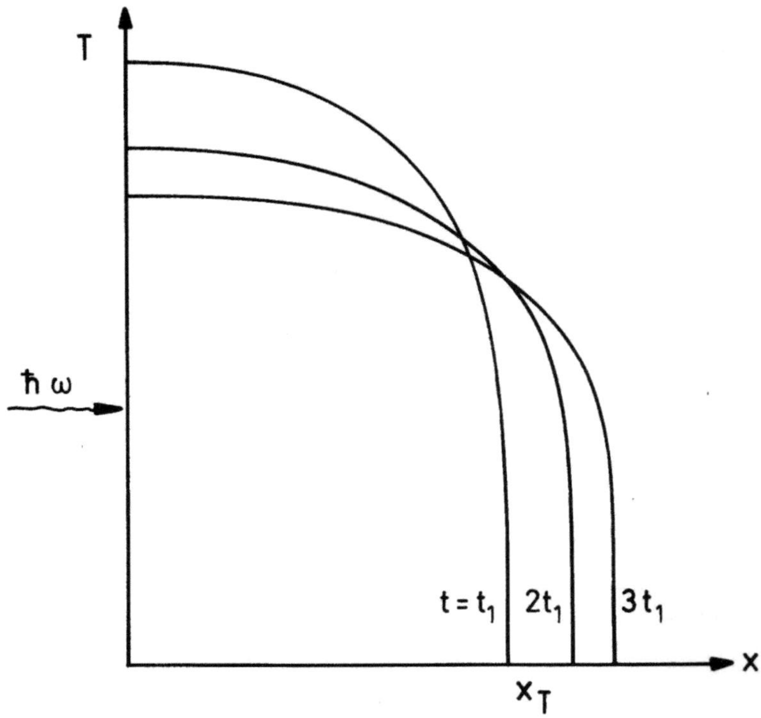

Fig. 1.12 Nonlinear electron heat wave at three equidistant time intervals $t_1, 2t_1, 3t_1$ in a medium at rest. The power of T for electron thermal conductivity is $\alpha = 5/2$. At $t = 0$ the diffusion speed is infinite; $x_T \sim t^{2/9}$

$$\Phi = \frac{I}{c_v} = Bt^{\gamma}.$$

If one looks for solutions of the form $T(x, t) = A(vt^{\beta} - x)^{\delta}$ one finds $\beta = 1,\ \gamma = \delta = 1/\alpha$. At the boundary $x = 0$ the flux and the temperature must obey $\Phi = -aT^{\alpha}\partial T/\partial x$. Inserting this in (1.63) yields

$$T(x, t) = \left[\frac{\alpha v}{a}(vt - x)\right]^{1/\alpha}, \quad v = \dot{x}_T = B^{\alpha/(\alpha+2)}\left(\frac{a}{\alpha}\right)^{1/(\alpha+2)}.$$

In the frame moving with x_T this heat wave is stationary.

Sudden release of heat from a point source. Another exact solution exists for a quantity $Q_0 = c_v q_0$ of heat suddenly released from a point into a half space filled with an isotropic homogeneous medium at rest. Since the procedure is the same as in the plane case only the final result is reported [79],

$$T(r,t) = T_0\left(1-\frac{r^2}{r_T^2}\right)^{1/\alpha}, \quad r_T = A'(aq_0^\alpha t)^{1/(3\alpha+2)},$$

$$A' = \left(2\frac{3\alpha+2}{\alpha\pi^\alpha}\right)^{1/(3\alpha+2)}\left\{\frac{\Gamma\left(\frac{5}{2}+\frac{1}{\alpha}\right)}{\Gamma\left(1+\frac{1}{\alpha}\right)\Gamma\left(\frac{3}{2}\right)}\right\}^{\alpha/(3\alpha+2)},$$

$$T_0 = \frac{4q_0}{\pi^{3/2}r_T^3}\frac{3\alpha+2}{2\alpha}\frac{\Gamma\left(\frac{3\alpha+2}{2\alpha}\right)}{\Gamma\left(\frac{1+\alpha}{\alpha}\right)}.$$

In particular with $\alpha = 5/2$,

$$T = T_0\left(1-\frac{r^2}{r_T^2}\right)^{2/5}, \quad r_T = q_0^{5/19}A'\left(\frac{\kappa_0}{c_v Z}t\right)^{2/19},$$

$$T_0 = 0.9893 q_0^{4/19}\left(\frac{c_v Z}{\kappa_0 t}\right)^{6/19}, \quad A' = 1.144.$$

The propagation velocity of the heat front $\dot{r}_T$ decreases extremely rapidly with time. The two solutions with sudden release of a heat quantity depend on one variable only instead on space and time separately; they belong to the class of *selfsimilar solutions*. In contrast, the steady state solution of constant heat front progression does not belong to the class because x and t are decoupled from each other.

1.3.1.6 Caruso's Model of Short Pulse-Overdense Matter Interaction

With respect to the achievement of dense, high temperature plasmas the most economic heating is realized when the energy input is so fast that the plasma is heated before any appreciable motion sets in and hence all the absorbed energy is converted into thermal energy. Let us assume that a laser pulse of sub-ns duration τ impinges onto the target. Let the absorbed energy per unit area, which is deposited at the surface of the target, be $q_0 = \int_\tau I_a dt$, I_a absorbed intensity. Then in plane geometry, with $T_e = T_i = T$ the interaction dynamics reduces to the heat diffusion equation (1.63) with $\alpha = 5/2$. The specific heat per unit volume is $c_v = 3n_0(Z+1)k_B/2$. The temperature $T(x,t)$ and its front $x_T(t)$ are given by (1.68). The value of $T_0/\langle T\rangle = 1.223$ and Fig. 1.12 show that the temperature profile is roughly approximated by a rectangle. In order to apply these results to heating with very short light pulses and to calculate the amount of heated matter we keep in mind that the lifetime τ_p of the solid state density plasma is approximately determined by the rarefaction wave which travels with the isothermal sound velocity $c_s(\langle T(t)\rangle) = \sqrt{(p_e+p_i)/\rho_0}$ into the heated material. According to an idea of A. Caruso the heat wave stops when the edge of the rarefaction wave reaches the front of the heat wave, i.e., when [81]

$$\int_0^{\tau_p} c_s dt = x_T(\tau_p).$$

This relation yields the following expressions for the lifetime τ_p, thickness of the heated slab $x_T(\tau_p)$, and maximum temperature $T(\tau_p)$ at the instant τ_p:

$$\tau_p = \frac{2}{3}\left(\frac{8}{9}\right)^{3/2} A^{9/4} k_B^{-7/4} \kappa_0^{1/2} \frac{m_i^{3/4}}{n_0 Z^{1/2}(Z+1)^{7/4}} q_0^{1/2}, \tag{1.69}$$

$$x_T(\tau_p) = \frac{2}{3}\left(\frac{8}{9}\right)^{1/3} A^{3/2} k_B^{-7/6} \kappa_0^{1/3} \frac{m_i^{1/6}}{n_0 Z^{1/3}(Z+1)^{7/6}} q_0^{2/3}, \tag{1.70}$$

$$T(\tau_p) = \frac{q_0}{c_v x_T(\tau_p)} = \left(\frac{8}{9}\right)^{1/3} A^{-3/2} k_B^{1/6} \kappa_0^{-1/3} \frac{Z^{1/3}(Z+1)^{1/6}}{m_i^{1/6}} q_0^{1/3}. \tag{1.71}$$

The constant A is given by (1.67). Note that $T(\tau_p)$ does almost not depend on the initial particle density n_0 and does depend only weakly on the absorbed energy per unit area q_0. $T(\tau_p)$ represents an upper limit for the temperature since at the instant τ_p a considerable fraction of the energy is already converted into kinetic energy: if an isothermal rarefaction wave is assumed (worst case) the ratio of kinetic energy to thermal energy at the instant τ_p is 2/3.

The above relations are valid for $T_i = T_e$. For the other limiting case of cold ions, $T_i = 0$, one has merely to substitute $Z + 1$ by Z in (1.69–1.71). In reality the ions are expected to be heated to a certain degree since the electron-ion relaxation time $\tau_{ei} = m_i/2m_e\nu_{ei}$ can be shown to be of the same order of magnitude as τ_p. In fact, with the relation $c_s\tau_p \approx x_F \approx (\kappa_e\tau_p/c_v)^{1/2}$ which follows from (1.69, 1.70) and $\kappa_e/c_v = \frac{1}{3}\lambda_e v_{\mathrm{th,e}}$ one obtains

$$\tau_p \approx \frac{1}{c_s^2}\frac{\kappa_e}{c_v} = \frac{1}{c_s^2}\frac{\lambda_e v_{\mathrm{th,e}}}{3} \approx \frac{m_i}{kT_e}\frac{v_{\mathrm{th,e}}^2}{3\nu_{ei}} = \frac{m_i}{2m_e}\frac{1}{\nu_{ei}} = \tau_{ei}.$$

Therefore, when effects are calculated for which the ion temperature enters very sensitively, e.g. neutron production, the ion temperature has to be evaluated accurately [82, 83]. The expansion of the plasma into vacuum for times $t \leq \tau_p$ and over distances Δx less than the focal radius are approximately described by the plane rarefaction wave.

1.3.1.7 Heat Front Versus Rarefaction Front Velocity

The foregoing model assumes the laser pulse length $\tau \ll \tau_p$. With long pulses it depends on the rise time of the laser intensity whether or not a thermal or a shock wave develops earlier. If the intensity rises too slowly heat conduction only influences the expanding plasma and not the undisturbed or shocked solid. The problem was solved by Anisimov [84]. Let the absorbed intensity $I_a(t)$ be such that $\langle T \rangle$ varies as

t^λ. Then

$$\dot{x}_T \simeq \frac{d}{dt}\left(aT^{5/2}t\right)^{1/2} \sim t^{(5\lambda-2)/4},$$

whereas the rarefaction wave travels at the speed of sound $c_s \sim \langle T\rangle^{1/2} \sim t^{\lambda/2}$. Comparing the two speeds leads to distinguishing the following situations:
(a) $\lambda > \frac{2}{3}$: For a certain time τ_p after the start of irradiation the sound velocity is greater than the penetration velocity of the thermal wave. Thus, the process of plasma production at the beginning will be governed by the dynamics of plasma expansion.
(b) $\lambda = \frac{2}{3}$: Both velocities $\dot{x}_T$ and c_s obey the same power law. The dominating energy transport mechanism, either mass flow or heat conduction, is determined by the coefficients A and B of the relations $v_T = At^{1/3}$ and $c_s = Bt^{1/3}$. If the heat conduction is the dominating process ($\dot{x}_T > c_s$), one can use the energy equation for a stationary medium without motion to calculate the time variation of the absorbed energy flux density $\Phi(t)$. From $I(t) \propto d[x_T(t)\langle T(t)\rangle]/dt$, it follows that the dependence is linear, $I_a(t) \sim t$. This is consistent with $\dot{x}_T > c_s$ since from (1.67) follows $B = 0.9A$.
(c) $\lambda < \frac{2}{3}$: This is just the inverse case to (a). Up to a time τ_p, heat conduction is the dominating energy transport mechanism. During this time the absorbed radiation intensity varies as t^α with $\alpha < 1$. The stationary heat wave (1.67) with $I_a \sim t^{2/9}$ belongs to this class.

The heat wave model was shown to be consistent with experiments performed with 10 ps Nd laser pulses of intensities up to 3×10^{15} Wcm^{-2} by Salzmann [85]. The measured electron temperatures in deuterium and carbon targets at 1 J pulse energy were 500 and 200 eV, respectively.

The model presented in this section is based on Spitzer–Braginskii's thermal conductivity following a power law $\kappa_e \sim T_e^\alpha$ [see (1.61)]. Subsequent to experiments showing the failure of such a simple local law [86] an avalanche-like enterprise started to investigate experimentally as well as theoretically phenomena of *heat flux inhibition*, *nonlocal*, and *anomalous transport*. It was discovered in experiments that heat diffusion remained at least by an order of magnitude (in some experiments by a factor of 10^{-2}, even 10^{-3}) below the *free streaming limit* (1.62).

Despite the enormous effort concentrated on electron (and radiative) heat transport the phenomenon is an open end problem. This situation raises the question on the correctness and usefulness of an analysis on the relation of heat diffusion to the dynamics induced by it according to [84]. The main motivation for reporting it is a methodological one. If power laws extracted from the experiment and supported by simple theoretical arguments are found in a certain intensity and time interval *dimensional analysis* yields powerful classification schemes, see for instance also [80]. Examples of this kind will be reported subsequently on various occasions. The reader may have asked himself also in connection with the instantaneous supply of an amount of heat to an overdense plasma how this can be accomplished. The answer to this very legitimate question will be given in connection with *ultrashort laser pulse*-target interaction.

1.3.2 Fully Developed Plasma Dynamics

In this section the dynamics of the hot laser plasma is treated. At moderate laser intensities, $I \leqslant 10^{14}$ Wcm^{-2}, collisional ionization prevails over multiphoton ionization as soon as a modest fraction of electrons has been set free. In the subsequent heating process a plasma from low Z material can be considered as fully ionized to a good approximation and the ionization energy can be disregarded in a first approach. The dynamics of the plasma is then well described by the two-fluid model, one for the electrons and the other one for the ions, with the absorption coefficient taken from the Drude model and the collision frequency from (1.30). The latter relation is only true if, as we assume here, collisional absorption dominates all other heating mechanisms. In simplifying one can go one step further by postulating $T_i = T_e$, or alternatively $T_i = 0$, to avoid the solution of the ion energy equation. But even then it is not possible to solve the remaining three balance equations analytically in plane or spherical geometry. The dynamics depends crucially on the time-dependence of the laser pulse. If the laser intensity at the surface of a solid target rises nearly instantaneously, say during a few ps, three different scales in time as well as in space have to be distinguished:

First, due to violent ionization an overdense plasma of solid state density is formed, the thickness of which is of the order of the skin depth $\delta_s = c/\omega_p \ll \lambda_{\mathrm{Nd}}$.

Secondly, an electron heat wave, see foregoing section, propagates into the undisturbed solid with a front coordinate $x_T(t) = \kappa_0^{2/9}(I^5/c_v^7)^{1/9}t^{7/9}$. This formula is easily recovered from dimensional considerations or from (1.68). For $t \to 0$, $\dot{x}_T$ tends to infinity which is characteristic of any continuum model of diffusion. From physical considerations it is clear that the real maximum speed is $\dot{x}_T \leqslant v_{\mathrm{th,e}} \simeq c_{se}$.

In the third stage the plasma rarefies towards the vacuum with ion sound speed $c_s \ll c_{\mathrm{se}}$ and, as the front of the heat wave slows down according to $\dot{x}_T \sim t^{-2/9}$ a shock wave propagating into the solid builds up as soon as $\dot{x}_T < c_s$ is fulfilled. From now on a significant amount of the absorbed laser energy originally stored in the electron fluid is transformed into kinetic energy of the plasma, i.e., of the ion fluid. For $t \to \infty$ all former plasma energy converts into kinetic ion energy.

1.3.2.1 Heating with Long Pulses

In order to get a more detailed picture of the laser target interaction, the one-fluid model has been solved numerically in plane and spherical geometry under the following conditions:

- The target material is solid hydrogen of particle density $n_0 = 4.5 \times 10^{22}$ and initial temperatures $T_e = T_i = 0$;
- constant laser intensity $I = I_0$ at Nd wavelength; the absorption is purely collisional;
- total reflection at the critical point;
- heat conduction according to Spitzer–Braginskii (1.61);
- absence of ponderomotive force, $\pi = 0$.

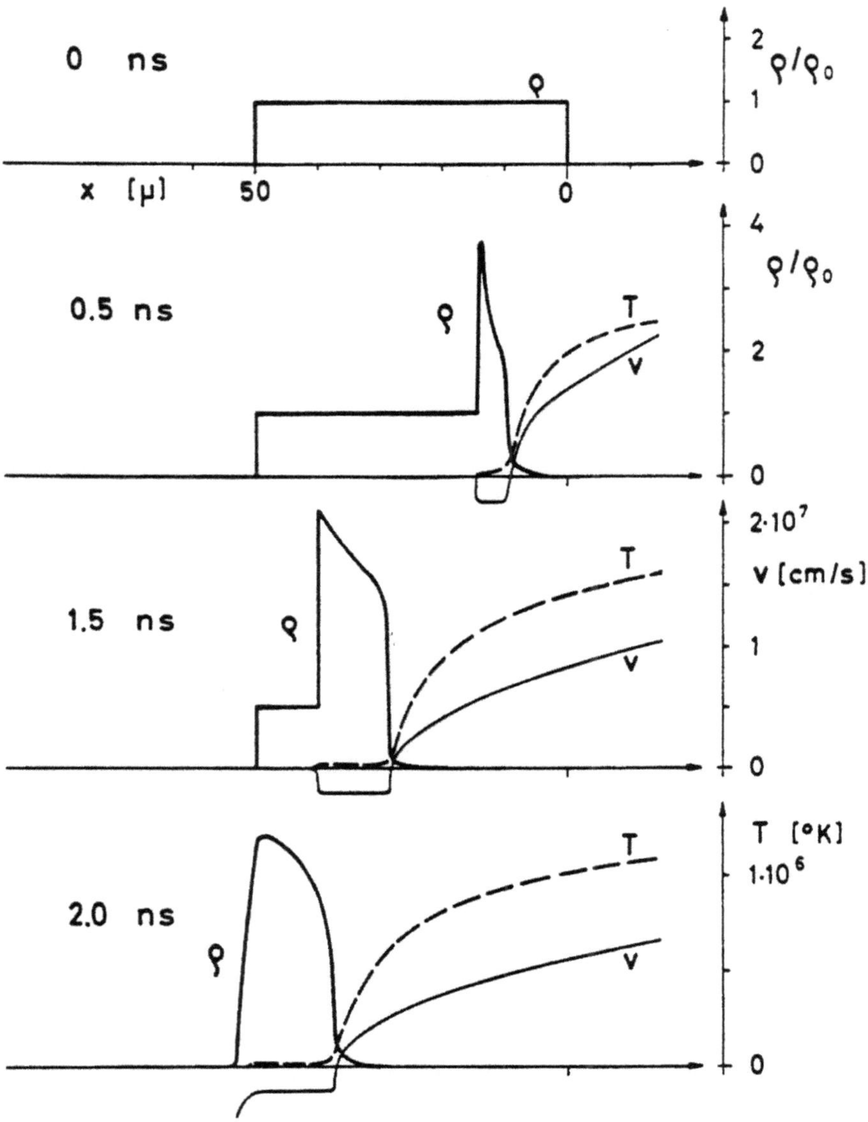

Fig. 1.13 Plane shock wave in a solid hydrogen foil (50μm thickness) at various times for a laser intensity $I_0 = 10^{12}$ Wcm^{-2}. T temperature, v flow velocity. The shock velocity is 2.7×10^6 cms^{-1}. Undisturbed foil (top) at $t = 0$ ns. Laser impinges from the right. ρ_0 initial density

Compression Wave

The dynamics of the high-density region in plane geometry is shown in Fig. 1.13 for a 50 μm thick hydrogen foil irradiated by the laser intensity $I_0 = 10^{12}$ Wcm^{-2}. For simplicity, in this special case $T_e = T_i = T$ is assumed. Starting from a low initial concentration of free electrons a strong absorbing layer forms at the front on a ps time scale. Owing to the high plasma pressure a shock builds up which travels at 2.7×10^6 cm s^{-1} into the foil, reaching the rear surface in less than 2 ns. Since the shock wave has to balance the momentum of the expanding plasma, and the densities in the two regions are very different, only a small fraction of the laser energy is transferred to the shock wave (in the present case 8%). Transfer of a large energy fraction to the compressed phase is possible in gas targets, where the plasma density is comparable to the density in the shocked material. The results of Fig. 1.13 were obtained by assuming the validity of the ideal gas law also for solid hydrogen. This may not be unreasonable because it is a very soft material; at only 40 kbar it is compressed by $\kappa = 2.7$ [87]. For comparison, to reach a comparable compression ratio of 2.2 in lead, 4 Mbar are needed.

As the amount of energy going into the shock wave is small and we are interested only in the properties of the hot plasma, the solid is treated as incompressible in the following and the interest is concentrated entirely on the plasma properties. The one-fluid conservation equations are solved in plane and spherical geometry. In the spherical case the radius of the pellet is chosen as $R_0 = 100$ μm. In order to make the two cases comparable and to show better the influence of purely geometrical effects, R_0 is kept constant by introducing a self-consistent matter source on the pellet surface which balances the amount of ablated material. The light flux F in spherical geometry is chosen to produce the intensity I_0 on the original pellet surface, i.e. $F = 4\pi R_0^2 I_0$.

Modest Laser Intensity

A calculation was performed for solid deuterium with a laser intensity $I_0 = 5 \times 10^{12}$ Wcm^{-2}. Results are shown in Fig. 1.14 for plane (upper figure) and spherical geometry (lower figure) after 5 and 10 ns. The distributions of the dynamical quantities $n_e (= n_i)$, T_e, T_i and of I and F are plotted as a function of x and r, respectively. Nearer the solid, the particle density drops rapidly and the plasma streams out with velocities of the order of 10^7cm/s. The incoming light and that reflected from the critical point are absorbed according to Beer's law and heat the plasma to considerable electron and ion temperatures. In plane geometry light is absorbed over a large distance on its way forward and backward so that the fraction of reflected light is very low (<2%). In the spherical case the plasma rarefies much more rapidly; accordingly, the absorption length is strongly reduced and the reflection is high (55%). The temperatures are considerably lower in the spherical case and the temperature maxima are shifted towards the target surface. A comparison of the temperature profiles for 5 ns with those for 10 ns and their position as a function of r shows immediately that in spherical geometry the plasma production process becomes stationary near the target.

The partition of the laser energy into internal and kinetic plasma energy strongly depends on the geometry. In the plane case 43% of the absorbed energy is stored as thermal energy of the electrons and ions; this ratio remains approximately constant in time. For the pellet a much more unfavorable ratio is expected. In fact, at 5 and 10 ns only 25 and 15%, of the absorbed energy is stored as thermal energy. At the intensity under consideration the influence of heat conduction turned out to be very small. Therefore matter is heated only in regions where light penetrates, and the

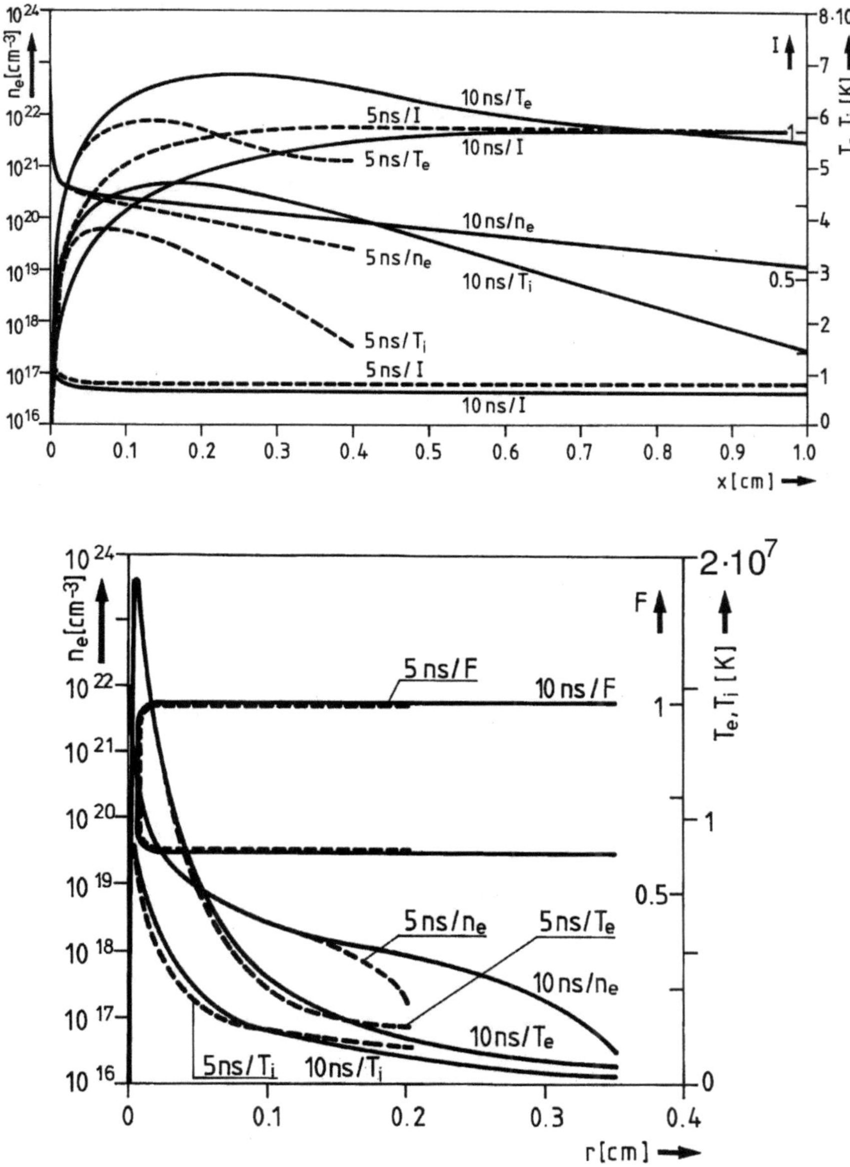

Fig. 1.14 Plasma production from solid deuterium in plane (upper figure) and spherical (lower figure) geometry with a Nd laser intensity $I_0 = 5 \times 10^{12}\,\mathrm{Wcm}^{-2}$ after 5 and 10 ns. n_e electron density, T_e electron temperature, T_i ion temperature, I laser intensity, $F = 4\pi r^2 I$ laser flux. Laser pulse impinges from the right. The ratio between thermal ion and electron energies in plane (spherical) geometry is 0.62 (0.47) at 5 ns and 0.66 (0.51) at 10 ns

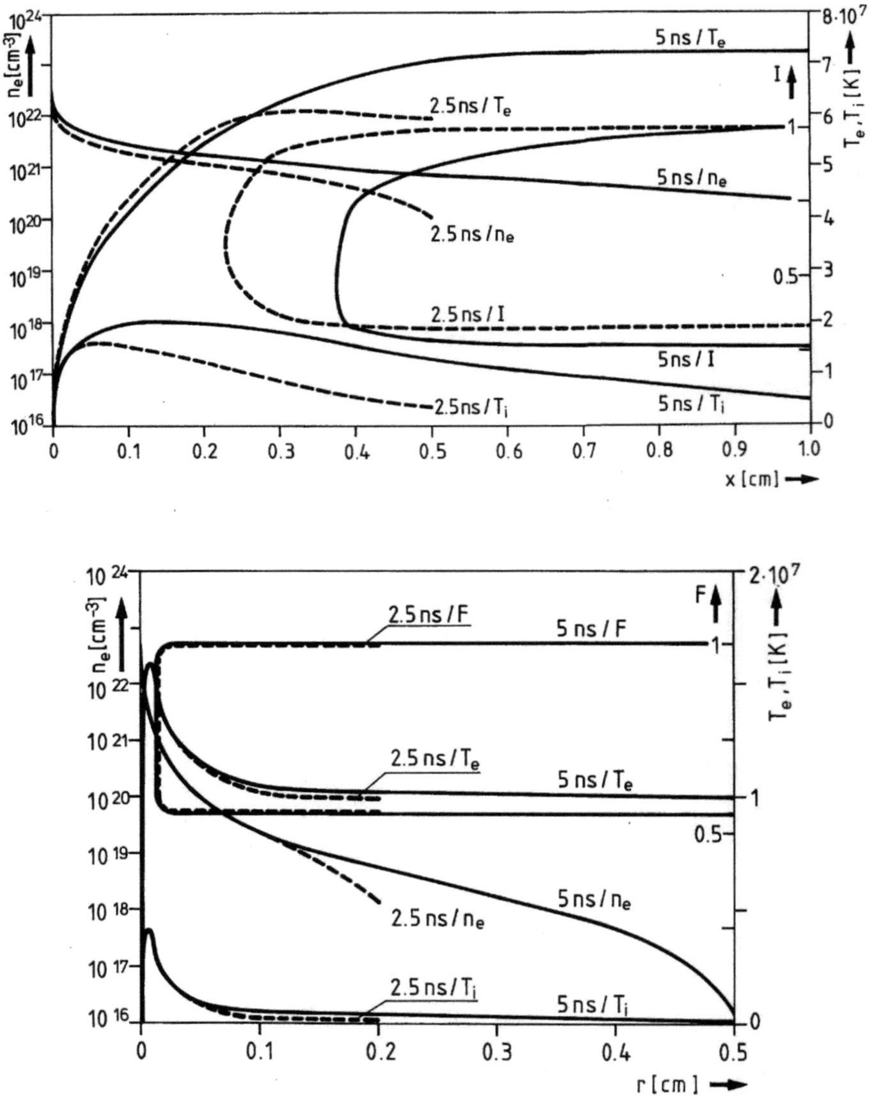

Fig. 1.15 Plasma production from solid deuterium in plane (upper figure) and spherical geometry (lower figure) with a Nd laser intensity $I_0 = 5 \times 10^{15}$ Wcm^{-2} after 2.5 and 5 ns. Symbols as in Fig. 1.14. The ratio between thermal ion and electron energies in plane (spherical) geometry is 0.31 (0.10) at 2.5 ns and 0.32 (0.06) at 5 ns. The target is irradiated from the right

plasma density, with the exception of a very small region at the target surface, is less than its critical value n_c.

High Laser Intensity

Corresponding calculations were performed with an intensity of $I_0 = 10^{15}\,\mathrm{Wcm}^{-2}$. The results are shown in Fig. 1.15 at 2.5 and 5 ns after the start of irradiation. In both geometries the electron temperature is 10 times higher, the ion temperature 4 times higher than in the previous case. At high temperatures the plasma production process is determined by light absorption and heat conduction. In contrast to light propagation, thermal diffusion is not limited by any cut-off density, and a large overdense region ($n_e > 10^{21}\mathrm{cm}^{-3}$) builds up in spherical and especially in plane geometry. Whereas at low intensities its thickness is about one wavelength and its structure is determined by light tunnelling at the reflection point, for intensities larger than about $I_0 = 10^{13}$ Wcm^{-2} its extension is given by heat conduction. From the figure in plane geometry one can see that the overdense region is growing in time: at 5 ns two thirds of the plasma produced is heated by thermal diffusion alone. In spherical geometry the heat conduction zone is much smaller and becomes stationary. In the plane case the ratio of thermal to total plasma energy is 47% at 5 ns and increases very slowly with time. Although heat conduction also plays a dominant role in spherical geometry, the partition of thermal to kinetic energy at $I_0 = 10^{15}$ Wcm^{-2} still favors the latter: at 2.5 ns only 23% and at 5 ns not more than 16% of the absorbed radiation appear as thermal energy.

With respect to laser light reflection we note that now, at the higher light intensity, the absorbing region has much larger dimensions; nevertheless the reflection is increased in both geometries as a result of the rise in the electron temperature (16% and 60% reflection).

Beer's law is a consequence of the WKB or optical approximation; it fails in the vicinity of the reflection point. By solving the wave equation for $I_0 = 10^{14}$ Wcm^{-2} in plane geometry one finds that about 15% of the total absorbed energy goes into the slab where the optical approximation breaks down; in the overdense region only 5% is absorbed. At higher intensities the absorption behind the reflection point is even lower because of the higher electron temperature. Thus the assumption of reflection at the critical density may be considered as approximately correct for calculating the overall dynamic behavior. For considerations in detail see Chaps. 6 and 7.

1.3.2.2 Multicomponent Plasma

Laser plasmas produced from high-Z elements represent mixtures of different ion charge states [88–91]. Such plasmas are of interest as sources of highly stripped ions for accelerators [92]. Optical and mass-spectroscopic investigations of the expanding plasma cloud on ns time scale show that the kinetic energy of the single species increases with its charge number Z [93, 95–99]. A first explanation of the energy spectra was based on electrostatic acceleration [94, 95, 100]. In the hot plasma cloud a quasi-static thermoelectric field $\mathbf{E}_s$ builds up which prevents the hot electrons from escaping. The various groups of ions should be accelerated in this E field corresponding to their individual charge state Z. As early as 1971 it was shown that in a fluid model 1% of the actual friction force an ion feels in moving through the plasma cloud is already strong enough to inhibit *ion separation* [101]. Similar independent theoretical investigations came to the same conclusion a decade later [102]. Meanwhile further investigations indicate that the observed ion separation is presumably a consequence of recombination [91, 103–105]. On such a background I. Kunz investigated the expansion dynamics of a multicomponent laser plasma numerically by solving coupled rate equations for a whole variety of ionization and recombination processes simultaneously with the expansion process [106]. Independently, similar dynamic calculations were performed in great detail by Granse et al. [107]. Very satisfactory agreement was found with the ion energy spectra most carefully measured by Dinger et al. [97]. Subsequently Rupp and Rohr [108] performed detailed measurements of the recombination dynamics of a plasma as a function of distance from the target. One of the remarkable and basic aspects of these investigations is the important role that *reheating* of the expanding plasma plays due to recombination: No charged particles would reach the ion collectors if the plasma had cooled down adiabatically. The sensitivity of the ion energies to reheating was shown also by an ingenious analytical investigation [109]. Summarizing, it can be concluded that in the intensity range $I \simeq 10^{11} - 10^{13}$ Wcm^{-2} the experimentally observed ion separation is very likely to be based mainly on the phenomenon of recombination and that electrostatic acceleration has to be excluded as predicted in the early stage of investigations on this subject. Meanwhile detailed measurements exist on the ion velocity distribution and anisotropy of kinetic ion temperature, angular emission distributions of neutrals and ions, and on the effect of recombination processes and ionization states [110, 111]. Binary systems show no directional segregation. The neutrals dominate at low laser intensities ($I < 10^{12}$ Wcm^{-2}) and large distances from the target. Their kinetic energy is generally low and smaller than that of the singly charged ions. All measurements underline the importance of recombination (e.g. three body recombination) up to distances of 1 m. The expansion dynamics shows also complex structures, i.e., the formation of localized groups of ions propagating at different velocities and densities. In such an environment groups of highly charged ions may also survive [112, 113].

Ion separation is studied more recently with intense short laser pulses in small liquid droplets and clusters. Here, the observed different energies of differently charged ions are generally attributed to thermoelectric acceleration. However, a quantitative proof has not been given yet. So, for instance in analogy to [101], the question of friction between the different ion species would have to be clarified for the new parameters associated with plasmas generated by intense sub ps laser pulses. A kinetic Vlasov treatment of the particle dynamics during the adiabatic expansion of a plasma bunch is presented in [114]. Here, the role of friction between different species has not been addressed either, and in later investigations on ion acceleration of different species by sub-ps ultraintense lasers, e.g. [115–117] the different configuration rendered the problem of friction irrelevant. The laser plasma from ns pulses is accessible to *similarity* considerations; they will be discussed in Chap. 3.

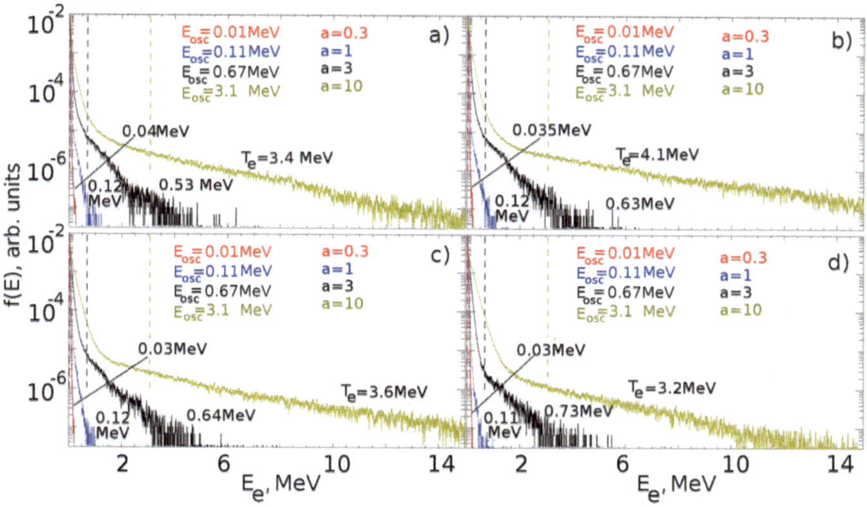

Fig. 1.16 Electron energy spectra $f(E)$ at (**a**) 30, (**b**) 35, (**c**) 40, and (**d**) 45 cycles after the beginning of the interaction. A $\sin^4$ laser pulse of peak amplitude $a = 0.3, 1, 3, 10$ and full width of 50 cycles impinges under 45° onto a hydrogen target with electron density n_{e0} such that $n_{e0}\lambda^2 = 9 \times 10^{22}\text{cm}^{-3}\mu\text{m}^2$. The pulses are identical in all four frames. Target thickness varies from 40 to 60 λ. Vertical dashed lines mark the mean oscillation energies. The hot electrons follow a Maxwellian distribution. The maximum mean energies $k_B T_e$ for $a \geq 1$ are by the factors 1.04, 1.09, 1.7 higher than the associated E_{os}. $k_B T_e$ increases during the evolution of the laser pulse for $a = 3$, for $a = 10$ it decreases. Power scaling $k_B T_e \sim I^\alpha$, $\alpha \geq 0$ not detected

1.4 Superintense Laser-Matter Interaction

Laser intensities from about 10^{17} to some 10^{22} Wcm^{-2} are addressed in this section. The interaction with matter plays in the relativistic domain. At the high intensities and high contrast prepulse-main pulse the laser field extracts bound electrons from atoms in less than a laser cycle by field ionization; the plasma formation process is nearly instantaneous. The most prominent signature is marked by the generation of high energy electrons and ions; for electrons see a 1D particle-in-cell (PIC) simulation in Fig. 1.16 with a Maxwellian tail (straight line of constant slope) in the upper energy interval [123]. At relativistic laser intensities it is convenient to express the irradiance $I\lambda^2$ by the dimensionless vector potential $\boldsymbol{a}$ which is $e\hat{\mathbf{A}}$ measured in units of the electron momentum $m_e c$. With the help of (1.2) the following relations hold

$$a = \frac{e|\hat{\mathbf{A}}|}{m_e c}, \ \hat{\mathbf{A}} = \frac{\hat{\mathbf{E}}}{\omega}, \ I = \frac{1}{2}\varepsilon_0 c\hat{\mathbf{E}}^2 \Rightarrow I\lambda^2[\text{Wcm}^{-2}\mu\text{m}^2] = 1.37 \times 10^{18}a^2. \tag{1.72}$$

For circular polarization the numerical factor is 2.74. The relativistic Lorentz factor of the free electron in the laser field is $\gamma = \sqrt{1 + \alpha a^2}$, $\alpha = 1/2$ for linear polarization and $\alpha = 1$ for circular polarization. The highest intensities are generally produced in the sub ps time interval, with a moderate energy output at the same time. This property allows the construction of small and flexible laser installations. At the same time another advantage is combined with the shortness of the pulses. For approximately 100 fs ion motion is not important; it lets the characteristic features of interaction appear in a genuine manner. An indispensable requirement in the experiment is a perfect contrast ratio between precursor and main laser pulse. Under such conditions experiments have shown, and 1D simulations have confirmed, that under oblique incidence absorption higher than 50% is possible [118]. It leads to the search for the effects that generate them in the absorption process of intense monochromatic light beams. Latest when the kinetic temperature reaches $10^3 Z^2$ eV in the plasma, collisional absorption is ineffective and other effects of non-collisional nature have to become active in order to ensure absorption. The best known non-collisional candidate so far is *resonance absorption* at oblique laser incidence [119]. It consists in the direct conversion of laser light into an electron plasma wave resonantly excited at the critical electron density where the laser frequency ω equals the plasma frequency ω_p. High intensity laser pulses in the intensity domain $I = 10^{16} - 10^{22}\,\mathrm{Wcm^{-2}}$ with good contrast ratio are so fast rising that there is no time to form an underdense preplasma in front of an irradiated solid sample that could couple to a resonantly excited plasma wave. Therefore the search began for new *collisionless absorption* processes. The first successful proposal was the so-called $\mathbf{j} \times \mathbf{B}$ heating [120]. The authors could show by particle-in-cell (PIC) simulations that at normal incidence the Lorentz force induces non-resonant electron oscillations at 2ω normal to the target surface which lead to appreciable absorption, target heating, and production of superthermal electrons at any density above critical. However, no attempt was made to explain how the observed absorption, i.e. *irreversibility*, comes into play. An extreme example of profile density steepening is presented in the following Fig. 1.17 from a particle-in-cell calculation with irradiance $I\lambda_0^2 = 5 \times 10^{21}\,\mathrm{Wcm^{-2}\mu m^2}$. At low laser intensities it may happen hat absorption is almost all collisional and nearly exhausted in the underdense plasma, i.e., $R \simeq 0$. In this situation the maximum of $\hat{E}^2$ is flat and no profile steepening develops, and the ablation pressure scales differently, as shown previously on plasma dynamics. The first experimental proof of profile steepening was reported in [74–76]. For a detailed theoretical study of profile structures and steepening see [77].

1.4.1 Collisionless Absorption

Let us consider the phenomenon of collisionless absorption of high-power laser beams from a more fundamental point of view. Under quasi-steady state conditions Poynting's theorem averaged over one laser cycle reduces to

$$\overline{\nabla \mathbf{S}} = -\overline{\mathbf{j}\mathbf{E}}. \tag{1.73}$$

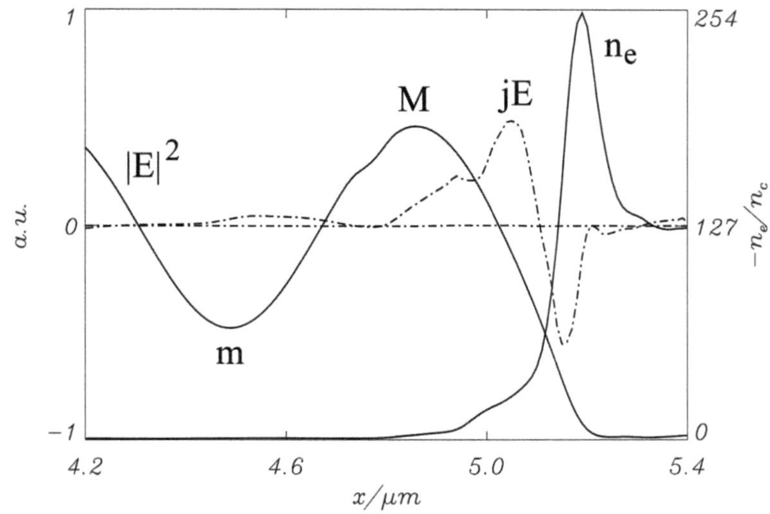

Fig. 1.17 Ponderomotive density profile steepening: Cycle-averaged distributions in space (x coordinate) of the field amplitude squared $|\mathbf{E}|^2$ with maximum M and minimum m closest to the target where it reduces to zero (arbitrary units); local absorption **jE** (left scale) and electron density n_e after 80 fs of irradiance $I\lambda^2 = 5\times10^{21}\,\mathrm{Wcm^{-2}\mu m^2}$ (right scale). Relativistic critical density $n_{\mathrm{cr}} = \gamma n_c = 43 n_c$. In the lab frame is $n_c = n_0/100$, n_0 target density. $x_{\mathrm{M}} - x_{\mathrm{m}} = \lambda_0/4$. Angle of incidence $\alpha = 45°$. Relativistic profile steepening is $L_{cr} = \lambda_0/34$

The equation describes all kinds of absorption, collisional, noncollisional, classical or quantized; in the latter case the current density and the electric field **E** are to be substituted by their operators acting on the corresponding state vector $|\psi\rangle$. In the intense laser field, despite the high particle densities involved, the classical picture is an excellent approximation. If the laser field evolves in time as $\mathbf{E} \sim \sin\omega t$ the current density follows as $\mathbf{j} \sim \cos(\omega t + \phi)$ and

$$\overline{\nabla \mathbf{S}} = -\overline{\mathbf{jE}} \sim -\overline{\cos(\omega t + \phi)\sin\omega t} = -\frac{1}{2}\sin\phi. \tag{1.74}$$

Dephasing between driver field and current determines the degree of absorption. A free electron in the vacuum starts oscillating with the arrival of the laser pulse; after the pulse is over the electron is at rest again, it has its oscillation energy given back to the wave (it has shifted only in position by a finite displacement owing to radiation pressure). Collisionless absorption reduces to the problem of finding out which effects lead to a finite phase shift in **j**. In collisional absorption it is the friction originating from the collisions between electrons and ions (see Drude model),

$$\overline{\mathbf{jE}} = \varepsilon_0 \omega_p^2 \frac{\nu_{ei}}{\omega^2 + \nu_{ei}^2} |\mathbf{E}|^2 > 0. \tag{1.75}$$

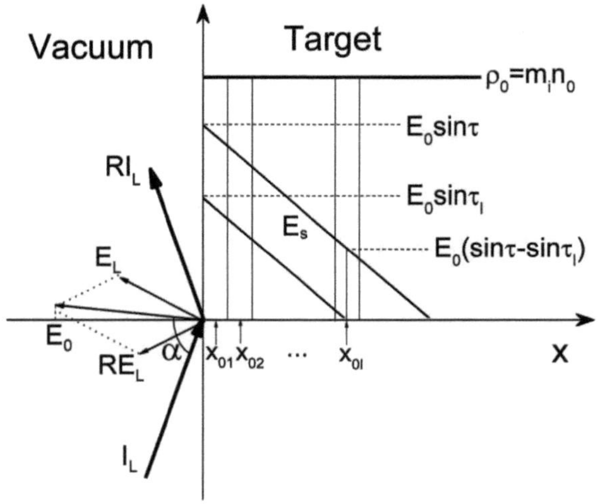

Fig. 1.18 Configuration of Brunel's absorption model. A p-polarized plane laser wave E_L (intensity I_L) impinges under angle α and is partially reflected as RE_L under α. At time instant $\tau_l = \omega t_l$ the resulting x component $E_0(\tau_l)$ penetrates up to the position x_{0l} given by (1.76). At the instant τ its magnitude there is $E_0(\sin\tau - \sin\tau_l)$ for $|\sin\tau| > |\sin\tau_l|$ and zero for $|\sin\tau| \leq |\sin\tau_l|$

At $\nu_{ei} = 0$ the collisional phase shift vanishes and any finite ϕ can only be of dynamic origin. This dynamic origin of absorption is found in the space charges induced by $\nabla\mathbf{u} \neq 0$ up to $I = 5\times10^{20} - 10^{21}\,\mathrm{Wcm}^{-2}$.

1.4.1.1 The Brunel Model

Not long after the $\mathbf{j}\times\mathbf{B}$ simulation [120] a remarkable step forward was made by F. Brunel in understanding high-power collisionless absorption. He could show after introducing a few modifications that the resonance absorption concept could be adapted to steep highly overdense plasma profiles, and significant absorption could be achieved under oblique incidence despite total absence of plasma resonance at $\omega = \omega_p$ ("not-so-resonant, resonant absorption" [121]) and no possibility for a plasma wave to propagate into a shallow *preplasma* in front of the target. Instead, now the energy imparted to the electrons is transported into the target and deposited there. The model is nonrelativistic, valid for $I_{\mathrm{Nd}} \lesssim 10^{18}\,\mathrm{Wcm}^{-2}$. For its analysis we follow [122].

A highly overdense target of constant electron density $n_e = n_0$ and zero temperature filling the half space $x \geq 0$ is given. From the vacuum the laser field component $E(\tau) = E_0\sin\tau$ is normally incident onto its surface, $\tau = \omega t$. From Poisson's law follows that the skin depth at the dimensionless time τ_l is

$$x_{0l} = \frac{\epsilon_0 E_0 \sin\tau_l}{en_0}, \tag{1.76}$$

see Fig. 1.18. With the amplitude of the oscillation velocity $v_0 = eE_0/(m_e\omega)$ in x direction the normalized velocity $w_l = v/v_0$ of the layer number l is given by

$$w_l(\tau) = (\cos\tau - \cos\tau_l) + (\tau - \tau_l)\sin\tau_l; \quad \tau \geq \tau_l. \tag{1.77}$$

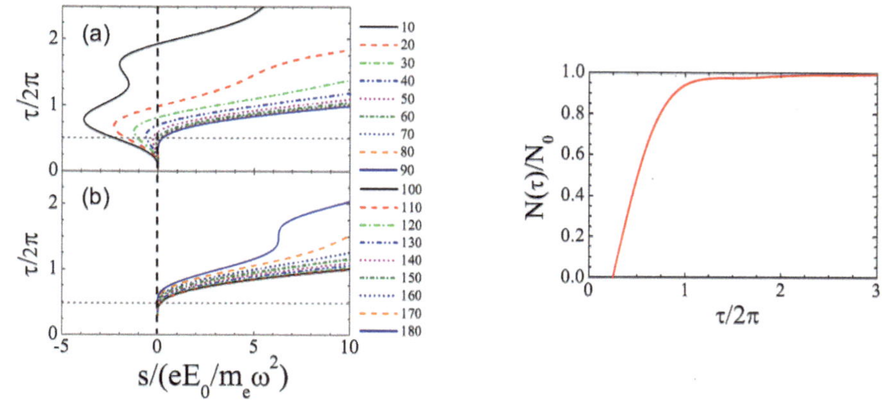

Fig. 1.19 Pictures (**a**), (**b**): Single layer displacement $s(\tau)$. The individual layer starts from position $x_{0l} \simeq 0$ at the time τ_l, indicated in degrees on the ordinate at the right ($10, 20, \ldots, 90 \Leftrightarrow \pi/2, 100, \ldots..$). Only displacements $x(\tau) \leq 0$ are physically real (**a**). In the region $x > 0$ the electric field $E_0(\tau)$ is zero at all times τ (**b**). Right picture: Number of particles per unit area $N(\tau)$ returned to the target in the interval $[0, \tau]$

Correspondingly, the normalized trajectory $s_l = x_l\omega/v_0$ results as

$$s_l(\tau) = (\sin\tau - \sin\tau_l) - (\tau - \tau_l)\cos\tau_l \\ + \frac{1}{2}(\tau - \tau_l)^2 \sin\tau_l + \frac{x_{0l}\omega}{v_0}; \quad \tau \geq \tau_l. \tag{1.78}$$

The number of particles per unit cross section that move during one cycle follows from (1.76) as $N_0 = n_0 x_0 = \epsilon_0 E_0/e$. Owing to the smallness of x_{0l} in a high density target in comparison to the oscillation amplitude v_0/ω it is set zero. When the lth particle is driven back to the target surface at the instant $\tau = \tau_f$ it enters the region $x > 0$ that is assumed to be field-free. Thus the electron has gained the kinetic energy $m_e v_0^2 w_l^2(\tau_f)/2$ from the laser field. For (1.77) and (1.78) to hold the orbits are not allowed to cross. A sufficient condition for this is that

$$\frac{\partial s_l}{\partial \tau_l} = \frac{1}{2}(\tau - \tau_l)^2 \cos\tau_l \tag{1.79}$$

does not change sign. In the interval $0 \leq \tau_l < \pi/2$ it is satisfied. In Fig. 1.19a, b representative orbits starting in the first interval $0 \leq \tau_l \leq 2\pi$ are shown. All of the N_0 electrons are pulled out into the vacuum during the first quarter time period $0 \leq \tau_l < \pi/2$ (see upper picture). In the second quarter period the two free fall terms $(\tau - \tau_l)\cos\tau_l$ and $(\tau - \tau_l)^2 \sin\tau_l/2$ in (1.78) prevail on the laser field term so that all s_l result positive (see lower picture) and have to be ignored (layers 91–180 in Fig. 1.19b). The change of sign in velocity w_l occurs exactly at $\tau_l = \pi/2$. In fact, for $\tau - \tau_l = \epsilon > 0$ and $\tau = \pi/2$ it follows $w_l = -\sin\epsilon + \epsilon\cos\epsilon = -\epsilon^2/2 < 0$. In contrast, $\tau_l = \pi/2$ yields $w_l = \epsilon^3/3! > 0$. In the second half period $\pi < \tau_l < 2\pi$ the

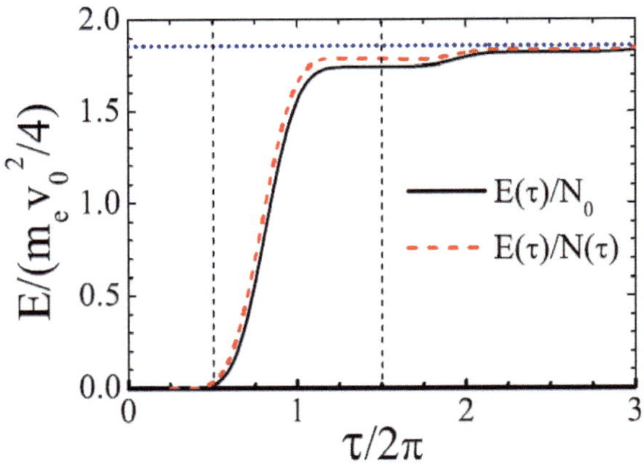

Fig. 1.20 Normalized average reentry energy as a function of reentry time τ. Bold curve: average over all particles pulled into vacuum during one laser cycle; dashed curve: average over $N(\tau)$ from Fig. 1.19

driving laser field is reversed and neutralized by the (infinitely) strong space charge field. Thus, when considering only particles with reentry times $\tau \leq 2\pi$ no crossing of orbits happens. Figure 1.19a shows that not all particles return during one period of τ. The front layer drifts away into the vacuum at the speed v_0 because of missing free fall term. The closer a layer is positioned to the front the longer is its reentry time. This gives rise to some orbit crossing for reentry times larger than one cycle. In the picture on the right the number of particles returning up to time τ is plotted in units of N_0. It shows that $N_0/2$ particles reenter in the interval $\pi/2 < \tau \leq \pi$ but only a small fraction (2.2%) comes back after one period 2π. At the starting of the next laser period the process described in Fig. 1.19 repeats with a new packet of N_0 electrons supplied by the slow return current which provides for quasineutrality.

The absorbed energy as a function of τ is of particular interest. The dashed curve in Fig. 1.20 shows the average reentry energy E in units of the mean oscillation energy $m_e v_0^2/4$ in the interval $[0, \tau]$. If normalization is done by the number of all particles N_0 that have been moved during one laser cycle, the solid line is obtained. The dashed line results from normalizing to $N(\tau)$. The results do not differ much from each other because the electrons $N_0/2$ reentering during the second quarter period contribute energetically by less than 1%. If summation is extended to all particles N_0 pulled out into the vacuum during one period the result is indicated by the horizontal dotted line. All N_0 electrons contribute to absorption. However the target is heated directly by the laser only by those coming back during the duration of the laser pulse. From Fig. 1.20 for the average energy $\eta_h m_e v_0^2/4$ gained per electron heating the target in the interval $[\pi, 3\pi]$ one reads $\eta_h = 1.74$ (Brunel: $\eta_h = 1.57$). The corresponding factor η_t of all electrons contributing to absorption is $\eta_t = 1.87$. One could think that the run away electrons could accumulate over several cycles and disturb the dynamics from (1.77) and (1.78). This may not happen because, in three dimensions, the space charge cloud rarefies rapidly and may also be quickly neutralized by a return current within the target.

Let us come now to the two most interesting questions, namely to the energy spectrum of the electrons and to the absorption coefficient in Brunel's model. The fastest electron of $w_f = 2.13$ finds back to the target at $\tau \simeq 2\pi$ corresponding to the cut off energy $E_f = (2w_f^2 = 9.07)\times\ m_e v_0^2/4$. The spectral distribution $f(E)$ normalized to unity is presented in Fig. 1.21a, b. Its peculiarity is the peak of f at the cut off energy exceeding the flat minimum by a factor of 7.5. If only non intersecting orbits are considered, i. e., $\tau \leq 5\pi/2$, almost no change is observed; in particular, cut off energy and the peak there in $f(E)$ are preserved. As the only free parameter is v_0 this spectrum is *self-similar* and universal. Equating the energy gain by the reentering electrons per period 2π to the energy supplied to the target yields

$$\frac{1}{4}N_0\eta_h m_e v_0^2 = \mathcal{A}I\cos\alpha\frac{2\pi}{\omega}, \quad \mathcal{A} = \frac{1}{2}\eta_h\frac{v_0}{c}\frac{E_0^2}{E_L^2}\frac{1}{2\pi\cos\alpha}. \tag{1.80}$$

α angle of incidence, $\mathcal{A}$ absorption coefficient, E_L laser field amplitude. With the help of the reflection coefficient R the field component E_0 follows from Fig. 1.18 as $E_0 = (1+\sqrt{R})E_L\sin\alpha$. Introducing the electron oscillation amplitude in vacuum $v_{os} = eE_L/m_e\omega$, $v_0 = v_{os}(1+\sqrt{R})\sin\alpha$, the heating and total absorption coefficients (indexes h, t) read

$$\mathcal{A} = \frac{1}{4\pi}\eta_{h,t}\frac{v_{os}}{c}(1+\sqrt{R})^3\frac{\sin^3\alpha}{\cos\alpha} \sim (I\lambda^2)^{1/2}; \ \frac{\eta_h}{4\pi} = 0.138,\ \frac{\eta_t}{4\pi} = 0.149. \tag{1.81}$$

$R = 1 - \mathcal{A}$ may be set and (1.80) may be solved for $v_{os}/c = 0.3$ as a function of α. The result is presented in the right picture of Fig. 1.21 for η_t. For $\alpha = 86°$ total absorption is reached and no solution exists beyond. At relativistic intensities an effective Lorentz factor γ has to be introduced. For $\alpha \leq 60°$ (1.81) yields only modest absorption in p-polarization in contrast to measurements [118]. Even setting $v_{os} = c$ (weakly relativistic case) and $R = 1$ produces collisionless absorption not exceeding $\mathcal{A} = 120\sin^3\alpha/\cos\alpha\ \%$ at moderate angles of incidence ($\alpha \leq 25°$). On the other hand, for $\alpha = 82°$ $\mathcal{A}$ exceeds unity at $v_{os}/c = 1$. In Brunel's model the absorbed energy scales like $(I\lambda^2)^{3/2}$ (!)

Contrary to a common believe that all electrons in one jet are pushed back by the inverted field, only half of them, lifted in the interval $(0, \pi/4)$, are in phase with the driver, the other half experience a weakened driver due to screening and fall back to the target, attracted by the immobile ions, before the laser field has changed direction. This leads quite naturally to a classification into energetic and less energetic electrons. With a view on Figs. 1.19–1.21, and bearing in mind what has been stated on particle dynamics in the combined space charge and laser fields, in particular on reentry, the decay of the electron energies into a slow and into a fast group has to be attributed to the following

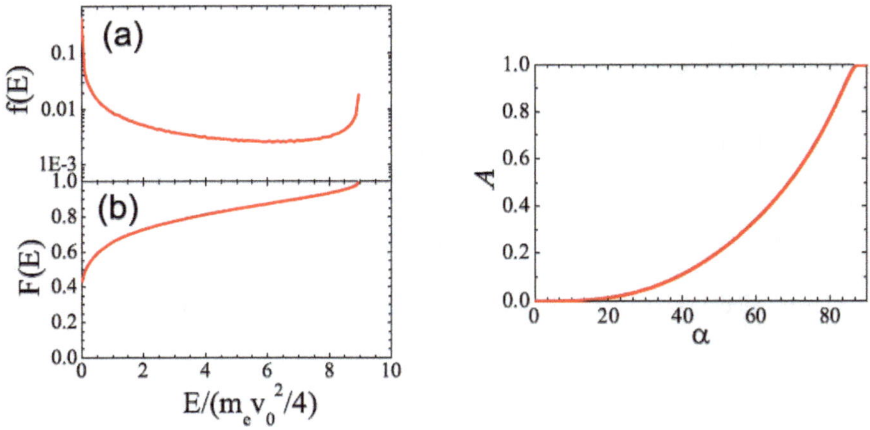

Fig. 1.21 Electron spectra in (**a**), (**b**) $f(E)$ and $F(E) = \int f(E')dE'$. Right picture: Absorption coefficient $\mathcal{A}$ as a function of angle of incidence α for normalized oscillation velocity $v_{os}/c = 0.3$ under condition $R = 1 - \mathcal{A}$; c speed of light

characteristics of Brunel's model. The particles returning to the target from vacuum during the second quarter cycle $(\pi/2, \pi)$ are those closer to the target. They feel a weakly screened space charge field and a weakened laser field. As a consequence, they are not pushed far out into to vacuum and return already when the two fields are still opposed to each other. This results into a low energy gain for half the total number N_0 of particles lifted into the vacuum (see Figs. 1.19, right picture, and 1.20). We give them the name *free falling particles*. For the $N_0/2$ particles reentering the target in the interval $(\pi, 2\pi)$ the situation is reversed; they are effectively accelerated back to the target by the two fields in phase with each other.

Brunel's model explains several important features of superintense laser beam-overdense target interaction: (i) *resonance absorption* concept adapted to targets overdense everywhere, (ii) origin of periodic jets of energetic electrons, (iii) acceleration in the vacuum, (iv) existence of groups of slow and of fast electrons, (v) *non-Maxwellian* subrelativistic electron spectrum around the energy cut off. The obvious deficiencies, e.g. wrong scaling of absorption, no crossing of electron orbits, are the price Brunel pays for oversimplification. The non-Maxwellian spectrum contrasts with Fig. 1.16. If we reserve the naming "Brunel" for those electrons which after having been pulled out into the vacuum cross the ion border again, spectra with non-Maxwellian tail are recovered also from PIC simulations [123], see Fig. 1.22. The arrival of single jets each laser cycle and the additional electron heating in the skin layer are illustrated by Fig. 1.23.

The essence of *collisionless absorption* in the overdense plasma is understood from Brunel's model. The laser forces the electron fluid to displace into the vacuum,

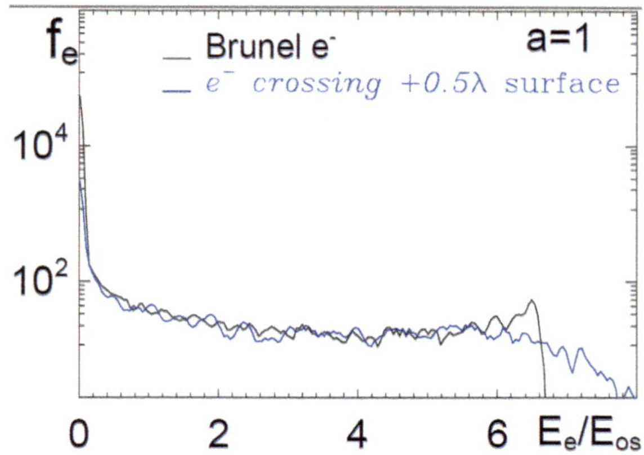

Fig. 1.22 Energy distributions $f(E)$ of Brunel electrons (black) and of electrons crossing positions $x = 0.5\,\lambda$ at intensity $a = 1$. The distributions are taken after 37 laser cycles when all electrons have returned to position $x = 2\,\lambda$. The non-Maxwellian structure of the Brunel electrons is preserved up to $a = 15$. The differences between "vacuum heating" and additional heating in the skin layer fade at higher intensities

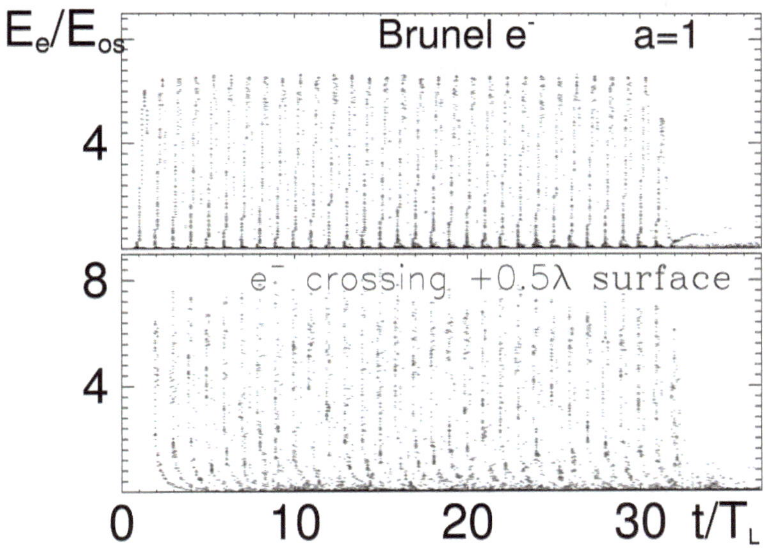

Fig. 1.23 Energy spectra of Brunel jets (upper picture) and jets at depth $\lambda/2$ (lower picture) as function of time (units in laser cycles) for $a = 1$. The additional heating in the skin layer is striking

either by the electric field component normal to the target at oblique incidence, or by the Lorentz force at normal incidence, in general by a superposition of both. The unbalanced space charge generates an electrostatic field that, superposed to the laser field, determines the electron motion and leads to the desired finite phase shift ϕ necessary for absorption in (1.74). The effect of charge separation can be seen most immediately considering a constant capacitor field $\mathbf{E_0}$ superposed to the oscillating laser field. It yields the amount of absorption per electron and unit time

$$\overline{\mathbf{jE}} = \frac{2\pi e^2 {\mathbf{E_0}}^2}{m_e \omega} > 0. \tag{1.82}$$

It is interesting to note and it can be formally shown, however it is also physically evident that $\overline{\mathbf{j}(\mathbf{E}_{\text{Laser}} + \mathbf{E_s})} = \overline{\mathbf{jE}}_{\text{Laser}}$; all work is done by the driver field, the space charge field $\mathbf{E_s}$ is inert, it provides for the phase shift only. This has its perfect analogy in the magnetic field, not doing any work, however leading to significant effects by changing the topology of the currents (e.g. Hall voltage). Some authors may attribute absorption to the Brunel like abrupt reduction of the laser wave amplitude in the skin layer. Due to this asymmetry the energy gained by an electron in the vacuum cannot be given back any more to the wave when entering the evanescent region. However, for this picture to work an electrostatic field component is needed too; transverse and longitudinal components cannot be isolated from each other. In the standard resonance absorption at the critical density it is the space charge field of the electron plasma wave that provides for collisionless absorption up to 49% through a phase shift $\phi \neq 0$.

Perhaps inspired from Brunel's mechanism of the electrons pulled out into the vacuum and then "pushed back" into the field-free target interior *vacuum heating* has been invoked for collisionless absorption to happen. If one wishes to use the term and to connect a well defined mechanism with it, it is most reasonable to identify it with Brunel's absorption model. For a more detailed analysis see [125].

1.4.2 Microstructured Targets

Absorption of fs pulses suffers from the disproportion of the electron density n_{e0} of the solid and the critical density n_c of the high power lasers, $n_c \ll n_{e0}$. The consequence is the heating of a thin surface layer to very high energy and to substantial reflection from this *plasma mirror*. Increased laser-solid coupling is obtained from *corrugated target* surfaces that allow the resonant excitation of surface plasmons [129, 130]. Increased coupling and volumetric heating are obtained from low den-

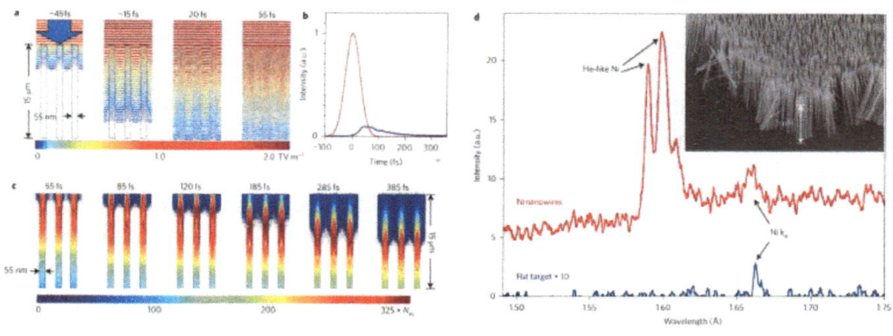

Fig. 1.24 Nanostructured target. (**a**): nanowire dimensions and 2ω TI:Sa laser intensity distribution at normal incidence, (**b**) reflected (blue) versus incident pulse (red), (**c**) plasma dynamics, (**d**) X-ray spectrum from structured (red) and from flat target (blue), inset: picture of nanowire array in Ni. Courtesy of [132]

sity foam targets, from clustered matter, from droplets and "smoked targets" due to optical path increase, broad variation of incident angles and polarizations (s and p polarization randomly distributed), local microshock generation, bubble collapse in foams, and inhibited plasma rarefaction due to expansion. Focusing of the laser beam onto the front end of a wire leads to collimation of the hot electrons and to distributed collisional heating along the wire axis [131]. The use of microstructured targets in the form of regular arrays of nanonwires turned out to be particularly efficient in producing volumetric hot dense plasmas [132]. In Fig. 1.24 the dimensions of a Ni nanowire array and the intensity distribution of the second harmonic Ti:Sa laser beam at normal incidence (a), almost total absorption (b), blue line, computed electron density evolution (c), and single shot X-ray spectra (red) from the microstructured target (inset) compared to the spectral output from the flat Ni target (blue) (d) are shown. In (a) and (c) the reader may recognize the enhanced Brunel-like electric field distribution perpendicular to the wires as well as their pinching under the action of the solenoidal magnetic field (Pukhov in [133]). The analysis of the X-ray spectra [as in (d)] showed that the enhanced emission stems from He-like Ni excited by thermal electrons in the 2–4 keV temperature range. Remember, collisional absorption increases with Z^2. The increase of the hot electron energy is by a factor of 2; they are responsible for the Ni K_α line. Owing to their small number the increase in output is modest, whereas in the 7–8 keV spectral domain the He-like line emission increased by more than a factor of 50. Highly charged ions from collisional ionization (e.g. Au^{+52}) and electron densities 300 times critical density have been obtained from sub-J fs laser pulses. By extrapolating to $I \simeq 10^{22}\,\mathrm{Wcm}^{-2}$ plasma energy densities and pressures in the GJ region and multi-Gbar domain seam to be feasible. For comparison, the pressure in the center of the sun amounts to 160 Gbar. Figure 1.24c illustrates the nanowire dynamics. The circumvention of the low critical density limit with its surprising escalations and benefits persists as long no overcritical plasma fills the interspaces.

1.5 Summary

Plasmas and hot matter created by high-power Nd and Ti:Sa lasers are considered in the intensity domain of I ranging from $(10^9 - 10^{10})$ to $10^{22}\,\mathrm{Wcm}^{-2}$. They are hot, with electron temperature T_e generally much higher than the ion temperature T_i. Their description is possible in terms of an ideal fully ionized plasma (ideality parameter $\Gamma < 0.1$). The laser light can penetrate up to the critical density n_c, at relativistic intensities, i.e. $a \gtrsim 1$, penetration up to $n_{cr} = \gamma n_c$ is possible, with the *local* Lorentz factor $\gamma = [1 + a^2/(2)]^{1/2}$ [126]. Plasma produced with long pulses in the ns regime is underdense except restricted high density layers penetrated by the electron heat front and a shock wave in front of it. At relativistic intensities ($I \gtrsim 10^{18}$ Wcm^{-2}) overdense preheating by fast electrons and return current takes place. The expanding hot plasma behind the critical point acts like a piston (rocket effect, ablation pressure) which drives a shock wave into the overdense target. The ablation pressure is a powerful instrument to generate states of compressed matter. The radiation pressure $p_L = (1 + R)I/c$, R reflection coefficient, helps to reinforce the plasma pressure at $I \gtrsim 10^{16}$ Wcm^{-2} and dominates it from relativistic intensities on. At moderate intensities p_L is weak, however it cannot be neglected because it leads to density profile steepening close to the critical density.

Pressure generation through laser ablation offers an instrument to generate dense ideal cold plasmas, similar to metallic conductors. It is the degenerate ideal quantum plasma which is described by the ideal Fermi gas. For matter to fall in this category the Fermi energy E_F must dominate the thermal electron energy, $E_F \gg k_B T_e$, and the deBroglie wavelength must exceed the mean interelectron distance, $\lambda_B > n_e^{-1/3}$.

The laser plasma is quasineutral if it is not subject to locally strong forces from outside, except at the borders over distances of the local Debye length. Quasineutrality means that the electron density n_e equals the ion density $n_i \times Z$, with Z the (average) ion charge. However, setting $n_e - Zn_i = 0$ is strictly forbidden unless it can be shown in the individual situation that the electrostatic fields due to small charge imbalance are negligible *and* are not subject to *unstable* behavior. The cold plasma exhibits a longitudinal eigen oscillation or resonance at the plasma frequency ω_p. Longitudinal and transverse electric waves of frequencies ω cannot propagate into overdense plasma. It is this important property whereof the critical density receives its significance. It is defined as the electron density at which equality $\omega_p = \omega$ is reached. At $\omega_p > \omega$ the square of the refractive index $\eta^2 = 1 - \omega_p^2/\omega^2$ assumes negative values, η becomes imaginary, direct plasma heating in depth by the laser is not possible.

The electron ion collision frequency ν_{ei} plays a key role in collisional plasma heating, in electron heat transport and in the thermalization processes of the electrons in themselves, determined by $\nu_{ee} \simeq \nu_{ei}$, the formation of an ion temperature T_i and the attainment of complete thermal equilibrium $T_i = T_e$. The time interval τ of the inverse of the collision frequency is the appropriate scale of the equilibration processes. This means (for low Z): for T_e it is $\tau_{ee} = 1/\nu_{ee}$, for T_i it is $\tau_{ii} = 1/\nu_{ii}$, $\nu_{ii} \simeq (m_e/m_i)^{1/2}\nu_{ee}$, and for $T_i = T_e$ it is $\tau_{\mathrm{eq}} = m_i/(m_e\nu_{ei})$. For hydrogen it results in the equality of ratios,

$$\tau_{ee} : \tau_{ii} : \tau_{\text{eq}}, = 1 : \left(\frac{m_e}{m_i}\right)^{1/2} : \frac{m_e}{m_i}.$$

Laser light absorption by electron-ion collisions follows from the Drude model. It consists in the introduction of a linear friction term $\nu_{ei}\mathbf{u}$ in the equation of motion of the average drift velocity $\mathbf{u}$. Its correctness for $|\mathbf{u}| \ll v_{\text{th,e}}$ is proven for hard sphere collisions, (1.19). With ν_{ei} properly defined its validity extends to realistic electron-ion Coulomb collisions, (1.30). The energy gained by one collision on the average over one laser cycle $\tau_L = 2\pi/\omega$ is twice the mean oscillation energy W of the free electron in the laser field.

Latest when T_e exceeds the limit $10^3 Z$ eV collisional heating is inefficient. It will be replaced by linear (and nonlinear) resonance absorption which consists in the resonant excitation of an electron plasma wave at the critical point [127]. A detailed description of the phenomenon will be presented in Chap. 6. For this mechanism to happen a critical density in the expanding plasma must develop during the evolution of the pulse to make linear resonance possible. The necessary power supply driving the rarefaction wave may happen by collisional absorption of a low intensity prepulse, by skin layer absorption [128], or by heat diffusion into the overdense target.

Almost no rarefaction wave builds up in the interaction of intense ultrashort laser pulses, further supported by profile steepening due to radiation pressure. In the early time of studies several noncollisional absorption mechanisms had been proposed among which Brunel's model of "resonant not so resonant absorption" emerged. By an electrostatic ansatz it was shown that the charge separation produced by the laser field at oblique angles and by the Lorentz force at perpendicular incidence leads to the compulsory dephasing between laser field and induced current density to provide for absorption. The resonance is never perfect, except particular cases; the residual fraction of laser power is reflected from the plasma. Substantial reduction of reflection is achieved from target microstructuring. Absorption up to 95% is achieved from Brunel effect, increased collisional absorption and excitation of surface waves.

Laser-matter interaction in the entire intensity domain faced so far is based on the existence of free electrons. For conductors and semiconductors, as well as weakly bound electrons in impurities, there exists no problem for them to be freed by the low laser photon energies. In clean dielectric targets (gases or solids) the first electrons are set free by multiphoton and field ionization. Which one prevails is decided by the *Keldish parameter* $\gamma_K = (E_I/2W)^{1/2}$. At $\gamma > 1$ multiphoton interaction dominates, at $\gamma_K < 1$ field ionisation wins. With an adequate number of electrons freed by the two processes ionization will continue to grow fast by collisional ionization. The subsequent avalanche-like electron multiplication process ends in the plasma breakdown from which the interaction processes described above can start.

1.6 Problems

▹ In an atom an electron is confined within $R \simeq 0.05$ nm. Calculate the maximum voltage difference over the distance $2R$ when a laser beam of intensity corresponding to $a = 1$ is impinging on it.

▹ Assume the electron in the neutral atom to be bound by the Coulomb potential $U \sim 1/r$. (a) What is the potential depression U_{m} by the laser from the foregoing problem? (b) What is the electron velocity in the undisturbed atom at this potential energy U_{m}? *Hint:* Use the virial theorem. Compare the "frequency" of the electron with the frequency of the laser of $\lambda = 1\,\mu$m.

▹ Show, γ_K from (1.11) and $\gamma_K' = \omega\tau_I$ are (nearly) identical. *Solution:* $\tau_I = x_{\mathrm{m}}/\overline{v} = x_{\mathrm{m}} m_e/(m_e E_I)^{1/2}$. Assume $E_I = eE_{\mathrm{m}} x_{\mathrm{m}} = \omega m_e \overline{v}(\overline{v}/\omega) = m_e\sqrt{2W} x_{\mathrm{m}}$; $\Rightarrow$ $(E_I/2W)^{1/2} = \omega x_{\mathrm{m}} E_I^{1/2}/2W$. From $E_I = 2W$ follows $\gamma_K = \omega x_{\mathrm{m}}/E_I^{1/2} = \omega\tau_I = \gamma_K'$.

▹ Show that a plasma sphere in vacuum oscillates at $\omega_0 = \omega_p/\sqrt{3}$. *Hint:* Calculate the restoring force in analogy to Fig. 1.6. (Do you remember the Clausius-Mossotti problem?)

▹ The pressure p has the dimension of an energy density $\varepsilon = nm\langle v^2\rangle/2$. Show that for the ideal gas of isotropic velocity $|v| = v_0$ holds $p = s\varepsilon$, $s = 3/2$. *Solution:* Chose arbitrary direction of x coordinate and box $V = L_x L_y L_z = L^3$. Particle impinges $\nu = v_x/2L$ times (*collision frequency*) onto surface L^2 perpendicular to x axis and exerts the force $f_x = 2mv_x\nu = mv_x^2/L$ on it. $\Rightarrow pL^2 = Nf_x = Nmv_x^2/L \Rightarrow p = 2(N/V)mv_x^2/2$. Isotropy: $\langle v_x^2\rangle = \langle v_y^2\rangle = \langle v_z^2\rangle = v_0^2/3$. Hence, $p=(2/3)\varepsilon$; $n=N/V$. *Remark:* Alternatively, $p = nf_x$ is the momentum flux density *through* the perpendicular surface L^2.

▹ Verify that $(\beta/\pi)^{3/2}$ in (1.18) is the correct factor of normalization to unity.

▹ Calculate the averages $\langle|\mathbf{v}|\rangle$, $(\langle\mathbf{v}^2\rangle)^{1/2}$, $\langle\mathbf{v}^{-2}\rangle$ for the Maxwellian distribution.

▹ Show that the differential collision cross section of a mass point colliding with an elastic hard sphere of radius R is $\sigma_\Omega = R^2/4$. *Solution:* Use the definition of σ_Ω from (2.148) and Fig. 1.5: Scattering angle is $\vartheta = 2(\pi/2 - \alpha)$. Collision parameter is $b = R\cos\theta \;\Rightarrow\; bdbd\varphi = -R^2\cos\theta\sin\theta d\varphi d\theta = -(R^2/4)\sin\vartheta d\varphi d\vartheta \;\Rightarrow\; \sigma_\Omega = R^2/4$.

▹ Show that (1.24) applied to the hard sphere model with Maxwellian f_0 reproduces (1.19). *Hint:* (i) Neglect B, (ii) take $\mathbf{v}_{\mathrm{os}}$ from (1.15), $|\mathbf{v}_{\mathrm{os}}| \ll v_{\mathrm{th}}$, and approximate $\sqrt{(\mathbf{v}+\mathbf{v}_{\mathrm{os}})^2} = v(1+2v_{\mathrm{os}}/v)^{1/2} = v[1+(v_{\mathrm{os}}/v)\cos(\mathbf{v},\mathbf{v}_{\mathrm{os}})]$.

▹ Derive the dc electric conductivity σ_0 from Drude's equation (1.24). *Answer:* $\sigma_0 = n_e e^2/m_e\nu_{ei} = \varepsilon_0\omega_p^2/\nu_{ei}$.

▹ Quasineutrality: Starting with $n_e = n_c$ express $\mathbf{E}$ as a function of the pressure gradient length L and show with the help of the Poisson equation that n_e and Zn_i differ significantly from each other for $L < \lambda_D$.

▹ If the plasma slab in Fig. 1.6 is excited by an external electric field $E(t) = \hat{E}\cos\omega_p t$

in the direction of δ the center of mass of the plasma slab does not move if it was at rest before (why?). (a) What does change with ω_p if momentum conservation is included now? *Answer:* m_e is substituted by the reduced mass of the ion-electron system $\mu = m_e m_i/(m_e + m_i)$. (b) What is the plasma frequency in an electron-positron plasma?

▹ In a cold plasma of common electron and ion density $n(x, t = 0) = n_0 + n_1 \sin kx$ in equilibrium the electron fluid is shifted by the small amount δ, $k\delta \ll 2\pi$ at $t + \varepsilon$ to $n_0 + n_1 \sin k(x + \delta)$. With a view on Fig. 1.6 show that the electron fluid oscillates as a whole at ω_p with amplitude δ. *Hint:* Expand $\sin k(x + \delta)$ in Taylor series at x and determine the local E field from the corresponding Maxwell equation.

▹ Use $\delta = \hat{\delta} \exp(ikx - i\omega t)$ for the longitudinal displacement to show that the cold plasma frequency is insensitive to the wave number k.

▹ Assume the static electron density modulation $n_1(x) = \hat{n}_1 \sin(kx)$ in the lab frame S. Which frequency ω' does the observer see in $S'(-v)$? *Answer:* $x = x' - vt \Rightarrow n_1(x') = \hat{n}_1 \sin(kx' - kvt) = \hat{n}_1 \sin(kx' - \omega' t) \Rightarrow \omega' = kv$ (nonrelativistic Doppler effect.)

▹ Verify (1.33), in particular, starting from protons convince yourself of the Z dependencies.

▹ Derive the vectorial expression of the thermoelectric field in (2.83).

▹ Derive (1.45): $p_P = \rho_0 v_S u_1$ for the strong shock.

▹ Show that $u_1 = v_S(1 - 1/\kappa)$ in Fig. 1.9. *Solution:* In unit time the shock has advanced by v_S and compressed the material $\rho_0 v_S$ to $\rho_1(v_S/\kappa)$ of thickness v_S/κ. Thus the compressed layer has advanced in unit time by $u_1 = v_S - v_S/\kappa$.

▹ Derive the conversion efficiency of the rocket under the assumption that the thrust p_p in the burn chamber is an ideal gas.

▹ Derive the piston pressure formula for a weak shock. *Hint:* p_0 must be added to the momentum balance.

▹ Derive the rocket equation (1.48).

▹ Apply the Rankine relation to a monoatomic ideal gas of low Z to determine the temperature after the shock as a function of compression κ and compare the result with the temperature obtained from the same adiabatic compression. *Hint:* In adiabatic compression holds $T\rho^{-2/3} = \text{const}$. Assume full ionization in region 1.

▹ A Ti:Sa laser pulse of $I = 10^{12}\,\text{Wcm}^{-2}$ is focused to a spot diameter of $d = 50\,\mu\text{m}$. Estimate the time τ until a steady plasma flow into vacuum is established. *Answer*: $\tau = d/c_s$, c_s sound velocity.

▹ Estimate $r = \hat{E}'^2/(k^2\hat{E}^2)$ in (1.55). *Answer:* $\hat{E}'^2 \simeq \hat{E}^2/(\lambda/2)^2 \Rightarrow r = 1/(\pi^2) = 0.1$.

▹ What is the change in the profile steepening length L_c if λ_0 is replaced by its local value $\lambda = \lambda_0/\eta$? *Hint:* Do not forget the swelling factor of $\hat{E}^2$ resulting from $I = c\eta\varepsilon_0\hat{E}^2/2$.

▹ Determine the position x/x_T from which on $\lambda_e < \epsilon L$, $L = T_e/|\nabla T_e|$ is violated for $\epsilon = 0.1$, $\epsilon = 0.01$ in Fig. 1.12.

▹ Show that $\nabla T = -\infty$ at the heat front if the initial temperature is zero.

▹ Derive an expression for the energy fraction r going into the shock wave from long laser pulses interacting with a target of initial density ρ_0. *Answer:* $r \simeq [(\kappa - 1)/\kappa]^{1/2}(\rho_c/\rho)^{1/2} < (\rho_c/\rho)^{1/2}$; κ compression ratio $\leqslant 4$.

▹ Give an estimate of the ablation pressure in Mbar in Fig. 1.13.

▹ (a) Calculate the radiation pressure of a laser beam of $I = 10^{22}\,\mathrm{Wcm}^{-2}$ in units of bar. (b) Imagine an ideal Fermi gas at $T_e = 0$ to be in equilibrium with the radiation pressure in (a). What is its electron density?

▹ Consider Fig. 1.18. Brunel's thickness of the skin layer (1.76) oscillates with frequency ω between zero and the maximum value of the laser field component normal to the target and, at constant finite target density, it falls off linearly. The electromagnetic skin layer thickness is constant, $\delta_s \simeq c/\omega_p$, and the laser field falls off exponentially. What is the reason for this different behaviour? *Answer:* In Brunel's model the contribution of the magnetic field is missing.

▹ According to Fig. 1.19 no particle is extracted into vacuum after the laser field has reached its maximum value at a quarter oscillation period, see lower picture (b). Give a physical explanation for this behaviour. *Hint:* Explain why the attractive force by the ions prevails on the outwards directed laser field.

1.7 Self-assessment

• The polarizing E field in Fig. 1.6 conserves total momentum of the ion and electron fluids; why? What does this imply for the plasma frequency? *Answer:* m_e in ω_p is to be replaced by the reduced mass $\mu = m_e m_i/(m_e + m_i) = m_e/(1 + m_e/m_i)$. What is the displacement of the ions?

• Do you know a physical quantity for which infinite heat conduction $\kappa_e = \infty$ may be assumed? *Answer:* Ion sound velocity $c_s = \sqrt{p_e/\rho}$ at high electron temperature T_e.

• What is the number of degrees of freedom f corresponding to infinite thermal conductivity $\kappa_e = \infty$? *Answer:* $f = \infty$. Explain why.

• (a) In which system of reference does the rocket equation (1.48) hold? (b) A rocket flies to the moon. Where ends the center of mass of the combined system (rocket + exhaust)?

• (a) Is thermodynamic efficiency $\eta_R = 1$ (all chemical energy converted into work) of a rocket driven by an ideal gas possible under the assumption of no lateral heat losses? (b) Why does $\eta_R = 1$ not violate the second law of thermodynamics?

• Which is the simplification introduced in (1.63) owing to which $\dot{x}_T = \infty$ at $t = 0$ and $x_T = \infty$ at $t = 0$ in case of constant heat diffusion? *Answer:* Electron inertia suppressed.

• Give the definition of the binary differential collision cross section σ_Ω in he center of mass system.

• The electron-ion collision frequency for momentum transfer of an electron of velocity $\mathbf{v}_e$ is given by $\nu_{ei} = n_i\sigma_t|\mathbf{v}_e|$. Its dimension is $[t^{-1}]$, as it has to be. (a) Justify the formula from the physical point of view (i) for hard spheres and (ii) for Coulomb interaction, in particular the appearance of the ion density n_i in it. How do you justify simultaneous multiple collisions? (b) Is ν_{ei} also correct if $\nu_{ei} \ll \omega$? *Answer:* Yes.. (c) Is ν_{ei} a temporal or ensemble average? *Answer:* ensemble.

• Why and where in deriving (1.19) does the limitation $|\hat{\mathbf{v}}| \ll v_{\text{th}}$ enter?

• Why in the soft Coulomb collision no change of position $\mathbf{x}$ is to be considered during the close encounter? Give an upper bound for $\Delta\mathbf{x}$ during the interaction time shorter than $T/3$.

• In the thermal plasma (1.17) for hard sphere collisions does not depend on the thermal motion. (a) Justify. *Hint:* Thermal motion cancels by symmetry when averaged in large ensemble. Reconsidering Fig. 1.5 may help. (b) Does (1.17) hold also in presence of a strong drift $|\mathbf{u}| \simeq v_{\text{th}}$? No.

• Prove in detail the correctness of (1.17) for the irreversible energy gain in a hard sphere collision from the high frequency electromagnetic field (for example microwave, Terahertz radiation, laser). In particular, justify the special choice of the reference system S' used. *Remark:* The problem is not trivial. The energy gain depends sensitively on the choice of S'. For the covariant, i.e. S' independent solution, see Problems in Chap. 2.

• Why do electron-electron collisions not contribute to plasma heating? *Answer:* In the tangent inertial frame $S'(\mathbf{v}_{\text{os}})$ co-moving with $\mathbf{v}_{\text{os}}$ the effect of the laser field disappears.

• Thermalization of the free electron cloud is accomplished by electron-electron and electron-ion collisions. In what do the two processes differ from each other? *Answer:* Electron-ion collisions are effective in isotropization of the electrons; their Gaussian energy distribution is due to the electron-electron energy exchange in the collision.

• The fully ionized plasma resembles very much a metallic conductor. In the plasma the electric conductivity is described by electron-ion collisions (Drude model); in the metal it is determined by electron-*phonon interaction*. Why two different pictures? *Hint:* Compare the corresponding deBroglie wavelengths. In the solid the electron does not "see" single ions.

• In the Fig. 1.6 the electrons on the left border of the layer shifted by δ experience the field $E = \sigma/2\varepsilon_0$, whereas the electrons on the right border of the layer "see" no field. Nevertheless the derivation of ω_p illustrated by the Figure is all right. Why? *Answer:* In the next half period the situation is reversed: the electrons of the former right border feel the full field. It is this alternating behavior which guarantees the slow evolution of the higher harmonics of ω_p.

• Why (and when) does the edge of the rarefaction wave not depend on geometry? *Answer:* as long as the thickness of the undisturbed layer is much smaller than its curvature radius.

• Show by dimensional analysis that in plane halfspace the rarefaction wave depends on the combination $\xi = x/c_s t$ only and not on $x,\ t$ separately: The plane rarefaction wave is *selfsimilar*.

• At which speed does a small density perturbation propagate in a homogeneous isothermal plasma of cold ions ($T_i = 0$) and very high thermal electron coefficient?

• Neglecting the shock wave of Fig. 1.13 in the energy balance of plasma production corresponds to the assumption of starting from initial target density $\rho_0 = \infty$ which is equivalent to assumption of *incompressible* target. When and why is the assumption allowed? *Hint:* For solution see estimate of energy fraction going into the shock in Problems.

• Why is the density ρ in the shockwave of Fig. 1.13 not constant, in contrast to ρ_1 in Fig. 1.9?

• Give a physical interpretation of the pressure term in (1.50).

• In the heat wave model by Caruso the ps laser pulse must be absorbed to a certain fraction. By which mechanisms may this be accomplished? Discuss various possibilities in dependence of the laser intensity.

• A powerful laser pulse is reflected by the low fraction R from the focusing lens. Why does the pulse with its high radiation pressure not "blow away" the lens? *Hint:* Estimate the momentum imparted to the lens by the laser pulse and discuss what happens.

• What is the meaning of "vacuum heating" (expression coined by Gibbon and Bell [124])?

• Describe the essence of collisionless absorption of an intense laser beam from overdense targets. *Hint:* Consider phase shift ϕ in (1.74). Which are the main forces on the electrons?

• List effects which may contribute to collisionless absorption. *Hint:* (anomalous) skin effect, radiation pressure, radiation losses, etc.

• Guess (with following calculation): Which is higher, the radiation pressure at $I = 10^{21}$ Wcm^{-2}, or the gas pressure in the center of the sun?

• Consider tunnelling through a rectangular potential barrier of hight $V = V_0$ and width d. What is the limiting value for $V_0 \to \infty$ under $V_0 d =$ const? Establish the analogy with optics.

• How is the Keldish parameter defined and which criterion is connected with it?

• In what do multiphoton and field ionization differ from each other? *Answer:* Multiphoton ionization is resonant excitation over many laser cycles; field ionization occurs within half a laser cycle.

• Why do orbits in the Brunel model not cross, and in PIC simulations they do?

• In the plasma oscillation and in the electron-ion collision total momentum of electrons and ions is conserved. If the small ion motion is taken into account what changes in the corresponding formulas of ω_p and ν_{ei}? *Answer:* m_e is replaced by the reduced mass $\mu = m_e m_i/(m_e + m_i)$.

1.8 Glossary

Constants:

Electron mass $m_e = 9.1 \times 10^{-31}$ kg,
Electron charge $e = 1.602 \times 10^{-19}$ C
Planck constant $\hbar = 1.055 \times 10^{-34}$ Js
Boltzmann constant $k_B = 1.3806 \times 10^{-23}$ JK $^{-1}$
Velocity of light $c = 3 \times 10^8$ ms^{-1}
Vacuum dielectric constant $\varepsilon_0 = 8.86 \times 10^{-12}$ CV^{-1}m^{-1} ; $4\pi\varepsilon_0 = 1.11 \times 10^{-10}$
T [eV] = 11600 T [K] $\simeq 10^4 T$ [K]
Laser intensity (plane wave)

$$\mathbf{I} = \varepsilon_0 c^2 \overline{\mathbf{E} \times \mathbf{B}} = \frac{1}{2}\varepsilon_0 c \mathbf{k}^0 \hat{\mathbf{E}}\hat{\mathbf{E}}^*; \quad \mathbf{k}^0 = \mathbf{k}/|\mathbf{k}|. \tag{1.2}$$

Electric field amplitude

$$\hat{E}\,[\mathrm{Vcm}^{-1}] = 27.5 \times (I\,[\mathrm{Wcm}^{-2}])^{\frac{1}{2}}. \tag{1.3}$$

Cycle-averaged oscillation energy

$$W = \frac{e^2}{4m_e\omega^2}\mathbf{E}\mathbf{E}^* \sim I\lambda^2. \tag{1.4}$$

Nonresonant multiphoton ionization probability

$$P_n \simeq \sigma_n I^n. \tag{1.6}$$

Field ionization, tunnelling probability

$$T = \frac{1}{1+\mathrm{e}^{-\zeta}}, \quad \zeta = 2\pi\left(\frac{m_e}{k}\right)^{1/2}\frac{\epsilon - eU_\mathrm{m}}{\hbar}, \quad k = -e\left(\frac{\partial^2 U}{\partial x^2}\right)_{x_\mathrm{m}} = -\alpha(\alpha+1)\frac{eU_0}{x_\mathrm{m}^2}. \tag{1.9}$$

Tunnelling time

$$\tau_i = \frac{1}{T}\left(\frac{m_e}{2|\epsilon|}\right)^{1/2} \times \min(d, 2x_\mathrm{m}). \tag{1.10}$$

Keldysh parameter

$$\gamma_K = \left(\frac{E_I}{2W}\right)^{1/2}. \tag{1.11}$$

Bohr radius

$$a_B = \frac{4\pi\varepsilon_0\hbar^2}{m_e e^2 Z} = \frac{0.0529}{Z}\ \mathrm{nm}.$$

Debye length

$$\lambda_D = \left(\frac{\varepsilon_0 k_B T_e}{n_e e^2}\right)^{1/2}, \quad \lambda_D[\text{cm}] = 6.9\left(\frac{\mathrm{T_e[K]}}{\mathrm{n_e[cm^{-3}}}\right)^{1/2} = 743\left(\frac{\mathrm{T_e[eV]}}{\mathrm{n_e[cm^{-3}]}}\right)^{1/2}. \tag{4.69}$$

Average electron energy gain per collision

$$\mathcal{E}_e = 2W = \frac{1}{2} m_e \hat{\mathbf{v}}^2. \tag{1.17}$$

Maxwell distribution

$$f_0 = \left(\frac{\beta}{\pi}\right)^{3/2} \mathrm{e}^{-\beta v_0^2}, \quad \beta = \frac{m_e}{2k_B T_e} = \frac{1}{2v_{\text{th}}^2}. \tag{1.18}$$

Drude's absorption model: friction $-m_e \nu \mathbf{u}$

$$\frac{\mathrm{d}\mathbf{u}}{\mathrm{d}t} + \nu \mathbf{u} = \frac{e}{m_e}(\mathbf{E} + \mathbf{u} \times \mathbf{B}); \quad \mathbf{u} = \int \mathbf{v} f_e(\mathbf{v}) \mathrm{d}\mathbf{v}. \tag{1.24}$$

Poynting's theorem (steady state): *dephasing* ϕ

$$\overline{\nabla \mathbf{S}} = -\overline{\mathbf{jE}} \sim -\overline{\cos(\omega t + \phi) \sin \omega t} = -\frac{1}{2} \sin \phi. \tag{1.74}$$

Plasma refractive index η:

$$\eta^2 = 1 + i\frac{\sigma}{\varepsilon_0 \omega} = 1 - \frac{\omega_p^2}{\omega^2} \frac{1}{1 + i\nu/\omega}. \tag{1.25}$$

Absorption coefficient

$$\alpha = 2k_0 \Im \eta. \tag{1.25}$$

Electron plasma frequency

$$\omega_p^2 = \frac{n_e e^2}{m_e \varepsilon_0}; \quad \omega_p\,[\mathrm{s^{-1}}] = 5.64 \times 10^4 (n_e\,[\mathrm{cm^{-3}}])^{1/2}. \tag{1.26}$$

Critical electron density

$$n_c = \frac{\varepsilon_0 m_e}{e^2}\omega^2 = 1.75 \times 10^{21} \left(\frac{\omega}{\omega_{\text{Ti:Sa}}}\right)^2 [\mathrm{cm^{-3}}]. \tag{1.27}$$

Solid hydrogen density

$$n_{\rm H} = 4.5 \times 10^{22}\,{\rm cm}^{-3} \;\Rightarrow\; \rho_{\rm H} = 0.075\,{\rm gcm}^{-3};\;\; \rho_{\rm DT} = 2.5\rho_{\rm H} = 1.9\,{\rm gcm}^{-3}.$$

Electron-ion collision frequency ν_{ei} [s^{-1}]:

$$\nu_{ei} = \frac{4}{3}(2\pi)^{1/2}\left(\frac{Ze^2}{4\pi\varepsilon_0 m_e}\right)^2\left(\frac{m_e}{k_B T_e}\right)^{3/2} n_i \ln\Lambda = 3.6 \times \frac{Zn_e[{\rm cm}^{-3}]}{(T_e[{\rm K}])^{3/2}}\ln\Lambda. \tag{1.30}$$

Fermi energy of ideal cold electron gas

$$E_F = (3\pi^2)^{2/3}\frac{\hbar^2}{2m_e}n_e^{2/3} = 3.65 \times 10^{-15}(n_e\,[{\rm cm}^{-3}])^{2/3}\,{\rm eV}. \tag{1.35}$$

Electron deBroglie wavelength

$$\lambda_B = \frac{\hbar}{m_e v} = \frac{0.185}{(E[{\rm eV}])^{1/2}}\,[{\rm nm}],\;\; \lambda_{\rm th} = \frac{\hbar}{m_e v_{\rm th}} = \frac{0.34}{(k_B T_e[{\rm eV}])^{1/2}}\,[{\rm nm}]. \tag{1.39}$$

Thermal electron velocity

$$v_{\rm th}\,[{\rm cms}^{-1}] = \left(\frac{k_B T_e}{m_e}\right)^{1/2} = 4.2 \times 10^7 (T_e[{\rm eV}])^{1/2}. \tag{1.40}$$

Confinement time of sphere

$$\tau = \frac{R}{c_s},\quad c_s = \left(\alpha\frac{p}{\rho}\right)^{1/2},\; \alpha \approx 1. \tag{1.43}$$

Piston pressure on shock

$$p_P = \rho_0 v_S u_1 = \rho_0 v_S^2\left(1 - \frac{1}{\kappa}\right);\qquad p_0 = 0;\quad \kappa = \frac{\rho_1}{\rho_0}. \tag{1.45}$$

Hugoniot relation

$$\varepsilon_1 - \varepsilon_0 = \frac{1}{2}(p_0 + p_1)\left(\frac{1}{\rho_0} - \frac{1}{\rho_1}\right). \tag{3.164}$$

Ablation pressure

$$P_a = p + \rho u^2 = 0.96\rho_c^{1/3}\left(\frac{I_a}{Z}\right)^{2/3} \sim \left(\frac{I_a}{\lambda Z}\right)^{2/3}. \tag{1.48}$$

Radiation pressure

$$p_L = (1 + R)\frac{I}{c}. \tag{1.52}$$

Ponderomotive force density on free electrons

$$\pi = n_e \mathbf{f}_p = -\varepsilon_0 \frac{\omega_p^2}{4\omega^2} \nabla |\mathbf{E}|^2, \ ; \ \mathbf{f}_p = -\nabla W. \tag{1.55}$$

Ponderomotive profile steepening $1/L_c = (1/\rho)|\partial_x \rho|$;

$$L_c = \frac{\rho_c c_s^2}{p_\pi \eta} \lambda_0 = 2\frac{n_c}{Z} k_B T_{\text{eff}} \frac{c\lambda_0}{(1+R)I\eta}; \quad T_{\text{eff}} \text{ see (1.50).} \tag{1.58}$$

Electron heat conduction coefficient

$$\mathbf{q}_e = -\kappa_e \nabla T_e; \quad \kappa_e = \eta_e \frac{p_e}{\nu_{ei}} \frac{k_B}{m_e} = \kappa_0 T_e^{5/2}; \ \eta_e = 3.16(Z=1) - 13(\text{high}Z); \tag{1.61}$$

$\kappa_0 = 1.8 \times 10^{-10}\ \text{Js}^{-1}\text{m}^{-1}\text{K}^{-7/2}$.
Plane heat wave

$$T = T_0 \left(1 - \frac{x^2}{x_T^2}\right)^{2/5}, \ x_T = 1.48\, q_0^{5/9} \left(\frac{\kappa_0}{c_v Z} t\right)^{2/9}, \ \frac{T_0}{\langle T \rangle} = 1.223. \tag{1.68}$$

Relativistic laser intensity, linear polarization:

$$a = e|\hat{\mathbf{A}}|/m_e c; \ I\lambda^2[\text{Wcm}^{-2}\mu\text{m}^2] = 1.37 \times 10^{18} a^2; \ \gamma = \sqrt{(1 + a^2/2)}. \tag{1.72}$$

Brunel's fraction of absorption

$$\mathcal{A} = \frac{1}{4\pi}\eta_{h,t}\frac{v_{os}}{c}(1+\sqrt{R})^3 \frac{\sin^3\alpha}{\cos\alpha} \sim (I\lambda^2)^{1/2}; \ \frac{\eta_h}{4\pi} = 0.138, \ \frac{\eta_t}{4\pi} = 0.149. \tag{1.81}$$

1.9 Further Readings

P. Mulser, D. Bauer, *High Power Laser-Matter Interaction*, Springer Tracts in Modern Physics 238 (Springer, Heidelberg, 2010), Chap. 7.
C.J. Joachain, N.J. Kylstra, R.M. Potvliege, *Atoms in Intense Laser Fields* (Cambridge University Press, Cambridge, 2012).
Ya.B. Zeldovich, Yu.P. Raizer, *Physics of Shock Waves and High-Temperature Hydrodynamic Phenomena* (Academic Press, New York, 1966), Chap. 10 (Thermal Waves).
S. Eliezer, *The Interaction of High-Power Lasers with Plasmas*, (Institute of Physics Publishing, Bristol, 2002).
A. Macchi, *A Superintense Laser-Plasma Interaction Theory Primer*, Springer Briefs in Physics, (Springer, Heidelberg, 2013).

References

1. V.E. Fortov, *Extreme States of Matter*, 2nd edn. (Springer, Hielberg, 2016)
2. P.D. Maker , R.W. Terhune, C.M. Savage, *Proceedings of the 3rd International Conference on Quantum Electronics*, ed. by P. Grivet, N. Bloembergen (Columbia U, New York, 1963), pp. 1559–1565
3. C.G. Morgan, Laser-induced breakdown of gases. Rep. Prog. Phys. **38**, 621 (1975). http://iopscience.iop.org/0034-4885/38/5/002
4. T.X. Phuoc, Optics Commun. **175**, 259 (2000)
5. F.H.M. Faisal, *Theory of Multiphoton Processes* (Plenum Press, New York, 1987)
6. H.R. Reiss, Phys. Rev. A **22**, 1786 (1980). H.G. Müller, A. Tip, M.J. van der Wiel, Phys. Rev. A **30**, 3039 (1984). S. Chu, J. Cooper, Phys. Rev. A **32**, 2769 (1985)
7. W. Becker et al., J. Opt. Soc. Am. B **4**, 743 (1987). M. Edwards et al., J. Phys. B **17**, L515 (1984)
8. P. Lambropoulos, X. Tang, J. Opt. Soc. Am. B **4**, 821 (1987)
9. S.L. Chin, P. Lambropoulos, *Multiphoton Ionisation of Atoms* (Academic Press, New York, 1984)
10. G.S. Voronov, N.B. Delone, Sov. Phys. JETP **23**, 54 (1966)
11. M.D. Perry, A. Szoke, O.L. Landen, E.M. Campbell, Phys. Rev. Lett. **60**, 1270 (1988)
12. F. Yergeau, S.L. Chin, P. Lavigne, J. Phys. B: At. Mol. Phys. **20**, 723 (1987)
13. A. L'Huillier, L.A. Lompré, G. Mainfray, C. Manus, J. Phys. B: At. Mol. Phys. **16**, 1363 (1983)
14. B. Walker, B. Sheehy, L.F. DiMauro et al., Phys. Rev. Lett. **73**, 1227 (1994)
15. M.R. Cervenan, N.R. Isenor, Opt. Commun. **13**, 175 (1975)
16. R. Poprawe, G. Herziger, IEEE. J. Quant. Electron. **22**, 590 (1986)
17. G. Bekefi, *Principles of Laser Plasmas* (Wiley, New York, 1976)
18. P. Clarke et al., Appl. Phys. A **69**, 117 (1999)
19. S. Izumida, K. Onishi, M. Saito, Jap. J. Appl. Phys. **1**(37), 2039 (1997)
20. M.F. Koldunov, A.A. Manenkov, I.L. Pokotilo, J. Opt. Technol. **63**, 128 (1996)
21. M. Hosada et al., Jap. J. Appl. Phys. **1**(38), 3567 (1999)
22. A.V. Fedonov et al., Appl. Surf. Sci. **197**, 45 (2002)
23. L.D. Landau, E.M. Lifshitz, *Quantum Mechanics*, vol. 3 (Pergamon Press, Oxford, 1965), p. 176
24. L.V. Keldysh, Sov. Phys. JETP **20**, 1307 (1965)
25. D. Bauer, Phys. Rev. A **55**, 2180 (1997)
26. P. Mulser, F. Cornolti, D. Bauer, Phys. Plasmas **5**, 4466 (1998)
27. N.H. Burnett, G.D. Enright, I.E.E.E. Quantum, Electron **26**, 1797 (1990)
28. D.C. Eder, E. Amendt, S.C. Wilks, Phys. Rev. A **65**, 6761 (1992)
29. S.J. McNaught, J.P. Knauer, D.D. Meyerhofer, Phys. Rev. Lett. **78**, 626 (1997)
30. W. Lotz, Z. Physik **216**, 241 (1968); Z: Physik **220**, 466 (1969)
31. R. Fabbro, B. Faral, J. Virmont, H. Pepin, F. Cottet, J.P. Romain, Laser Part. Beams **4**, 413 (1986)
32. R. Fabbro, B. Faral, J. Virmont, F. Cottet, J.P. Romain, H. Pepin, Phys. Fluids B **1**, 644 (1989)
33. B. Faral, R. Fabbro, J. Virmont, F. Cottet, J.P. Romain, Phys. Fluids B **2**, 371 (1990)
34. R. Cauble, D.W. Phillion, T.J. Hoover, N.C. Holmes, J.D. Kilkenny, R.W. Lee, Phys. Rev. Lett. **70**, 2102 (1993)
35. S.P. Regan et al., Phys. Rev. Lett. **89**, 085003 (2002)
36. T. Löwer, R. Sigel, K. Eidmann et al., Phys. Rev. Lett. **72**, 3186 (1994)
37. A.M. Evans et al., Laser Part Beams **14**, 113 (1996)
38. M. Koenig, F. Faral, J.M. Boudenne, Phys. Rev. E **50**, R3314 (1994)
39. A. Benuzzi et al., Phys. Rev. E **54**, 2162 (1996)
40. D. Batani et al., Laser Part Beams **21**, 481 (2003)
41. C.E. Max, C.F. McKee, W.C. Mead, Phys. Fluids **23**, 1620 (1980)
42. B. Ahlborn, M.H. Key, A.R. Bell, Phys. Fluids **25**, 541 (1982)

43. C. Yamanaka, Laser plasma and inertial confinement fusion, in *Handbook of Plasma Physics*, eds. by M.N. Rosenbluth, R.Z. Sagdeev, vol. 3 (North-Holland, Amsterdam, 1991), p. 3
44. R.K. Janev, L. P. Preniakov, V. P. Shevelko, *Physics of Highly Charged Ions* (Springer, Berlin, 1985)
45. P. Drude, Annalen der Physik **1**, 566 (1900); Annalen der Physik **3**, 369 (1900)
46. A.B. Langdon, Phys. Rev. Lett. **44**, 575 (1980)
47. R. Balescu, *Equilibrium and Nonequilibrium Statistical Mechanics* (Wiley Interscience, New York, 1975)
48. R. Sauerbrey et al., Phys. Plasmas **1**, 1635 (1994)
49. T. Feurer et al., Phys. Rev. E **56**, 4608 (1997)
50. K. Eidmann, R. Sigel, in *Laser interaction and related plasma phenomena*, eds. by H.J. Schwarz, H. Hora, vol. 3B (Plenum Press, New York, 1974)
51. S.I. Anisimov, A.M. Prokhorov, V.E. Fortov, Sov. Phys. Usp. **142**, 395 (1984)
52. K. Eidmann et al., Phys. Rev. A **30**, 2568 (1984)
53. J. Lindl, Phys. Plasmas **2**, 3933 (1995)
54. J. Meyer-ter-Vehn, Nucl. Fusion **22**, 561 (1982)
55. A. Caruso, Plasma Phys. Contr. Fusion **18**, 241 (1976)
56. M. Tabak et al., Phys. Plasmas **1**, 1626 (1994)
57. P. Mulser, S. Hain, F. Cornolti, Nucl. Instr. Meth. Phys. Res. A **415**, 165 (1998)
58. C.G.M. van Kessel, R. Sigel, Phys. Rev. Lett. **33**, 1020 (1974)
59. R. Evans et al., Phys. Rev. Lett. **77**, 3359 (1996)
60. S.P. Obenschain, J. Grun, B.H. Ripin, E.A. McLean, Phys. Rev. Lett. **46**, 1402 (1981)
61. M.C. Richardson, Direct drive fusion studies, in *Handbook of Plasma Physics*, eds. by, M.N. Rosenbluth, R.Z. Sagdeev, vol. 3 (North-Holland, Amsterdam, 1991), p. 199
62. J.L. Bobin, Phys. Fluids **14**, 2341 (1971)
63. E. Cojocaru, P. Mulser, Plasma Phys. **17**, 393 (1974)
64. S.J. Gitomer, R.L. Morse, B.S. Newberger, Phys. Fluids **20**, 234 (1977)
65. M.K. Matzen, R.L. Morse, Phys. Fluids **22**, 654 (1979)
66. J. Sanz, A. Liñán, M. Rodriguez, J.R. Sanmartin, Phys. Fluids **24**, 2098 (1981)
67. R. Fabbro, E. Claire, R. Max, E. Fabre, Phys. Fluids **28**, 1463 (1985)
68. S. Hain, P. Mulser, Laser Part. Beams **15**, 541 (1997)
69. J. Grun et al., Phys. Fluids **26**, 588 (1983)
70. W.M. Manheimer, D.G. Colombant, J.H. Gardner, Phys. Fluids **26**, 1644 (1982)
71. D. Kurilovich, M.M. Basko, D.A. Kim et al., Phys. Plasmas **25**, 012709 (2018)
72. L. Zhou et al., J. Appl. Physics **117**, 125904 (2015)
73. D.E. Fratanduono et al., J. Appl. Physics **110**, 073110 (2011)
74. R. Fedosejevs et al., Phys. Rev. Lett **39**, 932 (1977)
75. D.T. Attwood et al., Phys. Rev. Lett **40**, 184 (1978)
76. A. Raven, O. Willi, Phys. Rev. Lett **43**, 278 (1979)
77. S. Hüller, E. Ahedo, Phys. Plasmas **1**, 4061 (1994)
78. S.I. Braginskii, in *Review of Plasma Physics*, ed. by M.A. Leontovich, vol. 1 (Consultants Bureau, New York, 1966)
79. Y.B. Zeldovich, Y.P. Raizer, *Physics of Shock Waves and High-Temperature Hydrodynamic Phenomena* (Academic Press, New York, 1967)
80. R. Pakula, R. Sigel, Phys. Fluids **28**, 232 (1985); Phys. Fluids **29**, 1340 (1986)
81. A. Caruso, R. Gratton, Plasma Phys. **11**, 839 (1969)
82. J.W. Shearer, W.S. Barns, Phys. Rev. Lett. **24**, 92 (1970)
83. S.D. Zacharov et al., JETP Lett. **12**, 36 (1970)
84. S.I. Anisimov, JETP Lett. **12**, 287 (1970)
85. H. Salzmann, J. Appl. Phys. **44**, 113 (1973)
86. R.C. Malone, R.L. McCrory, R.L. Morse, Phys. Rev. Lett. **34**, 721 (1975)
87. M. van Thiel, B.J. Alder, Mol. Phys. **10**, 427 (1966)
88. H. Opower, W. Press, Z. Naturforsch. **21a**, 344 (1966)
89. D.W. Gregg, S.J. Thomas, J. Appl. Phys. **37**, 4313 (1966)

90. N.G. Basov et al., JETP Lett. **5**, 141 (1967)
91. M. Mattioli, D. Veron, Plasma Phys. **11**, 684 (1969); ibid. **13**, 19 (1971)
92. F.E. Irons, N.J. Peacock, J. Phys. B **7**, 2084 (1974)
93. N.C. Fenner, Phys. Lett. **22**, 421 (1966)
94. V. A. Gurevich et al., Sov. Phys. JETP **22**, 449 (1966); Sov. Phys. JETP **36**, 646 (1973)
95. B.E. Paton, N.R. Isenor, Can. J. Phys. **46**, 1237 (1968)
96. W. Demtröder, W. Jantz, Plasma Phys. **12**, 691 (1970)
97. R. Dinger, K. Rohr, H. Weber, J. Phys. D **13**, 2301 (1980); ibid. **17** 1707 (1984); Laser Part. Beams **4**, 239 (1986); ibid. **5**, 691 (1987)
98. K. Pitsch, K. Rohr, H. Weber, J. Phys. D **14**, L51 (1980)
99. K. Mann, K. Rohr, Laser Part. Beams **10**, 435 (1992)
100. B.C. Boland, F.E. Irons, R.W.P. McWhirter, J. Phys. B **1**, 1180 (1968)
101. P. Mulser, Plasma Phys. **13**, 1007 (1971)
102. F. Begay, D.W. Forslund, Phys. Fluids **25**, 1675 (1982)
103. R.R. Goforth, P. Hammerling, J. Appl. Phys. **49**, 3918 (1976)
104. M.K. Matzen, J.S. Pearlman, Phys. Fluids **22**, 449 (1979)
105. S.V. Latyshev, I.V. Rudskoi, Sov. J. Plasma Phys. **11**, 669 (1985)
106. I. Kunz, *Interpretation of the Ion Energy Spectra in Laser-produced Plasmas Based on the Recombination Model)*. Ph.D. thesis (in German), TH Darmstadt, (1990)
107. G. Granse, S. Völlmer, A. Lenk, A. Rupp, K. Rohr, Appl. Surf. Sci. **96–98**, 97 (1996)
108. A. Rupp, K. Rohr, J. Phys. D: Appl. Phys. **28**, 468 (1995)
109. A. Caruso, G. Gatti, C. Strangio, Nuovo Cimento **2**, 1213 (1983)
110. B.K. Sinha, A. Thum-Jäger, K.P. Rohr, J. Plasma Fusion Res. **2**, 406 (1999); ibid. 410 (1999)
111. A. Thum-Jäger, K.P. Rohr, J. Phys. D: Appl. Phys. **32**, 2827 (1999)
112. A. Thum-Jäger, B.K. Sinha, K.P. Rohr, Phys. Rev. E **61**, 3063 (2000); ibid. 016405 (2001)
113. T. Müller, B.K. Sinha, K.P. Rohr, Phys. Rev. E **67**, 026415 (2003)
114. V.F. Kovalev, V.Y. Bychenkov, V.T. Tykhonchuk, J. Exp. Theor. Phys. **95**, 226 (2002)
115. R.V. Volkov, V.M. Gordienko et al., JETP Lett. **85**, 23 (2007)
116. N. Naumova, T. Schlegel, V.T. Tikhonchuk et al., Phys. Rev. Lett. **102**, 025102 (2009)
117. A. Macchi, M. Borghesi, M. Passoni, Ion acceleration by superintense laser-plasma interaction. Rev. Mod. Phys. **85**, 751–793 (2013)
118. M. Cerchez, R. Jung, J. Osterholz, T. Toncian, O. Willi, P. Mulser, H. Ruhl, Phys. Rev. Lett. **100**, 245001 (2008)
119. A.D. Piliya, Sov. Phys. Tech. Phys. **11**, 609 (1966)
120. W.L. Kruer, K. Estabrook, Phys. Fluids **28**, 431 (1985)
121. F. Brunel, Phys. Rev. Lett. **59**, 52 (1987); Phys. Fluids **31**, 2714 (1988)
122. P. Mulser, S.M. Weng, T. Liseykina, Phys. Plasmas **19**, 043301 (2012)
123. T. Liseykina, P. Mulser, M. Murakami, Phys. Plasmas **22**, 033302 (2015)
124. P. Gibbon, A.R. Bell, Phys. Rev. Lett. **68**, 1535 (1992)
125. D. Bauer, P. Mulser, Phys. Plasmas **14**, 023301 (2007)
126. S.M. Weng, P. Mulser, Z.M. Sheng, Phys. Plasmas **19**, 022705 (2012)
127. N.G. Denisov, Sov. Phys. JETP **4**, 544 (1957)
128. W. Rozmus, V.T. Tikhonchuk, Phys. Rev. A **42**, 7401 (1990)
129. K. Miyazakia, G. Miyaji, J. Appl. Phys. **114**, 153108 (2013)
130. C. Shukla, A. Das, arXiv:1707.00480 [physics.plasm-ph] (2017)
131. J.J. Honrubia, Nat. Phys. **3**, 835 (2007)
132. M.A. Purvis et al., Nat. Photonics **7**, 796 (2013)
133. C. Bargsten et al., Sci. Adv. **3**, e1601558 (2017)

Chapter 2
Single Particle Motion

Motivation

Laser-matter interaction is a many body problem. In the classical realm already the three body problem is not solvable analytically. On the other hand understanding physics evolves along the analytic path. So, what is the solution to the dilemma? There is a twofold approach: development of collective concepts and models that are able to describe macroscopic aspects of matter, for example the many body phenomenon, to be summarized under the general concept of screening, with Debye screening as a special example. Alternatively, there is the possibility of reduction to single particle behavior. Reduction yields very satisfactory insight in the case the forces are known that act on the single particle. Unfortunately, this is generally not the case. Here, help comes from statistics. In the kinetic theory systematic procedures of reduction have been developed and applied with great success, for instance the BBGKY hierarchy, to be introduced in a further chapter.

We start with the description of the single particle in model situations. As our main concern is the behavior of a point particle in the laser field this kind of motion is studied in the non-relativistic and in the fully relativistic regime. It is characterized by rapid oscillations. In the single monochromatic wave it is quite simple to obtain a precise picture of the motion over a few periods as long as radiative effects can be neglected. In the near infrared frequency domain of high power lasers it covers the intensity interval from zero to $10^{22}\,\mathrm{Wcm}^{-2}$. In high power laser-matter interaction the classical single particle dynamics is generally believed to give satisfactory answers. Why this is so will be discussed with typical dynamical situations under quantum aspects. However, there are exceptions. No general rules for deviations exist so far; one aspect of the same motion satisfies classical behavior, others deviate from it.

Frequently one is not interested in the details of particle motion, rather is the orbit on large spacetime scales averaged over the single oscillation of primordial interest, the so called oscillation center motion. This faces us with the delicate question of correct averaging. Particular attention is dedicated to averaging in connection with the so-called ponderomotive force and related ponderomotive potential.

P. Mulser, *Hot Matter from High-Power Lasers*, Graduate Texts in Physics,
https://doi.org/10.1007/978-3-662-61181-4_2

Understanding single particle dynamics results most immediately from Newtonian mechanics. General laws, however, and powerful methods of calculation take advantage from Hamiltonian and Lagrangian formulation of mechanics. In view of their eminent role in all fields of dynamics, e.g., statistical physics and field theory, the essentials of classical mechanics are outlined in this chapter. Some aspects addressed in the context may be of interest also to the specialist.

There are two powerful means to test correct averaging in space-time: adiabatic invariants and change of the reference system. If forces and potentials become explicitly time dependent a Hamiltonian is no longer conservative. The question arises which quantity, if there is any, takes the place of the former energy conservation. In periodic motions free of hidden resonances there exists such a quantity under the condition that the Hamiltonian depends weakly on an external parameter; it is the adiabatic invariant. The particle bouncing forth and back between two elastic walls in motion against each other and the pendulum with slow and smooth variation in length are prominent examples. In both cases the cycle averaged energy times the oscillation period is conserved; the energy is floating.

The other powerful discriminator is the change of inertial system of reference. A correctly formulated force for instance has to satisfy well defined requirements, for instance to behave like a three vector in Galileian relativity, that is under slow motions, and like a four vector in genuine, i.e., Lorentzian relativity. In this context an important discovery is immediate: Maxwell's equations, the best theory we have got in physics, are incompatible with Galileian relativity, as is shown already by elementary examples. Throughout the book and, in particular, in the present chapter frequent use is made of reference system changes. In specific cases such a change may be all but simple.

2.1 Non-relativistic Regime

2.1.1 Electron in the Electromagnetic Wave

Many essential properties of the behavior of matter in the laser field are well described by the dynamics of a single electron in a monochromatic linearly polarized plane wave. The wave is characterized by the amplitude $\hat{\mathbf{E}}$, the wave vector $\mathbf{k}$, $|\mathbf{k}| = 2\pi/\lambda$, λ wavelength, and the frequency ω. In complex notation it reads in space-time $(\mathbf{x}, t)$

$$\mathbf{E}(\mathbf{x}, t) = \hat{\mathbf{E}}(\mathbf{x}, t)\mathrm{e}^{\mathrm{i}(\mathbf{k}\mathbf{x}-\omega t)}. \tag{2.1}$$

In order to be defined properly the amplitude $\hat{\mathbf{E}}(\mathbf{x}, t)$ must evolve on a distinctly slower space-time scale than the instantaneous field $\mathbf{E}(\mathbf{x}, t)$. From Maxwell's equation $\nabla \times \mathbf{E} = -\partial_t \mathbf{B}$ the magnetic field $\mathbf{B}$ associated with $\mathbf{E}$ follows,

$$\mathbf{B} = \frac{\mathbf{k}}{\omega} \times \mathbf{E}. \tag{2.2}$$

The fields **E** and **B** are intended as to refer to an inertial system of reference S, in our context frequently named laboratory (lab) frame. For the coupling of the fields to matter a force equation is needed. We define it in the system S' in which the particle is supposed to be momentarily at rest. A function of position **x**, velocity zero, and time t is a force $\mathbf{f}(\mathbf{x}, t)$ on a mass point of mass m if its change of momentum is described by the equality

$$m\frac{\mathrm{d}^2}{\mathrm{d}t^2}\mathbf{x} = \mathbf{f}(\mathbf{x}, t). \tag{2.3}$$

More specifically, it is named Newton force $\mathbf{f}_N$, $\mathbf{f}_N(\mathbf{x}, t) = \mathbf{f}(\mathbf{x}, \mathbf{v} = 0, t)$. The Newton force $\mathbf{f}_N$ on a point charge q is

$$\mathbf{f}_N(\mathbf{x}, t) = q\mathbf{E}(\mathbf{x}, t). \tag{2.4}$$

Forces of type $\mathbf{f}(\mathbf{x}, t)$ from (2.4) are called *impressed* forces, in contrast to the *inertial* forces expressing the identity $\mathbf{f} = m\ddot{\mathbf{x}}$ in a *tangent* inertial system, i.e., reference frame locally moving with **v** of the particle. The centrifugal and the Coriolis force are examples of inertial forces. In order to define the force on a particle moving at velocity **v** the transformation laws between the inertial systems S and S' must be specified. Here we do it in non-relativistic mechanics for velocities $v = |\mathbf{v}| \ll c$, c the velocity of light.

Galileian Relativity

In Galileian relativity there is no upper limit to velocity and there is a universal time t. Hence, an inertial system S' of coordinates $\mathbf{x}'$, t' moving with velocity **v** with respect to S is related to this by the Galilei transformation

$$\mathbf{x}' = \mathbf{x} - \mathbf{v}t, \quad t' = t; \qquad \mathbf{x} = \mathbf{x}' + \mathbf{v}t', \quad t = t'. \tag{2.5}$$

Two physical points P = $(\mathbf{x}, t)$ and P' = $(\mathbf{x}', t')$ are the same point P when their coordinates are linked together by a Galileian transformation (2.5). The set of points P is the *affine vector space* R^4. Relations (2.5) imply for the force **f** in S

$$\mathbf{f}(\mathbf{x}, t) = \mathbf{f}_N(\mathbf{x}', t') = q\mathbf{E}'(\mathbf{x}', t'). \tag{2.6}$$

if q is at rest in S'. In order to express **f** in terms of q and the electromagnetic field in S their transformation from S to S' must be established. The components of **E** and **B** parallel to **v** are not affected when a capacitor and an infinitely long coil are moved with their field lines parallel to **v**, hence $\mathbf{E}'_\parallel = \mathbf{E}_\parallel$ and $\mathbf{B}'_\parallel = \mathbf{B}_\parallel$. Concerning the perpendicular components Faraday's law of induction tells for any deformable moving loop L

$$\int_{L(t')} \mathbf{E}'\mathrm{d}\mathbf{s}' = -\frac{\mathrm{d}}{\mathrm{d}t}\int_{\boldsymbol{\Sigma}(t)} \mathbf{B}\mathrm{d}\boldsymbol{\Sigma} = -\int_{\boldsymbol{\Sigma}(t)} \frac{\partial}{\partial t}\mathbf{B}\mathrm{d}\boldsymbol{\Sigma} - \int_{L(t)} \mathbf{B}\mathbf{v}\times\mathrm{d}\mathbf{s} = \int_{L(t)} (\mathbf{E}+\mathbf{v}\times\mathbf{B})\mathrm{d}\mathbf{s}, \tag{2.7}$$

see Fig. 2.1. In Galileian relativity holds $\mathrm{d}\mathbf{s}' = \mathrm{d}\mathbf{s}$ and $t' = t$. The transformation of the perpendicular component of $\mathbf{E}$ from S to $S'(\mathbf{v})$ follows from an arbitrary loop $L(t)$ by letting shrink its diameter to zero along one direction,

$$\mathbf{E}'(\mathbf{x}', t') = \mathbf{E}(\mathbf{x}, t) + \mathbf{v}(\mathbf{x}, t) \times \mathbf{B}(\mathbf{x}, t), \qquad t' = t. \tag{2.8}$$

The LHS term in the first equality (2.7) is the voltage of E' along the moving and deforming loop at the time instant $t' = t$, the term on the RHS is the magnetic flux change in S and as such, owing to $t' = t$, the same in S' as well. The reader may convince himself that all following terms are to be associated with S. By the same procedure the transformation law of $\mathbf{B}$ follows from the Maxwell equation $\varepsilon_0 c^2 \nabla \times \mathbf{B} = \varepsilon_0 \partial_t \mathbf{E}$,

$$\mathbf{B}' = \mathbf{B} - \frac{\mathbf{v}}{c^2} \times \mathbf{E}. \tag{2.9}$$

From (2.6) and (2.8) the well known Lorentz force results and the momentum equation reads in the lab frame,

$$m\frac{\mathrm{d}^2\mathbf{x}}{\mathrm{d}t^2} = \mathbf{f}(\mathbf{x}, \mathbf{v}, t) = q[\mathbf{E}(\mathbf{x}, t) + \mathbf{v}(\mathbf{x}, t) \times \mathbf{B}(\mathbf{x}, t)]. \tag{2.10}$$

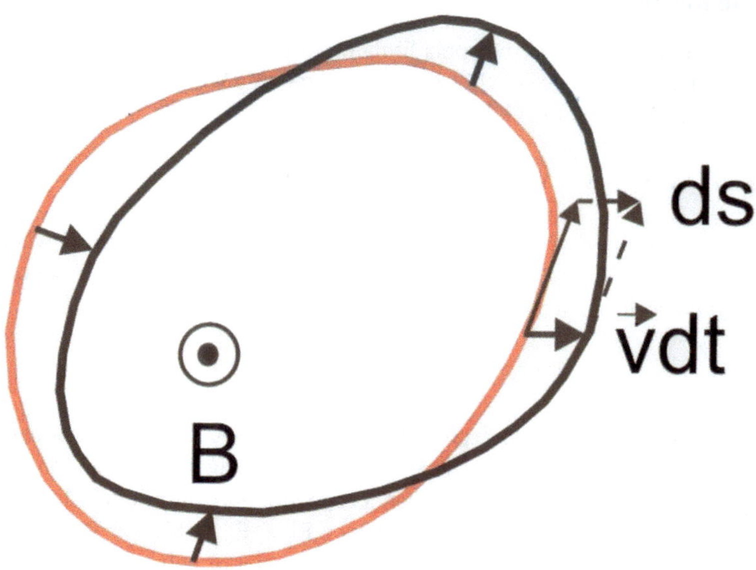

Fig. 2.1 Faraday's induction law applied to loop which deforms from red to black in the time interval $\mathrm{d}t$; $\mathrm{d}s$ line element. Magnetic field **B** points out of plane

The Lorentz force **f** must transform like **v** in R^3 and so do **v**, **E** and $\mathbf{v} \times \mathbf{B}$. This implies that under the inversion operation $\mathbf{x} \rightarrow -\mathbf{x}$, **B** is not affected: $\mathbf{B}(-\mathbf{x}) = \mathbf{B}(\mathbf{x})$.

The field transformations (2.8) and (2.9) follow from Maxwell equations in Galileian invariance of d**s** and dt. Nevertheless they are in contradiction with elementary optics. In fact, the transformation $\mathbf{B} \rightarrow \mathbf{B}'$ can be followed directly from (2.8) with the aid of (2.5) by transforming $\mathbf{E}'$ to $S''(-\mathbf{v})$ and observing the identity $S'' = S$. The stringent result is $\mathbf{B}' = \mathbf{B}$, in net contradiction to (2.9). Analogously, in Galileian relativity follows $\mathbf{E}' = \mathbf{E}$ from (2.9) and (2.5). Thus, (2.5) violates orthogonality between $\mathbf{E}'$, $\mathbf{B}'$ and $\mathbf{k}'$ of the plane wave. In other words, Maxwell's equations are incompatible with Galileian relativity. Lorentzian relativity will tell us later that (2.8) and (2.9) are the correct nonrelativistic limits of $\mathbf{E}'$ and $\mathbf{B}'$ but among themselves they do not transform Galilei-like.

Field theoretical view on Lorentz force. According to (2.10) the magnetic field influences the motion of a charge q when looking from the lab frame. In the frame comoving with the charge the particle sees only the electric field $\mathbf{E}'$ at its place and no magnetic field, see (2.3) and (2.8). This is in agreement with the absence of magnetic charges: The magnetic field does not couple to charged matter at rest. Frequently Ohms law is written as $\mathbf{j} = \sigma \mathbf{E}$. It must be stressed that this is true only if **E** is taken in the reference system comoving with the charge. Thus, in the lab frame Ohm's law reads $\mathbf{j} = \sigma(\mathbf{E} + \mathbf{v} \times \mathbf{B})$.

In the bare electromagnetic wave (for example, no static electric fields) the correction of (2.9) by the electric field term can be ignored at nonrelativistic speeds. Note, when making use of approximate equations combinations among them must be selected in a way as to minimize the deviations. This is a principle of all perturbation theory. Sometimes proceeding according to orders of perturbation, first order perturbation, second, third order, etc. may be alright. Attention has to be paid to small terms if they evolve unstable. Sometimes symmetry properties may help to choose the right combinations: In the nonrelativistic regime $\mathbf{E}'$ and $\mathbf{B}'$ from (2.8) and (2.9) go together because of the symmetry of Maxwell's equations of **E** and **B**, and not (2.8) combined with $\mathbf{B}' = \mathbf{B}$, although the latter follows from the Galileian relativity.

Oscillation Energy of the Free Electron

In a wide range of laser intensities and frequencies all properties of matter are determined by the dynamics of the electron as the second lightest massive stable particle. Its non-relativistic oscillatory motion is obtained from (2.10) by setting $q = -e$, $e = 1.6 \cdot 10^{-19}$ C, and $m = m_e$, $m_e = 9.1 \cdot 10^{-31}$ kg. The total derivative $\mathrm{d}/\mathrm{d}t = \partial/\partial t + (\mathbf{v}\nabla)$ is approximated by its linear part ∂_t. Formally, the approximation corresponds to the transformation to the comoving frame $S'(\mathbf{v})$ and hence the Lorentz term $\mathbf{v} \times \mathbf{B}$ and the convective term $(\mathbf{v}\nabla)$ as its counterpart have to be omitted simultaneously. This corresponds to determine the oscillatory motion of the electron in its fundamental mode ω,

$$\mathbf{v} = \hat{\mathbf{v}} e^{i(kx-\omega t)}, \qquad \boldsymbol{\delta} = \hat{\boldsymbol{\delta}} e^{i(kx-\omega t)}. \tag{2.11}$$

with the oscillation center $\boldsymbol{x}_0$ at rest. Strictly speaking, in S' ω is the Doppler shifted $\omega' = \omega - \mathbf{kv} \simeq \omega$. The amplitudes of velocity $\hat{\mathbf{v}}$, of displacement $\hat{\boldsymbol{\delta}}$, and of the mean oscillation energy $W = m_e(\Re\hat{\mathbf{v}})^2/4$ are given by

$$\hat{\mathbf{v}} = -i\frac{e}{m_e\omega}\hat{\mathbf{E}}, \qquad \hat{\boldsymbol{\delta}} = \frac{e}{m_e\omega^2}\hat{\mathbf{E}}, \qquad W = \frac{e^2}{4m_e\omega^2}\hat{\mathbf{E}}\hat{\mathbf{E}}^*. \tag{2.12}$$

The energy flux of the laser is characterized by the intensity $\mathbf{I}$. In the case of a plane wave it is to be identified with the time averaged Poynting vector $\boldsymbol{S}$ or, when convenient, with its modulus $I = |\boldsymbol{S}|$,

$$\mathbf{I} = \overline{\boldsymbol{S}} = \varepsilon_0 c^2 \overline{\boldsymbol{E} \times \boldsymbol{B}} = \frac{\mathbf{k}^0}{2} c\varepsilon_0 \hat{\mathbf{E}}\hat{\mathbf{E}}^*; \qquad \mathbf{k}^0 = \mathbf{k}/|\mathbf{k}|. \tag{2.13}$$

Numerically, wave amplitude and intensity are connected by

$$\hat{E}[\mathrm{Vcm}^{-1}] = 27.5 \times (I[\mathrm{Wcm}^{-2}])^{1/2}. \tag{2.14}$$

At $I = 3.5 \times 10^{16}\,\mathrm{Wcm}^{-2}$ the electric field amplitude equals the field strength $E = 5 \times 10^9\,\mathrm{Vcm}^{-1}$ of the electron in the ground state of the hydrogen atom. At the wavelength of $\lambda = 1\,\mu\mathrm{m}$ the associated velocity and oscillation amplitudes of the free electron and its quiver energy are $\hat{v} = 4.8 \times 10^7\,\mathrm{ms}^{-1}$, corresponding to $\beta = \hat{v}/c = 0.16$, $\hat{\delta} = 25.5\,\mathrm{nm}$, $W = 3.4\,\mathrm{keV}$. Let us compare these values with them of a free electron exposed to the sun light on earth, all assumed as concentrated at one micron wavelength: $I = 0.133\,\mathrm{Wcm}^{-2}$, $\hat{E} = 10\,\mathrm{Vcm}^{-1}$, $\hat{v} = 0.1\,\mathrm{ms}^{-1}$, $\hat{\delta} = 5 \times 10^{-8}\,\mathrm{nm}$, $W = 1.3 \times 10^{-14}\,\mathrm{eV}$.

Physical quantities are real. Often it may be convenient to use complex expressions for observable quantities, as for example the fields in (2.1) and (2.2). In *linear polarization* they stay either for their real or their imaginary parts. A complex solution of a linear equation with *real* coefficients is automatically a solution of its real and its imaginary part separately. The advantage is lost in nonlinear equations or when complex coefficients are involved. The use of complex amplitudes $\hat{A} = |\hat{A}| \exp i\phi$ is particularly indicated because they allow to express phase differences ϕ in a compact manner. Consequently, the physical amplitude is $|\hat{A}|$. In quadratic terms it appears as the product $\hat{A}\hat{A}^*$.

In circular polarization the electric field $\mathbf{E}$ is most conveniently described by taking its real and its imaginary part in the complex plane (y, iz) perpendicular to the direction of propagation $\mathbf{k}$ along x. Then, with real and imaginary parts taken both, expression (2.1) stands for circular polarization as well. At fixed position $\mathbf{x}$ in space right circular polarization is defined by the field vectors rotating *clockwise* when viewing along the propagation vector $\mathbf{k}$. Thus the exponent $\exp i(\mathbf{kx} - \omega t)$ in (2.1) defines right polarization. The same wave is also described by $\exp -i(\mathbf{kx} - \omega t)$ cor-

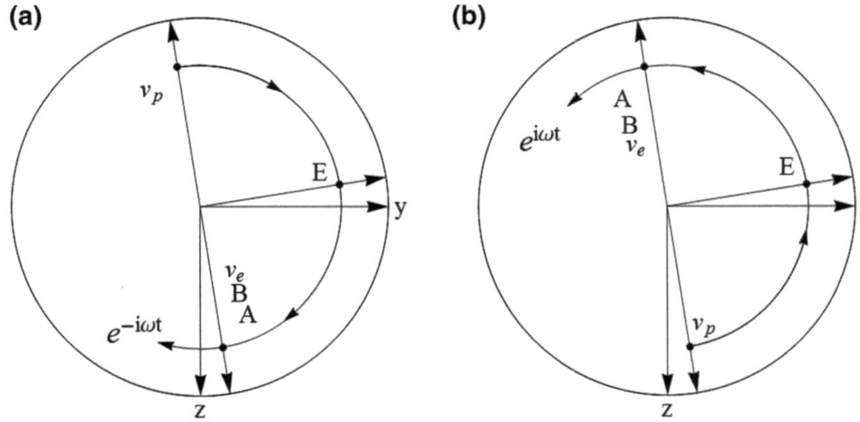

Fig. 2.2 **a**: Right hand circular polarization of an electromagnetic wave propagating into x direction, helicity −1: Polarization vector at fixed position **x** moves clockwise as in (2.1). Magnetic field **B**, vector potential **A**, and the velocity $\mathbf{v}_e$ of a negative charge are parallel to each other and orthogonal to the electric field **E**. **b**: Left hand circular polarization, helicity +1

responding to simultaneous space and time inversion $\mathbf{k} \to -\mathbf{k}, t \to -t$. At fixed time the field vector rotates in space in opposite direction, expressed by *helicity* $h = -1$. In left hand circular polarization the rotation of the field vectors at fixed position occurs *counter clockwise*, with helicity assigned $h = +1$. Left hand polarization is achieved by changing $\exp i(\mathbf{kx} - \omega t)$ into $\exp i(\mathbf{kx} + \omega t)$. According to our definition this is equivalent to $\exp -i(\mathbf{kx} + \omega t)$. In reflection of a light beam from a plane surface circular reflection and helicity are conserved at any angle of incidence and reflection. The motion of a charge in the single mode **k** lies entirely in the plane normal to **k**. For the electron the relative phases of the field quantities are shown in Fig. 2.2 for $h = -1$, right hand circular polarization (a), $h = +1$, left hand circular polarization (b), and **k** pointing along x: All quantities in (a) rotate clockwise in time at fixed position, **E** with phase delay of $\pi/2$ relative to **B**, **A**, and the velocity of a negative charge $\mathbf{v}_e$; in (b) they rotate counter clockwise. The expressions for the amplitudes $\hat{\mathbf{v}}$ and $\hat{\boldsymbol{\delta}}$ in (2.12) hold for linear as well as for circular polarization, the oscillation energy W, and the intensity I in circular polarization are twice their values of linear polarization. Hence, in circular polarization (2.14) is to be replaced by

$$\hat{E}[\mathrm{Vcm}^{-1}] = 27.5/\sqrt{2} \simeq 20 \times I[\,\mathrm{Wcm}^{-2}]^{1/2}. \tag{2.15}$$

The radius of rotation R_e of the electron in circular polarization is the modulus of the amplitude $\hat{\delta}$ in linear polarization, $R_e = |\hat{\delta}|$.

Electron with damping. The motion of a free electron embedded in an environment of other particles can be described in its simplest form by introducing a phenomenological linear damping to obtain the equation of motion under friction,

$$\frac{\partial \mathbf{v}}{\partial t} + \nu \mathbf{v} = -\frac{e}{m_e}\mathbf{E}. \tag{2.16}$$

The steady state solution is

$$\mathbf{v} = -i\frac{e}{m_e(\omega + i\nu)}\mathbf{E} = -i\frac{e}{m_e\omega}\frac{1 - i\nu/\omega}{1 + \nu^2/\omega^2}\mathbf{E}. \tag{2.17}$$

This and corresponding quantities differ now from (2.12) solely by the factor $(1 + i\nu/\omega)^{-1}$ for $\mathbf{E}$ and $(1 + \nu^2/\omega^2)^{-1}$ for W. In linear polarization the cycle averaged dissipation by friction is under steady state conditions

$$\dot{E} = m_e\nu\,\overline{(\Re\,\mathbf{v})^2} = -e\overline{\Re\,\mathbf{v}\,\Re\,\mathbf{E}} = -\frac{e^2}{m_e\varepsilon_0\omega^2}\frac{\nu}{1 + \nu^2/\omega^2}\frac{I}{c}. \tag{2.18}$$

This is the classical formula for collisional absorption of a laser beam by the single electron in the fully ionized plasma. For circular polarization the absorption is twice the last term on the right.

The friction coefficient ν has the dimension of an inverse time, $\nu = 1/\tau$. It has its origin in the deviations of the electron from a straight line by collisions with other electrons, with ions, and with neutral atoms or molecules. We expect that a week electric field (low laser intensity) does not influence the collision dynamics so that ν is an intrinsic property of matter. To give it a physical meaning it is therefore legitimate to consider (2.16) in the absence of an electric field,

$$\frac{\mathrm{d}\mathbf{v}}{\mathrm{d}t} = -\frac{\mathbf{v}}{\tau}; \qquad \nu = \frac{1}{\tau}. \tag{2.19}$$

This can be interpreted as follows. Each time interval τ the electron undergoes a deflection of $\pi/2$ by an elastic collision the loss of velocity in the original direction of $\mathbf{v}$ is given by $\mathrm{d}\mathbf{v}/\mathrm{d}t$. In collisions of hard spheres the average deviation from the direction of $\mathbf{v}$ in one collision is exactly $\pi/2$. Therefore τ is given the name *collision time* and consequently ν is the *collision frequency*. Collision time thereby is to be understood as the time between two consecutive collisions that occur instantaneously. In the kinetic chapter it will be shown that the simple model is close to reality if $\nu = 1/\tau$ is taken as an ensemble average. The relation (2.18) is true in the mean of an ensemble average, not for an individual electron.

General Harmonic Dynamics

In view of its eminent importance in all fields of physics let us briefly recall the *harmonic oscillator* and its properties. The equation of motion of a point charge of mass m in the one dimensional potential $V = m\omega_0^2x^2/2$, $\omega_0 = \text{const}$, is

$$\ddot{x} + \omega_0^2 x = 0. \tag{2.20}$$

Two linearly independent solutions are $\sin \omega_0 t$ and $\cos \omega_0 t$ or $\exp i\omega_0 t$ and $\exp -i\omega_0 t$ and any linear combination of them. The linearly damped or amplified harmonic oscillation is described by

$$\ddot{x} + \nu\dot{x} + \omega_0^2 x = 0; \quad \nu > 0: \text{ damping}, \quad \nu < 0: \text{ amplification}. \tag{2.21}$$

A solution is found by the ansatz $x = \exp \lambda t$ and

$$(\lambda^2 + \nu\lambda + \omega_0^2)x = 0 \;\Rightarrow\; \lambda_{1,2} = -\frac{\nu}{2} \pm \frac{1}{2}(\nu^2 - 4\omega_0^2)^{1/2}.$$

We distinguish
(a) $\nu^2 < 4\omega_0^2$ with $\lambda_{1,2} = -\nu/2 \pm i\omega_1,\ \omega_1 = (\omega_0^2 - \nu^2/4)^{1/2}$ and the general solution

$$x(t) = \mathrm{e}^{-(\nu/2)t}\big[C_1 \cos \omega_1 t + C_2 \sin \omega_1 t\big] = \mathrm{e}^{-(\nu/2)t} A \cos(\omega_1 t - \varphi),$$

$$A = (C_1^2 + C_2^2)^{1/2},\ \tan \varphi = \frac{C_2}{C_1}. \tag{2.22}$$

(b) $\nu^2 > 4\omega_0^2$ with $\lambda_{1,2} = -\nu/2 \pm \omega_1,\ \omega_1 = (\nu^2/4 - \omega_0^2)^{1/2}$ and the general solution

$$x = \mathrm{e}^{-(\nu/2)t}\big[C_1 \mathrm{e}^{\omega_1 t} + C_2 \mathrm{e}^{-\omega_1 t}\big]. \tag{2.23}$$

(c) $\lambda_{1,2} = -\nu/2$. The general solution for $\omega_1 = 0$ is

$$x = \mathrm{e}^{-(\nu/2)t}\big[C_1 t + C_2\big]. \tag{2.24}$$

as seen by inspection.
The driven harmonic oscillator excited by the harmonic driver $D \cos \omega t$ under condition (a) $\nu^2 < 4\omega_0^2$,

$$\ddot{x} + \nu\dot{x} + \omega_0^2 x = D \cos \omega t \tag{2.25}$$

is of high relevance. In presence of damping or amplification $\nu \neq 0$ a particular solution is $x = x_D$,

$$x_D = x_0 \cos(\omega t + \varphi);\ x_0 = \frac{D}{\big[(\omega^2 - \omega_0^2)^2 + \nu^2\omega^2\big]^{1/2}}, \quad \tan \varphi = \frac{\nu\omega}{\omega^2 - \omega_0^2}. \tag{2.26}$$

At $\omega = \omega_0$ the *dephasing* is $\varphi = \pm\pi/2$, driver and displacement x_D are orthogonal to each other. The displacement x reaches its maximum at the resonance ω_0',

$$\omega_0' = \omega_0 \left(1 - \frac{\nu^2}{2\omega_0^2}\right)^{1/2}. \tag{2.27}$$

Multiplication of (2.25) with $\dot{x}$ and time averaging over one oscillation period T yields the important relation

$$\int\left(\frac{\mathrm{d}}{\mathrm{d}t}\frac{\dot{x}^2}{2}+\nu\dot{x}^2+\omega_0^2\frac{\mathrm{d}}{\mathrm{d}t}\frac{x^2}{2}\right)\frac{\mathrm{d}t}{T}=\frac{1}{2}\nu\omega^2x_0^2=\frac{\nu\omega^2D^2}{2\left[(\omega^2-\omega_0^2)^2+\nu^2\omega^2\right]}. \tag{2.28}$$

The term $\nu\overline{\dot{x}^2}T=\int\dot{x}D\cos\omega t\mathrm{d}t$ is the dissipated work per cycle. Its maximum is reached at the unshifted resonance frequency $\omega=\omega_0$ for any $\nu>0:\max\nu\overline{\dot{x}^2}=D^2/2\nu$. In the absence of damping or amplification, $\nu=0$, no steady state solution of (2.25) exists at resonance $\omega=\omega_0$. The corresponding particular solution $x_D(t)$ is found by the intuitive ansatz $x_D=C(t)\sin\omega_0 t$ and confirmed as

$$x_D=\frac{D}{2\omega_0}t\sin\omega_0 t;\qquad \nu=0. \tag{2.29}$$

Alternatively, it can be obtained for $\nu=0$ in the limit $\omega\mapsto\omega_0$.

The quantity x is introduced as a displacement. The harmonic oscillator equation (2.25) is of nearly universal applicability in linear oscillation problems, like linear waves, their resonant interactions and stability, collisional and resonant absorption, and parametric growth (see the following chapters). Correspondingly, x is any quantity, velocity, density, field, operator. A second order linear wave equation turns into an oscillator equation after a Fourier transformation in space or in time. Accordingly, the derivatives $\dot{x}$, $\ddot{x}$ refer to time or to the space coordinates. An eminent example of the latter is the stationary electromagnetic wave equation of Chap. 5. Some significant properties of (2.25) are

- The parabolic potential is unique in so far as it is the only one with constant oscillation period $T=2\pi/\omega$ for all displacements x.
- Damping and growth lead to *dephasing* between driver D and displacement x, and to
- Detuning of the resonance frequency and broadening of the frequency band.

2.1.2 Lagrangian and Hamiltonian Description of Motion

In principle, all we can calculate in classical dynamics can be done in Newtonian formalism. The prototype equation is the equation of motion (2.3). Far more and immediate insight in the dynamics, as well as flexibility in performing calculations, is gained by the Lagrangian description, to mention in first place conservation laws and Noether's theorem, and the advantage of generalized coordinates. A further step towards symmetrization is achieved in the Hamiltonian formulation of mechanics. From cyclic coordinates and Poisson brackets conserved quantities and symmetries are seen at first glance. From the Poincaré–Cartan invariant Liouville's theorem and affine conservation laws follow that are at the basis of kinetic theory and statistical physics. Canonical perturbation theory is a major field of application in nonlinear dynamics. Finally, the Hamiltonian formalism allows for developing classical and quantum dynamics in parallel.

In short, the Hamiltonian description shows its power in establishing general laws, Newtonian representation is the basis of intuition and physical insight; for calculations the Lagrangian procedure is advised.

Hamilton's Principle of Least Action

Consider a free particle subject to the force $\mathbf{f}(\mathbf{x}, t)$. Under its action it moves in such a way that the impressed force $\mathbf{f}(\mathbf{x}, t)$ is balanced by its inertial force $m\ddot{\mathbf{x}}$, i.e., $\mathbf{f}(\mathbf{x}, t) - m\ddot{\mathbf{x}} = 0$. If we transform to the inertial frame $S'[\mathbf{v} = \mathbf{v}(t)]$ co-moving with the particle at the instant t, the so called *tangential* reference frame, there is no criterion that allows to distinguish between impressed and inertial forces. For instance, the centrifugal force is felt like a "real force". In the tangential reference frame the system is static and the principle of virtual displacements $\delta\mathbf{x}'$ is true by intuition: zero force generates zero work. Furthermore, as $\delta\mathbf{x}'$ is arbitrary, also arbitrarily short and the system is static for a time τ, the minimum time Δt_{min} required for the physical displacement $\Delta\mathbf{x} = \delta\mathbf{x}$ can be chosen sufficiently short, i.e. $\Delta t_{\mathrm{min}} \leq \tau$, and at the same time long enough to simultaneously minimize the inertia involved in the displacement process. As the mathematical procedure is playing on differentials, i.e., on the linear part of finite physical displacements $\Delta\mathbf{x}$, this means that any $\delta\mathbf{x}'$ can be assumed to occur instantaneously to simplify the mathematical procedure without physical consequences. After transforming back to the lab frame it follows that the property of virtual displacement $\delta\mathbf{x}$ is maintained because of the arbitrariness of $\delta\mathbf{x}'$; the classes $\{\delta\mathbf{x}'\}$ and $\{\delta\mathbf{x}\}$ are the same. Thus,

$$[m\ddot{\mathbf{x}} - \mathbf{f}(\mathbf{x}, t)]\delta\mathbf{x} = 0 \tag{2.30}$$

holds in all inertial reference systems. This is *D'Alembert's principle* of dynamics of a mass point. Once the equivalence between impressed and inertial forces is taken for true, as a consequence of the identity of classes $\{\delta\mathbf{x}'\}$ and $\{\delta\mathbf{x}\}$ D'Alembert's principle is identical with the principle of virtual work. This is particularly evident in the case of a mass point in the free falling elevator: it costs zero work to displace the mass point by a vanishing amount. D'Alembert's principle is generally considered as a fundamental principle of dynamics, not deducible from Newton's law. The reason for giving the preference to the formulation in terms of virtual work, i.e., by $\delta\mathbf{x}$ occurring instantaneously, instead of equating the sum of forces to zero is that a scalar quantity is easier to handle, in particular in presence of constraints imposed on a system from outside. However, even the scalar version (2.30) of Newton's law of motion (2.3) is still not easy enough to be practiced straightforwardly.

In the following it suffices to limit ourselves to potential forces. By definition they fulfill

$$\mathbf{f}(\mathbf{x},t) = -\nabla V(\mathbf{x},t) \Leftrightarrow \mathrm{d}V(\mathbf{x},t=\mathrm{const}) = \nabla V(\mathbf{x},t)\mathrm{d}\mathbf{x} = \nabla V(\mathbf{x},t)\delta\mathbf{x} = \delta V. \tag{2.31}$$

Then from (2.3) the existence of the kinetic energy $T = m\mathbf{v}^2/2$ follows as a unique function of position and time and (2.30) reads along all points of a physical orbit

$$\delta T(\mathbf{x},t) + \delta V(\mathbf{x},t) = 0$$

In the variations $\delta T, \delta V$ the time t is held fixed, as required by (2.31). This can be cast into the equivalent integral along an orbit passing through any couple of points $\mathbf{x}_0, \mathbf{x}_1$ at times t_0, t_1,

$$0 = \int_{\mathbf{x}_0}^{\mathbf{x}_1} \delta(T+V)\mathrm{d}t = \int_{\mathbf{x}_0}^{\mathbf{x}_1} \left(\frac{\partial T}{\partial \dot{\mathbf{x}}}\delta\dot{\mathbf{x}} + \frac{\partial V}{\partial \mathbf{x}}\delta\mathbf{x}\right)\mathrm{d}t. \tag{2.32}$$

To proceed further and to eliminate the dependence of $\delta\dot{\mathbf{x}}$ on the free displacement $\delta\mathbf{x}$ use is made of the property of $\delta\mathbf{x}$ to be virtual, i.e., taken at $t = \mathrm{const}$. It implies the very essential differential relation

$$\frac{\mathrm{d}}{\mathrm{d}t}\delta\mathbf{x} = \delta\frac{\mathrm{d}}{\mathrm{d}t}\mathbf{x} = \delta\dot{\mathbf{x}}, \tag{2.33}$$

where the differential $\mathrm{d}\delta\mathbf{x}$ is the difference of the isochronous shifts $\delta\mathbf{x}(t+\mathrm{d}t) - \delta\mathbf{x}(t) = \delta\mathbf{v}\mathrm{d}t$. The interchangeability of $\mathrm{d}/\mathrm{d}t$ with δ is a consequence of the variations $\delta T, \delta V$ performed in the class $\{\delta\mathbf{x}\}$ of the free virtual displacements .

It was recognized by Lagrange and Euler that the bounded variation $\delta\dot{\mathbf{x}}$ can be eliminated by passing to the integral in time of $\delta(T+V) = 0$ if the two endpoints of the orbit are kept constant; thus $\delta\mathbf{x}_0 = \delta\mathbf{x}_1 = 0$. The integrand on the derivative of T in (2.32) splits into a difference of two parts, $(\partial T/\partial\dot{\mathbf{x}})\delta\dot{\mathbf{x}} = \mathrm{d}[\delta\mathbf{x}(\partial T/\partial\dot{\mathbf{x}})]/\mathrm{d}t - \delta\mathbf{x}\mathrm{d}(\partial T/\partial\dot{\mathbf{x}})/\mathrm{d}t$, the first of them does not contribute to the integral owing to vanishing variation $\delta\mathbf{x}$ at the endpoints and we are left with the integral on the variation $\delta L(\mathbf{x},\dot{\mathbf{x}},t) = \delta T(\dot{\mathbf{x}}) - \delta V(\mathbf{x},t)$,

$$\delta S = \int_{\mathbf{x}_0}^{\mathbf{x}_1}\left(\frac{\mathrm{d}}{\mathrm{d}t}\frac{\partial T}{\partial\dot{\mathbf{x}}} + \frac{\partial V}{\partial\mathbf{x}}\right)\delta\mathbf{x}\mathrm{d}t = \int_{\mathbf{x}_0}^{\mathbf{x}_1}\left(\frac{\mathrm{d}}{\mathrm{d}t}\frac{\partial L}{\partial\dot{\mathbf{x}}} - \frac{\partial L}{\partial\mathbf{x}}\right)\delta\mathbf{x}\mathrm{d}t = \int_{\mathbf{x}_0}^{\mathbf{x}_1}\delta L\mathrm{d}t = 0. \tag{2.34}$$

By the equality $\delta S = \int \delta L\mathrm{d}t$ the variation of the *action* S of the free particle is defined. The variation $\delta S = \delta\int L\mathrm{d}t = 0$ is the *Hamilton principle* (*of least action*). Integration of $L(\mathbf{x},\dot{\mathbf{x}},t)$ along the orbit is an elegant way to account for the interdependence of $\delta\mathbf{x}$ and $\delta\dot{\mathbf{x}}$ according to equation (2.33). In this way $\delta\dot{\mathbf{x}}$ eliminates under the condition of the end points $P_0 = \mathbf{x}_0$, $P_1 = \mathbf{x}_1$ held fixed. In other words, by Hamilton's variational principle the bound variation of $\dot{\mathbf{x}}$ is converted into a free variation of $\mathbf{x}$. This is exactly the reason for the passage from the differentials of D'Alembert's principle (2.30) to the integrals on them in (2.32) and formulating the variational principle as the variation of the integral on the function $L(\mathbf{x}, \dot{\mathbf{x}}, t)$ between fixed positions. $L = T - V$ is the *Lagrangian* or *Lagrange function* of a mass point in a potential. The differential expression of L within the round brackets in (2.34) when equated to zero represents the Newtonian equation of motion of the mass point, hence, it follows that the variation of the action S along the true orbit is zero. Conversely, from the requirement that δL is piecewise continuous and $\delta S = 0$ Newton's law $m\ddot{\mathbf{x}} = \mathbf{f}(\mathbf{x}, t)$ follows, now in terms of the Lagrangian L,

$$\frac{\mathrm{d}}{\mathrm{d}t}\frac{\partial L}{\partial \dot{\mathbf{x}}} - \frac{\partial L}{\partial \mathbf{x}} = 0 \quad \Leftrightarrow \quad \frac{\mathrm{d}}{\mathrm{d}t}\frac{\partial L}{\partial \dot{x}_i} - \frac{\partial L}{\partial x_i}; \quad i = 1, 2, 3. \tag{2.35}$$

These are the celebrated one particle *Lagrangian equations* of motion. Whether the action of the physical orbit is a minimum (least action), a maximum or a saddle point depends on the potential V.

When no potential exists. D'Alembert's principle (2.30) and its integral

$$\int [m\ddot{\mathbf{x}} - \mathbf{f}(\mathbf{x}, t)]\delta\mathbf{x}\mathrm{d}t = 0$$

are equivalent for the class $\{\delta\mathbf{x}(t)\}$ continuous in (t), see pertinent exercise [for the reader who likes more rigor: continuous except a subset of t of measure zero, as for example on all rational values of t]. The inertial quantity $m\ddot{\mathbf{x}}\delta\mathbf{x}$ represents a non-integrable term in general. However, as $\delta\mathbf{x}$ is arbitrary D'Alembert's principle must hold also in direction tangential to the orbit, and in particular $\delta\mathbf{x}$ can be chosen equal to $\mathbf{v}\mathrm{d}t$, with $\mathbf{v}$ the local velocity vector. With this special choice the variation of the kinetic energy $\delta T = m\ddot{\mathbf{x}}\dot{\mathbf{x}}\mathrm{d}t = m\mathrm{d}\dot{\mathbf{x}}^2/2$ is restored and the variation of the action integral reads

$$\delta S = \int_{\mathbf{x}_0}^{\mathbf{x}_1} \left[\delta T + \mathbf{f}(\mathbf{x}, t)\delta\mathbf{x}\right]\mathrm{d}t = -\int_{\mathbf{x}_0}^{\mathbf{x}_1} \left[\frac{\mathrm{d}}{\mathrm{d}t}\frac{\partial T}{\partial \dot{\mathbf{x}}} - \mathbf{f}(\mathbf{x}, t)\right]\delta\mathbf{x}\mathrm{d}t \Rightarrow \frac{\mathrm{d}}{\mathrm{d}t}\frac{\partial T}{\partial \dot{\mathbf{x}}} = \mathbf{f}(\mathbf{x}, t).$$

This is again Newton's equation (2.3).

The concept of neighboring orbits. It is possible to introduce D'Alembert's principle and the concept of virtual displacement $\delta\mathbf{x}$ as two new principles of mechanics. Their success is then confirmed by the results: agreement of Lagrange equations with Newton's law and high flexibility of the variational method. On the other hand it has been discussed above that D'Alembert's principle is a consequence of Newton's law if no conceptual distinction is made between impressed and inertial forces. It is useful and elucidating further to introduce the concept of neighboring orbits and to apply on them Hamilton's principle of least action.

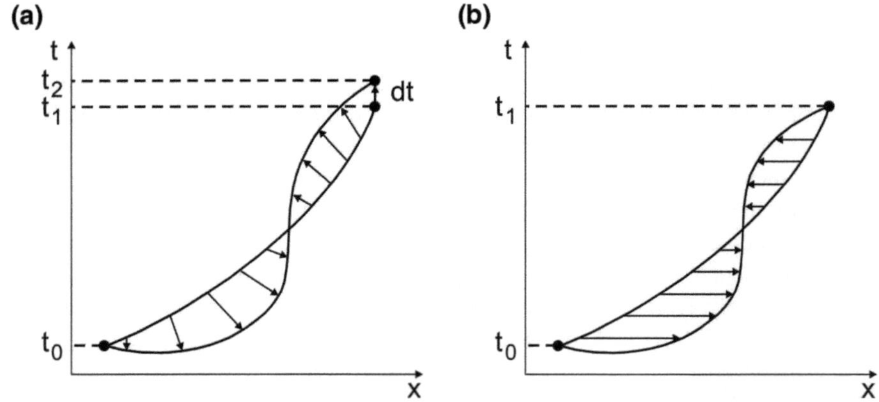

Fig. 2.3 True orbit $x(t)$ and neighboring trajectory $y = x(t) + h(t)$ connecting the fixed positions x_0, x_1. **a** $y(t')$ is generated by the physical displacement δy applied to $x(t)$: $x(t) + \delta y = y(t')$. Time elapsed along $y(t')$ is $dt = t_2 - t_1$. **b** The same trajectory is produced by virtual (= isochronous) displacements δx, except infinitesimal corrections in the neighborhood of (x_1, t_1)

Assume the true physical trajectory $\mathbf{x}(t)$ passing through the fixed points $\mathbf{x}_0, \mathbf{x}_1$ and consider an arbitrary orbit $\mathbf{y}(t') = \mathbf{x}(t) + \mathbf{h}(\mathbf{x}, t')$ in its neighborhood connecting the two endpoints. A mass point starting in $\mathbf{x}_0$ will describe the neighboring orbit at times t' different from t along $\mathbf{x}(t)$ [see Fig. 2.3a]. The time differences $\Delta t = t' - t$, in particular $\Delta t = t_2' - t_1$ at the endpoint $\mathbf{x}_1$ along the two action integrals, shrink to zero with $\mathbf{h}(\mathbf{x}, t') \to 0$ everywhere. For simplicity we assume the existence of a potential and approximate to first order $\mathbf{h}(\mathbf{x}, t') = \mathbf{h}(\mathbf{x}, t) + \partial_t \mathbf{h}(\mathbf{x}, t)\Delta t$. The isochronous shift $\mathbf{h}(\mathbf{x}, t)$ is arbitrary and free, it is identical with the virtual displacement $\delta\mathbf{x}$ from (2.30) [see Fig. 2.3b]. As T and V are differentiable functions the integrands of the action integral on $L = T - V$ from $\mathbf{x}_0$ to $\mathbf{x}_1$ along $\mathbf{y}(t') = \mathbf{x}(t) + \mathbf{h}(\mathbf{x}, t')$ and along $\mathbf{y}_1(t') = \mathbf{x}(t) + \delta\mathbf{x}$ differ by first or lower order and hence

$$S = \int_{t_0}^{t_2} L(\mathbf{y}(t'), \dot{\mathbf{y}}(t'), t')\mathrm{d}t' = \int_{t_0}^{t_1} \left[L(\mathbf{y}_1(t'), \dot{\mathbf{y}}_1(t'), t') + O(t' - t)\right]\mathrm{d}t'. \quad (2.36)$$

The variation of the second integral without the ignorable quantity $O(t' - t)$ is identical with the isochronous standard variation δS from (2.30)and (2.34), and Fig. 2.3b. Note that on the neighboring orbit $\mathbf{y}(t')$ the endpoint $\mathbf{x}_1$ is reached at $t = t_2 \neq t_1$ in general, however, with shrinking $\mathbf{h}$ it approaches t_1 [see vertical $\delta\mathbf{y} = \mathrm{d}t$ in Fig. 2.3a]. Apart from detailed insight in the physical nature of the variational procedure the concept of the neighboring orbit makes it clear why L is the difference of kinetic to potential energy.

Why $L = T - V$ and not something else? In the absence of a potential the quantity that is free to vary arbitrarily in (2.30) is the kinetic energy T. In presence of a potential $V(\mathbf{x})$, only the difference $T - V(\mathbf{x}) = L$ can be subject to a free variation irrespectively of whether $\mathbf{x}$ is a position on the true orbit $\mathbf{x}(t)$ or on the neighboring orbit $\mathbf{y}(t)$. As shown later in connection with the finding of the correct ponderomotive potential this fact will prove its significance.

The validity of (2.35) is guarantied for time dependent potentials as a consequence of the definition of the force $\mathbf{f} = -\nabla V(\mathbf{x}, t)$ taken at time instant held fixed. No general prediction on the existence of a Lagrange function is possible in the case V depends on velocity. A remarkable exception is made by the Lorentz force. The Lagrangian exists and assumes the simple (non covariant) form

$$L(\mathbf{x}, \mathbf{v}, t) = \frac{1}{2}m\mathbf{v}^2 + q\mathbf{v}\mathbf{A}(\mathbf{x}, t) - q\Phi(\mathbf{x}, t). \tag{2.37}$$

$\mathbf{A}$ is the vector potential, Φ the scalar potential. They are related to the electric and magnetic fields by

$$\mathbf{E} = -\frac{\partial}{\partial t}\mathbf{A} - \frac{1}{c}\nabla\Phi, \qquad \mathbf{B} = \nabla \times \mathbf{A}. \tag{2.38}$$

Three remarks concerning the variational principles are in order.

(i) The standard derivation starting from D'Alembert suggests that the true orbit and the neighboring trajectory generated by the isochronous, i.e., parallel $\delta\mathbf{x}$ are reached by the mass point isochronously and therefore frequently the upper time limit $t = t_1$ is taken as fixed during the process of variation. This is unnecessary and, even worse, mathematically wrong in the strict sense. In the derivation of the Lagrange equations use is made solely of the endpoints P_0 and P_1 being fixed; time is free. The derivation presented here shows that the upper limit must be allowed to vary in time by the infinitesimal amount $\mathrm{d}t = t_2 - t_1$ [see Fig. 2.3b]. The mathematical reason is that by the parallel shifts $\delta\mathbf{x}$ two neighboring time intervals lengthen or contract by an infinitesimal amount $\mathrm{d}t$. What has been shown by the argument related to the neighboring trajectory $\mathbf{y}(t') = \mathbf{x}(t) + \delta\mathbf{y}$ is that the $\mathrm{d}t$ are of higher order but in the integral they may sum up to to a finite time shift $\Delta t = t_2 - t_1$ at the end point $\mathbf{x}_1$. It is exactly by this fact that Hamilton's principle of least action contains as variants the principles of least time and shortest distance (e.g. of light rays). By fixing the times t_0, t_1 the two principles are excluded.

(ii) The concept of neighboring orbits shows the equivalence of Hamilton's and D'Alembert's principles. In addition, the equivalence of them with Newton's law has been shown also. If the equivalence of impressed and inertial forces is taken for granted the variational principles applied to point particles do not introduce any physics beyond Newton. Their superiority is all of methodological nature.

(iii) From an axiomatic point of view D'Alembert's principle (2.30) with isochronous displacements $\delta \mathbf{x}$ may be taken as an independent principle of Mechanics, but then also (2.33) is to be postulated as an additional axioma to show that Newton's law (2.3) follows from it.

The Lagrangian in generalized coordinates. The variational treatment is now extended to the dynamics of N interacting particles of identical mass m and charge q. In general the system is subject to constraints and exhibits a number of degrees of freedom $f < 3N$. For example, the motion of one particle may be bound to a line in space ($f = 1$) or confined to a surface ($f = 2$). The position of a solid is uniquely determined by the coordinates of its center of mass ($f = 3$) and the rotation angle ($f = 1$) around an axis ($f = 2$), hence $f = 6$ coordinates of position q_i determine its dynamics. The specific choice of the set of the q_i is governed by symmetries and intuition. We may write them as a f dimensional vector $\mathbf{q} = \{q_i\}$ in the Euclidian configuration space $\mathbf{R}^f$ with the positions and velocities of the individual particles given by

$$\mathbf{x}_k = \mathbf{x}_k(\mathbf{q}, t), \quad \mathbf{v}_k = \dot{\mathbf{x}}_k = \sum_i \frac{\partial \mathbf{x}_k}{\partial q_i}\dot{q}_i + \frac{\partial \mathbf{x}_k}{\partial t} = \frac{\partial \mathbf{x}_k}{\partial \mathbf{q}}\dot{\mathbf{q}} + \frac{\partial \mathbf{x}_k}{\partial t} \Rightarrow \frac{\partial \dot{\mathbf{x}}_k}{\partial \dot{q}_i} = \frac{\partial \mathbf{x}}{\partial q_i}. \tag{2.39}$$

The last step is the result of the partial derivative of $\dot{\mathbf{x}}_k$ with respect to $\dot{q}_i$. The variables q_i and $\dot{q}_i$ are the *generalized coordinates* and the *generalized!velocities*. Note, they constitute a set of $2f$ independent variables, in perfect analogy to the set of six variables $x_i, \dot{x}_i$ in the one particle Lagrangian. The total kinetic energy is given by

$$T(\mathbf{q}, \dot{\mathbf{q}}, t) = \sum_k \frac{1}{2} m \mathbf{v}_k^2 = \frac{1}{2} m \sum_k \left(\frac{\partial \mathbf{x}_k}{\partial \mathbf{q}}\dot{\mathbf{q}} + \frac{\partial \mathbf{x}_k}{\partial t} \right)^2. \tag{2.40}$$

For N particles D'Alembert's principle reads

$$\sum_{k=1}^{N} (m\ddot{\mathbf{x}}_k - \mathbf{f}_k)\delta \mathbf{x}_k = 0. \tag{2.41}$$

After substitution of the $\mathbf{x}_k$ by the generalized coordinates and identification of the set of all possible virtual displacements $\delta \mathbf{x}_k$ with the subset $\{\delta \mathbf{x}_k = \mathrm{d}\mathbf{x}_k = \mathbf{v}_k \mathrm{d}t\}$ it transforms into

$$\sum_{k=1}^{N}\left(m\frac{\mathrm{d}\dot{\mathbf{x}}_k}{\mathrm{d}t}\frac{\partial \mathbf{x}_k}{\partial \mathbf{q}}-\mathbf{f}_k\frac{\partial \mathbf{x}_k}{\partial \mathbf{q}}\right)\delta\mathbf{q}=\left(\sum_{k=1}^{N}m\frac{\mathrm{d}\dot{\mathbf{x}}_k}{\mathrm{d}t}\frac{\partial \dot{\mathbf{x}}_k}{\partial \dot{\mathbf{q}}}-\mathbf{Q}\right)\delta\mathbf{q}$$

$$=\left[\sum_{k=1}^{N}\left(\frac{\mathrm{d}}{\mathrm{d}t}\frac{\partial m\dot{\mathbf{x}}_k^2/2}{\partial \dot{\mathbf{q}}}-m\dot{\mathbf{x}}_k\frac{\mathrm{d}}{\mathrm{d}t}\frac{\partial \dot{\mathbf{x}}_k}{\partial \dot{\mathbf{q}}}\right)-\mathbf{Q}\right]\delta\mathbf{q}=\left(\frac{\mathrm{d}}{\mathrm{d}t}\frac{\partial T}{\partial \dot{\mathbf{q}}}-\frac{\partial T}{\partial \mathbf{q}}-\mathbf{Q}\right)\delta\mathbf{q}=0. \tag{2.42}$$

Thereby equality

$$m\dot{\mathbf{x}}_k\frac{\mathrm{d}}{\mathrm{d}t}\frac{\partial \dot{\mathbf{x}}_k}{\partial \dot{\mathbf{q}}}=m\dot{\mathbf{x}}_k\frac{\mathrm{d}}{\mathrm{d}t}\frac{\partial \mathbf{x}_k}{\partial \mathbf{q}}=m\frac{\partial T}{\partial \mathbf{q}}$$

follows from (2.39) and $\mathrm{d}_t\partial\mathbf{q}=\partial\mathbf{q}\mathrm{d}_t$. The f components Q_i of the vector $\mathbf{Q}$,

$$Q_i(q_j,\dot{q}_j,t)=\sum_{k}^{N}\mathbf{f}_k(\mathbf{x},\dot{\mathbf{q}},t)\frac{\partial \mathbf{x}_k}{\partial q_i};\qquad \mathbf{Q}=(Q_1,\ldots\ldots,Q_f)$$

are the *generalized forces*. The last equality in (2.42) expresses D'Alembert's principle in generalized coordinates. It is invariant with respect to point transformations. As a consequence of the constraints numerous forces $\mathbf{f}_k$ may add to zero, as for example the internal forces in a solid. In (2.41) only the irreducible forces count, the so called impressed forces, e.g. the force onto the center of mass of a solid, and contribute to the generalized force component Q_i.

Equality (2.42) is satisfied only if the bracket expressions vanish. In the form in terms of T and $\mathbf{Q}$ they are known as the Lagrange equations of motion of the first kind. Under the assumption that the forces are derivable from a potential of the form $V=V(\mathbf{q},t)$, i.e., $Q_i=-\partial V/\partial q_i$, $\partial V/\partial\dot{\mathbf{q}}=0$, they can be cast into the equivalent form of the Lagrange equations of the second kind

$$\frac{\mathrm{d}}{\mathrm{d}t}\frac{\partial L}{\partial \dot{\mathbf{q}}}-\frac{\partial L}{\partial \mathbf{q}}=0\quad\Leftrightarrow\quad\frac{\mathrm{d}}{\mathrm{d}t}\frac{\partial L}{\partial \dot{q}_i}-\frac{\partial L}{\partial q_i}=0;\qquad L=T(\mathbf{q},\dot{\mathbf{q}},t)-V(\mathbf{q},t). \tag{2.43}$$

Hamilton's variational principle has been shown to be equivalent to D'Alembert's principle for three degrees of freedom. All what was said there translates without limitation to f degrees of freedom, equivalence of the two principles included. Let us assume that the dynamics of a system is given in terms of the Lagrangian $L(\mathbf{q},\dot{\mathbf{q}},t)$. Condition (2.33) translates now into

$$\frac{\mathrm{d}}{\mathrm{d}t}\delta\mathbf{q}=\delta\dot{\mathbf{q}}. \tag{2.44}$$

We apply Hamilton's principle (2.34),

$$\delta S=\delta\int_{\mathbf{q}_0(t_0)}^{\mathbf{q}_1}L(\mathbf{q},\dot{\mathbf{q}},t)\mathrm{d}t=\int_{\mathbf{q}_0(t_0)}^{\mathbf{q}_1}\left(\frac{\mathrm{d}}{\mathrm{d}t}\frac{\partial L}{\partial \dot{\mathbf{q}}}-\frac{\partial L}{\partial \mathbf{q}}\right)\delta\mathbf{q}\delta t=0, \tag{2.45}$$

and are led straightforwardly to the Lagrange equations (2.43). As the property of S to be extremal does not depend on the special choice of coordinates the invariant structure of the Lagrange equations towards point transformations appears once more. As we shall see the variational formulation enlarges the field of applications in various directions, for instance to the dynamics of continuous media, to wave dynamics, and field theoretical modelling. The free choice of generalized coordinates means an enormous increase in flexibility and economics of description. By definition, the quantities $\mathbf{p} = \{p_i\}$,

$$\mathbf{p}(\mathbf{q}, \dot{\mathbf{q}}, t) = \frac{\partial L}{\partial \dot{\mathbf{q}}} \quad \Leftrightarrow \quad p_i(q_j, \dot{q}_j, t) = \frac{\partial L(q_j, \dot{q}_j, t)}{\partial \dot{q}_i}, \quad j = 1, \ldots\ldots, f. \tag{2.46}$$

are the *generalized* or *canonical momenta* . In case L does not depend on one coordinate q_k, from (2.43) follows for the associated generalized momentum $\mathrm{d}p_k/\mathrm{d}t = 0$, i.e., it is conserved, $p_k(q_{i\neq k}, \dot{q}_{i\neq k}, c_k, t) = c_k =$const and q_k and $\dot{q}_k$ can be eliminated from L. Such coordinates are named *cyclic* or *ignorable*. As an important application consider a plane wave of type (2.1) in a medium that is inhomogeneous only in propagation direction $\mathbf{k}^0$. Its Lagrangian does not depend in direction perpendicular to $\mathbf{k}^0$ and hence the canonical momentum $\mathbf{p}_\perp$ is conserved,

$$\mathbf{p}_\perp = m\mathbf{v} + q\mathbf{A} = \text{const}. \tag{2.47}$$

Frequent use will be made of this fact in the following to obtain analytical expressions in simplified geometry.

Consequences

Energy Conservation

The Lagrange equations admit a generalized energy conservation theorem as follows. We define the new function $H = \mathbf{p}\dot{\mathbf{q}} - L$ and take its derivative with respect to time,

$$\frac{\mathrm{d}}{\mathrm{d}t}H = \frac{\mathrm{d}}{\mathrm{d}t}\left(\mathbf{p}\dot{\mathbf{q}} - L\right) = \dot{\mathbf{q}}\frac{\mathrm{d}}{\mathrm{d}t}\frac{\partial L}{\partial \dot{\mathbf{q}}} + \frac{\partial L}{\partial \dot{\mathbf{q}}}\ddot{\mathbf{q}} - \frac{\partial L}{\partial \mathbf{q}}\dot{\mathbf{q}} - \frac{\partial L}{\partial \dot{\mathbf{q}}}\ddot{\mathbf{q}} - \frac{\partial L}{\partial t} = -\frac{\partial L}{\partial t}. \tag{2.48}$$

If L does not depend explicitly on time, H is conserved. In that case we conclude from Euler's theorem on homogeneous functions, or, in case of $\mathbf{p} = m\mathbf{v}$, directly from (2.40), $\mathbf{p}\dot{\mathbf{q}} = 2T$ and hence $H = T + V$ is the total energy.

By a Galilei boost a time independent potential $V(\mathbf{x})$ in S becomes time dependent in $S'(\mathbf{v})$. Time dependent potentials $V(\mathbf{x}, t)$ are of great interest in applications like heating or cooling of matter and acceleration of particles because energy is not conserved and may be varied according to appropriate schemes acting on time. In view of the so called ponderomotive force to be introduced later velocity dependent potentials of the form $V(\mathbf{x}, \mathbf{v})$ deserve special interest.

The Virial Theorem

The *virial* G is the quantity $G = \mathbf{pq} = \sum p_i q_i$. With $L(\mathbf{q}, \dot{\mathbf{q}})$ its time derivative yields

$$\frac{\mathrm{d}}{\mathrm{d}t}G = \mathbf{q}\frac{\mathrm{d}}{\mathrm{d}t}\frac{\partial L}{\partial \dot{\mathbf{q}}} + \mathbf{p}\dot{\mathbf{q}} = \frac{\partial L}{\partial \mathbf{q}}\mathbf{q} + 2T.$$

Let us now assume that G is bound by an upper limit for all times, then the time average $\overline{\mathrm{d}G/\mathrm{d}t}$ tends to zero with the time interval growing beyond all limits, and hence

$$\overline{\frac{\partial L}{\partial \mathbf{q}}\mathbf{q}} = -2\overline{T}. \tag{2.49}$$

The virial theorem has manifold applications, often of great practical relevance. For example, the virial of a fixed number of free particles is $G = \sum m\mathbf{v}_k\mathbf{x}_k$. In a time-independent potential (2.49) reads

$$2\overline{T} + \overline{\sum \mathbf{f}_i\mathbf{x}_i} = 2\overline{T} - \overline{\sum (\partial V/\partial \mathbf{x}_i)\mathbf{x}_i} = 0.$$

Application to a potential of the form $V = C\sum r_i^\alpha$ leads to

$$\overline{T} = \frac{\alpha}{2}\overline{V}; \quad i \geq 1. \tag{2.50}$$

With $\alpha = 2$ equipartition between average kinetic and average potential energy follows for the free harmonic oscillator. If the oscillator of eigenfrequency ω_0 is driven into a steady state at $\omega \neq \omega_0$ the ratio between the two averaged energies may assume any value, e.g., infinity in the limiting case of the free electron in the electromagnetic wave (2.1). In the Coulomb potential with $\alpha = -1$ the kinetic energy of a Kepler orbit is half its binding energy. Caution is indicated when the virial theorem is applied to unstable ensembles with fast particle losses, e.g. self gravitating open clusters of objects.

Noether's Theorem

It links continuous symmetries with conservation laws of Lagrangian dynamics. Each time a physical system exhibits a continuous or discrete symmetry there exists an associated quantity that is conserved. Conversely, it has become customary to speak of an underlying symmetry if a physical quantity of a system is conserved; hence, conservation laws and symmetries are treated as synonymous. Assume the point transformation and its unique inversion, differentiable with respect to the continuous parameter s,

$$\mathbf{r} = (\mathbf{q}, s) \quad \text{with} \quad \mathbf{r} = (\mathbf{q}, s = 0) = \mathbf{q}.$$

The theorem states: If the time independent Lagrangian $L(\mathbf{q}, \dot{\mathbf{q}})$, with $\mathbf{q}$ from the holonomic constraints on the $\mathbf{x}_k$ according to (2.39), is invariant under the above point transformation for all values of s, i.e.,

$$L(\mathbf{r}, \dot{\mathbf{r}}, s) = L(\mathbf{r}, \dot{\mathbf{r}}, s = 0) \tag{2.51}$$

then the quantity

$$I(\mathbf{q}, \dot{\mathbf{q}}) = \frac{\partial L(\mathbf{r}, \dot{\mathbf{r}})}{\partial \dot{\mathbf{r}}} \frac{\partial \mathbf{r}}{\partial s}\bigg|_{s=0} \tag{2.52}$$

is an integral of the Lagrangian equations of motion.

The proof is as simple as far reaching the consequences of the theorem are; it can hardly be overestimated. In fact, by making use of the invariance of the Lagrange equations relative to the point transformations above, the derivative of (2.51) with respect to s yields zero,

$$\frac{\partial L}{\partial \mathbf{r}} \frac{\partial \mathbf{r}}{\partial s} + \frac{\partial L}{\partial \dot{\mathbf{r}}} \frac{\partial \dot{\mathbf{r}}}{\partial s} = \frac{\partial \mathbf{r}}{\partial s} \frac{\mathrm{d}}{\mathrm{d}t} \frac{\partial L}{\partial \dot{\mathbf{r}}} + \frac{\partial L}{\partial \dot{\mathbf{r}}} \frac{\mathrm{d}}{\mathrm{d}t} \frac{\partial \mathbf{r}}{\partial s} = \frac{\mathrm{d}}{\mathrm{d}t} \left(\frac{\partial L}{\partial \dot{\mathbf{r}}} \frac{\partial \mathbf{r}}{\partial s}\bigg|_{s=0} \right) = 0.$$

From translational invariance in space and time momentum and energy conservation follow from Noether's theorem, and from invariance with respect to rotation, i.e., isotropy of space, follows the angular momentum conservation. Energy conservation is directly seen from identification of s with t and $\partial L/\partial t = 0$ in (2.48). In presence of external fields or forces at least one of spacetime symmetries is broken. For such reasons in realistic situations in plasma physics the number of symmetries is very restricted if not zero. On the other hand, plasma or fluid dynamics is accessible to analytical treatment just when use can be made of special symmetries. In a complete theory the external forces or fields have to become a part of the system and then all symmetries in spacetime are restored.

In this context one could ask the following question: Is it possible that energy conservation is violated on a fundamental level once in the future? Whenever in the past energy was not conserved a new kind of energy was found, e.g. the internal energy or heat, the energy of the escaping neutrino; or will be found, e.g. the dark energy in the universe. Hence, to violate energy conservation an existing theory must exhibit the two properties: (a) it is capable of explaining all phenomena pertaining to its realm and (b) at the same time the introduction of a new energy component is in contradiction to this theory. Intuitively, on the level of measurements (events "on shell") we believe that such a situation is to be excluded; disproof of the energy principle is impossible. In other words, the continuous symmetries of space-time are not laws of physics but postulates any complete theory must obey.

Reference Systems

If the Lagrange equations are correct in one reference system they hold in any other system of reference which leaves the accelerations invariant. Their totality represents the class of *inertial systems*. They move at constant velocity $\mathbf{v}$ relative to each other.

To find one representative of them it is sufficient to ascertain the validity of Newton's law (2.4) in it. Hence, the validity of the Lagrange equations is assured by the class of inertial systems. However, their range of validity is easily extended to any accelerated reference system by completing the Lagrangian with the corresponding apparent forces, e.g. of centrifugal and Coriolis type; as an example see the curvature drift (2.74) in a magnetic field.

The Hamiltonian

From $\mathbf{p}(q_i, \dot{q}_i, t) - \partial L(q_i, \dot{q}_i, t)/\partial\dot{\mathbf{q}} = 0$ the generalized velocities $\dot{\mathbf{q}}$ can be expressed as functions of the independent variables $\mathbf{p}$ and $\mathbf{q}$, $\{\dot{q}_i\} = \{q_i(p_k, q_k, t)\}$, and inserted in $L(\mathbf{q}, \dot{\mathbf{q}}, t) \to L(\mathbf{p}, \mathbf{q}, t)$. The energy function H from (2.48), now expressed in terms of $\mathbf{p}$ and $\mathbf{q}$,

$$H(\mathbf{p}, \mathbf{q}, t) = \mathbf{p}\dot{\mathbf{q}}(\mathbf{p}, \mathbf{q}, t) - L(\mathbf{p}, \mathbf{q}, t) \tag{2.53}$$

is the *Hamiltonian* of the dynamic system. From equating the differentials of $L = \mathbf{p}\dot{\mathbf{q}} - H$ in the two pairs of variables $(\mathbf{q}, \dot{\mathbf{q}})$, $(\mathbf{p}, \mathbf{q})$ and t,

$$\mathrm{d}L = \frac{\partial L}{\partial \mathbf{q}}\mathrm{d}\mathbf{q} + \frac{\partial L}{\partial \dot{\mathbf{q}}}\mathrm{d}\dot{\mathbf{q}} + \frac{\partial L}{\partial t}\mathrm{d}t = \mathbf{p}\mathrm{d}\dot{\mathbf{q}} + \dot{\mathbf{q}}\mathrm{d}\mathbf{p} - \frac{\partial H}{\partial \mathbf{p}}\mathrm{d}\mathbf{p} - \frac{\partial H}{\partial \mathbf{q}}\mathrm{d}\mathbf{q} - \frac{\partial H}{\partial t}\mathrm{d}t, \tag{2.54}$$

and substituting $\partial L/\partial\mathbf{q}$ from (2.43) the *Hamiltonian*, or *canonical*, equations of motion result,

$$\dot{\mathbf{p}} = -\frac{\partial H}{\partial \mathbf{q}},\ \dot{\mathbf{q}} = \frac{\partial H}{\partial \mathbf{p}} \quad \Leftrightarrow \quad \dot{p}_i = -\frac{\partial H}{\partial q_i},\ \dot{q}_i = \frac{\partial H}{\partial p_i},\quad i = 1, \ldots\ldots, f; \qquad \frac{\partial H}{\partial t} = -\frac{\partial L}{\partial t}. \tag{2.55}$$

Analogously to (2.48) the energy conservation now follows from the canonical equations,

$$\frac{\mathrm{d}H}{\mathrm{d}t} = \frac{\partial H}{\partial \mathbf{p}}\dot{\mathbf{p}} + \frac{\partial H}{\partial \mathbf{q}}\dot{\mathbf{q}} + \frac{\partial H}{\partial t} = \frac{\partial H}{\partial \mathbf{p}}\frac{\partial H}{\partial \mathbf{q}} - \frac{\partial H}{\partial \mathbf{q}}\frac{\partial H}{\partial \mathbf{p}} + \frac{\partial H}{\partial t} = \frac{\partial H}{\partial t}. \tag{2.56}$$

This is a special case of total derivative of a variable $b = b(\mathbf{p}, \mathbf{q}, t)$ in phase space with respect to t,

$$\frac{\mathrm{d}b}{\mathrm{d}t} = \dot{\mathbf{q}}\frac{\partial b}{\partial \mathbf{q}} + \dot{\mathbf{p}}\frac{\partial b}{\partial \mathbf{p}} + \frac{\partial b}{\partial t} = \{H, b\} + \frac{\partial b}{\partial t};\quad \{H, b\} = \frac{\partial H}{\partial \mathbf{p}}\frac{\partial b}{\partial \mathbf{q}} - \frac{\partial H}{\partial \mathbf{q}}\frac{\partial b}{\partial \mathbf{p}}. \tag{2.57}$$

{H, b} is the *Poisson bracket* of the classical mechanics. Under the condition of invertibility between the variable pairs $(\mathbf{q}, \dot{\mathbf{q}})$, $(\mathbf{p}, \mathbf{q})$ the Hamiltonian and Lagrange equations are equivalent.

From (2.37) and (2.53) the canonical momentum and the Hamiltonian of a point charge in the electromagnetic field is obtained in a straightforward way,

$$\mathbf{p} = m\dot{\mathbf{q}} + q\mathbf{A}, \qquad H(\mathbf{p}, \mathbf{q}, t) = \frac{1}{2m}(\mathbf{p} - q\mathbf{A})^2 + q\Phi(\mathbf{q}, t). \tag{2.58}$$

The quantity $\mathbf{p}_\mathrm{m} = \mathbf{p} - q\mathbf{A} = m\mathbf{v}$ is the mechanical momentum and consequently the numerical value of $H = T + V$ is the total energy of the single free particle.

Hamilton's extended variational principle. In (2.53) the Lagrangian $L = \mathbf{p}\dot{\mathbf{q}} - H$ may be thought as depending on $\mathbf{p}$, $\mathbf{q}$, and $\dot{\mathbf{q}}$. We impose the constraints $\delta\mathbf{p}_1 = \delta\mathbf{p}_0$, $\delta\mathbf{q}_1 = \delta\mathbf{q}_0$ onto the end points of a trajectory and equate the variation $\delta S(\mathbf{p}, \mathbf{q}, \dot{\mathbf{q}}, t)$ to zero under observing condition (2.44) and $\mathbf{p}\mathrm{d}\delta\mathbf{q}/\mathrm{d}t = \mathrm{d}(\mathbf{p}\delta\mathbf{q})/\mathrm{d}t - \dot{\mathbf{p}}\mathrm{d}\mathbf{q}$:

$$\delta S = \int_{\mathbf{p}_1,\mathbf{q}_1}^{\mathbf{p}_2,\mathbf{q}_2} \left[\left(-\dot{\mathbf{p}} - \frac{\partial H}{\partial \mathbf{q}}\right)\delta\mathbf{q} + \left(\dot{\mathbf{q}} - \frac{\partial H}{\partial \mathbf{p}}\right)\delta\mathbf{p}\right]\mathrm{d}t = 0. \tag{2.59}$$

This is *Hamiltonian's extended variational principle* in the free variables $\mathbf{p}$, $\mathbf{q}$ from which the canonical equations follow as necessary and sufficient conditions. Note, here no condition on the variation of $\mathbf{p}$ at the endpoints is required; only when switching to new canonical pairs of variables is undertaken it is needed to fix the one to one correspondence between them.

Phase Space and Liouville's Theorem

The variables $(\mathbf{p}, \mathbf{q})$ form an Euclidean space of $2f$ dimensions, the so called *phase space*. The Hamiltonian dynamics induces a continuous succession of mappings of the phase space onto itself, the so-called *phase flow*. It has the remarkable property that the volume element $\mathrm{d}\tau = (\mathrm{d}\mathbf{p}\mathrm{d}\mathbf{q})$ is an invariant of motion of the phase points $(\mathbf{p}(t), \mathbf{q}(t))$, or equivalently, with the Jacobian J is

$$\mathrm{d}\tau = J(t, t_0)\,\mathrm{d}\mathbf{p}_0\mathrm{d}\mathbf{q}_0 = \frac{\partial(\mathbf{p}(t), \mathbf{q}(t))}{\partial(\mathbf{p}_0, \mathbf{q}_0)}\,\mathrm{d}\mathbf{p}_0\mathrm{d}\mathbf{q}_0. \tag{2.60}$$

$\mathbf{p}_0, \mathbf{q}_0$ are the initial values at arbitrary time t_0.

Proof The elementary proof consists in showing that $\mathrm{d}J/\mathrm{d}t = 0$. The derivative of J is the sum of $2f$ matrices, hence

$$\frac{\mathrm{d}}{\mathrm{d}t}\tau = \frac{\mathrm{d}}{\mathrm{d}t}J(t, t_0)\,\mathrm{d}\mathbf{p}_0\mathrm{d}\mathbf{q}_0 = \sum_{i=1}^{f}\left(\frac{\mathrm{d}}{\mathrm{d}t}\frac{\partial p_i}{\partial p_{0i}} + \frac{\mathrm{d}}{\mathrm{d}t}\frac{\partial q_i}{\partial q_{0i}}\right)\mathrm{d}p_{0i}\mathrm{d}q_{0i}\prod_{i\neq j}^{f}\mathrm{d}p_j\mathrm{d}q_j$$

$$= \sum_{i=1}^{f}\left(\frac{\partial \dot{p}_i}{\partial p_{0i}} + \frac{\partial \dot{q}_i}{\partial q_{0i}}\right)\mathrm{d}p_{0i}\mathrm{d}q_{0i}\prod_{i\neq j}^{f}\mathrm{d}p_j\mathrm{d}q_j.$$

Note $\mathrm{d}_t\partial_{p_{0i}}p_i = \partial_{p_{0i}}\mathrm{d}_t p_i$ and $\mathrm{d}_t\partial_{q_{0i}}q_i = \partial_{q_{0i}}\mathrm{d}_t q_i$. The bracket expression is

$$-\frac{\partial^2 H}{\partial p_{0i}\partial q_i} + \frac{\partial^2 H}{\partial q_{0i}\partial p_i} \longrightarrow 0, \;\; i = 1, \dots, f \quad \Longrightarrow \quad \frac{\mathrm{d}}{\mathrm{d}t_0}J = 0. \tag{2.61}$$

for the time derivatives of $\mathbf{p}(t)$, $\mathbf{q}(t)$ taken at $t = t_0$, and t_0 chosen arbitrary. For $t = t_0$ the functional determinant $J = J(t_0, t_0)$ equals unity; thus $J(t_0, t) = 1$ for all times.

Corollary *An important variant is the Jacobian $J' = \partial(\mathbf{p}'(t), \mathbf{q}'(t))/\partial(\mathbf{p}_0, \mathbf{q}_0)$ with $\mathbf{p}' = \mathbf{p} + G(\mathbf{q}', t)$ and $\mathbf{q}' = \mathbf{q}$. The volume change of $d\tau'$ results as*

$$\frac{d}{dt}\tau' = \frac{d}{dt}J'(t, t_0)\, d\mathbf{p}_0 d\mathbf{q}_0 = \sum_{i=1}^{f}\left(\frac{\partial \dot{p}_i'}{\partial p_{0i}} + \frac{\partial \dot{q}_i}{\partial q_{0i}}\right) dp_{0i}dq_{0i}\prod_{i\neq j}^{f} dp_j dq_j .$$

Only the term within the bracket has changed. With the help of the new Hamiltonian $H'(\mathbf{p}', \mathbf{q}, t)$ it reads

$$-\frac{\partial^2 H'}{\partial p_{0i}\partial q_i} + \frac{\partial^2 H'}{\partial q_{0i}\partial p_i'} = -\sum_{l=1}^{f}\frac{\partial^2 H'}{\partial p_{0l}'\partial q_i}\frac{\partial p_{0l}'}{\partial p_{0i}} + \frac{\partial^2 H'}{\partial q_{0i}\partial p_i'} = -\frac{\partial^2 H'}{\partial p_{0i}'\partial q_i} + \frac{\partial^2 H'}{\partial q_{0i}\partial p_i'}$$

$$\lim_{t\mapsto t_0}\left(-\frac{\partial^2 H'}{\partial p_{0i}'\partial q_i} + \frac{\partial^2 H'}{\partial q_{0i}\partial p_i'}\right) = 0, \quad i = 1, \dots, f \quad \Longrightarrow \quad \frac{d}{dt_0}J' = 0. \quad (2.62)$$

The corollary has an important application to the electromagnetic Hamiltonian (2.58). From setting $G(\mathbf{x}, t) = -q\mathbf{A}(\mathbf{x}, \mathbf{t})$ and $\mathbf{p}' = \mathbf{p} + q\mathbf{A}(\mathbf{x}, \mathbf{t})$, $\mathbf{p} = m\mathbf{v}$, follows that electromagnetic interaction conserves the mechanical phase volume.

Remark on the microcanonical ensemble. The equilibrium statistical mechanics, i.e., thermostatistics, is derived from the very natural postulate that all states $(\mathbf{p}, \mathbf{q})$ of a Hamiltonian system $H(\mathbf{p}, \mathbf{q})$ that are accessible to the system have equal probability. The energy of an isolated system is conserved, $H(\mathbf{p}, \mathbf{q}) = E$; the accessible states are bound to a hypersurface Σ in phase space, for example to a hypersphere in the case of noninteracting point particles. At first glance one could be tempted to assign equal probability to all points on the hypersurface Σ. This is correct for independent point particles in standard description $(\mathbf{p} = \{m\dot{x}_i\}, \mathbf{x} = \{x_i\})$, and it is wrong for independent oscillators, like molecules. For this reason standard textbooks introduce an energy uncertainty ΔE. In such a hypershell of finite thickness the phase space volume is certainly conserved. On the other hand we make use of sharp energy constraints in other disciplines whenever E is conserved. This should be possible also with the microcanonical ensemble.

The solution to the problem is as follows. The Hamiltonian is not unique; a change to other canonical variables will conserve the phase space volume but the shape of the hypersurface $\Sigma = (H = E)$ and the measure $\mu(\Delta\Sigma)$ may change considerably. It is easily shown by simple examples that Liouville's theorem does not hold on Σ. However, an invariant measure $\mu(\Delta\Sigma)$ for the probabilities is obtained on Σ by setting

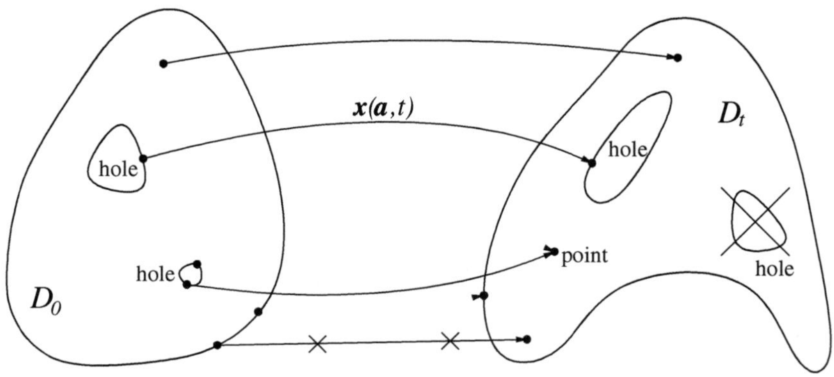

Fig. 2.4 Homeomorphism generated by the canonical equations in phase space. The compact domain D_0 is mapped onto the compact domain D_t at the time instant t. Inner points are mapped into inner points; holes can neither be created nor closed. When a hole shrinks to a point the phase space density is zero there. Orbits cannot cross

$$\mathrm{d}\mu(\Sigma) = \frac{\mathrm{d}\Sigma}{|\mathrm{grad}_{\mathbf{p},\mathbf{q}}(H = E)|}, \qquad \mathrm{grad}_{\mathbf{p},\mathbf{q}} H = \nabla H = \left(\frac{\partial H}{\partial \mathbf{p}}, \frac{\partial H}{\partial \mathbf{q}}\right). \tag{2.63}$$

The invariance of μ is seen from $\mathrm{d}\mu \Delta E = \mathrm{d}\tau$ in phase space (see $\mathrm{d}\mu$ for the harmonic oscillator in Sec. Problems). For ΔE shrinking to zero this goes over into $\delta(H(\mathbf{p}, \mathbf{q}) - E)$ with infinitesimal thickness varying along $\Sigma = (H - E = 0)$. Canonical transformations preserve the phase space volume.

Topological Properties of the Phase Space

It can be assumed that the mapping of the initial point $P_0 \to P_t$ by $(\mathbf{p}, \mathbf{q})(\mathbf{a}; t)$, $\mathbf{a} = (\mathbf{p}_0, \mathbf{q}_0)$, is continuous in the compact domain D_0. Then its image D_t at time t is also compact, and viceversa. The continuous mapping has the following remarkable properties, see Fig. 2.4:

- interior points end on interior points; the border is mapped onto the border
- a hole remains a hole, however it can shrink to a singular point
- the degree of connection (simply, twice, etc., connected) of D_0 is an invariant of the phase flow.

The continuous one to one mapping generates a *homeomorphism*. If orbits in configuration space, as for example in 3 space, are considered the same properties hold. At first glance these topological properties may appear rather abstract. As will be seen in further chapters they will reveal themselves of significant relevance.

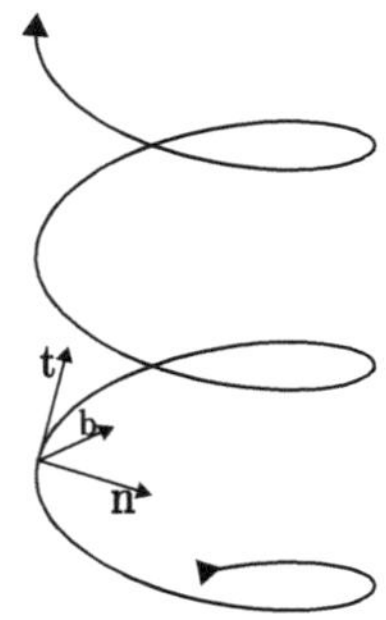

Fig. 2.5 The accompanying trihedron of tangent vector **t**, normal **n**, and binormal **b** is the natural intrinsic reference system to the curve

2.1.3 Charged Particle Motion in Crossed Static Fields

Frenet's Formulas

A natural way to describe the trajectory $\mathbf{r}(t)$ of a point particle is the introduction of a local orthogonal system, the *accompanying trihedron*, consisting of the three unit vectors: **t** *tangent vector* tangent to the orbit, the *principal normal* (vector) **n** in the tangent plane normal to **t**, and the *binormal* (vector) **b** orthogonal to **t** and **n** (see Fig. 2.5). When they are known as functions of the length s the trajectory is uniquely determined. Length and time are connected by $s = \int v\mathrm{d}t$. The curvature $\kappa = R^{-1}$, R curvature radius, and the torsion τ are connected by the Frenet formulas [1]

$$\frac{\mathrm{d}\mathbf{t}}{\mathrm{d}s} = \kappa\mathbf{n}, \qquad \frac{\mathrm{d}\mathbf{n}}{\mathrm{d}s} = -\kappa\mathbf{t} - \tau\mathbf{b}, \qquad \frac{\mathrm{d}\mathbf{b}}{\mathrm{d}s} = \tau\mathbf{n}. \tag{2.64}$$

The relations follow from the mutual orthogonality and the unit length of $(\mathbf{t}, \mathbf{n}, \mathbf{b})$. The three pairs of them fix the *osculating*, the *rectifying* and the *normal* plain. It is recognized by intuition from the Figure that the curvature vector is collinear with the principal normal, that its variation has the negative component κ in tangent direction and torsion τ in negative binormal direction.

Constant Magnetic Field

Identical equations exhibit identical solutions, Richard Feynman says. In the absence of an electric field the velocity component $\mathbf{v}_{\|}$ parallel to the magnetic field is not affected by **B**, hence, $\mathbf{v} = \mathbf{v}_{\|} + \mathbf{v}_{\perp}$ is the natural decomposition. The equation of motion in the magnetic field and that of a solid rotating with $\boldsymbol{\omega}$ around a fixed axis are mathematically identical,

$$\dot{\mathbf{v}}_{\perp} = \frac{q}{m}\mathbf{v}_{\perp} \times \mathbf{B}, \qquad \mathbf{v}_{\perp} = -\mathbf{r}_{\perp} \times \boldsymbol{\omega} \quad \Rightarrow \quad \dot{\mathbf{v}}_{\perp} = -\mathbf{v}_{\perp} \times \boldsymbol{\omega}. \tag{2.65}$$

Thus, the orbits $\mathbf{r}_{\perp}(t) = \mathbf{r}_G$ are circles. The gyrofrequency (or cyclotron frequency) $\omega_G = \omega$, velocity $\mathbf{v}_{\perp}$, and magnetic moment μ follow correspondingly,

$$\omega_G = -\frac{q}{m}\mathbf{B}, \qquad \mathbf{v}_\perp = \boldsymbol{\omega}_G \times \mathbf{r}_G, \qquad r_G = \frac{v_\perp}{\omega_G} = \frac{(2mE_\perp)^{1/2}}{qB},$$

$$\boldsymbol{\mu} = \frac{1}{2}q\mathbf{r}_\mathbf{G} \times \mathbf{v} = -\frac{q}{2}r_G^2\boldsymbol{\omega}_G \Rightarrow \mu = |\boldsymbol{\mu}| = \frac{q}{2}\frac{\mathbf{v}_\perp^2}{\omega_G} = \frac{E_\perp}{B}; \; E_\perp = \frac{1}{2}mv_\perp^2 \quad (2.66)$$

In presence of a magnetic field in the Hamiltonian of a particle the vector potential **A** appears. Again, for a spatially homogeneous magnetic field **A** follows from the analogy with the rotating solid,

$$\nabla \times (\boldsymbol{\omega} \times \mathbf{r}) = 2\boldsymbol{\omega}, \qquad \nabla \times \mathbf{A} = \mathbf{B} \quad \Rightarrow \quad \mathbf{A} = \frac{1}{2}\mathbf{B} \times \mathbf{r}_G. \qquad (2.67)$$

With this we calculate the integral $\oint \mathbf{pdq}$ along the circle,

$$\oint \mathbf{pdq} = \oint \left(m\mathbf{v}_\perp + \frac{q}{2}\mathbf{B} \times \mathbf{r}_G\right)\mathbf{v}_\perp \mathrm{d}t = \frac{2\pi}{\omega_G}\left(m\mathbf{v}_\perp^2 - \frac{1}{2}m\mathbf{v}_\perp^2\right) = 2\pi\frac{m}{q}\mu. \qquad (2.68)$$

The motion of the electron is right handed with respect to the magnetic field, the motion of the proton is left handed. The magnetic moment is antiparallel to **B**, irrespective of charge (diamagnetism: the induced magnetic field is parallel to $\boldsymbol{\mu}$) and opposite to its angular momentum **L**, see Fig. 2.6. Numerically, with B given in Tesla [T] gyrofrequency ω_G, gyroradius r_G, and magnetic moment μ of the electron read as follows,

$$\omega_G[\mathrm{s}^{-1}] = \frac{e}{m_e}B = 1.8 \times 10^{11}B[\mathrm{T}], \quad r_G[\mathrm{m}] = \frac{\sqrt{2m_e}}{e}\frac{\sqrt{E_\perp}}{B} = 3.4 \times 10^{-6}\frac{\sqrt{E_\perp[\mathrm{eV}]}}{B[\mathrm{T}]},$$

$$\mu[\mathrm{eVT}^{-1}] = 1.6 \times 10^{-19}\frac{E_\perp[\mathrm{eV}]}{B[\mathrm{T}]}; \; 1\,\mathrm{eV} = 1.602 \times 10^{-19}\mathrm{J}, \; 1\mathrm{Tesla} = 10^4\,\mathrm{Gauss}. \qquad (2.69)$$

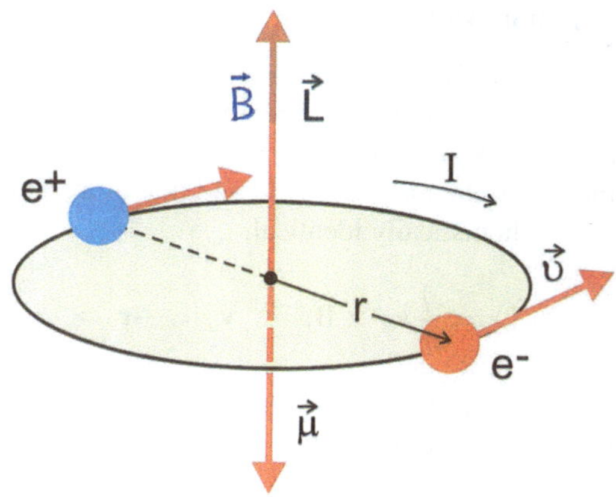

Fig. 2.6 Magnetic moment μ of a charge is opposite to its angular momentum **L**; $\mu = -(|q|/2m)\mathbf{L}$. The motion of the electron is righthanded with respect to **B**, the proton circulates lefthanded. $|\mu|$ is an adiabatic invariant

Gyrofrequency ω_G and gyroradius r_G of an ion of charge Z and mass m_i and the gyrovelocity of the electron change by the factor Zm_e/m_i and $(m_i/m_e)^{1/2}/Z$, respectively; the magnetic moment is by m_i/Zm_e larger.

Constant Crossed Electric and Magnetic Fields

The constant electric field $\mathbf{E}$ is assumed orthogonal to $\mathbf{B}$. Switching from the lab frame S to the drifting reference system $S'(\mathbf{v}_D)$ with

$$\mathbf{v}_D = \frac{\mathbf{E}\times\mathbf{B}}{B^2}; \qquad \mathbf{v} = \mathbf{v}_D + \mathbf{v}'. \tag{2.70}$$

yields

$$q(\mathbf{E}+\mathbf{v}\times\mathbf{B}) = q(\mathbf{E}+\mathbf{v}_D\times\mathbf{B}+\mathbf{v}'\times\mathbf{B}) = q\mathbf{v}'\times\mathbf{B}. \tag{2.71}$$

In S' the motion is a pure gyration; in the lab frame S the gyration undergoes the constant drift $\mathbf{v}_D = \mathbf{E}\times\mathbf{B}/B^2$. It does not depend on the sign of the charge. $\mathbf{B}' = \mathbf{B} - \mathbf{v}_D\times\mathbf{E}/c^2$ has been identified with $\mathbf{B}$. In the case the electric field is substituted by a constant gravity field $\mathbf{g}$, $\mathbf{v}_D$ is to be replaced by the gravity drift $\mathbf{v}_g$,

$$\mathbf{v}_g = \frac{m}{q}\frac{\mathbf{g}\times\mathbf{B}}{B^2}. \tag{2.72}$$

It induces charge separation.

Gradient Drift

An inhomogeneous magnetic field $\mathbf{B}(\mathbf{x})$ of constant direction and constant weak gradient $\nabla|\mathbf{B}(\mathbf{x})|$ is considered now. "Weak" means that the field variation over the largest gyroradius is small compared to the local value of B. In this special field configuration the local curvature radius $R_c(\varphi)$ to the particle orbit assumes the value [see Fig. 2.7a]

$$R_c(\varphi) = r_G\left(1 - r_G\frac{|\nabla B|}{B}\sin\varphi\right); \qquad \mathbf{B} = \mathbf{B}(\mathbf{x}_0), \qquad \nabla B = \nabla B(\mathbf{x}_0).$$

Position $\mathbf{x}_0$ is the gyrocenter. The curvature radii are (nearly) pointing towards $\mathbf{x}_0$; for higher order deviations see Sec. on adiabatic conservation of $\mu = \mathrm{const_a}$. Owing to $R_c(\varphi+\pi) > R_c(\varphi)$, $\varphi < \pi$, during one cycle the gyromotion exhibits the drift of the magnitude δ,

$$\delta = \oint R_c(\varphi)\sin\varphi \mathrm{d}\varphi = \pi B r_G^2\frac{|\nabla B|}{B}.$$

orthogonal to both, $\mathbf{B}$ and the gradient of B, see inset of Fig. 2.7. Correspondingly, the average drift speed is $v_D = \omega_G\delta/2\pi$,

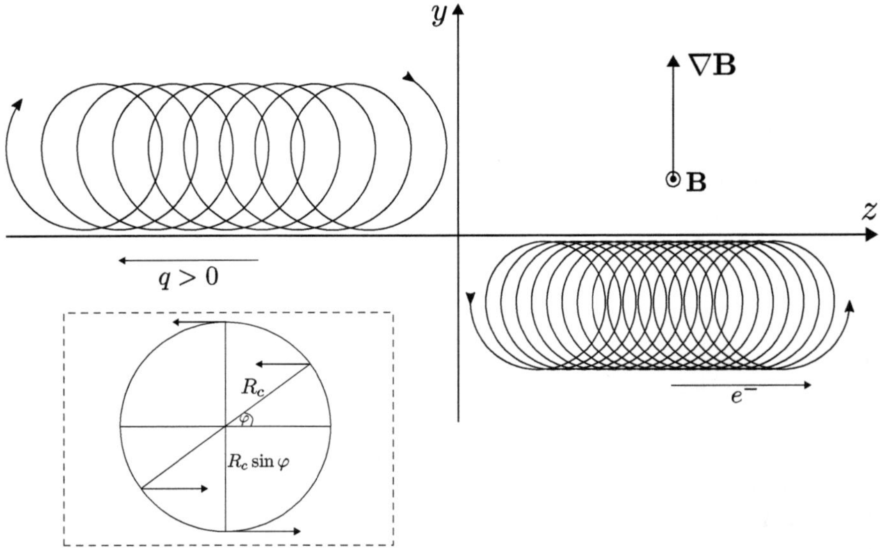

Fig. 2.7 Gradient drift leads to charge separation. Magnetic field **B** oriented in $-x$ direction, grad **B** points along y. Electron drifts to the right, ion to the left; electron velocity is assumed much smaller than ion velocity. The local curvature radius $R_c(\pi+\varphi)$ and its projection $R_c(\pi+\varphi)\sin(\pi+\varphi)$ along ∇B are larger than $R_c(\varphi)$ and projection $R_c(\varphi)\sin\varphi$. The projections are opposite to each other (see inset). The curvature radius $\mathbf{R}_c = R_c\mathbf{n}$ points towards the gyration center $\mathbf{x}_0$. The arrows in the inset are the local contributions to the drift along z

$$\mathbf{v}_D = \frac{q}{2m} r_G^2 \frac{\mathbf{B} \times \nabla B}{B}. \tag{2.73}$$

A quick view on Fig. 2.7 shows that the cross product reproduces the right drift direction of positive and negative charges. The gradient drift leads to charge separation.

Curvature Drift

A velocity $v_\parallel$ parallel to a magnetic field line of curvature radius $\mathbf{R}_c = R_c\mathbf{n}$ produces the centrifugal force $mv_\parallel^2/R_c$. According to (2.72) the concomitant drift velocity $\mathbf{v}_D$ produced by the magnetic field is

$$\mathbf{v}_D = \frac{mv_\parallel^2}{qB^2} \frac{\mathbf{R}_c \times \mathbf{B}}{R_c^2}. \tag{2.74}$$

2.1.4 Slow Motions and Adiabatic Invariants

Cycle-Averaged Motion

Often we are not interested in the microstructure of particle motion because it shows considerable regularity in its fine structure which we believe to be familiar with; we think we have a sufficiently detailed picture or imagination what happens on the fast time scale and on the short scale in space. We have encountered already prominent examples of such types of motion: particles gyrating in a static magnetic field and electrons oscillating in the electromagnetic wave of the laser. However, it may be of great interest to know particle flows, electric currents, momentum, and energy flows and their evolutions in time and space on large scales. Drift motions of the foregoing section may give answers on particle losses in magnetic devices. When studying new concepts of particle accelerators based on lasers the radiation pressure on the single electron, but not the details of the oscillatory motion, have to be known. Generally speaking, "coarse graining", i.e., desirable loss of information by switching from microscopic to macroscopic scales is successful with motions that exhibit oscillatory character in time and/or space. The general procedure is time- and space-averaging over the fine structures. A characteristic example of averaging is the gradient drift motion in the magnetic field. For the method to be successful the microscopic scales must be *clearly separable* from the resultant macroscopic scales.

It has been C. F. Gauss to recognize that periodic motions on the short and fast scale lead to non-periodic, unidirectional macroscopic motions and it seems that he has been the first to make use of it in perturbation theory. The averaging method has been successfully applied first to celestial mechanics, for example by Gauss on the asteroid Ceres in January 1801, to study slow sensible drifts on long time scales. Such non-periodic motions resulting from averaging over periodic orbits are called *secular motions*. According to our definition of force (2.3) *secular forces* are to be associated with secular accelerations.

In numerous cases averaging over one oscillation or rotation yields acceptable results. On the other hand there is no general prescription how averaging has to be done. A simple example is the averaged kinetic energy of a point particle in an electromagnetic wave. It yields the correct radiation pressure. However, when the oscillation energy of a particle with internal degrees of freedom is cycle-averaged a wrong result is obtained, and the same is true when averaging is done with a point particle without internal degrees of freedom in presence of a static magnetic or periodic electric field. In addition, the question about the precision of the results may be raised and one may ask what is the limiting accuracy of averaging when it is done in the most careful way.

The answer to these questions, correct averaging and highest precision, exists in the case the Hamiltonian of a *periodic system* depends weekly on a parameter λ, $H = H(\mathbf{p}, \mathbf{q}, t; \lambda)$. The parameter λ may be the amplitude of an electromagnetic wave, a pendulum of length changing slowly in time or an atom bouncing from the walls of a slowly shrinking box. The fast motion may be governed by a time-dependent Hamiltonian that disappears after averaging at λ held fixed. However, as λ changes, either in time or in space, or both, the translational and rotational symmetries are broken and, according to Noether's theorem (2.52), see also (2.48), (2.56), linear and angular momentum as well as energy are no longer conserved. They are replaced by the corresponding *adiabatic invariants* if the change of λ is slow *and* smooth; internal resonances may destroy the adiabatic conservation. The adiabatic procedure is the correct and most precise averaging we are looking for.

Adiabatic invariants are of great practical relevance; they may allow fast and precise numerical calculations. For example, in an interference experiment in an inhomogeneous medium phase and amplitude of monochromatic light may have to be known to high precision in a given position far away from the source, with thousands of wavelengths in between. Solving the wave equation brings the answer. This is time consuming and probably not precise enough because of the many rounding errors and because of numerical diffusion. Instead, use of the *Wentzel–Kramers–Brillouin (WKB)* or *eikonal approximation* solves the problem quickly with high precision. The shortcomings of the standard perturbation and the averaging method in relation to the adiabatic invariants will be illustrated in detail in connection with the ponderomotive force on the single electron and the radiation pressure on volume elements of matter.

Three Examples of Adiabatic Motion

Conservation of the Magnetic Moment

The magnetic moment of a charged particle is an adiabatic invariant, $\mu = \mathrm{const_a}$. The simplest proof is based on dimensional considerations. The temporal variation of μ depends only on the length r_G and the magnetic field B. Therefore

$$\frac{1}{\mu}\frac{\mathrm{d}\mu}{\mathrm{d}t} = \mathrm{const} \qquad \Rightarrow \qquad \mu = \mathrm{const_a} \tag{2.75}$$

by the following argument. The expression on the LHS of the first equality has the dimension of inverse time. As from a length and a Tesla no inverse time can be formed in (2.66) the expression can only be a constant for all values of B. For $B = \mathrm{const}$ its time variation is zero, hence $\mathrm{d}\mu/\mathrm{d}t = 0$ for all B. The situation of the gyration center $\mathbf{x}_0$ moving in space is reduced to time variation of B by switching to an inertial reference system co-moving with $\mathbf{x}_0$.

One could argue about choosing the derivative with respect to the dimensionless angle $\varphi = \omega_G t$ instead of t, but then nothing can be concluded because $\mathrm{d}\mu/\mathrm{d}\varphi = 0$ is trivial since under an adiabatic change of B the magnetic moment does not depend on the gyration angle. The success of dimensional analysis depends on the appropriate choice of independent variables.

An elegant, alternative proof (surprisingly not found in the standard literature) is based on Faraday's law and the $\mathbf{E} \times \mathbf{B}$ drift. By Faraday's law the electric field is induced parallel to the particle orbit and gives rise to the drift $\mathbf{v}_D$ towards to the center $\mathbf{x}_0$,

$$\mathbf{E} = \frac{1}{2}\mathbf{r}_G \times \dot{\mathbf{B}}, \qquad \mathbf{v}_D = \frac{\mathbf{E} \times \mathbf{B}}{B^2} = -\mathbf{r}_G \frac{\dot{B}}{2B},$$

or in scalar form

$$\dot{r}_G = -r_G \frac{\dot{B}}{2B} \qquad \Rightarrow \qquad \ln(r_G^2 B) = \mathrm{const_a}. \tag{2.76}$$

In both proofs the requirement of adiabatic change of **B** is hidden in the condition that the local curvature radius $\mathbf{R}_c$ (i) does not vary sensitively during one cycle and (ii) $\mathbf{R}_c$ points towards $\mathbf{x}_0$ in order to be identified with $-\mathbf{r}_G$. The local angular deviation is $\alpha = dr_\perp/ds = \dot{r}_\perp \mathrm{d}t/r_\perp \mathrm{d}\varphi = \dot{r}_\perp/r_\perp \omega_G$, thus

$$\alpha = \frac{\dot{r}_\perp}{r_\perp \omega_G} \simeq \frac{\Delta r_G}{2\pi r_G} = \frac{\Delta B}{4\pi B}.$$

Δr_G and ΔB indicate the changes over one circle. The smallness of α is very well satisfied and deviates very little from the principal normal **n**.

Guiding Center Motion

Often it is more important to know the trajectory of the guiding or gyration center than the detailed particle orbit. It is instructive and useful also for analogous situations, e.g. for the ponderomotive force to apply the cycle averaging method to derive its Hamiltonian. In a first step let **B** be static, $\mathbf{B} = \mathbf{B}(\mathbf{x})$. Then $H = C = \text{const}$ holds. Keeping in mind that $\mathbf{v}_\perp = \mathbf{v}_{0\perp} + \dot{\mathbf{r}}_G$ ($\mathbf{v}_{0\perp}$ stands for a drift across **B**), averaging the Hamiltonian (2.58) over one cycle yields

$$\bar{H} = \frac{1}{2}m\left(\mathbf{v}_0^2 + \dot{\mathbf{r}}_G^2\right) = \frac{\mathbf{p}_0^2}{2m} + \frac{m\dot{\mathbf{r}}_G^2}{2B}B = \frac{\mathbf{p}_0^2}{2m} + \mu B = C. \tag{2.77}$$

The gyroenergy $m\dot{\mathbf{r}}_G^2/2 = \mu B = -\boldsymbol{\mu}\mathbf{B}$ is a function of the gyrocenter position $\mathbf{x} = \mathbf{x}_0$ only. From $H = C = \text{const}$ follows that the averaged term $\mu B = -\boldsymbol{\mu}\mathbf{B}$ is the potential $V(\mathbf{x})$ governing the gyrocenter motion through the canonical equations

$$\dot{\mathbf{p}}_0 = -\frac{\partial \bar{H}}{\partial \mathbf{x}_0} = \nabla(\boldsymbol{\mu}\mathbf{B}) = (\boldsymbol{\mu}\nabla)\mathbf{B} + (\mathbf{B}\nabla)\boldsymbol{\mu}, \qquad \dot{\mathbf{x}}_0 = \frac{\partial \bar{H}}{\partial \mathbf{p}_0}. \tag{2.78}$$

It is known from Electrodynamics (e.g. Jackson, p. 185) that the Hamiltonian of a localized current distribution in an external field is identical with the cycle averaged Hamiltonian (2.77) (true even for time-varying fields) and owing to $\nabla \times \mathbf{B} = 0$ in vacuum and $\nabla \times \boldsymbol{\mu} = 0$ the force reduces to

$$\mathbf{f} = (\boldsymbol{\mu}\nabla)\mathbf{B}. \tag{2.79}$$

Let now **B** slowly depend on time. From $\boldsymbol{\mu} = \mathrm{const_a}$ follows again that (2.79) holds.

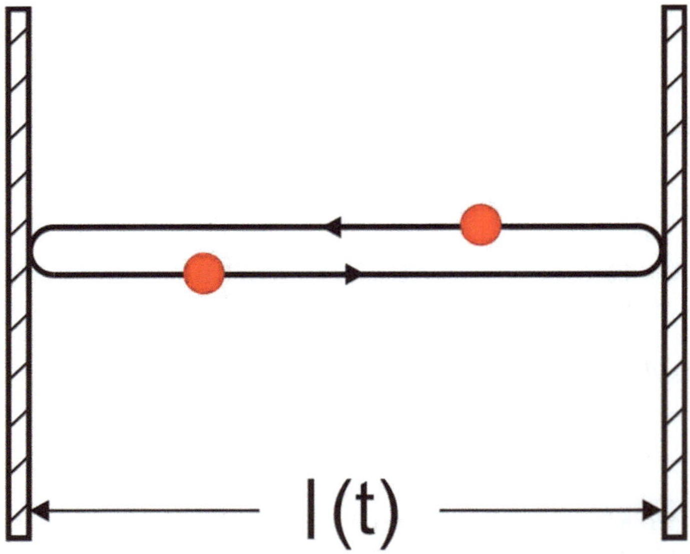

Fig. 2.8 Ball bouncing between slowly moving elastic walls at distance $l(t)$. Quantity mvl is adiabatically conserved

Generally magnetic field lines are not straight. If the curvature radius R_c is considerably larger than the local gyroradius r_G, $R_c \gg r_G$, the particles follow the local *direction* of the field lines with lateral displacements given by the gradient and the curvature drifts. In that sense a charged particle moves along a magnetic field line. This is also an aspect of adiabatic behavior. When its guiding center moves with initial momentum $\mathbf{p}_0$ into a region of increasing $\mathbf{B}$ it is slowed down and may come to rest eventually, and is reflected back as described by (2.79). This is the principle of the *magnetic mirror* and the static magnetic acceleration. Consider a magnetic coil of maximum field $B_{\max}$ on the axis of its plane and an ensemble of particles of energy $E = m(v_\parallel^2 + v_\perp^2)/2$ far out on the axis. Particles escape from the mirror if

$$\sin^2\alpha = \frac{v_\parallel^2}{v_\parallel^2 + v_\perp^2} < \frac{\mu B_{\max}}{E} = \sin^2\alpha_0. \tag{2.80}$$

The pitch angle α_0 below which the particle is lost defines the *loss cone* of the mirror in velocity space.

Particle Between Slowly Moving Walls

Let us consider a particle bouncing forth and back at speed v between two parallel elastic walls the distance l of which is slowly varying in time (see Fig. 2.8). By reflection from the moving wall the particle undergoes the momentum change $\Delta p = -2m\dot{l}$ each time interval $T = 2l/v$ and frequency $\omega = 2\pi/T$. With the averaged force per cycle $\dot{p} = \Delta p/T$ the particle dynamics follows the simple law

$$\dot{p} = -\frac{2m\dot{l}}{T} = -mv\frac{\dot{l}}{l} \quad \Rightarrow \quad \frac{\dot{p}}{p} + \frac{\dot{l}}{l} = 0 \quad \Rightarrow \quad pl = 4\pi\frac{E_{\mathrm{kin}}}{\omega} = \mathrm{const_a}. \tag{2.81}$$

Note, $\dot{p}$ is a smooth function, whereas Δp is not. The conservation of pl is a special case of

$$\oint p\mathrm{d}q = 2pl = \mathrm{const_a}. \tag{2.82}$$

$\oint p\mathrm{d}q$ reproduces the adiabatic invariant.

Application: Adiabatic change of state in ideal classical or Bose, or Fermi gas. The adiabatic change of state for one, two, and three degrees of freedom is easily derived from (2.81); we do it for $f = 3$. Starting from the initial values indicated by the index 0 for volume, particle density, temperature, pressure, and entropy V, n, T, p, S holds

$$pl = p_0 l_0, \quad E = \frac{\mathbf{p}^2}{2m} = E_0\left(\frac{l_0}{l}\right)^2 = E_0\left(\frac{V_0}{V}\right)^{2/3} = E_0\left(\frac{n}{n_0}\right)^{2/3} \Rightarrow Tn^{-2/3} = \mathrm{const_a}.$$

$$T\mathrm{d}S = \mathrm{d}E + p\mathrm{d}V = 0 \quad \Rightarrow \quad p = -\left(\frac{\partial E}{\partial V}\right)_S = \frac{2}{3}\frac{E_0}{V_0}\left(\frac{V_0}{V}\right)^{5/3} \quad \Rightarrow \quad pn^{-5/3} = \mathrm{const_a}. \tag{2.83}$$

In one dimension the particles may not equilibrate or only after a long time, e.g. an arbitrary one-dimensional rarefied particle distribution. The adiabatic invariance $pl = \mathrm{const_a}$ applies on such ensembles far from thermal equilibrium too.

Einstein's Pendulum of Variable Length

The question may arise what happens when a pendulum changes slowly its length during operation: Is there a quantity that is conserved? Lord Rayleigh gave the answer to this problem in 1902. Einstein, not aware of Rayleigh's result, reconfirmed the result on the Solvay Conference in 1911. Here, $\lambda = l$ is the parameter in the Hamiltonian of the pendulum. For simplicity we treat the problem for small displacements (angle α small). The quantity that is conserved is the *action* energy×time,

$$\frac{E_{\mathrm{os}}}{\omega} = \mathrm{const_a} \tag{2.84}$$

and not the oscillation energy E_{os} taken at fixed length l. The latter assumes the role of the parameter λ in the Hamiltonian of the plane pendulum.

Proof 1 (a) We consider E_{os} during one cycle $(t, t+T)$ with $l = l(t)$ fixed during T. The orbit in phase space is closed. For small displacements $l\alpha$ (for simplicity) holds

$$\ddot{\alpha} + \omega^2\alpha = 0 \quad \Rightarrow \quad E_{\mathrm{os}} = \frac{m}{2}l^2(\dot{\alpha}^2 + \omega^2\alpha^2); \quad \omega^2 = \frac{g}{l}.$$

Now repeat the same procedure, with l held fixed during one cycle, a time interval δt later during which l has changed from $l(t)$ to $l(t+\delta t)$. In correspondence E_{os}

and ω have changed from $E_{\text{os}}(t)$ and $\omega(t)$ to $E_{\text{os}}(t+\delta t)$ and $\omega(t+\delta t)$. With the displacement $l\alpha$ given initially, E_{os} is only a function of the slowly varying length or oscillation frequency ω (hint from [2]), i.e.

$$\lim_{\delta t\to 0}\frac{\Delta E_{\text{os}}}{\Delta\omega}=\frac{\mathrm{d}E_{\text{os}}}{\mathrm{d}\omega}=ml\alpha^2\omega=\frac{E_{\text{os}}}{\omega}\ \Rightarrow\ \frac{\mathrm{d}E_{\text{os}}}{E_{\text{os}}}=\frac{\mathrm{d}\omega}{\omega}\ \Rightarrow\ \frac{E_{\text{os}}}{\omega}=\text{const}_{\text{a}}.$$

(b) In reality l changes continuously with the consequence that the orbit is no longer closed in phase space. The reader may easily show that the open orbit deviates from the closed cycle by at most first order in ω and therefore the foregoing result remains valid. The arguments above allow the extension of proof 1 to arbitrary excursions α of the pendulum, see (2.129)–(2.131).

We consider the spherical pendulum. It has the advantage to reduce to simple circular motion with velocity $v=\text{const}$ and to make the underlying physics immediately evident. The Galileian pendulum above is obtained by projection of the motion on a vertical plane.

Proof 2 Gravity g does not affect the motion when l changes slowly because it is orthogonal to the circular motion. The same is true with the centrifugal force. The angular momentum $\mathbf{L}$ points along the symmetry axis of the cone and is conserved (hint from [2]),

$$\left|\frac{\mathrm{d}}{\mathrm{d}t}\mathbf{L}\right|=\frac{\mathrm{d}}{\mathrm{d}t}rmv=\frac{\mathrm{d}}{\mathrm{d}t}\frac{v}{\omega}mv=\frac{\mathrm{d}}{\mathrm{d}t}\frac{E_{\text{rot}}}{\omega}=0\ \Rightarrow\ \frac{E_{\text{rot}}}{\omega}=\text{const}_{\text{a}}.$$

Strictly speaking, orthogonality is not exactly obeyed with l changing in time. With ϵ indicating the deviation from orthogonality in $\mathbf{L}$ a correction term $\cos\epsilon$ appears. For small values of $1-\epsilon$ holds $1-\cos\epsilon=\epsilon^2/2+O(\epsilon^4)$; thus, adiabatic conservation is safe.

From Fig. 2.9 the role of the centrifugal force as an additional gravity force becomes clear,

$$g'=\left(g^2+\frac{v^2}{r}\right)^{1/2}\simeq g\left(1+\frac{1}{2}\alpha^2\right);\ \ \frac{v^2}{gr}\simeq\frac{r}{l}=\sin\alpha\simeq\alpha\ \Rightarrow v^2=g\alpha^2 l.$$

The effective potential energy is $-m(g'-g)l=-mg\alpha^2l/2$. The energy E_{rot} of the pendulum is according to (2.50) $E_{\text{rot}}=E_{\text{kin}}+E_{\text{pot}}=mv^2$.

Proof 3 This is a detailed proof based on the dynamics of the spherical pendulum. The energy conservation reads

$$\frac{\mathrm{d}}{\mathrm{d}t}mv^2=-\frac{1}{2}mg\alpha^2\dot{l}$$

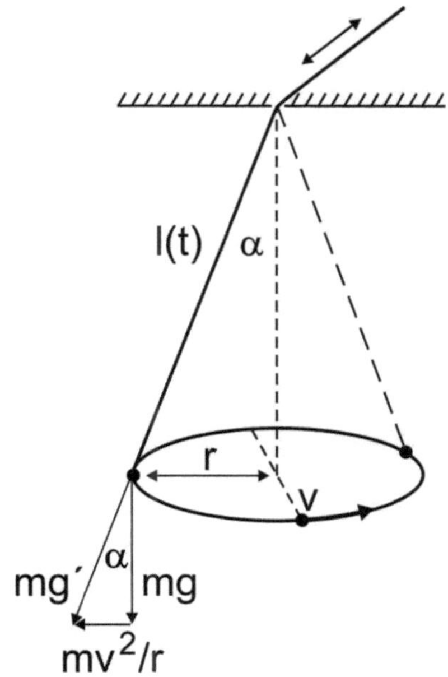

Fig. 2.9 Einstein's pendulum of length l slowly varying in time, here in spherical configuration as the result of two orthogonal plane Galileian pendula superposed. Note that the variation of l must be slow enough as to guarantee alignment of g' and l. The adiabatic invariant is $E_{\rm rot}/\omega$

$$\Rightarrow\quad 0=\frac{\mathrm{d}}{\mathrm{d}t}v^2+\frac{1}{2}g\alpha^2\dot{l}=g\left[\frac{\mathrm{d}}{\mathrm{d}t}(\alpha^2 l)+\alpha^2\dot{l}\right]=gl\alpha^2\left[\frac{\dot{(\alpha^2)}}{\alpha^2}+\frac{3}{2}\frac{\dot{l}}{l}\right].$$

From the integration of the last bracket expression and Galilei's formula $\omega=\sqrt{l/g}$ follows

$$\alpha^2 l^{3/2}=\mathrm{const_a}\quad\Rightarrow\quad\frac{E_{\rm rot}}{\omega}=l^2\alpha^2\omega=g^{1/2}\alpha^2 l^{3/2}=\mathrm{const_a}.$$

The last equality may be compared with our loop integral

$$\oint p\mathrm{d}q=2E_{\rm rot}T=4\pi\frac{E_{\rm rot}}{\omega}=\mathrm{const_a}. \tag{2.85}$$

We conclude that in all three examples considered the phase integral $\oint \mathbf{pdq}$ produces the correct adiabatic invariant. In the adiabatic process energy conservation is lost but in its place the action is conserved. The oscillation energy $E_{\rm os}$ of the linear pendulum is half the energy of the spherical pendulum. By inserting the local centrifugal force in the plane case one arrives at the same equation of motion and the identical conclusion (see exercises). The change from one dimension to the spherical case in two dimensions not only proves that sometimes, owing to an additional symmetry in higher dimensions, the treatment of a problem results easier but it gives also essential insight. In the particular case here it shows in what the adiabaticity consists physically: the resulting force of gravitation and centrifugal force and the tension along $l(t)$ must result collinear to second order in their deviation from the straight angle

Fig. 2.10 **a** Picture of swinging lady by Jean Honore Fragonard (1732–1806). Being trapped on the swing how does she manage to couple energy and momentum into the swing? The answer is given by (**b**) and Fig. 2.9 (see also Problems). **b** First and second region of stability and instability of the Mathieu map. a_0, a_1, and b_1 are Mathieu functions, μ is the characteristic exponent in stable regions and the second instability region. q and a are free parameters of the Mathieu equation. At q fixed instability increases with growing distance from the boundaries a_0, a_1, and b_1

π. Or in equivalent terms, the deviation $\angle\epsilon$ of the spiralling up or down motion due to the variation in length from the circular orbit at $l =$ const is of second order in ϵ. Historically, adiabatic invariances have plaid an important role in understanding non-spherical orbit quantization in the early phase of quantum mechanics.

Resonances Destroy the Adiabatic Invariance, or: **Theory of the Swing**

As a rule, slow change of length $l(t)$, or of the λ parameter in the Hamiltonian in general, does not guarantee adiabatic invariance. A quick view on the swing in action (Fig. 2.10a) tells the contrary, and already the simplest theory shows that in the absence of friction periodic length variation of vanishing amplitude superposed on its uniform slow increase or decrease can drive the pendulum unstable. Therefore, for safety reasons we must additionally postulate smoothness of length variation, i.e.,

$$\mathrm{d}^2 l/\mathrm{d}t^2 = 0 \tag{2.86}$$

as a sufficient condition. Everybody drives instinctively the swing unstable. However, ask anybody how he does it, specifically at what frequency, in 50% of cases you may get the wrong answer (author's experience). The general theory of the swing is very complex, similar to the theory of driving a bicycle, a simple stability analysis is feasible with the help of Mathieu's differential equation for the planar pendulum [3],

$$\ddot{x} + \omega^2\left(1 - 2\frac{q}{\omega^2}\cos 2\omega' t\right)x = 0 \;\Rightarrow\; \frac{\mathrm{d}^2 x}{\mathrm{d}\tau^2} + a\big(1 - b\cos 2\tau\big)x = 0, \tag{2.87}$$

$a = (\omega/\omega')^2$, q, $b = 2q/\omega^2$ parameters, $\tau = \omega' t$. The general solution assumes the structure

$$x(t) = C_1 e^{i\mu\omega' t} f(\omega t) + C_2 e^{i\mu\omega' t} g(\omega' t). \tag{2.88}$$

The functions f and g are real and of periodicity ω', μ is the characteristic exponent. The adiabatic constants C_1, C_2 must be chosen such as to fulfill the undisturbed adiabatic motion (2.85) for $q = 0$. Figure 2.10b shows the stability chart: dashed regions stability, white regions instability. Depending on the initial conditions (constants C_1, C_2) maximum amplification or damping happens at length modulations with twice the eigenfrequency ω and any positive amplitude $b > 0$ whatever its magnitude is. In the chart these points correspond to $(a, b) = (1, b\omega^2/2)$. Negative amplitudes $b < 0$, corresponding to changes of modulation phase by π, are stable. Arrows pointing obliquely into the instability zone represent detuning of frequency ω' from ω. They show weaker growth and damping and, finally, stability. Equation (2.87) for the pendulum of variable length l is the prototype of the class of the *parametric instabilities*. The modulation of the length may be prescribed from outside or generated intrinsically by the oscillator itself (e.g., stimulated Brillouin scattering). The Mathieu equation is also a model equation for the band structure (conduction band, valence band) in the periodic potential of solids; gaps between bands are unstable.

2.1.5 Poincaré–Cartan Invariant and the Adiabatic Theorem

A geometrical interpretation of the Hamiltonian mechanics and its visualization can be achieved if the $2n$ dimensional phase space of the $(\mathbf{p}, \mathbf{q})$ points is enlarged to the $2n + 1$ dimensional *extended phase space* of the points $(\mathbf{p}, \mathbf{q}, t)$ containing the time t as an additional coordinate. In the mathematical language R^{2n+1} is the Cartesian product $R^{2n} \times R_t$. To each point $(\mathbf{p}, \mathbf{q}, t)$ we associate a vector $\dot{\mathbf{z}} = (-H_q, H_p, 1)$, $H_q = \partial_q H$, $H_p = \partial_p H$. By the yield of directions $\dot{\mathbf{z}}$ bundles of non-crossing curves, the so-called *vortex lines*, are defined which map one by one the points of the area of a closed loop L_0 at $t = t_0$ onto the points of the closed loop L_1 at time $t = t_1$ (see Fig. 2.11). By applying now the multidimensional Stokes theorem to the vector field $\{\dot{\mathbf{z}}\}$ on the bundle and keeping in mind that the flux through the mantel of the Hamiltonian flow tube is zero the *Poincaré–Cartan integral invariant*

$$\oint_{L_0} \mathbf{p}\mathrm{d}\mathbf{q} = \oint_{L_1} \mathbf{p}\mathrm{d}\mathbf{q} = \oint_{L_2} (\mathbf{p}\mathrm{d}\mathbf{q} - H\mathrm{d}t) \tag{2.89}$$

follows as the circulation of the *Poincaré–Cartan* differential 1-form

$$\mathrm{d}F(\mathbf{p}, \mathbf{q}, t) = \mathbf{p}\mathrm{d}\mathbf{q} - H\mathrm{d}t. \tag{2.90}$$

The flux of the vortex lines of the bilinear Poincaré–Cartan form through closed loops on the arbitrary Hamiltonian flux tube is conserved and equals the circulation

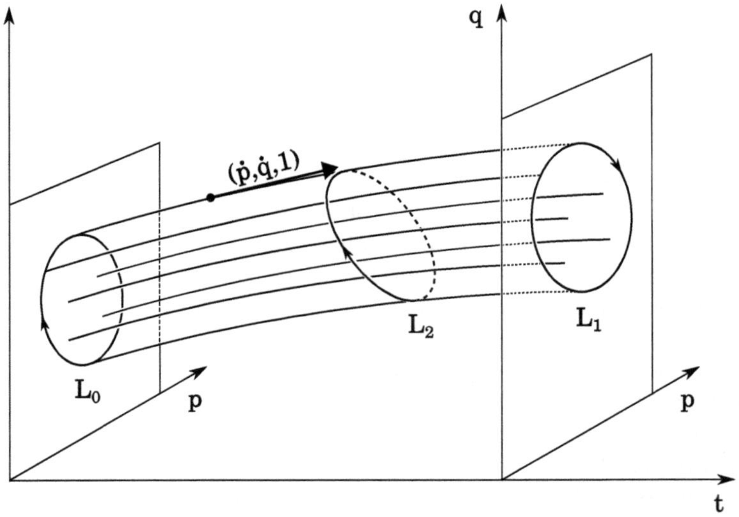

Fig. 2.11 Poincaré–Cartan integral invariant. The vectors $(\dot{\mathbf{p}}, \dot{\mathbf{q}}, 1)$ define a bundle of orbits emanating from a closed loop L_0 in the plane $t = t_0$ into the extended phase space $\{\mathbf{p}, \mathbf{q}, t\}$. By their projection onto the plane $t = t_1$ loop L_0 is mapped onto loop L_1 of the same area. In the case of a Hamiltonian $H(\mathbf{p}, \mathbf{q}, t; \lambda)$ strictly periodic in time $T = t_1 - t_0$ at λ held constant the fluxes of vortex lines through L_0 and L_1 are identical. In addition, an arbitrary cut across the flux tube generates the same flux $\oint(\mathbf{p}\mathrm{d}\mathbf{q} - H\mathrm{d}t)$ through the surface encircled by L_2. Under λ varying adiabatically within T the orbits deviate only by a small amount from strict periodicity and L_1 nearly reproduces L_0

$\oint \mathbf{p}\mathrm{d}\mathbf{q}$ at arbitrary $t = \text{const}$. It follows that the Poincaré–Cartan 1-form is an exact (or total) differential. The loops L_0, L_1 are bound to $t = 0$ and $t = t_1$, respectively. L_2 is a closed loop in the full extended phase space.

The elementary proof of (2.89) proceeds as follows. Select an arbitrary 3D vector $\dot{\mathbf{z}}_j = (-H_{q_j}, H_{p_j}, 1)$ and apply Stokes theorem to its t-component $\dot{z}_{j,t} = 1$ in the plane $\{(p_j, q_j, t = t_0)\}$,

$$\int_{L_{0j}} \mathrm{d}p_j \mathrm{d}q_j = \oint_{L_{0j}} p_j \mathrm{d}q_j = \oint_{L_{1j}} p_j \mathrm{d}q_j \Rightarrow \int_{L_1} \mathrm{d}\mathbf{p}\mathrm{d}\mathbf{q} = \sum_j \oint_{L_{1j}} p_j \mathrm{d}q_j = \oint_{L_1} \mathbf{p}\mathrm{d}\mathbf{q}.$$

Application of Stokes theorem to a loop in the oriented $\{(t, p_j, q_j = \text{const}\}$ plane and, analogously, to a loop in the oriented $\{(q_j, t, p_j = \text{const}\}$ plane yields

$$\int H_{p_j} \mathrm{d}p_j \mathrm{d}t = -\int \Delta_j H(p_j, t)\mathrm{d}t = -\oint H(p_j, t)\mathrm{d}t,$$
$$\int H_{q_j} \mathrm{d}q_j \mathrm{d}t = -\int \Delta_j H(q_j, t)\mathrm{d}t = -\oint H(q_j, t)\mathrm{d}t.$$

In the loop integrals on the RHS all coordinates in H are fixed except p_j and q_j, respectively. Summation over all partial loops results in the projection of loop L_2 onto the t-axis with the result $\sum_j \int [H_{p_j} \mathrm{d}p_j + H_{q_j} \mathrm{d}q_j]\mathrm{d}t = -\int_{L_2} H(\mathbf{p}(t), \mathbf{q}(t), t)\mathrm{d}t$.

Two immediate consequences of (2.89) are Liouville's conservation of the phase space volume owing to $\int \mathrm{d}\mathbf{p}\mathrm{d}\mathbf{q} = \oint \mathbf{p}\mathrm{d}\mathbf{q}$ encircling the same phase flux at arbitrary $t = \text{const}$, and the canonical equations of motion (2.55). The individual vortex line is completely described by its initial coordinates $(\mathbf{p}_0, \mathbf{q}_0)$ and its projection onto the t-axis, $\mathbf{p} = \mathbf{p}(\mathbf{p}_0, \mathbf{q}_0, t)$, $\mathbf{q} = \mathbf{q}(\mathbf{p}_0, \mathbf{q}_0, t)$. The differential of (2.90) is

$$\begin{aligned}\mathrm{d}\mathbf{p}\mathrm{d}\mathbf{q} - \mathrm{d}H\mathrm{d}t &= \sum_j [\mathrm{d}q_j(\mathbf{p}_0, \mathbf{q}_0, t)\mathrm{d}p_j - \mathrm{d}p_j(\mathbf{p}_0, \mathbf{q}_0, t)\mathrm{d}q_j] - \mathrm{d}H\mathrm{d}t \\ &= \left(\frac{\partial q_j(\mathbf{p}_0, \mathbf{q}_0, t)}{\partial t}\mathrm{d}p_j\mathrm{d}t - \frac{\partial p_j(\mathbf{p}_0, \mathbf{q}_0, t)}{\partial t}\mathrm{d}q_j\mathrm{d}t - \frac{\partial H}{\partial p_j}\mathrm{d}p_j\mathrm{d}t - \frac{\partial H}{\partial q_j}\mathrm{d}q_j\mathrm{d}t\right).\end{aligned}$$

In order to be exact it must fulfill

$$\dot{p}_j \equiv \frac{\partial p_j(\mathbf{p}_0, \mathbf{q}_0, t)}{\partial t} = -H_{q_j}, \quad \dot{q}_j \equiv \frac{\partial q_j(\mathbf{p}_0, \mathbf{q}_0, t)}{\partial t} = H_{p_j} \tag{2.91}$$

because $\mathrm{d}p_j, \mathrm{d}q_j$ are free. Equation (2.91) are the canonical equations of motion.

In view of forthcoming important applications it is useful to give an equivalent alternative formulation of Liouville's theorem on the conservation of the volume in the restricted $\{(\mathbf{p}, \mathbf{q})\}$ phase space. To this aim we consider a volume element $\mathrm{d}\tau = \mathrm{d}\mathbf{p}\mathrm{d}\mathbf{q}$ at the neighboring times t and $t + \mathrm{d}t$. With the Jacobian J it holds

$$\mathrm{d}\tau(t + \mathrm{d}t) = \mathrm{d}\tau(t)J(t + \mathrm{d}t) = \mathrm{d}\tau(t)\frac{\partial([\mathbf{p}(t) + \dot{\mathbf{p}}\mathrm{d}t], [\mathbf{q}(t) + \dot{\mathbf{q}}\mathrm{d}t])}{\partial(\mathbf{p}(t), \mathbf{q}(t))}.$$

The velocities $\dot{\mathbf{p}}$ and $\dot{\mathbf{q}}$ are taken at time t and depend each on all p_i, q_i. J consists of the product of all diagonal elements $\pm$ the products of all other combinations,

$$\prod_1^f (1 + \frac{\partial \dot{p}_i}{\partial p_i}) \prod_1^f (1 + \frac{\partial \dot{q}_i}{\partial q_i})\, \mathrm{d}t + o(\mathrm{d}t) = 1 + \sum_1^f \frac{\partial \dot{z}_i}{\partial z_i}\, \mathrm{d}t + o(\mathrm{d}t) = 1 + \operatorname{div} \dot{\mathbf{z}}\, \mathrm{d}t + o(\mathrm{d}t).$$

The non-diagonal products are all $o(\mathrm{d}t)$, i.e., of second and higher order up to $(\mathrm{d}t)^{2f}$. Thus, according to $\mathrm{d}J/\mathrm{d}t = 0$ from (2.61) follows

$$\frac{\mathrm{d}}{\mathrm{d}t}\mathrm{d}\tau = \mathrm{d}\tau \nabla \dot{\mathbf{z}} = 0 \quad \Rightarrow \quad \tau = \text{const} \quad \Leftrightarrow \quad \operatorname{div} \dot{\mathbf{z}} = 0; \qquad f \geq 1 \tag{2.92}$$

Incompressibility and divergence-free phase flow are equivalent.

Canonical Transformations

In dynamics the principal scope of a transformation of coordinates in general is to facilitate the integration of the equations of motion. Thereby the ultimate goal is to change to new coordinates $\mathbf{P} = \{P_j\}$, $\mathbf{Q} = \{Q_j\}$, s,

$$\mathbf{P} = \mathbf{P}(\mathbf{p}, \mathbf{q}, t), \qquad \mathbf{Q} = \mathbf{Q}(\mathbf{p}, \mathbf{q}, t), \qquad s = s(\mathbf{p}, \mathbf{q}, t). \tag{2.93}$$

in such a way as to get the solution $\mathbf{P} = \text{const}$, $\mathbf{Q} = \text{const}$. No such general integration procedure is known. So, as a more realistic goal one would like to keep all the advantages of the Hamiltonian dynamics, for instance Liouville's theorem which is basic for kinetic theory and thermostatistics. In other words, we aim at the most general transformation of kind (2.93) that preserves Hamilton's equations, with the new Hamiltonian $K = K(\mathbf{P}, \mathbf{Q}, s)$,

$$\dot{\mathbf{p}} = -\frac{\partial K}{\partial \mathbf{Q}} \quad \dot{\mathbf{Q}} = \frac{\partial K}{\partial \mathbf{P}}, \quad \frac{\mathrm{d}K}{\mathrm{d}s} = \frac{\partial K}{\partial s}; \qquad \dot{P}_i = \frac{\mathrm{d}P_i}{\mathrm{d}s}, \quad \text{etc.} \tag{2.94}$$

and all integral invariants of (2.90). Transformations with this property are called *canonical*. The class of (2.93) does not preserve the Hamiltonian structure in general, however the subclass of point transformations $\mathbf{Q} = \mathbf{Q}(\mathbf{q}, t)$ does it, as we have known already from Lagrangian dynamics. The associated canonical momenta follow by differentiation, $\dot{P}_i = \partial L(Q_j, \dot{Q}_j, s)/\partial \dot{Q}_j$. In the search for the widest class of canonical transformations the Poincaré–Cartan invariant plays the key role. The dependence on the canonical momenta has no influence on the circulation since integration of (2.90) acts only on $\mathbf{q}$ or on $\mathbf{Q}$. Whether a transformation (2.93) is canonical is regulated by the following

Theorem *The transformation (2.93) is canonical if and only if it obeys the equality*

$$\mathbf{p}d\mathbf{q} - H dt = \mathbf{P}d\mathbf{Q} - K ds + dS, \tag{2.95}$$

where dS is a total differential of $S(\mathbf{p}, \mathbf{q}, t)$ or $S(\mathbf{P}, \mathbf{Q}, s)$.

Proof Integration along any closed loop L with $s = \text{const}$ yields $\int \mathrm{d}S = 0$ and $\oint \mathbf{P}\mathrm{d}\mathbf{Q} = \oint \mathbf{p}\mathrm{d}\mathbf{q} = \text{const}$, thus from the multidimensional Stokes theorem follows (2.94), as well as the conservation of the integral invariants. Furthermore, the bundles generated by the two sets of orbits from $H(\mathbf{p}, \mathbf{q}, t)$ and $K(\mathbf{P}, \mathbf{Q}, s)$ are the same with a one to one correspondence of their orbits. This follows from the fact that at any phase point $\mathbf{z} = (\mathbf{p}, \mathbf{q}, t) = (\mathbf{P}, \mathbf{Q}, s)$ the projections of the tangent vectors $\dot{\mathbf{z}}(H)$ and $\dot{\mathbf{z}}(K)$ on t and s, respectively, are unique.

Corollary *If $s = t$ it follows*

$$\frac{d\mathbf{P}}{dt} = -\frac{\partial K}{\partial \mathbf{Q}} \quad \frac{d\mathbf{Q}}{dt} = \frac{\partial K}{\partial \mathbf{P}}, \qquad K(\mathbf{P}, \mathbf{Q}, t) = H(\mathbf{p}, \mathbf{q}, t). \tag{2.96}$$

It is an immediate consequence of (2.95), see (2.91), and $\oint dS = 0$. The corollary forbids scale transformations in general. The special transformation $\mathbf{P} = \alpha\mathbf{p}$, $\mathbf{Q} = \beta\mathbf{q}$ is canonical with $K(\mathbf{P}, \mathbf{Q}, t) = \alpha\beta H(\mathbf{p}/\alpha, \mathbf{q}/\beta, t)$ under the constraint $\alpha\beta = 1$. The latter guarantees also the invariant measure of volume elements, $d\mathbf{P}\, d\mathbf{Q} = d\mathbf{p}\, d\mathbf{q}$.

Generating Functions

For simplicity $s = t$ is set. Let us assume that in some region of the phase space $\mathbf{p}$ can be expressed as a single valued function $\mathbf{p} = \mathbf{p}(\mathbf{Q}, \mathbf{q})$ and hence $S(\mathbf{p}, \mathbf{q}, t) = S_1(\mathbf{Q}, \mathbf{q}, t)$. Then (2.95) reads

$$\mathbf{p}\mathrm{d}\mathbf{q} - H\mathrm{d}t = \mathbf{P}\mathrm{d}\mathbf{Q} - K\mathrm{d}t + \frac{\partial S_1}{\partial \mathbf{q}}\mathrm{d}\mathbf{q} + \frac{\partial S_1}{\partial \mathbf{Q}}\mathrm{d}\mathbf{Q} + \frac{\partial S_1}{\partial t}\mathrm{d}t$$

$$\Rightarrow \mathbf{p} = \frac{\partial S_1}{\partial \mathbf{q}}, \quad \mathbf{P} = -\frac{\partial S_1}{\partial \mathbf{Q}}, \quad H = K + \frac{\partial S_1}{\partial t}. \tag{2.97}$$

$S_1(\mathbf{Q}, \mathbf{q}, t)$ is a *generating function* for the transformation $(\mathbf{p}, \mathbf{q}) \to (\mathbf{P}, \mathbf{Q})$. It is easily seen that $S_1(\mathbf{Q}, \mathbf{q}, t)$ does not exhaust the class of canonical transformations. For instance it fails in the case of the identical transformation $(\mathbf{P}, \mathbf{Q}) = (\mathbf{p}, \mathbf{q})$ because $\mathbf{p} = \mathbf{p}(\mathbf{q})$ is no longer valid. Under the assumption of no constraints between $\mathbf{P}$ and $\mathbf{q}$ the generating function $S_2(\mathbf{P}, \mathbf{q}) = S_1 - \mathbf{P}\mathbf{Q}$, with S_1 expressed in $\mathbf{P}$ and $\mathbf{q}$ and no time dependence, yields

$$\mathbf{p}\mathrm{d}\mathbf{q} - H\mathrm{d}t = -\mathbf{Q}\mathrm{d}\mathbf{P} - K\mathrm{d}t + \frac{\partial S_1}{\partial \mathbf{q}}\mathrm{d}\mathbf{q} + \frac{\partial S_1}{\partial \mathbf{P}}\mathrm{d}\mathbf{P} \;\Rightarrow\; \mathbf{p} = \frac{\partial S_1}{\partial \mathbf{q}}, \quad \mathbf{Q} = \frac{\partial S_1}{\partial \mathbf{P}}, \quad K = H. \tag{2.98}$$

Several other combinations of old and new canonical variables are possible in the generating function S.

Extended and Reduced Phase Space

If H depends explicitly on time the energy is no longer conserved, see (2.56). In order to get $\mathrm{d}H/\mathrm{d}t = 0$ a new Hamiltonian $\mathcal{H}$ has to be looked for which numerically yields $\mathrm{d}\mathcal{H}/\mathrm{d}t = \mathrm{d}H/\mathrm{d}t - \partial_t H = 0$. This is accomplished in the extended phase space by defining

$$\mathcal{H} = \mathcal{H}(\mathbf{p}, \mathbf{q};\, p_t, q_t) = H(\mathbf{p}, \mathbf{q}, t) + p_t; \quad p_t = -H, \; q_t = t \tag{2.99}$$

$$\Rightarrow \quad \dot{p}_t = -\frac{\partial \mathcal{H}}{\partial q_t} = \frac{\partial H}{\partial t}, \quad \dot{q}_t = \frac{\partial \mathcal{H}}{\partial p_t} = \frac{\partial p_t}{\partial p_t} = 1; \quad \frac{\mathrm{d}\mathcal{H}}{\mathrm{d}t} = 0. \tag{2.100}$$

In the extended phase space $(\mathbf{p}, \mathbf{q}, t)$ the Hamiltonian $\mathcal{H}$ is energy conserving. The Poincaré–Cartan invariant reads $F = \oint \mathbf{p}\mathrm{d}\mathbf{q}$ and $\mathrm{d}F = \mathbf{p}\mathrm{d}\mathbf{q}$, respectively.

The $2f$ phase space *reduces* by two variables p_k, q_l if a conservation law $G = G(\mathbf{p}, \mathbf{q}) = G_0$ exists. g may stay for instance for energy conservation, $H(\mathbf{p}, \mathbf{q}) = E$. After elimination either of p_k or q_l from G the Hamiltonian H does no longer contain either p_k or q_l. Thus

$$\dot{q}_k = \frac{\partial H}{\partial p_k} = 0 \Rightarrow q_k = \text{const}; \quad \dot{p}_l = \frac{\partial H}{\partial q_l} = 0 \Rightarrow p_l = \text{const}. \tag{2.101}$$

It may happen that p_k or q_l do not appear in H from the very first. Such variables are named *cyclic*. Then, by the same reasoning as before the phase space reduces again from $2f$ to $2f - 2$ dimensions. In Lagrangian mechanics the set p_k, q_l corresponds to $q_k, \dot{q}_l$ with the same conclusion (2.101).

The Adiabatic Theorem

Quasiperiodic motions of mass points in the (2n + 1) dimensional phase space are of particular interest in interactions of charged particles with static magnetic fields and electric and electromagnetic waves. They can be described by a Hamiltonian $H(\mathbf{p}, \mathbf{q}, t; \lambda)$ where λ is a parameter, for instance, $\lambda = t$, with λ slowly changing in time. We formulate the adiabatic theorem in 7 dimensionsional phase space:

(i) $\lambda = \text{const}$ implies strict periodicity of motion, and time dependence, if any, fast with respect to the slow variation of $\lambda(t)$;
(ii) the variation of λ is such that the orbits are nearly closed, i.e., an oscillation/gyration center exists;
(iii) the variation of λ over several cycles is smooth.

Under these conditions holds

$$F = \oint (\mathbf{p}d\mathbf{q} - H\mathrm{d}t) = \text{const}_a. \tag{2.102}$$

Variations of this kind induced by λ are called *adiabatic* variations and indicated by the symbol const_a, defined by (2.103). If λ is a function of time, $\lambda = \lambda(t)$ in $H(\mathbf{p}, \mathbf{q}, t; \lambda)$, the time scale of the periodic motion must be fast in comparison to the time scale of λ. Hitherto we have encountered examples of slow variations where λ stood for a length (pendulum, distance between walls), magnetic field varying in space and/or time, electromagnetic field amplitude varying in space and/or time. The existence of a gyration or oscillation center implies the presence of two well separated space or time scales.

There are two modes of periodic motion: *libration*, if the motion is periodic in the restricted phase space, e.g. Galileian pendulum, electron trapped in the electrostatic wave, particle bouncing between moving walls; *rotation*, if the motion is open in the restricted phase space, e.g. rotating pendulum, detrapped electron. During one period T the oscillation/rotation center $\mathbf{x}_0$ shifts by the displacement $\mathbf{s}$, $\mathbf{x}_0(t + T) = \mathbf{x}_0(t) + \mathbf{s}$.

In order to give relation (2.102) a clear meaning a definition of adiabatic invariant is needed. If over one cycle T to start from t_0 the parameter $\lambda(t_0) = \lambda_0$ is kept constant the orbit is closed and $F(t_0)$ is well defined. At the time $t_0 + \delta t$ the parameter λ has assumed the new value $\lambda = \lambda_0 + \delta\lambda$ and the corresponding Poincaré–Cartan invariant is $F(t_0 + \delta t)$. For the *adiabatic change* $\Delta F = F(t_0 + \delta t) - F(t_0)$ we can give the following

Definition F is said *adiabatically conserved*, i.e $F = \text{const}_a$, if for the number of cycles N over which λ changes by the fixed amount $\Delta\lambda$

$$\lim_{N\to\infty} N\Delta F = 0 \tag{2.103}$$

is fulfilled [4, 5]. In words: Adiabatic invariance is guaranteed if the total deviation $N\Delta F$ reduces to zero for $N \mapsto \infty$ at the variation $\Delta\lambda$ held fixed over N cycles. The slower the variation of λ the more precise F becomes. All we have to show now is $\mathrm{d}F/\mathrm{d}t = 0$ in first order.

Proof For the difference $F(t_0 + \delta t) - F(t_0)$ of the periodic orbits starting from t_0 and $t_0 + \delta t$ holds

$$F(t_0 + \delta t) - F(t_0) \equiv F(t_0 + \delta t, \lambda_0 + \delta\lambda) - F(t_0, \lambda_0) =$$

$$[F(t_0 + \delta t, \lambda_0 + \delta\lambda) - F(t_0 + \delta t, \lambda_0)] + [F(t_0 + \delta t, \lambda_0) - F(t_0, \lambda_0)].$$

The second square bracket is zero because the Poincaré–Cartan invariant applies. The Jacobian $j_{t_0+\delta t}(\lambda_0, \lambda_0 + \delta\lambda)$ maps the first onto the second term in the first square bracket,

$$F(t_0 + \delta t, \lambda_0 + \delta\lambda) - F(t_0 + \delta t, \lambda_0)$$

$$= \oint \{[j_{t_0+\delta t}(\lambda_0, \lambda_0 + \delta\lambda) - 1](\mathbf{p}\mathrm{d}\mathbf{q} - H(\mathbf{p}, \mathbf{q}, \lambda_0))\mathrm{d}t\}.$$

The unit Jacobian $j = 1$ stands for $j_{t+\delta t}(\lambda_0, \lambda_0)$. In first order the Jacobian greatly simplifies, $j_{t+\delta t}(\lambda_0 + \delta\lambda) - 1 = \nabla\dot{\mathbf{z}}\,\delta t = 0$, $\dot{\mathbf{z}}$ taken at $t = t_0 + \delta t, \lambda = \lambda_0 + \delta\lambda$. Thus

$$j(t_0, t_0 + \delta t) - 1 = j(t_0, t_0 + [\partial\lambda/\partial t]\,\delta t) - 1 = \nabla\frac{\partial\dot{\mathbf{z}}}{\partial\lambda}\frac{\partial\lambda}{\partial t}(\delta t)^2$$

$$\Rightarrow \frac{\mathrm{d}F(t_0)}{\mathrm{d}t} = \lim_{\delta t\to 0}\frac{1}{\delta t}\oint \nabla\frac{\partial\dot{\mathbf{z}}}{\partial\lambda}\frac{\partial\lambda}{\partial t}(\delta t)^2(\mathbf{p}_0\mathrm{d}\mathbf{q}_0 - H\mathrm{d}t) = 0$$

$$\Rightarrow \oint \left[\mathbf{p}\mathrm{d}\mathbf{q} - H\mathrm{d}t\right] = \text{const}_a. \tag{2.104}$$

End of proof. The true orbit of the single mass point originates from the continuous change of λ all over T. It is open and differs from the closed orbit at λ fixed in at most first order, thus $\lim N\Delta F = 0$ persists.

In the literature only the adiabatic invariant $F = \oint \mathbf{p}\mathrm{d}\mathbf{q}$ restricted to *libration* of a time independent Hamiltonian $H(\mathbf{p}, \mathbf{q}; \lambda)$ seems to be common knowledge. In connection with the generalized longitudinal ponderomotive force we will show that the dynamic mode of rotation is only described correctly by F containing the additional integrand $H(\mathbf{p}(t), \mathbf{q}(t), \lambda = \mathrm{const})$. With the aid of $\mathcal{H}$ both dynamic modes are described by

$$F(t_0) = \oint_{\mathcal{H}, \lambda=\lambda(t_0)} \mathbf{p}\mathrm{d}\mathbf{q} = \mathrm{const_a}. \tag{2.105}$$

If the Hamiltonian depends on time energy conservation gets lost. It is replaced by the approximate conservation of action F in case the motion is (i) nearly periodic and (ii) the time variation is slow.

The nonrelativistic momentum of a point particle is $\mathbf{p} = m(\mathbf{v}_0 + \mathbf{w})$, $\mathbf{v}_0 = \mathbf{s}/T$. In the libration mode this translates (2.104) into

$$F = \oint \mathbf{p}\mathrm{d}\mathbf{q} = \int_T m(\mathbf{v}_0 + \mathbf{w})^2 \mathrm{d}t = m(\mathbf{v}_0^2 - \mathbf{v}_0^2)T + m \oint \mathbf{w}^2 \mathrm{d}t = 4E_{\mathrm{kin}}T = \mathrm{const_a} \tag{2.106}$$

Application of criterion (2.106) is immediate for the gyrating electron in the magnetic field varying in space and time. Compared with the individual analyses of the three examples presented in the foregoing subsection the theorem (2.104) means a significant step forward towards economy of thought.

A first qualitative criterion of adiabatic dynamics is that the kinetic energy increment per cycle is much less than the cycle averaged kinetic energy; in short, it is much smaller than the kinetic energy content of the particle. In general this condition is the more stringent the lower the rotation frequency ω becomes. Particular attention has to be paid to processes in which the frequency crosses zero. A priori it is unpredictable what happens to F there. Numerical studies have shown that $F = \mathrm{const_a}$ may be violated or may be not. If one wants to classify interactions of physical subsystems one observes that they all lie between the two extremes of smooth adiabatic exchanges and violent collisional interactions. Adiabatic transitions in general indicate continuous changes from one state into another state, into another shape or another property, see Fig. 2.12.

It is further observed that there is a close affinity between an adiabatic change and an alternating series. Consider a harmonic wave of slowly increasing amplitude. There is a net increase in area of $\sin(kx - \omega t) > 0$ over half a wavelength, and there is even a slightly higher negative contribution over the following half wave where $\sin(kx - \omega t) < 0$. The sum is an alternating series which converges if the single term tends to zero, whereas the sum of their moduli may diverge. As will be shown in Chap. 5 on waves in the plasma, in the *WKB approximation* waves follow Hamiltonian dynamics.

Fig. 2.12 Maurits Cornelis Escher (1898–1972): "Sky & Water I". Adiabatic transformation from fish to bird. Adiabatic transitions save entropy.

2.1.6 The Ponderomotive Force

In this section the concept of the *ponderomotive force* is developed. An electric wave, longitudinal or transverse, exerts a secular force on a single charge particle. The concept of the adiabatic invariance represents the natural basis along which this force is derived for various situations encountered in high power laser-matter interaction. Starting from the time-averaging concept expressions are derived for free and bound particles at intermediate intensities. As seen soon no difference of secular force appears between a transverse and a longitudinal wave. First the transverse wave is treated and extended into the relativistic domain. It will become apparent that at high intensities covariant cycle averaging extends to averaging in time *and* in space. As soon as the single particle dynamics in the longitudinal wave is followed close to the trapping limit the concept of adiabatic invariance will reveal a quite different behaviour in an electron plasma wave, characterized by the appearance of a cycle averaged electrostatic potential; it is absent in the transverse wave.

2.1.6.1 Transverse Wave

Light exerts a pressure on matter. It is a secular force. At an angle of incidence α towards the normal to the surface and a reflection coefficient R it is in vacuum

$$p_{\mathrm{L}} = (1 + R)\frac{I}{c}\cos^2\alpha. \tag{2.107}$$

The pressure is a second rank tensor and transforms therefore like $\mathbf{xx} = \{x_i x_j\}$. In contrast, the intensity $\mathbf{I}$ transforms like a position vector $\mathbf{x}$ with projection $\sim \cos\alpha$. Light pressure is the sum of forces on a certain number of electrons on the surface and is thus a proof of the existence of a secular, i.e., zero frequency force on the single particle. As we shall see in a further chapter the kinematic effect of an electromagnetic wave on protons and heavier nuclei in the optical and near infrared frequency domain is completely negligible up to laser intensities of 10^{22} Wcm^{-2}. The more our interest concentrates on the cycle averaged force of a monochromatic wave on the single electron. It has become customary to designate it *ponderomotive* force $\mathbf{f}_p$.

The ponderomotive force has been proven of central importance for the dynamics of warm dense matter and laser generated plasmas. Its importance becomes clear from a simple numerical example. Numerically, p_{L} from (2.107) reads

$$p_{\mathrm{L}}[\mathrm{bar}] = 3 \times 10^{-10}(1 + R)\,\mathrm{I}\,[\mathrm{Wcm}^{-2}]; \qquad 1\ \mathrm{bar} = 10^5\ \mathrm{Pa}.$$

A laser beam of 10^{21} Wcm^{-2} generates at least 300 Gbar; it is twice the gas pressure in the center of the sun. The ponderomotive force embraces a realm of physics: compression of matter, collisionless shock wave generation, novel schemes of particle acceleration, stimulated Brillouin and Raman scattering in solids, liquids, gases, and plasmas, ac Stark effect in atoms, and the Lamb shift. For the reason of universal appearance of this secular force, in the following it is derived and studied under various aspects. Essential properties of $\mathbf{f}_p$ for free and bound particles can be derived from simple energy considerations. A general theory will then be given in terms of the adiabatic theorem; it is the most natural and most general basis for it. Historically, the first derivation was based on perturbative methods in Netwon's mechanics.

Relevant electromagnetic field structures we have to consider, among others, are the following, expressed in the vector potential $\mathbf{A}$:

(a) Travelling wave with stationary amplitude in the lab frame, $\mathbf{A} = \hat{\mathbf{A}}(\mathbf{x})\,\mathrm{e}^{\mathrm{i}(\boldsymbol{k}\boldsymbol{x}-\omega t)}$ corresponding to a focused laser beam and no reflection
(b) Travelling monochromatic pulse, $\mathbf{A} = \hat{\mathbf{A}}(\mathbf{x} - \mathbf{v}_g t)$, $\mathbf{v}_g$ group velocity
(c) Standing wave, $\mathbf{A} = \hat{\mathbf{A}}(\mathbf{x}, t)\,\mathrm{e}^{-i\omega t}$

Ponderomotive Force and Potential from an Energy Principle

In case (a) the simplest derivation of $\mathbf{f}_p$ is obtained from an energy argument. In addition to simplicity the procedure has the advantage of validity beyond the perturbation level. The Hamiltonian of the charge q is according to (2.58)

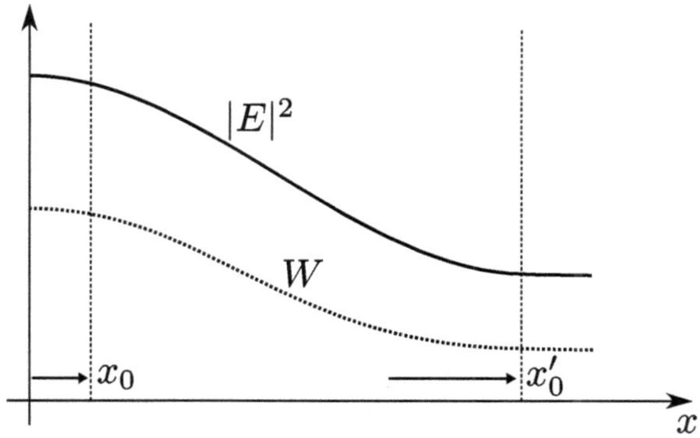

Fig. 2.13 Derivation of ponderomotive force from an energy principle in transverse wave of amplitude $\hat{\mathbf{E}}(\mathbf{x})$. At slow oscillation center motion the cycle averaged energy W is a function of position of the oscillation center $\mathbf{x}_0$. For particles streaming in from left at $\mathbf{x} = \mathbf{x}_0$ with velocity $\mathbf{v}_0$ a steady state of mass and energy flow builds up at arbitrary position $\mathbf{x} = \mathbf{x}_0'$, showing that the total cycle averaged energy $E = m\mathbf{v}_0^2/2 + W$ is conserved. W is the ponderomotive potential

$$H(\mathbf{p}, \mathbf{x}, t) = \frac{1}{2m}(\mathbf{p} - q\mathbf{A})^2. \tag{2.108}$$

The energy of the particle is not conserved since H varies rapidly in time, however the cycle-averaged oscillation energy W in the wave with stationary amplitude is only a function of position if the oscillation center $\mathbf{x}_0$ undergoes a modest velocity variation $\Delta\mathbf{v}_0$ only over the whole space interval, i.e., $|\mathbf{k}||\Delta\mathbf{v}_0| \ll \omega - \mathbf{k}\mathbf{v}_0$. When the particle moves from a region of high vector potential $\mathbf{A}$ to one of low vector potential, the question arises where the difference ΔW has been dissipated (Fig. 2.13). From a quantum point of view one is induced to argue that ΔW has been converted into kinetic energy of the oscillation center since a free particle cannot absorb photons and hence ΔW cannot be given back to the hf field. We show that classical arguments lead to the same conclusion and, consequently, the sum of kinetic energy of the oscillation center and of W is (nearly) conserved. As a consequence the force on the oscillation center is $\mathbf{f}_p = -\nabla W$. Therefore W is given the name *ponderomotive potential* and the symbol Φ_p. In linear polarization they are in the comoving field $\mathbf{E}'$

$$\Phi_\mathrm{p}(\mathbf{x}_0, \mathbf{v}_0) = W(\mathbf{x}_0, \mathbf{v}_0) = \frac{q^2}{4m(\omega - \mathbf{k}\mathbf{v})^2}\hat{\mathbf{E}}'\hat{\mathbf{E}}'^*, \qquad \mathbf{f}_p = -\nabla\Phi(\mathbf{x}_0, \mathbf{v}_0). \tag{2.109}$$

For circular polarization these expressions are to be multiplied by the factor 2.

The classical proof of our energy argument is as follows, see (Fig. 2.13). By injecting N particles per unit time at a position $\mathbf{x} = \mathbf{x}_0$ with the oscillation center velocity $\mathbf{v}_0$, stationary time-averaged mass and energy flows build up between $\mathbf{x} = \mathbf{x}_0$ and $\mathbf{x} = \mathbf{x}_0'$. The position $\mathbf{x}_0'$ is arbitrary; hence

$$N\mathbf{v}_0 = \text{const}, \qquad N\mathbf{v}_0\left(\frac{\mathbf{p}_0^2}{2m} + W(\mathbf{x}_0, \mathbf{v}_0)\right) = \text{const}$$

must hold along the oscillation center orbit $\mathbf{x}_0(t)$, or in particular,

$$H_0(\mathbf{p}_0, \mathbf{x}_0) = \frac{\mathbf{p}_0^2}{2m} + W(\mathbf{x}_0, \mathbf{v}_0) = \text{const} \tag{2.110}$$

thus showing that the cycle-averaged quantity from $\mathbf{p} - q\mathbf{A} = m\mathbf{v} = m(\mathbf{v}_0 + \mathbf{w})$,

$$H_0 = \frac{1}{T}\int_t^{t+T}\frac{1}{2m}(\mathbf{p} - q\mathbf{A})^2 \mathrm{d}t = \frac{1}{T}\int_t^{t+T}\frac{1}{2}m(\mathbf{v}_0 + \mathbf{w})^2 \mathrm{d}t = \frac{\mathbf{p}_0^2}{2m} + W(\mathbf{x}_0, \mathbf{v}_0). \tag{2.111}$$

is the Hamiltonian governing the oscillation center motion and $\Phi_p = W = m\overline{\mathbf{w}^2/2}$ is the ponderomotive potential of the monochromatic transverse wave (a) in linear polarization. It can easily be shown that $\mathbf{v}_0$ is the average that minimizes the sum of square deviations: $\oint[\mathbf{v}(t) - \mathbf{v}_0]^2\,\mathrm{d}t = \min$ (see exercise). Likewise, the energy argument applies to circular polarisation as well, with Φ_p and $\mathbf{f}_p$ twice the values of (2.109). So far no use has been made of the transverse character in the cycle averaging in (2.111): the second integral term remains the same for the longitudinal electron plasma wave of type (a). Hence, expressions (2.109) stands for a *transverse* and a *longitudinal* **E** field. The derivation is of nonperturbative character.

Let us now assume that in (a) the field amplitude depends on position and time, $\hat{\mathbf{A}} = \hat{\mathbf{A}}(\mathbf{x}, t)$. According to the definition of the infinitesimal potential increase as the work $\mathrm{d}W = -\mathbf{f}_\mathrm{p}\,\mathrm{d}\mathbf{x}$ at fixed time instant t the property of potential extends also to $\Phi_\mathrm{p}(\mathbf{x}_0, \mathbf{v}_0, t)$ with $\mathbf{f}_p = -\nabla\Phi$ pointwise. However, energy conservation on the large time and space scale no longer holds. Besides, on a large space scale energy conservation will be violated for time independent amplitude $\hat{\mathbf{A}}$ because of the difficulty to satisfy condition $|\mathbf{k}||\Delta\mathbf{v}_0| \ll \omega - \mathbf{k}\mathbf{v}_0$.

The travelling wave (b) leads to the same expressions for Φ_p and $\mathbf{f}_p$, if the nonrelativistic group velocity $\mathbf{v}_g$ is of the order of $\mathbf{v}_0$. If $|\mathbf{v}_g|$ is of the order of c and $|\mathbf{v}_0| \ll c$ generally the standard derivation of $\mathbf{f}_p$ applies as for wave (a). For relativistic $\mathbf{v}_0$ a general relativistic derivation of $\mathbf{f}_p$ for transverse waves will be presented in a following section. For the standing wave (c) Φ_p and $\mathbf{f}_p$ from (2.109) remain valid if $\mathbf{v}_0$ is at least by one order of magnitude smaller than the phase velocity $\omega/|\mathbf{k}|$ of the two counterpropagating wave components. Otherwise their Doppler shifted frequencies lead to interferences and possibly to chaotic trajectories, see [6].

Standard Derivation of Ponderomotive Potential and Force

Despite its inherent limitations this procedure may provide an alternative, detailed, physical insight in how the secular force $\mathbf{f}_p$ originates. The derivation starts from the Lorentz equation (2.10) of a charged particle in a monochromatic electromagnetic field and makes use of the decomposition of the trajectory $\mathbf{x}(t)$ into oscillation center and quiver motion,

$$\mathbf{x}(t) = \mathbf{x}_0(t) + \boldsymbol{\xi}(t), \qquad \mathbf{v}(t) = \mathbf{v}_0(t) + \mathbf{w}(t). \tag{2.112}$$

Under the restriction that the oscillation amplitude $\hat{\boldsymbol{\xi}}$ is much smaller than the local wavelength $\lambda(\mathbf{x}_0)$, i.e. $|\hat{\boldsymbol{\xi}}| \ll \lambda(\mathbf{x}_0)$, the Lorentz equation may be linearized in $\hat{\boldsymbol{\xi}}$ and solved separately,

$$m\frac{d\mathbf{v}}{dt} \simeq m\left[\frac{\partial \mathbf{w}}{\partial t} + (\mathbf{v}_0\nabla)\mathbf{w}\right] = q\left[\mathbf{E}(\mathbf{x}_0) + \mathbf{v}_0 \times \mathbf{B}(\mathbf{x}_0)\right] = q\mathbf{E}'(\mathbf{x}_0)$$

$$\mathbf{w}(t) = i\frac{q}{m(\omega - \mathbf{k}\mathbf{v}_0)}\hat{\mathbf{E}}'(\mathbf{x}_0)\,\mathrm{e}^{-i(\omega-\mathbf{k}\mathbf{v}_0)t} = \frac{q}{m}\hat{\mathbf{A}}'(\mathbf{x}_0)\,\mathrm{e}^{-i(\omega-\mathbf{k}\mathbf{v}_0)t}, \tag{2.113}$$

$\mathbf{E}' = -\partial_t\mathbf{A}'$, $\mathbf{A}'$ vector potential comoving with $\mathbf{v}_0$. The term $(\mathbf{v}_0\nabla)\mathbf{w}$ has been taken into account mainly for correctness reasons and in view of the Doppler effect to be treated later. Here, without loss of generality $\mathbf{v}_0 = 0$ can be set (comoving reference system). In the next order from the Lorentz equation $\mathbf{v}_0(t)$ is calculated, the time variation of which determines $\mathbf{f}_p$,

$$\mathbf{f}_p = m\frac{d\mathbf{v}_0}{dt} = q\left\{\mathbf{E}(\mathbf{x}_0 + \boldsymbol{\xi}(t)) + \mathbf{w}(t) \times \mathbf{B}(\mathbf{x}_0)\right\}_0.$$

The bracket contains terms varying with $2\,\omega$ and zero frequency since $\mathbf{E}(\mathbf{x}_0)$, $\mathbf{B}(\mathbf{x}_0)$, and $\boldsymbol{\xi}$, $\mathbf{w}$ both oscillate with ω. Therefore the subscript "0" has been added to indicate that in determining the secular force $\mathbf{f}_p$ the 2ω term has to be cut off. Taylor expansion of $\mathbf{E}$ in $\boldsymbol{\xi} = i\mathbf{w}/\omega$, using $\mathbf{B} = \nabla \times \mathbf{A}$, and observing that physical quantities have to be real leads directly to

$$q(\mathbf{E} + \mathbf{w} \times \mathbf{B})_0 = -\frac{q^2}{4m}\{(\hat{\mathbf{A}}^*\nabla)\hat{\mathbf{A}} + \hat{\mathbf{A}}^* \times \nabla \times \hat{\mathbf{A}} + \mathrm{cc}\} = -\frac{q^2}{4m}\nabla(\hat{\mathbf{A}}\hat{\mathbf{A}}^*). \tag{2.114}$$

Hence, $\mathbf{f}_p$ is the gradient of a quantity which can be interpreted as a potential Φ_p; it is identical with the ponderomotive potential given by (2.109). In a transverse plane wave $(\mathbf{w}\nabla)\mathbf{E}$ is zero; if $\mathbf{E}$ is longitudinal (e.g. an electron plasma wave) $\mathbf{B}$ is zero and $\mathbf{f}_p$ arises entirely from $(\mathbf{w}\nabla)\mathbf{E}$. In an inhomogeneous transverse wave, e.g., Gaussian beam, $(\mathbf{w}\nabla)\mathbf{E}$ also contributes to Φ_p.

Expression (2.109) for Φ_p is valid if damping can be ignored. In presence of a linear damping ν associated with $\mathbf{w}$ the following expression for $\mathbf{f}_p$ is found straightforwardly in terms of $\mathbf{E} = \mathbf{E}_r + \mathrm{i}\mathbf{E}_i$

$$\mathbf{f}_p = -\frac{q^2}{4m_e\omega^2(1 + \nu^2/\omega^2)}\left\{\nabla\hat{\mathbf{E}}\hat{\mathbf{E}}^* + 2\frac{\nu}{\omega}\left[\hat{\mathbf{E}}_i \times (\nabla \times \hat{\mathbf{E}}_r) - \hat{\mathbf{E}}_r \times (\nabla \times \mathbf{E}_i)\right]\right\}, \tag{2.115}$$

$\omega, \nu = \mathrm{const.}$ $\nabla \times \mathbf{f}_p$ differs from zero in this case; it has no potential.

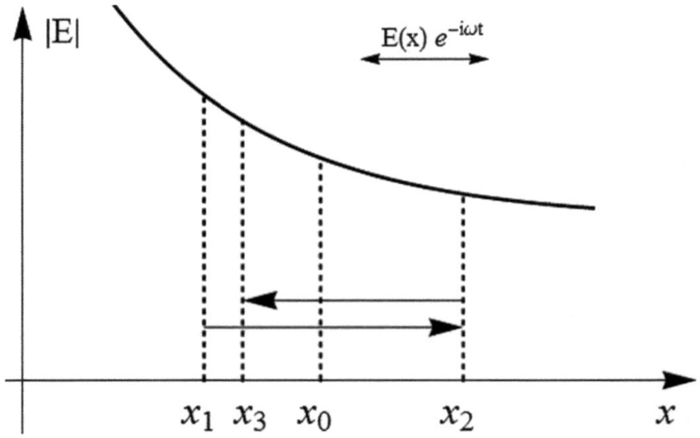

Fig. 2.14 In the inhomogeneous longitudinal electric field a free point charge experiences a drift into direction of decreasing amplitude $|\mathbf{E}|$. In the transverse plane wave the drift is produced by the grad $\hat{\mathbf{E}}$ term and the Lorentz force

In terms of the momentum equation the physical interpretation of the ponderomotive force $\mathbf{f}_p$ is as follows. Let us first assume a pure electrostatic wave in the x-direction of the form $\hat{E}(x)\,\mathrm{e}^{-i\omega t}$ with decreasing amplitude $\hat{E}(x)$ (Fig. 2.14). A free electron is shifted by the E-field from its original position x_0 to x_1. From there it is then accelerated to the right until it has passed x_0. From that moment on the electron is decelerated by the reversed E-field and is stopped at position x_2.

If x_0' designates the position in which the field is reversed ($x_0' > x_0$), the deceleration interval $x_2 - x_0'$ is larger than that of acceleration since on the right hand side of x_0 the E-field is weaker and therefore a longer distance is needed to take away the energy gained in the former quarter period of oscillation. On its way back the electron is stopped in the region of higher amplitude; the turning point is shifted from x_1 to x_3 into the direction of decreasing wave amplitude $\hat{E}(x)$. In an (inhomogeneous) medium this drift produces, by charge separation, a static E-field which transmits the force to the ions. Since this consideration is based only on energy and work, there was no need to specify the field direction with respect to the particle speed—it follows that the drift (or force) is independent of the sign of charge: positrons drift into the same direction. In the case of a plane electromagnetic wave the drift is caused by both, the gradient term $(\boldsymbol{\xi}\nabla)\hat{\mathbf{E}}$, and the $\mathbf{v} \times \mathbf{B}$ force which acts along the direction of propagation.

The comparison of the two types of derivations, energy principle versus first order perturbation method, clearly shows the superiority of our first method. In the perturbation approach the orbit $\mathbf{x}(t)$ must be calculated. In the first procedure the only requirement is the existence of an oscillation center and the knowledge of the quiver energy W, the shape of the orbit does not matter; inhomogeneities in $\hat{\mathbf{E}}$ in any direction are automatically included through the space dependence of the quiver energy $W(\mathbf{x}_0)$. Compare the analogous force on the guiding center $\mathbf{f} = \nabla(\boldsymbol{\mu}\mathbf{B}) = -\nabla m\mathbf{v}_\perp^2/2$ from (2.78) and (2.66).

The perturbation method reveals a dramatic deficiency when it is applied along the axis of a focused cylindrical laser beam. A view on Fig. 2.2 yields the result $\mathbf{f}_p = 0$ from the symmetry of motion entirely in the plane perpendicular to propagation; from the energy principle follows twice the value given in (2.109) as it should be.

Moderate Doppler Effect

Here it is shown that the non relativistic ponderomotive potential in the transverse wave is insensitive to the motion $\mathbf{v}_0$ of its oscillation center. A charged particle may move at velocity with respect to the lab frame that is no longer negligible, hitherto assumed $v \ll c$, now allowed to be of the order of c but still not relativistic, i.e. $v^2 \ll c^2$. For the electromagnetic wave we face a focused pulse of situation (a) and a propagating pulse (b). Due to the motion the charge "sees" a Doppler shifted frequency ω' and a Doppler shifted wave vector $\mathbf{k}'$. The nonrelativistic transformations from S to $S'(\mathbf{v})$ are obtained from the corresponding relativistic relations (see following section) by setting the Lorentz factor $\gamma = 1$,

$$\mathbf{k}' = \mathbf{k} - \frac{\mathbf{v}}{c}|\mathbf{k}|, \qquad \omega' = \omega - \mathbf{vk}. \tag{2.116}$$

The transformation of $\omega \to \omega'$ follows in Galileian relativity (2.5) from the invariance of phase $\phi = \mathbf{kx} - \omega t$ with respect to two fixed observers in S and $S'(\mathbf{v})$ because if S counts N wave crests passing by in the time interval t, S' must count the same number N during its time interval t'. In Galileian relativity t' equals the universal time t. Hence, $\phi = \mathbf{kx} - \omega t = \mathbf{k}'\mathbf{x}' - \omega' t = \mathbf{k}'\mathbf{x}' - (\omega - \mathbf{kv})t \to \omega' = \omega - \mathbf{vk}$. The invariance of phase applied to space yields $\mathbf{k} = \mathbf{k}'$ as a consequence of $c = \infty$, in contradiction to (2.116). This simple argument shows once more the incompatibility of the Galileian transformation with electrodynamics.

In the tangential reference system $S'(\mathbf{v}_0)$ the oscillation energy of the free charge in the plane electromagnetic wave $\mathbf{E}'(\mathbf{x}', t) = \hat{\mathbf{E}}'(\mathbf{x}') \exp \mathrm{i}(\boldsymbol{kx}' - \omega' t)$ in linear polarization is

$$W' = \Phi' = \frac{q^2}{4m}\hat{\mathbf{A}}'\hat{\mathbf{A}}'^* = \frac{q^2}{4m}\frac{\hat{\mathbf{E}}'(\mathbf{x}')\hat{\mathbf{E}}'^*(\mathbf{x}')}{(\omega - \mathbf{kv}_0)^2}; \quad \mathbf{E}' = -\partial_t \mathbf{A}'. \tag{2.117}$$

Transposed to the lab frame the energy W' changes into $E = \mathbf{p}_0^2/2m + W' = \text{const}_a$, as shown by (2.111). In terms of $\hat{\mathbf{E}}'$ in the lab frame, $\hat{\mathbf{E}}' = \hat{\mathbf{E}} + (\mathbf{v}_0 + \Re\mathbf{w}) \times \hat{\mathbf{B}}$, and $\hat{\mathbf{B}}$ from (2.9) follows with $v_{0\|} = \mathbf{v}_0\mathbf{E}/|\mathbf{E}|$, $\mathbf{k}^0 = \mathbf{k}/|\mathbf{k}|$

$$\mathbf{E}'\mathbf{E}'^* = \left[\left(1 - \mathbf{k}^0\frac{\mathbf{v}_0}{c}\right)^2 + \left(\frac{v_{0\|}}{c}\right)^2 + 2\frac{v_{0\|}}{c}|\mathbf{a}| + |\mathbf{a}|^2\right]\mathbf{E}\mathbf{E}^*; \qquad \mathbf{a} = \frac{q\mathbf{A}}{mc}.$$

For intensities $I < 10^{17}$ Wcm^{-2} at wavelength $\lambda \leq 1\mu$ the vector potential $|\mathbf{a}|$ normalized to m_e is less than 0.3; hence with $v_{0\|} < 0.3c$ the second and third term in the square bracket can be ignored and the ponderomotive potential appears *insensitive* to subrelativistic changes of the oscillation center motion $\mathbf{v}_0$,

$$\Phi(\mathbf{x}_0, \mathbf{v}_0) = \Phi(\mathbf{x}_0) = \frac{q^2}{4m}\mathbf{A}(\mathbf{x}_0)\mathbf{A}^*(\mathbf{x}_0) = \frac{q^2}{4m\omega^2}\hat{\mathbf{E}}(\mathbf{x}_0)\hat{\mathbf{E}}^*(\mathbf{x}_0). \tag{2.118}$$

The oscillation velocity $\mathbf{v}_0$ is to be measured in the system in which the wave amplitude is stationary or slowly varying in time. However, caution is advised. The ponderomotive force as the gradient is correctly given from $\Phi(\mathbf{x}_0)$ in (2.118). The potential itself shows a hysteresis if $\mathbf{v}_0$ changes over an extended path.

Historical remark. The ponderomotive force of a monochromatic electromagnetic wave on a point charge q of mass m has been derived at least eight times independently [7, 8], to begin with A. V. Gapunov and H. A. H. Boot in 1958. Since then a high number of papers by various authors has been published. Among them we mention as instructive examples [9, 10], and the recent paper [11] on higher order nonlocal corrections in the transverse field. It has been Heinrich Hora who stressed first the significance of ponderomotive action for its dc character in the plasma medium [7].

Ponderomotive Force on "Atoms"

For both, further insight on ponderomotive action as well as for applications, the slow time dynamics of particles with internal degrees of freedom is in order. First we consider the harmonic oscillator δ driven by the asymmetric harmonic field $\hat{\mathbf{E}}(\mathbf{x}, t)\,\mathrm{e}^{-i\omega t}$,

$$\ddot{\delta} + \omega_0^2\delta = \frac{q}{\mu}\hat{\mathbf{E}}(\mathbf{x})\,e^{-i\omega t}; \qquad \mu = \frac{m_1 m_2}{m_1 + m_2}. \tag{2.119}$$

The averaged oscillation energy is

$$W = \frac{1}{2}\mu\overline{\dot{\delta}^2} + E_{\text{pot}} = \frac{q^2}{4\mu}\,\frac{\omega^2 + \omega_0^2}{(\omega^2 - \omega_0^2)^2}\hat{\mathbf{E}}\hat{\mathbf{E}}^*.$$

In determining Φ_p one has to keep in mind that when the field is switched on the oscillator gains internal energy E_{in} also and that the forces producing this energy are internal forces and as such, by definition, cancel each other. Thus Φ_p is given by

$$\Phi_p = W - \overline{E}_{\text{in}} = W - E_{\text{pot,max}} = W - \frac{q^2}{4\mu}\frac{2\omega_0^2}{(\omega^2-\omega_0^2)^2}\hat{\mathbf{E}}\hat{\mathbf{E}}^*$$
$$= \frac{q^2}{4\mu(\omega^2-\omega_0^2)}\hat{\mathbf{E}}\hat{\mathbf{E}}^*. \tag{2.120}$$

In the driven harmonic oscillator E_{kin} differs from $\overline{E}_{\text{pot}}$. The same result is obtained by solving (2.119) up to the first order and determining its secular component. $\Phi_p = W - \overline{E}_{\text{in}}$ also holds for anharmonic **E** fields or when the Lorentz force is included. For $\omega_0 < \omega$ the oscillator behaves like a free particle, i.e. $\mathbf{f}_p$ tries to drive it into a region of decreasing field amplitude. For $\omega_0 > \omega$ it moves into the opposite direction. By introducing a damping term it is shown that at $\omega = \omega_0$ the force $\mathbf{f}_p$ reduces to zero (Fig. 2.15). If ω_0 is properly chosen and an effective oscillator strength is inserted the expression also applies to the radiation force on an atom with negligible degeneracy. Φ_p in this case is identical to the level shift ΔE of the linear dynamical Stark shift by the electric wave. If degeneracy is present, (2.120) changes accordingly. At first glance one might argue that (2.120) is not useful for a fully ionized plasma in which all electrons are free. However, it will become clear that an oscillating external field, for instance an electron plasma wave, resembles properties of a localized oscillator field. According to (2.4) and (2.10) a point charge interacts with fields only locally.

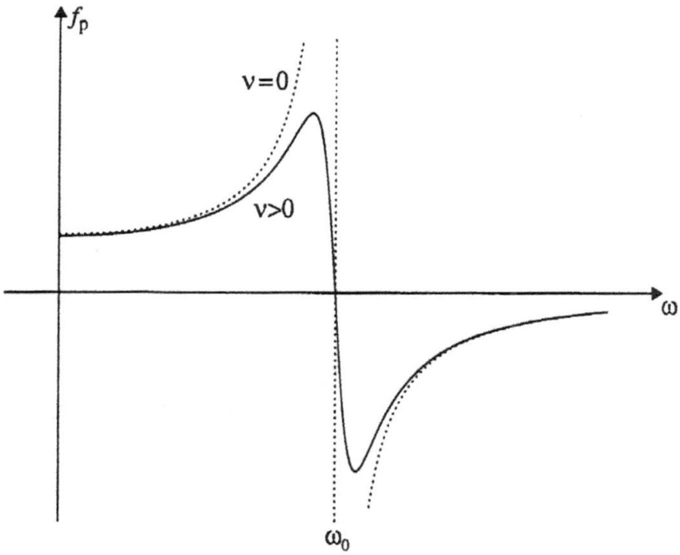

Fig. 2.15 A harmonic oscillator with eigenfrequency ω_0 and a charged point particle in a constant perpendicular magnetic field B_0 (cyclotron frequency ω_c) are shifted towards decreasing electric field amplitude when the driver frequency ω exceeds ω_0 and ω_c (special case: free particle with $\omega_0 = \omega_c = 0$); it moves into opposite direction if $\omega < (\omega_0,\ \omega_c)$ holds. At resonance $(\omega = \omega_0,\ \omega_c)$ $\mathbf{f}_p$ is zero. Solid curve: damped oscillator, dashed curve: oscillator without damping

Therefore the reaction of an electron on $\mathbf{E}$ does not distinguish between a localized and an extended field.

Next we consider the magnetic "atom": a charged point particle in a static magnetic field $\mathbf{B}_0$. There exist important applications of this type of situation in magnetically confined plasmas. With $\mathbf{E}(x,t)$ in the x-direction and $\mathbf{B}_0$ parallel to z one has to set

$$W = \frac{1}{2}m\overline{(v_x^2 + v_y^2)},\ \bar{V} = \mu B_0.$$

$\mu B_0 = AIB_0$ (μ magnetic moment, A cross section of the closed orbit, I current) is the "inner" energy of the orbiting particle configuration which has to be subtracted since, by definition, internal forces do not contribute to $\mathbf{f}_p$. Specializing to $\mathbf{E} = \mathbf{E}_x \sim \mathrm{e}^{-i\omega t}$ yields in analogy to (2.119) and (2.120)

$$\ddot{v}_x + \omega_c^2 v_x = -i\omega\frac{q}{m}E_x, \quad v_y = -i\frac{\omega_c}{\omega}v_x, \quad \omega_c = \frac{q}{m}B;$$

$$\begin{aligned}\Phi_p = W - \bar{V} &= \frac{q^2}{4m}\frac{\omega^2 + \omega_c^2}{(\omega^2 - \omega_c^2)^2}\hat{\mathbf{E}}\hat{\mathbf{E}}^* - \frac{q^2}{4m}\frac{2\omega_c^2}{(\omega^2 - \omega_c^2)^2}\hat{\mathbf{E}}\hat{\mathbf{E}}^*\\ &= \frac{q^2}{4m(\omega^2 - \omega_c^2)}\hat{\mathbf{E}}\hat{\mathbf{E}}^*.\end{aligned} \tag{2.121}$$

The cyclotron frequency ω_c takes the place of ω_0.

In the three cases of a free particle, a harmonic oscillator and a particle in a static magnetic field, considered here, the ponderomotive potential can be written with the help of the dipole moment $\mathbf{p} = q\boldsymbol{\delta}$ as

$$\Phi_p = E_{\mathrm{int}} = -\frac{1}{2}\overline{\mathbf{p}\mathbf{E}}. \tag{2.122}$$

If $\omega_0 > \omega$, $\mathbf{p}$ is parallel to $\mathbf{E}$ and the oscillator moves into the direction of increasing field; if $\omega_0 < \omega$, $\mathbf{p}$ is anti-parallel to $\mathbf{E}$; at resonance the phase shift is $\pi/2$.

2.1.6.2 Longitudinal Wave

a. Energy Conserving Wave Potential

In the derivations of the ponderomotive force presented so far nowhere the necessity appeared to distinguish between a transverse and a longitudinal wave. The question arises to what extent such an identity may be true because eventually there must be a difference between the two polarizations since with the longitudinal wave a potential $\Phi(x - \int v_\varphi \mathrm{d}t)$ is associated, see Fig. 2.16. We consider in a first step the slow motion in an electron plasma wave of constant frequency ω and constant phase

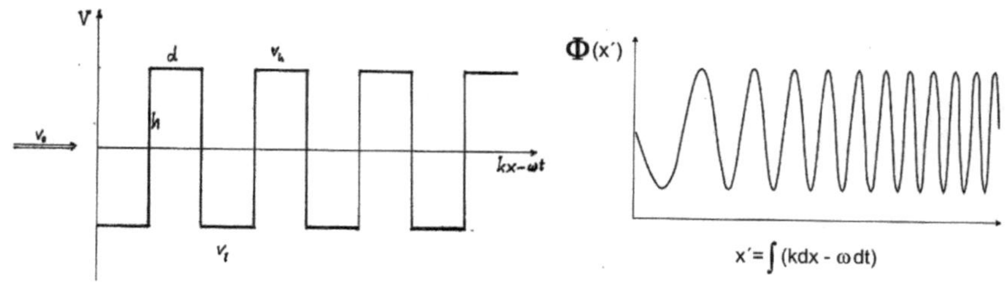

Fig. 2.16 LHS: Rectangular potential of longitudinal wave in comoving frame $x' = x - v_\varphi t$; wavelength $\lambda = 2d$, constant potential amplitude $V = h$. RHS: Potential Φ of longitudinal sinusoidal wave of constant amplitude, constant frequency, however variable wavelength

velocity $v_\varphi = \omega/k$, and no magnetic field present. The longitudinal electric wave exhibits a potential $\Phi(x - \int v_\varphi \mathrm{d}t)$, with v_φ the phase velocity. In the comoving reference system $S'(v_\varphi)$ of $x' = x - v_\varphi t$ ("*wave frame*") the wave structure is static, $\Phi = \Phi(x')$, the frequency ω' is zero.

Square Wave

To simplify further a rectangular profile of the potential per unit mass $V = \pm h$ with periodicity $\lambda = 2d$ is assumed, as sketched in the RHS picture of Fig. 2.16. It is exactly soluble in closed form and all essential features can be learned from it. At any position x_i' with $V(x_i') = 0$, or under the assumption that $V = 0$ for $x' \to -\infty$ the initial velocity of the mass point in the wave frame is fixed as v_i'. The local velocities v_1, v_2, the oscillation period T and the cycle averaged velocity v_0 are as follows,

$$v_{1,2}^2 = v_i'^2 \pm 2h = v_i'^2(1 \pm \kappa), \ \ \kappa = \frac{2h}{v_i'^2}; \quad T = \frac{d}{v_1} + \frac{d}{v_2} = \frac{2d}{v_0};$$

$$v_0 = v_1 \frac{t_1}{T} + v_2 \frac{t_2}{T} = 2v_i' \frac{(1-\kappa^2)^{1/2}}{(1+\kappa)^{1/2} + (1-\kappa)^{1/2}}; \quad v_1 t_1 = v_2 t_2 = d.$$

The average velocity v_0 is a function of the normalized amplitude κ but it is independent of the wavelength $2d$ as long as d does not depend on x'. For $\kappa \ll 1$ the dependence on the potential hump is weak, $v_0 = v_i'(1 - 3\kappa^2/8)$. The local velocities relative to the average translational speed v_0 and the centered averaged oscillation energy $W(x')$ are

$$v_1 - v_0 = v_1 \frac{v_1 - v_2}{v_1 + v_2}, \ \ v_2 - v_0 = v_2 \frac{v_2 - v_1}{v_1 + v_2}; \ \ W(x') = \frac{1}{2T}[(v_1 - v_0)^2 t_1 + (v_2 - v_0)^2 t_2]. \tag{2.123}$$

The oscillation amplitude is $\hat{\delta} = (v_1 - v_0)t_1/2 = (v_0 - v_2)t_2/2 = d(v_1 - v_2)/2(v_1 + v_2)$; maximum excursion $+\hat{\delta}$ and $-\hat{\delta}$ are reached after the times $t_+ = t_1/2 = \hat{\delta}/(v_1 - v_0)$ and $t_- = t_1 + t_2/2 = \hat{\delta}/(v_0 - v_2)$. Note, with respect to time the oscillation may be very asymmetric, $t_1 \ll t_2$. The energy balance in the rectangular potential reads with the aid of the formulas above

$$E = \frac{1}{2}v'^2_i = \frac{1}{2}\left[(v_1^2 - 2h)\frac{t_1}{T} + (v_2^2 + 2h)\frac{t_2}{T}\right] = \frac{1}{2}v_0^2 + W + \bar{V}; \quad \bar{V} = h\frac{t_2 - t_1}{t_1 + t_2}. \tag{2.124}$$

The terms $v_0^2/2$, W, and $\bar{V}$ can be expressed entirely as functions of the relative hump κ and the initial energy $mv'^2_i/2$,

$$\frac{m}{2}v_0^2 = 2m\frac{(v_1v_2)^2}{(v_1+v_2)^2} = mv'^2_i\frac{1-\kappa^2}{1+(1-\kappa^2)^{1/2}},$$

$$\bar{V} = \frac{m}{2}v'^2_i\kappa\frac{v_1 - v_2}{v_1 + v_2} = \frac{m}{2}v'^2_i\frac{\kappa^2}{1+(1-\kappa^2)^{1/2}}, \tag{2.125}$$

$$W = \frac{m}{2}v_1v_2\frac{(v_1-v_2)^2}{(v_1+v_2)^2} = \frac{m}{2}v'^2_i(1-\kappa^2)^{1/2}\frac{1-(1-\kappa^2)^{1/2}}{1+(1-\kappa^2)^{1/2}}.$$

The three components sum up to $mv'^2_i/2$, as to be expected from correct averaging. The total potential energy and the ratio $W/\bar{V}$ are

$$Z(\kappa) = W(\kappa) + \bar{V}(\kappa) = \frac{m}{2}v'^2_i\frac{2\kappa^2 + (1-\kappa^2)^{1/2} - 1}{1+(1-\kappa^2)^{1/2}}; \quad \frac{\bar{V}}{W} = \frac{\kappa^2}{(1-\kappa^2)^{1/2} - (1-\kappa^2)}. \tag{2.126}$$

In the lab frame the rectangular potential structure moves with phase velocity v_φ. The cycle averaged (*secular*) Hamiltonian of a point mass m with $p'_0 = mv_0$ in S' (static potential) is

$$H(p'_0, x'_0) = \frac{p'^2_0}{2m} + Z(x'_0); \tag{2.127}$$

The potential Z contains, in addition to the average oscillation energy, the time average of the potential $V = \pm h$ that arises from the asymmetry of the oscillation in time, $t_1 < t_2$. For $\kappa \to 0$ the contribution of $\bar{V}$ is twice as high as W, at $\kappa = 0.5$ it is slightly increased to $\bar{V}/W = 2.15$. Explosive growth $(1-\kappa^2)^{-1/2}$ of the ratio occurs for $\kappa \to 1$. Normalized velocity v_0, potential Z, and ratio $\bar{V}/W$ are shown in Fig. 2.17. For $\kappa^2 \ll 1$ the oscillation center moves at $v_0 = v'_i(1 - 3\kappa^2/4)^{1/2}$. Note, in the absence of $\mathbf{B}$ the longitudinal amplitude $\hat{\mathbf{E}}$ is invariant under reference system changes.

The rectangular model potential offers essential insight. A ponderomotive force and particle acceleration arise from any adiabatic change of the hump h, or κ. If the hump remains constant but its periodicity d changes (see Fig. 2.16, RHS) f_p changes, however no change in the particle energy follows because any increase (decrease) in f_p is neutralized by the shortening (lengthening) of the path. Translated to the lab frame it implies that changes of frequency ω and wave vector in space or time do not affect particle acceleration as long as the phase $\omega - \mathbf{k}\mathbf{v}_{\text{lab}}$ remains constant. A potential structure that is static, and thus is conservative, in general will become

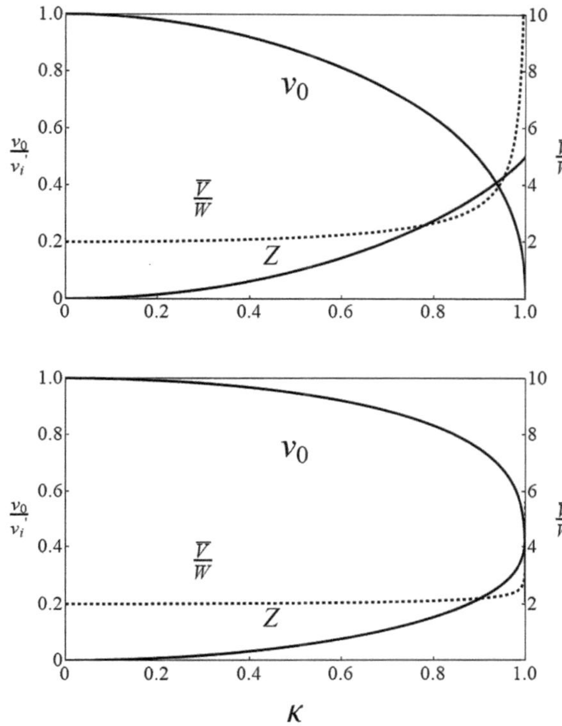

Fig. 2.17 Time averaged velocity v_0, potential $Z = W + \overline{V}$, (W cycle averaged oscillation energy, $\overline{V}$ cycle averaged potential energy), and ratio $\overline{V}/W$ are depicted as functions of $\kappa = 2\hat{V}/v_i'^2$, $v_i'^2/2$ particle injection energy; top: square wave, bottom: sinusoidal wave. The injection velocity v_i' refers to the wave frame. Note the dramatic decrease of $v_0 \to 0$ when κ approaches unity

explicitly time dependent in the lab frame S and no longer energy conserving owing to $\mathrm{d}H/\mathrm{d}t = \partial V/\partial t$.

In a periodic potential $V(x')$ of general shape and periodicity λ the oscillation period T and mean velocity v_0 are obtained from

$$T = \oint \frac{\mathrm{d}x'}{v'} = \oint \frac{\mathrm{d}x'}{\sqrt{\frac{2}{m}[E - V(x')]}}, \qquad v_0 = \frac{\lambda}{T}; \qquad E = \frac{1}{2}mv_i'^2. \tag{2.128}$$

Setting $y = \xi x'$, ξ scaling factor, leads to a change of the period T in (2.128) into $T' = \xi T$ and leaves v_0 unchanged owing to $T'v_0' = \xi\lambda$. If in addition the scaling factor changes slowly in space, $\lambda = \lambda(x)$, see RHS of Fig. 2.16, T results in a change of higher than first order in $|\nabla\lambda|/\lambda$, easily verified in the rectangular potential of the LHS in Fig. 2.16. As a consequence the ponderomotive force is zero at $\Phi(x') = \mathrm{const}$, v_0 does not change. If however, the wave structure changes its shape, for instance due to wave steepening, a non-zero force is expected to arise from (2.128) despite the amplitude of the wave potential Φ is constant.

Sinusoidal Wave

The period T for the *sinusoidal* wave in the wave frame is determined from the potential $V(x') = \hat{V}\cos kx'$ as follows. For simplicity the electron mass m_e is set to unity until stated differently. The local particle velocity is

$$v^2 = v_i'^2 + 2\hat{V}\cos\varphi = v_i'^2 - 2\hat{V}(1-\cos\varphi) + 2\hat{V} = v_i'^2 + 2\hat{V} - 4\hat{V}\sin^2\theta$$

$$\varphi = kx', \quad \theta = \varphi/2, \quad \cos\varphi = (1 - 2\sin^2\theta). \tag{2.129}$$

Use of (2.128) yields

$$T = \frac{2}{k}\left(\frac{\alpha}{\hat{V}}\right)^{1/2}\int_0^{\pi/2}\frac{d\theta}{(1-\alpha\sin^2\theta)^{1/2}} = \frac{2}{k}\left(\frac{\alpha}{\hat{V}}\right)^{1/2}\mathrm{K}(\alpha); \; \alpha = \frac{4\hat{V}}{v_i'^2 + 2\hat{V}} = \frac{2\kappa}{1+\kappa}. \tag{2.130}$$

$\mathrm{K}(\alpha) = \int_0^{\pi/2}(1-\alpha\sin^2\theta)^{-1/2}d\theta$ is the *complete elliptic integral* of the first kind. It diverges logarithmically for $\alpha = 1$. Likewise follow $v_0 = \lambda/T$, $E_0 = v_0^2/2$, $W + E_0 = \overline{v^2}/2$, and $\overline{V}$ as

$$v_0 = \frac{\lambda}{T} = \pi\frac{\hat{V}^{1/2}}{\alpha^{1/2}\mathrm{K}(\alpha)}, \quad \frac{1}{2}\overline{v^2} = 2\frac{\hat{V}}{\alpha}\frac{\mathrm{E}(\alpha)}{\mathrm{K}(\alpha)}, \quad \overline{V} = \frac{1}{2}v_i'^2\left\{1 - \frac{2\kappa}{\alpha}\frac{\mathrm{E}(\alpha)}{\mathrm{K}(\alpha)}\right\}. \tag{2.131}$$

$\mathrm{E}(\alpha) = \int_0^{\pi/2}(1-\alpha\sin^2\theta)^{1/2}d\theta$ is the *complete elliptic integral* of the second kind.

In first order the potential $V = \hat{V}\sin kx$ may be approximated by the substitution of the rectangles in Fig. 2.16 by *triangles* of basis d and hight h,

$$V = \frac{2h}{d}x \;\text{ in }\; 0 \le x \le d/2, \quad V = h - \frac{2h}{d}(x - d/2) \;\text{ in }\; d/2 \le x \le 3d/2$$
$$V = -h + \frac{2h}{d}(x - 3d/2) \;\text{ in }\; 3d/2 \le x \le 2d = \lambda. \tag{2.132}$$

Elementary integration of (2.128) with V from (2.132) yields

$$T = \frac{\lambda}{v_i'}\left(\frac{2}{1+(1-\kappa^2)^{1/2}}\right)^{1/2}, \quad \frac{1}{2}v_0^2 = \frac{1}{4}v_i'^2[1 + (1-\kappa^2)^{1/2}],$$

$$W = \frac{v_i'^2}{12}[1 - (1-\kappa^2)^{1/2}], \quad \overline{V} = \frac{v_i'^2}{6}[1 - (1-\kappa^2)^{1/2}] = 2W; \quad \kappa = \frac{2h}{v_i'^2}. \tag{2.133}$$

In summary, the three potential structures produce three different times $T = T_{\mathrm{sqw}}, T_{\mathrm{h}}, T_{\mathrm{t}}$, for crossing one wavelength λ of the meander, harmonic, and triangular waves,

$$T_{\rm sqw} = \frac{\lambda}{2v_i'}\frac{(1+\kappa)^{1/2}+(1-\kappa)^{1/2}}{(1-\kappa^2)^{1/2}},$$
$$T_{\rm h} = \frac{2\lambda}{\pi v_i'(1+\kappa)^{1/2}}{\rm K}(\alpha);\quad \alpha = \frac{2\kappa}{1+\kappa},\quad \kappa = \frac{2\hat{V}}{v_i'^2}, \tag{2.134}$$
$$T_{\rm t} = \frac{\lambda}{v_i'}\left(\frac{2}{1+(1-\kappa^2)^{1/2}}\right)^{1/2}.$$

$T_{\rm h}$ is identical with, for instance, the period of the pendulum in the rotation mode in the earth potential $V = mgl$. The period T and the concomitant average velocity v_0 depend very sensitively on the injection velocity v_i' at $\hat{V} = 0$. The dependence is strongest for $T_{\rm sqw}$ and weakest for $T_{\rm t}$. $T_{\rm sqw}$, $T_{\rm h}$ diverge for $\kappa \to 1$, $T_{\rm t}$ remains finite. For κ approaching unity follows

$$\kappa \to 1 \Rightarrow T_{\rm sqw} \sim (1-\kappa)^{-1/2},\ T_{\rm h} \sim \ln\left(\frac{32}{1-\kappa}\right)^{1/2} \to \infty,\ T_t = \sqrt{2}\lambda/v_i'. \tag{2.135}$$

The ponderomotive force is dominated by the contribution from the averaged potential $\overline{V}$. At vanishing wave amplitude it is twice the mean oscillatory energy. In contrast, from the energy principle, and also from the perturbative standard derivation (2.114) the same expression $\mathbf{f}_p$, up to a small Doppler effect, is obtained from (2.109) for the transverse and the longitudinal sinusoidal wave because no contribution from $\overline{V}$ is there. By the two cases two different situations are described. In the first case the amplitude is stationary in the lab frame (*generalized group velocity* $v_g = 0$, we refer to as the *standard* case), in the second case the pulse travels at phase velocity $v_g = v_\varphi$. In applying the energy principle to the standard situation $|\mathbf{v}_0| \ll v_\varphi$ has been tacitly assumed. The ratio $\overline{V}/W$ is highest for the rectangular wave. Therefore, for its evaluation in the standard case the expressions (2.125) have to be transformed solely from the frame comoving with the wave to the lab frame. W remains unchanged, however the relevant κ in $\bar{V}$ changes into $\kappa' = 2h^2/(v_\varphi - v_i')^2$. From $|v_0| \ll |v_\varphi|$ follows also $|v_i'| \ll |v_\varphi|$. Thus

$$\overline{V}_L = \frac{1}{T}\int V(x(t),t){\rm d}t \simeq \frac{1}{T}\int V\frac{{\rm d}x}{v_\varphi}{\rm d}x \simeq \frac{v_0}{v_\varphi}\overline{V} \approx 0. \tag{2.136}$$

In the standard configuration the time averaged potential is negligible as long as v_0, $v_i' \ll v_\varphi$, or the particle is far from being trapped.

b. Ponderomotive Force from the Adiabatic Theorem

The aim is to derive the longitudinal ponderomotive force for all electron injection velocities from $|v_i'| \ll v_\varphi$ to $|v_i| = v_\varphi$ to include particle trapping which is of ponderomotive character. In general the amplitude of an electric wave exhibits a slow variation in time and in space. In addition the frequency ω and the wave vector $\mathbf{k}$ may slowly change in time and/or space, and the wave may undergo variations of the shape along its path. Last but not least, in a general formula of $\mathbf{f}_p$ the group velocity v_g of a wave pulse envelope must be allowed to assume all values from zero or negative values up to the phase velocity v_φ or slightly above. In a wave focused onto a fixed position in space v_g is zero; due to selffocusing evolving in time v_g may assume negative values owing to recession of the focus. A wave pulse propagating in vacuum exhibits $v_g = v_\varphi$. In any of these situations elimination of the fast oscillation is achieved again by cycle averaging provided a clearly recognizable oscillation center exists. However it is essential to perform the averaging in a Lorentz or Galilei invariant manner. It is accomplished by choosing the integration for instance in the Lorentz (Galilei) invariant phase

$$\phi = \int (\mathbf{k}\mathrm{d}\mathbf{x} - \omega \mathrm{d}t), \tag{2.137}$$

see Sect. 1.2.1. It implies coupled averaging in time *and* in space, in contrast to the *standard ponderomotive force* (2.114) which is obtained by averaging in time only. Unfortunately, missing invariance of averaging merely in time led the authors of [11] to the rejection of the oscillation center concept, see appendix there.

In deriving the ponderomotive potential and force for this general situation no longer use can be made of energy conservation because of $\partial H_0/\partial t \neq 0$ of the secular Hamiltonian H_0. The only conserved quantity, though limited to first order, is the adiabatic invariant (2.102).

The decomposition of $\mathbf{v}(t)$ allows the calculation of the adiabatic invariants for libration and rotation from the two loops sketched in Fig. 2.18. Under $\lambda = \text{const}$ the straight orbits emanating from L_0 are all identical and coincide with the motion of the invariant oscillation center $OC(t_0) \rightarrow OC(t_0 + T)$ positioned at the virtual "center of mass". The true orbits of $\oint v_0 \mathrm{d}t = 0$ over period $2T$ are closed and wind around the bundle of straight orbits forth and back. The flux $F = \oint p\mathrm{d}q$ through them equals the area of the loop L_0. The open loops of the rotation mode wind around the same bundle L_0 of straight orbits and yield $\int (p\mathrm{d}q - H\mathrm{d}t)$ over the period T and can be closed by the trajectory $(p, q = \text{const})$ along $-[(t_0 + T) - t_0]$.

In order to separate between libration from rotation and at the same time to eliminate the fast time dependence in H averaging of the Hamiltonian $H(\mathbf{p}, \mathbf{q}, t; \lambda(t) = \text{const})$ is done in a Lorentz, at least Galilei invariant way, for instance along the invariant phase $\Phi = \oint (\mathbf{k}\mathrm{d}\mathbf{x} - \omega \mathrm{d}t)$. In the nonrelativistic longitudinal wave it suffices to time average $H'(p'.q')$ in the wave frame by decomposing $\mathbf{p}$ into $m(\mathbf{v}_0 + \mathbf{w}(t))$ in a way to minimize the mean oscillatory component $W = T^{-1} \oint m\mathbf{w}^2 \mathrm{d}t$, T time period. This yields in the longitudinal wave of wavelength $\lambda = 2\pi/k$ (with m set to unity) in the wave frame

$$\oint H'\mathrm{d}t = H_0(v_0, x')T = \left(\frac{1}{2}v_0^2 + W + \overline{V}\right) T = ET,$$
$$E = \text{const}, \ T = \int \frac{\mathrm{d}x}{v(x)}, \ v_0 = \frac{\lambda}{T}, \ \overline{V} = \int \frac{V(x)}{v(x)}\mathrm{d}x. \tag{2.138}$$

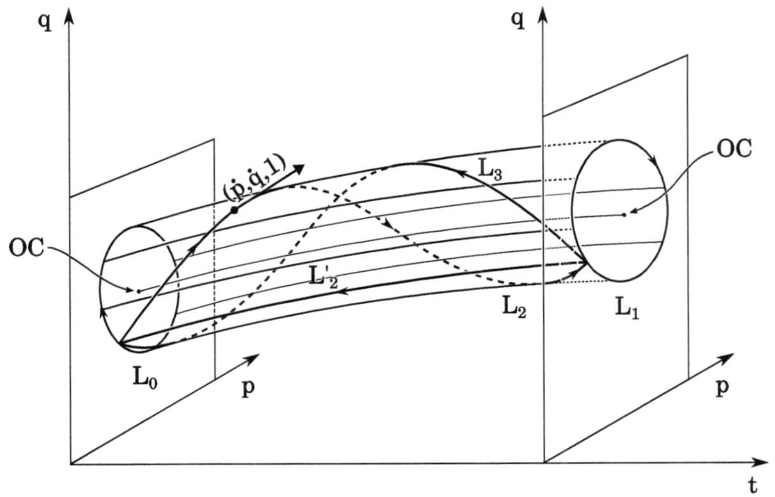

Fig. 2.18 The true particle orbits, here one in the rotation mode and another closed orbit indicating the separatrix towards trapping (libration mode), wind around the bundle of associated oscillation/gyration center trajectories. OC–OC marks the oscillation/gyration center. The libration mode consists of the forward loop L_2 and the backward loop L_3, each of them of periodicity T. Rotation mode: open loop L_2 ending at $T = t_0 + T$. If it is closed by L_2' along ($p =$ const, $q =$ const) the flux through $L_2 + L_2'$ equals half the flux through the libration loop $L_2 + L_3$ and L_0, here separatrix with $\text{const}_\text{a}' = \text{const}_\text{a}$ and peaked edges

Under adiabatic variation, $\lambda(t) \neq 0$ the two quantities F of libration and rotation in the electron plasma wave are adiabatically conserved. Hence, with $E_0 = v_0^2/2$

$$\text{libration}: \oint_{L_0} \mathbf{pdq} = 4T(E_0 + W) = 2\,\text{const}_\text{a};\ v_0 = 0\,, \tag{2.139}$$

$$\text{rotation}: \int_T \mathbf{pdq} - H\text{d}t = T(E_0 + W - \overline{V}) = \text{const}_\text{a}'\,. \tag{2.140}$$

If loop L_2 is closed by L_2' in Fig. 2.18 the flux through $L_2 + L_2'$ must equal $\int_T pdq = 2T(E_0 + W) = \text{const}_\text{a}'$. This is indeed the case because $\oint_{L_2+L_2'} H\text{d}t = 0$. The two relations (2.139) and (2.140) enable to determine the most correct ponderomotive force and the adiabatic transition from free to trapped particles, and viceversa. This is done now with a square wave and a harmonic wave potential, see Fig. 2.16. The potential at the left is chosen because all quantities are obtained by elementary algebra.

Comparison of (2.125) and (2.131) with (2.109) show that the standard cycle averaged Hamiltonian (2.111) is incomplete by the time averaged $\overline{V} \neq 0$, in contrast to space average $\langle V \rangle = 0$. Its weight becomes dramatic for the potential amplitude $\hat{V}$ approaching the injection energy $v_i'^2/2$, i.e. $\kappa \to 1$. The various quantities as functions of κ are shown in Fig. 2.17. The secular Hamiltonian in the wave frame is analogous to (2.127), $H_0(p_0, x') = p_0^2/2 + Z(x'),\ p_0 = v_0$ with $Z = W + \overline{V}$ chosen individually.

Extensive numerical runs show excellent agreement of the adiabatic invariant (2.140) in the wave frame with the solution of the nonrelativistic fully time dependent Lorentz equation of motion. Exceptions are encountered if v_0 approaches zero, i.e. in the vicinity of trapping. There it may happen that just above trapping the transmitted electron describes a smoother trajectory then that from the exact equation of motion; or even worse, the exact solution is that of particle reflection from $\hat{V}_{\max}$ in contrast to electron transmission from (2.140). A similar behavior has been observed in [11] for the transverse wave and in [6] at the onset of chaos. The general case of the longitudinal wave pulse $V(x,t) = \hat{V}(x - v_g t)\cos(kx - \omega t)$ propagating at group velocity $v_g \neq v_\varphi$ reads in the wave system $x' = x - v_\varphi t,\ t' = t$ as $V(x', t) = \hat{V}(x' - [v_g - v_\varphi]t, t)\cos kx'$. Here, for comparison we are interested in the free particle adiabatic invariant in the standard situation of $v_g = 0$ with particle injection from the left and and v_φ always taken positive,

$$T\left[\frac{1}{2}(v_L - v_\varphi)^2 + W - \overline{V}\right] = \text{const}_\text{a}'; \quad v_L - v_\varphi = v_0 \geq 0. \tag{2.141}$$

The *ponderomotive force* for constant phase velocity is given as

$$f_p = m\frac{\mathrm{d}}{\mathrm{d}t}v_0; \qquad v_\varphi = \text{const}. \tag{2.142}$$

The adiabatic invariant (2.140) of the free particle in the sinusoidal potential reads in the wave frame

$$\frac{2\kappa}{\alpha}E(\alpha) = K(\alpha). \tag{2.143}$$

In the standard situation of (2.109) the amplitude is stationary in the lab frame with $v_g = 0$, the potential $V(x)$ propagates to the right with v_φ in the lab frame and the injection velocity v_L is positive. The invariant (2.140) changes into

$$T\left[\frac{1}{2}(v_L - v_\varphi)^2 + W - \overline{V}\right] = \text{const}_\text{a}'; \quad v_L - v_\varphi \geq 0. \tag{2.144}$$

The adiabatic transition from the rotation (free particle) to the libration mode (trapped particle) can occur at $v_0 = 0$. There, however, the validity of $\text{const}_\text{a}' = \text{const}_\text{a}$ is not guaranteed a priori. Our extensive numerical tests show that the relative violation of const_a does not exceed 0.3% in a narrow interval of $T = \infty$ and then it continues with the same value as before for the free particle, in agreement with older studies on the pendulum [12]. The continuity of const_a' enables one to study the conditions for trapping and reflection by equating the two adiabatic invariants (2.139), (2.140) and the resulting change of the electron spectrum after one-time, i.e. non recurring, detrapping (multiple trapping is possible, however ends in a complex

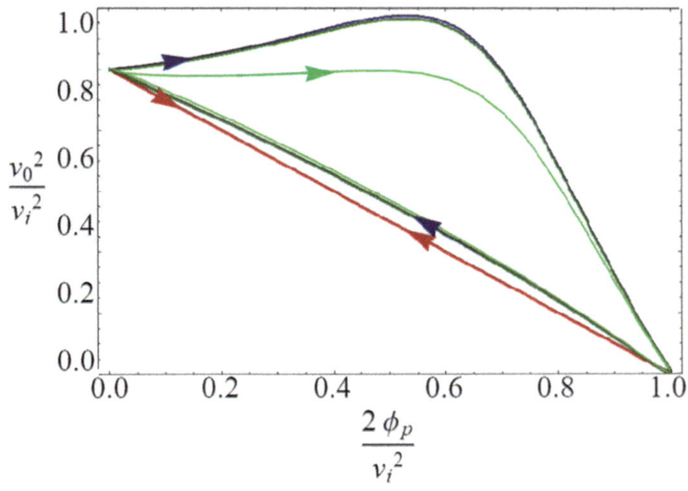

Fig. 2.19 Uphill acceleration and test of (2.143) in the lab frame. In the diagram of normalized initial velocity $v_i = v_i' + v_\varphi$ versus the normalized standard potential Φ_p a particle running against a fixed wave pulse ($v_g = 0,\ v_\varphi > 0$) and back after reflection according to (2.109) follows the straight line. The two proximate curves on top and the pair just above the straight line are the exact numerical solutions and the results from (2.143), respectively, over 1000 oscillations. The agreement is perfect. The curve below the upper pair, although having crossed 10–15 oscillations in the pulse of identical envelope over the same distance only with poor adiabatic behaviour, still indicates uphill acceleration. On the way back it is very close to the exact result owing to Doppler frequency up shift from $\omega = kv_\varphi$ to $\omega_r = \omega + kv_L$ after reflection

analysis). Note, no asymptotic energy change in the wave frame is possible without temporary trapping.

The most striking phenomenon of particle dynamics beyond the range of (2.109) is uphill acceleration from (2.143) with v_i' replaced by $v_i - v_\varphi$ in κ ad α, $v_i = v_i' + v_\varphi$; see Fig. 2.19. It takes place for initial velocities $v_i > (1/3)v_\varphi$. The hysteresis produced by (2.143) under reflection stems from the v_0 dependence of the "ponderomotive potential"; the particle sees a low frequency in direction of the wave (high $Z = W + \overline{V}$) and a high frequency (low Z) moving back after reflection. Although in the dashed line the particle undergoes not more than 10–15 oscillations in the uphill domain the adiabatic invariance reproduces already qualitatively quite well the phenomenon. At 100 times more oscillations in the same pulse (highest pair of curves and pair closest to the straight line) the agreement with the exact numerical solution is perfect. Note, the point of reflection is the same for (2.109) and (2.143).

A remark on uphill acceleration is in order here. Such kind of acceleration gives the impression of a wave potential dragging the particle behind. The picture of acceleration through surfing has been coined for it. Frequently the picture is applied to make Landau damping "understandable" with the effect to definitely obscuring the phenomenon. However, in terms of Newtonian mechanics there is no room for such a construction. Energy conservation holds only in the wave frame, in the lab frame $\partial H/\partial t \neq 0$. In fact, in the wave frame uphill acceleration reveals as "ponderomotive" deceleration; all dragging effect disappears.

A last remark is due to the concept of the standard ponderomotive potential $W(\mathbf{x}_0)$ in (2.109) and the vanishing of $\overline{V}$. In the way introduced in (2.125) and (2.131) $W(\mathbf{x}_0)$ is a Galileian invariant with respect to translation from wave to lab frame. It tells that if calculated from $V(x,t)$ of the longitudinal wave in the lab frame a first order correction has to take care of the Doppler effect, i.e. W depends on $\mathbf{x}_0$ and on the translation velocity $\mathbf{v}_0$, the latter disregarded usually. Regarding $\overline{V}$ under standard conditions of the derivation of (2.109) in the lab frame averaging in time of V for $v_0 \ll v_\varphi$ results as negligible,

$$\overline{V}_L = \frac{1}{T}\int V(x(t),t)\mathrm{d}t \simeq \frac{1}{T}\int V\frac{\mathrm{d}x}{v_\varphi}\mathrm{d}x \simeq \frac{v_0}{v_\varphi}\overline{V} \approx 0. \tag{2.145}$$

Under such limitations (2.140) reduces to (2.109), at a price of excluding the transition from rotation to libration, or from adiabatic particle deceleration to trapping.

2.1.7 *Particle Trapping*

There are three possibilities for a particle to interact ponderomotively with the longitudinal wave packet: transmission, reflection, and trapping by the wave. It follows from the adiabatic conservation of $F = \oint(\mathbf{p}\mathrm{d}\mathbf{q} - H\mathrm{d}t) = \oint \mathbf{p}\mathrm{d}\mathbf{q}$ that the asymptotic gain or loss of energy is zero if the oscillation period has a lower finite upper bound $T_0 < \infty$, $T \leq T_0$ because $mv_0^2 = \mathrm{const_a} - W \leq mv_i^2$. In other words, particle transmission and reflection conserve the energy asymptotically. For trapping the particle must cross the *separatrix* $L_2 + L_3$ with $\overline{V} = mv_i^2/2$ in Fig. 2.18. It is intuitively clear (but difficult to formulate in rigorous mathematical terms) that for an adiabatic transition from rotation to libration $\mathbf{p}_0$ must approach mv_φ and W must vanish. In a potential differentiable everywhere this is identical with T becoming unbound. Thus, the only possibility, if there is any, to acquire or to lose energy asymptotically is by trapping. For the non differentiable triangular wave profile the variation of $\oint \mathbf{p}\mathrm{d}\mathbf{q}$ is small but does not necessarily tend to zero. Once trapped, in any case the particle follows the adiabatic invariance $F = \int \mathbf{p}\mathrm{d}\mathbf{q} = 2\,\mathrm{const_a}$. At the transition *free* ⇆ *trapped* the period T becomes infinite and continuous transition $\mathrm{const}'_\mathrm{a} \leftrightarrows \mathrm{const}_\mathrm{a}$ is not guaranteed. The generation of chaos has been observed [6] and is well documented also elsewhere. On the other hand, extensive numerical tests by the author's research group have shown a relative local violation of $\mathrm{const}'_\mathrm{a} = \mathrm{const}_\mathrm{a}$ in a narrow interval $T = \infty$ does not exceed 0.3% and then it continuous with the same value as before for the free particle, in agreement with older studies on the pendulum [12].

2.1.8 Binary Collisions

The interaction of two billiard spheres represents the prototype of an event which is described by the word collision, more precisely *binary collision*: Two free particles move in against each other along straight orbits, interact within a finite region, and finally move out as free particles along straight lines again. The interaction occurs during a time that is short. It has to result from the context what short means in the specific case of interaction. Before and after the interaction the two collision partners are *asymptotically free*, in the interaction region the trajectories are bent each by a definite angle. The degree of bending depends on how close their approach will be. We may assume that the dynamics of the collision is described by a Hamiltonian $H(\mathbf{p}_1, \mathbf{p}_2, \mathbf{x}_1 - \mathbf{x}_2)$. The dependence on the relative vector $\mathbf{r} = \mathbf{x}_1 - \mathbf{x}_2$ is a consequence of the asymptotic freedom. Momentum and energy conservation before and after the collision (dash ') impose

$$\mathbf{p}'_1 + \mathbf{p}'_2 = \mathbf{p}_1 + \mathbf{p}_2 \quad \Leftrightarrow \quad m_1\mathbf{v}'_1 + m_2\mathbf{v}'_2 = m_1\mathbf{v}_1 + m_2\mathbf{v}_2$$

$$m_1\mathbf{v}'^2_1 + m_2\mathbf{v}'^2_2 = m_1\mathbf{v}^2_1 + m_2\mathbf{v}^2_2. \tag{2.146}$$

By the use of the center of mass and difference vectors, $\mathbf{V} = (m_1\mathbf{v}_1 + m_2\mathbf{v}_2)/(m_1 + m_2)$ and $\mathbf{w} = \mathbf{v}_1 - \mathbf{v}_2$, and the reduced mass $\mu = m_1 m_2/(m_1 + m_2)$ momentum and energy conservation translate into

$$\mathbf{V}' = \mathbf{V}, \quad |\mathbf{w}'| = |\mathbf{w}| = w; \qquad \mathbf{v}_1 = \mathbf{V} + \frac{\mu}{m_1}\mathbf{w}, \quad \mathbf{v}_2 = \mathbf{V} - \frac{\mu}{m_2}\mathbf{w} \tag{2.147}$$

and $\mathbf{v}'_1, \mathbf{v}'_2$ correspondingly (see Fig. 2.20). The angle by which $\mathbf{w}'$ is rotated with respect to $\mathbf{w}$ is the *scattering angle* $\vartheta = \angle(\mathbf{w}, \mathbf{w}')$. In general $\mathbf{w}$ and $\mathbf{w}'$ are not complanar. $(\mu/m_1)\mathbf{w}_1$ and $-(\mu/m_2)\mathbf{w}_2$ are the velocities of the collision partners in the center of mass frame. For an electron-ion pair with the ion at rest follows $\mu \simeq m_e$, $\mathbf{V} \simeq (m_e/m_i)\mathbf{v}_e$, $\mathbf{w} \simeq \mathbf{v}_e$, and $|\mathbf{w}'_e| \simeq |\mathbf{v}_e|$.

In the important case the interaction potential V in the Hamiltonian depends only on the relative distance $r = |\mathbf{x}_1 - \mathbf{x}_2|$ the relative velocities $\mathbf{w}, \mathbf{w}'$ are coplanar and scattering by the angle θ reduces to single particle dynamics of a point of mass μ in the central field $\pm\nabla V(r)$ fixed at the origin $\mathbf{r} = 0$, as sketched for an attractive and

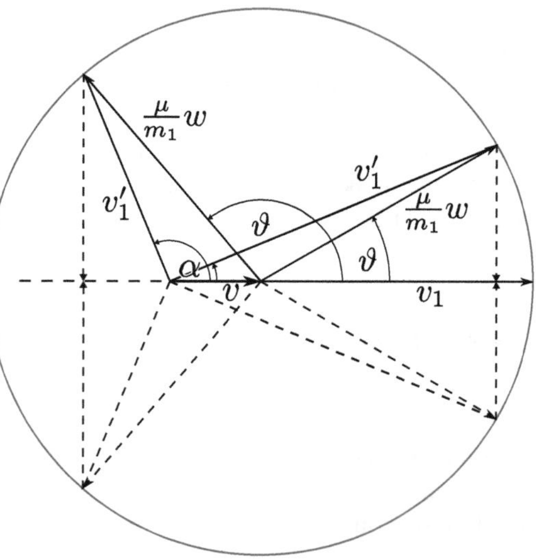

Fig. 2.20 Elastic collision of a particle of mass m_1 and velocity $\mathbf{v}_1$ with a particle of mass $m_2 = 2m_1$ at rest, $\mathbf{v}_2 = 0$. Center of mass velocity is $\mathbf{V} = \mathbf{v}_1/3$, relative velocity $\mathbf{w} = \mathbf{v}_1$. Velocity after collision $\mathbf{v}_1' = \mathbf{V} + 2\mathbf{w}'/3$, $|\mathbf{w}'| = |\mathbf{w}|$. Scattering for two different angles ϑ. Cylindrical symmetry around $\mathbf{v}_1$

a repulsive potential in Fig. 1.7. The total angular momentum is conserved and is $\mathbf{J} = \mathbf{r} \times \mu\mathbf{w}$ in the center of mass system. With b the *collision* or *impact parameter* the modulus of the angular momentum is $|\mathbf{J}| = \mu w b$. The scattering process exhibits cylindrical (rotational) symmetry; it does not depend on the azimuth angle φ. The scattering angle θ is completely determined once w and b are given.

Differential Collision Cross Section

Imagine the flux of noninteracting particles of intensity $I = nw = n(v_1 - v_2)$ entering in Fig. 1.7 from the left at $x = -\infty$ parallel to the axis. The number of particles entering through the infinitesimal area $\mathrm{d}F = b\mathrm{d}b\mathrm{d}\varphi$ per unit time is $\mathrm{d}N = I\mathrm{d}F$. After scattering at $x = 0$ they are found at $x = +\infty$ in the solid angle $\mathrm{d}\Omega = \sin\vartheta\mathrm{d}\varphi\mathrm{d}\vartheta$. As $\mathrm{d}N$ is proportional to the flux density I we can establish the equality

$$\mathrm{d}N = I\mathrm{d}F = I\sigma_\Omega\mathrm{d}\Omega \qquad \Rightarrow \qquad \sigma_\Omega = \frac{1}{I}\frac{\mathrm{d}N}{\mathrm{d}\Omega}. \tag{2.148}$$

The fraction of particles $\mathrm{d}N$ entering from the left has been defined as that number entering per second through the area $\mathrm{d}F = b\mathrm{d}b\mathrm{d}\varphi$. As it is the same number scattered into the solid angle $\mathrm{d}\Omega$ the definition of the proportionality constant σ_Ω in the second relation maintains its meaning also when no classical orbit or impact parameter can be defined. σ_Ω is the *differential cross section*. It expresses a property of how the two collision partners interact with each other. Its dimensions are $[\sigma_\Omega] = \text{area} \times \text{sterad}^{-1}$. For σ_Ω the symbols $\sigma(\varphi, \vartheta)$, $\sigma(\vartheta)$, and σ_ϑ are also in use.

The differential cross section σ_Ω is the probability for one particle in the center of mass system to be scattered into the solid angle $\mathrm{d}\Omega$. The meaning of $\mathrm{d}\Omega$ is that of direction $\mathrm{d}\boldsymbol{\Omega} \sim -\mathrm{d}\boldsymbol{\varphi} \times \mathrm{d}\boldsymbol{\vartheta}$ and magnitude $\sin\vartheta \mathrm{d}\varphi \mathrm{d}\theta$.

Differential Coulomb Cross Section

The classical *Coulomb* or *Rutherford differential cross section* $\sigma_\Omega = \sigma_C$ from the Coulomb potential $V(r) = q_1 q_2/4\pi\varepsilon_0 r$ is expressible in a compact way by the introduction of the collision parameter for perpendicular deflection $b_\perp$ and its connection with the scattering angle ϑ,

$$\sigma_C = \frac{b_\perp^2}{4\sin^4(\vartheta/2)}, \quad \tan\frac{\vartheta}{2} = \frac{b_\perp}{b}, \quad b_\perp = \frac{Ze^2}{4\pi\varepsilon_0 \mu \mathbf{w}^2} = 0.7\frac{Z}{E_r[\mathrm{eV}]}\,\mathrm{nm}. \tag{2.149}$$

The relative energy $E_r = \mu\mathbf{w}^2/2$ is the energy in the center of mass system. With Z the charge number these expressions hold for electron-electron as well as for electron-ion collisions. Note, in the first case $\mu = m_e/2$.

Collision Frequency of Momentum Transfer

The concept of *collision frequency* of momentum transfer is a key issue of microscopic transport theory. Collisional heating by laser is determined by the ratio ν_{ei}/ω. Consider a particle of momentum $m_1\mathbf{v}_1$ impinging on matter *at rest* of density n_2 and $\rho_2 = m_2 n_2$. The collision frequency $\nu_{12}(v_1)$ for momentum transfer by non-overlapping binary collisions must fulfill the equation of motion

$$m_1\frac{\mathrm{d}\mathbf{v}_1}{\mathrm{d}t} = -m_1\nu_{12}(v_1)\mathbf{v}_1. \tag{2.150}$$

If σ_t is the total collision cross section for momentum transfer, $\nu_{12}(v_1)$ is given by

$$\nu_{12}(v_1) = n_2\sigma_t v_1; \quad \sigma_t = 2\pi\int_{b=0}^{\infty}\frac{\mu}{m_1}\sigma_\vartheta(1-\cos\vartheta)\sin\vartheta \mathrm{d}\vartheta. \tag{2.151}$$

Consequently, $\lambda = 1/n_2\sigma_t$ is the *mean free path*. The correctness of expression $\nu_{12}(v_1)$ is shown with the help of Fig. 2.21 as follows. Consider the scattered particle m_1 moving along the x axis into an ensemble of a large number of N_2 scattering centers m_2 of density n_2. In the time interval Δt the projectile m_1 crosses the layer of cross section $A(x)$ and thickness $\Delta x = v_1\Delta t$ containing $\mathrm{d}N_2 = n_2 A(x)\Delta x$ centers. The thickness is chosen such that the scatterers do not overlap *or* the deflections are so weak that they can be treated as sequential events (see Chap. 7 on transport). The m_1 interacts with all scatterers $\mathrm{d}N_2(b) = 2\pi n_2 b\mathrm{d}b$ at the distance of the collision parameter b causing the deflection σ_ϑ. Total momentum loss $\Delta\mathbf{p}_1 = n_2\sigma_t m_1\mathbf{v}_1\Delta x$ into forward direction in the time interval Δt is by integration of b from 0 to the diameter of A. On the microscopic scale A is large so that b can be extended to infinity.

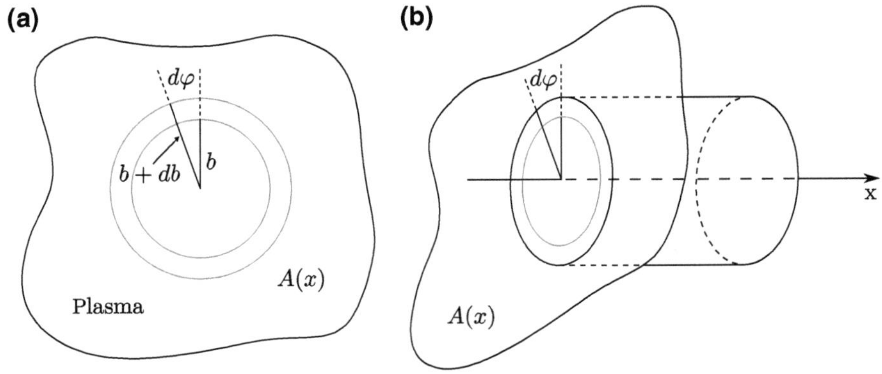

Fig. 2.21 Derivation of the mean collision frequency $\nu_{12}(v_1)$. (a) View along projectile velocity: $A(x)$ plasma cross section perpendicular to $\mathbf{v}_1$ at position x, b collision parameter, $d\varphi$ azimuthal angle element. (b) Side view: Slab of thickness $\Delta x = v_1 \Delta t$ and cross section $A(x)$ contains $dN_2 = n_2 A \Delta x$ scatterers. All of them act on the projectile on axis

Equivalently, ϑ varies from π to 0. The radial components cancel because of cylindrical symmetry. The axial component before the collision is $v_1 = V + (\mu/m_1)v_1$ and after the collision $v_1' \cos\alpha = V + (\mu/m_1)v_1 cos\vartheta$, see Fig. 2.20. Hence, (2.151) are the quantities that had to be proven, in particular that $\nu_{12}(v_1)$ is proportional to the density n_2. In the case of the electron-ion collision frequency ν_{ei} can be set $\mu = m_e$, $V = 0$, and $\mu/m_1 = 1$ in the integral of σ_t. Owing to the discreteness of matter the integration in b undergoes large fluctuations in the domain $b \lesssim b_\perp$. Therefore the collision frequency ν is to be intended as an ensemble average. In the general case of $\mathbf{v}_2 \neq 0$ in ν_{12} of (2.151) the velocity $\mathbf{v}_1$ is to be substituted by $\mathbf{w} = \mathbf{v}_1 - \mathbf{v}_2$. We note that the same result for ν_{12} is obtained from considering the scattering of m_1 with a single m_2 and then multiplying it with the density of the scattering centers n_2. This procedure with one electron scattered on one ion corresponds to what is known as the *test particle* method.

The proof given here is alright for short range interactions. In this respect the long range Coulomb potential is the worst that can occur. (1) Owing to its long range character a charged particle never interacts with one scatterer only. The number of *simultaneous collisions* is of the order of Λ, i.e., from 10 to a few hundred in laser plasmas and $10^4 - 10^7$ in a Tokamak plasma. (2) The charged particle interacts never with a single Coulomb scatterer, rather feels it the simultaneous influence of many other projectiles impinging onto the scatterer. Reduction to a sequence of binary collisions is possible if (1) momentum loss is caused by small angle deflections and (2) the Coulomb potential is substituted by an effective *screened potential*. Details are presented in Chap. 7 on transport.

Is ν_{ei} a time or an ensemble average? Owing to the discreteness of the ionic fluid in a given instant t the electron is scattered from

$$\sum_{N=1}^{N_2} \delta(\mathbf{x} - \mathbf{x}_N(t)).$$

This is a special configuration. All other positions are possible with equal likelihood, thus resulting into the uniform distribution $\int n_2 \times 2\pi b db$ which is the statistical ensemble average of the discrete sum above. Instead one could think of the mean collision frequency ν_{ei} as a time series of collisions of a single electron averaged in time, $\nu_{ei} = 1/\tau_{ei}$,

$$\tau_{ei} = \lim_{N\to\infty} \sum_{N=1}^{\infty} \tau_{ei,N}.$$

Apart from the question in which sense such a time limit exists there is the question of whether the two definitions are equal. The answer is given by Birkhoff's *ergodic theorem* in 1923 [13, 14]. Subsequently, and on the research related to the question, the modern understanding of statistical means is to be *ensemble averages.*

The model of colliding *hard ideal spheres* of radius r_1 and r_2 is useful in transport theory and equation of state considerations. The differential cross section has the interesting properties neither to depend on the relative energy E_r because of no surface friction, nor on the scattering angle ϑ or collision parameter b. With $r = r_1 + r_2$ it is, together with its total cross section σ_t, as follows,

$$\sigma_\Omega = \frac{r^2}{4}, \qquad \sigma_t = \int \sigma_\Omega(1 - \cos\vartheta)\mathrm{d}\varphi \mathrm{d}\cos\vartheta = \int \sigma_\Omega \mathrm{d}\varphi \mathrm{d}\cos\vartheta = \pi r^2. \tag{2.152}$$

The total scattering cross section of hard spheres for momentum transfer and for deflection only are identical and hence, the same holds for the collision frequency $\nu_{12}(v_1)$. The scattering angle ϑ refers to the center of mass system. For the transformation to fixed scattering center see Fig. 2.20.

2.2 Relativistic Regime

2.2.1 Essential Relativity

2.2.1.1 Lorentz Transformation

Already in Galileian relativity forces are the same in all inertial reference systems. Furthermore, there is no criterion that allows to distinguish between inertial and impressed forces. This suggests that laws of physics have to be covariant, i.e., there is no preferred reference system. Finally, it is a postulate that empty space has to be homogeneous and isotropic. How can these facts be reconciled with the postulates? It was the great step by Einstein to recognize that the synthesis is possible only in four dimensional space of elements $(\mathbf{x}, t)$, i.e., after revision of time.

From homogeneity and isotropy of empty tridimensional space follows that the transformation from a system of reference S to one S' moving at constant velocity $\mathbf{v}$ relative to S must be *linear* in $(\mathbf{x}, t)$. Further, if the existence of *physical states* or time independent configurations is postulated, the linear transformations from S to S' must form a group. These two requirements allow two classes of transformations only, Galileian (2.5) if no maximum speed exists, and Lorentzian if there exists a maximum finite speed. Einstein's argument for a maximum speed is basic and simple. Imagine an observer following a light front from behind at v slightly greater than c, slowing down to c when he has caught up with the front and than accelerating slowly away. Along such a travel he observes a finite, quasistatic electric field behind the light front and zero field ahead. In other words, he sees field lines ending in empty space. This is in contradiction to Maxwell's equation $\nabla\mathbf{E} = \rho_{el}/\epsilon_0$, which implies that static field lines must end in a charge or extend to infinity. With a limiting speed $c \geq v$ the only transformation from S to S' obeying linearity, homogeneity, and isotropy of empty space, and forming a group is Lorentzian:

$$\mathbf{x}' = \mathbf{x} + \frac{\gamma - 1}{v^2}(\mathbf{v}\mathbf{x})\mathbf{v} - \gamma\mathbf{v}t, \qquad t' = \gamma\left(t - \frac{\mathbf{v}\mathbf{x}}{c^2}\right); \qquad \gamma = \left(1 - \frac{v^2}{c^2}\right)^{-1/2}. \tag{2.153}$$

If $\mathbf{v}$ is chosen along x the transformation results in

$$x' = \gamma(x - vt), \quad t' = \gamma\left(t - \frac{v}{c^2}x\right); \quad y' = y, \quad z' = z. \tag{2.154}$$

The reverse is also true: From (2.154) follows (2.153) because they form a group. The inverse transformation $(\mathbf{x}', t) \rightarrow (\mathbf{x}, t)$ must show the same structure with the relative velocity $\mathbf{v}'$ of S with respect to S'. From the principle that all S are equivalent to each other follows $\mathbf{v}' = -\mathbf{v}$, a property which was taken for granted by Einstein and followers and has been proven only half a century later [15]. Thus

$$\mathbf{x} = \mathbf{x}' + \frac{\gamma - 1}{v^2}(\mathbf{v}\mathbf{x}')\mathbf{v} + \gamma \mathbf{v}t', \qquad t = \gamma\left(t' + \frac{\mathbf{v}\mathbf{x}'}{c^2}\right). \tag{2.155}$$

$$\mathbf{v} \parallel x', x \;\Rightarrow\; x = \gamma\big(x' + vt'\big), \quad t = \gamma\left(t' + \frac{v}{c^2}x'\right); \quad y = y', \quad z = z'. \tag{2.156}$$

From either (2.153) or (2.155) one deduces the invariant

$$\mathbf{x}'^2 - c^2t'^2 = \mathbf{x}^2 - c^2t^2. \tag{2.157}$$

With the *pseudo-euclidian metric* $g_{\alpha\beta} = \delta_{\alpha\beta}$ for $\alpha, \beta = 1, 2, 3$, and $g_{4\beta} = -\delta_{4\beta}$ with $\beta = 1, 2, 3, 4$ the invariant is recognized as the scalar product of the four vector $X = (\mathbf{x}, ct) = (x^1, x^2, x^3, x^4)$ with itself

$$X^2 = XX = g_{\alpha\beta}x^\alpha x^\beta = \mathbf{x}^2 - c^2t^2 = s^2. \tag{2.158}$$

The quantity $d = |s|$ is the length of the four vector X. In fact, $|s|$ satisfies the requirements of a *metric* inside and outside the *light cone*; on the light cone it is degenerate: (i) $d(X = 0) = 0$, (ii) $d(A, B) = d(B, A)$ (symmetry), (iii) $d(A, C) \leq d(A, B) + d(B, C)$ (triangle inequality). A, B, C are the end points of the corresponding four vectors $X, Y, X + Y$. The invariance of $|s|$ under a Lorentz transformation tells that a reference system change from S to $S'(\mathbf{v})$ in four space is a rotation by a certain angle around a fixed axis. In (2.158) summation is done over equal indices α and β in upper and lower position from 1 to 4 (Einstein's summation convention). Greek indices running from 1 to 4 are reserved to four quantities while latin indices $i, j, \ldots = 1, 2, 3$ are used to indicate the spatial components of the corresponding four quantities. Bold italic letters designate vectors in the physical space $\mathbb{R}^3$.

Einstein's clock. Imagine a mass point moving according to $\mathbf{x} = \mathbf{v}t$ in S. Owing to $s' = s =$ const in any inertial frame, in the comoving system $S'(\mathbf{v})$ holds $X' = (\mathbf{0}, ct)$ and

$$-c^2t'^2 = v^2t^2 - c^2t^2 \quad \Rightarrow \quad t' = \left(1 - \frac{v^2}{c^2}\right)^{1/2} t = \frac{t}{\gamma}. \tag{2.159}$$

The time t' in the system comoving with the particle is called the *proper time* τ. In differential form (2.158) reads

$$\mathrm{d}s^2 = \mathrm{d}\mathbf{x}^2 - c^2\mathrm{d}t^2 = -c^2\mathrm{d}\tau^2 \quad \Rightarrow \quad \mathrm{d}\tau = \frac{\mathrm{d}t}{\gamma}. \tag{2.160}$$

This is the formula for *time dilation* (e.g., of the lifetime of a moving excited atom). Conversely, it can be shown that from the invariance of $\mathrm{d}s$ alone follows the Lorentz transformation (2.153). The proof of this assertion and all formulas in this section can be found in Weinberg [16], Chap. 2 (or, better, can be derived as an exercise by the reader himself on the basis of the principles of homogeneity, isotropy, existence of states, and relativity principle).

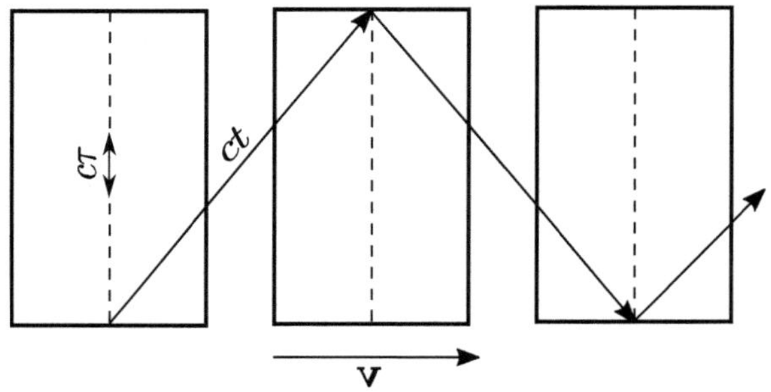

Fig. 2.22 Einstein's clock showing time dilation and Lorentz contraction

Proper time is best visualized by Einstein's clock in Fig. 2.22. An ultrashort laser pulse is bouncing forth and back between two mirrors at distance h from each other and at rest in $S'(v_\perp)$. The speed of light in both systems is the same. In proper time τ the light pulse covers the distance $h = c\tau$ in S' whereas the distance observed from the lab frame S is

$$h^2 + v_\perp^2 t^2 = c^2 t^2 \quad \Rightarrow \quad c^2 t^2 \left(1 - \frac{v_\perp^2}{c^2}\right) = h^2 = c^2 \tau^2 \quad \Rightarrow \quad t = \gamma\tau.$$

Next the *Lorentz contraction* of moving scale is deduced. A rod of length L' extends in S' from $x_1' = 0$ to x_2'. These end points must be measured as x_1 and x_2 both simultaneously at the instant t on the clock in S. This is achieved by sending two ultrashort laser pulses from x_1' and from x_2' into opposite directions in such a way that they meet at the center $L/2$ of the flying rod. By knowing the positions x_1, x_2 when they overlap with x_1', x_2' the length L is given by $L = x_2 - x_1$. This is equivalent to sending a light pulse from x_1 at time t_1 towards x_2, reflecting it there at time t and receiving it at time t_2 at x_1. The length of the rod in S is then given by $2L = c(t_2 - t_1)$. According to this latter procedure of measurement the end points x_2 right and x_1 left are reached at t_2 and t_1. For them the equalities hold

$$x_2 = L + v(t - t_1) = c(t - t_1), \quad x_1 = L - v(t_2 - t) = c(t_2 - t)$$

$$\Rightarrow \; t - t_1 = \frac{L}{c - v}, \quad t_2 - t_1 = \frac{L}{c + v}, \quad t_2 - t_1 = \frac{2L/c}{1 - v^2/c^2} = \frac{2L/c}{\gamma^2}.$$

In the time interval $t_2 - t_1$ in S, the pulse has travelled forth and back in S' in time $t_2' - t_1' = L'/c$. Owing to time deletion $t_2 - t_1 = \gamma(t_2' - t_1') = 2L'/c$, and hence the scale L' moving at velocity v in S is contracted according to

$$L = \frac{1}{2}c(t_2 - t_1) = \frac{L'}{\gamma}. \tag{2.161}$$

The speed of light is the same in S and S'. It requires

$$c = \frac{2L}{t_2 - t_1} = \frac{2L'}{t_2' - t_1'} = \frac{2L}{\gamma(t_2' - t_1')} \quad \Rightarrow \quad t_2 - t_1 = \gamma(t_2' - t_1'). \tag{2.162}$$

With $t = t_2 - t_1$, $\tau = t_2' - t_1'$ this is identical with the time dilation (2.159), this time Einstein's clock moved at velocity v parallel to the laser beam. It shows that time intervals measured with Einstein's clock are independent of the orientation of the laser beam with respect to its direction of motion.

Pedestrian's derivation of the Lorentz transformation. Assume (i) standard conditions: S and $S'(\mathbf{v})$ with axes x, y, z and x', y', z' oriented parallel and S' sliding along x with common origin at $t = t' = 0$; (ii) take Lorentz contraction (and time dilation) for granted. The position x' is mapped onto $x - vt$ for all times. Owing to length contraction of x there must hold, with the left end point $x(t)$ chosen to coincide with the origin of S,

$$x - vt - [x(t) = 0] = \frac{x'}{\gamma} \quad \Rightarrow \quad x' = \gamma(x - vt), \;\; y' = y, \;\; z' = z. \tag{2.163}$$

The time interval t' refers to the origin $x = 0$ held fixed. Then the distance $d' = ct'$ is mapped onto $d = ct - vt$, where $t = x/c$, hence $ct - vt = ct - vx/c$. Owing to the invariance of c with respect to all reference systems the transformation $t \to t'$ is recovered from dividing by c. The two distances covered by the light pulse are related by $d = d'/\gamma$ through the equality

$$\frac{ct'}{\gamma} = ct - \frac{vx}{c} \quad \Rightarrow \quad t' = \gamma\left(t - \frac{v}{c}\frac{x}{c}\right). \tag{2.164}$$

For $\gamma \to 1$ Galileian relativity $x' = x - vt$, $t' = t$ is recovered. The step from these expressions to the general transformations (2.153) is simple.

(i) Figure 2.22 shows that synchronization of two clocks is lost when they after synchronization move along different trajectories and then meet again (twin paradox is real!). Two events P_1, P_2 are the same if they transform into each other by a Lorentz transformation. (ii) Which of the Lorentz transformations to use, direct or inverse? If length is to be measured in S the events of the endpoints are simultaneous in S: $P_1 = (\mathbf{x}_1, t_1)$, $P_2 = (\mathbf{x}_2, t_2 = t_1)$; thus, the transformation (2.153) for $(\mathbf{x}, t) \to (\mathbf{x}', t')$ is in order. If the elapsed time is to be determined in S from the proper time registered by a clock at rest at x' in S' the inverse transformation (2.156) is to be used. (iii) Time dilation and Lorentz contraction are intimately connected. (iv) The Lorentz contraction cannot be "seen". Formula (2.161) refers to distances between simultaneous events P_1, P_2 in spacetime. Observations relate to simultaneous *arrivals* of light signals from P_1, $P_2(t_2 \neq t_1)$ in the observer's eye.

A first useful application of (2.160) and (2.161) is encountered in the transformation of the electromagnetic field from S to $S'(\mathbf{v})$. Assume an electric field $\mathbf{E}$ and a magnetic field $\mathbf{B}$ in S. Imagine an equivalent $\mathbf{E}$-field in a given point, generated by a capacitor at rest in S. An electron at rest in $S'(\mathbf{v})$ sees an invariant surface charge density σ when $\mathbf{v}$ is perpendicular to the plates, hence the field component parallel to $\mathbf{v}$ remains unaltered, $\mathbf{E}'_{\parallel} = \mathbf{E}_{\parallel}$. When the plates are parallel to $\mathbf{v}$ the charge density is increased by γ due to length contraction. Thus $\mathbf{E}'_{\perp} = \gamma\mathbf{E}_{\perp}$ in the absence of a magnetic field. The transformation of $\mathbf{E}_{\perp}$ in presence of $\mathbf{B}$ is obtained from Faraday's law in the form (2.7) with $\mathrm{d}\mathbf{s}'$ on its LHS taken at constant time instant t'. Hence, (2.153) applies to $\mathrm{d}\mathbf{s} = \gamma\mathrm{d}\mathbf{s}'$ on the RHS of (2.7). Alternatively one may argue on the relation between the time derivatives of the magnetic fluxes on LHS and RHS of (2.7). The magnetic flux counts the number of field lines through the loop and is therefore not affected from a change of the reference system. From Fig. 2.22 follows, with proper time $\tau = t'$ on the LHS, time in S is $t = \gamma\tau$ and consequently $\mathrm{d}/\mathrm{d}t = \gamma\mathrm{d}/\mathrm{d}\tau$. Note, this is also the result of applying (2.155). These relations, either $\mathrm{d}\mathbf{s}$ or $\mathrm{d}/\mathrm{d}t$, substituted in (2.7) generate the local transformation law of $\mathbf{E}_{\perp}$,

$$\mathbf{E}'_{\perp} = \gamma(\mathbf{E}_{\perp} + \mathbf{v} \times \mathbf{B}), \qquad \mathbf{E}'_{\parallel} = \mathbf{E}_{\parallel}. \tag{2.165}$$

The magnetic field generated by an infinitely long coil does not change when the coil moves with velocity $\mathbf{v} \parallel \mathbf{B}$ because cross section and magnetic field flux (number of field lines) remain unchanged. To find the transformation law of $\mathbf{B}_{\perp}$ the inverse transformation $\mathbf{E}'_{\perp} \to \mathbf{E}_{\perp} = \gamma(\mathbf{E}'_{\perp} - \mathbf{v} \times \mathbf{B}')$ can be used to calculate $\mathbf{v} \times \mathbf{E}_{\perp}$ with the help of (2.167),

$$\begin{aligned}\frac{1}{\gamma^2}\mathbf{v} \times \mathbf{E}_{\perp} &= \frac{1}{\gamma}\mathbf{v} \times (\mathbf{E}'_{\perp} - \mathbf{v} \times \mathbf{B}'_{\perp}) = \mathbf{v} \times (\mathbf{E}_{\perp} + \mathbf{v} \times \mathbf{B}_{\perp}) - \frac{1}{\gamma}\mathbf{v} \times (\mathbf{v} \times \mathbf{B}'_{\perp}) \\ &= \mathbf{v} \times \mathbf{E}_{\perp} - v^2\mathbf{B}_{\perp} + \frac{1}{\gamma}v^2\mathbf{B}'_{\perp}.\end{aligned} \tag{2.166}$$

Thus $\mathbf{B}$ transforms according to

$$\mathbf{B}'_{\perp} = \gamma\left(\mathbf{B}_{\perp} - \frac{1}{c^2}\mathbf{v} \times \mathbf{E}\right), \quad \mathbf{B}'_{\parallel} = \mathbf{B}_{\parallel}. \tag{2.167}$$

The second equation in (2.167) is directly deduced from 2.165 by making use of the analogy of $\nabla \times \mathbf{B} = \partial_t\mathbf{E}/\varepsilon_0 c^2$ with $\nabla \times \mathbf{E} = -\partial_t\mathbf{B}$ (once more Feynman's principle). From (2.165) and (2.167) one deduces

$$\mathbf{E}'^2 - c^2\mathbf{B}'^2 = \mathbf{E}^2 - c^2\mathbf{B}^2, \qquad \mathbf{E}'\mathbf{B}' = \mathbf{E}\mathbf{B}. \tag{2.168}$$

i.e., the two expressions are Lorentz scalars. The second relation says that $\mathbf{E}\perp\mathbf{B}$ is an invariant property with respect to a Lorentz transformation; for example the $\mathbf{k}$

vector of a plane electromagnetic wave may change direction, however, in contrast to Galileian relativity, **k**, **E**, **B** remain mutually orthogonal.

Four vectors. Any quantity Y which transforms like a position vector $X = (x, y, z, ct) = (x^1, x^2, x^3, x^4)$ is called a *four vector*. The following easily provable criterion is very useful.

Four vector criterion. If the scalar product $ZY = g_{\alpha\beta}z^\alpha y^\beta$ of a four vector Z with a four quantity Y remains unchanged under an arbitrary Lorentz transformation (2.153), Y is a four vector, and vice versa, i.e., the scalar product of two four vectors Z and Y is Lorentz invariant.

Using this criterion, the correct transformation laws of frequency ω and wave vector $\mathbf{k}$ of a light wave in vacuum is found. The phases $\phi = \mathbf{kx} - \omega t$ in S and $\phi' = \mathbf{k'x'} - \omega' t'$ in S' must be equal because they count the number of maxima, for example, which reach the observers in S and S'. Hence, $K = (\mathbf{k}, \omega/c)$ is a four vector according to the preceding criterion and transforms as

$$\mathbf{k}' = \mathbf{k} + \frac{\gamma - 1}{v^2}(\mathbf{vk})\mathbf{v} - \gamma\frac{\mathbf{v}}{c^2}\omega, \qquad \omega' = \gamma(\omega - \mathbf{kv}). \tag{2.169}$$

This is the relativistic Doppler effect of an electromagnetic wave in vacuum.

Unlimited frequency upconversion by reflection from a countermoving mirror is possible due to the Lorentz factor γ. Time dilation makes it possible. In Galileian relativity the maximum reachable frequency would be 2ω only. In the homogeneous medium at rest the vacuum light velocity in γ has to be replaced by the phase velocity $v_\varphi = c/\eta$, η refractive index.

Lorentz boost. An electromagnetic plane wave incident in the (x, y) plane under angle α relative to x represents a twodimensional problem, at normal incidence it would be onedimensional. This reduction is achieved by applying a Lorentz boost parallel to the unit vector $\mathbf{e}_y$ of velocity

$$\mathbf{v}_0 = c\mathbf{e}_y\frac{\mathbf{k}\mathbf{e}_y}{|\mathbf{k}|}, \qquad v_0 = c\sin\alpha, \qquad \alpha = \angle(\mathbf{k}, \mathbf{e}_x). \tag{2.170}$$

The wave vector $\mathbf{k}$ transforms in the reference system $S'(v_0)$ into

$$\mathbf{k}' = \mathbf{k} - \frac{\mathbf{v}_0}{v_0^2}(\mathbf{v}_0\mathbf{k}), \qquad |\mathbf{k}'| = k\cos\alpha = |\mathbf{k}'_x|; \qquad \mathbf{k}'\perp\mathbf{e}_y.$$

For the frequency in $S'(\mathbf{v}_0)$ (2.169) yields with $\gamma_0 = \left(1 - \sin^2\alpha\right)^{-1/2} = \cos^{-1}\alpha$

$$\omega' = \frac{\omega}{\gamma_0} = \omega\cos\alpha. \tag{2.171}$$

From (2.165) and (2.167) follows for the incident wave

in p-polarization (**E** field in the plane of incidence):

$$E'_y = E_y = \frac{|\mathbf{E}|}{\gamma_0}, \quad E'_x = 0, \quad B'_x = B'_y = 0, \quad B'_z = \frac{|\mathbf{E}|}{c\gamma_0} \tag{2.172}$$

in s-polarization (**B** field in the plane of incidence):

$$E'_x = E'_y = 0, \quad E'_z = \frac{E_z}{\gamma_0}, \quad B'_y = B_y = \frac{|\mathbf{B}|}{\gamma_0}, \quad B'_x = 0. \tag{2.173}$$

The intensity I' at normal incidence is obtained from the substitution of $\Re\mathbf{E}$ by the corresponding $\Re\mathbf{E}'$ component. As expected it is identical for both polarizations,

$$I' = \frac{1}{2}\varepsilon_0 c\,|\mathbf{E}'|^2 = \frac{I}{\gamma_0^2} = I\cos^2\alpha. \tag{2.174}$$

The reduction of oblique incidence to normal incidence is of great advantage in analytic theory as well as for numerical simulations. In the case of p-polarization it enables one to reduce an originally 2D2V problem to 1D2V one space dimension, two velocity components. In particle-in-cell (PIC) simulations such a Lorentz boost was used for the first time in [17] and in Vlasov simulations in [18].

(1) The transformation to $S'(v_0 = c\sin\alpha)$ in (2.170) is pure geometry, valid also for the obliquely falling rain in the wind; by the corresponding boost the wind disappears. Relativity enters with $\mathbf{E}' \perp \mathbf{k}'$, in contrast to Galileian relativity. (2) Note, one factor $\cos\alpha$ in (2.174) stems from the change of direction of the vector **I** by the angle α, the second factor comes from the Lorentz contraction of the unit area by γ_0.

It is evident that the following quantities with proper time τ and rest mass m are four vectors:

four velocity $\quad V = \mathrm{d}X/\mathrm{d}\tau = \gamma(\mathbf{v}, c), \quad V^2 = \gamma^2(v^2 - c^2) = -c^2,$

four momentum $\quad P = mV = (\mathbf{p}, \gamma mc), \quad \mathbf{p} = \gamma m\mathbf{v}, \quad P^2 = -m^2c^2,$

four acceleration $\quad A = \mathrm{d}V/\mathrm{d}\tau, \quad VA = \frac{1}{2}\mathrm{d}V^2/\mathrm{d}\tau = 0 \quad \Rightarrow \quad A \perp V$

(2.175)

$$A^2 = \mathbf{a}^2, \quad \mathbf{a} = \mathrm{d}\mathbf{v}/\mathrm{d}\tau.$$

A special criterion for a four quantity Y to be a four vector is that its square $Y^2 = g_{\alpha\beta}y^\alpha y^\beta$ is invariant against a Lorentz transformation. From the knowledge $V^2 = -c^2$, $P^2 = -m^2c^2$, $A^2 = \mathbf{a}^2$ alone follows that V, P, A are four vectors. At first glance it may surprise that V, P, A are constructed with the help of the position vector X in S and the proper time τ related to the comoving frame $S'(\mathbf{v})$. However, from (2.160) follows that $\mathrm{d}\tau$ is proportional to $\mathrm{d}s$, which, as an invariant, may be evaluated in S as well.

If a four quantity $F = F(\mathbf{x}, \mathbf{v})$ is related to the change of the four momentum P by

$$\frac{\mathrm{d}P}{\mathrm{d}\tau} = F. \tag{2.176}$$

F is called a four force or *Minkowski force*. If its spatial components are indicated by $\mathbf{f} = (f^1, f^2, f^3)$, it follows

$$\frac{\mathrm{d}\mathbf{p}}{\mathrm{d}\tau} = \mathbf{f} \quad \Leftrightarrow \quad \frac{\mathrm{d}\mathbf{p}}{\mathrm{d}t} = \mathbf{f}_\mathrm{E}, \qquad \mathbf{f}_\mathrm{E} = \mathbf{f}/\gamma. \tag{2.177}$$

For obvious reasons $\mathbf{f}_\mathrm{E}$ is given the name *Einstein force*. It is the relativistically correct three force describing the momentum change in an arbitrary inertial reference frame. In a reference frame S' comoving with the point particle, $\mathrm{d}\mathbf{p}/\mathrm{d}t$ reduces to $m\,\mathrm{d}\mathbf{v}/\mathrm{d}t$, and $\mathbf{f}_\mathrm{E}$ becomes equal to $\mathbf{f}$, which we give the index N, $\mathbf{f} = \mathbf{f}_\mathrm{N}$, and call $\mathbf{f}_\mathrm{N}$ a Newton force,

$$m\frac{\mathrm{d}\mathbf{v}}{\mathrm{d}t} = \mathbf{f}_\mathrm{N}. \tag{2.178}$$

In general it is correct only in the comoving inertial frame.

Owing to $\mathrm{d}p^4/\mathrm{d}t = \mathrm{d}p^4/\mathrm{d}\tau = 0$ in the comoving frame S', the Minkowski force is $F = (\mathbf{f}_\mathrm{N}, 0)$, and hence in an arbitrary reference system $S(\mathbf{w})$ the four force $F = (\mathbf{f}, f^4)$ is given through

$$\mathbf{f} = \mathbf{f}_\mathrm{N} + \frac{\gamma - 1}{w^2}(\mathbf{w}\mathbf{f}_\mathrm{N})\mathbf{w}, \qquad f^4 = -\gamma\frac{\mathbf{w}}{c}\mathbf{f}_\mathrm{N} = -\frac{\mathbf{w}}{c}\mathbf{f}. \tag{2.179}$$

Note that if the particle has a velocity $\mathbf{v}$ in S, owing to $\mathbf{v} = -\mathbf{w}$ one has $f^4 = \gamma\mathbf{v}\mathbf{f}_\mathrm{N}/c$. This relation follows also from $F \perp P$. In fact, $\mathrm{d}P^2/\mathrm{d}\tau = 2PF = 2\gamma m(\mathbf{v}\mathbf{f} - cf^4) = 0$. With the help of (2.179) the Einstein force on a point charge q is found. By definition of the electric field, in the rest frame S' of the charge $\mathbf{f}_\mathrm{N} = q\mathbf{E}' = q[\mathbf{E}_\parallel + \gamma(\mathbf{E}_\perp + \mathbf{v} \times \mathbf{B})]$ holds [see (2.165). In a frame S moving at $-\mathbf{v}$ relative to S' the transformation law (2.179) requires $F = q[\mathbf{E}_\parallel + \gamma(\mathbf{E}_\perp + \mathbf{v} \times \mathbf{B}) + (\gamma - 1)\mathbf{E}_\parallel, \gamma\mathbf{v}\mathbf{E}]$. Hence, $\mathbf{f}_\mathrm{E} = q(\mathbf{E} + \mathbf{v} \times \mathbf{B})$, and the correct relativistic equation of motion in three-space reads

$$\frac{\mathrm{d}}{\mathrm{d}t}(\gamma m\mathbf{v}) = q(\mathbf{E} + \mathbf{v} \times \mathbf{B}). \qquad \frac{\mathrm{d}}{\mathrm{d}t}m\gamma c^2 = q\mathbf{v}\mathbf{E}. \tag{2.180}$$

Sometimes the difference between the orbits from (2.177) or (2.180) and the corresponding nonrelativistic equation of motion is attributed to the "relativistic mass increase". Except bound systems, like atoms, nuclei, the gyrating or oscillating electron, that can be considered as separate entities this is not a good concept because the "mass increase" depends on the particular motion (e.g. "longitudinal" mass $m_{\parallel}$ differing from "transverse" mass $m_{\perp}$) and may obscure the real origin of (2.177) and (2.180) that is the Minkowski metric of space time.

Momentum. It shows that the kinetic energy of the particle is $E_{\mathrm{kin}} = \gamma mc^2 - mc^2 = E - E_0$ because $\int_0^t \mathbf{f}_{\mathrm{E}}\, \mathrm{d}\mathbf{x}$ is the work done by the Einstein force. Hence

$$E = E_0 + E_{\mathrm{kin}} = \gamma mc^2 \qquad \Leftrightarrow \qquad E^2 = m^2c^4 + p^2c^2. \tag{2.181}$$

The second relation is nothing but $P^2 = -m^2c^2 = -E_0^2/c^2$.

The Lagrangian of a particle of rest mass m moving in a time-independent potential $V = V(\mathbf{x})$ is

$$L(\mathbf{x}, \mathbf{v}) = T - V = -\frac{m}{\gamma}c^2 - V \tag{2.182}$$

because it obeys the equation of motion

$$\frac{\mathrm{d}}{\mathrm{d}t}\frac{\partial L}{\partial \mathbf{v}} - \frac{\partial L}{\partial \mathbf{x}} = \frac{\mathrm{d}}{\mathrm{d}t}(\gamma m\mathbf{v}) + \nabla V = 0,$$

if the force is correctly given by $-\nabla V$. Keeping in mind that it should be an Einstein force $\mathbf{f}_{\mathrm{E}}$, the weakness of a non covariant Lagrangian (2.182), i.e., a Lagrangian not expressed in four quantities, appears. For instance, when going from S to $S'(\mathbf{v})$, the potential may assume the wrong time dependence. If (2.182) is accepted the Hamiltonian is given by

$$H(\mathbf{p}, \mathbf{x}) = \frac{\partial L}{\partial \mathbf{v}}\mathbf{v} - L = (m^2c^4 + p^2c^2)^{1/2} + V(\mathbf{x}), \tag{2.183}$$

and the canonical equations of motion

$$\frac{\mathrm{d}}{\mathrm{d}t}\mathbf{p} = -\frac{\partial H}{\partial \mathbf{x}}, \qquad \frac{\mathrm{d}\mathbf{x}}{\mathrm{d}t} = \frac{\partial H}{\partial \mathbf{p}} \tag{2.184}$$

yield the equation of motion above and the connection between $\mathbf{v}$ and $\mathbf{p}$ for a point particle, respectively.

To find now the relativistic Lagrangian of a charge q in the electromagnetic field we compare (2.37) with (2.182) and suggest

$$L(\mathbf{x}, \mathbf{v}, t) = -\frac{mc^2}{\gamma} + q\mathbf{v}\,\mathbf{A} - q\Phi(\mathbf{x}, t). \tag{2.185}$$

It generates (2.180). In fact, (2.35) and (2.38) yield

$$\frac{\mathrm{d}}{\mathrm{d}t}\frac{\partial L}{\partial \mathbf{v}} - \frac{\partial L}{\partial \mathbf{x}} = \frac{\mathrm{d}}{\mathrm{d}t}\gamma m\mathbf{v} + q\partial_t\mathbf{A} + q(\mathbf{v}\,\nabla)\mathbf{A} - q\nabla(\mathbf{v}\,\mathbf{A}) + q\nabla\Phi$$

$$= q(\partial_t\mathbf{A} - \mathbf{v}\times\nabla\times\mathbf{A} + \nabla\Phi) = \frac{\mathrm{d}}{\mathrm{d}t}\gamma m\mathbf{v} - q(\mathbf{E} + \mathbf{v}\times\mathbf{B}) = 0.$$

The Hamiltonian results from (2.183) and (2.58) as follows,

$$H(\mathbf{p},\mathbf{q},t) = \gamma m\mathbf{v}^2 + \frac{mc^2}{\gamma} + q\Phi = \gamma mc^2 + q\Phi = [m^2c^4 + c^2(\mathbf{p} - q\mathbf{A})^2]^{1/2} + q\Phi(\mathbf{x},t) \tag{2.186}$$

It generates (2.180) through the canonical equations (2.184). The difference $\mathbf{p} - q\mathbf{A}$ is the kinetic momentum $\gamma m\mathbf{v}$. In the relativistic regime it is convenient to normalize the momenta to mc. Then the electromagnetic part of the Hamiltonian (2.186) reads

$$H(\boldsymbol{\pi},\mathbf{q},t) = mc^2[1 + (\boldsymbol{\pi} - \mathbf{a})^2]^{1/2}; \quad \boldsymbol{\pi} = \frac{\mathbf{p}}{mc}, \quad \mathbf{a} = \frac{q\mathbf{A}}{mc}, \quad \boldsymbol{\pi} - \mathbf{a} = \gamma\frac{\mathbf{v}}{c}.$$

The connection $\mathbf{a}$ with the laser intensity I in linear (circular) polarization is

$$I\,[\mathrm{Wcm}^{-2}] = 1.37\,(2.7)\times 10^{18}a^2; \qquad a = |\mathbf{a}|. \tag{2.187}$$

Addition theorem. Sometimes formulas relating the velocity $\mathbf{u}$ and the four momentum P measured in S to the corresponding quantities $\mathbf{u}'$ and P' measured in $S'(\mathbf{v})$ are useful. According to (2.155) holds for $\mathrm{d}\mathbf{x} = \mathbf{u}\,\mathrm{d}t$ and $\mathrm{d}\mathbf{x}' = \mathbf{u}'\,\mathrm{d}t'$

$$\mathbf{u}\,\mathrm{d}t = \left(\mathbf{u}' + \frac{\gamma - 1}{v^2}(\mathbf{v}\mathbf{u}')\mathbf{v} + \gamma\mathbf{v}\right)\mathrm{d}t', \quad \mathrm{d}t = \gamma\left(1 + \frac{\mathbf{v}\mathbf{u}'}{c^2}\right)\mathrm{d}t', \quad \gamma = \gamma(v).$$

Elimination of $\mathrm{d}t'$ leads to the so-called relativistic velocity addition theorem $\mathbf{v}$ and $\mathbf{u}'$,

$$\mathbf{u} = \frac{\mathbf{u}' + \mathbf{v}[\gamma(1 + \mathbf{u}'\mathbf{v}/v^2) - \mathbf{u}'\mathbf{v}/v^2]}{\gamma(1 + \mathbf{u}'\mathbf{v}/c^2)}. \tag{2.188}$$

If $u' < c$, $v < c$ is fulfilled, $u < c$ follows because (2.188) implies $(c^2 - u^2)/c^2 = (c^2 - u'^2)(c^2 - v^2)/(c^2 + \mathbf{u}'\mathbf{v})$. The inverse transformation for $\mathbf{u}'$ is obtained from (2.188) by interchanging $\mathbf{u}$ and $\mathbf{u}'$ and substituting $\mathbf{v}$ by $-\mathbf{v}$. The analogous formula for P and P' is accordingly

$$P = (\mathbf{p},\gamma mc) = \left(\frac{\mathbf{p}'/\gamma' + \mathbf{v}[\gamma(m + \mathbf{p}'\mathbf{v}/(\gamma'v^2)) - \mathbf{p}'\mathbf{v}/(\gamma'v^2)]}{1 + \mathbf{p}'\mathbf{v}/(\gamma'mc^2)}, \gamma mc\right) \tag{2.189}$$

with $\gamma' = \gamma(u')$.

Time like and space like four vectors. A four quantity Y is *time like* if $\mathbf{y}^2 - y^4 y_4 < 0$. It follows that a time like vector can be reduced to $Y = (0, 0, 0, y^4)$. The four velocity V and the four momentum P are time like four vectors. If $Y^2 > 0$, Y is a *space like* vector. In this case a Lorentz transformation exists which transforms Y into $Y = (y^1, y^2, y^3, 0)$. The four acceleration A and the Minkowski force F are space like vectors. There is no upper limit of force and acceleration in three dimensions. The position vector can be both.

2.2.1.2 Particle Motion in Intense Laser Wave

In order to explore the limits of particle acceleration by laser pulses the idealized situation of a plane, linearly polarized electromagnetic wave propagating in x-direction and acting on a point charge of rest mass m and charge q is considered. Let the wave be described by the vector potential $\mathbf{A}$ and $\Phi = 0$ (Coulomb gauge),

$$\mathbf{A} = \mathbf{e}_\perp A(\xi), \qquad \xi = x - ct, \qquad \mathbf{e}_\perp \mathbf{e}_x = 0, \qquad \mathbf{e}_\perp^2 = 1. \tag{2.190}$$

The equations of motion with $\mathbf{e}_\perp = \mathbf{e}_y$ are

$$\dot{p}_y = -\frac{\partial H}{\partial y} = 0, \qquad \dot{p}_x = -\frac{\partial H}{\partial x} = \frac{q}{m\gamma}(p_y - qA)\frac{\mathrm{d}A}{\mathrm{d}\xi}. \tag{2.191}$$

The first equation implies the conservation of the transverse canonical momentum

$$p_y = \gamma m v_y + qA = p_i = \text{const.} \tag{2.192}$$

From (2.191) and (2.192) follows

$$v_y(\xi) = \frac{p_i}{m\gamma} - \frac{q}{m\gamma}A(\xi), \qquad \frac{\mathrm{d}}{\mathrm{d}t}\gamma m v_x = \frac{q}{m\gamma}(p_i - qA)\frac{\mathrm{d}A}{\mathrm{d}\xi}. \tag{2.193}$$

By $\mathrm{d}/\mathrm{d}t = (\mathrm{d}\xi/\mathrm{d}t)\mathrm{d}/\mathrm{d}\xi = (v_x - c)\mathrm{d}/\mathrm{d}\xi$ the first of the two equations transforms into

$$\frac{\mathrm{d}y}{\mathrm{d}\xi} = \frac{p_i}{m\gamma(v_x - c)} - \frac{q}{m\gamma(v_x - c)}A(\xi) \tag{2.194}$$

Energy and momentum conservation require

$$\frac{\mathrm{d}}{\mathrm{d}t}m\gamma c^2 = qE_y v_y, \qquad \frac{\mathrm{d}}{\mathrm{d}t}m\gamma v_x = qv_y B_z$$

so that

$$\frac{\mathrm{d}}{\mathrm{d}t}m\gamma(v_x - c) = qv_y\left(\frac{1}{c}\frac{\partial A}{\partial t} + \frac{\partial A}{\partial x}\right) = 0 \Rightarrow m\gamma(v_x - c) = \text{const.} \tag{2.195}$$

Starting with $v_x = 0$, $x = 0$, and $y = 0$ at $t = 0$ yields

$$y(\xi) = \frac{q}{mc\gamma_i}\int_0^\xi A(\xi')\,\mathrm{d}\xi' - \frac{p_i}{mc\gamma_i}\xi, \qquad \gamma_i = \gamma(v_i), \qquad v_i = v_y(t = 0). \tag{2.196}$$

By integration under the initial conditions $v_x(t = 0) = 0, v_i = v_y(t = 0), \gamma_i = \gamma(v_i)$ the second equation of (2.193) becomes

$$\gamma m v_x = \frac{q^2}{2\gamma_i mc}[A^2(\xi) - A^2(\xi = 0)] - \frac{q}{\gamma_i mc}p_i[A(\xi) - A(\xi = 0)]. \tag{2.197}$$

Integrating once more $v_x = \mathrm{d}x/\mathrm{d}t$ of this expression in t under the initial conditions above one arrives at

$$x(\xi) = -\left(\frac{q}{mc\gamma_i}\right)^2 \frac{1}{2}\left[\int_0^\xi A^2(\xi')\,\mathrm{d}\xi' - A^2(\xi = 0)\xi\right] + \frac{q}{(mc\gamma_i)^2}p_i\left[\int_0^\xi A(\xi')\,\mathrm{d}\xi' - A(\xi = 0)\xi\right]. \tag{2.198}$$

The energy of the point charge follows from $E^2 = m^2c^4 + c^2m^2\gamma^2(v_x^2 + v_y^2)$. By substituting the formulas above for v_x, v_y this results in a lengthy expression. It shrinks to a compact size for the initial conditions $v_x(0) = v_y(0) = A(0) = 0$,

$$E = mc^2\left[1 + \frac{1}{2}\left(\frac{qA}{mc}\right)^2\right]. \tag{2.199}$$

Finally, $t(\xi)$ is recovered from $\xi = x - ct$,

$$t = \frac{1}{c}(x - \xi) = -\frac{\xi}{c} + \frac{x(\xi)}{c}, \tag{2.200}$$

with $x(\xi)$ according to (2.198). The trajectory of the particle is completely determined by (2.196) and (2.198) as a function of ξ. Owing to $x < ct$ follows $\xi < 0$ for increasing t and hence, a positive or negative charge initially at rest at $x = \xi = 0$ is pushed into the direction of the propagating wave, i.e., in positive x-direction here.

For an arbitrary plane wave holds:

(i) After the pulse has left the particle behind, v_x, v_y, and the energy E of the point charge return to their initial values before the arrival of the pulse. In particular, no net energy gain (acceleration) is possible in an interaction over a whole number of cycles. This result for a plane wave in vacuum is a special case of the Lawson–Woodward theorem which states that the net energy gain of a relativistic electron interacting with a continuous electromagnetic field in vacuum is zero [19]. In the photon picture this is a familiar statement: A free charge cannot absorb or emit real photons. The only net effect is a shift in position by Δx according to

(2.198) or (2.204). If the action of the particle back onto the wave is also taken into account the finite shift produces weakening of the pulse in its ascending part and strengthening of its decreasing flank.

(ii) During the laser pulse, maximum acceleration of a charge initially at rest is

$$\Delta E = \frac{1}{2}\hat{a}^2 mc^2. \tag{2.201}$$

Its lateral angular spread in a pulse of $A(\xi = 0) = 0$ is

$$\tan\alpha = \left|\frac{v_y}{v_x}\right| = \frac{2}{\hat{a}}. \tag{2.202}$$

The monochromatic plane wave with slowly varying amplitude $\hat{\mathbf{A}}(\varphi)$

$$\mathbf{A}(\mathbf{x}, t) = \mathbf{e}_y \hat{A}(\varphi)\sin\varphi, \qquad \varphi = kx - \omega t, \quad \omega = ck \tag{2.203}$$

is of particular interest. With the charge initially ($t = 0$) at rest at $\mathbf{x} = 0$ follows

$$\begin{aligned}
\gamma\frac{v_x}{c} &= \frac{1}{2}\hat{a}^2\sin^2\varphi, \quad \gamma\frac{v_y}{c} = -\hat{a}\sin\varphi, \quad \frac{v_z}{c} = 0, \quad \hat{a} = \frac{q\hat{A}}{mc},\\
x &= -\frac{1}{4k}\hat{a}^2\left(\varphi - \frac{1}{2}\sin(2\varphi)\right), \quad y = -\frac{1}{k}\hat{a}\cos\varphi, \quad z = 0,\\
t &= \frac{1}{\omega}(kx - \varphi), \quad E = mc^2\left(1 + \frac{1}{2}\hat{a}^2\sin^2\varphi\right)
\end{aligned} \tag{2.204}$$

In an intense laser field ($\hat{a} \gg 1$) a free charge is mainly accelerated in forward direction under the angle $\alpha = 2/\hat{a}$. It should be noted that the Lawson–Woodward theorem can be violated in the presence of additional **E** and **B** fields or in fields of finite extension.

The reference frame in which the oscillation center is at rest deserves special attention. The vector potential in this frame may be assumed as

$$\mathbf{A}(\varphi) = \mathbf{e}_y\hat{A}(\varphi)\cos\varphi, \qquad \varphi = k(x - ct).$$

From (2.193) and (2.196)–(2.198) follows

$$\begin{aligned}
\frac{v_y}{c} &= -\frac{\hat{a}}{\gamma(\varphi)}\cos\varphi, \qquad y = \frac{1}{k}\frac{\hat{a}}{\gamma_i}\sin\varphi,\\
\frac{v_x}{c} &= \frac{\hat{a}^2}{4\gamma_i\gamma(\varphi)}\cos(2\varphi), \qquad x = -\frac{\hat{a}^2}{8\gamma_i^2}\frac{1}{k}\sin(2\varphi).
\end{aligned} \tag{2.205}$$

Equation (2.205) describes a horizontal figure eight shaped trajectory in the plane (y, x), see Fig. 2.23. The constant γ_i is conveniently determined with the help of

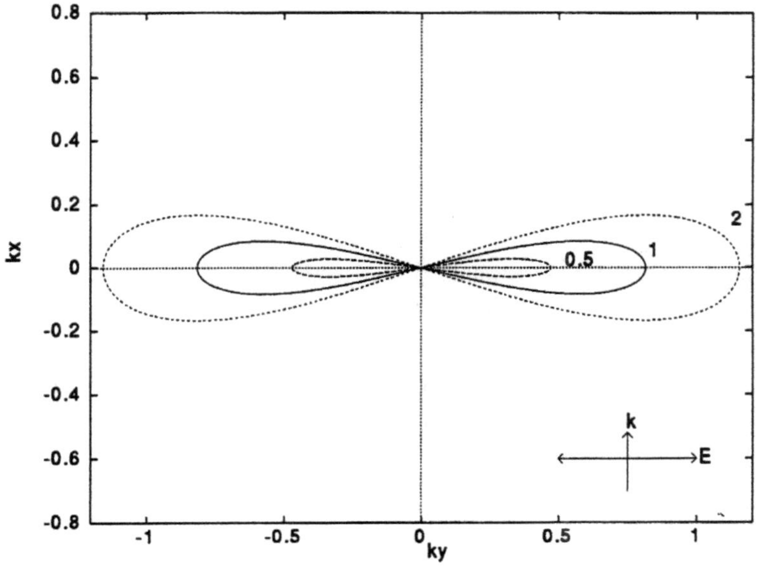

Fig. 2.23 Electron motion in the linearly polarized monochromatic laser wave of wave number k as a function of the field strength $\hat{a} = 0.5,\ 1,\ 2;\ \hat{a} = e\hat{A}/m_e c,\ \hat{A} = \hat{E}/\omega$

(2.195) for $\varphi = \pi/4$ since $v_x(\pi/4) = 0$ and $v_i = v_y(\pi/4) = -2^{-1/2}\hat{a}c/\gamma_i$. Hence,

$$\gamma_i = \left(1 - \frac{v_i^2}{c^2}\right)^{-1/2} = \left(1 + \frac{\hat{a}^2}{2}\right)^{1/2}. \tag{2.206}$$

The ratio of the excursion amplitudes is given by $\hat{x}/\hat{y} = \hat{a}/(8\gamma_i)$. For $I \to 0$ it behaves proportional to $I^{1/2}$ while for $I \to \infty$ it saturates at $\hat{x}/\hat{y} = 2^{-1/2}/4 = 0.1768$, i.e., the figure eight shape never becomes fat. With a monochromatic wave in the oscillation center frame the velocities $v_{y\,\max}$ and $v_{x\,\max}$ are reached at, e.g., $\varphi = 0$ and $\varphi = \pi/2$, respectively. For $\hat{a} \to \infty$, $v_{y\,\max}$ approaches c, as expected, whereas for $v_{x\,\max}$ from (2.195) and (2.205)

$$\frac{v_{x\,\max}}{c} = \frac{\hat{a}^2}{4(1 + 3\hat{a}^2/8)}, \tag{2.207}$$

results with the high intensity limit $v_{x\,\max} \to 2c/3$.

The mean oscillatory energy W in the reference frame $S' = S_{\mathrm{oc}}$ of the oscillation center may be regarded as an internal energy, $W = mc^2(\overline{\gamma} - 1)$. From the considerations on how to define an invariant center of mass it becomes clear that averaging has to be done over the invariant phase $\varphi = k(x - ct)$ which differs from averaging over time. By observing

$$\gamma\,\mathrm{d}\varphi = \gamma\frac{\mathrm{d}\varphi}{\mathrm{d}t}\mathrm{d}t = \gamma k(v_x - c)\,\mathrm{d}t = -\gamma_i\, d(\omega t) \tag{2.208}$$

the invariant oscillation energy results as

$$W = mc^2\left[\left(1+\frac{\hat{a}^2}{2}\right)^{1/2} - 1\right] = (\gamma_i - 1)mc^2. \tag{2.209}$$

For a circularly polarized laser field $\hat{a}^2/2$ has to be replaced by $\hat{a}^2$. Note, here $\hat{a}$ refers to S_{oc}, in (2.199) it is the normalized amplitude in the lab frame.

If the light pulse is not monochromatic but of the form (2.190) the averaging is meaningful if there exists a tangential reference frame in which the particle orbit is almost closed. If in this reference frame S_{oc} the averaged particle mass is $\gamma_i m$, its four momentum in the lab frame S is $P = \gamma_i m V$. Owing to $c^2P^2 = -\gamma_i^2 m^2 c^4$ the energy W averaged in ξ over one period is correctly given in the lab frame by

$$W = mc^2\left[\left(1+\overline{a^2(\xi)}\right)^{1/2} - 1\right] = \gamma(v_{\mathrm{L}})(\gamma_i - 1)mc^2. \tag{2.210}$$

Integrating (2.176) from $\tau = 0$ to $\tau > 0$ with the point particle initially at rest yields with the help of (2.177)

$$P(t) - P(0) = \left[\int_0^\tau \mathbf{f}\,\mathrm{d}\tau, \frac{1}{c}\int_0^\tau \mathbf{f}\,\mathbf{v}\,\mathrm{d}\tau\right] = \left[\int_0^t \mathbf{f}_{\mathrm{E}}\,\mathrm{d}t, \frac{1}{c}\int_0^t \mathbf{f}_{\mathrm{E}}\,\mathrm{d}\mathbf{x}\right] = [\mathbf{p}, (\gamma_{\mathrm{L}} - 1)mc]$$

with $\gamma_{\mathrm{L}} = \gamma(v_{\mathrm{L}} = v_{\mathrm{oc}})$ the Lorentz factor of the oscillation center velocity in the lab frame S.

To get a feeling for how efficient particle acceleration by laser in vacuum is, the energy and other relevant parameters for free electrons and protons in a Ti:Sa laser beam of intensities between $I = 10^{18}$ and $10^{24}\,\mathrm{Wcm}^{-2}$ are shown in Table 2.1. The front of the laser pulse is assumed to pass $x = 0$ at $t = 0$ and then to increase adiabatically to the steady state intensities given in the Table. Electron (index e) and proton (index p) are at rest at $t = 0$. Maximum energy E_{free} is gained during a quarter cycle ($\Delta\varphi = \pi/2$). If instead a pulse of the form $\hat{A}_y \cos\varphi$ is used with an amplitude extremely rapidly rising from $\hat{A}_y(t = 0) = 0$ to its constant full value, according to (2.199) follows the same maximum energy $mc^2\hat{a}^2/2$ if now acceleration occurs over half a period $\Delta\varphi = \pi$, i.e., $T' = \pi/\omega'$ in the co-moving frame. In the lab frame this "half period" T can be very long. For electrons the amplitude ratio $\hat{x}/\hat{y}$ of the quiver motion saturates already at $I = 10^{20}\,\mathrm{Wcm}^{-2}$. From $I = 10^{22}\,\mathrm{Wcm}^{-2}$ on proton acceleration E_{free} and quiver energy W are no longer negligible. It is further seen from Table 2.1 that in direct acceleration by $I = 10^{22}\,\mathrm{Wcm}^{-2}$ the electrons become "heavier" than the protons (1.2 GeV vs. 0.94 GeV). Comparing the two Lorentz factors, γ_{free} vs. γ_{oc}, it becomes evident that, in order to get maximum energy gain from the laser field, free particles gain higher energies than particles whose oscillation

Table 2.1 Maximum achievable acceleration E_{free} of electrons and protons in a fourth cycle of a plane Ti:Sa laser pulse (row 5) and corresponding quiver energy W (row 7) in the oscillation center frame. I laser intensity, $\hat{a}$ normalized vector potential amplitude, v_y/v_x ratio of velocities in field and pulse direction, γ_{free}, γ_{oc} Lorentz factors, $\Delta x/\lambda$ acceleration distance during a fourth cycle $\Delta\varphi = \pi/2$ in units of Ti:Sa wavelength ($\lambda = 800\,\text{nm}$) during a fourth cycle, $\hat{x}/\hat{y}$ ratio of oscillation amplitudes. First column for a given intensity yields the values for electrons, second column (where listed) the values for protons

I [Wcm^{-2}]	10^{18}	10^{20}	10^{22}		10^{24}	
$\hat{a}_e, \hat{a}_p$	0.69	6.87	68.67	3.7×10^{-2}	686.7	0.37
$\lvert v_y/v_x \rvert$	2.91	0.3	0.03	53.5	2.9×10^{-3}	5.35
$\gamma_{\text{free}} - 1$	0.24	23.6	2.36×10^3	7×10^{-4}	2.36×10^5	0.07
E_{free}	120.5 keV	12.0 MeV	1.2 GeV	656 keV	120 GeV	65.6 MeV
$\Delta x/\lambda$	0.03	2.95	295	8.7×10^{-5}	2.95×10^4	8.7×10^{-3}
$\gamma_{\text{oc}} - 1$	0.16	3.96	47.6	3.5×10^{-4}	484.6	0.034
W	81.8 keV	2.02 MeV	24.3 MeV	292 keV	247.6 MeV	32.3 MeV
$\hat{x}/\hat{y}$	0.0740	0.1730	0.1766	4.6×10^{-3}	0.1767	0.045

center is held fixed by reaction forces as is the case in overdense targets. For this reason highest electron energies are observed in tenuous plasmas. Although formulas (2.197) and (2.199) are valid for plane traveling waves only they represent a useful generalization of the ponderomotive concept. When the laser pulse starts sweeping over a slow particle it is ponderomotively accelerated into the direction of the light pulse. With decreasing frequency in the particle frame Φ_p increases faster than the intensity of the pulse. Finally, at sufficiently high laser field strength the particle is trapped in the wave and, from then on it gains energy by direct acceleration in the longitudinal field component $\mathbf{E}' = \gamma \mathbf{v}_y \times \mathbf{B}_z$ thereby maintaining its velocity $\mathbf{v}$ perpendicular to the total Lorentz force.

2.2.1.3 Relativistic Ponderomotive Force

Here, the ponderomotive force $\mathbf{f}_p$ of a point charge q is derived in the transverse electromagnetic field. Particular attention is devoted to the travelling monochromatic electromagnetic wave. With the ever growing laser intensities a real need arises for a covariant expression of $\mathbf{f}_p$. In addition to the practical requirement there is also the basic question on the dependence of the ponderomotive potential on the oscillation center velocity $\mathbf{v}_0$. The quantity that is conserved is the adiabatic invariant (2.104). When aiming at the most general expression of $\mathbf{f}_p$ one must start from there. Along this way also the connection of the energy considerations in (2.111) and Fig. 2.13 with (2.104) will become clear.

Assume that the charge is subject to a static potential $V = q\Phi(\mathbf{x})$ and the time-varying electromagnetic field $\mathbf{A}(\mathbf{x}, t)$. The Hamiltonian in this reference system is according to (2.186)

$$H(\mathbf{p},\mathbf{q},t) = [m^2c^4 + c^2(\mathbf{p} - q\mathbf{A})^2]^{1/2} + q\Phi(\mathbf{x}). \tag{2.211}$$

It reproduces correctly the motion of the particle but Φ is not in a covariant form; when changing to a different reference system Φ becomes time-dependent and follows a wave equation of type (5.10) from Chap. 5. The electromagnetic part with the *kinetic momentum* $\mathbf{p} - q\mathbf{A} = \gamma m\mathbf{v}$ is covariant. For $\mathbf{f}_p$ to make sense the particle must describe a quasiperiodic orbit with a clearly recognizable oscillation/gyration center. In a reference system change this point must transform according to Lorentz. It implies that cycle averaging $\oint \mathbf{p}\mathrm{d}\mathbf{q}$ has to be done in an invariant way, in contrast to merely time averaging in Sect. 2.1.6.1. The quasiperiodic motion makes the canonical transformation to action angle variables S and η possible. In the important case of a single monochromatic wave $\eta = \int(\mathbf{k}\mathrm{d}\mathbf{x} - \omega\mathrm{d}t)$ fulfils the requirement. In Fig. 2.18 the projection of the orbit $\{\mathbf{p}(\eta), \mathbf{q}(\eta)\}$ is an 8 shape like loop. The adiabatic invariant obtained from (2.211) is

$$\oint_{\text{cycle}} [\mathbf{p}\mathrm{d}\mathbf{q} - H(t)\mathrm{d}t] = \oint_{\eta(t)}^{\eta(t)+2\pi} \{[\gamma(\eta)\mathbf{v}(\eta) + q\mathbf{A}(\eta)]\mathbf{v}(\eta) - H(\eta)\}\frac{\mathrm{d}t}{\mathrm{d}\eta}\mathrm{d}\eta - \overline{V}(\mathbf{x}_0)\,T. \tag{2.212}$$

Thereby $\overline{V} = (1/T)\int V(x')dx'/v(x')$ along the wave and T is the duration of a closed cycle in the system in which $V = V(\mathbf{x})$ is at rest, $\mathbf{x}_0$ is the oscillation center, η is normalized to 2π. Analytic evaluation of this invariant is very limited. There exist conditions in which the static potential does not sensitively influence the orbit generated by the electromagnetic Hamiltonian component [20]. Unfortunately, the problem, even under this limitation, is not solvable in desired generality. A detailed study of the validity of the ponderomotive concept of free electrons in finite laser pulses is presented in [21]; for Gaussian beams and circular polarization see [22]. With a view to the fundamental question of reference system changes we limit ourselves here to the propagating wave type (b) of Sect. 2.1.6.1 and weak static potential V. In a first step $V = 0$ is set. In the system of the oscillation center at rest, $\mathbf{v}_0 = 0$, with the help of (2.209), (2.212) assumes

$$\oint_{\text{cycle}} [\mathbf{p}\mathrm{d}\mathbf{q} - H(t)\mathrm{d}t] = mc^2\left(1 + \frac{\hat{a}^2}{2}\right)^{1/2} T = \gamma_{\mathrm{os}}mc^2T = \mathrm{const}_{\mathrm{a}}. \tag{2.213}$$

The oscillation center coordinate $\mathbf{x}_0$ can be interpreted as the covariant center of mass, here center of orbit. It may be instructive to take a look at Sect. 3.2.1.1 to understand the importance of Lorentz invariant averaging. If the effective mass $m_{\mathrm{eff}} = \gamma_{\mathrm{os}}m$ is introduced it follows from (3.111) that in an arbitrary inertial system with m_{eff} moving at $\mathbf{v}_0$ the following quantities hold

$$\mathbf{p}_0 = \gamma_0 m_{\mathrm{eff}}\mathbf{v}_0, \quad H_0(\mathbf{x}_0, \mathbf{v}_0, t) = \gamma_0 m_{\mathrm{eff}}c^2; \quad \gamma_0 = \left(1 - \frac{\mathbf{v}_0^2}{c^2}\right)^{1/2}. \tag{2.214}$$

Correspondingly the associated adiabatic invariant and the ponderomotive force read

$$\oint_{\text{cycle}} [\mathbf{p}\mathrm{d}\mathbf{q} - H(t)\mathrm{d}t] = \gamma_0 m_{\text{eff}} c^2 T = \text{const}_\text{a}, \quad \mathbf{f}_p = -\frac{\partial H_0}{\partial \mathbf{x}_0} = -\gamma_0 c^2 \nabla m_{\text{eff}}. \tag{2.215}$$

The invariant can be cast into the equivalent form

$$\frac{1}{T}\oint_{\text{cycle}} [\mathbf{p}\mathrm{d}\mathbf{q} - H(t)\mathrm{d}t] = \gamma_0 m_{\text{eff}} c^2 = L_0. \tag{2.216}$$

The cle averaged Lagrangian L_0 is defined in (2.220). The period T depends on the reference system, $T_{\mathbf{v}_0} = \gamma_0 T_{(\mathbf{v}_0=0)}$.

Finally, there is still the question pending of whether the ponderomotive potential Φ_p is a function of the oscillation velocity $\mathbf{v}_0$ in addition to the space dependence $\mathbf{x}_0$. From the energy principle (2.109) it could not be answered. A partial, approximate, answer however could be given for Φ_p of an electromagnetic wave. The definite answer can be given now on the basis of the relativistic Hamiltonian (2.211) with the static potential $\Phi(\mathbf{x}) = 0$. In the oscillation center system Φ_p is $m_{\text{eff}}c^2$. Equation (2.214) shows that it transforms like an energy. The same happens with the ponderomotive force $\mathbf{f}_p$; it transforms like the space component of a four vector because $\nabla_{\text{lab}} = \gamma\nabla_{\text{os}}$. Hence, Φ_p and $\mathbf{f}_p$ are *independent* of $\mathbf{v}_0$. There is an essential difference in the transformation behaviour of the two ponderomotive quantities between the longitudinal and the transverse wave.

The Ponderomotive Potential and Force from a Lagrangian

It may be of interest to some readers to derive the ponderomotive action of a monochromatic electromagnetic wave in vacuum of type (a) and (b) from a Lagrangian [6],

$$L(\mathbf{x}, \mathbf{v}, t) = -\frac{mc^2}{\gamma}; \quad \dot{\mathbf{x}} = \mathbf{v}, \quad \gamma = (1 - v^2/c^2)^{-1/2}. \tag{2.217}$$

Again the existence of an oscillation/gyration center of motion is clearly recognizable to switch to action angle variables S and $\eta = \eta(\mathbf{x}, t)$. The motion is governed by Hamilton's principle

$$\delta S = \delta \int_{\eta_1}^{\eta_2} L(\mathbf{x}(\eta),\ \mathbf{v}(\eta),\ t(\eta))\frac{\mathrm{d}t}{\mathrm{d}\eta}\mathrm{d}\eta = \delta \int_{\eta_1}^{\eta_2} \mathcal{L}_0(\eta)\mathrm{d}\eta = 0. \tag{2.218}$$

From the Lorentz invariance of S and η it follows that the Lagrangian $\mathcal{L}(\eta) = L(\mathrm{d}\eta/\mathrm{d}t)^{-1}$ is invariant with respect to a change of the inertial reference system. Assuming that η is normalized to 2π for one full cycle or period of motion T, the cycle-averaged Lagrangian $\mathcal{L}_0$,

$$\mathcal{L}_0(\eta) = \frac{1}{2\pi}\int_{\eta}^{\eta+2\pi} \mathcal{L}(\eta')\mathrm{d}\eta'. \tag{2.219}$$

is defined. It depends only on the secular (i.e. oscillation center) coordinates $\mathbf{x}_0$ and $\mathbf{v}_0$ through η. The motion of the oscillation center is governed now by the Lagrange equation

$$\frac{\mathrm{d}}{\mathrm{d}t}\frac{\partial L_0}{\partial \mathbf{v}_0} - \frac{\partial L_0}{\partial \mathbf{x}_0} = 0, \quad L_0 = \mathcal{L}_0\frac{\mathrm{d}\eta}{\mathrm{d}t}. \tag{2.220}$$

If the effective mass $m_{\mathrm{eff}} = \mathcal{L}_0\gamma_0(\mathrm{d}\eta/\mathrm{d}t)/c^2$ is introduced, $L_0 = \mathcal{L}_0\mathrm{d}\eta/\mathrm{d}t$ shows that, in an arbitrary inertial frame in which the oscillation center moves at speed $\mathbf{v}_0$, L_0 determines the motion of a free particle with space and time dependent mass m_{eff},

$$L_0(\mathbf{x}_0, \mathbf{v}_0, t) = -\frac{m_{\mathrm{eff}}c^2}{\gamma_0}, \gamma_0 = \left(1 - \frac{v_0^2}{c^2}\right)^{-1/2}, \quad \mathbf{p}_0 = \gamma_0 m_{\mathrm{eff}}\mathbf{v}_0; \tag{2.221}$$

The mass m_{eff} is identical with the effective mass in (2.214) since $L = \int \mathbf{p}\mathrm{d}\mathbf{q} - H$. In the oscillation center system, i.e., in the inertial frame in which at the instant t $\mathbf{v}_0(t) = 0$ holds, the ponderomotive force follows from (2.220) and (2.221):

$$\mathbf{f}_p^{\mathrm{N}} \equiv \frac{\mathrm{d}\mathbf{p}_0}{\mathrm{d}t} = \frac{\partial L_0}{\partial \mathbf{x}_0} = -c^2\nabla m_{\mathrm{eff}}. \tag{2.222}$$

For the meaning of $\mathbf{f}_p^{\mathrm{N}}$ and its transformation to another reference system see (2.178).

Within the Lagrangian approach it remains to show that (2.220) is equivalent to the Hamiltonian adiabate (2.212). To demonstrate this assertion we prove the following theorem.

Theorem *The validity of (2.218) implies*

$$\delta\int_{\eta_i}^{\eta_f} \mathcal{L}_0(\eta)d\eta = o(N^{-1}), \tag{2.223}$$

where $N = (\eta_f - \eta_i)/2\pi$ is the number of cycles over which $\mathcal{L}_0$ undergoes an essential change. The symbol $o(N^{-1})$ means "vanishes at least with order $1/N$".

Proof Let the variation be an arbitrary piecewise continuous function $\Delta(\eta)$. The nth cycle starts at $\eta = \eta_n$, where for brevity we use the symbols $\Delta_n = \Delta(\eta_n)$, $\partial\mathcal{L}_0/\partial\eta_n = (\partial\mathcal{L}_0/\partial\eta)_{\eta=\eta_n}$. If the same quantities refer to an intermediate point $\eta \leq \eta_{\mathrm{a}} \leq \eta_n + 2\pi$ we write Δ_{a} and $\partial\mathcal{L}_0/\partial\eta_{\mathrm{a}}$ and omit the index n of the interval. To leading order the following holds:

$$
\begin{aligned}
\left|\delta \int_{\eta_i}^{\eta_f} \mathcal{L}_0 \mathrm{d}\eta\right| &= \left|\int_{\eta_i}^{\eta_f} \delta(\mathcal{L}_0 - \mathcal{L}) \mathrm{d}\eta\right| \\
&= \left|\int_{\eta_i}^{\eta_f} [\mathcal{L}_0(\eta + \Delta) - \mathcal{L}(\eta + \Delta)]\,\mathrm{d}\eta - \int_{\eta_i}^{\eta_f} [\mathcal{L}_0(\eta) - \mathcal{L}(\eta)]\,\mathrm{d}\eta\right|
\end{aligned}
$$

$$
\begin{aligned}
&\leq \sum_n \left|\int_{\eta_n}^{\eta_n+2\pi} \mathcal{L}_0(\eta + \Delta)\mathrm{d}\eta - 2\pi\mathcal{L}_0(\eta_n + \Delta_n) - \int_{\eta_n}^{\eta_n+2\pi} \mathcal{L}_0(\eta)\mathrm{d}\eta - 2\pi\mathcal{L}_0(\eta_n)\right| \\
&= \sum_n \left|\int_{\eta_n}^{\eta_n+2\pi} \frac{\partial \mathcal{L}_0}{\partial \eta_n}\Delta(\eta)\mathrm{d}\eta - 2\pi\frac{\partial \mathcal{L}_0}{\partial \eta_n}\Delta_n\right| = 2\pi\sum_n \left|\frac{\partial \mathcal{L}_0}{\partial \eta_a}\Delta_a - \frac{\partial \mathcal{L}_0}{\partial \eta_n}\Delta_n\right|.
\end{aligned}
$$

In the last step the mean value theorem is used. The function $\Delta(\eta)$ is arbitrary. Therefore at $\eta = \eta_n,\ \Delta_n = \Delta_a$ can be chosen without affecting Δ_a. With this substitution the leading order gives the result

$$
\left|\delta \int_{\eta_i}^{\eta_f} \mathcal{L}_0 \mathrm{d}\eta\right| \leq (2\pi)^2 \sum_n \left|\frac{\partial^2 \mathcal{L}_0}{\partial \eta_n^2}\Delta_a\right| \leq (2\pi)^2 N \max\left|\frac{\partial^2 \mathcal{L}_0}{\partial \eta_n^2}\right| \times \max|\Delta_a|.
$$

In this last step it is essential that $\partial\mathcal{L}_0/\partial\eta$ is a smooth function (in contrast to $\partial L/\partial\eta$, which is generally not). Now, $N = \min(1/2\pi)|\mathcal{L}_{0\,\max}/(\partial\mathcal{L}_0/\partial\eta_n)|$ is chosen; i.e., over N cycles $\mathcal{L}_{0\,\max}$ changes at most by $\mathcal{L}_0$. It follows that

$$
\left|\delta \int_{\eta_i}^{\eta_f} \mathcal{L}_0 \mathrm{d}\eta\right| \leq \frac{|\mathcal{L}_{0\,\max}|}{N} \times \max|\Delta|. \tag{2.224}
$$

Performing the variation of this inequality leads to (2.220) with the 0 replaced by a function f not larger than $|\mathcal{L}_{0\,\max}| \times \max|\Delta|/N^2$.

In order to understand what inequality (2.224) means let us specialize to a case of the averaged Lagrangian $\mathcal{L}_0$ not depending explicitly on time. Then the Hamiltonian $H_0 = \mathbf{p}_0\mathbf{v}_0 - L_0(\mathbf{x}_0, \mathbf{v}_0)$, where $L_0 = -\mathcal{L}_0(\omega - \mathbf{k}\mathbf{v}_0)$, owing to $\mathrm{d}H_0/\mathrm{d}t = \partial H_0/\partial t = -\partial L_0/\partial t = 0$, expresses energy conservation

$$
H_0 = E = \text{const.}
$$

A straightforward estimate shows that the uncertainty f in (2.220) leads to an energy uncertainty $\Delta H_0/H_0 \leq 2\pi/N$. This means that (2.220) is adiabatically zero and the total cycle-averaged energy is an adiabatic invariant in the rigorous mathematical sense in agreement with Arnold's definition [5]. For $N \to \infty$ (2.220) becomes exact. This is the mathematical proof of assertion (2.110) which was established there by physical arguments. Furthermore, physical arguments were used for expressions (2.120) and (2.121). By making use of (2.218) both expressions follow without using further arguments since nonrelativistically $L(\mathbf{x}, \mathbf{v}) = 1/2m\mathbf{v}^2 - V(\mathbf{x})$, $V(\mathbf{x})$ potential energy, and $L(\mathbf{x}, \mathbf{v}) = 1/2m\mathbf{v}^2 - V(\mathbf{x}) + q\mathbf{v}\mathbf{A}$ in the presence of a magnetic field.

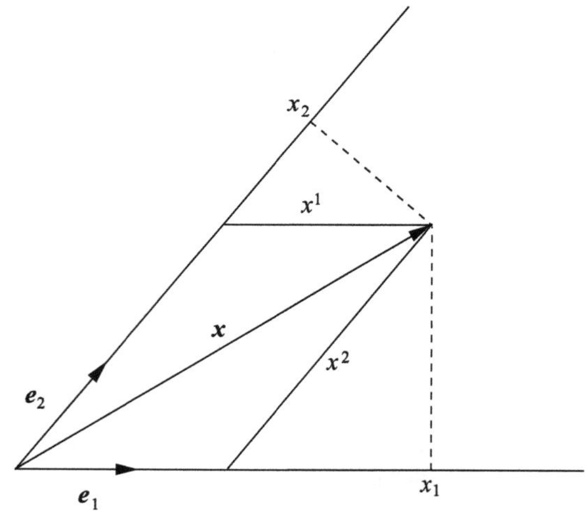

Fig. 2.24 Contravariant components x^1, x^2 and covariant components (projections) x_1 and x_2

2.2.2 Scalars, Contravariant, and Covariant Quantities

In the basis $\{\mathbf{e}_i\}$ the vector $\mathbf{x}$ is the unique linear combination $\mathbf{x} = x^i\mathbf{e}_i$, with x^i the *contravariant* components (Fig. 2.24). The same vector can also be expressed uniquely by its *covariant* components x_i, which are the projections of $\mathbf{x}$ onto the vectors $\mathbf{e}_i$, $x_i = \mathbf{x}\mathbf{e}_i$.

With the help of the Euclidean metric $g_{ik} = \mathbf{e}_i\mathbf{e}_k = g_{ki}$ and the inverse matrix $(g^{ik}) = (g_{ik})^{-1}$, the transformation from one set of components to the other is accomplished by

$$x_i = g_{ik}x^k, \qquad x^i = g^{ik}x_k. \tag{2.225}$$

The use of the two types of coordinates allows a compact representation of the scalar product,

$$\begin{aligned} \mathbf{xy} &= (x^i\mathbf{e}_i)\mathbf{y} = x^i(\mathbf{e}_i\mathbf{y}) = x^i y_i = g_{ik}x^i x^k, \\ \mathrm{d}s^2 &= \mathrm{d}\mathbf{x}\,\mathrm{d}\mathbf{x} = g_{ik}\,\mathrm{d}x^i\,\mathrm{d}x^k. \end{aligned} \tag{2.226}$$

The metric of special relativity for a Cartesian coordinate system in $\mathbb{R}^4$ is given by the $g_{\alpha\beta}$ of (2.158). It is identical with its inverse matrix, $g^{\alpha\beta} = g_{\alpha\beta}$.

A scalar quantity, or *Lorentz scalar*, is by definition a quantity which is invariant under a general Lorentz transformation. The rest mass m of a particle and its charge q are Lorentz scalars, but the density of particles n, of charge $\rho_{\mathrm{el}} = nq$, and of mass ρ are not. In fact, from (2.161) and (2.181) follows in $S'(\mathbf{v})$

$$n = \gamma n_0, \qquad \rho_{\mathrm{el}} = \gamma\rho_{\mathrm{el},0}, \qquad \rho = \gamma^2\rho_0, \qquad \gamma = (1 - v^2/c^2)^{-1/2}. \tag{2.227}$$

The index "0" refers to the rest frame S.

Any four quantity Y, which transforms like the position vector X, is a four vector, e.g., V, P, A, $K = (\mathbf{k}, \omega/c)$. For Y this is the case with an arbitrary X if $XY = x^\alpha y_\alpha = f$ is a Lorentz scalar. Any four quantity $T^{\alpha\beta}$, which transforms like the product $x^\alpha x^\beta$, is a four (or Lorentz) contravariant tensor of second rank, etc. In particular, any null four quantity is a four tensor of first (vector), second, etc., rank. The four gradient ∂_α,

$$\mathrm{grad}_\alpha = \partial_\alpha = \left(\frac{\partial}{\partial x^1}, \frac{\partial}{\partial x^2}, \frac{\partial}{\partial x^3}, \frac{\partial}{\partial x^4}\right) \tag{2.228}$$

when applied to a scalar f generates a four vector since

$$\mathrm{d}f = \frac{\partial}{\partial x^\alpha}\,\mathrm{d}x^\alpha = \mathrm{grad}_\alpha f\,\mathrm{d}X$$

is invariant under a Lorentz transformation, and $\mathrm{d}X$ is a four vector. According to the criterion above $\mathrm{grad}_\alpha f = \partial_\alpha f$ is a covariant four vector. Analogously, $\mathrm{grad}^\alpha f = \partial^\alpha f = \partial f/\partial x_\alpha$ is a contravariant four vector. Further, it is evident that $\mathrm{grad}_\alpha Y = \partial_\alpha y^\beta$ generates a mixed tensor of second rank T_α^β. The four divergence of a four vector $V = v^\alpha$, a four tensor $T = T^{\alpha\beta}$, etc.,

$$\mathrm{div}\,V = \partial_\alpha v^\alpha = \frac{\partial}{\partial x^\alpha} v^\alpha, \qquad \mathrm{div}\,T = \partial_\alpha T^{\alpha\beta} = \frac{\partial}{\partial x^\alpha} T^{\alpha\beta} \tag{2.229}$$

generates a Lorentz scalar, a four vector, etc. It is left as an exercise to the reader that the converse is also true.

In practical applications one is frequently faced with the problem of how the partial derivatives ∂_t and ∂_x of an entity $F(x, t)$ transform to $\partial_{t'}$ and $\partial_{x'}$ operating on $F'(x', t')$ in a system S' moving with velocity v along x. To this aim let us assume $x = g(x', t')$, $t = h(x', t')$. Equating the total differentials $\mathrm{d}F$ and $\mathrm{d}F'$,

$$\begin{aligned}\mathrm{d}F &= \partial_{t'} F'\,\mathrm{d}t' + \partial_{x'} F'\,\mathrm{d}x' = \partial_t F\,\mathrm{d}t + \partial_x F\,\mathrm{d}x\\ &= \partial_t F(h_{t'}\mathrm{d}t' + h_{x'}\mathrm{d}x') + \partial_x F(g_{t'}\mathrm{d}t' + g_{x'}\mathrm{d}x'),\end{aligned}$$

$g_{t'} = \partial_{t'} g$, etc., leads to

$$\partial_{t'} F' = \partial_t F\,h_{t'} + \partial_x F\,g_{t'}, \qquad \partial_{x'} F' = \partial_t F\,h_{x'} + \partial_x F\,g_{x'}. \tag{2.230}$$

In the special case of $g = \gamma(x' + vt')$, $h = \gamma(t' + vx'/c^2)$ results

$$\begin{aligned}&\partial_{t'} F' = \gamma(\partial_t F + v\partial_x F), \quad \partial_{x'} F' = \gamma\left(\frac{v}{c^2}\partial_t F + \partial_x F\right);\\ &\partial_t F = \gamma(\partial_{t'} F' - v\partial_{x'} F'), \quad \partial_x F = \gamma\left(-\frac{v}{c^2}\partial_{t'} F' + \partial_{x'} F'\right).\end{aligned} \tag{2.231}$$

They show the symmetry required by the relativity principle, i.e., the inverse transformation follows from interchanging F and F' and substituting v by $-v$. In Galileian relativity $x = x' + vt'$, $t = t'$ and

$$\partial_{t'} F' = \partial_t F + v\partial_x F, \qquad \partial_{x'} F' = \partial_x F. \tag{2.232}$$

2.3 Summary

Newtonian Mechanics. Single particle motion in the electromagnetic field give a first insight into the dynamics of laser generated plasmas. Slow motion with $|\mathbf{v}| \ll c$ can be studied in Galileian relativity. There, a reference system $S'(\mathbf{v})$ moving with constant velocity $\mathbf{v}$ with respect to S ("lab frame") is defined by $\mathbf{x}' = \mathbf{x} - \mathbf{v}t$ and the universal time $t' = t$. From Faraday's induction law the nonrelativistic transformations of the $\mathbf{E}$ and $\mathbf{B}$ and the nonrelativistic Lorentz force on a charged point particle $\mathbf{f} = q(\mathbf{E} + \mathbf{v} \times \mathbf{B})$ follow in the lab frame. The Lorentz force governs a whole variety of motions in the plasma: Electron oscillations in the laser field, gyration motion in the pure magnetic field, drift displacements of charges in crossed $\mathbf{E}$ and $\mathbf{B}$ fields, as well as gradient and curvature drifts in the inhomogeneous $\mathbf{B}$ field.

Lagrangian and Hamiltonian Mechanics. Gathering the physical content of dynamical processes is most immediate in Newtonian mechanics. General laws, symmetries, constants of motion and statistical properties are best studied in the phase space of canonical conjugate variables, for free particles preferentially in the coordinates of canonical momenta $\mathbf{p}$ and the conjugate positions $\mathbf{x} = \mathbf{q}$. Liouville's conservation of phase space volumes, conservation of canonical momenta, canonical transformations to appropriate variables, conservation of slowly varying quantities in time dependent systems, the so-called adiabatic invariants, and last but not least quantum systems have their natural basis in the Hamiltonian conception of mechanics. Specific problems are generally best handled in the Lagrange formalism. The concept of generalized coordinates confers it unexcelled flexibility. In the present and following chapters intense use is made of all above mentioned advantages. The nonrelativistic Hamilton and Lagrange functions of a free point charge q in the electromagnetic fields are

$$H(\mathbf{p}, \mathbf{q}) = \frac{\mathbf{p}^2}{2m} + V(\mathbf{q}, t), \;\; \mathbf{p} = m\mathbf{v} + q\mathbf{A}(\mathbf{q}, t);$$
$$L(\mathbf{q}, \dot{\mathbf{q}}, t) = T - V = \frac{1}{2}m\mathbf{v}^2 + q\mathbf{A}(\mathbf{q}, t) - V(\mathbf{q}, t).$$

$\mathbf{A}$ is the electromagnetic vector potential. In vacuum it is $\mathbf{A} = i\mathbf{E}/\omega$, in the constant magnetic field $\mathbf{A} = \mathbf{B} \times \mathbf{q}/2$.

Ponderomotive forces and adiabatic invariants. Hamiltonian mechanics as a whole follows from the Poincaré–Cartan integral invariant $\oint \left[\mathbf{p}d\mathbf{q} - Hdt\right] = 0$. The ponderomotive force (and potential) is a secular, i.e. zero frequency force induced by the charge oscillating at ω in the transverse laser field and in the longitudinal E field of the electron plasma wave. It is of particular interest in laser plasmas for fast electron generation, particle acceleration, plasma density profile steepening and a whole variety of parametric effects, like Brillouin and Raman scattering. The secular force is obtained from an averaging process over one oscillation. The way how averaging has to be accomplished is prescribed by the adiabatic theorem inferred in turn from the Poincaré–Cartan invariant. In the monochromatic transverse wave the ponderomotive potential of the free particle is identical with the invariant average of the quiver energy W; $\Phi_p = W$. In a compound particle ("atom") with internal degrees of freedom the internal energy has to be subtracted from W. In the longitudinal wave averaging is delicate: in addition to W the Φ_p there is a contribution from the time averaged wave potential $\overline{V}$ and the translational kinetic energy $m\mathbf{v}_0^2/2$ to be added. With these terms included adiabatic transition from the free particle to its trapping in the wave is possible. A simple extension of the underlying idea allows for the description of nonadiabatic trapping also.

Binary collisions. Adiabatic changes are smooth. Collisions between charged particles consist in an abrupt change of the related velocities and represent just the extreme opposite processes to adiabaticity. In energy conserving binary collisions the key concept is the differential collision cross section σ_Ω in the common center of mass system, defined as $\mathrm{d}I_{\boldsymbol{\Omega}} = I\sigma_\Omega \mathrm{d}\boldsymbol{\Omega}$; I incident parallel particle flux, $\mathrm{d}I_{\boldsymbol{\Omega}}$ particle flow scattered into direction $\boldsymbol{\Omega}$. Hard spheres are isotropic scatterers, charged particles scatter preferentially into forward direction. The cross section depends on the inverse of the relative energy squared and weakly on the Coulomb logarithm.

Relativistic dynamics. The velocity of light sets an upper limit to all speeds of material objects. It makes the revision of time and of the concept of simultaneous events necessary. The two reference systems S, $S'(\mathbf{v})$ are connected by the linear Lorentz transformation

$$\mathbf{x}' = \mathbf{x} + \frac{\gamma - 1}{v^2}(\mathbf{v}\mathbf{x})\mathbf{v} - \gamma\mathbf{v}t, \qquad t' = \gamma\left(t - \frac{\mathbf{v}\mathbf{x}}{c^2}\right); \qquad \gamma = \left(1 - \frac{v^2}{c^2}\right)^{-1/2}.$$

The invariance of the line element $\mathrm{d}s^2 = \mathrm{d}\mathbf{x}^2 - c^2\mathrm{d}t^2 = -c^2\mathrm{d}\tau^2 \Rightarrow \mathrm{d}\tau = \frac{\mathrm{d}t}{\gamma}$ with respect to changes from S to $S'(\mathbf{v})$ is a consequence of the Lorentz transformation. The proper time τ is the time that elapses on a clock comoving with the observer at rest in $S'(\mathbf{v})$. Four vectors, of position, velocity, acceleration, wave four vector $K = (\mathbf{k}, \omega/c)$, second rank tensors of energy-momentum, pressure, etc., are to be constructed as to respect the invariance of the invariant $\mathrm{d}s^2$. Four vectors are recognized by the invariance of their moduli in a transition from S to $S'(\mathbf{v})$. Physical laws formulated in four quantities are said *covariant*. The proof that Maxwell's equations transform according to Lorentz is straightforwardly shown in covariant formulation through the field tensors $F_{\alpha\beta}$, $F'^{\alpha\beta}$. However it is equally well accomplished (contrary to frequent opposite claim) in the conventional three vector form as shown in the chapter. To avoid running into contradictions it is a necessity to give up Galileian relativity and to reformulate Newtonian mechanics accordingly. The

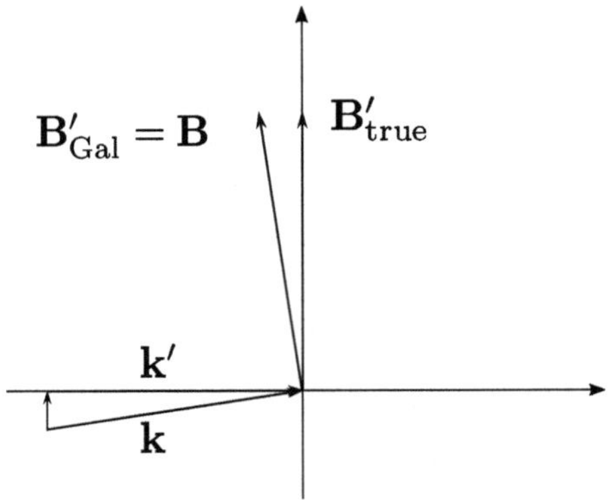

Fig. 2.25 In the lab frame a plane wave is impinging in the direction of the wave vector $\mathbf{k}$ onto an imaginary vertical plane. For an observer moving at speed $c \sin\alpha$ parallel to the plane the wave is incident normally along $\mathbf{k}'$ and the magnetic field $\mathbf{B}'$, indicated by $\mathbf{B}'_{true}$, is orthogonal to $\mathbf{k}'$. In contrast, Galileian relativity ($c = \infty$) yields $\mathbf{k}' = \mathbf{k}$ (!) from (2.116) and $\mathbf{B}' = \mathbf{B}'_{Gal} = \mathbf{B} \neq \mathbf{B}'_{true}$ from (2.8)

relativistic Hamiltonian and Lagrangian in the electromagnetic wave read

$$H(\mathbf{p}, \mathbf{q}, t) = [m^2c^4 + c^2(\mathbf{p} - q\mathbf{A})^2]^{1/2} + V(\mathbf{q}, t); \; L(\mathbf{q}, \dot{\mathbf{q}}) = -\frac{m}{\gamma}c^2 + q\mathbf{v}\mathbf{A} - V(\mathbf{q}, t).$$

The relativistic equation of motion is $\mathrm{d}\mathbf{p}/\mathrm{d}t = q(\mathbf{E} + \mathbf{v} \times \mathbf{B})$. The RHS is the Einstein force $\mathbf{f}_E$. Detailed application of H and L are made to describe relativistic electron motion in the running monochromatic wave and to obtain relativistic formulas of the ponderomotive force in the monochromatic transverse wave.

2.4 Problems

▷ Derive $\mathbf{B}' = \mathbf{B}$ from (2.8): Convince yourself that (2.7) is correct.
▷ Modify Maxwell's equations to fulfill (2.8). *Solution:* Drop the displacement current density $\partial_t(\varepsilon_0\mathbf{E})$.
▷ A plane s-polarized wave is crossing an imaginary surface under the small angle α (see Fig. 2.25). Apply a boost $v = c \sin\alpha$ in the plane of incidence and parallel to the surface such that the wave impinges normally to the surface. Show that according to (2.5) and (2.8) $\mathbf{B}$ is no longer normal to the new $\mathbf{k}'$ direction, in contradiction to $\nabla\mathbf{E}' = -\partial_t\mathbf{B}'$. *Solution:* Apply a special Galilei transformation twice.
▷ Verify (2.12), (2.13), (2.15).
▷ Give arguments for the conservation of helicity (= right/left polarization) in the reflection of a laser beam from a plane.
▷ What is the the laser intensity (linear polarization) corresponding (a) to the ground

state E field of hydrogen, (b) to the ground state E field of U^{+91}? *Solution (b):* $I = 6 \times 10^{24}$ Wcm^{-2}. $I \sim \gamma^2 Z^4$; γ Lorentz factor.

▹ (a) Show the last expression on the right of (2.18) equals $2W\nu$. (b) Rewrite (2.18) for circular polarization.

▹ Show that only the quadratic potential exhibits an oscillation frequency not depending on the degree of excitation. *Hint:* Use (2.128).

▹ Frequency detuning: Verify (2.27).

▹ Recover (2.29) from (2.25) under $\nu = 0$ and $\lim \omega = \omega_0$. *Hint:* Start from the undamped oscillator at rest: $x(t = 0) = \dot{x}(t = 0) = 0$.

▹ Free fall on earth after Fermat–Lagrange. Show the orbit under minimum time, i.e. $\delta \int \mathrm{d}t = 0$ from (2.34), is the trajectory of the free fall under $\delta S = 0$. *Hint:* $\mathrm{d}t = |d\mathbf{x}|/v$.

▹ What does Hamiltons principle of least action prescribe concerning the fixed end points: Do t_1, t_2 also remain fixed or must they be allowed to change in the variation process? Discuss it in the light of foregoing problem.

▹ Prove the following theorem that is basic for many applications in mathematical physics, e.g. variational calculus: (a) If $\int_A f(\tau)\mathrm{d}\tau = 0$ over an arbitrary measurable set $A \to f(\tau) = 0$ nearly everywhere. Under which conditions does the theorem hold? (b) Show the equivalence with (a) of the this version of the theorem: If $\int_A f(\tau)g(\tau)\mathrm{d}\tau = 0$ with an arbitrary function $g(\tau)$ over a fixed set $A \to f(\tau) = 0$ nearly everywhere on A. Specify the meaning of "nearly". *Possible answer to* (a): f is continuous except on a subset of measure zero. No conclusion can be drawn on its value there.

▹ In the variational calculus the existence of differentials [e.g., $\delta L(q, \dot{q}, t)$] plays a central role. (a) Give a sufficient condition for the existence $\delta f(x, y) = f_x \delta x + f_y \delta y$. (b) Give an example of non existing differential.

Possible solutions: (a) Sufficient condition for $\mathrm{d}f(x, y) = f_x \mathrm{d}x + f_y \mathrm{d}y$: f_x, f_y exist in a neighborhood of (x_0, y_0) and f_x is continuous at (x_0, y_0). (b) Choose $f(x, y) = xy$. $\Rightarrow f(0, y) = f(x, 0) = 0$, $f_x(0, 0) = f_y(0, 0) = 0 \Rightarrow f$ non differentiable in $(0, 0)$.

▹ Draw pictures that illustrate the meaning of relation (2.33) for neighboring orbits.

▹ Derive RHS expression in (2.43).

▹ Show in detail the validity of $L = T - V$ when V depends explicitly on time.

▹ Show that point transformations $q_i = q_i(r_j, t)$ preserve the invariance of the Lagrange equations. *Hint:* $\partial \dot{q}_i / \partial \dot{r}_j = \partial q_i / \partial r_j$.

▹ Do the freely falling elevators in a constant gravitational field form a class of inertial systems? *Answer:* yes.

▹ A potential step ΔV moves at velocity u into x direction. What is the minimum speed v of a mass point to overcome the potential difference?

▹ Deduce linear and angular momentum conservation from Noether's theorem.

▹ How do you verify that (2.37) is the correct Lagrange function?

▹ An electron at rest at position $x = 0$ is hit by a laser pulse (plane wave) propagating along the x axis. What can you say about position x and velocity v under the limitation $v \ll c$ of the electron after the beam has passed? *Hint:* Consider the conservation of the canonical momentum and calculate. After the pulse has passed,

the electron has undergone a finite displacement and is at rest again. If solution too difficult get inspired from the ponderomotive potential.
▹ Formulate sufficient conditions for the inversion of $\mathbf{p}(\mathbf{q}, \dot{\mathbf{q}}, t)$ to $\dot{\mathbf{q}}(\mathbf{p}, \mathbf{q}, t)$.
Solution: $\det(\partial^2 L/\partial q_i \partial q_j) \neq 0$ is a sufficient condition.
▹ Show that, given the existence of inversion between $(\mathbf{q}, \dot{\mathbf{q}})$ and $(\mathbf{p}, \mathbf{q})$, the Lagrange equations follow from Hamilton's equations (2.184).
▹ In (2.53) L is expressed by the independent variables $\mathbf{p}$ and $\mathbf{q}$. Why are you allowed to consider them as independent although in reality they are not?
▹ Derive (2.10) from (2.58). *Hint:* Inspection of (2.185) may be helpful.
▹ Prove Liouville's theorem for $f = 2$.
▹ Show that the invariant measure μ in phase space of the onedimensional harmonic oscillator $p^2 + (\alpha q)^2 = 2mE$ is

$$\mathrm{d}\mu(p) = \frac{\mathrm{d}\Sigma}{|\nabla H|_{H=E}} = \frac{\mathrm{d}s}{2\alpha(2mE - \frac{\alpha^2-1}{\alpha^2}p^2)^{1/2}}$$

$\mathrm{d}s$ line element on the ellipse in phase space, $\alpha = \sqrt{2}m\omega$. For $\alpha = 1$ the ellipse is a circle, thus $\mathrm{d}\mu(p) \sim \mathrm{d}s$.
▹ Consider a high number of noninteracting point particles of energies $E \leq E_0$ in the parabolic potential $V(x) = \kappa x^2$. Calculate the particle density $n(x)$ under the assumption that each energy value E is assumed with the same probability.
Answer: $n(x) \sim [1 - \kappa x^2/E_0]^{1/2}$.
Prove the three topological properties of phase flow.
▹ Verify the Frenet formulas (2.64).
▹ Solve the equation of motion for $\mathbf{E}$ not orthogonal to $\mathbf{B}$ in (2.70).
▹ Show that the local curvature radius in the magnetic field is $R_c(\mathbf{x}, t) = v_\perp/\omega_G|_{(\mathbf{x},t)} = mv_\perp/qB$.
▹ Sketch possible trajectories in the magnetic bottle (two opposite mirrors) and determine the loss cone angle α_0
▹ Two magnetic mirrors are moving adiabatically against each other and simultaneously the currents in the two coils change also slowly. Determine the adiabatic invariance of the charged particle motion along the axis. Hint: Apply (2.81).
▹ Consider the Galileian pendulum of length l varying slowly in time. How does the displacement of the small angle α vary with ω?
▹ Treat the planar pendulum in analogy to the spherical pendulum in the text.
▹ Newton's third law: Actio = reactio, and Fig. 2.10: How can the lady impart a tangential force to the swing? Explain the $2\,\omega$ frequency in Mathieu's equation.
▹ Derive (2.89) in three dimensions with $\nabla \times \dot{\mathbf{z}} = (\dot{v}, v, 1)$, $v = \dot{x}$.
▹ Show that (2.90) is an exact differential. *Hint:* Use Hamilton's equations of motion.
▹ Is the following transformation canonical? $P = \alpha p$, $Q = \beta q$, $s = t$; α, β arbitrary constants $\neq 1$. *Answer:* In general not. Can you find a special class of H for which it is canonical?
▹ Derive the generating function S of the Galileian canonical transformation.
▹ Reproduce the proof of $\oint[p\mathrm{d}q - H(\lambda)\mathrm{d}t] = \mathrm{const_a}$ in one dimension for nearly periodic motion. Alternatively, find your own proof of (2.104).

▹ Why and when is the integrand $-H\mathrm{d}t$ essential in (2.104)? Give the answer by an example. *Hint:* Take H time dependent.
▹ Show that the difference in $\oint \mathbf{p}\mathrm{d}\mathbf{q}$ for $\lambda = \text{const}$ in $H(\mathbf{p}, \mathbf{q}, \lambda)$ and λ slowly varying is of second or higher order.
Proof If the angle between the two line elements $\mathrm{d}\mathbf{q}$ along the two orbits is ϵ the difference $\Delta \oint \mathbf{p}\mathrm{d}\mathbf{q} = \oint \mathbf{p}(1 - \cos\epsilon)\mathrm{d}\mathbf{q} = \oint \mathbf{p}[\epsilon^2/2 + O(\epsilon^4)]\mathrm{d}\mathbf{q} = o(\oint \mathbf{p}\mathrm{d}\mathbf{q})$.
▹ Show that the oscillation center velocity $\mathbf{v}_0$ has the property of minimum square deviation, $\oint[\mathbf{v}(t) - \mathbf{v}_0]^2\,\mathrm{d}t = \min$.
Proof $\partial/\partial\mathbf{v}_0 \oint(\mathbf{v} - \mathbf{v}_0)^2\mathrm{d}t = -2\oint(\mathbf{v} - \mathbf{v}_0)\mathrm{d}t = 0,\ \partial^2/\partial\mathbf{v}_0^2 \oint(\mathbf{v} - \mathbf{v}_0)^2\mathrm{d}t = 2\oint \mathrm{d}t = 2T > 0$.
▹ Verify (2.115).
▹ Show that in presence of an electron plasma wave the standard ponderomotive potential is given by the "atomic" expression for Φ with ω_0 replaced by the Bohm–Gross frequency ω_{es}.
▹ Verify that $t_1 \neq t_2$ in (2.124) is a second order effect. Verify expressions in (2.125).
▹ Show that adiabatic stretching $\lambda(x)$ of a periodic wave structure leaves $Z = W + \overline{V}$ adiabatically constant (change is of order $o(|\nabla\lambda|\lambda)$.
▹ Calculate the oscillation period T of the pendulum in the libration and in the rotation mode. What is the oscillation period at the transition from libration to rotation? Compare with T_t from (2.134).
▹ Calculate $Z = W + \overline{V}$ from (2.131) for the sinusoidal wave.
▹ A charged particle either crosses a wave pulse or is reflected from it. Show that in these processes the particle does not gain or lose energy unless it is temporarily trapped.
▹ The angular momentum in the two-body central force field is $\mathbf{J} = \mathbf{R} \times (m_1 + m_2)\mathbf{V} + \mathbf{r} \times \mu\mathbf{w};\ \mathbf{R} = (m_1\mathbf{x}_1 + m_2\mathbf{x}_2)/(m_1 + m_2),\ \mathbf{V} = \dot{\mathbf{R}},\ \mathbf{r} = \mathbf{x}_1 - \mathbf{x}_2$.
▹ Calculate σ_Ω and σ_t for hard spheres. Solution for σ_Ω is found in Problems of Chap. 1. $\sigma_\Omega = R^2/4 = (r_1 + r_2)^2/4;\ r_1 = r_2 = r \Rightarrow \sigma_\Omega = r^2. \sigma_t = \int \sigma_\Omega \sin\vartheta\, \mathrm{d}\varphi\mathrm{d}\vartheta = 4\pi\sigma_\Omega = \pi R^2$.
▹ Derive the Coulomb cross section in (2.149) from the Kepler problem.
▹ Show the correctness of (2.151) for hard spheres. *Hint:* Start directly from the definition of σ_Ω in (2.148). Interpreter $\int \sigma_\Omega(1 - \cos\theta)\mathrm{d}\varphi\mathrm{d}\cos\theta = \int \sigma_\Omega\mathrm{d}\varphi\mathrm{d}\cos\theta = \pi r^2$ physically. Does the equality of these integrals also hold for the Debye scattering potential?
▹ Show that the *mean free path* of hard spheres is given by $\lambda = 1/(n_2\sigma_t)$. *Solution:* The probability $\mathrm{d}p$ of a particle of radius r_1 from a parallel beam of cross section A to collide with one of fixed centers of radius r_2 and density n_2 is $A\mathrm{d}p = \sigma_t dN = \sigma_t n_2 A\mathrm{d}x \Rightarrow \mathrm{d}p/\mathrm{d}x = n_2\sigma_t$. If $N_1(x)$ particles of the beam arrive at x, the fraction $\mathrm{d}N_1$ of them are eliminated in $\mathrm{d}x$, hence $\mathrm{d}N_1 = -N_1(x)\mathrm{d}p = -N_1(x)n_2\sigma_t\mathrm{d}x \Rightarrow N_1(x)/N_0 = \exp -x/n_2\sigma_t = p(x) = \exp -x/n_2\sigma_t$. This is the probability to survive without a collision over the distance x. The probability to cover exactly the distance x' is $p(x')$ of surviving times the probability to collide in $\mathrm{d}x'$, i.e. $p(x')\mathrm{d}p$. The *mean free path* is the average $\lambda = \int_0^\infty x' p(x')\mathrm{d}x'/n_2\sigma_t = 1/n_2\sigma_t$.

▹ Determine the differential collision cross section of hard spheres in the system in which the center of mass moves at velocity $\mathbf{V}$. *Hint:* Make use of Fig. 2.20.
▹ Verify (2.157).
▹ Show by direct verification that (2.186) is the correct Hamiltonian. *Hint:* Use the vector identity $\mathbf{a} \times \nabla \times \mathbf{b} = \nabla(\mathbf{a}_c\mathbf{b}) - (\mathbf{a}\nabla)\mathbf{b}$ in the canonical equations. Index "c" means that $\mathbf{a}$ is kept constant; special case: $\nabla \mathbf{a}^2 = 2[\mathbf{a} \times \nabla \times \mathbf{a} + (\mathbf{a}\nabla)\mathbf{a})]$.
▹ Set $\tau = ct$, $\boldsymbol{\beta} = \mathbf{v}/c$ and split $\mathbf{x}$ into a component $\mathbf{x}_\parallel$ parallel to $\mathbf{v}$ and a component $\mathbf{x}_\perp$ perpendicular to $\mathbf{v}$. Show that (2.153) is identical to $\mathbf{x}'_\parallel = \gamma(\mathbf{x} - \boldsymbol{\beta}\tau)$, $\tau' = \gamma(\tau - \boldsymbol{\beta}\mathbf{x}_\parallel)$ and the identity $\mathbf{x}'_\perp = \mathbf{x}_\perp$.
▹ Precision experiment of Lorentz contraction. A scale of length L is sliding at velocity v on a table parallel to its extension. From a superintense laser mounted on the sealing an ultrashort laser pulse (plane wave) is impinging vertically onto the table and leaves an imprint (shadow) of the moving scale. What is the length of the shadow? Why is it $L' = L/\gamma$ and not L?
Hint: Determine the angle of incidence of the laser in the moving frame.
▹ Reference system S' moves with v along x of S. The length L of a scale oriented along x in S transforms into $L' = \gamma L$ in S'. Owing to symmetry between S and S' and $\gamma(v) = \gamma(-v)$ holds $L = \gamma L'$. What is the value of L if you know L'; by which criterion do you decide between the two transformations?
▹ Prove the triangular inequality of four vectors lying inside of the light cone, and outside of it.
▹ Exercise relativity of simultaneous events: Park a 10 m long bus in your garage of 5 m length. What is the minimum velocity of the bus? ($v_{\text{bus}} = \sqrt{3}/2c$).
▹ The number π can be defined on the circle as the ratio perimeter/diameter. Consider a circle rotating in the plane of the observer at velocity v on its circumference. Calculate π. *Hint:* Approximate the circle by a polygon and go to the limes. Any number of π is possible between nearly zero and 3.141592....
▹ Show that Einstein's clock measures the same Δt for arbitrary orientation of the light pulse with respect to its motion $\mathbf{v}$, in particular $\mathbf{k} \parallel \mathbf{v}$ and $\mathbf{k} \perp \mathbf{v}$.
▹ Derive the general expressions (2.153) from (2.154).
▹ Show $A^2 = \mathbf{a}^2$. *Solution:* (a) V is a four vector $\Rightarrow$ $\mathrm{d}V/\mathrm{d}\tau$ is a four vector $\Rightarrow$ $A^2 = (\mathbf{a}, 0)^2 = \mathbf{a}^2$ owing to the invariance of the modulus under transformation to the comoving system $S(\mathbf{v})$. (b) A is a space like four vector with vanishing 4th component in appropriate S'.
▹ Express A^2 in covariant form. *Answer:* $A^2 = (1/m^2)[(\mathrm{d}\mathbf{p}/\mathrm{d}\tau)^2 - (\mathrm{d}E/\mathrm{d}\tau)^2/c^2] = (1/m^2)[(\mathrm{d}\mathbf{p}/\mathrm{d}\tau)^2 - (v/c)^2(\mathrm{d}|\mathbf{p}|/\mathrm{d}\tau)^2]$.
▹ Show that the relativistic gyroradius r_{Gr} and relativistic magnetic moment μ_r are $r_{Gr} = \gamma r_G$ and $\mu_r = \gamma\mu$, with r_G, μ the nonrelativistic expressions from (2.66).
▹ Show that (2.199) and (2.209) are connected by (2.210).
▹ Show that the vectors of the reciprocal lattice of cristallography $\mathbf{b}_i = \epsilon_{ijk}\mathbf{a}_j \times \mathbf{a}_k/(\mathbf{a}_1 \cdot \mathbf{a}_2 \times \mathbf{a}_3)$ are covariant (contravariant) with respect to the contravariant (covariant) vectors $\mathbf{a}_i$ in $\mathbf{R}^3$.
▹ Show that the div operator *decreases* the rank of a four tensor by one degree; the grad operator *increases* its degree by one.

▹ (a) Cast the Lorentz transformations (2.153) and (2.155) into matrix representation $\mathbf{x}' = \Lambda\mathbf{x} \Leftrightarrow x^{\alpha'} = \Lambda^{\alpha'}_{\beta} x^{\beta}$; $\mathbf{x} = \Lambda^{-1}\mathbf{x}' \Leftrightarrow x^{\beta} = \Lambda^{\beta}_{\alpha'} x^{\alpha'}$. (b) Show $\Lambda^{\alpha'}_{\beta}$ and $\Lambda^{\beta}_{\alpha'}$ are orthogonal. (c) Verify $\mathbf{e}_{\beta'} = \Lambda^{\alpha}_{\beta'}\mathbf{e}_{\alpha}$ and inverse transformation.

▹ Verify: $-mc^2/\gamma$ in the relativistic Lagrangian (2.185) is not the kinetic energy T.

▹ Derive the velocity addition theorem with the help of formulas (2.231).

2.5 Self-assessment

- (a) Characterize the following expressions by assigning them linear or right hand/left hand circular (elliptic) polarization and helicity: $\mathbf{E} = \hat{\mathbf{E}} \exp \pm\imath(\mathbf{kx} - \omega t)$, $\hat{\mathbf{E}} \exp -\imath(\mathbf{kx} + \omega t)$, $\hat{\mathbf{E}} \exp -\imath(-\mathbf{kx} - \omega t)$. (b) A right hand circularly (linearly) polarized wave is partially or totally reflected from a plane surface. How does the polarization change? *Answer:* Helicity is preserved.
- Explain why the Lagrangian L is $T - V$ and not the sum of the energies, or another composition of them.
- Is the relativistic T in L the kinetic energy? *Answer:* no.
- Justify in terms of physics that $(\mathrm{d}/\mathrm{d}t)\delta\mathbf{x} = \delta(\mathrm{d}\mathbf{x}/\mathrm{d}t)$. $\delta\mathbf{x}$ is the virtual displacement.
- The first term in D'Alembert's principle (2.30) may not be integrable in general; for instance because it depends explicitly on $\mathbf{v}$. Does it imply the nonexistence of a Lagrangian L? *Answer:* no. How is the extension to nonintegrable terms achieved? You find the answer in the foregoing text; perhaps read again.
- Does the virial theorem $\overline{T} = \overline{V}$ hold for the driven harmonic oscillator?
- Is $\mathbf{p} - q\mathbf{A}$ in H from (2.186) the canonical or the kinetic momentum? *Answer:* kinetic.
- Liouville's theorem holds in a space $\{\mathbf{p} \times \mathbf{q}\}$ of two canonical conjugate variables. In presence of a vector potential $\mathbf{A}$ the canonical momentum is $\mathbf{p} = \mathbf{p}_{\mathrm{kin}} + q\mathbf{A}$, $\mathbf{p}_{\mathrm{kin}} = \gamma m\mathbf{v}$. Does the theorem hold also in the mechanical phase space $\{\mathbf{p}_{\mathrm{kin}} \times \mathbf{q}\}$? (*Answer:* yes. For proof see text).
- Do (2.81) and (2.83) hold with soft walls, i.e., when the space in between is filled with a potential $V(\mathbf{x})$? *Answer:* no. Give an example.
- (a) Are point transformations $(\mathbf{p}, \mathbf{q}) \to (\mathbf{P}, \mathbf{Q})$ canonical? (b) Are the Lagrange equations invariant under point transformations? *Answer:* (a): no, apart from restrictions; (b): yes.
- How does the generating function of the Galileian transformation look like?
- Which class of motions induced by a time dependent Hamiltonian exhibits a conservation equation in place of energy conservation? *Hint:* See the introduction to slow motions.
- A is an adiabatic invariant. Which conditions does it fulfill? Is entropy an adiabatic invariant?
- Give at least 10 examples of adiabatic invariants.

- The *standard* ponderomotive potential Φ_p of the free electron is its mean oscillation energy W, for the harmonic oscillator it is not. Why? How does in the latter case Φ_p relate to $\oint \mathbf{p}d\mathbf{q}$?
- Show by an example that in the general case the criterion of adiabaticity is $J = \oint(\mathbf{p}d\mathbf{q} - Hdt)$. When is the term $\oint Hdt$ compulsory?
- Why is the ponderomotive potential Φ of a transverse wave not a potential in the strict sense? *Answer:* It depends on the *changes* of the initial drift velocity $\mathbf{v}_0$ during wave—particle interaction. What is the effect of the $\mathbf{v}_0$ dependence?
- The insensitivity of Φ_p to the initial oscillation (gyration) center velocity is a property of the transverse or the longitudinal electric wave? *Answer:* transverse wave.
- Solve the following paradox. An electron plasma wave of spatially constant amplitude $\hat{E}(t)$ and constant frequency ω can accelerate electrons although f_p is zero in the whole domain of $\nabla\hat{E}(t) = 0$. *Hint:* Consider a sphere rolling along on a billiard table in an elevator. What is its energy change with respect to the ground floor and where does it happen?
- Consider a 1D electron plasma wave of constant potential amplitude $\hat{\Phi} = \hat{E}/k$ and constant frequency ω but slowly varying wave number k in space. Does it exhibit a finite ponderomotive force f_p? You find the answer in the text.
- Consider *uphill acceleration* in Fig. 2.19. (a) Why do the end points $2\Phi_p/v_i^2$ coincide? (b) Do the positions of reflection coincide? *Answer:* no. Proof! (c) Which reflection point is closest to the foot point of the wave packet envelope?
- The mean energy gain of an electron from the laser in a hard sphere collision is twice the mean oscillation energy W, see (1.17). The result follows easily if it is calculated in the reference system in which the electron before the collision does not oscillate. However, already Galileian invariance tells that the result cannot depend on the choice of the reference system. You may find the solution with the help of the subsequent paradox and the observation that the laser field conserves the center of mass in the electron-ion collision.
- A car starts from zero and accelerates up to 100 miles/h. What is its fuel consumption? Easy answer: Absorbed energy (fuel) is $E_{\text{final}} - E_{\text{initial}} = mv_{\text{final}}^2/2$. An observer from a train travelling at 50 miles/h finds $E_{\text{final}} - E_{\text{initial}} = m(+50)^2/2 - m(-50)^2/2 = 0$!
  Invariant solution: The car is accelerated by *internal* forces between m of the car, velocity v_1, and M, velocity v_2, of the earth leaving the velocity v_C of the center of mass invariant. Thus

$$v_2 - v_C = -\frac{m}{M}(v_1 - v_C) \Rightarrow E = \frac{1}{2}m(v_1 - v_C)^2 + \frac{1}{2}M(v_2 - v_C)^2 = \frac{1}{2}\frac{m^2}{\mu}(v_1 - v_C)^2$$

The energy gain $\Delta E = E_{\text{final}} - E_{\text{initial}}$; μ is the reduced mass, $\mu \simeq m$. In the first case $v_C = 0$ and $\Delta E = mv_{1\text{final}}^2/2$. In the second case (from the train) is $v_C = v_{1\text{initial}} = -v_{1\text{final}}/2 \Rightarrow \Delta E = m(v_{1\text{final}}/2 + v_{1\text{final}}/2)^2/2$.

- In the expressions of collision frequency and mean free path according to (2.151) *simultaneous multiple collisions* are excluded; why? Give a criterion for the validity of the formulas. *Criterion:* $\lambda = 1/n_2\sigma_t \gtrapprox \sigma_t^{1/2} \Rightarrow n_2\sigma_t^{3/2} \lessapprox 1$.
- Why time dilation and Lorentz contraction of space and not time contraction and space dilation?
  Answer: LT in $\mathbf{R}^1$ are: $x' = \gamma(x - vt),\ t' = \gamma(t - vx/c^2),\ x = \gamma(x' + vt'),\ t = \gamma(t' + vx'/c^2)$. (a) Time dilation: Clock is at rest in $S \Rightarrow v = 0 \Rightarrow LT$ for x' to be used $\Rightarrow t' = \gamma t \geq t$. (b)Length contraction: Scale L extends from $x = 0$ to $x = L \Rightarrow L' = \gamma(L - vt),\ L = \gamma(L' + vt')$. If L is to be measured in moving $S'(v)$, endpoints are to be measured at the same time instant t' because end points move. $\Rightarrow$ Transformation to L' is to be used $\Rightarrow$, $L = L'/\gamma \leq L'$. The same result is obtained from S with end points to be measured simultaneously in S'.
- If in deriving (2.164) time dilation $t = \gamma t'$ is used γ would result replaced by γ^{-1}. Where is the conceptional mistake?
- Where do $y' = y,\ z' = z$ in (2.163) follow from? *Answer:* from requirement of linearity (e.g., plane wave remains plane under LT) and invariance with respect to rotation (i.e., isotropy). Try the proof.
- How do you distinguish a time like four vector from a space like four vector? Give examples of vectors belonging to the two classes.
- For which pairs of four vectors X, Y does an inequality $(X + Y)^2 \leq X^2 + Y^2$ hold, time like or space like, or both?
- The ponderomotive force $\mathbf{f}_p$ from (2.215) is an Einstein force. Why is it equal to $-\mathrm{d}\mathbf{p}_0/\mathrm{d}t$ with $\mathbf{p}_0$ from (2.215)?
- Difficult questions: (a) Does the ponderomotive potential Φ_p of the monochromatic travelling wave depend on the oscillation center motion $\mathbf{v}_0$? *Hint:* Answer is given by (2.214) and (2.215) in the text. (b) Does the electromagnetic ponderomotive potential exhibit hysteresis, like that of the electron plasma wave, or not?
- The relativistic ponderomotive force is *independent* of the oscillation center motion $\mathbf{v}_0$, whereas in the subrelativistic case, (2.117), it is only *insensitive* to $\mathbf{v}_0$? *Answer:* Newtonian mechanics is incompatible with Maxwell equations. Give an example for the latter.

2.6 Glossary

Free electron oscillation

$$\mathbf{v} = \hat{\mathbf{v}}\mathrm{e}^{\mathrm{i}(kx-\omega t)},\ \boldsymbol{\delta} = \hat{\boldsymbol{\delta}}\mathrm{e}^{\mathrm{i}(kx-\omega t)},\ \hat{\mathbf{v}} = -i\frac{e}{m_e\omega}\hat{\mathbf{E}},\ \hat{\boldsymbol{\delta}} = \frac{e}{m_e\omega^2}\hat{\mathbf{E}},\ W = \frac{e^2}{4m_e\omega^2}\hat{\mathbf{E}}\hat{\mathbf{E}}^*. \tag{2.11}$$

Laser intensity, **B** field, and $\hat{E}$ field in linear polarization

$$\mathbf{I} = \overline{S} = \varepsilon_0 c^2 \overline{\boldsymbol{E} \times \boldsymbol{B}} = \frac{\mathbf{k}^0}{2} c\varepsilon_0 \hat{\mathbf{E}}\hat{\mathbf{E}}^*,\ \mathbf{B} = \frac{\mathbf{k}}{\omega} \times \mathbf{E};\ \hat{E}[\mathrm{Vcm}^{-1}] = 27.5 \times (I[\mathrm{Wcm}^{-2}])^{1/2}. \tag{2.13}$$

Hamilton's principle of last action $\delta S = \delta \int L(\mathbf{x}, \dot{\mathbf{x}}, t)\mathrm{d}t = 0$
Lagrange equation in generalized coordinates $\mathbf{q}$, $\dot{\mathbf{q}}$

$$\frac{\mathrm{d}}{\mathrm{d}t}\frac{\partial L}{\partial \dot{\mathbf{q}}} - \frac{\partial L}{\partial \mathbf{q}} = 0 \quad \Leftrightarrow \quad \frac{\mathrm{d}}{\mathrm{d}t}\frac{\partial L}{\partial \dot{q}_i} - \frac{\partial L}{\partial q_i} = 0; \qquad L = T(\mathbf{q}, \dot{\mathbf{q}}, t) - V(\mathbf{q}, t). \tag{2.43}$$

Virial theorem

$$2\overline{T} + \overline{\sum \mathbf{f}_i \mathbf{x}_i} = 2\overline{T} - \overline{\sum (\partial V/\partial \mathbf{x}_i)\mathbf{x}_i} = 0;\ \ V = C\sum r_i^{\alpha} \Rightarrow \overline{T} = \frac{\alpha}{2}\overline{V};\ \ i \geq 1. \tag{2.50}$$

Noether's theorem

$$\frac{\mathrm{d}}{\mathrm{d}t}\left(\frac{\partial L}{\partial \dot{\mathbf{r}}}\frac{\partial \mathbf{r}}{\partial s}\Big|_{s=0}\right) = \frac{\partial L}{\partial \mathbf{r}}\frac{\partial \mathbf{r}}{\partial s} + \frac{\partial L}{\partial \dot{\mathbf{r}}}\frac{\partial \dot{\mathbf{r}}}{\partial s} = \frac{\partial \mathbf{r}}{\partial s}\frac{\mathrm{d}}{\mathrm{d}t}\frac{\partial L}{\partial \dot{\mathbf{r}}} + \frac{\partial L}{\partial \dot{\mathbf{r}}}\frac{\mathrm{d}}{\mathrm{d}t}\frac{\partial \mathbf{r}}{\partial s}\frac{\partial L}{\partial \mathbf{r}}\frac{\partial \mathbf{r}}{\partial s} + \frac{\partial L}{\partial \dot{\mathbf{r}}}\frac{\partial \dot{\mathbf{r}}}{\partial s} = 0. \tag{2.52}$$

Hamiltonian

$$H(\mathbf{p}, \mathbf{q}, t) = \mathbf{p}\dot{\mathbf{q}}(\mathbf{p}, \mathbf{q}, t) - L(\mathbf{p}, \mathbf{q}, t) \Rightarrow \dot{\mathbf{p}} = -\frac{\partial H}{\partial \mathbf{q}},\ \dot{\mathbf{q}} = \frac{\partial H}{\partial \mathbf{p}},\ \frac{\mathrm{d}H}{\mathrm{d}t} = \frac{\partial H}{\partial t}. \tag{2.3}$$

L and H in electromagnetic field, $\mathbf{p} = m\dot{\mathbf{q}} + q\mathbf{A}$ canononical momentum

$$L(\mathbf{x}, \mathbf{v}, t) = \frac{1}{2}m\mathbf{v}^2 + q\mathbf{v}\mathbf{A}(\mathbf{x}, t) - q\Phi(\mathbf{x}, t),\ \ H(\mathbf{p}, \mathbf{q}, t) = \frac{1}{2m}(\mathbf{p} - q\mathbf{A})^2 + q\Phi(\mathbf{q}, t). \tag{2.37, 2.58}$$

Vector potential $\mathbf{A}$:

$$\text{em. } \mathbf{A} = -i\mathbf{E}/\omega;\ \ \mathbf{B} = \text{const} \Rightarrow \mathbf{A} = \frac{1}{2}\mathbf{B} \times \mathbf{r}_G \tag{2.67}$$

trihedron $\mathbf{t}$, $\mathbf{n}$, $\mathbf{b}$:

$$\frac{\mathrm{d}\mathbf{t}}{\mathrm{d}s} = \kappa\mathbf{n}, \qquad \frac{\mathrm{d}\mathbf{n}}{\mathrm{d}s} = -\kappa\mathbf{t} - \tau\mathbf{b}, \qquad \frac{\mathrm{d}\mathbf{b}}{\mathrm{d}s} = \tau\mathbf{n}. \tag{2.64}$$

Gyrofrequency $\boldsymbol{\omega}_G$, gyroradius$\mathbf{r}_G$, magnetic moment $\boldsymbol{\mu}$

$$\boldsymbol{\omega}_G = -\frac{q}{m}\mathbf{B}, \qquad \mathbf{v}_\perp = \boldsymbol{\omega}_G \times \mathbf{r}_G, \qquad r_G = \frac{v_\perp}{\omega_G} = \frac{(2mE_\perp)^{1/2}}{qB},$$

$$\boldsymbol{\mu} = \frac{1}{2}q\mathbf{r}_\mathbf{G} \times \mathbf{v} = -\frac{q}{2}r_G^2\boldsymbol{\omega}_G \Rightarrow \mu = |\boldsymbol{\mu}| = \frac{q}{2}\frac{\mathbf{v}_\perp^2}{\omega_G} = \frac{E_\perp}{B};\ \ E_\perp = \frac{1}{2}mv_\perp^2. \tag{2.66}$$

Numerical

$$\omega_G[\mathrm{s}^{-1}] = \frac{e}{m_e}B = 1.8 \times 10^{11} B[\mathrm{T}], \; r_G[\mathrm{m}] = \frac{\sqrt{2m_e}}{e}\frac{\sqrt{E_\perp}}{B} = 3.4 \times 10^{-6}\frac{\sqrt{E_\perp[\mathrm{eV}]}}{B[\mathrm{T}]},$$

$$\mu[\mathrm{eVT}^{-1}] = 1.6 \times 10^{-19}\frac{E_\perp[\mathrm{eV}]}{B[\mathrm{T}]}; \; 1\,\mathrm{eV} = 1.602 \times 10^{-19}\mathrm{J}, \; 1\mathrm{Tesla} = 10^4\,\mathrm{Gauss}. \tag{2.69}$$

$\mathbf{E} \times \mathbf{B}$ drift and gravity drift

$$\mathbf{v}_D = \frac{\mathbf{E} \times \mathbf{B}}{B^2}; \qquad \mathbf{v}_g = \frac{m}{q}\frac{\mathbf{g} \times \mathbf{B}}{B^2}. \tag{2.70, 2.72}$$

Gradient drift and curvature drift

$$\mathbf{v}_D = \frac{q}{2m}r_G^2\frac{\mathbf{B} \times \nabla B}{B}, \qquad \mathbf{v}_D = \frac{mv_\parallel^2}{qB^2}\frac{\mathbf{R}_c \times \mathbf{B}}{R_c^2}. \tag{2.73, 2.74}$$

Adiabatic invariance after Poincaré–Cartan:

$$\text{closed orbit} \oint \mathbf{p}\mathrm{d}\mathbf{q} = \mathrm{const_a}; \quad \text{periodic orbit} \oint (\mathbf{p}\mathrm{d}\mathbf{q} - H\mathrm{d}t) = \mathrm{const_a}. \tag{2.104}$$

Lorentz transformations:

$$\mathbf{x}' = \mathbf{x} + \frac{\gamma - 1}{v^2}(\mathbf{vx})\mathbf{v} - \gamma\mathbf{v}t, \qquad t' = \gamma\left(t - \frac{\mathbf{vx}}{c^2}\right); \qquad \gamma = \left(1 - \frac{v^2}{c^2}\right)^{-1/2}, \tag{2.153}$$

$$\mathbf{x}'^2 - c^2t'^2 = \mathbf{x}^2 - c^2t^2. \tag{2.157}$$

$$\mathbf{k}' = \mathbf{k} + \frac{\gamma - 1}{v^2}(\mathbf{vk})\mathbf{v} - \gamma\frac{\mathbf{v}}{c^2}\omega, \qquad \omega' = \gamma(\omega - \mathbf{kv}). \tag{2.169}$$

$$\mathbf{E}'_\perp = \gamma(\mathbf{E}_\perp + \mathbf{v} \times \mathbf{B}), \qquad \mathbf{E}'_\parallel = \mathbf{E}_\parallel. \tag{2.165}$$

$$\mathbf{B}'_\perp = \gamma\left(\mathbf{B}_\perp - \frac{1}{c^2}\mathbf{v} \times \mathbf{E}\right), \; \mathbf{B}'_\parallel = \mathbf{B}_\parallel. \tag{2.167}$$

$$\text{four velocity} \quad V = \mathrm{d}X/\mathrm{d}\tau = \gamma(\mathbf{v}, c), \quad V^2 = \gamma^2(v^2 - c^2) = -c^2,$$

$$\text{four momentum} \quad P = mV = (\mathbf{p}, \gamma mc), \quad \mathbf{p} = \gamma m\mathbf{v}, \qquad P^2 = -m^2c^2,$$

$$\text{four acceleration} \quad A = \mathrm{d}V/\mathrm{d}\tau, \quad VA = \frac{1}{2}\mathrm{d}V^2/\mathrm{d}\tau = 0 \quad \Rightarrow \quad A \perp V. \tag{2.175}$$

$$A^2 = \mathbf{a}^2, \quad \mathbf{a} = \mathrm{d}\mathbf{v}/\mathrm{d}\tau.$$

Newton force $\mathbf{f}_\mathrm{N}$ and Einstein force $\mathbf{f}$:

$$\mathbf{f} = \mathbf{f}_{\mathrm{N}} + \frac{\gamma - 1}{w^2}(\mathbf{w}\mathbf{f}_{\mathrm{N}})\mathbf{w}, \qquad f^4 = -\gamma\frac{\mathbf{w}}{c}\mathbf{f}_{\mathrm{N}} = -\frac{\mathbf{w}}{c}\mathbf{f}. \tag{2.179}$$

$$\frac{\mathrm{d}}{\mathrm{d}t}(\gamma m\mathbf{v}) = q(\mathbf{E} + \mathbf{v}\times\mathbf{B}). \qquad \frac{\mathrm{d}}{\mathrm{d}t}m\gamma c^2 = q\mathbf{v}\mathbf{E}. \tag{2.180}$$

Particle energy E, Lagragian L, and Hamiltonian H:

$$E = E_0 + E_{\mathrm{kin}} = \gamma mc^2 \qquad \Leftrightarrow \qquad E^2 = m^2c^4 + p^2c^2. \tag{2.181}$$

$$L(\mathbf{x}, \mathbf{v}) = T - V = -\frac{m}{\gamma}c^2 - V. \tag{2.182}$$

$$H(\mathbf{p}, \mathbf{q}, t) = \gamma m\mathbf{v}^2 + \frac{mc^2}{\gamma} + q\Phi = \gamma mc^2 + q\Phi = [m^2c^4 + c^2(\mathbf{p} - q\mathbf{A})^2]^{1/2} + q\Phi(\mathbf{x}, t). \tag{2.186}$$

Addition theorems, velocities, and momenta:

$$\mathbf{u} = \frac{\mathbf{u}' + \mathbf{v}[\gamma(1 + \mathbf{u}'\mathbf{v}/v^2) - \mathbf{u}'\mathbf{v}/v^2]}{\gamma(1 + \mathbf{u}'\mathbf{v}/c^2)}. \tag{2.188}$$

$$P = (\mathbf{p}, \gamma mc) = \left(\frac{\mathbf{p}'/\gamma' + \mathbf{v}[\gamma(m + \mathbf{p}'\mathbf{v}/(\gamma' v^2)) - \mathbf{p}'\mathbf{v}/(\gamma' v^2)]}{1 + \mathbf{p}'\mathbf{v}/(\gamma' mc^2)}, \gamma mc\right). \tag{2.189}$$

Transverse wave, monochromatic or oscillation center exists, amplitude $\hat{\mathbf{A}}$ weakly space and time dependent:

$$H = \left[m^2c^4 + c^2(\mathbf{p} - q\mathbf{A})^2\right]^{1/2} + q\Phi; \quad \mathbf{p} - q\mathbf{A} = \gamma m\mathbf{v},$$
$$m_{\mathrm{osc}} = M_C = \frac{1}{T_C}\oint \gamma_C(v_C)m\mathrm{d}\tau; \ \text{ index C}: \mathbf{v}_0 = 0,$$

$$\oint[\mathbf{p}\mathrm{d}\mathbf{q} - H\mathrm{d}t] = [\gamma_0 m_{\mathrm{osc}}c^2]T = \mathrm{const_a}; \quad \gamma_0 = (1 - v_0^2/c^2)^{1/2}; \quad \mathbf{p}_0 = \gamma_0 m_{\mathrm{osc}}\mathbf{v}_0, \tag{2.215}$$
$$\varphi = \int(\mathbf{k}\mathrm{d}\mathbf{x} - \omega\mathrm{d}t) \text{ Lorentz invariant.}$$

Cycle averaged Hamiltonian $H_0(\mathbf{p}_0, \mathbf{x}_0)$; $\mathbf{p}_0, \mathbf{x}_0$ momentum and position of oscillation center, $\mathbf{f}_p$ ponderomotive force, lab frame:

$$H_0(\mathbf{p}_0, \mathbf{x}_0) = m_{\mathrm{osc}}c^2\left[1 + \left(\frac{\mathbf{p}_0}{m_{\mathrm{osc}}c}\right)^2\right]^{1/2}; \quad \mathbf{f}_p = -\frac{\partial H(\mathbf{p}_0, \mathbf{x}_0)}{\partial \mathbf{x}_0}. \tag{2.214}$$

Monochromatic $\Rightarrow m_{\mathrm{osc}} = m[1 + a^2/2]^{1/2}$
Poincaré–Cartan invariant of the longitudinal electron plasma wave, non relativistic:

$$T(E_0 + W - \overline{V}) = \text{const}'_a \quad \Leftrightarrow \quad E_0 + W - \overline{V} = \frac{1}{2} m_e v_i'^2. \tag{2.140}$$

For the harmonic wave $V = \hat{V} \cos kx$ in the wave frame it becomes

$$\frac{2\kappa}{\alpha} E(\alpha) = K(\alpha); \ \kappa = \frac{2\hat{V}}{m_e v_i'^2}, \ \alpha = \frac{2\kappa}{1+\kappa}. \tag{2.143}$$

2.7 Further Readings

A. Macchi, *A Superintense Laser-Plasma Interaction Theory Primer*, SpringerBriefs in Physics (Springer, Heidelberg, 2013), Chap. 2.
C. Lanczos, *The Variational Principles of Mechanics*, 4th edn. (Dover Publications, Inc., New York, 1970)
V.I. Arnold, *Mathematical Methods of Classical Mechanics* (Springer, New York, 1978)
J.D. Jackson, *Classical Electrodynamics*, 2nd Edn. (Wiley, New York, 1975)
H. Goldstein, *Classical Mechanics* (Addison, Wesley, Boston, 2002)
J. Mathews, R.L. Walker, *Mathematical Methods of Physics* (Benjamin/Cummings, Menlo Park, Ca., 1970)
S. Weinberg, *Gravitation and Cosmology* (Wiley, New York, 1972), Chap. 2

References

1. J. Mathews, R.L. Walker, *Mathematical Methods of Physics* (Benjamin, Menlo Park, 1970), Chap. 12-2
2. Private communication by Klaus Eidmann
3. M. Abramowitz, I.A. Stegun, *Handbook of Mathematical Functions* (Dover Publ. Inc., New York, 1970), Chap. 20
4. V.I. Arnold, *Mathematical Methods of Classical Mechanics* (Springer, New York, 1978), p. 297
5. V.I. Arnold, *Dynamical systems*, vol. 3 (Springer, New York, 1988), pp. 200–211
6. D. Bauer, P. Mulser, W.-H. Steeb, Phys. Rev. Lett. **75**, 4622 (1995)
7. H. Hora, D. Pfirsch, A. Schlüter, Z. Naturforsch. **22**, 278 (1967)
8. A.V. Gapunov and A.V. Miller, J. Exp. Theor. Phys. **7**, 168 (1958), H.A.H. Boot, S.A. Self, R.B.R. Shersby-Harvie, J. Electr. Control **4**, 434 (1958), E.S. Weibel, J. Electr. Control **5**, 435 (1958), L.D. Landau, E.M. Lifshitz, *Mechanics* (Pergamon Press, Oxford, 1976), p. 93, T.W.B. Kibble, Phys. Rev. **150**, 1060 (1966), F.A. Hopf, P. Meyestre, M.O. Scully, and W.H. Louisell, Phys. Rev. Lett. **37** 1342 (1976), R.W. Müller, GSI/Darmstadt (1984), private communication
9. M. Kono, M.M. Skoric, D. ter Haar, Phys. Rev. Lett. **45**, 1629 (1980); Phys. Fluids **27**, 1996 (1984)
10. M. Kono, M.M. Skoric, *Nonlinear Physics of Plasmas* (Springer, Berlin, 2010), Chap. 9
11. N. Iwata, Y. Kishimoto, Phys. Rev. Lett. **112**, 035002 (2014); Phys. Plasmas **8**, 1201160 (2013)
12. R.E. Aamodt, E.F. Jäger, Phys. Fluids **17**, 1386 (1974)
13. G.D. Birkhoff, Proc. Nat. Acad. Sci. U.S.A. **17**, 656 (1931)

14. M. Keane, K. Petersen, arXiv:math/0608251 [math.DS]. Accessed 10 Aug 2006
15. V. Berzi, V. Gorini, Math. Phys. **10**, 1518 (1969)
16. S. Weinberg, *Gravitation and Cosmology* (Wiley, New York, 1972), Chap. 2
17. P. Gibbon, A.R. Bell, Phys. Rev. Lett. **68**, 1535 (1992)
18. H. Ruhl, P. Mulser, Phys. Lett. A **205**, 388 (1995)
19. J.D. Lawson, IEEE Trans. Nucl. Sci. **NS-26**, 4217 (1979)
20. T.C. Pesch, H.-J. Kull, Phys. Plasmas **14**, 083103 (2007)
21. B. Quesnel, P. Mora, Phys. Rev. E **58**, 3719 (1998)
22. A.J. Castillo, V.P. Milantev, Tech. Phys. **59**, 1261 (2014)

Chapter 3
Laser Induced Fluid Dynamics

Characterization

In many respects the plasma may be considered as a medium that assumes any form provided by external boundaries and forces acting on it from outside. The main parameters are a local particle density $n(\mathbf{x}, t)$, mass density $\rho(\mathbf{x}, t)$, flow velocity $\mathbf{u}(\mathbf{x}, t)$, and internal energy density $\epsilon_{in}(\mathbf{x}, t)$. Microscopic order like crystalline structure or ordered micropatterns are of no interest. Such a medium is called a *fluid*. Examples of fluids are water, magma, sand, air, ideal gas, but also steel under high pressure in the form press, a neutron star, and the plasma. Steel and sand are *incompressible* under technical pressures. At sufficiently high pressure and strain all fluids turn into compressible media.

The forces acting on matter in a fluid element of volume ΔV mostly balance each other. Their nonzero resultant manifests itself as a *surface force* like *pressure*, shear and viscous forces, to be characterized as a class of *macroscopic forces*. If the forces in the fluid of finite extension exactly balance locally, like cohesion or surface tension, they are classified as *internal forces*. Internal forces have no influence on the dynamics of the fluid as a whole, they do not change the motion of its center of mass. However internal forces act microscopically through the border delimiting an arbitrary portion of the fluid. The depth of action of the surface forces may be characterized by the mean free path of the fluid carriers in neutral gases, or by the local Debye length in charged fluids like ideal plasmas, or by microscopic electrostatic forces between neighbours in dense fluids and nonideal plasmas. The second class of macroscopic forces are the *impressed* or *external* forces, like the Lorentz force or gravity. They are *volume forces* and are in their nature entirely different from surface forces; for instance, they do not change the entropy of the fluid.

In a specific sense the description of the plasma as a fluid is equivalent to the reduction of an unsolvable many body problem to a single particle problem via the mass element $\Delta M = \rho \Delta V$, $\rho = \rho(\mathbf{x}, t)$ local mass density. The reduction is achieved by coarse graining, the mass element $\Delta M = \rho \Delta V$ is to be chosen small on the macroscopic scale, and large and convex (to include the holes between the single particles) on the microscopic dimension. With n the particle density and λ the mean

P. Mulser, *Hot Matter from High-Power Lasers*, Graduate Texts in Physics,
https://doi.org/10.1007/978-3-662-61181-4_3

free path the extension $(\Delta V)^{1/3}$ of the fluid element has to obey the inequality

$$\{n^{-1/3}, \lambda\} \ll (\Delta V)^{1/3} \ll L = \frac{n}{|\nabla n|}. \tag{3.1}$$

For clarity, in the following the volume element is indicated either by the symbol Δ for difference or by d for the differential with the same significance since both are finite and differ only by higher order, $\Delta V = \mathrm{d}V + o(\mathrm{d}V)$. The surface of ΔV is indicated by $\Sigma(\Delta V)$. The surface element $\mathrm{d}\boldsymbol{\Sigma}$ is oriented outward by definition. The dynamics of the *rarefied* or *collisionless fluid* obeys the same conservation laws, however the dynamic quantities, like flow velocity, energy density, and pressure are defined by folding the one particle distribution function with the corresponding dynamic variable.

3.1 Conservation Laws

A fluid is characterized by a homogeneous composition, particle density $n(\mathbf{x}, t)$, mass density $\rho(\mathbf{x}, t) = mn(\mathbf{x}, t)$, electric charge density $\rho_{\mathrm{el}}(\mathbf{x}, t)$, flow velocity $\mathbf{u}(\mathbf{x}, t)$, pressure tensor $P(\mathbf{x}, t)$ and, in the case of *local thermodynamic equilibrium* (LTE), by $T(\mathbf{x}, t)$.

3.1.1 Particle and Mass Conservation

During the time $\mathrm{d}t$ the number of particles $\mathrm{d}N$ flow into the fixed volume V,

$$\mathrm{d}N = -\int_{\Sigma(V)} n(\mathbf{u}\mathrm{d}t)\mathrm{d}\,\boldsymbol{\Sigma} \quad \Rightarrow \quad \frac{\mathrm{d}N}{\mathrm{d}t} = -\int_{\Sigma(V)} n\mathbf{u}\mathrm{d}\,\boldsymbol{\Sigma},$$

see Fig. 3.1. With the help of Gauss' law follows

$$\frac{\mathrm{d}N}{\mathrm{d}t} = \frac{\mathrm{d}}{\mathrm{d}t}\int_V n\mathrm{d}V = \int_V \frac{\partial n}{\partial t}\mathrm{d}V = -\int_{\Sigma(V)} n\mathbf{u}\mathrm{d}\,\boldsymbol{\Sigma} = -\int_V \nabla(n\mathbf{u})\mathrm{d}V.$$

From V arbitrary and the integrands assumed continuous follow the particle and mass conservation laws

$$\frac{\partial n}{\partial t} + \nabla(n\mathbf{u}) = 0, \qquad \frac{\partial \rho}{\partial t} + \nabla(\rho\mathbf{u}) = 0. \tag{3.2}$$

The fluid is *incompressible* if $\mathrm{d}n/\mathrm{d}t = 0$. This translates to the equivalent condition $\operatorname{div}\mathbf{u} = 0$ since

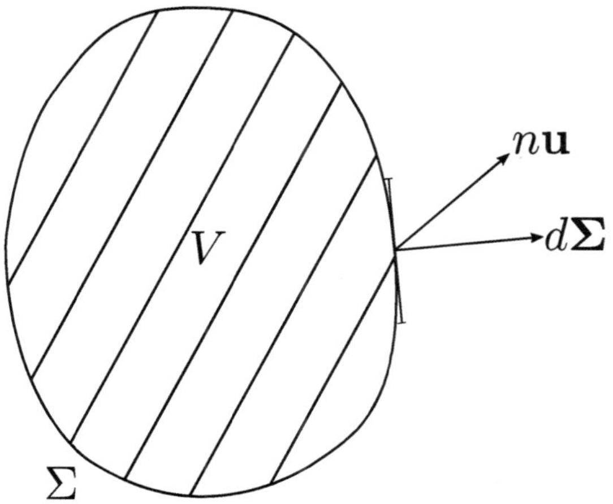

Fig. 3.1 Particle and mass conservation in arbitrary volume V fixed in space. During the time $\mathrm{d}t$ the number of particles $\mathrm{d}n = -n\mathbf{u}\mathrm{d}t\mathrm{d}\,\boldsymbol{\Sigma}$ cross the surface element $|\mathrm{d}\,\boldsymbol{\Sigma}|$. $\mathrm{d}n > 0$: particle increase, $\mathrm{d}n < 0$: particle loss in V

$$\frac{\mathrm{d}n}{\mathrm{d}t} = \frac{\partial n}{\partial t} + \mathbf{u}\nabla n = -n\nabla\mathbf{u} = 0.$$

The fluid flow is *stationary* if an inertial reference system can be found in which the partial derivative $\partial n/\partial t = 0$ holds. It implies that the flow is divergence-free, $\operatorname{div} n\mathbf{u} = 0$, or not explicitly time-dependent. If λ particles or μ mass per unit volume and unit time are created (3.2) is to be replaced by

$$\frac{\partial n}{\partial t} + \nabla(n\mathbf{u}) = \lambda, \qquad \frac{\partial \rho}{\partial t} + \nabla(\rho\mathbf{u}) = \mu. \tag{3.3}$$

Linearized particle conservation. Assume a uniform fluid at rest,

$$n = n_0 = \text{const}, \quad \mathbf{u} = 0.$$

and a small disturbance in particle density $n_1 \ll n_0$ and velocity $\mathbf{u}$. If only the first order terms are kept (3.2) reduces to the linear conservation equation

$$\frac{\partial n_1}{\partial t} + n_0\nabla\mathbf{u} = 0; \qquad \nabla(n_1\mathbf{u}) \ll \nabla(n_0\mathbf{u}). \tag{3.4}$$

A useful relation follows in terms of the fluid displacement $\boldsymbol{\delta}(\mathbf{x}, t)$. If $|(\boldsymbol{\delta}\nabla)\boldsymbol{\delta}(\mathbf{x}, t)| \ll |\boldsymbol{\delta}(\mathbf{x}, t)|$, as for example if $\boldsymbol{\delta}(\mathbf{x}, t)$ is small, the conservation equation becomes

$$0 = \frac{\partial n_1}{\partial t} + n_0\nabla\frac{\mathrm{d}\boldsymbol{\delta}}{\mathrm{d}t} = \frac{\partial n_1}{\partial t} + n_0\nabla\left[\frac{\partial\boldsymbol{\delta}}{\partial t} + (\mathbf{u}\nabla)\boldsymbol{\delta}\right] \simeq \frac{\partial n_1}{\partial t} + n_0\nabla\frac{\partial\boldsymbol{\delta}}{\partial t}.$$

Integration in time of the last sum leads to

$$n_1(\mathbf{x}, t) = -\nabla(n_0\boldsymbol{\delta}). \tag{3.5}$$

The formula relates the local density perturbation to the divergence of the local displacement $\boldsymbol{\delta}$.

3.1.2 Navier–Stokes Equation

Pressure-Viscosity Tensor

The viscosity is a shear force acting on the surface of the volume element $\mathrm{d}V$. Its origin derives from the momentum exchange by particle diffusion through the surface of $\mathrm{d}V$; it is illustrated by Fig. 3.2. The momentum flux of particles into the volume element through the top surface $\mathrm{d}\,\boldsymbol{\Sigma}_y = \mathbf{e}_y \mathrm{d}x\mathrm{d}z$ per unit time is $mn\langle u_{x1}\mathbf{v}\rangle \mathrm{d}\,\boldsymbol{\Sigma} = -mn[\langle u_{x1}v_y\rangle \mathrm{d}\,\boldsymbol{\Sigma}_y + \langle u_{x1}v_z\rangle \mathrm{d}\,\boldsymbol{\Sigma}_z]$ and is deposited inside within a mean free path λ_y and λ_z, respectively. At the same time the momentum loss by outflowing particles is $mn\langle u_{x2}\mathbf{v}\rangle \mathrm{d}\,\boldsymbol{\Sigma}$.

Note, $n\mathbf{v}$ is the same in both cases. In the one component fluid any difference between them gives a contribution to the mean flow $\rho\mathbf{u}$ and as such is classified as the separate phenomenon of *convection*. In the charged fluid, e.g., the plasma, charge separation sets in a very short time and stops convection.

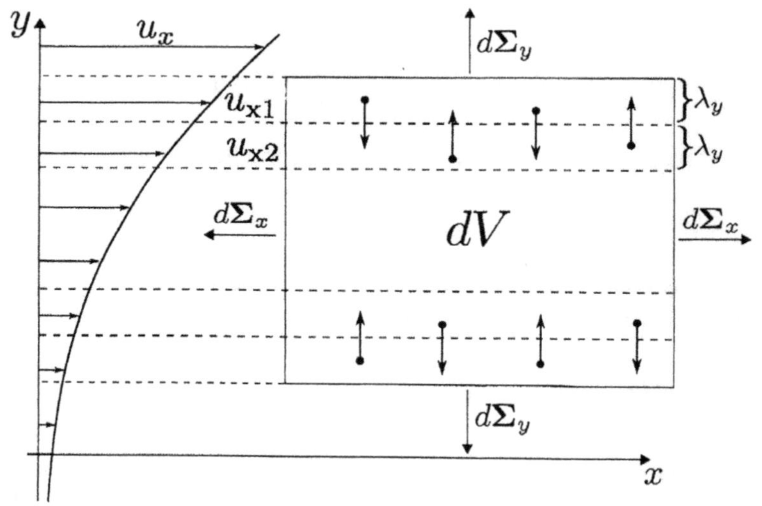

Fig. 3.2 Origin of the shear force acting on volume element $\mathrm{d}V$ in x-direction. Momentum mu_x and mu_z exchange in $\mathrm{d}\boldsymbol{\Sigma}_y$ direction occurs from a mean free path layer above the $\mathrm{d}x\mathrm{d}z$ surface and a mean free path layer underneath, indicated by vertical arrows. The single particle starts at a position y_1 in the upper layer with momentum mu_{x1} and deposits it after a mean free path λ_y in the layer underneath, and vice versa. The mean fluxes $n_1\langle v_{y1}\rangle$ from above and $n_2\langle v_{y2}\rangle$ from below are the same and equal to $n\langle v_y\rangle$. Momentum exchange of mu_z is analogous

With $u_{x1} - u_{x2} = (\partial u_x/\partial y)\lambda_y + (\partial u_x/\partial z)\lambda_z$, $\lambda_{y,z} = \langle v_{y,z}\rangle\tau$, $\tau = 1/\nu$ the mean collision time, the net force in x-direction $\mathrm{d} f_x$ = momentum gain per unit time flowing in laterally along $\mathrm{d}\,\boldsymbol{\Sigma}_y$ and $\mathrm{d}\,\boldsymbol{\Sigma}_z$ is

$$\mathrm{d} f_x = -P_{xy}\mathrm{d}\,\Sigma_y - P_{xz}\mathrm{d}\,\Sigma_z = -mv_y^2\tau\frac{\partial u_x}{\partial y}\,\mathrm{d}\,\Sigma_y - mv_z^2\tau\frac{\partial u_x}{\partial z}\,\mathrm{d}\,\Sigma_z.$$

The remaining two components follow accordingly as

$$\mathrm{d} f_y = -P_{yx}\mathrm{d}\,\Sigma_x - P_{yz}\mathrm{d}\,\Sigma_z, \quad \mathrm{d} f_z = -P_{zx}\mathrm{d}\,\Sigma_x - P_{zy}\mathrm{d}\,\Sigma_y.$$

The individual velocity components v_x, v_y, v_z entering linearly in the formulas above are the sum of flow velocity u_i and the relative kinetic (thermal) velocity component. In the differences between momentum gain and loss the flow components u_i automatically cancel; hence the v_i components may be considered as the kinetic components of the fluid at rest. In the expressions containing v_i^2 only the kinetic components enter with the fluid at rest.

From the construction of $P_{xy} = mn\langle v_y^2\rangle\tau\,\partial u_x/\partial y$ follows that the components $P_{ij},\ i, j \to x, y, z$, form a tensor P. In the isotropic fluid it is natural to extend isotropy to the velocity distribution, $v_x^2 = v_y^2 = v_z^2 = \mathbf{v}^2/3$. Furthermore, to avoid that adjacent volume elements do rotate against each other $\nabla \times \mathbf{u} = 0 \Leftrightarrow \partial u_i/\partial x_j = \partial u_j/\partial x_i$ must be imposed. Hence, the *viscosity tensor*, or *viscous stress tensor*, of shear is symmetric,

$$P = P_{ij} = mn\langle\mathbf{v}^2/3\rangle\tau\,\frac{\partial u_i}{\partial x_j} = 2\mu\frac{\partial u_i}{\partial x_j} = \mu\left(\frac{\partial u_i}{\partial x_j} + \frac{\partial u_j}{\partial x_i}\right) = P_{ji},\ i \neq j. \quad (3.6)$$

The shear viscosity coefficient μ of the dilute fluid is independent of particle density.

The force components acting perpendicularly onto the surfaces of $\mathrm{d}V$, i.e., antiparallel to $\mathrm{d}\,\boldsymbol{\Sigma}_i$, are the negative pressure components $-p_{xx}, -p_{yy}, -p_{zz}$. They do not generate shear, $P_{ii} = 0\ \ i = 1-3$. This is equivalent to the requirement of vanishing trace, $\mathrm{Tr}P = 0$, because the trace is an invariant of reference system rotations. Combining this property with P from above the complete shear viscosity tensor reads

$$P = P_{ij} = \mu\left(\frac{\partial u_i}{\partial x_j} + \frac{\partial u_j}{\partial x_i} - \frac{2}{3}\delta_{ij}\nabla\mathbf{u}\right); \qquad \mu = mn\langle\mathbf{v}^2/6\rangle\tau. \qquad (3.7)$$

If the isotropic fluid is exposed to weak external forces, or if it is highly collisional it can be assumed that isotropy is preserved everywhere throughout the fluid and the pressure remains isotropic too, $p_{xx} = p_{yy} = p_{zz} = p$.

When the volume element $\mathrm{d}V$ is compressed or expanded an internal friction force arises from momentum diffusion of the single particles over one mean free path just in exactly the same way as with the shear viscosity,

$$\mathrm{d}f_x = -m\langle v_x^2\rangle\tau_x'\frac{\partial u_x}{\partial x} = -\mu_x'\frac{\partial u_x}{\partial x}|\mathrm{d}\boldsymbol{\Sigma}_x|; \qquad \mathrm{d}f_i = -\mu_i'\frac{\partial u_i}{\partial x_i}|\mathrm{d}\,\boldsymbol{\Sigma}_i|$$

with τ_x' the mean collision or relaxation time. Under the condition of isotropy $\mu_i' = \mu' = mn\langle \mathbf{v}^2/3\rangle\tau'$ holds. As a rule, τ and $\tau' = \tau_x' = \tau_y' = \tau_z'$ may differ from each other. In contrast to shear, μ' is the coefficient of *volume viscosity* or *bulk viscosity*, or *second viscosity* coefficient. The total pressure-viscosity tensor Π of the isotropic fluid can be summarized as

$$\Pi = \Pi_{ij} = \left(p + \mu'\nabla\frac{\mathbf{u}}{3}\right)\delta_{ij} + P_{ij} \tag{3.8}$$

with P_{ij} from (3.7). The single force element acting on the surface in i-direction is

$$\mathrm{d}f_i = -\sum_{j=1}^{3}\Pi_{ij}\mathrm{d}\,\Sigma_j. \tag{3.9}$$

Volume Forces

In conducting fluids like the plasma the Lorentz force is of eminent significance. It is obtained by summing all individual forces (2.180) in the volume element. Owing to its restricted dimensions retardation effects can be ignored and Newton's 3rd principle, actio = reactio, holds. As the result of mutual compensations in leading order the force density $\pi_L' = \rho_{\mathrm{el}}(\mathbf{x}', t')\mathbf{E}'(\mathbf{x}', t')$ is obtained in the comoving reference system $S'(\mathbf{u})$. Thus, in the lab frame the relativistic force density is

$$\pi(\mathbf{x}, t) = \rho_{\mathrm{el}}(\mathbf{x}, t)[\mathbf{E}(\mathbf{x}, t) + \mathbf{u}(\mathbf{x}, t) \times \mathbf{B}(\mathbf{x}, t)] = \rho_{\mathrm{el}}\mathbf{E} + \mathbf{j} \times \mathbf{B}; \quad \mathbf{j} = \rho_{\mathrm{el}}\mathbf{u}. \tag{3.10}$$

In the two-component fluid of electrons and ions friction may play an important role. Introduced already by (2.16) for the single particle in Chap. 2 it generalizes straightforwardly to the electron fluid,

$$\boldsymbol{\pi}_f = -m_e n \nu_{ei}(\mathbf{u}_e - \mathbf{u}_i) \tag{3.11}$$

with ν_{ei} the mean electron-ion collision frequency. The collision frequency does not depend on the difference of the flow velocities as long as the electron kinetic velocity, in particular thermal velocity, is much larger than $|\mathbf{u}_e|$. Electron-electron and ion-ion collisions do not contribute because the flow velocities $\mathbf{u}_e$, $\mathbf{u}_i$ have no influence on collisions between partners of the same species. They may however indirectly influence ν_{ei} through their distribution functions (see Chap. 7 on Transport in Plasma).

A third kind of volume forces in laser-fluid interaction is the ponderomotive force density $\boldsymbol{\pi}_p$. In the fully ionized plasma it is given in its leading term by $\boldsymbol{\pi}_p = \boldsymbol{\pi}_0$ according to (2.118),

$$\boldsymbol{\pi}_0 = -n\nabla\Phi_p = -\frac{ne^2}{4m_e\omega^2}\nabla\mathbf{E}\mathbf{E}^*. \tag{3.12}$$

Momentum Conservation

Consider an arbitrary Volume V fixed in space. The total momentum flux $\mathbf{M}$ per unit time through its surface Σ is

$$M_i = -\int_{\Sigma(V)}\sum_j P_{ij}\mathrm{d}\boldsymbol{\Sigma}_j = -\int_V\sum_j\frac{\partial P_{ij}}{\partial x_j}\mathrm{d}V.$$

Including the volume force densities $\mathbf{f}$ for the mass element $\mathrm{d}M = \rho\mathrm{d}V$ this translates into,

$$\mathrm{d}M\frac{\mathrm{d}u_i}{\mathrm{d}t} = \rho\mathrm{d}V\frac{\mathrm{d}u_i}{\mathrm{d}t} = -\left(\sum_j\frac{\partial\Pi_{ij}}{\partial x_j}\right)\mathrm{d}V + f_i\mathrm{d}V;\ i = 1-3.$$

This is the *Navier–Stokes equation*,

$$\rho\left[\frac{\partial}{\partial t} + (\mathbf{u}\nabla)\right]\mathbf{u} = -\nabla\Pi + \boldsymbol{f}. \tag{3.13}$$

$$\nabla\Pi = \mathrm{div}\Pi = \sum_j\frac{\partial\Pi_{ij}}{\partial x_j},\ \ i = 1, 2, 3$$

$$\Pi_{ij} = \left(p + \mu'\nabla\frac{\mathbf{u}}{3}\right)\delta_{ij} + P_{ij}, \qquad P_{ij} = \mu\left(\frac{\partial u_i}{\partial x_j} + \frac{\partial u_j}{\partial x_i} - \frac{2}{3}\delta_{ij}\nabla\mathbf{u}\right). \tag{3.14}$$

In explicit form (3.13) reads

$$\rho[\partial_t + (\mathbf{u}\nabla)]u_i = -\partial_i\left(p + \mu'\nabla\frac{\mathbf{u}}{3}\right) - \mu\nabla^2 u_i + \frac{\mu}{3}\partial_i\nabla\mathbf{u} + f_i - R_i$$

$$R_i = \mathrm{grad}\mu(\mathrm{grad}u_i + \partial_i\nabla\mathbf{u}) - \frac{2}{3}\partial_i\mu\nabla\,\mathbf{u}.$$

Under the assumption of $\mu' = 2\mu$ and $|\partial_j\mu| \ll |\partial_j u_i|$ (3.13) simplifies to

$$\rho\left[\frac{\partial}{\partial t} + (\mathbf{u}\nabla)\right]\mathbf{u} = -\nabla p - \mu\Delta\mathbf{u} - \mu\mathrm{grad}\nabla\mathbf{u} + \mathbf{f}. \tag{3.15}$$

Often this form is understood as the proper Navier–Stokes equation. Under the additional assumption of incompressibility follows from (3.15) and $(\mathbf{u}\nabla\mathbf{u}) = \nabla\mathbf{u}^2/2 - \mathbf{u}\times(\nabla\times\mathbf{u})$

$$\frac{\partial}{\partial t}\mathbf{u} = -\frac{1}{\rho}\nabla\left(p+\frac{1}{2}\rho\mathbf{u}^2\right) + \mathbf{u}\times(\nabla\times\mathbf{u}) - \frac{\mu}{\rho}\Delta\mathbf{u} + \frac{1}{\rho}\mathbf{f}. \tag{3.16}$$

The Euler Equation

The *ideal fluid* is defined by the absence of shear and volume viscosity, $\mu = \mu' = 0$. Its dynamics underlies the *Euler equation*

$$\rho\left[\frac{\partial}{\partial t} + (\mathbf{u}\nabla)\right]\mathbf{u} = -\nabla p + \mathbf{f}. \tag{3.17}$$

In the anisotropic fluid, e.g., owing to a magnetic field, the gradient of the scalar pressure p is replaced by the divergence of the pressure tensor p_{ij}.

Bernoulli's Law

Let us chose an ideal fluid (i) in a steady state (ii) that is subject to a volume potential force per particle $\mathbf{f} = -\nabla\Phi$. The integration of the Euler equation along a stream line $\mathbf{s}$ yields owing to $\mathbf{u}\times(\nabla\times\mathbf{u}) \perp (\mathbf{u} \parallel d\mathbf{s})$

$$\int\frac{1}{\rho}\nabla p\, d\mathbf{s} + \frac{1}{2}\mathbf{u}^2 + \frac{1}{m}\Phi = \text{const.} \tag{3.18}$$

This is Bernoulli's law for the compressible ideal fluid. For its relevance to applications it is specialized to one dimensional flow. From $\rho\mathbf{u} = \text{const}$ inserted in Euler's equation (3.17) results in

$$p + \rho\mathbf{u}^2 + \frac{\rho}{m}\Phi = \text{const.} \tag{3.19}$$

In the incompressible fluid, $\rho = \text{const}$, Bernoulli's law in three dimensions (3D) assumes the familiar form

$$p + \frac{1}{2}\rho\mathbf{u}^2 + \frac{\rho}{m}\Phi = \text{const.} \tag{3.20}$$

Note, in one dimension the conserved quantities (3.19) and (3.20) differ by a factor of 2 in $\rho\mathbf{u}^2$.

For the isothermal atmosphere in equilibrium Bernoulli's law yields another well known relation,

$$\nabla p + mng = 0 \quad\Rightarrow\quad k_BT\frac{\partial n}{\partial z} + mng = 0 \quad\Rightarrow\quad n(z) = n_0\exp-\frac{mgz}{k_BT}.$$

In plasma equilibrium the gravitational potential is replaced by the electrostatic potential Φ to yield the electron density n_e in dependence of Φ,

$$n_e(\mathbf{x}) = n_0 \exp \frac{e\Phi(\mathbf{x})}{k_B T}. \tag{3.21}$$

n_0 is the electron density at position $\Phi(\mathbf{x}) = 0$.

3.1.3 Energy Conservation

Kinetic and Internal Energies

The volume forces, i.e., external forces applied from outside, impart kinetic energy of density $\epsilon_{\text{kin}} = \rho\mathbf{u}^2/2$, $\mathrm{d}E_{\text{kin}} = \epsilon_{\text{kin}}\mathrm{d}V$ to the fluid element $\mathrm{d}V$. *Internal energy* $\mathrm{d}E_{\text{in}} = \epsilon_{\text{in}}\mathrm{d}V$ is supplied by the work of internal forces: by friction of one fluid against the other fluid component, by viscosity, and by mechanical work of the pressure, by heat supply from outside through thermal conduction, by absorption and emission of radiation. Sometimes it may be a question of modelling whether internal energy is intended as to be fed by heat or by mechanical work. For example, Ohmic heating may be considered as a thermal process although in the microscopic picture energy is supplied to the electrons by the electric field. Subsequently the imparted kinetic energy is converted into heat by collisions of the electrons with the ions. The transformation of mechanical work into internal energy, and vice versa, is accounted for in the first principle of thermodynamics in which heat and mechanical work enter as a sum. Reduced to its very essence this principle ensures the existence of an internal energy in the absence of macroscopic fluid motion, irrespective of whether it is in thermal equilibrium or far from it.

Let $\mathbf{q}$ be the energy density flux without material convection and $\dot{h}$ the heating function, i.e., the supply of internal energy density per unit time mediated by the external sources, as for instance heating by laser irradiation. Then the energy balance of the volume element $\mathrm{d}V$ at rest in the tangent system $S'(\mathbf{u})$ reads as follows,

$$\frac{\partial}{\partial t'}(\epsilon_{\text{in}}\mathrm{d}V) = \dot{h}\mathrm{d}V - \int \mathbf{q}\mathrm{d}\,\boldsymbol{\Sigma} - \sum_{ij}\int \Pi_{ij}u'_j\mathrm{d}\,\boldsymbol{\Sigma}_j = \left[\dot{h} - \nabla\mathbf{q} - \sum_{ij}\partial_j(\Pi_{ij}u'_j)\right]\mathrm{d}V.$$

The balance is based on the following model, see Fig. 3.3. In case ϵ_{in} is thermalized the energy balance expresses the first law of thermodynamics. Owing to $\mathbf{u}' = 0$ in $S'(\mathbf{u})$ the last term simplifies to $\sum_{ij}\partial_j(\Pi_{ij}u'_j) = \sum_{ij}\Pi_{ij}\partial_j u_j$.

Energy Conservation

For compactness reasons from here on, except special occasions, $\mu = 0$ is set. Owing to the smallness of shear viscosity in the laser plasma such a setting is widely justified. With $\mathrm{d}M = \rho\mathrm{d}V = \text{const}$ with respect to time the balance relation above reads in

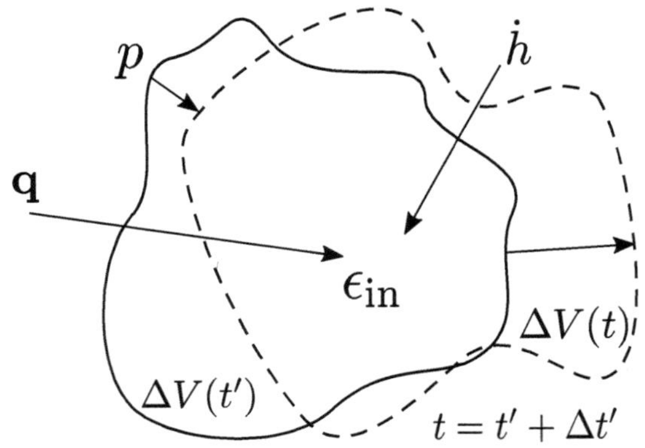

Fig. 3.3 Internal energy is supplied by $\dot{h}$ and thermal conduction $\mathbf{q}$ from outside to the comoving *rigid* volume element $\Delta V(t')$ at time t' and subsequently ΔV is deformed from $V(t')$ to $V(t)$ during the infinitesimal time interval $\Delta t'$ by the pressure-viscosity term, here p for simplicity

the lab frame as

$$\rho[\partial_t + (\mathbf{u}\nabla)]\frac{\epsilon_{\text{in}}}{\rho} = -\left(p + \mu'\nabla\frac{\mathbf{u}}{3}\right)\nabla\mathbf{u} - \nabla\mathbf{q} + \dot{h}. \tag{3.22}$$

Remark 3.1 Heat flux $\mathbf{q}$, pressure p, μ', and $\dot{h}$ have to be calculated in the comoving frame $S'(\mathbf{u})$ to keep these quantities well separated from convection.

Remark 3.2 $\rho \mathrm{d}V \mathrm{d}(\epsilon_{\text{in}}/\rho)/\mathrm{d}t$ is the change of internal energy per unit time of the number $\mathrm{d}N$ of particles held fixed. Hence, $\rho\mathrm{d}(\epsilon_{\text{in}}/\rho)/\mathrm{d}t$ is the corresponding quantity per unit volume. In classical terms the change is due to variation of the interparticle distances (interparticle potentials) and their kinetic energy (temperature); in the quantum picture it is caused by changes in the occupation numbers of the energy levels and the shift of the single energy levels, for instance due to compression.

Special Cases

(1) The fluid is in local thermodynamic equilibrium (**LTE**). It holds $\mathrm{d}(\epsilon_{\text{in}}\mathrm{d}V) = \mathrm{d}(c_V \mathrm{d}T\mathrm{d}V) = \mathrm{d}M\mathrm{d}[(c_V/\rho)\mathrm{d}T]$, c_V, c_V/ρ specific heat per unit volume and per unit mass, respectively.
(2) **Ideal classical gas**:

$$\epsilon_{\text{in}} = (f/2)nk_BT, \;\; p = nk_BT, \;\; c_V = (f/2)k_B, \;\; \mu' = 0. \tag{3.23}$$

f number of degrees of freedom.
(3) **Adiabatic changes**, i.e., entropy conserving processes: Set $\mathbf{q} = 0$, $\dot{h} = 0$. With $-\nabla\mathbf{u} = (\mathrm{d}n/\mathrm{d}t)/n$ (3.22) becomes for the ideal gas

$$\frac{f}{2}k_B\frac{\mathrm{d}T}{\mathrm{d}t} - k_B T\frac{\mathrm{d}n}{\mathrm{d}t} = 0 \;\Rightarrow\; \frac{\mathrm{d}}{\mathrm{d}t}\left(T^{f/2}n^{-1}\right) = 0$$

$$\Rightarrow\; Tn^{-2/f} = Tn^{-\gamma+1} = \text{const} \;\Leftrightarrow\; pn^{-\gamma} = \text{const}\;;\gamma = \frac{f+2}{f}, \tag{3.24}$$

"const" means constant with respect to any shift in space and time of a fixed volume element. The adiabatic coefficient γ is the nonrelativistic value. It is the same for the ideal classical as well as the ideal Fermi and Bose gases. The superrelativistic value is $\gamma_{\text{sr}} = (f+1)/f$, applicable for instance to photons.
(4) **Change of entropy**:

$$\mathrm{d}\epsilon_{\text{in}} = T\mathrm{d}\sigma + p\mathrm{d}\left(\frac{1}{n}\right) = T\mathrm{d}\sigma + \frac{f}{2}nk_B\mathrm{d}T \;\Rightarrow\; T\mathrm{d}\sigma = 0 \;\Rightarrow\; \sigma = \text{const}. \tag{3.25}$$

The quantity σ is the entropy per unit volume, $\sigma = \mathrm{d}S/\mathrm{d}V$. Conservation of entropy in the adiabatic process will be shown later in full generality by thermodynamic arguments (see next chapter).

3.1.4 *Two-Fluid Model of the Fully Ionized Plasma*

The description of a plasma in terms of two interpenetrating fluids coupled together by electrostatic forces and by friction has proved to be extremely successful. Owing to the different inertial response of the two fluids to external forces it is most natural to introduce two time scales to describe their dynamics. In first place electron currents and space charges build up as a reaction to the forces imposed from outside. They give rise to electric and magnetic fields which, in turn, act back onto the electron fluid as a whole. They build up during the time $\tau \sim 1/\omega_p$; ω_p electron plasma frequency. These fields tell the ions how to move. Owing to their much higher inertia the ion response takes place on the well distinct slow time scale. As the electrons are tied to the ions by quasineutrality they finally are the winners and tell the electrons how to move on the slow ion time scale. The electron-ion interplay generates the rich variety of *collective effects*. They may be considered as the signature of the plasma. Besides coupling by the collective fields interparticle momentum and energy exchanges provide for a second class of the two-fluid coupling. The most familiar examples of this kind are friction, e.g., Ohmic resistance, and exchange of internal energy, in particular two-temperature relaxation. The step from their microscopic origin to the macroscopic effects affecting the fluids is accomplished by the introduction of *transport coefficients*, derivable only from single particle kinetics.

Coupled Electron-Ion Fluids

The ion charge is assumed to be Z-fold, $q = Ze$, with Z constant in space and time. In the absence of external forces quasineutrality is assumed, $n_e(\mathbf{x}, t) = Zn_i(\mathbf{x}, t)$;

the charge density is $\rho_{\mathrm{el}} = e(Zn_i - n_e)$. The internal forces are the scalar pressures $p_{e,i}$ and the friction between the electron and ion fluids. Particle and momentum conservation are as follows,

$$\frac{\partial n_e}{\partial t} + \nabla(n_e \mathbf{u}_e) = 0, \qquad \frac{\partial n_i}{\partial t} + \nabla(n_i \mathbf{u}_i) = 0; \tag{3.26}$$

$$[\partial_t + (\mathbf{u}_e \nabla)]\mathbf{u}_e = -\frac{1}{m_e n_e}\nabla p_e - \frac{e}{m_e}(\mathbf{E} + \mathbf{u}_e \times \mathbf{B}) - \nu_{ei}(\mathbf{u}_e - \mathbf{u}_i). \tag{3.27}$$

$$[\partial_t + (\mathbf{u}_i \nabla)]\mathbf{u}_i = -\frac{1}{m_i n_i}\nabla p_i + \frac{Ze}{m_i}(\mathbf{E} + \mathbf{u}_i \times \mathbf{B}) + \frac{m_e}{m_i}\nu_{ei}(\mathbf{u}_e - \mathbf{u}_i). \tag{3.28}$$

Note, the friction force densities $\boldsymbol{\pi}_{fe,i}$ on electrons and ions are equal and opposite to each other, $\boldsymbol{\pi}_{fi} = -\boldsymbol{\pi}_{fe}$ [see (3.11)]. From strict application of actio = reactio follows that they have to be multiplied by μ/m_e and μ/m_i, respectively; $\mu = m_e m_i/(m_e + m_i)$. Volume and shear viscosity are ignored.

The overwhelming majority of heating mechanisms acts on the electrons, e.g., laser irradiation or ion stopping. Heating by shock waves and by resonances at ion frequencies are the exceptions. The internal energy densities of the electrons and the ions are the averages of the energy moments centered around their flow velocities, $\epsilon_{e,i} = m_{e,i} n_{e,i} \langle \mathbf{v}_{e,i}^2 \rangle/2$; in LTE they are the thermal motions. In general $\langle \mathbf{v}_{e,i}^2 \rangle \gg \mathbf{u}_{e,i}^2$ holds. In the electron-ion collision the electron transmits the momentum $\nu_{ei} m_e |\langle \mathbf{v}_e \rangle|$ per unit time to the ions. For the energy transmission to the ion in a collision this results in

$$\Delta \frac{\epsilon_i}{n_e} = \nu_{ei} \frac{1}{2} m_e \langle \mathbf{v}_e^2 \rangle \left\{ 1 - \left(1 - \frac{m_e}{m_i} \right)^2 \right\} = \nu_{ei} \frac{m_e}{m_i} \langle \mathbf{v}_e^2 \rangle m_e.$$

Analogously, the ion gives the amount of energy back to the electron per collision

$$\Delta \frac{\epsilon_i}{n_e} = \nu_{ei} \frac{m_e}{m_i} \langle \mathbf{v}_i^2 \rangle m_i.$$

Therefore the energy balances for the electron and the ion fluid read

$$n_e[\partial_t + (\mathbf{u}_e \nabla)]\frac{\epsilon_e}{n_e} = -p_e \nabla \mathbf{u}_e - \nabla \mathbf{q} - \nu_{ei} n_e \frac{m_e}{m_i}\left(m_e \langle \mathbf{v}_e^2 \rangle - m_i \langle \mathbf{v}_i^2 \rangle\right) + \dot{h}_e$$

$$n_i[\partial_t + (\mathbf{u}_i \nabla)]\frac{\epsilon_i}{n_i} = -p_i \nabla \mathbf{u}_i + \nu_{ei} n_e \frac{m_e}{m_i}\left(m_e \langle \mathbf{v}_e^2 \rangle - m_i \langle \mathbf{v}_i^2 \rangle\right) + \dot{h}_i. \tag{3.29}$$

Ion heat conduction $\mathbf{q}_i$ is ignored.

(a) **Two-Fluid Model of the Ideal Laser Plasma in LTE**

The dilute laser plasma in LTE with collisional heating $\dot{h}_e$ is characterized by

$$\frac{\epsilon_{e,i}}{n_{e,i}} = \frac{3}{2}k_B T_{e,i}, \qquad p_{e,i} = n_{e,i}k_B T_{e,i}, \qquad \dot{h}_e = \alpha(n_e, T_e)I,$$

$$\nu_{ei}n_e\frac{2m_e}{m_i}\left(m_e\langle\mathbf{v}_e^2\rangle - m_i\langle\mathbf{v}_i^2\rangle\right) = \nu_{ei}n_e\frac{m_e}{m_i} \times 3n_e k_B(T_e - T_i).$$

The collisional absorption coefficient is α and I is the laser intensity. We summarize the conservation equations under LTE for their practical relevance. With the assumptions above and no friction (except in α) (3.26), (3.28), (3.29) reduce to

$$\frac{\partial n_e}{\partial t} + \nabla(n_e\mathbf{u}_e) = 0, \qquad \frac{\partial n_i}{\partial t} + \nabla(n_i\mathbf{u}_i) = 0; \qquad n_e = Zn_i. \tag{3.30}$$

$$m_e n_e\frac{\mathrm{d}}{\mathrm{d}t}\mathbf{u}_e = -\nabla p_e - n_e e(\mathbf{E} + \mathbf{u}_e \times \mathbf{B}); \qquad m_i n_i\frac{\mathrm{d}}{\mathrm{d}t}\mathbf{u}_i = -\nabla p_i + Zn_i e(\mathbf{E} + \mathbf{u}_i \times \mathbf{B}). \tag{3.31}$$

$$\frac{3}{2}n_e k_B\frac{\mathrm{d}}{\mathrm{d}t}T_e = -p_e\nabla\mathbf{u}_e + \nabla(\kappa_e\nabla T_e) - 3k_B n_e\nu_{ei}(T_e - T_i) + \alpha I. \tag{3.32}$$

$$\frac{3}{2}n_i k_B\frac{\mathrm{d}}{\mathrm{d}t}T_i = -p_i\nabla\mathbf{u}_i + 3k_B n_e\nu_{ei}(T_e - T_i). \tag{3.33}$$

Remember the criterion for the plasma to be ideal:

The plasma is to be considered ideal if the ideality parameter Γ or the number of electrons in the Debye sphere N_D fulfill the conditions

$$\Gamma = \frac{\langle E_{\mathrm{pot}}\rangle}{\langle E_{\mathrm{kin}}\rangle} = \frac{0.2}{N_D^{2/3}} < 0.1 \quad \Leftrightarrow \quad N_D \geq 2-3; \qquad N_D = \frac{4\pi}{3}n_e\lambda_D^3 \tag{3.34}$$

N_D is the electron number in the Debye sphere of radius $r = \lambda_D$.

(b) **Change of Entropy in LTE**

According to the first law of Thermodynamics holds with $c_{ve} = 3n_e k_B/2$

$$n_{e,i}\,\mathrm{d}\frac{\epsilon_{e,i}}{n_{e,i}} = \mathrm{d}\,Q_{e,i} + \frac{p_{e,i}}{n_{e,i}}\mathrm{d}n_{e,i}; \qquad \mathrm{d}\,Q_{e,i} = -\nabla\mathbf{q}_{e,i}\mathrm{d}t \mp \nu_{ei}c_{ve}(T_e - T_i)\mathrm{d}t.$$

If all quantities involved are continuous in space and time, reversibility of $\mathrm{d}Q_{e,i}$ is ensured and with the entropy per unit volume $\sigma = \mathrm{d}S/\mathrm{d}V$ follows the entropy change per particle

$$\mathrm{d}\frac{\sigma_{e,i}}{n_{e,i}} = -\frac{1}{T_{e,i}}\{\nabla\mathbf{q}_{e,i} \pm \nu_{ei}c_{ve}(T_e - T_i) - \dot{h}_{e,i}\}\mathrm{d}t. \tag{3.35}$$

Heat production by friction between electrons and ions is included in the heating functions $\dot{h}_e$ and $\dot{h}_i$. In the dynamics of ideal fluids is $\mathbf{q}_{e,i} = 0$, $\dot{h}_{e,i} = 0$, $\nu_{ei} = 0$, hence the entropy per particle is conserved. This is in agreement with (3.32), (3.33), allowing only adiabatic changes of the ideal gas.

Reduction to the One-Fluid Model

The electron and the ion fluids can alternatively be described by the motion of the center of momentum and the electric current density,

$$\rho = m_e n_e + m_i n_i, \qquad \rho\mathbf{u} = \rho_e\mathbf{u}_e + \rho_i\mathbf{u}_i, \qquad \mathbf{j} = e(Zn_i\mathbf{u}_i - n_e\mathbf{u}_e). \tag{3.36}$$

Multiplying (3.27) by m_e and (3.28) by m_i and adding them yields

$$\rho\frac{\mathrm{d}_1\mathbf{u}}{\mathrm{d}t} = \mathbf{j}\times\mathbf{B} - \nabla(p_e + p_i); \quad \frac{\mathrm{d}_1\mathbf{u}}{\mathrm{d}t} = \frac{\mathrm{d}\mathbf{u}}{\mathrm{d}t} + \frac{\mu^2}{m_e m_i}\left(\frac{\mathbf{j}}{n_e}\nabla\right)\frac{\mathbf{j}}{n_e}, \quad \mu = \frac{m_e m_i}{m_e + m_i}. \tag{3.37}$$

μ is the *reduced mass*. Setting $Zn_i = n_e$ in (3.36) and subtracting (3.27) from (3.28) results into the *generalized Ohm's law*

$$\frac{1}{e^2 n_e}\mu\frac{\mathrm{d}_2\mathbf{j}}{\mathrm{d}t} + \frac{1}{e^2 n_e}\mu\nu_{ei} = \mathbf{E} + \mathbf{u}\times\mathbf{B} + \frac{1}{e\rho}\big[(m_e - m_i)\mathbf{j}\times\mathbf{B} + n_e m_i\nabla p_e - n_e m_e\nabla p_i\big]. \tag{3.38}$$

The first term in the square bracket is the *Hall term*, the difference of the two remaining terms is the *pressure diffusion term*. The total derivative $\mathrm{d}_2/\mathrm{d}t$ is defined by

$$\frac{1}{n_e}\frac{\mathrm{d}_2\mathbf{j}}{\mathrm{d}t} = \frac{\mathrm{d}(\mathbf{j}/n_e)}{\mathrm{d}t} + \left(\frac{\mathbf{j}}{n_e}\nabla\right)\mathbf{u} - \mu\left(\frac{\mathbf{j}}{n_e}\nabla\right)\frac{\mathbf{j}}{n_e}; \quad \frac{\mathrm{d}}{\mathrm{d}t} = \frac{\partial}{\partial t} + (\mathbf{u}\nabla). \tag{3.39}$$

Under the assumption of *quasineutrality* the two (3.37) and (3.38) are equivalent two the momentum equations (3.27), (3.28) of the individual fluids. Depending on the physical process the two complex equations fortunately can be greatly simplified. The first reduction is introduced by the separation of the fast from the slow time scale. For this aim it is more convenient to start directly from (3.27) and (3.28) and then to compare the result with (3.37) and (3.38).

The fields $\mathbf{E}$, $\mathbf{B}$ exhibit fast oscillating components $\mathbf{E}_1$, $\mathbf{B}_1$ on the time scale of $1/\omega$ due to the presence of laser irradiation and slow components $\mathbf{E}_0$, $\mathbf{B}_0$ of the ion time scale. The induced electron motion separates into two components, $\mathbf{u}_e = \mathbf{u}_{e1} + \mathbf{u}_{e0}$ and $n_e = n_{e1} + n_{e0}$. The electron momentum equation decays in leading order into the following components for the fast and the slow dynamics,

$$m_e n_{e0}[\partial_t + (\mathbf{u}_{e0}\nabla)]\mathbf{u}_{e1} = -\nabla p_{e1} - en_{e0}(\mathbf{E}_1 + \mathbf{u}_{e1}\times\mathbf{B}_{e0}) - \nu_{ei}m_e n_{e0}\mathbf{u}_{e1}. \tag{3.40}$$

$$m_e n_{e0}[\partial_t + (\mathbf{u}_{e0}\nabla)]\mathbf{u}_{e0} = -\nabla p_{e0} - en_{e0}(\mathbf{E}_{e0} + \mathbf{u}_{e0} \times \mathbf{B}_{e0}) + \pi_0 - \nu_{ei} m_e n_{e0}\mathbf{u}_{e0}. \tag{3.41}$$

$\pi_0 = n_{e0}\mathbf{f}_p$ is the ponderomotive force; it collects combinations of fast motion resulting into a force on the slow time scale. Here it is achieved by averaging quadratic fast terms over one laser cycle.

Fast components contribute to slow motion. The paradigmatic case of (3.41) deserves attention. It is best understood in terms of Fourier components. Owing to the nonlinearity of the momentum equation a periodic force of frequency ω generates all harmonics $\pm n\omega$ in the dynamics of the particle, here the velocities $u_{\pm n\omega}$ of the electron. In turn, the same nonlinearity ties all combinations together to produce all harmonics $\pm m\omega$ and the *zero frequency* u_0 through $u_\omega v_{-\omega} + u_{2\omega}u_{-2\omega} + \cdots + u_{n\omega}u_{-(n+1)\omega}n_{e,\omega} + \cdots$.

The ion motion on the fast time scale is negligible. If correctly taken into account, as done in (3.37) and (3.38), it leads to the replacement of m_e by the reduced mass $\mu = m_e m_i/(m_e + m_i)$. So, on the slow time scale $n_i = n_{i0}$, $\mathbf{u}_i = \mathbf{u}_{i0}$, $p_i = p_{i0}$ is set; the ion motion is entirely bound to

$$m_i n_i[\partial_t + (\mathbf{u}_i\nabla)]\mathbf{u}_i = -\nabla p_i + Zen_i(\mathbf{E}_0 + \mathbf{u}_i \times \mathbf{B}_0). \tag{3.42}$$

In the absence of strong electron ring currents applies $\mathrm{d}\mathbf{u}_e/\mathrm{d}t \simeq \mathrm{d}\mathbf{u}_{i0}/\mathrm{d}t$; (3.41) and (3.42) can be added,

$$m_i n_0[\partial_t + (\mathbf{u}\nabla)]\mathbf{u} = -\nabla(p_e + p_i) + \mathbf{j}_0 \times \mathbf{B}_0 + \pi_0. \tag{3.43}$$

The simplification $\mathbf{u}_i = \mathbf{u}_{i0} \simeq \mathbf{u}$ is made possible because of $\mathbf{u}_1 \simeq 0$. The slow electric component $\mathbf{E}_0' = \mathbf{E}_0 + \mathbf{u}_0 \times \mathbf{B}_0$ is the mediator for pressure p_{e0}, π_0 and friction force from the electrons to the ions. The left hand acceleration term in (3.41) is vanishingly small and hence

$$\mathbf{E}_0 + \mathbf{u}_0 \times \mathbf{B}_0 = -\frac{1}{en_{e0}}\nabla p_e - \frac{1}{e}\nabla\Phi_p - \frac{m_e}{e}\nu_{ei}(\mathbf{u}_{e0} - \mathbf{u}_{i0}). \tag{3.44}$$

This is a very useful formula. It tells that the electrons are held back by the inertia of the ions. The last term can be cast into is the slow current component $\mathbf{j}_0$ divided by the dc conductivity σ,

$$-\frac{m_e}{e}\nu_{ei}(\mathbf{u}_{e0} - \mathbf{u}_{i0}) = m_e\nu_{ei}\frac{\mathbf{j}_0}{n_{e0}e^2} = \frac{\mathbf{j}_0}{\sigma}; \qquad \sigma = \frac{n_{e0}e^2}{m_e\nu_{ei}}. \tag{3.45}$$

σ is the dc electric conductivity, see (4.137) for $\omega = 0$. The quasistatic field $\mathbf{E}_0$ exhibits a quasistatic potential $\Phi(\mathbf{x}, t)$ and is a measure of the imperfect plasma neutrality,

$$\mathbf{E}_0 = -\nabla\Phi, \qquad \nabla\mathbf{E}_0 = \frac{e}{\varepsilon_0}(Zn_i - n_{e0}).$$

Volume forces act on the kinetic energy and leave internal energy and entropy unaltered. Hence, internal energy conservation of the electron and the ion fluids happens on the slow time scale only. Equations (3.29) yield

$$n_{e0}\frac{\mathrm{d}}{\mathrm{d}t}\left(\frac{\epsilon_{e0}}{n_{e0}}\right) = -p_{e0}\nabla\mathbf{u} - 2\nu_{ei}c_{ve0}(T_e - T_i) + \alpha I - \nabla\mathbf{q}_{e0}. \tag{3.46}$$

$$n_{i0}\frac{\mathrm{d}}{\mathrm{d}t}\left(\frac{\epsilon_{i0}}{n_{i0}}\right) = -p_{i0}\nabla\mathbf{u} + 2\nu_{ei}c_{ve0}(T_e - T_i). \tag{3.47}$$

Summation of the two equations results in the monofluid energy balance

$$n\frac{\mathrm{d}}{\mathrm{d}t}\left(\frac{\epsilon}{n}\right) = -(p_{e0} + p_{i0})\nabla\mathbf{u} + \alpha I - \nabla\mathbf{q}_{e0}; \quad n_{e0} = Zn_{i0}, \ \epsilon = \epsilon_e + \epsilon_i. \tag{3.48}$$

On the ω time scale the electric current $\mathbf{j} = \mathbf{j}_1$ may be approximated by $\mathbf{j}_1 = -en_0\mathbf{u}_{e1}$. Inserted in (3.40) it yields the *generalized Ohms law* on the fast time scale in a slightly different approximation compared to (3.38)

$$[\partial_t + (\mathbf{u}_{e0}\nabla)]\mathbf{j}_1 = \frac{e}{m_e}\nabla p_{e1} + \varepsilon_0\,\omega_p^2\,\mathbf{E}_1 + \mathbf{j}_1 \times \mathbf{B}_{e0} - \nu_{ei}\mathbf{j}_1; \quad \omega_p^2 = \frac{n_{e0}e^2}{m_e\varepsilon_0}. \tag{3.49}$$

Ohm's law on the slow time scale is found from the obvious relations

$$\frac{\partial\mathbf{j}_0}{\partial t} \simeq 0, \ \ \mathbf{j}_0 = \sigma(\mathbf{E}_0 + \mathbf{u}_0 \times \mathbf{B}_0), \ \ \varepsilon_0 c^2\nabla \times \mathbf{B}_0 = \mathbf{j}_0, \ \ \nabla \times \mathbf{E}_0 = -\,\partial_t\mathbf{B}_0. \tag{3.50}$$

By eliminating $\mathbf{j}_0$ and taking the curl of $\mathbf{E}_0$ the two desired equations are obtained with the help of (3.44) and (3.45),

$$\mathbf{E}_0 = \frac{\mathbf{j}_0}{\sigma} - \mathbf{u}_0 \times \mathbf{B}_0 - \frac{1}{e}\left(\frac{1}{n_{e0}}\nabla p_{e0} + \nabla\Phi_p\right). \tag{3.51}$$

$$\frac{\partial\mathbf{B}_0}{\partial t} = \varepsilon_0 c^2\nabla^2\frac{\mathbf{B}_0}{\sigma} + \nabla \times (\mathbf{u}_0 \times \mathbf{B}_0) + \frac{1}{e}\nabla\frac{1}{n_{e0}} \times \nabla p_{e0}. \tag{3.52}$$

Whenever the gradients of the density and the electron pressure are not collinear their cross product is the generator of a quasistatic magnetic field. As seen later, in laser generated plasmas it is responsible for Megagauss magnetic fields.

3.1.5 Standard Form of the Conservation Equations

The standard form of a conserved quantity is obtained by the following reasoning. Let V be an arbitrary volume fixed in space and ϱ the quantity under investigation. It is conserved if its change in V per unit time is balanced by the flux per unit time into V or out of it through the surface Σ enclosing V (see Fig. 3.1 for $\rho = mn$). With $\mathrm{d}\boldsymbol{\Sigma}$ oriented outward it is in mathematical terms

$$\frac{\mathrm{d}}{\mathrm{d}t}\int_V \varrho \mathrm{d}V = -\int_{\Sigma(V)} (\varrho\mathbf{u})\mathrm{d}\boldsymbol{\Sigma} \quad \Leftrightarrow \quad \frac{\partial \varrho}{\partial t} + \mathrm{div}(\varrho\mathbf{u}) = 0. \tag{3.53}$$

Note, if $(\varrho\mathbf{u}, \varrho) = s = s^\alpha, \alpha = 1-4$, is defined as a four quantity and div is completed by the fourth partial derivative ∂_t the differential form above assumes the compact structure

$$\mathrm{div} s = 0. \tag{3.54}$$

In four form conservation of a physical quantity means that it is divergence free.

If ϱ is a scalar quantity, e.g., particle density n, its flux density is the vector $\varrho\mathbf{u}$ (tensor of first rank), if ϱ is a vector, e.g., momentum density $\boldsymbol{\pi} = \rho\mathbf{u} = \rho(u_x, u_y, u_z)$, the associated flux density (momentum flux density) is a second rank tensor $[\varrho\mathbf{u}]_{ij} = \varrho_i u_j$, e.g., $\rho u_i u_j$, $i, j = 1, 2, 3$, etc. It indicates that each of its components ϱ_i may be transported in any of the directions $\mathbf{e}_j$. Mass conservation (3.2) is already in the standard or *conservative form*, but momentum and energy conservation (3.13), resp. (3.17), and (3.22) are not. However, by adding appropriate zeros they are easily cast into the form of (3.53), or (3.54), respectively.

From mass conservation (3.2) follows

$$\mathbf{u}\frac{\mathrm{d}\rho}{\mathrm{d}t} = -\rho\mathbf{u}(\nabla\mathbf{u}), \qquad \left(\frac{\epsilon_{\mathrm{in}}}{\rho} + \frac{\mathbf{u}^2}{2}\right)\frac{\mathrm{d}\rho}{\mathrm{d}t} = -\left(\epsilon_{\mathrm{in}} + \frac{\rho\mathbf{u}^2}{2}\right)\nabla\mathbf{u}. \tag{3.55}$$

With the help of the first relation in (3.55) the total change in time of the momentum density $\rho\mathbf{u}$ results from (3.13) as

$$\frac{\mathrm{d}}{\mathrm{d}t}\rho\mathbf{u} = \rho\frac{\mathrm{d}\mathbf{u}}{\mathrm{d}t} + \mathbf{u}\frac{\mathrm{d}\rho}{\mathrm{d}t} = -\nabla\Pi + \mathbf{f} - \rho\mathbf{u}(\nabla\mathbf{u}) = -\nabla\big[\rho\mathbf{u}\mathbf{u} + \Pi\big] + \mathbf{f}.$$

This fits the general scheme of a conserved quantity,

$$\frac{\partial}{\partial t}\rho\mathbf{u} + \mathrm{div}[\rho\mathbf{u}\mathbf{u} + \Pi] = \mathbf{f}; \qquad [\rho\mathbf{u}\mathbf{u} + \Pi]_{ij} = \rho u_i u_j + \Pi_{ij}. \tag{3.56}$$

Analogously the conservation of the energy density $\epsilon_{\text{in}} + \rho\mathbf{u}^2/2$ follows from (3.22) with the help of the second expression in (3.55) and mass conservation (3.2)

$$\frac{\mathrm{d}}{\mathrm{d}t}\left(\epsilon_{\text{in}} + \frac{\rho\mathbf{u}^2}{2}\right) = \rho\frac{\mathrm{d}}{\mathrm{d}t}\left(\frac{\epsilon_{\text{in}}}{\rho} + \frac{\mathbf{u}^2}{2}\right) + \left(\frac{\epsilon_{\text{in}}}{\rho} + \frac{\mathbf{u}^2}{2}\right)\frac{\mathrm{d}}{\mathrm{d}t}\rho =$$

$$-\nabla(\Pi\mathbf{u} + \mathbf{q}) + \dot{h} + \mathbf{fu} - \left(\epsilon_{\text{in}} + \frac{\rho\mathbf{u}^2}{2}\right)\nabla\mathbf{u}. \tag{3.57}$$

Rearranged in standard form energy conservation reads

$$\frac{\partial}{\partial t}\left(\epsilon_{\text{in}} + \frac{\rho\mathbf{u}^2}{2}\right) + \nabla\left\{\mathbf{u}\left(\epsilon_{\text{in}} + \frac{\rho\mathbf{u}^2}{2} + \Pi\right) + \mathbf{q}\right\} = \dot{h} + \mathbf{fu}. \tag{3.58}$$

Note the energy density flux contains the work by the pressure-viscosity tensor $\Pi\mathbf{u}$. It represents the expense of energy per unit time to displace and deform the neighboring volume elements. For the ideal fluid $\dot{h} = \mathbf{q} = \mathbf{f} = 0,\ \Pi = p\delta_{ij}$ energy conservation reduces to

$$\frac{\partial}{\partial t}\left(\epsilon_{\text{in}} + \frac{\rho\mathbf{u}^2}{2}\right) + \nabla\mathbf{u}\left(\epsilon_{\text{in}} + \frac{\rho\mathbf{u}^2}{2} + p\right) = 0. \tag{3.59}$$

The reader may ask what the advantage of the standard form of conservation equations is relative to the forms (3.13), (3.17), and (3.22). Apart from the aesthetics and unifying aspect of the standard form there may be four arguments in favour of it. (i) If a quantity in its standard form can be extended to a relativistic four expression this is its relativistic conservation law; for an example see (3.100) of Sect. 3.2.1. (ii) Starting from the standard form of the conservation laws may frequently be advantageous in numerical simulations. (iii) If the conservation equations are of hyperbolic type the standard form can be given the structure of disturbances propagating along their *characteristics*; for example see Sect. 3.3.3 and Fig. 3.7. In stationary problems the partial time derivative vanishes and one is left with the spatial divergence of an expression which is immediately recognized as the correct conservation law. (iv) Finally, as seen from (3.3), (3.56), and (3.58) on the right hand side there are the source terms of particle generation λ, μ, external forces $\mathbf{f}$ and external work and heat supply $\mathbf{fu} + \dot{h}$. If read from right to left the fluxes appear as determined by the sources; conversely, sometimes it may be easier to determine the sources from the fluxes. An example for this situation is given in the following in connection with the correct collective ponderomotive force.

3.1.6 Collective Ponderomotive Force Density

The ponderomotive force considered so far is a bulk force. Its density is proportional to the electron (respectively particle) density and is indicated by π_0. A plasma in general is capable of excitation of collective electron and ion density fluctuations. These, in turn, are accompanied by collective longitudinal and transverse electric fields and superpose to the incident laser field. The secular component of this mixture gives rise to a ponderomotive force density π_t. As its origin lies in the plasma fluctuations it is different in its nature from the bulk force density $\boldsymbol{\pi}_0$.

Ponderomotive Force Density π_t from Fluctuations

The following considerations are based on momentum conservation. By definition the ponderomotive force density is a secular, i.e., low frequency force which originates from a high frequency transverse or longitudinal electric field and applies to the one-fluid model. Thus, when $\rho_0 = \langle \rho(t) \rangle$ and $\mathbf{u}_0 = \langle \mathbf{u}(t) \rangle$ are the density and velocity of a volume element averaged over the high frequency oscillations, $\boldsymbol{\pi}_0$ is defined by the equation of motion

$$\rho_0 \frac{\mathrm{d}\mathbf{u}_0}{\mathrm{d}t} = -\nabla(p_e + p_i) + \mathbf{f}_0 + \boldsymbol{\pi}_0. \tag{3.60}$$

where p_e, p_i are the electronic and ionic thermal pressures and f_0 is any low frequency force (electric and magnetic, gravitational, viscous, frictional, etc.). To apply the concept of the ponderomotive force it is important that the spectra of fast and slow motions in $\rho\mathbf{u} = \rho_i \mathbf{u}_i + \rho_e \mathbf{u}_e$ are well separated from each other. Including all forces $\mathbf{f}$, $\mathbf{f} = \mathbf{f}_0 + \mathbf{f}_h$, where $\mathbf{f}_h$ originates from the high frequency fields the general momentum conservation equation in the one-fluid model is

$$\frac{\partial}{\partial t}\rho\mathbf{u} + \nabla(\rho\mathbf{u}\mathbf{u} + p\mathbf{I}) = \mathbf{f}.$$

The momentum flow density $T = \rho\mathbf{u}\mathbf{u} + p\mathbf{I}$, $p = p_e + p_i$, is a second rank tensor with the components $T_{kl} = \rho v_k v_l + p\delta_{kl}$. Hence, by comparison with (3.60) π_t results as follows,

$$\pi_t = \left\langle \frac{\partial}{\partial t}\rho\mathbf{u} + \nabla(\rho\mathbf{u}\mathbf{u} + p\mathbf{I}) - \mathbf{f}_h \right\rangle - \rho_0 \frac{\mathrm{d}\mathbf{u}_0}{\mathrm{d}t}. \tag{3.61}$$

In the absence of ionic resonances (i.e., no static magnetic fields)

$$\rho_i \mathbf{u}_i = m_i n_{i0} \mathbf{u}_{i0}, \quad \mathbf{u}_{e0} = \mathbf{u}_{i0},$$

$$\rho_e \mathbf{u}_e = m_e(n_{e0} + n_{e1} + n_{e2} + \ldots..)(\mathbf{u}_{e0} + \mathbf{u}_{e1} + \mathbf{u}_{e2} + \ldots..); \tag{3.62}$$

holds. The indices 1, 2, etc. indicate the Fourier components of frequencies ω, 2ω, etc., induced by the high frequency electric or electromagnetic field

$$\mathbf{E}(\mathbf{x}, t) = \hat{\mathbf{E}}(\mathbf{x}, t)\mathrm{e}^{-i\omega t}. \tag{3.63}$$

$\hat{\mathbf{E}}(\mathbf{x}, t)$ is the amplitude slowly varying in space and time. Inserting the quantities (3.62) into (3.61) leads to

$$(m_i n_{i0} + m_e n_{e0})\langle\nabla(\mathbf{u}_{e1}\mathbf{u}_{e1} + \mathbf{u}_{e2}\mathbf{u}_{e2} + \ldots\ldots) - \mathbf{f}_h\rangle = \boldsymbol{\pi}_0,$$

where the brackets $\langle\rangle$ contain, in addition, all sums of products $\mathbf{u}_{ej}\mathbf{u}_{ek}$ with an even number $l = j + k$ and $j < k$. The first nonvanishing term contributing to the collective force $\boldsymbol{\pi}_t$ originates from $n_{e1}\mathbf{u}_{e1}$ in $\langle\partial\rho\mathbf{u}/\partial t\rangle$. Since this is also the leading term, to lowest order follows

$$\boldsymbol{\pi}_t = \left\langle\frac{\partial}{\partial t}(m_e n_{e1}\mathbf{u}_{e1})\right\rangle. \tag{3.64}$$

There is no corresponding contribution from the flux term since $\langle m_e n_{e1}\mathbf{u}_{e1}\mathbf{u}_{e1}\rangle$ is zero. With the help of the electron equation of motion,

$$\frac{\partial\mathbf{u}_{e1}}{\partial t} = -\frac{e}{m_e}\hat{\mathbf{E}}(\mathbf{x}, t)\mathrm{e}^{-i\omega t},$$

for $\mathbf{u}_{e1}$ and displacement $\boldsymbol{\delta}_{e1}$ one obtains

$$\mathbf{u}_{e1} = i\frac{e}{m_e\omega}\mathrm{e}^{-i\omega t}\left(1 - \frac{i}{\omega}\frac{\partial}{\partial t}\right)\hat{\mathbf{E}}(\mathbf{x}, t),$$

$$\boldsymbol{\delta}_{e1} = \frac{e}{m_e\omega^2}\mathrm{e}^{-i\omega t}\left(1 - \frac{2i}{\omega}\frac{\partial}{\partial t}\right)\hat{\mathbf{E}}(\mathbf{x}, t). \tag{3.65}$$

Analogously to (3.5), the density variation n_{e1} follows from linearization of (3.30):

$$n_{e1} + \nabla(n_{e0}\boldsymbol{\delta}_{e1}) = 0. \tag{3.66}$$

With these relations for $\boldsymbol{\delta}_{e1}$ and n_{e1}, $\boldsymbol{\pi}_t$ can be expressed in terms of the electric field amplitude of the high frequency wave as follows:

$$\boldsymbol{\pi}_t = i\frac{\epsilon_0}{4\omega}\frac{\partial}{\partial t}\left\{\hat{\mathbf{E}}\left(\nabla\frac{\omega_\mathrm{p}^2}{\omega^2}\hat{\mathbf{E}}^*\right) - \hat{\mathbf{E}}^*\left(\nabla\frac{\omega_\mathrm{p}^2}{\omega^2}\hat{\mathbf{E}}\right)\right\}. \tag{3.67}$$

The collective force density $\boldsymbol{\pi}_t$ plays a decisive role when an electrostatic fluctuation is modulated by the laser field to resonantly drive another electrostatic mode, as for example in the two-plasmon decay, see Chap. 6 on parametric instabilities. There is a whole variety of expressions for $\boldsymbol{\pi}_t$ in the literature differing from each other [1–3] and, when $\boldsymbol{\pi}_t$ is presented in terms of $\hat{\mathbf{E}}$ as in (3.67), it is not easy to recognize which of them are correct and which are not. However, the correctness of (3.67) is easily revealed through (3.64), since the latter is accessible to an immediate physical interpretation: Due to the variation of $\hat{\mathbf{E}}(\mathbf{x}, t)$ in time the secular component of the momentum density $\langle m_e n_{e1}\mathbf{u}_{e1}\rangle$ induced by the wave changes also. Since, on the other hand the associated momentum flux density term $\langle m_e n_{e1}\mathbf{u}_{e1}\mathbf{u}_{e1}\rangle$ is zero, the momentum change of a volume element appears as an external force applied to it.

This is the advantage of starting from the momentum equation in standard form we wanted to show.

In principle the contributions of p_i, p_e to π_0 as well as to π_t have to be considered also. However, by keeping in mind that kinetically p_e and p_i are defined each relative to their mean flow velocities $\mathbf{u}_e(\mathbf{x}, t)$ and $\mathbf{u}_i(\mathbf{x}, t)$ (and *not* relative to the common center of mass speed $\mathbf{u} = (\rho_i\,\mathbf{u}_i + \rho_e\,\mathbf{u}_e)/(\rho_i + \rho_e))$ there is no change in p_e and p_i due to the presence of a hf E-field as long as the particle distribution functions $f_i(\mathbf{x}, \mathbf{u}, t)$, $f_e(\mathbf{x}, \mathbf{u}, t)$ are not affected: In the systems co-moving with $\mathbf{u}_e$ and $\mathbf{u}_i$ the external field $\mathbf{E}$ vanishes. The expression π_t from (3.64) or (3.67) is a good approximation as long as the higher harmonics are weak. For a resonantly excited electron plasma wave close to the breaking limit this may no longer be the case, because all higher harmonics may then become important.

Global Momentum Conservation

It may be instructive to cast π into a different form. Let $\boldsymbol{\kappa}$ be the force density caused by a hf field. With the help of the Poynting vector $\mathbf{S}$ and Maxwell's stress tensor $\mathbf{T}$ it is expressed in conservation form as

$$\boldsymbol{\kappa} = -\frac{1}{c^2}\frac{\partial}{\partial t}\mathbf{S} - \nabla\mathbf{T}.$$

$\boldsymbol{\kappa}$ is the sink of electromagnetic momentum and must, therefore, appear as a mechanical force density for the electronic and the ionic fluids according to

$$\boldsymbol{\kappa} = \frac{\partial}{\partial t}(\rho_e\,\mathbf{u}_e + \rho_i\,\mathbf{u}_i) + \nabla(\rho_e\,\mathbf{u}_e\,\mathbf{u}_e + \rho_i\,\mathbf{u}_i\,\mathbf{u}_i + p_e\mathbf{I} + p_i\mathbf{I}). \tag{3.68}$$

since the momentum of the three fluids of electrons, ions, and photons, must be conserved. $\boldsymbol{\kappa}$ is rapidly oscillating. Thus, in order to formulate a law of momentum conservation for observable quantities on the slow time scale one has to pass to a one-fluid description for ρ and $\mathbf{u}$, as done above. After some algebraic manipulations (3.68) is conveniently expressed as follows,

$$\frac{\partial}{\partial t}\rho\mathbf{u} + \nabla(\rho\mathbf{u}\mathbf{u} + p\mathbf{I}) + \frac{1}{c^2}\frac{\partial}{\partial t}\mathbf{S} + \nabla\left(\mathbf{T} + \frac{\rho_e\rho_i}{\rho}\mathbf{w}\mathbf{w}\right) = 0, \tag{3.69}$$

where $\mathbf{w} = \mathbf{u}_i - \mathbf{u}_e$ [4]. When $\mathbf{E} = \hat{\mathbf{E}}e^{-i\omega t}$ reduces to zero π vanishes and the last two terms in (3.69) disappear. Hence, $\pi = \pi_0 + \pi_t$ is to be identified with

$$\pi = -\left\langle\frac{1}{c^2}\frac{\partial}{\partial t}\mathbf{S} + \nabla\mathbf{T} + \nabla\frac{\rho_e\rho_i}{\rho}\mathbf{w}\mathbf{w}\right\rangle. \tag{3.70}$$

The ponderomotive force π is more than the divergence of the Maxwellian stress tensor; $\nabla\rho_e\rho_i\mathbf{w}\mathbf{w}/\rho$ is of the same order of magnitude.

The two terms π_0 and π_t are qualitatively different: In the absence of dissipation π_0/n_{e0} disappears from (3.52) because derivable from a potential, whereas $\pi_{t'}/n_{e0}$ is not since $\nabla \times \pi_t/n_{e0} \neq 0$. The latter appears as a possible source term for magnetic field generation in (3.52),

$$\frac{\partial \mathbf{B}_0}{\partial t} = \varepsilon_0 c^2 \nabla^2 \frac{\mathbf{B}_0}{\sigma} + \nabla \times (\mathbf{u}_0 \times \mathbf{B}_0) + \frac{1}{e}\nabla \frac{1}{n_{e0}} \times \nabla p_{e0} - \frac{1}{e}\nabla \times \frac{\pi_t}{n_{e0}}. \tag{3.71}$$

The derivation of π presented in this section reveals once more that $\pi_0 = -n_{e0}\nabla\Phi_\mathrm{p}$ is correct and not $-\nabla(n_{e0}\Phi_\mathrm{p})$. A derivation of π_t in the presence of a static magnetic field is given by several authors [5–7].

3.1.7 The Lagrangian Picture of the Fluid

An arbitrary field F is considered to be known if it is given as a function of space coordinate $\mathbf{x}$ in a bounded or unbounded region to all times t of a finite or infinite interval: $F = F(\mathbf{x}, t)$. F may stand for any piecewise continuous distribution like $\rho, \mathbf{u}, p, \mathbf{E}, \mathbf{S}, \Pi$. Transport coefficients, like heat conduction κ, collision frequency ν_{ei} or viscosity μ are also field quantities in principle. As they are intrinsic properties of the fluid they do not explicitly depend on $\mathbf{x}$ and on t neither. Sometimes, however, it may be convenient to introduce a time dependence to express changes in the material properties, e.g., fatigue of elasticity. In this case one must be aware that the Galileian relativity principle is violated because such a description is incomplete.

Depending on the situation it may be advantageous to describe a field F by giving it the initial position $\mathbf{x} = \mathbf{a}$ at the arbitrary time instant $t = t_0$: $\mathbf{a} = \mathbf{x}(t_0) = \mathbf{x}_0$. The evolution of F is completely known in terms of $F(\mathbf{a}, t)$ if the *trajectory* $\mathbf{x}(\mathbf{a}, t)$ is known along which the point $F(\mathbf{a}, t_0)$ has moved. Actual and initial positions are connected by

$$\mathbf{x}(t) = \mathbf{a} + \int_{t_0}^{t} \mathbf{v}(t')\mathrm{d}t'. \tag{3.72}$$

If F is a property tied to matter, like density ρ or temperature T, $\mathbf{v}$ is the fluid velocity $\mathbf{u}$; if F describes a small pressure disturbance in a homogeneous streaming fluid $\mathbf{v}$ is $\mathbf{u} + \mathbf{c}_s$ because such a disturbance evidently propagates with fluid + sound velocity $\mathbf{c}_s$. The *Lagrangian picture* consists in the representation of all dynamical variables F in *Lagrangian coordinates* $(\mathbf{a}, t)$. The representation $F = F(\mathbf{x}, t)$ is the *Eulerian picture*. The Lagrangian representation is appropriate in describing free expansion of a confined fluid, e.g., hot laser plasma into vacuum. In the Euler picture initial density, flow velocity, pressure, etc., have to be declared over all space that will be occupied by the fluid in the time interval considered; in the Lagrange representation these quantities have to be fixed in the domain the fluid occupies at initial time t_0; all reachable positions are calculated from (3.72).

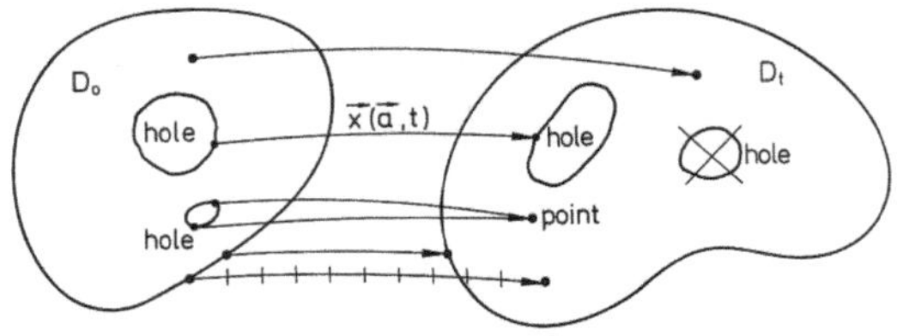

Fig. 3.4 Formally, fluid flow is a continuous mapping of one domain D_0 onto another domain D_t characterized by the parameter t (homeomorphism). In an ideal fluid a hole cannot be closed or opened (but can degenerate to a point) and a point on a boundary is always mapped onto the boundary. $\mathbf{a}$ initial position, $\mathbf{x}(\mathbf{a}, t)$ trajectory of a fluid element. The degree of connection of a domain is conserved

Topological properties of fluid flow. Formally, fluid flow is a continuous mapping (even a homeomorphism) of one region in space onto another one (Fig. 3.4): A volume element $\mathrm{d}V_0$, initially at position $\mathbf{x}(t=0) = \mathbf{a} = (a_1, a_2, a_3)$, is found at the point $\mathbf{x}(t)$ after time t, i.e., $\mathbf{x}(t)$ is the image of $\mathbf{a}$. The set of points $\mathbf{x}(\mathbf{a}, t)$, with $\mathbf{a}$ held fixed and $\mathbf{x}(\mathbf{a}, 0) = \mathbf{a}$, is the trajectory of the volume element starting from $\mathbf{a}$. It carries all dynamical variables from $\mathbf{x} = \mathbf{a}$ at $t = t_0$ to $\mathbf{x}(\mathbf{a}, t)$ at $t > t_0$. For instance, $\rho(\mathbf{a}, t)$ means $\rho(\mathbf{x}(\mathbf{a}, t), t) = \rho(\mathbf{x}, t)$. In contrast to the Eulerian picture in the Lagrangian description each fluid element tells us where it has come from, like the mass point with index i in Newton's Mechanics follows its trajectory $\mathbf{x}(t)$. The only difference of the trajectory of the fluid element $\mathrm{d}V$ is its continuous index $\mathbf{a}$, $\mathbf{x}_i(t) \rightarrow \mathbf{x}_\mathbf{a}(t) = \mathbf{x}(\mathbf{a}, t)$.

Conservation Laws in the Lagrangian Picture

Mass, momentum, and energy conservation are derived altogether by the same technique. We assume again, for the moment, that the fluid is sufficiently dense so that (i) the extension d of a volume element ΔV in which n, $\mathbf{u}$ etc. are reasonably constant is much larger than $\min\{\lambda, \lambda_D\}$, ($\lambda$ is the mean free path), and that (ii) Π from (3.8) is isotropic for simplicity, $\Pi = p\delta_{ij}$. Mass conservation in the Lagrangian representation is formulated as follows. The domains D_0, $D_t = D(t)$ in Fig. 3.4 and finite subdomains can be chosen arbitrary. They represent co-moving volumes $V_0 = V(t_0)$, $V_t = V(t)$ containing the same number of particles,

$$\int_{V_0} n(\mathbf{a}, t_0)\mathrm{d}V_0 = \int_{V_t} n(\mathbf{x}, t)\mathrm{d}V_t.$$

To compare the particle densities n at times t_0 and t the volumes must be identical and arbitrary. The transformation from V_t to V_0 is achieved with the help of the Jacobian J,

$$\int_{V_t} n(\mathbf{x}, t)\mathrm{d}V_t = \int_{V_0} n(\mathbf{a}, t)J\mathrm{d}V_0; \qquad J(\mathbf{a}, t) = \left|\frac{\partial(\mathbf{x}(\mathbf{a}, t))}{\partial(\mathbf{a})}\right|. \tag{3.73}$$

V_0 is arbitrary and hence

$$n(\mathbf{a}, 0) = n(\mathbf{a}, t)\left|\frac{\partial(\mathbf{x})}{\partial(\mathbf{a})}\right| \tag{3.74}$$

is the mass conservation in Lagrangian coordinates. At first glance it looks very different from (3.2), however it is the same. In fact, for the infinitesimal time $t = \mathrm{d}t$ and $\mathbf{x} = \mathbf{a} + \mathbf{u}\mathrm{d}t$ (3.74) reads

$$n(\mathbf{a}, 0) = \begin{vmatrix} \frac{\partial}{\partial a_1}(a_1 + u_1\mathrm{d}t) & \frac{\partial}{\partial a_1}(a_2 + u_2\mathrm{d}t) & \frac{\partial}{\partial a_1}(a_3 + u_3\mathrm{d}t) \\ \frac{\partial}{\partial a_2}(a_1 + u_1\mathrm{d}t) & \frac{\partial}{\partial a_2}(a_2 + u_2\mathrm{d}t) & \frac{\partial}{\partial a_2}(a_3 + u_3\mathrm{d}t) \\ \frac{\partial}{\partial a_3}(a_1 + u_1\mathrm{d}t) & \frac{\partial}{\partial a_3}(a_2 + u_2\mathrm{d}t) & \frac{\partial}{\partial a_3}(a_3 + u_3\mathrm{d}t) \end{vmatrix} n(\mathbf{a}, \mathrm{d}t).$$

The determinant is $1 + (\partial_{a_1}u_1 + \partial_{a_2}u_2 + \partial_{a_3}u_3)\mathrm{d}t = 1 + \nabla_{\mathbf{a}}\mathbf{u}\mathrm{d}t$, plus 15 terms of higher order in $\mathrm{d}t$. To leading order it follows that $n(\mathbf{a}, \mathrm{d}t) - n(\mathbf{a}, 0) = -n(\mathbf{a}, \mathrm{d}t)$ $\nabla_a\mathbf{u}\,\mathrm{d}t$ or, by continuity of n with respect to t,

$$\frac{\partial}{\partial t}n(\mathbf{a}, 0) = -n(\mathbf{a}, 0)\nabla_{\mathbf{a}}\mathbf{u}. \tag{3.75}$$

The temporal derivative is to be taken at constant $\mathbf{a}$ which means along the trajectory $\mathbf{x}(\mathbf{a}, t)$,

$$\begin{aligned} \frac{\partial}{\partial t}n(\mathbf{a}, 0) &= \lim_{\Delta t\to 0} \frac{n(\mathbf{x}(\mathbf{a}, \Delta t), \Delta t) - n(\mathbf{a}, 0)}{\Delta t} \\ &= \frac{\mathrm{d}}{\mathrm{d}t}n(\mathbf{x}, t = 0) = \left(\frac{\partial}{\partial t} + \mathbf{u}\nabla\right) n(\mathbf{x}, 0). \end{aligned} \tag{3.76}$$

Hence, the partial time derivative ∂_t in the Lagrangian representation is the total (*convective* or *substantial*) time derivative $\mathrm{d}_t = \partial_t + (\mathbf{u}\nabla)$ of the Eulerian picture. At $\mathbf{u} = 0$, $\partial_t^{\mathrm{Lag.}}$ becomes equal to $\partial_t^{\mathrm{Eul.}}$. In this sense the Lagrangian time derivative is the derivative in the (tangent) inertial system co-moving with the fluid. With the help of (3.76), mass conservation in the Eulerian picture (3.2) follows immediately from (3.75). In one dimension (3.74) is particularly intuitive and simple. The particles contained in the interval $\mathrm{d}a$ at $t = 0$ will occupy the interval $\mathrm{d}x$ at a later time, i.e., $n(a, 0)\mathrm{d}a = n(a, t)\mathrm{d}x$, or $n(a, 0) = n(a, t)|\partial x/\partial a| = n(a, t)J$. Correspondingly, in 3 dimensions J is the volume $\nabla x_1(\nabla x_2 \times \nabla x_3)$ of the parallelepiped formed by the three oblique vectors $\nabla_{\mathbf{a}}x_i$ in $\mathbf{a}$-space.

The same technique is adopted in the derivation of the momentum conservation. We start from a macroscopic volume $V(t)$ moving with the fluid. Condition (i) ensures that the pressure p acts on its surface Σ. Together with the external force density $\mathbf{f}$ the balance reads

$$\frac{\mathrm{d}}{\mathrm{d}t}\int_{V_t} \rho \mathbf{u} \mathrm{d}V = -\int_{\Sigma(V_t)} p\,\mathrm{d}\boldsymbol{\Sigma} + \int_{V_t} \mathbf{f}\mathrm{d}V$$

$$\begin{aligned}\frac{\mathrm{d}}{\mathrm{d}t}\int_{V(t)} \rho \mathbf{u} \mathrm{d}V &= \lim \frac{1}{\Delta t}\left(\int_{V'} \rho' \mathbf{u}' \mathrm{d}V' - \int_V \rho \mathbf{u} \mathrm{d}V\right) \\ &= \lim \frac{1}{\Delta t}\int_V \left(J'\rho'\mathbf{u}' - \rho\mathbf{u}\right)\mathrm{d}V \\ &= \lim \int_V \rho(t)\frac{\mathbf{u}(t') - \mathbf{u}(t)}{\Delta t}\mathrm{d}V = \int_V \rho\frac{\partial \mathbf{u}(\mathbf{a}, t)}{\partial t}\mathrm{d}V. \quad (3.77)\end{aligned}$$

with $t' = t + \Delta t$, and ρ', $\mathbf{u}'$, V', ρ, $\mathbf{u}$, V taken at t' and t, respectively. In one of the steps relation (3.74) has been used. The rest of the calculations leading to Euler's equation (3.17) are standard. However, for concluding from the integrals to the integrands assumption (i) has to be fulfilled. By the identical technique energy conservation is deduced from the existence of an energy density consisting of the sum of internal and kinetic energy densities.

3.1.8 Kinetic Foundation of Diluted Fluids

So far the laws governing the dynamics of fluids have been derived under the assumption that (i) the forces on a volume element $\mathrm{d}V$ exerted by the surrounding fluid (negative "pressure-viscosity tensor") act on its surface. It has been tacitly assumed further that (ii) along its trajectory $\mathbf{x}(t)$ particle diffusion, i.e., losses of particles to neighboring fluid elements and gain of particles from them, is inhibited. In the hot laser plasma none of the two conditions may be fulfilled at all and the question arises, whether, and if so, to what extent the governing equations have to be modified. The situation is even more serious when new particles are created or annihilated in $\mathrm{d}V$, for instance by ionization, pair creation or escape of photons. The doubts can be summarized by the question: What is the meaning of the trajectory $\mathbf{x}(t)$ of $\mathrm{d}V$ in such cases?

Closer inspection may bring partial clarification. It is true that the problem is serious for relations like (3.74) if applied to finite time intervals. Concerning particle exchange and interpretation of $\mathbf{x}(t)$ the validity of the fluid laws in differential form is guaranteed if condition (ii) above is fulfilled for infinitesimal times $\mathrm{d}t$ as in (3.77) and in all former derivatives of the context. The problem with condition (i) however still persists. Furthermore it is to be expected that particle creation and annihilation may lead to corrections of the fluid equations. In what follows it will be shown that the extension of the fluid equations to collisionless or weakly collisional fluids is achieved by a kinetic approach.

The Vlasov Equation

Consider the one particle phase space volume $\mathrm{d}\tau = \mathrm{d}\mathbf{p}\mathrm{d}\mathbf{q}$. The set of points enclosed in $\mathrm{d}\tau$ represents all *non-interacting* particles with momenta $\mathbf{p} = \gamma m\mathbf{v}$ between $\mathbf{p}$ and $\mathbf{p} + \mathrm{d}\mathbf{p}$ and concentrated in the space volume $\mathrm{d}\mathbf{x}$. In a given physical ensemble of non-interacting particles the number $\mathrm{d}N = f(\mathbf{x}, \mathbf{p}, t)\mathrm{d}\tau$ may occupy this phase space element $\mathrm{d}\tau$. The density $f(\mathbf{x}, \mathbf{p}, t)$ is named *one particle distribution function*. The total particle number ΔN at position $\mathbf{x}$ is obtained from integration over all momenta $\mathbf{p}$,

$$\Delta N = \left(\int_{\mathbf{p}} f(\mathbf{x}, \mathbf{p}, t)\mathrm{d}\mathbf{p}\right)\mathrm{d}\mathbf{x} \quad \Rightarrow \quad n(\mathbf{x}, t) = \int f(\mathbf{x}, \mathbf{p}, t)\mathrm{d}\mathbf{p}. \tag{3.78}$$

$n(\mathbf{x}, t)$ is the particle density in space. If f is normalized to unity, $f \to f/n$, it can be interpreted as the probability for an arbitrary particle out of $\mathrm{d}N$ finding itself at position $\mathbf{x}$ with momentum $\mathbf{p}$. The particle number $\mathrm{d}N$ of *noninteracting* particles is conserved in time because according to Fig. 3.4 no particle can cross the border of $\mathrm{d}\tau$ unless it undergoes a *collision* with another particle in $\mathrm{d}\mathbf{x}$. Thus, with the help of Liouville's theorem holds $\mathrm{d}\mathrm{d}\tau/\mathrm{d}t = 0$ and

$$\frac{\mathrm{d}\mathrm{d}N}{\mathrm{d}t} = \frac{\mathrm{d}(f\mathrm{d}\tau)}{\mathrm{d}t} = \frac{\mathrm{d}f}{\mathrm{d}t}\mathrm{d}\tau + f\frac{\mathrm{d}\mathrm{d}\tau}{\mathrm{d}t} = \frac{\mathrm{d}f(\mathbf{x}, \mathbf{p}, t)}{\mathrm{d}t}\mathrm{d}\tau = 0$$

$$\Rightarrow \quad \frac{\mathrm{d}f}{\mathrm{d}t} = \frac{\partial f}{\partial t} + \dot{\mathbf{x}}\frac{\partial f}{\partial \mathbf{x}} + \dot{\mathbf{p}}\frac{\partial f}{\partial \mathbf{p}} = 0.$$

With $\dot{\mathbf{x}} = \mathbf{p}/\gamma m$ and $\dot{\mathbf{p}} = q(\mathbf{E} + \mathbf{v} \times \mathbf{B})$ the mean field force at the position $\mathbf{x}$ the *relativistic Vlasov equation* becomes

$$\frac{\partial f}{\partial t} + \frac{\mathbf{p}}{\gamma m}\frac{\partial f}{\partial \mathbf{x}} + q(\mathbf{E} + \mathbf{v} \times \mathbf{B})\frac{\partial f}{\partial \mathbf{p}} = 0. \tag{3.79}$$

Trajectories. An arbitrary function $f(\phi)$ of a single particle trajectory ϕ,

$$\phi = \delta\big(\mathbf{x} - \mathbf{x}(t)\big)\delta\big(\mathbf{v} - \mathbf{v}(t)\big) \tag{3.80}$$

is a solution of the Vlasov equation. The dependence on t is expressed by $\mathbf{x}(t),\ \mathbf{v}(t)$. The proof is simple:

$$\frac{\partial \phi}{\partial t} = -\mathbf{v}(t)\delta\big(\mathbf{v} - \mathbf{v}(t)\big)\frac{\partial \delta}{\partial \mathbf{x}} - \dot{\mathbf{v}}(t)\delta\big(\mathbf{x} - \mathbf{x}(t)\big)\frac{\partial \delta}{\partial \mathbf{v}} = \mathbf{v}(t)\frac{\partial \phi}{\partial \mathbf{x}} + \dot{\mathbf{v}}(t)\frac{\partial \phi}{\partial \mathbf{v}}$$

$$\Rightarrow \quad \frac{\mathrm{d}\phi}{\mathrm{d}t} = 0 \quad \Rightarrow \quad \frac{\mathrm{d}}{\mathrm{d}t}f(\mathbf{x}, \mathbf{v}, t) = \frac{\mathrm{d}f}{\mathrm{d}\phi}\frac{\mathrm{d}\phi}{\mathrm{d}t} = 0.$$

In particular, the assertion holds for the single trajectory $f = \phi$.

Collisions. If a particle of momentum $\mathbf{p}$ at position $\mathbf{x}$ interacts by binary collision with another particle it will be lost from $\mathrm{d}\mathcal{T}(\mathbf{p})$. Conversely, a particle from $\mathrm{d}\mathcal{T}(\mathbf{p}')$ colliding with a particle $(\mathbf{p}_1')$ may be scattered into $\mathrm{d}\mathcal{T}(\mathbf{p})$ with finite probability per unit time (gain). This causes a change of the distribution function $f(\mathbf{x}, \mathbf{p}, t)$ per unit time. It can be described by a *binary collision term* $(\delta f/\delta t)_{\mathrm{coll}} = R(\mathbf{p}, \mathbf{p}_1; \mathbf{p}', \mathbf{p}_1')$ acting as a source in (3.79),

$$\frac{\partial f}{\partial t} + \frac{\mathbf{p}}{\gamma m}\frac{\partial f}{\partial \mathbf{x}} + q(\mathbf{E} + \mathbf{v} \times \mathbf{B})\frac{\partial f}{\partial \mathbf{p}} = R(\mathbf{p}, \mathbf{p}_1; \mathbf{p}', \mathbf{p}_1'). \tag{3.81}$$

Note, a binary collision term $R(\mathbf{p}, \mathbf{p}_1; \mathbf{p}', \mathbf{p}_1')$ of elastic or inelastic collisions does not change the normalization of f.

Throughout the text $f(\mathbf{x}, \mathbf{p}, t)$ is used either normalized to the particle density $n(\mathbf{x}, t)$ or normalized to unity; to which it refers will arise from the context.

Boltzmann's Binary Collision Integral

Let a rarefied fluid have the following properties: (i) The differential scattering cross section σ_Ω of two identical colliding particles does depend only on their relative velocity $\mathbf{w} = \mathbf{v}_1 - \mathbf{v}$ and on the scattering angle Ω (see Chap. 2, Sect. 2.1.8 on binary collisions); it implies time reversal:

$$\sigma_\Omega(-\mathbf{p}', -\mathbf{p}_1' \to -\mathbf{p}, -\mathbf{p}_1) = \sigma_\Omega(\mathbf{p}', \mathbf{p}_1' \to \mathbf{p}, \mathbf{p}_1), \tag{3.82}$$

(ii) the interaction length $\mathrm{d} = \sqrt{\sigma_t}$ is considerably shorter than the mean free path λ, e.g., $\mathrm{d} < \lambda/3$, σ_t total cross section. Collisions are assumed to occur instantaneously at their center of mass; a collision changes the momenta of both partners but both remain in $\mathrm{d}\mathbf{x}$ immediately after the encounter.

The number of lost particles from the volume element $\mathrm{d}\mathcal{T} = \mathrm{d}\mathbf{x}\mathrm{d}\mathbf{v}$ per unit time at position $\mathbf{x}$ is $f(\mathbf{v})f(\mathbf{v}_1)\,\sigma_\Omega|\mathbf{w}|\mathrm{d}\mathbf{x}\mathrm{d}\mathbf{v}\mathrm{d}\mathbf{v}_1$ integrated over the solid angle Ω and all velocities $\mathbf{v}_1$. σ_Ω is the differential collision cross section $\sigma(\{\mathbf{v}, \mathbf{v}_1\} \to \{\mathbf{v}', \mathbf{v}_1'\})$. Analogously, the gain of particles in $\mathrm{d}\mathcal{T} = \mathrm{d}\mathbf{x}\mathrm{d}\mathbf{v}$ from a scattering of two arbitrary particles $\mathbf{v}'$ and $\mathbf{v}_1'$ from $\mathrm{d}\mathcal{T}' = \mathrm{d}\mathbf{x}\mathrm{d}\mathbf{v}'$ with one of them ending in $\mathrm{d}\mathcal{T}$ is $f(\mathbf{v}')f(\mathbf{v}_1')\,\sigma_\Omega'|\mathbf{w}'|\mathrm{d}\mathcal{T}'$ integrated over the solid angle Ω and all velocities $\mathbf{v}_1'$. One particle out of the pair $\{\mathbf{v}', \mathbf{v}_1'\}$ ending in $\mathbf{v}\mathrm{d}\mathcal{T}$ means that either $\mathbf{v}'$ or $\mathbf{v}_1'$ depends uniquely on $\mathbf{v}$; it is not free owing to $\mathbf{V}' = \mathbf{V}$ and $|\mathbf{w}'| = |\mathbf{w}|$. In the case of classical particles it can be assumed that it is always $\mathbf{v}'$ that depends on $\mathbf{v}$.

Elastic binary collisions are such that, given the scattering angle Ω, the destination $\{\mathbf{v}', \mathbf{v}_1'\}$ is unique. If now the velocities $\{\mathbf{v}', \mathbf{v}_1'\}$ are inverted into $\{-\mathbf{v}', -\mathbf{v}_1'\}$ (time inversion) they collide and end in another pair of velocities. For point particles, or more generally, spherical particles without angular momentum or spin, it is quite immediate to assume that they end uniquely in $\{\mathbf{v}, \mathbf{v}_1\}$ if they encounter under the same solid angle $\Omega' = \Omega$. This means we postulate the equality of the cross sections

$\sigma(\{-\mathbf{v}, -\mathbf{v}_1\} \to \{-\mathbf{v}', -\mathbf{v}_1'\}) = \sigma(\{-\mathbf{v}', -\mathbf{v}_1'\} \to \{-\mathbf{v}, -\mathbf{v}_1\})$. By successive application of parity operation, i.e., mirroring in space follows the equality between the direct collision cross section σ_Ω and the indirect collision cross section σ'_Ω. As long as σ_Ω is of electromagnetic nature time reversal and parity are conserved. Additionally $|\mathbf{u}'| = |\mathbf{u}|$ holds. The Jacobian is $J(\mathbf{v}', \mathbf{v}_1'|\mathbf{v}, \mathbf{v}_1) = 1$ for particles of the same species because of $\mathrm{d}\mathbf{v}\mathrm{d}\mathbf{v}_1 = \mathrm{d}\mathbf{V}\mathrm{d}\mathbf{w} = \mathrm{d}\mathbf{V}'\mathrm{d}\mathbf{w} \to \mathrm{d}\mathbf{w} = \mathrm{d}\mathbf{w}' \to \mathrm{d}\mathbf{V}'\mathrm{d}\mathbf{w}' = \mathrm{d}\mathbf{V}\mathrm{d}\mathbf{w}$ for each value of σ_Ω fixed. Hence, the *Boltzmann equation* reads

$$\frac{\mathrm{d}f(\mathbf{x}, \mathbf{p}, t)}{\mathrm{d}t} = R(\mathbf{p}, \mathbf{p}_1; \mathbf{p}', \mathbf{p}_1') = \int \big(f(\mathbf{v}')f(\mathbf{v}_1') - f(\mathbf{v})f(\mathbf{v}_1)\big)\,|\mathbf{w}|\,\sigma_\Omega\,\mathrm{d}\Omega\,\mathrm{d}\mathbf{v}_1. \tag{3.83}$$

The left hand side of the Vlasov and the Boltzmann equation are the seven-dimensional divergence of the distribution function; they are in their standard, or conservative form. The Boltzmann collision integral on the right hand side is a source term analogous to λ and μ of particle and mass conservation (3.3). A source term in general indicates that there is some additional physics happening which is outside the possibility of description of the quantity under the div operation. Boltzmann achieves *closure* in $f(\mathbf{x}, \mathbf{p}, t)$ by the hypothesis that in the dilute fluid the particles are uncorrelated most of their path (*molecular chaos*) and are correlated, that means bound together, by the collision cross section σ_Ω when they come close to each other.

Maxwell Distribution

In the case of thermal equilibrium and the absence of external forces the left hand side of this equation must vanish because of no spatial gradients and no time dependence. As a consequence, the Boltzmann collision term must also vanish. The symmetry between direct and inverse collision (or time reversal) is such that even the integrand itself must vanish under the assumption that f is continuous and $\sigma_\Omega \geq 0$ everywhere; by simple arguments it can nowhere be negative. Thus, for thermal equilibrium follows

$$f(\mathbf{v}')f(\mathbf{v}_1') - f(\mathbf{v})f(\mathbf{v}_1) = 0 \;\Rightarrow\; \ln f(\mathbf{v}') + \ln f(\mathbf{v}_1') = \ln f(\mathbf{v}) + \ln f(\mathbf{v}_1) \tag{3.84}$$

for all pairs of velocities of binary collisions. This shows that $\ln f(\mathbf{v}) = \mathbf{C}m\mathbf{v}$, with $\mathbf{C}$ any constant vector, solves the functional equation because of momentum conservation in the collision. Summation over all velocities results in the projection of the fluid momentum onto $\mathbf{C}$. In the system of the fluid at rest the total momentum is zero, f results isotropic and can only depend on $|\mathbf{v}|$. However, among all powers of the modulus solely the energy conserving $m\mathbf{v}^2$ satisfies (3.84),

$$\ln f(\mathbf{v}) = \ln C - \beta\mathbf{v}^2 \quad \Rightarrow \quad f(\mathbf{v}) = C\mathrm{e}^{-\beta\mathbf{v}^2}.$$

Normalization to unity of f and equating the mean energy per particle ϵ_{in}/n to $3k_BT$ yields the celebrated and extremely useful *Maxwell distribution* $f = f_M$,

$$f_M(\mathbf{v}) = \left(\frac{\beta}{\pi}\right)^{3/2} \mathrm{e}^{-\beta \mathbf{v}^2}; \qquad \beta = \frac{m}{2k_BT}. \tag{3.85}$$

Moments of the Vlasov Equation

An infinite number of equations is obtained by multiplying (3.81) with all powers of $p_x^{r_x} p_y^{r_y} p_z^{r_z}$ or $v_x^{r_x} v_y^{r_y} v_z^{r_z}$ r_x, r_y, r_z nonnegative integers, and integrating over all momenta $\mathbf{p}$ or $\mathbf{v}$. To ensure the existence of the *moment* $\int p_x^{r_x} p_y^{r_y} p_z^{r_z} f(\mathbf{x}, \mathbf{p})\mathrm{d}\mathbf{p}$ of degree $r = \max\{r_x, r_y, r_z\}$, f must vanish as fast as $\exp(-|\mathbf{p}|)$ for $|\mathbf{p}| \to \infty$. The Maxwellian distribution f_M shows such a behavior. It can be proven that the resulting infinite set of *moment equations* is equivalent to the Vlasov equation (3.81). The zeroth and the first moments are

$$n(\mathbf{x}, t) = \int f(\mathbf{x}, \mathbf{v}, t)\mathrm{d}\mathbf{v}, \qquad \mathbf{u} = \frac{1}{n(\mathbf{x}, t)} \int \mathbf{v} f(\mathbf{x}, \mathbf{v}, t)\mathrm{d}\mathbf{v}. \tag{3.86}$$

The single particle velocity with respect to the fluid is $\mathbf{w} = \mathbf{v} - \mathbf{u}(\mathbf{x}, t)$. The *binary collisions* in (3.81) do not change to the moment equations since, as easily seen from symmetry

$$\int p_x^{r_x} p_y^{r_y} p_z^{r_z} R(\mathbf{p}, \mathbf{p}_1; \mathbf{p}', \mathbf{p}_1')\mathrm{d}\mathbf{p} = 0; \quad r_x, r_y, r_z \geq 0.$$

However, their conservation equations differ from the ideal fluid equations. The latter follow in the absence of collisions. Therefore we concentrate here on the Vlasov equation (3.79). Its zeroth moment is

$$\int \left\{ \frac{\partial f}{\partial t} + \mathbf{v}\frac{\partial f}{\partial \mathbf{x}} + \frac{q}{m}(\mathbf{E} + \mathbf{v} \times \mathbf{B})\frac{\partial f}{\partial \mathbf{v}} \right\} \mathrm{d}\mathbf{v} = \frac{\partial n}{\partial t} + \frac{\partial}{\partial \mathbf{x}} \int \mathbf{v} f \mathrm{d}\mathbf{v} \; + \frac{q}{m}(\mathbf{E} + \mathbf{v} \times \mathbf{B}) f \Big|_{\mathbf{v}=-\infty}^{\mathbf{v}=+\infty}$$

$$- \frac{q}{m} \int f \frac{\partial}{\partial \mathbf{v}}(\mathbf{E} + \mathbf{v} \times \mathbf{B})\mathrm{d}\mathbf{v} = \frac{\partial n}{\partial t} + \nabla(n\mathbf{u}) = 0. \tag{3.87}$$

This is the particle conservation. Integration by parts has been employed. For physical reasons $\mathbf{v} f = 0$ at infinities of $\mathbf{v}$, and so are the higher powers $v_x^{r_x} v_y^{r_y} v_z^{r_z} f = 0$ at the infinities of $\mathbf{v}$ for physical reasons. Differentiation by components shows $\nabla_{\mathbf{v}}(\mathbf{E} + \mathbf{v} \times \mathbf{B}) = 0$.

The first moment of $\mathbf{v}$, conveniently employed by components v_i, leads to momentum conservation as follows

$$\int \left\{ v_i \frac{\partial f}{\partial t} + v_i \sum_{j=1}^{3} v_j \partial_j f + \frac{q}{m} v_i \sum_{j=1}^{3} \left(E_j + \sum_{kl} \delta_{jkl} v_k B_l \right) \partial_{v_j} f \right\} \mathrm{d}\mathbf{v}$$

$$= \frac{\partial}{\partial t}(nu_i) + \sum_{j=1}^{3} \frac{\partial}{\partial x_j} \int (u_i + w_i)(u_j + w_j) f(\mathbf{w}) \mathrm{d}\mathbf{w}$$

$$+ \frac{q}{m} \sum_{j=1}^{3} \int v_i \frac{\partial}{\partial v_j} \left(E_j + \sum_{kl} \delta_{jkl} v_k B_l \right) f(\mathbf{v}) \mathrm{d}\mathbf{v} =$$

$$\frac{\partial}{\partial t}(nu_i) + \sum_{j=1}^{3} \frac{\partial}{\partial x_j} \int (u_i + w_i)(u_j + w_j) f(\mathbf{w}) \mathrm{d}\mathbf{w} - \frac{q}{m} \int \left(E_i + \sum_{kl} \delta_{jkl} v_k B_l \right) f(\mathbf{v}) \mathrm{d}\mathbf{v}$$

$$= \frac{\partial}{\partial t}(nu_i) + \sum_{j=1}^{3} \frac{\partial}{\partial x_j}(nu_i u_j) + \sum_{j=1}^{3} \frac{\partial}{\partial x_j} \int w_i w_j f \mathrm{d}\mathbf{w} - \frac{q}{m} n \big(\mathbf{E} + \mathbf{u} \times \mathbf{B}\big)_i = 0. \tag{3.88}$$

We define the pressure tensor P as

$$P = p_{ij} = m \int (v_i - u_i)(v_j - u_j) f \mathrm{d}\mathbf{v} \tag{3.89}$$

substitute it in (3.88) and obtain

$$\frac{\partial}{\partial t} \rho \mathbf{u} + \mathrm{div}\big(\rho \mathbf{u}\mathbf{u} + P\big) = n\big(\mathbf{E} + \mathbf{u} \times \mathbf{B}\big). \tag{3.90}$$

Compared to (3.56) the shear and volume viscosities are missing.

Multiplication of (3.79) with $m\mathbf{v}^2/2$ and integration over the velocity (or momentum) space yields the energy conservation equation straightforwardly with the help of integration by parts. On defining the *kinetic temperature* T_{kin},

$$\frac{3}{2} k_B T_{\mathrm{kin}} = \int \frac{1}{2} m (\mathbf{v} - \mathbf{u})^2 f \mathrm{d}\mathbf{v}. \tag{3.91}$$

follows $\epsilon_{\mathrm{in}} = (3/2) n k_B T_{\mathrm{kin}}$. Finally, there remains the term of heat flow density to be defined kinetically,

$$\mathbf{q} = \int \frac{1}{2} m (\mathbf{v} - \mathbf{u})^2 (\mathbf{v} - \mathbf{u}) f \mathrm{d}\mathbf{v}. \tag{3.92}$$

Now, all quantities appearing in the energy conservation equation in standard form (3.58) are determined from the one particle distribution function f, except dissipation by viscosity of shear and volume compression and heating from outside $\dot{h}$.

A kinetic equation of heat conduction $\mathbf{q}$ is obtained from the third order moment with $\mathbf{v}^2\mathbf{v}$. This, in turn introduces a forth order moment, etc. To stop at a definite order moment a *closure condition* is needed. One way is to go to moments of such an order r_0 that the moment of order $r_0 + 1$ is sufficiently small. Alternatively, one may look for an appropriate closure condition as for example by making use of the *Fourier ansatz* $\mathbf{q} = -\kappa\nabla T$ and the heat conduction coefficient taking from measurement or from additional microscopic arguments.

Collision frequency from the Boltzmann integral (3.83). As a useful application of the Boltzmann integral the electron-ion collision frequency ν of hard spheres is determined in the Lorentz approximation (ions at rest). The electron distribution function is assumed as $f_e(\mathbf{v}_e, t) = n_e\delta(\mathbf{v}_e - \mathbf{v}_0),\ f_i(\mathbf{v}_i) = n_i\delta(\mathbf{v}_i - 0);\ n_i = n_e$. Friction is connected with momentum loss, hence, the first moment of the Boltzmann equation has to be determined,

$$\int m_e\mathbf{v}\frac{\partial f_e}{\partial t}\mathrm{d}\mathbf{v} = m_e\int \left(f_e(\mathbf{v}_e')f_i(\mathbf{v}_i') - f_e(\mathbf{v}_e)f_i(\mathbf{v}_i)\right)|\mathbf{w}|\,\sigma_\Omega\,\mathrm{d}\Omega\,\mathrm{d}\mathbf{v}_i\mathrm{d}\mathbf{v}_e$$

$$\Rightarrow m_e n_e\frac{\partial\mathbf{v}_0}{\partial t} = m_e n_e n_i\int\left\{[\delta(\mathbf{v}_e' - \mathbf{v}_0) - \delta(\mathbf{v}_e - \mathbf{v}_0)]\frac{R^2}{4}\mathrm{d}\varphi\sin\theta\mathrm{d}\theta\right\}|\mathbf{v}_e|\mathrm{d}\mathbf{v}_e$$

$$= -m_e n_e n_i\frac{\pi}{2}R^2\int_0^{-1}\{[\cos\theta - 1]\mathrm{d}\cos\theta\}|\mathbf{v}_e|\delta(\mathbf{v}_e - \mathbf{v}_0)\mathrm{d}\mathbf{v}_e$$

$$\Rightarrow \frac{\partial\mathbf{v}_0}{\partial t} = -\pi R^2 n_i\mathbf{v}_0 = -\nu\mathbf{v}_0;\qquad \nu = n_i\sigma_t v_0. \tag{3.93}$$

The square bracket in the second line for each scattering angle θ stands for the difference $-(1 - \cos\theta)$ and expresses the loss of momentum per particle in forward direction.

Lagrange–Heisenberg Picture Versus Euler–Schrödinger Picture

The macroscopic conservation equations of fluids are obtained by recognizing combinations of moments as physical quantities, for example the total energy density ϵ per particle,

$$\epsilon(\mathbf{x}, t) = \int\frac{1}{2}m\mathbf{v}^2 f(\mathbf{x}, \mathbf{v}, t)\mathrm{d}\mathbf{v} = \frac{1}{2}m\mathbf{u}^2 + \int\frac{1}{2}m(\mathbf{v} - \mathbf{u})^2 f\mathrm{d}\mathbf{v} = \epsilon_{\text{kin}} + \epsilon_{\text{in}}. \tag{3.94}$$

This is the *Euler picture* of the expectation value of the single particle energy $\langle E(\mathbf{x}, t)\rangle = \epsilon(\mathbf{x}, t)$ as the result of folding the time and position independent quantity (operator) $\mathbf{v}^2$ with the normalized distribution function $f(\mathbf{x}, \mathbf{v}, t)$ depending on time (and position). In quantum systems it corresponds exactly to the *Schrödinger picture* in Hilbert space: The expectation value $\langle b\rangle$ of the operator b is

$$\langle b(t)\rangle = \int\psi_S(\mathbf{x}, t)^* b\,\psi_S(\mathbf{x}, t)\mathrm{d}\mathbf{x}. \tag{3.95}$$

Its evolution in time is accomplished through the time dependence of the state vector $|\psi(t)\rangle_S$ which follows from the time dependent Hamiltonian $H(\mathbf{p}, \mathbf{q}, t)$, i.e., from the time dependent Schrödinger equation. This contrasts with the equivalent *Heisenberg picture*: The state vector $|\psi\rangle_H$ is time independent, $\langle b(t)\rangle$ is by folding the time dependent operator $b(t)_H = \exp -(iH_S t/\hbar) b_S \exp i(H_S t/\hbar)$ (simplest case) with $|\psi\rangle_H$,

$$\langle b(t)\rangle = \int \psi_H^*(\mathbf{x}) b_H(t)\, \psi_H(\mathbf{x}) \mathrm{d}\mathbf{x}. \tag{3.96}$$

The equivalence of the two pictures rises the question whether such a correspondence exists for the classical Euler representation. In (3.94) b is to be identified with $\mathbf{v}^2$. In general, the dynamic quantity b will be a function of $\mathbf{p} = m\mathbf{v}$ and of position $\mathbf{q}$ in order to describe averages of *potential energies*. Hence, $b = b(\mathbf{p}, \mathbf{q}, \mathbf{x})$. The dependence on the macroscopic variable has the same meaning as in the Vlasov equation (3.79). In the Eulerian picture the expectation value of the time independent b is given by

$$\langle b\rangle = \langle b(\mathbf{x}, t)\rangle = \int b(\mathbf{p}, \mathbf{q}, \mathbf{x}) f(\mathbf{x}, \mathbf{p}, \mathbf{q}, t) \mathrm{d}\mathbf{p}\mathrm{d}\mathbf{q}. \tag{3.97}$$

There is a one to one correspondence of the single trajectory $(\mathbf{p}, \mathbf{q})$ to its initial values $(\mathbf{q} = \mathbf{a}, \mathbf{p} = m\mathbf{v}_0)$ at $t = 0 \Rightarrow (\mathbf{p}(t), \mathbf{q}(t)) = (\mathbf{p}(\mathbf{a}, \mathbf{v}_0, t), \mathbf{q}(\mathbf{a}, \mathbf{v}_0, t))$, the transformation is canonical; J relating $\mathrm{d}\mathbf{q}\mathrm{d}\mathbf{p}$ to $\mathrm{d}\mathbf{a}\mathrm{d}\mathbf{v}_0$ is the constant mass m. Thus

$$\langle b\rangle = \int b(\mathbf{a}, \mathbf{v}_0, t) f(\mathbf{x}, \mathbf{a}, \mathbf{v}_0, t) \mathrm{d}\mathbf{a}\mathrm{d}\mathbf{v}_0 = \int b(\mathbf{a}, \mathbf{v}_0, t) f_0(\mathbf{x}, \mathbf{a}, \mathbf{v}_0) \mathrm{d}\mathbf{a}\mathrm{d}\mathbf{v}_0. \tag{3.98}$$

The last term on the right is the Lagrangian representation of the expectation value $\langle b\rangle$. The equality $f(\mathbf{x}, \mathbf{p}, \mathbf{q}, t) = f_0(\mathbf{x}, \mathbf{a}, \mathbf{v}_0)$ follows from $\mathrm{d}f/\mathrm{d}t = 0$ as a consequence of Liouville's theorem. The evolution in time of b is given by the classical Hamiltonian $H(\mathbf{p}, \mathbf{q}, t)$. A formal proof of the equivalence between the Eulerian and the Lagrangian picture in 6N dimensions can be found in [8]. Generally the Eulerian representation is used in practice. Here we make use of the Lagrangian representation in connection with Landau damping in Chap. 5.

The equivalence of

$$\langle b(t)\rangle = \int b(\mathbf{p}, \mathbf{q}) f(\mathbf{x}, \mathbf{p}, \mathbf{q}, t) \mathrm{d}\mathbf{p}\mathrm{d}\mathbf{q} = \int b(\mathbf{p}_0, \mathbf{q}_0, t) f_0(\mathbf{x}, \mathbf{p}_0, \mathbf{q}_0) \mathrm{d}\mathbf{p}_0\mathrm{d}\mathbf{q}_0$$

is intuitively clear. The first integral can be interpreted as the expectation value $b(\mathbf{p}, \mathbf{q})$ at the time t because $f(\mathbf{x}, \mathbf{p}, \mathbf{q}, t)$ is the likelihood of b to assume the value $b(\mathbf{p}, \mathbf{q})$ at time t. The second integral represents the expectation value of $b(\mathbf{p}, \mathbf{q}, t)$ at $t = 0$. Since the particle involved are the same at any time $t > 0$, $\langle b(t)\rangle$ is determined by the operator b evolving in time.

Towards Non-standard Fluid Dynamics: The Case of Field Ionization

Kinetic theory allows to expand fluid theory from its phenomenological limits far down to the (nearly) collisionfree medium. Such an extension is of highest relevance for hot matter description like the laser generated and the cosmic plasma. At first glance it may be surprising that in the collisionless case the governing fluid equations maintain the same structure as known from the collision-dominated fluid. The trajectory of a dilute fluid element is no longer definable by the motion of its center of mass; the only statement that can be set is the trajectory as an indicator of the direction into which a localized ensemble of particles moves on the average during a restricted time interval Δt. At an appreciably larger time interval the trajectory $\mathbf{x}(t)$ is represented by an ensemble of new particles. Starting from a definition of the dynamic variables of the fluid as instantaneous quantities is helpful to find the correct equations of motion by intuition in simple cases. For instance, some further reasoning makes it clear that in the absence of particle creation and annihilation there is very little freedom to come to conservation equations differing from those in the previous sections.

Considerably more intuition is required in situations of the creation and annihilation of new particles, because new terms may be needed to describe a dilute, but also a dense fluid adequately. Kinetic theory is of great help also here. As a paradigm for numerous relevant cases we study the situation of field ionization by an intense laser beam.

Assume the electron density to be n_e, $n_b = n_0 - n_e/Z$ that of the neutral particles, with Z the ion charge, and n_0 the initial number of atoms. The probability of field ionization per unit time is λ_I. The electrons are released with the average initial velocity $\langle\mathbf{v}\rangle$ and average initial energy $m\langle\mathbf{v}^2\rangle/2$ at time instant t. From Vlasov's equation with λ_I as the ionization source term one derives the following moment equations

$$\frac{\partial n_e}{\partial t} + \nabla(n_e\mathbf{u}_e) = n_b\lambda_I,$$

$$\rho_e\frac{\mathrm{d}\mathbf{u}_e}{\mathrm{d}t} + n_b\lambda_I m_e(\mathbf{u}_e - \langle\mathbf{v}\rangle) + (\nu_{ei} + \nu_{en})\rho_e\mathbf{u}_e = -\nabla P_e - en_e\mathbf{E},$$

$$n_e\frac{\mathrm{d}}{\mathrm{d}t}\left(\frac{\epsilon_{\mathrm{in}}}{n_e} + \frac{m_e}{2}\mathbf{u}_e^2\right) + \left(\frac{m_e}{2}(\mathbf{u}_e^2 - \langle\mathbf{v}^2\rangle) + \frac{\epsilon_{\mathrm{in}}}{n_e}\right)n_b\lambda_I = -\nabla(\mathbf{u}_e P_e) - \nabla\mathbf{q}_e + \mathbf{j}_e\mathbf{E}. \tag{3.99}$$

The ideal electron pressure tensor P_e is to be calculated according to (3.89). The electron current density $\mathbf{j}_e$ is calculated from the momentum equation. The heating function is obtained by taking the cycle averaged work done by the acting electric field, $\dot{h} = \overline{\mathbf{j}_e\mathbf{E}}$. Averaging is necessary because the instantaneous value of $\mathbf{j}_e\mathbf{E}$ contains contributions of reversible oscillatory energy of the electrons. Only if $\mathbf{u}_e = \langle\mathbf{v}\rangle$ and $\mathbf{u}_e^2 = \langle\mathbf{v}^2\rangle$ the additional non-standard terms vanish although in that case new electrons may have been emitted. This fact sheds light on why the collisionless fluid is governed by equations of the structure of moderately dense fluids.

3.2 Relativistic Fluid Dynamics

Lorentz scalars, four vectors, and four tensors are powerful instruments for extending non-relativistic laws into the relativistic domain. One of the valuable strategies is to build a four quantity and to take its divergence. If this is zero in one reference system it remains zero in all systems $S(\mathbf{v})$. Any zero four quantity, scalar, vector, tensor, transforms into zero. It is the aim to look for a reference system in which the divergence of the four quantity of interest is zero. If it expresses a meaningful non-relativistic conservation law there, its relativistic extension is proven. As an example the procedure is applied to the ideal fluid.

3.2.1 *Ideal Fluid Dynamics*

Particle density at rest n_0 multiplied by the four velocity $U = \gamma(\mathbf{u}, c) = u^\alpha$ and by the four momentum $P = mV$ are the four vectors J_n and J_ρ of particle and momentum flow densities. As a consequence $\mathrm{div} J_n = \partial_\alpha j_n{}^\alpha$ and $\mathrm{div} J_\rho$ are Lorentz scalars. In the system at rest their divergence is zero, see (3.54), thus (3.2) are already the relativistically correct particle and mass conservation laws,

$$\partial_\alpha j_n{}^\alpha = 0 \quad \Leftrightarrow \quad \frac{\partial n}{\partial t} + \nabla(n\mathbf{u}) = 0; \quad \Rightarrow \quad n = \gamma n_0. \tag{3.100}$$

$$\partial_\alpha j_\rho{}^\alpha = 0 \quad \Leftrightarrow \quad \frac{\partial \rho}{\partial t} + \nabla(\rho\mathbf{u}) = 0; \quad \Rightarrow \quad \rho = \gamma^2 \rho_0. \tag{3.101}$$

Note $u^\alpha = \gamma u_i$ for the spatial components of U. The particle density transformation $n = \gamma n_0$ is a consequence of Lorentz contraction or time dilation. Consider now the tensor $T = T^{\alpha\beta} = \rho_0 u^\alpha u^\beta$. If in the momentum equation (3.56) pressure and external force density are set to zero for vanishing flow velocity holds

$$\frac{\partial}{\partial t}\rho\mathbf{u} + \mathrm{div}\rho\mathbf{u}\mathbf{u} = \partial_\alpha T^{\alpha\beta} = 0; \;\; \beta = 1 - 3; \;\; \partial_\alpha T^{\alpha 4} = c\{\partial_t \rho + \partial_i \rho u_i\} = 0. \tag{3.102}$$

So, div T is zero in any reference system and T as a four tensor is the relativistic energy-momentum conservation law of the pressure-less ideal fluid. $T^{\alpha 4}$ multiplied by c is the energy conservation in the absence of internal energy, or at zero temperature, respectively. In (3.101) $\rho = \gamma^2 \rho_0$ has been postulated from the mass increase $m(\mathbf{u}) = \gamma m$, in (3.102) it is a consequence of the four momentum conservation. $\mathrm{div} T = 0$ may be a model equation of dust, e.g., interplanetary or intergalactic dust.

In the presence of finite pressure p a four tensor $S^{\alpha\beta}$ must be added to $T = \rho_0 v^\alpha v^\beta$ to yield $\partial_i S^{ij} = \partial_i p\, \delta^{ij}$, $i, j = 1 - 3$. This is accomplished by setting

$$T^{\alpha\beta} = \rho_0 u^\alpha u^\beta + S^{\alpha\beta} = \left(\rho_0 + \frac{p}{c^2}\right) u^\alpha u^\beta + p g^{\alpha\beta}$$

with $g^{\alpha\beta}$ the metric tensor. $T^{\alpha\beta}$ is clearly a four tensor, and hence $\partial_\alpha T^{\alpha\beta}$ is a four vector F, which for vanishing velocity $\mathbf{u}$ must reduce to zero [see (3.56), $\mathbf{f} = 0$]. Therefore $F = 0$ must hold in any inertial frame, i.e., $\partial_\alpha T^{\alpha\beta} = 0$ are the relativistic conservation equations of energy (mass) and momentum. In $(\mathbf{x}, t)$ representation they read

$$\partial_t \left(\rho c^2 + \gamma^2 p \frac{\mathbf{u}^2}{c^2} \right) + \nabla(\rho c^2 + \gamma^2 p)\mathbf{u} = 0. \tag{3.103}$$

$$\partial_t \left(\rho + \gamma^2 \frac{p}{c^2} \right) \mathbf{u} + \nabla \left\{ \left(\rho + \gamma^2 \frac{p}{c^2} \right) \mathbf{u}\mathbf{u} + p\mathbf{I} \right\} = 0, \quad \mathbf{I} = (\delta^{ij}). \tag{3.104}$$

In contrast to (3.56) and (3.58) in the relativistic formulation the pressure seems to contribute to the internal and kinetic energy density ρc^2 and to the momentum density $\rho\mathbf{u}$. However, this would be an erroneous interpretation because the internal energy is defined in the rest frame of the fluid where the pressure term in (3.103) reduces to zero. Its appearance in the time derivatives of the relativistic conservation equations is a necessary consequence once the spatial derivatives of the pressure have to be included. In fact, the time derivative must be added to the spatial derivatives to form a four vector. The term $\nabla(\gamma^2 p\mathbf{u})$ in (3.103) stands for the work a volume element spends to remove the neighboring fluid elements. To make it relativistically invariant the four product containing the time coordinate is needed.

With the help of (3.103) momentum conservation (3.104) is reduced to the more intuitive relativistic Euler equation

$$\left(\rho + \gamma^2 \frac{p}{c^2} \right) \left(\frac{\partial \mathbf{u}}{\partial t} + (\mathbf{u}\nabla)\mathbf{u} \right) = -\left(\nabla p + \mathbf{u} \frac{\partial p}{\partial t} \right). \tag{3.105}$$

3.2.1.1 Digression on Particles with Zero Rest Mass (Photons)

No relation $n = \gamma n_0$ analogous to (3.100) can be defined. However, if the particle number is conserved, e.g., of photons in the absence of sources, the quantity $J = (\mathbf{j}, nc)$ is defined in S, and if it is a four vector, $\partial_\alpha j^\alpha = 0$ holds. $J = (\mathbf{S}, nc)$ with $\mathbf{S}$ the Poynting vector, or $\mathbf{j} = n\mathbf{c}$ for massless particles in general, is a four vector because $J^2 = n(\mathbf{c}^2 - c^2) = 0$. Hence, J transforms as

$$J' = (\mathbf{j}', n'c) = \left(\mathbf{j} + \frac{\gamma - 1}{v^2}(\mathbf{j}\mathbf{v})\mathbf{v} - \gamma \frac{\mathbf{v}}{c} nc,\ \gamma \left(nc - \frac{\mathbf{v}}{c}\mathbf{j} \right) \right). \tag{3.106}$$

In the special case of $\mathbf{j} \parallel \pm\mathbf{v}$ the massless particle density transforms like

$$n' = n\gamma\Big(1 \mp \frac{v}{c}\Big) = \Big(\frac{1 \mp \beta}{1 \pm \beta}\Big)^{1/2}. \tag{3.107}$$

3.2.1.2 Center of Momentum and Mass of Free Particles

The center of mass $\mathbf{x}_C$ is a physical entity, for example the center of mass of the earth. This implies that it must obey the Lorentz point transformation $X \to X' = x^{\alpha'} = \Lambda^{\alpha'}_{\beta} x^{\beta}$. The necessity of an invariant definition of $\mathbf{x}_C$ is illustrated by Fig. 3.5. The reader may imagine two massive blocks $M_1 = M_2 = M$ moving against each other at speeds close to light speed. An observer comoving with mass M_1 localizes the center of mass close to C_2, an observer moving with M_2 perceives it close to C_1.

An invariant definition of C is found with the help of the four momentum. To this aim consider an ensemble of N free point particles of rest masses $m_i \neq 0$ and four velocities V_i. The sum of momenta is the four momentum P,

$$P = \sum_{i=1}^{N} m_i V_i = \sum_{i=1}^{N} m_i \gamma_i \left(\mathbf{v}_i, c\right) = MV = M\gamma(v)\left(\mathbf{v}, c\right) \tag{3.108}$$

with $\gamma_i = \gamma(v_i)$. This definition implies $\gamma(v)M = \sum_i m_i\gamma_i$. It is evident that a reference system S_C exists in which $V = (0, c)$ in (3.108). Then

$$M_\mathrm{C} = \sum_{i=1}^{N} m_i \gamma_{i\mathrm{C}}, \qquad P = M_\mathrm{C} U = \gamma(\mathbf{u}) M_\mathrm{C}(\mathbf{u}, c). \tag{3.109}$$

$$\mathbf{x}_\mathrm{C} = \frac{1}{M_\mathrm{C}} \sum_{i=1}^{N} m_i \gamma_{i\mathrm{C}} \mathbf{x}_i, \qquad \gamma_{i\mathrm{C}} = \gamma_i(\mathbf{v}_{i\mathrm{C}}). \tag{3.110}$$

By $\mathbf{x}_\mathrm{C}$ the center of mass in the rest frame is defined. $X_\mathrm{C} = (\mathbf{x}_\mathrm{C}, ct)$ is a Lorentz vector, and with $\mathbf{x}_\mathrm{C}(t) = \mathbf{x}_j(t)$ also $\mathbf{x}'_\mathrm{C}(t') = \mathbf{x}'_j(t')$ holds. An additional advantage

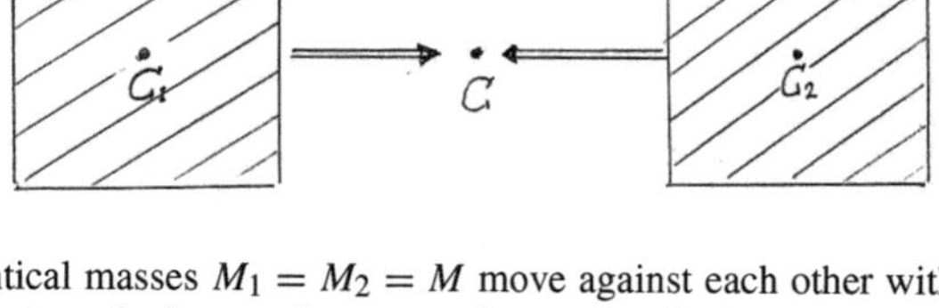

Fig. 3.5 Two identical masses $M_1 = M_2 = M$ move against each other with speeds close to c. Depending on the system of reference the center of mass C varies between C_1 and C_2. The invariant center coincides with the symmetry center C

of the special choice of the definition (3.110) is that $M_C c^2$ can be given the simple interpretation of the total internal energy in the case of noninteracting particles. The kinetic energy of the system of N point masses in the rest frame then reads $(M_C - \sum_i m_i)c^2$. Owing to

$$c^4(\sum m_i\gamma_i)^2 = M_C^2 c^4 + c^2\mathbf{p}_C^2, \quad vp_C = M_C\gamma_C\mathbf{v}_C. \tag{3.111}$$

the internal energy $M_C c^2$ is the minimum kinetic energy which the system can take under a Lorentz transformation.

For the interested reader a proof of the existence of S_C with $V = (0, c)$ is given. A four vector $Y = (\mathbf{y}, y^4)$ is *time-like* if $Y^2 = \mathbf{y}^2 - (y^4)^2 < 0$ and *space-like* if $Y^2 > 0$. By choosing a reference system $S(\mathbf{v}_t)$ with $\gamma_t\mathbf{v}_t = c\mathbf{y}/y^4$ the time-like Y reduces to $Y = (0, y_0^4)$ in $S(\mathbf{v}_t)$ whereas, when setting $\mathbf{v}_s = cy^4\mathbf{y}/\mathbf{y}^2$ a space-like Y becomes $Y = (\mathbf{y}_0, 0)$ in $S(\mathbf{v}_s)$. V_i and $P_i = mV_i$ are time-like. Hence, $P = P_1 + P_2 = (0, P_1^4) + (\mathbf{p}_2, P_2^4)$ is also time-like owing to $P^2 = -P_1^2 + \mathbf{p}_2^2 - (P_2^4)^2 - 2P_1^4P_2^4 < 0$ and $P_1^4, P_2^4 > 0$. Assertion $V = (0, c)$ follows by induction on P and from $V = P/M_C$, M_C from (3.110).

3.2.2 Moment Equations

The one-particle distribution function $f = \mathrm{d}N/\mathrm{d}\tau$ from (3.79) is a Lorentz scalar. To show it we observe that $\mathrm{d}N$ is an dimensionless pure number and as such a Lorentz scalar. The conservation of $\mathrm{d}\tau$ in t in a fixed reference system is guaranteed by Liouville's theorem (i). Its invariance with respect to a Lorentz transformation $S \to S'(\mathbf{v})$ is evidenced as follows. Suppose $\mathrm{d}\tau$ is at rest in S, $\mathrm{d}\tau = \mathrm{d}\mathbf{x}_0\mathrm{d}\mathbf{p}_0$. In $S'(\mathbf{v})$ it is $\mathrm{d}\tau' = \mathrm{d}\mathbf{x}'\mathrm{d}\mathbf{p}' = (\mathrm{d}\mathbf{x}_0/\gamma)(\gamma\mathrm{d}\mathbf{p}_0) = \mathrm{d}\tau$ as a consequence of the Lorentz contraction in $\mathbf{x}$ and time dilation in $\mathbf{p}$ (ii). Thus, f is an invariant scalar. The three-forms (3.79) and (3.81) are clearly Lorentz invariant since no use of a special reference system has been made in any step of their derivation.

Moment equations can be obtained from (3.81) by multiplying it with $t^{\alpha_1\alpha_2\ldots\alpha_s} = p^{\alpha_1}p^{\alpha_2}\ldots p^{\alpha_s}$ and integrating over the momentum space $\mathbb{R}_p^3$, just in the same way as in the nonrelativistic case. In order to preserve the four-character of $t^{\alpha_1\alpha_2\ldots\alpha_s}$ one has to integrate over an invariant volume element in momentum subspace, i.e., $\mathrm{d}\mathbf{p}/\gamma$ or $\mathrm{d}\mathbf{p}/p^4$, $p^4 = mu^4 = \gamma mc$. The integrals

$$T^{\alpha_1\alpha_2\ldots\alpha_s} = \int p^{\alpha_1}p^{\alpha_2}\ldots p^{\alpha_s} f(X, \mathbf{p})\frac{\mathrm{d}\mathbf{p}}{\gamma(\mathbf{p})mc} \tag{3.112}$$

clearly define a contravariant four tensor T of rank s because any finite sum over the four tensor $t^{\alpha_1\alpha_2\ldots\alpha_s}$ and hence also the limiting sum, i.e., the integral over the invariant volumes $\mathbf{dp}/(\gamma mc)$, are tensors of the same kind. The link with the fluid equations is established by introducing kinetic expressions for particle and charge density, internal energy, pressure, etc. and their fluxes. To this aim the global dynamics has to be split up into an inner component and a macroscopic outer flow component. In other words, the above developed relativistic concept of the center of momentum is to be used.

The number of particles in the volume element $\mathrm{d}V = \mathbf{dx}$ moving at velocity $\mathbf{u}_C$ is

$$\mathrm{d}N = n\mathbf{dx} = n_0\mathbf{dx}_0 \quad \Rightarrow \quad n = \gamma_C\, n_0; \quad \gamma_C = \gamma(\mathbf{u}_C) \tag{3.113}$$

owing to Lorentz contraction $\mathbf{dx} = \mathbf{dx}_0/\gamma_C$. For this relation to be valid it is essential that n_0 is at rest. It is also true by definition that

$$n\mathbf{dx} = \mathbf{dx}\int f(\mathbf{x},\mathbf{p},t)\mathbf{dp} = \mathbf{dx}\int \gamma(\mathbf{p}) f\,\mathrm{d}\frac{\mathbf{dp}}{\gamma(\mathbf{p})} = \gamma_C\mathbf{dx}\int f_0\,\mathrm{d}\frac{\mathbf{dp}}{\gamma(\mathbf{p})}.$$

$$\Rightarrow \quad n_0(\mathbf{x},t) = \int f(\mathbf{x},\mathbf{p},t)\,\mathrm{d}\frac{\mathbf{dp}}{\gamma(\mathbf{p})}. \tag{3.114}$$

owing to $f_0(\mathbf{x},\mathbf{p}=0,t) = f(\mathbf{x},\mathbf{p},t)$ and $\mathbf{dx}_0 = \gamma_C\mathbf{dx}$. $\gamma_C = \int \gamma\,\mathrm{d}(\mathbf{p}/\gamma)$ follows from (3.109). The kinetic definitions of the four current density J_n, particle density n, flow velocity $\mathbf{u}$, and the free particle energy density ϵ are accordingly,

$$\begin{aligned} J_\mathrm{n} &= c\int Pf\,\frac{\mathbf{dp}}{p^4} = n(\mathbf{u},c) = n_0 U, \qquad U = \gamma(\mathbf{u})(\mathbf{u},c),\\ n &= \int f\,\mathbf{dp}, \qquad \mathbf{u} = \frac{1}{n}\int \mathbf{v}f\,\mathbf{dp}, \qquad n = \gamma(\mathbf{u})n_0,\\ \epsilon &= c\int (p^4)^2 f\frac{\mathbf{dp}}{p^4} = mc^2\int \gamma(\mathbf{p})f\,\mathbf{dp} = \epsilon_\mathrm{in} + \epsilon_\mathrm{kin}. \end{aligned} \tag{3.115}$$

Index C is suppressed. The internal energy density ϵ_in is the energy density in the center of momentum system (3.109) with $\mathbf{u} = 0$. Note that the averaging procedure for particle density $n(\mathbf{x},t)$ and mean flow velocity $\mathbf{u}(\mathbf{x},t)$ in the relativistic and non-relativistic case are identical. The averaging procedure (3.115) can be justified on purely formal grounds: If a quantity under consideration coincides with its nonrelativistic expression in the rest frame (like n and j^α) it is relativistically correct in any frame $S'(\mathbf{v})$ because (3.112) generates four quantities out of four quantities.

Decomposition into intrinsic and dynamic components by means of the center of momentum concept (3.109) (Eckart's decomposition [9]) is appropriate for particles of finite rest mass. In contrast to the nonrelativistic case no simple addition theorem exists for the individual velocity three vectors. Decomposition is therefore done in four space. To this aim we observe first that in the non-relativistic regime the decomposition of the individual particle velocity $\mathbf{v}$ into the drift velocity $\mathbf{u}$ and the intrinsic velocity $\mathbf{w} = \mathbf{v} - \mathbf{u}$ can alternatively be expressed as $\mathbf{w} = \mathbf{w}_\parallel + \mathbf{w}_\perp$, where $\mathbf{w}_\parallel = (\mathbf{w}\mathbf{u}^0)\mathbf{u}^0$ is the projection of $\mathbf{w}$ onto the drift unit vector $\mathbf{u}^0$. With this in mind it is seen that

$$\int \mathbf{w} f \, \mathrm{d}\mathbf{w} = \int (\mathbf{v} - \mathbf{u}) f \, \mathrm{d}\mathbf{v} = \int [\mathbf{v} - (\mathbf{v}\mathbf{u}^0)\mathbf{u}^0] f \, \mathrm{d}\mathbf{v} = \int \mathbf{w}_\perp f \, \mathrm{d}\mathbf{w}_\perp. \tag{3.116}$$

This identity translates straightforwardly into four formalism: The individual particle momentum $P = (\mathbf{p}, p^4)$ is represented as the sum of a vector parallel to the mean flow four velocity U and a vector $W = (\mathbf{w}, w^4)$ perpendicular to it,

$$p^\alpha = m(gu^\alpha + w^\alpha), \qquad UW = u_\beta w^\beta = 0, \tag{3.117}$$

so that $w^4 = u_i w^i / u^4$. In the center of momentum system follows from (3.109) $\mathbf{w}_i = \gamma_{ic} \mathbf{v}_{ic}$ and $w_i^4 = 0$. W is a space-like vector. From $P^2 = m^2(gU + W)^2$, $U^2 = -c^2$, and $P^2 = -m^2c^2$ the coefficient $g = -UP/(mc^2)$ is obtained as $g = (1 + W^2/c^2)^{1/2}$. The average $\langle W \rangle$ is a measure of the internal energy per particle in units of the rest energy mc^2. In passing from $\mathbf{p}$ to $\mathbf{w}$ in (3.112)

$$\frac{\mathrm{d}\mathbf{p}}{p^4} = \frac{m^2}{g} \frac{\mathrm{d}\mathbf{w}}{u^4} \tag{3.118}$$

has to be used. The general validity of this relation can be verified by a lengthy explicit calculation of the Jacobian $|\partial(\mathbf{p})/\partial(\mathbf{w})|$. Alternatively it follows directly in the center of mass system ($w^4 = 0$) and hence in any inertial frame because $\mathrm{d}\mathbf{p}/p^4$ is an invariant measure. The decomposition (3.117) yields for the particle current density $J = j^\alpha$ and the energy-stress tensor $T = T^{\alpha\beta}$

$$j^\alpha = n_0 u^\alpha. \tag{3.119}$$

$$T^{\alpha\beta} = (\rho_0 + \epsilon/c^2) u^\alpha u^\beta + P^{\alpha\beta} + (u^\alpha q^\beta + u^\beta q^\alpha)/c, \tag{3.120}$$

where, with $h(X, \mathbf{w}) = m^3 c f(, \mathbf{p})$, the single terms are given by

$$\rho_0 = m \int h \frac{\mathrm{d}\mathbf{w}}{u^4}, \qquad \epsilon = mc^2 \int (g-1) h \frac{\mathrm{d}\mathbf{w}}{u^4},$$

$$P^{\alpha\beta} = m \int w^\alpha w^\beta h \frac{\mathrm{d}\mathbf{w}}{gu^4}, \qquad q^\alpha = mc^2 \int (g-1) w^\alpha h \frac{\mathrm{d}\mathbf{w}}{gu^4}. \tag{3.121}$$

In the local rest frame ($u^4 = c$, $w^4 = 0$) the mass density ρ_0, the internal energy ϵ, and the corresponding three vectors and three tensors read with $h' = h/c$

$$\rho_0 = m \int h' \,\mathbf{dw}, \quad \epsilon = mc^2 \int \left[\left(1 + \frac{\mathbf{w}^2}{c^2}\right)^{1/2} - 1 \right] h' \mathbf{dw}$$

$$P^{ik} = m \int \frac{w^i w^k}{(1+\mathbf{w}^2/c^2)^{1/2}} h' \mathbf{dw}, \quad q^i = mc^2 \int \left[1 - \frac{1}{(1+\mathbf{w}^2/c^2)^{1/2}}\right] w^i h' \,\mathbf{dw}. \tag{3.122}$$

The quantity q^i represents the local heat flow density. In the nonrelativistic limit q^i and the remaining quantities in (3.122) reduce to (3.92), (3.89), (3.91), and (3.86). The total energy density in the rest frame is $\rho_0 c^2 + \epsilon = T^{\alpha\beta} u_\alpha u_\beta$.

It is possible to identify U with the energy transport velocity and $J = J_m$ with $\rho_0 u + R$,

$$u^\alpha = (\rho_0 + \epsilon/c^2) u^\alpha u^\beta u_\beta = T^{\alpha\beta} u_\beta, \qquad R^\alpha = mc^2 \int w^\alpha h \frac{\mathbf{dw}}{g u^4} \tag{3.123}$$

in which case $q^\alpha = 0$ and $R^\alpha \neq 0$ holds (see Landau and Lifshitz [10]). With this decomposition the heat flux is hidden in J of (3.119). Such a choice may be advantageous in case of massless particles, e.g., photons.

The conservation laws (equations of motion) are obtained from the zeroth order and first order moments of $P = p^\alpha$,

$$\text{0th order}: \quad \int f(X,p)\,\mathbf{dp} = \int p^4 f(X,p) \frac{\mathbf{dp}}{p^4},$$

$$\text{1st order}: \quad \int p^\alpha f(X,p)\,\mathbf{dp} = \int p^\alpha p^4 f(X,p) \frac{\mathbf{dp}}{p^4}.$$

Folding the Vlasov equation (3.81) with p^4 by the invariant measure $\mathbf{dp}/p^4$,

$$\int p^4 \left[\partial_t f + \mathbf{v}\partial_\mathbf{x} f + q(\mathbf{E} + \mathbf{v} \times \mathbf{B})\partial_\mathbf{p} f\right] \frac{\mathbf{dp}}{p^4} = 0, \tag{3.124}$$

partial integration of the third term and comparison with (3.115) leads to the particle number or charge conservation $\partial_\alpha j_\mathrm{n}^\alpha = 0$ from (3.100):

$$\begin{aligned} 0 &= \int \partial_t f \,\mathbf{dp} + \int \mathbf{v}\partial_\mathbf{x} f \,\mathbf{dp} + q \int (\mathbf{E} + \mathbf{v} \times \mathbf{B})\partial_\mathbf{p} f \,\mathbf{dp} \\ &= \partial_t \int f \,\mathbf{dp} + \partial_\mathbf{x} \int \mathbf{v} f \,\mathbf{dp} + q \sum \int \left[(\mathbf{E} + \mathbf{v} \times \mathbf{B})^j f\right]_{p^j=-\infty}^{p^j=+\infty} \mathrm{d}p^k \,\mathrm{d}p^l \\ &= \partial_t n(\mathbf{x},t) + \nabla[n(\mathbf{x},t)\mathbf{u}(\mathbf{x},t)] = \partial_\alpha j_\mathrm{n}^\alpha. \end{aligned}$$

The third term vanishes owing to $f(p^i = \pm\infty) = 0$ and $\partial_{p^i} v_{j\neq i} = 0$. Momentum and energy conservation are obtained in the same way by folding (3.81) with $t^{\alpha 4} = p^\alpha p^4$ as the divergence of $T^{\alpha i}$ and $T^{\alpha 4}$,

$$\partial_\alpha T^{\alpha i} = nq(\mathbf{E} + \mathbf{u} \times \mathbf{B})^i, \qquad \partial_\alpha T^{\alpha 4} = nq\mathbf{uE}, \tag{3.125}$$

with $T^{\alpha i}$, $T^{\alpha 4}$ given by (3.120). Thereby use has been made of

$$\int (\mathbf{E} + \mathbf{v} \times \mathbf{B})^j p^i \frac{\partial f}{\partial p^j} \, \mathrm{d}\mathbf{p}$$

$$= \delta_{ij} \left\{ \int |\varepsilon_{jkl}| (\mathbf{E} + \mathbf{v} \times \mathbf{B})^j p^i f|_{p^j=-\infty}^{p^j=+\infty} \mathrm{d}p^k \mathrm{d}p^l - \int (\mathbf{E} + \mathbf{v} \times \mathbf{B})^j f \, \mathrm{d}\mathbf{p} \right\}, \quad (k < l)$$

and the relativistic single particle energy conservation for the fourth component of the Lorentz force.

For a gas of weakly interacting particles in local thermal equilibrium all thermodynamic properties (specific heat, pressure, etc.) are obtained from Maxwell's velocity distribution. From the canonical distribution for (distinguishable) particles the relativistic Maxwell distribution function is obtained by expressing the energy according to (2.181) with $\mathbf{p}$ taken in the local center of mass system,

$$f_{\mathrm{M}}(\mathbf{x}, \mathbf{p}, t) = n_0(\mathbf{x}, t) \mathcal{N} \exp \left[- \left(E_0^2 + p^2 c^2 \right)^{1/2} / (k_{\mathrm{B}} T) \right]. \tag{3.126}$$

$\mathcal{N}$ normalizes the exponential to unity, n_0 is the local particle density in the local rest frame and t is a time variation which is slow on the microscopic time scale. With f strictly Maxwellian (3.122) yields $q^i = 0$.

3.3 Similarity Solutions

In laser dynamics the diffusion of the absorbed energy by electronic heat conduction and by thermal radiation as well as subsequent expansion of the hot plasma into the vacuum or a surrounding low-density gas and plasma cooling by expansion are of particular interest. The dynamics studied in the laboratory are nonstationary and at best exhibit rotational symmetry around the normal to the target onto which the laser or ion beam is focused. In general these are typical cases for numerical computation. However, important aspects of laser and ion beam generated plasmas can be studied by the monofluid Euler equation. Here, a few idealized one-dimensional cases in simple geometry are even accessible to an analytical treatment. As a first example we study the plane and spherical adiabatic and isothermal rarefaction waves. They are basic for understanding the confinement of a hot plasma by its own inertia (*inertial confinement*) and the acceleration of ions from a sharp-edged plasma (TNSA: *target normal sheath acceleration*).

3.3.1 *Dimensional Analysis*

3.3.1.1 The Buckingham π-Theorem and Similarity

Nonlinear similarity solutions play such an important role in fluid dynamics that a brief outline of dimensional analysis and its theoretical basis is justified here. It was pointed out by Lord Rayleigh that sometimes physical laws can be found by dimensional considerations alone. His research in hydrodynamics is a good demonstration for this assertion [11, 12]. In 1914 the so called π-theorem was formulated by Buckingham which yields the formal basis for all dimensional and similarity considerations [13]. Roughly spoken, it states that any physical law must be expressible as an equation among dimensionless variables (so called π-variables). As a simple consequence it follows that a fundamental physical theory cannot be constructed without a minimum number of fundamental constants. However, this was explicitly expressed much later only by Sedov [14]. An elementary example may illustrate the fact. If one knows that a physical state decays exponentially in time t there must exist a constant τ with the significance of a mean life time to make the exponential dimensionless, $\exp(t/\tau)$.

Any physical law may be put into the form

$$y = f(x_1, x_2, \ldots, x_n), \qquad n \text{ integer} \tag{3.127}$$

with quantities y, x_i of certain dimensions. By dimension we mean the following: Given a system of k units M, L, T, ... (e.g., mass, length, time, ...) one has

$$[y] = \mathrm{M}^{\alpha}\mathrm{L}^{\beta}\mathrm{T}^{\gamma}\ldots, \quad [x_i] = \mathrm{M}^{\alpha_i}\mathrm{L}^{\beta_i}\mathrm{T}^{\gamma_i}\ldots.$$

The exponents $\alpha, \beta, \gamma, \ldots, \alpha_i, \beta_i, \gamma_i, \ldots$ constitute the complete dimensional matrix (Table 3.1). If its rank is indicated by r, then certainly $r \leq \min(n, k)$. If a new system of units is used, $\mathrm{M} = \mu \mathrm{m}$, $\mathrm{L} = \lambda \mathrm{l}$, $\mathrm{T} = \tau \mathrm{t}, \ldots$, the value of y transforms into $\bar{y} = \mu^{\alpha}\lambda^{\beta}\tau^{\gamma}\cdots =: Ky$ and correspondingly $\bar{x}_i = \mu^{\alpha_i}\lambda^{\beta_i}\tau^{\gamma_i}\cdots =: K_i x_i$. For (3.127) to be physically reasonable we postulate the following relations to be valid identically

$$\bar{y} = f(\bar{x}_1, \bar{x}_2, \ldots, \bar{x}_n) \text{ or } Kf(x_1, x_2, \ldots, x_n) = f(K_1 x_1, K_2 x_2, \ldots),$$

respectively. These relations express what we mean by dimensional homogeneity. $f(x_i)$ must contain at least all variables and constants on which y depends. For example, if y is the distance of the earth from the sun, f must contain the time t, the masses of earth and sun, total energy, angular momentum, and the gravitation constant (the position of the moon has also to be included if y must be known to a higher precision). With the help of linear algebra the following theorem can be demonstrated:

Table 3.1 Dimensional matrix consisting of n columns

	x_1	x_2	...	x_n
M	α_1	α_2	...	α_n
L	β_1	β_2	...	β_n
T	γ_1	γ_2	...	γ_n
$\vdots$	$\vdots$	$\vdots$		$\vdots$

Buckingham's π-Theorem. If $y = f(x_1, x_2, \ldots, x_n)$ is dimensionally homogeneous, (i) there exists a product $x_1^{s_1} x_2^{s_2} \ldots x_n^{s_n}$ of variables x_i from f of the same dimension as y, i.e.,

$$[y] = [x_1^{s_1} x_2^{s_2} \ldots x_n^{s_n}],$$

and (ii) $y = f(x_1, \ldots, x_n)$ can be put into the form

$$\pi = F(\pi_1, \pi_2, \ldots, \pi_{n-r}), \tag{3.128}$$

where $\pi, \pi_1, \ldots, \pi_{n-r}$ are $n - r$ independent dimensionless variables. The π_i's are power products of the x_i's and π may be taken as $y/(x_1^{s_1}\, x_2^{s_2} \ldots x_n^{s_n})$.
Corollary. If the rank r of the dimensional matrix is equal to n, F must be a dimensionless constant. This is the most interesting situation occurring with dimensional analysis since solving the problem under consideration reduces to determining this dimensionless constant from simple physical arguments, or from solving an ordinary differential equation, or from a single computer run, or from experiment.
Definitions. If from $\pi_1^{\sigma_1} \pi_2^{\sigma_2} \ldots \pi_{n-r}^{\sigma_{n-r}} = 1$ follows $\sigma_1 = \sigma_2 = \cdots = \sigma_{n-r} = 0$ the π_i's are called *independent*. If in a solution $y = f(x_1, x_2, \ldots, x_n)$ the number n of independent variables can be reduced at least by one we call it a *similarity solution*. $f(x_1, x_2, \ldots, x_n)$ is a *self similar solution* if the number of dimensionless variables π_i reduces to a single variable. (In the Russian literature all similarity solutions are often called self-similar. The author does not follow this convention.) For an elementary introduction to dimensional analysis and the π theorem see [15]. It may be helpful to illustrate the use of the π-theorem by a few simple examples.

Example 3.1 (*Electron-ion collision frequency* ν_{ei}) In principle it will depend on the electron velocity v_e, the size, i.e., area σ, of the ions, their number N_i, their volume V, and their velocity v_i. The ions are much slower than an average electron, $v_i \ll v_e$; so v_i can be neglected. Further, ν_{ei} is the same for V and a 10 times larger plasma, hence it depends only on the ratio $N_e/V = n_i$. We are left with $\nu_{ei} = f(n_i, \sigma, v_e)$ with the dimensions [cm^{-3}], [cm^2], [cm s^{-1}]. As a frequency, n_{ei} has the dimension [s^{-1}]. Thus, the unique combination is $\nu_{ei} = \alpha\, n_i \sigma v_e$, with α a dimensionless factor. Set now $\sigma = 1$ cm^2, $n_i = 1$ per cm^3, and $v_e = 1$ cm/s. Then, ν_{ei} is 1 and $\alpha = 1$ follows. Thus the electron-ion frequency is uniquely determined by $\nu_{ei} = n_i \sigma v_e$.

Example 3.2 (*Stokes law*) Assume a heavy sphere falling in deep water. What is the asymptotic velocity v of the sphere? One recognizes that a maximum number of variables is viscosity μ, weight G, and radius R. The dimensional matrix in the cgs system of units is presented in Table 3.2. Since its rank is $r = 3$, F from (3.128) is a constant, $F = C$, and v is uniquely determined by

$$v = C\frac{G}{\mu R} = C'\frac{R^2 g}{\eta}; \quad \eta = \frac{\mu}{\rho}.$$

This is Stokes' formula with $C = (6\pi)^{-1}$. It provides a counter example for the frequently expressed contention that C is always near unity in a "physical problem". If instead of this choice one takes μ, ρ, g, and R as independent variables the rank of the matrix is still $r = 3$ so that $4 - 3 = 1$ dimensionless quantity $\pi_1 = \rho^2 g R^2/\mu^2$ can be formed and v is no longer uniquely determined. The reason is that ρ, g, R appear in the problem only in the combinations R and $\rho R^3 g = 3G/4\pi$ and not as three independent quantities. The success of dimensional analysis depends crucially on the selection of the most restricted number of variables.

Example 3.3 (*Harmonic oscillator*) The motion of the harmonic oscillator $m\ddot{x} + \kappa x = 0$ is completely determined by the quantity κ/m and the oscillation amplitude $\hat{x}$. The dimensions of the two quantities are s^{-2} and cm, respectively. The oscillation period therefore is independent of $\hat{x}$ and is $T = C(m/\kappa)^{1/2}$. The unknown dimensionless constant C can be determined from passing to the spherical pendulum, see Sect. 2.1.4. It exhibits the same period T as the linear pendulum. The only velocity that can be constructed from $\hat{x}$ and κ/m is $v = \hat{x}(\kappa/m)^{1/2}$. This velocity equals $2\pi R/T$ from which $T = 2\pi(m/\kappa)^{1/2}$ follows. All Galileo would have had to do is to verify that the oscillation period of his pendulum becomes independent of amplitude when the excursion is small. Then $T = 2\pi/\omega = 2\pi(\ell/g)^{1/2}$ follows automatically. Incidentally, the invariance of T at small excursion angles α can be inferred from the centrifugal force $mR\omega^2 = mg\tan\alpha \Rightarrow \omega^2 = (g/R)\sin\alpha/\cos\alpha \sim 1/\cos\alpha = 1 + \alpha^2/2 \simeq 1$.

Example 3.4 (*Blast wave*) How secret can the TNT equivalent of a bomb be kept? Consider a homogeneous atmosphere of mass density ρ_0. In one very restricted region the energy W is supplied instantaneously ("point explosion"). As a consequence, a blast wave, i.e., a strong shock propagates out into the embedding material. Its radius $r_s(t)$ depends, under idealized conditions, on W/ρ_0 and t only. From $[W/\rho_0] = \text{cm}^5\ \text{s}^{-2}$ and $[r_s] = \text{cm}$ follows

Table 3.2 Dimensional matrix of the falling sphere; $r = 3$

	μ	G	R	v
g	1	1	0	0
cm	−1	1	1	1
s	−1	−2	0	−1

$$r_s(t) = C\left(\frac{W}{\rho_0}\right)^{1/5} t^{2/5}. \tag{3.129}$$

By taking pictures of the blast wave at different times W is recovered with good accuracy. Subsequently, corrections may have to be made for radiation losses and amplification of the shock front by absorption of radiation. For the constant C and variants of the model, for example supply of the energy W to a vacuum-solid interface see [14, 16]. In the kbar pressure range the time-dependence $\sim t^{2/5}$ in (3.129) has been confirmed recently in air [17].

Example 3.5 (*Rayleigh–Taylor instability*) Consider a fluid of density ρ_1 superposed to a fluid of density ρ_2 in a gravity field of acceleration g. Assume that the original horizontal interface extending in x-direction is subject to a vertical sinusoidal displacement $y = y_0 \sin kx$, $k = 2\pi/\lambda$. It is known from ocean waves in deep water that the vertical motion decays exponentially over the depth of a wavelength $\Delta y \simeq \lambda$. Hence, the mass per unit length involved in the acceleration is $(\rho_1 + \rho_2)(\lambda/s)$. The gravity force acting on this mass per unit length is $(\rho_1 - \rho_2)gy_0$. This results in the equation of motion

$$\ddot{y}_0 + (s/\lambda)Agy_0 = 0, \qquad A = (\rho_2 - \rho_1)/(\rho_1 + \rho_2). \tag{3.130}$$

A is the *Atwood number*. The complete analysis yields the dimensionless factor $s = 2\pi$ for small perturbations $y_0 \ll \lambda$. For $A > 0$ the liquid is stable fulfilling harmonic oscillations around $y = 0$. For $A < 0$ the interface becomes Rayleigh–Taylor unstable, $y_0(t) = \hat{y}_0 e^{\gamma t}$, with the linear growth rate

$$\gamma = \sqrt{kAg}, \qquad k = 2\pi/\lambda. \tag{3.131}$$

see Sect. 2.1.1. The exponential growth (and the harmonic oscillation) originates from the accelerated mass $\sim\lambda$.

3.3.1.2 Adiabatic and Isothermal Rarefaction Waves in Plane Geometry

In a first approach the expansion of a laser plasma into vacuum can be modelled as follows: Consider a half-space ($x \leq 0$) filled with an ideal gas of uniform density ρ_0, temperature T_0, and pressure p_0. Suppose the surface pressure at $x = 0$ is suddenly reduced to zero at $t = 0$. The subsequent dynamics is then governed by

$$\frac{\partial\rho}{\partial t} + \frac{\partial}{\partial x}(\rho u) = 0, \quad \rho\left(\frac{\partial u}{\partial t} + u\frac{\partial u}{\partial x}\right) = -c_{s0}^2\left(\frac{\rho}{\rho_0}\right)^{\gamma-1}\frac{\partial\rho}{\partial x}. \tag{3.132}$$

The adiabatic sound velocity is $c_s = (\gamma p/\rho)^{1/2} = c_{s0}(\rho/\rho_0)^{(\gamma-1)/2}$. The isothermal case $T = \text{const}$ is characterized by infinite thermal conductivity or an infinite number of degrees of freedom with $\gamma = 1$. Equations (3.132) show that ρ and u depend on ρ_0, c_{s0}, x, t. From the corresponding dimensional matrix we deduce one dimensionless variable $\pi = x/(c_{s0}t)$. Therefore the system above reduces to ordinary differential equations for $\rho = \rho_0 P(\pi),\ u = c_{s0}V(\pi)$,

$$P'(V-\pi) + PV' = 0, \quad V'(V-\pi) + P^{\gamma-2}P' = 0,$$

where the prime denotes $\mathrm{d}/\mathrm{d}\pi$. By eliminating P' one obtains

$$V'(V-\pi)^2 - P^{\gamma-1}V' = 0, \quad \text{or} \quad \left(u - \frac{x}{t}\right)^2 = c_s^2.$$

Using this relation in one of the dimensionless conservation equations it follows that $\mathrm{d}u = -c_s\mathrm{d}\rho/\rho$ and hence, for $\gamma \neq 1$ (adiabatic case)

$$\rho(x,t) = \rho_0\left[1 - \frac{\gamma-1}{2}\frac{u}{c_{s0}}\right]^{\frac{2}{\gamma-1}}, \quad u = \frac{2}{\gamma+1}\left(\frac{x}{t} + c_{s0}\right). \tag{3.133}$$

For γ approaching unity (isothermal case) ρ becomes

$$\rho(x,t) = \rho_0 \lim_{\gamma\to 1}\left(1 - \frac{\gamma-1}{2}\frac{u}{c_{s0}}\right)^{\frac{2}{\gamma-1}} = \rho_0\mathrm{e}^{-\frac{u}{c_{s0}}} = \rho_0\mathrm{e}^{-\left(1+\frac{x}{c_{s0}t}\right)}. \tag{3.134}$$

These solutions hold in the interval $-c_{s0}t \leq x \leq \frac{2}{\gamma-1}c_{s0}t$. The rarefaction wave propagates with the phase velocity $v_\varphi = c_{s0}$ into the undisturbed gas. For any instant $t > 0$ the front of the rarefaction wave x_F moves with constant speed $u_F = 2c_{s0}/(\gamma-1)$. For $\gamma = 5/3$ $u_F = 3c_{s0}$. For this case the rarefaction wave is shown in Fig. 3.6 before it collides with the wall at V_2.

It is interesting to calculate the heat flow density $q(x,t)$ which is needed for maintaining the temperature at the constant value T_0. From the energy equation

$$\frac{3}{2}nk_B\frac{\mathrm{d}T}{\mathrm{d}t} = -p\frac{\partial v}{\partial x} - \frac{\partial q}{\partial x} = 0,$$

we obtain

$$\begin{aligned} q(x,t) &= -\int_x^\infty p\frac{\partial u}{\partial x}\mathrm{d}x = -c_{s0}{}^3\rho_0\int_{u(x)/c_{s0}}^\infty \exp\left(-\frac{u}{c_{s0}}\right)\mathrm{d}\left(\frac{u}{c_{s0}}\right). \\ &= c_{s0}^3\rho = pc_{s0}. \end{aligned} \tag{3.135}$$

This is a useful formula for deducing a criterion for nearly isothermal plasma expansion in regions where the flow is planar. In fact, from (1.61) follows $pc_{s0} = \kappa_0 T_e^{5/2}|\partial T_e/\partial x|$. If the dimension of the plane region is L and Δp is the pressure difference at its end points, then the fractional temperature difference

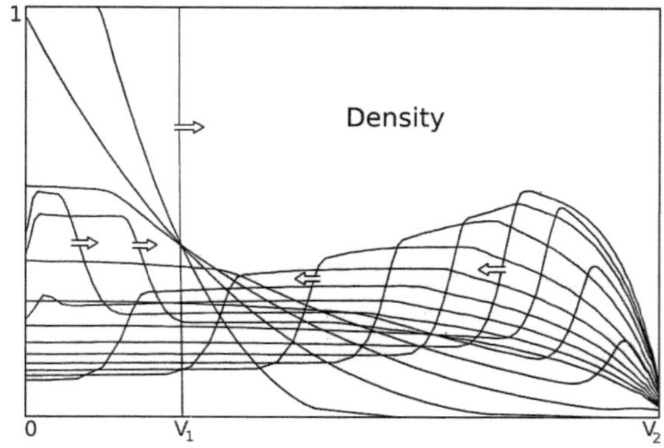

Fig. 3.6 Gay-Lussac experiment as an example of a rarefaction wave. A gas originally contained in the volume $(0, V_1)$ is suddenly set free to stream out into vacuum. After reflection of the rarefaction front at the left solid wall the density profile becomes surprisingly flat (see 3rd graph). When the density hits the solid wall a back running strong shock forms. At the wall the density remains zero for all times and the temperature diverges. The finite density in the figure is due to numerical diffusion

$$\frac{\Delta T_e}{T_e} \simeq L\frac{c_{s0}\Delta p}{\kappa_0 T_e^{7/2}} \tag{3.136}$$

must be much smaller than unity. In the adiabatic rarefaction wave q is zero and no entropy increase occurs by the expansion into vacuum.

The spherical rarefaction wave cannot be solved by dimensional analysis since the initial radius r_0 of the gas cloud enters as an additional independent variable. Only for $t = \mathrm{d}/s_0$, corresponding to distances d $\ll r_0$, is the expansion planar and it becomes clear that the front r_F moves at constant velocity u_F at all times which is the same as in planar geometry. However, with the aid of an additional assumption the authors of [18] were able to obtain a spherical self-similar solution of *nanocluster explosion*. Of particular relevance for various applications is a *demixing* and energy selection process occurring in spherical geometry. Clusters in the Coulomb explosion mode, i.e., when all free electrons have escaped from the cluster (‘*outer ionization*’ degree equals unity), are capable of generating *monoenergetic* beams of fast ions. A mixture of ions of different charge to mass ratio α is unstable against uniform mixing. Under appropriately chosen mixing ratios the component with higher α is accelerated to nearly monochromatic energies in the Coulomb field of the heavy ions. Efficient generation of quasimonoenergetic ions by Coulomb explosions of optimized nanostructured clusters is reported in [19].

Table 3.3 Dimensional matrix in cgsK units for solving (3.137). $r = 3$

	q_0	a	x	t
cm	1	2	1	0
s	0	−1	0	1
K	1	$-\alpha$	0	0

3.3.1.3 Nonlinear Heat and Radiation Waves

Here we solve the following nonlinear heat conduction and radiation transport problem by dimensional analysis [20]. A quantity of energy per unit area Q_0 is instantaneously released at $t = 0$ in a plane of a solid, liquid, or gas from which it diffuses according to the nonlinear equation into its interior

$$\frac{\partial T}{\partial t} = a\frac{\partial}{\partial x}\left(T^\alpha \frac{\partial T}{\partial x}\right); \qquad a = \frac{\kappa_0}{c_v}, \tag{3.137}$$

c_v specific heat. For heat conduction in a fully ionized plasma the exponent has to be taken $\alpha = \frac{5}{2}$; with other appropriate values of α it describes radiation transport. It is easy to show that for $\alpha > 0$ the heat front coordinate $x_T(t)$ is finite. For $\alpha = 0$ (constant thermal conductivity) $x_T = \infty$ for all $t > 0$. The temperature T will depend on $q_0 = Q_0/c_v$, x, and t. This leads to the conclusion that only one dimensionless variable

$$\pi = \frac{x}{(aq_0^\alpha t)^{1/(\alpha+2)}}$$

exists (see Table 3.3) and that $x_T(t)$ and the maximum temperature T_0 (at $x = 0$) are uniquely determined by

$$x_T(t) = A(aq_0^\alpha t)^{1/(\alpha+2)}, \quad T_0(t) = B\left(\frac{q_0^2}{at}\right)^{1/(\alpha+2)}.$$

With $BT(x,t) = T_0 f(\pi)$ reduces to the ordinary differential equation

$$(\alpha+2)\frac{\mathrm{d}}{\mathrm{d}\pi}\left(f^\alpha \frac{\mathrm{d}f}{\mathrm{d}\pi}\right) + \pi\frac{\mathrm{d}f}{\mathrm{d}\pi} + f = 0. \tag{3.138}$$

Several types of heat waves are treated in Chap. 1. Diffusion of black body radiation is postponed to a later chapter.

3.3.1.4 Huntley's Addition

In problems of dimensional analysis take the maximum possible number of independent units. For example, in some cases it is possible to distinguish between masses in the x and y directions or between "inertial" and "thermal" masses as separate physical entities. In this way problems were solved successfully by dimensional analysis that were not reducible a priori by the π-theorem. A necessary implication for the success of Huntley's method is that in the process under consideration there is no interaction between the quantities that are given different dimensions [21].

The interested reader may consult with profit several other textbooks on dimensional analysis in addition to Sedov [22–27]. The successful application of dimensional analysis and similarity considerations to continuum mechanics and hydrodynamics sometimes led to the conclusion that dimensional analysis would be appropriate mainly to these disciplines. Only rather recently has it become apparent that the application of Buckingham's π-theorem is not limited to such fields; it should also be important for example in quantum field theory and the theory of phase transitions [28]. Its successful application in biology was demonstrated also.

There are several solutions relevant for laser matter interaction which are not self-similar but become asymptotic solutions of this type. An example is the analytical treatment of homogeneous pellet compression for laser fusion with a behavior in time for ρ, u, and T according to Kidder [29]

$$\rho, u, T \sim \frac{1}{[1-(t/t_0)^2]^{3/2}} = f(t).$$

For instance, $\rho(R,t) = \rho_0(R/R_0)^2 f(t)$ is not self-similar. However, for t approaching the instant of collapse t_0 it comes arbitrarily close to

$$\rho = \frac{\rho_0}{2^{3/2}} \left(\frac{t_0^{1/2} R^{2/3}}{R_0^{2/3} \tau^{1/2}} \right)^3, \quad \tau = t_0 - t,$$

which is self-similar. Kidder's solutions are interesting also from the topological point of view: The void in a spherical shell only closes at $t = t_0$ when the whole pellet shrinks to a point, as is required from the topological considerations of Sect. 2.1.7.

In connection with converging shocks and compression waves Zeldovich [20] introduced a second "type of self-similar" solutions, a concept which was developed further by Barenblatt [28] with his formulation of intermediate asymptotics. Such a distinction may sometimes be justified from a practical point of view, but from a fundamental point of view no second class of similarity exists. In order to make sense for subdivision into similarities of the first and second kind all similarity problems should belong to these two classes. However, in view of certain solutions [30] a third class of similarity solutions would have to be introduced, and, it is not difficult to imagine that soon a fourth and fifth kind of similarity would have to be invented for other problems. Eventually some further progress beyond Buckingham's theorem may come from group theoretical methods and the determination of symmetry groups of differential equations [31, 32]. For the diffusion with $\alpha = -2$ (e.g., magnetic field diffusion) see in this context Euler et al. [33]. For symmetry properties of Euler-type plasma equations including magnetic fields see [34].

3.3.1.5 Similarities in Laser Generated Plasmas

The functional dependence of all variables of the laser plasma may be obtained from a large number of computer runs under systematic variation of the essential parameters. Such a procedure is generally time-consuming. Here dimensional analysis is much superior. However, if this method is applied to system (3.30)–(3.33) as it stands, more than one dimensionless variable π is found and the dimensional analysis is not conclusive. If instead T_i is assumed to be proportional to T_e, $T_i = \xi T_e$, $\mathbf{u}_e = \mathbf{u}_i = \mathbf{u}$, heat conduction is neglected and the average internal energy per unit mass $\varepsilon = 3k_B(ZT_e + T_i)/2m_i$ is introduced the system reduces in plane geometry to

$$\frac{\partial \rho}{\partial t} + \frac{\partial}{\partial x}\rho u = 0. \tag{3.139}$$

$$\frac{\partial v}{\partial t} + u\frac{\partial u}{\partial x} = -\frac{2}{3\rho}\frac{\partial}{\partial x}\rho\varepsilon. \tag{3.140}$$

$$\frac{\partial \varepsilon}{\partial t} + u\frac{\partial \varepsilon}{\partial x} = -\frac{2}{3}\varepsilon\frac{\partial u}{\partial x} + \frac{\alpha}{\rho}I(x). \tag{3.141}$$

$$\alpha = C\frac{\rho^2}{\varepsilon^{3/2}}, \quad I(x) = I_0 \exp\left(-\int_x^\infty \alpha \mathrm{d}x\right). \tag{3.142}$$

The initial values are $I_0 = 0,\ u = 0,\ \varepsilon = 0,\ \rho = \rho_0$ for $t \le 0$ and $I_0 = \text{const} > 0$ for $t > 0$. A look at (3.139)–(3.142) and the boundary/initial conditions reveal that ρ, u, ε depend on I_0, C, x, t, and ρ_0. However, as long as the average plasma density is much lower than ρ_0, almost all energy is fed into the plasma, only a small portion being coupled to the solid. The dependence on ρ_0 is weak and hence $\rho_0 = \infty$ can be set, so that ρ_0 is no longer a free variable [35]. The remaining dimensional matrix is given in Table 3.4. From this the existence of one dimensionless variable π is deduced,

$$\pi = \frac{x}{C^{1/8}I_0^{1/4}t^{9/8}}.$$

It is convenient to express all dimensional variables without using x; then they are uniquely determined from the π-theorem as follows,

$$\rho = I_0^{1/4}(Ct)^{-3/8}R(\pi), \quad u = I_0^{1/4}(Ct)^{1/8}V(\pi),$$

Table 3.4 Dimensional matrix of rank $r = 3$ for (3.139)–(3.142) with $\rho_0 = \infty$

	I_0	C	x	t
g	1	−2	0	0
cm	0	8	1	0
s	−3	−3	0	1

$$\varepsilon = I_0^{1/2}(Ct)^{1/4}E(\pi), \quad I(x) = I_0 J(\pi),$$

and (3.139)–(3.142) reduces to a system of ordinary differential equations (the prime corresponds to $\mathrm{d}/\mathrm{d}\pi$),

$$(8V - 9\pi)R' + 8RV' - 3R = 0$$

$$\left(8 - 9\frac{\pi}{V}\right) V' + \frac{16}{3}\frac{(RE)'}{RV} + 1 = 0$$

$$(8V - 9\pi)E' + \frac{16}{3}EV' + 2E - 8\frac{RJ}{E^{3/2}} = 0.$$

Instead of solving this system one can look at quantities not depending on x such as the mean values $\langle\rho\rangle$, $\langle u\rangle$, $\langle T_e\rangle \sim \langle\varepsilon_e\rangle$, and the amount of ablated matter per unit area $\mu(t)$, or at $u_{\max}(t)$ and $T_{e,\max}(t)$. They cannot depend on π and are therefore uniquely determined up to proportionality constants R_0, M_0, T_0,

$$\langle\rho\rangle = R_0 I_0^{1/4}(Ct)^{-3/8}, \quad \mu = M_0 I_0^{1/2}C^{-1/4}t^{3/4},$$

$$\langle T\rangle \sim T_{\max}(t) = T_0 I_0^{1/2}(Ct)^{1/4}. \tag{3.143}$$

These relations representing extremely useful formulae clearly show the power of dimensional analysis. The connection between R, V, E, etc. and R_0, M_0, T_0, etc. is found by integration, e.g.,

$$\mu(t) = \int_0^\infty \rho \mathrm{d}x = I_0^{1/2}C^{-1/4}t^{3/4}\int_0^\infty R(\pi)\mathrm{d}\pi \Rightarrow M_0 = \int_0^\infty R\mathrm{d}\pi. \tag{3.144}$$

The amount of plasma produced is proportional to $t^{3/4}$ and hence the process does not become stationary in one dimension. The physical reason is that with increasing $\mu(t)$ the radiation has to cross more absorbing plasma before reaching the solid. From (3.143) the ratio of thermal to kinetic energy turns out to be time-independent, in good agreement with the numerical results of the foregoing section. The same holds for the coefficient R_L of the light reflected from the plasma; in fact

$$\begin{aligned} R_L &= 1 - \exp\left(-2\int_0^\infty \alpha \mathrm{d}x\right) = 1 - \exp\left(-2C\int_0^\infty \frac{\rho^2}{\varepsilon^{3/2}}\mathrm{d}x\right) \\ &= 1 - \exp\left(-2\int_0^\infty \frac{R^2(\pi)}{E^{3/2}(\pi)}\mathrm{d}\pi\right) = \text{const.} \end{aligned}$$

Under the above conditions of $q_e = 0$ and $T_i \sim T_e$ the absorption in one dimension is self-regulating. This was the starting point for Caruso and Gratton [35] who first derived the relations (3.143) by using dimensional analysis. They also determined the constants R_0, M_0, T_0 with the help of simple physical arguments. To be precise, at finite R_L, I_0 has to be replaced by the absorbed intensity $I_a = (1 - R_L)I_0$. Considering $\langle\rho(t)\rangle$ for early times where it exceeds the critical density ρ_c it becomes clear that R_L depends also on ρ_c. In order to test (3.143) numerically $T_i = T_e = T$ was set and absorption was determined from $\mathrm{d}I/\mathrm{d}x = -\alpha I$ without taking into account that the beam cannot propagate in the overdense plasma; however, heat conduction was included. For solid hydrogen the comparison was more than satisfactory intensity interval $I_0 = 10^{11}$–$10^{13}\,\mathrm{Wcm}^{-2}$ [36]. Formulas (3.143) hold as long as the distance l from the plasma-vacuum boundary to the shock front in the solid is less than the focal radius r_0. For times $\tau_s \gtrsim r_0/v_{\mathrm{th}}$, l certainly exceeds r_0 and the plasma dynamics becomes stationary. For this case (3.143) transforms into the relations

$$\langle\rho\rangle = R_0'(Cr_0)^{-1/3}I_0^{1/3}, \quad \mu = M_0'C^{-1/3}r_0^{2/3}I_0^{1/3}t,$$

$$\langle T\rangle \sim T_{\max} = T_0'(Cr_0)^{2/9}I_0^{4/9}, \tag{3.145}$$

if t is replaced by $r_0/T_{\max}^{1/2}$. Alternatively one can write down the system (3.139)–(3.142) in spherical geometry,

$$\frac{\partial\rho}{\partial t} + \frac{1}{r^2}\frac{\partial}{\partial r}(r^2\rho u) = 0,$$

$$\frac{\partial u}{\partial t} + u\frac{\partial u}{\partial r} = -\frac{2}{3\rho}\frac{\partial}{\partial r}(\rho\varepsilon),$$

$$\frac{\partial\varepsilon}{\partial t} + u\frac{\partial\varepsilon}{\partial r} = -\frac{2}{3r^2}\varepsilon\frac{\partial}{\partial r}(r^2u) + C\frac{\rho}{\varepsilon^{3/2}}I.$$

The dynamics becomes stationary under the assumption $\rho_0 = \infty$ in a sphere of initial radius r_0. In this way the sphere becomes infinitely massive and its size r_0 is not affected by ablation. It is further clear that ρ, ε, v depend on C and $I_0(r_0)$ separately (with increasing C $T_{\max}$ rises at constant I_0). The dimensional matrix is given in Table 3.5. Its rank is $r = 3$. I_0, r_0, C, and t can be combined to form the dimensionless variable $\pi_1 = (r_0^8/I_0^2C)^{1/9}t$ which shows that $T_{\max}$, $\langle T\rangle$, $\langle\rho\rangle$, etc. are not stationary in a rigorous sense. This is not surprising since the plasma-vacuum interface is continuously increasing in size. Nevertheless it is easily seen in which sense the plasma expansion comes arbitrarily close to a steady state. In fact, far out in the "corona", say at $r = R$, the flow becomes highly supersonic. If a sink is applied there for the plasma such that $\rho(R) = 0$, the flow at $r < R$ is not altered

Table 3.5 Dimensional matrix for plasma production from a sphere. Possible dimensionless variables are $\pi_1 = \left(r_0^8/I_0^2C\right)^{1/9} t$, $\pi_2 = r/r_0$

	I_0	r_0	C	t
g	1	0	−2	0
cm	0	1	8	0
s	−3	0	−3	1

since no signal reaches this region from outside, and for large R the contribution of the region $r > R$ to the absorption of laser radiation becomes arbitrarily small. The overall effect of such a measure is that for $r_0 < r < R$ the flow becomes stationary after the time $\tau_s \simeq R/v_{\max}$. Then, combining I_0, r_0, and C to form the dimensional quantities, $\langle\rho\rangle$, μ, $T_{\max}$, $\langle T\rangle$ become unique and are given by (3.145). The relations can be tested by the numerical results. First of all the building up of a steady state is well confirmed in spherical geometry; e.g., see the constancy of the maxima of T_e, T_i and the fluxes F, and the linear increase of μ with time. For the ratio of $T_{\max}$ at $I_0 = 5 \times 10^{12}\,\mathrm{Wcm}^{-2}$ and $I_0 = 10^{15}\,\mathrm{Wcm}^{-2}$ one obtains 9.6 from the similarity model (note that the absorbed intensities I_a have to be used). In the numerical calculation this ratio amounts to 8.2; hence, there is satisfactory agreement. However, $\mu(t)$ does not scale according to $\mu \sim I_a^{1/3}$ since this quantity is very sensitive to heat conduction.

The dependence of the dynamic variables on laser wavelength λ and charge number Z and ion mass m_i enter only through the absorption constant C,

$$C \sim \lambda^2 Z^3 (Z+\xi)^{3/2} m_i^{-7/2},$$

(see (1.30) and (1.31)). With the approximation $Z + \xi \simeq Z$ follows in the plane case

$$\langle\rho\rangle \sim \lambda^{-3/4} Z^{-27/16} m_i^{21/16}, \quad \mu \sim \lambda^{-1/2} Z^{-9/8} m_i^{7/8},$$

$$\langle T\rangle \sim \lambda^{1/2} Z^{9/8} m_i^{-7/8}$$

and in the spherical case

$$\langle\rho\rangle \sim \mu \sim \lambda^{-2/3} Z^{-3/2} m_i^{7/6}, \quad \langle T\rangle \sim \lambda^{4/9} Z m_i^{-7/9}. \tag{3.146}$$

For fully stripped ions it may be assumed that $m_i \sim Z$ and the dependencies reduce to

$$\langle\rho\rangle \sim \lambda^{-3/4} Z^{-3/8}, \quad \mu \sim \lambda^{-1/2} Z^{-1/4}, \quad \langle T\rangle \sim \lambda^{1/2} Z^{1/4} \quad \text{(plane);}$$

$$\langle\rho\rangle \sim \mu \sim \lambda^{-2/3} Z^{-1/3}, \quad \langle T\rangle \sim \lambda^{4/9} Z^{2/9} \quad \text{(spherical).} \tag{3.147}$$

Numerical calculations give a detailed picture of the interaction and dynamics of a specific set of parameters. Similarity solutions are much less accurate, but a better overview and physical insight is gained from them. The functional dependencies given here are useful for designing experiments. For a Nd laser their validity range extends from $I_0 \gtrsim 10^{10}\,\mathrm{Wcm}^{-2}$ up to $I_0 \simeq 10^{14}\,\mathrm{Wcm}^{-2}$. As already seen, for some quantities the intensity interval extends even further. Strong thermal conduction puts a limit on the validity of (3.143) and (3.145). It is a general feature of nonlinear processes that, in contrast to linear ones, each dynamical variable obeys its own range. For some variables, for example ablation pressure or temperature and density distributions in the overdense plasma, more refined models are needed.

From the foregoing considerations the use of the π-theorem might appear more or less mechanical. In situations not involving any approximations or simplifications this is the case. In (3.139)–(3.142) one could decide for the cgs-Kelvin-system of units and for the particle density $n = n_i$ and the temperature $T = (ZT_e + T_i)/(Z+1)$ instead of ρ and ε. Then one has to deal with

$$\frac{\partial n}{\partial t} + \frac{\partial}{\partial x} nu = 0, \quad \frac{\mathrm{d}u}{\mathrm{d}t} = -(Z+1)\frac{k_B}{m_i}\frac{1}{n}\frac{\partial}{\partial x} nT,$$

$$\frac{\mathrm{d}T}{\mathrm{d}t} = -\frac{2}{3}T\frac{\partial u}{\partial x} + \frac{2}{3}\frac{\alpha}{k_B n(Z+1)} I,$$

where α is $C'Z^2n^2/T^{3/2}$. The independent variables are $I_0,\ C'/k_B,\ k_B/m_i,\ x, t$; the rank of the dimensional matrix is $r = 4$. $T_{\max}$ for instance is therefore uniquely determined, i.e., $T_{\max} \sim I_0^{1/4}(C'k_B)^{1/4}(k_B/m_i)^{-3/8}t^{-1/2}$. This differs from (3.143) and shows the wrong dependence on C' and t. Alternatively, the last term in the third equation can equally well be written as $\sim(\alpha/n)(I_0/k_B)$ and then $I_0/k_B,\ C',\ k_B/m_i,\ x,\ t$ appear to be independent variables. Since now no mass unit appears in $I_0/k_B,\ C'$, and k_B/m_i the rank of the matrix is $r = 3$ and nothing can be concluded on how $T_{\max}$ depends on $I_0/k_B,\ C',\ k_B/m_i$, and t; only the contradiction has disappeared. This digression illustrates that the use of the π-theorem generally requires physical intuition and it may stimulate the reader to find out the source of the wrong dependence of $T_{\max}$ given above.

3.3.2 Riemann Invariants

An alternative access to the solution of the adiabatic and isothermal rarefaction wave is afforded by the concept of the *Riemann invariants*. In addition this techniques allows the extension to more complex phenomena of rarefaction and compression of gas dynamics and offers additional physical insight.

The sound speed. We assume an *isentropic* fluid of constant density ρ_0 at rest and consider a small disturbance of the density $\rho_1(\mathbf{x}, t) \ll \rho_0$. Thereby isentropic means that the entropy per unit mass σ (or per particle, respectively) is constant in space and time. Then, the pressure p is a unique function of ρ and σ, $p = p(\rho, \sigma)$, and so is $\rho = \rho(p, \sigma)$. For the infinitesimal perturbation ρ_1 mass and momentum conservation read in first order

$$\partial_t \rho_1 + \rho_0 \nabla \mathbf{u} = 0, \qquad \rho_0 \partial_t \mathbf{u} = -\left(\frac{\partial p}{\partial \rho}\right)_{(\sigma,\rho_0)} \nabla \rho_1 = c_s^2 \nabla \rho_1. \tag{3.148}$$

The flow velocity $\mathbf{u}$ is eliminated by taking the partial time derivative of mass conservation and the ∇ operation of the momentum equation,

$$\frac{\partial^2}{\partial t^2}\rho_1 = c_s^2 \nabla^2 \rho_1; \qquad c_s = \sqrt{\left(\frac{\partial p}{\partial \rho}\right)_{(\sigma,\rho_0)}}. \tag{3.149}$$

This is the wave equation of the acoustic wave propagating at the adiabatic sound speed c_s. For the ideal gas $c_s = \sqrt{\gamma p/\rho}$. If instead of σ the temperature T is held constant the same wave equation is obtained with the propagation velocity now given by the isothermal sound speed $c_s = \sqrt{(\partial p/\partial \rho)_T}$. For the ideal gas this is $c_s = \sqrt{p/\rho}$. It is slower by the factor $\sqrt{\gamma}$. Formally it is obtained by setting $f = \infty$ in γ which in this limit becomes unity; physically this means infinite heat conduction.

The general solution of (3.149) is given by the Fourier integral $\rho_1(\mathbf{x}, t) = (2\pi)^{-3/2} \int \rho(\mathbf{k}) \exp i(\mathbf{kx} - \omega t)\mathrm{d}\mathbf{k}$. The single eigenmode $\rho(\mathbf{k})$ obeys the simplest possible dispersion relation $\omega = c_s k$, $k = |\mathbf{k}|$. As we shall see in connection with the ion acoustic waves it is valid in the long wavelength limit. In this limit the phase velocity $v_\varphi = \omega/k = c_s$ does not depend on the frequency and hence it is equal to the group velocity v_g and $\rho_1(\mathbf{x}, t)$ keeps its shape during propagation; it is *dispersion-free*. In reality, as a rule higher frequencies are more strongly damped than low frequencies.

The Riemann invariants. In a one dimensional fluid undergoing only adiabatic motions the change of the pressure $p = p(\rho, \sigma)$ and the flow velocity $\mathbf{u}(\mathbf{x}, t)$ of a volume element are given by

$$\frac{\mathrm{d}p}{\mathrm{d}t} = \left(\frac{\partial p}{\partial \rho}\right)_\sigma \frac{\mathrm{d}\rho}{\mathrm{d}t} + \left(\frac{\partial p}{\partial \sigma}\right)_\rho \frac{\mathrm{d}\sigma}{\mathrm{d}t} = -\left(\frac{\partial p}{\partial \rho}\right)_\sigma \rho \frac{\partial u}{\partial x} \Rightarrow \frac{1}{\rho c_s}\left(\frac{\partial p}{\partial t} + u\frac{\partial p}{\partial x}\right) + c_s \frac{\partial u}{\partial x} = 0$$

$$\frac{\partial u}{\partial t} + u\frac{\partial u}{\partial x} + \frac{1}{\rho}\frac{\partial p}{\partial x} = 0.$$

Adiabaticity of the single fluid element implies $\mathrm{d}\sigma/\mathrm{d}t = 0$. In the second equality of the first line mass conservation has been used. Taking the sum and the difference of the second equation in the first line and the momentum equation below

$$\frac{\partial u}{\partial t} + (u \pm c_s)\frac{\partial u}{\partial x} \pm \frac{1}{\rho c_s}\left(\frac{\partial p}{\partial t} + (u \pm c_s)\frac{\partial p}{\partial x}\right) = 0 \tag{3.150}$$

is obtained. Defining the two *characteristics* C^+ and C^-,

$$C^+ : \frac{\mathrm{d}x}{\mathrm{d}t} = u + c_s, \qquad C^- : \frac{\mathrm{d}x}{\mathrm{d}t} = u - c_s \tag{3.151}$$

from (7.64) the two differentials $\mathrm{d}J^+ = \mathrm{d}u + (1/\rho c_s)\mathrm{d}p$ and $\mathrm{d}J^- = \mathrm{d}u - (1/\rho c_s)\mathrm{d}p$ result zero, if $\mathrm{d}J^+$ is taken along C^+ and $\mathrm{d}J^-$ along C^-. Integrated along their characteristics they are

$$J^+ = u + \int \frac{\mathrm{d}p}{\rho c_s} = u + \int c_s \frac{\mathrm{d}\rho}{\rho} = \mathrm{const}^+ \text{ along } C^+,$$

$$J^- = u - \int \frac{\mathrm{d}p}{\rho c_s} = u - \int c_s \frac{\mathrm{d}\rho}{\rho} = \text{const}^- \text{ along } C^-. \tag{3.152}$$

J^+ and J^- are the *Riemann invariants* of gas dynamics. In the isentropic ideal gas with $\gamma > 1$ they are

$$\gamma > 1: \; J^\pm = u \pm \frac{2}{\gamma - 1} c_s; \qquad \gamma = 1: \; J^\pm = u \pm c_s \ln \rho. \tag{3.153}$$

In summary, small perturbations of density and pressure propagate at the local flow velocity plus and minus the local sound speed, in correspondence to the two directions sound can propagate away from the comoving fluid element. Owing to this property and the conservation of $\mathrm{d}J^\pm$ along $C^\pm$ the method of characteristics is advantageously used in the construction of difference schemes in numerics.

In spherical geometry the total time derivative preserves the structure of plane geometry, $\mathrm{d}_t = \partial_t + u\partial_r$, however $\nabla \mathbf{u} = \partial_r u + 2u/r$. It implies invariance of (3.151) for r in place of x, but (3.152) has to be replaced by

$$J^+ = u + \int c_s \frac{\mathrm{d}\rho}{\rho} = -2 \int c_s \frac{u}{r} \mathrm{d}t \text{ along } C^+,$$

$$J^- = u - \int c_s \frac{\mathrm{d}\rho}{\rho} = +2 \int c_s \frac{u}{r} \mathrm{d}t \text{ along } C^-. \tag{3.154}$$

Temperature and entropy are bound to matter; they propagate with the flow velocity $\mathbf{u}$. They constitute the *third species of characteristics*, indicated by C^0: $\mathrm{d}\mathbf{x}/\mathrm{d}t = \mathbf{u}$. Use of the Riemann invariants is now made to study rarefaction in one dimension $\tau = d/c_{s0}$.

The adiabatic and the isothermal rarefaction waves. An ideal gas of constant density ρ_0, temperature T_0 and pressure $p_0 = (\rho_0/m)k_B T_0$ at rest is assumed to fill the half space $x \le 0$ at times $t \le 0$. With $p\rho^{-\gamma} = p_0\rho_0^{-\gamma}$ the sound speed is $c_s = c_{s0}(\rho/\rho_0)^{(\gamma-1)/2}$. The governing equations are given by (3.132). At $t = 0$ the pressure at $x = 0$ is set to zero. For times $t > 0$ it holds along C^+

$$C^+: \frac{\mathrm{d}x}{\mathrm{d}t} = u + c_s, \qquad J^+ = u + \frac{2}{\gamma - 1} c_{s0} \left(\frac{\rho}{\rho_0}\right)^{\frac{\gamma-1}{2}} = \frac{2}{\gamma - 1} c_{s0}. \tag{3.155}$$

J^+ originates from the rarefaction edge where $u = 0$ and $\rho = \rho_0$. Zero pressure at the rarefaction front implies $\rho = 0$ and $c_s = 0$, and hence front velocity $u_F = 2/(\gamma - 1)c_{s0}$. For a fluid without internal degrees of freedom $\gamma = 5/3$ and $u_F = 3c_{s0}$. From $J^+ = \text{const}$ in (3.155) follows further that a point x of constant density moves at constant velocity u. Thus the characteristics C^+, C^- are straight lines and ρ and all other dynamic quantities depend on the dimensionless parameter $\xi = x/c_{s0}t$ only. Perturbations starting from the rarefaction region propagate inwards along $-x$. On C^- there holds: $x/t = u - c_{s0} = -c_{s0}$; the rarefaction edge moves into the undisturbed fluid with the unperturbed sound speed. As the rarefaction starts from $x = 0$ all C^+ characteristics pass through this point and consequently, the dynamical variables P, u, T keep their values constant in time; the rarefaction wave belongs to the class of *centered waves*, see Fig. 3.6 From (3.155) the density is obtained as a function of u. This is used then to express c_s in C^-: $x/t = u - c_s$ to obtain u as a function of $\xi = x/tc_{s0}$. The result is

$$\rho = \rho_0 \left(1 - \frac{\gamma - 1}{2}\frac{u}{c_{s0}}\right)^{\frac{2}{\gamma-1}}, \qquad \frac{u}{c_{s0}} = \frac{2}{\gamma + 1}(\xi + 1). \tag{3.156}$$

By the identical procedure the isothermal rarefaction wave follows from (3.153) for $\gamma = 1$,

$$\rho = \rho_0 \mathrm{e}^{-u/c_{s0}} = \rho_0 \mathrm{e}^{-(1+x/c_{s0}t)}; \qquad T = T_0. \tag{3.157}$$

Owing to infinite heat conduction arbitrarily high flow velocities are reached. This aspect makes this solution interesting for TNSA ion acceleration in hot laser plasmas.

The adiabatic and the isothermal rarefaction waves belong to the class of *self similar* solutions because they depend only on the dimensionless parameter $\xi = x/c_{s0}t$ in contrast to e.g., the travelling acoustic wave.

With the help of the rarefaction wave a measure is found for the containment time of a hot dense plasma cloud confined solely by its inertia. As the rarefaction edge runs into a plasma slab with sound velocity c_{s0} an indicator for confinement is the time τ in which the rarefaction edge covers the thickness d of a plasma slab:

$$\tau = \frac{\mathrm{d}}{c_{s0}}. \tag{3.158}$$

A high value of heat conduction helps to stretch the confinement time.

Adiabatic compression wave. The ideal gas law allows arbitrarily high densities by adiabatic compression at the minimum amount of work; as the entropy σ remains constant it is impossible to spend less work by any other process without cooling. Beyond pressures of several Megabar static compression fails; higher compression can only be realized dynamically. To this aim the time history of the compressing piston has to be programmed accordingly. For unlimited compression of a gas slab of thickness x_0 and constant density ρ_0 in the time interval τ correct programming of the piston is accessible to an analytical treatment. Achieving arbitrarily high density ρ requires that the characteristics emanating from the position $x_P(T)$ of the piston intersect all in x_0 at the instant τ, see Fig. 3.7. According to (3.152) the motion of the piston is governed by the two conditions on each C^+ characteristics

$$\dot{x}_P + c_{sP} = c_{s0}, \qquad \dot{x}_P + \frac{2}{\gamma-1}c_{sP} = \frac{2}{\gamma-1}c_{s0}.$$

Consider the density $\rho_P(x_P, t)$ adjacent to the piston. Its flow velocity is constant at all $t' \geq t$ owing to $J^+ = \text{const}$ along the characteristics $C^+(t)$ emanating from position $x_P(t)$. Thus, all C^+ characteristics converging towards (x_0, τ) are straight lines . With the slope of the first characteristics $\dot{x}_P = x_0/\tau$ at $t = 0$ the conditions above translate into

$$\dot{x}_P + c_s = \frac{x_0 - x_P}{\tau - t}, \quad \dot{x}_P = \frac{2}{\gamma-1}\left(\frac{x_0}{\tau} - \frac{x_0 - x_P}{\tau - t} + \dot{x}_P\right)$$

$$\Rightarrow (\gamma - 3)\frac{\mathrm{d}\xi}{\mathrm{d}\tau'} = 2\left(\frac{\xi}{\tau'} - \frac{x_0}{\tau}\right); \qquad \xi = x_0 - x,\ \tau' = \tau - t. \tag{3.159}$$

This is the equation governing the piston motion. Its solution is

$$x_P(t) = \frac{x_0}{\gamma-1}\left\{\gamma + 1 - 2\frac{t}{\tau} - (\gamma+1)\left(1 - \frac{t}{\tau}\right)^{2/(\gamma+1)}\right\}. \tag{3.160}$$

The density during compression is inhomogeneous, it is highest at the piston. In Fig. 3.7 the time history of the piston and selected values of compression $\kappa = \rho_P/\rho_0$ are depicted. At $t = t_0$ the piston velocity and the compression become infinite. The adiabatic compression wave is not selfsimilar.

3.3.3 *The Plane Shock Wave*

In the stable homogeneous fluid the propagation velocity of a density disturbance increases in proportion to its local value $\rho_1(\mathbf{x}, t)$ and hence leads to *profile steepening* in space. If damping, e.g., bulk viscosity, is too weak to stabilize it a discontinuity eventually forms, a so called *shock wave*. With respect to the unperturbed medium in front the shock moves at supersonic velocity; if not because of slowing down in some region, immediately a rarefaction wave starts from the edge and the discontinuity decays. The simplest situation of formation of a stationary shock is sketched in Fig. 1.9 in Chap. 1 in lab frame representation.

Rankine–Hugoniot relations. A piston moving at constant velocity v_P into a homogeneous fluid imparts the shocked region (in the figure region 1) the constant flow velocity $v_1 = v_P$. The unperturbed fluid ahead the discontinuity, here region 0, penetrates the shock front to a depth of a mean free path and stops; it has acquired the velocity v_1 and is compressed to the density ρ_1. In case the fluid is an ideal gas it can be shown that an arbitrarily low, uniform motion of the piston leads to formation of a shock. If the shock front S moves at v_S, mass conservation requires $\rho_0 v_S = \rho_1(v_S - v_1)$. Newton's law applied to the compressed volume between the

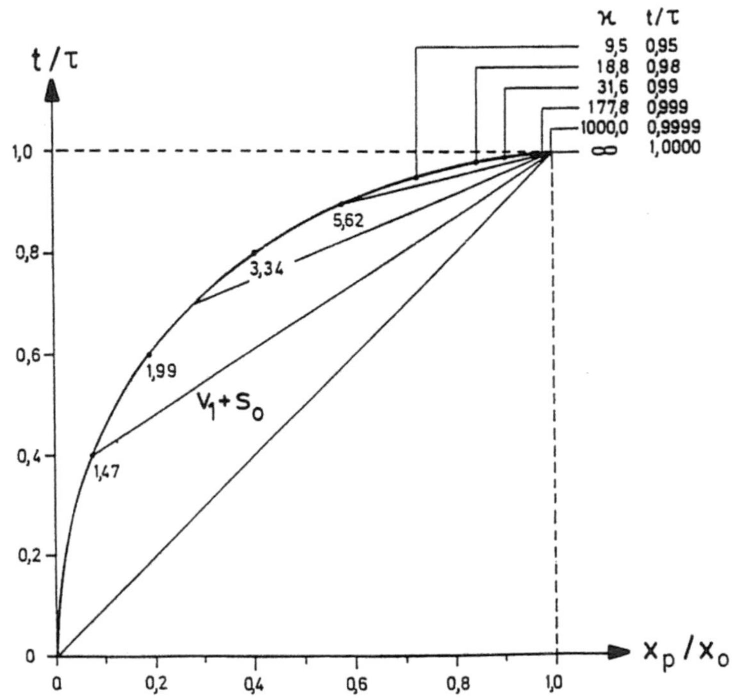

Fig. 3.7 Fast adiabatic compression in plane geometry. The piston (position x_P) starts from rest and moves as to keep all characteristics converging in (1, 1). The numbers entered along $x_P(t)$ represent some compression ratios $\kappa = \rho_P/\rho$

piston and the shock front at time t sets the balance between the momentum increase by unit time $\rho_0 v_S v_1$ and the force $p_1 - p_0$ acting on it. The mass $\rho_0 v_S$ stopped in the shock front per unit time undergoes the energy density increase per unit mass $\epsilon_1 - \epsilon_0 + v_1^2/2$ by the work done by the piston $p_1 v_1$, thus

$$\rho_0 v_S = \rho_1(v_S - v_1), \quad \rho_0 v_S v_1 = p_1 - p_0, \quad \rho_0 v_S(\epsilon_1 - \epsilon_0 + v_1^2/2) = p_1 v_1. \tag{3.161}$$

These are the conservation equations in the lab frame. From a measurement of v_S and v_1 the quantities ρ_1, p_1, and ϵ_1 are obtained. By switching to the system $S'(v_S)$ comoving with the shock front S the flow velocities are $u_0 = -v_S$ and $u_1 = v_1 - v_S$ and (3.161) transform into the symmetric scheme of the *Rankine–Hugoniot relations*,

$$\rho_1 u_1 = \rho_0 u_0;$$

$$\rho_1 u_1^2 + p_1 = \rho_0 u_0^2 + p_0;$$

$$\epsilon_1 + \frac{1}{2}u_1^2 + \frac{p_1}{\rho_1} = \epsilon_0 + \frac{1}{2}u_0^2 + \frac{p_0}{\rho_0}. \tag{3.162}$$

Their validity extends to any shock discontinuity as long as the shock width is much smaller than the curvature radius of the front. Since the shock wave is stationary in the system comoving with the shock front the conservation equations in their conservative version (3.53), (3.56), (3.59) apply with vanishing partial time derivatives; integration over a narrow zone including the shock front leads directly to (3.162). Thereby the transition from region 0 to region 1 can be of considerable finite extension as a consequence of complex physical processes going on in the shock, provided it is stationary. If shock induced phase transition comes into play it has to be taken into account in the momentum and energy balance because of possible changes of the equation of state and internal energy.

For specific calculations it is convenient to introduce the specific volume $\mathcal{V} = 1/\rho$. From the conservation of flow and the momentum in (3.162) the *Rayleigh equations* for the velocities relative to the shock front and the modulus of their difference Δu between ahead and behind it follow,

$$u_0^2 = \mathcal{V}_0^2 \frac{p_1 - p_0}{\mathcal{V}_0 - \mathcal{V}_1}, \quad u_1^2 = \mathcal{V}_1^2 \frac{p_1 - p_0}{\mathcal{V}_0 - \mathcal{V}_1}, \quad \Delta u = |u_0 - u_1| = [(p_1 - p_0)(\mathcal{V}_0 - \mathcal{V}_1)]^{1/2} \tag{3.163}$$

On eliminating the flow velocities u_0, u_1 in (3.162) one arrives at the *Hugoniot equation* of state that connects material properties only,

$$\epsilon_1 - \epsilon_0 = \frac{1}{2}(p_0 + p_1)\left(\frac{1}{\rho_0} - \frac{1}{\rho_1}\right). \tag{3.164}$$

It describes the change a mass element undergoes when it crosses the shock front. It is analogous to the adiabatic change of matter, however with the big difference of entropy conservation in the adiabatic transition and energy non-conservation in the shock transition. Stopping of the flow $\rho_0 u_0$ in the shock front transforms its kinetic energy into heat and possible other forms of energy, e.g., dissociation, ionization, or emission of radiation. Such phenomena represent cooling mechanisms. They may crucially affect the width of transition from the unperturbed region to the compressed region 1 in Fig. 1.9. As a rule, equilibration is fastest in the translational degrees of freedom, in particular between the ions and then with delay of the mass ratio m_e/m_i between ions and electrons [see (3.29)] for the plasma. Vibrational relaxation is considerably slower followed last by rotational excitation.

In the plasma the shock wave is one of the few non-resonant mechanism that heat the ions. Stopping of the electrons contributes by the negligible factor m_e/m_i and, to be precise, by inefficient adiabatic compression.

Shock compression of the ideal gas. At high energy densities matter tends to behave like an ideal gas. Pressure p and internal energy ϵ per unit mass are related by $\epsilon = p/[\rho(\gamma - 1)] = (f/2)p/\rho$ and $p = nk_BT, n = \rho/m$. They allow to recover the density and the pressure ratios explicitly from (3.164). To this aim we add $p/2\rho =$

$p\mathscr{V}/2$ to the internal energy ϵ,

$$\epsilon_{0,1} + \frac{p_{0,1}}{2\rho_{0,1}} = \frac{\gamma+1}{2(\gamma-1)} p_{0,1}\mathscr{V}_{0,1}$$

and add/subtract the same quantities on the RHS of (3.164) to eliminate cross products from the Hugoniot relation. Collection of the terms leads to the ratios

$$\frac{p_1}{p_0} = \frac{(\gamma+1)\mathscr{V}_0 - (\gamma-1)\mathscr{V}_1}{(\gamma+1)\mathscr{V}_1 - (\gamma-1)\mathscr{V}_0}, \quad \frac{\rho_0}{\rho_1} = \frac{\mathscr{V}_1}{\mathscr{V}_0} = \frac{(\gamma+1)p_0 + (\gamma-1)p_1}{(\gamma+1)p_1 + (\gamma-1)p_0}. \tag{3.165}$$

It is convenient to introduce the *Mach number* $M = v_S/c_{s0}$, $v_S = |u_0|$ as a measure of the relative shock speed and $\kappa = \rho_1/\rho_0$ for the compression. Then, after some straightforward elimination algebra the ratios p_1/p_0, κ, and T_1/T_0 follow from (3.163) and (3.165) as functions of the Mach number only. The results of compression, temperature, and entropy increase $\Delta\sigma$ per particle are as follows

$$\kappa = \frac{(\gamma+1)M^2}{(\gamma-1)M^2+2}, \quad \frac{T_1}{T_0} = \frac{2\gamma M^2 + 1 - \gamma}{(\gamma+1)\kappa}, \quad \Delta\sigma = k_B \ln\kappa \ln\left(\frac{T_1}{T_0}\right)^{f/2}. \tag{3.166}$$

The pressure ratio p_1/p_o is the product $\kappa \times T_1/T_0$. Downstream the shock front the flow is subsonic,

$$M_1 = \frac{|u_1|}{c_{s_1}} = \frac{1}{\kappa}\frac{u_0}{c_{s_0}}\frac{c_{s_0}}{c_{s_1}} = \frac{1}{\kappa^{1/2}}\left(\frac{\gamma+1}{2\gamma-(\gamma-1)/M^2}\right)^{1/2} < 1.$$

This is in agreement with the assumption of $\rho_1 = \text{const}$ between shock front and piston. A disturbance originating from the piston, as for example a small irregularity in its velocity, always catches up with the shock front, life time of the shock permitting of course. At $M = 1$ the shock velocity v_S equals c_{s_0}. As the Mach number increases the compression approaches finally the limiting value $\kappa = (\gamma+1)/(\gamma-1)$ of the so-called *strong shock*. It is obtained formally from setting $p_0 = 0$ in (3.162). For $p_0/p_1 \ll 1$ the assumption is justified. The finite asymptote of κ is a consequence of the dramatic increase of the temperature. In the strong shock the formulas above and the pressure simplify to

$$\kappa = \frac{\gamma+1}{\gamma-1}, \quad \frac{T_1}{T_0} = \frac{2\gamma}{(\gamma+1)\kappa}M^2, \quad \frac{p_1}{p_0} = \frac{2\gamma}{(\gamma+1)}M^2, \quad M_1 = \left(\frac{\gamma-1}{2\gamma}\right)^{1/2}. \tag{3.167}$$

For comparison, at the limiting κ value the adiabatic temperature increase amounts to the modest value $T_1/T_0 = \kappa^{2/(\kappa-1)}$. The flow velocity v_1 in the lab frame is connected with the shock velocity v_S through the compression ratio κ by the useful simple relation

$$v_1 = v_S\left(1 - \frac{1}{\kappa}\right). \tag{3.168}$$

Reflected shock. When the shock wave ρ_1 is stopped by a solid wall the flow velocity v_2 behind the shock front is zero. Giving the shocked quantities of the direct shock the index 1 and the reflected shock quantities the index 2 the flow velocities are with view on (3.168) as follows,

$$v_{S1} = M_1 c_{s0} = v_1 \frac{\kappa_1}{\kappa_1 - 1}, \; v_2 = 0, \; \kappa_2 = \frac{\rho_2}{\rho_1}, \; v_{S2} = M_2 c_{s1} = \frac{\kappa_2}{\kappa_2 - 1}; \; \Delta u_2 = \Delta u_1. \tag{3.169}$$

The velocities v refer to the lab frame, u to the frame moving at $v_{S1,2}$. The governing equations for the reflected shock are according (3.163) and (3.165)

$$\frac{p_2}{p_1} = \frac{(\gamma+1)\mathcal{V}_1 - (\gamma-1)\mathcal{V}_2}{(\gamma+1)\mathcal{V}_2 - (\gamma-1)\mathcal{V}_1}, \qquad \frac{\mathcal{V}_2}{\mathcal{V}_1} = \frac{(\gamma+1)p_1 + (\gamma-1)p_2}{(\gamma+1)p_2 + (\gamma-1)p_1},$$

$$\Delta u_2 = \Delta u_1 \Rightarrow [(p_2 - p_1)(\mathcal{V}_1 - \mathcal{V}_2)] = [(p_1 - p_0)(\mathcal{V}_0 - \mathcal{V}_1)]. \tag{3.170}$$

The relation describe shock reflection of arbitrary strength. The separation of variables is simple only for the most relevant situation in practice that the primary shock is strong. Then $p_0 = 0$ can be set and one obtains for κ_2 and the compression behind the reflected shock $\kappa = \kappa_1\kappa_2$,

$$\kappa_2 = \frac{\rho_2}{\rho_1} = \frac{\gamma}{\gamma - 1} \quad \Rightarrow \quad \kappa = \frac{\rho_2}{\rho_0} = \frac{\gamma}{\gamma-1}\frac{(\gamma+1)M^2}{(\gamma-1)M^2+2}. \tag{3.171}$$

It is worth noting that in a fluid with a polytropic equation of state $p\rho^{-\gamma}$ the reflected shock is never a strong shock, regardless of how strong the first shock is. For example, with $\gamma = 5/3$, $\kappa_2 = 5/2$ the limiting compression does not exceed $\kappa = 10$.

One may ask what happens when a second shock is launched and reaches the first shock front. In some sense this is the situation of reflection from a free end. From (3.162) and (3.168) follows that v_2 is subsonic and hence at the moment of shock front superposition a rarefaction wave starts running backwards that reduces the density gradually to the value of the primary shock with immediate reduction at the front.

The Hugoniot curve, in the literature also called sometimes *shock adiabate*. Suppose to know the *equation of state (EOS)* of a given material, explicitly $p = p(\epsilon, \mathcal{V})$ or implicitly $F(p, \epsilon, \mathcal{V}) = 0$. Substitution in the Hugoniot relation

$$\epsilon - \epsilon_0 = \frac{1}{2}(p_0 + p)(\mathcal{V}_0 - \mathcal{V}) \tag{3.172}$$

yields the pressure as a function of the specific volume $p = p(\mathcal{V})$ for any fluid. It describes the pressure undergoes in a shock from the initial state $(\mathcal{V}_0, p_0)$ to a general final state $(\mathcal{V}_1, p_1)$, see Fig. 3.8. In contrast to the adiabatic change of state occurring at constant entropy the Hugoniot curve exhibits an asymptote at non vanishing volume $\mathcal{V} > 0$. The general proof follows from (3.164) for the compression ratio $\kappa = \mathcal{V}_0/\mathcal{V}$

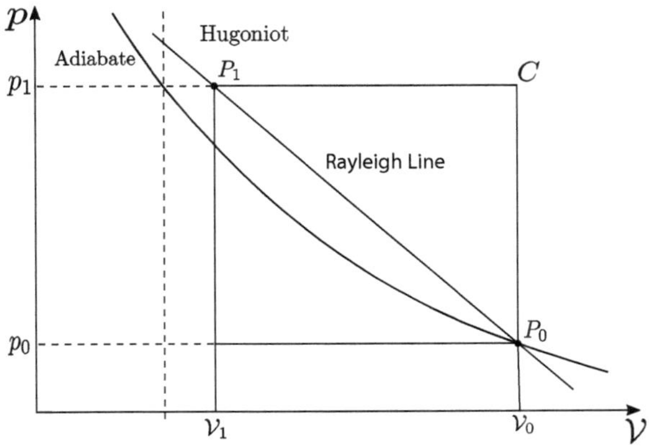

Fig. 3.8 Hugoniot diagram. Under fixed Mach number the fluid undergoes a transition from the thermodynamic state ρ_0, p_0, indicated by P_0, into the thermodynamic state P_1 of density and pressure ρ_1 and p_1. The Hugoniot curve exhibits a finite asymptote (vertical dashed line); in contrast, the adiabate allows arbitrarily high compression. In the limit $P_1 \to P_0$ the Rayleigh line becomes tangent to both, the Hugoniot curve and the adiabate through P_0. The areas of the rectangle $\mathcal{V}_1\mathcal{V}_0CP_1$ (triangles, rectangles, trapezium) visualize the different energies involved in shock compression

under the assumption that $\epsilon = p/((\gamma-1)\rho)$ is a function of the effective number $f = f(p)$ of degrees of freedom and that f remains finite. With $r = p/p_0$ follows

$$\lim_{r\to\infty} \kappa = \lim_{r\to\infty} \frac{(\gamma+1)r + (\gamma-1)}{\gamma+1+(\gamma-1)r} = \frac{\gamma+1}{\gamma-1} = 1 + f.$$

Only the isothermal case of no temperature increase, e.g., under strong cooling, is formally equivalent to infinite specific heat, i.e., $f = \infty$.

The Rayleigh equations allow a suggestive geometrical interpretation of Fig. 3.8: The area of the

(a) trapezium $\mathcal{V}_1\mathcal{V}_0P_0P_1\mathcal{V}_1 = \epsilon_1 - \epsilon_0 = \Delta\epsilon_{\text{in}}$,
(b) triangle $P_0CP_1 = v_1^2/2 = (v_1^2 - v_0^2)/2 = \epsilon_{\text{kin}}$ in lab frame,
(c) rectangle $\mathcal{V}_1\mathcal{V}_0CP_1 = \epsilon_{\text{in}} + \epsilon_{\text{kin}}$ in lab frame.

Weak shock. Consider the limit $\mathcal{V}_1 \to \mathcal{V}_0$ in (3.163). The slope of the Rayleigh line P_0P_1 with respect to the density assumes the value

$$\frac{\partial p}{\partial \rho}\Big|_{\rho_0} = -\mathcal{V}_0^2\frac{\partial p}{\partial \mathcal{V}}\Big|_{\mathcal{V}_0} = \gamma\frac{p_0}{\rho_0}. \tag{3.173}$$

This is the sound velocity squared c_{s0}^2 and shows that the shock velocity u_0 approaches the sound speed. At the same time the slope of the Hugoniot curve in P_0 coincides with the slope of the adiabate $p\rho^{-\gamma} =$ const. This has two consequences: (i) a succession of weak shocks comes arbitrarily close to the adiabate through P_0 and (ii) the pressure p_1 obtained from two successive shocks leads to higher compression κ than the corresponding single shock.

The coincidence of the shock velocity u_0 with the isentropic sound speed c_{s0} for vanishing shock compression suggests vanishing of the entropy increase per unit mass σ to higher order in Δp for weak compression $\Delta\kappa$. In fact, as we shall show now the entropy in the shock increases with third order increase in specific volume $\mathrm{d}\mathcal{V} = \mathcal{V}_0 - \mathcal{V}_1$ and pressure $\mathrm{d}p = p_1 - p_0$. In view of this result it is consistent to expand σ to first, p to second and ϵ to third order. The latter reads

$$\mathrm{d}\epsilon = T\mathrm{d}\sigma - p\mathrm{d}\mathcal{V} - \frac{1}{2}\left(\frac{\partial p}{\partial \mathcal{V}}\right)_{\sigma,0}\mathrm{d}^2\mathcal{V} - \frac{1}{6}\left(\frac{\partial^2 p}{\partial \mathcal{V}^2}\right)_{\sigma,0}\mathrm{d}^3\mathcal{V}.$$

Alternatively, from (3.164) follows

$$\mathrm{d}\epsilon = -\frac{1}{2}\left\{2p_0 + \left(\frac{\partial p}{\partial \mathcal{V}}\right)_{\sigma,0}\mathrm{d}\mathcal{V} + \frac{1}{2}\left(\frac{\partial^2 p}{\partial \mathcal{V}^2}\right)_{\sigma,0}\mathrm{d}^2\mathcal{V}\right\}\mathrm{d}\mathcal{V}.$$

Mixed derivatives of p with respect to $\mathcal{V}$ and σ can be droped as they contribute only beyond 3rd order. Equating the two terms of $\mathrm{d}\epsilon$ and realizing

$$\frac{\partial^2 \mathcal{V}}{\partial p^2} = \frac{\partial}{\partial p}\left(\frac{\partial p}{\partial \mathcal{V}}\right)^{-1} = -\left(\frac{\partial^2 p}{\partial \mathcal{V}^2}\right)\left(\frac{\partial p}{\partial \mathcal{V}}\right)^{-3}; \quad \left(\frac{\partial p}{\partial \mathcal{V}}\right)\mathrm{d}\mathcal{V} = -\mathrm{d}p$$

yields the desired result

$$T_0\mathrm{d}\sigma = \frac{1}{12}\left(\frac{\partial^2 p}{\partial \mathcal{V}^2}\right)_{\sigma,0}\mathrm{d}^3\mathcal{V} = \frac{1}{12}\left(\frac{\partial^2 \mathcal{V}}{\partial p^2}\right)_{\sigma,0}\mathrm{d}^3 p. \tag{3.174}$$

This important result is the a posteriori justification of the identification of the sound speed c_{s0} in (3.148) with its isentropic value in the absence of heat conduction.

The laser generated rarefaction shock. In the compression shock the undisturbed density flows into the shock front at supersonic velocity u_0 and streams up with subsonic speed u_1 at compressed density behind it, see Fig. 3.9a. In most fluids

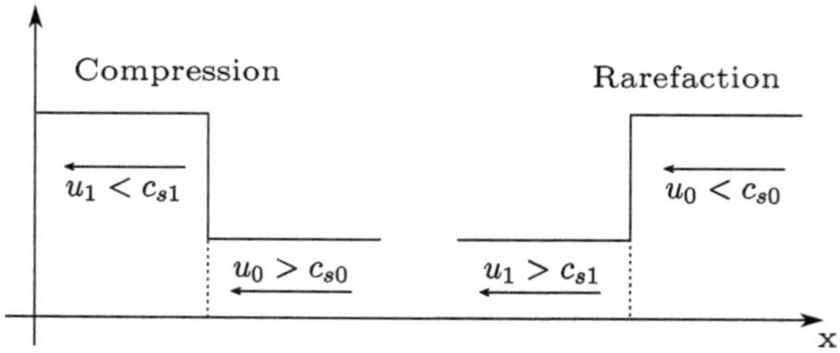

Fig. 3.9 Compression shock versus rarefaction shock. Flow velocities u_0, u_1 refer to the shock front

the pressure increases with temperature at constant volume or with entropy if the specific heat at constant volume C_V is positive. Under such circumstances follows $(\partial p/\partial \sigma)_{\mathcal{V}} > 0$. From (3.174) we deduce

$$\mathrm{d}p = \left(\frac{\partial p}{\partial \sigma}\right)_{\mathcal{V}} = \frac{1}{12T}\left(\frac{\partial p}{\partial \sigma}\right)_{\mathcal{V}}\left(\frac{\partial^2 p}{\partial \mathcal{V}^2}\right)_{\sigma,0}\mathrm{d}^3\mathcal{V}. \tag{3.175}$$

Hence, compression is ensured if $\left(\partial^2 p/\partial \mathcal{V}^2\right)_\sigma > 0$ is fulfilled along the Hugoniot curve.

The rarefaction shock is characterized by subsonic influx of the undisturbed fluid of density ρ_0 and supersonic downstream at the rarefied density ρ_1, see Fig. 3.9b. At positive specific heat c_v the fluid must exhibit anomalous material properties since $\left(\partial^2 p/\partial \mathcal{V}^2\right)_\sigma < 0$ holds now. Here, we do not discuss the implications for the equation of state, rather we give a very relevant example of rarefaction shock that arises by strong laser heating.

The interaction of a powerful ns laser pulse with a solid can be viewed in its simplest version just as a strong rarefaction shock. From the remarks in the introductory Chap. 1 we know that the laser beam is stopped at the critical density ρ_c and partially absorbed there in a narrow, at least spatially limited region. The fraction R is reflected. This implies strong heating $\dot{h} = (1-R)I$ in this region. From the study of the rarefaction wave (3.156) and (3.157) we know that downstream the hot plasma flow is supersonic. Assuming an ideal gas law for the plasma in perfect thermodynamic equilibrium $T_i = T_e = T$ and $Z = 1$ the Rankine–Hugoniot relations read

$$n_1 u_1 = n_0 u_0, \quad n_1(u_1^2 + c_{s1}^2) = n_0(u_0^2 + c_{s0}^2); \quad c_{s1}^2 = 2\gamma\frac{p_1}{\rho_1};$$

$$\frac{1}{2}m_i u_1^2 + \frac{2\gamma}{\gamma-1}k_B T_1 = \frac{1}{2}m_i u_0^2 + \frac{2\gamma}{\gamma-1}k_B T_0 + \frac{(1-R)I}{n_c c_{s1}}. \tag{3.176}$$

In the interaction region laser-solid around ρ_c the conditions for steady state 1D flow are fulfilled. From $\rho_c \ll \rho_{\mathrm{solid}}$ a simple estimate based on these conservation equations shows that (i) $u_0 < c_{s0}$ and can be neglected; (ii) the fraction of $\dot{h}$ transmitted to the dense region ρ_0 is at most $\sqrt{\rho/\rho_0}$, i.e., usually much less than unity. Setting $u_0 = 0$ is equivalent to assuming $\rho = \infty$. From the steady state flow also follows that the pressure p_0 can be evaluated at any position in region 1 where the flow is one dimensional. It represents the so called ablation pressure P_a, $P_a = p_0$, because it is generated by the outflowing hot plasma. The transition subsonic $\to$ supersonic has to happen in the absorption zone. Closer inspection in the following shows that the point of Mach $M = 1$ lies in the slightly underdense region $\rho|_{M=1} < \rho_c$ Here, with sufficient precision we may identify it with ρ_c. The Rankine–Hugoniot relations reduce to

$$\rho_1 u_1^2 + p_1 = \left(1+\frac{1}{\gamma}\right)\rho_c c_{sc}^2 = p_0, \qquad \left(\frac{1}{2}+\frac{1}{\gamma-1}\right)\rho_c c_{sc}^3 = (1-R)I; \quad \gamma \neq 1. \tag{3.177}$$

In the isothermal case, $\gamma = 1$, $T = T_e = T_i = \text{const}$ in space (and time), the bracket becomes $1/2 + 2 \times (3/2 + 1) = 11/2$. The quantities of interest are P_a and T. We solve them for the more relevant isothermal case, with $I_a = (1 - R)I$ for the absorbed power,

$$P_a = 2\left(\frac{2}{11}\right)^{2/3} \rho_c^{1/3} I_a^{2/3} = 0.64\rho_c^{1/3} I_a^{2/3},$$

$$k_B T = \frac{1}{2}\left(\frac{2}{11}\right)^{2/3} \left(\frac{m_i}{n_c^2}\right)^{1/3} I_a^{2/3} = 0.16 \left(\frac{m_i}{n_c^2}\right)^{1/3} I_a^{2/3}. \tag{3.178}$$

This may be compared with P_a from (1.51). Shock generation in solid and liquid matter by laser ablation is a powerful tool to study the equation of state (EOS) of compressed dense matter. A numerical example at $I_a = 10^{15}$ Wcm^{-2} Nd laser intensity on solid hydrogen, $T = T_e = T_i$, may illustrate the significance,

$$T = 5.5 \text{ keV}; \qquad P_a = 3.5 \text{ TPa} = 35 \text{ Mbar}; \qquad P_L = \frac{I_a}{c} = 3.3 \text{ Mbar}. \tag{3.179}$$

The light pressure P_L exerted by the laser beam under assumption of total absorption is also given for comparison,. Only at intensities three orders of magnitude higher it starts dominating the ablation pressure. At low intensities, to start at approximately $I = 10^{12}$ Wcm^{-2}, P_L leads to steepening and shortening of the density profile in the interaction region, see (1.58). The ablation pressure P_a drives a strong shock wave into the undisturbed target, most of the energy however remains in the plasma owing to the large amount of mass involved in region 0 compared to the rarefaction region 1 under the condition of equal momentum fluxes. The efficiency of ablative laser beam pressure generation appears remarkable if one bears in mind that the limit of static compression is at about 3–5 Mbar.

In view of the significance of P_a from (3.179) for its application in high pressure research it is advisable to summarize the conditions under which it is derived. The Hugoniot relations apply in their simplified version (3.177) under (i) steady state conditions in plane 1D flow of an ideal fully ionized plasma, (ii) heat supply in a characteristic interval, here the critical layer, of limited dimension, (iii) highly overdense target, $\rho_0 \gg \rho_c$, (iv) radiation losses from the plasma can be neglected. The extension to high-Z partially ionized plasmas with $T_e \neq T_i$ is straightforward; it is left as an exercise to the interested reader.

3.3.4 From Ablation to Radiation Pressure Under Heat Flow

Closed expressions for the ablation pressure can be developed only for quasi steady state flows. The expressions for the ablation pressure P_a found in the literature do not go substantially beyond the models leading to the simplified formulas (1.51) and

(3.178). Momentum equation and energy balance have to be modified in presence of electron heat conduction and radiation pressure from the laser. At high intensities P_a must undergo a continuous transition from the $I^{2/3}$ power law to a linear dependence on intensity because of the radiation pressure $p_L \sim I$. If the ablation pressure is determined by electron energy transport to the overdense plasma region sensitive deviations from an $I^{2/3}$ law are also expected.

3.3.4.1 The Critical Mach Number

First we study the plane 1D model. When a target with an initially flat surface is heated, the flow of evaporated matter is planar near the solid-gas or solid-plasma interface. As the distance increases the flow pattern becomes more and more divergent creating in this way a zone of a stationary flow field in front of the interface. The temperature increases from a low value in the target, reaches a maximum somewhere in the stationary zone and then decays because of cooling due to expansion (Fig. 3.10). Let us assume that the ablated gas or plasma follows a polytropic equation of state of the form $p\rho^{-\gamma} = \text{const},\ \gamma = \text{const}$, so that the sound velocity is $c_s = \sqrt{\gamma p/\rho}$. In the isothermal case (infinite heat conduction) $c_s = \sqrt{k_B(ZT_e + \gamma_i T_i)/m_i}$.

Absence of Radiation Pressure

Owing to its general relevance let us first consider the steady state 1D flow without radiation pressure. Mass flow obeys $\rho u = \text{const}$. The momentum flow is governed by

$$\frac{\partial}{\partial x}(\rho u^2 + p) = \rho u \frac{\partial}{\partial x}\left(u + \frac{p}{\rho u}\right) = \rho u \frac{\partial}{\partial x} s\left(M + \frac{1}{\gamma M}\right) = 0. \tag{3.180}$$

The momentum equation tells that

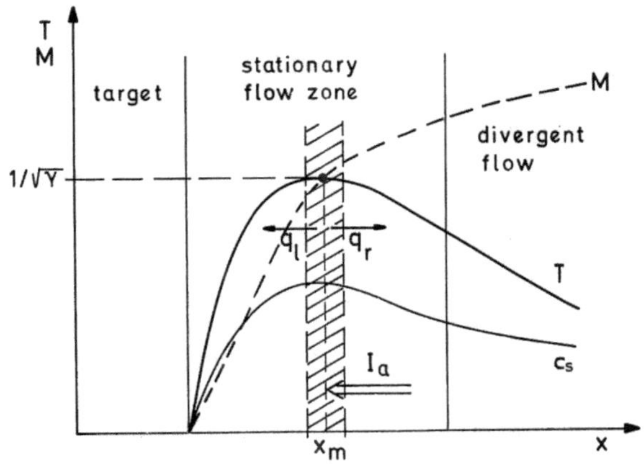

Fig. 3.10 Steady state one fluid 1D model. Distributions of temperature T, sound speed c_s and Mach number $M = u/c_s$ in a typical ablation flow into vacuum. I_a absorbed laser flux, q_l, q_r heat fluxes to left and to right; absorption region dashed

$$P_a = p + \rho u^2. \tag{1.48}$$

is invariant throughout the stationary flow region and represents the true ablation pressure. Differentiating the last expression of (3.180) leads to the relation

$$\left(M + \frac{1}{\gamma M}\right)\frac{1}{c_s}\frac{\partial c_s}{\partial x} + \left(M - \frac{1}{\gamma M}\right)\frac{1}{M}\frac{\partial M}{\partial x} = 0. \tag{3.181}$$

From this we conclude that at the position x_m of the maximum of temperature or sound speed

$$M = \frac{1}{\sqrt{\gamma}} \tag{3.182}$$

holds. The proof is as follows. As one moves from the target to the vacuum M increases from $M \ll 1$ to $M = \infty$. Let x_M be the smallest value at which (3.182) holds. $x_M < x_\mathrm{m}$ is not possible because the derivative of c_s is positive (for instance, there is no saddle point in c_s for $x < x_\mathrm{m}$). If $x_M \geq x_\mathrm{m}$ holds, $\partial c_s/\partial x$ must again be zero there. This implies either $x_M = x_\mathrm{m}$, or $\partial M/\partial x = 0$ at x_m, respectively. In the latter case differentiating (3.180) once more leads to the following relation at $x = x_\mathrm{m}$,

$$\left(M + \frac{1}{\gamma M}\right)\frac{1}{c_s}\frac{\partial^2 c_s}{\partial x^2} + \frac{c_s}{M}\left(M - \frac{1}{\gamma M}\right)\frac{\partial^2 M}{\partial x^2} = 0 \tag{3.183}$$

from which owing to $\partial^2 c_s/\partial x^2 < 0$ also $\partial^2 M/\partial x^2 < 0$ follows, i.e., u is a maximum. This is in contradiction to the physical assumption of the existence of only one maximum in T. In addition, (3.183) shows that $x = x_\mathrm{m}$ cannot be a saddle point for M. P_a can now be calculated from (1.48) if c_s and ρ are known at x_m. With resonance absorption dominating over collisional absorption (see Chap. 5), e.g., at $I\lambda^2 \gtrsim 10^{13}$ Wcm$^{-2}\mu$m^2, the energy deposition zone becomes centred around the critical point $x = x_c$ in a narrow region. Furthermore, in numerous experiments $\mathbf{q}_e$ seemed to be bound by a limit considerably lower than the classical value (1.61) (*heatflux inhibition* [37, 38]). This situation implies that x_m lies close to x_c and that the heat front x_T is sufficiently close to c_s to guarantee the flow to be approximately planar in between. To prove numerically whether in this case (3.182) is fulfilled or not, the fully time dependent initial value problem of target irradiation by a laser pulse of constant intensity was solved. The energy was deposited in a narrow zone around x_c and, in order to simulate the divergent flow, ρ was set equal to zero far out in the supersonic region. As Fig. 3.11a shows, the relation $M_c = 1/\sqrt{\gamma} = \sqrt{3/5} = 0.774$ is extremely well fulfilled. This is even more surprising if one takes into account that the maximum in T is very flat. A completely different picture with $M_c \ll 1/\sqrt{\gamma}$ is obtained when half of the incoming laser flux is deposited to the right of x_c of Fig. 3.11b. Nonetheless $M \simeq 1/\sqrt{\gamma}$ is again reached at the maximum of T which is now located outside the picture far away from the critical point. Several arguments have been given in the past to show that $M_c = 1$ should hold. It is evident from the analysis presented here that the validity of those arguments is questionable for $\gamma \neq 1$.

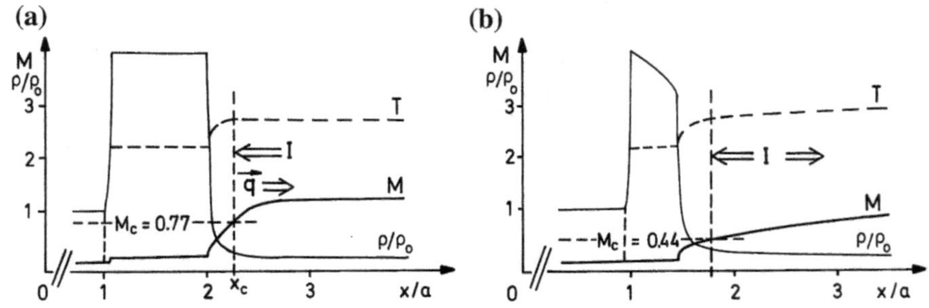

Fig. 3.11 **a** Steady state plane ablation. Energy input at $\rho_c/\rho_0 = 0.2$. $M_c = \sqrt{1/\gamma}$ is exactly confirmed for $\gamma = 5/3$. M_c does not depend on the value of ρ_c; heat conduction according (1.61) is included. **b** Plane nonsteady ablation due to energy deposition at x_c (50%) as well as in the ablated plasma. $M_c = 0.44$. The inhomogeneity of the compressed matter is due to decrease of $P_a(t)$ (see also Fig. 1.13 for the same reason)

It becomes further clear that a supersonic stationary flow at x_c [39] can only be due to deviations from plane geometry (e.g., divergence effects) regardless of how strong the heat flux in the overdense region is. In an unsteady flow such a restriction does not hold owing to the presence of additional, i.e., inertial terms.

Radiation Pressure Included

Mass and momentum conservation in plane 1D geometry with radiation pressure included follow from (3.53) and (3.56) in the steady state as

$$nu = \text{const}, \quad \rho u^2 + p - p_\pi = \text{const}, \quad p_\pi = -\int \frac{Zn}{4n_c}\frac{\partial}{\partial x}\left(\varepsilon_0 \hat{E}^2\right) \mathrm{d}x. \quad (3.184)$$

$$P_a = \rho u^2 + p - p_\pi. \quad (3.185)$$

$$\left(M + \frac{1}{\gamma M}\right)\frac{1}{c_s}\frac{\partial c_s}{\partial x} + \left(M - \frac{1}{\gamma M}\right)\frac{1}{M}\frac{\partial M}{\partial x} + \frac{\varepsilon_0}{4\rho_c c_s^2 M}\frac{\partial}{\partial x}\hat{E}^2 = 0. \quad (3.186)$$

The equations hold in the system in which the amplitude $\hat{E}^2$ is at rest. In a strict sense this means an inertial frame in which the ablation front stays in a fixed position. For thick targets such a requirement is generally fulfilled to a good approximation. Thin targets undergo an accelerated motion like a rocket as soon as the first shock wave has moved through. Here the correct reference system has to be readjusted at each instant of time; it is the *tangent inertial frame*.

In the following $\gamma = 1$ is set. We observe that $\hat{E}^2$ has the typical sharp maxima of a standing wave whereas that of $c_s(x > x_c)$ is flat, see Fig. 1.11. Therefore, in (3.186) the first term may be dropped since it cannot balance the two following expressions. Hence

$$\left(M - \frac{1}{M}\right)\frac{\partial M}{\partial x} + \frac{\varepsilon_0}{4\rho_c c_s^2}\frac{\partial}{\partial x}\hat{E}^2 = 0. \quad (3.187)$$

With profile steepening a steady state is reached during a time of the order of $\tau = \lambda/4c_s$, thus $\tau \lesssim 25$ ps. Under this assumption in the very overdense region u is subsonic ($M < 1$). In the corona u is supersonic ($M > 1$). Equation (3.187) states that the sonic point must coincide with the absolute maximum $\hat{E}^2_{\mathrm{m}}$ at x_{m}. Generally $\hat{E}^2_{\mathrm{m}}$ coincides with the first local maximum of $\hat{E}^2$ when counting them in positive x direction. In principle P_a can be evaluated at any point in the coaxial cylinder owing to $P_a =$ const throughout the steady state flow region. However, owing to the density dependence of p_{π} according to (3.184) it is convenient to evaluate P_a in a maximum or minimum of $\hat{E}^2$. We chose here the maximum at $x = x_{\mathrm{m}}$. At normal incidence the laser pulse amplitude follows the stationary wave equation

$$\frac{\partial^2}{\partial x^2}\hat{E} + k^2\left(1 - \frac{Zn}{n_c}\right)\hat{E} = 0; \;\; k = \omega/c. \tag{3.188}$$

Multiplying it by $\hat{E}' = \partial\hat{E}/\partial x$ yields at an arbitrary position

$$-\,p_{\pi} = \int \frac{Zn}{4n_c}\frac{\partial}{\partial x}\left(\varepsilon_0\hat{E}^2\right)\mathrm{d}x = \frac{\varepsilon_0}{4}\hat{E}^2 + \frac{\varepsilon_0}{4k^2}\hat{E}'^2. \tag{3.189}$$

In the maxima and minima $\hat{E}'$ vanishes and p_{π} simplifies. At x_{m} holds

$$P_a = 2\rho_{\mathrm{m}}c_s^2 + \frac{\varepsilon_0}{4}\hat{E}^2_{\mathrm{m}}; \quad \rho_{\mathrm{m}} = \rho_c M_c. \tag{3.190}$$

It is convenient to introduce the normalized quantity $\mathcal{E}^2 = \varepsilon_0\hat{E}^2/4p_c$. The critical Mach number M_c is obtained from integrating (3.187) between x_c and x_{m},

$$\mathcal{E}_c^2 = \mathcal{E}_{\mathrm{m}}^2 - \frac{1}{2}\left(M_c^2 - \ln M_c^2 - 1\right). \tag{3.191}$$

This equation together with (3.188) determines $\mathcal{E}_c$ uniquely if $\mathcal{E}^2_{\mathrm{m}}$ is assumed to be known. Although it is a simple system, it resisted to all attempts to solve it analytically in explicit form so far. The numerical analysis yields the following best fits

$$\mathcal{E}^2_{\mathrm{m}} < 1: \; M_c \simeq 1 - \sqrt{0.4\mathcal{E}^2_{\mathrm{m}}}; \quad \mathcal{E}^2_{\mathrm{m}} \geq 1: \; M_c \simeq \frac{\mathcal{E}^2_{\mathrm{m}}}{\exp(\mathcal{E}^2_{\mathrm{m}} + 0.5) - 2}. \tag{3.192}$$

Owing to the action of the radiation pressure M_c reduces monotonically with increasing $\mathcal{E}_{\mathrm{m}}$. At $\mathcal{E}^2_{\mathrm{m}} = 1$ the critical Mach number is reduced to $M_c = 0.38$. In the following $\mathcal{E}^2_{\mathrm{m}}$, together with P_a from (3.190), is evaluated as a function of the plasma pressure p_c at the critical point x_c for four relevant cases.

3.3.4.2 Perpendicular Incidence

Generally at densities below ρ_m at $x > x_m$ almost no collisional and collective absorption takes place; the reflection coefficient $R(x = x_m)$ is close to the overall reflection coefficient R and ρ is sufficiently smooth to apply the WKB approximation on $\hat{E}_m$,

$$\mathcal{E}_m^2 = \frac{(1+\sqrt{R})^2}{(1-\rho_m/\rho_c)^{1/2}} \frac{p_L}{2p_c}, \quad p_L = \frac{I}{c}. \tag{3.193}$$

For the WKB approximation see Chap. 5. In p_L the vacuum laser intensity I enters, possibly corrected by the collisional absorption in the corona $x > x_m$. The critical pressure p_c is to be linked to the incident laser intensity. For $L_c \simeq \lambda$ and shorter (5.180) will yield more precise values of $\hat{E}_m$. The transition to Fresnel's formula for strong profile steepening may be estimated from Table 5.1 and an Epstein transition layer [40]. For $\mathcal{E}_m^2 < 1$ this becomes with the help of (3.192)

$$\mathcal{E}_m^2 = 1.2\left(1+\sqrt{R}\right)^{8/5}\left(\frac{p_L}{2p_c}\right)^{4/5}. \tag{3.194}$$

The radiation pressure-corrected ablation pressure from (3.184) is given by

$$P_a = 2p_c\left(M_c + \frac{\mathcal{E}_m^2}{2}\right) = p_c\left(2M_c + \frac{(1+\sqrt{R})^2}{2(1-M_c)^{1/2}}\frac{p_L}{p_c}\right). \tag{3.195}$$

Contrary to what one may expect, for $\mathcal{E}_m^2 < 1.4$ radiation pressure leads to a reduction of P_a/p_c since at low ratios p_L/p_c the stagnation effect of the plasma flow prevails.

In Fig. 3.12 P_a/p_c is plotted as a function of $\mathcal{E}_m^2$. The dashed lines are obtained by making use of (3.192) for M_c. In the interval $0 \leq \mathcal{E}_m^2 \leq 1.41$ there is a depression of

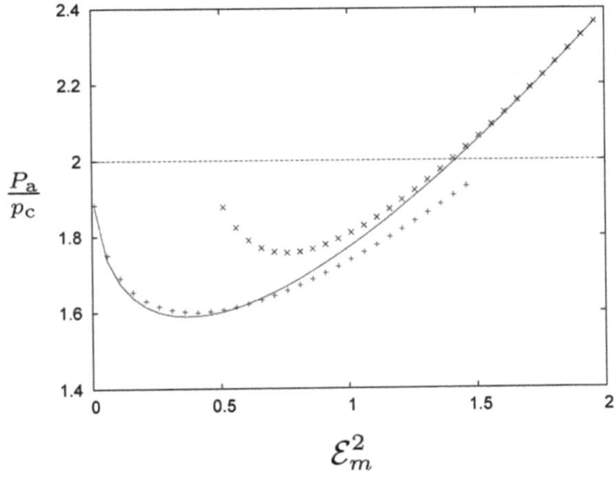

Fig. 3.12 Normalized ablation pressure P_a/p_c as a function of $\mathcal{E}_m^2$

P_a due to the stagnation effect of p_π on M_c, with a maximum reduction of 22% at $\mathcal{E}_\mathrm{m}^2$ as low as $\mathcal{E}_\mathrm{m}^2 = 0.35$. For $\mathcal{E}_\mathrm{m}^2 = 0.15$ the light pressure contributes to P_a by at most by 10% and can be neglected for $\mathcal{E}_\mathrm{m}^2 \leq 0.15$ and hence (3.190) yields $P_a \simeq 2p_cM_c$. In particular, at $\mathcal{E}_\mathrm{m}^2 = 0.15$ follows $P_a = 1.7\, p_c$. With ns laser pulses generally one moves around such low values of $\mathcal{E}_\mathrm{m}^2$ and it was correct to neglect p_π as an additional term in P_a of (3.184), as done by all authors so far. However, to neglect its stagnation effect on M_c would be incorrect. Flow inhibition at x_c becomes sensitive already at values as low as $\mathcal{E}_\mathrm{m}^2 = 0.02$, with a 10% reduction of P_a at temperature T held fixed. Only from $\mathcal{E}_\mathrm{m}^2 = 0.84$ on the radiation pressure starts dominating the ablative plasma pressure $2p_cM_c$. Stagnation and flow inhibition are synonymous with profile steepening [41, 42]. With the help of (3.187) its scale length L is given by

$$L = \frac{n}{|\nabla n|} = \frac{M}{|\nabla M|} = \frac{|1 - M^2|}{\nabla \mathcal{E}^2}.$$

At the critical point it becomes in units of vacuum wavelength λ

$$\frac{L_c}{\lambda} = \frac{1 - M_c^2}{2\lambda \mathcal{E}_c \partial \mathcal{E}/\partial x|_{x=x_c}}. \tag{3.196}$$

Multiplying (3.187) by $\rho/\rho_c = M_c/M$, then integrating it from x_c to x_m and substituting $p_{\pi m} - p_c$ from (3.189) yields

$$M_c\left\{2 - M_c - \frac{1}{M_c}\right\} + \mathcal{E}_\mathrm{m}^2 - \mathcal{E}_c^2 - \frac{1}{k^2}\mathcal{E}_c'^2 = 0.$$

Introducing $\mathcal{E}_\mathrm{m}^2$ and $\mathcal{E}_c^2$ from (3.191) leads to the desired expression for $\mathcal{E}_c'$ as a function of M_c,

$$\mathcal{E}_c'^2 = \left(2M_c - \ln M_c - \frac{1}{2}M_c^2 - \frac{3}{2}\right)k^2. \tag{3.197}$$

With this (3.196) becomes

$$\frac{L_c}{\lambda} = \frac{1 - M_c^2}{2\lambda k\{\mathcal{E}_\mathrm{m}^2 + (1 - M_c^2 + \ln M_c^2)/2\}^{1/2}\{2M_c - \ln M_c - M_c^2/2 - 3/2\}^{1/2}} \tag{3.198}$$

The evaluation of this equation proceeds as follows: $\mathcal{E}_\mathrm{m}$ is fixed first. Then the wave equation (3.188) is solved simultaneously with (3.187). In this way M_c is recovered to be used in (3.198). The result is shown in Fig. 3.13, solid line. If, instead, M_c is taken from (3.192) and used in (3.198) to determine L/λ, coincidence with the exact result is found (see Fig. 3.13, solid line, dots). For a quick estimate of L the derivative of $\mathcal{E}_c^2$ may be approximated by $\partial \mathcal{E}_c^2/\partial x \simeq 3k\mathcal{E}_\mathrm{m}^2[1 - 2M_c/(1 + M_c)]^{1/2}/4$, thus

$$\frac{L_c}{\lambda} \simeq \frac{2}{3\pi}(1 - M_c)^{1/2}(1 + M_c)^{3/2} = 0.17\left(\frac{2}{\mathcal{E}_\mathrm{m}} - 0.63\right)^{3/2}; \quad \mathcal{E}_\mathrm{m}^2 < 1. \tag{3.199}$$

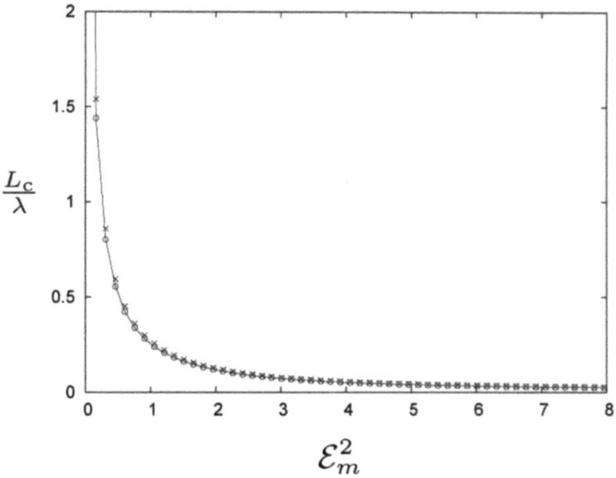

Fig. 3.13 Scale length at critical point L_c/λ as a function of $\mathcal{E}_{\mathrm{m}}^2$ from wave equation (solid), from (3.192) and (3.198) (dotted) and from (3.199) (crosses)

As the crosses in Fig. 3.13 show this simple formula is an excellent fit to (3.198).

3.3.4.3 Oblique Incidence: s-Polarization

The electric field follows the wave equation

$$\frac{\partial^2}{\partial x^2}\hat{E} + k^2\left(1 - \frac{Zn}{n_c} - \sin^2\alpha\right)\hat{E} = 0.$$

Proceeding in the same way as before one finds at the position x_{m}

$$p_\pi = \frac{\varepsilon_0}{4}\frac{\hat{E}_{\mathrm{m}}^2}{\cos^2\alpha}.$$

The density, pressure, and Mach number at the vertex, ρ_V, p_V, M_V,

$$\rho_V = \rho_c\cos^2\alpha,\ \ p_V = p_c\cos^2\alpha,\ \ \rho_{\mathrm{m}} = \rho_V M_V$$

assume the role of the former critical parameters ρ_c, p_c, and M_c With this step (3.194) transforms into

$$\mathcal{E}_{\mathrm{m}}^2 = \frac{(1+\sqrt{R})^2}{(1-\rho_{\mathrm{m}}/\rho_V)^{1/2}}\frac{p_L}{2p_c\cos^2\alpha} = \frac{(1+\sqrt{R})^2}{(1M_V)^{1/2}}\frac{p_L}{2p_V} \tag{3.200}$$

i.e., its structure remains invariant. Analogously, in (3.192) and (3.198) for the profile steepening, M_c is substituted by M_V and all calculations proceed in the same way as in the case of perpendicular incidence.

3.3.4.4 Oblique Incidence: p-Polarization

This is the situation of resonance absorption in a layered medium [43]. At electron temperatures $T_e \leq 10\,\text{keV}$ the electron plasma wave is emitted under the small angle $\alpha' = \beta \sin\alpha_0, \beta = c_{se}/c$ (see Chap. 6, Sect. 2.2.1: Inhomogeneous Stokes equation). In the region $x_c \leq x \leq x_\text{m}$ holds $|\hat{E}_x| \gg |\hat{E}_y|$ for all angles under which absorption is significant. Therefore p_π and $\mathcal{E}_\text{m}$ can be evaluated by using Piliya's equation (6.37) in the capacitor model approximation, i.e., neglecting E_y, $\beta^2 \sin^2\alpha_0$, and setting $1 - \beta^2 = 1$. The driving term containing B on the RHS of (6.37) is much smaller than $\hat{E}_{x,m}$ and (3.189) applies again with $p_{\pi,\text{m}} = \varepsilon_0/4\hat{E}^2_{x,\text{m}}$. Then (6.48) yields for $\mathcal{E}^2_\text{m} = \mathcal{E}^2_{x,m}$

$$\mathcal{E}^2_\text{m} = 1.1(1 - R)\left(\frac{kL}{\beta^4}\right)^{1/3} \frac{p_\text{L}}{2p_c}\cos\alpha_0. \tag{3.201}$$

Alternatively $\mathcal{E}^2_\text{m}$ can be calculated from the energy flux conservation (5.191) in combination with (5.193),

$$S_\text{es} = s_e\left(1 - \frac{\rho}{\rho_c}\right)^{1/2} \frac{\varepsilon_0}{2}\frac{\omega^2}{\omega_p^2}\hat{E}^2_\text{m} = (1 - R)cp_\text{L}\cos\alpha_0 \tag{3.202}$$

$$\Longrightarrow \qquad \mathcal{E}^2_\text{m} = \frac{1}{\beta}\frac{1 - R}{(1 - M_c)^{1/2}}\frac{p_\text{L}}{2p_c}\cos\alpha_0. \tag{3.203}$$

With (3.189)–(3.192) remaining valid the ablation pressure is given by

$$P_\text{a} = 2p_c\left(M_c + \frac{\mathcal{E}^2_\text{m}}{2}\right) \tag{3.204}$$

and L/λ follows from (3.198). For low intensities (3.202) may be correct. It has to be made sure that the resonance width d from (6.43) does not exceed the density scale length L_c of the Stokes equation. Expression (3.203) is more robust; however, it has also its limitations because S_es is based on linearized equations. On the other hand (6.55) on the limiting intensity I_max, Fig. 6.14 and the considerations on kinetic wave breaking in Sect. 6.3.3 may be helpful in the specific case.

3.3.4.5 Link to Laser Intensity

In general the energy supplied to the target may be deposited over a wide region in space as, for instance, is the case of collisionally absorbed short wavelength radiation. However, let us return to the better defined situation of Fig. 3.10 in which the energy absorption zone is centred in a narrow region of density as here (shaded region), and as in some deflagrations and detonations [44, 45]. In such cases the deposited energy flux density I_a may be thought of as deposited locally and split into the two heat fluxes q_l and q_r to the left and to the right, as indicated in the figure by the two arrows. In this model the slope of T is discontinuous at $x_{\rm m}$ in order to satisfy $q_l = \kappa(\partial T/\partial x)_{x_{\rm m}}$ and $q_r = -\kappa(\partial T/\partial x)_{x_{\rm m}}$. The energy flux into the target q_l covers the convection of enthalpy plus kinetic energy of the ablated material whereas q_r goes entirely into expansion work of the out flowing matter and does not contribute to the ablation pressure. The energy balance for the steady state is as follows,

$$q_l = I_a - q_r = \rho u\left(w + \frac{u^2}{2}\right) + q_e + P_a u_1, \quad w = \sigma + \varepsilon + \frac{p}{\rho}. \tag{3.205}$$

ε is the internal energy per unit mass and σ represents the heat of evaporation and/or ionization. The term p/ρ provides for the work done by the outflowing matter against the rarefying plasma in front of it. The quantity q_e is the longitudinal and transversal heat flux into the target which is not converted into steady state plasma outflow; $P_a u_1$ accounts for the work done by the laser to generate the shock wave travelling into the overdense material at shock speed v_S and matter velocity u_1. The heat flow to the right q_r is needed for the evaluation of the ablation pressure. In one-dimensional flow an upper limit for it is given by the isothermal rarefaction wave of $\gamma = 1$, $q_r = pc_s|_{x=x_{\rm m}}$. It is a good approximation at high electron temperature over a wide range in space. In the simplest case of $T_i = 0$, $\varepsilon = 3p/2\rho$, negligible ionization energy, and $q_e = 0$, $P_a u_1 = 0$ (moderate laser flux), one has

$$I_a = \rho u\left(\varepsilon + \frac{p}{\rho} + \frac{u^2}{2}\right)_{x=x_{\rm m}} + q_r = 4\rho_{\rm m}c_s^3|_{x=x_{\rm m}}, \quad q_l = 3\rho_{\rm m}c_s^3|_{x=x_{\rm m}}, \tag{3.206}$$

and hence, the maximum of the ratio of heat fluxes is $q_r/q_l = 1/3$. It indicates that q_r, in contrast to the common practice, should not be neglected. q_r is zero only when γ is identical with the adiabatic exponent.

The actual divergent flow in the outer corona leading to stationary conditions may be approximated by a spherical isothermal rarefaction wave which is the solution of $\partial_r \rho r^2 u = 0$ and $\partial_r u^2/2 = -c_s^2 \partial_r \rho/\rho$:

$$\rho = \rho_0 \mathrm{e}^{-(u^2-u_0^2)/2s^2}, \quad r^2 = r_0^2\frac{u_0}{u}\mathrm{e}^{(u^2-u_0^2)/2c_s^2}. \tag{3.207}$$

The maximum heat flux density q_r is calculated analogously to the plane case with the identifications $u_0 = c_s$, $\rho_0 = \rho_{\rm m}$,

$$\begin{aligned} q_r &= \int_{r_{\rm m}}^{\infty} \frac{p}{r^2}\frac{\partial}{\partial r}(r^2 u)\mathrm{d}r \\ &= \rho_{\rm m}c_s^3\left\{1 + \left(\frac{\pi \mathrm{e}}{2}\right)^{1/2}\left[1 - \mathrm{erf}\left(\frac{1}{\sqrt{2}}\right)\right]\right\} = 1.65\rho_{\rm m}c_s^3. \end{aligned} \tag{3.208}$$

It is higher by the factor 1.65 than for planar flow. The choice of the integration constants u_0, ρ_0 needs a short explanation. At $x = x_{\mathrm{m}}$, u is equal to the speed of sound for $\gamma = 1$; however the flow pattern, rather than being radial, looks as sketched in Fig. 3.14a for planar and Fig. 3.14b for concave targets; it is radial only for uniform illumination of spherical targets, e.g., fusion pellets for inertial fusion. Nevertheless, the heat flux $\mathbf{q}_e \mathrm{d}\boldsymbol{\Sigma}$ through any surface element $\mathrm{d}\boldsymbol{\Sigma}$ on $\Sigma(M = 1)$ is the same as for radial flow, provided $\mathbf{q}_e$ also follows the streamlines. In fact, under the condition $\rho \mathbf{v} \mathrm{d}\boldsymbol{\Sigma} = \mu = \mathrm{const}$ along a streamline, Bernoulli's equation indicates

$$\varepsilon + \frac{p}{\rho} + \frac{u^2}{2} + \frac{q_r}{\mu} = \mathrm{const.} \tag{3.209}$$

For $T = \mathrm{const}$ it simplifies to

$$\frac{u^2}{2} + \frac{q_r}{\mu} = \mathrm{const}'. \tag{3.210}$$

Since somewhere outside the target the flow becomes spherical this relation shows that q_r through $\Sigma(M = 1)$ is the same for all flow patterns owing to $u^2 = c_s^2$ there and $\mu =$ const along a flux tube. The assumption $\mathbf{q}_e \parallel \mathbf{u}$ is reasonable in the vicinity of the axis in many experiments for most of the interaction time. Strong lateral heat flow would invalidate (3.208). In particle-in-cell (PIC) calculations with $T_i = T_e = T$, for instance, the fraction of energy taken by q_r in the early stage of interaction could amount to 70% of the incident laser energy at $I = 10^{15}$ Wcm^{-2}; at later stages, when a large plasma cloud has formed this fraction reduced to an asymptotic value $q_r/I_a \simeq 20$–25%. From (3.208) the ratio $q_r/I_a = 1.65/(3 + 2 + 1/2 + 1.65) = 23\%$ is obtained. At such high laser intensities it is more realistic to assume $T_i = 0$ (see Fig. 1.14). Then, from (3.208) $q_r/q_l = 0.55$ follows. This shows once more that q_r should not be neglected.

The question arises whether the assumption of a steady state is a realistic hypothesis. As far as the overdense (ablative) region is concerned it is reasonable (for a detailed discussion see the instructive comments in [46, 47] and the references therein). For the underdense corona a more accurate evaluation of q_r may be based on an interesting discovery by Eidmann and Schmalz [48, 49]:

> The isothermal plasma ablation from a sphere of constant radius r_0 fills the surrounding space according to a stationary rarefaction wave (3.207) and then suddenly, at a finite time instant t_s passes over into a time-dependent solution of nearly linear velocity and nearly exponential density distribution in space, well approximated by an isothermal similarity solution. Such breaking up of the stationary solution is also observable in the numerical solutions, for instance in Fig. 1.14, spherical case. The position of breaking location $r(t_s)$ tends monotonically to ∞ with $t_s \to \infty$.

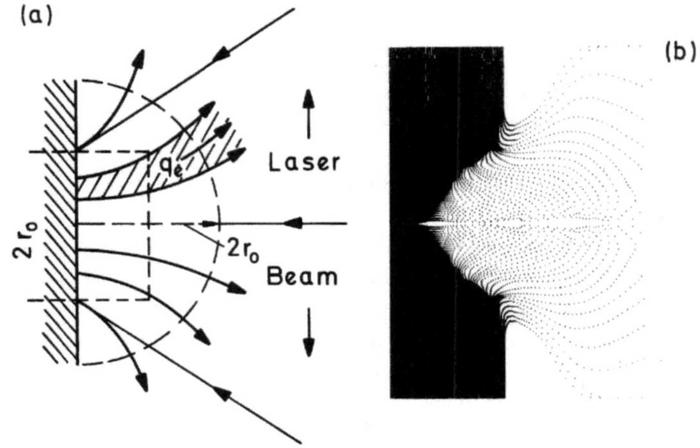

Fig. 3.14 **a** Streamlines originating from a plane target; **b** particle-in-cell (PIC) calculation with regular injection of particles with the parameters $n_0 = 5 \times 10^{22}\,\mathrm{cm}^{-3}$, $I = 10^{15}\,\mathrm{Wcm}^{-2}$, $\lambda = 1.06\,\mu\mathrm{m}$. The streamlines close to the shock follow the dots of the mass points. The shock remains attached to the surface of the crater

Under the assumption that the maximum of the electron temperature is located in the critical region and the flow there is essentially onedimensional the analysis of this section has shown that the effective laser intensity for generating ablation pressure I_a is

$$I_a = I_{\text{abs}} - q_r = q_l = 3\rho_{\text{m}}[c_s]^3_{\text{m}}, \quad I_{\text{abs}} \text{ absorbed intensity}. \tag{3.211}$$

Weakening of I_a due to ionization work is not considered. Elimination of c_s and assuming γ polytropic or isothermal the ablation pressure results as

$$P_a = 2\left[\frac{2(\gamma-1)}{(3\gamma-1)}\right]^{2/3} \rho_{\text{m}}^{1/3}(I_a - q_r)^{2/3} \quad \text{for} \quad \gamma \neq 1,$$

$$P_a = \frac{2}{3^{2/3}}\rho_{\text{m}}^{1/3}(I_a - q_r)^{2/3} \quad \text{for} \quad \gamma = 1. \tag{3.212}$$

Specifying for *planar* isothermal heat flow yields

$$P_a = \frac{2}{4^{2/3}}\rho_{\text{m}}^{1/3} I_a^{2/3} = 0.79\rho_{\text{m}}^{1/3} I_a^{2/3}; \qquad \gamma = 1, \tag{3.213}$$

and for *spherical* isothermal heat flow

$$P_a = \frac{2}{4.65^{2/3}}\rho_{\text{m}}^{1/3} I_a^{2/3} = 0.72\rho_{\text{m}}^{1/3} I_a^{2/3}; \qquad \gamma = 1. \tag{3.214}$$

If q_r is set equal to zero the numerical factor for the isothermal ablation pressure is $2/3^{2/3} = 0.96$, i.e., 33% higher, in agreement with (1.51). Scaling with the wavelength is $P_a \sim \lambda^{-2/3}$.

3.3.4.6 Perpendicular Incidence, Focused Beam

First it has to be made sure which of the absorption mechanisms dominates, collisional or resonance absorption. Such an estimate requires suitable averaging over all angles of incidence α_0 of the beam into the target crater. If resonance prevails, (3.204) and the angle-averaged (3.201) apply. If this is not the case no general recipe can be given here. Rather has there a combination between $\mathcal{E}_\mathrm{m}$ electromagnetic and $\mathcal{E}_\mathrm{m}$ electrostatic to be found. In the intensity regime $I_a = 10^{12}$–$10^{16}\,\mathrm{Wcm}^{-2}\ \mu\mathrm{m}^2$ and pulse durations ranging from several ns to ten ps $p_\pi \ll p_c$ is fulfilled. A similar inequality holds for the compression work going into the shock wave $P_a v_1$. With

$$P_a = \rho_0 v_S u_1 = g \rho_0 v_S^2, \quad u_1 = \left(g\frac{P_a}{\rho_0}\right)^{1/2}; \quad g = 1 - \frac{1}{\kappa} \tag{3.215}$$

follows

$$\frac{P_a u_1}{I} \leq \frac{P_a u_1}{I_a} \leq 2\left(M_c \frac{\rho_c}{\rho_0}\right)^{1/2}. \tag{3.216}$$

At short wavelengths in the UV range $P_a u_1$ must be taken into account. With $\mathcal{E}_\mathrm{m}$ as a free parameter P_a follows from (3.214) by replacing ρ_m by $\rho_c M_c$,

$$\begin{aligned} P_a &= 2\frac{(1-R)^{2/3}}{4.65^{2/3}} \rho_c^{1/3} M_c^{1/3} I^{2/3} \\ &= 3.3 \times 10^{-8} (1-R)^{2/3} M_c^{1/3} (\rho_c[\mathrm{gcm}^{-3}])^{1/3} \left(I[\mathrm{Wcm}^{-2}]\right)^{2/3} \mathrm{Mbar}. \end{aligned} \tag{3.217}$$

The critical Mach number M_c may be taken from (3.192). For plane isothermal heat flow into the corona the corresponding numerical factor is 3.6×10^{-8}.

Formula (3.217) holds for idealized conditions described in the previous analysis. It may be helpful in evaluating the single effects contributing to the ablation pressure in the individual experiment, like heat conduction, different kinds of laser energy absorption, radiation pressure and profile steepening, nonsteady state effects. Within the model presented here the most sensitive aspects are local and nonlocal energy deposition through direct laser light absorption, diffusive heat flow and delocalized preheat by fast electrons. Radiation pressure effects are important for profile steepening and its influence on energy absorption. For $\mathcal{E}_\mathrm{m}^2 \leq 1.4$, P_a is reduced as a consequence of the stagnation effect on the flow at critical density. The dependence of P_a on the laser intensity to the power of $2/3$ remains substantially unchanged as long as preheat and lateral heat losses can be ignored. In the experiment a whole variety of power dependences ranging between 0.15 and 2.5 are measured, e.g., 0.3 in [50]. Extensive experimental studies in the intensity range 10^{13}–10^{15} Wcm^{-2} and comparison with previous results have been presented by Dahmani [51]. The author finds a 2/3 power in I_a/λ, in agreement with (3.217), however, with a numerical factor slightly higher than 3.3×10^{-8}. From thorough acceleration studies of thin low-Z foils at the Asterix III iodine laser ($\lambda = 1.315\,\mu$m) by Eidmann et al. [38] in the intensity range 10^{11}–10^{16} Wcm^{-2} one extracts $P_a\,[\text{Mbar}] = 5.5 \times 10^{-9}\left(I[\text{Wcm}^{-2}]\right)^{0.65}$. In another series of experiments with the Asterix III laser at fundamental and third harmonic wavelength ($\lambda = 0.44\,\mu$m) the law $P_a = 2.8 \times 10^{-8} I_a^{0.6}\lambda^{-0.4}$ was found [52]. In a last example [53] 12 μm thick gold foils under Nd laser irradiation were analyzed with the result $P_a = 5.1 \times 10^{-9} I_a^{0.71}$. For additional comparisons with early experiments consider [54]. In a more recent experiment [55] the authors aimed at achieving 1D conditions without lateral energy losses for $\lambda = 0.44\,\mu$m up to intensities $I_a = 2 \times 10^{14}$ Wcm^{-2} on aluminum targets. They found a dependence of P_a on the target thickness d$^{-2/15}$ and an overall intensity dependence in "fair agreement with analytical models", e.g., [56]. In a numerical study by Evans et al. [57] at Nd wavelength with plane targets and no heat flux limit, and with spherical targets of radius 100 μm and heat flux limit $f = 0.03$ $P_a = 2.5 \times 10^{-9} I_a^{0.67}$ and $P_a = 1.9 \times 10^{-10} I_a^{0.767}$, respectively, was obtained in the interval 10^{12}–10^{16} Wcm^{-2}. To the authors' knowledge only very few contemporaneous investigations are available on the contribution of radiation pressure in this intensity regime, e.g., [58]. In recent years the ablation front structure and dynamics in strongly radiating plasmas has been treated as an eigenvalue problem by Basko et al. [59]. Power-law scaling of plasma pressure on laser-ablated tin microdroplets has been established by Kurilovich et al. [60]. They are supported by 2D hydrodynamic simulations.

At laser intensities well beyond 10^{17} Wcm^{-2} and on fs $\sim$ ps time scale the light pressure p_L clearly prevails on plasma pressure. The reason is that q_e of the fast and medium fast electrons nearly compensates I_a. To see this we consider heat diffusion in a solid target of $n_0 = 10^{23}$ cm^{-3} for $I_a = 10^{17}$ Wcm^{-2} irradiance at $t = 1$ ps. From (1.68) a penetration depth $x_T = 26(f/Z)^{2/9}\,\mu$m, f heat flux inhibition factor, and a plasma pressure $p = n_c k_B T = 2.5 \times 10^{12}(f/Z)^{-2/9}$ [cgs] are calculated. This has to be compared with the radiation pressure $p_L = I_a/c = 3 \times 10^{13}$ [cgs] which is an order of magnitude higher. In the transition region from $P_a \sim I_a^{2/3}$ to $P_a \sim I_a$ around $I = 10^{17}$ Wcm^{-2} accurate values of p/p_L are accessible only to numerical simulation.

3.3.4.7 Impulsive Loading

An alternative way to generate high pressure which is accessible to an analytical treatment is by impulsive loading. Suppose the laser pulse energy E is given. In view of the high pressure generation the question of the optimum pulse length is relevant. In general this is a complex problem. A partial answer can be given in the sense that the pressure ratio is between impulsive loading P_i and the steady state case of

P_a. For this purpose we observe that the shock wave cannot evolve before the heat diffusion speed $\dot{x}_T$ [see, e.g., (1.68)] has slowed down to the ion sound speed c_s. This means that the comparison of pressures has to be made at the time τ_p given by (1.69). Then the lowest value of P_i is determined at the edge of the rarefaction wave where $u = 0$ holds; hence $P_i \geq (Z+1)n_0 k_B T(\tau_p)$, whereas $P_a \leq p_\mathrm{m} + \rho_\mathrm{m} c_{s_\mathrm{m}}^2$. On approximating ρ_m by ρ_c the inequality $P_a \leq n_c\left(1+\frac{1}{Z}\right)k_B T(\tau_p)\left(1+\frac{1}{\sqrt{\gamma}}\right)$ results and hence, the ratio P_i/P_a fulfills

$$\frac{P_i}{P_a} \geq \frac{\sqrt{\gamma}}{1+\sqrt{\gamma}} Z \frac{n_0}{n_c}. \tag{3.218}$$

Generally this ratio is much larger than unity.

3.3.4.8 Is Critical Density Plateau Formation Possible?

Consider the stationary plasma flow in one dimension when it is dominated by the radiation pressure of the laser. The formation of possible density structures as a function of the Mach number obeys (3.187). Integrated once it reads

$$M^2 - \ln M^2 + \frac{\varepsilon_0}{2\rho_c c_s^2}\frac{\partial}{\partial x}|\hat{E}|^2 = \text{const.} \tag{3.219}$$

The dependence of $M^2 - \ln M^2$ is depicted as a function of M^2, or $|\hat{E}|^2$, see Fig. 3.15. In one dimensional (1D) *spherical geometry* extended steady sate structures are possible. In the pertinent equations (3.187) and (3.188) $\hat{E}$ is to be replaced by $r\hat{E}$, ∂_x becomes ∂_r and in (3.219) the term $4\ln r$ has to be added on its LHS. Possible density structures compatible with the steady state wave equation (3.188) in $r\hat{E}$ have been analyzed in [61]. In the subsonic region the familiar density steepening was found. If however the 1D plasma flow in the critical region in spherical geometry was assumed to be supersonic the formation of around the critical density had been shown to be the only possible solution. The investigation of the stability of plateau formation was outside the scope of the paper. Such an analysis has been undertaken one year later [62]. The authors "showed" that near-critical/subcritical plateaus are Brillouin unstable, and concluded that "an overdense pump detached in the underdense region is shown to be the only one-dimensionally stable solution" (Pellat). The scientific community seemed to share the conclusion since, except a few cases, e.g., [63], density plateaus in 1D have not been the subject of further investigations. The proof undertaken in [62] was highly idealized academic by performing a linear analysis with the ad hoc assumption of an initial perturbation in a homogeneous background. In reality, a whole variety of longitudinal instabilities are candidates for unstable behaviour in equal measure, Raman scattering for instance, and there are the transverse instabilities to be taken into consideration, first of all of Rayleigh–Taylor and Kelvin–Helmholtz type. After all profile steepening is subject to transverse instabil-

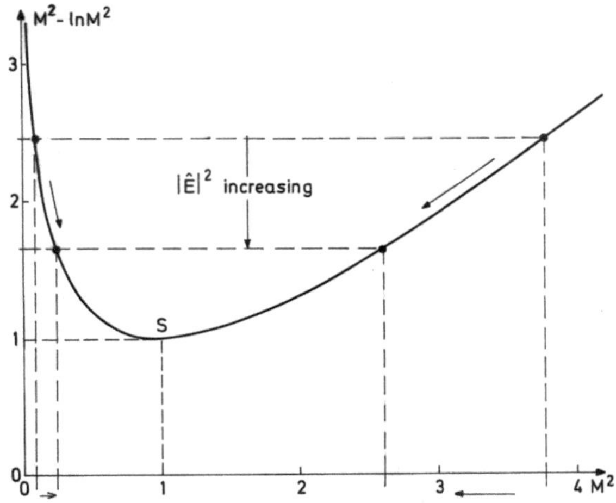

Fig. 3.15 The function $M^2 - \ln M^2$ exhibits a minimum, indicated by S, at the sonic point $M = 1$. Starting from a given value of the constant associated with a value of $|\hat{E}_0|^2$ its spatial decrease leads to a depression of the corresponding subsonic as well as supersonic Mach number. Conversely, starting from a local maximum of $|\hat{E}|^2$ at the sonic point its lowering leads either to a decrease of M and concomitant density increase in the subsonic domain or to an increase of M and decrease of the associated density in the supersonic region, corresponding to density step for $M < 1$ and, in onedimensional (1D) spherical geometry, to plateau formation for $M > 1$

ities, too. However, such general considerations are little conclusive because they do not exclude the occurrence of possible structures. To such an aim an analytical consideration is to be supported by numerical simulations. Such an analysis has been undertaken only more than two decades later.

Among the variety of ponderomotively induced density structures in laser produced plasmas in one dimension, it has been shown by particle-in-cell simulations (PIC) that the *stable plateau formation* at critical, subcritical, and above critical density is possible; stimulated Brillouin backscattering is suppressed with increasing laser intensity [64]. The PIC simulations have been performed in the intensity domain $a \simeq 1$–10. The onset of Brillouin/Raman scattering could be clearly discerned by the $\pi/2$ phase shift between driving field and plasma response, as it happens at resonance, see Chap. 6. The simulations have been performed in linear polarization and have been repeated afterwards also in circular polarization confirming essentially the linear results. The only noticeable difference was that for the survival of Brillouin scattering with increasing laser intensity is a little bit easier in circular polarization. This is in agreement with nearly noise-free Vlasov simulations of stimulated Raman scattering at high a values (see Sect. 6.4.6).

In [62] it has been shown that a homogeneous plasma close to or below critical density is subject to Brillouin backscattering in the absence of competing effects. The simulations presented in [64] show that efficient Brillouin (and Raman) suppression, either by a competing nonresonant ponderomotive effect in the underdense plasma or by the propagation cut off for nonrelativistic signals in the overdense domain between critical and relativistic critical density makes a stable density plateau formation in 1D possible. This fact is supported by analytical modelling on the Mathieu map Fig. 2.10b. As shown in [64] the region of stability is located in the dashed first stability zone left of the bold line $a = 1 - 2q$: Nonresonant ponderomotive structures enter into competition with the Brillouin and Raman growths; the observed suppression of these two resonant phenomena is not accidental. The research conducted in [64] with main results agrees in essential parts with findings by Kemp and Divol [65]: "Laser interaction.....extends as a density plateau just below n_c over several microns from the surface", see Fig. 3 there.

The main mechanism leading to plateau formation is this: light pressure pushes the overdense plasma of relativistic critical density ρ_{cr} back with velocity u_{cr}. Let us assume the plasma outflow underneath the density jump in the comoving reference system to be $\rho_P u_P$. When the equality $\rho_{cr} u_{cr} = \rho_P u_P$ is fulfilled in the laboratory frame, the instantaneously produced plasma is deposited at flow velocity $u = 0$ in the place created by target recession. The relevance of the plateau formation lies in the possibility of creating *hot dense matter* from *foams* with nowadays high-power lasers. As foams can be produced at arbitrary densities, from far below solid density up to hundreds of times higher, it may not be too difficult to fulfill the mass conservation relation. The density, length, and lifetime of the plateaus can be tailored to a high degree by choosing the starting foam density accurately.

In the subrelativistic intensity domain the probability to observe stable or slowly evolving and decaying plateaus is highest during the decaying phase of the laser pulse. There the divergent flow in the critical region takes advantage from the reducing ponderomotive pressure and the high thermal conduction. Both favour the formation of supersonic flow patterns at critical density. The experimental proof is still missing.

3.4 Summary

The hydrodynamic model of the plasma is capable of describing basic features of ionized matter. One of its prominent signatures is the dominance of *collective effects* in the plasma and in hot matter. Under the action of an intense laser beam they arise in the electron fluid from local macroscopic charge and current density perturbations on the *fast time scale*. The quasineutrality of the plasma favours the excitation of periodic features in the electrons which in turn give rise to *secular forces* by cycle averaging over the fast oscillations. These "zero frequency" or *ponderomotive forces* couple to the ionic and neutral fluid components and leave their characteristic imprints in matter by modifying the fluid dynamics as a whole. Density profile steepening and shock waves are prominent examples of such footprints of the high power laser in matter. The aim of the chapter is the formulation of the basic laws governing the fluid dynamic aspect of the laser-hot matter interplay.

Starting from a microscopically large and macroscopically small volume element dV or ΔV *impressed forces*, e.g., $q\mathbf{E}$, and forces acting on its surface, e.g., pressure p, viscosity μ, momentum conservation equations are derived for the electron and the ion fluid components. The surface forces are a peculiarity of the fluid model; on a microscopic level they do not exist. An analogous feature is encountered in the energy conservation. To describe it properly, besides the kinetic energy $(\rho\mathbf{u}^2/2)dV$ of the local mass center the internal energy $\epsilon_{in}dV$ as the sum of all microscopic energies relative to the collective motion $\mathbf{u}$ has to be considered. In local thermal equilibrium (LTE) $\epsilon_{in}dV$ is the *heat* content of dV. If divided by dV the quantities result as *force densities* and *energy densities*.

The conservation equations can be expressed as depending on space and time $(\mathbf{x}, t)$; this is the so called *Eulerian representation*. If one starts from Newtonian Mechanics of discrete indices of the single particles, in the fluid continuum it is evident that the indices have to be substituted by a continuous index as, for example, the initial position coordinate $\mathbf{a} = \mathbf{x}(t = t_0)$, or a *mass coordinate* $\rho_0\,\mathbf{a},\ \rho_0 = \rho(\mathbf{a}, t_0)$. This is the *Lagrangian representation*. The *Jacobian* $J(\mathbf{a}; \mathbf{x}(t))$ assumes the role of particle and mass conservation. In expansion problems it may be advantageous to make use of the Lagrangian representation because there is no need to define the initial distributions of the dynamical quantities in the whole domain reached finally by the fluid.

Conservation equations may be formulated in a spare version, i.e., for example referring to the internal energy only in the conservation of energy, see (3.22). Alternatively there is a systematic scheme suggested by particle conservation: The change in time dt of a physical quantity in dV equals its influx through the surface $d\Sigma$ during dt. After conversion of the surface integral into a volume integral the conservation assumes the structure of

$$\frac{\partial}{\partial t}\rho + \operatorname{div} F = 0.$$

If ρ is a scalar quantity, e.g., particle density n, energy density ϵ, F is a vector, e.g., $n\mathbf{u}$, $(\epsilon + p)\mathbf{u}$, p pressure. If ρ is a vector, F is a second rank tensor. If there are sources of ρ the zero on the RHS is replaced by a source term λ. This is the *standard* or *conservative form* of the conservation equations. Use of it may be advantageous in the relativistic domain and in the steady state case. In the latter $\partial\rho/\partial t = 0$ and one is left with $\operatorname{div} F = 0$ or $\operatorname{div} F = \lambda$, respectively. For an example see the *Rankine–Hugoniot relations*.

The phenomenological derivation of the fluid equations is based on the assumption that the fluid velocity $\mathbf{u}$ is identical with the center of mass motion of the fluid element dV. With decreasing density the mean free path of the single particle may result much larger than the linear dimension of dV. This is the case of the *collisionless fluid* in which the life time of a volume element, marked by a certain number of particles, is very short and the question arises on how the conservation equations have to read correctly. The problem is solved by the *moment equations* of the *Vlasov equation* once the single particle distribution function $f(\mathbf{x}, \mathbf{v}, t)$ is known. Particle density n, mean velocity $\mathbf{u}$, energy density ϵ, and pressure tensor $P = p_{ij}$ are defined through the first three moments of $f(\mathbf{x}, \mathbf{v}, t)$ and interlinked through the Vlasov equation. The resulting conservation equations assume the same structure as the macroscopic fluid equations. The difference now is that the dynamic quantities, pressure for instance, (i) result from kinetic definitions and (ii) form an infinite set of equations where the mth moment is linked to the unknown next higher moment of order $m + 1$. In order to stop with energy conservation a *closure relation* for the *heat flow density* $\mathbf{q}$ must be introduced, for example by the Fourier assumption $\mathbf{q} \sim -\nabla T$. Modifications of the conservation relations in a rarefied fluid are obtained from the *Boltzmann binary collision integral* as a source term of the Vlasov equation. Collisions lead to an increase of entropy.

The fluid equations are nonlinear and do not admit, except very special cases, analytic solutions. However, if reduced to the leading terms, they may show *similarities*. A straightforward method to find some of them is the *dimensional analysis*, formalized in the *Buckingham theorem*. A *similarity* exists if by the theorem the number of variables entering the set of conservation equations is reduced by at least one variable. The problem is *selfsimilar* if the number of variables reduces to one single variable. Examples of selfsimilarity are the adiabatic and isothermal rarefaction waves, the nonlinear heat and radiation waves in the fully ionized homogeneous plasma, the adiabatic plane compression wave and the *Caruso model* of laser generated plasma by long pulses. Simple, non similar applications of great relevance accessible to analytic treatments are shock generation and density profile steepening by laser. Steady state onedimensional hydrodynamic shocks are governed by the *Rankine–Hugoniot relations*. The transition from the undisturbed matter to its shocked state is governed by the *Hugoniot shock adiabate*; it is of general fluid dynamic validity. Laser generated shock waves by matter ablation are a powerful tool to study equations of state in the multi-Megabar-Gigabar pressure domain not accessible by the diamond pressure cell. Ablation pressure formulas are derived under various conditions in steady state flows.

3.5 Problems

▹ From (3.6) follows $\nabla \times \mathbf{u} = 0$. In a fluid rotating around a fixed axis with angular velocity $\boldsymbol{\omega}$ holds $\mathbf{u} = \boldsymbol{\omega} \times \mathbf{r}$ and $\nabla \times \mathbf{u} = 2\boldsymbol{\omega}$. How is this possible?
▹ How does volume viscosity arise between hard spheres?
▹ Derive (3.19)–(3.21) from Euler's equation (3.17).
▹ Why does μ' of the volume viscosity differ in general from the shear viscosity coefficient μ? Model an example. *Hint.* Cohesive forces may make the difference.
▹ Show that the torque induced by f_x on the volume element $\mathrm{d}V$ in Fig. 3.2 must be balanced by the torque of f_y since, in the case of imbalance, the angular acceleration $\dot{\omega}_z$ results proportional to $(\mathrm{d}V)^{-2/3}$; it becomes arbitrarily large for $\mathrm{d}V \to 0$.

Symmetry of P guarantees angular momentum balance to the single infinitesimal volume element dV. *Hint:* Use the equation of rotational motion in terms of angular velocity and moment of inertia.

▹ Show that (3.10) is relativistically correct. Give an explicit relativistic expression for $\mathbf{j}$.

▹ Linear screening. (a) An infinite plane carrying the uniform surface charge density σ is immersed in a homogeneous isothermal ideal plasma. Calculate $n(x)$ and the Debye screening length λ_D under the assumption $e\Phi \ll k_BT$. *Hint:* Use Poisson's equation. (b) Replace the charged plane by an ion of charge q and determine Debye screening length and Debye potential Φ_D under the assumption that the potential is weak. *Answer.* $\Phi_D(r) = \frac{q}{4\pi\varepsilon_0 r}\exp -\frac{r}{\lambda_D}$; $\lambda_D = \left(\frac{\varepsilon_0 k_B T}{n_e e^2}\right)^{1/2}$. (c) Find a criterion for the validity of assumption $e\Phi \ll k_BT$. *Hint:* Estimate the number of particles with $e\Phi \geq k_BT$ contributing to screening?

▹ Show $\gamma_{sr} = (f+1)/f$ for photons; $f = 3$ Why does polarization not count?

▹ Justify the corrections of π_f by μ/m_e and μ/m_i in (3.27), (3.28).

▹ Two heat baths of temperatures $T_2 > T_1$ are connected and exchange the amount of heat Q. (i) Calculate the entropy increase. (2) How does the entropy change if Q is given back again to the hotter heath bath? (3) Calculate the entropy increase by thermalization of two equal massive blocks of temperatures T_1 and T_2 brought into contact with each other.

▹ Calculate the ratio $|\mathbf{q}_i|/|\mathbf{q}_e|$ of the heat fluxes in a thermal plasma.

▹ Why is the work done by the electric field absent in (3.32) and (3.33)? *Answer:* Macroscopic fields $\mathbf{E}$, $\mathbf{B}$, depending only on one space variable $\mathbf{x}$, can only accelerate or decelerate a fluid element. Viscosity, friction, and heat exchange originate from collisions, i.e., from microfields depending on sets of at least *pairs* of variables $(\mathbf{x}_1, \mathbf{x}_2)$ in space.

▹ (a) Derive approximate expressions of $\dot{h}_e$ and $\dot{h}_i$ due to the friction of the electrons with the ions in a hf electric field for (3.32), (3.33), and (3.35). (b) Relate friction to the laser absorption coefficient α. *Hint for* (b): You may use Drude's model.

▹ Complete (3.40) by the next order terms contributing to frequency ω.

▹ Deduce the exact expression for $\rho d\mathbf{u}/dt$ from definitions (3.36) and show

$$\dot{\rho} + \nabla\rho\mathbf{u} = 0, \qquad \rho\frac{d\mathbf{u}}{dt} \simeq \rho_e\frac{d\mathbf{u}_e}{dt} + \rho_i\frac{d\mathbf{u}_i}{dt} \simeq \rho_i\frac{d\mathbf{u}_i}{dt}.$$

▹ Cast the two-fluid momentum and energy equations (3.31)–(3.33) into conservative form.

▹ Derive the energy conservation (3.22) from an arbitrarily moving volume V_t. *Hint:* Do not forget to include the kinetic energy density. How does it vanish in the final result of (3.22)?

▹ Think Lagrangian. For this purpose derive mass, momentum and energy conservation in 1D plane and 1D spherical symmetry.

▹ Vlasov equation: Why can the canonical momentum $\mathbf{p}_{can} = \mathbf{p} + q\mathbf{A}$ be replaced by the mechanical momentum $\mathbf{p} = \gamma m\mathbf{v}$ while Liouville's theorem holds in $\{(\mathbf{x}, \mathbf{p}_{can})\}$ phase space? *Hint:* You find the answer under the proof of Liouville's theorem.

▹ Vlasov kinetics: What is the probability p to find two free particles of velocities $\mathbf{v}_1$, $\mathbf{v}_2$ in position $\mathbf{x}$ at the time t? *Answer:* $p = f(\mathbf{x}, \mathbf{v}_1, t) f(\mathbf{x}, \mathbf{v}_2, t)$ because they are independent of each other.
▹ Define the 6-dimensional velocity vector $V = (\mathbf{v}, \mathbf{a})$, $\mathbf{a}$ vector of acceleration. Show that from Liouville's theorem follows

$$\frac{\partial f}{\partial t} + \mathrm{div}(fV) = 0$$

and that this is identical with the Vlasov equation (3.79).
▹ Boltzmann equation: What is the probability p that one particle of velocity $\mathbf{v}$ at $(\mathbf{x}, t)$ collides with any other particle at the same position $\mathbf{x}$?
Answer: $p = f(\mathbf{x}, \mathbf{v}, t) \int f(\mathbf{x}, \mathbf{v}', t)\sigma_\Omega \mathrm{d}\Omega \mathrm{d}\mathbf{v}'$.
▹ What is the previous probability per *unit time*?
Answer: $p = f(\mathbf{x}, \mathbf{v}, t) \int f(\mathbf{x}, \mathbf{v}', t)\sigma_\Omega \mathrm{d}\Omega |\mathbf{w}| \mathrm{d}\mathbf{v}'$.
▹ (a) Repeat the calculation of ν with the screened Coulomb cross section from the Boltzmann collision term (3.83) in (3.93). (b) Show that the differential collision cross section for momentum transfer is $(1 - \cos\theta)\sigma_\theta$. *Hint (b):* See (3.93) for hard spheres.
▹ Boltzmann equation: Does an equation of type (3.81) hold for dense fluids, too? *Answer:* No. Reason? However, is often used after appropriate modifications.
▹ Show that the Boltzmann collision term for momentum exchange between hard spheres in a gas mixture of particle masses m_1, m_2 results in $-\nu_{12}(\mathbf{u}_1 - \mathbf{u}_2)$ for the m_1 particles.
▹ Prove $J(\mathbf{v}', \mathbf{v}_1'|\mathbf{v}, \mathbf{v}_1) = 1$; J is the Jacobian.
Answer: $\mathrm{d}\mathbf{v}'\mathrm{d}\mathbf{v}_1' = J(\mathbf{v}'\mathbf{v}_1'|\mathbf{v}\mathbf{v}_1)\mathrm{d}\mathbf{v}\mathrm{d}\mathbf{v}_1 = \mathrm{d}\mathbf{v}\mathrm{d}\mathbf{v}_1$ because

$$J = \left|\frac{\partial \mathbf{v}' \partial \mathbf{v}_1'}{\partial \mathbf{v} \partial \mathbf{v}_1}\right| = -\begin{vmatrix} \mathbf{1} & \frac{\mu}{m_1}\mathbf{1} \\ \mathbf{1} & -\frac{\mu}{m_2}\mathbf{1} \end{vmatrix} = \mu\left(\frac{1}{m_1} + \frac{1}{m_2}\right) = 1$$

at $\Omega = \mathrm{const.}$
▹ Boltzmann equation: Prove that in the steady state $f(\mathbf{v}')f(\mathbf{v}_1') - f(\mathbf{v})f(\mathbf{v}_1)$ must vanish everywhere. *Hint:* Use the time reversal argument for collisions and for f to be continuous. *Alternatively:* $f(\mathbf{v}) = f(|\mathbf{v}|)$ because of isotropy. Rotate $\mathbf{v}, \mathbf{v}_1 \to \mathbf{v}', \mathbf{v}_1' \Rightarrow f(\mathbf{v})f(\mathbf{v}_1) - f(\mathbf{v}')f(\mathbf{v}_1') = f(\mathbf{v}')f(\mathbf{v}_1') - f(\mathbf{v})f(\mathbf{v}_1)$.
▹ Maxwell distribution: Verify the correctness of the steps from (3.84) to (6.18).
▹ Write down the Boltzmann integral for light particles of f_1 colliding with heavy (nearly immobile) particles of f_2. *Answer:* $\int (f_1' f_2' - f_1 f_2)\sigma_\Omega \mathrm{d}\Omega |\mathbf{w}| \mathrm{d}\mathbf{v}_2 = n_2 \int [f_1(\mathbf{v}') - f_1(\mathbf{v})]\sigma_\Omega \mathrm{d}\Omega |\mathbf{v}|$. This approximation is the *Lorentz transport model.* A gas (plasma) to which it applies is a *Lorentz gas (Lorentz plasma).*
▹ The equilibrium distribution of a plasma at rest in 1D is $f(v)$. An inertial system S' moves with velocity v_0; $v' = v - v_0, x' = x - v_0 t$. (a) How does the distribution function $f(v)$ read in S'? (b) Express the flux nu in S'. *Answer:* (a) $f'(v') = f(v' + v_0)$. (b) $nu = \int v' f(v' + v_0)\mathrm{d}v' = \int (v - v_0) f(v)\mathrm{d}v = -nv_0$; $\mathrm{d}v' = \mathrm{d}v$.
▹ Moments: Derive $p = nk_B T$ from (6.18).

▹ Maxwell distribution: Generalize (6.18) into the relativistic domain. How does the normalization function in place of C look like?
▹ Deduce (3.58) from Vlasov's equation without heating term and viscous dissipation .
▹ Field ionization: What is $\mathbf{E}$, the laser field or the laser field + binding atomic field in (3.99)? *Answer:* the laser field.
▹ Use the Drude model from Chap. 2 to obtain an explicit expression of $\dot{h} = \langle \mathbf{j}_e \mathbf{E} \rangle$.
▹ Derive (3.105) from (3.103) and (3.104).
▹ The following definition of mass center is compatible with (3.108):

$$\tilde{X} = \frac{1}{\gamma(\mathbf{v})M}\sum_{i=1}^{N} m_i\gamma_i X_i \qquad \Leftrightarrow \qquad \tilde{\mathbf{x}} = \frac{1}{\gamma(\mathbf{v})M}\sum_{i=1}^{N} m_i\gamma_i \mathbf{x}_i$$

as long as no external forces act on the particles. (a) Show the compatibility with (3.108). *Hint:* Differentiate with respect to t. (b) $\tilde{\mathbf{x}}$ depends on the choice of the reference system and M is not a minimum in general. (c) In the system of $\mathbf{x}_C$ all centers of mass $\tilde{\mathbf{x}}$ are also at rest.
▹ Show that the nonrelativistic (conventional) center of mass of the earth wanders around its geometrical center within a sphere of radius 10 m during its motion around the sun.
▹ Compatibility proof: Prove n_0 from (3.114) is the limit of (3.109) for $N/V \to \infty$ at $V = \text{const}$.
▹ Show the validity of (3.116).
▹ Derive (4.88) from the Jacobian $|\partial(\mathbf{p})/\partial(\mathbf{w})|$ by explicit calculation in an arbitrary reference system $S'(\mathbf{v})$.
▹ The non-relativistic velocity distribution is given as $\mathbf{v}^2 = (v_x - u)^2 + v_y^2 + v_z^2 = \text{const}$, $\mathbf{u} = (u, 0, 0)$. Find the corresponding relativistic decomposition $V = gU + W$.
▹ Calculate the quantities (3.122) for V from the preceding exercise and discuss the symmetries of P^{ik} and q^i.
▹ Verify that the fraction of intensity I_a absorbed by the strong shock in the target does not exceed $\sqrt{\rho/\rho_0}$. *Answer:* See text.
▹ Verify that the terms omitted in (3.148) are of second order.
▹ Multicomponent plasma. Calculate the ion sound wave of a mixture of ions of different charge and population under the assumption of common flow velocity u (no ion separation). *Hint:* Total pressure is the sum of the partial pressures.
▹ Derive the Rankine–Hugoniot relations from the standard form of the conservation equations (3.53), (3.56), (3.59).
▹ Derive the compression ratio κ in (3.166) from appropriate formulas in the text.
▹ Problem of practical relevance, e.g., in numerics: Imagine a piston starting at $v_P = 0$ moves smoothly up to $v_P = \text{const}$ after the finite time $t = t_1$ in an ideal gas. Describe the formation of a steady shock as a function of t and x, as shown in Fig. 1.9. The shock obeys the Rankine–Hugoniot relations.
▹ Verify κ_2 of the reflected shock in (3.171).

▹ Derive an expression for the ablation pressure P_a for a Z-fold ionized plasma with $T = T_e$ and $T_i = 0$.
▹ Calculate the average rarefaction factor of a hot plasma slab of thickness d after the time $\tau/2$.
▹ Show that solution (3.157) is the limiting case of the adiabatic rarefaction wave for $\gamma \to 1$.
▹ *Gas gun:* Calculate the velocity $v(t)$ of a foil of areal mass density μ.
▹ The characteristics in the isentropic compression of a spherical pellet are no longer straight lines. Sketch their shape and give an estimate from which radius on the maximum density deviates more than 10% from the plane case.
▹ Solve (3.159), plane geometry, and calculate the density distribution $\rho(x, t')$ for a time instant $0 < t' < \tau$. *Hint:* Solve first the homogeneous equation of $\dot{x}_P$.
▹ Justification of (3.177). Derive the first order correction to P_a for moderate density ratio $r = \rho_l\rho_c \simeq 5$, $Z_0 = Z_1$, and $T_i = 0$.
▹ In the nonideal fluid the potential energy density ϵ_{pot} is a significant part of ϵ_{in}. Discuss qualitatively its impact on shock velocity v_S and compression ratio κ.
▹ Assume uniform heat deposition $\dot{h}$ in the interval $|x - x_c| \leq \Delta$ at the critical point x_c of an ideal plasma of density and temperature $\rho_0 \gg \rho_c$, T_0. Calculate the 1D flow and the density scale length L_c.
▹ Consider the stationary spherical isothermal rarefaction wave (3.208). Calculate the heat flux at the sonic point $q_r(r = r_{\text{m}})$ needed to keep the rarefaction wave isothermal. *Solution:*

$$\begin{aligned} q_r &= \int_{r_{\text{m}}}^{\infty} \frac{p}{r^2}\frac{\partial}{\partial r}(r^2 u)\mathrm{d}r = \rho_{\text{m}} c_s^2 \mathrm{e}^{1/2} \int_{u=c_s}^{\infty} \exp\left(-\left(\frac{u}{c_s}\right)^2\right) u \,\mathrm{d}\exp\left(\frac{u^2}{2c_s^2}\right) \\ &= \sqrt{2\mathrm{e}}\rho_{\text{m}} c_s^3 \int_{1/\sqrt{2}}^{\infty} 2w^2 \mathrm{e}^{-w^2}\mathrm{d}w = \sqrt{2\mathrm{e}}\rho_{\text{m}} c_s^3 \left\{\frac{1}{\sqrt{2\mathrm{e}}} + \int_{1/\sqrt{2}}^{\infty} \mathrm{e}^{-w^2}\mathrm{d}w\right\} \\ &= \rho_{\text{m}} c_s^3 \left\{1 + \left(\frac{\pi \mathrm{e}}{2}\right)^{1/2}\left[1 - \operatorname{erf}\left(\frac{1}{\sqrt{2}}\right)\right]\right\} = 1.65\rho_{\text{m}} c_s^3. \end{aligned} \tag{3.220}$$

3.6 Self-assessment

- In a fluid domain $\mathcal{D}$ containing two isolated closed islands holds $\nabla \times \mathbf{u} = 0$ everywhere. Nothing is known about $\mathbf{u}$ on the islands. What is the circulation $\oint \mathbf{u}\mathrm{d}\mathbf{s}$ along an arbitrary path in $\mathcal{D}$?
- The viscosity tensor P_{ij} is symmetric and does not generate vorticity. In reality vortices (turbulence) exist. Where do they originate from?
- Under which conditions do (3.49) and (3.50) follow from (3.38)?
- LTE: (a) Which criterion decides on the applicability of LTE in the two fluid model? *Answer:* collision times $\tau_{ee}, \tau_{ii}, \tau_{ei}$ short compared with characteristic plasma evolution time T. (b) Complete the equation of ratios $\tau_{ee} : \tau_{ii} : \tau_{ei} = 1 :$? :?.

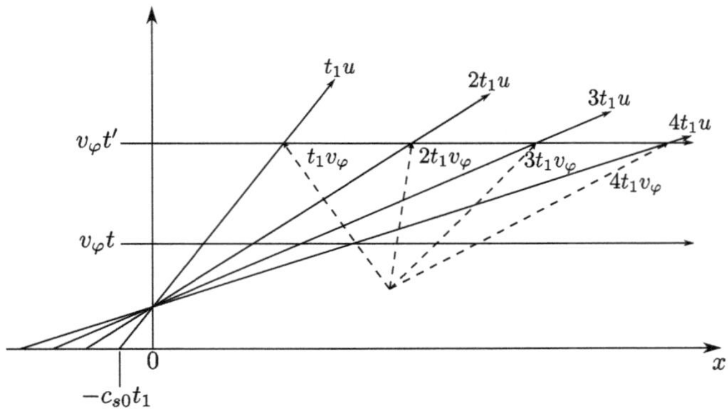

Fig. 3.16 Centered isothermal rarefaction wave: the characteristics $(u - c_{s0})t$ are straight lines. The phase velocity is a function of density; it increases with decreasing density

- LTE: In a plasma of $n_e = 10^{21}$ cm^{-3} the kinetic electron temperature is 1 keV. Is it Maxwellian and is p_e isotropic on ns, ps, fs time scale of T?
- Indicate the dependence of the viscosity coefficient μ on temperature and particle mass for $T_i = T_e$.
- Do the Eulerian and Lagrangian pictures have equivalents in another field of physics? *Answer: Schrödinger* and *Heisenberg* pictures! What is the correspondence?
- Show that the single orbit from Newton's equation is also a solution of the Vlasov equation.
- Moments: $n_0\mathbf{u} = \int \mathbf{v}f(\mathbf{v})\mathrm{d}\mathbf{v}$; $n_0 = \int f(\mathbf{v})\mathrm{d}\mathbf{v}$. If $f(\mathbf{w})$ is the thermal distribution function and you know that the fluid is streaming at velocity $\mathbf{u}$, how does the corresponding moment read? *Answer:* Set $\mathbf{v} = \mathbf{u} + \mathbf{w} \Rightarrow \mathrm{d}\mathbf{v} = \mathrm{d}\mathbf{w}$; $f(\mathbf{v}) = f(\mathbf{v} - \mathbf{u}) \Rightarrow \int \mathbf{v}f(\mathbf{v} - \mathbf{u})\mathrm{d}\mathbf{v} = \int(\mathbf{u} + \mathbf{w})f(\mathbf{w})\mathrm{d}\mathbf{w} = n_0\mathbf{u}$.
- Starting from the isothermal rarefaction wave $\rho = \rho_0 \exp -(1 + x/c_{s0}t)$ calculate the *substantial velocity* $u = u(x, t)$ and the *phase velocity* $v_\varphi = v(\rho = \text{const})$. *Solution [centered rarefaction wave]:* All $J^+ = 0$ in (3.153), $\gamma = 1$. From (3.153) or (3.156) follows $u = c_{s0} + x/t$. From $\rho =$ const follows $x/(c_{s0}t) = -\ln(\rho/\rho_0) - 1 = \quad \text{const} \quad \Rightarrow v_\varphi = \mathrm{d}x/\mathrm{d}t|_{\rho=\text{const}} - \text{const} \times c_{s0} = 0 \Rightarrow v_\varphi = \text{const} \times c_{s0} = -[\ln(\rho/\rho_0) + 1]c_{s0}$. See Fig. 3.16.
- Why does Boltzmann's equation (collisions) not change the normalization of $f(\mathbf{x}, \mathbf{v}, t)$ from Vlasov equation?
- Does Boltzmann's collision term conserve (a) the particle number? (b) Does it conserve entropy? *Answer:* (a) yes. (b) no.
- Which transport coefficients are not derivable from the Vlasov equation and which, however, follow from the Boltzmann collision term?
- Boltzmann's collision term does not apply to long range forces, e.g., bare binary Coulomb interactions. Why then the electron-ion collision frequency ν_{ei} derived from Boltzmann's equation is correct? *Hint:* Remember Debye screening from multiple collisions.

- A dynamical quantity $b(t)$, e.g., heat flow density, evolves in time according its Hamiltonian. At $t = 0$ its distribution function is $f_0(\mathbf{p}, \mathbf{q})$. If this latter evolves in time according to the Vlasov equation into $f(\mathbf{p}, \mathbf{q}, t)$ and the dynamical quantity is the time independent $b(\mathbf{p}, \mathbf{q})$ does the following equality hold?:

$$\int b(t) f_0(\mathbf{p}, \mathbf{q})\mathrm{d}\mathbf{p}\mathrm{d}\mathbf{q} = \int b(\mathbf{p}, \mathbf{q}) f(\mathbf{p}, \mathbf{q}, t)\mathrm{d}\mathbf{p}\mathrm{d}\mathbf{q}.$$

  Give an example.
- Boltzmann equation: Derive the relativistic version of Boltzmann's collision integral (3.83).
- Find the missing physics in (3.90) in comparison to (3.56). Discuss the discrepancy in the light of Fig. 3.2.
- Does the kinetic temperature T_{kin} obey the second law of thermodynamics? *Hint:* Not necessarily. Find a counterexample to the second law.
- Explain in physical terms why the non-collisional fluid is governed by equations of the same structure as those of the dense fluid. *Hint:* Start from particle density n and flow velocity $\mathbf{u}$.
- In a dense fluid the trajectory $\mathbf{x}(\mathbf{a}, t) = \int \mathbf{u}(\mathbf{a}, t)\mathrm{d}t$ of a volume element $\mathrm{d}V$, $\mathbf{a}$ Lagrange coordinate, has a clear meaning: it is the trajectory of the center of mass of $\mathrm{d}V$. In the rarefied plasma the mean free path λ is much longer than $(\mathrm{d}V)^{1/3}$ and the life time of $\mathrm{d}V$ is very short. What is the meaning of a trajectory $\mathbf{x}(\mathbf{a}, t)$ here? *Answer:* At each instant of time t the tangent to the instantaneous flow velocity $\mathbf{u}(t)$ represents the acceleration and results proportional to the sum of the instantaneously acting forces on the volume element. This is equivalent to assume that in the representative time interval Δt each escaping particle of momentum $\mathbf{p}$ is substituted by a particle of the same momentum $\mathbf{p}$ entering the volume element, or in turn $\Delta n/n = 1/\sqrt{n} \ll 1$. Thereby $n = N\mathrm{d}V/V$ with N the number of particles in V supposed reasonably homogeneous.
- Give a physical interpretation of the terms $\gamma^2 p/c^2$ and $\mathbf{u}\,\partial_t p$ in (3.103), (3.104).
- Where is the term $\rho\mathbf{u}^2/2$ from (3.58) hidden in (3.105)? Derive an explicit expression for ρ_0 of (3.102) to include internal and kinetic energy ϵ_{in} and ϵ_{kin}.
- To what extent does the formula for the confinement time also apply to a spherical plasma cloud?
- Riemann invariants. How is the connection of the flow velocity u with the phase velocity $v_\varphi = v(\rho = \mathrm{const})$ in the plane adiabatic compression wave of Fig. 3.7?
- Calculate the heat flux density $q(x, t)$ needed to keep the temperature $T(x)$ constant in the plane isothermal rarefaction wave. *Answer:* $q(x, t) = p\, c_{s0}$
- (a) What is the maximum speed a foil of mass μ/cm^2 applied at the vacuum-gas interface can be accelerated to at a given gas temperature? Does it depend on the sort of gas and on its density? (b) Can you describe $v(t)$?
- Knowing the adiabatic coefficient γ of the material indicate the flow velocity v_1 in dependence of the shock speed v_S.
- How does the pressure p_1 in the shocked material depend on the shock speed v_S in the strong shock?

- Give a quantitative criterion of when a shock is strong.
- What is the entropy increase in a strong shock?
- Why does $\Delta u_2 = \Delta u_1$ hold in shock reflection?
- What is the limiting value of compression κ in the reflected shock of the ideal monoatomic gas? Is the reflected shock a strong shock? *Answer:* no. What is the reason?
- Indicate the power law on laser intensity and cut off density in ablative pressure generation.
- Transcribe (3.187), (3.188), (3.219) from 1D plane geometry to 1D spherical geometry and indicate their limits of validity.

3.7 Glossary

Conservation laws

$$\frac{\partial n}{\partial t} + \nabla(n\mathbf{u}) = \lambda, \qquad \frac{\partial \rho}{\partial t} + \nabla(\rho\mathbf{u}) = \mu. \tag{3.3}$$

Navier–Stokes equation

$$\rho\Big[\frac{\partial}{\partial t} + (\mathbf{u}\nabla)\Big]\mathbf{u} = -\nabla p - \mu\Delta\mathbf{u} - \mu\mathrm{grad}\nabla\mathbf{u} + \mathbf{f}. \tag{3.15}$$

Euler equation

$$\rho\left[\frac{\partial}{\partial t} + (\mathbf{u}\nabla)\right]\mathbf{u} = -\nabla p + \mathbf{f}. \tag{3.17}$$

Energy balance

$$\rho[\partial_t + (\mathbf{u}\nabla)]\frac{\epsilon_{\mathrm{in}}}{\rho} = -\left(p + \mu'\nabla\frac{\mathbf{u}}{3}\right)\nabla\mathbf{u} - \nabla\mathbf{q} + \dot{h}. \tag{3.22}$$

Entropy balance

$$\mathrm{d}\epsilon_{\mathrm{in}} = T\mathrm{d}\sigma + p\mathrm{d}\left(\frac{1}{n}\right) = T\mathrm{d}\sigma + \frac{f}{2}nk_B\mathrm{d}T. \tag{3.25}$$

Bernoulli compressible

$$\int \frac{1}{\rho}\nabla p\,\mathrm{d}\mathbf{s} + \frac{1}{2}\mathbf{u}^2 + \frac{1}{m}\Phi = \mathrm{const.} \tag{3.18}$$

Boltzmann law

$$n_e(\mathbf{x}) = n_0 \exp\frac{e\Phi(\mathbf{x})}{k_B T}. \tag{3.21}$$

Two-fluid model in LTE

$$\frac{\partial n_e}{\partial t} + \nabla(n_e \mathbf{u}_e) = 0, \qquad \frac{\partial n_i}{\partial t} + \nabla(n_i \mathbf{u}_i) = 0; \qquad n_e = Zn_i. \tag{3.30}$$

$$m_e n_e \frac{\mathrm{d}}{\mathrm{d}t}\mathbf{u}_e = -\nabla p_e - n_e e(\mathbf{E} + \mathbf{u}_e \times \mathbf{B}); \quad m_i n_i \frac{\mathrm{d}}{\mathrm{d}t}\mathbf{u}_i = -\nabla p_i + Zn_i e(\mathbf{E} + \mathbf{u}_i \times \mathbf{B}). \tag{3.31}$$

$$\frac{3}{2} n_e k_B \frac{\mathrm{d}}{\mathrm{d}t} T_e = -p_e \nabla \mathbf{u}_e + \nabla(\kappa_e \nabla T_e) - 3k_B n_e \nu_{ei}(T_e - T_i) + \alpha I. \tag{3.32}$$

$$\frac{3}{2} n_i k_B \frac{\mathrm{d}}{\mathrm{d}t} T_i = -p_i \nabla \mathbf{u}_i + 3k_B n_e \nu_{ei}(T_e - T_i). \tag{3.33}$$

Entropy change per particle

$$\mathrm{d}\frac{\sigma_{e,i}}{n_{e,i}} = -\frac{1}{T_{e,i}}\{\nabla \mathbf{q}_{e,i} \pm \nu_{ei} c_{ve}(T_e - T_i) - \dot{h}_{e,i}\}\mathrm{d}t. \tag{3.35}$$

Generalized Ohm's law

$$\frac{1}{e^2 n_e}\mu \frac{\mathrm{d}_2 \mathbf{j}}{\mathrm{d}t} + \frac{1}{e^2 n_e}\mu \nu_{ei} = \mathbf{E} + \mathbf{u} \times \mathbf{B} + \frac{1}{e\rho}\left[(m_e - m_i)\mathbf{j} \times \mathbf{B} + n_e m_i \nabla p_e - n_e m_e \nabla p_i\right] \tag{3.38}$$

Slow motion

$$\mathbf{E}_0 = \frac{\mathbf{j}_0}{\sigma} - \mathbf{u}_0 \times \mathbf{B}_0 - \frac{1}{e}\left(\frac{1}{n_{e0}}\nabla p_{e0} + \nabla \Phi_p\right). \tag{3.51}$$

$$\frac{\partial \mathbf{B}_0}{\partial t} = \varepsilon_0 c^2 \nabla^2 \frac{\mathbf{B}_0}{\sigma} + \nabla \times (\mathbf{u}_0 \times \mathbf{B}_0) + \frac{1}{e}\nabla \frac{1}{n_{e0}} \times \nabla p_{e0} \tag{3.52}$$

Standard balance equations

$$\frac{\partial}{\partial t}\rho \mathbf{u} + \mathrm{div}[\rho \mathbf{u}\mathbf{u} + \Pi] = \mathbf{f}; \qquad [\rho \mathbf{u}\mathbf{u} + \Pi]_{ij} = \rho u_i u_j + \Pi_{ij}. \tag{3.56}$$

$$\frac{\partial}{\partial t}\left(\epsilon_{\mathrm{in}} + \frac{\rho \mathbf{u}^2}{2}\right) + \nabla\left\{\mathbf{u}\left(\epsilon_{\mathrm{in}} + \frac{\rho \mathbf{u}^2}{2} + \Pi\right) + \mathbf{q}\right\} = \dot{h} + \mathbf{f}\mathbf{u}. \tag{3.58}$$

Collective ponderomotive force

$$\boldsymbol{\pi}_t = \left\langle \frac{\partial}{\partial t}(m_e n_{e1} \mathbf{u}_{e1}) \right\rangle. \tag{3.64}$$

Magnetic field generation

$$\frac{\partial \mathbf{B}_0}{\partial t} = \varepsilon_0 c^2 \nabla^2 \frac{\mathbf{B}_0}{\sigma} + \nabla \times (\mathbf{u}_0 \times \mathbf{B}_0) + \frac{1}{e}\nabla \frac{1}{n_{e0}} \times \nabla p_{e0} - \frac{1}{e}\nabla \times \frac{\pi_t}{n_{e0}}. \tag{3.71}$$

Vlasov equation

$$\frac{\partial f}{\partial t} + \frac{\mathbf{p}}{\gamma m}\frac{\partial f}{\partial \mathbf{x}} + q(\mathbf{E} + \mathbf{v} \times \mathbf{B})\frac{\partial f}{\partial \mathbf{p}} = 0. \tag{3.79}$$

Boltzmann equation

$$\frac{\mathrm{d}f(\mathbf{x}, \mathbf{p}, t)}{\mathrm{d}t} = R(\mathbf{p}, \mathbf{p}_1; \mathbf{p}', \mathbf{p}_1') = \int \left(f(\mathbf{v}')f(\mathbf{v}_1') - f(\mathbf{v})f(\mathbf{v}_1)\right)|\mathbf{w}|\,\sigma_\Omega\,\mathrm{d}\Omega\,\mathrm{d}\mathbf{v}_1. \tag{3.83}$$

Fluid flow velocity **u**

$$n(\mathbf{x}, t) = \int f(\mathbf{x}, \mathbf{v}, t)\mathrm{d}\mathbf{v}, \qquad \mathbf{u}(\mathbf{x}, t) = \frac{1}{n(\mathbf{x}, t)}\int \mathbf{v} f(\mathbf{x}, \mathbf{v}, t)\mathrm{d}\mathbf{v}. \tag{3.86}$$

Vlasov moments:

pressure tensor

$$P = p_{ij} = m\int (v_i - u_i)(v_j - u_j) f \mathrm{d}\mathbf{v}. \tag{3.89}$$

kinetic temperature

$$\frac{3}{2}k_B T_{\mathrm{kin}} = \int \frac{1}{2}m(\mathbf{v} - \mathbf{u})^2 f \mathrm{d}\mathbf{v}. \tag{3.91}$$

heat flow

$$\mathbf{q} = \int \frac{1}{2}m(\mathbf{v} - \mathbf{u})^2(\mathbf{v} - \mathbf{u}) f \mathrm{d}\mathbf{v}. \tag{3.92}$$

Relativistic particle conservation

$$\partial_\alpha j_n{}^\alpha = 0 \quad \Leftrightarrow \quad \frac{\partial n}{\partial t} + \nabla(n\mathbf{u}) = 0; \quad \Rightarrow \quad n = \gamma n_0. \tag{3.100}$$

Relativistic energy conservation

$$\partial_t\left(\rho c^2 + \gamma^2 p \frac{\mathbf{u}^2}{c^2}\right) + \nabla(\rho c^2 + \gamma^2 p)\mathbf{u} = 0. \tag{3.103}$$

Relativistic momentum conservation

$$\partial_t\left(\rho + \gamma^2 \frac{p}{c^2}\right)\mathbf{u} + \nabla\left\{\left(\rho + \gamma^2 \frac{p}{c^2}\right)\mathbf{u}\mathbf{u} + p\mathbf{I}\right\} = 0, \quad \mathbf{I} = (\delta^{ij}). \tag{3.104}$$

Sum of four momenta

$$P = \sum_{i=1}^{N} m_i V_i = \sum_{i=1}^{N} m_i \gamma_i \left(\mathbf{v}_i, c\right) = MV = M\gamma(v)\left(\mathbf{v}, c\right). \tag{3.108}$$

Energy-stress tensor

$$T^{\alpha\beta} = (\rho_0 + \epsilon/c^2)u^\alpha u^\beta + P^{\alpha\beta} + (u^\alpha q^\beta + u^\beta q^\alpha)/c. \tag{3.120}$$

Relativistic Maxwellian

$$f_{\mathrm{M}}(\mathbf{x}, \mathbf{p}, t) = n_0(\mathbf{x}, t)\mathcal{N} \exp\left[-\left(E_0^2 + p^2c^2\right)^{1/2}/(k_{\mathrm{B}}T)\right]. \tag{3.126}$$

Buckingham's π-theorem

$$\pi = F(\pi_1, \pi_2, \ldots, \pi_{n-r}). \tag{3.128}$$

Adiabatic slab compression

$$x_P(t) = \frac{x_0}{\gamma - 1}\left\{\gamma + 1 - 2\frac{t}{\tau} - (\gamma + 1)\left(1 - \frac{t}{\tau}\right)^{2/(\gamma+1)}\right\}. \tag{3.160}$$

Hugoniot equation

$$\epsilon_1 - \epsilon_0 = \frac{1}{2}(p_0 + p_1)\left(\frac{1}{\rho_0} - \frac{1}{\rho_1}\right). \tag{3.164}$$

Ideal gas shock

$$\kappa = \frac{(\gamma + 1)M^2}{(\gamma - 1)M^2 + 2}, \quad \frac{T_1}{T_0} = \frac{2\gamma M^2 + 1 - \gamma}{(\gamma + 1)\kappa}, \quad \Delta\sigma = k_B \ln\kappa \ln\left(\frac{T_1}{T_0}\right)^{f/2}. \tag{3.166}$$

Shock reflection

$$\kappa_2 = \frac{\rho_2}{\rho_1} = \frac{\gamma}{\gamma - 1} \Rightarrow \kappa = \frac{\rho_2}{\rho_0} = \frac{\gamma}{\gamma - 1}\frac{(\gamma + 1)M^2}{(\gamma - 1)M^2 + 2}. \tag{3.171}$$

Steady state ablation pressure

$$P_a = \rho u^2 + p - p_\pi. \tag{3.185}$$

Profile steepening

$$\frac{L_c}{\lambda} \simeq \frac{2}{3\pi}(1 - M_c)^{1/2}(1 + M_c)^{3/2} = 0.17\left(\frac{2}{\mathcal{E}_{\mathrm{m}}} - 0.63\right)^{3/2}; \ \mathcal{E}_{\mathrm{m}}^2 < 1. \tag{3.199}$$

Ablation pressure, spherical

$$P_a = \frac{2}{4.65^{2/3}} \rho_{\rm m}^{1/3} I_a^{2/3} = 0.72 \rho_{\rm m}^{1/3} I_a^{2/3}; \qquad \gamma = 1. \tag{3.214}$$

P_a from focused laser beam, $I[\mathrm{Wcm}^{-2}]$, $\rho_c[\mathrm{gcm}^{-3}]$:

$$P_a = 2\frac{(1-R)^{2/3}}{4.65^{2/3}} \rho_c^{1/3} M_c^{1/3} I^{2/3} \Rightarrow P_a[\mathrm{Mbar}] = 3.3 \times 10^{-8}(1-\mathrm{R})^{2/3}\mathrm{M}_\mathrm{c}^{1/3}\text{æ}_\mathrm{c}^{1/3}\mathrm{I}^{2/3}. \tag{3.217}$$

3.8 Further Readings

Ya.B. Zeldovich, Yu.P. Raizer, *Physics of Shock Waves and High-Temperature Hydrodynamic Phenomena* (Academic, New York, 1967). Chap. I.
G.W. Bluman, V.D. Cole, *Similarity Methods for Differential Equations* (Springer, New York, 1974).
R. Courant, K.O. Friedrichs, *Supersonic Flow and Shock Waves* (Springer, Heidelberg, 1976).
S. Chandrasekhar, *Hydrodynamic and Hydromagnetic Stability* (Oxford University Press, Oxford, 1961)

References

1. A. Zeidler, H. Schnabl, P. Mulser, Phys. Fluids **28**, 372 (1985)
2. N.C. Lee, G.K. Parks, Phys. Fluids **26**, 724 (1983)
3. G.W. Kentwell, D.A. Jones, Phys. Rep. **145**, 319 (1987)
4. P. Mulser, J. Opt. Soc. Am. B **2**, 1814 (1985)
5. Ph.L. Similon, A.N. Kaufman, D.D. Holm, Phys. Fluids **29**, 1900 (1986)
6. G. Stratham, D. ter Haar, Plasma Phys. **25**, 681 (1983)
7. M.M. Skoric, M. Kono, Phys. Fluids **31**, 418 (1988)
8. R. Balescu, *Equilibrium and Nonequilibrium Statistical Mechanics* (Wiley, New York, 1975). Sect. 2.2
9. C. Eckart, Phys. Rev. **58**, 919 (1940)
10. L.D. Landau, E.M. Lifshitz, *Fluid Mechanics* (Pergamon Press, Oxford, 1959). Chap. XV
11. L. Rayleigh, Philos. Mag. **34**, 59 (1892); Proc. R. Soc. **L XIV**, 68 (1899)
12. L. Rayleigh, Nature **95**, 66, 591, 644 (1915)
13. E. Buckingham, Phys. Rev. **4**, 345 (1914)
14. L.I. Sedov, *Similarity and Dimensional Methods* (Academic, New York, 1959)
15. J.C. Gibbings, *Dimensional Analysis* (Springer, Heidelberg, 2011), 297 pp
16. C. De Izarra, J. Caillard, O. Valle, Mod. Phys. Lett. B **16**, 69 (2002)
17. Z. Zheng et al., Optoelectron. Lett. **3**, 394 (2007)
18. M. Murakami, M. Tanaka, Phys. Plasmas **15**, 082702 (2008)
19. M. Murakami, K. Mima, Phys. Plasmas **16**, 103108 (2009)
20. Ya.B. Zeldovich, Yu.P. Raizer, *Physics of Shock Waves and High-Temperature Hydrodynamic Phenomena* (Academic, New York, 1967)
21. H.E. Huntley, *Dimensional Analysis* (McDonald, London, 1953)

22. P.W. Bridgman, *Dimensional Analysis*, Revised edn. (Yale University Press, New Haven, 1931)
23. W.J. Duncan, *Physical Similarity and Dimensional Analysis* (Arnold, London, 1953)
24. G. Birkhoff, *Hydrodynamics; A Study in Logic, Fact and Similitude* (Princeton University Press, Princeton, 1960)
25. H.L. Langhaar, *Dimensional Analysis and Theory of Models* (Wiley, New York, 1964)
26. H. Görtler, *Dimensionsanalyse* (Springer, New York, 1975) (in German)
27. H.J. Spurk, *Dimensionsanalyse in der Strömungslehre* (Springer, Berlin, 1992) (in German)
28. G.I. Barenblatt, *Similarity, Self-similarity, and Intermediate Asymptotics* (Consultants Bureau, New York, 1979)
29. R.E. Kidder, Nucl. Fusion **16**, 405 (1976)
30. G. Wilks, Dyn. Probl. Math. Phys. **26**, 151 (1983)
31. G.W. Bluman, V.D. Cole, *Similarity Methods for Differential Equations* (Springer, New York, 1974)
32. P.G. Glockner, M.C. Singh (eds.), *Symmetry, Similarity and Group Theoretical Methods in Mechanics* (The University of Calgary, Calgary, 1974)
33. N. Euler, W.H. Steeb, P. Mulser, J. Phys. Soc. Jpn. **60**, 1132 (1991)
34. F. Ceccherini, G. Cicogna, F. Pegoraro, J. Phys. A: Math. Gen. **38**, 4597 (2005)
35. A. Caruso, R. Gratton, Plasma Phys. **10**, 867 (1968)
36. P. Mulser, Z. Naturforsch. **25a**, 282 (1970)
37. R. Fabbro, C. Max, E. Fabre, Phys. Fluids **28**, 1463 (1985)
38. K. Eidmann et al., Phys. Rev. A **30**, 2568 (1984)
39. M.K. Matzen, J.S. Pearlman, Phys. Fluids **22**, 449 (1979)
40. K. Rawer, Ann. Phys. **35**, 385; **42**, 294 (1942)
41. D.T. Attwood et al., Phys. Rev. Lett. **40**, 184 (1978)
42. A. Raven, O. Willi, Phys. Rev. Lett. **43**, 278 (1979)
43. J.D. Lindl, P.K. Kaw, Phys. Fluids **14**, 371 (1971)
44. G.J. Pert, Plasma Phys. **29**, 415 (1983)
45. A.N. Drenim, *Toward Detonation Theory* (Springer, Berlin, 1999)
46. J.R. Sanmartin, J.L. Montanes, A. Barrero, Phys. Fluids **26**, 2754 (1983)
47. W.M. Manheimer, D.G. Colombant, J.H. Gardner, Phys. Fluids **26**, 2755 (1983)
48. R.F. Schmalz, K. Eidmann, Phys. Fluids **29**, 3483 (1986)
49. R.F. Schmalz, Phys. Fluids **29**, 1389 (1986)
50. M.H. Key et al., Phys. Fluids **29**, 2011 (1983)
51. F. Dahmani, J. Appl. Phys. **74**, 622 (1993); Phys. Fluids B **4**, 1585 (1992)
52. A.G.M. Maaswinkel, K. Eidmann, R. Sigel, S. Witkowski, Opt. Commun. **51**, 255 (1984)
53. B.K. Godwal, T.S. Shirsat, H.C. Pant, J. Appl. Phys. **65**, 4608 (1989)
54. B. Ahlborn, M.H. Key, A.R. Bell, Phys. Fluids **25**, 541 (1982)
55. D. Batani et al., Phys. Rev. E **68**, 067403 (2003)
56. P. Mora, Phys. Fluids **25**, 1051 (1982)
57. R.G. Evans, A.R. Bell, B.J. MacGowan, J. Phys. D: Appl. Phys. **15**, 711 (1982)
58. P. Mulser, S. Hain, F. Cornolti, Nucl. Instrum. Methods Phys. Res. A **415**, 165 (1998)
59. M.M. Basko et al., Phys. Plasmas **22**, 053111 (2015)
60. D. Kurilovich et al., Phys. Plasmas **25**, 012709 (2018)
61. P. Mulser, C. van Kessel, Phys. Rev. Lett. **38**, 902 (1977)
62. J. Virmont, R. Pellat, P. Mora, Phys. Fluids **21**, 567 (1978)
63. V.D. Shapiro, V.I. Shevchenko, in *Handbook of Plasma Physics*, vol. 2, ed. by M.N. Rosenbluth, R.Z. Sagdeev (North-Holland, Amsterdam, 1984), p. 168
64. P. Mulser, S.M. Weng, Phys. Plasmas **17**, 102707 (2010)
65. A.J. Kemp, L. Divol, Phys. Rev. Lett. **109**, 195005 (2012)

Chapter 4
Hot Matter in Thermal Equilibrium

Introduction

The model of matter as a fluid provided us with conservation laws of mass, momentum, and energy. They hold for all kinds of fluids independently of their shape, their density, and their chemical composition. We have applied them successfully to describe neutral fluids like water or dilute gases, or plasma. The constituents of the plasma are electrically charged and couple strongly to electric fields and are, as a consequence, largely dominated by collective effects. For such phenomena modelling the plasma as a fluid turns out to be very appropriate.

In the description of matter as a fluid the concept of *internal energy* has to be introduced to fulfill energy conservation. A fluid macroscopically at rest in all its parts, flow velocity $\mathbf{u} = 0$, in special cases is capable of absorbing energy and yet to remain at rest in all its parts. In a generic process of energy absorption nothing can be said about the partition of the amount of supplied energy among the particles, molucules, ions, electrons. In concomitance with fast absorption groups of particles may be accelerated to high energies, like fast electron production by laser. Alternatively, energy absorption may also solely lead to an increase of temperature in the Maxwellian velocity distribution of the fluid particles or, in contrast, may lead to heating of the plasma ions only, as for example in the strong shock. Heating of the electrons separately is an additional important variance; it is the case in collisional laser beam absorption. Subsequently on longer time scale the ions may share internal energy with the electrons to an extent that will strongly depend on the confinement time of the plasma. In yet another situation the internal energy increment may preferentially lead to fast dissociation and ionization.

Considerations like these stimulate the question whether, under limiting conditions, there exists an additional quantity that permits to determine the evolution of the fluid composition and its dynamics. In this context possible chemical reactions, degrees of ionization, partition of internal energy between electrons and ions may be at the center of interest. The absorption and emission of radiation or the transport of energy in general are other relevant processes that may perhaps be governed by common rules.

P. Mulser, *Hot Matter from High-Power Lasers*, Graduate Texts in Physics,
https://doi.org/10.1007/978-3-662-61181-4_4

The *entropy* is the quantity we look for. It is a general law of nature that any isolated material system comes to rest in all its macroscopic parameters y_i after due time, i.e., $\mathrm{d}y_i/\mathrm{d}t \to 0$. In particular, $\mathrm{d}\mathbf{u}/\mathrm{d}t \to 0$ in all its parts. Once this condition is reached we say the system is in *perfect thermal equilibrium*. As a consequence, the internal energy $\mathcal{E}_{\mathrm{in}}$ becomes a unique function of the macroscopic parameters y_i characterizing the system, as for example its volume V, particle number N, temperature T, electric field $\mathbf{E}$, and perhaps additional parameters. This defines the *state variables* of the system. In thermal equilibrium the internal energy $\mathcal{E}_{\mathrm{in}} = \mathcal{E}_{\mathrm{in}}(n, \rho, T, \mathbf{E}, ...)$ is a state variable. A given value a state variable assumes is entirely independent of how and under which constraints it has been reached.

In this chapter it will be shown that the entropy S is a state variable and that it governs all processes in thermal equilibrium. The science of entropy is the contents of *Thermodynamics*. Its strength is that processes in thermodynamic equilibrium can be calculated without going through the detailed history of evolution to equilibrium. The theory of thermal systems is very general and finds application from extremely dilute systems like the intergalactic gas of 100 particles per cm^3 to extremely dense nuclear matter and quark-gluon plasma up to black holes.

There are two roots to thermodynamics, the *phenomenological* and the *statistical approach*. The statistical approach is more powerful, provides microscopic insight and allows the absolute determination of entropy and estimates of how fast equilibrium will be reached. The phenomenological method has its own beauty and strength, in particular because of its vicinity to intuition, in particular in cases where a rigorous statistical treatment fails owing to the complexity of calculation. After all, the statistical definitions of thermodynamic quantities like temperature, specific heat, thermodynamic potentials, are inspired from their phenomenological meaning and tested with it.

The basic laws of thermal equilibrium are exposed in the homogeneous medium. Along this way a *thermostatics* is developed which soon turns out to be of very restricted value. Fortunately macroscopic systems exhibit the fundamental property to assume thermal equilibrium in the restricted domain of typically the order of a mean free path on a short time scale impressed by the electromagnetic interaction of the microscopic constituents. This fact allows the application of thermal concepts like temperature, internal energy, entropy, to apply locally and to describe the dynamics in the large, i.e. extended regions and large time scales, by transport equations, like heat conduction, radiation transport or gasdynamic processes. The transition to *local thermal equilibrium* (LTE) makes of thermostatics a powerful and versatile dynamics, the *thermodynamics* of inhomogeneous systems.

Starting from the three fundamental laws their statistical interpretation is developed. The thermodynamic potentials are introduced and explicitly calculated for the ideal Bose, Fermi, and Boltzmann gas. Applications of them to black body radiation, chemical equilibrium in dilute systems and ionization equilibrium by means of the Saha equation are to be presented. The last section is devoted to warm dense and hot matter, with detailed discussion of the Thomas–Fermi model for ionized matter. The first deviations from ideality of neutral matter with increasing density are studied in connection with the semiempirical van der Waals equation of state.

4.1 Phenomenological Approach to Entropy

4.1.1 The Fundamental Laws of Thermodynamics

In the following sections, if not stated differently, finite systems are considered that are homogeneous and at rest in all their parts, in particular flow velocity $\mathbf{u}(\mathbf{x}) = 0$. When two systems are brought into contact with each other and at the same time as a whole they are isolated from the rest of the world all parameters y_i will assume new stationary values. The compound system has reached thermal equilibrium. Experience has shown that four fundamental laws are sufficient to arrive at the concept of entropy as a state variable and to deduce its basic properties.

Zeroth Law of Thermodynamics

Consider three systems A, B, C. The zeroth law is generally formulated in this way:

If A is in equilibrium with B and A is in equilibrium with C, then B is in equilibrium with C.

In the case A, B, C consist of the same composition it can be given a less formal meaning because it states that the internal energy per unit volume $\epsilon_{\text{in}} = \mathcal{E}_{\text{in}}/V$ (or unit mass, respectively) does not depend on the shape of the system, or equivalently, the surface energy is negligibly small compared to its volume energy. To see this it is sufficient to think system A being subdivided into pieces that are in contact with each other. Mathematically the zeroth law obeys the requirements of an equivalence, namely of identity, symmetry, and transitivity.

First Law of Thermodynamics

Frequently this law is considered as the proof for the existence of an internal energy $\mathcal{E}_{\text{in}}$. A fluid far from thermal equilibrium may exhibit a well defined internal energy, for instance a dilute gas with an arbitrary distribution function $f(\mathbf{x}, \mathbf{v}, t)$, see the preceding chapter. In reality the law tells much more. We know that in thermal equilibrium the distribution function f is a Maxwellian in the specific example; it is a unique function of temperature and number of particles N only. The power of thermodynamics consists in the limitation imposed by the equilibrium. We define the first law as follows:

> In thermal equilibrium the internal energy $\mathcal{E}_{\text{in}}$ becomes a unique function of a few macroscopic parameters y_i, $\mathcal{E}_{\text{in}} = \mathcal{E}_{\text{in}}(y_i)$.

The law allows the definition of *heat* Q supplied to a system, and of *warm* and *cold*, by definition identified also with *high temperature* and *low temperature*. The existence of the quantity *entropy* as a state variable can be deduced from the property that free flow of heat occurs from hot to cold, but not the opposite direction, see Clausius'

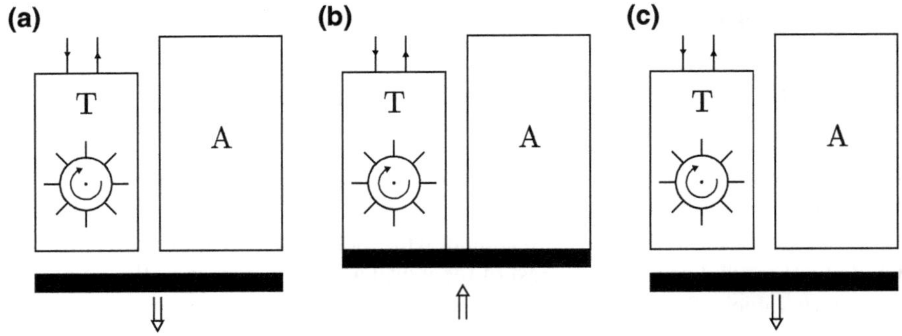

Fig. 4.1 Joule's wheel $\mathcal{T}$ in action to define "cold" and "warm": **a** Test system $\mathcal{T}$ and system A are each in its own equilibrium. **b** They are brought into thermal contact until a new equilibrium is built up. **c** $\mathcal{T}$ is separated from A and the turbine is activated to bring $\mathcal{T}$ into the state it was in **a**, if possible

assertion below. In order to give a meaning to his principle a definition of "cold" and "warm" is needed first.

To this aim we chose a test system $\mathcal{T}$ consisting of a fluid in which an electrically driven turbine is immersed, see Fig. 4.1. $\mathcal{T}$ is in thermal equilibrium. We bring it in contact with a system A in an isolated environment. After some time $\mathcal{T}$ and A will be arbitrarily close to a new equilibrium. We say they have assumed equal temperature. Then we separate the systems and activate the turbine of $\mathcal{T}$ and measure the electrical energy supplied to the fluid during the equilibration process. Two possibilities can happen:

(1) By the supply of the amount of electrical energy $Q > 0$ system $\mathcal{T}$ has reached the state it had before getting into contact with A. By the first law Q is the difference of internal energy $\Delta\mathcal{E}_{\text{in}}$ transferred from $\mathcal{T}$ to A and is defined as *heat*. The kind of contact of $\mathcal{T}$ with A which happened without any supply of energy from outside is defined as *thermal contact*. The positive transfer of heat Q is expressed by the statement that, before contact, A was at lower temperature than the test system $\mathcal{T}$ or, equivalently, A was colder than the test system $\mathcal{T}$.

(2) The original state of $\mathcal{T}$ cannot be established by electrical energy supply of any positive amount Q to it. Then we say that A was at higher temperature than $\mathcal{T}$ before thermal contact.

Heat exchange by thermal contact of the system A with the test system $\mathcal{T}$ is possible also without temperature change, i.e., with the two systems in thermal equilibrium all the time by feeding the heat δQ given to A as electrical energy $\mathrm{d}W$ in each moment to $\mathcal{T}$. This is heat transfer at constant temperature; we say system $\mathcal{T}$ works in the *heat reservoir* mode. Test system $\mathcal{T}$ can work as a heat reservoir also without activating the turbine. The internal energy of a homogeneous system $\mathscr{R}$ is proportional to its mass. For $M \to \infty$ it follows that $\mathcal{E}_{\text{in}}(y_i)$ does not change in any of its parameters y_i if a finite amount of Q is detracted from it and transferred to A. The massive $\mathscr{R}$ is a heat reservoir.

As a result, the internal energy $\mathcal{E}_{\text{in}}$ of a system A of a given amount of mass M_A and a fixed number of particles N_A can be changed by supplying an amount of heat Q to it and by doing work W on it of any kind, mechanical, electrical, gravitational. The work W can be supplied from both sides, from inside and from outside. Thus, the first law can be given the structure

$$\mathrm{d}\mathcal{E}_{\text{in}} = \delta Q + \mathrm{d}W. \tag{4.1}$$

We recall that volume forces do not affect the internal energy. The most frequently encountered kind of work on a fluid is mediated by the fluid pressure tensor Π. For simplicity we choose it isotropic and viscosity-free as the pressure p. Then, the work is expressed as $\mathrm{d}W = -p\mathrm{d}V$ and the internal energy of mass M and the internal energy per particle read

$$\mathrm{d}\mathcal{E}_{\text{in}} = \delta Q - p\mathrm{d}V, \qquad \mathrm{d}\epsilon_{\text{in}} = \mathrm{d}q + p\frac{\mathrm{d}n}{n^2}; \qquad q = \frac{Q}{N}, \quad n = \frac{N}{V}. \tag{4.2}$$

Heat Q is *not* a state variable; it depends on the heat transfer process. Proof by example: (i) Let He in the Gay-Lussac experiment (see Fig. 3.6) undergo a sudden expansion by twice the original volume. The result is $Q = 0$ because $\mathcal{E}_{\text{He}}$ is (almost) independent of V at room temperature and $\Delta W = 0$. (ii) Bring the He gas in thermal contact with a very massive test system $\mathcal{T}$ and let it expand adiabatically from V to $2V$. From $Q - \int p\mathrm{d}V = \Delta\mathcal{E}_{\text{He}} = 0$ follows $Q = Nk_BT\ln 2 > 0$. Note: If the gas is adiabatically compressed in contact with $\mathcal{T}$ the original state is reestablished without any changes in the two systems; the process of adiabatic expansion is perfectly *reversible*. To bring the system back to the original volume V in case (i) compression work must be done of the amount$Nk_BT\ln 2$. The total system is in a different state from before; the process (i) is *irreversible*. Note, here adiabatic means compression has to occur so slowly that thermal equilibrium is maintained throughout the entire volume.

Second Law of Thermodynamics

The second law may be formulated as follows:
In thermal equilibrium there exists a state function $S = S(y_i)$ of the system variables y_i, called *entropy*, with the following properties:
(a) S is an extensive quantity, i.e., S is proportional to its mass M;
(b) S is additive, i.e., if S_1, S_2 are the entropies of two systems the total entropy is the sum of both, $S = S_1 + S_2$;
(c) in a thermally isolated system, i.e., in a system that is protected against heat exchange with other systems, S cannot decrease, $\mathrm{d}S \geq 0$.
(d) The entropy of a thermally isolated system in thermal equilibrium is a maximum.

Property (d) determines the ionization equilibrium of the plasma. It is the basis of the *Saha equation*. Property (a) follows from the assumptions of homogeneity and vanishing surface energy compared to $\mathcal{E}_{\text{in}}$. Property (b) is quite natural in the case of homogeneous noninteracting systems with vanishing surface entropy.

There exist various equivalent formulations of the second law. Because of their simplicity we present two of them:

Assertion C after Clausius: *Spontaneous* transfer of heat from lower to higher temperature is impossible. "Spontaneous" thereby means that such a transition is the *only* change in the transition process.

Assertion K after *Kelvin*: No thermodynamic change of state exists with the *only* effect of extracting an amount of heat Q from a reservoir and transforming it entirely into work W.

Conversion of an amount of Q entirely into work W is possible. For instance, let a cylinder filled with gas isothermally expand by doing the work $W = -\int p\mathrm{d}V$. In this process, however, $Q \rightarrow W$ is not the only change; here, there is also the change in volume of the gas. Hence, there is no contradiction with assertion K. From each of assertions C and K the existence of entropy S follows with the properties (a)–(c) indicated above. For property (d) statistical insight is necessary.

The perfect heat engine. It consists of two heat reservoirs $\mathcal{R}_1, \mathcal{R}_2$ at temperatures $T_1 > T_2$ and an arbitrary working substance. The engine operates in three steps and only in these three steps:

(1) taking the amount of heat Q_1 from $\mathcal{R}_1$,

(2) giving off the amount of heat Q_2 to $\mathcal{R}_2$,

(3) doing the work $W = Q_1 - Q_2$.

The definition implies that the perfect heat engine is periodic and reversible. In particular, the working substance is again at temperature T_1 after step (3). *Reversible* means that a well defined process starting from a fixed state $\mathcal{S}$ can be reversed to return infinitesimally close to $\mathcal{S}$ without leaving any other changes. The arbitrarily small difference must be admitted to tell the working substance in thermal contact into which direction an amount of heat Q has to go. The perfect heat engine working between two fixed temperatures T_1, T_2 is a special type of the *Carnot engine* .

The amount of heat Q_2 is always positive. $Q_2 \leq 0$ violates assertion C. In the mode of $Q_2 > Q_1$, $W = Q_1 - Q_2 < 0$ the Carnot engine works in the cooling mode. It follows further from assertion C that there is no more efficient heat engine working between two fixed temperatures T_1, T_2 than the Carnot engine. In terms of efficiency η it means

$$\eta = \frac{W}{Q_1} = 1 - \frac{Q_2}{Q_1}. \tag{4.3}$$

The ratio Q_2/Q_1 is a dimensionless quantity. It does not depend on the working substance.

Note, so far nowhere use has been made of temperature. It has been used only as a word to characterize a heat bath as 'hotter' or 'colder'. The Carnot engine allows a universal definition of temperature and how it can be quantified.

The absolute temperature. Heat Q is internal energy. Thus, because of homogeneity of $\mathcal{E}_{\text{in}}$ it is guaranteed by the first law that the ratio of heats Q_2/Q_1 does not depend on any extensive parameter y_i like mass M, volume V, or number of particles N. On the other hand, if it depends on more than one intensive parameter they are proportional to each other and hence they are reducible to one single universal parameter by rescaling. We call it the *absolute temperature* T and define it by setting

$$\frac{T_2}{T_1} = \frac{Q_2}{Q_1} = 1 - \eta \quad \Leftrightarrow \quad \frac{Q_1}{T_1} - \frac{Q_2}{T_2} = 0. \tag{4.4}$$

T can be given in arbitrarily fine units by putting s Carnot engines in series between the heat reservoirs $\mathscr{R}_1, \mathscr{R}_2$ with $Q'_{\sigma+1} = (1-\eta)Q'_\sigma$, fixing T_1 arbitrarily positive and renaming $Q_2 \to Q_s$

$$\frac{T_{\sigma+1}}{T_\sigma} = \frac{Q'_{\sigma+1}}{Q'_\sigma} = 1 - \eta, \quad Q'_1 = Q_1, \quad Q_s = (1-\eta)Q'_{s-1}s.$$

Let Q_2 approach zero, $Q_2 = \varepsilon \to 0,\ T_s = C_V\varepsilon$, where C_V is merely a constant factor for dimensionality reasons, then the absolute temperature can be extended arbitrarily close to $T = +0$. However $T_s = 0$ is forbidden by the *third law of thermodynamics*. The absolute temperature T is always non-negative because Q_s is positive for assertion K and the Carnot engine is in the working mode, $\Delta W > 0$. By convenient choice of T_1 and the number of Carnot engines s the unit of T becomes 1 K (Kelvin). The absolute temperature defined by (4.4) is identical with T of the ideal gas law.

The state function entropy S. Consider the cyclic process from state A to B and back in such a way that the temperature T is defined at all intermediate steps. For an arbitrary process of this kind the *Clausius theorem* holds,

$$\int_A^B \frac{\delta Q}{T} + \int_B^A \frac{\delta Q}{T} = \oint \frac{\delta Q}{T} \le 0. \tag{4.5}$$

The equality sign holds if the cylce is reversible in all its steps from A to B and back.

Proof Let a Carnot engine operate between a fixed temperature $T_0 > \max T$ and the individual step temperature T. According to (4.4) holds $\delta Q/T = \delta Q_0/T_0$. Hence

$$\oint \frac{\delta Q}{T} = \frac{1}{T_0}\oint \delta Q_0 = \frac{Q_0}{T_0}.$$

If Q_0 is positive this quantity has entirely been converted into work $\Rightarrow$ contradiction to assertion K. However, no contradiction arises if Q_0 is negative. In the case of global reversibility
(a) along the same path holds $\int_B^A = -\int_A^B$ because the Carnot engine works both ways under inversion of sign, i.e. $\delta Q_{AB} = -\delta Q_{BA} \Rightarrow \oint \delta Q/T = 0$;
(b) in case of path BA different from path AB, say $(AB)'$, both reversible in themselves, leave the whole system unchanged, too $\Rightarrow \oint \delta Q/T = 0$.

This proves that the entropy S defined as

$$S(B) = S(A) + \int_A^B \frac{\delta Q_{\text{rev}}}{T} \quad \Leftrightarrow \quad \mathrm{d}S = \frac{\mathrm{d}Q_{\text{rev}}}{T} \tag{4.6}$$

is independent of how state B is reached by *reversible* supply of heat $\delta Q_{\text{rev}} \Rightarrow$ The entropy S is a *state function* or *state variable*.

S obeys the properties (a), (b), (c). The first is a consequence of negligible surface contributions, the second tells that interaction between the two systems is absent. Property (c) tells that in the thermally isolated system the entropy cannot decrease, $dS \geq 0$, because of $Q_0 = 0$ by definition.

Third Law of Thermodynamics.

The third law has been formulated first by Walther Nernst in 1906 and is referred to also as Nernst's theorem or postulate. We present it in our context merely for completeness because for our purposes it plays no role. One possible formulation of it is as follows:

In an arbitrary system it is impossible to reach the absolute zero $T = 0$ by a finite number of operations of any kind.

One consequence of the assertion for the specific heats at constant pressure p and at constant volume V is $\lim C_p \to C_V \to 0$ for $T \to 0$. Another popular formulation is:

For T approaching zero the entropy S approaches zero, too.

Without further specifications such a formulation is at least misleading. In full generality it is false or inapplicable. We come briefly back to this second formulation of the third law in the following section on thermostatistics.

Nomenclature. Sometimes it may be useful to distinguish between *specific heat* per unit mass or volume c_v and c_p and *heat capacity* for a given amount of matter C_V and C_p. Here, throughout the book, we do not follow such a distinction because it should become clear from the context which of the two quantities is addressed. We use generally c_v, c_p for both, except for clarity it is advisable to use C_V and C_p.

4.1.2 Properties and Applications of Entropy

The significance of heat and internal energy. According to (4.1) the state function of internal energy $\mathcal{E}_{\text{in}}$ is the sum of heat Q and work W,

$$\Delta\mathcal{E}_{\text{in}} = \Delta Q + \Delta W \Rightarrow \Delta Q = \Delta\mathcal{E}_{\text{in}} - \Delta W. \tag{4.7}$$

As long as ΔW is not specified the amount of heat supplied to the system is not defined. It can assume any value between zero in the adiabatic process and $\Delta\mathcal{E}_{\text{in}}$ in pure heat transfer by thermal contact. During the heat transfer process the system may even be far from equilibrium with no existing temperature. Solely after termination of the transfer $\mathcal{E}_{\text{in}}$ will assume the status of a state function again with a new well defined temperature. If the temperature of the system exists all the time along any reversible path the amount of heat ΔQ_{rev} is a well defined quantity and results according to the second law (4.6) as

$$\Delta Q_{\text{rev}} = \int T\mathrm{d}S \quad \Leftrightarrow \quad \mathcal{E}_{\text{in}} = \int T\mathrm{d}S + \int \mathrm{d}W_{\text{rev}}. \tag{4.8}$$

In summary, heat Q is an arbitrary quantity in general; only Q_{rev} has a precise meaning and is calculable from the entropy S or from the internal energy and the work reversibly done to the system. The second law allows the elimination of the concept of heat from thermodynamics. The reason why the poorly defined term heat is still in use all the time is a practical one and it has remained the same since the invention of the Carnot engine and the steam engine. Consider for instance the isothermal transfer of the amount of heat ΔQ from a heat reservoir as the only process. Then we know immediately $\Delta\mathcal{E}_{\text{in}} = \Delta Q$. It may be a difficult task to do the calculation of the internal energy change according to (4.8) because during the heat transfer process the system may expand or contract and, in case of a plasma, change its ionization degree. Entropy conserving thermodynamic processes are called *adiabatic*. They are reversible.

Heat transfer and the Clausius inequality. Consider a system that takes the internal energy $\mathcal{E}_{\text{in}}$ from a heat reservoir at temperature T_1 and transmits it to a second heat reservoir on temperature T_2. Extraction and transmission can be performed by direct thermal contact but also by a process of any kind, also arbitrarily far from any thermodynamic equilibrium. What is the entropy change of the whole system?
(**a**) Use a Carnot engine for doing the transfer of $Q = \mathcal{E}_{\text{in}}$ in a reversible way. Extraction at T_1 reduces the entropy by Q/T_1. By reversible work of the Carnot engine the temperature is brought down to T_2. Transfer of Q to the second reservoir at this temperature leads to the entropy increase Q/T_2. Hence, the very useful formula results,

$$\Delta S = Q\left(\frac{1}{T_2} - \frac{1}{T_1}\right). \tag{4.9}$$

The maximum work that Q is capable of doing between T_1 and T_2 is $W_{\text{max}} = QT_1/(T_1 - T_2)$ under transferring $Q_2 = QT_2/T_1$. By direct transfer of Q between the two temperature levels the amount $Q - Q_2$ is wasted under the aspect of energy production at the constraint of zero entropy change.

In order to extract the amount of Q from a heat bath of temperature T, according to assertion C the temperature of the system in thermal contact with the heat bath must have a little bit lower temperature $T' = T - \varepsilon$ to indicate Q in which direction to go. How does the difference ε affect the entropy? Subdivide the temperature into $N - 1$ equal levels, hence $\varepsilon = (T_1 - T_2)/N$. By indicating with Q' the heat given to level $T - \varepsilon$ and by $\mathrm{d}S'$ the entropy gain from the transfer of Q' between the two levels there holds

$$\frac{Q}{T} - \frac{Q'}{T-\varepsilon} = 0 \Rightarrow Q' = Q\left(1 - \frac{\varepsilon}{T}\right), \quad \mathrm{d}S = Q\frac{\varepsilon}{T^2}, \quad \mathrm{d}S' = Q'\frac{\varepsilon}{T^2} \Rightarrow \mathrm{d}S - \mathrm{d}S' = \frac{\varepsilon}{T}\mathrm{d}S. \tag{4.10}$$

The consequences are twice:
(α) Entropy production by direct heat transfer between two adjacent temperature levels vanishes in first order with their difference.
(β) The entropy production of N Carnot engines in series between the two fixed levels $T - 1 > T_2$ vanishes in first order of N^{-1}.

As a result of (α) together with (β) the difference of heat Q_1 absorbed from level T_1 and the heat Q_2 transferred to level T_2 shrinks to zero to first order in N^{-1}. This leads to a second method to determine ΔS in heat transfer:

(b) Once T_1 and T_2 are connected by a continuous temperature function, i.e., if in (4.5) T is defined along the whole path from A to B only equality exists; the process can be inverted as BA or $(BA)'$ with the Carnot machine. Result (4.9) immediately follows from $N \to \infty$ because the difference between Q_1 absorbed by the first Carnot engine at T_1 and Q_2 transferred by the last Carnot engine in series at T_2 vanishes with N increasing.

(c) Property (d) of the entropy to assume a maximum value compatible with the outer constraints y_i of the thermally isolated system cannot be shown within the phenomenological frame. What can be stated with certainty is that if the thermally isolated system is in a maximum of its entropy its state is stable.

(d) Suppose system 1 has got the entropy S_1 at temperature T_1 and system 2 has got the entropy S_2 at temperature $T_2 \neq T_1$. The total entropy is $S_i = S_1 + S_2$. After they have brought into thermal contact a new equilibrium is reached with S_f at T_f. The result is that the entropy has increased, $S_f > S_i$. In fact, equilibrium implies that according to Clausius' assertion C exchange of heat $Q > 0$ has spontaneously happened between the two systems with the result of (4.9).

Clausius' inequality. Imagine that some of the intermediate steps of the transition AB are leaking because of heat conduction. According to Clausius' assertion C this means additional heat transfer $\delta Q'$ from a higher temperature level T_i to its neighboring level $T_i - \delta T_i$. As a consequence $\int \delta Q/T = \int \delta Q'/T < 0$ and, by the equivalence of C and K, this special path is *irreversible*. Thereby it is tacitly assumed that the thermalization process is muck faster than the heat conduction process. Thus, in the closed circle *Clausius' inequality* reads as

$$\oint \frac{\delta Q}{T} \leq 0. \tag{4.11}$$

Thereby the equality sign refers to all partial processes being *reversible*.

The quantities $\mathrm{d}\mathcal{E}_{\mathrm{in}}$ and $\mathrm{d}S$ are exact differentials of the variables y_i. The differential of a function $F(y_i)$ may be given as

$$\mathrm{d}F = \sum_{1}^{k} A_j(y_i)\mathrm{d}y_j \tag{4.12}$$

with the A_j arbitrary continuous functions of the y_i variables. If F is a state function with continuous partial derivatives its total or exact differential exists and takes the form

$$\mathrm{d}F = \sum_{1}^{k} \frac{\partial F}{\partial y_j}\mathrm{d}y_j \quad \Leftrightarrow \quad A_j = \frac{\partial F}{\partial y_j}.$$

Suppose now that $\nabla \times \mathbf{A}$ of the vector $\mathbf{A} = \{A_j(y_i)\}$ exists and is zero then (4.12) is a total differential and F is a state function with the A_j its partial derivatives $\partial_j F$.

Let the differential work $\mathrm{d}W$ be given by $\mathrm{d}W = -p\mathrm{d}V$, then by the identification $\delta Q = T\mathrm{d}S$ in (4.1), $\mathrm{d}\mathcal{E}_{\mathrm{in}}$ = TdS - pdV becomes a total differential of S and V because the property of state function for $\mathcal{E}_{\mathrm{in}}$ is guaranteed by the first law. It entails the fulfilment of

$$\mathrm{d}\mathcal{E}_{\mathrm{in}} = T\mathrm{d}S - p\mathrm{d}V \Rightarrow \left(\frac{\partial \mathcal{E}_{\mathrm{in}}}{\partial S}\right)_V = T, \ \left(\frac{\partial \mathcal{E}_{\mathrm{in}}}{\partial V}\right)_S = -p; \ \left(\frac{\partial T}{\partial V}\right)_S = -\left(\frac{\partial p}{\partial S}\right)_V. \tag{4.13}$$

The differential form of $\mathcal{E}_{\mathrm{in}}$ shows that its natural variables are S and V, $\mathcal{E}_{\mathrm{in}} = \mathcal{E}_{\mathrm{in}}(S, V)$. With this choice T and p are fixed as the related derivatives of $\mathcal{E}_{\mathrm{in}}$. For this analogy with the forces in mechanics $\mathcal{E}_{\mathrm{in}}$ is referred to as a *thermodynamic potential*. $\mathcal{E}_{\mathrm{in}}$ may depend on further variables y_i, like particle number N, polarization $\mathbf{P}$, magnetization $\mathbf{M}$, mixing ratio of components, etc., all of them *independent from each other*. Its differential reads

$$\mathrm{d}\mathcal{E}_{\mathrm{in}} = T\mathrm{d}S - p\mathrm{d}V + \sum_j \frac{\partial \mathcal{E}_{\mathrm{in}}}{\partial y_j} dy_j. \tag{4.14}$$

Again, by analogy the partial derivatives of the thermodynamic potential $\mathcal{E}_{\mathrm{in}}$ are given the name of *generalized forces* or *thermo[dynamic] forces*. For example, $\partial \mathcal{E}_{\mathrm{in}}/\partial \mathbf{P} = -\mathbf{E}$. The entropy is determined by (4.13) as a function of $\mathcal{E}_{\mathrm{in}}$ and V, $T\mathrm{d}S = \mathrm{d}\mathcal{E}_{\mathrm{in}} + p\mathrm{d}V$. The internal energy is a strongly monotonic function of the entropy at fixed volume owing to $T > 0$ and can be uniquely inverted to obtain the state function S as a function of $\mathcal{E}_{\mathrm{in}}$, and vice versa. Its differential is

$$\mathrm{d}S = \frac{1}{T}\left[\mathrm{d}\mathcal{E}_{\mathrm{in}} + p\mathrm{d}V\right] \Rightarrow \left(\frac{\partial S}{\partial \mathcal{E}_{\mathrm{in}}}\right)_V = \frac{1}{T}, \ \left(\frac{\partial S}{\partial V}\right)_{\mathcal{E}_{\mathrm{in}}} = \frac{p}{T}. \tag{4.15}$$

$S(\mathcal{E}_{\text{in}}, V)$ is a thermodynamic potential for T and p. As a function of energy it is not easily accessible to a measurement.

Entropy from Measurable Qantities

We recall the following measurable quantities

$$\begin{array}{ll} C_V = \left(\frac{\partial \mathcal{E}_{\text{in}}}{\partial T}\right)_V = T\left(\frac{\partial S}{\partial T}\right)_V & \text{specific heat at constant volume} \\ C_p = \left(\frac{\partial \mathcal{E}_{\text{in}}}{\partial T}\right)_p + p\left(\frac{\partial V}{\partial T}\right)_p = T\left(\frac{\partial S}{\partial T}\right)_p & \text{specific heat at constant pressure} \\ \alpha = \frac{1}{V}\left(\frac{\partial V}{\partial T}\right)_p = -\frac{1}{\rho}\left(\frac{\partial \rho}{\partial T}\right)_p & \text{thermal expansion coefficient} \\ \kappa_T = -\frac{1}{V}\left(\frac{\partial V}{\partial p}\right)_T = \frac{1}{\rho}\left(\frac{\partial \rho}{\partial p}\right)_T & \text{isothermal compressibility} \\ \kappa_S = -\frac{1}{V}\left(\frac{\partial V}{\partial p}\right)_S = \frac{1}{\rho}\left(\frac{\partial \rho}{\partial p}\right)_S & \text{adiabatic compressibility} \end{array} \tag{4.16}$$

According to (4.13) or (4.15) $\mathcal{E}_{\text{in}}$ and S can be uniquely expressed either in the independent variables T and V or in the independent variables p and V, if $p > 0$ in the whole domain under consideration. Uniqueness is guaranteed by the properties of the *Legendre transformation*. In both pairs of variables dS is a total differential; the requirement implies the equalities

$$\begin{aligned} \left(\frac{\partial \mathcal{E}_{\text{in}}}{\partial V}\right)_T + p = T\left(\frac{\partial p}{\partial T}\right)_V &= -T\left(\frac{\partial p}{\partial V}\right)_T\left(\frac{\partial V}{\partial T}\right)_p = \frac{\alpha T}{\kappa_T} \\ \left(\frac{\partial \mathcal{E}_{\text{in}}}{\partial p}\right)_T + p\left(\frac{\partial V}{\partial p}\right)_T &= -T\left(\frac{\partial V}{\partial T}\right)_p = \alpha T V. \end{aligned} \tag{4.17}$$

With the help of these relations between the various partial derivatives follow the so called $T\mathrm{d}S$ *relations* in (T, V),

$$T\mathrm{d}S = \left(\frac{\partial \mathcal{E}_{\text{in}}}{\partial T}\right)_V \mathrm{d}T + \left[\left(\frac{\partial \mathcal{E}_{\text{in}}}{\partial V}\right)_T + p\right]\mathrm{d}V = C_V\mathrm{d}T + \frac{\alpha T}{\kappa_T}\mathrm{d}V \tag{4.18}$$

and in (T, p),

$$T\mathrm{d}S = \left[\left(\frac{\partial \mathcal{E}_{\text{in}}}{\partial T}\right)_p + p\left(\frac{\partial V}{\partial T}\right)_p\right]\mathrm{d}T - T\left(\frac{\partial V}{\partial T}\right)_p \mathrm{d}p = C_p\mathrm{d}T - \alpha T V\mathrm{d}p. \tag{4.19}$$

Both forms allow the determination of entropy differences from measurable quantities. From their difference

$$(C_p - C_V)\mathrm{d}T - \frac{\alpha T}{\kappa_T}\mathrm{d}V - \alpha T V\mathrm{d}p = 0$$

follows under isochoric conditions $\mathrm{d}V = 0$

$$C_p - C_V = \frac{\alpha^2 T V}{\kappa_T}. \tag{4.20}$$

In the majority of cases this is a positive difference. Only under the anomaly of $\kappa_T < 0$ it becomes negative. In such a case the system is *mechanically unstable.*

A first application of the TdS forms is obtained by setting $dS = 0$ in (4.18) and (4.19),

$$T\mathrm{d}S = 0 \;\Rightarrow\; C_V = -T\left(\frac{\partial p}{\partial T}\right)_V \left(\frac{\partial V}{\partial T}\right)_S, \quad C_p = T\left(\frac{\partial V}{\partial T}\right)_p \left(\frac{\partial p}{\partial T}\right)_S$$

$$\Rightarrow\; \frac{C_p}{C_V} = -\frac{\left(\frac{\partial p}{\partial T}\right)_V \left(\frac{\partial V}{\partial T}\right)_S}{\left(\frac{\partial V}{\partial T}\right)_p \left(\left(\frac{\partial p}{\partial T}\right)_S\right.} = \frac{\left(\frac{\partial p}{\partial V}\right)_S}{\left(\frac{\partial p}{\partial V}\right)_T} = \frac{\kappa_S}{\kappa_T}. \tag{4.21}$$

The last step towards this remarkable and yet intuitive result follows from

$$\mathrm{d}p = \left(\frac{\partial p}{\partial V}\right)_T \mathrm{d}V + \left(\frac{\partial p}{\partial T}\right)_V \mathrm{d}T.$$

by setting once $dp = 0$ and then $\mathrm{d}T = 0$. The specific heats follow now as

$$C_V = \frac{\alpha^2 \kappa_S}{\kappa_T - \kappa_S} T V, \qquad C_p = \frac{\alpha^2}{\kappa_T - \kappa_S} T V. \tag{4.22}$$

The compressibility is related to the sound speed by $c_S^2 = \kappa_S \rho$ for the adiabatic sound speed c_S and $c_T^2 = \kappa_T \rho$ for the isothermal sound speed c_T. A general adiabatic change of state is expressible in the form

$$pV^\gamma = \text{const} \qquad \Leftrightarrow \qquad p\rho^{-\gamma} = \text{const} \tag{4.23}$$

The *adiabatic exponent* γ rather than to be a constant will in general be a function of the thermodynamic state, e.g. of (S, V). Differentiation of this equation of state with respect to V leads to the definition of γ as

$$\gamma = -\frac{V}{p}\left(\frac{\partial p}{\partial V}\right)_S = \frac{1}{p\kappa_S} = \frac{\rho}{p} c_S^2. \tag{4.24}$$

From the second defining form γ may be determined experimentally. In the case of the ideal gas holds $\gamma = C_p / C_V$.

For its practical as well as theoretical relevance the *Grüneisen parameter* Γ is to be introduced here,

$$\Gamma = -\frac{V}{T}\frac{\partial^2 \mathcal{E}_{\text{in}}}{\partial V \partial S} = \frac{V}{C_V}\left(\frac{\partial p}{\partial T}\right)_V = \frac{\alpha V}{C_V \kappa_T}. \tag{4.25}$$

If Γ is a constant the last term set constant is the so called *Grüneisen equation of state*. A system is *thermally stable* if $C_V > 0$ and $\kappa_T > 0$. Then from (4.25) follows that α and Γ have the same sign. The Grüneisen parameter of the ideal gas is $\Gamma = \gamma - 1$. This is for the nonrelativistic monoatomic gas $\gamma = 5/3$, $\Gamma = 2/3$. For the extremely relativistic gas, e.g. photons, it is $\gamma = 4/3$, $\Gamma = 1/3$. Note, in order to determine the quantities $\mathcal{E}_{\text{in}}$, S, C_V with the help of the equation of state for p one of them must be known or measured. For the non relativistic monoatomic ideal gas the ratio $\mathcal{E}_{\text{in}}/p = 3/2$ is obtained from simple reasoning. With this and $p = nkT$ follows $\mathcal{E}_{\text{in}}$, C_V, and S from one of the $T\mathrm{d}S$ relations,

$$\mathcal{E}_{\text{in}} = \frac{3}{2}Nk_BT, \quad C_V = \frac{3}{2}Nk_B, \quad S = S_0 + Nk_B \ln\left(\frac{T^{3/2}V}{T_0^{3/2}V_0}\right). \tag{4.26}$$

The Stefan–Boltzmann Law

To get an additional feeling for the power of the entropy concept the *Stefan–Boltzmann law* of thermal radiation in equilibrium is derived. The photons represent the most perfect gas known so far, a fact guaranteed by the linearity of Maxwell's equations. At a given temperature T the energy is given by $\mathcal{E}_{\text{in}} = \epsilon(T)V$ as a consequence of homogeneity of the radiation field. Further, the photons are an extremely relativistic gas of rest mass zero and the single momentum $\mathbf{p}$ related to its energy by $E = c\mathbf{p}^2$, in contrast to $E = \mathbf{p}^2/2m$ for the nonrelativistic massive particle. Together with the property of isotropy of ϵ the radiation pressure is $p = \epsilon/3$. The first of the relations (4.17) requires

$$\frac{4}{3}\epsilon = \frac{1}{3}T\frac{\partial \epsilon}{\partial T} \quad \Rightarrow \quad \frac{d\epsilon}{\epsilon} = 4\frac{\mathrm{d}T}{T} \quad \Rightarrow \quad \epsilon = C_{\mathrm{B}}T^4 \quad \Rightarrow \quad \mathrm{d}I = \frac{\epsilon}{4\pi}c\,\mathrm{d}\Omega. \tag{4.27}$$

This is the Stefan–Boltzmann law of thermal radiation in equilibrium as derived by Boltzmann from the second law. The connection of the radiation field ϵ with the black body intensity I follows from its isotropy. The entropy of the radiation field is obtained from one of the $T\mathrm{d}S$ relations, e.g., from (4.18),

$$\mathrm{d}S = [4VT^2\mathrm{d}T + \frac{4}{3}T^3\mathrm{d}V]C_{\mathrm{B}} \Rightarrow S = \frac{8}{3}C_{\mathrm{B}}T^3V = \frac{8}{3}\frac{\epsilon}{T}V. \tag{4.28}$$

The constant C_B contains the Boltzmann constant k_B and the Planck constant $\hbar$ and cannot be derived in the realm of phenomenological thermodynamics. Its value is

$$C_{\mathrm{B}} = \frac{\pi^2 k_B^4}{15\hbar^3 c^3} = 7.55 \times 10^{-16}\,[\,\mathrm{Jm^{-3}K^{-4}}]. \tag{4.29}$$

4.1.3 Thermodynamic Potentials

From $\mathcal{E}_{\text{in}}$ as a function of entropy and volume, and additional parameters y_i, all other thermodynamic quantities result as partial derivatives with respect to these variables. Often, however, other combinations of variables are easier to be handled, for example temperature and volume, or pressure and volume. At the same time one would like to keep the advantage of potentials.

4.1.3.1 The Enthalpy H

It is evident that the *enthalpy* $H = \mathcal{E}_{\text{in}} + pV$ is a state variable. To see in which variables it assumes the property of a potential we form its total differential

$$\mathrm{d}H = \mathrm{d}\mathcal{E}_{\text{in}} + \mathrm{d}(pV) = T\mathrm{d}S + V\mathrm{d}p \;\Rightarrow\; T = \left(\frac{\partial H}{\partial S}\right)_p, \quad V = \left(\frac{\partial H}{\partial p}\right)_S. \tag{4.30}$$

Piecewise uniqueness of the substitution of V by p is ensured by (4.13). The expression of the specific heat under constant pressure simplifies now as $C_p = (\partial H/\partial T)_p$. H may be the appropriate quantity for processes that evolve under constant pressure. However, it still depends on the entropy and is therefore difficult to control. The passage to the easily measurable variables T and V is reached by the

4.1.3.2 The Helmholtz Free Energy F

The *Helmholtz free energy* is $F = \mathcal{E}_{\text{in}} - TS$. Its property as a very useful potential results from its differential form

$$\mathrm{d}F = \mathrm{d}\mathcal{E}_{\text{in}} - \mathrm{d}(TS) = -S\mathrm{d}T - p\mathrm{d}V \;\Rightarrow\; \left(\frac{\partial F}{\partial T}\right)_V = -S, \; \left(\frac{\partial F}{\partial V}\right)_T = -p. \tag{4.31}$$

F is the maximum energy that can be converted into reversible work in an isothermal process,

$$W = \int p\mathrm{d}V\big|_{T=\text{const}} = -\int \left(\frac{\partial F}{\partial V}\right)_T \mathrm{d}V = -\Delta F\big|_T \;\Rightarrow\; W + \Delta F = \text{const.}$$

From Clausius theorem (4.11) follows: In the thermally isolated system the free energy cannot increase. If F is a relative or absolute minimum the thermally isolated system is in a state of equilibrium. The knowledge of the free energy of a given system enables one to calculate all equilibrium parameters as derivatives of F. Analogous behavior at constant pressure is expected from the *free enthalpy* G

4.1.3.3 The Free Enthalpy or Gibbs Potential G

The free enthalpy $G = H - TS$ is the state function alternative to F. In its natural variables T and p it is a thermodynamic potential,

$$dG = dH - d(TS) = -SrmdT + Vdp \Rightarrow \left(\frac{\partial G}{\partial T}\right)_p = -S, \; \left(\frac{\partial G}{\partial p}\right)_T = V. \tag{4.32}$$

At constant pressure and constant temperature the maximum energy extracted reversibly from a system is

$$W = \int p\mathrm{d}V\big|_{p,T} = p\Delta V = -\Delta F$$

and

$$\Delta G \leq 0. \tag{4.33}$$

The last inequality follows from $\mathrm{d}G = \mathrm{d}F + \mathrm{d}(pV) \leq S\mathrm{d}T + V\mathrm{d}p = 0$ and $dG = \mathrm{d}F + p\mathrm{d}V \leq 0$.

Maxwell relations. It is well known that if the partial derivatives f_x, f_y of a function $f(x, y)$ exist, and in addition one of the mixed derivatives exists, say f_{xy}, *and* is continuous then it holds $f_{yx} = f_{xy}$. Application of the interchangeability of the second partial derivatives to the thermodynamic potentials yields the 4 *Maxwell relations*, for example,

$$\left(\frac{\partial T}{\partial V}\right)_S = -\left(\frac{\partial p}{\partial S}\right)_V \tag{4.34}$$

from (4.13), etc. The state functions $\mathcal{E}_{\mathrm{in}}$, H, F, G are the most common thermodynamic potentials. However, for particular processes different potentials may be introduced by Legendre transforms, just in analogy to the foregoing procedures.

4.2 LTE: The Local Thermodynamic Equilibrium

4.2.1 Evolution to Thermal Equilibrium

Conditions imposed by the laws of thermodynamics are very stringent and enable one to describe complex phenomena, like the degree of ionization of a plasma, the partition of energies, or the emission of thermal radiation, from the sun for instance. The general laws are applicable to a system if it is in or close to thermal equilibrium. It may take a long time to reach such conditions. For example, below 13.4 °C the white metallic tin (β tin) transforms into a gray non metallic powder (α tin); the mutation process may extend over centuries. Unstable systems may never reach equilibrium, as for example uranium U^{235}. In general, an open system under oscillating supply of energy will never reach global equilibrium. The planet earth does never evolve into a state of thermal equilibrium - unless it will be expelled from the solar system. And yet, in all cited cases experience has shown that a definite temperature can be assigned. It is an assignment dictated by the practice, so to the metals tin and uranium, as well as to each *local* region of the earth. In general the equilibration time of a system is the longer the bigger its extension and the more intense its exchange with an unsteady environment is.

Fortunately there are two characteristics which help to overcome the difficulty with the perfect and global equilibrium. They will allow an enormous extension of applicability of thermodynamic arguments. They are the limitation to (i) *imperfect thermal equilibrium* by restricting the equilibrium to the components of main interest energetically, and (ii) the restriction to *local thermodynamic equilibrium.* The criterion for the applicability of the restricted equilibrium follows from the first and second law of thermodynamics:

> A system is in thermal equilibrium if an absolute temperature T can be assigned to its specific components, and vice versa.

The Imperfect Equilibrium

Consider a rarefied fluid at a given temperature T. To be specific, let us take a hydrogen plasma of density $n = 10^{21}\mathrm{cm}^{-3}$ and temperature of 1 keV. Its thermal electron energy density is $\epsilon_{\mathrm{in}} = 3nk_BT = 4.8 \times 10^5$ Jcm^{-3}. In perfect equilibrium its radiation energy density according to (4.27) is $\epsilon_{bb} = \rho(T) = CT^4 = 1.4 \times 10^7$ $\mathrm{Jcm}^{-3} \gg \epsilon_{\mathrm{in}}$. If the plasma is not enclosed in a *hohlraum*, it is optically thin and all radiation escapes as *bremsstrahlung* of power $P_{bs} \sim n^2Z^2T^{1/2} \simeq 9 \times 10^{13}$ Wcm^{-3}. Owing to the radiation loss the plasma cannot be in perfect thermal equilibrium, no radiation temperature T_{rad} can be assigned to the photonic component. It is useful to compare the data:

$$n = 10^{21}\mathrm{cm}^3,\, T = 1\mathrm{keV} \Rightarrow \epsilon_{\mathrm{in}} = 4.8 \times 10^5,\, \epsilon_{bb} = 1.4 \times 10^7,\, P_{bs} = 9 \times 10^{13}\mathrm{Wcm}^{-3}.$$

The energy loss by bremsstrahlung in the unit volume is $P_{bb}/\nu_{ei} = 9 \times 10^{13}/5 \times 10^{11} \simeq 200$ Jcm^{-3} per electron-ion collision. This means that the number of collisions N to irradiate the thermal energy ϵ_{in} at fixed temperature is $N = \nu_{ei}(\epsilon_{\mathrm{in}}/P_{bs}) = 2.5 \times 10^3$. It is the number an electron must collide with an ion to lose $3k_BT$. Correspondingly, it loses the small fraction $\Delta\epsilon_e/\epsilon_{\mathrm{in}} = 4 \times 10^{-4}$ of energy in one electron-ion encounter. The typical time for the evolution into a Maxwellian is of the order of $\tau \simeq 1/\nu_{ei}$. To be more scrupulous let us assume $\tau = 10/\nu_{ei}$. Then the loss is still as low as 0.004, i.e., negligible with respect to the fractional energy exchange of order unity in an electron-electron encounter. Hence, it is perfectly justified to assume that the electrons are in a partial, *kinetic*, equilibrium, with a Maxwellian velocity distribution of temperature $T_e = T_i = T = 1$ keV. The number N is independent of T_e and n_e. We conclude

For the electron component of a plasma to be in thermal equilibrium it is necessary and sufficient to be in *kinetic* equilibrium with a distribution function of the free electrons close to a Maxwellian.

Maxwell Distribution from a Sufficient Condition

Assume now that locally each gas or fluid particle undergoes from several to many collisions during the characteristic time of change of density, $\tau = n(\mathbf{x})/\partial_t n(\mathbf{x}) = N\tau_{\text{coll}}$, then one can assume that (i) all correlations between the velocity components v_x, v_y, v_z and (ii) any memory of directions are lost. Let $\mathrm{d}p_x = f(v_x)\mathrm{d}v_x$, $\mathrm{d}p_y = f(v_y)\mathrm{d}v_y$, $\mathrm{d}p_z = f(v_z)\mathrm{d}v_z$ be the probability to find the velocity component of a single particle in the interval $\mathrm{d}v_x, \mathrm{d}v_y, \mathrm{d}v_z$. f does not depend on direction, hence owing to (ii) $f(v_i) = f(|v_i|) = f(v)$, where $f(v)$ is any function of the modulus of $|\mathbf{v}|$, for instance $g(\mathbf{v}^2) = g(v_x^2 + v_y^2 + v_z^2)$. According to (i) of independent probabilities holds

$$f(v)\mathrm{d}\mathbf{v} = g(v_x^2 + v_y^2 + v_z^2)\mathrm{d}\mathbf{v} = g(v_x^2)g(v_y^2)g(v_z^2)\mathrm{d}v_x\mathrm{d}v_y\mathrm{d}v_z$$

$$\Rightarrow g(v_x^2 + v_y^2) = g(v_x^2)g(v_y^2)g(v_z^2 = 0); \quad g(0)g(0) = g(0)^2 \Rightarrow g(0) = 1. \quad (4.35)$$

The only continuous function exhibiting property (4.35) is the exponential function $f(\mathbf{v}) = \exp -\beta(\mathbf{x})\mathbf{v}^2$. Thus, the probability distribution $f(\mathbf{v})$, normalized to unity is the Maxwell distribution (1.18) for dilute fluids. For the proof see Sect. 4.7. The identification of $\beta(\mathbf{x})$ with a local temperature $T(\mathbf{x}) = m/2k_B\beta(\mathbf{x})$ follows from the moment of the internal energy (3.91).

A caveat is in order here insofar as the preceding estimates on losses, or gains, are based on averaged quantities. A thorough discussion will concentrate on the different function of individual groups of electrons in a physical process, e.g. weakly bound electrons, or electrons involved in heating; they may differ from electron groups responsible for heat diffusion. For the bound electrons in the plasma to be assigned a temperature means that the electron energy levels are thermally occupied. In presence of a radiation field LTE may e heavily disturbed and recurrence must be made to involved and demanding rate equations [1]. It is well known and understandable that the thermal distribution between two levels is the more difficult to be achieved the bigger their energy difference is. There must be enough energetic electrons available in the Maxwell distribution of the free electrons to guarantee a sufficiently high collision frequency of all electrons energetically above the excitation level. In high-Z plasmas this is generally not the case. A successful recipe is to fix a collisional limit in energy above which the bound electrons are in thermal equilibrium and to calculate the occupation of the levels below by rate equations.

Non-Maxwellian Distribution by Strong Energy Supply

To be on the safe side with a Maxwellian electron distribution high losses as well as intense supply of energy must be avoided. In high-Z plasmas of the parameters chosen above the bremsstrahlung losses may seriously disturb the thermal equilibrium of the electron component if the ion charge exceeds $Z = 50$ and the loss increases to $\Delta\epsilon_e/\epsilon_e = 1$.

Strong heating of matter is best achieved by laser irradiation. In the laser field the free electrons gain oscillatory energy from the laser field. By collisions with the ions this ordered energy is deviated into different directions and cannot be given back to the laser field; the transfer process is irreversible. The electron-electron collisions do not participate in the absorption process because for $k\hat{\delta} \ll 1$ their collision dynamics is unaffected by the $\mathbf{E}$ field (k wave number, $\hat{\delta}$ oscillation amplitude). To see this it is sufficient to go into a reference system comoving with the oscillating electron. However, ν_{ee} is the crucial quantity that determines finally the time of relaxation from isotropy to a Maxwellian. Isotropization is reached first by both, electron - ion and electron-electron collisions; in the slow final diffusion of the electron velocity $|\mathbf{v}_e|$ the ions do not participate owing to their heavy mass. According to (1.17) the electron gets twice the mean oscillation energy W from (1.1) per electron-ion collision. With a laser of $\lambda \simeq 1\mu$ and intensity $I \leq 10^{15}$ Wcm^{-2} the inequality $W \ll k_B T$ holds. Under these limitations the collision frequencies, the heating rate $\dot{h}$ and the diffusion rate $\dot{D}$ per Coulomb collisions are

$$\nu_{ee} = \frac{C}{v_{\mathrm{th}}^3}, \quad \nu_{ei} = \frac{CZ^2}{v_{\mathrm{th}}^3}, \quad \dot{h} \simeq 2W\nu_{ei} \sim \frac{Z^2 v_{\mathrm{os}}^2}{v_{\mathrm{th}}^3}, \quad v_{\mathrm{os}} = \frac{e\hat{E}}{m_e\omega}, \quad v_{\mathrm{th}}^2 = \frac{k_B T}{m_e},$$

$$\dot{D} \leq \nu_{ee} m_e v_{\mathrm{th}}^2 \sim \frac{v_{\mathrm{th}}^2}{v_{\mathrm{th}}^3} \Rightarrow \frac{\dot{h}}{\dot{D}} = \frac{Z^2 v_{\mathrm{os}}^2}{v_{\mathrm{th}}^2}. \tag{4.36}$$

Here is to say that $\dot{h}$ heats preferentially the slow electrons whereas $\dot{D}$ is strongest beyond v_{th} of the velocity spectrum. Depending on Z and laser intensity the ratio $\dot{h}/\dot{D}$ may easily exceed unity at moderate laser intensities with the result that the electron velocity distribution flattens around $\mathbf{v} = 0$ and does no longer relax to a Maxwellian. The formation of such *Supergaussian* distributions has been predicted by Bruce Langdon [2]. From the solution of the Vlasov equation (3.81) he has obtained distributions as indicated in Fig. 4.2.

Transition Regime Between Equilibria: LTE

Consider a rarefaction wave and the Gay-Lussac experiment, Fig. 3.6. As the pressure on one side is suddenly released motion sets in, the system becomes inhomogeneous. Each volume element loses internal energy on the expense of kinetic energy. Once the gas hits the opposite wall its stagnation and the back running strong hot shock lead to reconversion of the expansion energy into internal energy. After a few reflections the shock attenuates. Finally the gas fills twice the original volume homogeneously with a constant temperature. In the ideal gas the temperature is the same as before expansion.

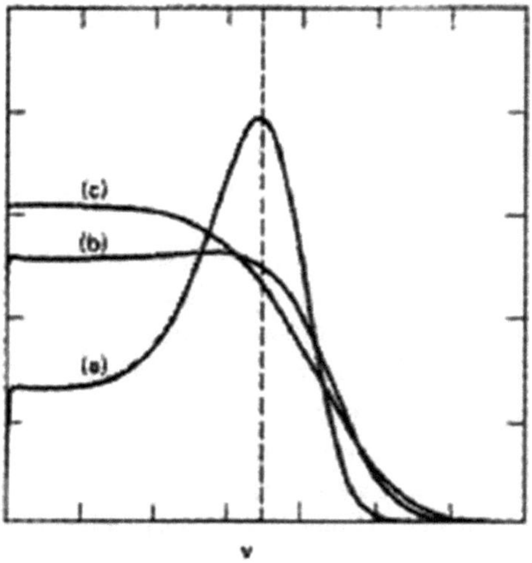

Fig. 4.2 Evolution of an initially monochromatic velocity (dashed line, units in v_{th}, see text) into a supergaussian distribution f_0 in presence of laser heating. At first, electrons diffuse both up and down in balance so that f_0 gains only 1% in energy [curve **a**]. Thereafter, the slowest electrons cannot lose more energy, while others continue to diffuse upward, resulting in net absorption. When the energy has increased by 10% [curve **b**] f_0 is close to the selfsimilar stationary solution (**c**): $f_0 \sim u^{-3} \exp -v^5/Cv_0^2 t$ [same normalization in energy as (**b**)]. Courtesy of [2]

The Maxwellian has been obtained under very reasonable conditions. It cannot be exact if $\nabla T \neq 0$ because (5.77) tells that energy (heat) flow $\mathbf{q}$ is nonzero. On the other hand, with a Maxwellian $\mathbf{q} = 0$ results from (3.92) because of symmetry. It is worth to examine this discrepancy. Assume an ideal fully ionized plasma with a constant temperature gradient ∇T_e over the length L in static equilibrium. According to (3.41) the thermoelectric field must balance the electron pressure to fulfil $\mathbf{u} = 0$ (less stringent $\dot{\mathbf{u}} \simeq 0$),

$$E = -\frac{k_B}{en_e}\partial_x(n_e T_e). \tag{4.37}$$

For mild gradients of T_e it can be assumed that the correct solution f of the steady state Vlasov equation differs only slightly from the Maxwellian f_0. Thus, $f_1 = f - f_0$ satisfies the linearized Vlasov equation,

$$v\partial_x f_1 - \frac{e}{m_e}E\partial_v f_0 = 0,\ (4.37) \Rightarrow\ v\left[\partial_x f_1 - \frac{\partial_x(n_e T_e)}{n_e T_e} f_0\right] = 0$$
$$\Rightarrow f_1(\mathbf{v}, x) \sim f_0(\mathbf{v}) p_e(x). \tag{4.38}$$

The result merely tells that the long range collective fields do not act locally and neither help to thermalize nor to destroy local thermal equilibrium, in contrast to the short range fields of the charged or polarized particles of the fluid. It shows once more that external forces do not change the entropy in the absence of discontinuities.

A criterion for the validity of local thermodynamic equilibrium can be extracted from the analysis of the heat flow density $\mathbf{q}$. For this purpose let a negative temperature gradient $\nabla T = -T(\mathbf{x})/L$ exist along $\mathbf{x}$ oriented from left to right. The steady state heat flux $\mathbf{q}$ is the net energy flux difference $\mathbf{q}_+ - \mathbf{q}_-$ between more energetic particles coming from $\mathbf{x}'$ left of $\mathbf{x}$ and less energetic particles coming from $\mathbf{x}''$ right of $\mathbf{x}$. Both deposit their energies $m\mathbf{v}'^2/2$ and $m\mathbf{v}''^2/2$ at $\mathbf{x}$ after having travelled a mean free path $\boldsymbol{\lambda}(\mathbf{v}') = \mathbf{v}'\tau(\mathbf{v}') = \mathbf{x} - \mathbf{x}'$ and $\boldsymbol{\lambda}(\mathbf{v}'') = \mathbf{v}''\tau(\mathbf{v}'') = \mathbf{x} - \mathbf{x}''$. The flux is

$$\begin{aligned}\mathbf{q} &= \frac{m_e}{2}\int \mathbf{v}^2\mathbf{v}f(\mathbf{v})d\mathbf{v} = \frac{m_e}{2}\int \mathbf{v}'^2\frac{\boldsymbol{\lambda}\nabla T}{|\boldsymbol{\lambda}\nabla T|}\mathbf{v}(\mathbf{x})f(\mathbf{v})d\mathbf{v} = q_+ - \mathbf{q}_- \\ &\simeq -\frac{3}{2}k_B(T_+ - T_-)n(\mathbf{x})\mathbf{v}_{\text{th}}d\tau d\mathbf{v} = -\frac{3}{2}\overline{\lambda}n|\mathbf{v}_{\text{th}}|\nabla k_B T = -\kappa\nabla T. \quad (4.39)\end{aligned}$$

The symbol $\overline{\lambda}$ is the mean free path averaged over all $\boldsymbol{\lambda}(\mathbf{v}')$; it is parallel to ∇T. The expressions of $\mathbf{q}$ in the first line are capable of describing *nonlocal* heat transport as well. For details see chapter on transport in plasma. The limitation to a local flux has occurred in the second line with the introduction of Fourier's approximation ∇T. It holds if $\overline{\lambda} \ll L$ is fulfilled. It turns out to be equivalent to the criterion for the existence of LTE. The assumption of LTE is guaranteed if the difference of the two energy flux densities $\mathbf{q}_+$ and $\mathbf{q}_-$ is small,

$$\frac{|\mathbf{q}|}{|\mathbf{q}_+| + |\mathbf{q}_-|} = \frac{|\mathbf{q}_+ - \mathbf{q}_-|}{|\mathbf{q}_+| + |\mathbf{q}_-|} \ll 1. \quad (4.40)$$

In this case the coefficient of heat conduction κ is a local quantity.

Inequality (4.40) for the electron component is the criterion for the existence of a local electron temperature $T_e(\mathbf{x}, t)$ and electron LTE. Ion temperature T_i and ion LTE obey the same criterion if $\mathbf{q}$ stands for the ion energy flow density relative to the plasma moving at $\mathbf{u} = 0$. Finally, the existence of a local radiation temperature $T_{\text{rad}}(\mathbf{x}, t)$ is also expressed by (4.40) with $\mathbf{q}$ the radiant flux density.

In principle, macroscopic conservation equations in terms of particle density n, flow velocity $\mathbf{u}$, temperature T and pressure tensor $P = P_{ij}$, or $P = p\delta_{ij}$ respectively, can be obtained from a Vlasov–Boltzmann equation with an arbitrary one particle distribution function $f(\mathbf{x}, \mathbf{v}, t)$. However, there is a fundamental difference between the so obtained conservation equations and those from *local thermodynamic equilibrium*. In the latter case (i) $f(\mathbf{x}, \mathbf{v}, t)$ is close to a Maxwellian with a deviation $|f_1| \ll f_M$ and (ii) entropy, internal energy, T and pressure P, p obey the fundamental laws of equilibrium thermodynamics *locally*. With an arbitrary distribution function $f(\mathbf{x}, \mathbf{v}, t)$ they may not. So, entropy is no longer defined in terms of n and T, and T_{kin} may violate assertion C after Clausius.

4.3 Essentials of Thermostatistics: Classical Systems

4.3.1 *The Fundamental Principle of Equilibrium Thermodynamics*

Any classical system is defined by its Hamiltonian $H(\mathbf{p}, \mathbf{q}, t)$. As two simple examples, for the ideal gas of N isotropic harmonic oscillators of frequency ω and the ideal fully ionized hydrogen plasma in an external potential V the Hamiltonians read

$$H(\mathbf{p}, \mathbf{q}) = \sum_{j=1}^{N} \frac{1}{2m}\mathbf{p}_j^2 + \frac{m\omega^2}{2}\mathbf{q}_j^2, \quad H(\mathbf{p}, \mathbf{q}) = \sum_{j=1}^{N} \left[\frac{\mathbf{p}_{ij}^2}{2m_i} + \frac{\mathbf{p}_{ej}^2}{2m_e} + V(\mathbf{q}_{ij}) - V(\mathbf{q}_{ej}) \right] \quad (4.41)$$

in the $6N$ dimensional phase space Γ. Any such position is allowed provided it stays on the $3N$ dimensional hypersurface $\Sigma_{H=E}$ because its energy E is conserved. The other constraint that limits the set of points on Σ is that all $\mathbf{q}_j$ of $\mathbf{q} = \{\mathbf{q}_j\}$ are confined to the volume V. There may be other external parameters that lead to further restriction of the accessible phase space on $\Sigma_{H=E}$. In a nearly ideal gas of arbitrarily weak interactions between the particles it is quite natural to assume that the system will occupy every arbitrarily small element on the accessible part of the hypersphere Σ because all initial conditions respecting the constraints are possible states the system can assume. If all masses are equal it must be followed further from the symmetry of the Hamiltonian $H = \sum \mathbf{p}_j^2/2m$ that all states are equally likely. The idea of symmetry is susceptible of the extension of equal probability to all accessible states of any Hamiltonian system. Theoretical conclusions and the experiment confirm the validity of the following basic postulate:

A Hamiltonian system in thermodynamic equilibrium fills the total accessible hypersurface $\Sigma(E) = \Sigma(H = E)$ with equal probability. ⇒ The probability density is given by the measure

$$\mathrm{d}\mu(\Sigma) = \frac{\mathrm{d}\Sigma}{|\mathrm{grad}_{(\mathbf{p},\mathbf{q})}(H = E)|\Omega} = \text{const} \quad (4.42)$$

on Σ and $\mathrm{d}\mu = 0$ outside, see 2.63 of Chap. 2. The normalization factor Ω and the gradient are given by

$$\Omega = \int_{\Sigma(E)} \mathrm{d}\mu(\Sigma); \quad \mathrm{d}\Gamma_{(H=E)} = \Omega\, \mathrm{d}E; \quad \mathrm{grad}_{(\mathbf{p},\mathbf{q})} H = \nabla H = \left(\frac{\partial H}{\partial \mathbf{p}}, \frac{\partial H}{\partial \mathbf{q}} \right).$$

Note, it is the surface element $\mathrm{d}\Sigma$ times the local thickness dE in phase space that is the measure of equal probability on Σ. The volume of the phase space enclosed by $\Sigma(H = E)$ is indicated by $\Gamma(E) = \int_{E' \leq E} \Omega \mathrm{d}E'$. The set of states $(\mathbf{p}, \mathbf{q})$ within $\Sigma(E)$ is the *microcanonical ensemble*.

The probability density is invariant with respect to an arbitrary canonical transformation. Since all possible states of the system are equivalent to each other, in thermodynamic equilibrium all macroscopic quantities of the system, for instance the pressure, are *ensemble averages* $\langle f \rangle$ of functions $f = f(\mathbf{p}, \mathbf{q})$ on $\Sigma(E)$,

$$\langle f \rangle = \frac{1}{\Omega} \int_{\Sigma(E)} f(\mathbf{p}, \mathbf{q}) \frac{\mathrm{d}\mathbf{p}\mathrm{d}\mathbf{q}}{|\mathrm{grad}_{\mathbf{p},\mathbf{q}}(H = E)|} = \frac{1}{\Omega} \frac{\partial}{\partial E} \int_{H \leq E} f(\mathbf{p}, \mathbf{q}) \mathrm{d}\mathbf{p}\mathrm{d}\mathbf{q}. \quad (4.43)$$

4.3.1.1 Entropy from the Microcanonical Ensemble

Let us consider the classical ideal gas of point particles and calculate the phase space volume $\Gamma(E) = \int_{H \leq E} \mathrm{d}\mathbf{p}\mathrm{d}\mathbf{q}$. The integral in momentum space is the $3N$ dimensional sphere with radius $R = \sqrt{2mE} = (\sum \mathbf{p}_j^2)^{1/2}$, hence

$$\Gamma(E) = \int_{H \leq E} \mathrm{d}\mathbf{p}\mathrm{d}\mathbf{q} = V^N \times C_N R^{3N}; \quad C_N = \frac{\pi^{3N/2}}{\Gamma(3N/2 + 1)}. \quad (4.44)$$

Γ with integers in the argument stands for the Gamma function,

$$\Gamma(n + 1) = n!, \qquad \Gamma(n + 1/2) = \frac{(2n)!}{n!4^n} \sqrt{\pi}.$$

Stirling's formula is very useful in combinatorics. For large arguments of the factorial a satisfactory approximation is Stirling's formula in the following version,

$$n! \simeq (2\pi n)^{1/2} \left(\frac{n}{\mathrm{e}}\right)^n. \quad (4.45)$$

For $n = 5$ the relative deviation is 1.6%, for $n = 10$ it shrinks to 0.8%. In Stirling's approximation $\Gamma(E)$ becomes

$$\Gamma(E) = V^N \times \frac{1}{\sqrt{3N\pi}} \left(\frac{2\pi\mathrm{e}}{3N}\right)^{3N/2} (2mE)^{3N/2}. \quad (4.46)$$

This is an extremely rapidly increasing function of energy. The fraction s of the whole volume of the momentum sphere lying in the layer below the surface of depth δ is given by

$$\delta = \left(1 - \mathrm{e}^{\ln(1/s)/3N}\right) R \simeq \frac{|\ln(1/s)|}{3N} R; \quad s = \frac{1}{2} \Rightarrow \delta = 0.2 \frac{R}{N} \sim N^{-1/2}. \quad (4.47)$$

i.e., all states lie just below the surface. $R = (2mE)^{1/2}$ is the radius of the hypersphere from (4.46). The property discloses two implications that are basic for the success of statistical thermodynamics:
(a) The ensemble average $\langle f \rangle$ results the same whether taken over all states of $\Gamma(E)$, or over a thin shell $\partial_E \Gamma(E)\Delta E$, or over Ω of the set of states $\{H = E\}$.
(b) The *most likely value* of f is the value of f taken at the highest density of states. The second implication is that the mean square deviation $(f - \langle f \rangle)^2/\langle f \rangle^2 \ll 1$, i.e., is very small. In short, nearly all states exhibit the same value $f = \langle f \rangle$.

In systems of weakly interacting particles the shape of $\Gamma(E, V)$ is close to Γ of (4.44). In general, in stable systems, including strongly coupled matter like nonideal plasmas and warm dense matter, the momentum phase space will still be convex and a non decreasing function of energy and volume, with an effective radius of the order of $(2mE)^{1/2}$. In addition we can assume that surface energies are negligible relative to the volume energy E. For simplicity such standard systems are named "convex".

Convex systems behave like (4.47).

In the microcanonical ensemble the entropy S of a convex system is defined by

$$S(E, V) = C \ln \Gamma(E, V) \sim C \ln \Omega(E, V) \sim C \ln[\Gamma(E, V) - \Gamma(E - \Delta E, V)]. \tag{4.48}$$

C is a constant to be fixed later. The three expressions for the entropy are all the same up to an insignificant difference of order $\ln N$. In the ideal gas follows from (4.44)

$$\begin{aligned} S(E, V) &= \ln C + N \ln V + \frac{3N}{2} \ln \frac{2\pi e}{3N} + \frac{3N}{2} \ln(2mE) \\ &= NK \ln \left[V \left(\frac{2\pi m E}{N} \right)^{3/2} \right] + \frac{3}{2} NK + \text{const.} \end{aligned} \tag{4.49}$$

K is an arbitrary dimensional constant so far. We show that the statistically defined entropy (4.48) fulfills the properties (a)–(c) of the phenomenological entropy plus property (d).
(a) $S(E, V)$ is an extensive quantity. In a homogeneous system the particle number N is proportional to its mass M. Hence, for convex systems $S \sim N \sim M$ is fulfilled.
(b) Imagine $\Gamma(E, V)$ subdivided into two systems S_1, S_2 with $N = N_1 + N_2$, both, N_1, N_2 macroscopic. Under the assumption of negligible surface energy holds

$$\Gamma(E, V) = \int_{E_1=0}^{E_1=E} \Omega(E - E_1)\Omega(E_1)dE_1. \tag{4.50}$$

In the subsystems S_1, S_2 assumed convex Γ_1 increases like $E_1^{3N_1/2}$ with E_1 and $\Omega(E - E_1)$ at the same time decreases like $E_2^{3(N_2-1)/2}$ from $\Gamma(E)$ to zero. The integrand reaches a sharply peaked maximum at E_1^* of width $\delta \sim \sqrt{(}2mE_1^*)/N_1 + \sqrt{(}2mE_2^*)/N_2$. According to the considerations on convex systems above follows

$$\begin{aligned}\ln \Gamma(E) &= \ln[\Omega(E - E_1^*)\Gamma(E_1^*)dE_1] = \ln[\Omega(E_2^*)dE_1] + \ln \Gamma(E_1^*)\\ &= \ln \Gamma(E_1^*) + \ln \Gamma(E_2^*).\end{aligned} \tag{4.51}$$

to an excellent approximation $\Rightarrow S = S_1 + S_2$ if the two systems are in equilibrium with each other. If each system is in its own but isolated thermal equilibrium additivity follows from property (a).

Statistical Temperature and Pressure

At the maximum of $\Omega(E - E_1)\Gamma(E_1)_{E_1=E_1^*}$ holds $\delta[\ln \Gamma(E_1^*) - \delta \ln \Gamma(E_2^*)] = 0$; in explicit form

$$\left(\frac{\partial S(E_1^*, V)}{\partial E_1}\right)_V = \left(\frac{\partial S(E_2^*, V)}{\partial E_2}\right)_V.$$

This equality between two arbitrary subsystems allows to define the absolute temperature T,

$$\frac{1}{CT} = \left(\frac{\partial S(E_1^*, V)}{\partial E_1}\right)_V \tag{4.52}$$

in formal agreement with (4.15). The definition of the thermodynamic equilibrium implies the existence of an absolute temperature T, in accordance with the phenomenological findings. Obviously T is a non negative quantity. In agreement with (4.15) we introduce further the equilibrium pressure p as

$$p = T\left(\frac{\partial S(E_1^*, V)}{\partial V}\right)_{E_1^*}. \tag{4.53}$$

Equation (4.52) guaranties the resolution with respect to E_1^*,

$$T\mathrm{d}S = \mathrm{d}E_1^* + p\mathrm{d}V \qquad \Leftrightarrow \qquad \mathrm{d}E_1^* = T\mathrm{d}S - p\mathrm{d}V. \tag{4.54}$$

This is the second law of thermodynamics, derived from statistics; E_1^*, *or* E, respectively, is the former internal energy $\mathcal{E}_{\text{in}}$.

(c) Thermal isolation means energy can be supplied only through equilibrium pressure p (non equilibrium pressure would make the system inhomogeneous, i.e. non local, to be excluded here). In this case $\delta Q = \delta Q_{rev} = 0 \Rightarrow \mathrm{d}S = 0 \Leftrightarrow \mathrm{d}S \not< 0$. In any case $\mathrm{d}S/\mathrm{d}E \geq 0$ at *fixed volume* is evident. Let now the two systems each be in its own equilibrium at different temperatures in a thermally isolated environment and then brought in contact for global equilibrium. According to the maximum property of (4.50) the system $S_1 + S_2$ occupied initially a position in phase space

$\Omega(E - E_1')\Gamma(E_1') < \Omega(E - E_1^*)\Gamma(E_1^*)$ and hence, $\Delta S > 0$ in the new equilibrium. (d) With E and V fixed, the equilibrium state occupies an Ω set of states that, by the principle (4.42), is not smaller than any nonequilibrium set of states $\Omega'(E, V)$, thus $S(\Omega) = \max$.

In the following the statistical approach to thermodynamics will reveal its full power, conceptually as well as in modelling thermodynamic processes. It is conceptually much simpler than the phenomenological approach. We shall see that entropy and reversibility are accessible to an exact and extremely simple picture on the microscopic level whereas in the macroscopic approach they are subtle and, sometimes, remain vague.

Equipartition: The Virial Theorem

Let x_j be either p_j or q_j. The aim is to calculate the ensemble average of $x_j \partial_k H$:

$$\begin{aligned}
\langle x_j \frac{\partial H}{\partial x_k} \rangle &= \frac{1}{\mu(\Omega)} \int_{\Omega} x_j \frac{\partial H}{\partial x_k} \mathbf{dpdq} = \frac{1}{\mu(\Omega)} \frac{\partial}{\partial E} \int_{H \le E} x_j \frac{\partial H}{\partial x_k} \mathbf{dpdq} \\
&= \frac{1}{\mu(\Omega)} \frac{\partial}{\partial E} \left\{ \int_{H \le E} \frac{\partial}{\partial x_k} [x_j (H - E)] \mathbf{dpdq} - \delta_{jk} \int_{H \le E} (H - E) \right\} \mathbf{dpdq} \\
&= \frac{\delta_{jk}}{\mu(\Omega)} \frac{\partial}{\partial E} \int_{H \le E} (E - H) \mathbf{dpdq} = \frac{\delta_{jk}}{\mu(\Omega)} \int_{H \le E} \mathbf{dpdq}.
\end{aligned}$$

The first integral in the second line vanishes because $H - E = 0$ on the surface $\Sigma(E)$. The last integral is easily evaluated by keeping in mind $\mu(\Omega) = \partial_E \Gamma(E)$. Thus

$$\langle x_j \frac{\partial H}{\partial x_k} \rangle = \delta_{jk} \left(\frac{\partial}{\partial E} \ln \Gamma(E) \right)^{-1} = \delta_{jk} \frac{C}{\partial_E S(E)} = \delta_{jk} CT \tag{4.55}$$

$$\Rightarrow \langle p_j \frac{\partial H}{\partial p_j} \rangle = CT; \quad \langle q_j \frac{\partial H}{\partial q_j} \rangle = CT \tag{4.56}$$

$$\Rightarrow \sum_1^{3N} \langle q_j \dot{p}_j \rangle = -3NCT. \tag{4.57}$$

$\sum_1^{3N} \langle q_j \dot{p}_j \rangle$ is the virial, introduced by (2.49) in Chap. 2. A frequently encountered Hamiltonion for noninteracting particles (ideal gas) or pseudoparticles (plane waves, phonon, plasmons, photons) is reducible to the structure

$$H = \sum_j A_j p_j^2 + \sum_j B_j q_j^2 \quad \Rightarrow \quad \sum_j \left(p_j \frac{\partial H}{\partial p_j} + q_j \frac{\partial H}{\partial q_j} \right) = 2H \tag{4.58}$$

with A_j and B_j constants. If their number of terms is f (*degrees of freedom*) (4.56) and (4.57) state

$$\langle H \rangle = \frac{f}{2} CT. \tag{4.59}$$

This is the equipartition theorem for harmonic oscillators. It is the same for translations and rotations ($B_j = 0, j \leq f$). Comparison with the ideal gas shows that C is the Boltzmann constant k_B. $\langle H \rangle$ is the internal energy $\mathcal{E}_{\mathrm{in}}$. The specific heat at constant volume is $C_V = (f/2)Nk_B$. The entropy expression (4.49) for the ideal gas with $K = k_B$ reduces to

$$S(E, V) = Nk_B \ln \left[V \left(\frac{2\pi m E}{N} \right)^{3/2} \right] + \frac{3}{2} N k_B + \text{const.} \tag{4.60}$$

Internal Energy, Temperature, and Heat

In the classical ideal gas holds

$$\mathrm{d}\mathcal{E}_{\mathrm{in}} = C_V \mathrm{d}T = \mathrm{d}Q_{\mathrm{rev}} \quad \Leftrightarrow \quad \Delta Q_{\mathrm{rev}} = \Delta \mathcal{E}_{\mathrm{in}}, \quad \mathcal{E}_{\mathrm{in}} \sim T.$$

It is perhaps this special relations that sometimes one is tempted to identify heat with internal energy and internal energy with absolute temperature. It is true that the existence of an absolute temperature T is ensured with the existence of internal energy $\mathcal{E}_{\mathrm{in}}$ and entropy s as state functions in thermal equilibrium. The temperature, however, that is reached depends on the individual substance; it is proportional to the average internal potential plus internal kinetic energy. Their ratio may be almost arbitrary in general. Only in the case of the harmonic oscillator (4.58) it is unity, or it is zero as in the ideal monoatomic gas. Proportionality between $\mathcal{E}_{\mathrm{in}}$ and T holds in a restricted temperature/energy interval, like $\mathrm{d}\mathcal{E}_{\mathrm{in}} = C_V \mathrm{d}T$, $\mathrm{d}\mathcal{E}_{\mathrm{in}} = C_p \mathrm{d}T$, $\mathrm{d}\mathcal{E}_{\mathrm{in}} = C_S \mathrm{d}T$. Over finite intervals the specific heat may vary considerably, in dense matter because of interparticle potential variations, in plasmas owing to ionization. After all, the origin of the energy may be fixed arbitrarily in a particular gauge, however the temperature T is always a positive quantity.

Owing to $\Delta Q_{\mathrm{irr}} < \Delta Q_{\mathrm{rev}}$ and also $\Delta Q_{\mathrm{rev}}|_{V=\mathrm{const}} < \Delta Q_{\mathrm{rev}}|_{p=\mathrm{const}}$ one must conclude that while heat contributes to the internal energy its amount is only defined as soon as the process is known. In the adiabatic process the internal energy varies but the heat supplied is zero. Finally, the entropy is only loosely connected with temperature and energy. In the adiabatic process energy and temperature increase or decrease but the entropy remains constant.

Internal energy, heat, entropy, and temperature are independent quantities.

Discrimination between these four quantities may decide on success or failure in establishing similarity laws and in dimensional analysis.

Range of Validity of the Maxwell Velocity Distribution

It is instructive to derive the Maxwellian velocity distribution from the microcanonical ensemble. As the whole thermodynamics is derivable from it the Maxwell distribution must follow as a special case. Given a system of N particles we ask for the probability $w(\mathbf{v})$ of any arbitrary particle to assume a velocity in the interval $[\mathbf{v}, \mathbf{v}+\mathrm{d}\mathbf{v}]$. According to the fundamental law of thermostatistics all states that are reachable energetically occur with equal likelihood. If one particle is in this velocity interval of $m\mathbf{v}^2/2 = E_1$ all states of the remaining $N-1$ particles are on the energy shell $\Omega(E-E_1)$, hence

$$w(\mathbf{v}) \sim N \times \mu[\Omega(E-E_1)] \sim N \times \Gamma(E-E_1). \tag{4.61}$$

E_1 is vanishingly small in comparison to $E-E_1$, hence this can be expanded in Taylor series around E. When expanding the question arises which is the most favorable expansion parameter. As a rule, expansion of an expression at a point of strong curvature is favorably transformed to a quantity that is almost flat at such a position. In our case of $w(\mathbf{v})$ we expand its logarithm:

$$\begin{aligned}\ln w(\mathbf{v}) &= \ln C_1 + \ln \Gamma(E-E_1) = \ln C_1 + \frac{1}{k_B}S(E-E_1) = \ln C - E_1\frac{1}{k_B}\frac{\partial S}{\partial E}\\ &= \ln C - \frac{E_1}{k_B T} \Rightarrow f_M(\mathbf{v}) = C\mathrm{e}^{-m\mathbf{v}^2/2k_BT};\ C = \left(\frac{m}{2\pi k_B T}\right)^{1/2}.\end{aligned} \tag{4.62}$$

$C = C_1\Gamma(E)$. In this derivation no use of the coupling strength between particles has been made. It follows that f_M is correct as long as the velocity $\mathbf{v}$ of a particle is meaningful. With increasing particle density the quantum effect of gradual delocalization becomes important. This puts a limit onto the validity of the classical Maxwellian. Two extensions are important:

(a) If the fluid is a mixture of particles with various masses m_α, e.g., electrons and ions, the velocity $\mathbf{v}$ has to be substituted by the momenta $\mathbf{p}_\alpha = m_\alpha \mathbf{v}_\alpha$, and f_M remains valid for mixtures, like macromolecules and dust.

(b) In the relativistic domain the energy shell of massive particles is defined by the Hamiltonian

$$\begin{aligned}H(\mathbf{p},\mathbf{q}) &= \sum_i \left[\left(m_i^2c^4 + c^2\mathbf{p}_i^2\right)^{1/2} + V(\mathbf{q}_i)\right] = E\\ &\Rightarrow f_M(\mathbf{p}) = C(k_BT)\mathrm{e}^{-\left[\sqrt{m_i^2c^4+c^2\mathbf{p}^2}+V(\mathbf{q})\right]/k_BT}.\end{aligned} \tag{4.63}$$

The normalization factor is no longer a simple function of temperature. An expression for the particle density $n(\mathbf{x})$ is obtained by integrating over all momenta,

$$n(\mathbf{x}) = n_0 \mathrm{e}^{-V(\mathbf{x})/k_B T}. \tag{4.64}$$

If $V(\mathbf{x} \to \infty) = 0$, $n_0 = n_\infty$ is the density at infinity.

Debye Screening in the Thermal Plasma

Quasineutrality. By definition the extension d of a plasma is several times larger than its Debye length λ_D. This implies the fundamental property of a plasma to be *quasineutral* in its interior if only quasistatic electric fields act on it: Zn_i can be replaced by n_e, $n_e = Zn_i$; however, the net charge density $n_e - Zn_i \neq 0$ in general; it couples to the electric field. To see how screening comes about we consider Fig. 4.3. The surface charge density σ of an infinitely extended plate in vacuum generates a constant electric field of strength $E = \sigma/2\varepsilon_0$. When the plate is immersed in a fluid of constant electron and ion temperature $T = T_e = T_i$ and the undisturbed electron and ion densities are $n_e = Zn_i = n_\infty = \mathrm{const}$, the free charges arrange according to (4.64)

$$n_e(x) = n_\infty \exp\left(\frac{e\Phi}{k_B T}\right), \quad Zn_i(x) = n_\infty \exp\left(\frac{-Ze\Phi}{k_B T}\right); \quad E = -\frac{\partial \Phi}{\partial x}. \tag{4.65}$$

Poisson's law states

$$\frac{\partial^2}{\partial x^2}\Phi = \frac{en_\infty}{\varepsilon_0}\left[\exp\left(\frac{e\Phi}{k_B T}\right) - \exp\left(-\frac{Ze\Phi}{k_B T}\right)\right]. \tag{4.66}$$

For $Z > 1$ no solution in closed form is known. However, expansion of the exponential under the condition $Ze\Phi/k_B T \ll 1$ reduces to

$$\frac{\partial^2}{\partial x^2}\Phi = \frac{n_\infty e^2 (Z+1)}{\varepsilon_0 k_B T}\Phi \quad \Rightarrow \quad \Phi(x) = \Phi_0 \exp\left(-\frac{x}{\lambda_D}\right). \tag{4.67}$$

It shows that σ and its field E are screened by 63% at the distance of a Debye length λ_D,

$$\lambda_D = \left(\frac{\varepsilon_0 k_B T}{n_\infty e^2 (1+Z)}\right)^{1/2}, \tag{4.68}$$

$$\lambda_D\,[\mathrm{cm}] = 6.9\left(\frac{T\ [\mathrm{K}]}{(1+Z) n_\infty\ [\mathrm{cm}^{-3}]}\right)^{1/2} = 743\left(\frac{T\ [\mathrm{eV}]}{(1+Z) n_\infty\ [\mathrm{cm}^{-3}]}\right)^{1/2}. \tag{4.69}$$

A thermal fluid of minimum extension $d \gg \lambda_D$ is quasineutral. For λ_D one frequently uses the expression

$$\lambda_D = \left(\frac{\varepsilon_0 k_B T_e}{n_\infty e^2}\right)^{1/2} = \frac{v_{\mathrm{eth}}}{\omega_p}, \ \lambda_D\,[\mathrm{cm}] = 7\left(\frac{T_e[\mathrm{K}]}{n_\infty[\mathrm{cm}^{-3}]}\right)^{1/2} = 743\left(\frac{T\ [\mathrm{eV}]}{n_\infty\ [\mathrm{cm}^{-3}]}\right)^{1/2}. \tag{4.70}$$

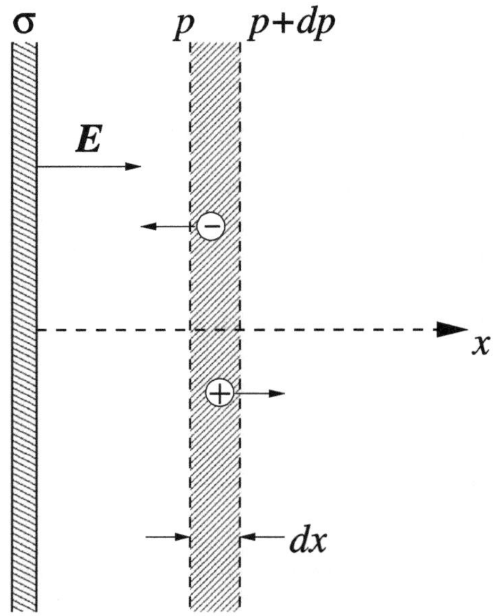

Fig. 4.3 Screening of an electric charge σ immersed in an isothermal plasma. Potential Φ and field E fall off as $\exp(-x/\lambda_D)$ with the Debye length λ_D. p is the thermal pressure

Expression (4.70) is the correct screening length in case the ions do not have time to redistribute. For $Z = 1$ and ions in equilibrium (4.68) becomes $\lambda_D\,[\mathrm{cm}] = 5(T[\mathrm{K}]/n_e[\mathrm{cm}^{-3}])^{1/2}$, if $n_e = n_\infty$ is used for the undisturbed electron density. In laser plasmas the following situation may occur: n_1 electrons have temperature T_1 while $n_2 = n_e - n_1$ have temperature T_2. Then

$$\lambda_D = 6.9\,\{T_1T_2/[(\xi_1T_2 + \xi_2T_1)n_e]\}^{1/2}; \quad \xi_{1,2} = n_{1,2}/n_e. \tag{4.71}$$

is the correct screening length of the electrons. If ξ_1 and ξ_2 are of the same order and $T_1 \gg T_2$ screening is dominated by the cold electrons.

On the Domain of Validity of Boltzmann's Formula (4.66) **in One Dimension**

Assume the electron distribution function of interest, e.g. hot electrons only, given by $f_e(x, v)$. The kinetic electron temperature T_e and the electron pressure p_e are given by

$$k_BT_e(x) = \frac{1}{n_e(x)}\int m_e(v-u)^2 f_e(x,v)\mathrm{d}v, \quad p_e(x) = \int m_e(v-u)^2 f_e(x,v)\mathrm{d}v.$$

Hence, $p_e = n_e k_B T_e$. The electron and the cold ion fluids are governed by

$$\rho_e\frac{\mathrm{d}u_e}{\mathrm{d}t} = -\partial_x p_e - en_eE, \quad \rho_i\frac{\mathrm{d}u_i}{\mathrm{d}t} = Zen_iE. \tag{4.72}$$

As long as the slow component $m_e n_e u_e$ is much less than $m_i n_i u_i$ and $|Zn_i - n_e| \lesssim Zn_i$ the electron balance is entirely determined by the electric field and the electron pressure, $\partial_x p_e = -en_e E$. If T_e is isothermal it can be integrated in space to yield the generalized Boltzmann formula

$$n_e(x) = n_{e0} \exp[e\Phi(x)/k_B T_e]. \tag{4.73}$$

The expression of $n_e(x)$ is valid for an arbitrary electron distribution function $f_e(x, v)$ as for example by two opposite monoenergetic electron beams. Thermal equilibrium and quasineutrality are not required. It can be extended to three dimensions if the propotionality factor between p_e and $n_e k_B T_e$ is known. In thermal equilibrium it is 2/3.

The Screened Coulomb Potential of an Ion

A test charge $q = Ze$ at rest in a plasma is subject to the same screening process as described for the charged wall. In the Poisson equation (4.66) ∂_{xx} is to be substituted now by $\nabla^2 = (1/r)\partial_{rr} r$ of spherical geometry,

$$\frac{1}{r}\frac{\partial^2}{\partial r^2} r\Phi = \frac{en_\infty}{\varepsilon_0}\left[1 - \exp\left(-\frac{Ze\Phi}{k_B T}\right)\right].$$

Expansion of the exponential and setting $\Psi = r\Phi$ results in

$$\frac{\partial^2}{\partial x^2}\Psi = \lambda_D^{-2}\Psi, \quad \lambda_D = \left(\frac{\varepsilon_0 k_B T_e}{n_\infty e^2}\right)^{1/2} \Rightarrow \Phi(r) = \frac{q}{4\pi\varepsilon_0 r}\exp\left(-\frac{r}{\lambda_D}\right). \tag{4.74}$$

For small radii r^* the number N of shielding charges may become very small,

$$N = \frac{4\pi}{3} r^{*3} n_e \ll 1 \tag{4.75}$$

to make the inner screening inefficient and to justify linearization of (4.66). However, it may also happen that the criterion is no longer fulfilled for r* as large as λ_D. This is the case of *nonideal, strongly coupled plasma*, with an effective potential differing from Φ_D. From (4.70) one deduces for the ideal plasma that the spherical potential Φ_D holds for charge motions relative to the electron plasma fluid not exceeding the sound velocity v_{eth}. For the derivation of the Debye length (4.74) from an arbitrary isotropic distribution function $f(v)$ see Chap. 7 on transport. There it is shown explicitly that shielding originates from the attraction of the electrons by the ions. The Debye potential Φ_D is the statistical average of the rapidly fluctuating electron charge around the single ion.

Debye screening is by the rearrangement of the *free* electrons around the ion and perhaps by the repulsion of the other ions in its neighbourhood. Relation (4.65) is not limited to thermal equilibrium. With a view at Fig. 4.3 one deduces

$$\mathrm{d}p_e = k_B T \mathrm{d}n_e = -en_e E\mathrm{d}x, \quad \Rightarrow \quad \frac{1}{n_e}\frac{\partial n_e}{\partial x} = \frac{e}{k_B T}\frac{\partial \Phi}{\partial x}$$

All that is needed is the knowledge of pressure p_e (and possibly also p_i) and the kinetic temperature T_e (possibly also T_i). The only change is that p_e/n_e is no longer connected with T_e by the factor 1 but by another proportionality, generally not differing much from unity.

Validity of the Test Particle Potential

Φ_D has been obtained from the linearized Poisson equation of *one single particle*. In contrast, to calculate screening at an arbitrary position $\mathbf{x}$ in the plasma Poisson's equation refers to the potential $\Phi(\mathbf{x})$ which results from the equilibrium of all electrons in the Coulomb fields of the N/Z plasma ions. The ions undergo slow motions in comparison to the electrons and can be assumed here as fixed at positions $\mathbf{x}_i$. We have to discern between three classes of electron-ion interactions. (A) Electrons passing at a distance $d \lesssim n_i^{-1/3}/3$ from the ion: They feel essentially the influence of this single ion along a bent trajectory. (B) Electrons streaking two or more ions simultaneously at a distance $d \simeq n_i^{-1/3}$ following a curved or twisted orbit. (C) Electrons describing nearly straight orbits because of being subject to small local deviations δ_i. In the case of ideal plasma class C prevails by far over the "nonlinear" classes A and B and provides for shielding. It is the situation of $\lambda_D \gg n_i^{-1/3}$. In this case, and only in it, $\Phi(\mathbf{x})$ is the superposition of the self consistent single electron potentials $\Phi_C(\mathbf{x}-\mathbf{x}_i)$ produced by the single ion at position $\mathbf{x}_i$. Correspondingly Poisson's equation reads

$$\Phi(\mathbf{x}) = \sum_{i=1}^{N/Z} \Phi_C(\mathbf{x}-\mathbf{x}_i) \Rightarrow \nabla^2\Phi(\mathbf{x}) = \frac{e}{\varepsilon_0} n_\infty \left(\exp\frac{e\Phi(\mathbf{x})}{k_B T} - 1\right).$$

Expansion of the exponent yields

$$\nabla^2\Phi(\mathbf{x}) = \sum_{i=1}^{N/Z} \nabla^2\Phi_C(\mathbf{x}-\mathbf{x}_i) = \sum_{i=1}^{N/Z} \frac{e^2 n_\infty}{\varepsilon_0 k_B T_e}\Phi_C(\mathbf{x}-\mathbf{x}_i) = \sum_{i=1}^{N/Z} \frac{1}{\lambda_D^2}\Phi_C(\mathbf{x}-\mathbf{x}_i). \tag{4.76}$$

From the equation it is to be concluded:

the validity of the test particle model is intimately connected with the superposition principle, i.e linearity, of the perturbation by the single ions. It is achieved by the exclusion of the orbits of classes A and B causing high frequency fluctuations. The Debye length λ_D is the result of a *mean field approach.*

4.3.1.2 The Canonical Ensemble

The natural variables of entropy in the microcanonical ensemble are the internal energy and the volume, and perhaps some additional external parameters y_i of the system. Again it is desirable to transform to easily controllable and measurable variables like temperature for the internal energy. The way according to which such a change is accomplished can be guessed from the derivation of the Maxwell velocity distribution above. Instead of one single particle there now a system of an arbitrary number of particles N and energy E in the heat bath of temperature T is faced. *Heat bath* either means that its number of particles $N_0 - N$ and its energy $E_0 - E$ are much bigger than N and the most probable $\langle E\rangle$ of the system, or a system under continuous energy supply to keep $T = \text{const}$. The coupling Hamiltonian of heat bath and system is strong enough to establish thermal equilibrium and at the same time so weak as to be ignored energetically. We ask for the probability $w(E, V)$ to find the system in an energy state E between $E = 0$ and $E = E_0$ of the bath. The probability is proportional to its phase space volume $\Gamma(E)$ and the heat bath phase space $\Gamma(E_0 - E)$,

$$
\begin{aligned}
w(E) &= C_1\Gamma(E) \times \Gamma(E_0 - E); \quad \Rightarrow \quad \ln w(E) = \ln C_1 + \ln \Gamma(E) + \ln \Gamma(E_0 - E) \\
&= \ln C - E\frac{1}{k_B}\frac{\partial S_0}{\partial E}\bigg|_{E_0} = \ln C - \frac{E}{k_B T}.
\end{aligned} \tag{4.77}
$$

$C = C_1\Gamma(E)\Gamma(E_0)$. The probability $w(E)$ is a strongly peaked function of E/k_BT, similar to that in the decomposition of the microcanonical ensemble into two subensembles. The system in the heat bath contains all states $(\mathbf{p}, \mathbf{q})$ on the energy shell $H(\mathbf{p}, \mathbf{q}) = E$. According to the fundamental axioma of thermostatistics all of these states have equal probability. Hence $\rho(\mathbf{p}, \mathbf{q}) = w(E)/\mu[\Omega(E)]$ is the *canonical density of states* in the heat bath, their sum $Q_N(V, T)$ is the *canonical partition function.*

Partition Function and Free Energy

The internal energy $\mathcal{E}_{\text{in}}$ expressed in terms of the Helmholtz free energy $F(V, T)$ is

$$
\mathcal{E}_{\text{in}} = F - T\left(\frac{\partial F}{\partial T}\right)_V = \langle H(\mathbf{p}, \mathbf{q})\rangle \Rightarrow \int_{\Omega(E_0)} [H - F + T(\partial_T F)_V]\mathrm{e}^{-\beta H}\mathbf{dpdq} = 0.
\tag{4.78}
$$

The integral can be rewritten as

$$\int [H - F + T(\partial_T F)_V] \mathrm{e}^{-\beta H} \mathbf{dpdq} = -\frac{\partial}{\partial \beta} \int \mathrm{e}^{-\beta[H - F(V,T)]} \mathbf{dpdq} = 0$$

$$\Rightarrow \quad Q_N(V,T) = \mathrm{e}^{-\beta F(V,T)} \quad \Leftrightarrow \quad \ln Q_N(V,T) = -\frac{F(V,T)}{k_B T}. \qquad (4.79)$$

The logarithm of the partition function in the variables of volume V and temperature T is a thermodynamic potential from which all equilibrium parameters follow by partial derivation, either from $F(V,T)$ in (4.31), or from the integral form $Q_N(V,T) = \int \exp[-H(\mathbf{p},\mathbf{q})/k_B T] \mathbf{dpdq}$:

$$\mathcal{E}_{\text{in}}(V,T) = k_B T^2 \left(\frac{\partial}{\partial T} \ln Q_N \right)_V, \quad S(V,T) = k_B \left(\ln Q_N + T \frac{\partial}{\partial T} \ln Q_N \right)_V$$

$$p(V,T) = k_B T \left(\frac{\partial}{\partial V} \ln Q_N \right)_T. \qquad (4.80)$$

In Q_N the integration can be extended to $\mathbf{p}, \mathbf{q} \to \infty$ owing to the fast decaying exponential. So far it has not been specified how the correct enumeration of the states $(\mathbf{p}, \mathbf{q})$ is to be done and in what units they have to be measured. Both questions can only be answered on the basis of quantum mechanical considerations in the next section.

The microcanonical and the canonical ensembles must be equivalent. With a view to the properties of the structure function in the microcanonical ensemble it is to be expected that in the canonical ensemble the most likely energy is close to the average $\mathcal{E}_{\text{in}}$. This can be shown by calculating the mean square deviation $(H - \langle H \rangle)^2 = \langle H^2 \rangle - \langle H \rangle^2$. Its evaluation is done by deriving the following identity and making use of (4.79):

$$\begin{aligned} 0 &= \frac{\partial}{\partial \beta} \int (H - \langle H \rangle) \mathrm{e}^{-\beta[H - F(V,T)]} \mathbf{dpdq} \\ &= -\frac{\partial \mathcal{E}_{\text{in}}}{\partial \beta} - \int (H - \langle H \rangle)[H - F + T(\partial_T F)_V] \mathrm{e}^{-\beta(H-F)} \mathbf{dpdq} \\ \Rightarrow -\frac{\partial \mathcal{E}_{\text{in}}}{\partial \beta} &= \int (H - \langle H \rangle)^2 \mathrm{e}^{-\beta(H-F)} \mathbf{dpdq} = \langle H^2 \rangle - \langle H \rangle^2 \end{aligned}$$

$$\Rightarrow \quad \langle H^2 \rangle - \langle H \rangle^2 = k_B T^2 C_V; \qquad \frac{\partial}{\partial \beta} = -k_B T^2 \frac{\partial}{\partial T}. \qquad (4.81)$$

For an ideal fluid component (gas, plasma) with f degrees of freedom the mean square deviation in energy $\langle \Delta \mathcal{E}_{\text{in}} \rangle$ becomes

$$\frac{\langle \Delta \mathcal{E}_{\text{in}} \rangle}{\mathcal{E}_{\text{in}}} = \frac{\langle \Delta T \rangle}{T} = \frac{1}{\sqrt{(f/2)N}}. \tag{4.82}$$

It proves that nearly all macrosystems have energies close to the ensemble average and that this is close to the most likely energy. In this sense the microcanonical and the canonical distributions are equivalent.

The Grand Canonical Ensemble

Imagine a system occupying a volume V in a heat bath of the same substance and volume $V_0 \gg V$ that is capable of exchanging freely particles, for instance through porous walls. In analogy to the canonical system where a fixed number of particles is free to exchange energy with the bath, now we ask for the probability $w(E, V, N)$ to find N particles in V with total energy E. Evidently it is given by

$$w(E, V, N) = C_N \Gamma(E, V, N) \times \Gamma(E_0 - E, V_0, N_0 - N)$$

$$\Rightarrow \ln w(E, V, N) = \ln C_N + \ln \Gamma(E, V, N) + \ln \Gamma(E_0, V_0, N_0) - \frac{E}{k_B T} + \frac{\mu N}{k_B T}$$

$$\Rightarrow \rho(\mathbf{p}, \mathbf{q}) = C' \mathrm{e}^{\beta(\mu N - H_N)}; \quad \beta\mu = -\frac{\partial}{\partial N} \ln \Gamma(E_0, V_0, N)\big|_{N=N_0}, \; \beta = \frac{1}{k_B T}. \tag{4.83}$$

$\rho(\mathbf{p}, \mathbf{q})$ is the density of states of the *grand canonical* ensemble. The new quantity μ is the chemical potential. It is a property of the heat bath. The concomitant state functions of the grand canonical partition function $\mathcal{Z}$ and its thermodynamic potential J are defined in analogy to the canonical quantities,

$$\mathcal{Z}(V, T, \mu) = \sum_{N=0}^{N=N_0} \int \mathrm{e}^{\beta(\mu N - H_N(\mathbf{p}, \mathbf{q}))} d\mathbf{p} d\mathbf{q} = \mathrm{e}^{-\beta J(V, T, \mu)} \tag{4.84}$$

$$\Rightarrow \rho(\mathbf{p}, \mathbf{q}) = \frac{1}{\mathcal{Z}(V, T, \mu)} \mathrm{e}^{\beta(\mu N - H_N)}. \tag{4.85}$$

For the quantities $\exp(\mu N - H_N)$ and $J(V, T, \mu)$ the names *Gibbs factor* and *grand potential* are also in use. The only question not faced yet is the relation of J with the free energy in the grand canonical ensemble; it is answered now.

Let $\overline{E}$ and $\overline{N}$ be the values at which $w(E, V, T)$ reaches its maximum. From (4.81) it follows that the maximum is strongly peaked and lies inside $\Gamma(E_0, V_0, N_0)$. It holds

$$\begin{aligned}
\ln \mathcal{Z}_N &= \ln \int \mu[\Omega(E, N)] \mathrm{e}^{\beta(\mu' N - E)} dE \\
&= \ln \Gamma(\overline{E}, \overline{N}) - \beta \overline{E} + \beta \mu' \overline{N} = -\beta F(V, T, \overline{N}) + \beta \mu' \overline{N} = -\beta J \\
&\Rightarrow J(V, T, \mu') = F(V, T, \overline{N}) - \mu' \overline{N}.
\end{aligned}$$

Note, the chemical potential μ' is associated with the energy $\overline{E}$. We compare the differentials

$$\mathrm{d}J = \left(\frac{\partial J}{\partial T}\right)_{V\mu'} \mathrm{d}T + \left(\frac{\partial J}{\partial V}\right)_{T\mu'} \mathrm{d}V + \left(\frac{\partial J}{\partial \mu'}\right)_{VT} \mathrm{d}\mu' = -S\mathrm{d}T - p\mathrm{d}V - \overline{N}\mathrm{d}\mu'.$$

For convex systems we can assume $\overline{N}$ is very close to $\langle N\rangle$. Then, from the comparison of the previous differential with that of J in (4.84) follows $\mu' = \mu$ and

$$S = -\left(\frac{\partial J}{\partial T}\right)_{V\mu}, \quad p = -\left(\frac{\partial J}{\partial V}\right)_{T\mu}, \quad \overline{N} = -\left(\frac{\partial J}{\partial \mu}\right)_{VT} \tag{4.86}$$

These relations connect the grand potential J with the thermodynamics.

The Variance of the Particle Number N

We derive a measure for the deviation of $\overline{N}$ from $\langle N\rangle$ as follows:

$$\frac{\partial}{\partial(\beta\mu)} \ln \mathcal{Z} = \langle N\rangle, \quad \frac{\partial^2}{\partial(\beta^2\mu^2)} \ln \mathcal{Z} = \langle N^2\rangle - \langle N\rangle^2 = \langle \Delta N^2\rangle. \tag{4.87}$$

$$\frac{\partial^2}{\partial(\beta^2\mu^2)} \ln \mathcal{Z} = k_B T \frac{\partial\langle N\rangle}{\partial\mu} \quad \Rightarrow \quad \frac{\langle \Delta N^2\rangle}{\langle N\rangle^2} = \frac{k_B T}{\langle N\rangle^2}\left(\frac{\partial\langle N\rangle}{\partial\mu}\right)_{VT}. \tag{4.88}$$

$$\left(\frac{\partial V}{\partial p}\right)_{T\overline{N}} \left(\frac{\partial p}{\partial \overline{N}}\right)_{VT} \left(\frac{\partial \overline{N}}{\partial V}\right)_{Tp} = -1 = -V\kappa_T \left(\frac{\partial \mu}{\partial V}\right)_{T\overline{N}} \left(\frac{\partial \overline{N}}{\partial V}\right)_{Tp}. \tag{4.89}$$

In the last step $\partial p/\partial\overline{N} = -\partial\mu/\partial V$ has been used. The equality follows from the total differential $\mathrm{d}F = S\mathrm{d}T - p\mathrm{d}V + \mu d\overline{N} = 0$ in thermal equilibrium at T held constant. The last relation can be further reduced by observing

$$\overline{N} = V\overline{n}(T,p) \quad \Rightarrow \quad \left(\frac{\partial \overline{N}}{\partial V}\right)_{Tp} = \overline{n}$$

$$\mu = f(\overline{n}, T) \quad \Rightarrow \quad \left(\frac{\partial \mu}{\partial V}\right)_{T\overline{N}} = \left(\frac{\partial \mu}{\partial \overline{n}}\right)_{T} \left(\frac{\partial \overline{n}}{\partial V}\right)_{T} = -\overline{n}\left(\frac{\partial \mu}{\partial \overline{N}}\right)_{VT}$$

$$\Rightarrow \quad \left(\frac{\partial \overline{N}}{\partial \mu}\right)_{VT} = -\overline{n}\left(\frac{\partial \mu}{\partial V}\right)^{-1}_{T\overline{N}} = \kappa_T V \overline{n}^2. \tag{4.90}$$

Substitution of (4.90 c) in (4.88) leads to the desired result

$$\frac{\langle \Delta N^2\rangle}{\langle N\rangle} = \frac{\langle \Delta n^2\rangle}{\langle n\rangle} V = \kappa_T \overline{n} k_B T. \tag{4.91}$$

This is a remarkable outcome, intuitively clear a posteriori. Consider a dilute fluid, for instance the fully ionized plasma with Maxwellian velocity distribution. A density fluctuation means diffusion in itself. Thus the fluctuation level $\langle \Delta N^2 \rangle$ will be proportional to the average thermal energy per particle $k_B T$ and to the particle density squared; but there is also the counteracting potential of the pressure p that inhibits arbitrarily high accumulation of particles. In the plasma $\kappa_T = p^{-1} = [\langle n \rangle k_B T]^{-1}$ and hence

$$\frac{\sqrt{\langle \Delta n^2 \rangle}}{\langle n \rangle} = \frac{\sqrt{\overline{n}}}{V^{1/2} \langle n \rangle} \simeq \frac{1}{\sqrt{\langle n \rangle V}}. \tag{4.92}$$

This is the result known of free particles moving without collisions. In most situations (4.91) yields very small values. In dense fluids the isothermal compressibility κ_T will be low so that the identification of the most likely number of particles $\overline{N}$ with their average number $\langle \Delta N \rangle$ in the fixed volume V is highly justified. All depends on the stiffness κ_T of matter. From Rayleigh-Thomson scattering the phenomenon of *critical opalescence* is well known: Transparent liquids appear cloudy owing to large density fluctuations in the region of second order phase transitions. There the compressibility κ_T may assume very large values.

The microcanonical, the canonical and the grand canonical ensembles are equivalent to each other because at fixed temperature and volume the overwhelming number of states in phase space assumes values of energy E close to its ensemble average $\langle E \rangle$ and particle numbers N close to the ensemble average $\langle N \rangle$. As the density of states in convex systems is strongly peaked the averages coincide with the most likely values $\overline{E}$ and $\overline{N}$. For these properties the more flexible grand canonical distribution may advantageously be used in situations where energy and particle number are fixed. With respect to $\langle N \rangle \gg \sqrt{\langle N^2 \rangle}$ critical phenomena make an exception.

Shortcomings of Classical Entropy from (4.49): The Gibbs Paradox

Consider a box of volume $2V$ separated in the middle by a wall and the two chambers filled one with He^3 and the other chamber filled with He^4, both at room temperature and equal pressure. The total entropy is S_0. Remove the wall and wait until a new equilibrium is established. According to (4.60) the entropy increase in this process is $\Delta S = S_0 \ln 2$. Now substitute He^4 by He^3 and let the two identical gases mix again as before. The entropy increase is again $\Delta S = S_0 \ln 2$, however, in the case of identical gases in the chambers the state before mixing and after do not differ in their macroscopic parameters. Thus, ΔS must be zero. The solution of the paradox is due to a correction given by Sackur-Tetrode of the formula (4.49). It refers to the correct counting of states. Physics of particles of the same species, e.g. electrons, does not depend on *order of numbering* j of the particles in the Hamiltonian (4.41). Two classical states $(\mathbf{p}_1, \mathbf{q}_1; \mathbf{p}_2, \mathbf{q}_2)$ and $(\mathbf{p}_2, \mathbf{q}_2; \mathbf{p}_1, \mathbf{q}_1)$ of two identical particles are two points in phase space. However, as we know for example the classical one particle distribution function $f(\mathbf{x}, \mathbf{v}, t)$ is obtained from the two particle distribution function $f_2(\mathbf{x}_1, \mathbf{v}_1; \mathbf{x}_2, \mathbf{v}_2)$ by integration either of $\mathbf{x}_2, \mathbf{v}_2$ or of $\mathbf{x}_1, \mathbf{v}_1$ over its particular phase space,

$$f(\mathbf{x}, \mathbf{v}, t) = \int f_2(\mathbf{x}, \mathbf{v}; \mathbf{x}_2, \mathbf{v}_2)\mathrm{d}\mathbf{x}_2\mathrm{d}\mathbf{v}_2 = \int f_2(\mathbf{x}_1, \mathbf{v}_1; \mathbf{x}, \mathbf{v})\mathrm{d}\mathbf{x}_1\mathrm{d}\mathbf{v}_1.$$

Applying this idea of symmetry to the counting of physical states on the energy shell $\Omega(E)$ one finds that the total number of dynamic states $(\mathbf{p}, \mathbf{q})$ of the He^3 gas has to be divided by the number of permutations $N! \simeq N \ln N$. Any permutation of the N particles represents an individual point on Σ, however as physical states, whether classical or quantum mechanical, all permutations are identical. The measure $d\mu$ in (4.42) is too large by the factor $N!$. As a consequence $\Gamma(E)$ of the hypersphere (4.44) is to be substituted by the volume $\Gamma(E)/N!$. With this change (4.60) goes over into

$$S(E, V) = Nk_B \ln\left[\frac{V}{N}\left(\frac{2\pi m E}{N}\right)^{3/2}\right] + \frac{3}{2}Nk_B + \text{const.} \tag{4.93}$$

It avoids Gibbs paradox because of $2V/2N$ after mixing.

And yet, another gedankenexperiment. Imagine chamber one filled with He^3 atoms of nuclear spin orientation (+) relative to an external homogeneous magnetic field B and chamber two filled with He^3 atoms of negative orientation (−) with respect to B. Before mixing the B field is shut down abruptly so that the two orientations survive mixing for a while. An apparatus able to discern between the two species measures an entropy increase by the factor ln 2 after mixing whereas for the chemist the two species are identical and the Gibbs paradox persists. We conclude: The value of the entropy depends on the degree of *coarse graining*, sometimes expressed also as degree of knowledge. Put into other words, depending on the set of phenomena to be considered it may be appropriate to introduce a scale of "disorder" or lack of knowledge and to associate to each degree its own entropy. See for further readings [3], Sects. 1, 2.

4.4 Essentials of Thermostatistics: Quantum Systems

4.4.1 The Density Matrix

Experience has shown that all systems obey the principles of quantum theory. Any classical description, i.e., equations in which the Planck constant $\hbar$ is absent, must follow from quantum mechanics (we use Dirac's quantity in place of Planck's $h = 2\pi\hbar$).

So far such a derivation is feasible for classes of problems, not in full generality. Often we make use of a classical description, as for instance of the isothermal rarefaction wave or the quark-gluon plasma, just because we are convinced that such a description represents an acceptable approximation. Although a classical model maybe simpler and more intuitive the problem of replacement still persists on the fundamental level. For a more detailed discussion of quantum vs classical systems see Chap. 8.

Quantum theory imposes three changes to classical thermostatistics:

- Classical energies $E = H(\mathbf{p}, \mathbf{q})$ are continuous with smooth energy density $\rho(\mathbf{p}, \mathbf{q})$; quantized energy levels are irregularly spaced with gaps in between
- In addition to classical symmetry there is the distinction between Bosons and Fermions; it is of pure quantum nature.
- Appearance of the new constant $\hbar$. Within classical theory an absolute determination of entropy is not possible.

In contrast to the number of classical states $\mu(\Omega)/N!$ forming a continuum, in a quantum system there may be a gap of states just at a particular value of E. In not too small systems this is not a problem. The macroscopic energy states $|\psi_E\rangle$ are highly degenerate in any energy interval dE and the actual irregular δ-like distribution can be replaced by a smooth density of states $\rho(E)$. A necessary condition for such a procedure however is $E_{j+1} - E_j \ll k_B T$.

The Average of an Operator

If a definite quantum state $|\psi\rangle$ of a system is known every physical quantity *(observable)* O is the average of a Hermitian operator $\hat{O}$, $\langle O\rangle = \langle\psi|\hat{O}|\psi\rangle$. We recall that an operator is defined by its matrix elements. The *pure* state $|\psi\rangle$ is a linear combination of a complete set of orthogonal states $|S\rangle$, hence

$$\langle O\rangle = \langle\psi|\hat{O}|\psi\rangle = \langle\psi|\sum_S |S\rangle\langle S|\hat{O}|\psi\rangle = \sum_S \langle S|\hat{O}|\psi\rangle\langle\psi|S\rangle; \quad \sum_S |S\rangle\langle S| = 1.$$

With the help of the density operator $\hat{\rho}$ the average can be cast in the compact form

$$\hat{\rho} = |\psi\rangle\langle\psi| \quad \Rightarrow \quad \langle O\rangle = \sum_S \langle S|\hat{O}\hat{\rho}|S\rangle = \mathrm{Tr}(\hat{O}\hat{\rho}).$$

The average $\langle O\rangle$ is invariant with respect to a change of basis from $|S\rangle$ to $|T\rangle$, however its diagonal form may be lost:

$$\begin{aligned}\hat{\rho} &= \sum_S \langle S|\hat{O}\hat{\rho}S\rangle = \sum_{S,T,T'} \langle S|T'\rangle\langle T'|\hat{O}\hat{\rho}|T\rangle\langle T|S\rangle \\ &= \sum_{S,T,T'} \langle T|S\rangle\langle S|T'\rangle\langle T'|\hat{O}\hat{\rho}|T\rangle = \sum_T \langle T|\hat{O}\hat{\rho}|T\rangle.\end{aligned}$$

Thus, the average can be evaluated in any orthonormal system $\{|T\rangle\}$, $\sum |T\rangle\langle T| = 1$. $\hat{\rho}$ is a Hermitian operator. It holds $(\hat{O}\hat{\rho})^\dagger = \hat{\rho}^\dagger\hat{O}^\dagger = \hat{\rho}\hat{O}$.

Often we do not know the state vector of a system, and we are not interested in it either. All we can know of a state vector $|\psi_r\rangle$ is the likelihood P_r to be in the state $|\psi_r\rangle$. If there are $s \leq \infty$ such states $|\psi_r\rangle$ occupied with probabilities P_r the average of the operator $\hat{O}$ is evidently given by

$$\langle O\rangle = \sum_{r=1}^{s} P_r\langle\psi_r|\sum_S |S\rangle\langle S|\hat{O}|\psi_r\rangle = \sum_S \langle S|\hat{O}\sum_{r=1}^{s} P_r|\psi_r\rangle\langle\psi_r|S\rangle; \quad \sum_{r=1}^{s} P_r = 1.$$

$\langle O\rangle$ is the average of a *mixed state* $|\psi\rangle = \sum c_r|\psi_r\rangle$ where only the time independent $|c_r|^2 = P_r$ are known. Defining the density operator $\hat{\rho}$ for the statistical mixture,

$$\hat{\rho} = |\psi\rangle\langle\psi| = \sum_{r=1}^{s} P_r|\psi_r\rangle\langle\psi_r| \tag{4.94}$$

the average $\langle O\rangle$ reduces to the compact of the pure state,

$$\langle O\rangle = \mathrm{Tr}(\hat{O}\hat{\rho}). \tag{4.95}$$

The formula holds for orthogonal as well as for non-orthogonal mixtures of states $|\psi_r\rangle$. $\hat{\rho}$ from (4.94) is Hermitian and normalized to unity,

$$\langle\psi|\psi\rangle = \sum_r \langle\psi|\psi_r\rangle\langle\psi_r|\psi\rangle = \sum_r |c_r|^2 = \sum_r P_r = 1. \tag{4.96}$$

It is easily shown that only the pure state $|\psi\rangle$ satisfies $\hat{\rho}^2 = \hat{\rho} \Rightarrow \hat{\rho} = 1$. In the Schrödinger picture the density operator evolves in time according to

$$\imath\hbar\partial_t|\psi\rangle = H|\psi\rangle, \quad -\imath\hbar\langle\psi|\partial_t = \langle\psi|H \quad \Rightarrow \quad \frac{\partial}{\partial t}\hat{\rho} = \frac{1}{\imath\hbar}[H, \hat{\rho}] \tag{4.97}$$

In general the quantum state $|\psi\rangle$ of a system to the energy eigenvalue E is a superposition of eigenstates $|\psi_r\rangle$ to E, $|\psi\rangle = \sum c_l|\psi_l\rangle$. The average of an observable O is given by

$$\langle O\rangle = \sum_r \langle\psi_r|O\sum_l c_l|\psi_l\rangle\langle c_l|\psi_l|\psi_r\rangle = \sum_{i,r} c_l c_r^*\langle\psi_r|O|\psi_l\rangle. \tag{4.98}$$

In the averaging procedure interference terms $c_l c_r^*$ appear. We expect that they are associated with randomly fluctuating phases, like the products of E-field modes of ordinary light of intensity I, not observable on time scales of changes of macroscopic variables, like volume, particle density, pressure, etc. Therefore it is legitimate to postulate:

$$c_l c_r^* = 0 \text{ for } l \neq r, \quad c_l c_r^* = |c_r|^2 \text{ for } l = r. \tag{4.99}$$

The postulate is in perfect analogy to the assumption on the classical states $(\mathbf{p}, \mathbf{q})$. In the quantum statistics they are replaced by the the number of eigenstates to the energy E. As a consequence of the setting (4.99) the average (4.98) needs to be normalized,

$$\langle O \rangle = \frac{|c_r|^2 \langle \psi_r | O | \psi_l \rangle}{\sum_r |c_r|^2}. \tag{4.100}$$

If the energy states are highly degenerate to form a convex system all arguments on classical systems apply in an identical manner. By the relations (4.99) the quantum mechanical version of the *microcanonical ensemble* (4.100) is introduced. The entropy is now given by

$$S(E, V) = k_B \ln W. \tag{4.101}$$

W is the number of all eigenstates to the energy E accessible to the system. The absolute temperature T and the pressure p follow by partial derivation.

4.4.1.1 The Canonical Ensemble

for a system of volume V and absolute temperature T held fixed in the heat bath follows from the microcanonical ensemble by the same arguments as for the classical system. Replacement of the classical states $\{(\mathbf{p}, \mathbf{q})\}$ by the energy eigenstates $\{E_l\}$ results in the probability of the individual state $w(E_l)$ and the partition function $Q_N(V, T)$ as follows,

$$w(E_l) = \frac{\mathrm{e}^{-\beta E_l}}{Q_N(V, T)} = \rho_{ll}; \quad Q_N(V, T) = \sum_l \mathrm{e}^{-\beta E_l}. \tag{4.102}$$

Analogously, the mean energy is obtained as

$$\mathcal{E}_{\text{in}} = \langle E \rangle = \frac{\sum_l E_l \exp(-\beta E_l)}{Q_N(V, T)} = -\frac{\partial \ln Q_N(V, T)}{\partial \beta}.$$

Here each state is counted seperately: $E_1 \le E_2 \le E_3 \le \ldots\ldots \le E_l \le \ldots\ldots$ Introducing the degree of degeneracy g_k counting is by $E_1 < E_2 < E_3 < \ldots\ldots < E_k < \ldots\ldots$ and the foregoing quantities read obviously

$$w(E_k) = g_k \frac{\mathrm{e}^{-\beta E_k}}{Q_N(V, T)} = g_k \rho_{kk}; \quad Q_N(V, T) = \sum_k g_k \mathrm{e}^{-\beta E_k}.$$

In the representation-free form of operators the expressions can be written as

$$\hat{\rho} = \frac{\mathrm{e}^{-\beta H}}{Q_N(V, T)}; \quad Q_N(V, T) = \mathrm{Tr}(\mathrm{e}^{-\beta H}); \quad \mathcal{E}_{\text{in}} = \langle H \rangle = -\frac{\partial \ln \mathrm{Tr}(\mathrm{e}^{-\beta H})}{\partial \beta}. \tag{4.103}$$

The link of the partition function with the thermodynamics is established by the free energy $F(V,T)$,

$$F(V,T) = -\beta \ln Q_N(V,T). \tag{4.104}$$

The entropy $S(V,T)$ follows from (4.80). Alternatively it is

$$\begin{aligned} S(V,T) &= k_B\beta(\mathcal{E}_{\text{in}} - F) = k_B\{\beta \sum_l w_l E_l + \ln Q_N\} \\ &= k_B\{-\sum_l w_l \ln(Q_N w_l) + \ln Q_N\} = -k_B \sum_l w_l \ln w_l; \end{aligned}$$

owing to $\sum_l w_l = 1$. In an arbitrary basis this reads

$$S(V,T) = -k_B \mathrm{Tr}(\hat{\rho} \ln \hat{\rho}). \tag{4.105}$$

4.4.1.2 The Grand Canonical Ensemble

In analogy to the classical system the partition function $\mathcal{Z}(V,T,\mu)$ is to be defined by

$$\mathcal{Z}(V,T,\mu) = \sum_{N=0}^{N=N_0} \mathrm{e}^{\beta\mu N}\, Q_N(V,T,N) = \sum_{N,l} \mathrm{e}^{\beta(\mu N - E_{N,l})} = \sum_N \mathrm{Tr}\ \mathrm{e}^{(\mu N - H_N)}. \tag{4.106}$$

The probability, i.e., *fractional population*, for an energy eigenstate with eigenvalue $E_{N,l}$ and N particles in the volume V, $w(V,T,N,E_{N,l})$ is

$$w(V,T,N,E_{N,l}) = \frac{\mathrm{e}^{\beta(\mu N - E_{N,l})}}{\mathcal{Z}(V,T,\mu)} = [\rho_N]_{ll}; \quad \hat{\rho} = \frac{\sum_N \exp[\beta(\mu N - H_N)]}{\mathcal{Z}(V,T,\mu)}. \tag{4.107}$$

$[\rho_N]_{ll}$ is the normalized density matrix element, $\hat{\rho}$ is the complete normalized density operator. By setting

$$J(V,T,\mu) = -k_B T \ln \mathcal{Z}(V,T,\mu), \tag{4.108}$$

the connection with the thermodynamic quantities is established. All operations are identical with those for the classical grand partition function. For example, in terms of $\hat{\rho}$ of (4.107) the entropy is given by (4.105).

4.4.2 Ideal Systems

The energy eigenvalue $E_l^{(N)}$ of a set of N noninteracting particles is the sum of N single particle energies E_k,

$$E_l^{(N)} = \sum_{k=0}^{k=s(N)} n_k E_k; \quad N = n_0 + n_1 + \cdots + n_s, \tag{4.109}$$

$\{n_k\}$, $n_k \geq 0$, is the set of possible *occupation numbers* of the individual energy levels. The probability at which a particular array is realized is given by

$$w(E_l, N, V) = \frac{e^{\beta(\mu N - E_l^{(N)})}}{\mathcal{Z}} = \frac{1}{\mathcal{Z}} \prod_{k=0}^{\infty} e^{\beta(\mu - E_k)n_k}; \quad k > s(N) \to n_k = 0.$$

In the evaluation of $\mathcal{Z}$ the grand canonical concept shows its power. By choosing the heat bath sufficiently large $N_0 = \infty$ can be set. It holds

$$\mathcal{Z}(V, T, \mu) = \sum_{N, E_l^{(N)}} e^{\beta(\mu N - E_l^{(N)})} = \sum_{\{n_k\}} \prod_{k=0}^{\infty} e^{\beta(\mu - E_k)n_k} = \prod_{k=0}^{\infty} \sum_{\{n_k\}} e^{\beta(\mu - E_k)n_k}. \tag{4.110}$$

The probability to find the jth energy level E_j occupied by n_j particles is

$$w(n_j) = \frac{1}{\mathcal{Z}} e^{\beta(\mu - E_j)n_j} \sum_{\{n_k\}}{}' \prod_{k \neq j}^{\infty} e^{\beta(\mu - E_k)n_k} = -\frac{\partial}{\partial(\beta n_j E_j)} \ln \mathcal{Z}. \tag{4.111}$$

4.4.2.1 The Ideal Bose Gas

The Bose–Einstein Distribution

The wave vector of particles with even spin is symmetric with respect to index exchanges of two particles. The individual occupation number n_j can be as high as N. The average occupation number $\langle n_j \rangle$ of level E_j is

$$\langle n_j \rangle = \sum_{n_j=0}^{\infty} n_j w(n_j) = -\sum_{n_j=0}^{\infty} n_j \frac{\partial}{\partial(\beta n_j E_j)} \ln \mathcal{Z} = -\frac{\partial}{\partial(\beta E_j)} \ln \mathcal{Z}$$

$$\mathcal{Z}(T, V, \mu) = \prod_{k=0}^{\infty} \sum_{n_k=0}^{\infty} e^{\beta(\mu - E_k)n_k} = \prod_{k \geq 0} \frac{1}{1 - e^{\beta(\mu - E_k)}}. \tag{4.112}$$

$$\langle n_j \rangle = -\frac{\partial}{\partial(\beta E_j)} \ln \mathcal{Z} = \frac{1}{e^{\beta(E_j - \mu)} - 1}. \tag{4.113}$$

This is the *Bose–Einstein distribution* of noninteracting Bose particles in thermal equilibrium in the fixed volume V. The total mean energy $E = \mathcal{E}_{\text{in}}$ and the mean number of particles are

$$E = \sum_{N,l} E_l^{(N)} w(N, E_l^{(N)}) = \langle \sum_k n_k E_k \rangle = \sum_k \frac{E_k}{e^{\beta(E_k-\mu)} - 1};$$
$$\langle N \rangle = \langle \sum_{k \geq 0} n_k \rangle = \sum_k \frac{1}{e^{\beta(E_k-\mu)} - 1}. \tag{4.114}$$

Thus, by eliminating μ from $E = E(V, T, \mu)$ and $\langle N \rangle = \langle N \rangle(V, T, \mu)$ the internal energy $\mathcal{E}_{\text{in}} = E(V, T, \langle N \rangle)$ is recovered.

Properties of the Bose–Einstein Distribution

(1) Owing to the dependence of the distribution on the difference $E - \mu$ one is free in the choice of the zero of E.
(2) If $T \geq 0$ together with $\lim T = 0$ and E_0 is the lowest energy level $E_0 = \min(E)$ then follows $\mu \leq E_0$ because $\langle n \rangle \geq 0$.
(3) *Bose–Einstein condensation.* Assume $\langle N \rangle \gg 1,\ k_B T \ll E_1 - E_0$ and the degeneracy factors $g_1 \simeq g_0$. It follows

$$\frac{1}{e^{\beta(E_1-\mu)} - 1} \geq \frac{1}{e^{\beta(E_1-E_0)} - 1}; \quad \frac{1}{e^{\beta(E_1-\mu)} - 1} < \frac{k_B T}{E_1 - E_0} \ll 1 \tag{4.115}$$

for $T \longrightarrow 0$, i.e., nearly all particles are in the ground state to form a Bose–Einstein condensate.
(4) Entropy S, pressure p and specific heat capacity C_V for the ideal Bose system are given by

$$S = -\left(\frac{\partial J}{\partial T}\right)_{V\mu} = \frac{\partial}{\partial T}\left(\frac{1}{\beta}\ln \mathcal{Z}\right) = k_B \sum_{k\geq 0}\left[\frac{e^{\beta(E_k-\mu)}}{e^{\beta(E_k-\mu)} - 1} - \ln\left(1 - e^{-\beta(E_k-\mu)}\right)\right],$$
$$p = -\left(\frac{\partial J}{\partial V}\right)_{T\mu} = \frac{\partial}{\partial V}\left(\frac{1}{\beta}\ln \mathcal{Z}\right) = -\sum_{k\geq 0}\frac{\partial E_k/\partial V}{e^{\beta(E_k-\mu)} - 1} = -\sum_{k\geq 0}\langle n_k\rangle \frac{\partial E_k}{\partial V},$$
$$C_V = \left(\frac{\partial E}{\partial T}\right)_V = \sum_{k\geq 0} E_k \frac{\partial \langle n_k\rangle}{\partial T}. \tag{4.116}$$

Entropy changes are due to changes in the occupation numbers of the single levels, *adiabatic* changes of pressure are due to shifts of the individual energy levels $E_k(V)$, but no changes in the occupation numbers occur and, consequently, S remains invariant. At this point compare with the adiabatic invariants (2.81) and (2.83). The interpretation of C_V is evident.

Fluctuations

$$\langle n_l^2 \rangle = \sum_{n_l \geq 0} n_l^2 w(n_l) = \frac{1}{\mathcal{Z}}\frac{\partial^2}{\partial(\beta E_l)^2}\mathcal{Z} = \frac{1}{\beta^2}\left\{\frac{\partial}{\partial E_l}\frac{1}{\mathcal{Z}}\frac{\partial \mathcal{Z}}{\partial E_l} + \frac{1}{\mathcal{Z}^2}\left(\frac{\partial \mathcal{Z}}{\partial E_l}\right)^2\right\}$$
$$= -\frac{1}{\beta}\frac{\partial \langle n_l \rangle}{\partial E_l} + \langle n_l \rangle^2 \Rightarrow \langle \Delta n^2 \rangle = \langle n_l^2 \rangle + \langle n_l \rangle.$$

Thus, the relative fluctuation squared results as large as its mean value $\langle n_l \rangle^2$,

$$\frac{\langle \Delta n_l^2 \rangle}{\langle n_l \rangle^2} = 1 + \frac{1}{\langle n_l \rangle}. \tag{4.117}$$

Compared to the fluctuation of classical free particles the ideal Bose gas tends to "cluster", as researchers occasionally say. For critique see last question in Self-assessment.

Photons in Thermal Equilibrium

The photons are the quanta of the electromagnetic radiation field. A particular field configuration is described by the number of photons $n_{\mathbf{k}}$ of momenta $\hbar\mathbf{k}$ and polarization vector $\epsilon_{\mathbf{k}}$ of a possible *eigenmode* of the field, i.e., n_1 photons in the first mode of propagation direction $\mathbf{k}_1$, n_2 photons in the mode $\mathbf{k}_2$, etc., and concomitant polarizations. In compact form we write $\{|n_{\mathbf{k}}\rangle\}$ for the field configuration (state vector) $\{|n_{\mathbf{k}}\rangle\} = |n_1, n_2, ..., n_{\mathbf{k}}, ...\rangle$ and suppress the polarization index. For field quantization the interested reader may have a look at Chap. 7. In vacuum the energy H of the radiation field of the **k**th mode is given by the number of photons $n_{\mathbf{k}}$,

$$H = \hbar\omega_{\mathbf{k}}(n_{\mathbf{k}} + \frac{1}{2}); \quad \omega^2 = c^2\mathbf{k}^2, \quad \mathbf{k} = \frac{2\pi}{L}\mathbf{n}. \tag{4.118}$$

By fixing volume and temperature the free energy is completely determined, $F = F(V;T)$, hence $\langle N(V,T)\rangle = \langle N\rangle(V,T)$ and $\mu = -(\partial J/\partial\langle N\rangle)_{V,T} = 0 \Rightarrow J = F$. Owing to the factor 1/2 in the Hamiltonian (4.118) the zero point energy, i.e., the energy of the vacuum, is infinite. Fortunately, in nearly all experiments only energy differences are measured so that E can be rescaled by subtracting the infinity E_0 from $E \Rightarrow E' = E - E_0$ and by setting $\mathcal{E}_{\text{in}} = \langle E'\rangle$. The reader may convince himself that all thermodynamic quantities come out right by the replacement

$$\mathcal{Z} = \sum_{\{n_{\mathbf{k}}\}} \prod e^{-\beta\hbar\omega_{\mathbf{k}}(n_{\mathbf{k}}+1/2)} \Rightarrow \mathcal{Z}' = \sum_{\{n_{\mathbf{k}}\}} \prod e^{-\beta\hbar\omega_{\mathbf{k}} n_{\mathbf{k}}};$$

$$w'(n_{\mathbf{k}}) = \frac{1}{\mathcal{Z}'} e^{-\beta\hbar\omega_{\mathbf{k}} n_{\mathbf{k}}} \sum_{\{n_{\mathbf{k}'}\}} \prod_{\mathbf{k}'\neq\mathbf{k}} e^{-\beta\hbar\omega_{\mathbf{k}'} n_{\mathbf{k}}}.$$

4.4.2.2 Black Body Radiation

The *Planck distribution* for the mode **k** results as

$$\langle n_{\mathbf{k}}\rangle = \sum_{n_{\mathbf{k}}\geq 0} n_{\mathbf{k}} w(n_{\mathbf{k}}) = \sum_{n_{\mathbf{k}}\geq 1} n_{\mathbf{k}} w'(n_{\mathbf{k}}) = -\frac{\partial}{\partial(\beta\hbar\omega_{\mathbf{k}})} \ln(\mathcal{Z}') = \frac{1}{\exp(\beta\hbar\omega_{\mathbf{k}}) - 1}. \tag{4.119}$$

In the cube of length L the number of modes $d\mathcal{N}$ per polarization ϵ is

$$d\mathcal{N} = 4\pi\mathbf{n}^2 d|\mathbf{n}| = 4\pi\frac{L^2}{4\pi^2}\mathbf{k}^2\frac{L}{2\pi}d|\mathbf{k}| = \frac{L^3\omega^2}{2\pi^2c^3}\mathrm{d}\omega \Rightarrow g(\omega) = 2\frac{\partial^2\mathcal{N}}{\partial V\partial\omega} \tag{4.120}$$

$g(\omega)$ is the *degeneracy factor* for each mode of the two polarization directions, $\mathbf{n}$ is the position vector of the Bravais lattice (n_1, n_2, n_3) of integers. Relation (4.120) is encountered whenever directional quantization is asked for. The spectral energy density $\rho(\omega, T)$ and its intensity $I(\omega, T)$ for both polarizations are

$$\rho(\omega, T) = \frac{\partial\varepsilon_{\mathrm{in}}(T)}{\partial\omega} = \hbar\omega\langle n_\omega\rangle = \frac{\hbar\omega^3}{\pi^2c^3}\frac{1}{\exp(\beta\hbar\omega) - 1}. \tag{4.121}$$

$$I(\omega, T) = \frac{c}{4\pi}\rho(\omega, T) = \frac{\hbar\omega^3}{4\pi^3c^2}\frac{1}{\exp(\beta\hbar\omega) - 1}. \tag{4.122}$$

The dimensions of $\rho(\omega, T)$ and $I(\omega, T)$ are $[\mathrm{Js/m^3}]$ and $[\mathrm{J/m^2}]$, respectively. $\rho(\omega, T)\mathrm{d}\omega$ is the energy per unit volume within the frequency interval $\mathrm{d}\omega = 2\pi\,\mathrm{d}v$. $I(\omega, T)\;\mathrm{d}f\;\mathrm{d}w\;\mathrm{d}\Omega$ is the energy emitted per unit time from an area $\mathrm{d}f$ within the frequency interval $\mathrm{d}\omega = 2\pi\,\mathrm{d}v$ into a solid angle element $\mathrm{d}\Omega$. Owing to isotropy it is independent of direction.

The relation between the energy density $\rho(\omega, T)$ and the intensity $I(\omega, T)$ is fundamental. Consider n particles per unit volume, each of them moving with $\mathbf{v} = v\boldsymbol{\Omega}$ into its individual direction $\boldsymbol{\Omega}$ with equal likelihood for all $\boldsymbol{\Omega}$. Then the number of particles dn moving into the solid angle $d\boldsymbol{\Omega}$ per unit time is the fraction $nd(\boldsymbol{\Omega}/4\pi)$. Correspondingly, the energy flux density is $d\mathbf{I} = \mathbf{I}\mathrm{d}\Omega = I\boldsymbol{\Omega}\mathrm{d}\Omega = \epsilon n\mathbf{v}\mathrm{d}(\boldsymbol{\Omega}/4\pi) = \rho v\boldsymbol{\Omega}\mathrm{d}(\boldsymbol{\Omega}/4\pi)$. Thus, with photons $v = c$ and

$$I = \frac{c}{4\pi}\rho \tag{4.123}$$

under the condition of isotropy.

Internal energy $\mathcal{E}_{\mathrm{in}}$, entropy S, radiation pressure p, and specific heat C_V per unit frequency interval result from (4.116) as

$$\begin{aligned}
\mathcal{E}_{\mathrm{in}}(\omega, V, T) &= \rho(\omega, T)V = \frac{\hbar\omega^3 V}{\pi^2c^3}\frac{1}{\exp(\beta\hbar\omega) - 1},\\
S(\omega, V, T) &= k_B V\frac{\omega^2}{\pi^2c^3}\left\{\frac{\beta\hbar\omega}{\exp(\beta\hbar\omega) - 1} - \ln[1 - \exp(-\beta\hbar\omega)]\right\},\\
p(\omega, T) &= -\left(\frac{\partial\mathcal{E}_{\mathrm{in}}}{\partial V}\right)_S = -\frac{\partial g(\omega)V}{\exp(\beta\hbar\omega) - 1}\left(\frac{\partial\hbar\omega}{\partial V}\right)_S = \frac{1}{3}\rho(\omega, T),\\
C_V(\omega, T) &= V\frac{\partial\rho(\omega, T)}{\partial T} = \frac{\rho(\omega, T)}{T}\frac{\hbar\omega}{k_BT}\frac{\exp(\beta\hbar\omega)}{\exp(\beta\hbar\omega) - 1}.
\end{aligned} \tag{4.124}$$

In determining the pressure p the number of modes $\mathscr{N} = g(\omega)V$ is to be kept constant. Further, at constant S holds $\omega V^{-1/3} = \mathrm{const}_a$ under adiabatic changes of the volume V. This is in agreement with the adiabatic coefficient $\gamma = 4/3$ for photons. Under a general adiabatic change of size and shape of the volume V enclosing the thermal radiation field within perfectly reflecting walls we can assume that thermal equilibrium is established again after a while. It implies for all frequencies $\beta\omega =$ const, or for the temperature T and the linear dimension R of V

$$T V^{1/3} \sim T R = \mathrm{const} \tag{4.125}$$

in agreement with (4.28). Under changes of constant entropy the number of photons in each $\mathbf{k}$ mode is conserved, as is the case with the quantum levels of massive particles. For the relevance of relation (4.125) see [4], Sect. 15.5 on the cosmic microwave radiation background.

The Stefan–Boltzmann Law

Is the total radiant energy density $\rho(T) = \varepsilon_{\mathrm{in}}(T)$, or the total radiated power per unit volume $I(T)$, respectively. With $\xi = \beta\hbar\omega$ holds

$$\rho(\omega, T)\mathrm{d}\omega = \frac{\hbar}{\pi^2 c^3}\left(\frac{k_B}{\hbar}\right)^4 \frac{\xi^3 \mathrm{d}\xi}{\exp\xi - 1}; \quad \int \frac{\xi^3}{\exp\xi - 1}\mathrm{d}\xi = \frac{\pi^4}{15}$$

$$\Rightarrow \rho(T) = \frac{\pi^2 k_B^4}{15 c^3 \hbar^3} T^4 = C T^4; \quad C = 7.55 \times 10^{-16}[\mathrm{Jm}^{-3}\mathrm{K}^4]. \tag{4.126}$$

$$I(T) = \frac{c}{4\pi} 2\pi\rho(T) \int_{\pi/2}^{0} \cos\theta \sin\theta \mathrm{d}\theta = \frac{c}{4}\rho(T) = \sigma T^4. \tag{4.127}$$

$$\sigma = \frac{\pi^2 k_B^4}{60 c^2 \hbar^3} = 5.66 \times 10^{-8}[\mathrm{Js}^{-1}\mathrm{m}^{-2}\mathrm{K}^{-4}].$$

In the region of soft photons, $\xi = \hbar\omega/k_B T \ll 1$ the exponential can be expanded and one arrives at the

Rayleigh–Jeans Limit

$$\rho(\omega, T) = \frac{\omega^2}{\pi^2 c^3} k_B T = g(\omega) k_B T \quad \Rightarrow \quad \langle \varepsilon_{\mathbf{k}} \rangle = k_B T. \tag{4.128}$$

This is a purely classical result obtained under the assumption of equipartition, i.e., that each linearly polarized mode $\mathbf{k}$ of photons $\hbar\omega_{\mathbf{k}}$ contributes by $k_B T/2$ to the mean energy $\langle \varepsilon_{\mathbf{k}} \rangle = k_B T$. Planck's constant is absent. The opposite limiting case for hard photons, $\xi = \hbar\omega/k_B T \gg 1$, leads to

Kirchhoff's Law:

$$\rho(\omega, T) = g(\omega)\hbar\omega \exp(-\frac{\hbar\omega}{k_B T}); \quad \hbar\omega/k_B T \gg 1. \tag{4.129}$$

Wien's Displacement Law

Expresses a simple property of Planck's law that is of practical relevance for quick estimates in radiation physics. The maximum of the Planck curve $\rho(\omega, T)$ lies at

$$\xi e^{\xi} = 3(e^{\xi} - 1) \quad \Rightarrow \quad \xi = \frac{\hbar\omega_{\max}}{k_B T} = 2.82 \quad \Rightarrow \quad \omega_{\max} = 3.69 \times 10^{11} T[\mathrm{s}^{-1}\mathrm{K}]. \tag{4.130}$$

From $\rho(\lambda, T)\mathrm{d}\lambda = -\rho(\omega, T)\mathrm{d}\omega = 2\pi c/\lambda^2 \rho(\omega, T)d\lambda$ the maximum of $\rho(\lambda, T)$ is found at

$$\xi e^{\xi} = 5(e^{\xi} - 1) \quad \Rightarrow \quad \xi = \frac{\hbar\omega_{\max}}{k_B T} = 4.965 \quad \Rightarrow \quad \lambda_{\max} = 2.89 \times 10^{6} T \ \ [\mathrm{nmK}]. \tag{4.131}$$

4.4.2.3 Principle of Detailed Balance

Consider an isolated two level system of states $|1\rangle$ and $|2\rangle$ (atom, molecule, ion, or other system) in thermal equilibrium with the radiation field (Fig. 4.4). With the coefficients A_{12} for *spontaneous* emission and B_{12}, B_{21} for *induced* emission and absorption of a photon holds for the occupation numbers n_1, n_2 (Einstein 1917)

$$\frac{dn_1}{\mathrm{d}t} = n_2[A_{12} + B_{12}\rho(\omega, T)] - n_1 B_{21}\rho(\omega, T) = -\frac{dn_2}{\mathrm{d}t} = 0. \tag{4.132}$$

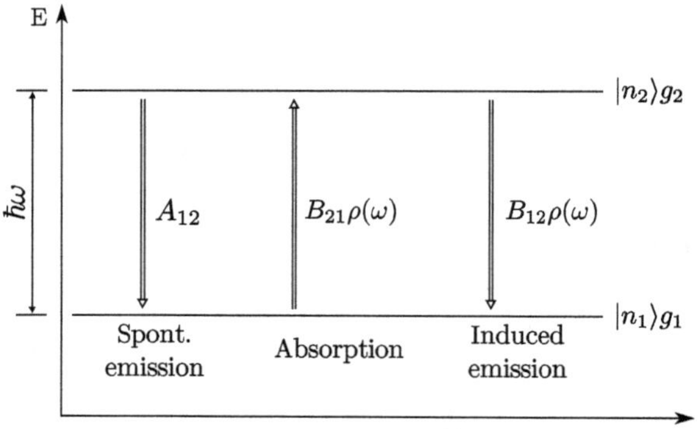

Fig. 4.4 Assume two eigenstates of matter $|n_1\rangle$, $|n_2\rangle$ of degeneracy g_1, g_2 and separated energetically by $\hbar\omega$ from each other. The spectral radiation density is $\rho(\omega)$. The population of the upper state $|n_2\rangle$ per unit time is $g_1 B_{21}\rho(\omega)$. Its depopulation per unit time is by spontaneous emission $g_2 A_{12}$ and by stimulated emission $g_2 B_{12}\rho(\omega)$. In thermal equilibrium between radiation and matter the two coefficients $B_{21}(\omega)$ and $B_{12}(\omega)$ become equal

The canonical distribution of the states $|1\rangle$ and $|2\rangle$ requires $n_2 = n_1 e^{-\beta\hbar\omega}$. For $T \to \infty \to n_2 \Rightarrow n_1$ and A_{12} can be neglected with the consequence of $B_{21} = B_{12}$. The Einstein coefficients do not depend on temperature, the equality of B_{21} with B_{12} is of general validity in thermal equlibrium and (4.132) results in

$$B_{21}\rho(\omega, T) = [A_{12} + B_{12}\rho(\omega, T)]\, e^{-\beta\hbar\omega}, \quad B_{21} = B_{12} \;\Rightarrow\; A_{12} = \frac{\omega^2}{\pi^2 c^3}\, B_{12}\hbar\omega. \tag{4.133}$$

It explains why hard X-ray lasing is difficult to achieve: Induced emission of (monomode) lasing is directed into one mode $\mathbf{k}$ whereas spontaneous emission occurs isotropically and with equal probability into all possible modes of number $\omega^2/\pi^2 c^3$ per unit volume and unit frequency interval. With the number of photons $\langle n_\mathbf{k}\rangle$ in each mode (4.133) can also be written as

$$B_{12}\rho(\omega, T) = A_{12}\langle n_\mathbf{k}\rangle, \quad B_{12}\rho(\omega, T) + A_{12} = A_{12}(\langle n_\mathbf{k}\rangle + 1). \tag{4.134}$$

For $\hbar\omega \gg k_B T$ spontaneous emission dominates.

4.4.2.4 Kirchhoff's Law of Detailed Radiation Balance

Consider a hohlraum, i.e., Planckian radiator, of spectral intensity $I(\omega, T)$ in perfect thermal equilibrium with matter of spectral emissivity ε_ω and spectral absorption coefficient α_ω; see Fig. 4.5.
The principle of detailed balance requires for the spectral absorptivity

$$\varepsilon_\omega = \alpha_\omega \rho(\omega, T) \quad \Rightarrow \quad \frac{\varepsilon_\omega}{\alpha_\omega} = \rho(\omega, T) = \frac{4\pi}{c} I(\omega, T). \tag{4.135}$$

This is *Kirchhoff's law* of radiation balance. As it is formulated in terms of intensities in this form it is based on the validity of the geometrical optics (see related exercise).

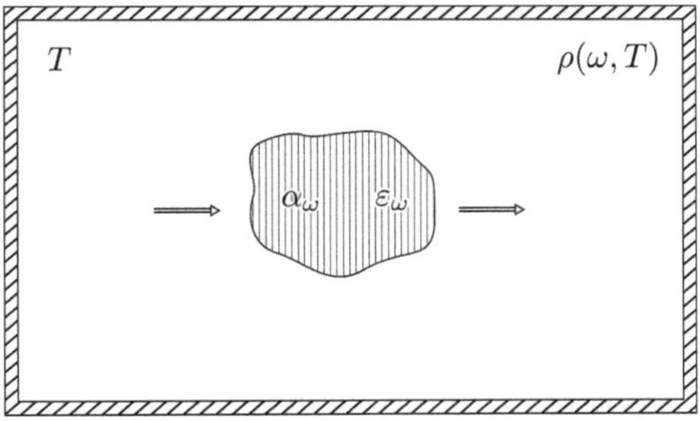

Fig. 4.5 A hohlraum with walls at temperature T emits black body radiation of uniform and isotropic intensity $I(\omega, T)$. Matter immersed in the cavity is in radiative equilibrium with the Planckian radiation field $\rho(\omega, T)$

According to (4.133) the *net spectral absorption coefficient* α_ω in (4.135) is the *total spectral absorption* coefficient κ_ω *minus* the coefficient of emission induced by the embedding radiation field. With a view on Fig. 4.4 the two quantities α_ω and κ_ω are connected by

$$\alpha_\omega = \left(\frac{n_1}{g_1}B_{21}g_1 - \frac{n_2}{g_2}B_{12}g_2\right) = n_1 B_{21}\left(1 - \frac{n_2/g_2}{n_1/g_1}\right),$$

$$\kappa_\omega = n_1 B_{21} \quad \Rightarrow \quad \alpha_\omega = \kappa_\omega\left(1 - \frac{n_2/g_2}{n_1/g_1}\right). \tag{4.136}$$

If α_ω is referred to $I(\omega, P)$, i.e. $\varepsilon_\omega = \alpha_\omega \times I(\omega, T)$, the above expression is to be multiplied by $4\pi/c$.

The expressions for κ_ω and α_ω are of general validity (see below). In *thermal equilibrium* of the two levels $|n_1\rangle$, $|n_2\rangle$ the net absorption is reduced by the *re-emission* factor $1 - \exp(-\beta\hbar\omega)$. In our formulation of Kirchhoff's law (4.135) the emission ε_ω and total absorption κ_ω are properties of the material (atoms, ions) and are independent of the embedding radiation field as long as it does not influence the energy spectrum of matter (e.g. atomic level shifts by ac Stark effect). Therefore the total absorption coefficient κ_ω and ε_ω are accessible to calculations or a measurement, whereas α_ω is not, except the surrounding radiation field is negligible.

In the optically thin plasma, and in a plasma at sufficiently low temperature in general, ε_ω is energetically negligible, i.e. $\rho(T) \ll nk_BT$. Nevertheless relations (4.135) and (4.136) apply if the electron-electron collision frequency is high enough to guarantee that the energy levels under consideration are thermally occupied. In the worst case Kirchhoff's balance applies to one or a few spectral lines only.

Einstein's relation (4.132) is of general validity, in and out of thermal equilibrium. In fact, in the quantized electromagnetic field (see Chap. 7) $a_\mathbf{k}$, $a_\mathbf{k}^\dagger$ when applied to a photon number state $|n_\mathbf{k}\rangle$ containing $n_\mathbf{k}$ photons reduce this by one photon, $a_\mathbf{k}|n_\mathbf{k}\rangle = n_\mathbf{k}^{1/2}|n_\mathbf{k} - 1\rangle$ (absorption), and augment it by one photon, $a_\mathbf{k}^\dagger|n_\mathbf{k}\rangle = (n_\mathbf{k} + 1)^{1/2}|n_\mathbf{k} + 1\rangle$ (emission). The Hermitian operator $a_\mathbf{k}^\dagger a_\mathbf{k}$ counts the photons, $a_\mathbf{k}^\dagger a_\mathbf{k}|n_\mathbf{k}\rangle = n_\mathbf{k}|n_\mathbf{k}\rangle$. With these rules in mind for the quantized field follows

$$\begin{aligned}
\rho_\omega &= \frac{\hbar\omega^3}{\pi^2 c^3} n_{\mathbf{k}} = \frac{\hbar\omega^3}{\pi^2 c^3} \langle n_{\mathbf{k}} | a^\dagger a | n_{\mathbf{k}} \rangle, \\
\text{Emission} &= C |\langle n_{\mathbf{k}+1} | a^\dagger | n_{\mathbf{k}} \rangle|^2 = (1 + n_{\mathbf{k}}) C = \left(1 + \rho_\omega \frac{\pi^2 c^3}{\hbar\omega^3}\right) C \\
= C + \frac{\pi^2 c^3}{\hbar\omega^3} C \rho_\omega &= A_{12} + B_{12}\rho_\omega \Rightarrow A_{12} = C, \quad B_{12} = \frac{\pi^2 c^3}{\hbar\omega^3} A_{12}
\end{aligned} \tag{4.137}$$

In the semiclassical theory of radiation (atom quantized, radiation field classical) with the interaction Hamiltonian in dipole approximation $H_\text{I} = e\mathbf{x}\mathbf{E}$ holds

$$B_{12} = K|\langle 1|\mathbf{x}|2\rangle|^2 = K|\langle 2|\mathbf{x}|1\rangle|^2 = B_{21}; \quad g_1, g_2 = 1, \quad K = \text{const}$$

At this level of description spontaneous emission is out of range. Not all degenerate levels of multiplicities g_1, g_2 may be in thermal equilibrium. In that case $B_{12}g_2$ may differ from $B_{21}g_1$. For its explanation the full dynamics of transitions beyond rate equations must be considered, see Chap. 7.

The Planck Hohlraum Filled with Matter

Radiation interacting with matter changes the vacuum wave vector $\mathbf{k}_0$ into $\mathbf{k} = \mathbf{k}_0\eta$. For a plasma with the dispersion relation $\omega^2 = \omega_p^2 + c^2\mathbf{k}^2$ this means that $\rho(\omega, T)$ from (4.121) is zero for $\omega < \omega_p$.

4.4.2.5 The Ideal Fermi Gas

Spin 1/2 particles follow the *Fermi-Dirac* statistics. The probability $w(E_l, N, V)$ for N independent particles to occupy the energy level E_l and the average occupation number $\langle n_j \rangle$ of the single particle level E_k are again the same as for the ideal Bose gas

$$w(E_l, N, V) = \frac{e^{\beta(\mu N - E_l^{(N)})}}{\mathcal{Z}} = \frac{1}{\mathcal{Z}} \prod_{k=0}^{\infty} e^{\beta(\mu - E_k) n_k}; \quad \langle n_j \rangle = -\frac{\partial}{\partial(\beta n_j E_j)} \ln \mathcal{Z}.$$

For Fermions the single n_k take the values 0 and 1. Thus

$$\mathcal{Z}(V, T, \mu) = \sum_{\{n_k\}} \prod_{k=0}^{\infty} e^{\beta(\mu - E_k) n_k} = \prod_{k=0}^{\infty} \sum_{n_k=0}^{1} e^{\beta(\mu - E_k) n_k} = \prod_{k=0}^{\infty} \left(1 + e^{\beta(\mu - E_k)}\right). \tag{4.138}$$

This yields the *Fermi distribution*

$$\langle n_j \rangle = -\frac{\partial}{\partial(\beta E_j)} \ln \mathcal{Z} = \frac{1}{e^{\beta(E_j-\mu)}+1}. \tag{4.139}$$

The mean energy $E = \mathcal{E}_{\text{in}}$ and the mean particle number $\langle N \rangle$ are

$$E = \Big\langle \sum_{k=0}^{\infty} n_k E_k \Big\rangle = \sum_{k=0}^{\infty} \frac{E_k}{e^{\beta(E_k-\mu)}+1} = E(V, T, \mu),$$
$$\langle N \rangle = \Big\langle \sum_{k=0}^{\infty} n_k \Big\rangle = \sum_{k=0}^{\infty} \frac{1}{e^{\beta(E_k-\mu)}+1} = \langle N(V, T, \mu) \rangle. \tag{4.140}$$

By elimination of μ from $E(V, T, \mu)$ and $\langle N(V, T, \mu) \rangle$ the internal energy $\mathcal{E}_{\text{in}}(V, T)$ is recovered. Entropy S, pressure p, and specific heat C_V for the ideal Fermi gas are given by

$$S = -\left(\frac{\partial J}{\partial T}\right)_{V\mu} = \frac{\partial}{\partial T}\left(\frac{1}{\beta} \ln \mathcal{Z}\right) = k_B \sum_{k \geq 0} \left[\frac{e^{\beta(E_k-\mu)}}{e^{\beta(E_k-\mu)}+1} + \ln\left(1 + e^{\beta(\mu-E_k)}\right)\right],$$
$$p = -\left(\frac{\partial J}{\partial V}\right)_{T\mu} = \frac{\partial}{\partial V}\left(\frac{1}{\beta} \ln \mathcal{Z}\right) = -\sum_{k \geq 0} \frac{\partial E_k/\partial V}{e^{\beta(E_k-\mu)}+1} = -\sum_{k \geq 0} \langle n_k \rangle \frac{\partial E_k}{\partial V},$$
$$C_V = \left(\frac{\partial E}{\partial T}\right)_V = \sum_{k \geq 0} E_k \frac{\partial \langle n_k \rangle}{\partial T}. \tag{4.141}$$

At $T = 0$ the chemical potential equals the *Fermi edge* or *Fermi energy* E_F, $\mu = E_N = E_F$, N number of particles. For $k_B T \ll \mu$ the Taylor expansion of $\langle n_j \rangle$ in $\xi = \beta(E - \mu)$ yields

$$\langle n \rangle = \frac{1}{2}\left(1 - \frac{\xi}{2}\right).$$

At the Fermi edge $\langle n \rangle = 1/2$ holds for all temperatures.

Flucuations

Like the Bose gas $\langle n_l^2 \rangle$ is given by $\langle n_l \rangle^2 - k_B T \partial \langle n_l \rangle / \partial E_l$. Thus

$$\frac{\langle \Delta n_l^2 \rangle}{\langle n_l \rangle^2} = \frac{1}{\langle n_l \rangle} - 1. \tag{4.142}$$

At zero temperature the fluctuation results zero.

4.4.2.6 The Boltzmann Limit

Let us set the scale for the single particle energies E_j to zero for the ground state E_1. One may ask when there is no need to distinguish between spin zero or one half. This is certainly the case if all single particle occupations fulfill the inequality $\langle n_j \rangle \ll 1$. It tells that the likelihood for one energy level j to be occupied by more than one particle is vanishingly small. For (4.113) and (4.139) it is satisfied if

$$\mathrm{e}^{-\beta\mu} \gg 1 \tag{4.143}$$

because one concludes

$$\mathrm{e}^{\beta(E_j-\mu)} \geqslant \mathrm{e}^{-\beta\mu} \quad \Rightarrow \quad \langle n_j \rangle = \mathrm{e}^{-\beta(\mu-E_j)} \quad \forall j. \tag{4.144}$$

This is the *Boltzmann* or *Maxwell-Boltzmann statistics* for average single energy level occupation. Summing (4.144) over all states j results in the total mean particle number and single particle occupations

$$\langle N \rangle = \mathrm{e}^{\beta\mu} \sum_k \mathrm{e}^{-\beta E_k} = \mathrm{e}^{\beta\mu} \times Q_1(V,T) \;\Rightarrow\; \langle n_j \rangle = \frac{\langle N \rangle}{Q_1(V,T)} \mathrm{e}^{-\beta E_j}. \tag{4.145}$$

$Q_1(V,T)$ is the one particle partition function. The single particle Hamiltonian splits into $H_{\text{trans}} = \mathbf{p}^2/2m + H_{\text{int}}$ where H_{int} stands for the rotational and vibrational motions. In perfect analogy to (4.120) it is calculated as

$$Q_1 = \int_{|\mathbf{p}|\geqslant 0} \exp -\beta(\mathbf{p}^2/2m)\,\mathrm{d}\mathbf{p} \times Q_{1,\text{int}} = \left(\frac{2\pi m k_B T}{h}\right)^{3/2} V = \lambda_{\text{th}}^{-3} V.$$

The internal partition function results zero owing to the scaling $E_1 = 0$ above for point particles. Its use in (4.145) together with (4.143) yields

$$\exp -\beta\mu = \frac{Q_1(V,T)}{\langle N \rangle} = (n\lambda_{\text{th}}^3)^{-1} \quad \Leftrightarrow \quad d = n^{-1/3} \gtrsim 2\lambda_{\text{th}}. \tag{4.146}$$

Boltzmann statistics is correct in gases of mean particle distance d twice as or greater than the reduced *thermal deBroglie wavelength* $\lambda_{\text{th}} = (\hbar/mk_BT)^{1/2}$.

Flucuations

From the general expression for $\langle n_l^2 \rangle$ above in (4.87), but also from elementary probability arguments, the relative occupation variance follows for independent classical particles as

$$\frac{\langle \Delta n_l^2 \rangle}{\langle n_l \rangle^2} = \frac{1}{\langle n_l \rangle}. \tag{4.147}$$

4.4.2.7 The Single Particle Heat Bath

The probability $w(n_j)$ of n_j independent particles to occupy the energy level E_j is given by (4.111). It does not depend on the occupation of any other energy level

$E_k, k \neq j$, and reduces to

$$w(n_j) = \frac{1}{\mathcal{Z}_1} e^{\beta(\mu - E_j)n_j}, \quad \mathcal{Z}_1 = \sum_{n_l} = e^{\beta(\mu - E_l)n_l}. \tag{4.148}$$

Interpretation. For independent particles the jth energy level is one of the possible energy levels E_l in the heat bath of the single particle; $w(n_j)$ is the normalized probability; and $\mathcal{Z}_1$ is the one particle grand canonical partition function.

4.4.2.8 The Partition Function in the Boltzmann Limit

For its relevance in applications and for the simplicity of analytical formulas it is worth to present explicit expressions of the partition function of the ideal gas and plasma in the Boltzmann limit. If besides ideality condition (4.143) is fulfilled we refer to such a system also as *perfect classical gas or plasma*. If the energy levels of the single particles (electrons, protons, neutrals, ions) are labelled as $E_1 \leqslant E_2 \leqslant E_3 .... \leqslant E_r \leqslant ...$ the partition function for the single particle in the volume V is given by

$$Z_1(V, T) = \sum_r e^{-\beta E_r}. \tag{4.149}$$

From the mutual independence follows $Z(V, T, N)$ of N particles as

$$Z(V, T, N) = \frac{1}{N!} \left(\sum_r e^{-\beta E_r} \right)^N = \frac{1}{N!} Z_1(V, T)^N. \tag{4.150}$$

The reader may convince himself that $Z(V, T, N)$ is in agreement with the last expression in (4.110). The factor $N!$ in the denominnator makes Z independent of the order of numbering the particles: $\exp -\beta E_l \exp -\beta E_m = \exp -\beta E_m \exp -\beta E_l$, and there is only a vanishingly small number of terms with $l = m$; see also the Gibbs paradox. The phase space of the single particle separates into translation expressed by the momenta of the center of mass, $H_t = \mathbf{p}^2/2m$, and the innerparticle dynamics of rotation and vibration, and perhaps of innerionic excitations, H_{in}. From $H = H_t + H_{\text{in}}$ the energy E_r is the sum $E_r = E_{r,t} + E_{r,\text{in}}$ and

$$Z_1(V, T) = \sum_r e^{-\beta(E_{r,t} + E_{r,\text{in}})} = \sum_{r,t} \left\{ \sum_{r,\text{in}} e^{-\beta(E_{r,t} + E_{r,\text{in}})} \right\} = Z_{1t} Z_{1\text{in}} \tag{4.151}$$

factorizes into the *translational partition function* Z_{1t} and the *internal partition function* $Z_{1\text{in}}$. The internal partition function does not depend on V. Apart from the mass m, Z_{1t} is the same for all particle species. According to (4.120) and (4.164) the number of modes per polarization (spin state either + or −) is

$$d\mathcal{N} = \frac{4\pi \mathbf{p}^2 d|\mathbf{p}|}{h^3} V.$$

Hence

$$Z_{1t}(V,T) = V\int_0^\infty \frac{4\pi \mathbf{p}^2 d|\mathbf{p}|}{h^3} \mathrm{e}^{-\beta \mathbf{p}^2/2m} = \left(\frac{2\pi m k_B T}{h^2}\right)^{3/2} V. \tag{4.152}$$

Making use of Stirling's formula for $N! = (N/\mathrm{e})^N$ the Helmholtz free energy and entropy are

$$F(V,T,N) = -Nk_BT \ln\left[\frac{\mathrm{e}V}{N}\left(\frac{2\pi m k_B T}{h^2}\right)^{3/2} Z_{1\text{in}}\right]; \quad S(V,T,N) = \frac{\mathcal{E}_{\text{in}} - F}{T}. \tag{4.153}$$

The translational entropy component S_t is given by (4.93).

4.4.2.9 Chemical Equilibrium and Law of Mass Action

Chemical equilibrium stands for a chemical reaction, as for example

$$2H_2 + O_2 \leftrightarrows 2H_2O, \quad N_2 + O_2 \leftrightarrows 2NO, \quad N_2 + 2O_2 \leftrightarrows 2NO_2 \tag{4.154}$$

but also for phase transition liquid ⇆ vapor, dissociation of molecules, and ionization of atoms or ions. If the single component is indicated by A_k the reaction can be written as

$$k_1A_1 + k_2A_2 + \ldots\ldots k_sA_s = 0 \tag{4.155}$$

with the assignment for the first example $k_1 = 2$, $A_1 = H_2$, $k_2 = 1$, $A_2 = O_2$, $k_3 = -2$, $A_3 = H_2O$, etc. The free energy F is assumed to be the sum of the single components, $F = F(V, T, N_1, N_2, \ldots\ldots, N_s) = \sum_k F(V, T, N_k)$, $N = N_1 + N_2 + \cdots N_s$. In equilibrium F is a minimum in the particle numbers N_k,

$$\mathrm{d}F = \sum_1^s \left(\frac{\partial F}{\partial N_k}\right)_{V,T} dN_k = \sum_1^s \mu_k dN_k = 0. \tag{4.156}$$

$$\mu_k = \left(\frac{\partial F}{\partial N_k}\right)_{V,T,N_j(j\neq k)} = -k_BT\{\ln Z_{1k} - \ln N_k\}. \tag{4.157}$$

The individual chemical reaction (4.155) provides the prescription how the changes dN_k in (4.156) are linked together. Obviously they can vary only in proportion to the individual numbers k_i of particles involved in the reaction, i.e. $dN_k^i \sim k_i$. Setting $f_k(T) = Z_{1k}/V$ and the partial particle densities $n_k = N_k/V$ the chemical equilibrium imposes on the μ_k

$$\sum_1^s k\mu_k = 0; \quad \mu_k = k_B T\{\ln n_k - \ln f_k(T)\} \Rightarrow \sum_1^s kn_k = \sum_1^s kf_k(T). \quad (4.158)$$

The last equality is generally written in exponential form

$$\prod_1^s n_k^k = \prod_1^s f_k(T)^k = K(T). \quad (4.159)$$

It is known as the *mass action law*. The product on the RHS depends only on temperature. $K(T)$ is the *equilibrium constant* of the reaction. The individual f_k contain the temperature dependent internal partition functions $Z_{1k,\mathrm{in}}$. The energies E_r in the partition functions must all start from the same energy ground level.

4.4.2.10 The Saha Equation

In the dilute gas and plasma the mass action law can directly be applied to calculate the various equilibrium ionization degrees in the Boltzmann limit. For temperatures above 0.1 eV we may assume that all molecules are dissociated into neutral atoms and ions. We consider the equilibrium between the $(z+1)$ fold ionized neutrals of concentration $\eta_{z+1} = n_{z+1}/N$ and ionization energy E_{z+1} and the z fold ionized component of η_z and E_z according to the scheme

$$A^{z+} \leftrightarrows A^{z+1} + e^-, \quad z \geq 0. \quad (4.160)$$

The concentrations refer to the total number of atoms/ions N per unit volume. The mass action law (4.159) requires

$$\frac{\eta_e \eta_{z+1}}{N\eta_z} = 2\left(\frac{2\pi m_e k_B T}{h^2}\right)^{3/2} \frac{Z_{1\mathrm{in}}^{z+1}}{Z_{1\mathrm{in}}^z} \mathrm{e}^{-\beta(E_{z+1}-E_z)}; \quad z \geq 0. \quad (4.161)$$

If instead of considering the ionization degree at constant volume V, and so in general the mass action law, one wants the equilibrium condition under constant pressure p and constant temperature T one has merely to substitute F by the Gibbs potential $G = F + pV$ in the preceding derivations. Equations (4.161) form a complete set of algebraic equations which generally have to be solved numerically. In practical applications great simplifications are often possible, see Ya. B. Zeldovich, Yu. P. Raizer in FR, pp. 192–206. At low concentrations and high temperatures generally electronic excitation do not play a major role. Ionization prevails on excitation owing to the overwhelming number of free electron states.

The Saha equilibrium requires that all significant components (electrons, radiation in first place) contribute to the free energy or Gibbs enthalpy. Hot tenuous plasmas may be largely transparent to radiation and photoionization is almost absent. Among the various models considering partial equilibrium the *coronal model* has shown its relevance in laboratory plasmas and in stellar atmospheres. It considers the equilibrium of *collisional ionization* from the ground state and *radiative recombination* from high ionization states [5–7]. For the opposite situation of photoionization dominated plasmas see [8]. If strong radiation fields compete with collisional excitation and, in addition strong spatial density and temperature gradients are present LTE cannot be expected. To handle such complex situations the interested reader may consult appropriate literature [9].

4.5 From Warm Dense Matter to Hot Dense Plasma

The laser generated plasmas so far considered are hot and inhomogeneous, and expand rapidly. The Fermi energy is far below the mean kinetic energy and the ideality paramter Γ is less than 0.1. As a consequence their theoretical description is that of a dilute fluid by classical means. More precisely, such a treatment is adequate to the subcritical portion of samples directly heated by the laser and the indirectly heated supercritical region powered by electronic energy diffusion; it extends from the critical density to the electron heat front. Ideality condition in this latter part may not be so well fulfilled. With increasing laser intensity and pulse length a third region behind the heat front becomes increasingly important with rising laser intensity. It is moderately heated by shock compression and by energy deposition of the hot electrons. It is the region of *warm dense matter*: solid state density and above, and temperatures ranging from 0.1 to 100 eV. Succinctly it can be termed the *hot solid*. Intense ion beams are an additional tool to get solids and liquids warm.

Dense matter heated beyond temperatures of $T \simeq 100$ eV may be addressed as *hot dense plasma* because its properties will be largely determined by the characteristics of fully ionized plasmas. It is difficult to bring a sample with flat surfaces into this domain owing to the limiting critical density, unless one focuses at highly relativistic laser intensities where the critical density is $n_{\mathrm{cr}} = \gamma n_c$. At $I = 10^{20}\,\mathrm{Wcm}^{-2}$ the laser penetrates up to 100 fold the critical density, i.e. $n_e \simeq 10^{23}\ \mathrm{cm}^{-3}$, at most. The limit can be overcome by giving the sample a special *microstructure*, see Fig. 1.24. In this way energy densities in the GJ and pressures in the Gbar order of magnitude are reached, in the laboratory comparable only to compressed fusion pellets from the National Ignition Facility (NIF) and comparable installations.

The pressures achieved with static methods to compress matter is limited: 3.6 Mbar with the standard diamond cell, 6.4 Mbar in the nanodiamond cell. The study of matter under superhigh pressures is made possible in the shock wave from laser ablation and from the laser radiation pressure. Combined with similarity methods they make the study of astrophysical conditions feasible in the laboratory [10]. The Universe is a unique showcase of dense matter, to start with the "high vacuum" intergalactic gas of 100 particles per cm^{-3} and to end with neutron stars of 4×10^{14} gcm^{-3}, corresponding to particle density $n \simeq 4 \times 10^{41}$ cm^{-3}, for matter without an event horizon; this is higher than the 3×10^{14} gcm^{-3} of the atomic nucleus. Values in between show the center of the Earth, 10–20 gcm^{-3}, of Jupiter, 30 gcm^{-3}, of the Sun, 150 gcm^{-3}. The concomitant pressures and temperatures amount to: Earth, 4 Mbar and 0.5 eV, Jupiter, 50 Mbar and 2 eV, Sun 160–250 Gbar and 1.5 keV, neutron star, 10^{23} Mbar and 100 keV.

It is exciting to look for the equation of state (EOS), the energy density and, possibly, phase transitions for such a wide span of densities and temperatures. The investigation of the most likely thermodynamic properties will certainly get decisive input from 100 year solid state physics below about 0.3 eV temperature. At superhigh pressures the question of stability of matter arises. Up to the stability limit the only counteracting pressure against gravitation originates from the Pauli exclusion principle of the Fermions. As a guideline, with progressing density and pressure solid matter transforms into a liquid, or above the *critical point* directly into a dense gas and, finally, into a cold dense Fermi fluid, and perhaps again into a solid-like crystalline structure. Under strong heating the solid state may transform directly into the plasma state as we are faced with all the time in high power laser matter interaction. With intense ultrashort laser pulses a new kind of phase transition, the so called *nonthermal melting*, is possible [11]. In ordinary melting electrons couple to phonons and induce them to move by tunneling. Instead, in nonthermal melting the energy absorbed by the electrons is directly transmitted to the interatomic potentials on the fast electron time scale. Subsequently, on the slow atomic time scale, the potential changes lead to relocations of the ions.

4.5.1 The Equation of State of Dense Matter

So far matter has been considered as to be in a plasma state, more precisely, a classical fluid of free electrons and ions, disordered on the scale of atomic size. This is opposite to what the solid state physicist encounters in the temperature interval from zero to typically 0.2–0.3 eV. The density of atoms/ions at zero pressure is 10^{23} particles per cm^3, the electron density is $(1-5) \times 10^{24}$ cm^{-3}. Some elements are insulators, others are conductors with typically 10^{23} free electrons per cm^3. At such densities the Fermi energy E_F ranges from 2 to 10 eV and therefore Fermi-Dirac statistics applies to their thermal description at temperatures not distinctly above E_F. At solid state density matter is in a crystalline structure, i.e. long range order, or in a liquid state of short range order only. There is an enormous knowledge about the solid state and still rather poor knowledge about liquids. Both have in common high resistance to compression and similar cohesive force. It is the Fermionic character of the electrons which opposes to overlapping in compression. Under high pressure and shrinking volume their only choice is to escape into high energy states at vanishing temperature. The phenomenon of *pressure ionization* is a universal feature of highly compressed matter. This indicates that in an equation of state, if it has to hold within a wide pressure range, the description of the electron component will occupy the first place. The Fermi energy of free ions is by the ion/electron mass ratio lower and thus not significant in our context.

There is no partially empty conduction bend in the insulator; all electrons are tightly bound to the nuclei. And yet we have to distinguish between different types of insulators according to the manner how the electrons are distributed between the positive nuclei. The covalent crystal, diamond for example, is not dissimilar to metallic conductors but with the delocalized electrons accumulated in places between the nuclei rather than nearly uniformly distributed as in the conductor. In the molecular crystals the electrons are overwhelmingly bond to their nuclei and only a vanishing fraction is delocalized; therefore the low adhesion energy and the low boiling point. The best example of this type are the noble gases from Ne to Rd with their closed atomic shells. The ionic crystals, like the alkali halides NaCl, KCl, are the combination of a metallic and a nonmetallic element. They are characterized by strong electric polarization from where they take mainly their cohesion.

In view of plasmas and highly compressed matter produced from intense and superintense lasers thermodynamic properties valid for a wide range of parameters are required. This is essential in particular for the formulation of the equation of state. In the following the basic elements will be presented on which such EOS can be constructed. Thereby we are guided by a few papers more or less inspired by the work of More et al. [12, 13].

It is convenient to start from the canonical partition function $Q(\rho, T_e, T_i)$, or $F(\rho, T_e, T_i)$ as a function of the mass density ρ and the temperatures T_e and T_i for the three components F_e for the electrons, F_i for the ions and F_b for chemical bonding effects of the solid state and other quantum corrections,

$$F(\rho, T_e, T_i) = F_e(\rho, T_e) + F_i(\rho, T_i) + F_b(\rho, T_e). \tag{4.162}$$

No mutual interaction term between the fluids is considered. The pressures of the single components are obtained from (4.80) and (4.104),

$$p_e = \rho^2 \frac{\partial F_e}{\partial \rho}, \quad p_i = \rho^2 \frac{\partial F_i}{\partial \rho}, \quad p_b = \rho^2 \frac{\partial F_b}{\partial \rho}. \tag{4.163}$$

The relations are valid for an amount of mass or number of nuclei held fixed.

4.5.1.1 The Electron Component

The electron fluid is now described by the semiclassical *Thomas–Fermi model* (TF). The single electron is free and is subject to the Coulomb potential of the nucleus of charge number Z and the collective potential of the electrons bound to the nucleus. The atom or ion is assumed to be confined within a sphere of radius r_0. If the atomic mass number is A, m_p the proton mass, and ρ the plasma mass density r_0 follows from $4\pi r_0^3/3 = Am_p/\rho$. The number of electrons in the sphere is N. The Hamiltonian of the single electron is $E = \mathbf{p}^2/2m_e - eV(r)$ in the potential $V(r)$ of nucleus and bound electrons. The potential $V(r)$ is given by Poisson's equation

$$\varepsilon_0 \nabla^2 V(r) = -Ze\delta(\mathbf{r}) + en_e(r)$$

if the electron energy levels are closely spaced. The TF model consists in applying the Fermi-Dirac distribution (4.139) for free electrons *locally* to the potential $V(r)$. This is a major step and needs a special reflection.

Consider Fig. 4.6, first picture left. Exactly in this situation of a constant potential $V = V_0$, (4.139) multiplied with the degeneracy factor $g(|\mathbf{k}|)$ tells the number of available states $dn(k)$ in the the interval $dk = d|\mathbf{k}|$. The determination of g is analogous to the photon counting in the Planck formula. In the box of extension L the number of electron states is the shell $2 \times 4\pi k^2 dk$ divided by the volume of the unit cell $(\Delta k)^3 = (2\pi/L)^3$. This results in the number

$$d\mathcal{N} = \frac{1}{\pi^2} L^3 k^2 dk \Rightarrow dn(k) = \frac{1}{\pi^2} k^2 dk = \frac{1}{\pi^2 \hbar^3} \mathbf{p}^2 d|\mathbf{p}| \Rightarrow g(|\mathbf{p}|) = \frac{1}{\pi^2 \hbar^3}. \tag{4.164}$$

The factor 2 accounts for the two spin states. The Fermi energy E_F and the Fermi momentum $\mathbf{p}_F$ at the *Fermi edge* are related by $E_F = \hbar^2 \mathbf{k}_F^2/2m_e$. The total energy of an electron at the Fermi edge is $E = E_F - eV_0$ if the bottom of the box is at potential V_0.

In a second step chose a potential with a sink, picture 2 from left. Its effect on the states is a lowering of all energies, from the highest amount for the ground state to monotonous decrease of the higher levels. This is because the curvature of the eigenfunctions is an increasing distance of E to the potential V. Move now to the third picture with several potential boxes superposed on their ground states V_{0j}. The box number j undergoes an energy lowering of all its states $E > -eV_{0j}$ there owing

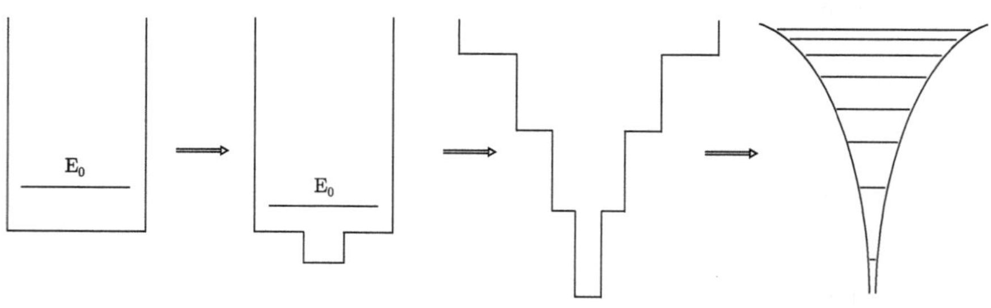

Fig. 4.6 From $V =$ const to $V(r)$. Fermi-Dirac distribution of free electrons applies to picture at left with potential at bottom $V = V_0$. Slight depression of energy levels by the potential sink in the following picture. Generalization by multistep potential in the subsequent picture. For the impact of non constant potential on the energy level distribution see text. Picture at the right is the limiting case of the multistep potential

to the sinks below $V = V_{0j}$. The same happens with the states in the box number $j+1$, but to a minor degree because it is larger. However, the key point is that it is the difference of states lost on the bottom to box $j-1$ and states gained at the top from box $j+1$. The corrections sum up to an alternating series of decreasing differences. An adjuvant effect of higher quantum states comes from the WKW like increase of the wave functions near the turning points; it helps further to mitigate the effect of the potential depression by the positive nucleus. Again we encounter an adiabatic behaviour of the overall sum of deviations in a spatially varying quantity with respect to its uniformity, as in Chap. 2 with a Hamiltonian depending on a slowly varying parameter. This makes the application of (4.139) possible for nonconstant V_0. We may take the limit of the multistep potential to the continuous potential $V(r)$ of the picture on the right provided there are enough states in each local interval $\mathrm{d}V(r)$. The reflection on the applicability of the Fermi model presented here may help to explain its success although a Coulomb like potential is far from a square well shape. It is the adiabatic behaviour of level shifts which turns the deviations into an alternating series in contrast to a rapidly swelling series of similar terms all positive.

The limiting consideration tells how to apply (4.139) to each local box: At each position r counting starts from $V(r)$. Hence, the electron density at r is

$$n(r) = \int n(k)dk = \int_{\{2m_e[eV(r)+\mu]\}^{1/2}}^{\infty} \frac{g(|\mathbf{p}|)|\mathbf{p}|^2 d|\mathbf{p}|}{\mathrm{e}^{\beta[\mathbf{p}^2/2m_e - eV(r)-\mu]}+1}. \tag{4.165}$$

The number N of electrons in the sphere of radius r_0 is

$$N = 4\pi \int_0^{r_0} n(r) r^2 dr. \tag{4.166}$$

The TF Model at $T_e = 0$

At zero electron temperature all available states are filled up to the Fermi edge. Beyond $k = k_F$ the $g(k)$ is zero and $n(r)$ results from $[\mathbf{p}^2/2m_e = eV(r)+\mu]$ in (4.165)

$$n(r) = \frac{1}{3\pi^2\hbar^3}\{2m_e[eV(r)+\mu]\}^{3/2}. \tag{4.167}$$

It is advantageous to regularize the Coulomb potential by introducing the new dimensionless function $\chi(r)$,

$$\frac{Ze^2}{r}\chi(r) = 4\pi\varepsilon_0[eV(r)+\mu]. \tag{4.168}$$

$Ze\chi(r)$ is the screened nuclear charge at distance r. It is $\chi(r) \Longrightarrow 1$ for $r \Longrightarrow 0$. Substitution of (4.167) in the Poisson equation leads to

$$\frac{d^2\chi}{dr^2} = \frac{(2m_e e^2)^{3/2}}{3\pi^2\hbar^3(4\pi\varepsilon_0^3)^{1/2}}\frac{Z^{1/2}}{r^{1/2}}\chi^{3/2}. \tag{4.169}$$

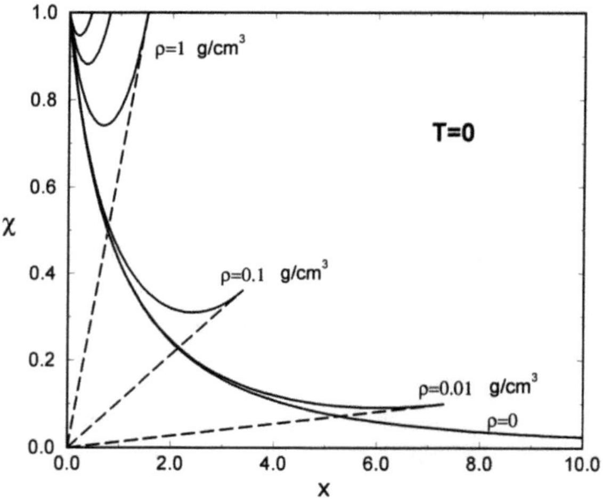

Fig. 4.7 Solution of the Thomas–Fermi equation for various densities and $Z = 1$ at zero temperature. The number of confined electrons is determined by the χ-value at which the tangent to $\chi(x)$ passes through the origin according to (4.172). Pressure ionization inceases rapidly with rising density. Courtesy of [13]

Rescaling with the hydrogen Bohr radius a_B,

$$x = \frac{r}{a_0}, \quad a_0 = \frac{1}{2}\left(\frac{3}{4}\pi\right)^{2/3}\frac{a_B}{Z^{1/3}} = 0.885\frac{a_B}{Z^{1/3}}; \quad a_B = \frac{4\pi\varepsilon_0\hbar^2}{m_e e^2} = 0.0529 \text{ nm} \tag{4.170}$$

yields the cold *Thomas–Fermi equation*

$$\frac{d^2}{dx^2}\chi(x) = \frac{1}{x^{1/2}}\chi(x)^{3/2}. \tag{4.171}$$

At the cell boundary $x = 1$ the electric field $E = -\partial_r V(r)|_{r=r-0}$ must vanish for translational symmetry reasons. Equation (4.168) translates this into the boundary condition

$$\frac{d\chi}{dx} = \frac{\chi}{x}\Big|_{x=x_0}. \tag{4.172}$$

In Fig. 4.7 χ is shown as a function of distance x for a selection of densities. The tangent at the border of the cell passes through the origin. This determines the degree of *pressure ionization*. It increases with increasing density. The Figure also shows the infinite extension of the neutral atom, a well-known defect of the Thomas–Fermi model.

The TF Model at Finite Temperature

At finite temperature the electron density $n(r)$ is given by expression (4.165) and must be evaluated numerically. In dealing with the TF thermodynamics integrals of the form

$$F_\nu(y) = \int_0^\infty \frac{x^\nu \mathrm{d}x}{1 + \exp(x - y)}$$

appear. Then, in terms of x and y the following identifications follow from (4.165)

$$\beta\left(\frac{\mathbf{p}^2}{2m_e}\right)=x,\quad y=\beta(eV+\mu),\quad p^2\mathrm{d}p=\frac{1}{2}\left(\frac{2m_e}{\beta}\right)^{3/2}x^{1/2}\mathrm{d}x,\quad \nu=\frac{1}{2}$$

and

$$n(r)=\frac{1}{2\pi^2}(\frac{2m_e}{\hbar^2})^{3/2}(k_BT_e)^{3/2}F_{1/2}\left[\frac{eV(r)+\mu}{k_BT_e}\right];\quad [n(r)]=\mathrm{cm}^{-3}. \tag{4.173}$$

The function $F_{1/2}(y)$ is known as the *Fermi-Dirac integral*. For $T_e\to 0$ expression (4.173) goes over into (4.167). The chemical potential follows from $\int n(r)\mathrm{d}\mathbf{r}=Z$ because the sphere of radius r_0 is neutral, with the electric field vanishing on the border of the cell. After setting

$$\xi=\frac{r}{r_0},\quad \Psi(\xi)=\xi\frac{eV+\mu}{k_BT_e},\quad a=\frac{m_ee^2r_0^2(2m_ek_BT_e)^{1/2}}{\pi\varepsilon_0\hbar^3}$$

with a dimensionless, the Thomas–Fermi equation for finite temperature reads

$$\frac{d^2}{\mathrm{d}\xi^2}\Psi(\xi)=a\xi F_{1/2}\left[\frac{\Psi(\xi)}{\xi}\right];\quad \Psi(0)=\frac{Ze^2}{4\pi\varepsilon_0r_0k_BT_e},\quad \frac{d\Psi(\xi=1)}{\mathrm{d}\xi}=\Psi(1). \tag{4.174}$$

Radius r_0 and temperature T_e scale with Z as $Z^{1/3}$ and $Z^{4/3}$. For (4.174) see also [14, 15].

The internal energy $\mathcal{E}_{\mathrm{in}}$ in the cell is the sum of the kinetic energy K, the potential energy of attraction nucleus-electron cloud $U_{ne}<0$, and the repulsive electron-electron energy $U_{ee}>0$,

$$\begin{aligned}K&=\int_{\mathrm{cell}}\mathrm{d}\mathbf{r}\int\frac{\mathbf{p}^2}{2m_e}\frac{2}{h^3}f(r,\mathbf{p})\mathrm{d}\mathbf{p}\\&=\frac{1}{2\pi^2}(\frac{2m_e}{\hbar^2})^{3/2}(k_BT_e)^{5/2}\int_{\mathrm{cell}}F_{3/2}\left[\frac{eV(r)+\mu}{k_BT_e}\right]\mathrm{d}\mathbf{r},\\U_{ne}&=-\int_{\mathrm{cell}}\frac{Ze^2}{4\pi\varepsilon_0r}n(r)\mathrm{d}\mathbf{r},\quad U_{ee}=\frac{1}{2}\int_{\mathrm{cell}}\frac{n(r)n(r')e^2}{4\pi\varepsilon_0|\mathbf{r}-\mathbf{r}'|}\mathrm{d}\mathbf{r}.\end{aligned} \tag{4.175}$$

The free energy F is obtained from integrating relation (4.78),

$$\mathcal{E}_{\mathrm{in}}=F-T_e\left(\frac{\partial F}{\partial T_e}\right)_V=\frac{\partial\beta F_e}{\partial\beta};\quad \beta=\frac{1}{k_BT_e},\quad \mathcal{E}_{\mathrm{in}}=K+U_{ne}+U_{ee}.$$

It yields

$$F_e=Z\mu-\frac{2}{3}K-U_{ee},\quad S_e=\frac{1}{T_e}(\frac{5}{3}K-Z\mu+U_{ne}+2U_{ee}),$$

$$p_e = \frac{1}{3\pi^2}(\frac{2m_e}{\hbar^2})^{3/2}(k_B T_e)^{5/2} F_{3/2}\left(\frac{\mu}{k_B T_e}\right). \tag{4.176}$$

Ionization degree with TF. At the border of the cell the electric field is zero, the electrons with higher energies are uniformly distributed as a requirement of translational symmetry. The potential $V(r_0)$ within the TF model has to be set to zero there. The number of these free electrons is given by (index I for ionization)

$$N_I = \int_{\text{cell}} \left(\int_{E>0} \frac{2}{h^3} f(r, p) \mathrm{d}\mathbf{p} \right) \mathrm{d}\mathbf{r} = \frac{4\pi}{3} r_0^3 n(r_0). \tag{4.177}$$

In reality, the ion potential for positive electron energies is not flat but oscillating in space because also *free* electrons feel the presence of the localized nuclei. In the TF model the bound electrons compensate exactly the attraction of the ions; the free N_I electrons provide for the Fermi pressure. At $T_e = 0$ the pressure ionization of aluminum at solid density for instance yields a pressure of 1 Mbar from the free electrons in TF instead $p = 0$. This is because of the absence of the long range Coulomb forces of the ions acting on each other. They must be taken into account when the cohesion energy or compressibility of ionic crystals is to be calculated, in a first approach for instance within the *Madelung model*, see [16], p. 403ff.

Z scaling of TF. The TF model developed previously has the great advantage that it scales with the nuclear charge Z. By substituting $Z = Z_1 = 1$ in the equations by the real value Z, e.g. $Z = 13$ for Al, and $\rho = AZ\rho_1$, $T = Z^{4/3}T_1$ one arrives at the following Z-scaling for the thermodynamic quantities [12]

$$\begin{aligned} N_I^Z(\rho, T) &= Z N_1(\rho_1, T_1), \\ E(\rho, T) &= Z^{7/3} E_1(\rho_1, T_1), \\ \mu(\rho, T) &= Z^{4/3} \mu_1(\rho_1, T_1), \\ p(\rho, T) &= Z^{10/3} p_1(\rho_1, T_1), \\ S(\rho, T) &= (Z/A) S_1(\rho_1, T_1), \\ F(\rho, T) &= Z^{7/3} F_1(\rho_1, T_1). \end{aligned} \tag{4.178}$$

A whole sequence of corrections have been devised to achieve higher accuracy of the TF model, like the Thomas–Fermi–Dirac (TFD) scheme which introduces the so called exchange energy of the electrons, well known for example from the Hartree-Fock equation, the Thomas–Fermi–Kirshnitz (TFK) scheme by implementing a gradient correction in first perturbation limit, quantum modifications taking into account shell structures, and relativistic effects for $p_F \gtrsim m_e c$. The latter become relevant for ρ above 2×10^6 gcm^{-3}. Quantum and gradient corrections may contribute each by 10% [12]. An alternative approach leading to systematic corrections to the Thomas–Fermi model without a gradient expansion is presented in [17]. With any of such corrections the Z scaling gets lost. Internal energy E (left) and pressure P (middle) from the TF model of aluminum from $T_e = 0$ to $T_e = 10^5$ eV as well as the ioniza-

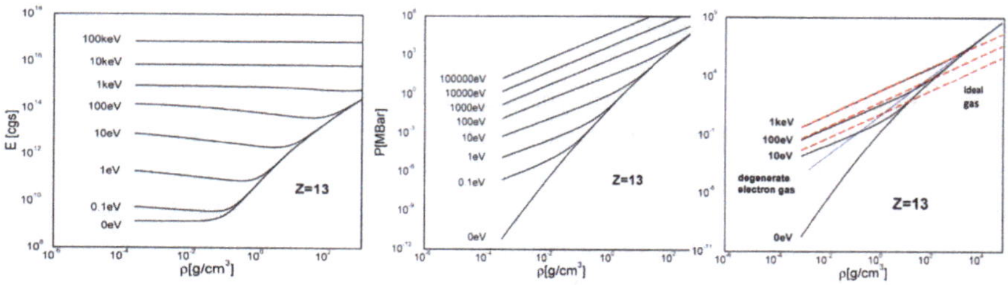

Fig. 4.8 Internal energy E (left) and pressure P (middle) from the TF model of aluminum for temperatures from $T_e = 0$ eV to $T_e = 10^5$ eV. The minimum in E vanishes with the number of bound electrons reducing to zero; it is equivalent to vanishing U_{ne} term and ionization $I_I^Z = Z$. Comparison of pressure P in aluminum as a function of density for temperatures from zero to 1 keV (right picture): confluent TF (solid) and ideal Fermi gas (dotted) at $T_e = 0$ eV, TF from $T_e = 10$ eV to $T_e = 1$ keV compared with ideal Fermi gas and ideal Boltzmann gas (dashed). Courtesy of [13]

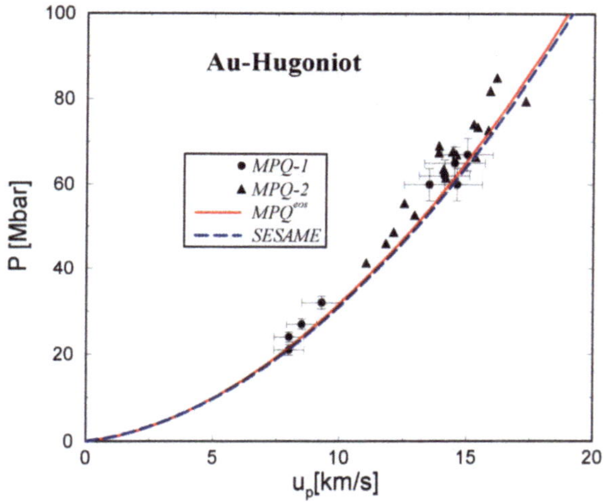

Fig. 4.9 Comparison of the equation of state from [13] (red solid) with shock compression measurements in gold from the Asterix laser in Garching/Munich and with SESAME (black dashed) from Los Alamos National Laboratory (LANL). The experimental points have been obtained from indirect drive in hohlraums as well as from direct drive by the laser [18]. Up to 80 Mbars have been achieved as a function of the shock velocity u_p, reported in units of km/s. Courtesy of [13]

tion of aluminum from $T_e = 0$ to $T_e = 1$ keV are reported in Fig. 4.8. An interesting comparison of experimental values for gold with the MPQ equation of state (red solid curve) are shown in the Hugoniot curve of Fig. 4.9. The red graph is the result of various corrections to the Thomas–Fermi model described in [13].

4.5.1.2 The Ion Component

Low temperature. The ion thermodynamic potentials can be treated as a nondegenerate classical fluid. Its Fermi energy at equal particle density is by the factor m_e/m_i smaller than the electron Fermi energy. At low temperature, say room temperature and below, account must be taken of the structure of matter as forming a crystalline structure or as being in a state of a dense fluid. When modelling the cohesive energy of a solid distinction must be made between the previously mentioned types of bondings. Nobel gas type crystals may be described by *Lennard-Jones type* intermolecular six-twelve potentials

$$\Phi(r) = -\frac{A}{r^6} + \frac{B}{r^{12}} \tag{4.179}$$

with r the intermolecular distance and $a,\ B$ constants. The electron clouds assume a dipole structure by their attraction each from the other nucleus with a power law p_{el}/r^3. As the nuclei get closer by compression the repulsive force prevails with an inverse power frequently modelled by r^{-12}. The repulsive component stems from the Fermionic character of the surrounding electrons: The number of available free states shrinks, entanglement with higher states takes place which, in turn, leads to fluctuating polarization. Now, after this quantum effect of mixing with excited states, from classical electrodynamics follows that the mutually induced dipole, each electron cloud repelled by the other one, goes with power r^{-6} from where the repulsive power 12 results finally. This mechanism is present also in other types of crystals, generally however it is weak.

The cohesive energy of ionic type crystals and fluids is determined in first place by the mutual long range attractive Coulomb potential energy $-\alpha e^2/r$, with α the *Madelung constant*, and to a weaker degree by the repulsive potential C/r^m, from experiments m ranging typically between 5 and 10. Summing the alternative series of Coulomb terms represents an interesting mathematical problem because the result depends on the order of summing up owing to the lacking absolute convergence of the series. There are numerous recipes how to do ordering motivated from physics; see e.g. [16], p. 403–8. The constants α and C depend on electron temperature and density, establishing in this way coupling of F_i with F_e in (4.162). Finally, for higher accuracy one has to take account also of the zero point ion oscillations, the more the lighter the ions are. For the the equations of state of the covalent crystals and metals we refer to the specialized literature.

High temperature. Material properties like specific heat, compressibility and transport coefficients of insulators and metals change, sometimes drastically, with temperature. It suggests that lattice vibrations in the crystal, and fluctuation in the fluid state cannot be ignored. In the crystal at finite temperature the atoms/ions fulfil bound motions around their equilibrium positions. We have to distinguish between the Bravais lattice at each point of which there is sitting a single species of atoms/ions (*monoatomic Bravais lattice*) and the situation where 2, generally p, particles are sitting in the unit cell, for instance the two ions Na^+, Cl^- or two Ge atoms in the unit cell. For small excursions ξ from the equilibrium positions each ξ is the linear superposition of $3p$ independent harmonic oscillators, the so called *normal oscillations* of the unit cell. In the crystal of N particles they propagate as the $3pN$ *normal modes*, i.e. plane waves $\boldsymbol{\xi} \sim \exp i(\mathbf{kx} - \omega t)$, with the dispersion relations $\omega = \omega(\mathbf{k})$. They are orthogonal to each other. In the monoatomic Bravais lattice their number is $3N$, N longitudinal and $2N$ transverse modes for the two polarizations. In the low frequency, long wavelength interval they are known as sound waves and therefore in the whole domain they are addressed as the *acoustic branches*. Note, they correspond to the $3N$ degrees of freedom of N particles coupled together by their interparticle forces. In the case of NaCl and Ge the 3 acoustic oscillations are characterized by displacements which leave the mutual distances Na^+ - Cl^- and Ge–Ge unaltered, both species oscillate in pairs, with frequency ω starting from close to zero for $\mathbf{k} \to 0$ and ending with the shortest possible wavelength imposed by the Bravais lattice constants. These find their exact correspondence in the longitudinal modes encountered in the plasma as the ion acoustic waves of dispersion $\omega = c_s k$; transverse modes in gases do not exist. In the crystal there exists the second type of modes where the oscillations between the pairs, or $(p-1)$ fold in general, are excited. There exist $3pN - 3N$ such modes, $(p-1)N$ of them are longitudinal and $2(p-1)N$ are transverse modes. They start from finite ω at $\mathbf{k} = 0$ and are high frequency modes. In the plasma they find the exact analogue in the high frequency electron plasma waves with $\omega(k=0) = \omega_p > 0$. All linear modes are quantized in the same manner as shown for the electromagnetic waves in Chap. 7 and then, in analogy to photons, they are called *phonons* and one distinguishes between *acoustic phonons* for the low frequency branch, and *optical phonons* for the high frequency branches. The reason for the distinction is that the optical phonons couple to photons through the *Raman effect*, like the electron plasma waves, called *plasmons* if quantized, undergo resonance absorption and stimulated Raman effect. Phonons are bosons and as such they follow the Bose–Einstein distribution (4.113) with the chemical potential $\mu = 0$. They contribute to the specific heat and to free energy and entropy in the same way as photons do, provided the distribution of modes in space is known. This is an all but easy calculation. Instead, depending on temperature, the simplified *Debye model* and the oversimplified *Einstein model* of the specific heat may serve as first approaches to obtain acceptable relations also for compressed matter. At high temperature both end up in the classical law of *Dulong-Petit*. Beyond a temperature $T \simeq 10$ eV the contribution of the atom/ion component to the free energy is less than 10% [12].

A nonideal gas. What the hydrogen atom meant for the development of quantum mechanics was the ideal gas for the development of statistical physics. So let us begin to consider what happens first if the classical ideal gas increases in density at a fixed temperature. In the ideal gas the particles are point like, their real size occupies a vanishingly small fraction of the volume V they fill in and they do not interact with each other (more precisely, the interaction is short range and so weak as to establish thermal equilibrium in a time which is short compared to the time imposed by changes of the constraints V and T). Imagine now the gas consisting of N particles of size $v_0 = 4\pi r_0^3/3$ each in the volume V. For the ith particle the volume $V_0 = \sum_j' 4\pi(r_i + r_j)^3/3 = 8(N-1)v_0$ is not accessible. This holds with equal probability for the N particles i in a wide range of density $n = N/V$. Thus the non accessible volume by the hard spheres is $V_0 = 8N(N-1)v_0/2 \simeq 4N^2v_0$ altogether. As long as three body collisions do not essentially contribute it is natural to write $V_0 = Nb$,

$$b = 4Nv_0, \quad V' = V - Nb \Rightarrow pV' = Nk_BT, \tag{4.180}$$

V' is the accessible volume. The equation of state to the right is a van der Waals equation of state for hard spheres. The van der Waals attraction of molecules by induced dipole forces can be calculated from the interaction potential $U(r1, ..., rN) = \sum U(r_i, r_j) = Ui, j$, split into pairs, and its related partition function

$$\begin{aligned} Q(V,T) &= \frac{C}{N!}\int \exp -\beta\left[\sum_i \mathbf{p}_i^2/2m + (1/2)\sum_{i,j} U_{i,j}\right]\prod(\mathrm{d}\mathbf{p}_i\mathrm{d}\mathbf{q}_i) \\ &= \frac{C}{N!}\left(\frac{2\pi m}{\beta}\right)^{3N/2}\prod_{i<j}\exp -\beta U_{i,j}\mathrm{d}\mathbf{r}_i. \end{aligned} \tag{4.181}$$

For further evaluation of the product $\prod_i \exp -\beta U_{i,j}\mathrm{d}\mathbf{r}_i$ is replaced by the mean value $(V - V_0)\exp -\beta\overline{U}$, taken as independent of j, to result in

$$Q(V,T) = \frac{C}{N!}\left[\left(\frac{2\pi m}{\beta}\right)^{3/2}(V - V_0)\exp -\beta\overline{U}\right]^N.$$

The pressure follows from (4.80),

$$p(V,T) = k_BT\left(\frac{\partial}{\partial V}\ln Q(V,T)\right)_T = \frac{k_BTN}{V - bN} - N\frac{\partial\overline{U}}{\partial V}.$$

The mean value $\overline{U}$ is an average of $N-1 \simeq N$ attractive pairs divided by the accessible volume $(V - V_0)$ for which can be set $\overline{U} - Na/V$. Insertion in $p(V;T)$ yields the *van der Waals equation of state*

$$p(VT) = \frac{k_BTN}{V - bN} - a\frac{N^2}{V^2} \Rightarrow \left(p + a\frac{N^2}{V^2}\right)(V - V_0) = Nk_BT. \tag{4.182}$$

By piecewise adjustment of the parameters a and b the equation can be used to describe phase transitions from the liquid to the gaseous state.In such transitions each particle comes close to several neighbours simultaneously. The derivation given here takes solely binary encounters into consideration and cannot be considered as a proof of (4.182). It merely indicates what kind the deviations from the ideal equation of state are when the gas density increases: Finite particle size leads to the restriction of the available volume from V to $(V - V_0)$ and to pressure increment. The mutual attraction of the particles by the weak *van der Waals forces* results in its lowering.

The corrections V_0 and a/V^2 represent the lowest order of a *virial expansion*. The van der Waals equation is of third order in volume. In a suitable temperature interval there exist three real roots in V for $p(V) = \text{const}$. At sufficiently high temperature the gas becomes early ideal and one is left with one real root V only. By simultaneous lowering of the temperature one arrives at a point T_c where the additional roots merge, just for continuity reasons, into one real value V_c. It is the *critical point* (V_c, T_c, p_c) defined by the obvious conditions on first and second derivatives of the pressure

$$\left(\frac{\partial p}{\partial V}\right)_T = 0; \quad \left(\frac{\partial^2 p}{\partial V^2}\right)_T = 0 \;\Rightarrow\; V_c = 3b, \quad k_B T_c = \frac{8a}{27b}, \quad p_c = \frac{a}{27b^2}. \tag{4.183}$$

At $T < T_c$ the liquid and the gaseous phases coexist.

4.6 Summary

The equilibrium of a homogeneous fluid macroscopically at rest, $\mathbf{u} = 0$, is characterized by the existence of an absolute temperature $T \geq 0$. Equivalent to it is the existence of a state variable called entropy $S \geq 0$. Both quantities are functions of macroscopic parameters like volume V, internal energy $\mathcal{E}$, macroscopic fields $\mathbf{E}$, $\mathbf{B}$, gravitation. In the phenomenological approach the theory of thermal equilibrium, i.e. a state that does not change in time as long as the macroscopic parameters are fixed, is based on three fundamental laws. The *first law* guarantees the existence of the internal energy $\mathcal{E}$ as a state variable. The existence of the state variables entropy and temperature follows from the *second law* of thermodynamics. In Kelvin's formulation it states that internal energy never flows *spontaneously* from a colder system in thermal equilibrium to a hotter thermal equilibrium system. The internal energy can be changed by external work to the system and by supply of heat Q, $d\mathcal{E} = \delta Q - p dV$. In a general process with $d\mathcal{E}$ fixed, δQ is not uniquely defined; it can vary from $\delta Q = 0$ to $\delta Q = \mathcal{E}$, the difference is compensated by the work of pressure p (or also by work of the macroscopic fields). In the general case the pressure will not remain homogeneous throughout the volume V, as for instance in fast gas dynamic compression. Such a process is called *irreversible*. It is *reversible* if the system remains homogeneous throughout the whole volume V. Only in such a *quasistatic process* all variables involved pass through a sequence of equilibrium states. Then, heat is uniquely defined by

$$\delta Q = T dS = \delta Q_{\text{rev}}, \qquad S = \int \frac{\delta Q_{\text{rev}}}{T} + S_0.$$

The quantity δQ_{rev} serves as the definition of the entropy S in the phenomenological approach of thermodynamics. By Clausius' theorem irreversibility is characterized by $\delta Q_{\text{irr}} < \delta Q_{\text{rev}}$.

In the thermally isolated system the entropy cannot decrease. This implies that in a homogeneous system composed of several components, for example ions, electrons, and neutrals, or consisting of two or more coexisting phases, like liquid and vapor, at thermal equilibrium the entropy is a maximum. For the converse to be true it must be postulated that the local maxima of entropy (i.e., local in the mathematical sense) are equilibrium states; it does not follow from Kelvin's (or Clausius') formulation of the second law of thermodynamics. The internal energy $\mathcal{E}(S, V, ...)$, the entropy $S(\mathcal{E}, V, ...)$, the Helmholtz free energy $F(V, T, ...) = \mathcal{E} - TS$, and the Gibbs free enthalpy $G(V, p, ...) = F - pV$ are *thermodynamic potentials* from each of which all other state variables follow by derivation.

The theory of thermal equilibrium, if limited to homogeneous finite systems, is of narrow applicability. By the passage from global to *local* thermal equilibrium thermo*statics* transmutes into true thermo*dynamics* and signifies an enormous enlargement of applicability. The underlying concept is the different time scales to assume a true local temperature $T(\mathbf{x}, t)$ and a global, spatially uniform temperature $T(t)$. For $T(\mathbf{x}, t)$, $T_e(\mathbf{x}, t)$, $T_i(\mathbf{x}, t)$, and any other temperature of a single component to hold the specific local interaction (collision) time τ_{int} must be short compared to the adiabatic change in time,

$$\tau_{\text{int}} \ll T/|\partial_t T(\mathbf{x}, t)| \quad \Leftrightarrow \quad \lambda \ll \frac{T(\mathbf{x}, t)}{|\nabla T(\mathbf{x}, t)|}; \qquad \lambda \text{ mean free path.}$$

Once the condition is satisfied, either in time (left inequality) or in space (right relation), all kinds of thermal transport, e.g., heat flow, radiation transport, can be described in terms of *local* state variables connected by transport coefficients with local fluxes. Fluxes of density, charge, momentum, energy, and entropy are described in terms of the local gradients. The various local transport quantities are linked together by the *Onsager relations*. The determination of the concomitant transport coefficients is the object of the kinetic theory and, in complex cases, of numerical simulations and the experiment.

The statistical treatment of thermodynamics is based on the principle of equal probability of all states of the system allowed by the external parameters, like volume, total energy, pressure etc. It leads to unexpected microscopic insight. The entropy becomes an expression of the averaged occupation numbers of the internal (quasi)energy levels. As such its variation is a measure of the deviation of a system from the adiabatic behavior. Adiabatic processes, volume compression for example, lead to changes of the individual energy levels but do not change their occupation numbers, they form the class of *isentropic*, i.e., *reversible* processes. The statistical foundation of entropy on the principle of equal likelihood of all energy states on the energy shell $E = \mathcal{E}$ allows a particularly lucid formulation of the thermodynamics. In the *microcanonical* ensemble (MC) the key quantity is the number $W(E)$ of *quantum* states on the energy shell $E(N, V, ...)$ of an isolated system with fixed number N of particles. The entropy is then defined as a function of E, V, and additional macroscopic parameters, like concentrations of components or ionization degrees. Alternatively, in the *canonical* ensemble (CE) the system of N particles and volume V is allowed to exchange energy with the weakly interacting heat bath of temperature

T. Finally, in the *grand canonical* ensemble (GE) volume and temperature are fixed, but the system is allowed to freely exchange particles with the heat bath, in addition to energy. In these three statistical ensembles the entropy is defined as

$$\text{MC}: \; S(N, E, V, ...) = k_B \ln W(N, E, V, ...); \quad \frac{1}{k_B T} = \frac{1}{\beta} = \frac{\partial W}{\partial E},$$

$$\text{CE}: \; S(N, V, T, ...) = -k_B \sum_l w_l \ln w_l; \; w_l = \frac{e^{-\beta E_l}}{Q_N(V, T, ...)}, \; Q_N = \sum_l e^{-\beta E_l},$$

$$\text{GE}: \; S(\mu, V, T, ...) = -k_B \sum_{l,N \geq 0} w_l(N) \ln w_l(N); \; w_l(N) = \frac{e^{-\beta E_{l(N)}}}{\mathcal{Z}(V, T, \mu, ...)},$$

$$\text{GE}: \; \mathcal{Z}(\mu, V, T, ...) = \sum_{l,N \geq 0} e^{\beta(\mu N - E_l(N))}, \quad \mu(V, T, ...) = k_B T \frac{\partial \mathcal{Z}}{\partial \langle N \rangle}.$$

The chemical potential is given by the state function $\mu(T, V, ...)$. In the statistical approach to thermal equilibrium the existence of entropy together with the identification of its maxima with the possible thermal equilibria is the content of the *second principle* of thermodynamics. For convex systems, in practice all stable macroscopic systems, the equivalence of the three statistical ensembles can be shown. As a consequence, the three definitions of S above are equivalent to each other. From the statistical definition of S *heat* is uniquely defined by $\delta Q_{\text{rev}} = T \mathrm{d}S$. With the determination of the canonical and the grand canonical partition functions Q_N and $\mathcal{Z}$ all knowledgeable quantities follow as partial derivatives from S or the thermodynamic potential of the Helmholtz free energy F,

$$F(\langle N \rangle, T, V, ...) = -\beta \ln Q_N, \qquad F(\langle N \rangle, T, V, ...) = -\beta \ln \mathcal{Z}(T, V, \mu, ...) + \mu \langle N \rangle.$$

The entropy tends to its absolute minimum for $T \to 0$. For a nondegenerate ground state, which is the case with any *condensed system*, the minimum is $S = 0$. This is the content of the *third* principle of thermodynamics.

The *ideal gas* is the hydrogen atom of thermostatistics. The interparticle potential is strong enough to drive the system into thermal equilibrium, and weak enough not to perturb sensitively the energy levels E_k of the single particle. Density effects, like ionization lowering, may be taken into account by the introduction of *effective* energy levels E'_k and the neglect of the interaction potential in the Hamiltonian. They become functions of the particle densities. In the following expressions E_k stands for the more general E'_k. The macroscopic energies E_l split into sums $E_l = \sum_k n_{lk} E_k$; the partition functions factorize. Here the grand canonical partition function $\mathcal{Z}$ shows its power. The mean occupation numbers $n_k = \langle n_{k,l} \rangle$ of the single particles result in the straightforward way as

$$n_k = \frac{1}{e^{\beta(E_k - \mu)} \pm 1}.$$

This is the *Fermi-Dirac* distribution for sign + and the *Bose–Einstein* distribution for sign −. The sign + stands for half integer spin particles (*Fermions*), the sign − for integer spin (*Bosons*). For $E_k \gg k_B T + \mu$ both distributions merge to the classical Boltzmann distribution

$$n_k = \mathrm{e}^{-\beta(E_k-\mu)}.$$

The set of levels E_k consists in general of a discrete and continuous spectrum. Correct normalization of the distribution is provided by the chemical potential μ. The distributions of n_k contain the *principle of detailed balance* between two levels,

$$\frac{n_j}{n_k} = \frac{\mathrm{e}^{\beta(E_k-\mu)} \pm 1}{\mathrm{e}^{\beta(E_j-\mu)} \pm 1}; \quad k_B T + \mu \ll E_{j,k} \quad \Rightarrow \quad \frac{n_j}{n_k} = {}^{-\beta(E_j-E_k)}.$$

The simplified expression on the right is the standard form of detailed balance. It is of general validity for ideal gases (neutrals, charged particles) if μ is vanishingly small, i.e. in the *Boltzmann limit.*

The *classical ideal gas (plasma)*, i.e. the Boltzmann limit of the ideal gas, is of considerable utility when the system is sufficiently diluted because the single particle partition function Z_1 splits into the product of a translational contribution depending on volume, and into an internal component describing rotation, vibration, and excitation states, independent of the volume. The total partition function Z of N particles is the Nth power of Z_1 divided by the essential factorial $N!$. In the frame of the classical ideal gas the chemical equilibrium is determined from the standard *mass action law*. The latter follows from minimizing the free energy or the Gibbs enthalpy. The ionization equilibrium is given by the *Saha equation* as a special application of the mass action law.

The *Thomas–Fermi model* is the prototype among the equations of state of dense matter. It applies if the pressure is determined by the quasi free electrons. For the metallic state this is the case. It is simple enough to serve as a guide to modelling matter under continuously increasing compression. The cold Thomas–Fermi model offers the additional advantage of scaling with the charge number Z. Finally, the degree of ionization can be determined from it. In some sense it is the Saha equation for compressed matter. Last but not least it is accessible to treat nonlinear and linear screening in Fermi fluids; for the latter see [19]. The Thomas–Fermi model applies with success to interacting systems of electrons. This may be surprising at first glance because it starts from the distribution of noninteracting fermions. Closer inspection reveals that it is applicable to potentials whenever the concomitant energy level shifts undergo solely adiabatic changes with respect to the free levels.

■ The relaxation time τ to an equilibrium distribution is the main criterion for the existence of local LTE in the plasma, $\tau_{ee} = 1/\nu_{ee}$ for the electrons to approach a Maxwellian with electron temperature T_e, $\tau_{ii} = 1/\nu_{ii}$ for the ions to approach a Maxwellian with ion temperature T_i, $\tau_{ei} = (m_e/m_i)\nu_{ei}$ for local $T_e = T_i$. For the ionization equilibrium the inverse of the ionizing collision frequency is the relevant quantity.

■ The principle of detailed balance applies to (quasi)ideal systems (weak particle-particle interaction). Its standard Boltzmann-like version requires at least that the chemical potential is small.
■ The definition of states and their correct counting is the exclusive result of quantum mechanics. The entropy never turns over into a classical variable, see e.g.

$$S(E, V) = k_B \ln W = \frac{1}{N!(2\pi\hbar)^{3N}} \int_{E<\mathcal{E}\leq E+\Delta E} \mathrm{d}^{3N}\mathbf{p}\ \mathrm{d}^{3N}\mathbf{q}.$$

Expressions of S not containing $\hbar$ either are differences ΔS between two thermodynamic states or ratios S/S_0.
■ The absolute temperature T is an entity on its own, different from energy $\mathcal{E}_{\text{in}}$. Consequence: the Boltzmann constant k_B cannot be eliminated from thermodynamics, in contrast to $\hbar$ which is absent in classical physics. This may become relevant for dimensional analysis in special cases, as for example in selfsimilar radiation waves.
■ The general thermodynamic laws apply to individual components, neutrals, electrons, ions ..., if for each of them an individual absolute temperature $T_\sigma(\mathbf{x}, t)$, $\sigma = n, e, i, ...$, can be defined locally.
■ Fluctuations are subject to the laws of quantum mechanics, dramatically evidenced by the equality of the mean occupation $\langle n \rangle_j$ of an energy level with its variance $\sqrt{\Delta \langle n_j^2 \rangle}$ in the *Bose statistics*. In contrast, *Fermions* show narrowing of the variance and perfect order with temperature shrinking much below the Fermi temperature. Only in the Boltzmann limit independent particles behave according to the classical rules of probability.
■ The quantum statistical approach to thermodynamics offers the maximum possible microscopic insight into the structure of a system in local or global thermal equilibrium. The phenomenological approach to thermodynamics represents a powerful alternative to the statistical procedure for extracting basic thermodynamic relations in complex systems. Moreover, the quantum statistical formulation of transport in such cases may be a hopeless task without the aid of classical modelling. As an example take volume viscosity of oil, flow of glaciers and lava, or viscous dynamics of the fission of an uranium nucleus.

4.7 Problems

▷ Show with the help of second law of thermodynamics that the internal energy of the ideal gas does not depend on the volume. *Hint:* dS is a total differential.
▷ Verify assertion on (4.12) to represent a total differential.
▷ Show the equivalence of assertions C and K. *Hint:* Show, if C is false follows K is false, too, and viceversa.
▷ Show that (i) $Q_2 = 0$ violates Clausius' assertion C (use the property of reversibility

of the Carnot engine); (ii) no other engine works with higher efficiency η between two temperatures than the Carnot engine.
▹ List processes of laser-matter interaction under 'reversible' and 'irreversible'.
▹ Show that the Grüneisen parameter of the ideal gas (particles, photons, phonons, plasmons) is $\Gamma = \gamma - 1$.
▹ Prove the statement: If the thermally isolated system is in a maximum (local or absolute) of entropy under fixed constraints $\Rightarrow$ it is in thermal equilibrium.
▹ Two metallic blocks with specific heats C_{p1} and C_{p2} are each in thermal equilibrium at temperatures T_1 and T_2. Calculate the equilibrium temperature and entropy increase after they are brought into thermal contact.
▹ *Shock tube.* A tank of volume V_1 is filled with hydrogen at $p = 150$ bar. By a membrane it is connected with a closed end shock tube of volume $V_2 \ll V_1$ filled with Ar gas at $p = 1$ bar. The thermally isolated facility is at room temperature. By a minimum increase of p in the tank the membrane breaks and the hydrogen gas is released to drive a strong shock into the Ar gas. (a) Calculate the temperature of the shock wave before reflection. (b) Determine temperature and pressures in hydrogen and argon after thermal equilibrium is restored. Assume hydrogen and Argon to be ideal gases. *Solution to (b):* No change of total volume, internal energy of gas does not depend on volume $\Rightarrow$ Both gases are at room temperature, $p_{H_2} = p_{Ar}$.
▹ (a) How does the internal energy of the van der Waals gas, $p = RT/(V-b) - a/V^2$, depend on the volume? Interpret the result. *Answer:* $\partial_V \mathcal{E}_{in}(V,T) = a/V^2$; pressure lowering and internal energy increase at constant temperature due to mutual particle attraction is provided by $-a/V^2$. (b) Show for the gas in (a) that C_V does depend on temperature only through b. *Answer:* $\partial_V C_V(V,T) = \partial_T[RT/(V-b)] = R/(V-b)$. (c) Calculate the infinitesimal change of entropy dS and internal energy $d\mathcal{E}_{in}(V,T)$ for the gas in (a). *Answer:* $T\mathrm{d}S = C_V\mathrm{d}T + [\partial_V\mathcal{E}_{in}(V,T) + p] \Rightarrow \mathrm{d}S = (C_V/T)\mathrm{d}T + [R/(V-b)]\mathrm{d}V$; $\mathrm{d}\mathcal{E}_{in}(V,T) = T\mathrm{d}S - p\mathrm{d}V = C_V\mathrm{d}T + a/V^2\mathrm{d}V$.
▹ (a) Prove (4.21). (b) Verify the validity of (4.20),–(4.22) for the ideal fully ionized plasma. (c) Convince yourself by intuition of the correctness of (4.21).
▹ Determine the adiabatic coefficient γ for (a) the ideal fully ionized plasma, (b) the photon gas, (c) the ideal cold Fermi gas.
▹ (a) Verify the last step in (4.25). (b) Find the relation between Γ and γ for the ideal gas.
▹ (a) Derive the four Maxwell relations from the thermodynamic potentials. (b) The entropy is a state function. Does an additional Maxwell relation follow from this fact?
▹ Prove: T from (4.4) coincides with T from the ideal gas law. *Proof:* Consider the Carnot cycle: (1) Isothermal expansion $V_0 \to V_1$ at $T = T_1$, (2) adiabatic expansion $V_1 \to V_2$, (3) isothermal compression $V_2 \to V_3$ at $T = T_2$, (4) adiabatic compression $V_3 \to V_0$.

$$Q_1 = RT_1 \ln\frac{V_1}{V_0},\quad Q_2 = RT_2 \ln\frac{V_2}{V_3},\quad V_2T_2^{\gamma-1} = V_1T_1^{\gamma-1},\quad V_3T_2^{\gamma-1} = V_0T_1^{\gamma-1}$$

$$\Rightarrow \frac{V_2}{V_3} = \frac{V_1}{V_0}; \quad 1 - \eta = \frac{Q_2}{Q_1} = \frac{T_2 \ln(V_2/V_3)}{T_1 \ln(V_1/V_0} = \frac{T_2}{T_1}.$$

▹ Compare the entropy of the Van der Waals equation of state with that of the ideal gas. Are there any anomalies?
▹ Verify that the radiation pressure in thermal equilibrium is $p = \epsilon/3$ (a) in the photon picture (use $\mathbf{p} = \hbar\mathbf{k}$; integrate over all angles) and (b) by starting from the adiabatic law $\omega V^{1/3} = \text{const}_a$ (see Chap. 2).
▹ Calculate the entropy production per unit time and unit length $\dot{\sigma}(\mathbf{x}, t)$ from local $\mathbf{q}(\mathbf{x}, t)$. *Answer:* $\dot{\sigma}(\mathbf{x}, t) = |\mathbf{q}||\nabla T|/T^2$; Fourier: $\Rightarrow \dot{\sigma}(\mathbf{x}, t) = \kappa(\nabla T)^2/T^2$.
▹ Prove: From $F(x + y) =: F(u) = F(x)F(y) \Rightarrow F(u) = A \exp Cu$, A, C constants. *Hint:* Take the partial derivative with respect to $x \Rightarrow F_x = F'(u)$, $F_x = F'(x)F(y) \Rightarrow F'(u)/F(u) = F'(x)/F(x) \;\; \vee\, y, \Rightarrow F'/F = C$.
▹ Why is setting $f(|\mathbf{v}|) = g(\mathbf{v}^2)$ unique and, for instance, not a supergaussian, like $f(|\mathbf{v}|) = g(|\mathbf{v}|^\alpha)$, $\alpha > 2$? *Hint:* Remember: Energy conservation must be fulfilled.
▹ Construct a distribution function $f(\mathbf{x}, \mathbf{v}, t)$ which violates Clausius' assertion of equilibrium thermodynamics.
▹ From Chap. 2–4 you have been familiar with four independent procedures to derive the Maxwell distribution function. Reproduce all and discuss the assumptions underlying to each of them.
▹ The structure function $\Omega(E_0)$ and its associated phase space volume $\Gamma(E_0)$ of the sum of two noninteracting systems are given by

$$\Omega(E_0) = \int_{E<E_0} \Omega(E)\Omega(E_0 - E)\mathrm{d}E, \quad \Gamma(E_0) = \int_0^{E_0} \mathrm{d}E_0' \int_{E<E_0'} \Omega(E)\Omega(E_0' - E)\mathrm{d}E.$$

Justify the replacement of $\ln \Omega$ by $\ln \Gamma$ in (4.77) and (4.83).
▹ In (4.79) from $\partial \int ...\mathrm{d}\mathbf{p}\mathrm{d}\mathbf{q}/\partial\beta = 0$ follows $\int ...\mathrm{d}\mathbf{p}\mathrm{d}\mathbf{q} = K = \text{const}$. Show that $K = 1$.
▹ Derive internal energy, entropy, and equation of state of the ideal gas from the virial theorem, e.g. in the form of (4.56).
▹ Apply the virial theorem to the CO_2 and H_2O molecules in thermal equilibrium.
▹ Evaluate the error by omitting the quadratic expansion term in (4.62) if $T = T(\mathbf{x})$.
▹ Thermal equilibrium of N_0 free particles in a large box of volume V_0 may be tested by choosing a small (but not too small) box of volume V and counting the particles in it at various positions $\mathbf{x}$ in V_0. What is the result for $\langle n \rangle$ and $\langle \Delta n^2 \rangle$ if $n(\mathbf{x})$ is their number in the box V?
▹ A thermal plasma of constant temperature T is enclosed in a box of volume V to which a constant electric field is applied from outside. (a) Calculate the distribution function $f(\mathbf{x}, \mathbf{v})$. (b) Evaluate the induced thermoelectric field.
▹ If the operator of unity is applied between $\hat{O}$ and $|\psi_r\rangle$ the result is $\hat{\rho}\hat{O}$. What is the difference between $\hat{O}\hat{\rho}$ and $\hat{\rho}\hat{O}$? *Answer:* $\hat{\rho}\hat{O} = (\hat{O}\hat{\rho})^\dagger = \hat{O}\hat{\rho}$, as expected.
▹ Show $\mathrm{Tr}(A)$ of an operator A is invariant with respect to basis changes.
▹ Show, $\rho^2 = \rho$ implies ρ acts on a pure state $\Rightarrow \rho = 1$. Is this result true also for $\rho = \sum P_r |R_r\rangle\langle R_r|$ with $\langle R_{r'}|R_r\rangle \neq 0$? *Answer:* Yes. Prove!

▹ Prove the correctness of the last expression for $\mathcal{Z}$ in (4.110). *Hint:* $(\sum_i A_i)(\sum_j B_j) = \sum_{ij} A_i B_j$.
▹ Show that (4.150) is in agreement with the last expression in (4.110).
▹ A hohlraum is filled uniformly with matter of refractive index η. Show

$$\rho(\omega, T) = \left(n + \omega\frac{\partial \eta}{\partial \omega}\right)\rho_0(\omega, T); \quad I(\omega, T) = \eta^2 I_0(\omega, T). \tag{4.184}$$

The index "0" stands for vacuum. Sketch the two quantities for a plasma-filled hohlraum of $\eta^2 = 1 - n_e/n_c$.
▹ Convince yourself that for $\lambda \ll V^{1/3}$ the density of modes in the Planck hohlraum does not depend on the shape of V. Give an estimate of the error.
▹ The ratio of spontaneous to stimulated emission is

$$\frac{A_{12}}{B_{12}\rho(\omega, T)} = \exp(\beta\hbar\omega) - 1 = \frac{1}{\langle n\rangle}. \tag{4.185}$$

For $T = 300$ [K] spontaneous emission begins to prevail already for λ shorter than 50μm. Determine this limit for a gas of $T_e = 1$ eV.
▹ Explain why the canonical distribution function of states is a strongly peaked function of the system energy states E. Give a measure of its width.
▹ Formulate Kirchhoff's law of radiation balance for (a) the single free electron, (b) the oscillating dipole of eigenfrequency ω_0, (c) a single atom (ion). How are resonances ($\omega = \omega_0$) to be treated?
▹ Calculate the entropy of a *single* plane wave in the hohlraum V of temperature T.
▹ Formulate the system of Saha equations for the equilibrium ionization of carbon. Discuss the simplifications for $k_B T = 5$ eV.
▹ Derive the Debye potential for a one component plasma of Z-fold ionized ions.
▹ Calculate the chemical potential for the Maxwellian velocity distribution.
▹ Derive (4.147) for independent classical point particles from first principles.
▹ Show for $k_B T_e \to 0$ (4.173) goes over into the cold Fermi charge density (4.167).
▹ Show the scaling of (4.174) with $Z^{-1/3}$ of the cell radius and $Z^{4/3}$ of the electron temperature.
▹ Verify (4.176) and interpret.

4.8 Self-assessment

- Consider rarefied He^4 and molecular D_2 gases at room temperature. The number of degrees of freedom is $f_{He} = 3$ and $f_{D_2} = 5$ (no vibrations). (1) Do rotations (and vibrations, if excited) contribute to pressure $p = p(\rho, T)$? (2) If not, why then the ratio of the sound velocities according to (4.24) is $[(5/3)/(7/5)]^{1/2} = 1.09$? (3) Do vibrations contribute to pressure in dense matter, e.g. solids? *Answer:* yes.
- Question related to (4.123): Is nc compatible with $n\mathbf{v} = \gamma n_0 \mathbf{v}$?

- Give reasons for $\mu = 0$ of photons in vacuum. *Hint:* (a) experimental proof of Planck's law; (b) photon number $N_\omega = f(V, T, \omega)$; black walls of heat bath emit one photon for one absorbed photon; ...
- Can photons form a Bose–Einstein condensate? *Answer:* Massless photons cannot; however, see Jan Klaers et al., Nature **468**, 545 (2010), and others.
- What is the value of the Grüneisen parameter of the photon gas?
- If the constant C in the entropy (4.49) for the ideal gas results $C = k_B$, why then this identification holds for all systems? *Hint:* Think of a Carnot cycle.
- What distinguishes T from $\mathcal{E}$, and S from $\mathcal{E}/T$? *Possible answer:* No proportionality or other universal relation between the two quantities. S allows to distinguish between mechanical work and heat.
- Show the equivalence of $\delta Q_{\text{irr}} < \delta Q_{\text{rev}}$ with the definition of *irreversibility* given in the summary.
- Kelvin's formulation of the second principle of thermodynamics is based on "colder" and "hotter". How do you define them without making recurrence to the concept of absolute temperature T?
- Show: External work by continuous forces does not change the entropy. How is the statement consistent with Ohmic heating by the applied electric field? *Answer:* Heating is produced by discontinuous (more precise: nonadiabatic) forces in the collisions.
- (a) Sketch the way along which follows from the Clausius inequality (4.11) that the free energy in a thermally isolated system cannot increase. (b) Under which conditions are thermodynamic equilibria determined from the equality in (4.33)?
- (a) List all thermodynamic potentials you know and indicate on which variables each of them depends. (b) Invent your own thermodynamic potential.
- Homogeneous vs inhomogeneous system: In the ideal gas a sudden volume change occurs from V to $2V$. When does the entropy start to increase and what is its amount in the new equilibrium?
- Local thermodynamic equilibrium (LTE) requires that the one particle distribution function is close to a Maxwellian? Why? *Hint:* The macroscopic thermodynamic quantities of LTE obey the fundamental laws of equilibrium thermodynamics.
- Which of the following forces do increase the entropy and which do not: shear viscosity, volume viscosity, electric field applied from outside, heat conduction, electron-ion friction, Lorentz force, gravitational potential? *Answer:* Forces involving collisions (two-body, many-body forces) produce entropy, forces depending on one spatial coordinate (and not containing gradients) do not.
- Justify the relativistic extension (4.63) of $f_M(\mathbf{v})$.
- Control the correctness of (4.81).
- Explain why the canonical distribution function of states is a strongly peaked function of the total energy E of the system. Give a measure of its width.
- Which are the underlying physical properties from which follows the equivalence of the microcanonical, the canonical and the grand canonical ensembles?
- Consider the fully ionized hydrogen plasma. Each ion is surrounded by its Debye cloud. How is this possible as there is only one electron per ion on the average?

- Convince yourself of the correctness of the first equality in (4.89).
- Show that the sum of all products equals the product of all possible sums in (4.110).
- The chemical potential μ in the grand canonical ensemble is defined as the change of energy by adding an additional particle to the system. It appears again in the probability w for the system occupying a given energy level $E_{l,N}$ in a system of N particles in the heat bath. Since w is a probability it must be normalizable to unity. Show that μ provides for the correct normalization of w in the Fermi distribution of the ideal Fermi gas.
- Calculate $\mu(V, T)$ of the Fermi-Dirac distribution as a function of T. How does μ decrease with T increasing?
- Formulate a condition for the validity of the Einstein relation between the coefficients A and B of spontaneous and induced emission of thermal radiation. Does the principle of detailed balance hold?
- The principle of detailed balance for the *single* ion is generally used to derive the Debye screening potential for *all* ions, i.e., for an arbitrary ion in the plasma. The derivation from the grand canonical probabilities refers to a single, i.e., a well defined ion. Under what conditions is the procedure correct? What is the limit of validity? *Hint:* Superposition principle requires linearity in density perturbation.
- Why is the chemical potential for massless particles zero?
- In Fig. 4.9 the internal energy E at $T_e = 0$ eV shows no minimum in the TF approximation. Explain. *Answer:* Kinetic energy $K = 0$ and Coulomb energies U_{ne} and U_{ee} compensate each other.
- Give a simple explanation for the Fermi energy $E_F \sim n_e^{2/3}$ and Fermi pressure $p_e(T_e = 0) \sim n_e^{5/3}$. This is the same dependence for the ideal classical gas at finite temperature. Accidental coincidence? *Answer:* no. Both are applications of the adiabatic invariance $vl = \mathrm{const_a}$ (see Chap. 2).
- List some effects contributing to an improvement of the finite temperature Thomas–Fermi model (4.174).
- Explain to the non specialist the physics leading to the Lennard-Jones potential. *Hint:* Use a semiclassical model, i.e., combine quantum mechanics with classical electrostatics.
- At sufficiently high temperature and/or when dealing with massive objects it is common believe that transition from quantum to classical behaviour occurs. Why entropy, black body radiation, electron partition function never get classical? *Hint:* You may find a criterion in the last section of Chap. 8.
- "Clustering" of Bosons? Consider clustering of degree $m = N/n$ of classical particles: From (4.147) follows $\sqrt{\langle \Delta n^2 \rangle}/\langle n \rangle = \sqrt{m/N}$. This is larger than $1/\sqrt{N}$ but far from unity for Bosons.

4.9 Glossary

First law of thermodynamics:

$$\exists\ \mathcal{E}_{\text{in}} \ \Leftrightarrow\ \mathrm{d}\mathcal{E}_{\text{in}} = \delta Q + \mathrm{d}W. \tag{4.1}$$

Clausius' inequality

$$\oint \frac{\delta Q}{T} \leq 0; \quad \oint \frac{\delta Q_{\text{rev}}}{T} = 0. \tag{4.5}$$

Second law of thermodynamics:

$$\exists\ S \ \Leftrightarrow\ \mathrm{d}S = \frac{\mathrm{d}Q_{\text{rev}}}{T}. \tag{4.6}$$

Entropy increase by heat transfer Q from T_1 to $T_2 < T_1$,

$$\Delta S = Q\left(\frac{1}{T_2} - \frac{1}{T_1}\right). \tag{4.9}$$

Absolute temperature T and pressure p

$$_{\text{in}} = T\mathrm{d}S - p\mathrm{d}V \Rightarrow T = \left(\frac{\partial \mathcal{E}_{\text{in}}}{\partial S}\right)_V; \quad p = -\left(\frac{\partial \mathcal{E}_{\text{in}}}{\partial V}\right)_S. \tag{4.13}$$

TdS relations

$$T\mathrm{d}S = \left(\frac{\partial \mathcal{E}_{\text{in}}}{\partial T}\right)_V \mathrm{d}T + \left[\left(\frac{\partial \mathcal{E}_{\text{in}}}{\partial V}\right)_T + p\right]\mathrm{d}V = C_V\mathrm{d}T + \frac{\alpha T}{\kappa_T}\mathrm{d}V. \tag{4.18}$$

$$T\mathrm{d}S = \left[\left(\frac{\partial \mathcal{E}_{\text{in}}}{\partial T}\right)_p + p\left(\frac{\partial V}{\partial T}\right)_p\right]\mathrm{d}T - T\left(\frac{\partial V}{\partial T}\right)_p \mathrm{d}p = C_p\mathrm{d}T - \alpha T V\mathrm{d}p. \tag{4.19}$$

Specific heats C_V, C_p

$$C_p - C_V = \frac{\alpha^2 T V}{\kappa_T}, \quad C_V = \frac{\alpha^2 \kappa_S}{\kappa_T - \kappa_S} T V, \qquad \frac{\alpha^2}{\kappa_T - \kappa_S} T V. \tag{4.20}$$

Adiabatic exponent γ and Grüneisen parameter Γ

$$\gamma = \frac{1}{p\kappa_S} = \frac{\rho}{p}c_S^2; \quad \Gamma = \frac{V}{C_V}\left(\frac{\partial p}{\partial T}\right)_V = \frac{\alpha V}{C_V \kappa_T}. \tag{4.25}$$

Thermodynamic potentials $\mathcal{E}_{\text{in}}$, S, $H = \mathcal{E}_{\text{in}} + pV$, $F = \mathcal{E}_{\text{in}} - TS$, $G = H - TS$:

$$\mathcal{E}_{\text{in}} = \mathcal{E}_{\text{in}}(S, V, ...),\ \mathrm{d}\mathcal{E}_{\text{in}} = T\mathrm{d}S - p\mathrm{d}V;\ S = S(\mathcal{E}_{\text{in}}, V, ...),\ \mathrm{d}S = \frac{1}{T}[\mathrm{d}\mathcal{E}_{\text{in}} + p\mathrm{d}V],$$

$$H = H(S, p, ...),\ \mathrm{d}H = T\mathrm{d}S + V\mathrm{d}p,\ F = F(T, V, ...),\ \mathrm{d}F = -S\mathrm{d}T - p\mathrm{d}V,$$

$$G = G(T, p, ...),\ \mathrm{d}G = -S\mathrm{d}T + V\mathrm{d}p. \tag{4.32}$$

Maxwell relations (interchanging second derivatives)

$$\left(\frac{\partial T}{\partial p}\right)_S = \left(\frac{\partial V}{\partial S}\right)_p,\quad \left(\frac{\partial T}{\partial V}\right)_S = -\left(\frac{\partial p}{\partial S}\right)_V,$$

$$\left(\frac{\partial S}{\partial V}\right)_T = \left(\frac{\partial p}{\partial T}\right)_V,\quad \left(\frac{\partial S}{\partial p}\right)_T = -\left(\frac{\partial V}{\partial T}\right)_p. \tag{4.34}$$

Microcanonical ensemble ($\Leftrightarrow$ = equivalence), $E = \mathcal{E}_{\text{in}}$

$$\Omega = \int_{\Sigma(E)} \frac{\mathrm{d}\Sigma}{|\mathrm{grad}_{(\mathbf{p},\mathbf{q})}(H = E)|} \Leftrightarrow \Gamma(E) = \int_{E' \leq E} \Omega \mathrm{d}E'. \tag{4.42}$$

Phase average $\langle f \rangle$ of phase function $f(\mathbf{p}, \mathbf{q})$:

$$\langle f \rangle = \frac{1}{\Omega}\frac{\partial}{\partial E}\int_{H \leq E} f(\mathbf{p}, \mathbf{q})\mathrm{d}\mathbf{p}\mathrm{d}\mathbf{q}. \tag{4.43}$$

Stirling's formula

$$n! \simeq (2\pi n)^{1/2}\left(\frac{n}{\mathrm{e}}\right)^n. \tag{4.45}$$

Microcanonical entropy

$$S(E, V) = k_B \ln \Omega(E, V) \Leftrightarrow k_B \ln \Gamma(E, V);\quad E = \mathcal{E}_{\text{in.}} \tag{4.48}$$

Absolute temperature

$$\frac{1}{T} = \left(\frac{\partial S}{\partial E}\right)_V. \tag{4.52}$$

Virial theorem

$$\left\langle p_j \frac{\partial H}{\partial p_j}\right\rangle = k_B T;\quad \left\langle q_j \frac{\partial H}{\partial q_j}\right\rangle = k_B T. \tag{4.56}$$

Equipartition theorem of $H = \sum_j A_j p_j^2 + \sum_j B_j q_j^2$ is $\langle H \rangle = \frac{f}{2} k_B T$.
f total number of terms.
Entropy of classical ideal gas

$$S(E,V)=Nk_B\ln\left[\frac{V}{N}\left(\frac{2\pi mE}{N}\right)^{3/2}\right]+\frac{3}{2}Nk_B+\text{const.} \tag{4.93}$$

Classical Canonical Ensemble

Probability $w(E)$ in heat bath and density of states $\rho(\mathbf{p},\mathbf{q})$

$$w(E)=\mu[\Omega(E,V)]\exp-\frac{E}{k_BT};\quad \rho(\mathbf{p},\mathbf{q})=\frac{w(E)}{\mu[\Omega(E,V)]}. \tag{4.77}$$

Partition function $Q_N(V,T)=\int\mu[\Omega(E,V)]\exp-\frac{E}{k_BT}\,\mathrm{d}E$
Connection with free energy

$$Q_N(V,T)=\exp-\frac{F(V,T)}{k_BT}. \tag{4.79}$$

Further relations

$$\mathcal{E}_{\text{in}}(V,T)=k_BT^2\left(\frac{\partial}{\partial T}\ln Q_N\right)_V,\quad S(V,T)=k_B\left(\ln Q_N+T\frac{\partial}{\partial T}\ln Q_N\right)_V$$

$$p(V,T)=k_BT\left(\frac{\partial}{\partial V}\ln Q_N\right)_T. \tag{4.80}$$

Mean square deviations

$$\frac{\langle\Delta\mathcal{E}_{\text{in}}\rangle}{\mathcal{E}_{\text{in}}}=\frac{\langle\Delta T\rangle}{T}=\frac{1}{\sqrt{(f/2)N}}. \tag{4.82}$$

Classical Grand Canonical Ensemble $\mathcal{Z}(V,T,\mu)$

$$\mathcal{Z}(V,T,\mu)=\sum_{N=0}^{N=N_0}\int \mathrm{e}^{\beta(\mu N-H_N(\mathbf{p},\mathbf{q}))}\mathbf{dpdq}=\mathrm{e}^{-\beta J(V,T,\mu)}, \tag{4.84}$$

$$\Rightarrow\rho(\mathbf{p},\mathbf{q})=\frac{1}{\mathcal{Z}(V,T,\mu)}\mathrm{e}^{\beta(\mu N-H_N)}. \tag{4.85}$$

Chemical potential $\mu=\mu(V,T,N_0)$ Mean particle fluctuations

$$\frac{\langle\Delta N^2\rangle}{\langle N\rangle}=\frac{\langle\Delta n^2\rangle}{\langle n\rangle}V=\kappa_T\bar{n}k_BT. \tag{4.91}$$

Ideal plasma

$$\kappa_T = p^{-1} = [(\langle n\rangle k_B T]^{-1} \;\Rightarrow\; \frac{\sqrt{\langle\Delta n^2\rangle}}{\langle n\rangle} = \frac{\sqrt{\bar{n}}}{\langle n\rangle} \simeq \frac{1}{\sqrt{\langle n\rangle}}. \tag{4.92}$$

Grand potential $J = J(V, T, \mu)$

$$S = -\left(\frac{\partial J}{\partial T}\right)_{V\mu}, \quad p = -\left(\frac{\partial J}{\partial V}\right)_{T\mu}, \quad \overline{N} = -\left(\frac{\partial J}{\partial \mu}\right)_{VT}. \tag{4.86}$$

Quantum Systems

Density matrix $\hat{\rho}$ and average of operator O

$$\hat{\rho} = |\psi\rangle\langle\psi| = \sum_{r=1}^{s} P_r|\psi_r\rangle\langle\psi_r|, \quad \sum_{r=1}^{s} P_r = 1; \quad \langle O\rangle = \mathrm{Tr}(\hat{O}\hat{\rho}). \tag{4.95}$$

Average of O:

$$\langle O\rangle = \sum_{i,r} c_l c_r^* \langle\psi_r|O|\psi_l\rangle; \quad c_l c_r^* = |c_r|^2\delta_{lr}. \tag{4.98}$$

Quantum microcanonical ensemble

$$c_l c_r^* = |c_r|^2\delta_{lr} \quad \Rightarrow \quad \langle O\rangle = \frac{|c_r|^2\langle\psi_r|O|\psi_l\rangle}{\sum_r |c_r|^2}, \tag{4.100}$$

von Neumann equation (quantum Liouville equation)

$$\imath\hbar\partial_t|\psi\rangle = H|\psi\rangle, \quad -\imath\hbar\langle\psi|\partial_t = \langle\psi|H \quad \Rightarrow \quad \frac{\partial}{\partial t}\hat{\rho} = \frac{1}{\imath\hbar}[H, \hat{\rho}]. \tag{4.97}$$

Quantum canonical ensemble $Q_N(T; V)$: $E_1 \le E_2 \le E_3 \le \cdots \le E_l \le \cdots$

$$w(E_l) = \frac{\mathrm{e}^{-\beta E_l}}{Q_N(V, T)} = \rho_{ll}; \quad Q_N(V, T) = \sum_l \mathrm{e}^{-\beta E_l}. \tag{4.102}$$

Mean energy

$$\langle E\rangle = \frac{\sum_l E_l \exp(-\beta E_l)}{Q_N(V, T)} = -\frac{\partial \ln Q_N(V, T)}{\partial\beta}.$$

Degree of degeneracy g_k: $E_1 < E_2 < E_3 < \cdots < E_k < \cdots$

$$w(E_k) = g_k\frac{\mathrm{e}^{-\beta E_k}}{Q_N(V, T)} = g_k\rho_{kk}; \quad Q_N(V, T) = \sum_k g_k \mathrm{e}^{-\beta E_k}. \tag{4.102}$$

Canonical free energy $F(V, T)$ and entropy $S(V, T)$ are

$$F(V,T) = -\beta \ln Q_N(V,T); \quad S(V,T) = -k_B \sum_l w_l \ln w_l = -k_B \mathrm{Tr}(\hat{\rho} \ln \hat{\rho}). \tag{4.105}$$

Grand canonical ensemble

$$\mathcal{Z}(V,T,\mu) = \sum_{N=0}^{N=N_0} \mathrm{e}^{\beta\mu N}\, Q_N(V,T,N) = \sum_{N,l} \mathrm{e}^{\beta(\mu N - E_{N,l})} = \sum_N \mathrm{Tr}\ \mathrm{e}^{(\mu N - H_N)}. \tag{4.106}$$

Connection with thermodynamics: $J(V,T,\mu) = -k_B T \ln \mathcal{Z}(V,T,\mu)$

Ideal systems: $E_l^{(N)} = \sum_{k=0}^{k=s(N)} n_k E_k; \quad N = n_0 + n_1 + \cdots + n_j + \cdots + n_s$

Bose–Einstein distribution

$$\langle n_j \rangle = -\frac{\partial}{\partial(\beta E_j)} \ln \mathcal{Z} = \frac{1}{\mathrm{e}^{\beta(E_j-\mu)} - 1}. \tag{4.113}$$

Variance

$$\frac{\langle \Delta n_j^2 \rangle}{\langle n_j \rangle^2} = 1 + \frac{1}{\langle n_j \rangle}. \tag{4.117}$$

Fermi-Dirac distribution

$$\langle n_j \rangle = -\frac{\partial}{\partial(\beta E_j)} \ln \mathcal{Z} = \frac{1}{\mathrm{e}^{\beta(E_j-\mu)} + 1}. \tag{4.139}$$

Variance

$$\frac{\langle \Delta n_j^2 \rangle}{\langle n_j \rangle^2} = \frac{1}{\langle n_j \rangle} - 1. \tag{4.142}$$

Boltzmann distribution, $E_j - \mu \gg k_B T$

$$\langle n_j \rangle = \mathrm{e}^{-\beta(E_j-\mu)}; \quad \frac{\langle \Delta n_j^2 \rangle}{\langle n_j \rangle^2} = \frac{1}{\langle n_j \rangle}. \tag{4.147}$$

Black Body Radiation

Planck formula in vacuum

$$I(\omega,T) = \frac{c}{4\pi}\rho(\omega,T) = \frac{\hbar\omega^3}{4\pi^3 c^2} \frac{1}{\exp(\beta\hbar\omega) - 1}. \tag{4.122}$$

Stefan–Boltzmann law

$$I(T) = \frac{c}{4}\rho(T) = \sigma T^4; \quad \sigma = \frac{\pi^2 k_B^4}{60 c^2 \hbar^3} = 5.66 \times 10^{-8} [\mathrm{Js^{-1}m^{-2}K^{-4}}]. \tag{4.127}$$

Rayleigh-Jeans limit, $\hbar\omega \ll k_B T$

$$\rho(\omega, T) = \frac{\omega^2}{\pi^2 c^3} k_B T = g(\omega) k_B T \quad \Rightarrow \quad \langle \varepsilon_{\mathbf{k}} \rangle = k_B T. \tag{4.128}$$

Kirchhoff's law, $\hbar\omega \gg k_B T$

$$\rho(\omega, T) = g(\omega)\hbar\omega \exp\left(-\frac{\hbar\omega}{k_B T}\right). \tag{7.9}$$

Wien's displacement law

$$\omega_{\max} = 3.69 \times 10^{11} T[\mathrm{s}^{-1}\mathrm{K}]; \quad \lambda_{\max} = 2.89 \times 10^6 T \ \ [\mathrm{nmK}]. \tag{4.131}$$

Einstein's emission coefficients

$$A_{12} = \frac{\omega^2}{\pi^2 c^3} \left(B_{12}\hbar\omega\right). \tag{9.7}$$

Kirchhoff's law of total and net absorption coefficients κ_ω and α_ω

$$\kappa_\omega = \frac{c}{4\pi} n_1 B_{21}, \quad \alpha_\omega = \kappa_\omega \left(1 - \frac{n_2/g_2}{n_1/g_1}\right) = \left(1 - \mathrm{e}^{-\beta\hbar\omega}\right). \tag{4.136}$$

Perfect classical gas (Boltzmann limit)

$$F(V, T, N) = -N k_B T \ln\left[\frac{\mathrm{e}V}{N}\left(\frac{2\pi m k_B T}{h^2}\right)^{3/2} Z_{1\mathrm{in}}\right]. \tag{4.153}$$

Chemical equilibrium

$$k_1 A_1 + k_2 A_2 + \ldots\ldots.. k_s A_s = 0. \tag{4.155}$$

Mass action law

$$\prod_1^s n_k^k = \prod_1^s f_k(T)^k = K(T). \tag{4.159}$$

Saha equation

$$\frac{\eta_e \eta_{z+1}}{N \eta_z} = 2\left(\frac{2\pi m_e k_B T}{h^2}\right)^{3/2} \frac{Z_{1\mathrm{in}}^{z+1}}{Z_{1\mathrm{in}}^{z}} \mathrm{e}^{-\beta(E_{z+1}-E_z)}; \qquad z \geq 0. \tag{4.161}$$

Thomas–Fermi Equation, at T = 0

$$\frac{d^2}{\mathrm{d}x^2}\chi(x) = \frac{1}{x^{1/2}}\chi(x)^{3/2}. \tag{4.171}$$

finite temperature

$$\xi = \frac{r}{r_0}, \quad \Psi(\xi) = \xi \frac{eV + \mu}{k_B T_e}, \quad a = \frac{4\pi m_e e^2 r_0^2 (2m_e k_B T_e)^{1/2}}{\pi \hbar^3}$$

$$\frac{d^2}{d\xi^2}\Psi(\xi) = a\xi F_{1/2}\left[\frac{\Psi(\xi)}{\xi}\right]; \quad \Psi(0) = \frac{Ze^2}{k_B T_e r_0}, \quad \frac{d\Psi(\xi = 1)}{d\xi} = \Psi(1). \quad (4.174)$$

Thomas–Fermi pressure

$$p_e[\text{Mbar}] = \frac{1}{3\pi^2}\left(\frac{2m_e}{\hbar^2}\right)^{3/2}(k_B T_e)^{5/2} F_{3/2}\left(\frac{\mu}{k_B T_e}\right) = 3.26 \times 10^{-3} F_{3/2}(T_e[\text{eV}])^{5/2}. \quad (4.176)$$

Semiempirical van der Waals equation

$$\left(p + a\frac{N^2}{V^2}\right)(V - V_0) = N k_B T. \quad (4.182)$$

4.10 Further Readings

R. Balescu, *Equilibrium and Nonequilibrium Statistical Mechanics* (Wiley, New York, 1975)

F. Mandl, *Statistical Physics*, (Wiley, New York, 1997)

Y.B. Zeldovich, Y.P. Raizer, *Physics of Shock Waves and High-Temperature Hydrodynamic Phenomena*, (Academic, New York, 1966)

V.E. Fortov, *Extreme States of Matter*, (Springer International Publishing Switzerland, 2016)

P. Drake, *High Energy Density Physics: Foundations, Inertial Fusion, and Experimental Astrophysics*, (Springer, Berlin, 2006)

J. Larsen, *Foundations of High-Energy-Density Physics: Physical processes of matter at extreme conditions*, (Cambridge University Press, Cambridge, 2017)

S.J. Rose, *The effect of a radiation field on excitation and ionization in non-LTE high energy density plasmas*, *Laser-Plasma Interactions and Applications* ed. by P. McKenna et al. (eds), (Springer, Heidelberg, 2013), Chap. 4

N.W. Ashcroft, N. David Mermin, *Solid State Physics*, (Saunders College, Philadelphia, 1976)

B.-G. Englert, *Semiclassical Theory of Atoms*, (Springer, Berlin, 1988)

References

1. S.J. Rose, High Energy Density Phys. **5**, 26 (2009)
2. A.B. Langdon, Phys. Rev. Lett. **44**, 575 (1980)
3. H. Grad, The Many Faces of Entropy. Commun. Pure Appl. Math. XIV **323–354** (1961)
4. S. Weinberg, *Gravitation and Cosmology* (Wiley, New York, 1972)
5. D. Mosher, Phys. Rev. A **10**, 2330 (1974)
6. M.A. Dopita, R.S. Sutherland, *Astrophysics of the Diffuse Universe* (Springer, Heidelberg, 2003). Chapt. 5
7. S.F. Garanin, E.M. Palagina, Plasma Phys. Rep. **33**, 684 (2007)
8. F.P. Keenan, S.J. Rose, Astron. Geophys. **45**, 6.18 (2004)
9. E. Hill, S.J. Rose, Phys. Plasmas **17**, 103301 (2010)
10. B.A. Remington, R.P. Drake, H. Takabe, D. Arnett, *A Review of Astrophysics Experiments on Intense Lasers*, LLNL, Contract No.W-7405-Eng-48 (2000); Science **284**(5419), 1488–1493 (1999)
11. N. Medvedev, Z. Li, B. Ziaja, Phys. Rev. B **91**, 054113 (2015)
12. R.M. More, K.H. Warren, D.A. Young, G.B. Zimmerman, Phys. Plasmas **31**, 3059 (1988)
13. A.J. Kemp, J. Meyer-ter-Vehn, Nucl. Inst. Methods Phys. Res. A **415**, 674 (1988)
14. R. Feynman, N. Metropolis, E. Teller, Phys. Rev. **75**, 1561 (1949)
15. R. Cowan, J. Ashkin, Phys. Rev. **105**, 144 (1957)
16. N.W. Ashcroft, N.D. Mermin, *Solid State Physics* (Saunders College, Philadelphia, 1976)
17. T.T. Chau, J.H. Hue, M.-I. Trappe, B.-G. Englert, New. J. Phys. **20**, 073003 (2018)
18. A. Benuzzi, T. Löwer, M. Koenig, B. Faral, D. Batani, D. Beretta, C. Danson, D. Pepler, Phys. Rev. E **54**, 2162 (1996)
19. N.W. Ashcroft, N. David Mermin, *Solid State Physics*, (Saunders College, Philadelphia, 1976), p. 340

Chapter 5
Waves in the Ideal Plasma

Introduction

The wave is one of the fundamental phenomena occurring in physics. Locally acting forces produce perturbations that propagate in space and time. Their dynamics shows features that allow general classifications of a wide variety of phenomena and, in the linear regime, general relations among the dynamic variables can be formulated. Limiting to electromagnetic interaction the forces driving the dynamics of matter are the electric and magnetic fields coupled to matter by electric charges and currents. The coupling can be viewed as the interaction of *two fluids*, i.e., the electromagnetic fluid of fields with the matter as a fluid. The result are *field equations* that connect forces and momenta locally. Their speed of propagation is the result of the interplay of local inertia with local forces. Correct relativistic treatment limits their propagation to finite velocities, in this way automatically excluding non causal interaction at distance. The plasma as a special state of matter is described as a fluid in 3D in real space, see Chap. 3, or as a fluid in phase space by a Vlasov-like dynamic equation. In this chapter both procedures will be used.

The electromagnetic "fluid" is described by Maxwell's equations. They are linear; in contrast the fluid equations of matter, either in real or in phase space, are highly nonlinear. This fact confines a general analytic treatment to small deviations of the material fluid from static equilibrium and, in concomitance poses limits on the field strengths. To obtain general laws *linearization* is a necessity. Only in sporadic cases fully nonlinear solutions are known, as for example rarefaction, compression, shock, and heat waves under special initial and boundary conditions (see Chap. 3). The net forces on a fluid element are differences, expressed by gradients and divergences, the infinitesimal changes in time result in partial derivatives with respect to time. In the homogeneous medium these partial derivatives in space and time are converted

P. Mulser, *Hot Matter from High-Power Lasers*, Graduate Texts in Physics,
https://doi.org/10.1007/978-3-662-61181-4_5

into linear algebraic relations by the *Fourier–Laplace* integral transform. The resulting linear waves are characterized by the wave vectors **k** and frequencies ω linked together by *dispersion relations*. Owing to these properties the wave equations are homogeneous, and are referred to as *eigenmodes*. They are orthogonal to each other. A general linear perturbation is a superposition of eigenmodes.

In the following the electromagnetic eigenmodes in the homogeneous unmagnetized and magnetized plasma background are analyzed. The description of the main types of waves relevant to the laser plasma is done in the ideal two fluid model. Kinetic aspects are elaborated in the Vlasov model. Extension of the description into moderately inhomogeneous plasma is achieved in *WKB approximation* (after Wentzel, Kramers, Brillouin). The approximation builds a bridge to conservation of action and to a Hamiltonian description of the *classical photon*. The concept marks the limits to which any *quasiparticle* concept is subject, the quantum concept of the photon included. Electron plasma waves can resonantly couple to the laser field and may be driven into a highly nonlinear regime, and may eventually break. Adequate attention is dedicated to this aspect within the fluid description, in particular to self interaction of modes, *wave breaking* and self-annihilation. Finally, analytic expressions of *collisional absorption* in characteristic plasma density profiles and of energy densities and fluxes carried by the standard types of waves are derived. The derivation of the linear *Landau damping* is simplified with respect to Landau's approach.

The various wave equations of the modes encountered in the chapter reveal similarities in their structures which in turn constitute links among the phenomena they describe. As an example it will be shown that identical structures in the dispersion relations lead to identical structures in their energy conservations. There exists a one to one correspondence between the linearized wave equations of the plasma, and of linear optics in general, and the Schrödinger wave equation. Correspondingly, a variety of solutions in one field are mapped onto the affine area, like diffraction—scattering, evanescence—tunnelling, dispersion—quantum diffusion, Landau echoes—quantum revivals.

5.1 The Maxwell Equations

In 3D representation $(\mathbf{x}, t)$ the local quantities of the electric field $\mathbf{E}(\mathbf{x}, t)$, the magnetic field $\mathbf{B}(\mathbf{x}, t)$, the electric current density $\mathbf{j}(\mathbf{x}, t)$ and the total local electric charge density $\rho_{\mathrm{el}}(\mathbf{x}, t)$ are connected by *Maxwell's equations* in the following way

$$\varepsilon_0 c^2 \nabla \times \mathbf{B} = \varepsilon_0 \dot{\mathbf{E}} + \mathbf{j}. \tag{5.1}$$

$$\nabla \times \mathbf{E} = -\dot{\mathbf{B}}. \tag{5.2}$$

$$\nabla \mathbf{E} = \frac{\rho_{\mathrm{el}}}{\varepsilon_0}. \tag{5.3}$$

$$\nabla \mathbf{B} = 0. \tag{5.4}$$

The electric charge and current densities in the two fluid model are

$$\rho_{\text{el}} = e(Zn_i - n_e); \qquad \mathbf{j} = e(Zn_i\mathbf{u}_i - n_e\mathbf{u}_e). \tag{5.5}$$

The constants are $\varepsilon_0 = 8.854 \times 10^{-12}[\text{AsV}^{-1}\text{m}^{-1}]$, $\varepsilon_0 c^2 = \mu_0^{-1} = 7.97 \times 10^5[\text{VsA}^{-1}\text{m}^{-1}]$. By convention, the dot on field quantities stands for the partial time derivative. The quantities $n_{e,i}$ are the electron and ion densities, $\mathbf{u}_{e,i}$ are their flow velocities, and Z is the ion charge number. The particle densities are the Lorentz contracted values as measured at speeds $|\mathbf{u}_e|$ and $|\mathbf{u}_i|$, respectively, i.e., $n_{e,i} = \gamma_{e,i} n_{0e,i}$. $\gamma_{e,i}$ are the Lorentz factors and $n_{0e,i}$ the densities at rest. As shown in Chap. 2 (5.1)–(5.5) are fully relativistic, although not presented in manifest covariant form.

The first of Maxwell's equations (5.1) results from electric charge conservation, the most precise conservation law in physics,

$$\dot{\rho_{\text{el}}} + \nabla\mathbf{j} = 0. \tag{5.6}$$

By defining a vector field **E** through

$$\nabla\varepsilon_0\mathbf{E}(\mathbf{x}, t) = \rho_{\text{el}}(\mathbf{x}, t) \tag{5.7}$$

(5.6) reads

$$\nabla(\varepsilon_0\dot{\mathbf{E}} + \mathbf{j}) = 0.$$

The general solution div**F** = 0 is the curl of an arbitrary vector field **B**, thus

$$\varepsilon_0 c^2 \nabla \times \mathbf{B} = \varepsilon_0\dot{\mathbf{E}} + \mathbf{j}.$$

By applying the two fields **E** and **B** to special configurations they are recognized in these particular cases as the familiar electric and magnetic fields. Their general validity then follows from the fact that no experiment has ever been found that could disprove charge conservation (5.6). The same procedure holds for the definition of **E** by (5.7).

In simple configurations the experiment tells that the electric charges are the sources of the electric fields. It can be shown that if $\nabla\mathbf{E} = \rho_{\text{el}}/\varepsilon_0$ is satisfied at one time instant t_0 it is satisfied for all times $t \gtrless t_0$. The second of Maxwell's equations is Faraday's induction law applied to an arbitrary loop fixed in space. The forth of Maxwell's equations, $\nabla\mathbf{B} = 0$ expresses the fact that no magnetic charges have ever been found so far. In other words, the magnetic field does not couple to matter through magnetic charges as sources and sinks of **B**.

5.2 Wave Equations

It is convenient to introduce the vector potential $\mathbf{A}(\mathbf{x}, t)$ by

$$\mathbf{B} = \nabla \times \mathbf{A}. \tag{5.8}$$

Then (5.2) reads

$$\nabla \times (\mathbf{E} + \dot{\mathbf{A}}) = 0.$$

The general solution of this equation is the gradient of an arbitrary scalar function $-\Phi(\mathbf{x}, t)$:

$$\mathbf{E} + \dot{\mathbf{A}} = -\nabla\Phi. \tag{5.9}$$

In the potentials $\mathbf{A}$ and Φ the Maxwell equations reduce to coupled inhomogeneous wave equations. By imposing the *Lorentz gauge* $\nabla\mathbf{A} + \dot{\Phi}/c^2 = 0$ the coupled system decouples into two inhomogeneous wave equations, one for $\mathbf{A}$ and the other one for Φ,

$$\nabla^2\mathbf{A} - \frac{1}{c^2}\frac{\partial^2\mathbf{A}}{\partial t^2} = -\frac{\mathbf{j}}{\varepsilon_0 c^2}; \quad \nabla^2\Phi - \frac{1}{c^2}\frac{\partial^2\Phi}{\partial t^2} = -\frac{\rho_{\mathrm{el}}}{\varepsilon_0}; \quad \nabla\mathbf{A} + \frac{1}{c^2}\frac{\partial\Phi}{\partial t} = 0. \tag{5.10}$$

Observation of the Lorentz gauge (3rd relation) is always possible without any restrictions of the physical reality; it is a question of convenience, see Sect. 2.3.3. In the source terms all kinds of electric currents and all kinds of charges enter. If, as in dielectric media, the polarization $\mathbf{P}$ and the dielectric displacement $\mathbf{D} = \varepsilon_0\mathbf{E} + \mathbf{P}$ are introduced the polarization current and charge densities $\mathbf{j}_P$ and ρ_P, given by

$$\mathbf{j}_P = \frac{\partial\mathbf{P}}{\partial t}; \quad \rho_P = -\nabla\mathbf{P} \quad \Rightarrow \quad \nabla\mathbf{D} = \nabla\varepsilon_0\mathbf{E} - \rho_P = \rho_{\mathrm{free}}, \tag{5.11}$$

the current and charge densities in (5.10) are the sums $\mathbf{j} = \mathbf{j}_P + \mathbf{j}_{\mathrm{free}}$ and $\rho = \rho_P + \rho_{\mathrm{free}}$. The index "free" stands for the free charges.

The solutions of the wave equations are given by the well known time retarded integrals

$$\mathbf{A}(\mathbf{x}, t) = \mathbf{A}_0(\mathbf{x}, t) + \frac{1}{4\pi\varepsilon_0 c^2}\int\limits_V \frac{\mathbf{j}(\mathbf{x}', t')}{|\mathbf{x} - \mathbf{x}'|}\mathrm{d}\mathbf{x}'; \tag{5.12}$$

$$\Phi(\mathbf{x}, t) = \Phi_0(\mathbf{x}, t) + \frac{1}{4\pi\varepsilon_0}\int\limits_V \frac{\rho_{\mathrm{el}}(\mathbf{x}', t')}{|\mathbf{x} - \mathbf{v}\mathbf{x}'|}\mathrm{d}\mathbf{x}'; \quad t' = t - \frac{|\mathbf{x} - \mathbf{x}'|}{c}, \tag{5.13}$$

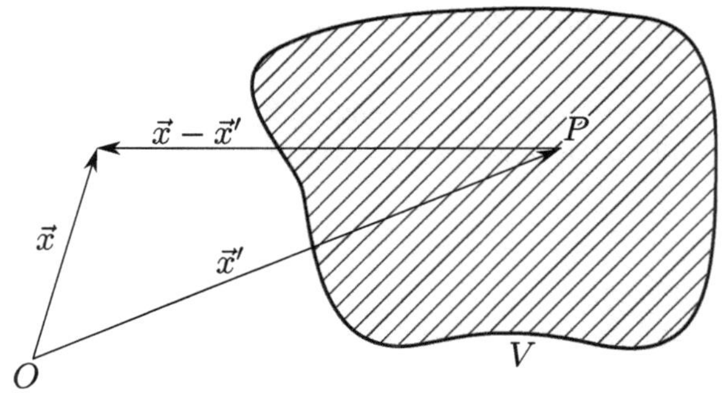

Fig. 5.1 Orientation of the vectors to obtain the retarded potentials $\mathbf{A}(\mathbf{x},t)$, $\Phi(\mathbf{x},t)$ from integrals on the corresponding sources $\mathbf{j}(\mathbf{x}',\mathbf{t}')$, $\rho_{el}(\mathbf{x}',t')$ in the volume V

where $\mathbf{A}_0(\mathbf{x},t)$, $\Phi_0(\mathbf{x},t)$ are arbitrary solutions of the corresponding homogeneous wave equations. For the orientation of the position vectors see Fig. 5.1. Specialized to a point charge $q\delta(\mathbf{x}')$, $\mathbf{j}=q\mathbf{v}\delta(\mathbf{x}')$, the integrals reduce to the familiar *Lienard–Wiechert potentials*

$$\mathbf{A}(\mathbf{x},t)=\frac{q}{4\pi\varepsilon_0c^2}\frac{\mathbf{v}(t')}{R-\mathbf{Rv}/c};\qquad \Phi(\mathbf{x},t)=\frac{q}{4\pi\varepsilon_0}\frac{1}{R-\mathbf{Rv}/c}. \tag{5.14}$$

The vector $\mathbf{R}=\mathbf{x}-\mathbf{x}_0(t')$ has its origin at the retarded position $\mathbf{x}_0(t')$ of the charge q and points to position $\mathbf{x}$ of the observer; $t'=t-R/c$, $R=|\mathbf{R}|$. In calculating the fields $\mathbf{E}$ and $\mathbf{B}$ from (5.9) the derivative with respect to the retarded time t' is important; it confers the waves transverse character and a fall off like $1/R$ at large distances R.

A gauge free wave equation is obtained for $\mathbf{E}$ by taking the partial time derivative of $\mathbf{A}$ and the gradient of Φ in the wave equations (5.10). Observing (5.9) one arrives at the wave equation for the electric field

$$\nabla\times\nabla\times\mathbf{E}+\frac{1}{c^2}\frac{\partial^2}{\partial t^2}\mathbf{E}=-\frac{1}{\epsilon_0c^2}\frac{\partial\mathbf{j}}{\partial t}. \tag{5.15}$$

Correspondingly, application of the curl operation to the wave equation for $\mathbf{A}$ leads to the gauge free wave equation for the magnetic field,

$$\nabla^2\mathbf{B}-\frac{1}{c^2}\frac{\partial^2}{\partial t^2}\mathbf{B}=-\frac{1}{\epsilon_0c^2}\nabla\times\mathbf{j}. \tag{5.16}$$

Note, the differential operators interchange. Direct derivation of the two equations from Maxwell's equations (5.1)–(5.4) leads to the same result, of course.

5.2.1 The Poynting Theorem

Energy conservation is deduced by taking the scalar products of **E** with (5.1), $-\mathbf{B}$ with (5.2) and adding the two expressions to obtain

$$\frac{\partial}{\partial t}\frac{1}{2}\varepsilon_0(\mathbf{E}^2 + c^2\mathbf{B}^2) + \nabla\mathbf{S} = -\mathbf{jE}; \qquad \mathbf{S} = \varepsilon_0 c^2 \mathbf{E}\times\mathbf{B}. \tag{5.17}$$

The term $\epsilon_{em} = \frac{1}{2}\varepsilon_0(\mathbf{E}^2 + c^2\mathbf{B}^2)$ is the electromagnetic energy density, the energy flux density is expressed by the *Poynting vector* **S**. The term **jE** represents the coupling to matter; it expresses the work of the electromagnetic field on it. The absence of an analogous term containing the magnetic field **B** is a consequence of zero magnetic charge. The magnetic field does not couple to matter directly. The momentum conservation of the electromagnetic field is expressed in terms of Poynting's vector **S** and stress tensor T.

5.2.2 Maxwell's Stress Tensor T

Substituting ρ_{el} and **j** from Maxwell's equations in the Lorentz force density π,

$$\pi = \rho_{el}\mathbf{E} + \mathbf{j}\times\mathbf{B} \tag{5.18}$$

and adding $\varepsilon_0 c^2\mathbf{B}(\nabla\mathbf{B}) = 0$, $\varepsilon_0\mathbf{E}\times(\dot{\mathbf{B}} + \nabla\times\mathbf{E}) = 0$ yields the electromagnetic momentum conservation

$$\frac{1}{c^2}\frac{\partial}{\partial t}\mathbf{S} + \nabla\mathbf{T} = -\pi; \quad \mathbf{T} = T_{ij} = \varepsilon_0\left[E_iE_j + c^2B_iB_j - \frac{1}{2}\delta_{ij}(\mathbf{E}^2 + c^2\mathbf{B}^2)\right]. \tag{5.19}$$

The stress tensor **T** is symmetric. Its divergence means $\nabla\mathbf{T} = \sum_j \partial_j T_{ij}$. The time derivative of **S**, added by Lorentz subsequently to Maxwell, is essential for exerting zero force onto the vacuum. There are no discussions on the expression presented here with charges and currents understood as taken on the background of the vacuum. The situation is very different in the case Maxwell's equations are formulated inside matter. Several forms of the stress tensor have been proposed in the past and new variants are under discussion at present. The interested reader may consult the enlightening article by Shevchenko and Hoenders [1]. The authors present a microscopic derivation of the macroscopic terms.

Equations (5.10) are capable of describing all kinds of linear and nonlinear waves in the plasma without and with magnetic fields in the frame of the fluid model. Furthermore, their solutions are obtained by the retarded integrals (5.12) and (5.13), provided the self consistent source terms of charges and currents are known. This is the crucial point. The source terms are the result of simultaneously solving Maxwell's equations coupled to the nonlinear two fluid equations of Chap. 3. Apart from special cases, analytical solutions are attainable only under linearization. This is essentially the program of the present chapter.

5.3 Covariant Electrodynamics

5.3.1 The Electromagnetic Field Tensor

By defining the four quantity

$$A = (c\mathbf{A}, \Phi) = (A^1, A^2, A^3, A^4 = \Phi) = (A_1, A_2, A_3, -A_4 = -\Phi) \tag{5.20}$$

and supposing for the moment that A is a four vector, the quantity

$$\frac{F_{\alpha\beta}}{c} = \partial_\alpha A_\beta - \partial_\beta A_\alpha. \tag{5.21}$$

is a covariant, antisymmetric four tensor (see criterion of Chap. 2). The relations (5.9) are linked to $F_{\alpha\beta}$ by

$$\begin{aligned} cB_x &= \partial_2 A_3 - \partial_3 A_2 = F_{23}/c, & E_x &= \partial_1 A_4 - \partial_4 A_1 = F_{14}/c, \\ cB_y &= \partial_3 A_1 - \partial_1 A_3 = F_{31}/c, & E_y &= \partial_2 A_4 - \partial_4 A_2 = F_{24}/c, \\ cB_z &= \partial_1 A_2 - \partial_2 A_1 = F_{12}/c, & E_z &= \partial_3 A_4 - \partial_4 A_3 = F_{34}/c. \end{aligned} \tag{5.22}$$

Hence

$$F_{\alpha\beta} = c\begin{pmatrix} 0 & cB_z & -cB_y & E_x \\ -cB_z & 0 & cB_x & E_y \\ cB_y & -cB_x & 0 & E_z \\ -E_x & -E_y & -E_z & 0 \end{pmatrix}, \quad F^{\alpha\beta} = c\begin{pmatrix} 0 & cB_z & -cB_y & -E_x \\ -cB_z & 0 & cB_x & -E_y \\ cB_y & -cB_x & 0 & -E_z \\ E_x & E_y & E_z & 0 \end{pmatrix}.$$

In a general Euclidean metric g holds $F^{\alpha\beta} = g^{\alpha\gamma} g^{\beta\delta} F_{\gamma\delta}$. With the four velocity vector V and the electric charge density at rest $\rho_{\mathrm{el},0}$ the four current density is given by

$$J = \rho_{\mathrm{el},0} V, \qquad \partial_\alpha j^\alpha = 0. \tag{5.23}$$

The four current is divergence-free, i.e., the electric charge is conserved pointwise. With the help of (5.23) it can be shown that the total charge Q is both conserved and a Lorentz scalar (see Weinberg, Sect. 6). The first and third of Maxwell equations read in terms of $F^{\alpha\beta}$

$$\partial_\alpha F^{\alpha\beta} = \frac{j^\beta}{\varepsilon_0} \quad \Leftrightarrow \quad \operatorname{div} F = \frac{J}{\varepsilon_0}. \tag{5.24}$$

Since J is a four vector $F^{\alpha\beta}$ is a four tensor, and consequently A from (5.20) is a four vector potential. By comparing the second and fourth of Maxwell equations

$$\nabla \times \mathbf{E} + \dot{\mathbf{B}} = \kappa \mathbf{j}_{\mathrm{mag}}, \qquad \nabla \mathbf{B} = \kappa \rho_{\mathrm{mag}}, \quad J_{\mathrm{mag}} = (\mathbf{j}_{\mathrm{mag}}, c\rho_{\mathrm{mag}}) = 0$$

with the former pair of Maxwell equations the corresponding field tensor

$$F'^{\alpha\beta} = \varepsilon^{\alpha\beta\gamma\delta} F'_{\alpha\beta} = c \begin{pmatrix} 0 & E_z & -E_y & cB_x \\ -E_z & 0 & E_x & cB_y \\ E_y & -E_x & 0 & cB_z \\ -cB_x & -cB_y & -cB_z & 0 \end{pmatrix} \tag{5.25}$$

satisfying

$$\mathrm{div}\, F' = 0 \quad \Leftrightarrow \partial_\alpha F'^{\alpha\beta} = 0 \tag{5.26}$$

is obtained. κ is a proportionality constant. Equation (5.26) is identical with the second and forth of Maxwell's equations (5.2) and (5.4). This completes the proof of covariance of the Maxwell equations.

Covariant Lorentz Force

The relativistic momentum and energy conservation of a point charge assume the familiar structure in three form

$$\frac{\mathrm{d}}{\mathrm{d}t} m\gamma\mathbf{v} = q(\mathbf{E} + \mathbf{v} \times \mathbf{B}), \qquad \frac{\mathrm{d}}{\mathrm{d}t} m\gamma c^2 = q\mathbf{v}\mathbf{E}.$$

In terms of the covariant field tensor $F_{\alpha\beta}$ they are transcribed into manifestly covariant form as

$$m\frac{\mathrm{d}}{\mathrm{d}\tau} V = -\frac{q}{c^2} F V, \ V = v^\alpha = \gamma(\mathbf{v}, c) \quad \Leftrightarrow \quad m\frac{\mathrm{d}v_\alpha}{\mathrm{d}t} = \frac{q}{c^2} F_{\alpha\beta} v^\beta. \tag{5.27}$$

5.3.2 Lorentz Scalars and Lorentz Invariant Operators

There are the two simple, nontrivial Lorentz scalars

$$F_{\alpha\beta} F^{\alpha\beta} = 2c^2(c^2\mathbf{B}^2 - \mathbf{E}^2), \qquad \frac{1}{8c^2} \varepsilon_{\alpha\beta\gamma\delta} F^{\alpha\beta} F^{\gamma\delta} = \mathbf{EB}, \tag{5.28}$$

where $\varepsilon_{\alpha\beta\gamma\delta}$ is the totally antisymmetric *Levi-Civita tensor*. The invariance of **EB** against a Lorentz transformation also follows explicitly using (2.165) and (2.167) from Chap. 2,

$$\mathbf{E}'\mathbf{B}' = \mathbf{E}'_{\parallel}\mathbf{B}'_{\parallel} + \mathbf{E}'_{\perp}\mathbf{B}'_{\perp} = \mathbf{E}_{\parallel}\mathbf{B}_{\parallel} + \gamma^2 \left[\mathbf{E}_{\perp}\mathbf{B}_{\perp} - \frac{1}{c^2}(\mathbf{v}\times\mathbf{B}_{\perp})(\mathbf{v}\times\mathbf{E}_{\perp})\right]$$

$$= \mathbf{E}_{\parallel}\mathbf{B}_{\parallel} + \mathbf{E}_{\perp}\mathbf{B}_{\perp} = \mathbf{E}\mathbf{B},$$

or from the transformation law of the electromagnetic field tensors F and F'. In fact, the Lorentz transformations (2.153) and (2.155) are linear and hence can be written in the form

$$x'^{\beta} = \Lambda^{\beta}_{\alpha}x^{\alpha}, \quad x^{\alpha} = \Lambda^{\alpha}_{\beta}x'^{\beta}, \quad \Lambda^{\gamma}_{\beta}\Lambda^{\beta}_{\alpha} = \delta^{\gamma}_{\alpha} \Leftrightarrow (\Lambda^{\beta}{}_{\alpha}) = (\Lambda^{\alpha}{}_{\beta})^{-1} \tag{5.29}$$

to obtain $F'^{\alpha\beta}$. (It is more convenient to associate the dash $'$ with the indices, i.e., to write $x^{\beta'}$ for x'^{β} and $F^{\alpha'\beta'}$ for $F'^{\alpha\beta}$). Hence $F^{\alpha'\beta'} = \Lambda^{\alpha'}_{\alpha}\Lambda^{\beta'}_{\beta}F^{\alpha\beta}$. From this equality formulas (2.165) and (2.167) are found again. The matrix Λ describes a so-called *Lorentz boost*.

The already familiar div operator ∂_{α} and the *d'Alembertian* $\Box$,

$$\Box = \partial^{\alpha}_{\alpha} = \partial^{\alpha}\partial_{\alpha} = g^{\alpha\beta}\frac{\partial}{\partial x^{\alpha}}\frac{\partial}{\partial x^{\beta}} = \nabla^2 - \frac{1}{c^2}\frac{\partial^2}{\partial t^2} \tag{5.30}$$

are Lorentz covariant and Lorentz invariant operators, respectively. The invariance of the d'Alembertian is evident from the scalar product $\Box = \partial_{\alpha}\partial^{\alpha}$. To show the covariance of the four divergence consider $\partial_{\alpha}T^{\alpha\beta}$ where $T^{\alpha\beta}$ is an arbitrary four tensor. For $x^{\alpha'} = \Lambda^{\alpha'}_{\alpha}x^{\alpha}$, $T^{\alpha'\beta} = \Lambda^{\alpha'}_{\kappa}T^{\kappa\beta}$

$$\partial_{\alpha'}T^{\alpha'\beta} = \frac{\partial x^{\alpha}}{\partial x^{\alpha'}}\partial_{\alpha}\Lambda^{\alpha'}_{\kappa}T^{\kappa\beta} = \Lambda^{\alpha}_{\alpha'}\Lambda^{\alpha'}_{\kappa}\partial_{\alpha}T^{\kappa\beta} = \delta^{\alpha}_{\kappa}\partial_{\alpha}T^{\kappa\beta} = \partial_{\alpha}T^{\alpha\beta}$$

holds, demonstrating the proper transformation behavior of ∂_{α} as a four vector. Moreover, if $\partial_{\alpha}T^{\alpha\beta} = t^{\beta}$ with t^{β} a four vector, $T^{\alpha\beta}$ is a four tensor (and vice versa). Use of this criterion was made above after (5.24) for the electromagnetic field tensor.

5.3.3 *Gauge Transformations*

The four vector potential leading to a given field configuration $F^{\alpha\beta}$ is not unique. Assume $\mathbf{A}_0(\mathbf{x}, t)$ and $\Phi_0(\mathbf{x}, t)$ to be two particular solutions of (5.9). Then, an arbitrary solution

$$\mathbf{A}(\mathbf{x}, t) = \mathbf{A}_0(\mathbf{x}, t) + \mathbf{A}_1(\mathbf{x}, t), \quad \Phi(\mathbf{x}, t) = \Phi_0(\mathbf{x}, t) + \Phi_1(\mathbf{x}, t)$$

of (5.9) must satisfy

$$\nabla\times\mathbf{A} - \mathbf{B} = \nabla\times\mathbf{A}_1 = 0; \quad \dot{\mathbf{A}} + \nabla\Phi + \mathbf{E} = \dot{\mathbf{A}}_1 + \nabla\Phi_1 = 0. \tag{5.31}$$

The general solution of the first equation, $\nabla \times \mathbf{A}_1 = 0$, is $\mathbf{A}_1(\mathbf{x}, t) = \nabla\Psi(\mathbf{x}, t)$. Then, the second equation requires $\nabla(\dot{\Psi} + \Phi_1) = 0$. Hence, the general *Lorentz gauge* with arbitrary $\Psi(\mathbf{x}, t)$,

$$\mathbf{A}' = \mathbf{A} + \nabla\Psi, \quad \Phi' = \Phi - \dot{\Psi} \quad \text{(General Lorentz gauge)}, \tag{5.32}$$

leaves the fields **E** and **B** invariant. This remains true even if $\Psi + \Phi_1 = f(t)$ with $f(t)$ an arbitrary function of time only.

Restricted Lorentz Gauge

The current and charge density are linked to the fields **E** and **B** by

$$\mathbf{j} = \varepsilon_0(c^2\nabla \times \mathbf{B} - \partial_t\mathbf{E}), \qquad \rho_{\mathrm{el}} = \varepsilon_0\nabla\mathbf{E}.$$

Substitution of them by the potentials of (5.9) results in two coupled equations for the potentials which separate into the symmetric inhomogeneous wave equations (5.10) if the *restricted Lorentz gauge*

$$\nabla c\mathbf{A} + \frac{\partial}{\partial ct}\Phi = 0 \qquad \text{(Lorentz gauge)} \tag{5.33}$$

is imposed. In manifestly covariant form (5.10) reads

$$\Box A = -\frac{J}{\varepsilon_0 c^2}; \qquad \partial_\alpha A^\alpha = 0. \tag{5.34}$$

The Coulomb Gauge

The Coulomb or *transverse* gauge may be convenient in applications to separate the transverse from the longitudinal fields. The starting point is

$$\mathrm{div}\mathbf{A} = 0. \tag{5.35}$$

Such a choice is guaranteed by (5.32). The electric field is written as the sum of a longitudinal and a transverse component $\mathbf{E}_l$ and $\mathbf{E}_t$ in such a way that they fulfil

$$\mathbf{E} = \mathbf{E}_l + \mathbf{E}_t; \quad \varepsilon_0\nabla\mathbf{E}_l = \rho_{\mathrm{el}}, \quad \Rightarrow \quad \nabla\mathbf{E}_t = 0. \tag{5.36}$$

From (5.9) and (5.3) follows $\rho_{\mathrm{el}}/\varepsilon_0 = \nabla\mathbf{E}_l = -\nabla^2\Phi$. It has the particular (but not unique) solution

$$\Phi(\mathbf{x}, t) = \frac{1}{4\pi\varepsilon_0}\int\frac{\rho_{\mathrm{el}}(\mathbf{x}, t)}{|\mathbf{x} - \mathbf{x}'|}\mathrm{d}\mathbf{x}'; \quad \mathbf{E}_l = -\frac{1}{4\pi\varepsilon_0}\int\frac{\rho_{\mathrm{el}}(\mathbf{x}, t)(\mathbf{x} - \mathbf{x}')}{|\mathbf{x} - \mathbf{x}'|^3}\mathrm{d}\mathbf{x}'. \tag{5.37}$$

This expression for $\mathbf{E}_l$ inserted in the second of (5.9) leads to $\mathbf{E}_t = -\dot{\mathbf{A}}$. Decomposing $\mathbf{j} = \mathbf{j}_l + \mathbf{j}_t$ such that $\mathbf{j}_l = -\dot{\mathbf{E}}_l$ yields

$$\varepsilon_0 c^2 \nabla \times \mathbf{B} = \varepsilon_0 \dot{\mathbf{E}} + \mathbf{j} = \varepsilon_0 \dot{\mathbf{E}}_t + \mathbf{j}_t \quad \Rightarrow \quad \nabla^2 \mathbf{A} - \frac{1}{c^2}\frac{\partial^2}{\partial t^2}\mathbf{A} = -\frac{1}{\varepsilon_0 c^2}\mathbf{j}_t. \tag{5.38}$$

The equation of the transverse electric field $\mathbf{E}_t$ follows from (5.15) or (5.38), or from Maxwell's equation directly,

$$\nabla^2 \mathbf{E}_t - \frac{1}{c^2}\frac{\partial^2}{\partial t^2}\mathbf{E}_t = \frac{1}{\epsilon_0 c^2}\frac{\partial}{\partial t}\mathbf{j}_t. \tag{5.39}$$

The particular solution of the wave equation (5.38) with the transverse current density $\mathbf{j}_t$ as the source term is

$$\mathbf{A}(\mathbf{x}, t) = \frac{1}{4\pi\varepsilon_0 c^2}\int_V \frac{\mathbf{j}_t(\mathbf{x}', t')}{|\mathbf{x} - \mathbf{x}'|}\mathrm{d}\mathbf{x}'; \quad t' = t - \frac{|\mathbf{x} - \mathbf{x}'|}{c}. \tag{5.40}$$

In general, however, the source term $\mathbf{j}_t$ is not known a priori.

$\mathbf{A}(\mathbf{x}, t)$ together with $\Phi(\mathbf{x}, t)$ from (5.37) are the potentials in the Coulomb gauge. The non retarded solution $\Phi(\mathbf{x}, t)$ taken as a single expression violates causality. However, $\Phi(\mathbf{x}, t)$ is not an observable quantity; it has to be considered together with $\mathbf{A}(\mathbf{x}, t)$ driven by the transverse current component $\mathbf{j}_t$ in which a term $\mathbf{E}_l = -\nabla\Phi$ cancels with $\mathbf{E}_l$ from (5.37), see for example [2]. The Coulomb gauge separates the transverse radiation field from the longitudinal space charge fields.

5.4 Eigenmodes of the Uniform Plasma

5.4.1 *The Unmagnetized Fully Ionized Plasma*

5.4.1.1 The Electromagnetic Wave

The electromagnetic wave is transverse, expressed by the condition $\mathrm{div}\mathbf{E} = 0$. As a consequence the scalar potential $\Phi(\mathbf{x}, t) = 0$ can be set. The Lorentz gauge (5.33) and the Coulomb gauge (5.35) together with the wave equation (5.10)

$$\nabla\mathbf{A} = 0 \quad \Rightarrow \quad \nabla^2\mathbf{A} - \frac{1}{c^2}\frac{\partial^2}{\partial t^2}\mathbf{A} = -\frac{1}{\varepsilon_0 c^2}\mathbf{j}, \quad \mathbf{j} = \mathbf{j}_t; \quad \mathbf{E} = \mathbf{E}_t = -\frac{\partial}{\partial t}\mathbf{A} \tag{5.41}$$

are simultaneously fulfilled. The wave equation is solved by the Fourier ansatz

$$\mathbf{A}(\mathbf{x}, t) \mid \mathbf{j}(\mathbf{x}, t) = \frac{1}{(2\pi)^4}\int [\mathbf{A}(\mathbf{k}', \omega') \mid \mathbf{j}(\mathbf{k}', \omega')]\mathrm{e}^{\imath(\mathbf{k}'\mathbf{x} - \omega' t)}\mathrm{d}\mathbf{k}'\mathrm{d}\omega'.$$

Substitution in (5.41), then multiplication by $e^{-i\mathbf{kx}}$ and integration in $\mathbf{x}$ yields $\sim \delta(\mathbf{k}-\mathbf{k}')$; subsequent integration in $\mathbf{k}'$ leads to

$$\frac{1}{(2\pi)}\int\left\{\left[-\mathbf{k}^2+\frac{\omega'^2}{c^2}\right]\mathbf{A}(\mathbf{k},\omega')+\frac{1}{\varepsilon_0 c^2}\mathbf{j}(\mathbf{k},\omega')\right\}e^{-i\omega' t}d\omega'=0.$$

Repetition of the same procedure with ω' and t yields an algebraic relation in the single Fourier components

$$\left\{\left[-\mathbf{k}^2+\frac{\omega^2}{c^2}\right]\mathbf{A}(\mathbf{k},\omega)+\frac{1}{\varepsilon_0 c^2}\mathbf{j}(\mathbf{k},\omega)\right\}=0. \tag{5.42}$$

It proves that the linear wave equation with constant coefficients must be fulfilled by the single Fourier components $\mathbf{A}(\mathbf{k},\omega), \mathbf{j}(\mathbf{k},\omega)$ of the same wave vector $\mathbf{k}$ and the same frequency ω as a consequence of orthogonality. In the plasma at rest, $\mathbf{u}_i=\mathbf{u}_{e0}=0$, the linear current density is linked to the electric field by the high frequency (hf) *Ohm's* law

$$\mathbf{j}(\mathbf{k},\omega)=\sigma(\mathbf{k},\omega)\,\mathbf{E}(\mathbf{k},\omega)=i\,\omega\,\sigma(\mathbf{k},\omega)\mathbf{A}(\mathbf{k},\omega). \tag{5.43}$$

The *electric conductivity* $\sigma(\mathbf{k},\omega)$ is easily determined from the linearized electron fluid equation (3.28)

$$\partial_t\mathbf{u}_e=-\frac{e}{m_e}\mathbf{E}=\frac{e}{m_e}\partial_t\mathbf{A}\quad\Rightarrow\quad\mathbf{u}_e=\frac{e}{m_e}\mathbf{A}. \tag{5.44}$$

The contribution to the current density by the ions is by the factor $m_e/m_i\simeq 10^{-3}-10^{-5}$ smaller. Equation (5.44) must be fulfilled in the single Fourier components, thus

$$\mathbf{j}(\mathbf{k},\omega)=-en_e\mathbf{u}_e j(\mathbf{k},\omega)=-\frac{n_e e^2}{m_e}\mathbf{A}(\mathbf{k},\omega)\Rightarrow\sigma(\mathbf{k},\omega)=i\frac{n_e e^2}{m_e\omega}. \tag{5.45}$$

Note, only in Fourier space proportionality between $\mathbf{j}$ and $\mathbf{E}$ holds. Superposition of two modes for example results in a conductivity modulated in time. Elimination of $\mathbf{j}(\mathbf{k},\omega)$ from (5.42) with the help of Ohm's law (5.43) ends in the condition for the eigenmode

$$\left\{\omega^2-c^2\mathbf{k}_0^2\varepsilon(\mathbf{k},\omega)\right\}\mathbf{A}(\mathbf{k},\omega)=0;\;\;\varepsilon(\mathbf{k},\omega)=1+i\frac{\sigma(\mathbf{k},\omega)}{\varepsilon_0\omega}=1-\frac{\omega_p^2}{\omega^2};\;\;\omega_p^2=\frac{n_e e^2}{m_e\varepsilon_0}. \tag{5.46}$$

Thereby $\mathbf{k}_0^2=\omega^2/c^2$ for the wave vector in the vacuum has been set. The first equality expresses the dispersion relation of the electromagnetic wave in the fully ionized plasma. For the single Fourier mode of frequency ω to propagate

$$\omega^2 = c^2\mathbf{k}_0^2\varepsilon(\mathbf{k},\omega) = c^2\mathbf{k}_0^2\left(1 - \frac{\omega_p^2}{\omega^2}\right) \quad \Leftrightarrow \quad \omega^2 = \omega_p^2 + c^2\mathbf{k}^2 \tag{5.47}$$

must be fulfilled. The quantity ε is the *transverse dielectric constant*.

Note, σ and ε are given in the *long wavelength limit* $\mathbf{k} \to 0$ for a cold plasma; no $\mathbf{k}$ dependence appears in (5.45) and (5.46). In the absence of resonances (e.g., no static magnetic field) $v_{\mathrm{os}}^2 \ll v_\varphi^2$ always for subrelativistic intensities. Thermal motion in general leads to *Doppler broadening* in frequency ω of light propagation and scattering into an arbitrary direction. In the normal mode $(\mathbf{k},\omega)$ the single electron "sees" the Doppler-shifted mode $(\mathbf{k}',\omega')$ according to (2.169) and emits it in its frame at ω' into direction $\mathbf{k}'$ which in the lab frame is $\mathbf{k}$. As a consequence, ω' transforms back to ω of the lab frame irrespective of the thermal velocity component. Thermal motion or, as a special case, constant flow $\mathbf{u}_e$ of the electron fluid do not affect light propagation into forward direction. They only act on the phase of the mode (see *Fresnel effect* in Chap. 6).

The term ω_p is assigned the name *plasma frequency* for reasons that will become clear in connection with the longitudinal plasma modes. The electron density at which ω_p becomes equal to the wave frequency defines the so called *critical plasma density* n_c,

$$\omega^2 = \omega_p^2 = \frac{n_e e^2}{m_e\varepsilon_0} \quad \Leftrightarrow \quad n_c = \frac{\varepsilon_0 m_e}{e^2}\omega^2; \quad n_c[\mathrm{cm}^{-3}] = \frac{10^{21}}{(\lambda\,[\text{¯m}])^2}, \quad \lambda = \frac{2\pi c}{\omega}. \tag{5.48}$$

For $n_e = 0$ the wave vector $\mathbf{k}$ is $\mathbf{k}_0$ of the vacuum. At the critical density, $n_e = n_c$, follows $\mathbf{k} = 0 \Leftrightarrow \lambda = \infty$. For $n_e > n_c$ the wave vector becomes purely imaginary, the associated mode behaves like $\exp\pm(|\mathbf{k}|x)$. Such solutions can only exist in regions of finite extent under appropriate boundary conditions. Dispersion relation (5.47) is the simplest possible relation in matter; it is characterized by the absence of resonances of the carriers, e.g. ions, atoms, molecules. The *de Broglie waves* of matter follow a dispersion relation of identical structure: $(E/\hbar)^2 = (mc^2/\hbar)^2 + c^2\mathbf{k}^2$.

Phase and group velocities. A *general dispersion relation* between wave vector $\mathbf{k}$ and frequency ω is defined by an implicit relation of the form

$$D(\mathbf{k},\omega) = 0. \tag{5.49}$$

Resolved piecewise for ω as a function of $\mathbf{k}$ it may decay into different branches $\omega_r(\mathbf{k})$. In any case $\mathbf{k}$ and ω of an eigenmode in general are linked together by a dispersion relation characterizing the type of wave. Owing to this link the single mode is perfectly defined by one of the quantities

$$\mathbf{A}(\mathbf{x},t)\,|\,\mathbf{E}(\mathbf{x},t)\,|\,\mathbf{B}(\mathbf{x},t) = [\hat{\mathbf{A}}_\mathbf{k}\,|\,\hat{\mathbf{E}}_\mathbf{k}\,|\,\hat{\mathbf{B}}_\mathbf{k}]\mathrm{e}^{i(\mathbf{kx}-\omega t)};\;\; \hat{\mathbf{E}}_\mathbf{k} = i\omega\hat{\mathbf{A}}_\mathbf{k},\;\; \hat{\mathbf{B}}_\mathbf{k} = i\mathbf{k}\times\hat{\mathbf{A}}_\mathbf{k} \tag{5.50}$$

with $\hat{\mathbf{A}}_\mathbf{k}, \hat{\mathbf{E}}_\mathbf{k}, \hat{\mathbf{B}}_\mathbf{k}$ the amplitudes and $\mathbf{kx}-\omega t = \Phi(\mathbf{x},t)$ the phase. The velocity at which $\Phi(\mathbf{x},t) = \mathbf{kx}-\omega t = \text{const}$ propagates is the phase velocity $v_\varphi = \omega/|\mathbf{k}|$. In general it is a function of $\mathbf{k}$.

As a consequence of dispersion $\omega(\mathbf{k})$, a wave packet $\mathbf{A}(\mathbf{x}, t=0)$ will change its shape during propagation because in its Fourier decomposition

$$\mathbf{A}(\mathbf{x},t) = \frac{1}{(2\pi)^{3/2}}\int \mathbf{A}_\mathbf{k}\mathrm{e}^{i[\mathbf{kx}-\omega(\mathbf{k})t]}\mathrm{d}\mathbf{k} \tag{5.51}$$

the single components propagate in time each at its individual phase speed. For a local field maximum centered at $\mathbf{k}=\mathbf{k}_0$, $\omega_0=\omega(\mathbf{k}_0)$ the deformation leads to a propagation velocity different from $v_\varphi(\mathbf{k}_0)$. First order Taylor expansion of ω yields

$$\omega(\mathbf{k}) = \omega_0 + \mathbf{K}\frac{\partial\omega}{\partial\mathbf{k}}\Big|_{\omega=\omega_0} \Rightarrow \mathbf{kx}-\omega(\mathbf{k})t = \mathbf{k}_0\mathbf{x}-\omega_0 t + \mathbf{K}\left(\mathbf{x}-\frac{\partial\omega}{\partial\mathbf{k}}t\right),\;\; \mathbf{K}=\mathbf{k}-\mathbf{k}_0. \tag{5.52}$$

Its substitution in (5.51),

$$\mathbf{A}(\mathbf{x},t) = \frac{1}{(2\pi)^{3/2}}\mathrm{e}^{i(\mathbf{k}_0\mathbf{x}-\omega_0 t)}\int \mathbf{A}_\mathbf{K}\mathrm{e}^{i\mathbf{K}(\mathbf{x}-\partial\omega/\partial\mathbf{k}\,t)}\mathrm{d}\mathbf{K} \tag{5.53}$$

shows a plane wave of phase $\Phi_0 = \mathbf{k}_0\mathbf{x}-\omega_0 t$ and an integral as amplitude propagating in space and simultaneously deforming in time.

Let us assume that the deformation, i.e. *dispersion*, is weak and the local maximum is a stable Gaussian wave packet in space,

$$\mathbf{A}(\mathbf{x},t) = \mathbf{A}_0\mathrm{e}^{-(\mathbf{x}-\mathbf{v}_g t)^2/2d^2}\times\mathrm{e}^{i(\mathbf{k}_0\mathbf{x}-\omega_0 t)};\quad \mathbf{v}_g = \frac{\partial\omega}{\partial\mathbf{k}}\Big|_{\omega=\omega_0}\quad v_g = |\mathbf{v}_g|. \tag{5.54}$$

Its connection with the Fourier components $\mathbf{A}_\mathbf{K}$ is

$$\mathbf{A}_\mathbf{K} = \mathbf{A}_0 d^3\,\mathrm{e}^{-\mathbf{K}^2 d^2/2}. \tag{5.55}$$

The Gaussian envelope propagates at *group velocity* $\mathbf{v}_g$. After differentiation of the electromagnetic dispersion relation (5.47) with respect to $\mathbf{k}$ the relation

$$v_\varphi\, v_g = c^2 \tag{5.56}$$

follows. At the critical density the monochromatic electromagnetic wave degenerates: $\mathbf{k}$ approaches zero, v_φ tends to infinity and the monochromatic pulse comes to rest. Under linear dispersion (5.52) it maintains its Gaussian shape.

At high frequencies beyond atomic and ionic resonances, in the X-ray regime for example, dispersion approaches the plasma dispersion (5.47)

$$\omega = \omega_p\left(1+\frac{c^2k^2}{\omega_p^2}\right)^{1/2} \simeq \omega_p + \frac{c^2k^2}{2\omega_p} = \omega_p + \frac{C}{4}k^2. \tag{5.57}$$

If the Gaussian pulse $\mathbf{A}(\mathbf{x}, t)$ is considered around $\mathbf{k} = 0$ where the dispersion is strongest (curvature in Fig. 5.10 is highest) Fourier decomposition follows immediately by substituting $2d^2$ in the exponent by $d^2 + iCt$,

$$\mathbf{A}(\mathbf{x}, t) = \frac{\mathbf{A}_0 d}{\sqrt{d^2+iCt}}\exp -\frac{\mathbf{x}^2}{d^2+iCt} - i\omega_p t = Re^{i\psi},$$

$$R(\mathbf{x}, t) = \frac{\mathbf{A}_0 d}{(d^4+C^2t^2)^{1/4}}\exp -\frac{\mathbf{x}^2 d^2}{d^2+C^2t^2}. \tag{5.58}$$

The real amplitude $R(\mathbf{x}, t)$ of the wave packet reduces in time and its extension increases simultaneously; it diffuses in time. The situation is typical of nonlinear dispersion.

Refractive index η. The wave vector in the plasma is $\mathbf{k} = \mathbf{k}_0\,\varepsilon^{1/2}$. Accordingly, the phase velocity and the refractive index η are

$$v_\varphi = \frac{\omega}{|\mathbf{k}|} = \frac{\omega}{|\mathbf{k}_0|\varepsilon^{1/2}} = \frac{c}{\varepsilon^{1/2}} \Rightarrow \eta = \eta_\perp = \varepsilon^{1/2},\ \omega = \frac{c}{\eta}|\mathbf{k}|,\ v_g = \frac{c}{\eta + \omega(\partial\eta/\partial\omega)}. \tag{5.59}$$

The term $\epsilon_{em} = \frac{1}{2}\varepsilon_0(\mathbf{E}^2 + c^2\mathbf{B}^2)$ is the electromagnetic energy density, the energy flux density is expressed by the *Poynting vector* $\mathbf{S}$. The term $\mathbf{jE}$ represents the coupling to matter; it expresses the work of the electromagnetic field on matter. The magnetic field does not couple to matter directly. The momentum conservation of the electromagnetic field is expressed in terms of Poynting's vector $\mathbf{S}$ and stress tensor T. It shows that the refractive index η is the square root of the dielectric constant ε. The relations (5.59) are of general validity for all transverse waves. In the fully ionized unmagnetized plasma holds

$$\eta = \left(1-\frac{\omega_p^2}{\omega^2}\right)^{1/2};\quad v_\varphi = \frac{c}{\eta},\quad v_g = c\eta \tag{5.60}$$

At the sharp interface between vacuum (index 1) and the homogeneous plasma (index 2) the parallel and perpendicular components of $\mathbf{E}$ and $\mathbf{B}$ obey the continuity relations

$$\mathbf{E}_{1\parallel} = \mathbf{E}_{2\parallel},\quad \mathbf{E}_{1\perp} = \varepsilon_2\mathbf{E}_{2\perp};\qquad \mathbf{B}_{1\parallel} = \mathbf{B}_{2\parallel},\quad \mathbf{B}_{1\perp} = \mathbf{B}_{2\perp}. \tag{5.61}$$

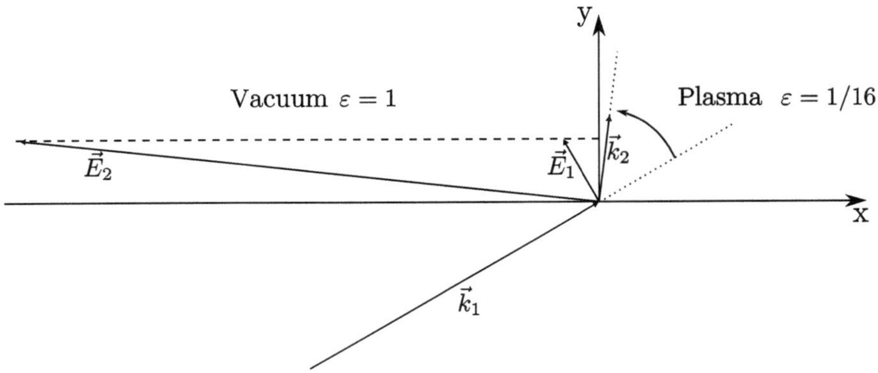

Fig. 5.2 Refraction of electric field and wave vector at the interface vacuum/plasma not far from critical density, $\varepsilon = 1/16$. $\mathbf{E}_{2\perp} = \mathbf{E}_{1\perp}/\varepsilon,\ k_2 = k_1\varepsilon^{1/2}$

Each of the monochromatic fields is the sum of an incident and of a reflected component. In a plasma not far from the critical density an incident wave undergoes a strong refraction towards the surface normal as sketched in Fig. 5.2.

5.4.1.2 Longitudinal Waves

In the absence of a static magnetic field the plasma is capable of supporting two types of eigenmodes with the electric field into direction of propagation: the high frequency *electron plasma wave* and the low frequency *ion acoustic wave*. The longitudinal structure of the **E** field implies div$\mathbf{E} \neq 0$. As a consequence, the electron and ion densities and their pressures appear modulated, too. The driving forces are a combination of electric field strength and fluid dynamic pressure. We shall see now how this interplay of electric with thermal forces acts as a function of frequency.

The High Frequency Electron Plasma Wave

The two-component plasma is capable of rapid charge density oscillations. Owing to their large mass the ions can be considered as a uniform background at rest, whereas for the electrons (3.26) and (3.27) apply. For short time intervals heat flow among electrons $\mathbf{q}_e$ and heat flow to the ions need not be considered and, in the absence of external heating, the energy equation simplifies to the isentropic state relation, see (3.24) or (4.23),

$$T_e n_e^{-(\gamma_e - 1)} = \text{const}, \quad p_e n_e^{-\gamma_e} = \text{const}; \quad \gamma_e = \frac{5}{3}. \tag{5.62}$$

However, it must be pointed out that the adiabatic coefficient $\gamma_e = 5/3$ is correct only under the assumption that there are enough collisions during a change of n_e to redistribute the compression work isotropically. In contrast, in a fast one-dimensional compression occurring during a time shorter than the collision time τ_{ee} the compression (or expansion) work acts only on one of f degrees of freedom so that $f/2 = 3/2$ in (4.23) or (4.58) has to be replaced by $f/2 = 1/2$ and the adiabatic coefficient becomes $\gamma_e = (f+2)/f = 3$ in (5.62), in agreement with the kinetic result from the Vlasov equation (3.79). Under the above assumptions the electronic fluid decouples from the ions and its dynamics is determined by the equations

$$\frac{\partial n_e}{\partial t} + \nabla(n_e \mathbf{u}_e) = 0, \quad n_e \left(\frac{\partial}{\partial t} + \mathbf{u}_e \nabla \right) \mathbf{u}_e = -\frac{1}{m} \nabla p_e - n_e \frac{e}{m_e} \mathbf{E} - n_e \nu_{ei} \mathbf{u}_e,$$

$$\nabla \mathbf{E} = -\frac{e}{\varepsilon_0}(n_e - Z n_0), \quad p_e n_e^{-\gamma_e} = \text{const.} \tag{5.63}$$

In general this system is still "intractable" [19]. However, it becomes solvable in the case of small disturbances $n_1 = n_e - Zn_0$ from equilibrium $n_{e0} = n_e - n_1 = Zn_0 = \text{const}$, $\mathbf{u}_{e0} = 0$, $T_{e0} = T_e - T_1 = \text{const}$, $|n_1| \ll n_e$. In this case the system can be linearized by omitting all higher order terms,

$$\frac{\partial n_1}{\partial t} + n_{e0} \nabla \mathbf{u}_e = 0, \quad n_{e0} \frac{\partial \mathbf{u}_e}{\partial t} = -c_{\text{se}}^2 \nabla n_1 - n_{e0} \frac{e}{m_e} \mathbf{E} - n_{e0} \nu_{ei} \mathbf{u}_e; \tag{5.64}$$

$$c_{\text{se}}^2 = \gamma_e \frac{k_B T_{e0}}{m_e}. \tag{5.65}$$

Sometimes it may be useful to cast mass conservation into an alternative form with the help of the displacement $\boldsymbol{\delta}_e(\mathbf{x}, t)$,

$$0 = \frac{\partial n_1}{\partial t} + \nabla \left(n_e \frac{\mathrm{d}\boldsymbol{\delta}_e}{\mathrm{d}t} \right) \simeq \frac{\partial n_1}{\partial t} + \nabla \left(n_{e0} \frac{\partial \boldsymbol{\delta}_e}{\partial t} \right).$$

Integration in time connects the density perturbation with the local displacement,

$$n_1(\mathbf{x}, t) = -\nabla(n_{e0} \boldsymbol{\delta}_e). \tag{5.66}$$

The quantities $\mathbf{u}_e$ and $\mathbf{E}$ are removed by differentiating the first equation of (5.64) with respect to time and taking the divergence of the second equality, and then using (5.3) for $\nabla \mathbf{E}$. The straightforward result is

$$c_{\text{se}}^2 \nabla^2 n_1 - \frac{\partial^2}{\partial t^2} n_1 - \nu_{ei} \frac{\partial}{\partial t} n_1 - (\omega_p^2 - c_{\text{se}}^2 \nabla^2) n_1 = 0. \tag{5.67}$$

Here we encounter again the *plasma frequency* ω_p from (5.46),

$$\omega_p^2 = \frac{n_{e0}e^2}{\varepsilon_0 m_e} \tag{5.68}$$

and understand why this name for the quantity (5.68). The general solution of the linear wave equation (5.67) shows the identical structure of the electromagnetic wave equation (5.38_2) with constant coefficients not depending on n_1. It is again obtained from the Fourier decomposition into normal modes $n_{e\mathbf{k}}(\mathbf{x}, t)$,

$$n_{e\mathbf{k}}(\mathbf{x}, t) = \hat{n}_{e\mathbf{k}} e^{i(\mathbf{k}_e\mathbf{x}-\omega t)}, \qquad n_1(\mathbf{x}, t) = \frac{1}{(2\pi)^{3/2}} \int n_{e\mathbf{k}}(\mathbf{x}, t)\, \mathrm{d}\mathbf{k} \tag{5.69}$$

together with the *Bohm–Gross* dispersion relation

$$\omega^2 = \omega_p^2 + c_{\mathrm{se}}^2 \mathbf{k}_e^2 - i\nu_{ei}\omega. \tag{5.70}$$

The longitudinal electron plasma wave and the electromagnetic wave follow dispersion relations of the same structure. Therefore, what has been said about the group and phase velocities and the refractive index $\eta = \eta_\perp$ for the transverse eigenmodes holds for (5.70) with $\nu_{ei} = 0$ also,

$$v_\varphi\, v_g = c_{\mathrm{se}}^2; \quad \eta = \eta_\| = \frac{c_{\mathrm{se}}}{v_\varphi} = \frac{c_{\mathrm{se}}|\mathbf{k}_e|}{\omega} = \left(1 - \frac{\omega_p^2}{\omega^2}\right)^{1/2} = \eta_\perp. \tag{5.71}$$

In particular, Snell's law of refraction and the critical density are the same for both waves. The longitudinal electron wave cannot penetrate into an *overdense* plasma of $n_{e0} > n_c$.

In the cold plasma ($T_e = 0$) the restoring force is entirely of electrostatic nature, the group velocity is zero and an arbitrary, small charge perturbation n_1 oscillates at the plasma frequency ω_p defined by the expression of (5.68). This is why ω_p deserves this name, see also Fig. 1.6. In the warm plasma ($T_e > 0$) there is the additional thermal pressure contribution $c_{\mathrm{se}}^2\mathbf{k}_e^2$ to ω_p^2 of the electrostatic wave and v_g is finite; ω_p is the lower limit of ω for the wavelength λ tending to infinity. The wave is *cut-off* at the critical electron density n_c. Normalized to the Titanium-saphire laser for example it is given by

$$n_c = \frac{\varepsilon_0 m}{e^2}\omega^2 = 1.75 \times 10^{21} \left(\frac{\omega}{\omega_{\mathrm{Ti:Sa}}}\right)^2 \; [\mathrm{cm}^{-3}]. \tag{5.72}$$

Formally, for λ tending to zero ω approaches the other asymptote $\omega = c_{se}k_e$ and $v_\varphi = v_g = c_{se}$. For this reason c_{se} is called the *electron sound velocity*.

It is rather surprising at first glance that the same ω_p appears in the dispersion of the transverse wave also. However, the long wavelength limit $\mathbf{k} \to 0$ in (5.47) is equivalent to $\mathbf{B} \to 0$ in Maxwell's equation (5.2) and, accordingly, of (5.1) shrinking to charge conservation $\varepsilon_0 \dot{\mathbf{E}} + \mathbf{j} = 0$. The latter is obeyed by the longitudinal as well as the transverse wave in the limit $\mathbf{k} \to 0$. Furthermore, if one takes the resonant interaction of a transverse with a longitudinal wave at oblique incidence onto an inhomogeneous plasma for granted the presence of ω_p and its identity in both dispersion relations is stringent, see Sect. 6.2.

In the linearized version of system (5.63) all quantities $\mathbf{E}, \mathbf{u}_e, n_1, T_1,\ p_1 = n_1 k_B T_{e0}$, except one of them, can be eliminated to obtain a wave equation equivalent to (5.67). This is typical of linear waves in homogeneous media. For the normal modes this means that they are all proportional to each other. In the electron plasma wave, named also *electrostatic* or *Langmuir* wave, the dependences of the quantities and their amplitudes are

$$E = i\frac{e}{\varepsilon_0 k_e} n_1,\ \Phi = i\frac{E}{k_e} = -\frac{e}{\varepsilon_0 k_e^2} n_1,\ u_e = v_\varphi \frac{n_1}{n_{e0}} = -i\omega\delta,\ T_1 = (\gamma_e - 1) T_{e0} \frac{n_1}{n_{e0}},$$

$$\hat{E} = \frac{e}{\varepsilon_0 k_e}\hat{n}_1,\ \hat{\Phi} = \frac{\hat{E}}{k_e} = \frac{e}{\varepsilon_0 k_e^2}\hat{n}_1,\ \hat{u}_e = v_\varphi \frac{\hat{n}_1}{n_{e0}},\ \hat{T}_1 = (\gamma_e - 1) T_{e0} \frac{\hat{n}_1}{n_{e0}}, \tag{5.73}$$

Φ is the electrostatic wave potential. The proportionality of the potential energy V and the density fluctuation n_1

$$V = -e\Phi = \frac{e^2}{\varepsilon_0 k_e^2} n_1 = m_e \frac{\omega_p^2}{k_e^2} \frac{n_1}{n_0} \tag{5.74}$$

may be useful. It shows once more the electrostatic origin of the ω_p^2 term in the dispersion relation (5.70). The relation of the flow velocity $u_e = |\mathbf{u}_e|$ to the phase velocity v_φ in (5.73) is of practical importance. It is capable of an intuitive interpretation when written as $n_{e0} u_e = n_1 v_\varphi$, see Fig. 5.3: The longitudinal wave is the product of total density $n_e \simeq n_{e0}$ with the oscillatory particle velocity u_e. On the other hand the same wave can be thought as produced by the density perturbation n_1 sliding with the phase speed v_φ over n_{e0}.

Collisional damping of the electron plasma wave , with $\Im\omega = -\nu_{ei}/2$, in the hot laser plasma is generally small. For example $\nu_{ei}/\omega_{\mathrm{Nd}} = 3 \times 10^{-5} Z$ at $n_{e0} = 10^{20}\ \mathrm{cm}^{-3}$ and $T_e = 1$ keV. However, there is an important *collisionless* damping mechanism. For k_e approaching the order of magnitude of $k_D = 1/\lambda_D$ the electrostatic field $E_\mathbf{k}$ and its potential $\Phi_\mathbf{k}$ increase at constant density amplitude $\hat{n_1}$ and lead to deep modulation of electrons with velocities close to the thermal speed $v_{\mathrm{th},e}$:

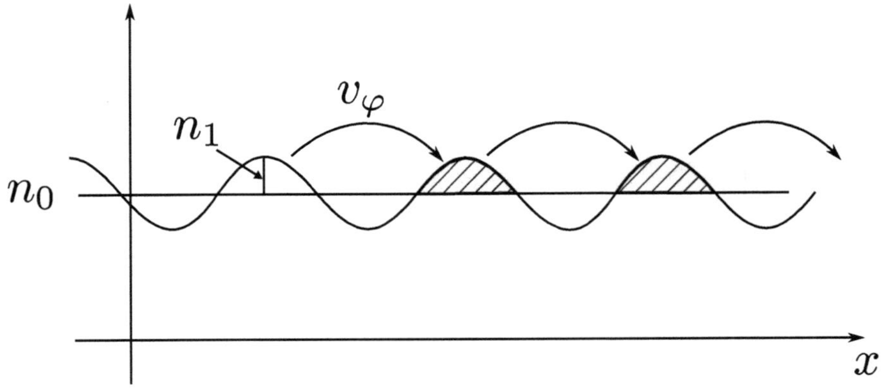

Fig. 5.3 Geometrical interpretation of the longitudinal wave dynamics. The Figure tells that the wave n_1 can be imagined as produced by the shift of the density perturbation n_1 at its phase velocity v_φ and the main fluid at rest; alternatively, the total density moves with fluid velocity $\mathbf{u}$. This is what linearized mass conservation imposes, i.e. $n_0\mathbf{u} = n_1\mathbf{v}_\varphi$. It establishes the connection of flow velocity u with density

the particle nature of the electron fluid manifests itself through the phenomenon of *Landau damping*. It has to be discussed in detail separately.

Longitudinal dielectric constant $\varepsilon_\parallel$. Historically a dielectric *permissivity* $\mathbf{D} = \varepsilon\varepsilon_0\mathbf{E}$,

$$\mathbf{D_k} = \varepsilon_\mathbf{k}\,\varepsilon_0\,\mathbf{E_k} = \varepsilon_0\mathbf{E_k} + \mathbf{P} \tag{5.75}$$

with $\mathbf{P}$ the dielectric polarization, has been introduced to distinguish "external" electric charges from "bound" charges. Since the homogeneous plasma does not contain external charges it is expected that the electron plasma wave satisfies $\mathbf{D_k} = 0$. In linear approximation the polarization is $\mathbf{P_k} = -n_{e0}e\boldsymbol{\delta}_e$, with $\boldsymbol{\delta}_e$ the electron displacement. From (5.64) follows

$$\mathbf{u}_e = -i\,\frac{e}{m_e\omega}\left(1 + \frac{c_{\rm se}^2\mathbf{k}_e^2}{\omega_p^2}\right)\mathbf{E}, \quad \boldsymbol{\delta}_e = i\,\mathbf{u}_e/\omega \;\Rightarrow\; \mathbf{P_k} = -\varepsilon_0\frac{\omega_p^2}{\omega^2}\left(1 + \frac{c_{\rm se}^2\mathbf{k}_e^2}{\omega_p^2}\right)\mathbf{E}$$

$$\Rightarrow\; \mathbf{D_k} = \varepsilon_0\left(1 - \frac{\omega_p^2 + c_{\rm se}^2\mathbf{k}_e^2}{\omega^2}\right)\mathbf{E_k} = 0; \quad \varepsilon_\parallel(\mathbf{k},\omega) = 1 - \frac{\omega_p^2 + c_{\rm se}^2\mathbf{k}_e^2}{\omega^2}. \tag{5.76}$$

Thus, $\mathbf{D_k} = 0$ is a consequence of the dispersion relation. Conversely, if the function $\varepsilon_\parallel(\mathbf{k},\omega)$ is known the dispersion relation follows from setting $\varepsilon_\parallel(\mathbf{k},\omega) = 0$. Its connection with the electric conductivity $\sigma(\mathbf{k},\omega)$ results from $\partial_t\mathbf{P} = \mathbf{j} = \sigma\mathbf{E}$ as

$$\sigma = i\,\frac{\varepsilon_0}{\omega}(\omega_p^2 + c_{\rm se}^2\mathbf{k}_e^2) \quad\Rightarrow\quad \varepsilon_\parallel(\mathbf{k},\omega) = 1 + i\,\frac{\sigma(\mathbf{k},\omega)}{\varepsilon_0\omega}. \tag{5.77}$$

This is the same dependence as for the transverse wave in (5.46). The difference lies in the electrical conductivity. In longitudinal direction there is the thermal pressure contribution to the electric restoring force, in transverse direction it is missing.

For both waves, *transverse* and *longitudinal* $\mathbf{E}_\mathbf{k}$ is the measured electric field, e.g. in a narrow parallel vacuum slit, at position and time $(\mathbf{x}, t)$ in presence of the polarization $\mathbf{P}_\mathbf{k}$. It is the sum of the vacuum field and the field generated by the polarization $\mathbf{P}_\mathbf{k} = \varepsilon_0 \chi_\mathbf{k} \mathbf{E}_\mathbf{k}$, with $\chi_\mathbf{k}$ the *dielectric susceptibility*. A plasma sphere in vacuum oscillates at its lowest frequency with $\omega_0 = \omega_p/\sqrt{3}$.

The electron plasma wave must follow from the general wave equation (5.15). For $\mathbf{E} \parallel \mathbf{k}_e$, $|\mathbf{E}| = E$ it reads

$$-\frac{1}{c^2}\frac{\partial^2}{\partial t^2}\mathbf{E} = \frac{1}{\varepsilon_0 c^2}\frac{\partial \mathbf{j}}{\partial t}; \quad \mathbf{E}, \mathbf{j} \sim \mathrm{e}^{i(\mathbf{k}_e\mathbf{x}-\omega t)} \quad \Rightarrow \quad E = -\frac{i}{\varepsilon_0 \omega} j.$$

The current density of the single mode is $j = \sigma E$, with σ from (5.77). Hence,

$$\left(\omega^2 + i\frac{\omega}{\varepsilon_0}\sigma\right)E = \left(\omega^2 - \omega_p^2 - c_{\mathrm{se}}^2 k_e^2\right)E = 0; \;\; E \neq 0 \;\Rightarrow\; \omega^2 = \omega_p^2 + c_{\mathrm{se}}^2 k_e^2.$$

The longitudinal wave is characterized by div$\mathbf{E} \neq 0$. It leads to the additional contribution from the electron pressure term in the conductivity σ.

Cut offs and resonances. It is the electron current density $\mathbf{j}$ induced by the $\mathbf{E}$ field of the wave that determines its propagation and stopping. The *cut off* of a mode occurs if the refractive index goes to zero and the phase velocity tends to infinity. The incident wave generates a current density which, in turn, generates a magnetic field of sign *opposite* to the vacuum magnetic field. The general situation may be exemplified by the motion of an oscillator of a single eigenfrequency ω_0, driven by the vacuum field $\mathbf{E}_v = \mathbf{E}_0 \exp -i(\mathbf{k}_0\mathbf{x} - \omega t)$,

$$\ddot{\delta} + \omega_0^2 \delta = -\frac{e}{m_e}\mathbf{E}_v. \tag{5.78}$$

The vacuum magnetic field $\mathbf{B}_v$, the induced current density $\mathbf{j}_v = -en_1\dot{\delta}$ from (5.78 with the low electron density $n_1 \ll n_c$ and the magnetic field $\mathbf{B}_1$ generated by $\mathbf{j}_v$ result as

$$\mathbf{B}_v = \frac{\mathbf{k}_0}{\omega} \times \mathbf{E}_v; \quad \mathbf{j}_v = i\omega\epsilon_0 \frac{\omega_{p1}^2}{\omega^2 - \omega_0^2}\mathbf{E}_v; \quad \mathbf{B}_1 = -\frac{\omega_{p1}^2}{\omega^2 - \omega_0^2}\frac{\mathbf{k}_1}{\omega} \times \mathbf{E}_v,$$

$$\omega_{p1}^2 = n_1 e^2/\epsilon_0 m_e, \quad \mathbf{k}_1 = \mathbf{k}_0 \eta_1, \quad \eta_1 = \left(1 - \frac{\omega_{p1}^2}{\omega^2}\right)^{1/2} \quad \Rightarrow |\mathbf{k}_1| < |\mathbf{k}_0|. \tag{5.79}$$

The induced magnetic field $\mathbf{B}_1$ follows from (5.1). In the ideal plasma the eigenfrequency is $\omega_0 = 0$; comparison with $\mathbf{B}_v$ shows indeed that (i) $\mathbf{B}_1$ weakens the vacuum field $\mathbf{B}_v$ and (ii) the wavelength in the tenuous plasma is lengthened. The same arguments apply to an arbitrary electron density $n_e + n_1$ with the identical conclusions (i) and (ii). They show why the total magnetic field $\mathbf{B}$ and the wave vector $\mathbf{k}$ reduce to zero at the cut off in the case of plasma. In addition, they make clear why a hump in $\mathbf{E}$ and $\mathbf{E}$ show a recession into direction opposite to v_φ; in other words, the group velocity decreases with increasing plasma density. At the critical point the induced current $\mathbf{j}$ is such as to cancel completely the magnetic field. A transverse wave with $\mathbf{B} = 0$ cannot propagate further into the overdense plasma. At the critical point the vacuum magnetic energy $\varepsilon_0\mathbf{B}^2/2$ is transformed into oscillatory energy $m_e\mathbf{v}_{os}^2/2$. A consideration analogous to (5.78) and (5.79) applies to $\mathbf{v}_{os}$ which beyond the critical point with $\omega_= = \omega_p > \omega$ tends to cancel the electric field. A similar argument leads to the cut off of the electron plasma wave.

In the case $\omega < \omega_0$ the induced current takes on the opposite phase of plasma, the wave propagates without cut off; it is the case of transparent glass. Only much beyond resonance, $\omega \gg \omega_0$ the plasma behaviour with cut off is re-established; propagation is forbidden.

In case the electrons go through a *resonance* $\omega = \omega_0$ it follows from (5.78), and it is well known from classical optics, that $\mathbf{j}$ and its conductivity σ tend to infinity. In reality friction prevents all physical quantities involved from unlimited growth. The wave energy is totally absorbed or absorbed in part and partially reflected. A second possibility is the total or partial conversion into another type of wave, for instance resonant excitation of an electron plasma wave by the laser beam. This is the so called *resonance absorption*. From (5.46) together with (5.59) follows that resonance in the ideal plasma is recognized by the refractive index tending to infinity and the phase velocity v_φ reducing to zero; the wave cannot propagate. The two phenomena of cut off and resonance separate the homogeneous plasma into density regions of propagation and regions where wave propagation is not possible or strong absorption takes place.

5.4.1.3 Kinetic Treatment of the Linear Electron Plasma Wave

Dispersion Relation

We start from the Vlasov equation 3.79 for f with $\mathbf{B}$ set to zero. In the homogeneous plasma at rest and no electric field present the equilibrium distribution function $f = f_0(\mathbf{v})$ satisfies the Vlasov equation $df_0/dt = 0$. Throughout the book f is normalized either to unity or to the local particle density n for convenience. A small perturbation in the distribution function is described by $f_1 = f_1(\mathbf{x}, \mathbf{v}, t)$, $|f_1| \ll |f_0|$. It will produce an electrostatic field $\mathbf{E}(\mathbf{x}, t)$. The distribution function $f = f_0 + f_1$ obeys the linearized Vlasov/Poisson system

$$\frac{\partial f_1}{\partial t} + \mathbf{v}\frac{\partial f_1}{\partial \mathbf{x}} - \frac{e}{m_e}\mathbf{E}\frac{\partial f_0}{\partial \mathbf{v}} = 0; \quad \varepsilon_0\nabla\mathbf{E} = -e\int_{-\infty}^{+\infty} f_1 d\mathbf{v}. \tag{5.80}$$

Now let us look for the existence of a harmonic wave $f_1 = \hat{f}_1 \exp i(\mathbf{kx} - \omega t)$. From Poissons equation in (5.80) follows the same harmonic dependence on $\mathbf{x}, t$ and

$$i(\mathbf{kv} - \omega) f_1 - \frac{e}{m_e}\mathbf{E}\frac{\partial f_0}{\partial \mathbf{v}} = 0; \; i\varepsilon_0 \mathbf{kE} = -\int f_1 \mathrm{d}\mathbf{v}.$$

Substitution of $\mathbf{E}$ and integration in $\mathbf{v}$ leads to the dispersion equation $\varepsilon(\mathbf{k}, \omega) = 0$ of the electron plasma wave,

$$i\left\{\mathbf{kv} - \omega - \frac{\omega_p^2}{k^2}\mathbf{k}\frac{\partial f_0}{\partial \mathbf{v}}\right\}\int f_1 \mathrm{d}\mathbf{v} = 0 \Rightarrow 1 - \frac{\omega_p^2}{k^2}\int_{-\infty}^{+\infty}\frac{\mathbf{k}\partial f_0/\partial \mathbf{v}}{\mathbf{kv} - \omega} = 0. \quad (5.81)$$

The distribution function f_0 factorizes into $f_0 = f_{0x} f_{0y} f_{0z}$. With $\mathbf{k}$ along x the dispersion assumes the scalar form

$$\varepsilon(k, \omega) = 1 - \frac{\omega_p^2}{k^2}\int_{-\infty}^{+\infty}\frac{\partial f_{0x}/\partial v}{v - \omega/k}\mathrm{d}v = 0 \Leftrightarrow 1 + \frac{\omega_p^2}{k^2}\int_{-\infty}^{+\infty}\frac{f_{0x}}{(kv - \omega)^2}\mathrm{d}v = 0. \quad (5.82)$$

The last version follows from partial integration in v. The evaluation of the dispersion relation $\varepsilon(k, \omega) = 0$ is of relevance under the constraint of either k taken real and $\omega = \omega(k)$ satisfying the dispersion (5.82) and $\Re\omega > 0$, or ω real and positive and $k = k(\omega)$ obeying (5.82). The first case is relevant if the wavelength is fixed by the experiment, for instance by the dimensions of the experimental apparatus or the resonator; the second case is realized if the electron plasma wave is excited from outside at a fixed frequency $\omega > 0$.

Apart from special forms of the equilibrium distribution function f_0 (in the following we drop the index x), (5.82) can only have solutions of either complex $\omega(k)$ or of complex $k(\omega)$. For both k and ω real the integrand becomes infinite when v equals the phase velocity $v_\varphi = \omega/k$; no solution to (5.82) exists. However, it is a matter of basic calculus to show that for smooth and symmetric $f_0(v)$ in v at least one solution exists for k real and ω complex, and for ω real and k complex. For k real and $\omega = \omega_r + i\gamma$, $\omega_r > 0$ the real and imaginary parts of (5.82) satisfy the continuous functions $F(\omega_r, \gamma) = 0$ and $G(\omega_r, \gamma) = 0$ in the infinite domain $(\omega_r > 0, |\gamma| > 0)$. Under the previous conditions on f_0 they are resolvable as continuous functions $\gamma = \gamma_F(\omega_r)$ and $\gamma_G(\omega_r)$, $\gamma_F(\omega_r) \not\propto \gamma_G(\omega_r)$, hence $\gamma_F(\omega_r) = \gamma_G(\omega_r)$ yields a solution ω of (5.82). The case of real ω is analogous. For real k Landau has solved the dispersion equation explicitly for $\omega = \omega_r + i\gamma$ and has found a negative value of γ, that means *damping* without collisions, for stable f_0 to the astonishment of the scientific community [29]. His treatment is purely formal, the result is obtained from conditions of the existence of the inverse Fourier transform. It represents one of the great triumphs of mathematical physics.

Landau Damping

"Landau damping may be the single most famous mystery of classical plasma physics." With this statement Clément Mouhot and Cédric Villani [30], Fields laureat

of 2013, express a widely diffused feeling among scientists about Landau damping altogether. Their assertion is still true if the word "plasma" is dropped. And it remains true, to minor extent, if in addition the adjective "classical" is taken out of the statement also. It must be said that still after 70 years the majority of plasma physicists feel the statement not misplaced at all, even in case it is limited solely to linear Landau damping.

Landau damping is a universal phenomenon, encountered in excitation of atoms, e.g., revivals, spin waves, classical and quantum echoes, nonlinear optics, e.g. free induction decay, see Chap. 7. The authors of [30] give most significant contributions to the study of nonlinear Landau damping and on its long time, i.e., asymptotic, behaviour. The paper represents major advancements in the rigorous mathematical foundation of this subject which during seventy years has evolved into enormous complexity. They show in addition, together with other contributors, its affinity to weak turbulence and transport theory, gravity waves in astrophysics, to the KAM (Kolmogorov– Arnold–Moser) theory, e.g. stability of the planetary system, and even to the formation of traffic jam on highways. Despite the rigorous mathematical basis the paper is surprisingly accessible to the interested physicist.

In a seminal paper of 1946 Landau found that, in the absence of collisions, the longitudinal plasma wave decays in time exponentially as $\exp(\gamma t)$ [29], with the damping coefficient

$$\gamma = \frac{\pi}{2}\frac{\omega_r\omega_p^2}{k^2}\left(\frac{\partial f_0}{\partial v}\right)_{v=v_\varphi}. \tag{5.83}$$

The expression holds under the conditions (i) $\gamma \ll \omega$ and (ii) $(k\lambda_D)^2 \ll 1$ for the eigenmode (ω, k) of wave number $k = 2\pi/\lambda$. The real part ω_r of the complex frequency obeys the Bohm–Gross dispersion relation

$$\omega^2 = \omega_p^2 + 3v_{\mathrm{th}}^2 k^2. \tag{5.84}$$

The wavelength λ exhibits an upper limit at $k\lambda_D \simeq 1$. With the Maxwellian distribution function $f_0(v) = \sqrt{\pi/\beta}\exp -\beta v^2$, $\beta = m_e/2k_BT_e$, this results in the familiar expression

$$\begin{aligned}\gamma &= -\left(\frac{\pi}{8}\right)^{1/2}\frac{\omega_p}{(k\lambda_D)^3}\exp\left(-\frac{1}{2(k\lambda)^2}-\frac{3}{2}\right)\\ &= -0.14\frac{\omega_p}{(k\lambda_D)^3}\exp-\left(\frac{1}{2(k\lambda)^2}\right);\quad \lambda_D = \frac{v_{\mathrm{th}}}{\omega_p}.\end{aligned} \tag{5.85}$$

The passage from (5.83) to (5.85) is by simple algebra. The "astonishing" finding of Landau [31] was that, strictly speaking, no linear electron plasma eigenmode exists, all modes are exponentially damped. An example may illustrate the efficiency of linear Landau damping. At $(k\lambda_D)^2 = 1/6$ dispersion (5.84) holds and restriction (ii) is satisfied, $\lambda = 15\lambda_D$, $\omega = 1.2\omega_p$. The damping is $|\gamma/\omega| = 0.08$; the wave survives

$N = \Delta t/\tau = \omega/2\pi|\gamma| = 2$ periods $T_0 = 2\pi/\omega$. In contrast, at twice the wavelength, $\lambda = 30\lambda_D$, the wave undergoes nearly 1000 oscillations before reducing to 1/e.

Landau derived (5.83) from the nonlinear eigenvalue equation (5.82) in the complex frequency ω for real k. The derivation is purely formal, the physics behind is hidden. A frequently encountered physical argument is this: Particles moving more slowly than the wave gain energy from the wave, particles moving faster than the wave loose energy (by way of example see [32], *"Landau damping"*, perhaps also [33] for the references therein, [34, 35]). Hence, if f_0 shows negative slope damping by these "resonant particles" is the consequence.The argument, though present throughout all pertinent literature, is vague; energy gain and loss depend on the system of reference. In addition, back action of the single particle onto the plasma wave is outside of a linear treatment. The term "resonant particles" contributed to a good portion to obscure the phenomenon and to make a mystery of Landau damping. After all, linear Landau damping starts from perturbing linear orbits, a feature common to all linear theory, and excludes particle trapping. An alternative picture invoked to explain damping is based on the *phase mixing* concept (see for example [36], Chap. 8.6 and Appendix C, also [30]). This may be a useful concept in interacting modes with broad band spectrum, as weak plasma turbulence for example. In linear Landau damping over short time intervals it is not a helpful model for it is too complex. It was John Dawson fifteen years after Landau who presented a derivation of (5.83) by calculating the total work done by the electric field on the electron fluid [37]. The most lucid and transparent treatment of this type is perhaps that found in [38]. It shows the compatibility of Landau's formula with the energy conservation by expanding the perturbation of orbits to second order in the electric field.

Consider the linear wave $n_1(x,t) = \hat{n}_1 \exp i(kx - \omega t)$ on the undisturbed back ground n_0 propagating with phase velocity v_φ to the right in the lab frame S. Once wave number and frequency ($k > 0,\ \omega > 0$) are given the electron plasma mode is uniquely determined by one of the amplitudes of electron density n_1, electric field E, potential energy V, oscillation velocity $v_{\text{os}} = u_e$, and oscillatory displacement δ_{os}, see (5.73) and (5.74).

This property of a linear electron plasma wave propagating to the right is preserved if the wave is transformed from the lab frame S to the comoving frame ("wave frame") $S'(v_\varphi) : x' = x - v_\varphi t$. The wave becomes static with zero frequency, $n_1(x') = \hat{n}_1 \cos kx'$, and the plasma streams to the left with $-v_\varphi$, see Fig. 5.4. No damping means that the same *number* of particles (not the same particles) $n_1 dx'$ from position x' finds itself at position $x' - \lambda$ after the time $T_0 = 2\pi/\omega$. The argument can even be strengthened. By symmetry, and the principle of detailed balance in force-free space, in any equilibrium distribution $f_0(\mathbf{v})$ each $-\mathbf{v}$ corresponds to $+\mathbf{v}$ with equal probability. It is a consequence of the linearized term $n_0 v_{\text{os}}$ used under the divergence for $(n_0 + n_1)v_{\text{os}}$: All fluid elements undergo the same shift per unit time to the left with respect to the wave potential. In terms of the undisturbed distribution function it is defined by the flux

$$\int v_i' f_0(v_i)\mathrm{d}v_i = -n_0 v_\varphi; \ \ v_i' = v_i - v_\varphi; \ \ \int v_i f_0(v_i)\mathrm{d}v_i = 0. \tag{5.86}$$

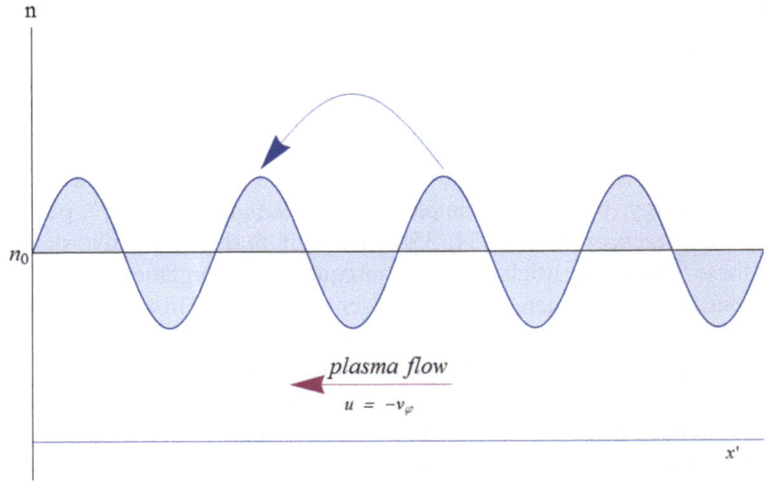

Fig. 5.4 Undamped Vlasov mode (ω, k) in the reference frame $S'(v_\varphi)$ ("wave frame"): electron density n_1 and potential energy V are static; the plasma flows with $u = -(v_\varphi)$ to the left

Under the assumption of $|\gamma| \ll \omega_r$ the flow is quasistationary. The derivative $\dot{n}_1$ can be expressed as the change of n_1 over one wavelength which in turn is the difference of the fluxes

$$\dot{n}_1 = \frac{n_1(x' - \lambda) - n_1(x')}{T_0} = -\int v_i' f_0(v_i)\mathrm{d}v_i \Big|_{x'-\lambda}^{x'} = 0. \tag{5.87}$$

Damping means $\dot{n}_1 < 0$. Here the difference of fluxes is zero because in the undamped wave the flux shows the periodicity of λ.

Imagine now that each v_i' is reduced by the individual amount $\Delta(|v_i'|) > 0$ in a narrow neighbourhood around $v_i' = 0$. As a consequence a reduced number of particles arrive at (x', t) which have started from $(x' - \lambda, t - T_0)$. Thus (5.87) is replaced now by

$$\dot{n}_1 = \int \Delta(v_i') f_0(v_i)\mathrm{d}v_i = \int \Delta(|v_i'|) v_i' \frac{\partial f_0}{\partial v_i}\Big|_{v_i = v_\varphi} \mathrm{d}v_i. \tag{5.88}$$

Under the assumption of negative derivative of f_0 at v_φ (more slow than fast particles) from (5.88) the wave n_1 results *damped*, in the opposite case it is amplified, no damping follows from zero slope of f_0 at $v_i = v_\varphi$. In presence of the wave potential $V(x')$ from (5.73) the single particle is retarded over the distance of λ relative to its free flow. In fact,

$$T(|v_i'|) = \int_\lambda \frac{\mathrm{d}x'}{v(x')} = \sqrt{\frac{m_e}{2}} \int_\lambda \frac{\mathrm{d}x'}{[m_e v_i'^2/2 - V(x')]^{1/2}} > \frac{\lambda}{|v_i'|}. \tag{5.89}$$

In the wave frame $v_i' = v_i - v_\varphi$ is the velocity far out of the wave train with adiabatically increasing and decreasing amplitude. In the following m_e is set to unity. The ratio $\kappa(x') = 2\hat{V}(x')/v_i'^2$ is to be chosen less than 1 because Landau damping (5.83) refers to free particles only [36]. $T(v_i')$ is of second order in κ. For a sinusoidal potential the period results as

$$T = \frac{2\lambda \mathrm{K}(\alpha)}{\pi v_i'(1+\kappa)^{1/2}}; \quad \alpha = 2\kappa/(1+\kappa),$$

$\mathrm{K}(\alpha)$ is the elliptic function of the first kind. For $\kappa \to 0$ follows $T = \lambda(1 + 3\kappa^2/16)/v_i$. The presence of a potential introduces dispersion in $T(v_i')$. This is exactly what is behind the phenomenon of linear Landau damping. By intuition it becomes clear that dispersion must lead to decay of ordered structures ("collisionless damping"), directly evident from transcribing Landau's damping formula (5.83) for $\omega \simeq \omega_p$ into

$$\gamma = \frac{\pi}{2}\frac{\omega\omega_p^2}{k^2}\frac{\partial f_0}{\partial v_\varphi} = \frac{\pi^2}{T_0}v_\varphi^2\frac{\partial f_0}{\partial v_\varphi} \iff j = -Cv_{\mathrm{th}}^2\frac{\partial n_0}{\partial x}. \quad (5.90)$$

The second expression of γ shows the same structure as Fick's law for particle (or heat) diffusion j (expression on the right) when a local gradient exists. C is a material constant. Landau damping is diffusion in velocity (or momentum) space. For special distribution functions dispersion in velocity leads to reappearance of a pulse somewhere in space (see Landau plasma *echo* (8.40) in Sect. 7.8.1, see also [36]). From Fig. 5.4 follows that when starting from an equilibrium distribution $f_0(v_i)$ (i) the result $T(|v_i'|) > \lambda/|v_i'|$ leads to slowing down of *all* particles (they arrive late) , and (ii) damping and amplification of n_1 depend on negative or positive slope of $f_0(v_i)$ in the vicinity of $v = v_\varphi$; there the dilation of T is highest. Zero slope keeps the wave undamped.

Landau's pioneering achievement has consisted in finding entirely by formal arguments that the wave is damped, or equivalently, indicating the correct path of integration in the complex plane. There the integration is not straightforward because the technique with the semicircle vanishing at $\pm\infty$ does not work. Here we may simplify Landau by choosing the path of integration as sketched in Fig. 5.5. The dispersion relation (5.82) has a pole at $\omega = \omega + i\gamma$ with $\gamma < 0$ in order to satisfy it for real k, see Fig. 5.5. The denominator can be split into two parts to make Cauchy's integral formula applicable,

$$2i\pi f(z) = \oint \frac{f(\zeta)d\zeta}{(\zeta - z)} = \oint \left\{ \frac{f(\zeta)d\zeta}{2(\zeta - z)} + \frac{f(\zeta)d\zeta}{2(\zeta - z^*)} \right\}. \quad (5.91)$$

As easily seen for small γ the two integrals are nearly equal and become identical in the limit $\gamma \to 0$. The integration path is taken, as sketched in the Figure, on the lower half plane along the real axis from $-\infty$ to $v = -\epsilon$, $\epsilon = (2\hat{V})^{1/2}$, then encircling the pole $\omega = \omega_r + i\gamma$ and from $+\epsilon$ to $+\infty$, from there on the upper half plane along

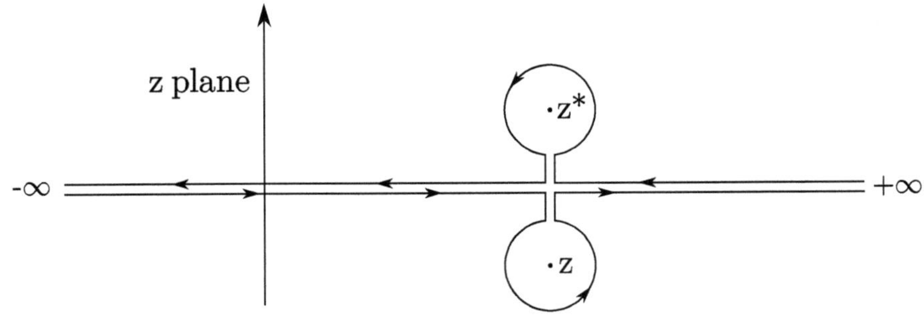

Fig. 5.5 Integration path in the complex z plane. Integration of (5.91) is taken around z (first term) from $-\infty$ to $+\infty$ and around z^* (second term) from $+\infty$ to $-\infty$

the real axis back to ϵ, encircling the pole $\omega^* = \omega_r - i\gamma$ and from $-\epsilon$ to $-\infty$. In the gap $[-\epsilon, +\epsilon]$ the particles are trapped ($\kappa > 1$!). Application of (5.91) yields for $|\gamma| \ll \omega_r$, with ϵ indicating the gap,

$$2\int_{-\infty-\epsilon}^{+\epsilon+\infty} \frac{\partial f_0(v)/\partial v}{v - \omega/k}\mathrm{d}v \simeq 2i\pi \frac{\partial f_0(v)}{\partial v}\bigg|_{\omega=\omega_r+i\gamma}. \tag{5.92}$$

Thus, we have reproduced Landau's residue $i\pi\partial f_0(v_\varphi)/\partial v$ on a semicircle and his dispersion function ,

$$\frac{\omega_p^2}{k^2}\left\{\mathrm{Pr}\int \frac{\partial f_0(v)/\partial v}{v - \omega/k}\mathrm{d}v + i\pi\frac{\partial f_0(v)}{\partial v}\right\}_{\omega_r+i\gamma} = 1. \tag{5.93}$$

Note, to motivate integration on a semicircle a symmetry argument is to be used (without symmetry or additional condition instead other values for the residue, e.g. $2i\pi$, are mathematically equally stringent). The contour from $-\infty$ to $+\infty$ with semicircle below the pole ω/k is unique.

The trapped particles in the gap must be treated separately unless their number is small, as assumed in [29], for instance if v_φ is much larger than the average electron velocity, see Fig. 5.6. Negligible number means that the gap is nearly zero and the passage from (5.92) to (5.93) is possible with the principal value (Pr $\int$). The passage from (5.93) to Landau's expression (5.83) is straightforward algebra.

A second transcription of the damping coefficient γ from (5.83) can be given with the help of the ratio $V/n_1 = \hat{V}/\hat{n}_1 = e^2/\varepsilon_0 k^2 = \mathrm{const}$,

$$\gamma = \frac{\dot{n}_1}{n_1} = \frac{\pi^2}{T_0} n_0 \frac{V}{m_e n_1} \frac{\partial f_0}{\partial v_\varphi}. \tag{5.94}$$

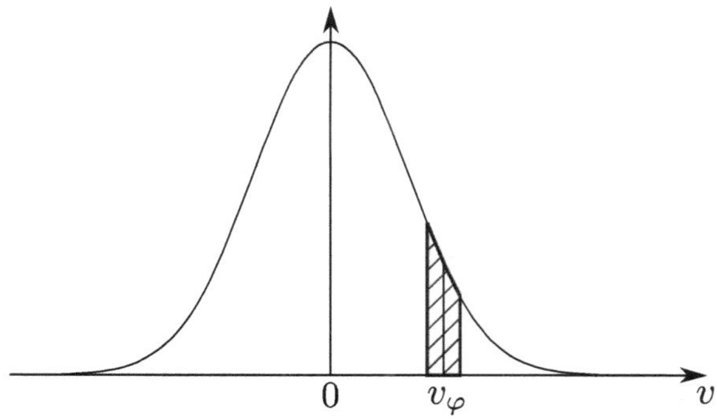

Fig. 5.6 Distribution function $f_0(v_i)$ in the lab frame S and $f(v_i') = f_0(v_\varphi + v_i')$ in the wave frame$S'(v_\varphi)$.Trapped particles ($\kappa = 2\hat{V}/v_i'^2 > 1$) in the narrow dashed band around v_φ of width 2ϵ are neglected in linear Landau damping

It is the direct proof of the wave potential $V(x')$ to be the generator of Landau damping. The first equality in this equation is the definition of damping in the linear regime. The second equality shows that it is independent of the magnitude of $\hat{V}$, it is *selfsimilar* in V/n_1; only the ratio $\hat{V}/\hat{n}_1$ figures. This may explain why Vlasov overlooked the significance of the tiny linear wave potential in the linear treatment.

The analysis presented here shows that Landau damping is diffusion in momentum space. The presence of the electrostatic potential introduces dispersion in the equilibrium distribution of the charged particles. The essence of the phenomenon is governed by the single equation (5.89). It expresses the individual delay each particle undergoes in the periodic potential structure. In the reference system comoving with the wave strict energy conservation holds; the single particle undergoes an exact periodic, quasi steady motion without suffering neither acceleration nor deceleration from one period T_0 to the next. In a linear treatment no room is left for back action of free (detrapped) particles onto the wave. In the lab frame the potential is time dependent and not conservative; $H =$ const is forbidden by the fundamental Hamiltonian relation $dH/dt = \partial H/\partial t = \partial V/\partial t$. The standard, misleading, interpretations (surfing, resonances) stem from a mixture of energy arguments applied to a domain where energy conservation does not hold.

A far higher degree of difficulty is connected with the treatment of the long term asymptotic decay of an electron plasma wave. It is no longer exponential, see [30]. On large time scales nonlinear fluid like terms become significant. Our treatment may give incentives for extensions of Landau damping into the nonlinear regime.

High Amplitude Electron Plasma Wave. Hydrodynamic Wave Breaking

A periodic electron plasma wave is considered again in the homogeneous plasma of undisturbed density n_0. In a reference system moving with local phase velocity v_φ holds for the relative flow velocity

$$w_e = u_e - v_\varphi, \quad \frac{\partial}{\partial x} n_e w_e = 0 \Rightarrow n_e w_e = -n_0 v_\varphi, \quad n_e = \frac{n_0}{1 - u_e/v_\varphi},$$

$$\frac{1}{2} m_e \frac{\partial w_e^2}{\partial x} = -m_e \frac{c_{\text{se}}^2}{\gamma_e - 1} \frac{\partial}{\partial x} \left(\frac{n_e}{n_0} \right)^{\gamma_e - 1} - eE, \quad \gamma_e \neq 1.$$

By eliminating E and w_e the following relations are obtained for $\xi = n_e/n_0$,

$$\gamma > 1 : \frac{\partial^2}{\partial x^2} \left\{ \xi^{\gamma_e - 1} + \frac{\mu}{\xi^2} \right\} = b(\xi - 1), \quad \mu = \frac{\gamma_e - 1}{2} \left(\frac{v_\varphi}{c_{\text{se}}^2} \right)^2,$$
$$b = (\gamma - 1) \frac{\omega_p^2}{c_{\text{se}}^2},$$

$$\gamma_e = 1 : \frac{\partial^2}{\partial x^2} \left\{ \ln \xi + \frac{\mu}{\xi^2} \right\} = b(\xi - 1), \quad \mu = \frac{1}{2} \left(\frac{v_\varphi}{c_{\text{se}}^2} \right)^2, \quad b = \frac{\omega_p^2}{c_{\text{se}}^2}.$$

If these equations are integrated once, from a position with $\xi = \min$ to an appropriate x-value,

$$\left(\frac{c_{\text{se}}^2}{v_\varphi^2} \xi^{\gamma_e - 2} - \xi^{-3} \right) \xi' = \frac{b}{2\mu} \int_{x:\xi=\min}^{x} (\xi - 1) \mathrm{d}x, \tag{5.95}$$

it can be seen that the solution $\xi(x)$ is symmetric with respect to its minima and maxima for

$$\frac{n_e}{n_0} \leq \xi_0 = \left(\frac{v_\varphi}{c_{\text{se}}^2} \right)^{2/(\gamma+1)}. \tag{5.96}$$

However, there is no permanent wave for $\xi > \xi_0$. $\xi = \xi_0$ is a branch point with a periodic wave of zero radius of curvature in the maxima and an aperiodic solution: The wave *breaks* beyond $\xi = \xi_0$. Inequality (5.96) is the fluid dynamic wave breaking criterion by Coffey [12]. (Note that in a periodic wave with $\xi > \xi_0$ intervals could be found in which the LHS and RHS of (5.95) would have different signs!). Below the breaking limit the wave crests exhibit finite curvature, at the limit their curvature is zero, the crests become acute. In the cold plasma, with $c_{\text{se}}^2 = 0$ this happens at infinite density n_e. The physical reason for the existence of $\xi_0 < \infty$ is that for $\xi > \xi_0$ the compression work increases in such a way that the energy balance can no longer be satisfied. In the derivation of the criterion the integration is done over half a wavelength; hence, it is insensitive with respect to mild inhomogeneities.

It is worthwhile to study wave breaking under a periodic driver in more detail in mildly inhomogeneous plasma in the capacitor approximation of resonance absorption (see Chap. 6). A plasma of finite temperature at rest is assumed. In the capacitor approximation, i.e., $\mathbf{E} = (E, 0, 0)$, $\nabla \mathbf{E} = \partial E/\partial x$, $\nabla \times \mathbf{B} = (ik_y B, 0, 0)$, Maxwell's equations and the equation of motion are

$$\frac{\partial E}{\partial t} = ik_y c^2 B + \frac{e}{\varepsilon_0} n_e u_e, \quad \frac{\partial E}{\partial x} = \frac{e}{\varepsilon_0}(n_0 - n_e),$$

$$\frac{\mathrm{d}u_e}{\mathrm{d}t} = -\frac{c_{\mathrm{se}}^2}{n_e}\frac{\partial n_e}{\partial x} - \frac{e}{m_e}E. \tag{5.97}$$

Elimination of E leads to

$$\frac{\mathrm{d}^2 u_e}{dt^2} + \frac{\mathrm{d}}{\mathrm{d}t}\left(c_{\mathrm{se}}^2 \frac{1}{n_e}\frac{\partial n_e}{\partial x}\right) + \omega_p^2 u_e = -i\frac{ec^2}{m_e}k_y B. \tag{5.98}$$

Advantageously, here ω_p contains the ion background density n_0/Z

$$\omega_p^2 = \frac{e^2 n_0(x,t)}{\varepsilon_0 m_e},$$

rather than the instantaneous electron density n_e. Replacing (x, t) by the Lagrangian coordinates (a, t), in which a is the initial position of the electron and ion fluid elements, leads to

$$x_i = a, \quad x_e(a,t) = a + \delta_x(a,t), \quad \delta_x(a,t) = \int_0^t u_e(a,t)dt, \quad n_e = \frac{n_0}{1 + \partial\delta_x/\partial a}. \tag{5.99}$$

In the new coordinates (5.98) becomes

$$\frac{\partial^2}{\partial t^2}u_e + \frac{c_{\mathrm{se}}}{\gamma n_0^\gamma}\frac{\partial^2}{\partial t \partial a}\left(\frac{n_0}{1 + \partial\delta_x/\partial a}\right)^{\gamma_e} + \omega_p^2(a,t)u_e = \frac{1}{2}\omega^2 v_d \mathrm{e}^{-i\omega t} + \mathrm{cc} \tag{5.100}$$

The oscillation center approximation was assumed to obtain the term $\omega_p^2(a,t)u_e$. This is justified for inhomogeneity lengths L and electrostatic wavelength λ_e if $L > 2\lambda_e$. The electron fluid is assumed to obey an adiabatic law, $p_e = \mathrm{const} \times n_e^\gamma$, the driver v_d stands for $v_d = -iec^2 k_y B(0)/m_e\omega^2$. For $L > 2\lambda_e$, one may further transform (5.100),

$$\frac{\partial}{\partial t}\left\{\frac{\partial^2}{\partial t^2}\delta_x + \frac{c_{\mathrm{se}}^2}{\gamma_e}\frac{\partial}{\partial a}\left(1 + \frac{\partial}{\partial a}\delta_x\right)^{-\gamma_e} + \omega_p^2\delta_x\right\} = \frac{1}{2}\omega^2 v_d \mathrm{e}^{-i\omega t} + \mathrm{cc},$$

and integrate it once to obtain

$$\frac{\partial^2}{\partial t^2}\delta_x + \frac{c_{\mathrm{se}}^2}{\gamma}\frac{\partial}{\partial a}\left(1 + \frac{\partial}{\partial a}\delta_x\right)^{-\gamma} + \omega_p^2\delta_x = \frac{i}{2}\omega v_d \mathrm{e}^{-i\omega t} + \mathrm{cc} \tag{5.101}$$

For small amplitudes $(1 + \partial\delta_x/\partial a)^{-\gamma_e} \simeq 1 - \gamma_e\partial\delta_x/\partial a$ and, with the help of $\delta_e = \delta_x$ from (2.11), (5.101) reduces to the linear capacitor model (6.51), since then $\partial/\partial_a \simeq \partial/\partial_x$. For a cold plasma ($c_{\mathrm{se}}^2 = 0$) it leads to the Koch–Albritton model [13] of uncoupled harmonic oscillators from which, through (5.99), the electron density is recovered. Although there is strong coupling between the single fluid elements, u_e and δ_x are sinusoidal in the cold plasma, but n_e is not, in contrast to a linear wave which is sinusoidal in all variables. By setting

$$\delta_x = \delta_1 + \frac{1}{2}\sum_{\sigma\geq 2}\{\delta_\sigma e^{-i\sigma\omega t} + \delta_\sigma^* e^{i\sigma\omega t}\}$$

$$g = 1 + \frac{1}{4}\binom{\gamma+2}{2}k_1^2|\delta_1|^2 + \frac{1}{8}\binom{\gamma+4}{4}k_1^4|\delta_1|^4 + \ldots.$$

in the warm plasma the amplitude-dependent dispersion relation for the fundamental wave δ_1,

$$\omega^2 = \omega_p^2\left\{1 + g\left(\frac{c_{se}k_1}{\omega_p}\right)^2\right\}, \tag{5.102}$$

and the amplitude ratios for the higher harmonics are obtained [14],

$$\frac{\hat{\delta}_2}{\hat{\delta}_1} = \frac{\gamma+1}{2}\frac{\eta^{3/2}g^{-1/2}\omega^2}{3\omega_p^2 g + 4(g-1)\eta^2\omega^2}\frac{\hat{u}_1}{c_{se}}, \quad u_1 = \dot{\delta}_1,$$

$$\frac{\hat{\delta}_3}{\hat{\delta}_1} = \frac{\gamma+1}{4}\frac{\eta^3\omega^2}{g^2}\left(\frac{\hat{u}_1}{c_{se}}\right)^2\frac{\frac{3(\gamma+1)\omega^2}{3\omega_p^2 g+4(g-1)\eta^2\omega^2} + \frac{\gamma+2}{2}\eta}{8\omega_p^2 + 9(g-1)\eta^2\omega^2}.$$

In the resonance region $\eta^2 \ll 1,\ (c_{se}k_1/\omega_p)^2 \ll 1,\ g = 1,\ \omega_p^2 = \omega^2$ holds and the relations simplify to

$$\frac{\hat{\delta}_2}{\hat{\delta}_1} \simeq \frac{\gamma+1}{6}\eta^{3/2}\frac{u_1}{c_{se}}, \quad \frac{\hat{\delta}_3}{\hat{\delta}_1} \simeq \frac{1}{2}\left(\frac{\gamma+1}{4}\right)^2\eta^3\left(\frac{u_1}{c_{se}}\right)^2 \tag{5.103}$$

showing the interaction of modes: As the excited wave moves away from the resonance region towards lower densities the refractive index η increases from zero to one, thus leading to an increase of higher order modes. The modified dispersion (5.102) is a consequence of *self-interaction* of the fundamental mode δ_1. Due to the excitation of higher harmonics at finite electron temperature u_e and δ_x no longer remain sinusoidal. This behavior is clearly seen in the numerical solution of (5.100) in Fig. 5.7. The maxima of u_e are more peaked than its minima. As a result, δ_x has a tendency to assume a sawtooth shape. The spike-like character is especially pronounced in the electron density n_e. Such structures of *bubbles and spikes* are very characteristic of large amplitude electrostatic waves and are encountered in the nonlinear evolution of the Rayleigh–Taylor instability, see [15].

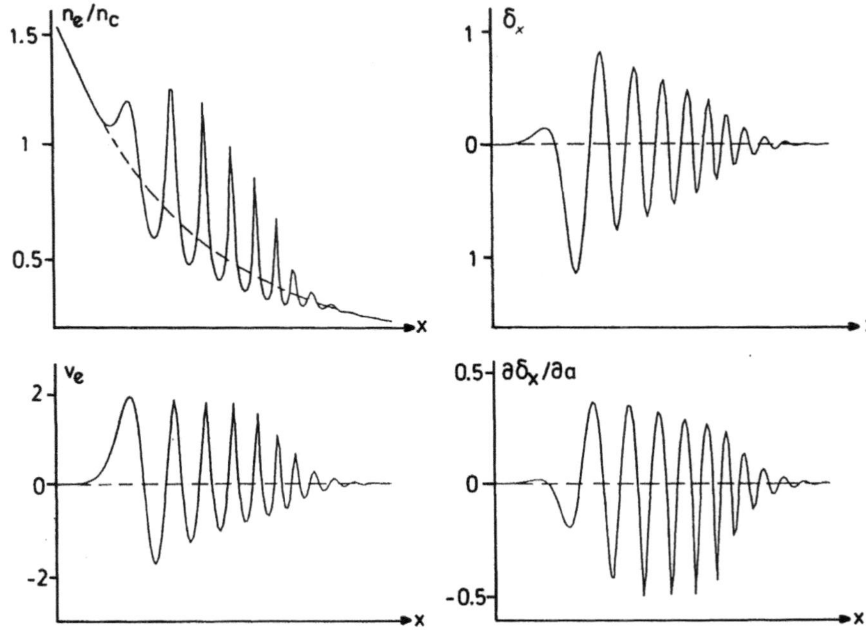

Fig. 5.7 Nonlinear electron plasma wave. Normalized electron density n_e, electron quiver velocity u_e, excursion $\delta_x = \int u_e \mathrm{d}t$ and $y = \partial\delta_x/\partial a$ as functions of x. Nearly the same picture results if these quantities are plotted as functions of the Lagrange coordinate a

The main contribution to peaking of n_e comes from mass conservation of (5.99) and is best understood in the cold plasma case; there

$$\delta_x = \hat{\delta}\cos\omega_p t, \quad n_e = \frac{n_0}{1 + (\partial\hat{\delta}/\partial a)\cos\omega_p t} \tag{5.104}$$

holds and n_e oscillates between $n_0/2$ and ∞. In the warm plasma n_e remains above $n_0/2$ in all situations [16].

Resonant fluiddynamic wave breaking. Out of resonance there is a limiting amplitude beyond which no periodic wave exists; it is given by the Coffey criterion (5.96). Additional light is shed on breaking by studying the phenomenon in a cold streaming plasma resonantly driven at fixed frequency ω. The problem is reducible to the solution of a resonantly driven harmonic oscillator with time varying eigenfrequency. Energy transport out of the resonance region is accomplished by convection at flow velocity u_0. Introducing this in (5.99) by setting $x_i = a + \int u_0(a,t)\mathrm{d}t$ and $x_e = x_i + \int u_e \mathrm{d}t$ and $c_{se} = 0$ one arrives at the harmonic oscillator equation for the individual fluid element of position a at $t = 0$,

$$\frac{\partial^2}{\partial t^2}u_e + \omega_p^2(a,t)u_e = \frac{1}{2}\omega^2 v_d \mathrm{e}^{-i\omega t} + \mathrm{cc} \tag{5.105}$$

with resonance frequency ω_p varying in time: A volume element starting in the overdense region is off resonance owing to $\omega_p \gg \omega$; it then approaches resonance $\omega_p = \omega$ and, rarefying further, in the underdense region ($\omega_p < \omega$) it decouples again from the driver. The solution of (5.105) is found from the general formula for the inhomogeneous solution of $y'' + a(x)y = f(x)$,

$$y(x) = h(x)\int^x \frac{g(\xi)f(\xi)}{gh' - g'h}\mathrm{d}\xi - g(x)\int^x \frac{h(\xi)f(\xi)}{gh' - g'h}\mathrm{d}\xi \tag{5.106}$$

with g, h two linearly independent solutions of the homogeneous equation. At $t = 0$ the fluid element starts at $u_e = 0$ in the highly overdense region and evolves in $u = u_e - u_0$ according to

$$u = i\frac{\omega^2 v_d}{2\omega_p^{1/2}}\mathrm{e}^{-i\varphi}\int_0^t \frac{\mathrm{e}^{i(\omega t' - \varphi')}}{\omega_p^{1/2}}\mathrm{d}t'; \quad \varphi(a, t) = \int_0^t \omega_p(a, t')\mathrm{d}t', \tag{5.107}$$

if the flow velocity obeys the inequality $u_0 \leq L\omega/\pi$. With L the density scale length at resonance a self-consistent density profile is $n_0 = n_c/(1 + a/2L)^2$. Inserting n_0 in (5.107) yields in the resonance region

$$u(a, t) = i\frac{L^{1/2}\omega^{1/2}v_d\mathrm{e}^{i\omega a/u_0 + \sigma^2}}{u_0^{1/2}}\int_{-\infty}^{\sigma} \mathrm{e}^{-i\sigma'^2}\mathrm{d}\sigma', \quad \sigma = \frac{1}{2}t\left(\frac{wu_0}{L}\right)^{1/2}. \tag{5.108}$$

The modulus of $\int_{-\infty}^{\sigma} \mathrm{e}^{-i\sigma'^2} d\sigma'$ is the familiar Cornu spiral.. It shows that secular growth of $|u(a, t)|$ occurs around resonance (Fig. 5.8). With σ running from $-\pi/2$ to $+\pi/2$ the resonance width Δ is deduced to be

$$\Delta = 2\pi\left(\frac{u_0 L}{\omega}\right)^{1/2}. \tag{5.109}$$

Wave breaking occurs when in (5.99) $\partial\delta_x/\partial a = -1$. For this to occur at the edge of resonance ($\sigma = \pi/2$).

$$\frac{v_d}{u_0} = \left|\frac{1}{2} + i\frac{\pi}{2}\mathrm{e}^{i(\pi/2)^2}\int_{-\infty}^{\pi/2} \mathrm{e}^{-i\sigma^2}\mathrm{d}\sigma\right|^{-1} = 0.36 \tag{5.110}$$

must be fulfilled [17]. At resonance cold wave breaking is due to the overlapping of two originally distinct volume elements as a result of their different phase shifts in the neighbourhood of the resonance point. Due to such shifts a Langmuir wave in a cold streaming plasma may break as soon as $u_{\mathrm{os}} > 0.75\ v_\varphi$ is satisfied. This is slightly lower than the first breaking criterion $u_{\mathrm{os}} > v_\varphi$ which was obtained by J. M. Dawson on the basis of a different model [18]. For breaking limits under various conditions

Fig. 5.8 The value of $|w| = \left|\int_{-\infty}^{\sigma} \exp -i\sigma'^2 \mathrm{d}\sigma'\right|$ is the length of the vector extending from the center of the lower Cornu spiral to a point of the double spiral determined by the parameter σ, which is proportional to the length of the arc, i.e., time. Maximum growth occurs around the resonance $\sigma = 0$. The maximum of $|w| = 2.074$ is reached at $\sigma = 1.53$ (compare $\pi/2 = 1.57$)

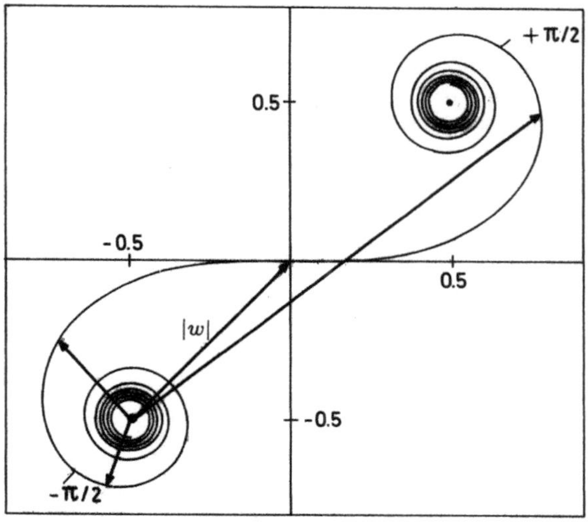

and limits of validity of (5.105) and its solution (5.108) see [17]. A detailed breaking scenario of a resonantly driven electron plasma wave is presented in Fig. 5.9.

Electron wave breaking is a complex phenomenon, a whole variety of scenarios inducing it are possible. So far only hydrodynamic breaking has been faced for two situations. One is breaking of an unboundedly growing wave out of resonance. The growth of the amplitude may happen under WKB like conditions in a inhomogeneous plasma background (see Sect. 5.4.2) or by geometrical focusing. The other scenario is breaking in the critical plasma region by resonant mode conversion. Here it has been described in terms of a harmonic oscillator with time varying eigenfrequency under the limitation of zero electron temperature. In the warm plasma dispersion provides for additional energy transport out of the critical zone and, consequently, for an increase of the breaking limit (5.110). A quantitative analysis has to be achieved from the nonlinear partial equation (5.101), defined as "untractable" [19]. A hydrodynamic approach to wave breaking in some cases may be appropriate, in other cases it may be not owing to the phenomenon of electron trapping in the high amplitude wave. This is the domain of the kinetic approach, to be treated in Sect. 6.2.3.

The Ion Acoustic Wave

One could think of treating the ion acoustic wave formally by scaling down the frequencies ω and ω_p by the ratio $(Zm_e/m_i)^{1/2}$. However, in such a slow motion the electrons have time to follow adiabatically any ion density modulation and to neutralize it. Under slow motion the momentum equation for the electron fluid reduces to (3.41), in the context here with $\pi = 0$, $\mathbf{B}_s = 0$. Neglecting viscosity the ions obey (3.43). For small density disturbances $n_1 = n_i - n_0$, $n_{e1} = n_e - Zn_0$ it reads as follows,

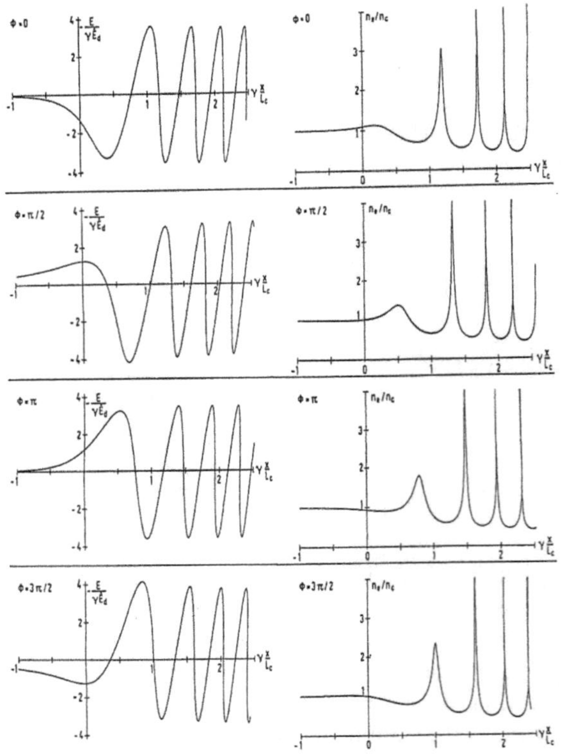

Fig. 5.9 Resonant excitation and breaking of a could electron plasma wave in the critical region of plasma streaming at velocity u_0. Left: Electrostatic wave around resonance point as a function of the Eulerian coordinate x at four different times. Right: Corresponding electron density n_e normalized to the critical density n_c. $|v_d/u_0| = 0.16$ has been chosen. Breaking occurs in the third density spike

$$n_0 \frac{\partial \mathbf{u}}{\partial t} = -\frac{1}{m_i} \nabla p_i - \frac{1}{m_i} \nabla p_e = -c_{\mathrm{si}}^2 \nabla n_1 - \frac{m_e}{m_i} c_{\mathrm{se}}^2 \nabla n_{e1}; \quad c_{\mathrm{si}}^2 = \gamma_i \frac{k_B T_i}{m_i}. \quad (5.111)$$

In the absence of heating the ions satisfy the adiabatic law $p_i n_i^{-\gamma_i} = \mathrm{const}$ from which c_{si} results as specified. From (3.44) and the Poisson equation,

$$\mathbf{E} = -\frac{m_e}{n_{e0} e} c_{\mathrm{se}}^2 \nabla n_{e1}, \quad \nabla \mathbf{E} = \frac{e}{\varepsilon_0} (Z n_1 - n_{e1}).$$

n_{e1} is determined as a function of n_1 through elimination of $\mathbf{E}$,

$$-c_{\mathrm{se}}^2 \nabla^2 n_{e1} = \omega_p^2 (Z n_1 - n_{e1}) \Rightarrow (1 - \gamma_e \lambda_D^2 \nabla^2) n_{e1} = Z n_1$$

with λ_D from (4.70). Elimination of $\mathbf{u}$ from the linearized momentum equation above with the help of $\partial_t n_1 + n_0 \nabla \mathbf{u} = 0$ yields

$$\frac{\partial^2 n_1}{\partial t^2} = c_{\mathrm{si}}^2 \nabla^2 n_1 + \frac{m_e}{m_i} c_{\mathrm{se}}^2 \nabla^2 n_{e1} = \left[c_{\mathrm{si}}^2 + \frac{Z \frac{m_e}{m_i} c_{\mathrm{se}}^2}{1 - \frac{1}{n_{e1}} \gamma_e \lambda_D^2 \nabla^2 n_{e1}} \right] \nabla^2 n_1. \quad (5.112)$$

Fourier analysis shows that the plane ion acoustic waves

$$n_1 = \hat{n}e^{i(\mathbf{k}_a\mathbf{x}-\omega_a t)}.$$

form a complete set of solutions to (5.112) if ω_a satisfies the dispersion relation

$$\omega_a = c_s k_a, k_a = |\mathbf{k}_a,| \quad c_s^2 = \frac{k_B}{m_i}\left(\gamma_i T_i + \gamma_e \frac{ZT_e}{1+\gamma_e(k_a\lambda_D)^2}\right). \tag{5.113}$$

Quasineutrality requires that n_{e1} is modulated with the same k_a. The quantity c_s is the *ion sound velocity*. It remains finite when T_i approaches zero owing to the electron pressure which in the ion sound wave is transmitted to the ions by the electrostatic field of type (3.44) rather than by collisions. The condition however for this to happen is $k_a\lambda_D \ll 1$. The ion plasma frequency ω_{pi} does not appear in (5.113) since the slow ionic charge disturbance is nearly completely screened by the fast electrons. Again, if y stands for any of the perturbations of particle densities n_e, n_i, temperatures T_e, T_i, velocities $\mathbf{u}_e$, $\mathbf{u}_i$, electric field $\mathbf{E}$ or potential Φ, each of them obeys the linear acoustic wave equation

$$\frac{\partial^2}{\partial t^2}y - c_s^2\nabla^2 y = 0.$$

They are related by

$$T_{\alpha 1} = (\gamma_\alpha - 1)T_{\alpha 0}\frac{n_{\alpha 1}}{n_{\alpha 0}}, \quad v_\alpha = c_s\frac{n_{\alpha 1}}{n_{\alpha 0}}, \quad n_{e1} = \frac{Zn_1}{1+\gamma_e(k_a\lambda_D)^2},$$

$$E = -i\,\frac{e}{\varepsilon_0}\gamma_e(k_a\lambda_D)^2 n_{e1}, \quad \Phi = i\,\frac{E}{k_a}. \tag{5.114}$$

In the long-wavelength limit c_s is almost dispersion free owing to nearly perfect screening; $\mathbf{u}_e$ is very close to $\mathbf{u}_i = \mathbf{u}$ and frictional damping plays no role. Generally the electrons have time to transmit their compression energy by thermal conduction to their neighborhood during an oscillation whereas the ions do not. Hence, $\gamma_e = 1$, i.e., the electrons behave isothermally and $\gamma_i = 3$ holds. If however, $\nu_{ii} > \omega/2\pi$ is fulfilled, γ_i has to be set equal to $5/3$.

The ion plasma frequency ω_{pi}. In laser generated plasmas $T_e \gg T_i$ may be easily reached. If λ_D is much larger than the wavelength λ_i of a periodic perturbation of n_i the electrons form a uniform homogeneous negative background. This is what the vanishing ratio of n_{e1} to n_1 in (5.114) tells. The ion charge imbalance $n_{i1} = n_1$ is no longer screened at all, its dynamics is driven by the ion pressure gradient and the accompanying collective electric field of n_{i1}. The ions oscillate according to a dispersion relation of Bohm–Gross structure (5.70) with all electron quantities (index e) replaced by those for the ions,

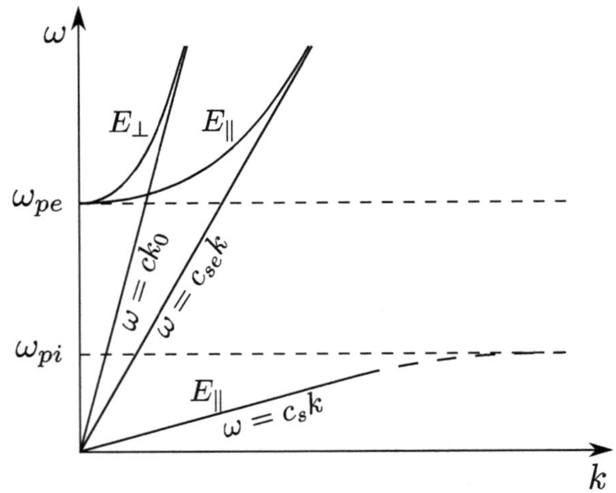

Fig. 5.10 The dispersion relations (5.47), (5.70), $\nu = 0$, and (5.113) of the three free modes in the unmagnetized ideal plasma are sketched as functions of the wave number k. The velocities of the asymptotes are: c light velocity in vacuum, c_{se} electron sound velocity, c_s sound velocity of the ion acoustic wave

$$\omega_a^2 = \omega_{pi}^2 + c_{\mathrm{si}}^2 \mathbf{k}_a^2, \qquad \omega_{pi}^2 = \frac{n_i Z^2 e^2}{\varepsilon_0 m_i}. \tag{5.115}$$

The dispersion relations of the three unmagnetized wave types, electromagnetic, electron plasma, ion acoustic mode, are sketched in Fig. 5.10 under standard conditions as functions of the modulus k of the wave vector $\mathbf{k}$.

The Shortest Wavelength of a Normal Mode

It is well known that ultrasound generation in neutral gas becomes inefficient with increasing frequency. The driving force is the pressure gradient which in the wave must be transmitted to the neighbouring fluid element over the distance of the mean free path d_f less than or approximately equal to $\lambda/4$. Hence, the wavelength and frequency limits are determined by the condition $|\mathbf{k}|d_f \leq 1$. In the electron plasma or ion sound wave the Debye length takes on the role of the limiting quantity for wavelength and frequency as soon as $d_f > \lambda_D$ is fulfilled. In the ideal plasma this is always the case. In the transverse wave of a given wavelength λ the requirement is that there must be enough particles to mark unambiguously a current modulation over λ. This implies $|\mathbf{k}|n_e^{-1/3} < 1$. Thus, in the plasma holds

$$\begin{aligned} &\text{Longitudinal plasma wave}: |\mathbf{k}|\lambda_D \leqslant 1,\\ &\text{Transverse plasma wave}: |\mathbf{k}|n_e^{-1/3} \leqslant 1. \end{aligned} \tag{5.116}$$

The problem of the shortest existing eigenmode can be formulated in an alternative way that emphasizes its connection with the Doppler effect and the scattering of light. To this aim let us express the particle density in terms of delta functions and subject them to a Fourier transformation,

$$n(\mathbf{x},t) = \sum_{j \in N} \delta(\mathbf{x} - \mathbf{x}_j(t)) \quad \Rightarrow \quad n(\mathbf{k},t) = \frac{1}{(2\pi)^{3/2}} \int n(\mathbf{x},t) \mathrm{e}^{-i\mathbf{kx}} \mathrm{d}\mathbf{x}. \tag{5.117}$$

The sum extends to all N particles of the single plasma components, the Fourier transform is taken over the whole 3D space. By the subsequent Fourier transform in time,

$$n(\mathbf{k},\omega) = \frac{1}{(2\pi)^{1/2}} \int n(\mathbf{k},t) \mathrm{e}^{i\omega t} \mathrm{d}t = \frac{1}{(2\pi)^2} \int n(\mathbf{x},t) \mathrm{e}^{-i(\mathbf{kx}-\omega t)} \mathrm{d}\mathbf{k}\mathrm{d}t \tag{5.118}$$

the decomposition into orthogonal eigenmodes $n(\mathbf{k},\omega)$ is completed for all wave vectors $\mathbf{k}$ and all frequencies ω. Substitution of the inverse transformations in the corresponding wave equations ends up in an infinite set of dispersion relations of these modes. As $\mathbf{k} \rightarrow \infty$ there is a $\mathbf{k}_{\text{lim}}$ such that $|\mathbf{k}_{\text{lim}}| n(\mathbf{k}_{\text{lim}})^{-1/3} < 1$. This means that the fluctuation in $n(\mathbf{k}_{\text{lim}})$ exceeds this value. The subsequent Fourier decomposition in frequency, although it exists mathematically, it does not lead to a meaningful dispersion relation $\omega(\mathbf{k})$ because the particles in the mode are uncorrelated. Applied to light scattering it tells that the signal from the individual scattering centers adds in intensities instead of $\mathbf{E}$ fields. Applied to *Thomson scattering* it leads to the distinction of *incoherent* from *coherent* (collective) light scattering.

5.4.2 The Magnetized Fully Ionized Plasma

The Ideal Magnetohydrodynamics

The dynamics of the ionic plasma component is described by the one fluid model (see Chap. 3, Sect. Two fluid model). It is quasi static compared with the time scale of the electronic plasma component. The governing equations of interest here are special cases of (3.50) and (3.52) in the following sense:

- Absence of the ponderomotive potential Φ_p
- Ideality: electron-ion collision frequency is almost negligible $\Longleftrightarrow$ electric conductivity is very high (formally: $\sigma \rightarrow \infty$, $\omega \cong 0$).

From energy considerations easily follows that the current density $\mathbf{j}_0$ remains finite in the stable plasma equilibrium. Alternatively, at low frequencies the electrons follow the ion motion quasistatically, and this in turn is governed by the ion inertia at any finite frequency, Thus, (3.50) remains unchanged,

$$\frac{\partial \mathbf{j}_0}{\partial t} \simeq 0, \quad \mathbf{j}_0 = \sigma(\mathbf{E}_0 + \mathbf{u}_0 \times \mathbf{B}_0), \quad \varepsilon_0 c^2 \nabla \times \mathbf{B}_0 = \mathbf{j}_0, \quad \nabla \times \mathbf{E}_0 = -\partial_t \mathbf{B}_0 \tag{5.119}$$

and (3.52) becomes

$$\mathbf{E}_0 = -\mathbf{u}_0 \times \mathbf{B}_0 - \frac{1}{en_{e0}}\nabla p_{e0}, \qquad \frac{\partial \mathbf{B}_0}{\partial t} = \nabla \times (\mathbf{u}_0 \times \mathbf{B}_0) + \nabla \frac{1}{en_{e0}} \times \nabla p_{e0}. \tag{5.120}$$

The electric field stems in part from the electron pressure gradient and in part it is compensated by the Lorentz force $\mathbf{u}_0 \times \mathbf{B}_0$ in such a way as to keep $\mathbf{j}_0$ finite.

Frozen magnetic field lines. Under the assumption of uniform electron pressure, $p_e = \text{const}$, (5.120) reads

$$\mathbf{E}_0 = -\mathbf{u}_0 \times \mathbf{B}_0; \qquad \frac{\partial \mathbf{B}_0}{\partial t} = \nabla \times (\mathbf{u}_0 \times \mathbf{B}_0) \tag{5.121}$$

and agrees with Faraday's law, $\nabla \times \mathbf{E}_0 = -\partial_t \mathbf{B}_0$. Consider the change of magnetic flux through an arbitrary oriented surface $\boldsymbol{\Sigma}(t = \text{const})$ spanned over an arbitrary deformable closed line $L(t)$ and use Stokes' theorem on (5.121)

$$\int_{\boldsymbol{\Sigma}} \frac{\partial}{\partial t}\mathbf{B}_0 d\boldsymbol{\Sigma} = \oint_{L(t)} (\mathbf{u}_0 \times \mathbf{B}_0)\mathrm{d}\mathbf{s} = -\oint_{L(t)} \mathbf{B}_0(\mathbf{u}_0 \times \mathrm{d}\mathbf{s}) \iff \frac{\mathrm{d}}{\mathrm{d}t}\int_{\boldsymbol{\Sigma}(t)} \mathbf{B}_0 d\boldsymbol{\Sigma} = 0, \tag{5.122}$$

see Fig. 2.1. The last equality is Alfvén's theorem on the *frozen magnetic flux* of ideal magnetofluid dynamics:

The number of magnetic field lines through any oriented surface $\boldsymbol{\Sigma}(t)$ comoving with the plasma is conserved.

Owing to the approximate equality for $\mathbf{E}_0$ in (5.119) (displacement current $\varepsilon_0\dot{\mathbf{E}}$ is suppressed) it is advisable to check the degree of ideality in the individual case. It is essential to bear in mind that $\mathbf{E}_0 = -\mathbf{u}_0 \times \mathbf{B}_0$ from (5.122) refers to Ohm's law for the total current density $\mathbf{j}_0$ and tells that the electric field is nearly, for $\sigma = \infty$ exactly, compensated by the Lorentz term $\mathbf{u}_0 \times \mathbf{B}_0$, and thus it is legitimate to substitute it in Faraday's law from (5.119) by $\mathbf{u}_0 \times \mathbf{B}_0$. In the Galileian relativity of the one fluid model what figures is the Lorentz force $\rho\mathbf{f} = \mathbf{j}_0 \times \mathbf{B}_0$.

Equilibrium in ideal magnetofluid dynamics. The governing equations are

$$\nabla \times \mathbf{B}_0 = \mu_0 \mathbf{j}_0, \quad \nabla \mathbf{B}_0 = 0, \quad \nabla \times \mathbf{E}_0 = -\frac{\partial \mathbf{B}_0}{\partial t}, \quad \mathbf{E}_0 = -\mathbf{u}_0 \times \mathbf{B}_0 \tag{5.123}$$

the last relation to be used properly. We evaluate the magnetic pressure p_{m}. For the equilibrium in the plasma must hold

$$-\nabla p + \mathbf{j}_0 \times \mathbf{B}_0 = -\nabla p + \frac{1}{\mu_0}(\nabla \times \mathbf{B}_0) \times \mathbf{B}_0 = -\nabla\left(p + \frac{\mathbf{B}_0^2}{2\mu_0}\right) + \frac{1}{\mu_0}(\mathbf{B}_0\nabla)\mathbf{B}_0 = 0. \tag{5.124}$$

After adding the zero term $\mathbf{B}_0\mathrm{div}\mathbf{B}_0$ this appears as the divergence of the stress tensor $-\mathbf{T} = (B_i B_j - \delta_{ij}\mathbf{B}^2/2)/\mu_0$ from (5.19) and can be defined as the magnetic pressure tensor p_m. In a Cartesian coordinate system with the z axis oriented along $\mathbf{B}_0$ the total pressure on the plasma becomes diagonal with the non-zero components equated to zero (*principal axes* of stress tensor),

$$\partial_x\big(p + \frac{\mathbf{B}_0^2}{2\mu_0}\big) = 0, \quad \partial_y\big(p + \frac{\mathbf{B}_0^2}{2\mu_0}\big) = 0, \quad \partial_z\big(p - \frac{\mathbf{B}_0^2}{2\mu_0}\big) = 0. \tag{5.125}$$

The negative magnetic stress τ_m along the field lines is the sum of $-\mathbf{B}_0^2/\mu_0$ along the field lines and the positive isotropic magnetic pressure $\mathbf{B}_0^2/2\mu_0$.

Consider the surfaces $p(\mathbf{x}) = \mathrm{const}$ and $\mathbf{B}_0^2 = \mathrm{const}$. From the equilibrium condition (5.124) follows

$$\mathbf{B}_0\nabla p = 0, \quad \mathbf{j}\nabla p = 0, \quad \mathbf{j}\nabla\mathbf{B}_0^2 = 0. \tag{5.126}$$

The magnetic field lines lie on surfaces which are complanar with the surfaces of constant pressure, the electric current flows along these surfaces.

Eigenmodes of the Ideal Magnetized Fluid

The monofluid plasma is subject to a magnetic stress in addition to the thermal pressure of the unmagnetized plasma. Therefore eigenmodes similar to the ion acoustic wave are to be expected. Owing to the tensorial character of the magnetic pressure transverse excitation of the plasma should be possible also. In the homogeneous plasma in equilibrium we set

$$n, \rho, p, \mathbf{B} = n_0, \rho_0, p_0, \mathbf{B}_0 = \mathrm{const}, \quad \mathbf{u} = \mathbf{u}_0 = 0; \quad p\rho^{-\gamma} = \mathrm{const}.$$

The compession Alfvén wave. Consider a density compression $\rho_0 + \rho_1(x)$, $\rho_1(x) \ll \rho_0$, with the concomitant velocity $u(x)$ along the x axis and the frozen magnetic field $\mathbf{B} = \mathbf{B}_0 + \mathbf{B}_1$ in z direction. Owing to $|\mathbf{B}| \sim \rho$ the magnetic pressure $p_\mathrm{m} = \mathbf{B}^2/2\mu_0 \sim \rho^2$ follows the adiabate $p_\mathrm{m}\rho^{-\gamma_\mathrm{m}} = \mathrm{const}$, with $\gamma_\mathrm{m} = 2$. The gradient of the total pressure $P = p + p_\mathrm{m}$ is

$$\frac{\partial P}{\partial x} = \big(\gamma\frac{p}{\rho} + \gamma_\mathrm{m}\frac{p}{\rho}\big)\frac{\partial \rho_1}{\partial x}.$$

Explicit calculation to first order yields

$$\gamma\frac{p}{\rho} + \gamma_\mathrm{m}\frac{p}{\rho} = \gamma\frac{p_0}{\rho_0}\big[1 + (\gamma - 1)\frac{\rho_1}{\rho_0}\big] + \gamma_\mathrm{m}\frac{p_{\mathrm{m}0}}{\rho_0}\big[1 + (\gamma_\mathrm{m} - 1)\frac{\rho_1}{\rho_0}\big] \simeq \gamma\frac{p_0}{\rho_0} + \frac{\mathbf{B}_0^2}{\mu_0\rho_0}$$

and after introducing the definitions of c_s and c_A the total linearized pressure gradient results as

$$\frac{\partial P}{\partial x} = (c_s^2 + c_A^2)\frac{\partial \rho_1}{\partial x}; \qquad \gamma\frac{p_0}{\rho_0} = c_s^2, \quad \frac{\mathbf{B}_0^2}{\mu_0\rho_0} = c_A^2. \tag{5.127}$$

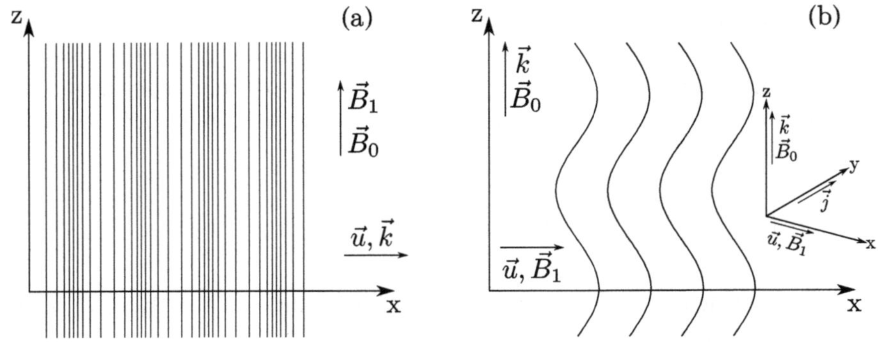

Fig. 5.11 **a** Compression Alfvén wave $\mathbf{B}_1$ propagating across the constant magnetic field $\mathbf{B}_0$, wave is longitudinal, $\mathbf{B}_1 \| \mathbf{k}$, magnetic pressure p_m is proportional to ρ^2. **b** Shear Alfvén wave $\mathbf{B}_1$ propagating along $\mathbf{B}_0$, wave is transverse, $\mathbf{B}_1 \perp \mathbf{k}$. Tension along $\mathbf{B} = \mathbf{B}_0 + \mathbf{B}_1$ is proportional to length increase of magnetic field lines. Compression wave is faster

The dynamics of the perturbation n_1 is governed by the linearized mass and momentum conservation equations

$$\partial_t \rho_1 + \rho_0 \partial_x u = 0, \qquad \rho_0 \partial_t u + (c_s^2 + c_A^2)\partial_x \rho_1 = 0.$$

Partial derivation of the first equation with respect to t and of the second equation with respect to x and elimination of the mixed derivative terms yields the simple undamped wave equation for the *compression Alfvén mode*,

$$\partial_{xx}\rho_1 - (c_s^2 + c_A^2)^{-1}\partial_{tt}\rho_1 = 0 \;\Rightarrow n_1(x,t) = \hat{n}_1 \mathrm{e}^{i(kx \pm \omega t)}, \;\; \omega^2 = (c_s^2 + c_A^2)k^2. \tag{5.128}$$

The wave is longitudinal, $\mathbf{k}$ is parallel to $\mathbf{u}$ and perpendicular to $\mathbf{B}_0$, as illustrated in Fig. 5.11(a). $c_A = \sqrt{\mathbf{B}_0^2/\mu_0\rho_0}$ is the *Alfvén velocity*.

The shearAlfvén wave. It belongs to the class of transverse (electromagnetic) waves. The wave vector $\mathbf{k}$ is assumed parallel to $\mathbf{B}_0$ and $\mathbf{u}$ perpendicular to it. Owing to $\nabla\mathbf{u} = 0$ the equilibrium density ρ_0 remains unchanged. The field dynamics follows from the linearized Faraday law (5.121) and the linearized momentum equation

$$\frac{\partial \mathbf{B}_1}{\partial t} = \nabla \times (\mathbf{u} \times \mathbf{B}_0), \qquad \rho_0 \frac{\partial \mathbf{u}}{\partial t} = \mathbf{j} \times \mathbf{B}_0 = \frac{1}{\mu_0}(\nabla \times \mathbf{B}_1) \times \mathbf{B}_0.$$

Substitution of $\partial_t \mathbf{B}_1$ yields

$$\rho_0 \frac{\partial^2 \mathbf{u}}{\partial t^2} = \frac{1}{\mu_0}(\nabla \times \dot{\mathbf{B}}_1) \times \mathbf{B}_0 = -\frac{1}{\mu_0}\mathbf{B}_0 \times \nabla \times [\nabla \times (\mathbf{u} \times \mathbf{B}_0)]. \tag{5.129}$$

Fourier decomposition

$$\mathbf{u},\ \mathbf{j},\ \mathbf{B}_1 = \hat{\mathbf{u}},\ \hat{\mathbf{j}},\ \hat{\mathbf{B}}_1 e^{i(\mathbf{kx}-\omega t)}, \qquad \nabla\times = i\mathbf{k}\times$$

shows that $\mathbf{B}_1$ is parallel to $\mathbf{u}$ and $\mathbf{j}$ is perpendicular to $\mathbf{B}_0$ and $\mathbf{u}$. With this ansatz the right term in (5.129) reduces to

$$\frac{1}{\mu_0}\mathbf{B}_0 \times \mathbf{k} \times [\mathbf{k} \times (\mathbf{u} \times \mathbf{B}_0)] = -\frac{1}{\mu_0}(\mathbf{kB}_0)^2\mathbf{u}$$

and $\mathbf{u}$ obeys the dispersion relation of the *shear Alfvén wave*

$$[\rho_0\omega^2 - \frac{\mathbf{B}_0^2}{\mu_0}\mathbf{k}^2]\mathbf{u} = 0 \ \Rightarrow\ \omega^2 = c_A^2\mathbf{k}^2; \quad c_A = \left(\frac{\mathbf{B}_0^2}{\mu_0\rho_0}\right)^{1/2}. \tag{5.130}$$

Let $\mathbf{u} = \mathbf{u}(z)$ lie along $\mathbf{e}_x$. Then, $\mathbf{B}_1 = \hat{\mathbf{B}_1} \exp i(kz - \omega t)$ points also along $\mathbf{e}_x$, whereas $\mathbf{j}$ is oriented in $\mathbf{e}_y$ direction; compare Fig. 5.11(b). The shear Alfvén wave is driven by the magnetic tension along the field lines. To see this compare with the transverse wave of excursion $x(z,t)$ from equilibrium on the elastic string under tension τ. The length of a string element is $dl = dz/\cos\alpha$, $\cos\alpha = \partial z/\partial l$. For small angles $\alpha \Rightarrow \cos\alpha = 1 - \alpha^2/2$ and dl simplifies to $dl = dz(1 + \alpha^2/2)$, thus the force per unit length is $-\tau\partial^2 l/\partial z^2$ and the equation of motion reads

$$\rho_0\partial_{tt}x + \tau\partial_{zz}x = 0; \quad \Rightarrow \quad \omega^2 = v_\varphi^2 k^2, \quad v_\varphi = \left(\frac{\tau}{\rho_0}\right)^{1/2}.$$

Comparison with (5.130) shows that the tension along a frozen magnetic field line is $\tau = \tau_m = \mathbf{B}_0^2/\mu_0$.

Magnetoacoustic waves. The constant magnetic field $\mathbf{B}_0$ may be oriented along the z axis. The wave vector $\mathbf{k}$ is now pointing into an arbitrary direction and splits into $\mathbf{k} = \mathbf{k}_\parallel + \mathbf{k}_\perp$ parallel and perpendicular to $\mathbf{B}_0$. The linearized equations governing the dynamics of the wave are the same as for the shear Alfvén wave above, complemented now by the pressure term $-\nabla p$,

$$\rho_0\frac{\partial\mathbf{u}}{\partial t} = -\nabla p + \mathbf{j}\times\mathbf{B}_0, \quad \mathbf{E} = -\mathbf{u}\times\mathbf{B}_0, \quad \nabla\times\mathbf{B}_1 = \mu_0\mathbf{j}, \quad \nabla\times\mathbf{E} = -\dot{\mathbf{B}}_1. \tag{5.131}$$

Elimination of $\mathbf{j}$, $\mathbf{B}_1$, $\mathbf{E}$ from the partial time derivative of the first equation is straightforward with the help of $\nabla\dot{p} = -\rho_0\nabla(\nabla\mathbf{u}) = -\rho_0\nabla\times(\nabla\times\mathbf{u}) + \nabla^2\mathbf{u}$:

$$\frac{\partial^2\mathbf{u}}{\partial t^2} = -c_s^2[\nabla\times(\nabla\times\mathbf{u}) + \nabla^2\mathbf{u}] - c_A^2\frac{\mathbf{B}_0}{B_0}\times\nabla\times\left[\nabla\times\left(\mathbf{u}\times\frac{\mathbf{B}_0}{B_0}\right)\right].$$

The dispersion relation follows from setting $\mathbf{u} = \hat{\mathbf{u}} \exp i(\mathbf{kx} - \omega t)$,

$$\omega^2 \mathbf{u} - c_s^2[\mathbf{k} \times (\mathbf{k} \times \mathbf{u}) + \mathbf{k}^2\mathbf{u}] - c_A^2 \frac{\mathbf{B}_0}{B_0} \times \mathbf{k} \times \left[\mathbf{k} \times \left(\mathbf{u} \times \frac{\mathbf{B}_0}{B_0}\right)\right] = 0. \quad (5.132)$$

If $\mathbf{u}\|\mathbf{k}$ is along the z axis the dispersion of the ordinary ion sound wave $\omega^2 - c_s^2\mathbf{k}^2 = 0$ results. For the general case $\mathbf{k} = \mathbf{k}_\| + \mathbf{k}_\perp$ and $\mathbf{u} = (u_x, u_y, u_z)$ (5.132) yields

$$(\omega^2 - k_\|^2 c_A^2)u_x = (\omega^2 - k_\perp c_s^2 - k^2 v_A^2)u_y - k_\perp k_\| c_s^2 u_z = k_\perp k_\| c_s^2 u_y - (\omega^2 - c_s^2 k^2)u_z = 0$$

Finite velocity $\mathbf{u}$ or, equivalently, finite $\mathbf{B}_1, \mathbf{j}, \mathbf{E}$ requires vanishing determinant of this system,

$$D(\omega, \mathbf{k}) = (\omega^2 - k_\|^2 c_A^2)[\omega^4 - k^2(c_s^2 + c_A^2)\omega^2 + k^2 k_\|^2 c_s^2 c_A^2] = 0. \quad (5.133)$$

The dispersion relation is satisfied either by the vanishing of the first bracket which describes the shear Alfvén wave, or of the square bracket. The latter is a biquadratic equation in ω with the solution

$$\omega^2 = \frac{1}{2}k^2(c_s^2 + c_A^2)\left\{1 \pm \left[1 - 4\frac{k_\|^2 c_s^2 c_A^2}{k^2(c_s^2 + c_A^2)}\right]^{1/2}\right\}. \quad (5.134)$$

The term in the square bracket varies between zero and 1, hence the linear magnetoacoustic waves are undamped stable eigenmodes of the homogeneous plasma. They are named *magnetoacoustic waves* because of their compressional contribution c_s to the propagation speed, except in the pure shear mode. The two brunches $\omega(\mathbf{k})$ stand for the *fast* and the *slow* mode. The maximum and the minimum of ω are reached for $k_\| = 0$ with $\omega_{\max}^2 = \mathbf{k}^2(c_s^2 + c_A^2)$ for the compressional Alfvén wave, and $\omega_{\min}^2 = 0$ for the non propagating slow mode in this special case. We conclude, the presence of a static magnetic field leads to an increase of the frequency of the ion acoustic wave according to (5.128) versus (5.113),

$$\frac{c_A^2}{c_s^2} = \frac{B_0^2}{\mu_0 \gamma_a p} = 6.33 \times 10^{11}\frac{(B_0[\mathrm{T}])^2}{\gamma_a p[\mathrm{Pa}]} = 6.2 \times 10^6 \frac{(B_0[\mathrm{T}])^2}{\gamma_a p[\mathrm{bar}]}.$$

$$\gamma_a p = \gamma_e p_e + \gamma_i p_i; \;\; 1\mathrm{T(Tesla)} = 10^4\,\mathrm{Gauss}, \;\; 1[\mathrm{bar}] = 1.02 \times 10^5\,\mathrm{Pa}. \quad (5.135)$$

High frequency modes in the magnetic field. One of the severe restrictions on the interaction of matter with the high power laser beam lies in the low cut off of propagation. Overdense matter, $n_e > n_c$ cannot directly interact with the light beam. Heating of dense matter by infrared or visible radiation must rely on electron heat conduction and on shock compression. Fortunately, in the relativistic regime there is an increase of cut off as a consequence of the relativistic electron mass increase to be treated later. Here we investigate the possibility of a critical density increase by applying a strong magnetic field $\mathbf{B}_0$ on non relativistic intensities. For this purpose we examine the linear dispersion relation of a transverse wave in the magnetized homogeneous plasma in the most relevant configuration.

The idea is the following. The left hand circularly polarized laser wave forces the single electron into an anti-clockwise circular motion as illustrated by Fig. 2.2. A constant magnetic field $\mathbf{B}_0$ oriented along the $\mathbf{k}$ vector of the wave imposes, according to (2.66), a circular motion into clockwise sense to the electron. The superposition of both is expected to lead to lowering of the induced current and, in concomitance, to an increase of the critical density.

The propagation of the wave is determined from the wave equation (5.15) for the transverse $\mathbf{E}$ vector. All to be calculated is the current density $\mathbf{j} = -en_e\mathbf{v}_\perp$. The static magnetic field $\mathbf{B}_0$ induces a rotation of $\mathbf{v}_\perp$ perpendicularly to its direction according to (2.65), $\dot{\mathbf{v}}_\perp = \boldsymbol{\omega}_G \times \mathbf{v}_\perp$. The variation of the modulus into radial direction is accomplished by the $\mathbf{E}$ vector. We observe that the wave equation (5.15) holds for circular polarization as well, provided the field vector is chosen complex, $\mathbf{E} = \mathbf{E}_y + i\mathbf{E}_z$, and all coefficients are real valued. The latter property is assured by linearization of $\mathbf{j} = \sigma\mathbf{E}$. Keeping in mind this picture the linearized equation of motion of a right hand polarized Fourier component $\mathbf{E} \sim \exp i(\mathbf{kx} - \omega t)$ and the current density follow as

$$\dot{\mathbf{v}}_\perp = -i\omega\mathbf{v}_\perp = -\frac{e}{m_e}(\mathbf{E} + \mathbf{v}_\perp \times \mathbf{B}_0); \ \boldsymbol{\omega}_G = \frac{e}{m_e}\mathbf{B}_0, \ \boldsymbol{\omega}_G \times \mathbf{v}_\perp = -i\mathbf{v}_\perp$$

$$\Rightarrow \qquad \mathbf{v}_\perp = \frac{e}{m_e\omega(\omega - \omega_G)}\mathbf{E}, \quad \mathbf{j} = -\varepsilon_0\frac{\omega_p^2}{\omega^2\left(1 - \frac{\omega_G}{\omega}\right)}\mathbf{E}.$$

In connection with wave propagation it is customary to rename the gyrofrequency ω_G by the *cyclotron frequency* $\omega_c = \omega_G$. The current density for the left hand circularly polarized wave is obtained from changing $1 - \omega_c/\omega$ into $1 + \omega_c/\omega$. Keeping in mind $\mathbf{k} = \mathbf{k}^0\eta$, η refractive index the wave equation (5.15) yields for the right handed circularly polarized wave R and its left handed counterpart L

$$\mathrm{R}: \ \eta^2 = \frac{c^2\mathbf{k}^2}{\omega^2} = 1 - \frac{\omega_p^2}{\omega^2}\frac{1}{1 - \omega_c/\omega}; \quad \mathrm{L}: \ \eta^2 = \frac{c^2\mathbf{k}^2}{\omega^2} = 1 - \frac{\omega_p^2}{\omega^2}\frac{1}{1 + \omega_c/\omega}. \tag{5.136}$$

In our context the L wave is of interest. The presence of an axial static B field leads to an increased critical density n_{ec} relative to the critical density n_c at $B_0 = 0$,

$$n_{ec} = n_c\left(1 + \frac{\omega_c}{\omega}\right); \quad n_{ec} = n_c\left(1 + \frac{1.8 \times 10^{11} B_0[\mathrm{T}]}{2 \times 10^{15}}\right) \cong n_c(1 + 10^{-4} B_0[\mathrm{T}]). \tag{5.137}$$

The use of a static magnetic field is not very efficient at a typical infrared laser frequency of 2×10^{15} [s^{-1}] between Nd and Ti:Sa.

Faraday effect. The different phase velocities of the R and L waves can be used as a diagnostic tool to measure either plasma densities or static magnetic fields. In the magnetic field the linearly polarized wave separates into an R and L component. The polarization of each mode is transported with its phase velocity v_φ. After the length L the difference in angle of orientation $\Delta\varphi$ is with $\varepsilon = \omega_c/\omega$

$$\Delta\varphi = [k(\mathrm{R}) - k(\mathrm{L})]L = \frac{\omega}{c}\Big[\Big(1 - \frac{\omega_p^2}{\omega^2}\frac{1}{1-\varepsilon}\Big)^{1/2} - \Big(1 - \frac{\omega_p^2}{\omega^2}\frac{1}{1+\varepsilon}\Big)^{1/2}\Big]L. \quad (5.138)$$

The precision of a measurement depends crucially on the degree of homogeneity of n_e and B_0.

5.5 Waves in the Inhomogeneous Plasma

Two Basic Considerations

Local amplitude. How to describe a wave in the inhomogeneous medium? In principle one has to formulate the charges and currents point by point and to solve Maxwell's equations with these quantities in a self-consistent manner. It means to calculate the expressions (5.12) and (5.13), or the wave equations (5.10) in the Lorentz gauge, or equivalent equations in the apposite gauge. In general this is a business for the computer, not for human brains. Fortunately, there is a whole class of situations in which all we have learned in the homogeneous medium applies, with supplements, to media with properties varying in space and time. It is the class of solutions where the variations of the sources ρ_{el} and $\mathbf{j}$ are smooth and slowly varying over one wavelength λ or one oscillation period $T = 2\pi/\omega$. Let a component of the actual source be indicated by $\sigma_l(\mathbf{x}, t), l = 1, 2, 3$. Then, *slow* means

$$\left|\frac{\lambda\, \mathrm{grad}\sigma_l(\mathbf{x}, t)}{\sigma_l(\mathbf{x}, t)}\right| \ll 1 \Leftrightarrow \left|\frac{\mathrm{grad}\sigma_l(\mathbf{x}, t)}{k\sigma_l(\mathbf{x}, t)}\right| < 1 \Leftrightarrow kL > 1; \quad L = \left|\frac{\sigma_l}{\nabla\sigma_l}\right|. \quad (5.139)$$

The corresponding inequality for slow time variation reads

$$T\left|\frac{\dot{\sigma}_l(\mathbf{x}, t)}{\sigma_l(\mathbf{x}, t)}\right| \ll 1 \Leftrightarrow \left|\frac{\dot{\sigma}_l(\mathbf{x}, t)}{\omega\sigma_l(\mathbf{x}, t)}\right| < 1. \quad (5.140)$$

The definition of the *gradient length L* results obvious; examples are Figs. 1.17 and 5.7. Under the constraints of smoothness above the local forces producing the wave motion do almost not feel the inhomogeneities and hence the dispersion relation $D(\mathbf{k}(\mathbf{x}, t), \omega(\mathbf{x}, t)) = 0$ will be only weakly affected, i.e., it remains invariant in zeroth order.

For this purpose imagine the wave field distribution at a fixed time over one wave-

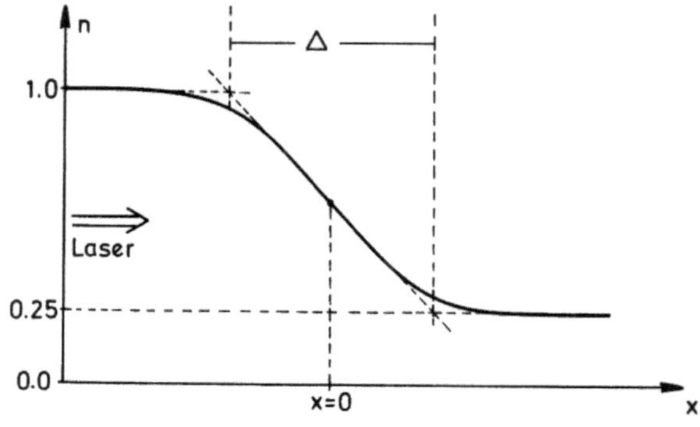

Fig. 5.12 Epstein transition layer of width Δ for the square of refractive index. The wave is incident from the left. Parameter $P = -3/4$ (plasma case) is assumed

length $\lambda < L(\mathbf{x})$ in a medium with smoothly decaying refractive index η, as shown in Fig. 5.12 for η^2. The difference $\Delta\eta$ with respect to the center of the wave at $\lambda/2$, integrated over the first half wavelength compensates in first order the corresponding term over the second half wavelength. As a result the impact of the inhomogeneity is of second or higher order. The phase $\Psi(\mathbf{k}, \omega, \mathbf{x}, t) = \int [\mathbf{k}(\mathbf{x}, t)\mathrm{d}\mathbf{x} - \omega(\mathbf{x}, t)\mathrm{d}t]$ that replaces now $\Psi(\mathbf{k}, \omega) = \mathbf{kx} - \omega t$ behaves like an adiabatic invariant (see the explanation of adiabatic invariant in Chap. 2, in particular its character of an alternating series). As a consequence, only the amplitude A of the wave under consideration as a first order quantity has to be adjusted locally, i.e., $A = \text{const} \Longrightarrow A(\mathbf{x}, t)$. Should A not obey the inequalities (5.139) and (5.140) in some region of space the splitting of the field quantity into amplitude and phase will lose its meaning. Such a violation always occurs at the cut offs $\mathbf{k} = 0, \omega = 0$. The procedure resulting from the slowness-smoothness conditions (5.139), (5.140) has been named *WKB approximation* or *optical approximation*. In regions where it fails recurrence has to be made to the general wave equations, for instance (5.10).

Smooth medium does not reflect. As will be seen in the following the flows of quantities like energy and momentum, action in general, associated with the wave in the smooth medium happen along rays without *reflection* so that useful conservation relations can be established along them.

The statement of no reflection can be quantified for the perpendicular incidence of a laser beam from vacuum onto an *Epstein transition layer* of the following form [3, 4]

$$\eta^2(x) = 1 + P\frac{\exp(kx/s)}{1 + \exp(kx/s)}, \tag{5.141}$$

where P and s are free parameters. In Fig. 5.12 $\eta^2(x)$, with $P = -3/4$, is sketched. The transition width Δ is determined through s in the following way:

$$\Delta = \frac{4s}{k}, \quad k = 2\pi/\lambda.$$

Table 5.1 Normalized reflection coefficient r for the layer of Fig. 5.12 ($P = -3/4$ for plasma) and $P = +3/4$ as a function of the transition width Δ.

s	2	1	1/2	1/4	1/8	1/16
Δ/λ	1.27	0.64	0.32	0.16	0.08	0.04
$P = +3/4$	6.1×10^{-10}	1.36×10^{-4}	3.92×10^{-2}	0.37	0.77	0.93
$P = -3/4$	3.1×10^{-5}	1.5×10^{-2}	0.25	0.78	0.91	0.98

For a light beam normally incident from the left-hand side the reflection coefficient R is given by [4]

$$R = \frac{\sinh^2[\pi s(1 - \sqrt{1+P})]}{\sinh^2[\pi s(1 + \sqrt{1+P})]}. \tag{5.142}$$

In Table 5.1 R normalized to the Fresnel value of the reflection coefficient $R_0 = (\eta - 1)^2/(\eta + 1)^2$ for a refractive index step, $r = R/R_0$, is shown for $P = \pm 3/4$ and different values of s. It may be surprising how fast reflection drops from its Fresnel value $r = 1$ as soon as Δ becomes larger than $\lambda/4$; at $P < -1$ total reflection occurs for all values of s [4]. As expected convergence towards nonreflecting behaviour is faster for $P > 0$ than for $P < 0$. In the latter case already the vicinity of the reflecting cut off at $\eta = 0$ becomes noticeable. As soon as a WKB condition is fulfilled, local reflection becomes extremely low, and the radiation field can be uniquely split into incident and reflected waves. The formulas equally apply to a Langmuir wave incident from a plasma of constant density n_{e0} on the left. When the optical approximation fails, a local reflection coefficient $R(x)$ is no longer uniquely determined by the bare difference $\Delta\eta$.

5.5.1 *From the Transverse Wave to the Classical Photon*

The gauge free wave equation

$$\nabla \times \nabla \times \mathbf{E} + \frac{1}{c^2}\frac{\partial^2}{\partial t^2}\mathbf{E} = -\frac{1}{\epsilon_0 c^2}\frac{\partial \mathbf{j}}{\partial t} \tag{5.15}$$

is of general validity, for transverse as well as for longitudinal waves in an arbitrarily inhomogeneous medium in motion. Thereby $\mathbf{j} = -n_e e\mathbf{u}_e + Zn_i e\mathbf{u}_i$ is the total current. Its connection with the electric field is given by the generalized Ohm's law from fluid theory. Traditionally, in optics of dielectric media $\mathbf{j}$ is replaced by the polarization $\mathbf{P}$, but this is a pure convention if their relation

$$\mathbf{j} = \frac{\partial}{\partial t}\mathbf{P} + \nabla \times \left(\mathbf{P} \times \frac{\mathbf{p}}{\rho}\right), \quad \mathbf{p} = m_i\mathbf{u}_i + m_e\mathbf{u}_e$$

is observed [5]. It can be regarded as the definition of **P**. Plasma motion is not very significant in the high frequency domain of the laser and can generally be ignored, see the estimate below. Hence, in the smooth medium Ohm's law reduces to the simple form of $\mathbf{j} = -n_{e0}\mathbf{u}_{e1}$ of the fast electron motion. Then, in presence of a linear friction $\nu\mathbf{u}_{e1}$ the transverse refraction index η for the ω mode is given by

$$\eta^2 = 1 - \frac{\omega_p^2}{\omega^2}\frac{1}{1+\nu^2/\omega^2} + i\frac{\nu}{\omega}\frac{\omega_p^2}{\omega^2}\frac{1}{1+\nu^2/\omega^2}. \tag{5.143}$$

If ν is the electron-ion collision frequency ν_{ei} the current density **j** is identical with that from the Drude model of Chap. 1. In the inhomogeneous plasma, and in dielectrics also, *resonant coupling* between the electromagnetic and the longitudinal electron plasma wave can occur. It is therefore necessary to include the possibility of electron density modulations in Ohm's law. From the electron fluid equations for the fast component and the Maxwell relation $\varepsilon_0\nabla\mathbf{E} = -en_{e1}$ its linearized form follows straightforwardly as

$$\frac{\partial \mathbf{j}}{\partial t} = \varepsilon_0\omega_p^2\mathbf{E} - \varepsilon_0 c_{se}^2\nabla(\nabla\mathbf{E}) - \nu\mathbf{j}. \tag{5.144}$$

Then, for a wave of frequency ω (5.15) reduces to the linear wave equation of plasma optics,

$$\nabla\times\nabla\times\mathbf{E} - k_0^2\eta^2\mathbf{E} = \beta^2\nabla(\nabla\mathbf{E}), \quad \beta = c_{se}/c. \tag{5.145}$$

For the pure transverse wave ($\nabla\mathbf{E} = 0$), with no coupling to an electron plasma wave, it simplifies further to the *stationary wave equation*

$$\nabla^2\mathbf{E} + \mathbf{k}_0^2\eta^2(\mathbf{x})\mathbf{E} = 0. \tag{5.146}$$

Identical equations yield identical solutions, Richard Feynman says. Excitation of a two level atomic system is isomorphous to the dynamics of a single spin. This realization enabled him to construct the extremely useful and pictorial *optical Bloch model* [27]. Another useful parallelism exists between the wave equation (5.146) and the *Schrödinger equation* of a particle in the potential $V(\mathbf{x})$,

$$-\frac{\hbar^2}{2m}\nabla^2\psi(\mathbf{x}) + V(\mathbf{x})\psi(\mathbf{x}) = E\psi(\mathbf{x}); \qquad E \text{ energy eigenvalue.} \tag{5.147}$$

After identification $1 - V(\mathbf{x})/E = \eta(\mathbf{x})^2$ and $2mE/\hbar^2 = \mathbf{k}_0^2$ the Schrödinger equation transforms (apart from polarization) into (5.146). Methodically and with regard to contents classical optics can benefit from the formal affinity to quantum theory, as for example in tunnelling and trapping of electromagnetic radiation.

Nonlinear Ohm's law with plasma flow included*. Only the electron motion has to be considered because at high frequency ion motion is insignificant, or more precisely, is taken into account by replacing m_e by the reduced mass μ of the two species. The current density $\mathbf{j}$ follows from the combination of the momentum and the particle conservations of the electron fluid,

$$\frac{\partial \mathbf{j}}{\partial t} = \varepsilon_0\omega_p^2\mathbf{E} - \varepsilon_0 c_{\mathrm{se}}^2\nabla(\nabla\mathbf{E}) - \nu\mathbf{j} - \\ \frac{e}{m_e}\mathbf{j}\times\mathbf{B} + \frac{e^2}{m_e}(n_e - n_0)\mathbf{E} - (\mathbf{j}\nabla)\mathbf{u}_e - \mathbf{u}_e(\nabla\mathbf{j}). \tag{5.148}$$

The quantity ν is a linear collisional or noncollisional damping coefficient. The Ohm law is nonlinear. For subrelativistic intensities the oscillation amplitude $\hat{\boldsymbol{\delta}}_e$ fulfils $|\hat{\boldsymbol{\delta}}_e| \ll \lambda$. Under the additional constraint of $|\hat{\boldsymbol{\delta}}_e| \ll L$, the plasma frequency $\omega_p = (n_0e^2/\varepsilon_0 m_e)^{1/2}$ is defined locally and no higher harmonics of the laser frequency ω have to be considered in (5.148). Thus, if the flow velocity is $\mathbf{u}_0(\mathbf{x},t)$ and $\mathbf{j} = \mathbf{j}_{\mathrm{os}} = -en_0(\mathbf{u}_e - \mathbf{u}_0)$ Ohm's law can be written as follows:

$$\frac{\partial \mathbf{j}}{\partial t} = \varepsilon_0\omega_p^2(\mathbf{E} + \mathbf{u}_0\times\mathbf{B}) - \varepsilon_0 c_{\mathrm{se}}^2\nabla(\nabla\mathbf{E}) - \nu\mathbf{j} - (\mathbf{j}\nabla)\mathbf{u}_0 - \mathbf{u}_0(\nabla\mathbf{j}). \tag{5.149}$$

With the help of the quantities $k_0 = \omega/c,\ \beta = c_{\mathrm{se}}/c,\ v_\varphi v_g = c^2$ for phase and group velocities of the electromagnetic wave and $v_\varphi v_g = c_{\mathrm{se}}^2$ for the electrostatic wave one estimates

$$\left|\frac{c_{\mathrm{se}}^2\nabla(\nabla\mathbf{E})}{\omega_p^2\mathbf{E}}\right| = \gamma_e(\lambda_D k_e)^2, \quad \left|\frac{(\mathbf{j}\nabla)\mathbf{u}_0}{\varepsilon_0 c_{\mathrm{se}}^2\nabla(\nabla\mathbf{E})}\right| \le \frac{u_0}{c}\frac{\beta^2}{k_0 L(\lambda_D k_e)^2},$$

$$\left|\frac{\mathbf{u}_0(\nabla\mathbf{j})}{\varepsilon_0 c_{\mathrm{se}}^2\nabla(\nabla\mathbf{E})}\right| \le \frac{u_0}{v_g}. \tag{5.150}$$

As a consequence, plasma flow is generally irrelevant in laser plasma optics and one may set $\mathbf{u}_0 = 0$. The term $\mathbf{E}' = \mathbf{E} + \mathbf{u}_0\times\mathbf{B}$ is the electric field seen by an observer comoving nonrelativistically with the plasma and can be identified with $\mathbf{E}$; hence, (5.149) reduces to (5.144) as expected. The nonlinear terms of (5.148) give rise to the ponderomotive force and to most of nonlinear optics phenomena encountered in laser plasmas (see Chap. 6).

Rays, Phases, and Ray Tracing

Amplitude $\hat{\mathbf{E}}(\mathbf{x},t)$ and phase $\Psi(\mathbf{x},t)$ are complementary, either both are meaningful simultaneously or both can no longer be defined locally. Let us assume that such a splitting of the wave field $\mathbf{E}(\mathbf{x},t)$, longitudinal or transverse, is appropriate,

$$\mathbf{E}(\mathbf{x},t) = \hat{\mathbf{E}}(\mathbf{x},t)\mathrm{e}^{i\Psi(\mathbf{x},t)}. \tag{5.151}$$

The definition of a local wave vector and a local frequency consistent with the splitting (5.151) is

$$\mathbf{k} = \nabla\Psi, \quad \omega = -\frac{\partial}{\partial t}\Psi, \tag{5.152}$$

(see e.g. [6]), Sect. 7.7). The phase element $d\Psi$ is a total differential and the phase $\Psi(\mathbf{x}, t)$ is independent of the path of integration between two points P_1, P_2,

$$\Psi(\mathbf{x}, t) = \int_{P_1}^{P_2} (\mathbf{k}\mathrm{d}\mathbf{s} - \omega \mathrm{d}t). \tag{5.153}$$

Use of (5.151) in (5.145) under neglect of second order terms results in

$$(\nabla\Psi)^2\mathbf{E} - \nabla\Psi(\mathbf{E}\nabla\Psi) + \beta^2\nabla\Psi(\mathbf{E}\nabla\Psi) = k_0^2\eta^2\mathbf{E}.$$

By splitting $\mathbf{E}$ into $\mathbf{E}_\perp + \mathbf{E}_\parallel$ with respect to the wave vector $\mathbf{k} = \nabla\Psi$, the two *eikonal equations* and dispersion relations follow for the electromagnetic and the electrostatic components,

$$\mathbf{E}_\perp : \ (\nabla\Psi)^2 = \mathbf{k}_0^2\eta^2, \ \omega^2 = \omega_p^2 + c^2\mathbf{k}^2, \ k_0 = \frac{\omega}{c}, \ k = k_0\eta, \ \eta^2 = 1 - \frac{\omega_p^2}{\omega^2},$$

$$\mathbf{E}_\parallel : \beta^2(\nabla\Psi)^2 = k_0^2\eta^2, \ \omega^2 = \omega_p^2 + c_{se}^2\mathbf{k}_e^2, \ k_{0e} = \frac{\omega}{c_{se}}, \ k_e = k_{0e}\eta, \ \eta^2 = 1 - \frac{\omega_p^2}{\omega^2}. \tag{5.154}$$

Here and in the rest of this section we take $\nu = 0$. In the collisionless plasma the refractive index $\eta = c/v_{\varphi,\mathrm{em}}$ or $\eta = c_{se}/v_{\varphi,\mathrm{em}}$, respectively, is the same for both waves and, in an isothermal plasma, both propagate along identical trajectories. These trajectories or rays are orthogonal to the surfaces of constant phase, $\Psi(\mathbf{x}, t = \mathrm{const}) = \mathrm{const}$ (Fig. 5.13). In a stationary plasma with Ψ not depending on time the phase difference between two points P_1, P_2 at a given time is uniquely determined by

$$\Psi(P_2) - \Psi(P_1) = \int_{P_1}^{P_2} |\mathbf{k}|\mathrm{d}s$$

along a ray segment C. Along an arbitrary path C' from P_1 to P_2 one has

$$\Psi(P_2) - \Psi(P_1) = \int_{C'} \mathbf{k}\mathrm{d}\mathbf{s} \leq \int_{C'} k\mathrm{d}s. \tag{5.155}$$

This expresses *Fermat's principle*. In the homogeneous medium it is the principle of the shortest distance or the shortest time between the fixed points P_1, P_2.

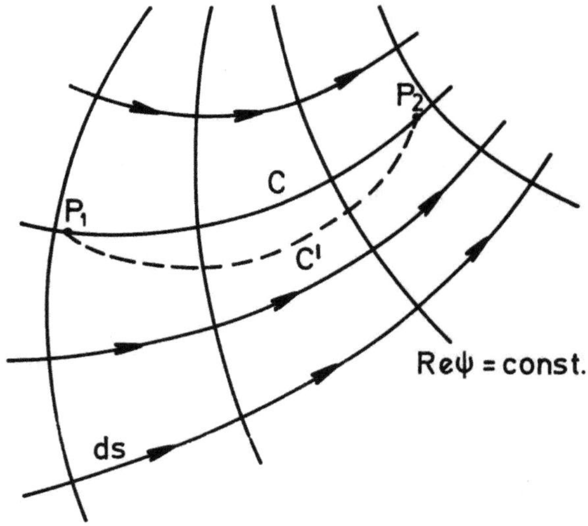

Fig. 5.13 The rays $\{\mathbf{k} \parallel \mathrm{d}\mathbf{s}\}$ form a manifold of curves which are perpendicular to the surfaces of constant phase $\Psi(\mathbf{x}, t = \text{const}) = \text{const}$

Ray equation. From

$$\frac{\mathrm{d}}{\mathrm{d}s}(\mathbf{k}_0\eta) = \frac{1}{k_0\eta}(\mathbf{k}_0\eta\nabla)(\mathbf{k}_0\eta) = \frac{1}{2k_0\eta}\nabla(\mathbf{k}_0\eta)^2 - \frac{\mathbf{k}_0}{k_0} \times \nabla \times \mathbf{k}_0\eta$$

$$\nabla \times \mathbf{k}_0\eta = \nabla \times (\nabla\Psi) = 0$$

follows the *ray equation*

$$\frac{\mathrm{d}}{\mathrm{d}s}(\mathbf{k}_0\eta) = k_0\nabla\eta. \tag{5.156}$$

Thereby use has been made of the vector identity $\mathbf{a} \times (\nabla \times \mathbf{a}) = \nabla(\mathbf{a}^2/2) - (\mathbf{a}\nabla)\mathbf{a}$. Equation (5.156) is the basis of *raytracing*. From it η is again recognized as the refractive index. In the fully ionized plasma it is the same for Langmuir waves as for electromagnetic waves. For the electron plasma wave c_{se} plays the same role as the speed of light c does for the laser wave.

In general, (5.156) has to be solved numerically. Only in plane and spherical geometry (*stratified* or *layered* medium) it is easily integrated. With the constant vector $\mathbf{n} = \nabla\eta/|\nabla\eta|$ in plane geometry one has

$$\frac{\mathrm{d}}{\mathrm{d}s}(\mathbf{n} \times \mathbf{k}_0\eta) = \mathbf{n} \times \frac{\mathrm{d}}{\mathrm{d}s}(\mathbf{k}_0\eta) = \mathbf{n} \times k_0\nabla\eta = 0.$$

Similarly, in spherical geometry a vector $\mathbf{r}$ having its origin in the center of symmetry of $\eta(r)$ satisfies

$$\frac{\mathrm{d}}{\mathrm{d}s}(\mathbf{r} \times \mathbf{k}_0\eta) = \frac{\mathbf{k}_0}{k_0} \times \mathbf{k}_0\eta + \mathbf{r} \times \frac{\mathrm{d}}{\mathrm{d}s}(\mathbf{k}_0\eta) = \mathbf{r} \times k_0\nabla\eta = 0$$

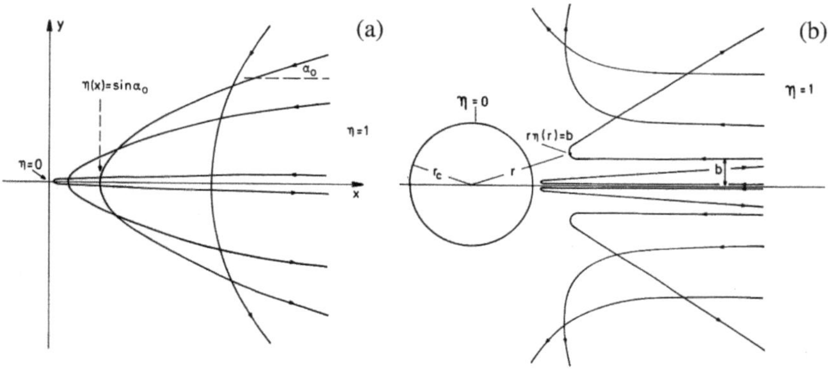

Fig. 5.14 Raytracing. **a** Plane layered medium. A ray incident from vacuum at the angle α_0 has its turning point at $\eta(x) = \sin\alpha_0$. The y-component $\hbar k_y = \hbar k_0 \eta \cos\alpha = \hbar k_0 \cos\alpha_0$ of the photon momentum is conserved in the stationary medium. **b** Deflection of a parallel beam by a spherical plasma cloud. In the stationary medium the angular momenta $\hbar k_0 \eta r \sin\alpha = \hbar k_0 b$ of the photons are conserved along their trajectories

since $\nabla\eta$ is parallel to $\mathbf{r}$. If α is the angle between the local wave vector and $\nabla\eta$, the two relations can be expressed in integrated form

$$\eta(x)\sin\alpha = \text{const} \quad (\text{plane}), \quad r\eta(r)\sin\alpha = \text{const} \quad (\text{spherical}). \tag{5.157}$$

Two examples of ray distributions for plane and spherical geometry are sketched in Fig. 5.14. In the plane stratified medium the ray directions are such that the y-component of $\mathbf{k}_0\eta$ is conserved. In spherical geometry the momentum $\mathbf{r} \times \mathbf{k}_0\eta$ relative to the center of symmetry remains constant. It follows from (5.154) that Langmuir and electromagnetic waves cannot propagate beyond the critical density given by (5.72). At oblique or nonradial incidence from the vacuum at the angle α_0 or distance b from the radius, respectively, the turning point of a ray is located at

$$\eta(x) = \sin\alpha_0 \quad (\text{plane}), \quad r\eta(r) = b \quad (\text{spherical}). \tag{5.158}$$

Hamiltonian equations of motion. The phase Ψ is a function of $\mathbf{x}$ and t. From the definition (5.152) of the wave vector it is clear that $\mathbf{k} = \mathbf{k}(\mathbf{x}, t)$; hence, owing to the dispersion relations (5.154), the functional dependence of the frequency is $\omega = \omega(\mathbf{k}, \mathbf{x}, t)$. The explicit dependence on $\mathbf{x}$ enters through $\omega_p^2(\mathbf{x}, t)$ and $c_{\text{se}}^2(\mathbf{x}, t)$. From (5.152) and $\nabla \times \mathbf{k} = \nabla \times \nabla\Psi = 0$ it follows that

$$\frac{\partial k_i}{\partial t} + \left(\frac{\mathrm{d}\omega}{\mathrm{d}x_i}\right)_{t=\text{const}} = 0, \qquad \frac{\partial k_i}{\partial x_j} = \frac{\partial k_j}{\partial x_i}.$$

With the help of the second relation the first transforms into

$$\frac{\partial k_i}{\partial t} + \frac{\partial \omega}{\partial k_j}\frac{\partial k_j}{\partial x_i} + \frac{\partial \omega}{\partial x_i} = \frac{\partial k_i}{\partial t} + \frac{\partial \omega}{\partial k_j}\frac{\partial k_i}{\partial x_j} + \frac{\partial \omega}{\partial x_i}$$

$$= \left(\frac{\partial}{\partial t} + \frac{\partial \omega}{\partial k_j}\frac{\partial}{\partial x_j}\right) k_i + \frac{\partial \omega}{\partial x_i} = 0; \quad \Leftrightarrow \quad \left(\frac{\partial}{\partial t} + \mathbf{v}_g \nabla\right) k_i = -\frac{\partial \omega}{\partial x_i}.$$

Hence, the set of equations governing an electromagnetic or electrostatic wave, i.e.,

$$\frac{\mathrm{d}\mathbf{x}}{\mathrm{d}t} = \frac{\partial \omega}{\partial \mathbf{k}}, \quad \frac{\mathrm{d}\mathbf{k}}{\mathrm{d}t} = -\frac{\partial \omega}{\partial \mathbf{x}}, \quad \frac{\mathrm{d}\omega}{\mathrm{d}t} = \frac{\partial \omega}{\partial t} \tag{5.159}$$

are Hamilton's canonical equations if we take $H(\mathbf{p}, \mathbf{q}, t) = \omega(\mathbf{k}, \mathbf{x}, t)$ with $p_i = k_i$, $q_i = x_i$. We observe that $\partial \omega / \partial \mathbf{k}$, generally defined as the *group velocity* of a wave packet, appears to be the transport velocity of the $\mathbf{k}$ vector. The last of (5.159) is a consequence of the preceding canonical equations of motion. From (5.152) follows also the *Hamilton–Jacobi equation*:

$$\frac{\partial \Psi}{\partial t} + \omega\left(\frac{\partial \Psi}{\partial \mathbf{x}}, \mathbf{x}, t\right) = 0 \tag{5.160}$$

thus showing that the phase Ψ is an action variable. The first of the canonical equations of motion (5.159) enables one to calculate the propagation of a pulse of narrow frequency band $\omega(\mathbf{k}, \mathbf{x}, t)$ more accurately than by $\mathbf{v}_g$ from (5.54). The propagation velocity $\mathbf{v}_\mathbf{k}$ and its path $\mathbf{x}_\mathbf{k}(t)$ are determined for the individual Fourier intensity component $I(\mathbf{k}, \mathbf{x}, t) = f(\mathbf{k}, \mathbf{x}, t)v_\varphi$ and then recomposed to yield the pulse $\int I(\mathbf{k}, t)\mathrm{d}\mathbf{k}$. The quantity $f(\mathbf{k}, \mathbf{x}, t)$ is the photon number density distribution in the one photon phase space $\Gamma(\mathbf{k}, \mathbf{x})$ and $I(\mathbf{k}, \mathbf{x}, t)$ is its flux density. The Vlasov equation of $f(\mathbf{k}, \mathbf{x}, t)$ reads

$$\frac{\partial f}{\partial t} + \frac{\partial \omega}{\partial \mathbf{k}}\frac{\partial f}{\partial \mathbf{x}} - \frac{\partial \omega}{\partial \mathbf{x}}\frac{\partial f}{\partial \mathbf{k}} = \lambda(\mathbf{x}, \mathbf{k}, t) - \alpha(\mathbf{x}, \mathbf{k}, t) f. \tag{5.161}$$

For fixed $(\mathbf{x}, t)$ the term $-\alpha f$ expresses the fact that there is an equal probability to be absorbed, and equal probability to be re-emitted, for all photons $\mathbf{k}$; $\alpha(\mathbf{x}, \mathbf{k}, t)$ is the *net absorption* coefficient. Equation (5.161) is the basic equation of *radiation transport*. In the case of the refractive index $\eta = 1$ it reduces to the familiar relation

$$\frac{\partial f}{\partial t} + \mathbf{c}\frac{\partial f}{\partial \mathbf{x}} = \lambda(\mathbf{x}, \mathbf{k}, t) - \alpha(\mathbf{x}, \mathbf{k}, t) f. \tag{5.162}$$

It is valid for short wavelength radiation. From (5.161) it is easily seen how it modifies for a homogeneous medium with $\eta = \text{const}$. In the Poisson bracket formalism [4] the LHS of (5.161) can also be written as $\partial f / \partial t + \{\omega, f\}$.

The classical frequency $\omega(\mathbf{k}, \mathbf{x}, t)$ assumes the role of the Hamilton function H of a photon, interacting collectively with matter through the refractive index η, however non-interacting with another photon. In the case of $\partial \omega / \partial t = 0$ the frequency is conserved. If ω is multiplied by a constant K of the dimension of an action the canonical equations (5.159) remain unchanged if the same proportionality constant K is introduced in $\mathbf{k}$. The new Hamiltonian $K\omega$ assumes the dimension of an energy. The experiment with microscopic entities tells that K is to be identified with $\hbar$. In the approximation of geometrical optics (WKB approximation) the cycle-averaged electromagnetic energy flux density $\overline{\mathbf{S}}$ of frequency ω, identified with the laser intensity $\mathbf{I}_\omega$ in (2.13), propagates along the rays of Fig. 5.13. In the absence of emission and absorption f is conserved, $f(\mathbf{k}, \mathbf{x}, t) = C = \text{const}$. Thus

$$I_\omega \mathrm{d}\omega = v_g f|\mathrm{d}\mathbf{k}| = c\eta f \eta |\mathrm{d}\mathbf{k}_0|, \quad \mathrm{d}\omega = c|\mathrm{d}\mathbf{k}_0| \Rightarrow \frac{I_\omega}{\eta^2} = C = \text{const}, \tag{5.163}$$

$I_\omega = |\overline{\mathbf{I}}_\omega|$. This relation is valid in the stratified medium where (5.157) holds. In the anisotropic medium η is to be substituted by a considerably more sophisticated expression, derived for example in [7]. The identification of I_ω with the photon number density $f(\mathbf{x}, \mathbf{k}, t)$ is made through the setting

$$f\frac{\partial\omega}{\partial\mathbf{k}} = \frac{1}{2}\frac{\varepsilon_0}{\hbar}\frac{\mathbf{E}\mathbf{E}^*}{\omega}\mathbf{v}_g. \tag{5.164}$$

Equation (5.161) is the Vlasov equation of incoherent monochromatic photon transport. It describes the photon field in terms of intensities and ignores mutual interferences. The substitution of the divergence of the Ponting vector of (5.17) by the gradient in (5.162) expresses the fact that in optical approximation changes of intensity occur along the rays only. From (5.164) and (5.162) follows that in the absence of sources and sinks the action density $(\varepsilon_0/2)\mathbf{E}\mathbf{E}^*/\omega$ is conserved rather than the energy density $(\varepsilon_0/2)\mathbf{E}\mathbf{E}^*$. This is in perfect agreement with the adiabatic theorem (2.104). Analogous Vlasov conservation equations hold for plasmons and phonons, and for quasiparticles in general [8]. The more general case of $\mathbf{I}_\omega$ propagating obliquely to the rays is considered in connection with radiation transport in Chap. 7.

5.5.2 *Wave Amplitudes*

WKB Approximation

Splitting of a wave field into amplitude and phase according to (5.151) implies the transport of wave energy along the characteristics $\mathbf{x}(t)$ of photons. Now, let the medium be stationary, i.e., $\partial_t\eta = 0,\ \eta = \eta(\mathbf{x})$. Then the characteristics $\mathbf{x}(t)$ coincide with the rays and the phase reduces to the familiar expression $\Psi(\mathbf{x}, t) = \pm k_0 \int \eta ds - \omega t$. By designating the local cross section of a narrow ray bundle by $Q(s)$, in the eikonal region $\mathbf{x} = \mathbf{x}(s)$ the solutions of (5.145) for $\mathbf{E} = \mathbf{E}_\perp$ and $\mathbf{E} = \mathbf{E}_\parallel$ are

$$\mathbf{E}(s) = \left(\frac{Q(s_0)}{\eta(s)Q(s)}\right)^{1/2} \left\{\hat{\mathbf{E}}_+ \mathrm{e}^{ik_0\int\eta ds} + \hat{\mathbf{E}}_- \mathrm{e}^{-ik_0\int\eta ds}\right\} \mathrm{e}^{-i\omega t}. \tag{5.165}$$

These are the WKB solutions with $\eta(s_0) = 1$. They simply express the fact that the energy flows with the group velocity along the rays without being attenuated by gradual reflection. The propagating wave $\mathbf{E}_+$ and counterpropagating wave $\mathbf{E}_-$ are normal or parallel, respectively, to the rays and do not interact with each other: no crossing of rays from different origin. In the plane layered medium holds $\mathbf{k}_0\eta\mathrm{d}\mathbf{x} = k_0\eta\mathrm{d}x/\cos\alpha$ and $Q(x) \sim \mathrm{d}y\cos\alpha$. The angle α along a ray is given by (5.158). Thus

$$\begin{aligned}\mathbf{E}(\mathbf{x},t) = &\frac{\hat{\mathbf{E}}_+(\mathbf{x}_0, t=0)}{(\eta^2 - \sin^2\alpha_0)^{1/4}} \mathrm{e}^{ik_0\int_{x_0}^{x}\sqrt{(\eta^2-\sin^2\alpha_0)}\mathrm{d}x' - i\omega t} \\ &+ \frac{\hat{\mathbf{E}}_-(\mathbf{x}_0, t=0)}{(\eta^2 - \sin^2\alpha_0)^{1/4}} \mathrm{e}^{-ik_0\int_{x_0}^{x}\sqrt{(\eta^2-\sin^2\alpha_0)}\mathrm{d}x' - i\omega t}. \end{aligned} \tag{5.166}$$

In the neighborhood of a critical point the condition $\lambda(\mathbf{x}) = 2\pi/k_0\eta(\mathbf{x}) \ll L(\mathbf{x})$ of the optical approximation is never fulfilled and the eikonal approach with its consequences (5.156)–(5.161) becomes meaningless. In general, the limitation holds also at the turning points. In (5.165) and (5.166) the amplitudes diverge there. At normal incidence the swelling factor of the amplitudes is $\eta^{-1/2}$. The longitudinal Langmuir wave follows the same refraction index. Therefore (5.165) and (5.166) hold for it if k_0 is replaced by $k_{0e} = k_0/\beta$, $\beta = c_{se}/c$, $c_{se} = \mathrm{const}$.

A serious difficulty arises with the expressions (5.156) and (5.159) in an absorbing medium, in particular when the absorption is strong, since then the imaginary parts of $\mathbf{k}$ and $\mathbf{v}_g$ can no longer be ignored. Several solutions have been proposed to overcome this difficulty. We remark that an interesting reinterpretation of the group velocity in terms of a time-dependent combination of $\Re(\partial\omega/\partial\mathbf{k})$ and $\Im(\partial\omega/\partial\mathbf{k})$ was presented and appropriate ray-tracing equations were derived in [9]. For more recent developments on this subject the interested reader may consult [10, 11]. In the case of $\nu \neq 0$ and η complex the concept of rays still makes sense provided $\Im\eta \ll \Re\eta$. Then it does not matter much whether in the ray equation (5.156) and in (5.166) η is substituted by $\eta_r = \Re\eta$ or by $|\eta| = (\eta_r^2 + \eta_i^2)^{1/2} \simeq \eta_r(1 + \eta_i^2/2\eta_r^2) \simeq \eta_r$, $(\eta_i = \Im\eta)$. If it is again postulated that the energy flow is along the rays, i.e. $\overline{\mathbf{S}}\|\mathrm{d}\mathbf{s}$, then η is to be replaced by η_r in these equations. It is also in accordance with Ginzburg's suggestion (see [20], p. 249).

Wave Structure in the Critical Region

In the experiments of laser-plasma interaction two questions are of main importance: the maximum field amplitude and the contribution to collisional absorption in the evanescent region. The distribution of the electric field $\mathbf{E}(\mathbf{x})$ around $\eta(\mathbf{x}) = 0$ and turning points can be investigated without too much effort in the layered medium. It is onedimensional, $\eta = \eta(x)$, with the reflection (or turning) point given by $\eta^2(x) - \sin^2\alpha_0 = 0$ for the electromagnetic as well as the electron plasma wave. The transverse wave is assumed to be in the so called s-*polarized configuration*, with $E = E_z(x, y)$ perpendicular to the plane of incidence (x, y). Fourier decomposition of $E_z(x, y)$ along y yields $E_z(x, k_y) = E_z(x)\exp(ik_y(x) - i\omega t)$. From (5.158) follows $k_y(x) = k_0\sin\alpha_0 = \mathrm{const}$ and (5.145) simplifies to

$$\text{s wave}: \quad \frac{\partial^2 E_z}{\partial x^2} + k_0^2[\eta^2(x) - \sin^2\alpha_0]E_z = 0; \;\; k_{0e} = \omega/c. \tag{5.167}$$

$$\text{long. wave}: \quad \frac{\partial^2 E_l}{\partial x^2} + k_{0e}^2[\eta^2(x) - \sin^2\alpha_0]E_l = 0; \;\; k_{0e} = \omega/c_{se}. \tag{5.168}$$

The electron plasma wave E_l is curl-free, $\nabla \times \mathbf{E} = 0$. Its Fourier decomposition is identical because it follows the same refraction law if k_0 is replaced by $k_{0e} = k_0/\beta$. In s configuration $\partial E_z/\partial z = 0$, the two waves do not couple. In the critical region η^2 may be bridged by a linear profile of slope q around the turning point $x_0 = x_c + \sin^2\alpha_0/q$ provided the curvature of the profile is not too large. With the new coordinate ξ,

$$\xi = (\kappa^2 q)^{1/3}(x - x_0), \quad x_0 = x_c + \frac{\sin^2\alpha_0}{q}, \quad \kappa = k_0, \text{ and } \kappa = k_{0e} \tag{5.169}$$

x_c the critical point, (5.167) and (5.168) transform into the homogeneous *Stokes equation*

$$y'' + \xi y = 0; \qquad y = E_z. \tag{5.170}$$

A solution is found by Fourier-transforming y,

$$-k^2\varphi(k) + i\frac{\mathrm{d}\varphi}{dk} = 0 \quad \Rightarrow \quad \varphi(k) = \mathrm{e}^{-ik^3/3}$$

and inverting φ along a suitable contour C,

$$y(\xi) = y_0 \int_C \exp i\left(k\xi - \frac{k^3}{3}\right) \mathrm{d}k. \tag{5.171}$$

$\Re y$ and $\Im y$ represent two linearly independent solutions of (5.170). When a light beam is incident from the right the solution must become evanescent in the overdense halfspace $\xi < 0$. This is accomplished by combining $\Re y$ and $\Im y$ in a suitable way; the resulting solution is the Airy function $\mathrm{Ai}(\xi)$, if $y_0 = 1/(2\pi)$:

$$\mathrm{Ai}(\xi) = \frac{f(\xi)}{3^{2/3}(2/3)} - \frac{g(\xi)}{3^{1/3}(1/3)};$$

$$f(\xi) = 1 - \frac{1}{3!}\xi^3 + \frac{1\cdot 4}{6!}\xi^6 - \frac{1\cdot 4\cdot 7}{9!}\xi^9 + \frac{1\cdot 4\cdot 7\cdot 10}{12!}\xi^{12} \mp \ldots,$$

$$g(\xi) = -\xi + \frac{2}{4!}\xi^4 - \frac{2\cdot 5}{7!}\xi^7 + \frac{2\cdot 5\cdot 8}{10!}\xi^{10} \mp \ldots \tag{5.172}$$

([21], §10.4 with $\xi = -z$). The second, linearly independent solution is chosen such that it diverges monotonically,

$$\mathrm{Bi}(\xi) = \frac{f(\xi)}{3^{1/6}(2/3)} + \frac{g(\xi)}{3^{-1/6}(1/3)},$$

Ai and Bi are shown in Fig. 5.15a. The power series representation is suitable for small values of $|\xi|$. For large $|\xi|$, $\mathrm{Ai}(\xi)$ is expected to be well approximated by its asymptotic forms. With $\zeta = \int_0 \eta d\xi \sim \int_0 \xi^{1/2} d\xi = \frac{2}{3}\xi^{3/2}$ and $\arg \xi^{3/2} = \frac{3}{2}\arg \xi$, $\arg \xi^{1/4} = \frac{1}{4}\arg \xi$ they are

$$\begin{aligned}
\mathrm{Ai}(\xi) &= \mathrm{Ai}_1(\xi) = \frac{1}{2\sqrt{\pi}\xi^{1/4}} e^{-i\zeta}, \quad \text{for } \frac{\pi}{3} \le \arg \xi \le \frac{5}{3}\pi, \\
\mathrm{Ai}(\xi) &= \mathrm{Ai}_2(\xi) = \frac{1}{2\sqrt{\pi}\xi^{1/4}} \left\{ \exp i(\zeta - \frac{\pi}{4}) + \exp\left[-i\left(\zeta - \frac{\pi}{4}\right)\right]\right\}, \\
&\text{for } -\frac{\pi}{3} \le \arg \xi \le \frac{\pi}{3}.
\end{aligned} \tag{5.173}$$

The rays emanating from the origin under the angles $\delta = \pm\pi/3, \ -\pi$ are the so-called anti-Stokes lines. They play an important role for the determination of the correct WKB solutions connecting different regions in the complex ξ-plane (see, e.g., [22], Sect. 1–4). In Fig. 5.15b a comparison is made between $|\mathrm{Ai}|^2$ and its asymptotic expansion $|\mathrm{Ai}_2|^2$ for real ξ. As soon as $|\xi| \ge 1.3$ the relative error $|\mathrm{Ai} - \mathrm{Ai}_2|/|\mathrm{Ai}|$ becomes less than 1.5%. Even at the maximum of Ai which lies at $\xi = 1.045$ the deviation is only 4%. The deviation starts to become exponentially large when $|\xi|$ is less than 0.5. It is typical for the WKB approximation in general that even the maximum E-field and its location are approximately reproduced. The accuracy of (5.173) can be estimated analytically by casting the exact solution into the form

$$y(\xi) = C_+ W_+ + C_- W_-; \quad W_\pm = \xi^{-1/4}\exp(\pm i\zeta), \quad \xi > 0.$$

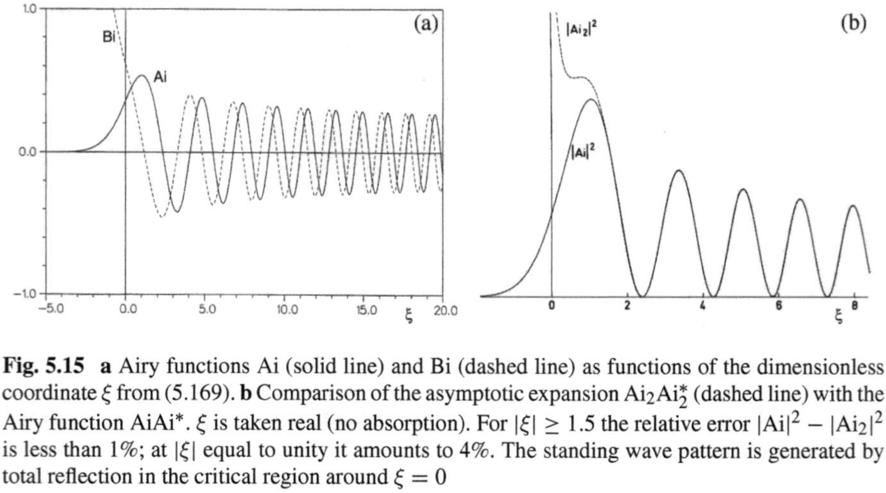

Fig. 5.15 **a** Airy functions Ai (solid line) and Bi (dashed line) as functions of the dimensionless coordinate ξ from (5.169). **b** Comparison of the asymptotic expansion $\mathrm{Ai}_2\mathrm{Ai}_2^*$ (dashed line) with the Airy function AiAi^*. ξ is taken real (no absorption). For $|\xi| \ge 1.5$ the relative error $|\mathrm{Ai}|^2 - |\mathrm{Ai}_2|^2$ is less than 1%; at $|\xi|$ equal to unity it amounts to 4%. The standing wave pattern is generated by total reflection in the critical region around $\xi = 0$

The deviations of $C_\pm$ from unity are a suitable measure of the accuracy of $W_\pm$. In general (see, e.g., [22], (1)–(106))

$$|\Delta C_\pm| \lesssim \frac{5}{48|\xi|}. \tag{5.174}$$

For large $|\xi|$ the WKB form becomes extremely precise. Absorption and reflection coefficients can be calculated for arbitrary, smooth density profiles by bridging the critical point with the help of Stokes' equation. To this end let us assume a wave of unit amplitude incident from vacuum and let the amplitude of the reflected wave be s. Then from the requirement that the electric field must be continuous one obtains at x, with the associated $|\xi|$ lying in the WKB region,

$$\eta^{-1/2}\exp(-ik_0\int_x^\infty \eta \mathrm{d}x) + s\eta^{-1/2}\exp(ik_0\int_x^\infty \eta \mathrm{d}x)$$
$$= C\xi^{-1/4}\left(\mathrm{e}^{\mathrm{i}(\zeta-\pi/4)} + \mathrm{e}^{-i(\zeta-\pi/4)}\right). \tag{5.175}$$

In order to satisfy this relation the incident wave at the RHS must be equal to the incident wave at the LHS, and the same must be true for the reflected wave since, within the limit of the WKB approximation the two waves do not interact. By eliminating the factor C one obtains for the amplitude ratio s

$$s = \exp\left[-2i(k_0\int_x^\infty \eta \mathrm{d}x - \zeta + \pi/4)\right].$$

The reflection coefficient R for the intensity is given by

$$R = ss^* \exp\left[-4(\Im\zeta + k_0\int_x^\infty |\Im\eta| \mathrm{d}x)\right]. \tag{5.176}$$

The position x is somewhat arbitrary. On the one hand $\Delta x = x - x_c$ must not be too small for the validity of the WKB approximation; on the other hand it should not be too large in order to keep the difference between η^2 and its linear approximation small. This difference can be evaluated by taking the quadratic term in the expansion of η^2 into account. The coefficient s would be independent of x if the WKB solution were exact. In order to be a good approximation, $|(1/s)(\mathrm{d}s/\mathrm{d}x)|\Delta x \ll 1$ has to be fulfilled. Straightforward algebra yields

$$\left|\frac{1}{s}\frac{\mathrm{d}s}{\mathrm{d}x}\right|\Delta x < \left|\frac{1}{s}\frac{\mathrm{d}s}{\mathrm{d}x}\right|(x-x_0) = \frac{1}{2}\left|\frac{q'}{(k_0q^2)^{2/3}\xi^{5/2}}\right| \ll 1;\ q' = \left.\frac{\mathrm{d}q}{\mathrm{d}x}\right|_{x_0}. \tag{5.177}$$

The inequality is certainly fulfilled if $|q'/(k_0 q^2)^{2/3}| \ll 1$ holds. The formula (5.176) for determining R is rather general and can be cast into the more convenient form

$$R = \exp(-4k_0 \int_{x_0}^{\infty} |\Im \eta| \, \mathrm{d}x) = \exp(-2 \int_{x_0}^{\infty} \alpha \mathrm{d}x), \tag{5.178}$$

provided that

(i) in the region around the turning point x_0, η can be approximated by a linear function and
(ii) the position ξ connecting the linear function with the smooth but otherwise arbitrary function $\eta^2(x)$ lies in the Stokes region of Ai_2.

In the form (5.178) for R all direct evidence of the use of Stokes' equation has disappeared; it merely served to fix the correct lower limit of integration x_0. Collisional absorption occurring in the overdense (tunnelling) region is already included when starting the integration in (5.178) from the turning point x_0. The factor 4 comes from $\alpha = 2k_0|\Im \eta|$ in (5.189) and from the fact that the fraction of attenuation of the laser beam is the same on its way into the plasma and out of it after reflection around x_0.

Spherical wave. The homogeneous Stokes equation applies equally well to spherical geometry and normal incidence. Here $\nabla^2 E = \partial_r(r^2 \partial_r E)/r^2 = \partial_{rr}(rE)/r$. Hence, by setting $y = rE$ the stationary electromagnetic wave equation reads

$$\frac{\partial^2}{\partial r^2} y + k_0^2 \eta^2 y = 0. \tag{5.179}$$

Equation (5.170) is the special case for linear η^2. There is an important topological difference between solutions of (5.179) for longitudinal and transverse waves. If y is polarized along r (e.g., acoustic or electron plasma wave) spherically symmetric solutions of the form $y = C\mathrm{e}^{i\phi}/r$, $C = \mathrm{const}$, exist with infinite amplitude at the origin. No such solution exists for a transverse wave. In the latter case C is always a function of the polar angles φ and ϑ as well; the wave shows diffraction at the origin and the amplitude remains finite [23].

Uniform illumination of a sphere by a monochromatic light wave is impossible.

However, as long as the dominant variation of the electric field in the wave equation (5.146) is into r direction a focused laser beam is well described by (5.179). In the case of small scale beam filamentation [24, 25] it may fail.

Maximum Electric Field Amplitude

Near the critical or turning point the electric field amplitude may increase considerably. The factor f by which this happens is the ratio between the maximum E-field in the plasma $\hat{\mathbf{E}}_P$ and the vacuum field amplitude $\hat{\mathbf{E}}_V$ at the same position,

$$f = \frac{|\hat{\mathbf{E}}_P|}{|\hat{\mathbf{E}}_V|} = \frac{|y_P|}{|y_V|}.$$

Let the incident wave have an amplitude of unity. Then at a suitable position x or r, (5.175) holds, from which the scaling factor C is determined,

$$|C| = \frac{\xi_r^{1/4}}{\eta_r^{1/2}} \exp\left[-\left(\Im\xi + k_0 \int_x^\infty |\Im\eta|\,\mathrm{d}x\right)\right] = \left(\frac{k_0}{|q|}\right)^{1/6} R^{1/4}.$$

Keeping in mind that

$$f = |\hat{\mathbf{E}}_P| = 2\sqrt{\pi}|C||\mathrm{Ai}|_{\max},$$

one obtains for the field increase

$$f = 2 \times 0.54\sqrt{\pi}|k_0 q^{-1}|^{1/6} R^{1/4} = 1.90(k_0 L)^{1/6} R^{1/4}. \qquad (5.180)$$

This formula is valid for an arbitrary smooth electron density distribution. Its limitation consists only in the approximation of (5.167) and (5.179) by the Stokes equation (5.170) around the turning point, the validity of which is guaranteed by condition (5.177). In the case of considerable density profile distortions or when the maximum lies in front of the linear region the formula fails. The dependence of f on R is very weak. It reflects the experience that in numerical calculations, even with considerable absorption, the field still increases towards the reflection point. In fact, when, for example, 90% of light is absorbed ($R = 0.1$), f reduces by a factor of only about 2 with respect to the nonabsorbing case. The knowledge of the f factor is useful for calculating threshold intensities of parametric instabilities or local light pressure effects.

Swelling factor in the transition layer. The radiation pressure of the laser beam remains below the ablation pressure of the downstreaming plasma in the subrelativistic intensity range $a < 1$. However, orders of magnitude below it leads already to plasma density profile steepening in the critical region as soon as the light pressure exceeds a few percent of the plasma pressure. As soon as the density gradient length L_c shrinks below a laser wavelength λ the swelling factor f may be determined more appropriately from the transition layer (5.141). The factor f is obtained from the smooth connection of the solution from [4] with the WKB solution.

A Reflection-Free Density Profile

Instead of starting from the energy flux conservation to derive the WKB solution (5.166) one can alternatively look for an approximate solution to the stationary wave equation (5.146) which, for simplicity, is used now for normal incidence only (the extension to $\alpha_0 \neq 0$ is straightforward),

$$E'' + k_0^2\eta^2(x)E = 0. \tag{5.181}$$

Differentiating twice the eikonal expression $E = C_1 \exp(\pm ik_0 \int \eta \mathrm{d}x)$ yields

$$E'' = -k_0^2 E + \left\{\pm ik_0(\eta' C_1 + 2\eta C_1') + C_1''\right\} \exp(\pm ik_0 \int \eta \mathrm{d}x).$$

Inserting this in (5.181) and dropping C_1'' results in the constraint $\eta'/\eta + 2C_1'/C_1 = 0$, or when integrated, $C_1 = \mathrm{const}/\eta^{1/2}$ in agreement with (5.166). Setting $C_1 = C_2(x)/\eta^{1/2}$ and repeating the previous procedure generates the second correction, etc.. However, it is a general experience that if the first order correction, generally named *the* WKB expression, is not sufficiently accurate, inclusion of the next higher orders does not substantially alter the situation. The first approximation with $C_2 = \mathrm{const}$ is an exact solution of (5.181) for a medium with the refractive index n given by

$$n^2 = \eta^2 + \frac{1}{4k_0^2}\left\{2\frac{\eta''}{\eta} - 3\left(\frac{\eta'}{\eta}\right)^2\right\} = \eta^2 + \eta_1^2, \tag{5.182}$$

as one may verify by differentiating it twice. The general solution of (5.181) with n^2 for the refractive index squared and $y = E$ (plane case) is the sum of two noninteracting terms,

$$E = C_+\eta^{-1/2}\mathrm{e}^{+ik_0\int \eta \mathrm{d}x} + C_-\eta^{-1/2}\mathrm{e}^{-ik_0\int \eta \mathrm{d}x} = E_+ + E_-.$$

Consequently, $E = E_+$ with $C_- = 0$ is also an exact solution. This leads to an interesting conclusion [26]: Any refractive index profile n constructed according to (5.182) on an arbitrary twice differentiable function $\eta(x)$ is *reflection-free*. The η_1^2 term acts in such a way that the reflected wave is cancelled by interference with the incident wave. For the WKB approximation to hold η_1^2 must be small in comparison with η^2. When the collision frequency is high and η^2 very smooth, $\eta_1^2 \ll \eta^2$ may hold even at x_c. Consequently the overall reflection coefficient R is low, in agreement with one would expect since in such a case the density scale length L is large and the laser beam is nearly absorbed before reaching x_c. In the opposite case of ν_{ei} small and/or high density gradient around x_c, η_1^2 is large and no longer monotonic; hence, (5.178) no longer applies to such *resonant* (reflection-free) profiles and no conflict arises. For $\nu_{ei} = 0$ a nonsingular reflection-free profile around x_c does not exist.

5.5.3 High Frequency Energy Fluxes

Energy balance of the different modes and energy fluxes between them deserve particular attention in view of mode coupling and excitation of unstable modes in the laser plasma. Energy conservation is expressed by Poynting's theorem (5.17)

$$\frac{1}{2}\varepsilon_0\frac{\partial}{\partial t}(\mathbf{E}^2 + c^2\mathbf{B}^2) + \nabla\mathbf{S} = -\mathbf{jE}.$$

The term in parentheses contains the energy density of purely electromagnetic nature. It can vary in time by irradiation, expressed by the divergence of the Poynting vector $\mathbf{S}$, and by doing mechanical work on matter through the coupling term $\mathbf{jE}$. To interpret this source and sink term the nonrelativistic Ohm's law (5.149) in a tangent inertial frame comoving with the plasma may again be used, with $(\mathbf{j}\nabla)\mathbf{u}_0$ set equal to zero,

$$\mathbf{E} = \frac{1}{\varepsilon_0\omega_p^2}\left(\frac{\partial\mathbf{j}}{\partial t} + \nu\mathbf{j} + \varepsilon_0 c_{se}^2\nabla(\nabla\mathbf{E})\right).$$

Now let $\mathbf{E}$ be purely transverse, $\mathbf{E} = \mathbf{E}_\perp$. Multiplying by $\mathbf{j}$ and averaging over one period leads to $\overline{\Re(\mathbf{jE})} = \partial_t(n_e W) + 2\nu n_e W$, where W is the mean oscillation energy of the electron. Hence, the cycle-averaged relation (5.17) can be written as

$$\frac{\partial}{\partial t}\left\{\frac{\varepsilon_0}{4}\hat{\mathbf{E}}\hat{\mathbf{E}}^* + \frac{\varepsilon_0 c^2}{4}\hat{\mathbf{B}}\hat{\mathbf{B}}^* + \frac{1}{2}n_e W\right\} + \nabla\overline{\mathbf{S}} = -2\nu n_e W. \tag{5.183}$$

If there is no dissipation ($\nu = 0$) the only energy "absorption" $\nabla\overline{\mathbf{S}} < 0$ is accomplished by increasing the electric, magnetic, or kinetic energy densities; they are reversible. When the wave field as well as the plasma density are both stationary, $\nabla\overline{\mathbf{S}} = 0$ must hold. As the wave penetrates a fully ionized plasma with an electron density slowly varying in space the magnetic energy density decreases according to $\varepsilon_0 c^2\hat{\mathbf{B}}\hat{\mathbf{B}}^*/4 = \eta^2\varepsilon_0\hat{\mathbf{E}}\hat{\mathbf{E}}^*/4$, whereas the kinetic energy density increases by the same amount. In fact, the total density $\mathcal{E}$ is

$$\begin{aligned}\mathcal{E} &= \frac{\varepsilon_0}{4}\left(\hat{\mathbf{E}}\hat{\mathbf{E}}^* + c^2\hat{\mathbf{B}}\hat{\mathbf{B}}^*\right) + \frac{1}{2}n_e W \\ &= \frac{\varepsilon_0}{4}\hat{\mathbf{E}}\hat{\mathbf{E}}^*\left(1 + \eta^2 + \frac{\omega_p^2}{\omega^2}\right) = \frac{\varepsilon_0}{2}\hat{\mathbf{E}}\hat{\mathbf{E}}^*\end{aligned} \tag{5.184}$$

thus revealing that the magnetic energy gradually transforms into oscillation energy. At the critical point $B = \eta E/c \sim \eta^{1/2}$ tends to zero and the oscillation energy is exactly equal to the electric energy. Equations (5.183) and (5.184) are in perfect agreement with the WKB solution (5.166). It is interesting to note that, without dissipation, (5.183) yields with the aid of (2.1), (2.2), and (5.184)

$$\frac{\partial}{\partial t}\mathcal{E} + \nabla\overline{\mathbf{S}} = \frac{\partial}{\partial t}\mathcal{E} + \nabla\left(\frac{\varepsilon_0}{2}c\eta\hat{\mathbf{E}}\hat{\mathbf{E}}^*\right) = \frac{\partial}{\partial t}\mathcal{E} + \nabla(\mathbf{v}_g\mathcal{E}) = 0$$

for the fully ionized plasma. In a general dispersive medium the last version of energy conservation also holds, but its derivation is more subtle. In the presence of dissipation ($\nu > 0$) the pure absorptive case with $\partial\mathcal{E}/\partial t = 0$ is of greatest relevance,

$$\nabla\overline{\mathbf{S}} = -2\nu n_e W, \quad n_e W = \frac{1}{2}\left(1 - \Re\eta^2\right)\mathcal{E}. \tag{5.185}$$

In the dispersive medium light rays are no longer parallel, see Fig. 5.13. A narrow light bundle changes its cross section Q along the ray path s. Introducing the flux $J(s) = Q(s)I(s)$ (5.185) transforms with the aid of Gauss' law into

$$\frac{\mathrm{d}J}{\mathrm{d}s} = -\frac{\nu}{c\eta_r}\frac{\omega_p^2}{\omega^2 + \nu^2}J = -\alpha J \Leftrightarrow J(s) = J(s_0)\mathrm{e}^{-\int_{s_0}^{s}\alpha\mathrm{d}s}. \tag{5.186}$$

The integrated version on the RHS is recognized as *Beer's law*. Alternatively, $I = |\overline{\mathbf{S}}|$ may be calculated directly from (5.165) with the undamped local amplitude $\hat{\mathbf{E}}(s) = \hat{\mathbf{E}}_+[Q(s_0)/\eta(s)Q(s)]^{1/2}$

$$I(s) = \varepsilon_0 c^2|\overline{\mathbf{E}\times\mathbf{B}}| = \frac{1}{2}\varepsilon_0 c\eta_r\hat{\mathbf{E}}\hat{\mathbf{E}}^*\mathrm{e}^{-2k_0\int^s\Im\eta\mathrm{d}s},$$

$$\alpha = 2k_0\Im\eta. \tag{5.187}$$

From $A = \Re\eta^2$ and $B = \Im\eta^2$ in (5.143) one deduces

$$\eta_r = \Re\eta = \left\{\frac{1}{2}\left(\sqrt{A^2 + B^2} + A\right)\right\}^{1/2},$$

$$\eta_i = \Im\eta = \left\{\frac{1}{2}\left(\sqrt{A^2 + B^2} - A\right)\right\}^{1/2}. \tag{5.188}$$

Expanding η for $B^2 \ll A^2$ leads to

$$\eta_r = \left(1 - \frac{\omega_p^2}{\omega^2 + \nu^2}\right)^{1/2}, \quad \eta_i = \frac{B}{2\eta_r}, \quad \alpha = 2k_0\Im\eta = \frac{\nu}{c\eta_r}\frac{\omega_p^2}{\omega^2 + \nu^2}, \tag{5.189}$$

i.e., the absorption coefficients from (5.186) and (5.187) become identical, as expected. The refractive index η_r in the denominator is due to the increase of $\mathbf{E}$ and W with increasing plasma density. If the eikonal approximation is no longer valid, no simple relation exists between $\mathcal{E}$ and $\overline{\mathbf{S}}$ and the absorption has to be calculated from (5.183) and the wave equation (5.145), or a simplified version of it, e.g. $\Delta\mathbf{E} + k_0^2\eta^2\mathbf{E} = 0$.

For the longitudinal component $\mathbf{E} = \mathbf{E}_\parallel$, $\mathbf{S} = 0$ holds and the energy is carried by the thermal motion of the electrons. On the other hand, one would argue that dispersion relations of identical structures lead to energy conservation equations of the same structure. Hence, in the dissipation-free case, $\partial_t \mathcal{E} + \nabla(\mathbf{v}_g \mathcal{E}) = 0$ should hold again with $\mathcal{E} = f \varepsilon_0 \hat{\mathbf{E}}_\parallel \hat{\mathbf{E}}_\parallel^* / 2$, where f is an eventual proportionality constant and $v_g = c_{\mathrm{se}}^2 / v_\varphi$. To see whether such an argument holds in the general case one may start from the linearized momentum equations (5.64) and (5.111) for electrons and ions without dissipation,

$$m_e n_{e0} \partial_t \mathbf{u}_e = -m_e c_{\mathrm{se}}^2 \nabla n_{e1} - n_{e0} e \mathbf{E},$$

$$m_i n_0 \partial_t \mathbf{u}_i = -m_i c_{\mathrm{si}}^2 \nabla n_{i1} + n_0 Z e \mathbf{E}.$$

Multiplying the first equation by $\mathbf{u}_e$ and the second equation by $\mathbf{u}_i$ and adding them yields

$$\partial_t \left(\frac{1}{2} n_{e0} m_e \mathbf{u}_e^2 + \frac{1}{2} n_0 m_i \mathbf{u}_i^2 \right) - \mathbf{jE} = -m_e c_{\mathrm{se}}^2 \mathbf{u}_e \nabla n_{e1} - m_i c_s^2 \mathbf{u}_i \nabla n_{i1}$$

$$= -m_e c_{\mathrm{se}}^2 \big(\nabla (n_{e1} \mathbf{u}_e) - n_{e1} \nabla \mathbf{u}_e \big) - m_i c_s^2 \big(\nabla (n_{i1} \mathbf{u}_i) - n_{i1} \nabla \mathbf{u}_i \big).$$

The linearized mass conservation equations imply

$$n_{\alpha 1} \nabla \mathbf{u}_\alpha = \frac{n_{\alpha 1}}{n_{\alpha 0}} \nabla n_{\alpha 0} \mathbf{u}_\alpha = -\frac{n_{\alpha 1}}{n_{\alpha 0}} \partial_t n_{\alpha 1} = -\frac{1}{2} n_{\alpha 0} \partial_t \left(\frac{n_{\alpha 1}}{n_{\alpha 0}} \right)^2, \quad \alpha = e, i,$$

and, when substituted in the foregoing equation,

$$\partial_t \left(\frac{1}{2} n_{e0} m_e \mathbf{u}_e^2 + \frac{1}{2} n_0 m_i \mathbf{u}_i^2 \right) + \partial_t \left[\frac{1}{2} n_{e0} m_e c_{\mathrm{se}}^2 \left(\frac{n_{e1}}{n_{e0}} \right)^2 + \frac{1}{2} n_0 m_i c_s^2 \left(\frac{n_{i1}}{n_0} \right)^2 \right]$$

$$+ 2 \nabla \left(\frac{1}{2} \mathbf{u}_{g,e} n_{e0} m_e \mathbf{u}_e^2 + \frac{1}{2} \mathbf{u}_{g,i} n_0 m_i \mathbf{u}_i^2 \right) - \mathbf{jE} = 0 \tag{5.190}$$

is obtained, or

$$\frac{\partial}{\partial t} \left(\mathcal{E}_{e,\mathrm{kin}} + \mathcal{E}_{i,\mathrm{kin}} + \mathcal{E}_{e,\mathrm{pot}} + \mathcal{E}_{i,\mathrm{pot}} \right) + 2 \nabla \left(\mathbf{u}_{g,e} \mathcal{E}_{e,\mathrm{kin}} + \mathbf{u}_{g,i} \mathcal{E}_{i,\mathrm{kin}} \right) - \mathbf{jE} = 0,$$

$v_{g,\alpha}$ group velocity. The divergence term was obtained with the help of $\mathbf{u}_\alpha = \mathbf{u}_{\varphi,\alpha} n_{\alpha 1} / n_{\alpha 0}$ from (5.73) and (5.114). When $\mathbf{jE}$ is substituted from (5.17) at every time instant the wave energy balance is given by

$$\frac{\partial}{\partial t}\left(\mathcal{E}_{\mathrm{e}}+\mathcal{E}_{e,\mathrm{kin}}+\mathcal{E}_{i,\mathrm{kin}}+\mathcal{E}_{e,\mathrm{pot}}+\mathcal{E}_{i,\mathrm{pot}}\right)+\nabla\left(\mathbf{S}+2\mathbf{S}_{e,\mathrm{kin}}+2\mathbf{S}_{i,\mathrm{kin}}\right)=0;$$

$$S_{e,\mathrm{kin}}=\frac{1}{2}\frac{c_{\mathrm{se}}^2}{v_\varphi}n_{e0}m\mathbf{u}_{e\parallel}^2;\quad S_{i,\mathrm{kin}}=\frac{1}{2}c_s n_0 m_i \mathbf{u}_{i\parallel}^2. \tag{5.191}$$

Specializing to a monochromatic electron-plasma wave we have, with the help of (5.73) and after cycle-averaging,

$$\overline{\mathcal{E}}_{e,\mathrm{kin}}+\overline{\mathcal{E}}_{e,\mathrm{pot}}=\frac{1}{2}m_e n_{e0}\left(1+\frac{c_{\mathrm{se}}^2}{v_\varphi^2}\right)\overline{\mathbf{u}_e^2}=\frac{\varepsilon_0}{4}\frac{\omega^2+c_{\mathrm{se}}^2\mathbf{k}_e^2}{\omega_p^2}\hat{\mathbf{E}}\hat{\mathbf{E}}^*,$$

$$\overline{S}_{e,\mathrm{kin}}=\frac{\varepsilon_0}{4}v_g\frac{\omega^2}{\omega_p^2}\hat{\mathbf{E}}\hat{\mathbf{E}}^*. \tag{5.192}$$

Hence, the total cycle-averaged energy density $\mathcal{E}_{\mathrm{s}}$ and flux density $I=2\overline{S}_{e,\mathrm{kin}}$ are

$$\mathcal{E}_{\mathrm{s}}=\mathcal{E}_{\mathrm{e}}+\mathcal{E}_{e,\mathrm{kin}}+\mathcal{E}_{e,\mathrm{pot}}=\frac{\varepsilon_0}{2}\frac{\omega^2}{\omega_p^2}\hat{\mathbf{E}}\hat{\mathbf{E}}^*,\quad I=v_g\mathcal{E}_{\mathrm{s}}. \tag{5.193}$$

This shows that the idea based on the dispersion relations was correct with $f=\omega^2/\omega_p^2$. When an electron-plasma wave is resonantly excited by the laser, $\omega=\omega_p$, $f=1$ holds, since $k_e=0$ at resonance (see Chap. 6). Energy conservation (5.191) becomes simplest for the ion acoustic wave,

$$\partial_t\mathcal{E}_i+\nabla\mathbf{I}_i=0;$$

$$\mathcal{E}_i=\overline{\mathcal{E}}_{i,\mathrm{kin}}+\overline{\mathcal{E}}_{i,\mathrm{pot}}=2\overline{\mathcal{E}}_{i,\mathrm{kin}}=\frac{1}{2}n_0 m_i\hat{\mathbf{u}}_i^2,\quad I_i=2\overline{S}_{i,\mathrm{kin}}=c_s\mathcal{E}_i. \tag{5.194}$$

The derivation of a relation analogous to (5.191) for dielectric media is more involved. It is interesting to note that the group velocity enters in the energy flux density only when the total energy density $\mathcal{E}$ is considered. If, instead, one includes in $\mathcal{E}$ solely the energy density $\mathcal{E}_c$ which is carried by the wave and transmitted from point to point v_g has to be replaced by another speed. Thus, for instance, for an electromagnetic wave in the plasma $\mathcal{E}_c=\mathcal{E}_{\mathrm{e}}-\mathcal{E}_{e,\mathrm{kin}}=\varepsilon_0\eta^2\hat{\mathbf{E}}\hat{\mathbf{E}}^*/4$ and the energy conservation $\partial_t\mathcal{E}+\nabla(\mathbf{v}_g\mathcal{E})=0$ is replaced by the equivalent expression

$$\partial_t\mathcal{E}_c+\nabla\overline{\mathbf{S}}=\partial_t\mathcal{E}_c+\nabla(\mathbf{v}_\varphi\mathcal{E}_c)=0;$$

$\mathcal{E}_c$ is always carried at the phase velocity $v_\varphi=c/\eta$. The energy balance presented above is strictly valid for waves of infinitesimal amplitude in a homogeneous plasma. Its applicability to nonstationary inhomogeneous plasmas is guaranteed as long as $\lambda/L\ll 1$ and $2\pi/\omega T\ll 1$, with L and T the characteristic length and time scales are fulfilled. More precisely, the validity of the above analysis is intimately connected with the adiabatic behavior of an action integral. Therefore, in

***Variational Treatment**

As outlined in Chap. 2 Lagrangian Mechanics is the most general and powerful formalism for solving problems of point dynamics, once the Lagrangian $L = L(q_i, \dot{q}_i, t)$ of generalized point coordinates q_i and their time derivatives $\dot{q}_i$ is known. The familiar equations of motion $d_t L_{\dot{q}_i} - L_{q_i} = 0$ follow from the principle of least action,

$$\delta \int_{t_1}^{t_2} L(q_i, \dot{q}_i, t) \mathrm{d}t = 0.$$

When passing from discrete mass points (δ-functions) to the continuous distribution of fields (see Sect. on Lagrangian in Chap. 3) the number of degrees of freedom becomes infinite and the coordinates $q_i, \dot{q}_i$ become continuous variables $q(\mathbf{x}, t)$ of the field densities and their derivatives $\dot{q} = \partial_t q + (\mathbf{v}\nabla)q = f(q_t, q_{x_j})$. Depending on the problem, q may represent a matter density $\rho(\mathbf{x}, t)$, a velocity field $\mathbf{u}(\mathbf{x}, t)$ or, as in this section, an electric (or magnetic) field. With $q = \mathbf{E}(\mathbf{x}, t)$ the principle of least action reads

$$\delta \int_{t_1}^{t_2} \int_R L(E_i, \partial E_i/\partial t, \partial E_i/\partial x_j, t) \mathrm{d}t \mathrm{d}\mathbf{x} = 0; \quad i = 1, 2, 3. \tag{5.195}$$

By introducing the arbitrary variations $h_i(\mathbf{x}, t)$ for

$$E_i(\mathbf{x}, t) = E_{i0}(\mathbf{x}, t) + h_i(\mathbf{x}, t),$$

and after integrating by parts to eliminate $\partial_t h_i$ and $\partial_j h_i = \partial h_i/\partial x_j$, the Lagrange equations of motion

$$\frac{\partial}{\partial t} L_{E_{it}} + \frac{\partial}{\partial x_j} L_{E_{ixj}} - L_{E_i} = 0; \quad L_{E_{it}} = \frac{\partial L}{\partial E_{it}}, \quad E_{it} = \frac{\partial E_i}{\partial t}, \text{ etc}, \tag{5.196}$$

are obtained in the standard way (e.g., see also [6], p. 391ff). It is seen by inspection that all Lagrangians proportional to

$$L_\perp = \frac{1}{2} \left\{ \frac{\omega_p^2}{c^2} E_i^2 + \sum_j (\partial_j E_i)^2 - \frac{1}{c^2} (\partial_t E_i)^2 \right\} \text{ for } \mathbf{E}_\perp, \quad i = 1, 2, 3,$$

$$L_\parallel = \frac{1}{2} \left\{ \frac{\omega_p^2}{c^2} E_i^2 + \beta^2 \sum_j (\partial_i E_j)^2 - \frac{1}{c^2} (\partial_t E_i)^2 \right\} \text{ for } \mathbf{E}_\parallel, \ i = 1, 2, 3, \tag{5.197}$$

reproduce the wave equation (5.15) with $\partial_t \mathbf{j} = \varepsilon_0 \omega_p^2 \mathbf{E} - \varepsilon_0 \omega_p^2 \nabla(\nabla \mathbf{E})$. For $\mathbf{E}_\perp$ and $\mathbf{E}_\parallel$ separately, these become

$$\nabla^2 \mathbf{E}_\perp - \frac{1}{c^2}\frac{\partial^2}{\partial t^2}\mathbf{E}_\perp - \frac{\omega_p^2}{c^2}\mathbf{E}_\perp = 0,$$

$$\beta^2 \nabla(\nabla \mathbf{E}_\parallel) - \frac{1}{c^2}\frac{\partial^2}{\partial t^2}\mathbf{E}_\parallel - \frac{\omega_p^2}{c^2}\mathbf{E}_\parallel = 0. \tag{5.198}$$

Under the assumption that $\mathbf{E}$, longitudinal or transverse, is described by the eikonal approximation, i.e., $\mathbf{E} = \hat{\mathbf{E}}(\mathbf{x}, t)\exp(i\Psi(\mathbf{x}, t))$, with $\hat{\mathbf{E}}$ and $\mathbf{k} = \nabla\Psi$ changing slowly in space and time, (5.195) transforms into equations containing only $\hat{\mathbf{E}}$, $\Psi_t = -\omega$, $\Psi_{x_j} = k_j$, which are all slowly varying variables. Therefore, instead of using $L = L(\mathbf{E}, \omega, \mathbf{k}, \mathbf{x}, t)$ it is appropriate to introduce the phase-averaged Lagrangian $\mathcal{L} = \mathcal{L}(\hat{\mathbf{E}}, \omega, \mathbf{k}, \mathbf{x}, t)$,

$$\mathcal{L}(\hat{\mathbf{E}}, \omega, \mathbf{k}, \mathbf{x}, t) = \frac{1}{2\pi}\int L(\mathbf{E}, \omega, \mathbf{k}, \mathbf{x}, t)\mathrm{d}\Psi, \tag{5.199}$$

which no longer exhibits any fast dependence on $\mathbf{x}$ and t. By observing that the averaging procedure is interchangeable with the $\mathbf{x}$ and t-integration up to higher orders in the phase averaged $h_i(\mathbf{x}, t)$,

$$\delta\int \mathcal{L}(\hat{\mathbf{E}}, \omega, \mathbf{k}, \mathbf{x}, t)\mathrm{d}t\mathrm{d}\mathbf{x} = \frac{1}{2\pi}\int \mathrm{d}\Psi[\delta\int L(\mathbf{E}, \omega, \mathbf{k}, vx, t)\mathrm{d}t\mathrm{d}\mathbf{x}]$$

the "averaged" principle of least action ([6], Chap. 14),

$$\delta\int \mathcal{L}(\hat{\mathbf{E}}, \omega, \mathbf{k}, \mathbf{x}, t)\mathrm{d}t\mathrm{d}\mathbf{x} = 0 \tag{5.200}$$

results. An explicit proof of interchangeability is given in Chap. 2, Sect. 1.2. Varying in $\hat{\mathbf{E}}, \omega, \mathbf{k}$ as before leads to

$$\mathcal{L}_{\hat{\mathbf{E}}} = 0, \quad \frac{\partial}{\partial t}\mathcal{L}_\omega - \frac{\partial}{\partial \mathbf{x}}\mathcal{L}_\mathbf{k} = 0; \quad (\mathcal{L}_\omega = -\mathcal{L}_{\Psi_t}). \tag{5.201}$$

In the WKB approximation the Lagrangians (5.197) transform into

$$\mathcal{L}_\perp = \frac{1}{4}\left\{\frac{\omega_p^2}{c^2} + \mathbf{k}^2 - \frac{\omega^2}{c^2}\right\}\hat{E}_\perp^2; \quad \mathcal{L}_\parallel = \frac{1}{4}\left\{\frac{\omega_p^2}{c^2} + \beta^2\mathbf{k}^2 - \frac{\omega^2}{c^2}\right\}\hat{\mathbf{E}}_\parallel^2, \tag{5.202}$$

$\mathcal{L}_{\hat{\mathbf{E}}} = 0$ is proportional to the dispersion relation $D(\omega, \mathbf{k}) = 0$; thus $\mathcal{L}_{\hat{\mathbf{E}}} = \hat{\mathbf{E}}^2 D(\omega, \mathbf{k})$, $\hat{\mathbf{E}} = \hat{\mathbf{E}}_\perp$ or $\hat{\mathbf{E}}_\parallel$. With this the second of (5.197) can be written as

$$\frac{\partial}{\partial t}(D_\omega \hat{\mathbf{E}}^2) - \frac{\partial}{\partial \mathbf{x}}(\hat{\mathbf{E}}^2 D_\mathbf{k}) = 0.$$

From $D = 0$ follows with $\omega = \omega(\mathbf{k}, \mathbf{x}, t)$ and

$$D_\omega \mathbf{v}_g + D_\mathbf{k} = 0 \quad \Rightarrow \quad \mathbf{v}_g = -\frac{D_\mathbf{k}}{D_\omega} = -\frac{\mathbf{f}(\mathbf{k}, \mathbf{x}, t)}{g(\mathbf{k}, \mathbf{x}, t)}, \quad D_\mathbf{k} = -g(\mathbf{k}, \mathbf{x}, t)\mathbf{v}_g.$$

Assume now that there is no explicit time dependence. Then, replacing D_ω and $D_\mathbf{k}$ by $g(\mathbf{k}, \mathbf{x})$ and $g(\mathbf{k}, \mathbf{x})\mathbf{v}_g$ in the above equation leads to

$$\frac{\partial}{\partial t}\left(g(\mathbf{k}, \mathbf{x})\hat{\mathbf{E}}^2\right) + \frac{\partial}{\partial \mathbf{x}}\left(g(\mathbf{k}, \mathbf{x})\mathbf{v}_g \hat{\mathbf{E}}^2\right) = g(\mathbf{k}, \mathbf{x})\left\{\frac{\partial}{\partial t}\hat{\mathbf{E}}^2 + \nabla(\mathbf{v}_g \hat{\mathbf{E}}^2)\right\} = 0. \tag{5.203}$$

In the last step $\nabla \times \mathbf{k} = 0$ has been used. This proves that dispersion relations of the same structure lead to analogous energy conservation relations. In addition, it shows that the total energy density in linear dielectrics also propagates with the group velocity.

5.5.4 Collisional Absorption in Special Density Profiles

For practical purposes it is very useful to have explicit formulas at hand for the reflectivity R and absorption $A = 1 - R$ in special density profiles. To keep them simple layered isothermal plasmas are considered only.

Linear Density Profile

As long as $\nu_{ei}^2/\omega^2 \ll 1$ the refractive index can be approximated by

$$\eta^2 = 1 - \frac{n_e}{n_c} + i\frac{\nu_{ei}}{\omega}\frac{n_e}{n_c} = 1 - \frac{n_e}{n_c} + i\mu\frac{n_e^2}{n_c^2}; \quad \mu = \frac{\nu_{ei,c}}{\omega}.$$

By $\nu_{ei,c}$ the collision frequency at the critical point is meant. With $n_e/n_c = 1 - x/L = 1 - u$, L profile length, and η_i approximated by (5.189),

$$\eta_i \simeq \frac{\mu}{2}u^{-1/2}(1 - u)^2,$$

the integral in (5.178) becomes

$$\int_{u=0}^{1} |\Im\eta| \mathrm{d}x = \frac{\mu L}{2} \int_{u=u_0>0}^{1} u^{-1/2}(1-u)^2 \mathrm{d}u$$
$$= \frac{\mu L}{2}\left(\frac{16}{15} - 2u_0^{1/2} + \frac{4}{3}u_0^{3/2} - \frac{2}{5}u_0^{5/2}\right). \tag{5.204}$$

The Taylor expansion of η_i overestimates its true value (5.188) around the critical point (it even diverges there); therefore the exact value is obtained when the lower limit of the integration is taken to be some positive u_0 instead of zero. Since, on the other hand the expansion is satisfactory from $u = 2\mu(1-u)^2 \simeq 2\mu$ upwards, the relative deviation from the leading term is of the order of $2.5\mu^{1/2}$. Hence, with $\mu \ll 1$, the reflection coefficient for a linear profile becomes

$$R = R_L \simeq \exp(-\frac{32}{15}k_0 L\mu); \quad \mu = \frac{\nu_{ei,c}}{\omega} \ll 1. \tag{5.205}$$

For high collision frequencies a more accurate value is obtained from (5.204) by determining u_0 or integrating (5.178) numerically.

The Profile $n_e = n_c(x_c/x)$

$$\int_0^{\infty} \eta_i \mathrm{d}x = \frac{\mu x_c}{2} \int_{u=0}^{1} \frac{\mathrm{d}u}{(1-u)^{1/2}} = \mu x_c; \quad u = x_c/x.$$

$$R = R_1 = \mathrm{e}^{-4k_0 x_c \mu}; \quad \mu \ll 1. \tag{5.206}$$

5.5.4.1 The Profile $n_e = n_c(x_c/x)^2$

$$\int_0^{\infty} \eta_i \mathrm{d}x = \frac{\mu x_c}{2} \int_{u=0}^{1} \frac{u^2 \mathrm{d}u}{(1-u^2)^{1/2}} = \frac{\mu x_c}{4} \arcsin 1 = \frac{\mu x_c}{4}\frac{\pi}{2}.$$

$$R = R_2 = \exp(-\frac{\pi}{2}k_0 x_c \mu); \quad \mu \ll 1. \tag{5.207}$$

This profile is a satisfactory approximation of a spherical isothermal rarefaction wave.

5.5.4.2 The Profile $n_e = n_c(x_c/x)^3$

The integral is evaluated with the help of the gamma function,

$$I = \frac{2}{x_c\mu}\int_{x>x_c}^{\infty}\eta_i \mathrm{d}x \simeq \int_0^1 \frac{u^4\mathrm{d}u}{(1-u^3)^{1/2}} = \frac{\sqrt{\pi}}{3}\frac{\Gamma(5/3)}{\Gamma(13/6)} = 0.4928; \quad \mu \ll 1.$$

$$R = R_3 = \exp(-4k_0\frac{x_c\mu}{2}I) = \exp(-0.98\, k_0 x_c\mu). \tag{5.208}$$

This profile represents an asymptotically spherical rarefaction wave. The wave equation (5.179) shows that formulas (5.206)–(5.208) are equally valid for spherical targets if x and x_c are substituted by the radii r and r_c. The scale length L of a profile of the form $n_e = n_c(r_c/r)^\alpha$ is connected with r_c by $L = L_\alpha = r_c/\alpha$ and the above reflection formulas read as follows,

$$R_L = \exp(-\frac{32}{15}k_0 L\mu),\ R_1 = \exp(-4k_0 L_1\mu),\ R_2 = \exp(-\pi k_0 L_2\mu),$$

$$R_3 = \exp(-2.2\, k_0 L_3\mu). \tag{5.209}$$

If the absorption is low, i.e., $k_0 r_c\mu$ is small, all four formulas give reasonably accurate results. In the case of strong absorption the formulas show that the form of the density profile very much influences the result. The parameter characterizing absorption is the product $k_0 r_c\mu$. From this expression the wavelength dependence can easily be studied under various conditions. If, for example, the critical radius is kept constant the absorption is independent of λ for a given density profile. In fact k_0 varies as $1/\lambda$ and $\mu = \nu_{ei,c}/\omega$ is proportional to λ. On the other hand, from (5.209) a very strong dependence of R on the ion charge number Z should be expected. Since $A \sim Z$ holds for the atomic mass A, under otherwise identical conditions, the reflection coefficients for deuterium and a high-Z material are related by the power law

$$R_Z = R_{D_2}^Z.$$

If heat conduction plays a role this relation is drastically modified [28].

5.5.5 *Summary*

Laser and intense ion beam generated plasmas show to a great extent *collective response* to the electric and magnetic fields **E** and **B**. The dynamics is adequately described by the interaction of two “fluids”, i.e, the ionized matter plasma generates local distributions of electric charges $\rho_{\mathrm{el}}(\mathbf{x}, t)$ and currents $\mathbf{j}(\mathbf{x}, t)$ which couple to the electromagnetic “fluid” $\{\mathbf{E}, \mathbf{B}\}(\mathbf{x}, t)$. The plasma as a fluid is described in Chap. 3, its linear interaction with the fields is the subject of the present chapter. The fields are governed by the Maxwell equations in which the material response enters in the form of source terms $\{\rho_{\mathrm{el}}, \mathbf{j}\}(\mathbf{x}, t)$. The dynamics of the plasma generates the fields and the fields, in turn, tell the plasma how to move. This is most elegantly described by the wave equations (5.10) in terms of the vector potential $\mathbf{A}(\mathbf{x}, t)$ and the scalar potential $\Phi(\mathbf{x}, t)$ in the Lorentz gauge,

$$\nabla^2\mathbf{A} - \frac{1}{c^2}\frac{\partial^2\mathbf{A}}{\partial t^2} = -\frac{\mathbf{j}}{\varepsilon_0 c^2}; \quad \nabla^2\Phi - \frac{1}{c^2}\frac{\partial^2\Phi}{\partial t^2} = -\frac{\rho_{\mathrm{el}}}{\varepsilon_0}; \quad \nabla\mathbf{A} + \frac{\partial\Phi}{c^2\partial t} = 0$$

with the retarded solutions (5.12) and (5.13). Although the field equations are linear, and hence guaranteeing the superposition of ion and electron contributions, the equations describing this self consistent mutual interaction of the two "fluids" result highly nonlinear in general. Only by the *linearization technique* general laws and rules can be found analytically. The procedure corresponds to the limitation to small signals, except one situation that is given by the propagation of a monochromatic transverse wave in circular polarization. In this case the electron motion is entirely confined to the plane perpendicular to the propagation direction of the plane wave and no coupling to longitudinal modes occurs.

Relativity. In the cold plasma the motion of the electrons becomes relativistic from intensities I corresponding to the normalized momentum $a = e\hat{A}/m_e c \geq 1$ on, $I[\mathrm{Wcm}^{-2}] = 1.37 \times 10^{18} a^2(\lambda[\mu\mathrm{m}])^{-2}$, linear polarization. The relativistic covariance of the Maxwell equations, the energy-momentum tensor, and the invariance of $c^2\mathbf{B}^2 - \mathbf{E}^2$ and $\mathbf{EB}$ is seen from their four dimensional formulation but follows, as shown (see Chap. 2), equally well from their 3D representations. Both representations have their advantages and are used in this chapter. Maxwell's equations combined with the relativistic two fluid model from Chap. 3 constitute the simplest fully relativistic tool to describe the laser-plasma dynamics in the high energy domain.

Free modes in the unmagnetized plasma. In the infinitely extended homogeneous plasma there exist two *transverse eigenmodes*, corresponding to the two possible polarizations (linear: parallel, orthogonal $\Longleftrightarrow$ circular: right, left) with the dispersion relation (5.47) and phase vs group velocity (5.56)

$$\omega^2 = \mathbf{k}_0^2\varepsilon(\mathbf{k}, \omega) = \mathbf{k}_0^2\left(1 - \frac{\omega_p^2}{\omega^2}\right) \quad \Leftrightarrow \quad \omega^2 = \omega_p^2 + c^2\mathbf{k}^2, \qquad v_\varphi v_g = c^2.$$

They have a *cut off* at the electron plasma frequency, $\omega = \omega_p$. Above the critical density $n_c = \varepsilon_0 m_e \omega^2/e^2 = 10^{21}[\mathrm{cm}^{-3}]/(\lambda[\mu\mathrm{m}])^2$ no laser beam propagation is possible. It implies that plasmas generated by high power lasers are hot and close to ideality. The critical density of the circularly polarized wave in the relativistic regime is increased by the Lorentz factor γ, $n_{cr} = \gamma n_c$. No similar simple relation exists for linear polarization (although $n_{cr} = \gamma n_c$ is widely used in linear polarization also). In addition to electromagnetic field energy and electron oscillatory energy it carries electron thermal energy, see (5.193).

The high frequency *electron plasma wave*, named also *electrostatic* or *Langmuir wave*, obeys the *Bohm–Gross dispersion relation*

$$\omega^2 = \omega_p^2 + c_{\rm se}^2 \mathbf{k}^2, \qquad v_\varphi v_g = c_{\rm se}^2; \qquad c_{\rm se}^2 = 3\frac{k_B T}{m_e}.$$

It exhibits the same cut off as the electromagnetic wave. In the nonlinear regime the electron plasma wave shows significant new features: *steepening, peaking, breaking, particle trapping, particle reflection.*

At low frequency a perturbation in the ion density is nearly completely neutralized by the quick electrons. The residual electric field merely transmits the electron thermal pressure to the ions. This has two consequences: (1) absence of a corresponding ion plasma frequency ω_{pi} and (2) a restoring force from ion as well as from the electron thermal pressure. The dispersion relation of the *ion acoustic wave* reads as (5.113)

$$\omega = s|\mathbf{k}|, \quad c_s^2 = \frac{\gamma_e Z k_B T_e + \gamma_i k_B T_i}{m_i}; \quad k\lambda_D \ll 1; \quad v_g = v_\varphi.$$

The ion acoustic wave is *dispersion free* because of $v_g = v_\varphi$. The three types of linear modes have one property in common, i.e., once ω or $\mathbf{k}$ is fixed it is sufficient to know one additional parameter, for instance its amplitude, to describe all properties of the mode, like charge density, thermal pressure (both absent in the transverse wave), electric field, potential, and quiver motion.

The shortest wavelength. The electromagnetic wave shows no finite limit in λ extending to zero because it exists also in the empty space with a continuous transition from plasma dispersion to $\omega = c|\mathbf{k}|$ in vacuum. For the longitudinal wave the situation is completely different because it is supported by the discrete matter. In the neutral fluid the acoustic wave is the only longitudinal mode with a limiting $\lambda \cong 4\Lambda$, with Λ the mean free path. Below no pressure transmission to the adjacent fluid element is possible in a coherent way. This is well known from ultrasound generation.

In the electron plasma wave the restoring forces from the charge accumulations and the thermal pressure are transmitted by the electric field. Therefore for coherence the wavelength limit is given by $k\lambda_D \leq 1$. The same limit holds for the ion acoustic wave, except if the electron temperature is so high that the electrons form a uniform background for the ions. In this case λ_D is to be replaced by the ion Debye length λ_{Di}.

Landau damping may put severe limits on the existence of short wave Langmuir waves long before the above criterion is reached. In the derivation of the damping coefficient Landau solved a nonlinear eigenvalue equation from which the underlying physics is not seen. In the context here an alternative derivation is given by transcribing Landau's formula for damping γ as follows,

$$\gamma = \frac{\pi}{2}\frac{\omega\omega_p^2}{k^2}\frac{\partial f_0}{\partial v_\varphi} = \frac{\pi^2}{T_0}v_\varphi^2\frac{\partial f_0}{\partial v_\varphi} = \frac{\pi^2}{T_0}n_0\frac{\hat{V}}{\hat{n}_1}\frac{\partial f_0}{\partial v_\varphi} = \frac{\dot{\hat{n}}_1}{\hat{n}_1}.$$

Landau's expression is given by the first term on the left. The term in the middle with $T_0 = 2\pi/\omega$ shows that damping is due to diffusion, whereas the third expression shows that the origin of diffusion is the potential energy $\hat{V}$ of the wave. The closer the velocity of the free electron is to the phase velocity the stronger is its diffusion out of the coherent motion of the density disturbance $\hat{n}_1$.

Modes in the static magnetic field $\mathbf{B}_0$. The magnetic field breaks the isotropy of space. As a consequence the thermal pressure shows its tensor character and the **k** vector depends on direction and the mutual orientation of **E** and $\mathbf{B}_0$. A whole variety of new waves is born and it is difficult to classify them, and the more so, as there arise remarkable differences between a wave in the cold or in the warm plasma. Fortunately, in laser matter interaction besides the low frequency *magnetoacoustic waves* the high frequency modes are not very important.

Low frequency modes. At low frequency the description greatly simplifies in the approximation of "infinite" electric conductivity and due to the omission of the displacement current $\varepsilon_0\dot{\mathbf{E}}$ in Mawell's equations. Infinite conductivity means no B field diffusion; the *frozen magnetic field* moves with the plasma; it rarefies and it is compressed with matter: The *magnetic pressure* $p_\mathrm{m} = \mathbf{B}^2/2\mu_0$ follows the adiabate of the ideal gas, $p_\mathrm{m}\rho^{-\gamma_\mathrm{m}} = \text{const}$, with $\gamma_\mathrm{m} = 2$. Bearing this in mind it is intuitive that the longitudinal *compression Alfvén wave*, with fluid motion $\mathbf{u}||\mathbf{k}$ and $\mathbf{k}\perp\mathbf{B}_0$, exhibits the dispersion relation (5.128),

$$\omega^2 = (c_s^2 + c_A^2)k^2; \qquad c_A = \left(\frac{\mathbf{B}_0^2}{\mu_0\rho_0}\right)^{1/2},$$

c_A is the Alfvén velocity. If the displacement **u** of the plasma is perpendicular to **k** and $\mathbf{k}||\mathbf{B}_0$ follows $\mathrm{div}\mathbf{u} = 0$ and no density variation occurs. The only restoring force is due to lengthening of the magnetic field lines which are under "tension" $\tau = \mathbf{B}_0^2/\mu_0$ and hence, the dispersion of the *shear Alfvén wave* is given by (5.130)

$$\omega^2 = c_A^2\mathbf{k}^2; \quad c_A = \left(\frac{\mathbf{B}_0^2}{\mu_0\rho_0}\right)^{1/2}.$$

The compression Alfvén wave is always faster than the shear Alfvén wave.

Arbitrary orientation of the wave vector **k** with respect to $\mathbf{B}_0$ leads to the family of the so called *magnetosonic waves*. It contains the foregoing two Alfvén waves as special cases. Their general dispersion relation is given by (5.133),

$$D(\omega, \mathbf{k}) = (\omega^2 - k_\parallel^2 c_A^2)[\omega^4 - k^2(c_s^2 + c_A^2)\omega^2 + k^2 k_\parallel^2 c_s^2 c_A^2] = 0.$$

From the vanishing of the square bracket the dispersion of a new wave follows,

$$\omega^2 = \frac{1}{2}k^2(c_s^2 + c_A^2)\left\{1 \pm \left[1 - 4\frac{k_\parallel^2 c_s^2 c_A^2}{k^2(c_s^2 + c_A^2)}\right]^{1/2}\right\}.$$

The term in the square bracket varies between zero and 1, hence the linear magnetoacoustic waves are undamped stable eigenmodes of the homogeneous plasma. The magnetoacoustic waves play a role in the expansion of an underdense hot plasma into vacuum under a static magnetic field $\mathbf{B}_0$.

High frequency modes. The electromagnetic wave with its propagation direction along the magnetic field $\mathbf{B}_0$ is of interest because it can lead to a *critical density increase* and it offers a diagnostic tool through the *Faraday effect* of polarization plane rotation. The increased critical density n_{ec} by a strong field results according to (5.137) as

$$n_{ec} = n_c\left(1 + \frac{\omega_c}{\omega}\right); \quad n_{ec} = n_c\left(1 + \frac{1.8 \times 10^{11} B_0[\mathrm{T}]}{2 \times 10^{15}}\right) \cong n_c(1 + 10^{-4} B_0[\mathrm{T}]).$$

The rotation of the polarization plane by the coaxial field $\mathbf{B}_0$ is determined from (5.138),

$$\Delta\varphi = [k(\mathrm{R}) - k(\mathrm{L})]L = \frac{\omega}{c}\left[\left(1 - \frac{\omega_p^2}{\omega^2}\frac{1}{1-\varepsilon}\right)^{1/2} - \left(1 - \frac{\omega_p^2}{\omega^2}\frac{1}{1+\varepsilon}\right)^{1/2}\right]L.$$

The hf transverse wave with $\mathbf{k}$ and $\mathbf{E}$ perpendicular to $\mathbf{B}_0$ is of interest because the *extraordinary wave* shows an absorbing resonance at the electron cyclotron frequency ω_c from where it does not propagate further. Here, the *cut off* occurs at zero phase velocity and zero wavelength.

Wave propagation in the inhomogeneous plasma. General laws can be given in the slowly varying medium with scale lengths $L(\mathbf{x}, t)$ much longer than the local wavelength $\lambda(\mathbf{x}, t)$. Then the *WKB or optical approximation* applies where the amplitude becomes a slowly varying quantity in space and time and the phase $\Phi = \mathbf{kx} - \omega t$ is replaced by the integral $\Phi = \int(\mathbf{k}\mathrm{d}\mathbf{x} - \omega \mathrm{d}t)$. The main consequences of the WKB ansatz are as follows:

- The energy flows along the *rays s* determined from the *ray equation*

  $$\frac{\mathrm{d}}{\mathrm{d}s}(\mathbf{k}_0\eta) = k_0\nabla\eta, \quad \eta \text{ refractive index.}$$

  In the absence of collisions ($\nu = 0$) the power flux along a ray bundle is conserved.
- The electromagnetic and the electron plasma wave obey the same *ray tracing* law because their refractive indexes exhibit the same structure (but their dielectric functions differ!).

- Translational symmetry in one direction implies the conservation of the wave vector component parallel to it. The angular momentum $\mathbf{x} \times \mathbf{k}$ is conserved in rotational symmetry of $\eta(\mathbf{x}) = \eta(r)$.
- Under WKB conditions there is almost no reflection.
- Appreciable reflection occurs
  - at transition layers of refractive index variations on lengths below $\lambda/2$, or
  - from resonant periodic structures even of vanishing amplitude, e.g., *stimulated Brillouin scattering*
  - from cutoffs. WKB never holds there.
- In the optical approximation the *classical photon* obeys the canonical equations of motion. With the Hamiltonian $\omega(\mathbf{x}, \mathbf{k}, t)$ holds

$$\frac{\mathrm{d}\mathbf{x}}{\mathrm{d}t} = \frac{\partial\omega}{\partial\mathbf{k}}, \quad \frac{\mathrm{d}\mathbf{k}}{\mathrm{d}t} = -\frac{\partial\omega}{\partial\mathbf{x}}, \quad \frac{\mathrm{d}\omega}{\mathrm{d}t} = \frac{\partial\omega}{\partial t}.$$

The second equation shows that photon momentum $\mathbf{k}$ is transported with group velocity v_g. The canonical equations form the basis for radiation transport and *photon acceleration.*

5.6 Problems

▹ Show that $\mathbf{j} = \rho_{\mathrm{el}}\mathbf{u}$ is the relativistic current density. *Hint:* $\rho_{\mathrm{e0,i0}}V$ is a four vector. How does ρ_{el} change with $\mathbf{u}$?
▹ Express $\mathbf{u}$ as a function of $\mathbf{u}_e$ and $\mathbf{u}_i$.
▹ Show: If $\nabla\varepsilon_0\mathbf{E} = \rho_{\mathrm{el}}$ for one time instant it holds true for all times.
▹ Verify (5.12), (5.13), and (5.14).
▹ Show that $\mathbf{A}$ and $\mathbf{B}$ from (5.12) and (5.13) fulfil the Lorentz gauge.
▹ Derive $\mathbf{E} = -\dot{\mathbf{A}} - \nabla\Phi$.
▹ Verify (5.17) and (5.19)
▹ Show that the total charge $Q = \int \rho_{\mathrm{el}}\mathrm{d}\mathbf{x}$ is a Lorentz scalar.
▹ Verify the Lorentz invariance of $c^2\mathbf{B}^2 - \mathbf{E}^2$ with the use of (2.165) and (2.167).
▹ Write the Lienard–Wiechert potentials in covariant form.
▹ Derive (5.15) from (5.34).
▹ In Fourier inversion of $g(k) = (2\pi)^{-1/2}\int f(x)\exp(-ikx)\mathrm{d}x$ to $f(x)$ the inversion of integrations

$$\int\{\int f(x')\exp(-ikx')\mathrm{d}x'\}\exp(ikx)\mathrm{d}k = \int\{\int f(x')\exp ik(x-x')\mathrm{d}k\}\mathrm{d}x'$$

is needed. Give a proof for the inversion of integrations. *Hint:* Make use of the convergence generator $\exp -\lambda|x|$ and make $\lambda > 0$ sufficiently small.
▹ Verify (5.58).

▹ Derive Snell's law of refraction from (5.61).
▹ (a) Derive Fresnel's formulas for reflection and transmission under perpendicular incidence onto the interface between two homogeneous plasmas. (b) Control the Poynting vectors.
▹ Show, in the reflection from the vacuum—overdense plasma boundary the parallel electric field components subtract from each other, the perpendicular components add to each other.
▹ Show, the critical density is that density at which the self induced field becomes as intense as the incident field.
▹ Show in detail why the induced electron oscillations of the Langmuir wave lead to an increase of **E** towards rising plasma density in the underdense region and to a cut off beyond the critical point. *Hint:* In the nonabsorptive medium the energy flux is conserved.
▹ Perform the Fourier transform of (5.54) to obtain (5.55).
▹ $\varepsilon(k, \omega)$ is derived in (5.82) by eliminating E. Derive it by eliminating f_1. Why is $f_0(v = \pm\infty) = 0$?
▹ Derive (5.85) from (5.83).
▹ Derive (5.91).
▹ Derive (5.89). *Hint:* Have a look at Chap. 2, ponderomotive force.
▹ Explain with the help of $T > T_0$ why linear Landau damping is proportional to $\partial f_0/\partial v$.
▹ Justify the appearance of Cauchy's principal value in (5.93).
▹ Give an estimate up to which limit of $k\lambda_D \ll 1$ (5.85) is an acceptable approximation.
▹ Determine $\varepsilon_\parallel$ of the ion acoustic wave.
▹ From which frequencies on the displacement current can no longer be neglected in (5.119)?
▹ Ion velocity **u** and current density $\mathbf{j}_1$ of the shear Alfv'en wave are orthogonal to each other. Explain why.
▹ (a) Sketch the range of $\omega(\mathbf{k})$ as a function of magnitude and orientation of **k** relative to the $\mathbf{B}_0$ field in (5.134). (b) Determine the orientation of **j** and **u** in the individual modes.
▹ Derive the dispersion relation for $\mathbf{k}\|\mathbf{e}_x$ and $\mathbf{B}_0\|\mathbf{e}_z$. (a) *Ordinary wave*: If **E** is parallel to $\mathbf{B}_0$ the wave is not affected by the $\mathbf{B}_0$ field $\Longrightarrow \omega^2 = \omega_p^2 + c^2\mathbf{k}^2$. (b) *Extraordinary wave*: $\mathbf{E}\|\mathbf{e}_y$. Show that the electron motion lies on an ellipse, i.e. it oscillates in transverse and longitudinal direction. Derive the dispersion relation.
▹ Write R from (5.142) for the electron plasma wave incident from left.
▹ Show that the high frequency ion motion in the laser field leads to the replacement of m_e by the reduced mass μ.
▹ Derive (5.144) from the fast fluid component in Chap. 3.
▹ Deduce the law of reflection and Snell's refraction law from Fermat's principle.
▹ Write (5.161) for constant refractive index. How does the number of modes change in a cavity?
▹ The photon in (5.164) is a *quasiparticle*. Determine its rest mass.
▹ Derive (5.166) from (5.153) and (5.165).

▹ Derive relations (5.169) and check the correctness of (5.177).
▹ Derive Beer's law (5.186).

5.7 Self-assessment

- Write the correct expression for current density $\mathbf{j}$ and charge density ρ_{el} in (5.1) and (5.3) (*Hint:* Is there a Lorentz factor γ or not?)
- Poynting's theorem (5.17) and the stress tensor (5.19) express electromagnetic energy and momentum conservation in standard form. Interpret the source terms $\mathbf{jE}$ and $-\boldsymbol{\pi}$. How would (5.17) read in presence of magnetic charges $\rho_{\text{mag}} = c\varepsilon_0\nabla\mathbf{B}$?
- Under which conditions do Lorentz and Coulomb gauge coincide? What is their striking difference?
- How do you define an eigenmode?
- What is the influence of thermal motion of the electrons on the dispersion relation (5.47)? *Answer:* none.
- Given the refractive index η, calculate the group velocity as a function of the wavelength. *Answer:* $v_g = v_\varphi[1 + (\lambda/\eta)(\partial\eta/\partial\lambda)]$.
- Discuss electromagnetic pulse propagation close to the critical density in the fully ionized plasma. Which energy fluxes do contribute to the wave energy transport?
- Why is glass transparent to the Nd laser whereas iron is not, although in both the electron densities are approximately the same?
- From which physical facts does follow that the cut offs of the transverse and the longitudinal electron wave must be the same ω_p? *Answer:* $\mathbf{k} \Rightarrow 0$ + comment on ω_p in the text.
- Show, in the highly overdense plasma ($\varepsilon \ll 0$) the electric field lines in the plasma are nearly parallel to the interface vacuum-plasma for almost all angles of incidence.
- Knowing the amplitude $\hat{n}_1$ of a longitudinal wave determine E, V, v_{os}.
- How are the longitudinal and transverse dispersion relations in an electron—positron plasma? *Hint:* Show that the contribution of an equal number of positive carriers to the electron current leads to the substitution of m_e by the reduced mass μ. How does the electron sound speed change? *Answer:* The same way.
- In what do the dielectric functions for the longitudinal and the transverse wave differ from each other and why?
- Derive expression (5.85) from (5.83) under conditions (i) and (ii). *Hint:* Use $\lambda_D = v_{\text{e,th}}/\omega_p$.
- The Landau contour is the lower graph extending from $-\infty$ to $+\infty$ with a semicircle around the pole approaching the real axis of ω from below. Show its uniqueness by physical arguments only. *Hint:* (a) Show that (5.89) together with negative slope of f_0 at $v_i = v_\varphi$ leads to damping $\Rightarrow \gamma < 0$. (b) Integrals along upper and lower contour in Fig. 5.5 become equal for $\gamma \Rightarrow 0$, i.e. they are nearly equal for sufficiently small $|\gamma|$. (c) Cauchy's integral formula is equivalent to semicircle for small $|\gamma|$ (symmetry argument). (a)+(b)+(c) $\Rightarrow$ (5.83).

- In the linear regime Landau damping γ does not depend on the amplitude of the electron plasma wave. Why then the residue cannot be calculated forwards along the Landau path and back on the real axis from $+\infty$ to $-\infty$? *Hint:* Lower and upper path are not symmetric.
- Under standard conditions the ion plasma frequency ω_{pi} plays no role because it is screened by the electrons. Under which conditions does it appear in the ion dispersion relation? You find the answer in the text.
- Where has the electric field "disappeared" in the expressions of the magnetic pressure (5.124) and (5.125)? *Hint:* Express the plasma equilibrium in the two fluid model for electrons and ions.
- The magnetic pressure is $p_{\mathrm{m}} = \mathbf{B}^2/2\mu_o$. Why then is $c_A^2 = \mathbf{B}^2/(\mu_0\rho_0)$? If you do not know the answer go back to the text. For alternative explanation see below.
- Explain in terms of physics the reduction of ω to zero for the slow magnetoacoustic wave.
- How can the critical density be increased by a magnetic field $\mathbf{B}_0$?
- Make a guess: What is the static magnetic field strength to increase the nonrelativistic critical density n_c for Nd by the factor of 2? *Answer:* 10^4 T = ? Gauss. Insert here the right figure.
- When does the group velocity of a laser pulse (approximately) coincide with the transport velocity $v_g = \partial\omega/\partial k$ of the $\mathbf{k}$ vector?
(*Hint*: Use Fourier decomposition).
- The adiabatic coefficient of the electron plasma wave is $\gamma_e = 3$, the adiabatic coefficient of $\mathbf{B}_0$ is $\gamma_{\mathrm{m}} = 2$. What is the rule behind and why the difference? *(Formal) Solution*: Nonrel.: $\gamma = (f+2)/f$, superrelativistic.: $\gamma = (f+1)/f$ for zero rest mass (photons); **B** is made of photons.
- The current induced by the electromagnetic wave in the plasma generates a magnetic field of opposite direction to the incident field. Show that just at the critical density they cancel each other.
- Where does the magnetic energy $\varepsilon_0 c^2\mathbf{B}^2/2$ go as the laser wave, incident from vacuum, approaches gradually the critical point? *Hint*: into electron oscillation energy. Show it.
- Deduce the energy balance of the transverse wave in the plasma in geometrical approximation from Poynting's theorem.
- Try the same for the Langmuir wave. *Hint*: Do not forget the thermal energy.
- Why so far no use of angular momentum conservation of the combined system plasma plus electromagnetic field has been made? Does it impose any new constraints on plasma-field dynamics? *Answer:* In the unbounded plasma it does not.
- Explain Landau damping on physical grounds to an undergraduate student. If you have difficulties, read again the explanation given here. Avoid the concept of "resonant particles"; it is not appropriate because it is in disagreement with energy conservation of the single particle.
- Why has c to be replaced by v_g in (5.164) for the plasma? Does it hold for any linear optical medium? (*Answer:* No).
- Phase and amplitude in (5.165) are adiabatic invariants. Indicate the slowness parameters for $\mathbf{k}$ and ω.

- Note, the plane Stokes equation (5.170) is selfsimilar. Indicate the similarity parameter.
- Beer's law: What is the limit of validity of $I(s) = \varepsilon_0 c^2 |\overline{\mathbf{E} \times \mathbf{B}}|$ between intensity and Poynting vector? *Answer:* Validity is bound to WKB (or optical) approximation.

5.8 Glossary

Wave equation

$$\nabla \times \nabla \times \mathbf{E} + \frac{1}{c^2}\frac{\partial^2}{\partial t^2}\mathbf{E} = -\frac{1}{\epsilon_0 c^2}\frac{\partial \mathbf{j}}{\partial t}. \tag{5.15}$$

Liénard–Wiechert potentials

$$\mathbf{A}(\mathbf{x}, t) = \frac{q}{4\pi\varepsilon_0 c^2}\frac{\mathbf{v}t'}{R - \mathbf{R}\mathbf{v}/c}; \qquad \Phi(\mathbf{x}, t) = \frac{q}{4\pi\varepsilon_0}\frac{1}{R - \mathbf{R}\mathbf{v}/c}. \tag{5.14}$$

Poynting's theorem

$$\frac{\partial}{\partial t}\frac{1}{2}\varepsilon_0(\mathbf{E}^2 + c^2\mathbf{B}^2) + \nabla\mathbf{S} = -\mathbf{j}\mathbf{E}; \quad \mathbf{S} = \varepsilon_0 c^2 \mathbf{E} \times \mathbf{B}. \tag{5.17}$$

Maxwell's stress tensor $\mathbf{T}$; $\quad \boldsymbol{\pi} = \rho_{\mathrm{el}}\mathbf{E} + \mathbf{j} \times \mathbf{B}$

$$\frac{1}{c^2}\frac{\partial}{\partial t}\mathbf{S} + \nabla\mathbf{T} = -\boldsymbol{\pi}; \quad \mathbf{T} = T_{ij} = \varepsilon_0[E_i E_j + c^2 B_i B_j - \frac{1}{2}\delta_{ij}(\mathbf{E}^2 + c^2\mathbf{B}^2)]. \tag{5.19}$$

Relativistic Lorentz force $\left[\frac{\mathrm{d}}{\mathrm{d}t}m\gamma\mathbf{v} = q(\mathbf{E} + \mathbf{v} \times \mathbf{B}), \quad \frac{\mathrm{d}}{\mathrm{d}t}m\gamma c^2 = q\mathbf{v}\mathbf{E}\right]$

$$\Leftrightarrow \quad m\frac{\mathrm{d}}{\mathrm{d}\tau}V = -\frac{q}{c^2}FV, \; V = v^\alpha = \gamma(\mathbf{v}, c) \quad \Leftrightarrow \quad m\frac{\mathrm{d}v_\alpha}{\mathrm{d}t} = \frac{q}{c^2}F_{\alpha\beta}v^\beta. \tag{5.27}$$

Lorentz gauge

$$\nabla c\mathbf{A} + \frac{\partial}{\partial ct}\Phi = 0 \quad \Leftrightarrow \quad \partial_\alpha A^\alpha = 0, \quad A = (c\mathbf{A}, \Phi). \tag{5.34}$$

Monochromatic *electromagnetic wave* $\nabla^2\mathbf{E} + \mathbf{k}^2\mathbf{E} = 0; \; \mathbf{k} = \mathbf{k}_0\eta, \; |\mathbf{k}_0| = \omega/c$
Fully ionized plasma dispersion: $\omega^2 = \omega_p^2 + c^2\mathbf{k}^2, \quad \omega_p^2 = n_e e^2/\varepsilon_0 m_e$

$$\eta^2 = 1 - \frac{\omega_p^2}{\omega^2}\frac{1}{1 + \nu^2/\omega^2} + i\frac{\nu}{\omega}\frac{\omega_p^2}{\omega^2}\frac{1}{1 + \nu^2/\omega^2} = \varepsilon(\mathbf{k} = 0, \omega). \tag{5.59}$$

Phase and group velocities v_φ, v_g:

$$v_\varphi = \omega/|\mathbf{k}| = c/\eta;\ \ v_g = |\partial\omega/\partial\mathbf{k}| = c/[\eta + \omega(\partial\eta/\partial\omega)]. \tag{5.59}$$

Fully ionized plasma: $v_\varphi v_g = c^2;\ \ v_\varphi = c/\eta \Rightarrow v_g = c\eta$
Field amplitude from intensity of linearly polarized plane wave

$$I = |\bar{\mathbf{S}}| = \frac{1}{2}\varepsilon_0 c\hat{\mathbf{E}}^2 \ \Rightarrow\ \hat{\mathbf{E}}[\mathrm{Vcm}^{-1}] = 27.5 \times (I[\mathrm{Wcm}^{-2}])^{1/2}. \tag{2.14}$$

Electron plasma wave: $\mathbf{E} \parallel \mathbf{k}$

$$(5.15) \Rightarrow -\frac{1}{c^2}\frac{\partial^2}{\partial t^2}\mathbf{E} = \frac{1}{\varepsilon_0 c^2}\frac{\partial \mathbf{j}}{\partial t}, \mathbf{j} = \sigma\mathbf{E} \Rightarrow \varepsilon_\parallel(\mathbf{k},\omega) = 1 - \frac{\omega_p^2 + c_{\mathrm{se}}^2\mathbf{k}_e^2}{\omega^2} = 0 \Rightarrow \tag{5.76}$$

Bohm–Gross dispersion: $\omega^2 = \omega_p^2 + c_{\mathrm{se}}^2\mathbf{k}^2,\ v_\varphi v_g = c_{\mathrm{se}}^2;\ c_{\mathrm{se}}^2 = \gamma_e k_B T_e/m_e$
Refractive index: $\eta_\parallel^2 = 1 - \omega^2/\omega_p^2 = \eta_\perp^2;\ \varepsilon_{\parallel,\perp}^2 = 1 + i\sigma_{\parallel,\perp}/\varepsilon_0\omega;\ \sigma_\parallel \neq \sigma_\perp$
Cut off *critical density* n_c

$$\omega^2 = \omega_p^2 \Rightarrow n_c = \frac{\varepsilon_0 m}{e^2}\omega^2 = 1.75 \times 10^{21}\left(\frac{\omega}{\omega_{\mathrm{Ti:Sa}}}\right)^2 [\mathrm{cm}^{-3}]. \tag{5.72}$$

$$\text{Polarization}\quad n_1(\mathbf{x},t) = -\nabla(n_{e0}\boldsymbol{\delta}_e),\quad \boldsymbol{\delta}_e = i\frac{\mathbf{u}_e}{\omega}. \tag{5.66}$$

Relations between dynamic quantities in linearized wave

$$E = i\frac{e}{\varepsilon_0 k_e}n_1,\ \Phi = i\frac{E}{k_e} = -\frac{e}{\varepsilon_0 k_e^2}n_1,\ u_e = v_\varphi\frac{n_1}{n_{e0}},\ T_1 = (\gamma_e - 1)T_{e0}\frac{n_1}{n_{e0}},$$

$$\hat{E} = \frac{e}{\varepsilon_0 k_e}\hat{n}_1,\ \hat{\Phi} = \frac{\hat{E}}{k_e} = \frac{e}{\varepsilon_0 k_e^2}\hat{n}_1,\ \hat{u}_e = v_\varphi\frac{\hat{n}_1}{n_{e0}},\ \hat{T}_1 = (\gamma_e - 1)T_{e0}\frac{\hat{n}_1}{n_{e0}}. \tag{5.73}$$

Linear *Landau damping*:

$$\gamma = \frac{\pi}{2}\frac{\omega\omega_p^2}{k^2}\frac{\partial f_0}{\partial v_\varphi};\quad f_0 \text{ distribution function.} \tag{5.83}$$

Ion acoustic wave

$$\omega_a = c_s k_a, k_a = |\mathbf{k}_a, |\ \ c_s^2 = \frac{k_B}{m_i}\left(\gamma_i T_i + \gamma_e\frac{ZT_e}{1 + \gamma_e(k_a\lambda_D)^2}\right). \tag{5.113}$$

Ideal Magnetohydrodynamics

Isotropic magnetic pressure $p_{\rm m} = \mathbf{B}^2/2\mu_0$
Magnetic tension along field lines $\tau_{\rm m} = -\mathbf{B}^2/\mu_0$
Alfvén velocity $c_A = (|\tau_{\rm m}|/\rho)^{1/2}$
Dispersion of *compression* Alfvén wave $\omega^2 = (c_s^2 + c_A^2)\mathbf{k}^2$
Dispersion of *shear* Alfvén wave $\omega = c_A|\mathbf{k}|$
Critical density increase by axial static **B** field

$$n_{ec} = n_c\left(1 + \frac{1.8\times 10^{11} B_0[\mathrm{T}]}{2\times 10^{15}}\right) \cong n_c(1 + 10^{-4}B_0[\mathrm{T}]). \tag{5.137}$$

Faraday effect: Polarization rotation over plasma length L :

$$\Delta\varphi = [k(\mathrm{R}) - k(\mathrm{L})]L = \frac{\omega}{c}\Big[\Big(1 - \frac{\omega_p^2}{\omega^2}\frac{1}{1-\varepsilon}\Big)^{1/2} - \Big(1 - \frac{\omega_p^2}{\omega^2}\frac{1}{1+\varepsilon}\Big)^{1/2}\Big]L, \tag{5.138}$$

$\omega_c[\mathrm{s}] = 1.8\times 10^{11} B\,[\mathrm{T}]$.

Inhomogeneous Plasma in Optical Approximation

$$\text{Phase of wave}: \ \Psi(\mathbf{x}, t) = \int_{P_1}^{P_2} (\mathbf{k}\mathrm{d}\mathbf{s} - \omega \mathrm{d}t); \quad \mathbf{k} = \nabla\Psi, \quad \omega = -\frac{\partial}{\partial t}\Psi. \tag{5.153}$$

Ray tracing

$$\frac{\mathrm{d}}{\mathrm{d}s}(\mathbf{k}_0\eta) = k_0\nabla\eta; \ \text{ plane}: \eta(x)\sin\alpha = \text{const}; \ \text{ spherical}: \ r\eta(r)\sin\alpha = \text{const}. \tag{5.156}$$

Wave in a ray bundle, **E** along ray s:

$$\mathbf{E}(s) = \left(\frac{Q(s_0)}{\eta(s)Q(s)}\right)^{1/2}\left\{\hat{\mathbf{E}}_+ \mathrm{e}^{ik_0\int\eta ds} + \hat{\mathbf{E}}_- \mathrm{e}^{-ik_0\int\eta ds}\right\}\mathrm{e}^{-i\omega t}. \tag{5.165}$$

If $\eta^2 = A + iB \ \Rightarrow \ \eta = \eta_r + i\eta_i$

$$\eta_r = \left\{\frac{1}{2}\left(\sqrt{A^2+B^2} + A\right)\right\}^{1/2}, \ \eta_i = \left\{\frac{1}{2}\left(\sqrt{A^2+B^2} - A\right)\right\}^{1/2}. \tag{5.188}$$

Oblique incidence of plane wave, **E**-field:

$$\mathbf{E}(\mathbf{x}, t) = \frac{\hat{\mathbf{E}}_+(\mathbf{x}_0, t=0)}{(\eta^2 - \sin^2\alpha_0)^{1/4}}\mathrm{e}^{ik_0\int_{\mathbf{x}_0}^{\mathbf{x}}\eta \mathrm{d}x' - i\omega t} + \frac{\hat{\mathbf{E}}_-(\mathbf{x}_0, t=0)}{(\eta^2 - \sin^2\alpha_0)^{1/4}}\mathrm{e}^{-ik_0\int_{\mathbf{x}_0}^{\mathbf{x}}\eta \mathrm{d}x' - i\omega t}. \tag{5.166}$$

Collisional absorption $A = 1 - R$ under normal incidence into layered and spherical density profiles $n_e/n_c = (x_c/x)^\alpha$, linear, and spherical $n_e/n_c = (r_c/r)^\alpha$, $\alpha = 1, 2, 3$ results from reflection R in Sect. 2.5.4.

5.9 Further Readings

M. Lazar, On Retardation, Radiation and Liénard–Wiechert Type Potentials in Electrodynamics and Elastodynamics. Wave Motion, Elsevier **50**, 1161–1174 (2013).
V.L. Ginzburg, *The Propagation of Electromagnetic Waves in Plasmas* (Pergamon Press , Oxford, 1964).
T.J.M. Boyd, J.J. Sanderson, *The Physics of Plasmas* (Cambridge University Press, Cambridge, 2005). Chap. 4: Ideal Magnetohydrodynamics.
L.M. Brekhovskikh, *Waves in Layered Media* (Academic, London, 1960). J.D. Jackson, *Classical Electrodynamics* (John Wiley, New York, 1975).

References

1. A. Shevchenko, B.J. Hoenders, New J. Phys. **12**, 053020 (2010)
2. O.L. Brill, B. Goodman, Am. J. Phys. **35**, 832 (1967)
3. P. S., Proc. Natl. Acad. Sci. U.S. **16**, 627 (1930)
4. K. Rawer, Ann. Phys. **35**, 385 (1939); **42**, 294 (1942)
5. P. Penfield, H.A. Haus, *Electrodynamics of Moving Media* (MIT Press, Cambridge, 1967), Chap. 1
6. G.B. Whitham, *Linear and Nonlinear Waves* (Wiley, New York, 1974)
7. G. Bekefi, *Radiation Processes in Plasmas* (John Wiley, New York, 1966), Sect. 1.6
8. J.T. Mendonca, Eur. Phys. J. D **68**, 79 (2014)
9. L. Muschietti, C.T. Dum, Phys. Fluids B **5**, 1383 (1993)
10. E. Sonnenschein, I. Rutkevich, D. Censor, Phys. Rev. E **57**, 1005 (1998)
11. B. Nordland, Phys. Rev. E **55**, 3647 (1997)
12. T.P. Coffey, Phys. Fluids **14**, 1402 (1971)
13. P. Koch, J. Albritton, Phys. Rev. Lett. **32**, 1420 (1974)
14. P. Mulser, W. Schneider, B. McNamara (eds.), *Excitation of Nonlinear Electron Plasma Waves and Particle Acceleration By Laser. Twenty Years of Plasma Physics* (World Scientific, Singapore, 1985), p. 264
15. J.-J. Kull, Phys. Rep. **206**, 197 (1991)
16. A. Bergmann, H. Schnabl, Phys. Fluids **31**, 3266 (1988)
17. P. Mulser, H. Takabe, K. Mima, Z. Naturforsch. **37a**, 201 (1982)
18. J.M. Dawson, Phys. Rev. **113**, 383 (1959)
19. C.R. Davidson, *Methods in Nonlinear Plasma Theory* (Academic , New York, 1972), Chap. 4
20. V.L. Ginzburg, *The Propagation of Electromagnetic Waves in Plasmas* (Pergamon Press, Oxford, 1964)
21. M. Abramowitz, I.A. Stegun, *Pocketbook of Mathematical Functions* (Harris Deutsch, Frankfurt, 1984)
22. J. Mathews, R.L. Walker, *Mathematical Methods of Physics* (Benjamin/Cummings, Menlo Park, 1964)
23. P. Mulser, C. van Kessel, J. Phys. D Appl. Phys. **11**, 1085 (1978)
24. O. Willi, P.T. Rumsby, Opt. Comm. **37**, 45 (1981)
25. R.D. Jones, W.C. Mead, S.V. Coggeshall et al., Phys. Fluids **31**, 1249 (1988)
26. W. Kofink, Ann. Physik **1**, 119 (1947)
27. R.P. Feynman, F.L. Vernon, R.W. Hellwarth, J. Appl. Phys. **28**, 49 (1957)
28. E. Cojocaru, P. Mulser, Plasma Phys. **17**, 393 (1974)
29. L. D. Landau, J. Phys. (U.S.S.R.) **10**, 25 (1946)

30. C. Mouhot, C. Villani, Acta Math. **207**, 29 (2011); 173 pages. Fields Medal awarded to C. Villani in 2013 for his contributions to Landau damping
31. T.H. Stix, *Waves in Plasmas* (American Institute of Physics, New York, 1992)
32. Wikipedia, *Landau damping*
33. A. Piel, *Plasma Physics* (Springer, Berlin, 2010), Chap. 9.3.3
34. F.F. Chen, *Plasma Physics and Controlled Fusion*, vol. 1 (Plenum Press, New York, 1984)
35. T.J.M. Boyd, J.J. Sanderson, *The Physics of Plasmas* (Cambridge University Press, Cambridge, 2005)
36. C.R. Davidson, *Methods in Nonlinear Plasma Theory* (Academic, New York, 1972)
37. J.M. Dawson, Phys. Fluids **4**, 869 (1961)
38. W.L. Kruer, *The Physics of Laser Plasma interactions* (Addison-Wesley, Redwood City, 1988), Chap. 9

Chapter 6
Unstable Fluids and Plasmas

6.1 Fluid Dynamic Instabilities and Unstable Waves

Richness of instabilities is one of continuous surprises of plasmas. Instabilities confer an originally uniform plasma unexpected structures: density and temperature modulations; turbulent shocks, violent bursts, and extremely energetic jets in cosmic plasmas; islands, disruptions, and anomalous transport in magnetic and inertial fusion devices; down and up conversion of modes, fast particle jets, Terahertz and hard X-ray photons in laser plasmas. The characteristic unstable equilibrium scenario is this: an arbitrarily small perturbation of the equilibrium produces a force that grows proportional to the perturbation. The growth is always accompanied by an increase of kinetic energy of the system. The unstable growth may either continue undefinitely or run into a saturation limit. If the instability is undesired, and most of them in fluid dynamics are, we may say a wrong channel has opened into which profitable energy from the original configuration is dissipated.

6.1.1 *Basic Unstable Phenomena*

The simplest example for an undesired energy sink is the

Rayleigh-Taylor Instability

Consider an incompressible fluid of density ρ_1 horizontally superposed on an incompressible fluid of density ρ_2 and both exposed to constant gravitational acceleration $-g$, see Fig. 6.1. This is a possible equilibrium, stable if $\rho_1 < \rho_2$, unstable if $\rho_1 > \rho_2$. If the interface is vertically perturbed by an arbitrary amount $h(\mathbf{x}) > 0$ to produce the volume difference V, (a), the system is exposed to the down force $(\rho_1 - \rho_2)Vg$ if $\rho_1 < \rho_2$. The force is proportional to V and counteracts to the perturbation; the system is stable. In the reversed case of the heavier fluid superposed on the lighter fluid

P. Mulser, *Hot Matter from High-Power Lasers*, Graduate Texts in Physics,
https://doi.org/10.1007/978-3-662-61181-4_6

the force $(\rho_1 - \rho_2)Vg$ is reversed into the upward direction and causes unlimited growth of the perturbation. The only possible channel the work by the displacement goes into is kinetic energy of the system. Without intervention from outside the undesired energy flow into the wrong channel continues growing proportional to the displacement volume; the system is Rayleigh–Taylor unstable. The force that drives the *Rayleigh–Taylor instability* (RTI) originates from the lowering of the center of mass in the gravity field $-g$ of the earth, or any massive object, or in a reference system undergoing the constant upward acceleration $+g$.

For a quantitative analysis of the growth the perturbation $h(\mathbf{x})$ is subject to a Fourier decomposition. For simplicity the infinitesimal initial perturbation of the interface is assumed vertical and sinusoidal, $y_i = h_0(t = 0)\sin kx$, Fig. 6.1b. The growth of the instability $\dot{h}(t)$ is calculated from Newton's law provided the involved accelerated mass is known. To this aim the decay of the vertical perturbation $h(y + h_0(t)\sin kx)$ at a generic position $y + h_0(t)\sin kx$ of the fluid must be known. Owing to the smallness of the interface excursion $h_0(t = 0)$ the approximation $y + h_0(t)\sin kx = y$ is allowed. It implies that $h(y)$ starts from the unperturbed interface $y = 0$ instead from $h_0(t)$. The velocity field $\mathbf{u} = u_x, u_y$ can be derived from a potential $\Phi(x, y, t)$, $u_x = -\partial_x\Phi$, $u_y = -\partial_y\Phi$, because $\nabla \times \mathbf{u} = 0$. It is intuitive to set for $y > 0$,

$$\Phi = -h_0(t)h(y)\sin kx \Rightarrow u_x = h_0(t)kh(y)\cos kx,\ u_y = h_0(t)\partial_y h(y)\sin kx. \tag{6.1}$$

The vertical velocity u_y is highest in the maxima of $h_0(t)$ in agreement with (6.1). Incompressibility of the fluids imposes $\nabla\mathbf{u} = 0$, thus

$$\frac{\partial u_x}{\partial x} + \frac{\partial u_y}{\partial y} = -h_0k^2h\sin kx + \partial_{yy}hh_0\sin kx = 0 \Rightarrow \frac{\mathrm{d}^2h}{\mathrm{d}y^2} = k^2h,\ h(y) = \mathrm{e}^{-k|y|}. \tag{6.2}$$

because at $y = \pm\infty$ the perturbation must vanish. The perturbation of the interface decays exponentially in depth, just like flat ocean waves in deep water. It is obvious that the horizontal velocity components $|u_x|$ are highest and opposed to each other at the nodes, see arrows in Fig. 6.1. This is accomplished by setting $\Phi_2 = -\Phi_1 = \Phi$. The stability analysis can be continued formally but it is more immediate just to apply Newton's second law. The formal approach will be presented later in connection with the Kelvin–Helmholtz instability. The total involved mass M and down force F per length dx at the maximum excursion are

$$M = \mathrm{d}x\left[\int_{-\infty}^{0}\rho_2\mathrm{e}^{ky}\mathrm{d}y + \int_{0}^{+\infty}\rho_1\mathrm{e}^{ky}\mathrm{d}y\right] = \frac{\mathrm{d}x}{k}(\rho_1 + \rho_2),\quad F = (\rho_1 - \rho_2)h_0(t)g. \tag{6.3}$$

The force F is best determined from a vertical virtual shift $\delta h_0 < 0$. By the shift the ρ_1 column of length $h_0(t)$ does the positive work $\rho_1 h_0(t)g\mathrm{d}x|\delta h_0|$. The fraction $\rho_2 h_0(t)g\mathrm{d}x|\delta h_0|$ of it is spent to lift the lighter column. The difference is the free energy $F\mathrm{d}x|\delta h_0| = (\rho_1 - \rho_2)h_0(t)g\mathrm{d}x|\delta h_0|$ going into kinetic energy increase. No other parts of the two liquids are affected by δh_0. The equation of motion of the

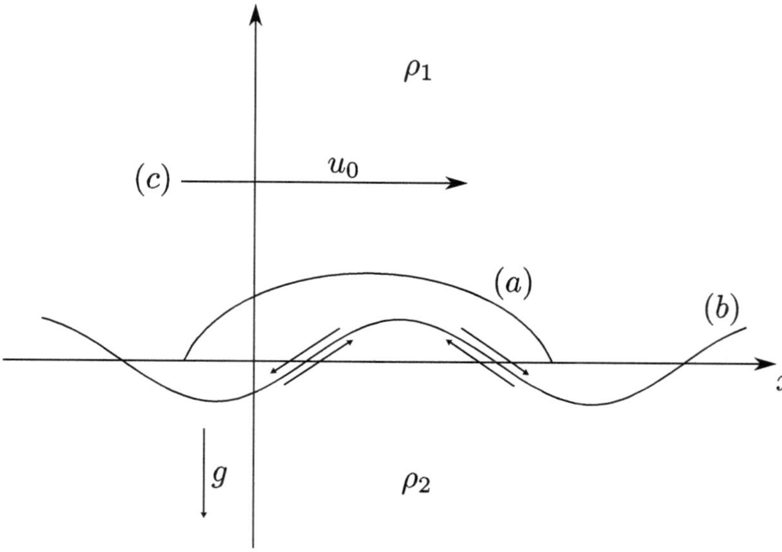

Fig. 6.1 The Rayleigh–Taylor interface instability is driven by the lowering of the potential energy of two superposed heavy fluids of densities ρ_1 and ρ_2. **a** arbitrary vertical perturbation of the interface, **b** sinusoidal interface perturbation, **c** Kelvin–Helmholtz instability from the upper fluid moving at velocity $\mathbf{u}_0$ and the lower fluid at rest

displacement and its solution are

$$M\frac{\mathrm{d}^2h_0}{\mathrm{d}t^2} = F \Rightarrow \ddot{h}_0 = \frac{\rho_1-\rho_2}{\rho_1+\rho_2}kgh_0 \Rightarrow h_0(t) = y_i\,\mathrm{e}^{(Akg)^{1/2}t};\ A = \frac{\rho_1-\rho_2}{\rho_1+\rho_2}. \tag{6.4}$$

The quantity A is the *Atwood number*; $\gamma_{\mathrm{RTI}} = (Akg)^{1/2}$ is S. Chandrasekhar's formula for the growth rate of the Rayleigh–Taylor instability in the ideal incompressible fluid [1]. The effective mass M to be accelerated grows in proportion to the wavelength of the perturbation. It is the reason for the square root of Akg.

An instructive application of the Rayleigh–Taylor instability is the

Levitation of a plasma by a magnetic field. A box filled with plasma in a horizontal magnetic field $\mathbf{B}$ should keep its position because if there are no collisions the electrons stick to the field lines, each of them winding around its own line. However, according to (2.72) the electrons and the ions undergo gravity drift motions $\mathbf{v}_g \sim q^{-1}\mathbf{g}\times\mathbf{B}$, each into opposite direction. Let $\mathbf{B}$ be directed along $\mathbf{e}_x$ and $\mathbf{g}$ along $\mathbf{e}_y$. The electrons drift along $\mathbf{e}_z$, the ions along $-\mathbf{e}_z$; the induced electric field lies along $\mathbf{e}_z$ and hence the induced $\mathbf{E}\times\mathbf{B}$ drift points along $\mathbf{g}$, the box falls, levitation fails.

Stability of a strip. In applications the thickness d of the upper fluid is finite. As long as $kd \gg 1$ the involved accelerated mass M is as determined in (6.3) and γ_{RTI} applies to the layer of thickness d. For $kd \ll 1$ the involved mass extends over the whole thickness and the growth is given by

$$\gamma_d = \Big(\frac{\rho_2}{\rho_1} - 1\Big) gkd. \tag{6.5}$$

For $kd = 0.1,\ \rho_1/\rho_2 = 2$ the growth reduces to $\gamma_d/\gamma_{\rm RTI} = 0.16$. With $\rho_1/\rho_2 \mapsto \infty$ the classical $\gamma_{\rm RTI}$ is recovered. Growth formula γ_d for $kd \ll 1$ and its limit $\gamma_{\rm RTI}$ are understood from the involved masses M in (6.3). Another relevant case consists in the separation of ρ_1 from ρ_2 by a linear density profile of width d which connects the two constant densities extending to $\pm\infty$. Again, from (6.3) one deduces $\gamma_d = \gamma_{\rm RTI}$ for $kd \ll 1$ and $\gamma_d \mapsto 0$ for $kd \gg 1$. Exact growth for all k values can be found for an exponential transition layer of width d in [2], Sect. 6.3.1-5.

RTI in inertial fusion. For the inertial fusion to work efficiently the fuel must be compressed, less with heavy ions as a driver, more with lasers, in this latter case at least one thousand fold solid compression is required. The compression is achieved by mass ablation from a spherical shell of radius R and shell thickness ΔR. In the compression phase the accelerated outer pellet surface is Rayleigh–Taylor unstable. Before the highest compression κ is reached in the stagnation phase the inner surface undergoes violent deceleration and becomes RT unstable. From the aspect of maximum kinetic energy input and at the same time mild acceleration and deceleration the use of very thin shells of *aspect ratio* $R/\Delta R$ is favorable. Stability requirements at the same time aim at a low product of acceleration g, compression time t and representative eigenmode k. In the spherical compression the shell is essentially freely accelerated over half the initial radius R, afterwards ΔR thickens due to convergence, and deceleration becomes noticeable. The assumption of constant acceleration translates into $gt^2/2 = R/2$. According to the foregoing considerations on a strip there is a maximum unstable k. Numerical simulations suggest $k \simeq 2/\Delta r$. Use of these values in (6.4) [or (6.5)] results in

$$\gamma_{\rm RTI} = \Big(2A\frac{R}{\Delta R}\Big)^{1/2} \Rightarrow \max\Big(\frac{R}{\Delta R}\Big) = \frac{1}{2A}\Big(\ln\frac{\kappa^{1/3}}{y_i}\Big)^2; \quad \kappa = \frac{3R^2\Delta R}{R_f^3}. \tag{6.6}$$

There is a maximum aspect ratio by setting the final radius R_f equal to the initial perturbation amplitude y_i of R. This is a severe limit either on the aspect ratio $R/\Delta R$, i.e on the rocket efficiency of compression, or on the sphericity and smoothness y_i of the fusion pellet. In direct drive the Atwood number is close to unity, in indirect drive it is between 1 and $1/2$. RTI not only sets limits to economic compression it even shows that acceleration increased by the factor f over $R/2$ results in a reduction of $\gamma_{\rm RTI}$ by $f^{1/2}$.

When the amplitude grows to a size of order two, substantial deviations from the linear theory are observed and the RTI evolves into a nonlinear stage. The light fluid rises into the heavy fluid in the form of *bubbles* and the heavy fluid falls into the light fluid in the form of *spikes*. The resulting structure turns out to be quite similar to the reverse shape of a spiking high amplitude electron plasma wave in Figs. 5.7 and

5.9. The bubble rise dynamics accompanied by numerical examples, as well as their competition is studied analytically in [2], Sect. 5. Collapse of bubbles and increased upward motion is observed. In the final stage, the RTI may evolve into turbulent or chaotic mixing. So far, in the linear regime only constant acceleration has been considered here. There is a rich and well established literature on the RTI under various conditions of acceleration. Impulsive acceleration of an interface separating two different densities is by passing of a shock wave. The unstable growth is described by the *Richtmyer–Meshkov instability* [4, 5]. An excellent introduction to the linear and nonlinear analytic theory of the RTI under a variety of conditions, accompanied by numerical evaluations, is presented by H.-J. Kull, see Further Readings.

Search for stabilization. Inertial fusion pellet design has forced the search for stabilizing effects and for the most dangerous perturbing modes. In the plane linear theory γ_{RTI} increases indefinitely with vanishing perturbing wavelength. Apart from geometrical effects there is a maximum growth at finite k and stabilization somewhere beyond owing to viscosity, mass ablation, and heat conduction. Here the "Takabe formula" [6],

$$\gamma = \alpha(kg)^{1/2} - \beta k V_a \tag{6.7}$$

with $\alpha \sim 1$ and V_a the ablation velocity across the ablation front has become popular. For various structures of the ablating plasma the growth rate γ is found to agree well with recent twodimensional simulations in a classical transport regime if $\alpha = 0.9$ and $\beta = 3 - 4$ is set [7]. At small k values growth is controlled by (6.4) but at large wave numbers an ablative cut off has been obtained. Owing to the ablative stabilization short wavelength lasers inducing enhanced mass ablation are suggested to be advantageous to stable implosion. The maximum growth emerged approximately at half of the classical free surface mode. Subsequently in an isobaric ablation model Takabe's expression has been generalized in [2] Sect. 9.2.2. Optimum fit of (6.7) for different flat targets driven by a laser pulse with a 1-ns linear ramp followed by a flat top pulse has yielded $\alpha = 0.94$, $\beta = 2.7$ in solid DT targets [8]. The authors conclude suggesting that (6.7) holds for relatively large *Froude numbers* between 0.1 and 5 and electronic heat conduction of $\kappa \sim T^{5/2}$, but it fails for small Froude numbers and radiative materials.

Experimental verification of formula (6.7) has been undertaken in [9] in plane polystyrene (CH) targets of 40 μ thickness. The formula can be rewritten in the approximate form

$$\gamma = \left\{\frac{kg}{1+kL}\right\}^{1/2} - \beta k \frac{\dot{m}}{\rho_a}. \tag{6.8}$$

Here $\dot{m}$ is the mass ablation rate, ρ_a the peak target density. With the exception of the scale length L, taken from the simulation, β was evaluated by measuring all quantities to amount to 1.2 ± 0.7. This is in reasonable agreement with the theoretical prediction 1.7 found in [8] for CH targets. For longer wavelengths perturbation a higher value of β has been found. In a very recent paper [10] the nonlinear multimode bubble front evolution of ablatively driven targets is investigated in two and three dimensions.

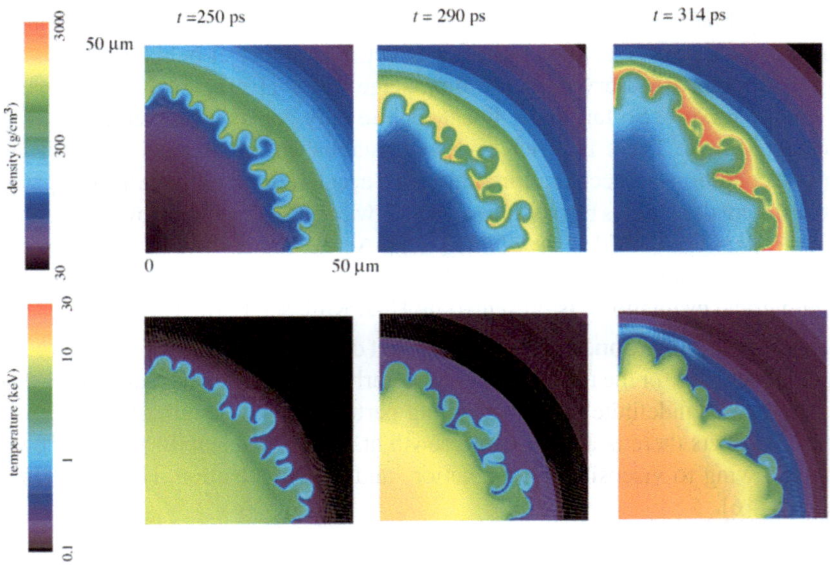

Fig. 6.2 Simulation of the nonlinear stage of a multimode initial perturbation ($l = 4 - 72$, white spectrum) inside a fusion capsule prior and slightly after ignition at 300 ps. Mode interaction and the formation of a highly distorted shell-gas interface is observed in the deceleration stage. The persistence of small scale structures in density and temperature indicates that the evolution into a stage of intense mixing is prevented by the outrunning thermonuclear burn wave. Courtesy of Atzeni et al. [3]

Ablation-driven vorticity accelerates the bubble velocity and prevents the transition from the bubble competition to the bubble merger regime at large initial amplitudes leading to higher α than in the classical case. Because of the dependence of the multimode bubble penetration on the initial perturbation and of vorticity generation, ablative stabilization of the nonlinear RTI is not as effective as previously anticipated for large initial perturbations (Fig. 6.2).

Whereas both examples of plasma instabilities presented, falling matter and magnetic levitation, grow out of *static equilibrium* situations a whole class of instabilities has its origin in a *dynamic* perturbation of the electron fluid on the fast time scale, or of the entire plasma on the ion time scale. This instability scenario is illustrated by the following two simple devices.

The Two-Stream Instability

This is the first simple example of a dynamically unstable plasma equilibrium. Imagine a homogeneous plasma of density $2n_0$ as the superposition of two counter streaming flows n_0u and $-n_0u$ along x and a small sinusoidal density perturbation $2n_1(x) = 2\hat{n}_1 \exp ikx$ impressed. This perturbation will influence both jets in the same way. Assume the jet to the right to be stationary, $(n_0 + n_1)(u_0 + u_1) = n_0u_0 =$ const. In first order holds $n_0u_1 = -n_1u_0$. The small disturbance (n_1, u_1) produces the electric potential $\Phi/2$ from Poisson's equation,

$$\varepsilon_0 \partial_{xx} \Phi = 2en_1 \quad \Rightarrow \quad k^2 \varepsilon_0 \Phi = -2en_1(x) \quad \Rightarrow \quad e\Phi = \frac{2\omega_p^2}{u_0^2 k^2} m_e u_0 u_1. \tag{6.9}$$

The plasma frequency ω_p refers to the single beam of density n_0. Energy conservation requires for the single electron

$$\frac{1}{2} m_e (u_0 + u_1)^2 - e\Phi = \frac{1}{2} m_e u_0^2 \Rightarrow m_e u_0 u_1 - e\Phi = 0. \tag{6.10}$$

The two expressions (6.9) and (6.10) are contradictory except for $2\omega_p^2/u_0^2 k_0^2 = 1$, i.e., for $k_0 = \sqrt{2}\omega_p/u_0$. The phase velocity of the disturbance relative to each of the two beams is $u_0 = \omega_p/k_0$. With respect to one plasma stream at rest this is a cold plasma wave of frequency $\omega = \omega_p$ and phase velocity $\pm u_0$. For all other k values a stable stationary flow is impossible: for $k < k_0$ there is the free energy difference $[2\omega_p^2/u_0^2 k_0^2 - 1] m_e u_0 u_1$ left to increase the wave kinetic and potential energy. This is achieved through the increase of u_1 and n_1; hence, for $k < k_0$ the wave is *unstable*. In the reference frame of the single beam at rest n_1 is a *negative energy wave*, see [11], Sect. 4.5. The energy set free by the wave drives the two stream instability in the cold plasma. Increase of the beam velocity u_0 has a stabilizing effect on the mode k. At $k > k_0$ there is a deficit left of free energy, thus equilibrium is possible only for $n_1 = 0$; the equilibrium is stable.

The stability analysis in the cold plasma proceeds as follows.

$$d_\pm n_1 + n_0 \partial_x n_1 = 0, \; m_e d_\pm u_1 = -eE, \; \varepsilon_0 \partial_x E = -en_1; \quad d_\pm = \partial_t \pm u_0 \partial_x. \tag{6.11}$$

Fourier transform , or the ansatz $n_1, u_1, E \sim \exp i(kx - \omega t)$, transforms the differential relations into an algebraic system of equations. The further reduction to a homogeneous equation of one unknown quantity proceeds by substitution in the standard way. If E is kept the result is $\varepsilon(k,\omega)E = 0$ with the dielectric function $\varepsilon(k,\omega)$ to be zero,

$$\varepsilon(k,\omega) = 1 - \frac{\omega_p^2}{(\omega - ku_0)^2} - \frac{\omega_p^2}{(\omega + ku_0)^2} = 0. \tag{6.12}$$

The system is unstable when the biquadratic dispersion relation admits complex roots $\omega = \omega_r + i\gamma$. It is the case when the minimum $\varepsilon(\omega = 0) = 1 - 2\omega_p^2/k^2 u_0^2$ becomes negative. Maximum growth occurs for $ku_0 \approx \omega_p$. The twostream instability is of high relevance in a system of a relativistic electron jet in a plasma and its return current. For oblique orientation of the **k** vector to the beam the unstable mode is of electromagnetic type. By varying the angle of **k** from zero degree in the twostream configuration to $\pi/2$ of filamentation a close interconnection of the two types with the *Weibel instability* is discovered [13].

In a general treatment of the two-stream instability it is advantageous to chose the reference system in which the total current density is zero. Otherwise the magnetic fields have to be taken into account.

The Kelvin–Helmholtz Instability

Is a dynamically unstable equilibrium. Consider again the situation of Fig. 6.1 for RTI of two superposed fluids with the difference now that the upper fluid slides with velocity u_0 over the lower fluid at rest, Fig. 6.1c. If the interface is assumed again to be perturbed initially by $y = h_0(t = 0) \sin kx$ the further evolution is described by a velocity potential $\Phi = \Phi_2(x, y)$ for the lower fluid which is identical with that of the RTI for ρ_2. The upper fluid differs by the constant motion u_0. It is accounted for by the additional term $-u_0 x$ to $\Phi = \Phi_1(x, y)$ in (6.1). Hence, with $h(y) = \exp -k|y|$ from (6.2) we set

$$\Phi_1 = C_1 h_0(t) \mathrm{e}^{-k|y|-ikx} - u_0 x, \quad \Phi_2 = -C_2 h_0(t) \mathrm{e}^{-k|y|-ikx}; \; C_1, C_2 = \text{const.} \tag{6.13}$$

From the correct application of Bernoulli's law on the perturbed interface in Fig. 6.1, $\rho u^2/2 + p = \text{const}$, it results that unstable behavior is to be expected because the maxima of perturbation experience a lifting force owing to a pressure sink there. The force increases with growing u_0 and drives the system into the *Kelvin–Helmholtz instability* (KHI). Each fluid obeys the equations of motion

$$\rho\left[\frac{\partial u_x}{\partial t} + u_0 \frac{\partial u_x}{\partial x}\right] + \frac{\partial p}{\partial x} = 0, \quad \rho\left[\frac{\partial u_y}{\partial t} + u_0 \frac{\partial u_y}{\partial x}\right] + \frac{\partial p}{\partial x} + \rho g = 0. \tag{6.14}$$

if for ρ_2 is obviously set $u_0 = 0$. In terms of the velocity potentials they transform into

$$\rho \frac{\partial}{\partial x}\left[\frac{\partial \Phi}{\partial t} + u_0 \frac{\partial \Phi}{\partial x} - \frac{p}{\rho}\right] = 0, \quad \rho \frac{\partial}{\partial y}\left[\frac{\partial \Phi}{\partial t} + u_0 \frac{\partial \Phi}{\partial x} - \frac{p}{\rho} - gy\right] = 0. \tag{6.15}$$

For the upper fluid, and the lower fluid accordingly, there are the first integrals $K_1(y)$ for x arbitrary and $y = \text{const}$, and $K_1'(x)$ for y arbitrary and $x = \text{const}$,

$$p_1 - \rho_1\left[\frac{\partial \Phi_1}{\partial t} + u_0 \frac{\partial \Phi_1}{\partial x}\right] = K_1(y), \quad p_1 - \rho_1\left[\frac{\partial \Phi_1}{\partial t} + u_0 \frac{\partial \Phi_1}{\partial x} - gy\right] = K_1'(x).$$

Their combination yields

$$K_1(y) - \rho_1 g y = K_1'(x) \;\Rightarrow\; K_1' = K_{01} = \text{const.}$$

An identical relation $K_2'(x) = K_{02} = \text{const}$ follows from (6.15) for the lower fluid with u_0 set to zero. For the pressures p_1, p_2 the two conditions translate into

$$p_1 = K_{01} - \rho_1 g y + \rho_1\left[\frac{\partial\Phi_1}{\partial t} + u_0\frac{\partial\Phi_1}{\partial x}\right], \quad p_2 = K_{02} - \rho_2 g y + \rho_2\frac{\partial\Phi_2}{\partial t}. \quad (6.16)$$

At the interface $y(x,t) = h_0(t)\exp -ikx$ the vertical fluid velocities $-\partial_y\Phi_1(y = 0)$, $-\partial_y\Phi_2(y = 0)$ must be equal to each other to keep the two fluids adjacent all time and equate $\mp \mathrm{d}y(x,t)/\mathrm{d}t$. For the velocities the condition reads in first order

$$C_1 k h_0 \mathrm{e}^{-ikx} = \frac{\partial y}{\partial t} + u_0\frac{\partial y}{\partial x}\Big|_{y(x)=0} = \left[\frac{\dot{h}_0}{h_0} - iku_0\right]\mathrm{e}^{-ikx} \Rightarrow C_1 = \frac{\dot{h}_0}{kh_0} - iu_0$$

$$-C_2 k h_0 \mathrm{e}^{-ikx} = \frac{\partial y}{\partial t} = \dot{h}_0 \mathrm{e}^{-ikx} \Rightarrow C_2 = -\frac{\dot{h}_0}{kh_0}. \quad (6.17)$$

The pressure must be continuous across the interface $y(x,t)$. In total equilibrium, $u_x, u_y = 0$, $y(x,t) = 0$, this imposes on (6.16)

$$K_{01} - \rho_1 u_0^2 = K_{02}. \quad (6.18)$$

If a small initial interface deformation $y_0 = \varepsilon \exp ikx$, $k|y_0| \ll 1$ is present (6.18) still holds in first order because K_{01}, K_{01}, u_0 are first order constants. Therefore the interface dynamics induced by y_0 obeys in first order (6.16),

$$\rho_1 g y(x,t) - \rho_1\left[\frac{\partial\Phi_1}{\partial t} + u_0\frac{\partial\Phi_1}{\partial x}\right] = \rho_2 g y(x,t) - \rho_2\frac{\partial\Phi_2}{\partial t}.$$

We substitute the velocity potentials from (6.13) and C_1, C_2 from (6.17) and set for abbreviation in full generality $\dot{h}_0/h_0 = -i\omega(t)$ to obtain

$$(\omega - iku_0)^2\rho_1 + \omega^2\rho_2 - gk(\rho_1 - \rho_2) = 0; \quad k > 0. \quad (6.19)$$

Thus, ω is a constant with respect to time. For $u_0 = 0$ this is the Rayleigh–Taylor instability from (6.4) with $\omega = (gkA)^{1/2} = \gamma_{\mathrm{RTI}}$. The general solution of (6.19) is

$$\omega = iku_0\frac{\rho_1}{\rho_1 + \rho_2} \pm \left\{k^2u_0^2\frac{\rho_1\rho_2}{(\rho_1 + \rho_2)^2} + gkA\right\}^{1/2}. \quad (6.20)$$

The first term on the RHS tells that the sinusoidal interface perturbation propagates with the velocity of the center of mass of the two fluids. The term in the bracket is always positive for positive Atwood number A and reinforces KHI. The stable RTI configuration $A < 0$ can always be driven unstable as soon as $u_0^2 > g(\rho_2^2 - \rho_1^2)/k\rho_1\rho_2$. In the absence of a gravity force the KHI occurs at arbitrarily low motion

u_0. If, in addition, the densities are equal the dispersion relation is particularly simple,

$$\omega = \frac{1}{2} k u_0 (i \pm 1). \tag{6.21}$$

Surface tension T provides for a stabilizing effect,

$$\omega = iku_0 \frac{\rho_1}{\rho_1 + \rho_2} \pm \left\{ k^2 u_0^2 \frac{\rho_1 \rho_2}{(\rho_1 + \rho_2)^2} + gkA - \frac{k^{3/2} T}{\rho_1 + \rho_2} \right\}^{1/2}. \tag{6.22}$$

A constant magnetic field $\mathbf{B}_0$ acts like a tension $\tau = \mathbf{B}_0^2/\mu_0$ along the field lines which counteract against their lengthening by a transverse perturbation, see (5.130). Its stabilizing effect under general orientation with respect to the interface has been studied recently in [14]. In the nonlinear stage of evolution of the RTI into spikes shear flow along a spike becomes significant and evolves into KHI-like perturbations.

Bernoulli's Role in KHI

For simplicity consider the situation of (6.21) for equal densities. Above a strip of width $y \simeq \lambda/2$ the flow is almost $u_0 = \text{const}$ everywhere. However, inside the strip where the sinusoidal excursion is positive it is squeezed and hence the flow velocity $u(x)$ along a streamline close to the separatrix $y(x)$ follows the variation of the latter. In the steady state the pressure follows Bernoulli's law $p(x) + \rho u^2(x) = \text{const}$ with a variation proportional to $-y(x)$. This results into the local force $-\partial_x p$ with maxima in the nodes and alternating direction such as to squeeze the fluid towards increasing amplitude of the separatrix. The pressure gradient is out of phase by $\pi/2$ with respect to $y(x)$ and $u(x) \sim y(x)$. In the reference system $S'(u_0)$ comoving with the upper fluid the separatrix is a sinusoidal wave $y(x') \sim \sin(kx' + \Im\omega t)$ with phase velocity $v_\varphi = -u_0/2$. The displacement of a volume element at fixed x' represents a harmonic oscillator driven at resonance because dephased by $\pi/2$ with respect to the driving force $-\partial_{x'} p$. In the ideal fluid there is no finite *threshold* for the onset of the KHI; it starts from arbitrarily low perturbation y_0. This is a consequence of resonant drive. The most immediate access to the KHI is by the centrifugal force. Any deviation from a straight flow means curvature and a concomitant force directed outward. Bernoulli's law and the growth rate (6.21) are directly calculated from it.

All further instabilities of relevance to laser-plasma dynamics are microinstabilities or of the type of resonant wave-wave coupling, like resonance absorption, stimulated Raman and Brillouin scattering, or two-plasmon decay. In all cases in which more than two modes are involved the instability is driven by the secular ponderomotive force. Stimulated scattering of an electromagnetic wave from electron density fluctuations induced by acoustic and Langmuir waves plays an important role for understanding the plasma dynamics in detail, processes of fast electron generation and plasma heating, as well as for laser plasma applications. In first approximation the longitudinal electric wave far from trapping obeys a wave equation of the same structure as the transverse electromagnetic wave. The same is true for the *standard* ponderomotive force. Hence, both types of waves are subject, damping rates permitting, to the same instability dynamics.

6.1.2 Summary: Plasma Modes, Energy Densities, and Fluxes

To elaborate the governing equations it may be useful to summarize the necessary main laws of energy transport by waves derived in the foregoing chapters. In this approach let us assume a fully ionized ideal plasma of constant density n_0, temperature T_0, and flow velocity $\mathbf{u}_0$ on which a small static electron density variation n_1, possibly accompanied by a small temperature variation T_1 is superimposed.

Wave Equations

It is convenient to split the current density $\mathbf{j} = -en_e(\mathbf{u} - \mathbf{u}_0)$ of (5.15) into the two components

$$\mathbf{j} = \mathbf{j}_0 + \mathbf{j}_1, \quad \mathbf{j}_0 = -en_0(\mathbf{u} - \mathbf{u}_0) = -en_0\mathbf{u}_e, \quad \mathbf{j}_1 = -en_1\mathbf{u}_e.$$

The wave equation for the transverse and parallel field components $\mathbf{E}_{\perp(\mathbf{x},\omega)}$ and $\mathbf{E}_{\parallel(\mathbf{x},\omega)}$ in the homogeneous medium ($n_1 = 0$) is obtained from Ohm's law (5.148) in its linearized version, $\mathbf{j}_0(\omega) = \sigma(\omega)\mathbf{E}(\omega, \mathbf{x}, t)$,

$$\nabla^2\mathbf{E}_{\perp,\parallel} - \frac{1}{v_\varphi^2}\frac{\partial^2}{\partial t^2}\mathbf{E}_{\perp,\parallel} = 0, \quad \nabla\mathbf{E}_\perp = 0, \quad \nabla\times\mathbf{E}_\parallel = 0. \tag{6.23}$$

The individual Fourier components $\mathbf{E}_\perp(\mathbf{k}, \omega)$ and $\mathbf{E}_\parallel(\mathbf{k}, \omega)$ obey the same dispersion relations, i.e., refraction law and ray equation, but propagate with different phase velocities v_φ,

$$\begin{aligned}\mathbf{E}_\perp(\mathbf{k}, \omega):\quad & v_\varphi = \frac{c}{\eta}, \quad \eta^2 = 1 - \frac{\omega_p^2}{\omega^2} = 1 - \frac{n_0}{n_c}, \\ & \omega^2 = \omega_p^2 + c^2k^2, \quad \mathbf{k} = \mathbf{k}_0\eta, \quad k_0 = \frac{\omega}{c}.\end{aligned} \tag{6.24}$$

$$\begin{aligned}\mathbf{E}_\parallel(\mathbf{k}, \omega):\quad & v_\varphi = \frac{c_{se}}{\eta}, \quad \eta^2 = 1 - \frac{\omega_p^2}{\omega^2} = 1 - \frac{n_0}{n_c}, \\ & \omega^2 = \omega_p^2 + c_{se}^2k^2, \quad \mathbf{k} = \mathbf{k}_0\eta, \quad k_0 = \frac{\omega}{c_{se}}.\end{aligned} \tag{6.25}$$

η refractive index, c_{se} electron sound speed. The density perturbation $n(\mathbf{k}, \omega)$ of the ion acoustic wave obeys

$$\nabla^2 n - \frac{1}{c_s^2}\frac{\partial^2}{\partial t^2}n = 0; \quad \eta = 1, \quad \omega = c_s k. \tag{6.26}$$

In general plasmas are inhomogeneous and nonstationary. In the WKB limit in space and time all relations (6.23)–(6.26) remain valid locally. In presence of $n_1 \neq 0$ in leading order n_0 is to be replaced by $n_e = n_0 + n_1$ in $\sigma(\omega)$, ω_p and η; $\mathbf{j}_0 = -en_e\mathbf{v}_{\text{os}}$ with $\mathbf{v}_{\text{os}}$ the electron quiver velocity. Together with these changes the wave equation, complemented by $\mathbf{j}_1$, reads now

$$\nabla^2\mathbf{E}_{\perp,\parallel} - \frac{1}{v_\varphi^2}\frac{\partial^2}{\partial t^2}\mathbf{E}_{\perp,\parallel} = -\frac{1}{\varepsilon_0 c^2\beta^2}\frac{\partial en_1\mathbf{u}_e}{\partial t}, \quad \mathbf{E}_\perp : \beta = 1, \quad \mathbf{E}_\parallel : \beta = c_{se}/c. \tag{6.27}$$

As we shall see resonant excitation of a density modulation $n_1 \ll n_0$ occurs when n_1 represents a plasma eigenmode, that is in the unmagnetized case an ion acoustic or Langmuir mode propagating with its phase velocity v_φ. The current density $\mathbf{j}_1 = -en_1\mathbf{u}_e$ is parallel to $\mathbf{E}_{\perp,\parallel}$. By changing to a co-moving reference system the perturbation n_1 becomes static; wave vector $\mathbf{k}$ and frequency ω of the pump wave are Doppler-shifted according to (6.83). In the co-moving frame the correctly transformed electric field $\mathbf{E}'(\mathbf{k}', \omega')$ obeys (6.23) with v_φ depending on the angle $(\mathbf{k}, \mathbf{u}_{\text{flow}})$. In the eigenmode system the steady state flow condition $n_e\mathbf{u}_e = n_0\mathbf{u}_\varphi$ holds.

Energy Transport

A short summary of Sect. 5.5.3 of the total energy densities of the electromagnetic, the electrostatic, and the ion acoustic wave $\mathcal{E}_{\text{em}}$, $\mathcal{E}_{\text{es}}$, $\mathcal{E}_{\text{a}}$ and their conservation equations is as follows,

$$\mathcal{E}_{\text{em}} = \mathcal{E}_{\text{e}} + \mathcal{E}_{\text{m}} + \mathcal{E}_{\text{os}} = \frac{1}{2}\varepsilon_0\mathbf{E}\mathbf{E}^*; \qquad \partial_t\mathcal{E}_{\text{em}} + \nabla\mathbf{v}_g\mathcal{E}_{\text{em}} = 0, \quad v_g = \frac{c^2}{v_\varphi}, \tag{6.28}$$

$$\mathcal{E}_{\text{es}} = \mathcal{E}_{\text{e}} + \mathcal{E}_{\text{os}} + \mathcal{E}_{\text{pot}} = \frac{1}{2}\varepsilon_0\frac{\omega^2}{\omega_p^2}\mathbf{E}\mathbf{E}^*; \quad \partial_t\mathcal{E}_{\text{es}} + \nabla\mathbf{v}_g\mathcal{E}_{\text{es}} = 0, \quad v_g = \frac{c_{se}^2}{v_\varphi}, \tag{6.29}$$

$$\mathcal{E}_{\text{a}} = \mathcal{E}_{\text{i,kin}} + \mathcal{E}_{\text{i,pot}} = n_0 m_i v_{\text{os}}^2; \quad \partial_t\mathcal{E}_{\text{a}} + \nabla\mathbf{v}_g\mathcal{E}_{\text{a}} = 0; \quad v_g = v_\varphi = c_s. \tag{6.30}$$

The indices os, kin, pot indicate the oscillatory, kinetic, and potential particle energy densities; $\mathcal{E}_{\text{e}}$ and $\mathcal{E}_{\text{m}}$ are the electric and magnetic contributions. The group velocity $\mathbf{v}_g = \partial\omega/\partial\mathbf{k}$ is the energy transport velocity of a narrow wave packet. More precisely, as apparent from the first of canonical equations (5.159) $\mathbf{v}_g$ is the velocity at which a constant $\mathbf{k}$ vector propagates. In a stationary plasma the *energy flux* of frequency ω along a ray bundle of cross section $S(\mathbf{x})$ must be constant, $v_g\mathcal{E}_{\text{em,es}}S = \text{const}$; hence, the WKB result $\mathcal{E}_{\text{em,es}}\eta^{1/2} = \text{const}$ in space follows in one dimension. Along the way towards the critical density the magnetic energy of $\mathbf{E}_\perp(\mathbf{k}, \omega)$ and the potential energy of $\mathbf{E}_\parallel(\mathbf{k}, \omega)$ transform gradually into oscillatory energy of the electrons. At the critical point the magnetic energy reduces to zero.

Imagine an electric wave with a fixed $\mathbf{k}$ vector in a homogeneous plasma, for instance a standing wave between two neighbouring fixed boundaries, the density of which changes slowly in time. Then

$$E^2\eta = \frac{E^2}{\omega}(\omega^2 - \omega_p^2)^{1/2} = \frac{E^2}{\omega}ck = \text{const} \Rightarrow \frac{E^2}{\omega} = \text{const} \tag{6.31}$$

Thus the *action* E^2/ω is an adiabatic invariant. This is in perfect analogy to the pendulum of slowly varying length where the energy $\mathcal{E} \sim \omega$. It shows that under an adiabatic change the photon (plasmon) number density $N = \mathcal{E}_{\mathrm{em,es}}/\hbar\omega \sim E^2/\omega$ is conserved.

6.2 Mode Conversion: Resonance Absorption

The general understanding of *absorption of light* is the annihilation of a photon (by the field operator a) and the creation of another energy carrier (by the creation operator b^+), like a fast electron or a plasmon. Under the aspect of efficiency the question may arise whether direct conversion of the transverse laser wave into an electron plasma wave is possible without loss of energy. Such a direct mode conversion must fulfil the resonance conditions in momentum, i.e., **k**, and energy, i.e., ω:

$$\text{Resonance absorption} \quad \Leftrightarrow \quad \mathbf{k}_{\mathrm{em}} = \mathbf{k}_{\mathrm{es}}, \quad \omega_{\mathrm{em}} = \omega_{\mathrm{es}}. \tag{6.32}$$

In the photon picture this process of *resonance absorption* is a two wave process: one photon converts into one plasmon. Inspection of the dispersion relations (6.24) and (6.25) reveals that owing to $c \gg c_{\mathrm{es}}$ resonance is possible only with $\mathbf{k}_{\mathrm{em}} = \mathbf{k}_{\mathrm{es}} = 0$, both in the critical plasma region owing to $\eta = 0$ only there. Additionally, conversion is possible only in the inhomogeneous plasma in order to give access to the laser beam through an underdense region. Thus, the presence of a density gradient at the critical point is essential. As we know the density gradient is steep for several reasons: hydrodynamic rarefaction of the hot plasma, ponderomotive profile steepening, inhomogeneities induced by parametric instabilities. Correspondingly, resonance occurs in a very narrow region around the critical surface only. These various conditions find their natural foundation in the excitation mechanism illustrated by the following Fig. 6.3. Under arbitrary incidence of the laser wave its electric field will form a finite angle with the excited electrostatic field and, despite resonance is fulfilled in $(\mathbf{k}, \omega)$, conversion cannot occur one to one and the laser wave will be partially reflected. One could be induced to think that the process of conversion could grow into a highly nonlinear regime owing to its resonant drive. However, there is the phenomenon of *convection* of wave energy out of the resonance zone which delimits the conversion against nonlinearity very frequently in the experiment.

The phenomenon of resonance absorption was first described by Denisov [15] for cold plasma and later by Piliya [16] for warm plasma. Only later was its relevance for collisionless laser plasma heating recognized [17]. Subsequently, accurate conversion rates as a function of density scale length L and angle of incidence α_0 were determined numerically [18]. The most accurate and complete study of the phenomenon in the linear regime was undertaken by Kull [19]. From (5.161) and (5.164) follows

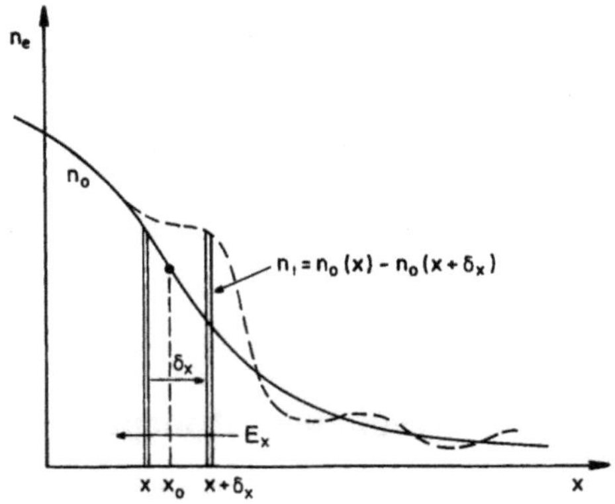

Fig. 6.3 Excitation of an electron plasma wave along a plasma density gradient. Electrons with equilibrium density $n_0(x) = Zn_i$ are shifted to $x + \delta_x$ by the electromagnetic E_x-field component. Since there the density of the nearly immobile ions is $n_i(x + \delta_x)$ a charge density imbalance $\rho_{el} = -en_1 = -e[n_0(x) - n_0(x + \delta_x)]$, oscillating at frequency ω, is produced. In the picture $Z = 1$

that in the absence of sources the action density $(\varepsilon_0/2)\hat{\mathbf{E}}\hat{\mathbf{E}}^*/\omega$ is conserved rather than the energy density $(\varepsilon_0/2)\hat{\mathbf{E}}\hat{\mathbf{E}}^*$. In the neighborhood of a critical point the optical approximation $\lambda(\mathbf{x}) = 2\pi/k_0\eta \ll L$ is never fulfilled and the eikonal approach with its consequences (5.156)–(5.161) becomes meaningless. In general, this is also true at the turning points. The simplest model in which this can be seen in detail analytically is by restriction to the layered medium in which the plasma inhomogeneity depends on the x component only, see Fig. 5.14. In this coordinate system the exact wave equation (5.145) may be split into its components. In the coordinate system of Fig. 5.14a the fact that η is independent of y and z, a Fourier decomposition in the planes $x =$ const is convenient, with k_y not depending on x. This is in agreement with the ray equation (5.156), $k_y = k_0\eta \sin\alpha = k_0 \sin\alpha_0 =$ const, and means that the momentum of a photon perpendicular to the density gradient (parallel to the y axis in the Figure) is conserved along its geometrical path. Hence,

$$\mathbf{E}(\mathbf{x}, t) = \left\{E_x(x),\ E_y(x),\ E_z(x)\right\} \mathrm{e}^{ik_y y - i\omega t}. \tag{6.33}$$

The individual components obey the three equations ($' = \mathrm{d}/\mathrm{d}x$)

$$E_x'' + \frac{k_0^2}{\beta^2}(\eta^2 - \sin^2\alpha_0)E_x = i\frac{k_0}{\beta^2}(1 - \beta^2)\sin\alpha_0 E_y', \tag{6.34}$$

$$E_y'' + k_0^2(\eta^2 - \beta^2\sin^2\alpha_0)E_y = ik_0(1 - \beta^2)\sin\alpha_0 E_x', \tag{6.35}$$

$$E_z'' + k_0^2(\eta^2 - \sin^2\alpha_0)E_z = 0. \tag{6.36}$$

E_z (s-polarized light) decouples from E_x and E_y, and is of purely electromagnetic nature. At its turning point $k_x = k_0(\eta^2 - \sin^2\alpha_0)^{1/2}$ vanishes and the approximations leading to the dispersion relations (5.154) are no longer justified. For the LHS of (6.34) to be resonantly excited by its RHS "driver", which in laser irradiation is an electromagnetic wave, in addition to $\omega_{es} = \omega_{em}$ both terms must have the same spatial dependence. Outside the turning point region (5.154) holds with $k_{x,em}$ very different from $k_{xes} = k_{x,em}/\beta$ for nonrelativistic temperatures. As a consequence, there the transverse and longitudinal components decouple and propagate without affecting each other. At the turning points resonant coupling is possible and is generally very effective.

A wave with **E** perpendicular to the plane of incidence is called s *polarized* (from german "senkrecht" = perpendicular) $\Rightarrow$ No resonant coupling.

A wave with **E** parallel to the plane of incidence is called p *polarized* $\Rightarrow$ Resonant coupling

A serious difficulty arises with the expressions (5.156) and (5.159) in an absorbing medium, in particular when the absorption is strong, since then the imaginary parts of $\mathbf{k}$ and $\mathbf{v}_g$ can no longer be ignored. Several solutions have been proposed to overcome this difficulty. We remark that an interesting reinterpretation of the group velocity in terms of a time-dependent combination of $\Re(\partial\omega/\partial\mathbf{k})$ and $\Im(\partial\omega/\partial\mathbf{k})$ was presented and appropriate ray-tracing equations were derived in [20]. For more recent developments on this subject the interested reader may consult [21, 22].

6.2.1 Inhomogeneous Stokes Equation

In contrast to E_z which propagates freely according to (6.36) the E_x component in (6.34) appears as a wave driven by the electromagnetic field component E_y. If there is no incident E_z-wave from the vacuum, E_z remains zero everywhere in the layered medium. E_z decouples from the other field components. This is not the case for E_x. Under oblique incidence E_y is different from zero and produces an E_x-field through (6.34). To see more clearly the structure of this field it is convenient to transform the driver with the help of Faraday's law,

$$(\nabla \times \nabla \times \mathbf{E})_x = ik_y E_y' + k_y^2 E_x = i\omega(\nabla \times \mathbf{B})_x = -k_0 k_y c B_z$$

to obtain Piliya's equation [16], with B for B_z,

$$E_x'' + \frac{k_0^2}{\beta^2}(\eta^2 - \beta^2 \sin^2 \alpha_0) E_x = -\frac{k_0 k_y c}{\beta^2}(1 - \beta^2) B(x).$$

At nonrelativistic temperatures, with $\beta^2 \ll 1$ and $k^2/\beta^2 \gg k_0^2$, the equation further simplifies to

$$E_x'' + \frac{k_0^2}{\beta^2}(\eta^2 - \beta^2 \sin^2 \alpha_0) E_x = -\frac{k_0 k_y c}{\beta^2} B(x). \tag{6.37}$$

This shows that the magnetic field or, equivalently, the E_x-component forces the electrons to oscillate along the density gradient and hence, produces a coherent charge separation or, under dispersion in the warm plasma, an electron plasma wave. Its longitudinal character is revealed by the local wave number $k/\beta = k_0\eta/\beta$ in (6.37) which is identical to that given in (5.154). Under the assumption of a linear density profile, with $\eta_i = 0$ and setting

$$\xi = \left(\frac{k_0^2}{\beta^2 L}\right)^{1/3}(x - x_0), \quad x_0 = L\beta^2 \sin^2\alpha_0,$$
$$w = \left(\frac{\beta}{k_0 L}\right)^{2/3}\frac{E_x}{cB(0)\sin\alpha_0}, \quad b = -\frac{B(x)}{B(0)}, \tag{6.38}$$

Equation (6.37) transforms into the dimensionless inhomogeneous Stokes-Airy equation

$$w'' + \xi w = b(\xi). \tag{6.39}$$

As apparent already from Fig. 5.14a effective excitation of an electron plasma wave puts certain constraints on the spatial structure of the driver b. Resonant coupling at frequency ω can occur over an extended region if the driver wavelength matches the local electrostatic wavelength; otherwise resonance is limited to a space region that is narrow compared with λ. The latter situation is most easily realized at the turning point owing to the reduction of $k_e = k_0\eta/\beta$ there. In Fig. 6.4 numerical solutions of (6.39) are shown for a driver of Gaussian structure acting at the turning point of the electron plasma wave, $b = \exp(-\xi^2/\delta^2)$, with $\delta^2 = 2,\ 8.33,\ 20$, and ∞. It is clearly seen that the maximum amplitude of the wave no longer increases much as soon as the driver halfwidth reaches half a local wavelength λ_e (compare (b), (c), (d)). The reason for this saturation lies in the constructive and destructive spatial interferences, alternating and nearly canceling each other when the driver is extended and smooth. As a consequence the electron plasma wave amplitude becomes modulated [see Fig. 6.4 (d). Starting from Ai(ξ) and Bi(ξ) the general solution of (6.39) is constructed in the familiar way, i.e., using the Green's function for (6.39)

$$w(\xi) = C_1\mathrm{Ai}(\xi)(\xi) + C_2\mathrm{Bi}(\xi) + \int^{\xi} b(z)\frac{\mathrm{Ai}(z)\mathrm{Bi}(\xi) - \mathrm{Ai}(\xi)\mathrm{Ai}(z)}{W}\mathrm{d}z \tag{6.40}$$

with the Wronskian $W = \mathrm{AiBi}' - \mathrm{Ai}'\mathrm{Bi} = 1/\pi$. For a spatially constant driver, $b = -1$, the inhomogeneous solution becomes the special function $\mathrm{Hi}(\xi) - 2\mathrm{Hi}(\xi)/3$. The correct outgoing electron plasma wave is found from the asymptotic representation of Ai, Bi, and Hi and its evanescent behavior for $\xi < 0$,

$$w = \pi(\mathrm{Hi} - \mathrm{Bi} - i\mathrm{Ai}) = \hat{w}e^{i\Theta}.$$

Here $\hat{w}$ is the amplitude of w, and Θ the phase. Around the turning point $\xi = 0$ the square of the amplitude $\hat{w}^2$ admits the power series expansion

$$\hat{w}^2(\xi) = 1.658 + 1.209\,\xi + 0.238\,\xi^2 - 0.0839\,\xi^3$$
$$-0.0523\,\xi^4 - 7.40 \times 10^{-3}\,\xi^5 + \ldots.$$

The maximum of $\hat{w} = 1.88$ occurs at $\xi_m = 1.8$. Figure 6.4 shows that the WKB solution of (6.39) deviates very little from the exact solutions for $\xi > \xi_{\mathrm{m}}$. The decay of $\hat{w}$ is slowest for $b = 1$ and behaves like $1/\xi^{1/4}$ for $\xi \gtrsim \xi_{\mathrm{m}}$. On the other hand, since saturation of $\hat{w}_{\mathrm{max}}$ occurs approximately for drivers of halfwidth ξ_{m}, ξ_{m} is the correct measure of the width of the resonance $x_m - x_0$. It increases as $L^{1/3}$. The lower curves in Fig. 6.4 indicate the phase velocity v_φ at which a point ξ satisfying $w(\xi, t) = 0$ propagates. In contrast to the Bohm–Gross expression the true v_φ does not diverge at the turning point.

In the laser generated plasma the incident electromagnetic wave acts as a resonant driver for the electron plasma wave propagating down the density gradient. Conversion depends on the angle of incidence: At normal incidence the laser field has no component along the density gradient; at grazing incidence the field has the right direction; however, the laser beam is reflected from the plasma too far away from the critical point. At an intermediate optimum angle a conversion factor of 0.5 is reached in flat density profiles. In steep plasma profiles as well as at relativistic temperatures the conversion rate and the optimum angle both increase and conversion can rise up to 100% [19]. The excited electron plasma wave is not converted back into electromagnetic waves and is, consequently, absorbed by the plasma. At low temperatures and flat density gradients significant damping can occur by electron-ion collisions. At intermediate laser intensities there may exist conditions for linear Landau damping to play a more important role [23]. At flux densities above $10^{14}(\omega/\omega_{\mathrm{Nd}})^2$ Wcm^{-2} the electrostatic wave becomes nonlinear in its amplitude and the electron density modulation assumes a very pronounced spike-like shape due to self-interaction of modes. In the potential wells associated with the wave, effective electron trapping and acceleration occurs and nonlinear Landau damping dominates. At even higher laser fluxes the Langmuir wave becomes aperiodic; *wavebreaking* occurs. In the following all these phenomena, starting from linear mode conversion, are treated extensively and the conversion rates are calculated.

Intense high contrast laser pulses exceeding $I = 10^{16}$Wcm^{-2} by orders of magnitude are standard. Their length can be chosen such that all interaction takes place before the ionic rarefaction wave starts. The observation of an angular dependence of *collisionless absorption* very similar to classical resonance absorption under p-polarization raises the question on the nature of the absorption process in this case and on the possible transition from resonant absorption to Fresnel formulas for discontinuous changes of the refractive index. Under certain conditions such a smooth transition from resonance absorption to ordinary linear optics at interfaces can be constructed. Whether such a coincidence has a real physical basis or whether it is purely accidental is to be discussed also. The question is of primary interest for a correct interpretation of the *collisionless absorption* mechanism measured in experiments and confirmed by simulations, however hitherto still not understood completely.

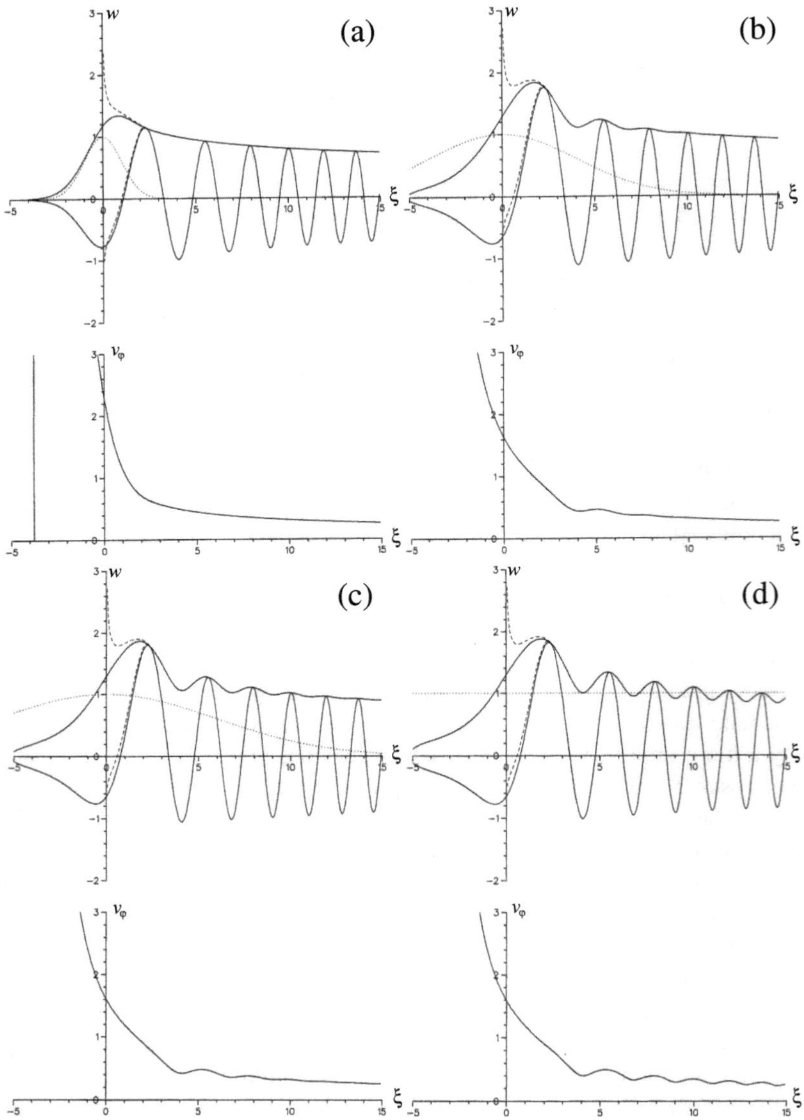

Fig. 6.4 Solution of the inhomogeneous Stokes equation $w'' + \xi w = -\exp(-\xi^2/\delta^2)$ for $\delta^2 = 2$ (**a**), 8.33 (**b**), 20 (**c**), and ∞ (**d**). Solid lines: numerical results of normalized field w and its amplitude $\hat{w}$ (upper curves), and phase velocity v_φ of the electron plasma wave (lower curves); dashed lines: WKB approximation of w and $\hat{w}$; dotted lines: driver $\exp(-\xi^2/\delta^2)$. For $\xi < 0$, w and $\hat{w}$ decay exponentially

6.2.2 Linear Resonance Absorption

An analytical and semianalytical treatment of resonance absorption is feasible in the 1D layered medium. Treatments in 2D and 3D are reserved to particle-in cell (PIC [25]) and Vlasov simulations. Fortunately all essential aspects of mode conversion and conversion rates can be studied to a satisfactory degree in a plane layered medium. In such a geometry the linear phenomenon is governed by the system of (6.34) and (6.35), or by (6.37) and (6.35), respectively. If the magnetic field is known the conversion can be determined numerically from (6.37), or Piliya's equation alone. Even for the layered medium a complete analytic solution of the coupled equations for E_x and B is not available so far. The field distributions are uniquely determined as soon as three independent parameters, e.g., α_0, β, and $q = (k_0 L)^{2/3} \sin^2 \alpha_0$ are given, see similarity parameters in (6.38) for the inhomogeneous Stokes equation. In the WKB region the electron plasma wave propagates nearly parallel to the density gradient since, as a consequence of the ansatz $\hat{\mathbf{E}}(x, y) = \hat{\mathbf{E}}(x) \exp(ik_y y)$ (momentum conservation), for its angle α_1 results

$$\sin \alpha_1 = \beta \sin \alpha_0, \tag{6.41}$$

and $E_{\mathrm{es}} \simeq E_{x,\mathrm{es}}$ holds. In the overdense region E_x as well as E_y become evanescent with a decay which is determined by the two periodicities k/β and k for the electron plasma wave and the electromagnetic mode. Typical distributions of $|E_x|^2$ and $|B|^2$ in a linear density profile are shown in Fig. 6.5 as functions of ξ for $q = 0.5$ and $\beta = 0.1$, the latter corresponding to a temperature of $T_e = 1.8$ keV. They represent the envelopes of $E_x = E_{x,\mathrm{em}} + E_{x,\mathrm{es}}$ and of B. The decrease of $|B|$ according to the WKB approximation (5.166),

$$|B(x)| = B_{\mathrm{inc}}(\eta^2 - \sin^2 \alpha_0)^{1/4} \tag{6.42}$$

its flat local maximum at $x = 0$, and the modulation of $E_{x,\mathrm{es}}$ due to the superposition of $E_{x,\mathrm{em}}$, i.e., $|E_x|^2 = |E_{x,\mathrm{es}} + E_{x,\mathrm{em}}|^2$ are characteristic of resonance absorption. Owing to the flat maximum, the driver B exhibits in the resonance region, the inhomogeneous Airy-Stokes equation (6.39) yields an excellent approximation to E_x with a constant driver $b = -1$, see the dashed curve close to resonance in the Figure. As outlined in the foregoing section for a spatially constant driver B, resonance terminates at ξ_{m} where $|E_x|^2$ reaches its maximum; thus the resonance width in dimensionless form is given by $2\xi_{\mathrm{m}}$. With x_m indicating the position of the maximum $|E_x|^2$ and with $x_0 = L\beta^2 \sin^2 \alpha_0$ this becomes in real space

$$d = 2x_{\mathrm{m}} = 2\left(\frac{\beta^2 L}{k_0^2}\right)^{1/3} \xi_{\mathrm{m}}, \quad \xi_{\mathrm{m}} = 1.8, \tag{6.43}$$

provided $x_0/(x_{\mathrm{m}} - x_0) = (k_0 L \beta^2)^{2/3} \sin^2 \alpha_0 / \xi_{\mathrm{m}} \ll 1$ is fulfilled. With $\beta = 0.1$ and $k_0 L \leq 50$ it is well obeyed for all angles of incidence α_0 The resonance curve is

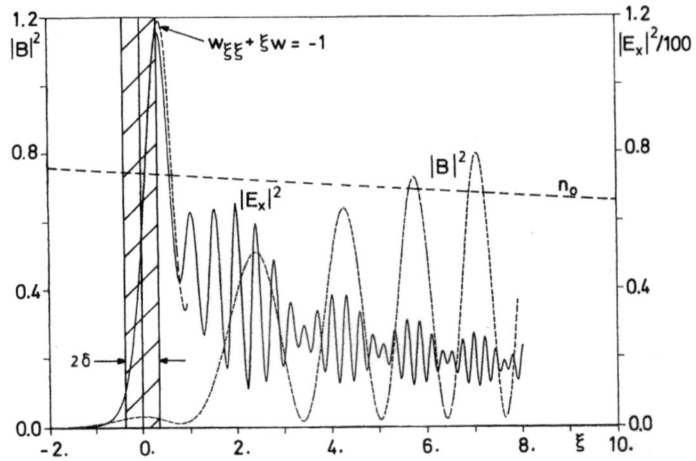

Fig. 6.5 Resonance absorption: Distribution of $|E_x|^2$ and $|B|^2$ over the flat plasma density profile n_0 as a function of ξ. Dashed region shows the resonance width $2\xi_m$ (ξ_m position of maximum $|E_x|^2$). Stokes equation reproduces well the resonance (dashed curve close to resonance)

nearly symmetric, hence d scales as $(k_1 L)^{1/3}/k_1$, $k_1 = k_0/\beta$. Although the density profile in Fig. 6.5 is flat, d is very small (dashed region). As a consequence of the driver not being exactly constant, in the exact solution of the coupled system of (6.34), (6.35) $\xi_m = 1.7$ is obtained from Fig. 6.5. When L is fixed at small angles of incidence the electromagnetic turning point is close to the critical point and much intensity reaches the coupling region, but the driver in (6.37), owing to the factor $\sin\alpha_0$ in k_y, is small. As α_0 increases less intensity tunnels to the critical surface; however, this may be overcompensated by the factor $\sin\alpha_0$, and resonant excitation of E_{es} reaches a maximum. With α_0 increasing further the exponential decrease of intensity tunneling to the critical point prevails and resonance absorption becomes ineffective. This situation is illustrated in Fig. 6.6. Resonance is strongest for $q \simeq 0.4$ (3rd picture).

Let us now determine the conversion rate $A = |S_{x,es}/S_{x,em}|$ for steady state resonance absorption. The electromagnetic and electrostatic flux densities in x-direction are (with I_0 the vacuum flux density of amplitude E_0)

$$\begin{aligned} S_{x,em} &= \frac{1}{2} v_{g,x} \varepsilon_0 |\hat{E}_{em}(x)|^2 = \frac{1}{2} c\eta (1 - \sin^2\alpha_0)^{1/2} \varepsilon_0 |\hat{E}_{em}(x)|^2 \\ &= (1 - \sin^2\alpha_0)^{1/2} I_0. \\ S_{x,es} &= \frac{1}{2} v_{g,x} \varepsilon_0 |\hat{E}_{es}(x)|^2 = \frac{1}{2} c_{se} \eta (1 - \sin^2\alpha_1)^{1/2} \varepsilon_0 |\hat{E}_{es}(x)|^2 \simeq I_{es}. \end{aligned}$$

By making use of the asymptotic expansion of (6.39) with $b = -1$ (see [27], p. 162 ff),

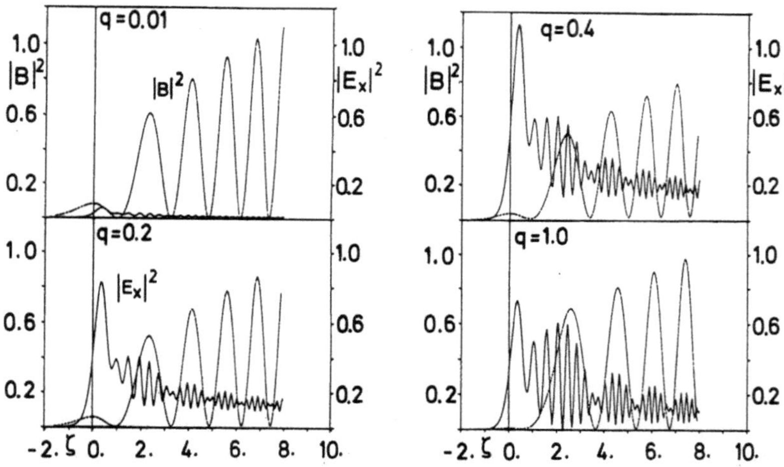

Fig. 6.6 Resonance absorption. The same distributions as in Fig. 6.5 for different angles of incidence, $q = (k_0 L)^{2/3} \sin^2 \alpha_0$

$$w = -\frac{\pi^{1/2}}{\xi^{1/4}} \exp i \left(\frac{2}{3} \xi^{3/2} + \frac{\pi}{4} \right); \quad \frac{\eta}{\xi^{1/2}} = \left(\frac{\beta}{kL} \right)^{1/3}.$$

From the definition of w in the foregoing subsection follows

$$S_{x,\mathrm{es}} = \frac{\pi}{2} c^3 \varepsilon_0 k L |B(0)|^2 \sin^2 \alpha_0. \tag{6.44}$$

Hence, the conversion factor of resonance absorption is

$$A = \pi k_0 L \frac{\sin^2 \alpha_0}{(1 - \sin^2 \alpha_0)^{1/2}} \left| \frac{B(0)}{B_{\mathrm{inc}}} \right|^2. \tag{6.45}$$

As soon as A is known this relation may be used to determine the "driver" $B(0)$. With I_0 in Wcm^{-2},

$$\begin{aligned} |B(0)| \sin \alpha_0 &= B_{\mathrm{inc}} (1 - \sin^2 \alpha_0)^{1/4} \left(\frac{A}{\pi} \right)^{1/2} (k_0 L)^{-1/2} \\ &= 5.16 \times 10^{-6} \left(\frac{A I_0}{k_0 L} \right)^{1/2} (1 - \sin^2 \alpha_0)^{1/4} \ [\mathrm{Tesla}]. \end{aligned} \tag{6.46}$$

Expressions (6.45), (6.46) are based on the assumption of $|B| = $ const over the whole resonance interval; for smooth density distributions with no significant curvature and $k_0 L \geq 2\pi$ this simplification is justified. In Fig. 6.7 the conversion factor A is shown as a function of q for different electron temperatures and $k_0 L \leq 2\pi$. As expected from (6.39) there is almost no difference between cold and warm plasma up to temperatures

$T_e = 20\,\text{keV}$ corresponding to $\beta = 0.1$. At $T_e = 0$ the maximum conversion rate is $A = 0.49$. For $\xi < 2$ the local electric field amplitude is given by

$$\left|\frac{\hat{E}_x}{E_0}\right|^2 = (1 - \sin^2\alpha_0)^{1/2}(k_0L)^{1/3}\frac{A}{\pi\beta^{4/3}}|w|^2. \tag{6.47}$$

Solving for its maximum value $|\hat{E}_{x,\max}|$ yields

$$\left|\frac{\hat{E}_{x,\max}}{E_0}\right|^2 = 1.1(1 - \sin^2\alpha_0)^{1/2}(k_0L)^{1/3}A\beta^{-4/3}. \tag{6.48}$$

This may be compared with the maximum amplitude $E_{z,\max}$ of s-polarized light not exhibiting resonance, (5.180). For $T_e = 0$ analytical conversion rates A have been given by several authors [28–30]. For finite T_e but with $B = \text{const}$ the authors of [31] have found

$$A = \frac{q}{|1 - i\pi^2 q\Gamma|^2}|2\pi\text{Ai}'|^2; \quad q = (k_0L)^{2/3}\sin^2\alpha_0, \quad \Gamma = \text{Ai}'(\text{Bi}' + i\text{Ai}'). \tag{6.49}$$

Its numerical evaluation is the dashed line in Fig. 6.7. The deviations from the result in [19] are due to the simplification $B = \text{const}$. For $\beta \leq 0.1$ the power series solution in [28] is much more accurate and practically coincides with the numerical result in [19]. For subrelativistic temperatures the conversion A peaks at

$$\sin\alpha_0 = \frac{(0.4 - 0.5)}{(kL)^{1/3}}. \tag{6.50}$$

As T_e becomes relativistic and c_{se} approaches c resonant phase matching occurs over a wide range and the conversion factor A approaches unity at $q \simeq 0.7 - 0.9$ [19]. The shift of the maximum of A towards higher q-values is a consequence of x_0 increasing with β in (6.38) and hence favoring resonance in lower density regions.

6.2.2.1 The Capacitor Model of Resonance Absorption

The following brief considerations may be useful in so far as to complete the picture of resonance absorption from the intuitive physical point of view. It may open rapid access to simple estimates. Owing to the $\exp(-i\omega t)$ time dependence of E_x and B, (6.37) is equivalent to the equation of a resonantly driven electron plasma wave parallel to ∇n_0,

$$\frac{\partial^2}{\partial t^2}E_x - c_{se}^2\frac{\partial^2}{\partial x^2}E_x + \omega_p^2 E_x = c\omega^2 B\sin\alpha_0 \Rightarrow \frac{\partial^2}{\partial x^2}E_x + \frac{k_0^2}{\beta^2}\eta^2 E_x = \frac{k_0^2}{\beta^2}cB\sin\alpha_0. \tag{6.51}$$

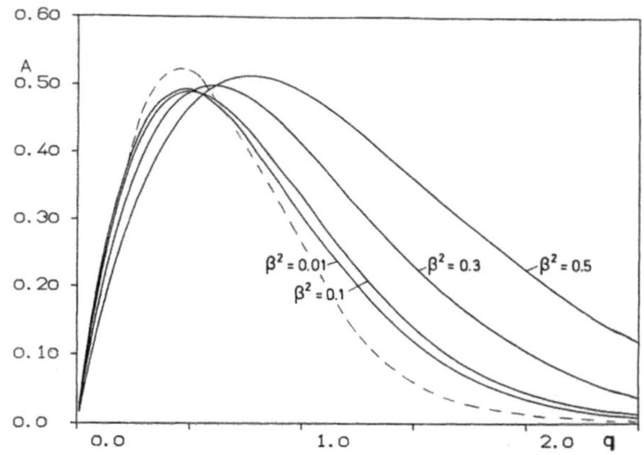

Fig. 6.7 Temperature dependence of resonance absorption. Conversion factor A as a function of $q = (k_0 L)^{2/3} \sin^2 \alpha_0$. Dashed line: Hinkel-Lipsker [31]

Let x_c be the position of the critical point and let us integrate (6.35) over an interval enclosing x_c under the restriction $\lambda \leq L$,

$$ik_{0y} \int_{x<x_c}^{x>x_c} E'_x \mathrm{d}x = ik_{0y} E_x(x) = \int_{x<x_c}^{x>x_c} (E''_y + k_0^2 \eta^2 E_y) \mathrm{d}x \simeq E'_y(x).$$

The inequality $\lambda \leq L$ enters in the last step; in steep density gradients η becomes large already in the vicinity of x_c. From $E'_y \simeq ik_{0y} E_x$ and $|E'_y| \simeq k_0 \eta |E_y|/\beta$ in the resonance interval, we have

$$|E_y| \simeq |E_x| \frac{\beta}{\eta} \sin \alpha_0 \ll |E_x|; \quad \beta/\eta << 1.$$

This is the main result of a more involved estimate [29]. It shows that $\nabla \mathbf{E} = \partial E_x / \partial x$ is a good approximation in not too steep density gradients and the displacement $\boldsymbol{\delta}_e$ of an electronic fluid element is nearly $\delta_{e,x}$. The latter is the basic assumption of the so-called capacitor model (6.51) [17, 38]. High laser intensities and plasma density steepening impose limits of applicability, at least in the red/near infrared wavelength domain. So far, starting from a rather general wave equation in the layered plasma, a few simplified assumptions had been introduced among which the linearity of the current density $\mathbf{j} = \sigma \mathbf{E}$ and, less stringent, a spatially constant driver $B = B(0)$ are most severe.

When curvature of the ion density profile in the resonance zone does not play a decisive role (this is the case in expanding plasmas produced by smooth laser pulses) the wave number k_e^* of the electron plasma wave at the edge of resonance is deduced from $\xi_\mathrm{m} = 1.8$ in (6.43),

$$k_e^* = \frac{\omega}{s_e} \left\{ \beta^2 \sin^2 \alpha_0 + \xi_\mathrm{m} \left(\frac{\beta}{k_0 L} \right)^{2/3} \right\}^{1/2} \simeq 1.8^{1/2} \left(\frac{k_0^2}{\beta^2 L} \right)^{1/3}. \tag{6.52}$$

The second expression holds when x_0 can be ignored. As the electron plasma wave moves out into the lower density corona, k_{es} increases and Landau damping sets in; it can also be determined in the framework of the capacitor model [23]. This has to compete with the WKB reduction of $\hat{E}_{es} \sim \eta^{-1/2}$. The corresponding dependence of the electron amplitude $\hat{n}_1 = \hat{n}_e - n_0,\ \ n_0 = \langle n_e \rangle$, shows the opposite behavior; thus

$$\hat{E}(x) = \left(\frac{\eta(x_1)}{\eta(x)}\right)^{1/2} \hat{E}(x_1), \quad \hat{n}_1(x) = \left(\frac{\eta(x)}{\eta(x_1)}\right)^{1/2} \hat{n}_1(x_1) \tag{6.53}$$

if x_1 is an arbitrary position out of resonance where the WKB approximation is valid.

6.2.2.2 Steep Density Gradients and Fresnel Formula

Total conversion at low β. The capacitor model just exposed may fail either when the density profile curvatures become significant in the resonance interval or/and when L becomes of the order of the electron displacement λ_e or less. The latter effect has been studied in the literature by adopting Epstein type density transition layers of the form [19]

$$\eta^2(x) = \frac{1}{2} - \frac{1}{1 + \exp(-x/\Delta)}, \qquad L = 4\Delta.$$

It was found, and it is shown in Fig. 6.8 that (i) the conversion A deviates from the curves of Fig. 6.7 as soon as $k\Delta$ becomes smaller than 2, (ii) at a given temperature the maximum of A increases with decreasing transition width Δ, and (iii) the local angle α at the resonant layer at which maximum conversion occurs is shifted towards larger values, in contrast to the vacuum angle α_0 which is reduced. A particularly interesting and clear result is obtained for an infinite plasma density step with $\Delta \to 0$. In this case total mode conversion, i.e., $A = 1$, is found for all temperatures when the electron plasma wave is emitted into the direction of the incident radiation field E_0 [19]. It requires the existence of an underdense plasma shelf in front of the discontinuity, generated for example by a laser prepulse. Resonance absorption of intense electromagnetic fields at the surface of solid targets plays an important role in the interaction of ultrashort laser pulses with matter (i.e., $\lesssim 10^{-13}$ s) since during such times almost no rarefaction wave builds up and absorption occurs at a density discontinuity. It has been argued that degenerate resonance absorption takes place here which is characterized by nonresonant, extremely anharmonic forced oscillations of the first electronic layer at the surface [32, 33], and that absorption occurs at rather large angles of incidence. This kind of absorption has been treated more extensively in the context of ultrashort laser pulse-matter interaction in Chap. 1 and its correct physical interpretation is given there.

Fresnel's formula as a limiting case? With Δ shrinking to zero one could expect the absorption A to approach Fresnel's formula for p-polarized light shining on a

Fig. 6.8 Conversion curves A as functions of the optical layer thickness $\bar{k}\Delta$ and angle of incidence. $q = (k_0 L)^{2/3} \sin^2 \alpha_0, \ \bar{k} = k_0/2, \ \beta^2 = 0.1$

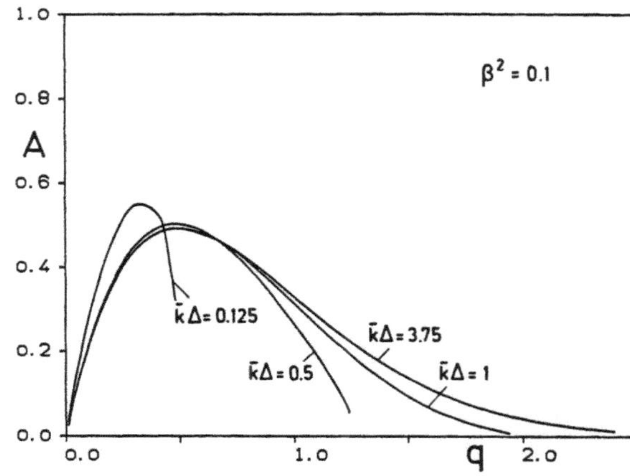

metal under oblique incidence. This is indeed the case for ratios ν_c/ω of collision frequency ν_c to laser frequency ω suitably chosen as shown explicitly in [34] by solving the exact Helmholtz equation in an exponential isothermal density profile of variable scale length L and in [35, 36] by applying a layered plasma model to various density (e.g., Epstein transition) profiles. These latter authors have found "qualitative" agreement between resonance absorption of 250 fs long pulses in short scale length plasmas and Fresnel optics. The finding may suggest that there is such a correspondence between resonance absorption and classical optics. In particular, the insensitivity of the reflection coefficient R towards ν_c/ω over a wide range of values and for $kL < 1$ [34] is understood from the following simple model for a sharp refractive index transition (see for instance [37], p. 215, [30], p. 42). The dielectric displacement component D_x is continuous across the transition, i.e., $\varepsilon E_x = E_{\mathrm{inc}}[1 + \sqrt{R}\exp(i\alpha)]$, α real. In a linear density profile of scale length L and a damping coefficient ν normalized to ω, εE_x translates into $E_x = E_{\mathrm{inc}}[1 + \sqrt{R}\exp(i\alpha)]/(\xi + i\nu)$, $\xi = (x_c - x)/L$, x_c critical point. The damping stands either for a collision frequency or a loss due to wave dispersion at finite electron temperature, i.e., energy outflow from the critical region. Then the absorption coefficient $A = 1 - R$ reads

$$A = \int \langle j_x E_x \rangle \, \mathrm{d}x / E_{\mathrm{inc}} \sim [1 + \sqrt{R}\mathrm{e}^{i\alpha}]\nu \int |E_x|^2 \, \mathrm{d}x \sim [1 + \sqrt{R}\mathrm{e}^{i\alpha}] \int \frac{\nu}{\xi^2 + \nu^2} \, \mathrm{d}\xi. \tag{6.54}$$

For $\nu \lesssim 0.3$ the integral can be extended from $-\infty$ to $+\infty$ to yield the value $1/\pi$. It shows that A becomes insensitive to damping ν and scale length L. On the other hand one must keep in mind that the dependence of R on small damping values is also weak and its value remains close to unity for electron densities of the order of 100 times critical density. In fact, at 80° of incidence and $L/\lambda < 0.1$ the reflection for $\nu = 0.01, 0.1$, and 1.0 is $R = 0.99, 0.92$, and 0.5, i.e., the penetrating field is very small and consequently A, despite the constant value of the integral close to $1/\pi$, vanishes for $L \to 0$, except for unrealistically high collision frequencies. There is an additional restriction. In order for the above expression for A to hold with ν reducing to zero the incident laser field must also reduce to zero to avoid *wave breaking* owing to $E_{x,\max} \sim 1/\nu$. After all, Fresnel's formulas are restricted to linear optics. From such a requirement one deduces for the limiting intensity $I_{\max}$ for $L \to 0$ [42] [see also (6.56) below]

$$I_{\max} = \frac{1}{2}\left(\frac{4}{3\pi}\right)^{2/3}\left(\frac{m}{e}\right)^2 (c^8\omega^7)^{1/3}(\epsilon L^{1/3})^2 \tag{6.55}$$

$$\to \quad I_{\max}[\mathrm{Wcm}^{-2}] = 2.9 \times 10^{18}(\epsilon \times L[\mathrm{m}]^{1/3})^2.$$

The tolerable limit for linearity of the electron plasma wave n_1 is indicated by $\epsilon = n_1/n_c$. With $\epsilon = 0.1$ and $L = 10\,\mathrm{nm}$ linear resonance absorption is at its limit for 6×10^{13} Wcm−2. The $L^{2/3}$-dependence reflects the property of the resonance width increasing with $L^{1/3}$. Particular experimental effort has been dedicated to the collisionless absorption of sub-ps laser pulses in short scale length plasmas from solid targets in [34–36, 44]. In [34] it is shown how resonance absorption may be used as an elegant and very reliable diagnostic tool for determining the laser scale length L In [44] the reader may find an interesting comparison between fluid models and collisionfree PIC simulations. Both, although fundamentally different from each other, fit well to the experimental results, at a price, however, of introducing an artificial viscosity and, in concomitance, strong artificial damping in the fluid program to avoid wave breaking. In presence of artificial viscosity it is necessary to check the absorption $\overline{\mathbf{jE}}$ over the whole density transition profile. Only if $\int \overline{\mathbf{jE}}\,\mathrm{d}x$ is essentially localized around the critical point and in the evanescent region the result perhaps may be reliable. In this context a short discussion of the correct boundary conditions at a vacuum-matter interface may shed additional light on the question whether resonance absorption for $\Delta \to 0$ goes over into the laws of linear optics. Thereby it must be kept in mind that such an analysis, owing to its linear character, is valid, if at all, only for very low intensities.

Extension of Frenel's boundary conditions. Fresnel's reflection formula for $R = 1 - A$ resembles one aspect of resonance absorption, namely the Brewster minimum of R which is a consequence of the fact that an oscillating electron does not radiate into its oscillation direction. This makes also result [19] of total conversion into an electron plasma wave emanating into the lower density shelf understandable. Fresnel's formula is derived from the conditions that the tangential component of the electric field E_t and the normal component of the electric displacement D_n are continuous. Now, to obtain Fresnel's situation the transition width Δ, together with the density of the lower shelf and the electron temperature T_e has to reduce to zero. From Maxwell's equations (5.1), (5.2) when applied to the infinitesimal area $\mathbf{dF} = \mathbf{e}_y \mathrm{d}x\mathrm{d}z$, follows for $\mathbf{B} = \mathbf{B}_t$

$$\varepsilon_0 c^2(\nabla \times \mathbf{B})\mathrm{d}\mathbf{F} = \left(j_y - i\omega\varepsilon_0 E_y\right)\mathrm{d}x\mathrm{d}z = (B_{z,\mathrm{vac}} - B_{z,\mathrm{plasma}})\mathrm{d}z.$$

For $\Delta \to 0$, E_y and j_y both diverge logarithmically at $T_e = 0$, $E_y, j_y \sim \ln(1/\Delta)$, and hence $\mathbf{B}_t = B_z\mathbf{e}_z$ is continuous across the vacuum-plasma interface. As a consequence the Poynting vector $\mathbf{S} = S_x\mathbf{e}_x$ is also continuous across the surface and resonance absorption reduces to zero. Continuity of $\mathbf{S}$ requires total reflection from the overdense plasma in perfect agreement with Fresnel's result. In order to be in agreement with the foregoing absorption formula $\sqrt{R}\exp(i\alpha) = -1$ must be set in (6.54).

However, this is not the whole story. From $\mathbf{D}_n$ being continuous a discontinuity of $\mathbf{E}_n$, i.e., the induction of an oscillating surface charge follows. The oscillation leads to a delocalization of this charge and to a finite transition width, as considered in [32]. Such an argument led to the boundary condition of $\mathbf{E}_n$ being also continuous at the interface and with it to a correction of Fresnel's reflection formula for p-polarized light incident onto a metal surface with $\omega_p > \omega$ [45]. For a quantitative discussion of this situation with a weak incident field the interested reader may consult the monograph of Forstmann and Gerhardts [46].

In conclusion, only within a restricted domain of parameters ν, L, and laser intensity I the Fresnel formula for absorption of p-polarized waves may appear as the limiting case of linear resonance absorption. On physical grounds, however, there is no basis for such an identification over the whole range of the density scale length $L \to 0$. For high laser intensities anharmonic resonance, not contained in Fresnel's formula at all, will play a decisive role.

6.2.3 Comparison with Experiments

Resonance absorption with ns laser pulses was experimentally shown to occur by numerous research groups, as for example [47–52], and to be in agreement with linear theory at Nd laser intensities up to $I = 10^{14}$ Wcm^{-2}. A clear difference between s- and p-absorption was found at all angles of incidence $\alpha_0 < 60^o$. The first experimental proof was given by J. E. Balmer and T. P. Donaldson by measuring the electron plasma temperature from the interaction of a 35 ps Nd laser pulse of intensity $I \simeq 2 \times 10^{13}$ Wcm^{-2} [47]. A pronounced distinction between s and p polarization was found. The angular dependence of T_e showed the characteristic shape of Fig. 6.7. The most accurate and complete measurements in the intensity range from 10^{13} to 10^{14} Wcm^{-2} were performed in the past in [53] on plexiglass targets with a Nd-YAG laser of 35 ps pulse length. The absorption was detected simultaneously in five parallel channels of observation: reflection of (1) the fundamental wave and generation of the second harmonic (2), see Fig. 6.9, charge (3) and velocity (4) distribution of the expanding plasma cloud, and relative ablation pressure (5) on the target. All five channels gave equivalent results and the absorption of p-polarization exceeded that

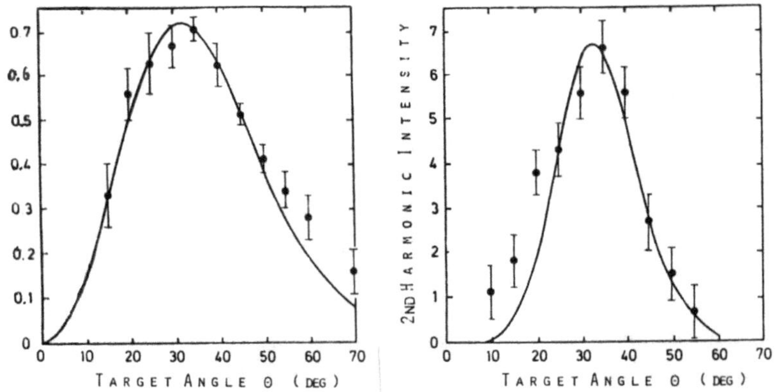

Fig. 6.9 Experimental verification of resonance absorption by Dinger et al. [53] as a function of angle of incidence θ. Left: Ratio $(R_s - R_p)/R_s$. R_s R_p reflected intensities in s and p polarization. Right: Intensity of 2ω emission. Courtesy of [53]

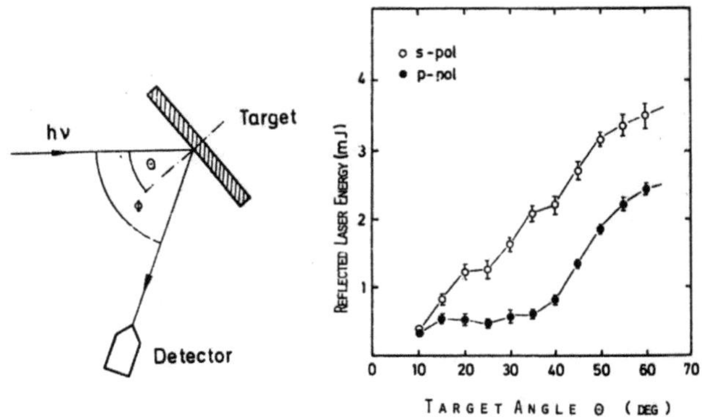

Fig. 6.10 Experimental verification of resonance absorption by Dinger et al. [53] as a function of angle of incidence θ for $s-$ and $p-$ polarization; reflected energy in units of [mJ] [53]

for s-polarization by about a factor of 2 on the average, see Fig. 6.10. The excited Langmuir wave can interact nonlinearly with the incident laser light, thus creating the second harmonic, $2\omega_{\mathrm{Nd}}$ with an intensity proportional to A. Some characteristic deviations between experiments and theory of resonance absorption may have to be attributed to fluctuations of the plasma density profile. Theoretically the influence of stochastic density oscillations was studied by several authors [55–57] for long pulses. It was found that the excitation level of the Langmuir wave decreases monotonically with increasing amplitude of the density fluctuations.

Resonance absorption by excitation of a plasma wave at twice the laser frequency is possible at normal or nearly normal incidence when the Lorentz force $-e\mathbf{v}_{\mathrm{os}} \times \mathbf{B}$ competes with or prevails over the driving force $-e\mathbf{E}$ at the fundamental frequency. The possibility increases with density profile steepening. In fact, the first studies into this direction have been realized in ultrashort laser pulse interaction in step like solid density profiles and the term $\mathbf{j} \times \mathbf{B}$ *heating* has been coined for the mechanism [54]. Its enhancement is based on *anharmonic resonance.*

6.3 Nonlinear Resonance Absorption

The theory of resonance absorption exposed in the preceding section is valid as long as the ratio $\hat{n}_1/n_0$ is small compared to unity in the resonance interval. From (2.11) and from δ in (5.66), we have $n_1 \simeq -in_0k_eeE/m - e\omega^2$. At the maximum of $\hat{E}$, one must take $k_e = k_e^*$ from (6.52), and (6.48) may be used to relate the amplitude $\hat{n}_1$ to the incident laser intensity I_0. After some transformations the following relation is obtained,

$$\frac{\hat{n}_1}{n_0} = \sqrt{2}\xi_{\mathrm{m}}^{1/2}\frac{e}{m_e(c\varepsilon_0)^{1/2}}(k_0L)^{5/6}A^{1/2}\frac{1}{\omega L}\left(\frac{I_0}{\omega^2}\right)^{1/2}. \tag{6.56}$$

At the Nd laser intensity $I_0 = 10^{14}$ W/cm^2 and $L = \lambda_{\mathrm{Nd}}$ results from there

$$\frac{\hat{n}_1}{n_0} = 0.28\ A^{1/2}. \tag{6.57}$$

It reveals that for these parameters the linear description of the electron plasma wave in the resonance interval is close to its limit. At higher intensities a nonlinear theory of mode conversion must be developed. In general, a complete description of this phenomenon can only be obtained by including Maxwell's equations and employing numerical methods. However, in order to describe essential features of the nonlinear behavior and to make analytical predictions, models of resonant electron plasma wave excitation can be used. The simplest choice is again the capacitor model. In the following it will be used to study some nonlinear resonance absorption characteristics. Vlasov simulations of the capacitor model are then used to test the hydrodynamic approach and to study electron trapping and acceleration. An entirely different model will be introduced in the following chapter to study collisionless absorption of ultrashort laser pulses on solid surfaces. The underlying idea is that with a strong driver the resonance frequency becomes a function of the oscillation amplitude, i.e., of driver strength. By this effect an originally nonresonant volume element may undergo resonance with the laser frequency if the beam intensity is sufficiently high (see Sect. 7.3.2 on anharmonic resonance).

6.3.1 *Wave Breaking*

In the foregoing chapter the notion of *wave breaking* has been introduced in connection with the high amplitude electron plasma wave. Studied first in 1959 it has gained increasing interest in the following decades accompanied by a significant amount of literature. A somehow intuitive knowledge of "irregularity" was associated with the concept of *breaking* of an originally "regular" wave. The simplest reason for the appearance of irregularity has been seen in the loss of uniqueness of flow. It happens when the Jacobian $J = |\partial \mathbf{x}/\partial \mathbf{a}|$, **a** Lagrangian coordinate, becomes infinite because two adjacent fluid volume elements superpose and cross. In the regular wave ordering of fluid elements remains the same all the time; the sequence of an element ahead of another element behind persists indefinitely. The Jacobian-based criterion is correct for the cold plasma. In a warm stationary plasma it has to be substituted by a criterion which in addition to the flux of matter accounts also for the shift of the phase of the wave, see Chap. 5, there formulated for an ideal thermal plasma. The reason for breaking was found in an excessive increase of the electrostatic potential of the wave and its enthalpy to no longer satisfy the momentum balance [41]. In connection with resonance absorption new criteria appeared, for instance in terms of a harmonic oscillator with time dependent eigenfrequency [38, 39]. These criteria have all to be referred to as *hydrodynamic wave breaking* in cold and warm fluids, out of resonance, and at resonance. In this context *geometric wave breaking* has to be cited [40]. Although, intuitively it may appear as the most natural braking scenario it was found late in 2D and 3D PIC simulations of the wake of an intense ultrashort laser pulse travelling through an underdense plasma. When the wave fronts off axis travel faster than the central part they may collapse on axis, interpenetrate, and destroy the wave.

New aspects of wave breaking came along with the discovery of hot electrons and electron acceleration by trapping. The discovery was that kinetic phenomena greatly affect the structure of a Langmuir wave and its damping. Not seldom, subsequently and even till present, wave breaking and particle acceleration have become synonyms. Sentences in the published literature appeared like "when the wave breaks electrons become fast". In concomitance, wave breaking is often invoked as responsible for collisionless absorption of superintense laser pulses. It is sufficient to think of particle accelerator technique to understand that for efficient acceleration wave breaking has to be avoided. It may be argued also from what has been outlined on self interaction of modes of a Langmuir wave in the nonlinear regime in the previous chapter and from Vlasov simulations in the following sections that wave breaking is detrimental to particle acceleration. Two main questions may be formulated to make further steps towards clarification:

(i) What is a meaningful definition of wave breaking in a kinetic approach, and
(ii) What is the appropriate breaking criterion if kinetic effects dominate.

For example, one may ask whether such a phenomenon like *kinetic wave breaking* occurs at all when one keeps in mind how strongly a Langmuir wave is damped by particle trapping and acceleration. There is only one early clear statement on this problem [70]: "Thus we see that there is no one-to-one correspondence between hot electron production and wave breaking as is sometimes suggested." The deceptive correlation still persists and creates confusion.

In the following sections 1D Vlasov simulations are used to clarify some of the questions just addressed: the role of damping, the significance and value of hydrodynamic breaking criteria, fast electron generation by electron trapping in the resonantly excited Langmuir wave, their non-Maxwellian distribution in the first trapping stage, and their final Maxwellian spectrum, well confirmed by the experiment, by PIC simulations (see Fig. 1.16), and the Vasov simulations presented here. In the last section the problem of kinetic wave breaking is tackled. A general definition of wave breaking, applicable to fluid dynamic as well as kinetic description, is proposed and trapping of entire bunches of electrons and their coalescence is recognized as one important scenario of breaking. The discussion is mainly based on a still topical (not much known) paper by Bergmann [41].

6.3.1.1 Resonance Absorption by Nonlinear Electron Plasma Waves

With increasing laser intensity or at low electron temperature the electron plasma wave may exceed the linear limit and may even break. Furthermore, particle-in-cell (PIC) simulations have shown that the particle aspect appears more and more as the Langmuir wave potential increases and particle trapping and acceleration dominates [69]. As we shall see in the following these phenomena lead to strong collisionless damping of the Langmuir wave. It is therefore quite natural to extend the traditional, somewhat vague definition of *nonlinear Landau damping* (see for example [11, 59], Sect. 5.3.1) to include such processes. This makes clear that at high intensities a kinetic description of resonance absorption is needed. With particle trapping the following questions arise: What structure does a strongly nonlinear Langmuir wave assume, and to what extent does the much simpler fluid description still give a satisfactory answer? How is the mode conversion efficiency, i.e., the absorption coefficient A, affected by the nonlinear evolution of the plasma wave? These questions can be answered properly by numerically solving the Vlasov equation. The evolution of the one-dimensional electron distribution function $f(x, v, t)$ under the action of an external driver field E_d and the induced self-consistent field $E(x, t)$ in such a fixed ion background of charge density $en_0(x)$ is governed by the Vlasov and Poisson equations,

$$\frac{\partial f}{\partial t} + v\frac{\partial f}{\partial x} - \frac{e}{m_e}(E + E_d)\frac{\partial f}{\partial v} = 0, \quad \frac{\partial E}{\partial x} = \frac{e}{\varepsilon_0}\left(n_0 - n_c\int_{-\infty}^{\infty} f\mathrm{d}v\right). \tag{6.58}$$

This procedure is much more involved than corresponding PIC simulations; however, as a compensation, it generates almost no numerical noise and leads to finer resolution in regions of low density in phase space. Starting with a Maxwellian distribution function

$$f_0(x, v) = \frac{1}{(2\pi)^{1/2}v_{\mathrm{th}}}\exp\left(-\frac{x}{L}\right)\exp\left(-\frac{v^2}{2v_{\mathrm{th}}^2}\right),$$

the result of the Vlasov calculation of the Langmuir wave is compared with hydrodynamic solutions obtained from (5.100) [43]. In Fig. 6.11a the driving field amplitude, measured in units of the thermal electric field $E_{\mathrm{th}} = k_BT_e/e\lambda_D = (m_ek_BT_e)^{1/2}\omega/e$, is $\hat{E}_d = 0.02$, and in (b)–(d) it is twice as strong, $\hat{E}_d = 0.04$. All pictures are taken at the same time $\omega t = 120\pi$. The first two, (a) and (b), are the hydrodynamic results

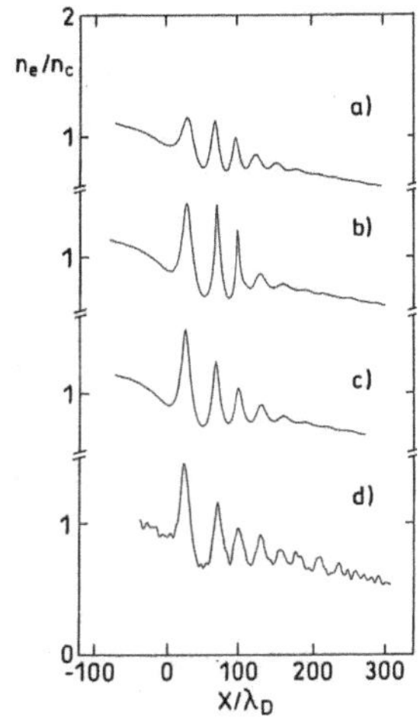

Fig. 6.11 High-amplitude electron plasma wave in the capacitor model with driving field amplitude **a** $\hat{E}_d = 0.02$ and **b–d** $\hat{E}_d = 0.04$ and $L = 600\lambda_D$ at time $\omega t = 120\pi$. The critical density is located at $x = 0$. Normalized electron density n_e/n_c **a, b** from the fluid model, **c** from the Vlasov simulation, and **d** from a PIC calculation with $6 \cdot 10^4$ particles (large numerical noise). Jitter in the phase between **a, b, c,** and **d** is due to small time step differences between the different numerical procedures

and show the characteristic peaking as presented in Figs. 5.7 and 5.9. The analogous Vlasov result is shown in (c). The agreement in the general wave structure is excellent. For comparison, in picture (d) the corresponding result of a PIC calculation with $6 \cdot 10^4$ particles, applied to a nonlinear density profile with the same gradient n_c/L at the resonance point, is plotted. The advantage of the 1D Vlasov simulation is evident. The amount of particles trapped and accelerated (detrapped) by the wave of Fig. 6.11 can be estimated from the distribution function in Fig. 6.12a: trapped particles are recognized from their closed loops in the potential wells, detrapped particles form the undulating contours leaning to the right. For comparison, in Fig. 6.12b the distribution function from the PIC calculation of the same wave is presented. Here, in contrast to the Vlasov code, the evaluation of the distribution function is not possible in those regions where almost no simulation particles are present. How strongly *nonlinear Landau damping* exceeds linear Landau damping can be deduced from Fig. 6.13 for the two driver strengths $\hat{E}_d = 0.04$ and $\hat{E}_d = 0.08$. It shows that, except at very low $\hat{E}_d$ values, linear Landau damping (dotted curves) is totally dominated by damping due to particle trapping and acceleration (at low driver strength the comparison with linear Landau damping was used to test the accuracy of the numerical procedure employed here). More quantitatively, Vlasov simulations show that *linear Landau damping* holds up to the following laser intensity limit,

$$I_0 = 2 \times 10^{13} \left(\frac{\omega_{\mathrm{Nd}}}{\omega}\right)^2 \left(\frac{300\lambda_D}{L}\right)^{1/3} \ \mathrm{Wcm}^{-2}. \tag{6.59}$$

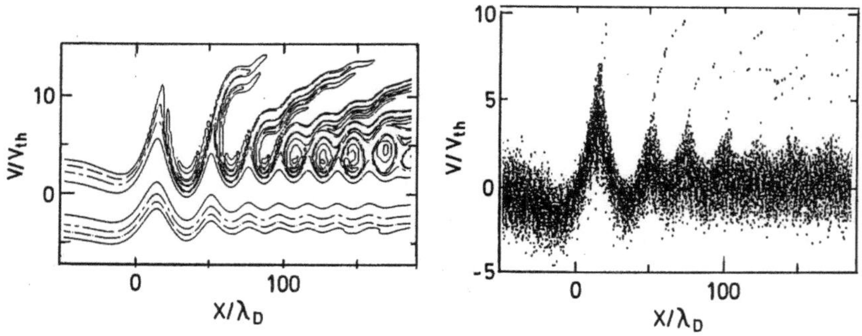

Fig. 6.12 Electron distribution in the wave in Fig. 6.11b–d after 60 cycles. **a** Vlasov simulation: contour lines of the distribution function for $f = 10^{-4},\ 10^{-3},\ 10^{-2}$, and 10^{-1} in phase space. **b** PIC simulation with $v_{\text{th}} = 0.05\,c$; a fraction of the particles are plotted as dots in phase space. Fast electrons escaping to the right are clearly seen

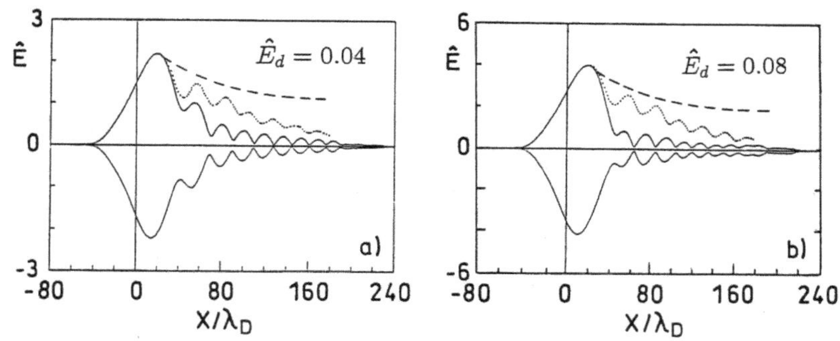

Fig. 6.13 Envelopes of the electric field for $L = 300\lambda_D$ and **a** $\hat{E}_d = 0.04$ and **b** $\hat{E}_d = 0.08$. Dashed curves: amplitude maxima from the fluid theory (no damping, $\hat{E} = \text{const}/\{1 - \exp[-x/L]\}^{1/2}$); dotted curves: linear Landau damping after [23]; solid curves: true nonlinear Landau damping. Amplitude modulations are due to $\hat{E}_d$ taken constant in space

The first wave crest is the only resonantly driven maximum. All other maxima which are outside the resonance region stop growing due to nonlinear Landau damping at $\hat{E}_d \approx 0.04$; the additional absorbed energy is transferred to fast electrons. The first field maximum increases with growing driver strength as plotted in Fig. 6.14. For $\hat{E}_d \gtrsim 0.15$, the Vlasov results deviate substantially from the linear fluid theory (6.55), (6.57), as indicated by the dots in the figure. If the driver strength is further increased (e.g., beyond $\hat{E}_d = 0.18$ for $L = 100\lambda_D$), all maxima except the resonant peak start fluctuating due to trapping of entire bunches of electrons, and out of resonance the Langmuir wave completely loses its periodic structure, i.e., it *breaks*. For the fraction A of resonantly absorbed intensity one would expect strong deviations from its linear value (6.45). At the latest, this should happen as soon as the highly nonlinear phenomenon of intense phase mixing sets in. However, surprisingly enough, the Vlasov simulations indicate that for moderately steep density profiles ($k_0 L \gtrsim 2$), at

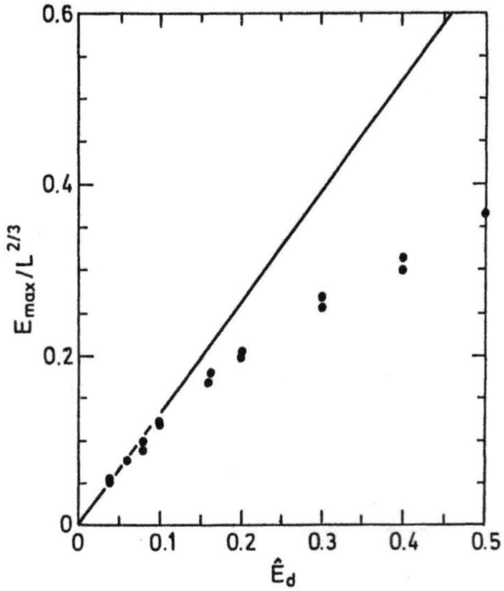

Fig. 6.14 Comparison of maximum electric-field amplitude from Vlasov simulations (dots) and from the linear theory (straight line). The latter scales as $\hat{E}_{\max} = 1.28 \cdot (L/\lambda_D)^{2/3}\hat{E}_d$ [23]

least up to $L\hat{E}_d^2 = 30$, the absorbed power in the capacitor model

$$\overline{P}_{\text{abs}} = AI_0 \cos\alpha_0 = \langle \int jE_d \mathrm{d}x\rangle_t,$$

j current density, is given by the linear theory with the maximum $A(\alpha_0)_{\text{opt}} = 0.49$ (see Fig. 6.7). Converting the $L\hat{E}_d$-values for a plasma of $T_e = 1$keV into laser intensities I_0 according to (6.45) and (6.46),

$$\hat{E}_d = \left(\frac{2AI_0\cos\alpha_0}{\pi\varepsilon_0\omega L}\right)^{1/2} \tag{6.60}$$

the intensity scale is indicated at the top of Fig. 6.15. It shows that the absorption A is correctly calculated from the linear theory at least up to

$$I_0 = 2 \times 10^{16}\ \text{Wcm}^{-2}/(\lambda_L/\mu\text{m})^2). \tag{6.61}$$

This is 200 times higher than the limit given by (6.57). It can be explained as follows. In the linear theory wave conversion takes place only within the resonance zone of width d from (6.43) given above. Outside this region the plasma wave decouples from the driver and becomes a freely propagating wave. Vlasov simulations show that in the nonlinear regime these characteristics still persist and almost no energy is coupled into the Langmuir wave outside the resonance region. In the nonlinear case the most remarkable difference is the generation of fast electrons. However, a large fraction of the energy transferred to the accelerated electrons is taken from

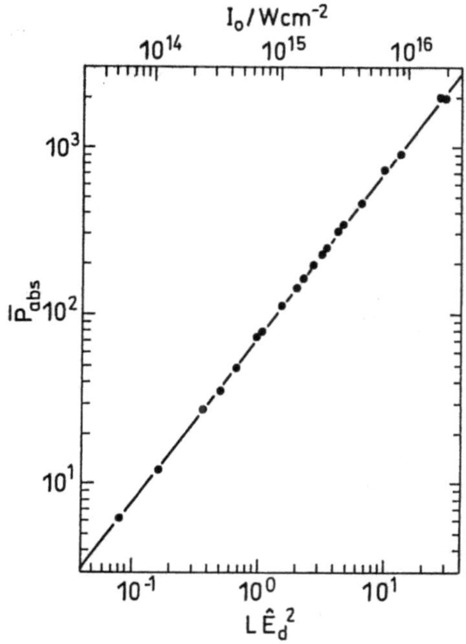

Fig. 6.15 Validity of linear resonance absorption: mean absorbed power $\overline{P}_{\text{abs}}$ as a function of the driving field strength $\hat{E}_d$ (*bottom scale*) and laser intensity for $\lambda = 1.06$ m and $T_e = 1$ keV in case of maximum absorption (*top scale*). *Dots: Vlasov simulations*; *straight line*: linear theory

the induced electrostatic wave field which is still almost sinusoidal in the resonance zone. Outside this region only thermalization of the plasma waves and of the fast electrons takes place. In other words, the accelerated electrons or fluid elements, respectively, have already acquired nearly their total energy just in the moment of breaking.

In order to show clearly the structure of the nonlinear plasma wave and its damping, in Figs. 6.11, 6.12, and 6.13 moderately steep density profiles have been chosen. Figures 6.14 and 6.15 contain data for density profiles with L as short as $20\lambda_D$ which in Nd-laser plasmas is typically $L \approx \lambda/5$; this models ponderomotive *profile steepening* which ranges from $L \simeq \lambda$ to $L \approx \lambda/(10-30)$ (see Sect. 3.3.4). Detailed calculations of the conversion efficiency in linear resonance absorption showed that only for $k_{\text{em}}L < 1$ the conversion exceeds the value $A = 0.49$ under special conditions [19].

Caviton and *soliton formation* with subsequent collapse occur in plasmas under favourable conditions. Just as interesting in themselves they have attracted the attention of theoreticians and have been extensively studied. For an introduction and overview [11], Chap. 6 is recommended. For detailed studies in connection with laser interactions contributions like [60] deserve interest. In models with fixed ions or smooth plasma background, as considered in our context, they are not observed. On the other hand, in laser plasmas produced by smooth ns and ps pulses from solid targets they represent transient phenomena and do not play a major role energetically, e.g. for absorption, since the plasma flow has a strong smoothing effect on any eventual ion density fluctuation generated in the resonance region. As the laser pulses become shorter and very intense, e.g., $I \gtrapprox 10^{19}$ Wcm^{-2}, long-lived, macroscopic bubble like structures have been detected via the electrostatic deflection of proton probe beams. The interpretation is supported by PIC simulations and other diagnostics [61].

6.3.2 Hot Electron Generation

Collisional, or more generally, stochastic laser (and particle beam) interaction with the plasma leads to a Maxwellian distribution of the free electrons to which a universal temperature T_e can be assigned with all properties of the fundamental laws of thermodynamics. A resonantly excited wave represents a deviation from thermodynamic equilibrium and may disturb the electron spectrum through the associated electric field. Wave-particle interaction in its mildest form is represented by linear Landau damping. Below a certain amplitude the traveling potential of an electron plasma wave is not strong enough to trap a significant fraction of the electrons; instead they follow straight orbits and remain free. With increasing amplitude, the potential V becomes sufficiently large to capture first the particles having velocities not far from the phase velocity v_φ of the wave. As an electron is trapped in a potential well it oscillates with the *bounce frequency* ω_B around its minimum and is carried into regions where v_φ and the density disturbance n_1 are increased or decreased. In the first case effective particle acceleration takes place. In the second case an electron may remain trapped and the difference of its kinetic energy is given back to the wave, or it becomes free again and carries away the energy imparted to it by the wave. Such *detrapped* electrons are seen on the wave crests in Fig. 6.12b. The bounce frequency of an electron oscillating close to the minimum of a sinusoidal potential $V = -(e/\varepsilon_0 k_e^2)n_1$ is easily determined. Expansion of V around its maximum x_0 yields

$$V = -\frac{e^2\hat{n}_1}{\varepsilon_0}\frac{(x-x_0)^2}{2}.$$

Hence, the bounce frequency is

$$\omega_B = \left(\frac{e^2\hat{n}_1}{\varepsilon_0 m_e}\right)^{1/2}, \qquad (6.62)$$

ω_B is the plasma frequency of the electron density amplitude $\hat{n}_1$. In the adiabatic transition free $\Rightarrow$ trapped the cycle averaged potential $\overline{V}$ plays a decisive role (see Sect. 2.1.6.2). When trapped an electron undergoes a momentum/energy change under the deformation of the wave potential along the propagation that is regulated by the adiabatic invariant $\oint pdq = \text{const}_\text{a}$. In the fixed sinusoidal potential the bounce frequency decreases with increasing bounce energy, the spectrum of the trapped electrons broadens. One may think of a whole variety of interactions, particle-particle, particle-wave, particle-fluctuations, which introduce a stochastic element that contributes to Maxwellization of their spectrum. In fact, *superthermal electrons* were observed first as early as in 1971 [62]. Since their energy distribution was not far from a Maxwellian a *hot temperature* T_h could be assigned to them, in contrast to the thermal electrons whose temperature T_e was given the name *cold electrons* although typical values of their *cold temperature* $T_c = T_e$ in laser produced plasmas on ns time scales range from 0.5 to 5 keV. In plasmas produced by high-intensity CO_2 lasers considerably higher values of T_c are reached. The hypothesis that the main sources for hot electron generation in the ns and ps regimes are resonance absorption at the critical density and stimulated Raman scattering as well as two-plasmon decay at a quarter of the critical density was successively confirmed by particle simulations [64–66]. Electron acceleration and, in concomitance, acceleration of ions up to several 10 MeV has become an important phenomenon when superintense laser pulses interact with gas and solid targets. Erroneously and repeatedly fast electrons are attributed to wave breaking (see next section). Currently, it is a main question and an unsolved problem what is the functional dependence of the hot temperature T_h and the fraction of hot electrons on the absorbed laser intensity ("hot electron scaling").

6.3.2.1 Particle Acceleration by a Discontinuous Langmuir Wave

Figure 6.5 suggests to model collisionless hot electron production by a suddenly rising wave structure at the critical point as sketched by Fig. 6.16. The wave is assumed sinusoidal with constant phase velocity v_φ and smoothly decaying amplitude $\hat{V}$ of Epstein type,

$$\hat{V}(x) = \frac{4w_0 \mathrm{e}^{\gamma k_e x}}{1 + \mathrm{e}^{\gamma k_e x}} v_\varphi^2. \tag{6.63}$$

In the wave frame $S' : x' = x - v_\varphi t$ the wave potential is static with respect to its fast oscillation and slowly time dependent in the amplitude, $V = \hat{V}(x' + v_\varphi t) \cos k_e x'$. For an electron to become trapped its oscillation center velocity must reduce to zero somewhere in the wave. There are three classes of electrons: A$\{v_i' < 0\}$ with negative initial velocities $v_i' < 0$, B$\{0 < v_i' < v_\varphi\}$ and C$\{v_i' \geq v_\varphi\}$. B and C enter the wave from left. The energy of a particle just after entering the wave depends not only on the energy $m_e v_i^2/2$ of the electron in free space but also on the edge of the potential V of width $2\hat{V}(x_c) = 4w_0 v_\varphi^2$. All electrons are assumed to enter with equal probability for modulus and phase of v_i. By solving numerically the exact equation of motion of 2400 electrons under the conditions above for $w_0 = 0.4$ and damping $\gamma = 0.003$ one arrives at the transition matrix $v_i \Rightarrow v_f$ depicted in Fig. 6.17 [67]. The solid contours of two symmetric spoons are the analytical result from the adiabatic invariants outlined in Sect. 2.1.6. Adiabaticity is well observed for γ not exceeding the value 0.08: Almost all points stay inside the analytic contours.

Three salient features have to be evidenced. (i) In contrast to a continuous wave envelope in the resonantly excited wave electrons of arbitrarily low velocity v_i can be trapped and accelerated. In the smooth continuous wave envelope, for instance in the envelope resulting from completing Fig. 6.16 by its mirror image $\hat{E}(-x + x_c) = \hat{E}(x - x_c)$, all particles with $|v_i|$ not exceeding a critical velocity v_c,

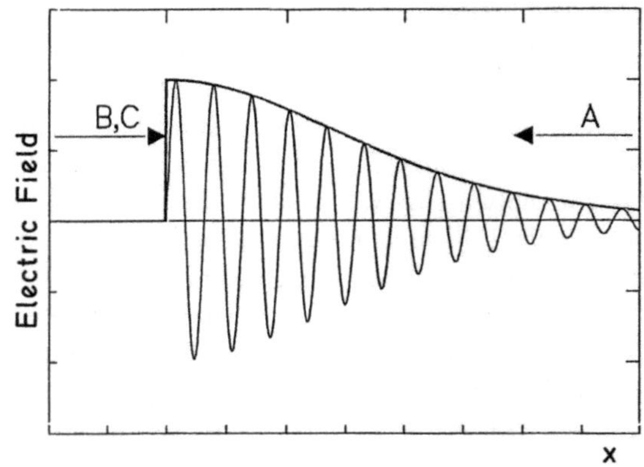

Fig. 6.16 Electric field of a wave with discontinuously rising envelope at $x_c . E(x, t) = \hat{E} \sin[k_e(x - x_c) - \omega t] = \partial_x V/e$. Potential $\hat{V}$ is taken from (6.63)

$$\frac{v_c}{v_\varphi} = \left[1 - \frac{8}{\pi^2}\right]^{1/2} \simeq 0.435 \tag{6.64}$$

are either reflected or pass through. Reflection occurs for $w_0 > 1.27$. Another essential result of the continuous envelope is that

In the wave of continuous envelope a particle can gain energy only if it is temporarily trapped.

(ii) The energy spectrum of the Figure is far from a Maxwellian; folding with a Maxwellian in a suitable velocity interval does not bridge the gap. (iii) Particles entering the field region from the left with initial velocities below the phase velocity can be accelerated to a maximum speed which is independent of v_i, but which increases with w_0, i.e., with $\hat{E}_{\max}$:

$$v_{\max} = v_\varphi \left[1 + 2\frac{\hat{E}_{\max}}{v_\varphi} + \frac{4}{\pi}\left(\frac{2\hat{E}_{\max}}{v_\varphi}\right)^{1/2}\right]^{1/2}. \tag{6.65}$$

This important formula shows that, owing to the time dependence of $\hat{E}$ in the wave frame, the maximum energy gained by a particle from the wave in the wave frame is *not* $\Delta\mathcal{E} = 2 \times \frac{1}{2}m(v_i - v_\varphi)^2$, although this expression is frequently found in the literature as a result from the *standard ponderomotive formula.*

6.3.2.2 Vlasov Simulations and Experiments

Extensive particle simulations [58, 64–66, 68, 69, 71] have shown that the spectrum of the hot electrons is close to Maxwellian and is sometimes attributed to multiple acceleration by Langmuir waves [69]. Subsequent calculations based on the Vlasov equation clearly confirmed the Maxwellian energy distribution of the trapped electrons and thus the existence of a temperature T_h. In Fig. 6.18 the space averaged distribution functions $f(v) = \langle f(x, v)\rangle$ for driver strengths of $\hat{E}_d = 0.04$ (a) and $\hat{E}_d = 0.08$ (b) in a fixed ion density profile of $L_c = 300\lambda_D$ are shown and the values of $T_h = 16$ keV and $T_h = 36$ keV are calculated. In addition to the much reduced noise in the Vlasov simulations the significant result which is found is that the Maxwellian character of $f(x, v)$ is already established at the right edge of the resonance zone of Fig. 6.5. In a run with $\hat{E}_d = 0.05$ and $L_c = 300\lambda_D$ a modest variation of $f(x, v)$ was only found when x was varied from $x = x_{\mathrm{m}}$ (end of resonance) to $x = L_c$, L_c critical scale length. This sheds new light on the different behaviour of Figs. 6.14 and 6.15 relative to their linear limits. However, it leads us back to arguments which may explain the discrepancy between Figs. 6.17 and 6.18. Figure 6.17 is the result of particles injected from left at x_c and from far right where the wave amplitude is zero. Apart from the fact of weak spectral widening due to the finite cold background temperature T_c there is also the possibility of electrons entering the wave laterally everywhere. Thus, there is a high probability for efficient Mawellization to occur by lateral electron inflow and repeated crossing the resonant wave.

The Maxwellian distribution suddenly breaks up at a maximum velocity v_{max}. A comparison of the "experimental" values from the simulations with formulas from the literature can be made, in particular with W. Schneider's expression (6.65) and another simple expression, which in normalized quantities reads [72]

$$v_{\mathrm{max}} = \left(8\hat{E}_d L_c\right)^{1/2}.$$

The result of a comparison with seven Vlasov runs is summarized in Table 6.1. Thereby v_φ was taken at $x = x_m$. The agreement with (6.65) is excellent whereas the simpler formula above, though leading to too low values, may serve for a rough estimate. Its agreement with simulations is expected to improve with decreasing scale length L_c. According to both formulae v_{max}, and the corresponding T_h increase with phase velocity v_φ. This means that more energetic electrons are to be expected from electron plasma waves excited in flat density profiles, e.g., from stimulated Raman scattering and two-plasmon decay; its confirmation by particle simulations and experiments is well established [64, 65, 73–75] (see also Chap. 7 on fs laser interaction). For practical purposes scaling laws for T_h and the number density n_h of hot electrons have been looked for. For the construction of such functional dependences from simple models in the ns laser pulse regime the reader may consult an early review by Haines [76]. In the long pulse regime all authors seem to agree on a combined $(I_0\lambda^2)^\delta$-dependence of T_h, and most of them (Vlasov simulations included) find δ-values not far from 1/3. According to [65],

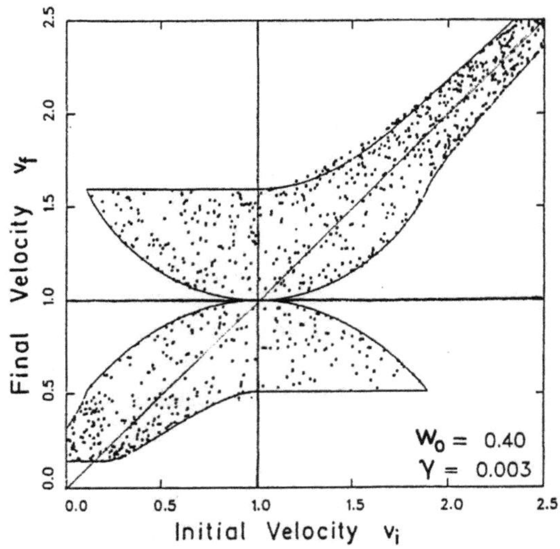

Fig. 6.17 "*Schneider's spoons*" [67]. Spectrum of final velocities v_f of 2400 particles accelerated in a Langmuir wave of suddenly rising amplitude after (6.63)

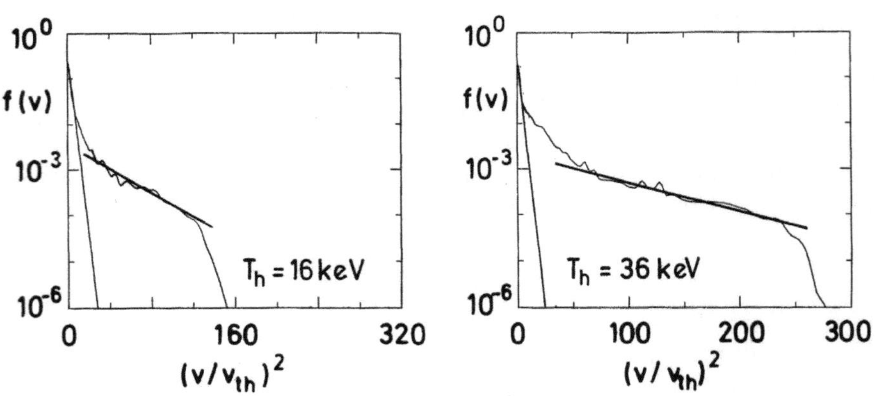

Fig. 6.18 Electron distribution function averaged over space, $f(v) = \langle f(x, v) \rangle$ for a normalized driver $\hat{E}_d = 0.04$ (**a**) and $\hat{E}_d = 0.08$ (**b**). The ion density scale length is $L_c = 300\lambda_D$ in both cases. T_h temperature of the fast electrons, T_c "cold" bulk temperature (steep straight lines). [After A. Bergmann (1990)]

$$T_h \simeq 14(I_0\lambda^2)^{1/3} T_c^{1/3}, \tag{6.66}$$

when T_h, T_c are measured in keV, I_0 in units of 10^{16} Wcm^{-2} and λ in µm. Their best fit to experiments at $(I_0\lambda^2) > 10^{15}$ Wµm^{2}cm^{-2} was $\delta = 0.25$. In this context the reader interested in a comparison with other simulations may consult [71].

Electron trapping and acceleration is one root of anomalous electron heat flux density $\mathbf{q}_e$ out of the critical layer [26, 43, 67]. In Fig. 6.19a the spatial distribution

Table 6.1 Comparison of the maximum electron velocity from a Vlasov simulation of resonance absorption with analytical formulae

$\hat{E}_d$	$L = 100\ \lambda_D$			$L = 300\ \lambda_D$			
	0.06	0.1	0.2	0.04	0.08	0.12	0.16
Vlasov simulation	≈10	≈12	≈15	≈12	≈17	19.5	≈22
Equation (6.65)	10.7	11.9	14.3	13	17.6	19.5	21.2
$v_{\max} = (8\hat{E}_d L_c)^{1/2}$	6.9	8.9	12.6	9.8	13.9	17	19.6

of q_e from Vlasov simulation at $\tau = 60\pi$ (dashed line) and $\tau = 80\pi$ is presented for $E_d = 0.04$ and $L_c = 300\lambda_D$. The first isolated peak of q_e reaches its maximum exactly at the border x_m of the resonance zone. The divergence of q_e manifests itself in large deviations of single volume elements from adiabatic behavior. In Fig. 6.19b the dependence of the kinetic pressure $p_e = m\int(v - u_e)^2 f\mathrm{d}v$ [see (3.92), relativistic: (3.121)] on the electron density n_e is shown in a log-log plot for a driver $E_d = 0.16$. In the absence of heat flow each volume element can move only along the diagonal corresponding to $\gamma = 3$ (nonrelativistic). Owing to the heat flux of the fast electrons a single volume element, when compressed in a density spike of the wave, at first remains in the neighborhood of the $\gamma = 3$ line (see lower arrow), but then, in the valley of the wave, a jet of fast electrons and with them a high value of q_e have been built up. The latter is responsible for the trajectory in the diagram far above the diagonal (follow upper arrow).

6.3.3 *Kinetic Theory of Wave Breaking*

A kinetic wave description is needed to answer questions (i) and (ii). It becomes further clear that in the case of $v_{\mathrm{os}} \ll v_{\mathrm{th}}$ the answers can hardly be found on the basis of PIC simulations owing to their inherent large noise. Therefore, again the system (6.58) in combination with the capacitor model is used and the driving field $\hat{E}_d$ is measured in units of the thermal field E_{th} as in the foregoing section. The ions are treated as a fixed smooth neutralizing background of inhomogeneity length L, the latter being chosen as a free parameter. The initial velocity distribution is Maxwellian, $f_0 = (2\pi)^{-1/2}v_{\mathrm{th}}^{-1}\exp(-x/L)\exp(-v^2/2v_{\mathrm{th}}^2)$. Figure 6.20 shows the evolution of f in terms of contour lines $f = \mathrm{const}$ after 20, 25, and 40 periods for $L = 300\ \lambda_D$ and $\hat{E}_d = 0.04$ with $E_{\mathrm{th}} = kT_e/e\lambda_D$ and Debye length at resonance, $\lambda_D = v_{\mathrm{th}}/\omega$. After a much higher number of cycles f looks very similar and does not show any new aspects, thus indicating that a quasi-steady state is reached at the times considered here. Closed loops in the contour plots refer to trapped particles. The contour lines extending to high velocities and leaning to the right are to be assigned to accelerated and detrapped particles. They are modulated by the periodic wave potential and their inclination increases as is characteristic for free streaming particles. The electron density is plotted in Fig. 6.21. The dashed smooth curve is the Coffey limit [77]. Owing to strong nonlinear damping the Langmuir wave never reaches this limit,

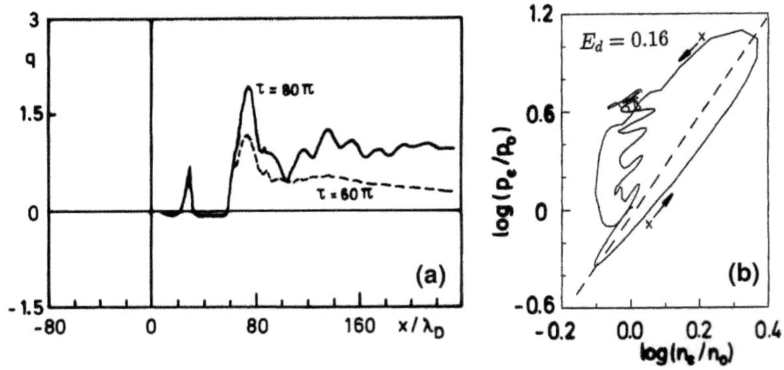

Fig. 6.19 **a** Anomalous heat current density $q = (m/2) \int (v - u_e)^3 f dv$ at $\tau = 60\pi$ (dashed) and 80π (solid) for a driver $E_d = 0.04$. **b** $\log n_e$- $\log p_e$- plot of an electron fluid element of particle density n_e and pressure p_e in the highly nonlinear electron plasma wave. The deviation of the trajectory from the straight line is a measure of the time-integrated divergence of the local heat flow q. Driver strength is $E_d = 0.16$

except for the first maximum, even at much higher driver strengths $\hat{E}_d$. This behavior is confirmed for three other L-values in Fig. 6.21. The Coffey criterion does not apply to the first (resonant) density maximum. For instance, in (a) it is clearly higher than the Coffey limit, but there is no indication of breaking (see also the first smooth and regular maximum in Fig. 6.22. For $\hat{E}_d \geq 0.04$ all maxima of n_e except the first do not increase further; but the wave remains periodic and smooth. As a first result, we can clearly see that, although fast electrons are generated by trapping, the wave is still regular and periodic in the sense mentioned above. In conclusion, fast electron generation alone does not indicate *wavebreaking*. However, above a certain threshold $\hat{E}_d^*$ the distribution function undergoes a *qualitative* change: the Langmuir wave becomes irregular, it breaks, as illustrated by Fig. 6.22 for n_e and E at $\hat{E}_d = 0.2$ in a density profile of scale length $L = 300\lambda_D$. Inspection of the corresponding evolution of the distribution function (see Fig. 6.23) reveals the reason for such a behaviour. The mean oscillatory velocitiy in the resonance region becomes so large that trapping of whole *bunches* of rather slow electrons occurs. These bunches of coherently moving electrons are partly *coalescing* (see black zones in the Figure) and remain trapped for at least several wavelengths, thus creating an additional autonomous macroscopic electric field. From a fluid point of view the phenomenon is similar to what is called intense mixing of volume elements exhibiting different oscillation phases. (A Grassberger-Procaccia analysis of the electric field and the electron density at fixed x reveals the transition from a quasiperiodic to a *chaotic attractor* [42]). The electrons in these bunches contribute to the heat conduction, so that the heat flux q increases suddenly with wavebreaking. For example at $x = 50\,\lambda_D$ one obtains (η refractive index, q electron heat flow density).

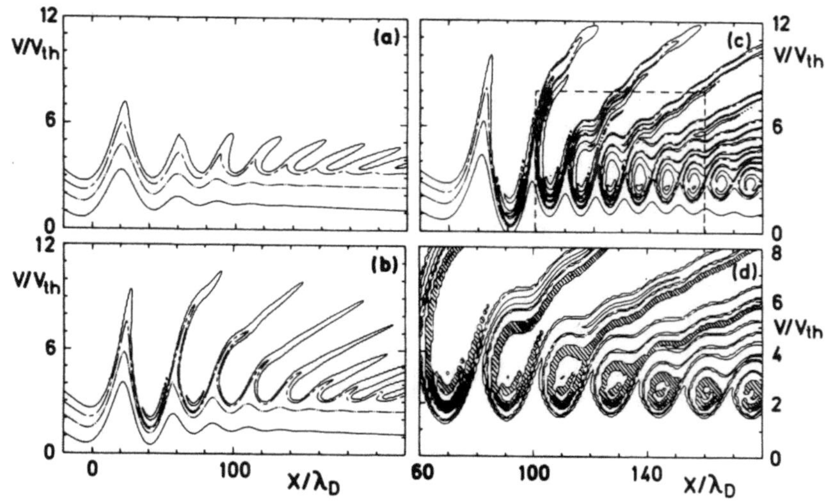

Fig. 6.20 Distribution function evolving from the initial distribution $f_0(x, v) = (2\pi)^{-1/2} \exp(-x/L) \exp(-v^2/2v_{\text{th}}^2)$ with $L = 300\ \lambda_D$ and $E_d = 0.04$ after **a** 20, **b** 25, and **c** 40 periods; **d** enlarged plot of area indicated by the dashed frame in **c** after 50 periods. The contour lines refer to $f = 10^{-1}$, 10^{-2}, 10^{-3}, and 10^{-4}, the critical density with $\omega_p = \omega$ is located at $x = 0$. Trapped electrons form closed loops (**d**); the velocity modulation of the detrapped electrons is due to the periodic electric field of the Langmuir wave. $\hat{E}_d$ is normalized to $E_{\text{th}} = kT_e/e\lambda_D$, $\lambda_D = v_{\text{th}}/\omega$

$$\begin{array}{llll} \eta: & 0.06 & 0.09 & 0.12 \\ q: & 0.31 & 0.32 & 0.49 \end{array}$$

In contrast to hot electron generation the bunches are not accelerated to high energies.

As a consequence of these numerical studies, the following definition can be given:
Wave breaking consists in the loss of periodicity of at least one of the macroscopically observable quantities [41].

This definition extends to both hydrodynamic and kinetic descriptions. In contrast to a linear wave where an irregularity occurs in all variables simultaneously, breaking may appear to a different degree in the various quantities, e.g., n_e and E in Fig. 6.22. The resonant density maximum, the first in the Figure, may satisfy Coffey's inequality or exceed this limit, even when the wave does not break (cf. Figs. 6.21, 6.22).

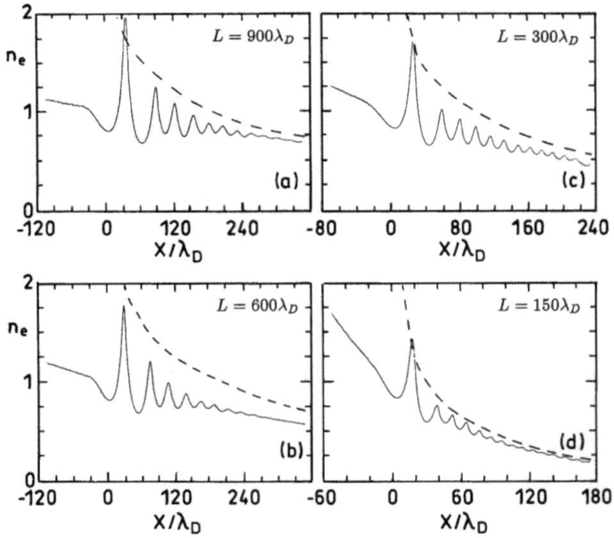

Fig. 6.21 Resonantly excited electron plasma waves in ion density profiles with different scale lengths L. Electron density n_e and breaking limit after Coffey (dashed curves); off resonance this limit is never reached owing to nonlinear Landau damping. **a** $\hat{E}_d = 0.06, \omega t = 120\ \pi$; **b** $\hat{E}_d = 0.06, \omega t = 80\ \pi$; **c** $\hat{E}_d = 0.08, \omega t = 100\ \pi$; **d** $\hat{E}_d = 0.1, \omega t = 70\ \pi$

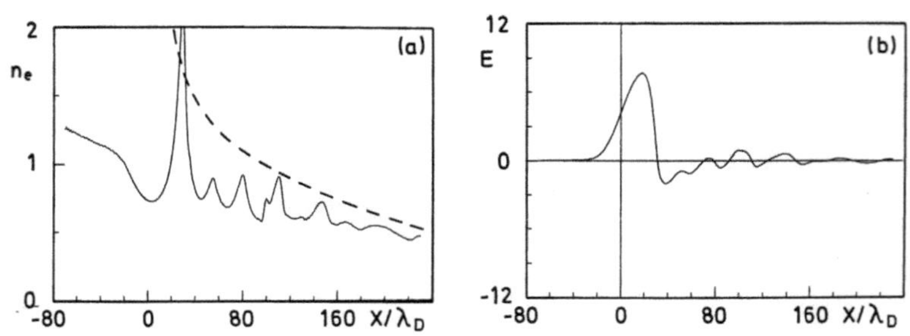

Fig. 6.22 Langmuir wave excited by driver of strength $\hat{E}_d = 0.2$ in a plasma of scale length $L = 300\ \lambda_D$: the irregular shape indicates breaking. **a** Electron density n_e; dashed curve: Coffey limit. **b** Electric field E as a function of space

Starting from the breaking condition for the streaming cold plasma, we replace the streaming velocity by the group velocity of the plasma wave, because the latter is now the speed of energy transport (at least approximately in a nonlinear wave). From the fluid theory of resonance absorption we obtain for the group velocity at the end of the resonance zone $v_g = s_e^2/v_\varphi \approx 2(L/\lambda_D)^{-1/3} v_{\text{th}}$. Inserting this into (5.110) at the place of v_0 yields

$$\hat{E}_d > \hat{E}_d^* = 0.72(L/\lambda_D)^{-1/3}. \tag{6.67}$$

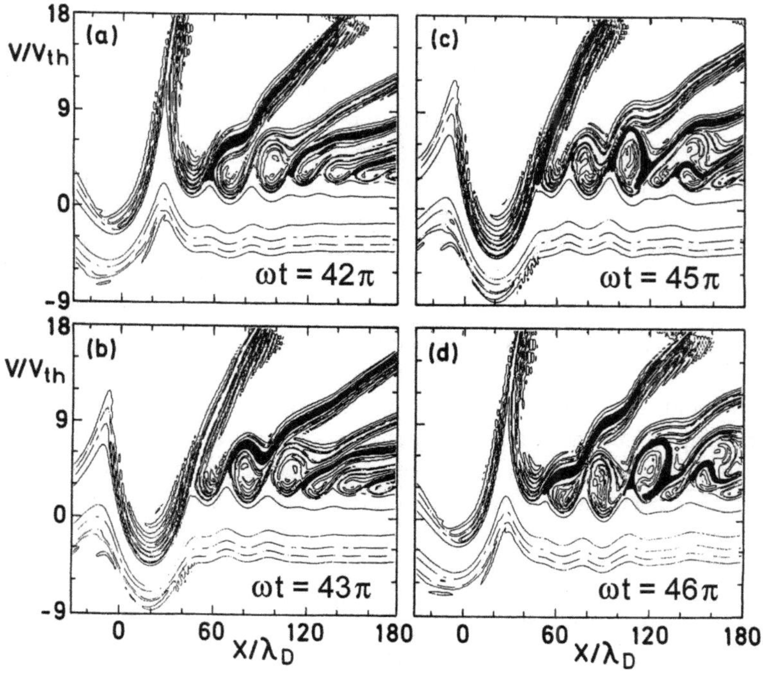

Fig. 6.23 Route to wavebreaking. The distribution function for $L = 300\ \lambda_D$ and $\hat{E}_d = 0.2$ after **a** 21, **b** 21.5, **c** 22.5, and **d** 23 periods. Contour lines for $f = 10^{-1},\ 10^{-2},\ 10^{-3}$, and 10^{-4}. Trapping and coalescence of entire electron *bunches* is evident (see black colored streaks)

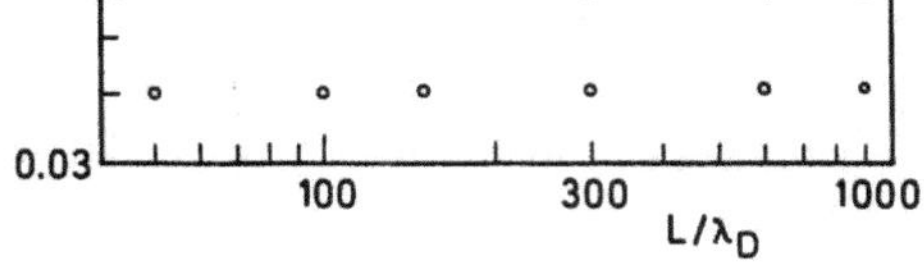

Fig. 6.24 Driver strength threshold $\hat{E}_d^*$ for wavebreaking as a function of scale length L. Straight line: (5.110), (6.67). The circles in the $(\hat{E}_d, L)$-plane represent Vlasov simulations with wavebreaking (solid) and without wavebreaking (blank)

This breaking threshold is in very good agreement with the results of the Vlasov simulations, which are displayed in Fig. 6.24. The reason for the validity of inequality (6.67) at finite electron temperature is that in the resonance region the electronic oscillatory motion v_e is mainly determined by the total electric field $E = \hat{E}_d + E_{\text{wave}}$, and is only slightly affected by the much smaller force due to the electron pressure gradient.

Next, the threshold intensity for wavebreaking is calculated. Making use of the scaling $L/\lambda_D \sim I_L^{-1/2}$ from [66], where I_L is the vacuum laser intensity, (6.67) translates to $I_L^* = 2 \cdot 10^{15}$ W/cm^2 for the Nd-laser if a degree of absorption of 25% is assumed. Finally it is mentioned that the kinetic analysis of wave breaking of freely propagating Langmuir waves in a *homogeneous* plasma shows that the same phenomenon also occurs here and that one scenario of breaking (others may also exist) is again trapping of *entire bunches* of electrons. However, the limit at which the wave breaks is much higher than the Coffey criterion indicates. It must also be mentioned that in the light of recent PIC and Vlasov simulations (for example [25]) it appears doubtful whether the hydrodynamic breaking limit becomes ever relevant because of the sudden increase of collective noise produced by kinetic effects. It is an open question how the two criteria correlate.

We conclude that wave breaking is a phenomenon on its own, to be distinguished from trapping of more or less uncorrelated electrons. The definition of wave breaking presented here has a clear meaning and is also applicable to a kinetic description. Further, it has been shown that a criterion for wave breaking in smooth ion density profiles can be deduced from the model of a cold streaming plasma. In addition, the Coffey criterion is not applicable to resonance absorption.

In general, when adjacent electron fluid elements are driven by a spatially inhomogeneous intense electromagnetic field they easily happen to cross and to mix up together, i.e., the regular motion becomes disordered. In particular this is believed to happen in the so-called Brunel effect at the plasma-vacuum interface [32, 33] and in "$\mathbf{j} \times \mathbf{B}$ heating" in the skin layer [54], and in collisionless absorption by collective interaction [78]. In the light of deeper insight into the nature of collisionless absorption at high laser intensities (wave) breaking is a consequence of anharmonic resonance. Wave breaking and breaking of flow will lead to an increased plasma fluctuation level, spectral broadening of reflected laser light, temporal pulsations, Maxwellian-like spectra of accelerated particles, and eventually to lower saturation levels of stimulated Raman and Brillouin scattering at high laser intensities. It seems to be a widely accepted belief that wave breaking is an efficient mechanism for electron acceleration. Closer inspection, in particular analysis of *resonant breaking*, leads to the conclusion that rather the opposite is true, i.e., first electrons are accelerated and then breaking sets in. Breaking is an extremely versatile phenomenon, like turbulence, and there may exist as many scenarios for its onset as for the onset of turbulence.

6.4 Resonant Three Wave Interactions

This is the subject of unstable simultaneous excitation of two eigenmodes in the unmagnetized homogeneous plasma by the laser wave. In the linear theory of resonance absorption one photon "decays" into one plasmon. Each fluid element is a harmonic oscillator directly coupled to the laser field. For the conversion to occur efficiently the single oscillator must be driven at resonance. The requirement leads to the matching conditions (6.32) of a two wave process for frequency ω (energy) and **k** (momentum). In a *three wave process* a fraction of the laser beam, or a longitudinal mode, decays into *two* modes of longitudinal and/or transverse polarization. In the case of *stimulated Brillouin back scattering* it will be explicitly shown that in linear theory each fluid element is a (i) harmonic oscillator which is (ii) driven resonantly by the (iii) ponderomotive force of the laser ("pump") beam. The third mode originates from the (iv) scattering of the pump mode off the plasma inhomogeneity created ponderomotively. Scattering in general originates from the electrons. Therefore the same mechanisms (i)–(iv) apply identically to the high frequency electron plasma mode to produce the *Raman instability*. The first goal is to show here that all possible three wave processes are resonantly driven by the standard ponderomotive force (2.109) and (2.120). By switching to the reference system $S'(v_\varphi)$ in which the electron modulation is static the scattered wave turns out to be unshifted reflection from the electron inhomogeneity. Subsequent transformation back to the lab frame S will lead to the standard representation of an electron mode and of a frequency shifted wave resonantly scattered off it.

6.4.1 Overview and Physical Picture

Streaming Plasma in Standing Wave

In order to see how an electromagnetic wave amplitude modulation acts on a steady state plasma flow let us consider a partially standing wave $E(x,t)$ of an incident and a reflected component E_0 and,

$$E(x,t) = \hat{E}_0 \mathrm{e}^{i(kx-\omega t)} + \hat{E}_r \mathrm{e}^{i(-kx-\omega t)} = (\hat{E}_0 - \hat{E}_r)\mathrm{e}^{i(kx-\omega t)} + 2\hat{E}_r \mathrm{e}^{-i\omega t}\cos kx \tag{6.68}$$

acting on a stationary isothermal plasma flow parallel to the x-axis. The flow is determined by

$$\rho u = \rho_0 u_0 = \text{const}, \quad \rho u \frac{\partial u}{\partial x} = -c_s^2 \frac{\partial \rho}{\partial x} - \frac{\varepsilon_0}{4\rho_c}\rho \frac{\partial}{\partial x}|E_0 + E_r|_0^2,$$

$$\frac{\partial}{\partial x}|E_0 + E_r|_0^2 = \frac{\partial}{\partial x}(E_0 E_r^* + E_0^* E_r) = \frac{\partial}{\partial x}\left(|\hat{E}_0 \hat{E}_r^*|\mathrm{e}^{2ikx} + |\hat{E}_0^* \hat{E}_r|\mathrm{e}^{-2ikx}\right).$$

The terms $|E_0^2|$ and $|E_r^2|$ can be ignored in the low signal case; in addition they are not resonant. With the help of the Mach number $M = u/c_s$ and the amplitudes $\hat{E}_0$, $\hat{E}_r$ both real, for $\rho_0, \rho \ll \rho_c$ these relations lead to

$$\left(1 - M^2\right)\frac{\partial \rho}{\partial x} = \frac{\varepsilon_0}{c_s^2}\frac{\rho}{\rho_c} k \hat{E}_0 \hat{E}_r \sin 2kx \Rightarrow \ln\frac{\rho}{\rho_0} = -\frac{\varepsilon_0 \hat{E}_0 \hat{E}_r}{2\rho_c c_s^2 (1 - M^2)}\cos 2kx,$$

$$M < 1 \Rightarrow \rho_1 \sim -\cos(2kx) = \cos(2kx + \pi), \quad M > 1 \Rightarrow \rho_1 = \rho - \rho_0 \sim \cos(2kx). \tag{6.69}$$

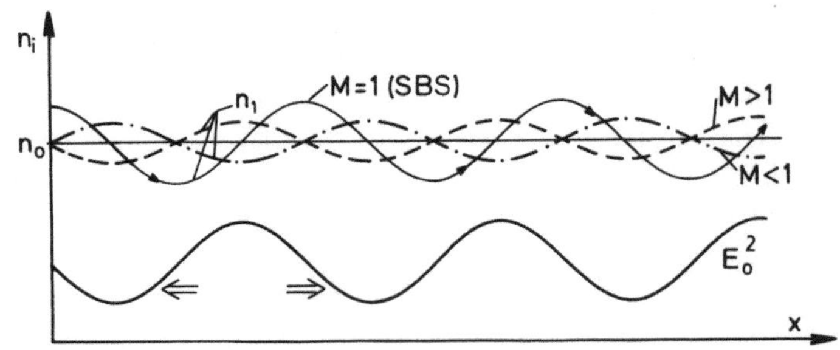

Fig. 6.25 A homogeneous plasma is periodically modulated by the pressure of a partially standing wave. $|E|$ electric field amplitude, n_0 average plasma density, $n_i = n_0 + n_1(x)$ local plasma density, $M = v/c_s$ Mach number. $M > 1$: maxima of n_1 in phase with maxima of $|E|^2$; $M < 1$: n_1 dephased by π; $M = 1$: n_1 dephased by $\pi/2$ (resonance: stimulated Brillouin scattering, SBS; $f_p \sim -\partial_x |E|^2$ in phase with n_i, see Fig. 6.27); lower arrows: direction of ponderomotive force

The ponderomotive density modulation is as follows. If M is less than unity everywhere a wave field of periodically modulated amplitude produces a stationary density modulation the maxima of which coincide with the minima of $|E|^2$ and the modulation amplitude increases with increasing Mach number; ρ_0 is the density at position $\cos 2kx = 0$. If the flow is supersonic in the whole region the density modulation is in phase with $|E|^2$, if it is subsonic the density and $|E|^2$ are out of phase by π, see Fig. 6.25. For a fixed modulation depth of $|E|$ the density modulation increases monotonically with the flow velocity when this approaches the sound speed from both sides; it is lowest for the plasma at rest ($M = M_0 = 0$), grows with M increasing and tends to zero with $M \to \infty$. The modulation is *static*. For $M \to 1$ the modulation tends to infinity and it violates the condition $\rho \ll \rho_c$. No steady state exists for $M = 1$, an indicator of a hidden instability. Evidently, at resonance we are not allowed to ignore inertia, or equivalently, the kinetic energy. At $M = 1$ the streaming perturbation is a sound wave driven at resonance. In analogy to the linear oscillator we also guess that at $M = 1$ the phase between density modulation and ponderomotive potential is $\pi/2$ now, like the phase of the resonantly driven harmonic oscillator (2.29). We show that the guess about resonance is correct. Let ρ_{10} be a small sinusoidal initial disturbance of ρ_0. The steady state equations above now have to be substituted by the linearized equations.

$$d_t\rho_1 + \rho_0\partial_x u = 0, \quad d_t u = -c_s^2\partial_x - \frac{\varepsilon_0}{4\rho_c}\partial_x|E_0 + E_r|^2; \quad d_t = \partial_t + c_s\partial_x. \quad (6.70)$$

The reflected wave E_r is assumed to be a weak signal, $|E_r| \ll |E_0|$, produced for instance by reflection of the incident wave E_0 from the plasma inhomogeneity of the acoustic mode. In the approximation of no *pump depletion* $E_0 = \text{const}$ can be assumed. Elimination of u by cross differentiation with respect to the interchangeable d_t, ∂_x yields directly the equation of a driven harmonic oscillator of eigenfrequency $\omega_a = c_s k_a$, $k_a = 2k$,

$$d_{tt}\rho_1 + 4c_s^2k^2\rho_1 = \varepsilon_0\frac{\rho_0}{\rho_c}k^2\hat{E}_0\hat{E}_r\cos 2kx;\quad d_{tt} = (d_t^2) \simeq c_s^2\partial_{xx} + 2c_s\partial_t\partial_x,$$

ρ_1 is the quasistatic amplitude. In the frame comoving with ρ_1 its secular growth is

$$\partial_t\rho_1(x,t) = -\varepsilon_0\frac{k^2}{2\omega_a}\frac{\rho_0}{\rho_c}\hat{E}_0\hat{E}_r\sin k_a x;\quad -\sin k_a x = \cos(k_a x + \pi/2),\quad k_a = 2k. \tag{6.71}$$

The spatial modulation of the density $\rho = \rho_0 + \rho_1$ at $t =$ const is identical with $\partial_t\rho_1(x, t = \text{const})$; hence, at $M = 1$ the phase shift of the density modulation is half the value of (6.69) for subsonic flow $M < 1$, see Fig. 6.25. It will turn out soon that the resonantly driven density modulation is the *stimulated Brillouin back scattering* of laser light. If E_r is proportional to ρ_1 the growth of ρ_1 is exponential. The analysis of the two collinear waves E_0 and E_r given here is straight forwardly generalized to arbitrary angles between the two waves. For the laser wave scattered from an electron plasma wave and for the scattering of an electron plasma wave from a wave of the same kind analogous considerations hold. Correspondingly, they are manifestations of *stimulated Raman back or side scattering*. In the frame $S'(v_\varphi)$ in which the modulation of the refractive index η by the electron modulation is static Brillouin and Raman scattering are classical reflection from refractive inhomogeneities with equal frequencies. The partial contributions reflected from the inhomogeneity of a single wavelength superpose constructively if the wave vectors add up. Thus, for *resonant*, i.e., constructive interaction in the comoving system $S'(v_\varphi)$ the matching conditions read

$$\mathbf{k}' = \mathbf{k}'_r + \mathbf{k}'_M,\qquad \omega' = \omega'_r + \omega'_M,\quad \omega'_M = 0. \tag{6.72}$$

The quantities referring to matter are marked by the index M.

6.4.2 The Doppler Effect

6.4.2.1 Doppler Effect in Vacuum

When transforming from one inertial system S to another inertial reference system $S'(\mathbf{v})$ in *vacuum* wave vectors $\mathbf{k}$ and frequencies ω change, however the phase $\Psi = KX = \mathbf{kx} - \omega t$ remains invariant, it is a Lorentz scalar,

$$\Psi = KX = K'X' \;\Leftrightarrow\; \mathbf{kx} - \omega t = \mathbf{k}'\mathbf{x}' - \omega' t' = \text{const}(\mathbf{v}). \tag{6.73}$$

As a consequence $K = (\mathbf{k}, \omega/c)$ is a four vector, and as such it transforms like $X = (\mathbf{x}, ct)$,

$$\mathbf{k}' = \mathbf{k} + \frac{\gamma - 1}{\mathbf{v}^2}(\mathbf{vk})\mathbf{v} - \gamma\omega\frac{\mathbf{v}}{c^2},\qquad \omega' = \gamma(\omega - \mathbf{kv}). \tag{6.74}$$

If a (partially reflecting) mirror is moving at velocity $\mathbf{v}$ and it is oriented in such a way that the wave vector of the reflected light is $\mathbf{k}_r$ in the lab frame, its frequency ω_r' in S' is $\omega_r' = \gamma(\omega_r - \mathbf{k}_r\mathbf{v})$. At the same time $\omega' = \gamma(\omega - \mathbf{k}\mathbf{v})$ equals ω_r' because of reflection. Resolution of the equality with respect to ω_r yields the basic Doppler formula in two different versions,

$$\omega_r = \omega - (\mathbf{k} - \mathbf{k}_r)\mathbf{v}. \tag{6.75}$$

$$\omega_r = \omega\frac{1 - \mathbf{k}^0\mathbf{v}/c}{1 - \mathbf{k}_r{}^0\mathbf{v}/c}; \quad \mathbf{k}^0 = \frac{\mathbf{k}}{|\mathbf{k}|}, \quad \mathbf{k}_r{}^0 = \frac{\mathbf{k}_r}{|\mathbf{k}_r|}. \tag{6.76}$$

Note, no Lorentz factor γ is involved in these formulas. The orientation of the mirror in vacuum follows from the condition of $\mathbf{k}_r'$ to be the specularly reflected vector $\mathbf{k}'$, i.e., $\mathbf{k}_r' \times \mathbf{n}' = \mathbf{k}' \times \mathbf{n}'$, $\mathbf{n}'$ normal vector. In forward direction $\mathbf{k}_r = \mathbf{k}$ the Doppler shift is zero, in backward direction it is

$$\Delta\omega = \omega\left(1 - \frac{1 \mp \beta}{1 \pm \beta}\right); \qquad \beta = \frac{v}{c}. \tag{6.77}$$

Only for $\beta \ll 1$ the approximation $k_r = 2k\sin(\vartheta/2)$ is legitimate for the modulus of $\mathbf{k}_r$; $\vartheta = \angle(\mathbf{k}, \mathbf{k}_r)$.

From the Doppler shift of a signal its velocity $\mathbf{v}$ can be inferred provided the emitting region is of limited extension or of special geometric structure. For the purpose of the latter consider an array of parallel partially reflecting mirrors moving at velocity $\mathbf{v}$ represent a periodic structure of wave vector $\mathbf{k}_M \uparrow\uparrow \mathbf{v}$ and frequency ω_M. For constructive interference to occur between the single contributions of reflection to the reflected wave $(\mathbf{k}_r, \omega_r)$ the array $(\mathbf{k}_M, \omega_M)$ must fulfill (6.75) and (6.76),

$$\mathbf{k}_M = \mathbf{k} - \mathbf{k}_r, \quad \omega_M = \omega - \omega_r \quad \Leftrightarrow \quad \mathbf{k} = \mathbf{k}_r + \mathbf{k}_M, \quad \omega = \omega_r + \omega_M. \tag{6.78}$$

If the array is made of an eigenmode in the homogeneous background plasma, e.g., an acoustic or electron plasma fluctuation, partial reflection occurs from the single oscillation. Although now the periodic structure propagates through the plasma medium with velocity $\mathbf{v} = \mathbf{v}_\varphi$ that is no longer isotropic because of plasma flow in $S'(\mathbf{v}_\varphi)$ the resonance conditions for constructive interference (6.78) still hold (see Fig. 6.26). From the reflected wave $(\mathbf{k}_r, \omega_r)$ the phase velocity of the extended periodic structure is determined. The single material velocity of an ion $\mathbf{v}_{os} = \mathbf{v}_\varphi n_1/n_0 \ll \mathbf{v}_\varphi$ cannot be "seen" by the incident wave.

Fulfilment of (6.78) in the medium follows from the general wave equation (5.15). It holds in any reference system S. This, although intuitively evident, is most immediately seen formally by writing $\mathbf{k}$ and ω on the RHS to obtaining a zero four vector on the LHS. The zero vector is invariant under a Lorentz transformation. The incident wave $\mathbf{E}(\mathbf{k}, \omega)$, transverse or longitudinal, produces a current density $\mathbf{j} = -en_e(\mathbf{x}, t)\mathbf{v}_{os}$, with $\mathbf{v}_{os} = -i(e/m_e\omega)\mathbf{E}\exp(i\mathbf{k}\mathbf{x} - i\omega t)$. The scattered field $\mathbf{E}_s$ obeys also (5.15). Expressing $n_e(\mathbf{x}, t)$ in Fourier–Laplace components $n_e(\mathbf{K}, \Omega)$, $\mathbf{E}_s$

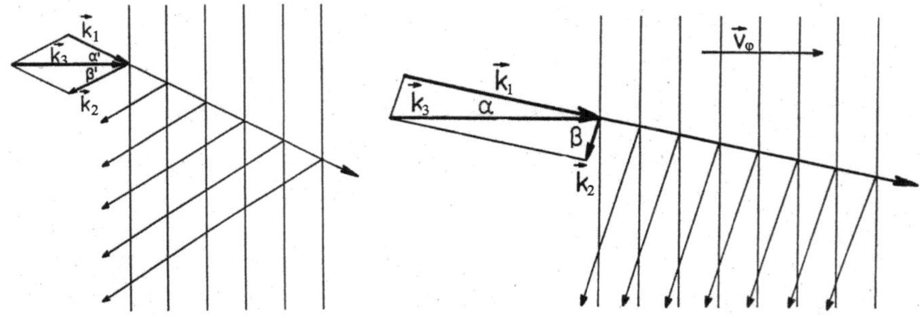

Fig. 6.26 Induced scattering of an intense pump wave $\mathbf{k}_1$ from a refractive index modulation $\mathbf{k}_3$; $\mathbf{k}_2$ scattered wave. **a** Co-moving frame, sketched for slow mode $\mathbf{k}_3$, e.g., Brillouin scattering, $\alpha' \simeq \beta'$, **b** lab frame (for clarity v_φ is assumed large)

and **j** read

$$\mathbf{E}_s(\mathbf{x},t) = \frac{1}{(2\pi)^2}\int \mathbf{E}(\mathbf{k}_s,\omega_s)\mathrm{e}^{i(\mathbf{k}_s\mathbf{x}-\omega_s t)}\,\mathrm{d}\mathbf{k}_s\mathrm{d}\omega_s,$$

$$\mathbf{j}(\mathbf{x},t) = -\frac{e^2\mathbf{E}}{(2\pi)^2 m_e\omega}\int\Big\{n_e(\mathbf{K},\Omega)\mathrm{e}^{i[(\mathbf{k}+\mathbf{K})\mathbf{x}-(\omega+\Omega)t]} + n_e^*(\mathbf{K},\Omega)\mathrm{e}^{i[(\mathbf{k}-\mathbf{K})\mathbf{x}-(\omega-\Omega)t]}\Big\}\,\mathrm{d}\mathbf{K}\,\mathrm{d}\Omega,$$

$\{\mathbf{k}_s,\omega_s\}$ represents an orthogonal basis and hence $\mathbf{E}_s$ and $\mathbf{j}$ must fulfil the wave equation component-wise. Thus the

$$\text{Stokes condition:}\qquad \mathbf{k}_s = \mathbf{k}-\mathbf{K},\qquad \omega_s = \omega-\Omega,$$

$$\text{and}\tag{6.79}$$

$$\text{anti-Stokes condition:}\qquad \mathbf{k}_s = \mathbf{k}+\mathbf{K},\qquad \omega_s = \omega+\Omega,$$

must be fulfilled. In spectroscopy the *Stokes* and *anti-Stokes* lines appear with frequencies $\omega_s = \omega - \Omega$ and $\omega_s = \omega + \Omega$, respectively. The fluctuating density mode $(\mathbf{K},\Omega)$ propagates at phase velocity $\mathbf{v}_\varphi = \mathbf{K}^0\Omega/|\mathbf{K}|$, $\mathbf{K}^0$ unit vector. Hence from $\Omega = \mathbf{v}_\varphi\mathbf{K}$ inserted in (6.4) relation (6.75) of the combined Doppler effect, $\omega_s = \omega - \mathbf{v}_\varphi(\mathbf{k}-\mathbf{k}_s)$, is recovered for both, Stokes and anti-Stokes frequencies. Use of (6.75) is legitimate owing to isotropy of propagation of incident and scattered mode in the plasma at rest. The relevant velocity responsible for the Doppler shift is the phase velocity $\mathbf{v}_\varphi$ of the refractive index modulation. Depending on the type of the modulating mode the velocity producing a Doppler shift may be a material velocity as in the case of the array above. Another example of material velocity is an entropy fluctuation in the absence of heat conduction because it is bound to the material flow. In the contrary, in scattering off a normal mode it is the immaterial *phase velocity* that produces the Doppler shift.

6.4.2.2 Doppler Effect in the Homogeneous Isotropic Medium

Consider a homogeneous isotropic medium at rest. A plane wave $(\mathbf{k}_0, \omega)$ entering the medium from the vacuum remains a plane wave of unchanged frequency ω. The wave vector $\mathbf{k}_0$ transforms into $\mathbf{k} = \mathbf{k}_0\eta$, the phase velocity is $v_\varphi = \omega/|\mathbf{k}|$ and hence the refractive index is $\eta = c/v_\varphi$, recall (5.59). The phase $\Psi = \mathbf{k}\mathbf{x} - \omega t$ in the medium is a Lorentz scalar for the same reason as in the vacuum. In fact, $\mathbf{k}$ and ω are defined as

$$\mathbf{k} = \nabla\Psi, \quad \omega = -\frac{\partial\Psi}{\partial t} \quad \Leftrightarrow \quad K = k_\alpha = \partial_\alpha\Psi \quad \Psi = KX = k_\alpha x^\alpha. \tag{6.80}$$

As a consequence (6.73) applies in the medium as well, with the four vector $K = (\mathbf{k}, \omega/v_\varphi)$. At fixed position $\mathbf{x}$ the phase Ψ counts the number of phase crests passing by. We recall that (6.74) from S to $S'(\mathbf{v})$ in the homogeneous and isotropic space, with γ unspecified, follows from the transformation requirements of (i) being linear and (ii) forming a group, see for example [63] or Sect. 2.2.1.1 in this book. The same requirements have to be applied to the homogeneous isotropic medium. We proceed at the determination of the Lorentz transformation in the medium by observing first that the Lorentz factor γ will change.

To this aim consider in vacuum $\mathbf{k}_{0\|}$ parallel to $\mathbf{v}$ in the moving system $S'(v)$ and $\mathbf{k}_{0\perp}$ perpendicular to $\mathbf{v}$. The resulting Doppler shifted frequencies are

$$\omega' = \gamma(\omega - \mathbf{k}_0\mathbf{v}) = \gamma(\omega - k_0 v), \qquad \omega' = \gamma\omega.$$

The term in the brackets is the result of Galileian relativity, γ is the *time dilation* factor of $T = 2\pi/\omega$ moving at $-\mathbf{v}$ with respect to S', see the Einstein clock Fig. 2.22. For illustration, an electromagnetic wave becomes static in the reference system $S'(v_\varphi)$ comoving with its phase velocity v_{phase} owing to $\omega - kv_\varphi = 0$. In the medium moving at $v \leq v_\varphi$ the wave maintains its frequency ω' when penetrating the vacuum-medium interface owing to $\mathrm{d}\omega'/\mathrm{d}t = \partial\omega'/\partial t = 0$. The Doppler formula applied to vacuum and to the moving medium yields

$$\omega' = \gamma(\omega - \mathbf{k}_0\mathbf{v}) = \gamma_\varphi[\omega - (\mathbf{k}/\eta)\mathbf{v}] = \gamma_\varphi[\omega - \mathbf{k}(\mathbf{v}/\eta)].$$

The wave vectors $\mathbf{k}_0$ and $\mathbf{k}$ obey Snell's refraction law at a discontinuity and (5.156) in the inhomogeneous medium. Under v/η approaching v_φ the following setting for v_φ is stringent:

$$\gamma = \left(1 - \frac{v^2}{c^2}\right)^{-1/2} = \left(1 - \frac{v^2/\eta^2}{c^2/\eta^2}\right)^{-1/2} = \left(1 - \frac{v^2/\eta^2}{v_\varphi^2}\right)^{-1/2} = \gamma_\varphi. \tag{6.81}$$

The Lorentz transformation is the product of a time corrected Galilei transform $S'(v)$ of the component $\mathbf{k}_{\|}$ parallel to $\mathbf{v}$ and the identity transformation of $\mathbf{k}_{\perp} = \mathbf{k} - \mathbf{k}_{\|}$ to $\mathbf{k}'_{\perp}$, i.e., $\mathbf{k}'_{\perp} = \mathbf{k}_{\perp}$. There is a one to one mapping of the electromagnetic mode $(\mathbf{k} = \mathbf{k}_0\eta, \omega)$ and changes direction. The transformation of the vacuum mode component $\mathbf{k}_{0\|}$ to $\mathbf{k}'_{0\|}$ in the special reference system $S'(\mathbf{w} = \mathbf{v}\eta)$ reads

$$\mathbf{k}'_{0\|} = \gamma\left(\mathbf{k}_{0\|} - \frac{\omega}{c^2}\mathbf{v}\eta\right) = \gamma\left(\left[\mathbf{k}_0\frac{\mathbf{v}}{|\mathbf{v}|}\right]\frac{\mathbf{v}}{|\mathbf{v}|} - \frac{\omega}{c}\frac{\mathbf{v}\eta}{c}\right) = \frac{\gamma}{\mathbf{v}^2}(\mathbf{k}_0\mathbf{v})\mathbf{v} - \gamma\frac{\omega}{c}\frac{\mathbf{v}}{v_\varphi}. \tag{6.82}$$

Multiplication of the RHS by η and addition of $\mathbf{k}\perp = \mathbf{k}_{0\perp}\eta$ yields

$$\mathbf{k}' = \mathbf{k} + \frac{\gamma - 1}{\mathbf{v}^2}(\mathbf{kv})\mathbf{v} - \gamma\omega\frac{\mathbf{v}}{v_\varphi^2}.$$

The Lorentz factor is $\gamma = (1 - v^2\eta^2/c^2)^{-1/2} = (1 - v^2/v_\varphi^2)^{-1/2}$ and is indicated by γ_φ. The transformation of the frequency remains the same as in the vacuum owing to $\mathbf{k}_0\mathbf{v}\eta = \mathbf{kv}$. Hence, the Lorentz transformations in the homogeneous isotropic medium differ from those in vacuum by c to be replaced by v_φ and the γ_φ factor,

$$\mathbf{k}' = \mathbf{k} + \frac{\gamma_\varphi - 1}{\mathbf{v}^2}(\mathbf{kv})\mathbf{v} - \gamma_\varphi\omega\frac{\mathbf{v}}{v_\varphi^2}, \quad \omega' = \gamma_\varphi(\omega - \mathbf{kv}); \quad \gamma_\varphi = \left(1 - \frac{v^2}{v_\varphi^2}\right)^{-1/2}. \tag{6.83}$$

This means for the Doppler shifted frequency ω_r in vacuum that the $(\mathbf{k}_0, \omega)$ wave incident onto a (partially reflecting) mirror moving with velocity $\mathbf{v}$ through the medium is given by

$$\omega_r = \omega - (\mathbf{k} - \mathbf{k}_r)\mathbf{v} = \omega\frac{1 - \mathbf{k}^0\mathbf{v}/v_\varphi}{1 - \mathbf{k}_r{}^0\mathbf{v}/v_\varphi}; \quad \mathbf{k}^0 = \frac{\mathbf{k}}{|\mathbf{k}|}, \quad \mathbf{k}_r{}^0 = \frac{\mathbf{k}_r}{|\mathbf{k}_r|}. \tag{6.84}$$

The velocity addition theorem in the medium follows from applying twice a Lorentz transformation and from the postulate that the Lorentz transformations must form a group. If $\mathbf{v}$ depends on space and time, for instance in presence of a plasma flow $\mathbf{v} = \mathbf{u}(\mathbf{x}, t)$, the phase is the integral $\Psi = \int \mathbf{k}d\mathbf{x} - \omega dt$.

Let $|\mathbf{v}| = w_0 < v_\varphi$ hold. In a plasma close to the critical density $\mathbf{k} = \mathbf{k}_0\eta \rightarrow 0$, $\mathbf{k}_r = \mathbf{k}_{0r}\eta_r \rightarrow 0$, thus the Doppler shift $\omega - \omega_r$ is strongly reduced with respect to reflection from a mirror moving at the same speed in vacuum. A second important property following from (6.84) is that a spatially constant flow of the medium does not affect the frequency of the electromagnetic wave; in other words, the frequency is invariant with respect to matter flow $\mathbf{u} = \mathbf{v}$. The proof is simple. The moving mirror sees the frequency $\omega' = \gamma_\varphi(\omega - \mathbf{kv})$ which after reflection into forward direction is unchanged because of $(\mathbf{k} - \mathbf{k}_r) = 0$ according to (6.84). In the reference system of the mirror at rest ω' is shifted back to ω but the plasma flows at $\mathbf{u} = -\mathbf{v}$. If the reader prefers a microscopic picture, in the streaming plasma the individual electron oscillates at $\omega'' = \gamma_\varphi(\omega - \mathbf{k}[\mathbf{v} + \mathbf{v}_{\text{th}}])$ and emits this frequency into forward direction, shifted additionally by its thermal velocity. In the lab frame ω'' is Doppler shifted back to ω.

The periodic array $(\mathbf{k}_M, \omega_M)$ is a partially reflecting eigenmode propagating at phase velocity $\mathbf{v}_M = \omega_M \mathbf{k}_M^0/|\mathbf{k}_M|$. When transforming to the frame $S'(\mathbf{v}_M)$ of the eigenmode at rest the phase velocities of the incident and reflected electromagnetic waves will depend on the angles between $\mathbf{k}$, $\mathbf{k}_r$, and $\mathbf{v}_M$. Furthermore, in the case of an electrostatic mode in the nearly critical plasma $\mathbf{v}_M$ may exceed c but not v_φ. In resonant scattering which is constructive reflection in $S'(v_\varphi)$ the phase matching condition (6.72) is fulfilled. In three wave interaction it may be convenient to assign the indices $1, 2, 3$ to the incident, the scattered, and the plasma mode. In the lab frame S in which matter is at rest with this convention (6.72) translates into the basic relation

$$\mathbf{k}_1 = \mathbf{k}_2 + \mathbf{k}_3, \qquad \omega_1 = \omega_2 + \omega_3. \tag{6.85}$$

The motion of a material point is generally studied by summing up all fields acting on the particle, those produced by neighboring matter, e.g., collisions, included. In such a treatment the Doppler effect of a single mass point is governed by the Lorentz γ of the vacuum and not by γ_φ because it corresponds to the fields taken in the vacuum and the influence of matter described by the true charge and current densities in empty space.

Medium in Motion (Fresnel–Fizeau Effect)

The flow of the homogeneous medium $\mathbf{v}$ destroys the isotropy of wave propagation. The velocity addition theorem (2.188) yields for the collinear $v_{\varphi\upuparrows}$ and $v_{\varphi\perp}$ orthogonal to $\mathbf{v}$ in the lab system

$$v_{\varphi\upuparrows} = \frac{v_\varphi + v}{(1 + v_\varphi v/c^2)}, \quad v_{\varphi\perp} = \frac{v_\varphi}{\gamma}. \tag{6.86}$$

The reduced component $v_{\varphi\perp}$ is an example of time dilation (compare Fig. 2.22). If $v \leqslant v_\varphi$ holds $v_{\varphi\upuparrows} > v_{\varphi\perp}$, as expected. The mode (k, ω) in the comoving system is assumed to be known. The question is how it transforms into (k_v, ω_v) of the lab frame moving at velocity $-v$. The refractive index in the two reference systems is

$$\eta = \frac{ck}{\omega}, \quad k = k_0\eta, \quad k_0 c = \omega; \qquad \eta_v = \frac{ck_v}{\omega_v}, \quad k_v = k_0\eta_v. \tag{6.87}$$

With $\beta = v/c$ expressions

$$ck_v = \gamma_\varphi kc\left(1 + \frac{\beta}{\eta}\right), \quad \omega_v = \gamma_\varphi \omega(1 + \beta\eta) \quad \Rightarrow \quad \eta_v = \frac{\eta + \beta}{1 + \beta\eta} \tag{6.88}$$

follow from (6.83. Note, $-v$ corresponds to $\mathbf{v}_M$ of (6.83); it is the material velocity. For $\beta \ll 1$ follows to first order in β

$$\eta_v = \eta(1 - \beta\eta)(\eta + \beta) \simeq \eta + \beta(1 - \eta^2). \tag{6.89}$$

In the fully ionized plasma at rest $\eta^2 = 1 - \omega_p^2/\omega^2$. The frequency ω is down shifted in the refractive index η from (6.87)–(6.89). Hence, the frequency corrected refractive index η_c entering in (6.87), (6.88), and (6.89) is given by

$$\eta_c^2 = 1 - \frac{\omega_p^2}{\omega^2} \frac{1}{\gamma_\varphi^2(1 - \beta\eta)^2} \simeq 1 - \frac{\omega_p^2}{\omega^2} \frac{1 + 2\beta\eta}{\gamma_\varphi^2}. \tag{6.90}$$

The influence of motion collinear to the propagation of the laser wave is quantified by (6.89) and (6.90). For an arbitrary orientation of the $\mathbf{k}$ vector to $\mathbf{v}$ the orthogonal Fresnel–Fizeau effect has to be evaluated in an analogous manner. As this is to be expected to be weaker we may conclude that, unless high precision is asked for (e.g., Doppler shift measurements from moving plasma), in plasma dynamics Doppler shifts can be ignored a long as $\beta \ll 1$ holds. There is one exception from this general rule, that is the anomalous dispersion region, i.e., if η is in the vicinity of atomic resonant frequencies.

6.4.2.3 Classification of Resonant Three Wave Interactions

By assigning the indexes em, es, and a to the different types of modes, electromagnetic, electrostatic, and acoustic, as the carriers of photons $\hbar\omega_{\text{em}}$, plasmons $\hbar\omega_{\text{es}}$, and phonons $\hbar\omega_{\text{a}}$, we conclude from the foregoing section that the following decay instabilities are possible and underlie the same physical principle:

1.	Brillouin instability (SBS):	$\omega_{\text{em}} \to \omega'_{\text{em}} + \omega_{\text{a}}$	
2.	Raman instability (SRS):	$\omega_{\text{em}} \to \omega'_{\text{em}} + \omega_{\text{es}}$	
3.	Two plasmon decay (TPD):	$\omega_{\text{em}} \to \omega'_{\text{es}} + \omega_{\text{es}}$	
4.	Parametric decay instability (PDI):	$\omega_{\text{em}} \to \omega_{\text{es}} + \omega_{\text{a}}$	
5.	Oscillating two-stream instability (OTSI):	$\omega_{\text{em}} \to \omega_{\text{es}} + \omega_{\text{a}},$	$\omega_{\text{es}} > \omega_{\text{em}}$
6.	Langmuir decay instability (LDI):	$\omega_{\text{es}} \to \omega'_{\text{es}} + \omega_{\text{a}},$	$\omega'_{\text{es}} \to \cdots$
7.	Two phonon decay of phonon (TPhD):	$\omega_{\text{a}} \to \omega'_{\text{a}} + \omega''_{\text{a}}$	

Stimulated decay of a phonon into two phonons is possible. In high power-laser plasma interaction it plays at most a secondary role and will not be pursued here. The opposite is true for stimulated *Brillouin* [79] or *Brillouin–Mandelstam scattering* (SBS) [80]. It consists in the stimulated amplification of an electromagnetic wave scattered off a low frequency ion acoustic mode satisfying the resonance conditions (6.85). In appositely prepared plasmas it can lead to 100% back reflection of the incident laser light [81, 82], in this way preventing the plasma from effective heating. The acoustic mode is a low frequency wave, $\omega_a \ll \omega_{em}$, and hence $\omega'_{em} \simeq \omega_{em}$ and $|\mathbf{k}_a| \simeq 2|\mathbf{k}_{em}|\sin(\vartheta/2)$, $\vartheta = \angle(\mathbf{k}_{em}, \mathbf{k}'_{em})$. Owing to the virtual high reflectivity the decay process has been intensively investigated experimentally and theoretically in connection with controlled inertial fusion over more than three decades [83–87]. The repeated decay of a Langmuir wave into another Langmuir wave and an ion acoustic mode (cascading) is called Langmuir decay instability (LDI). It is the electrostatic equivalent to SBS; the underlying coupling physics is the same as for SBS [88, 89]. The process tends to degenerate into nearly indistinguishable ion modes by multiple cascading. In a limited parameter regime LDI appears to be a secondary instability that can saturate the Langmuir wave generated by the process of stimulated Raman scattering [90, 91].

The stimulated Raman scattering (SRS) in plasma has been investigated for at least half a century [92]. It is the resonant decay of light into an electron plasma wave and a light wave of longer wavelength. The physical mechanism is the same as for Brillouin scattering. Owing to the much higher frequency of the plasma wave the scattered Stokes line undergoes a remarkable red shift, in perfect analogy to the Raman effect in solids from optically active crystal vibrations (optical phonons) producing similar large red shifts. SRS exhibits fast growth rates and high saturation levels in plasmas, at least under ideal conditions [93], and fast electron generation due to particle trapping in the Langmuir wave. The latter is of relevance to indirect drive of inertial fusion pellets [119]. From the inequality $\omega'_{em} + \omega_{es} \geq 2\omega_p$ follows that the instability is limited to the density domain $n \leq n_c/4$. At comparatively low density and high electron temperature there is another limit due to strong Landau damping of the electron plasma wave. It takes place in the domain $k_{es}\lambda_D \gtrsim 0.35$. On the other hand, kinetic effects induced by SRS show a fast onset which can lead to a strong modification of the electron distribution function and thus to phenomena like nonlinear enhancement of the instability ("inflation regime" [95]) and to the excitation of new modes, so called (electron) beam acoustic modes (BAM) [96].

The *two plasmon decay* (TPD) is excited in the vicinity of $n_c/4$ because of strong Landau damping outside $\omega_{es} \simeq \omega_p$. Like SRS, it is a potential candidate for fast electron generation [97] owing to the high phase velocity of the daughter waves limited only by the finite density scale length due to profile steepening at $n_c/4$ (compare Fig. 6.4). Langmuir cascading into two electron plasma waves, $\omega_{es} \rightarrow \omega'_{es} + \omega''_{es}$ is not possible because the resonance conditions (6.85) cannot be fulfilled in the entire density domain. For the same reason resonant transverse wave cascading into two electromagnetic waves does not occur either in the fully ionized plasma.

Perhaps the first light pressure driven instability in the plasma has been described by V. P. Silin in 1965 [98]. Imagine an ion acoustic wave n_a with its wave vector perpendicular to $\mathbf{k}_{em}$ of the laser beam. The laser field polarizes the plasma at frequency ω_{em} by forcing the single electron to oscillate with $\boldsymbol{\delta}_{os}$. The periodic perturbation proportional to $n_a\boldsymbol{\delta}_{os}$ excites an electron plasma wave symmetrically to the left and to the right along $\mathbf{k}_a$ with wave vectors $\mathbf{k}_{es} = \pm\mathbf{k}_a$. Their superposition generates a ponderomotive force in phase with n_a and forces it to grow. Its growth, in turn, reinforces the two Langmuir modes with the same rate as n_a. This is the *parametric decay* of a photon into a plasmon and a phonon. It is localized in the underdense plasma region close to n_c. In 1968 K. Nishikawa recognized that it can also work above n_c if $\omega_1 = \omega_{em} < \omega_2 = \omega_{es}$, and ω_3 degenerates to the quasimode $\omega_a = 0$ [99]. This is the case of the so-called *oscillating two stream instability* OTSI. Nishikawa's discovery had a major impact on the proper treatment of other parametric instabilities under the influence of a strong driver. If the dephasing ("mismatch") becomes larger than the frequency of one of the modes the contribution of the anti-Stokes component of the faster mode must also be taken into account. Pertinent examples will be discussed.

Linearized Ponderomotive Force for Resonant Drive The standard ponderomotive force derived in Chap. 2 refers to the reference system S' in which (i) the oscillation center is at rest and (ii) the wave E consists of monochromatic components only. The standard ponderomotive force density $\pi_0 = n_3\mathbf{f}_p$ is applicable to static electron density n_3. In order to describe its resonant coupling with a travelling electron mode of frequency ω_3 in the lab frame it is advisable to transform π_0 into this system. In the linear regime the resonant part of three wave coupling is assumed to be weak. As a consequence $|\mathbf{E}_2| \ll |\mathbf{E}_1|$ and $|\mathbf{E}_1| = \text{const}$ is approximated. *Pump depletion*, i.e., back action of the scattered wave onto the driver, is not considered. The reference system $S'(v_\varphi)$ comoving with the phase of the electron mode may be superior for the insight into the mechanism driving the instabilities.

Transformation of π_0 to the Lab Frame

Application of the perturbative derivation (2.113) to the resonant terms in the comoving system yields for the secular component

$$\pi_0' = -n_0'e\{(\xi_1'\nabla)\mathbf{E}_2' + (\xi_2'\nabla)\mathbf{E}_1' + \mathbf{w}_1' \times \mathbf{B}_2' + \mathbf{w}_2' \times \mathbf{B}_1'\}_0. \tag{6.91}$$

The quantities $\mathbf{w}_{1,2}'$, $\xi_{1,2}'$ are the oscillation velocities and the displacements. In the lab frame they oscillate at ω_1 for the pump and at ω_2 for the scattered wave. Substitution of $\mathbf{w}_{1,2} = -ie/2m_e\omega_{1,2}(\mathbf{E}_{1,2} - \mathbf{E}_{1,2}^*)$, $\xi_{1,2} = e/2m_e\omega_{1,2}^2(\mathbf{E}_{1,2} + \mathbf{E}_{1,2}^*)$, $\mathbf{B}_{1,2} = -i/2\omega_{1,2}\nabla \times (\mathbf{E}_{1,2} - \mathbf{E}_{1,2}^*)$ yields the resonant ponderomotive component

$$\pi_0 = -\frac{\varepsilon_0\omega_p^2}{2\omega_1\omega_2}\left\{\nabla(\mathbf{E}_1\mathbf{E}_2^{*c}) + \nabla(\mathbf{E}_2^*\mathbf{E}_1^c) + \frac{\omega_1 - \omega_2}{2}\left[\frac{(\mathbf{E}_2^*\nabla)\mathbf{E}_1}{\omega_2} - \frac{(\mathbf{E}_1\nabla)\mathbf{E}_2^*}{\omega_1}\right]\right\}. \tag{6.92}$$

in the lab frame with $\omega_3 = \omega_1 - \omega_2$ (c: field component kept constant). For $\mathbf{E}_1\nabla n_3 = 0$ the term in the square bracket vanishes; hence

$$\pi_0 = -i\frac{\varepsilon_0\omega_p^2}{2\omega_1\omega_2}(\mathbf{k}_1 - \mathbf{k}_2)(\hat{E}_1\hat{E}_2^*)e^{i[(\mathbf{k}_1-\mathbf{k}_2)\mathbf{x}-(\omega_1-\omega_2)t]}. \tag{6.93}$$

The plasma frequency ω_p is Lorentz-invariant because electron density and electron mass observe the same transformation law. Compared to $\boldsymbol{\pi}_0 = n_0\mathbf{f}_p$ from (3.60) in the co-moving frame, the new expression (6.93) differs only by the product of the Doppler shifted frequencies ω_1 and ω_2 in the denominator and the lab frame representation of the Lorentz-invariant phase Φ. The field components $\mathbf{E}_1$, $\mathbf{E}_2$ may stand for two transverse or two longitudinal waves or for a combination of a transverse with a longitudinal wave. In all cases $\boldsymbol{\pi}_0$ is the same formula (6.93). In S' assume a pump wave $\mathbf{E}_1(\mathbf{k}, \omega)$ incident onto a homogeneous plasma in which a much weaker plane wave $\mathbf{E}_2(\mathbf{k}', \omega)$ is present. If one of the waves or both are longitudinal in the lab frame S the secular term $\boldsymbol{\pi}_t = \partial_t\langle m_e n_1 \mathbf{u}_e\rangle$ has to be added to $\boldsymbol{\pi}_0$ [see 3.64]. It plays a role, for instance, in the two plasmon decay. In case $\mathbf{E}$ is the sum of a transverse and a longitudinal component, $\mathbf{E} = \mathbf{E}_\perp + \mathbf{E}_\parallel$, in (6.92) follows $\nabla\mathbf{E}\mathbf{E}^* = \nabla(\mathbf{E}_\perp\mathbf{E}_\perp^* + \mathbf{E}_\parallel\mathbf{E}_\parallel^*)$ only under the condition $\mathbf{E}\nabla n_e = 0$. The general case is more complex.

6.4.2.4 Physical Picture of Brillouin Backscattering and Parametric Decay

Physical intuition tells that the displacement δ of the harmonic oscillator (2.25) driven at $\omega < \omega_0$ is in phase with its driver $D = -(e/m_e)E_d$, driven at $\omega > \omega_0$ it is out of phase by $+\pi$. Maximum work by the driver is done at the detuned resonance $\omega_0' = \sqrt{\omega_0^2 - \nu^2/2}$ with the driver in phase with $v_{os} = \dot{\delta}$, see (2.26). Translated to the picture of the ion acoustic mode $n_3 = n_a$ at resonance the driving ponderomotive force f_p is in phase with $v_{os} = c_s(n_1/n_0) \sim n_a$, see (6.71 and Fig. 6.25.

Brillouin instability. This is exactly the picture of Fig. 6.27 in the comoving system, with the acoustic mode n_a identified with n_1 and the ponderomotive force density identified with $\pi_0 = n_0 f_p$ for resonant driving. What remains to show is that the

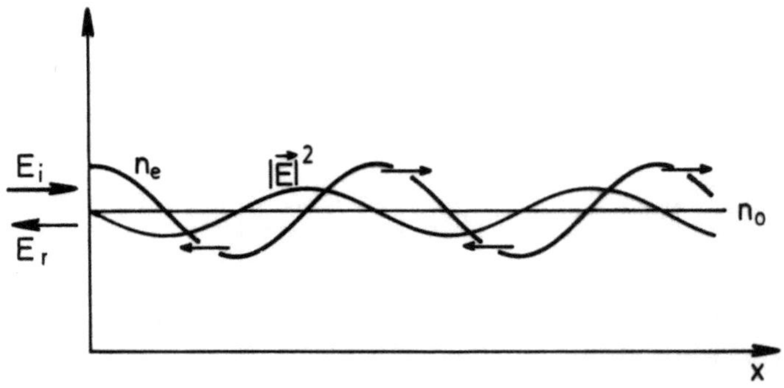

Fig. 6.27 A static electron density disturbance n_1 in the co-moving frame is resonantly driven by a modulated electric field amplitude $|\mathbf{E}|$ if the plasma moves at velocity $-v_\varphi = -\omega_3/k_3$. The arrows indicate the maxima and minima of the ponderomotive force π_0. Maximum flow inhibition π_{max} coincides with max n_1

phase of the electromagnetic wave reflected from n_a relative to the incident wave is such as to generate the partially standing wave $|E|^2 \simeq |E_1|^2 + E_1E_2^* + E_1^*E_2$ with phase delay of $\pi/2$ relative to n_a. In the comoving system (6.27) tells $E_2' \sim n_a'E_1'$ because n_a' is static, and hence $E_1'E_2'^* \sim n_a'|E_1'|^2$ is constant in time. The phase delay of $\pi/2$ results from the following argument for k', ω' in $S'(-v_\varphi)$. For constructive interference the reflected wave E_2' moves a quarter wavelength λ' backward during one period, or $\lambda'/4 = \lambda_a'/2$ because k_a' is twice k'. Simultaneously, the incident wave E_1' moves $\lambda'/2$ forward to add to the length of periodicity λ'. The result is a dephasing between $|E'|^2$ and n_a' as depicted in Fig. 6.27 for unstable growth. Transformation back to the lab frame confirms the matching conditions (6.85). This is the pictorial verification of the formal proof by (6.70) and (6.71).

In the system comoving with the ion mode Brillouin scattering is the geometrical reflection of an incident light wave from a periodic density perturbation. If the perturbation extends over many wavelengths only the resonantly driven scattered waves obeying **k** and ω matching survive. Resonance guarantees the fastest growth. Consequently, over long times the resonant mode growing out from an initial broad band perturbation (e.g., thermal noise) dominates. Feedback (coupling) is by the ponderomotive force. All three wave interactions listed above follow the same scheme.

In case of subsonic and supersonic flow with respect to the standing wave structure $|E|^2$ the wave reflected from the single density hump interferes destructively with the incident wave and has little effect on the standing wave in the homogeneous plasma, as confirmed by numerical simulations. The destructive interference is the reason for the existence of steady states at Mach $M < 1$ and $M > 1$.

Parametric decay is the scattering of a Langmuir wave ($E_{es} = E_2, \delta_2$) off an acoustic perturbation $n_a = n_3$ resonantly driven by the laser wave ($E_{em} = E_1, \delta_1$). The resonance conditions (6.85) are to be fulfilled under (i) $c_{se}^2 \ll c^2$, $\omega_3 \ll \omega_1$, and (ii) $\omega_p \simeq \omega_2$ because of strong Landau damping otherwise. From (i) follows $\omega_2 \simeq \omega_1$ and in combination with (ii) $|\mathbf{k}_2| \gg |\mathbf{k}_1|$. These inequalities are well fulfilled under the following idealized geometry in the lab frame:

$$\mathbf{k}_1 \perp \mathbf{k}_2, \mathbf{k}_3, \quad \mathbf{k}_1 \to 0, \quad \mathbf{k}_2 = -\mathbf{k}_3; \quad \omega_3 \to 0. \tag{6.94}$$

In the system S' comoving with the acoustic mode n_3 the field quantities can be assumed unaltered owing to the very low value of $v_{\varphi,3}$. The spatially uniform laser field $E_1 = \hat{E}_1 e^{-i\omega_1 t}$ polarized in x direction impinges along z on a static plasma density modulation $n_3 = \hat{n}_3 e^{ik_3 x}$. It forces the electrons to shift in $\pm x$-direction by the amount

$$\delta_1 = \Re \frac{e}{m_e\omega_1^2}\left(1 + i\frac{\nu_{ei}}{\omega_1}\right)\hat{E}_1 e^{-i\omega_1 t}. \tag{6.95}$$

In the presence of the ion density modulation n_3 it gives rise to an induced electric field E_d obtained from Poisson's equation,

$$\frac{\partial E_d}{\partial x} = -\frac{e}{\varepsilon_0}[n_3(x-\delta_1) - n_3(x)] = \frac{e}{\varepsilon_0}\delta_1\frac{\partial n_3}{\partial x} \quad \Rightarrow \quad E_d = \frac{e}{\varepsilon_0}\delta_1 n_3. \tag{6.96}$$

The further analysis is continued after [100] The sum of $|E_1 + E_d|^2$ is *in* phase with n_3 and hence stability would be the conclusion, if this were the total electric field involved. Then, and only then $\pi \sim -\nabla|\mathbf{E}|^2$. However, in S' the driver excites two electron plasma waves $E_2' = \hat{E}_2 \mathrm{e}^{i(\pm k_3 x - \omega_1 t)}$. Owing to the convection $u_2 = -c_s$ and $\partial_t \pm u\partial_x'$ their displacements $\delta_2 = \delta_+, \delta_-$ obey the harmonic oscillator equations

$$\frac{\mathrm{d}^2\delta_\pm}{\mathrm{d}t^2} + \nu\frac{\partial\delta_\pm}{\partial t} + \omega_2^2\delta_\pm = -\frac{e}{m_e}(\hat{E}_1 + \hat{E}_d)\mathrm{e}^{-i\omega_1 t}; \quad \omega_2^2 = \omega_p^2 + c_{se}^2 k_2^2, \quad k_2 = k_3 \tag{6.97}$$

with the true eigenfrequencies in S'

$$\omega_+ = \omega_2 + \omega_a, \quad \omega_- = \omega_2 - \omega_a$$

of the electron wave n_2 propagating to the left and to the right, respectively (k_3, ω_3 taken positive). Equation (6.97) is a transcription of n_e in (5.67) to the displacement δ_e from (5.73). The damping coefficient ν is of Landau or collisional type ν_{ei}. The amplitude $\hat{\delta}_2$ results modulated with periodicity k_3 from the polarization induced by the laser in n_3. The reason why a resonance frequency appears in (6.97) but not in (6.95) is because E_2' is not divergence free whereas $\nabla E_1 = 0$. The ponderomotive force of the two waves is given by the total field $E = E_1 + E_d + E_2$ or, equivalently, by $E_1 + E_d$ as a driver of the plasma oscillations. Thus, $\pi_\pm$ is

$$\pi_\pm = -\frac{\omega^2 - \omega_\pm^2}{(\omega^2 - \omega_\pm^2)^2 + \nu^2\omega^2}\frac{e^2\omega_p^2|E_1|^2}{4m_e(\omega^2 + \nu_{ei}^2)}k_3^2 n_3. \tag{6.98}$$

The wave pressure is in phase for $\omega < \omega_\pm$ and drives the plasma unstable, in agreement with Fig. 6.27 and with the rather lengthy treatments [98, 99]. The ratio of the amplitudes $\hat{n}_2/\hat{n}_3$ is determined by (6.97) and (6.96). In the absence of damping or growth the exact resonance occurs at $\omega\pm$, the driver $\pi_\pm$ vanishes. It may surprise at fist glance. A view to Fig. 2.15 clarifies the paradox. The singularity at $\nu = 0$ goes over into a continuous behaviour of the phase shift between driver and excursion with its maximum at the detuned resonance $\omega_0' = \omega_0\sqrt{1 - \nu^2/\omega_0^2}$.

Let ν be zero. Instability can occur for $\omega_2 < \omega < \omega_2(1 + 2\omega_3/\omega_2)$ with the ω_+ mode, but not with ω_-. The growth generates detuning and the effective resonance ω_2' may shift to the unstable ω_- branch. For this reason K. Nishikawa concluded that the contribution of both branches must be considered simultaneously. It led him to the discovery of the

Oscillating two-stream instability (OTSI). Let us go back to Fig. 6.25 and reconsider a sinusoidal ion density perturbation at rest in the lab frame, for instance as a consequence of strong detuning. Now ω_+ and ω_- become equal. As shown, for Mach $M = 0$ the density perturbation should decay, and it does indeed if the laser propagates and partially reflects along $\mathbf{k}_3$. In the geometry of PDI there is again the excitation of the additional field E_2 originating from the laser induced polarization δ_1 and acting like a harmonic oscillator of eigenfrequency ω_{es}, owing to Landau damping close to ω_p. For $\omega > \omega_p$ the ponderomotive force density $\boldsymbol{\pi}$ changes sign and acts like $M > 1$. In the ion acoustic eigenmode the plasma pressure is balanced by the inertia of the single fluid element. In the static "quasimode" n_3 the plasma pressure must be balanced by $\boldsymbol{\pi}$. Only beyond this threshold unstable growth sets in. This will be shown quantitatively later. The resonant PDI is capable of arbitrarily low threshold, the nonresonant OTSI shows a *finite threshold* for its unstable growth. Formally there is a one to one correspondence of the OTSI with the beam-plasma instability of two counter streaming electron beams.

Two plasmon decay (TPD). To give an intuitive physical picture of the TPD is more complex. It is true that also in this case the instability arises from the reflection of the pump wave $\mathbf{k}_1$ from the Langmuir wave $\mathbf{k}_2$ ($\mathbf{k}_3$) to result in a ponderomotive force in phase with $\mathbf{k}_3$ ($\mathbf{k}_2$, respectively). However, the geometry for maximum growth is peculiar. Starting from the configuration in the lab frame with the pump $\mathbf{k}_1$ incident under the arbitrary angle α onto the plasma mode $\mathbf{k}_2$ we observe that in good approximation holds

$$\omega_2 = \omega_3 = \omega_p = \omega_1/2, \qquad \eta_1 = \sqrt{3}/2, \qquad c_\varphi = v_{\varphi,1} = \omega_1/k_1, \tag{6.99}$$

$$k_1 = k_0\eta_1, \qquad k_2 = k_3 \simeq (c/2s_e)k_0 \gg k_1, \qquad k_{2,3} = \omega/(2v_{\varphi 2,3}).$$

It follows from $k_{2,3} \gg k_1$ that the two plasma waves are nearly collinear (see Fig. 6.28). The growth of the two daughter waves $k_{2,3}$ implies an increase of the momentum densities $m_e n_{2,3} v_1$ which, in turn, manifest themselves in a ponderomotive force like an inverse rocket effect. Thus, the ponderomotive force $\boldsymbol{\pi}_{2,3}$ driving TPD unstable is given by (3.67) or (3.64). It is physically obvious and confirmed by the quantitative analysis in the next section that it acts along $\mathbf{k}_1$ and exhibits the proportionality

$$\boldsymbol{\pi}_{2,3} \sim \mathbf{k}_1 |\mathbf{E}_1 \mathbf{E}_{2,3}^*|. \tag{6.100}$$

The growth rate γ is proportional to the projection of $\boldsymbol{\pi}_{2,3}$ in direction $\mathbf{k}_{2,3}$, i.e., $|\mathbf{k}_1 \mathbf{k}_{2,3}| \sim \cos\alpha$. On the other hand, the modulus $|\boldsymbol{\pi}_{2,3}| \sim |\mathbf{E}_1 \mathbf{E}_{2,3}^*| \sim \sin\alpha$; hence $\gamma \sim \sin\alpha\cos\alpha$, with maximum growth under $\alpha = \pi/4$. In linear theory one configuration and its mirror image with respect to $\mathbf{k}_1$ exhibit identical growth rates γ.

Excitation of TPD under perpendicular incidence, $\alpha = \pi/2$ is not possible as a consequence of $\gamma \sim \sin\alpha\cos\alpha$. This contrasts with PDI where $\boldsymbol{\pi}_t$ is nearly zero and a symmetric composition of the $\mathbf{k}$-diagram can be realized also in the lab frame. The

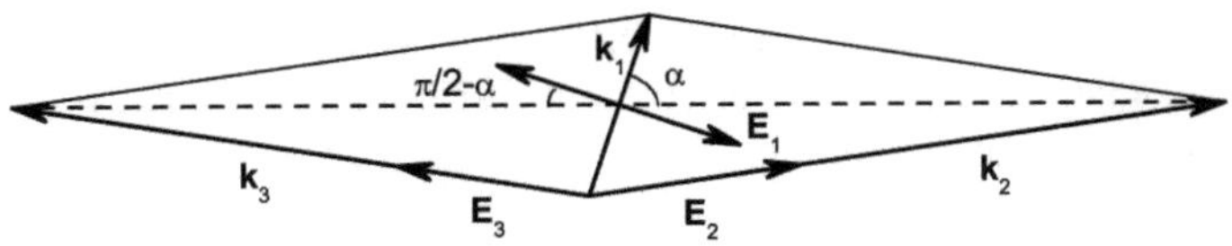

Fig. 6.28 Stimulated two-plasmon decay; $\mathbf{k}_1 = \mathbf{k}_2 - \mathbf{k}_3, \mathbf{k}_1 \ll \mathbf{k}_{2,3}$. Its mirror image with respect to $\mathbf{k}_1$ shows equal growth

difference in geometry originates from a slow (acoustic) mode, e.g., n_3, from which in the comoving frame two fast modes $n_e = n_2$ are excited by the laser in almost opposite directions. In the co-moving system of one of two fast modes, as in TPD, the left-right symmetry is destroyed.

6.4.3 *Growth Rates*

6.4.3.1 Basics About Normal Mode Analysis

Normal mode analysis is the simplest way to decide whether a linear system is stable or unstable. It is based on the Fourier–Laplace transform of small perturbations. The dynamics of them can be approximated by linear equations. In the actual and foregoing chapters repeatedly the method of *linearization* has been applied successfully. There exist two preferred model situations that greatly facilitate the normal mode analysis of linear instability. It seems quite intuitive to describe spatial growth for real frequencies ω fixed from outside, and temporal growth for real wave vectors $\mathbf{k}$ in the infinitely extended homogeneous medium or wave vector $\mathbf{k}$ fixed by the boundaries of a finite system. All kinds of unstable growth (and decay) not faster than exponential in space or time are encompassed by Fourier–Laplace analysis; *catastrophic growth*, i.e., growth with finite asymptote is excluded. The knowledge of one unstable mode does not tell what the long space-time behaviour of a system will be. However, it is physically reasonable to assume that there exists a finite neighbourhood of unstable modes around one mode, classified as unstable, and hence, from Fourier–Laplace inversion in the unstable domain follows that instability will happen also in direct space $(\mathbf{x}, t)$ for a given time interval. The time interval over which instability extends in real space may be finite or infinite. In the first case the instability is *convective*, in the second case it is *absolute*. The difference is illustrated by Fig. 6.29. A convective unstable pulse grows and then decays at each arbitrary position of interest (a); an absolutely unstable quantity grows indefinitely in time everywhere (b). The distinction is dictated by experimental requirements. If the group velocity of the pulse maximum in Fig. 6.29 is high enough the pulse may decay very quickly before growing to a dangerous height. The classification convective and absolute is not covariant; it depends on the relative motion of the system of reference. The interested reader may

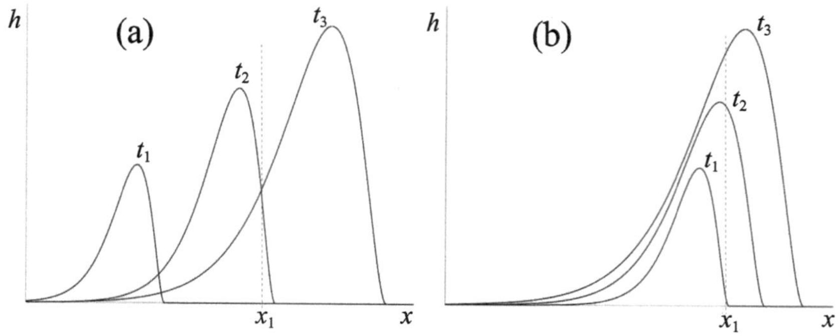

Fig. 6.29 *Convective* (**a**) and *absolute* instability (**b**). The convective unstable quantity $h(x)$ grows in time up to a maximum at an arbitrary position x_1 and then decays there. The absolutely unstable quantity $h(x)$ shows indefinite growth in time at any position x_1

find an excellent introduction to the theory of absolute and convective instabilities with significant case studies in Abraham Bers' handbook article in [101].

Complex stability analysis of real physical quantities. Fourier–Laplace transform in the complex domain is very convenient. Its physical interpretation is based on the property that the real and the imaginary parts of a complex solution are themselves linearly independent solutions of an equation, provided it is (i) linear and its coefficients are (ii) real constants. However, physical practice goes a step further by introducing also complex physical coefficients, like a complex refractive index η, see e.g., (5.188). In this and similar situations the original differential equations are of second order with real constant coefficients. With linear damping or amplification, i.e., negative damping, a first order derivative is introduced leading to (i) exponential damping (amplification) by the factor $\exp\gamma t$ and to (ii) an additional phase shift between field variable and driver. Both effects are described by allowing for a complex amplitude $\hat{\mathbf{x}} = \Re\hat{\mathbf{x}} + i\Im\hat{\mathbf{x}}$ of the complex $\mathbf{x}$ and a complex frequency $\Omega = \omega + i\gamma$ to yield

$$\mathbf{x} = \hat{\mathbf{x}}e^{-\Omega t} = (\Re\hat{\mathbf{x}} + i\Im\hat{\mathbf{x}})e^{-i(\omega+i\gamma)t} \Rightarrow \Re\mathbf{x} = e^{\gamma t}[\Re\hat{\mathbf{x}}\cos(\omega t) + \Im\hat{\mathbf{x}}\sin(\omega t)]$$

$$\Rightarrow \Re\mathbf{x} = \mathbf{x}^0\sqrt{(\Re^2\hat{\mathbf{x}} + \Im^2\hat{\mathbf{x}})}\,e^{\gamma t}\cos(\omega t + \phi), \quad \cos\phi = \frac{\Re\hat{\mathbf{x}}}{\sqrt{(\Re^2\hat{\mathbf{x}} + \Im^2\hat{\mathbf{x}})}}. \quad (6.101)$$

The symbol is the unit vector $\mathbf{x}^0$ in direction of $\mathbf{x}$. In the specific case of complex η from (1.75) $\mathbf{v}$ is to be identified with $\dot{\mathbf{x}}$. The expression $\Re\mathbf{x}$ is identical with the solution of the original equation with all quantities real. Negative γ in the complex Ω describes exponential damping, positive γ indicates unstable exponential growth. In summary, the complex calculus is a kind of short hand writing of linear physics with originally real quantities. In this sense the rule given above extends to linear equations with complex constant coefficients, too.

Laplace or Fourier transform? Generally an unstable mode grows out from noise containing a frequency component that is resonant according to (6.79). This implies that the Laplace transform is the appropriate procedure. Accordingly, parametric three wave interaction in its simplest form results in an inhomogeneous linear system of two coupled algebraic equations containing the initial values on the right, $A\mathbf{x} = \mathbf{x}_i$. As will be seen from the following case studies the system is a hidden homogeneous system with matrix A_c,

$$A\mathbf{x} = \mathbf{x}_i \quad \Leftrightarrow \quad A\mathbf{x} - \mathrm{I}\mathbf{x}_i = 0 \quad \Leftrightarrow \quad A_c\mathbf{x} = 0; \quad \text{I unit matrix.} \tag{6.102}$$

A non trivial solution of the last equation exists only if the determinant of A_c is zero. For $\mathbf{x}_i \to 0$ the two matrices A and A_c become identical. In conclusion, the unstable growth rates are obtained from a Fourier transform of the linear equations and setting their system determinant equal to zero.

Correct use of the Fourier transform. If Fourier analysis is performed in the complex domain completeness in $\{\mathbf{k}, \omega\}$ is reached by allowing for real wave vectors and real frequencies from $-\infty$ to $+\infty$. Orthogonality in the infinite space is expressed by

$$\int_{-\infty}^{+\infty} \mathrm{e}^{i(\mathbf{k}-\mathbf{k}')\mathbf{x}} d\mathbf{x} = (2\pi)^3 \delta(\mathbf{k}, \mathbf{k}'), \quad \int_{-\infty}^{+\infty} \mathrm{e}^{i(\omega-\omega')t} dt = 2\pi\delta(\omega, \omega').$$

Orthogonality holds for the real parts of $\mathbf{k}$ and ω. Whenever stability is analyzed in time by normal mode decomposition this has to be performed for the whole sequence of $\mathbf{k}$ vectors $\mathbf{k} \pm m\mathbf{k}_1$ and all complex $\Omega \pm m\omega_1$ with $(\mathbf{k}_1, \omega_1)$ impressed from outside real and positive (both always negative is equally possible), and m running through all non negative integers, $m = 0, 1, 2, \ldots..$ The physics behind is this: In presence of the laser beam any mode in the plasma is modulated by $\pm\mathbf{k}_1\mathbf{x}$ in space and by $\omega_1 t$ in time. Due to the nonlinearity of the equations of motion they interact in turn with $\mathbf{k}_1, \omega_1$, and so forth, to generate the infinite sequence above in m. Consequently, the Fourier analysis of the linearized fluid equations generates an infinite sequence of equations among the Fourier components. The corresponding dispersion relations deciding on stability or instability are obtained by the *continuous fraction method*. The alternative simpler method, used here, is to cut the system after an $m = m_{\max}$ as low as possible where no resonant modes contribute anymore to the dynamics.

Stimulated Brillouin Scattering

The transverse pump wave $\mathbf{E}_1(\mathbf{k}_1, \omega_1)$ impinges onto a homogeneous plasma of density n_0 which is modulated by the ion acoustic mode $n_3 = \hat{n}_3 \mathrm{e}^{i(\mathbf{k}_3\mathbf{x}-\omega_3 t)}$. The polarization of $\mathbf{E}_1$ is assumed to fulfil $\mathbf{k}_3\mathbf{E}_1 = 0$. The scattered wave $\mathbf{E}_2(\mathbf{k}_2, \omega_2)$ obeys (6.27); the acoustic density perturbation n_3 couples to the ponderomotive pressure. In leading order they follow the coupled equations with real coefficients

$$\nabla^2\mathbf{E}_2 - \frac{1}{c_{\varphi 2}^2}\frac{\partial^2\mathbf{E}_2}{\partial t^2} = -\frac{1}{\varepsilon_0 c^2}\frac{\partial}{\partial t}(en_3\mathbf{v}_{\mathrm{os},1}), \quad c_{\varphi 2} = \frac{c}{\eta_2}, \tag{6.103}$$

$$\nabla^2 n_3 - \frac{1}{c_s^2}\frac{\partial^2 n_3}{\partial t^2} = \frac{1}{m_i c_s^2}\nabla\pi_0, \quad \pi_0 = -\varepsilon_0 \frac{\omega_p^2}{\omega_1\omega_2}\nabla(\Re\mathbf{E}_1\Re\mathbf{E}_2). \tag{6.104}$$

We assume $|\hat{\mathbf{E}}_1| \gg |\hat{\mathbf{E}}_2|$ and $\hat{n}_3 \ll n_0$, and we consider growth in the small signal amplification domain and therefore do not need a coupled equation of $\mathbf{E}_1$. To take advantage of the complex Fourier analysis of these two equations in ($\mathbf{k}$, Ω) one of the quantities must be taken real. We choose $\Re\mathbf{E}_1 = (\mathbf{E}_1 + \mathbf{E}_1^*)/2$ and obtain the linked set of equations from the requirement of orthogonality

$$(\Omega^2 - v_{\varphi 2}^2 k_2^2)\mathbf{E}_2^{(\Omega)} = \frac{\Omega\omega_p^2}{2\eta_2^2 n_{e0}\omega_1}\left(\mathbf{E}_1 n_3^{(\Omega-\omega_1)} - \mathbf{E}_1^* n_3^{(\Omega+\omega_1)}\right), \tag{6.105}$$

$$(\Omega^2 - c_s^2 k^2)n_3^{(\Omega)} = -\frac{\varepsilon_0 k^2}{2m_i}\frac{\omega_p^2}{\omega_1\omega_2}\left(\mathbf{E}_1\mathbf{E}_2^{(\Omega-\omega_1)} + \mathbf{E}_1^*\mathbf{E}_2^{(\Omega+\omega_1)}\right). \tag{6.106}$$

On the RHS of (6.106) $\omega_2 = \omega_1 - \omega_3$ is associated with the Stokes component of $\mathbf{E}_2$ and $\omega_2 = \omega_1 + \omega_3$ refers to the anti-Stokes component of $\mathbf{E}_2$. We express $\mathbf{E}_2^{(\Omega-\omega_1)}$ and $\mathbf{E}_2^{(\Omega+\omega_1)}$ from (6.106) with the help of (6.105),

$$[(\Omega - \omega_1)^2 - v_{\varphi 2}^2(\mathbf{k} - \mathbf{k}_1)^2]\mathbf{E}_2^{(\Omega-\omega_1)} = \frac{\omega_p^2}{2\eta_2^2 n_{e0}}\mathbf{E}_1^* n_3^{(\Omega)}, \tag{6.107}$$

$$[(\Omega + \omega_1)^2 - v_{\varphi 2}^2(\mathbf{k} + \mathbf{k}_1)^2]\mathbf{E}_2^{(\Omega+\omega_1)} = \frac{\omega_p^2}{2\eta_2^2 n_{e0}}\mathbf{E}_1 n_3^{(\Omega)}, \tag{6.108}$$

and eliminate them from (6.106) to obtain

$$\left[(\Omega^2 - c_s^2 k^2) - \frac{\varepsilon_0}{4}\frac{k^2\omega_p^2}{m_i\eta_2^2 n_{e0}}\mathbf{E}_1\mathbf{E}_1^*\left(\frac{1}{D^{(\Omega-\omega_1)}} + \frac{1}{D^{(\Omega+\omega_1)}}\right)\right] n_3^{(\Omega)} = 0 \tag{6.109}$$

The Stokes and anti-Stokes ω_2 in (6.107) and (6.108) results from setting $\Omega = \omega_3$ in the RHS terms. The electromagnetic dispersion functions $(\Omega \pm \omega_1)^2 - v_{\varphi 2}^2(\mathbf{k} \pm \mathbf{k}_1)^2$ are abbreviated by the symbols $D^{(\Omega\pm\omega_1)}$. The infinite chain of (6.105) has been cut at $n_3^{(\Omega\pm\omega_1)}$ because owing to $|\Omega| \ll \omega_1$ these and all following terms are very small in comparison to the only resonant mode $n_3^{(\omega)}$. Furthermore, from (6.85) follows, even in the case of strong mismatch under a strong driver, $\omega_2 \simeq \omega_1$, $|\mathbf{k}_2| \simeq |\mathbf{k}_1|$, and hence $\mathbf{k}_1 \simeq \mathbf{k}_2 + \mathbf{k}_3$. It imposes $k_3 \simeq 2k_1 \sin(\vartheta/2)$, with the scattering angle $\vartheta = \angle(\mathbf{k}_1, \mathbf{k}_2)$. For backscattering, i.e., $\vartheta = \pi$, this implies $\mathbf{k} = \mathbf{k}_3 \simeq \mathbf{k}_1 - \mathbf{k}_2 \simeq 2\mathbf{k}_1$. As the intensity of the pump wave tends to zero the mismatch decreases also to zero. For mild backscattering $\gamma^2 \ll \omega_3^2 \simeq \omega_a^2$ holds, and from (6.85) follows more precisely $\mathbf{k}_3 = 2\mathbf{k}_1[1 - \eta_1 c_s/c]$. Hence, in (6.109) only the Stokes component $\mathbf{E}_2^{(\Omega-\omega_1)}$ is resonant; the anti-Stokes term $D^{(\Omega+\omega_1)}$ can be disregarded. This is the *weak coupling* limit [93]. The dispersion relation (6.109) shrinks to

$$(2i\gamma\omega_3 - \gamma^2)[(-2i\gamma\omega_1 - 2\omega_1\omega_3) - v_{\varphi 2}^2(k_3^2 - 2\mathbf{k}_1\mathbf{k}_3)] = \frac{\varepsilon_0 k_3^2 \omega_p^2}{4m_i\eta_2^2 n_{e0}}\mathbf{E}_1\mathbf{E}_1^* \quad (6.110)$$

With $\mathbf{k}_3$ from above for backscattering the sum of the real terms in the second bracket yields ω_3^2 and can be set to zero compared with $2\omega_1\omega_3$. Hence, one is left with the maximum growth rate for exact backscattering

$$\gamma_{\text{max}} = \frac{1}{2}\frac{\omega_{\text{pi}}}{c(m_e\eta_1^{3/2}n_{e0}c_s)^{1/2}}I^{1/2} = \frac{1}{2^{3/2}\eta_1}\omega_{\text{pi}}\frac{\hat{\mathbf{v}}_{\text{os}}}{(cc_s)^{1/2}}. \quad (6.111)$$

$I = \varepsilon_0 c\eta_1 \mathbf{E}_1\mathbf{E}_1^*/2$, $\omega_{\text{pi}}^2 = (m_e/m_i)\omega_{\text{pe}}^2$, $\eta_2 \simeq \eta_1$; $\hat{\mathbf{v}}_{\text{os}}$ is taken locally. In side scattering ($\vartheta < \pi/2$) the term

$$-2\omega_1\omega_3 - v_{\varphi 2}^2(k_3^2 - 2\mathbf{k}_1\mathbf{k}_3) = -4\omega_1 k_1 \sin\frac{\vartheta}{2}\left[c_s + \frac{c}{\eta_2}\left(\sin\frac{\vartheta}{2} - \cos\frac{\vartheta}{2}\right)\right]$$

is positive and contributes to reduce γ. The reduction is, primarily, a consequence of $|\mathbf{E}_1\mathbf{E}_2| = |\mathbf{E}_1||\mathbf{E}_2|\cos\vartheta$. A short remark may be in order: In (6.103) the scattered wave $\mathbf{E}_2$ depends linearly on the pump wave $\mathbf{E}_1$. As in side scattering the two vectors are not parallel to each other in general ($\vartheta \neq 0$) additional terms appear in $\mathbf{k}_3\mathbf{E}_1 \neq 0$ polarization. The dependence of γ on η_1 is due to shrinking of the ion sound wave vector in the plasma, $\boldsymbol{\pi}_0 \sim \mathbf{k}_3 \sim \mathbf{k}_0\eta_1$. In the so called *strong coupling* regime [54, 93], i e., when $|\Omega|^2 \gg k_3^2 c_s^2$ holds under the strong driver, in back scattering the approximations

$$\begin{aligned} D^{(\Omega-\omega_1)} &= \Omega^2 - 2\Omega\omega_1 + \omega_1^2 - \omega_p^2 - c^2k_1^2(1 - 2\eta_1 c_s/c)^2 \\ &\simeq -2\Omega\omega_1 + 4\eta_1^2\omega_1 c_s k_1 \simeq -2\Omega\omega_1, \quad \Omega^2 - c_s^2k^2 \simeq \Omega^2 \end{aligned}$$

are consistent and from (6.109), ignoring the anti-Stokes term, follows Ω as the complex cubic root

$$\Omega = \frac{1}{2^{4/3}\eta_1^{1/3}}\left[\omega_1\omega_{\text{pi}}^2\left(\frac{\hat{\mathbf{v}}_{\text{os}}}{c}\right)^2\right]^{1/3}(1 + i\sqrt{3}). \quad (6.112)$$

Unstable growth is particularly favoured in presence of a strong counter propagating electromagnetic wave, for example, when the incident wave reaches the critical density before sufficient attenuation. In such a case γ may exceed ω_a several times. In stationary plasmas with less than critical density a steady state solution of SBS usually establishes. In situations when light is reflected from the critical surface with frequency differing from the red-shifted $\mathbf{E}_2^{(\Omega-\omega_1)}$ mode SBS becomes nonstationary for a sufficiently high pump wave and the scattered spectrum undergoes strong broadening [103]. Latest in such a situation the scattered anti-Stokes field component $\mathbf{E}_2^{(\Omega+\omega_1)}$ can no longer be ignored a priori in the dispersion relation term $1/D^{(\Omega-\omega_1)} + 1/D^{(\Omega+\omega_1)}$ in (6.109) [104].

The anti-Stokes wave is close to resonance for the matching condition

$$\omega_2 = \omega_1 + \omega_3, \qquad \mathbf{k}_2 = \mathbf{k}_1 + \mathbf{k}_3. \tag{6.113}$$

It travels into the forward direction as the driver $\mathbf{E}_1(\omega_1, \mathbf{k}_1)$. This is most immediately seen in the co-moving frame. Under the influence of the driver $\mathbf{E}_1$ from each position $\mathbf{x}$ an elementary electromagnetic disturbance $\mathbf{E}_2^{(\omega_1')}$ propagates into opposite directions with strengths the difference of which reduces to zero with flow velocity $-c_s$ approaching zero. In the lab frame ω_1' becomes the anti-Stokes frequency $\omega_1 = \gamma_\varphi(\omega_1' + \mathbf{k}_2'\mathbf{c}_s) = \omega_1 + \omega_3$ for $\mathbf{k}_2\mathbf{k}_3 \geq 0$, see (6.83). When $\mathbf{k}_1$ and $\mathbf{k}_2$ are approximately collinear, $|\mathbf{k}_3| \simeq \eta_1\omega_3/c \simeq k_0 s/c$ is very small and $D^{(\Omega+\omega_1)}$ is close to resonant coupling,

$$D^{(\Omega+\omega_1)} = \Omega^2 + 2\Omega\omega_1 + \omega_1^2 - \omega_p^2 - c^2k_1^2\left(1 + \frac{2\eta_1 s}{c}\right)^2 \simeq \Omega^2 + 2\Omega\omega_1 - 4\eta_1^2\omega_1 s k_1$$

On the other hand, in nearly forward direction the Stokes $|\mathbf{k}_3|$ can assume arbitrarily small values; anti-Stokes and Stokes frequencies $\omega_2 = \omega_1 \pm s|\mathbf{k}_3|$ are very close to ω_1 and with increasing pump wave intensity their bandwidths begin to overlap in a $|\mathbf{k}_3|$ interval around the dispersion intersection points (see Fig. 6.30). This leads to a redistribution of the energy fluxes and to back reaction onto the (ω_3, k_3) mode that rather to be a free mode will convert into a forced quasi-mode at downshifted frequency ω_a. Nishikawas's oscillating two-stream instability with $\omega_a = 0$ for finite $|\mathbf{k}_2| = |\mathbf{k}_3|$ is perhaps the most instructive example for the Stokes-anti-Stokes interplay and concomitant frequency shift. The instability results in a long wavelength modulation of the amplitudes $\hat{n}_3$ and $|\mathbf{E}| = |\mathbf{E}_1 + \mathbf{E}_2|$. As a special case of SBS the filamentary instability is obtained from (6.109 by setting $\mathbf{k}_3\mathbf{k}_1 = 0$ and $\Omega = i\gamma$. Then, $D^{(\Omega\pm\omega_1)} = \pm 2i\gamma\omega_1 - c^2k_3^2$ leads to the equation of growth γ,

$$(\gamma^2 + c_s^2k^2)\frac{1}{4\gamma^2\omega_1^2 + c^4k^4} = \frac{\omega_{\mathrm{pi}}^2}{8}\left(\frac{\hat{\mathbf{v}}_{\mathrm{os}}}{c}\right)^2; \qquad \mathbf{k} \perp \mathbf{k}_1. \tag{6.114}$$

Here, as throughout the chapter, $\mathbf{E}_1\nabla n_3 = 0$ is assumed. As long as the pump wave does not drive plasma resonances, for instance owing to $\omega_1 > \omega_p$, polarization effects can be disregarded [103, 105]. Stokes and anti-Stokes electromagnetic components both act to drive the ion acoustic and the electron plasma mode whenever the detuning due to growth exceeds the Stokes shift $\Delta\omega$. It is advisable in general, also with other types of unstable modes, to include both components in the analysis and to compare with the result obtained from the resonant Stokes mode alone. After all it must be kept in mind that parametric systems of type (6.105), (6.106) represent a recurrence scheme of an infinite number of equations. For the PDI and OTSI sequences closure and impact of the anti-Stokes component are investigated quantitatively in [106]. The arguments for closure may vary from case to case. So for example numerical studies

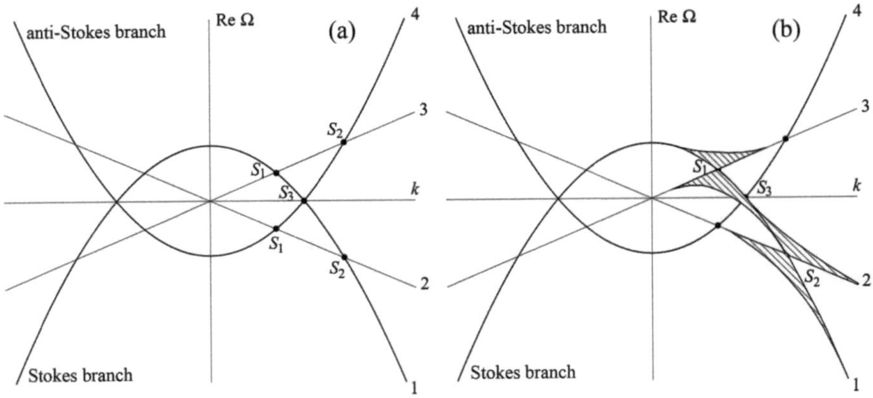

Fig. 6.30 **a** Brillouin diagram of wave-wave coupling in SBS and PDI. Dispersion diagrams of electromagnetic ($\omega_{\rm em}$: 1,4), electrostatic ($\omega_{\rm es}$: 1,4), and acoustic (ω_a: 2,3) modes. Resonant interaction (coupling) is strongest in the neighborhood of the intersection points S_1, S_2 and S_3. S_1, S_3 are unstable, see (6.96) for S_1, and(6.114) for S_3. From (6.95) with $1/D^{(\Omega-\omega_1)} = 0$ follows that S_2 is a stable interaction point in the approximation of (6.96). S_2 is not energy conserving. **b** Excitation of a quasi-mode (ω_3, k_3) from overlapping of mismatch zones around S_1 and S_2. For laser intensity $I \to 0$ the graphs delimiting the shadowed regions contract to the intersection points S_1 and S_2

have shown that ion acoustic harmonics can couple to SBS and lead to reduction of reflectivity, spatial decorrelation and to temporal chaotic plasma dynamics [85]. In a very recent investigation intense interplay of Stokes and anti-Stokes and higher order scattering and rescattering has been revealed in simulations, and bursts of reflectivity have been observed [107].

Stimulated Brillouin backscattering contains a relevant practical aspect; it can operate as a phase conjugated plasma mirror. The linear phase conjugation operator "$*$" acts on the spatial part of a plane wave $Ae^{i\mathbf{kx}-i\omega t}$ transforming it into its complex conjugate expression, $(Ae^{i\mathbf{kx}})^*e^{-i\omega t} = A^*e^{-i\mathbf{kx}-i\omega t}$. This is exactly what happens in Brillouin backscattering. The inversion of the wave vector **k** means that a wave on its way back crosses again the same space domain in opposite direction correcting thereby all undesired distortions of the wave front (phase) from its first passage (up to small differences originating from the difference between η_1 and η_2). The distortions of an originally plane wave are eliminated in the phase conjugated reflection. Specular reflection of a beam having for instance the cross section of a question mark produces its mirror image and to primary phase distortions it adds additional distortions on its way back, whereas the back scattered beam reproduces the original shape of the question mark (in laser-plasma interaction first discovered as the *Fragezeicheneffekt* in 1972 [108]).

In long scale length plasmas stimulated Brillouin back and side scattering from a coherent laser beam is an instability that grows rapidly in time to high levels of reflectivity [81, 82]. The acoustic wave has low frequency and, according to (6.30), low energy density $\mathcal{E}_{\rm a}$ and does almost no photon energy deposit in the medium, with dramatic consequences for plasma heating in general and inertial confinement fusion and X-ray sources in particular. Therefore much interest has been concentrated in the past three decades on experiments to clarify the saturation level of light reflection R in long scale length underdense plasmas. In early SBS studies with CO_2 lasers interacting with a smooth plasma [109] and subsequent similar experiments with gas jets [110] showed surprisingly low saturation levels of $\hat{n}_3/n_0 \simeq 10$–20%, corresponding to $R < 10\%$. Further drastic reduction of this values of R from SBS has been achieved by applying average laser beam smoothing through induced spatial incoherence (ISI: broad band laser beam is first divided into many independent beamlets, with mutual time shifts exceeding the coherence time, and then overlapped on the target). With ISI applied to a 2 ns Nd laser beam of peak intensity $I = 10^{14}$ Wcm^{-2} on CH disk targets R resulted below 1% [111]. These were good news but they resisted for nearly two decades to a satisfactory theoretical explanation. The studies focused on the discrepancy between theoretical expectations and experimental results mostly concentrated on saturation effects of the ion acoustic modes due to steepening and wavebreaking, generation of harmonics, particle trapping, and decay into subharmonics of $\hat{n}_3$.

In retrospect, however, the low backscatter levels measured in the past have not been analyzed with enough temporal resolution and therefore indicate merely the ratio of back scattered to incident laser energy, generally ignoring transient high backscatter levels that could potentially occur during a fraction of the laser pulse. The ability of combining fluid dynamic and kinetic SBS modelling with laser pulse shaping leads to a more realistic interpretation of the experiments. While often high backscatter levels occur temporarily, in agreement with theory, the average level over a longer time scale, e.g., tens of ps or entire laser pulses, may result moderate or low compared to the incident laser intensity. However, this should not lead to the hasty conclusion that SBS is not of concern under constraints imposed by specific goals, as for example in laser fusion schemes. Independently of all nonlinear effects examined with the attempt to explain saturated SBS levels, laser beam smoothing in plasmas can itself lead to reduction of backscattering: self-focusing of individual laser speckles of spatially smoothed beams (e.g., by random phase plate technique) contribute in this way to an additional plasma-induced smoothing as the beam propagates further into the plasma target.

In more recent time a breakthrough in understanding SBS suppression in long scale length plasmas has been achieved by Hüller et al. [112]. Sophisticated numerical simulations have shown a quite universal behavior applicable to laser beams containing high-intensity hot spots that are subject to self-focusing. Although the growth rate of SBS is higher than the one of self-focusing, SBS starts from a significantly lower noise level. Henceforth it grows primarily in intense self focused hot spots, giving rise to transient high backscatter flashes. Nevertheless, due to subsequent ponderomotive density depletion disruption of the SBS signal takes place [113] and, eventually, backscattering, averaged over the whole pulse length, remains on a moderate level. It is important to mention that the region behind the self-focusing hot spots does not contribute significantly to SBS due to plasma induced smoothing, see [112]. In this paper the features of SBS were studied for a mono-speckle laser beam at Nd peak intensity $I = 3 \times 10^{14}$ Wcm^{-2} including self-focusing and nonlocal energy transport self consistently. In the linearly rising laser pulse with its maximum at 200 ps SBS starts growing fast as predicted by (6.111) and reaches its maximum of $R \simeq 10\%$ already at 50 ps and decays after 100 ps so that averaged over the whole pulse of 1 ns $\overline{\langle R\rangle} \simeq 1$–$1.5\%$ results with nonlocal transport included, and $\overline{\langle R\rangle} \simeq 4\%$ without. The results have been found in agreement with the concomitant experiment. On the other hand, when self-focusing was suppressed in the simulation, maximum values of $R = 50\%$ were reached in coincidence with the maximum of the laser pulse at 200 ps. The conclusion is twofold: (1) self-focusing, i.e., local intensity concentration, is responsible for the fast growth; (2) the sudden disruption of SBS after a short time is triggered by an effect localized in spots of high laser intensity. The simulations allow the identification of (2) with the resonant filament instability [114, 115]. Due to the ponderomotive force a density well forms along a filament which is capable of supporting electromagnetic eigenmodes like a waveguide. When the threshold intensity for self-focusing is reached they grow rapidly unstable and lead to lateral bending of the filament and eventually to its complete breakup. After some time it forms again and disrupts. The process leads to increased lateral scattering of the laser beam and to correlation length reduction to 10–15 μm in laser plasma interaction regions of mm-scale. Very good agreement between experiment and the large scale simulations suggest the identification of the resonant filament instability with the low level SBS reflectivity. With the finding the two decades old problem of Brillouin anomaly in high power laser interaction seems to be solved, or to be very close to its definite solution. It should be stressed that the outlined scenario appears quite universal with beams containing hot spots at power densities above the self-focusing threshold [116]. In a more recent paper 3D simulations have been successfully performed on effective reduction of stimulated Brillouin scattering from an expanding plasma [117]. It is observed that, after a short transient stage, SBS reaches a significant level (i) in a single self-focused speckle located in the low-density front part of the plasma only (ii) as long as the incident laser pulse is increasing in amplitude. Using a variational approach, the model employed reproduces the position and the peak intensity of the self-focusing hot spot in the front part of the plasma as well as the local density depletion in the hot spot. The knowledge of these parameters makes it possible to estimate the spatial amplification of SBS as a function of the laser beam power, and consequently, to explain the experimentally observed SBS reflectivity, considerably reduced with respect to standard theory in the regime of large laser power.

Stimulated Raman Scattering (SRS)

Formally, stimulated Raman scattering in the linear fluid regime obeys the same equations as SBS. All difference in physics derives from the resonant excitation of the high frequency electron plasma wave which replaces the low frequency acoustic mode n_3 in (6.104). Thus, the governing equations are

$$\nabla^2 \mathbf{E}_2 - \frac{1}{v_{\varphi 2}^2}\frac{\partial^2 \mathbf{E}_2}{\partial t^2} = -\frac{1}{\varepsilon_0 c^2}\frac{\partial}{\partial t}(en_3 \mathbf{v}_{\mathrm{os},1}), \qquad c_{\varphi 2} = \frac{c}{\eta_2}. \tag{6.115}$$

$$\nabla^2 n_3 - \frac{1}{v_{\varphi 3}^2}\frac{\partial^2 n_3}{\partial t^2} = \frac{1}{m_e c_{se}^2}\nabla \boldsymbol{\pi}_0, \qquad \nabla \boldsymbol{\pi}_0 = -\varepsilon_0 \frac{\omega_p^2}{\omega_1 \omega_2}\nabla(\Re \mathbf{E}_1 \Re \mathbf{E}_2). \tag{6.116}$$

Limiting ourselves again to the small signal amplification the Fourier analysis reproduces (6.105)–(6.108), with c_s and m_i replaced by $v_{\varphi 3}$ from (6.25) and m_e. It leads to the dispersion equation

$$(\Omega^2 - v_{\varphi 3}^2 k^2) - \frac{\varepsilon_0}{4}\frac{k^2 \omega_p^2}{m_e n_{e0}}\mathbf{E}_1 \mathbf{E}_1^* \left(\frac{1}{D^{(\Omega - \omega_1)}} + \frac{1}{D^{(\Omega + \omega_1)}}\right) = 0. \tag{6.117}$$

For a moderately strong driver the matching conditions (6.85) are well satisfied. Th scattered frequency and wave number range from $(\omega_2, \mathbf{k}_2) = (\omega_1, \mathbf{k}_1)$ to $(\omega_1/2, 0)$. Since ω_3 is close to ω_p, for the Stokes component follows from $\omega_2 \simeq \omega_1 - \omega_p$ that in back or side scattering the anti-Stokes frequency $\omega_2 \simeq \omega_1 + \omega_p$ is nonresonant almost everywhere and $1/D^{(\Omega + \omega_1)}$ can be neglected. The dispersion relation reduces to

$$(\omega_3^2 + 2i\omega_3\gamma - \omega_p^2 c_{se}^2 k_3^2)[(\omega_3 - \omega_1)^2 + 2i\gamma(\omega_3 - \omega_1) - \omega_p^2 - c^2(k_3 - k_1)^2]$$
$$= \frac{\varepsilon_0}{4}\frac{k_3^2 \omega_p^2}{m_e n_{e0}}\mathbf{E}_1 \mathbf{E}_1^*. \tag{6.118}$$

Maximum growth occurs close to resonance of ω_2 and ω_3; hence

$$2i\gamma_{\max}\omega_3[2i\gamma_{\max}(\omega_3 - \omega_1)] = \frac{\varepsilon_0}{4}\frac{k_3^2 \omega_p^2}{m_e n_{e0}}\mathbf{E}_1 \mathbf{E}_1^* \tag{6.119}$$

$$\gamma_{\max} = \frac{k_3}{4}\left[\omega_p^2(\omega_2\omega_3)\frac{e^2}{m_2^2\omega_1^2}\mathbf{E}_1\mathbf{E}_1^*\right]^{1/2} = \frac{k_3}{4}\hat{v}_{\mathrm{os}}\left[\frac{\omega_p^2}{\omega_2\omega_3}\right]^{1/2}. \tag{6.120}$$

For exact back scattering follows from (6.85) and (6.25) $k_2 = k_0(1 - 2\omega_p/\omega_1)^{1/2}$, $k_3 = |k_3| = k_1 + k_2$. The comparison of $\gamma_{\max}$ for SRS and SBS for $c_s \simeq 10^{-3}c$ and $m_e/m_i \simeq 10^{-4}$ shows that they are of the same order of magnitude. SRS into exact forward direction is also possible. Its relevance lies in the possibility of exciting an electron plasma wave of particularly high phase velocity in the low density plasma of $\omega_p \ll \omega_1$, e.g., hohlraum plasma in indirect inertial fusion drive, and subsequent preheat by trapped fast electrons [94, 118, 120, 121]. In regions of low plasma density $\omega_3 \simeq \omega_p, \eta_1 = \eta \simeq 1$, and in forward direction the matching condition implies for the anti-Stokes and Stokes lines $k_3 = |\mathbf{k}_1 - \mathbf{k}_2| = (\omega_1 \pm \omega_p)/c \simeq \omega_1/c$. The dispersion functions $D^{(\Omega \pm \omega_1)} = D^{(\omega_3 + i\gamma \pm \omega_1)}$ simplify to

$$D^{(\Omega-\omega_1)} = 2i\gamma(\omega_p - \omega_1) \simeq -2i\gamma\omega_1, \qquad D^{(\Omega+\omega_1)} = 2i\gamma(\omega_p + \omega_1) \simeq +2i\gamma\omega_1,$$

$$D^{(\Omega-\omega_1)} + D^{(\Omega+\omega_1)} = 4i\gamma\omega_p, \qquad D^{(\Omega-\omega_1)}D^{(\Omega+\omega_1)} = 4\gamma^2\omega_1^2.$$

Both terms are nearly resonant and must be used in (6.117) to yield the correct growth rate

$$\gamma = \frac{1}{2^{3/2}}\frac{\omega_p^2}{\omega_1}\frac{\hat{v}_{os}}{c} \tag{6.121}$$

Two Plasmon Decay

Short inspection of (6.115) and (6.116) shows that the equations governing the two plasmon decay are obtained by replacing $\mathbf{E}_{\perp 2}$ by $\mathbf{E}_{\parallel 2}$ of a second plasma wave or associated electron density perturbation n_2,

$$\nabla^2 n_2 - \frac{1}{v_{\varphi 2}^2}\frac{\partial^2 n_2}{\partial t^2} = \frac{1}{m v_{\varphi 2}^2}\nabla\pi_2, \quad \nabla^2 n_3 - \frac{1}{v_{\varphi 3}^2}\frac{\partial^2 n_3}{\partial t^2} = \frac{1}{m v_{\varphi 3}^2}\nabla\pi_3. \tag{6.122}$$

The wave pressure term now is given by the resonant terms of $\pi = \pi_0 + \pi_t$, i.e.,

$$\pi_{2,3} = (\pi_0 + \pi_t)_{2,3} = -n_0 m\nabla(\mathbf{v}_1\mathbf{v}_{2,3}) + m\frac{\partial n_{2,3}\mathbf{v}_1}{\partial t} \tag{6.123}$$

with $\mathbf{v}_1 = (\mathbf{v}_{os} + \mathbf{v}_{os}^*)$ induced by the laser. We substitute $\mathbf{v}_{2,3}$ by $n_{2,3}$ with the help of (5.70), Fourier-analyze (6.122) and keep only the resonant terms to obtain

$$(-\Omega^2 + \omega_2^2)n_2^{(\Omega)} = -(\mathbf{k} - \mathbf{k}_1)\frac{\mathbf{v}_{os}}{2}\left[\frac{k^2(\Omega - \omega_1)}{(\mathbf{k} - \mathbf{k}_1)^2} + \Omega\right]n_3^{(\Omega-\omega_1)}$$

$$[-(\Omega - \omega_1)^2 + (\omega_2 - \omega_1)^2]n_3^{(\Omega-\omega_1)} = -\frac{\mathbf{k}\mathbf{v}_{os}^*}{2}\left[(\mathbf{k} - \mathbf{k}_1)^2\frac{\Omega}{k^2} + (\Omega - \omega_1)\right]n_2^{(\Omega)}.$$

Elimination of $n_3^{(\Omega-\omega_1)}$ and approximating $\omega_2 = \omega_3 = \omega_p = \omega_1/2$, yields the dispersion equation

$$(\Omega^2 - \omega_2^2)[(\Omega - \omega_1)^2 - (\omega_2 - \omega_1)^2] - \frac{1}{4}(\mathbf{k} - \mathbf{k}_1)\hat{\mathbf{v}}_{os}(\mathbf{k}\hat{\mathbf{v}}*_{os})\omega_p^2\frac{[(\mathbf{k} - \mathbf{k}_1)^2 - k^2]^2}{k^2(\mathbf{k} - \mathbf{k}_1)^2} \tag{6.124}$$

Setting $\Omega = \omega_p + i\gamma$ and $\mathbf{k} - \mathbf{k}_1 \simeq \mathbf{k}$ determines the growth rate γ,

$$\gamma = \mathbf{k}\hat{\mathbf{v}}_{os}\left|\frac{(\mathbf{k} - \mathbf{k}_1)^2 - k^2}{k|\mathbf{k} - \mathbf{k}_1|}\right|. \tag{6.125}$$

With the angle between the pump $\mathbf{k}_1$ and $\mathbf{k} = \mathbf{k}_2$ of n_2, $\alpha = \angle(\mathbf{k}_1, \mathbf{k}_2)$, and $(\mathbf{k} - \mathbf{k}_1)^2 - k^2 = -2\mathbf{k}_1\mathbf{k} + k_1^2 \simeq -2\mathbf{k}_1\mathbf{k}$ the growth γ results proportional to $|\mathbf{k}_2\hat{\mathbf{v}}_{os}||$

$\mathbf{k}_1\mathbf{k}| \sim \sin\alpha\cos\alpha$, in agreement with the elementary reasoning. For $\alpha = \pi/4$ it assumes its maximum value,

$$\gamma = \frac{\omega_1}{4\sqrt{2}}\frac{\hat{\mathbf{v}}_{os}}{c}. \tag{6.126}$$

A more elaborated analysis of the TPDI permits the determination of the electron temperature through dispersion modifications by the thermal pressure [122, 123]. From experiments with multiple overlapping laser beams in spherical and planar geometry it is concluded that the TPD instability is the main source of superthermal electron generation in the plasma corona with nonvanishing density scale length at n_c. The experiments show that the total overlapped intensity governs their scaling rather than the number of overlapped beams [124, 126]. Theoretical investigations and experiments confirm once more the production of hot electrons [127, 128] and show the transition of the SRS to the TPD when increasing the plasma density scale lengths [125]. An additional important feature is the emission of $1/2\,\omega_1$and $3/2\,\omega_1$ radiation [126] as a consequence of the current density component $\mathbf{j}_{3\omega/2} = -en_{e,\omega/2}\mathbf{v}_{os}$ in the wave equation of the driver $\mathbf{E}_1$.

Oscillating Two-Stream Instability

The growth rate γ of the OTSI is easily found with the help of π_0 from (6.98). Combining it with the ion momentum equation and particle conservation after linearization leads to

$$\frac{\partial^2}{\partial t^2}n_3 = -\left(\frac{\pi_0}{m_e n_0} + c_s^2\right)k_3^2 n_3. \tag{6.127}$$

The interesting unstable domain for ω_1 is close to $\omega_2 \simeq \omega_p$. For $\nu_{ei} \ll \omega_1$ and $\omega_1^2 - \omega_2^2 = 2\omega\varDelta$ follows

$$\frac{\gamma^2}{k_3^2} = -\frac{\varDelta}{\varDelta^2 + \nu^2/4}\frac{m_e}{m_i}\frac{\hat{v}_1^2}{4} - c_s^2. \tag{6.128}$$

Since n_3 is quasi-static and not propagating the instability is absolute and, typical for it, even in the absence of any dissipation, exhibits a finite threshold for the pump intensity. The reason is obvious: growth can start when the ponderomotive force ($\sim |\mathbf{E}_1|^2 n_3$) prevails on the thermal plasma pressure gradient ($\sim c_s^2 n_3$).

Parametric Decay Instability

In contrast, the PDI is the resonant decay of a photon into a plasmon and a propagating phonon. An observer co-moving with n_3 at speed $c_s > 0$ sees a static modulation to which the former analysis applies, with the difference however that the plasmons emanating from n_3 are Doppler-shifted by $\varDelta\omega_2 = \pm\omega_3$ whereas ω_1, apart from an insignificant transverse Doppler shift from time dilation, is not. Introducing $\omega_\pm = \omega_2 \pm \omega_3$, $d_t = \partial_t - c_s\partial_x$ momentum conservation and (6.98) combine to

$$\frac{\mathrm{d}^2}{\mathrm{d}t^2}n_3 = -\left\{\left[\frac{\omega_1^2-\omega_+^2}{(\omega_1^2-\omega_+^2)^2+\nu^2\omega_1^2}+\frac{\omega_1^2-\omega_-^2}{(\omega_1^2-\omega_-^2)^2+\nu^2\omega_1^2}\right]\mathrm{e}^{-i\pi/2}\right.$$
$$\left.\times\frac{e^2\omega_p^2|\mathbf{E}_1|^2}{4m_e m_i(\omega^2+\nu^2)}\right\}k_3^2 n_3. \qquad (6.129)$$

This is in agreement with the somewhat lengthy treatments [98, 99]. The phase change by $\Delta\varphi = -\pi/2$ comes from (6.69) at resonance, as extensively discussed in the first section of this chapter. In the denominator there appears the factor 4 now instead of 2 in (6.98) since, pictorially spoken, half the electrons have anti-Stokes eigenfrequency ω_+ and the other half are resonant at the Stokes frequency ω_- owing to the flow $v = -c_s$. The growth rate just above threshold follows from $d_t^2 = (\partial_t - c_s\partial_x)^2 \simeq -(2i\gamma\omega_3 + c_s^2k_3^2)$

$$\gamma = -\left\{\frac{\Delta_+}{\Delta_+^2+\nu^2/4}+\frac{\Delta_-}{\Delta_-^2+\nu^2/4}\right\}\frac{m_e\omega}{m_i\omega_3}\frac{\hat{v}_1}{16}; \qquad \Delta_\pm = \omega_+ - \omega_\pm. \qquad (6.130)$$

For vanishingly small damping it reduces to

$$\gamma = -\left\{\frac{1}{\Delta_+}+\frac{1}{\Delta_-}\right\}\frac{m_e\omega}{m_i\omega_3}\frac{\hat{v}_1^2}{16} \qquad (6.131)$$

thus indicating that at low driver intensity the PDI is convective. Both terms in the bracket are negative for $\omega_1 - \omega_- = \Delta + \omega_3 < 0$. Growth also occurs with $0 > \Delta > -\omega_3$. At exact resonance $\Delta = 0$ neither the OTSI nor the PDI can grow.

Impact of Dissipation and Inhomogeneities

At resonance an undamped harmonic oscillator $\delta(t)$ is excited to arbitrarily high amplitude by an arbitrarily weak driver E_d; the onset of growth starts at threshold zero. In presence of linear damping $\nu\dot{\delta}$ growth can start when $-e\overline{E_d\dot{\delta}} > m\nu\overline{\dot{\delta}^2}$ is fulfilled. Applied to the parametric three wave process with dispersion equation of type (6.109) this reads in the weak coupling limit at optimum matching, consistent with weak damping $\nu_{2,3} \ll \omega_{2,3}$,

$$\nu_2\delta_2^2 < C_2E_1^*\delta_3\delta_2, \qquad \nu_3\delta_3^2 < C_3E_1\delta_2\delta_3 \quad \Rightarrow \nu_2\nu_3 < C|E_1|^2, \qquad (6.132)$$

$C = C_2C_3 = \text{const.}$ $|E_{\text{thres}}|^2 = \nu_2\nu_3/C = \gamma_{\text{thres}}^2$ is the threshold pump intensity in presence of dissipation. By intuition $|E_{\text{thres}}|^2$ follows from the undamped growth rate

$$\gamma = \gamma_{\text{thres}} = \sqrt{\nu_2\nu_3}; \qquad C = \frac{\nu_2\nu_3}{|E_{\text{thres}}|^2}. \qquad (6.133)$$

Alternatively, inserting the damping terms $i\nu_2(\omega_3 - \omega_1)$ and $i\nu_3\omega_3$ in (6.108), (6.109), at exact resonance (6.132) becomes

$$(2\gamma + \nu_2)(2\gamma + \nu_3) - C|E_1|^2 = 0.$$

The solution for growth is

$$\gamma = -\frac{\nu_2 + \nu_3}{4} + \frac{1}{2}\left\{\frac{(\nu_2 + \nu_3)^2}{4} - (\nu_2\nu_3 - C|E_1|^2)\right\}^{1/2}$$

with (6.133) for $\gamma = 0$ at the threshold. Hence, in presence of weak linear dissipation the growth is given by

$$\gamma = -\frac{\nu_2 + \nu_3}{4} + \frac{1}{2}\left\{\frac{(\nu_2 - \nu_3)^2}{4} + C|E_1|^2\right\}^{1/2}. \tag{6.134}$$

If instead of ν the coefficient for kinetic energy damping ν_E is introduced $\nu_E = \nu/2$ has to be kept in mind. Damping may be either of collisional nature or of linear Landau type. Perhaps less intuitive, if one of the $\nu_{2,3}$ vanishes $|E_{\text{thres}}|$ is again zero.

As a rule laser interaction creates inhomogeneous plasmas. In a region of validity of the optical WKB approximation the phase mismatch $\psi(\mathbf{x})$ of the wave vectors is given by $\psi = \int (\mathbf{k}_1 - \mathbf{k}_2 - \mathbf{k}_3)\, d\mathbf{x}$. It is independent of the path of integration from $\mathbf{x}_0$ to $\mathbf{x}$. Let us assume that at position $\mathbf{x}_0$ there is perfect matching. Spatial growth will stop when $\psi = \pi$ is reached. Owing to $\nabla \times \mathbf{k}_{1,2,3} = 0$ holds $(d\mathbf{x}\nabla)\check{} = \nabla(d\mathbf{x}\check{}) = \kappa'^{\check{}0} d\mathbf{x}$, with $\check{} = \mathbf{k}_1 - \mathbf{k}_2 - \mathbf{k}_3$ and $\kappa' = \nabla|\check{}|$ pointing along $\check{}$. Under gentle electron density variation Taylor expansion in $\mathbf{x}_0$ along $\check{}$ yields $\psi = \int_{\mathbf{x}_0} (\check{})\, d\mathbf{x} = \kappa'(\mathbf{x}_0)(x - x_0)^2/2$. Thus the amplification length is $x - x_0 = \sqrt{2\pi/\kappa'}$. At threshold the energy fed into the unstable modes 2, 3 in the intervals $\Delta_{2,3} = (x - x_0)/\cos\alpha_{2,3}$, $\alpha_{2,3} = \angle(\check{}, \mathbf{k}_{2,3})$, is convectively carried out of the interaction region, $v_{g2,3}E_{2,3} = \nu_{2,3}\Delta_{2,3}E_{2,3}$. The damping coefficients inserted in (6.133) yield

$$\frac{C|E_1|^2}{\nu_2\nu_3} = 2\pi C|E_{\text{thres}}|^2 \frac{\cos\alpha_2 \cos\alpha_3}{|\kappa' v_{g2} v_{g3}|} = 1. \tag{6.135}$$

This is the celebrated Piliya-Rosenbluth criterion for $|E_{\text{thres}}|$ in the inhomogeneous plasma [130, 131]. As the emphasis in this chapter is primarily on the basic physical effects driving the plasma unstable and on the basic wave dynamics the influence of inhomogeneities on the individual growth rates is not pursued further. An extensive introduction to parametric instabilities in inhomogeneous plasmas may be found in [132] and a list of thresholds and linear growth rates may be found in [133].

6.4.4 *Parametric Amplification of Pulses*

The foregoing section was devoted to the basic mechanisms of resonant three wave interactions in the homogeneous plasma in situations where the constant amplitude approximation is justified. It represents the purest, i.e., the most idealized, case of wave-wave coupling by the wave pressure. To give this theory "a touch of reality" (Boyd/Sanderson [134], Sect. 10.3) in the following the coupled three wave equations for slowly varying amplitudes are formulated. At the same time such an extension will exhaust the limits of the Hamiltonian description of wave dynamics in terms of action angle variables and preserve the number of quasi-particles introduced by (5.164) in the Sect. on the classical photon, and finally, it will yield additional insight by revealing the perfect symmetry between the three coupled waves.

6.4.4.1 Slowly Varying Amplitudes

The waves considered in this subsection exhibit the more general structure $\mathbf{A}(\mathbf{x}, t) = \hat{\mathbf{A}}(\mathbf{x}, t)\exp[i\Phi(\mathbf{x}, t)]$ with amplitudes $\hat{\mathbf{A}}(\mathbf{x}, t)$ slowly varying in space according to the WKB approximation, and wave vectors $\mathbf{k}$ and frequencies ω defined by $\mathbf{k} = \nabla\Phi$, $\omega = -\partial_t\Phi$. We begin with the laser pump wave $\mathbf{E}_1$ by expanding (6.27) up to first order,

$$(-c_{\varphi 1}^2\mathbf{k}_1^2 + \omega_1^2)\mathrm{e}^{i\Phi_1}\hat{\mathbf{E}}_1 + 2i[c^2(\mathbf{k}\nabla)\hat{\mathbf{E}}_1 + \omega\partial_t\mathbf{E}_1]\mathrm{e}^{i\Phi_1} = \frac{e^2\omega_1}{2\varepsilon_0 m\omega_2}\mathrm{e}^{i(\Phi_2+\Phi_3)}\hat{\mathbf{E}}_2\hat{n}_3$$

$$= \frac{-ie\omega_1}{2m\omega_2}\mathrm{e}^{i(\Phi_2+\Phi_3)}\hat{\mathbf{E}}_2(\mathbf{k}_3\hat{\mathbf{E}}_3); \quad \mathbf{E}_3 = i\frac{e}{\varepsilon_0}\mathbf{k}_3 n_3. \tag{6.136}$$

Note that c in the second term is the light speed in vacuum. The first term vanishes by definition. Recalling $c_g c_\varphi = c^2$ one is led to

$$(\mathbf{c}_{g1}\nabla)\hat{\mathbf{E}}_1 + \partial_t\hat{\mathbf{E}}_1 = -\frac{1}{4}\frac{ek_3\omega_1}{m\omega_1\omega_2}\hat{\mathbf{E}}_2\hat{E}_3\mathrm{e}^{i(\Phi_2+\Phi_3-\Phi_1)}. \tag{6.137}$$

We chose the stimulated Raman effect as the most illustrative example in the context here. The scattered transverse wave $\mathbf{E}_2$ is subject to the same procedure as $\mathbf{E}_1$; it reads

$$(\mathbf{c}_{g2}\nabla)\hat{\mathbf{E}}_2 + \partial_t\hat{\mathbf{E}}_2 = \frac{1}{4}\frac{ek_3\omega_2}{m\omega_1\omega_2}\hat{\mathbf{E}}_1\hat{E}_3^*\mathrm{e}^{-i(\Phi_2+\Phi_3-\Phi_1)}. \tag{6.138}$$

Only the $\hat{n}_3$ and $\hat{\mathbf{E}}_3^*$ components fulfil the resonance condition. Equation (6.116) is a consequence of (6.27) because only the velocity component of $\mathbf{v}_e$ parallel to $\mathbf{E}_3$ and $\mathbf{k}_3$ contributes. Analogously to (6.138) with $v_g v_\varphi = c_{\mathrm{se}}^2$ we obtain

$$(\mathbf{c}_{g3}\nabla)\hat{\mathbf{E}}_3 + \partial_t\hat{\mathbf{E}}_3 = \frac{1}{4}\frac{e\mathbf{k}_3\omega_p^2}{m\omega_1\omega_2\omega_3}\hat{\mathbf{E}}_1\hat{\mathbf{E}}_2^*\mathrm{e}^{-i(\Phi_2+\Phi_3-\Phi_1)}. \tag{6.139}$$

Introducing the symbol $d_i = \partial_t + (\mathbf{v}_{gi}\nabla)$ for the total, i.e., convective, derivative, multiplying $\mathbf{E}_i$ by its cc $\mathbf{E}_i^*$ in the last three equations above and setting $\Delta\Phi = \Phi_2 +$

$\Phi_3 - \Phi_1$ for the *dephasing* ("mismatch") of the modes and $C = \varepsilon_0 e k_3/8m\omega_1\omega_2$, (6.137)–(6.139) read

$$\frac{1}{\omega_1} d_1 \left(\frac{\varepsilon_0}{2}\hat{\mathbf{E}}_1\hat{\mathbf{E}}_1^*\right) = -C\left(\hat{E}_1^*\hat{E}_2\hat{E}_3 e^{i\Delta\Phi} + \hat{E}_1\hat{E}_2^*\hat{E}_3^* e^{-i\Delta\Phi}\right). \quad (6.140)$$

$$\frac{1}{\omega_2} d_2 \left(\frac{\varepsilon_0}{2}\hat{\mathbf{E}}_2\hat{\mathbf{E}}_2^*\right) = C\left(\hat{E}_1^*\hat{E}_2\hat{E}_3 e^{i\Delta\Phi} + \hat{E}_1\hat{E}_2^*\hat{E}_3^* e^{-i\Delta\Phi}\right). \quad (6.141)$$

$$\frac{\omega_3}{\omega_p^2} d_3 \left(\frac{\varepsilon_0}{2}\hat{\mathbf{E}}_3\hat{\mathbf{E}}_3^*\right) = C\left(\hat{E}_1^*\hat{E}_2\hat{E}_3 e^{i\Delta\Phi} + \hat{E}_1\hat{E}_2^*\hat{E}_3^* e^{-i\Delta\Phi}\right). \quad (6.142)$$

The expressions on the LHS of (6.140)–(6.142) represent the convective derivatives of the energy densities ε_i divided by their frequencies ω_i, $i = 1,\ 2,\ 3$. Throughout the chapter we have used the electric fields $\mathbf{E}_i$ with frequencies $\omega_i = \text{const}$ for convenience. In the case of slight inhomogeneities in space and time the ω_i and $\mathbf{k}_i$ both are subject to slow (secular) changes that have to be taken into account in deriving the energy conservation relations. In Sect. 5.5.1 it has been shown that in the absence of source terms the Hamiltonian concept leads to the conservation of $\varepsilon_0\mathbf{E}_i\mathbf{E}_i^*/2\omega_i$ for transverse waves [see (5.164)] rather than of the energy densities $\varepsilon_0\mathbf{E}_i\mathbf{E}_i^*/2$. That means that when expressing the quiver velocity by the vector potential, $\mathbf{v}_{os} = -e\mathbf{A}/m$, and keeping the derivatives of the amplitudes one arrives at the correct expressions $d_i(\varepsilon_0\mathbf{E}_i\mathbf{E}_i^*/2\omega_i)$. For the longitudinal modes the same follows from the assertion that dispersion relations of identical structure lead to analogous conservation relations, proved in Sect. 5.5.3, (5.202)ff. Introducing $K = ek_3\omega_p/(8m_e^2\omega_1\omega_2\omega_3)^{1/2}$, $a_{1,2} = (\varepsilon_0/2\omega_{1,2})^{1/2}\mathbf{E}_{1,2}$, and $a_3 = (\varepsilon_0\omega_3/2\omega_p^2)^{1/2}\mathbf{E}_3$, (6.140)–(6.142) transform into

$$\begin{aligned} d_1|a_1|^2 &= -K(a_1^*a_2a_3 + \text{cc}), \\ d_1|a_2|^2 &= +K(a_1^*a_2a_3 + \text{cc}), \\ d_1|a_3|^2 &= +K(a_1^*a_2a_3 + \text{cc}). \end{aligned} \quad (6.143)$$

These relations show the perfect equivalence of coupling of mode i by the two modes j and k. It is straightforward to show it also for the acoustic mode. So far it has been evident from (6.116) and (6.122) that the longitudinal modes couple ponderomotively to transverse modes. The symmetry of (6.122) reveals the identical nature of coupling due to wave pressure for all modes. However, the coupling coefficients for $\mathbf{E}_\perp$ differ from those for $\mathbf{E}_\parallel$. When two modes drive a velocity field $\mathbf{v}$ with $div\,\mathbf{v} = 0$ the resulting unstable mode is an electromagnetic plane wave.

6.4.5 Quasi-particle Conservation and Manley-Rowe Relations

According to (5.164) the LHS of (6.140)–(6.142) are the total derivatives of the quasi-particle densities $f_{1,2,3}$. As the creation and annihilation rates on the RHS are identical the classical so-called Manley-Rowe relations hold [135, 137],

$$-\frac{\mathrm{d}f_1}{\mathrm{d}t} = \frac{\mathrm{d}f_2}{\mathrm{d}t} = \frac{\mathrm{d}f_3}{\mathrm{d}t}. \tag{6.144}$$

These relations yield the justification for characterizing the parametric instabilities in Sect. 6.4.1 in terms of *particle decay processes*. The extension of the operator d/dt to linear damping and diffusion is very easy in the framework of classical theory. It may be surprising at first glance that for the expectation values a quantum concept holds exactly with the same range of validity in a purely classical field. Only for counting the real number of quasi-particles additional information is needed, not available in classical physics. This new element is Planck's constant $\hbar$. The surprise vanishes after some simple reflections on what classical and quantum theory have in common in the homogeneous and isotropic environment. Free particles can be represented by plane waves and as such they are characterized by the parameters $\mathbf{k}$ (momentum) and ω (energy); furthermore, the complex exponentials $\mathrm{e}^{i\Phi}$ are eigenstates of the operators $\partial_\mathbf{k}$ and ∂_t of the wave operator $\nabla^2 - \partial^2_{v_g t}$. Finally, the steady state wave equation $\nabla^2\mathbf{E} + k_0^2\eta^2\mathbf{E} = 0$ is identical with the Schrödinger equation if the identifications $k_0^2 = 2mE/\hbar^2$ and $\eta^2 = 1 - V/E$ with potential and total energies V and E are made. When (6.144) is integrated over a finite volume V fixed in space the balance reads

$$\frac{\mathrm{d}}{\mathrm{d}t}\int_V (f_2 + f_3 - f_1)\mathrm{d}V + \int_{\Sigma(V)} (f_2 v_{g2} + f_3 v_{g3} - f_1 v_{g1})\mathrm{d}\Sigma = 0. \tag{6.145}$$

This is the conservation of quasi-particles in its standard form; the total loss through the surface $\Sigma(V)$ equals their production in the volume. As seen from (6.140)–(6.142) the production rates depend on the dephasing angle $\Delta\Phi$. If in the inhomogeneous plasma matching is perfect in one position in space it will, depending on the dispersion relations, deteriorate in its neighborhood and may eventually stop unstable growth.

6.4.6 *Light Scattering at Relativistic Intensities*

Stimulated Raman and Brillouin scattering from a pump wave in the relativistic regime, typically $I \geq 10^{18}$ W/cm^2, have been investigated analytically and semi-analytically by several authors, as for example [138–141]. By taking self-focusing, ponderomotive nonlinearities and pump depletion into account the latter author finds that reflection of the backscattered Brillouin wave at normalized incident intensity in circular polarization $a^2 = (e\hat{E}/mc\omega)^2 = 4.0$ does not exceed 15%. When considering the effect of Landau damping on Raman forward and backward scattering in a moderately relativistic regime one should be aware that the Landau damping coefficients for plasma parameters relevant to controlled thermonuclear fusion have to be revisited; they may be by several orders of magnitude lower than the nonrelativistic values [142]. A new and remarkable variant under several aspects of stimulated forward Raman scattering is reported in [143]: The Raman radiation is downshifted by half the plasma frequency; the growth is not exponential but of *explosive instability* type, i.e., it diverges after a finite time. Also, interestingly, a quasi-static electric field moving along with the laser pulse is found.

A first impression of unstable growth at high laser intensities may still be gained from the linear formulas of growth from Sect. 6.4.2, now taken for circular polarization for simplicity, by substituting

$$\left(\frac{\hat{v}_{os}}{c}\right)^2 \rightarrow \frac{\hat{a}^2}{1+\hat{a}^2}, \quad m \rightarrow \frac{m}{(1+\hat{a}^2)^{1/2}}; \quad \hat{a} = \frac{e\hat{E}}{mc\omega}. \tag{6.146}$$

The inverse *growth rates* for SRS, OTSI, and filamentation at $I = 10^{18}$ W/cm^2 are 2, 5, 26, 136 fs for Nd and 1.5, 4, 15, 117 fs for KrF. At $I = 10^{20}$ W/cm^2 they shorten to 13, 1.2, 11, 36 fs for Nd and 0.3, 0.6, 4, 18 fs for KrF, hence ranging from sub-fs to $\sim$ 100 fs. Growth is fastest for SRS and is expected to saturate quickly. To get insight into the dynamics of SRS growth and its saturation level, recurrence must be made to numerical simulations. In the following the results obtained by Hain [144] in a plasma of $Z = 1$ and $T_e = 1$ keV are summarized. First simulations from a 1D2V two-fluid model with a linearly polarized Nd laser beam in the intensity range of 10^{18} Wcm^{-2} are presented with parameters given in Fig. 6.31. Maxwell's equations together with the relativistic electron fluid equations of motion

$$\partial_t n_e + \nabla n_e v_e = 0, \quad \frac{\mathrm{d}\gamma\mathbf{v}_e}{\mathrm{d}t} = \frac{1}{m}\nabla p_e - e(\mathbf{E} + \mathbf{v}_e \times \mathbf{B}) \tag{6.147}$$

are solved. Typical results of Raman backward scattering from flat and perturbed ion background are presented in Fig. 6.31 First the instability builds up with a growth rate similar to (6.121) corrected by (6.146). At sub-relativistic intensity in (a) the electron mode shows the spiky structure well-known from Fig. 5.7 which then goes over into the broken structure of (b). As maximum growth occurs around $n_c/4$ the reflected wave manifests itself as a standing wave of wavelength of the extension of the active plasma volume, best seen in (a) from the low frequency modulation of E_y. In (b) strong trapping of incident light at $n_c/4$ up to 50% is observed with following *detrapping* into forward direction. The saturated reflection level does not exceed 10%. Backscattering from a stochastic ion background becomes significant ($\sim$10%) only from a Brillouin-matched spatial modulation of $2k_{\mathrm{Laser}}$. In the fluid dynamic simulation local wavebreaking must be bridged somehow artificially each time over the instant of breaking. Alternatively, SRS is simulated kinetically by solving the Vlasov equation. This is a 3D problem. However, if $T_e \neq 0$ is assumed only for the thermal v_x velocity component and the laser field is approximated by a plane wave $E_y = -\partial_t A_y$, owing to the *canonical momentum conservation* $mv_y = eA_y$ collinear scattering is reducible to a 1D1V problem for the electron distribution function $f(x, p_x, p_y, t) = f(x, p_x, t)\delta(p_y - eA_y)$. Including the Lorentz force component (ponderomotive term) the relativistic Vlasov equation (3.79) reads

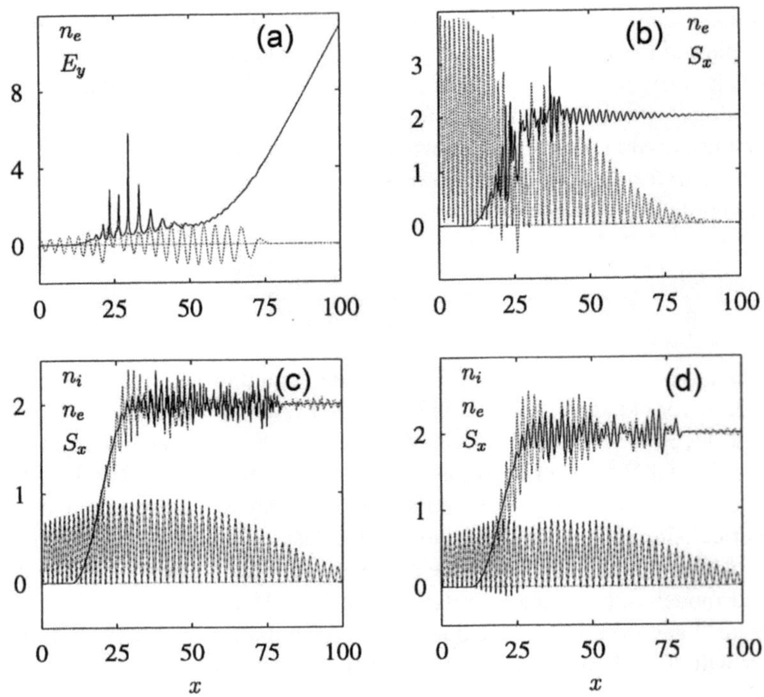

Fig. 6.31 1D2V fluid simulation of stimulated Raman scattering (SRS) at λ_{Nd}. Electron and ion densities n_e, n_i (solid and dotted, resp.) in units of 10^{21} cm^{-3}, critical density $n_c = 4 \times 10^{21}$ cm^{-3}, electric field E_y in units of 3×10^{10} Vcm^{-1}, Poynting flux S_x in units of 2.5×10^{18} Wcm^{-2} (lower graphs, dotted), space coordinate x in units λ_{Nd}/π. **a** Incident intensity from left $I = 3.5 \times 10^{17}$; electron plasma wave strongly nonlinear, regular (solid), electric field E_y (dotted); *laser beam trapping* around $n_c/4$, negligible reflection R. **b** $I = 1.4 \times 10^{18}$, electron plasma wave broken; strong trapping around $n_c/4$, relativistically increased by factor 1.5–1.6; $R = 10\%$. **c** SRS from stochastic ion density profile, static (solid), $\langle k \rangle > 2k_{\text{Laser}}$, $I = 3.5 \times 10^{17}$, n_e (dotted), $R =$ few %. **d** SRS from stochastic ion density profile (solid), $\langle k \rangle = 2k_{\text{Laser}}$, strong light trapping around rel. $n_c/4$ with subsequent detrapping; $R = 10\%$

$$\frac{\partial f}{\partial_t} + \frac{p_x}{\gamma m}\frac{\partial f}{\partial x} - e(E_x + \frac{e}{2\gamma m}\partial_x A_y^2)\frac{\partial f}{\partial p_y} = 0. \tag{6.148}$$

In circular polarization A_y, E_y are to be substituted by $A_\perp = (A_y^2 + A_z^2)^{1/2}$, $E_\perp = -\partial_t A_\perp$. The ion dynamics is calculated in the nonrelativistic fluid approach with an ion temperature $T_i = 100$ eV. For the electrons a "one-dimensional" temperature $T_e = 1$ keV is chosen. Simulations are made in linear and circular polarization with a Ti:Sa beam of constant 10^{18} Wcm^{-2} and rise time 30 fs that is impinging on ~10 wavelengths thick targets of constant densities $n_e = n_i = n_c/10$ and $n_c/4$. The results after different numbers of light periods are presented in Figs. 6.32, 6.33, and 6.34. We start with SRS from a beam in linear polarization propagating through $n_c/10$ density. In Fig. 6.32 $f(x, p_x)/f_{\max}$, normalized Poynting flux S_x and n_i (1:

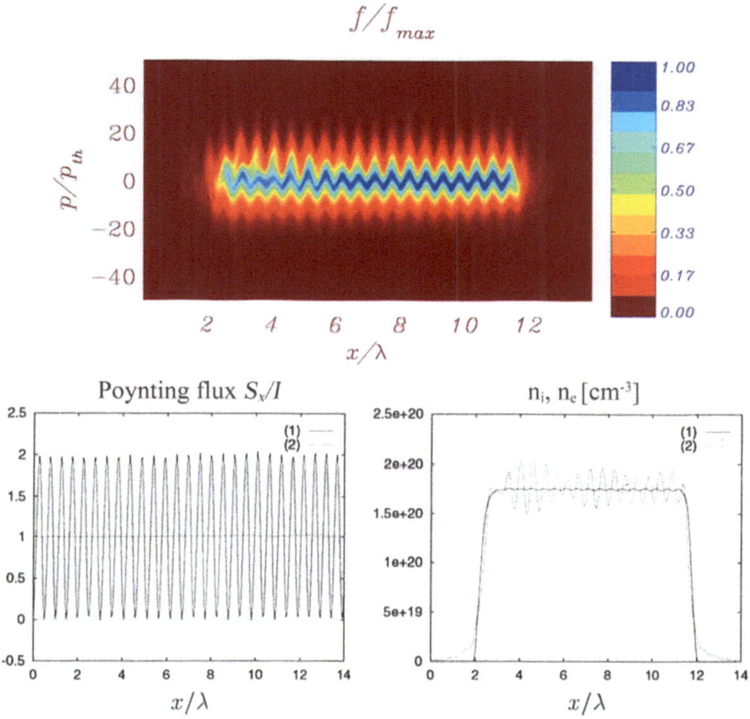

Fig. 6.32 SRS: Linearly polarized Ti:Sa laser pulse of $I = 10^{18}$ Wcm^{-2} impinges onto 10λ thick $n_c/10$ plasma from LHS. Electron distribution function $f(x, p_x)$, $p_{th} = (m_e k_B T_e)^{1/2}$, Poynting flux $S_x(t)$ (1: solid) and cycle-averaged S_x (2: dotted) and n_i (1: solid), n_e (2: dotted) after 40 laser cycles. Electron wave is perfectly regular; almost no back reflection. $\hat{E}_x \simeq 0.02\hat{E}_1$(pump)

solid) and n_e are shown after 40 laser cycles. The electron distribution function and n_e, n_i are perfectly regular and saturate at a very low level, $\hat{E}_x \approx 0.02\hat{E}_1$. The modulation in $\hat{E}_x$ originates from the Raman anti-Stokes component. A completely different picture of the electron dynamics is obtained from two counter-propagating pump beams (standing wave, e.g., total reflection from critical point) after 40 cycles (Fig. 6.33). When the two beams overlap the electrons start moving chaotically and back reflection saturates, under a low level again as inferred from the cycle-averaged Poynting flux S_x (dotted line). The ions behave perfectly regular. Stronger Raman back scattering and increased chaotic dynamics is to be expected from $n_c/4$ density. The latter is true as Fig. 6.34 shows, however, S_x/I (L) indicates that back reflection remains low in this case again. SRS of a circularly polarized traveling pump wave of $I = 10^{18}$ Wcm^{-2} from an $n_c/10$ plasma is presented in Fig. 6.35 after 40 cycles. The time for the instability to grow is 4 cycles only. At early times regular structures in n_i and n_e similar to those in Fig. 6.31 build up, however following onset of breaking destroys them (lower right picture) and stabilizes back reflection at 2–5%, see S_x/I in the inset of lower left picture of E_x. It has been a general observation

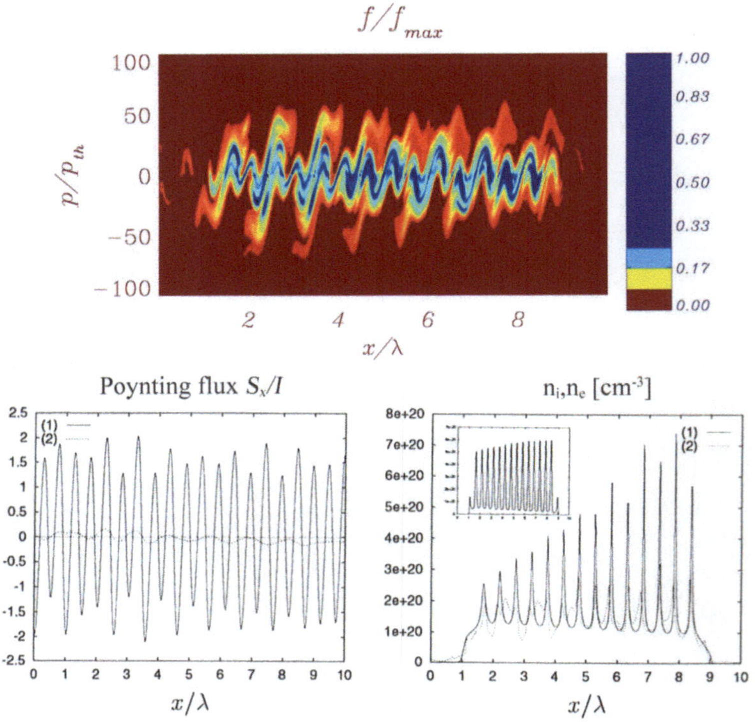

Fig. 6.33 SRS: Linearly polarized standing pump pulse. Parameters and symbols as in Fig. 6.31. Electron density becomes chaotic when the two pump beams start overlapping. The ponderomotively driven ion density profile remains regular; amplitudes are reduced due to electron chaos, compare inset on ion density profile from hydrodynamic simulation. Back reflection level remains very low despite $\hat{E}_x \approx 0.2\hat{E}_1$

through all simulations that circular polarization shows the tendency towards higher degree of chaotic dynamics and at the same time towards increased back reflection. With increasing background plasma density the difference tends to vanish. For the parameters of Fig. 6.34 there is almost no difference between linear and circular polarization. It can be concluded that at relativistic intensities SRS saturates, mainly owing to breaking, at an insignificant low level, in qualitative agreement with hydro simulations where the latter overestimate scattering for obvious reasons. In the context here it may be interesting to note that in a very early paper [145] it has been stated on pure physical grounds that at relativistic intensities parametric instabilities are not important. The assertion is in agreement with the simulations presented here.

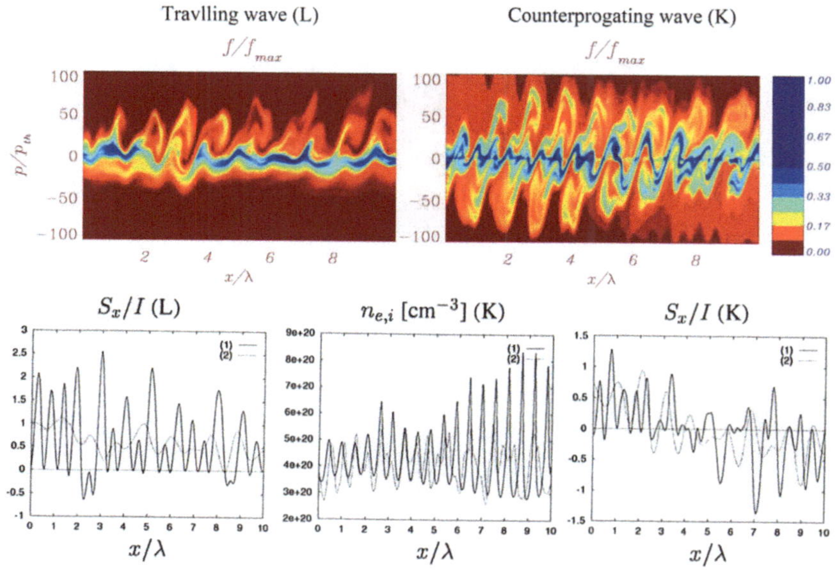

Fig. 6.34 SRS after 40 cycles from a linearly polarized traveling pump pulse (L), $I = 10^{18}$ Wcm^{-2}, and two counter-propagating pulses (K) in $n_c/4$ plasma, same intensity 10^{18} Wcm^{-2} both directions. Symbols and rest of parameters as in Fig. 6.32. In both cases, L and K, the electron plasma wave breaks and stabilizes back scattering; the ion density remains regular. $\hat{E}_x \approx 0.2\hat{E}_1$ (L) and $\hat{E}_x \approx 0.3\hat{E}_1$ (K). In the scattered low-frequency wave $\hat{E}_2$ trapping of $\hat{E}_1$ occurs up to 50% in intensity

6.4.7 *Self Focusing and Filamentation*

A uniform laser beam incident onto a homogeneous plasma may contract in its diameter, it undergoes *self focusing*; or it may split into a number of beamlets which themselves focus, it underlies *filamentation*, see Fig. 6.36. Both phenomena appear under a whole variety of forms. Controlled self focusing is a powerful mean to increase the intensity by a multiple of its original value. Under the interaction of the laser beam with the uniform plasma the refractive index η_0 becomes nonuniform and may be expressed in a power series of even powers of the laser amplitude $\hat{\mathbf{E}}$

$$\eta = \eta_0 + \eta_2(\hat{\mathbf{E}}^2) + \eta_4(\hat{\mathbf{E}}^4) + ...$$

Self focusing originates from the *ponderomotive force* of the inhomogeneous laser beam. The plasma is expelled away from the beam axis which leads to a positive term η_2. As a consequence the wave fronts propagate at lower phase velocity relative to the lower intensity borders, the rays, which are perpendicular to the phase fronts, are bent towards the axis according the ray equation (5.156). Beam contraction may continue up to dimensions of a wavelength where diffraction starts prevailing. This may set an end to the process of self focusing. However, it can also happen that the beam or a portion of it assumes an oscillating behaviour between focusing and expansion

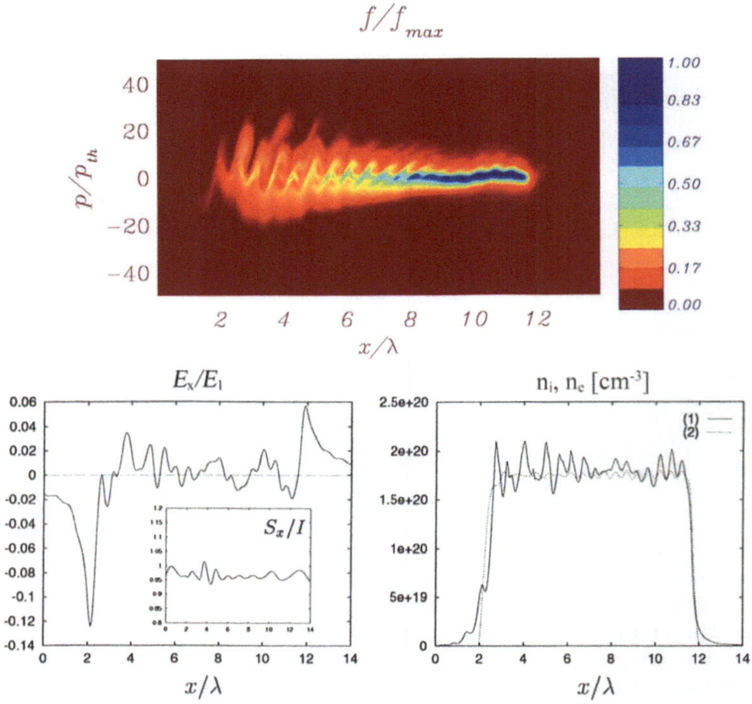

Fig. 6.35 SRS after 40 cycles from a linearly polarized traveling pump pulse in $n_c/10$ plasma density, $I = 10^{18}$ Wcm^{-2}. Symbols and parameters as in Fig. 6.32. Owing to breaking of n_e, see $f(x, p_x)$ and $E_x \approx 0.02\hat{E}_1$, reflection saturates at 2–5%, see Poynting flux S_x/I in the inset

[146, 147]. The interplay between focusing and diffraction determines shape and fine structure of self focused and filamentary patterns. In collisional interaction heating is proportional to the beam intensity and leads to *thermal self focusing* due to channel formation and to positive η_2. Whole beam focusing was reported for the first time in [148] and was attributed to local heating. In the relativistic intensity domain an increase of η is induced by the relativistic increase of the electron mass in the plasma frequency; it leads to *relativistic self focusing* [149].

Regular filamentary structures in ns beam plasma interactions were reported for the first time by O. Willi and P.T. Rumsby in 1981 [150] to the surprise of the scientific community. At the fundamental, 2nd, and 3rd harmonic Nd laser frequency periodic plasma density perturbations parallel to the incident laser beams had been diagnosed at intensities $I \simeq 10^{12} - 10^{15}$ Wcm^{-2}. In a representative number of shots the spatial periodicity was ranging from 10 to 18 μm. In burn-through experiments with flat solid targets the back action of the density modulations onto the intensity profile of the laser could be demonstrated. The plasma flow velocity under different geometries, oblique incidence on flat targets, radial irradiation of spherical targets, had no noticeable influence on the phenomenon. Filamentary structures with periodicity orthogonal to

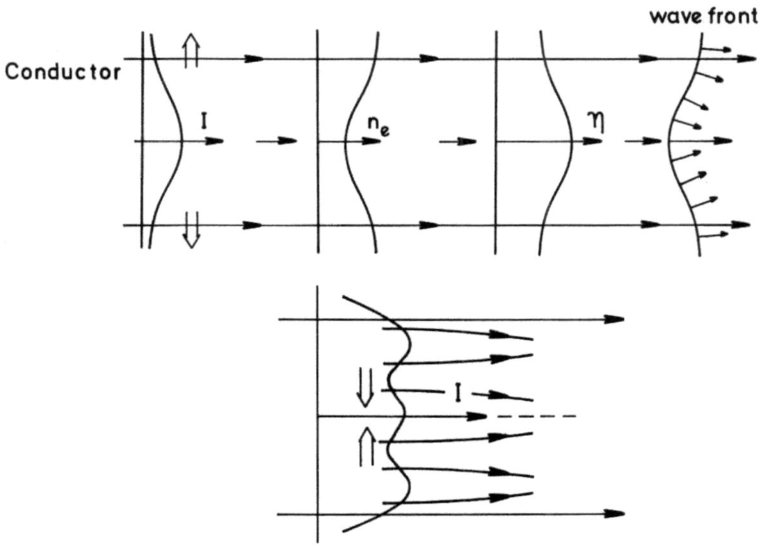

Fig. 6.36 Upper picture: Whole beam self focusing in electron density depression n_e causing refractive index increase η along the beam axis. Depression in n_e is either caused by the beam ponderomotive force or by local overheating of the plasma. Lower picture: Ponderomotive or thermal filamentation from intensity ripples in laser beam profile. Single beamlet underlies to the same laws as single beam self focusing

the laser beam direction are known to be driven also by ponderomotive, thermal, and relativistic effects. The *thermal instability* arises in the resistive plasma from local collisional overheating by an intensity spike in the laser beam and subsequent plasma expansion. The concomitant increase in the refractive index leads to bending of the rays in the single filament according to (5.156), lateral motion (*hosing instability* [152]), and to filament self-focusing in the underdense plasma.

Ponderomotive filamentation is very simple in principle. Assume a periodically modulated laser beam entering the homogeneous plasma in x-direction with a periodic field modulation $k_\perp$ perpendicular to it, say in z-direction. If the stationary motion of the isothermal plasma is also in z-direction equilibrium between thermal and ponderomotive pressure for our intensity modulation fixed previously would be established according to (6.69). Initially the ponderomotive pressure is dominating and as the beam progresses along x a density modulation ρ_1 along z is impressed, either out of phase by π from the intensity (subsonic) or in phase with it (supersonic). ρ_1 acts back on the laser beam through η_2,

$$\eta_2(\hat{\mathbf{E}}^2) = -\frac{1}{2\eta_0}\frac{\rho_1}{\rho_c}. \tag{6.149}$$

For $k >$ several $k_\perp$ the ray equation (5.156) applies. It describes bending and compression of the rays towards density minima of ρ which, in turn, reinforces plasma expulsion by increased ponderomotive action.

Alternatively, when the plasma density modulation $\rho_1[\exp(ik_\perp z) + \exp(-ik_\perp z)]$ is small compared with the unperturbed density ρ_0 a linearized ansatz for the laser wave $\mathbf{E} = \mathbf{E}_0 + \mathbf{E}_1 = \hat{\mathbf{E}}_0 \exp(ikx) + \hat{\mathbf{E}}_1 \exp(ikx + ik_\perp z)$ with $\mathbf{E}_0 \perp k\mathbf{e}_x$ and $\mathbf{E}_1 \perp (k\mathbf{e}_x + k_\perp \mathbf{e}_z)$ is appropriate. Its spatial distribution must fulfil the wave equation (5.146), i.e., $\nabla^2\mathbf{E} + k_0^2[1 - (\rho_0 + \rho_1)/\rho_c]\mathbf{E} = 0$. Ordering according to $\mathbf{k} = k\mathbf{e}_x$ and $\mathbf{k}' = k\mathbf{e}_x \pm k_\perp \mathbf{e}_z$, and by observing that $\mathbf{E}_0$ and $\mathbf{E}_1$ obey the dispersion relation $k^2 = k_0^2(1 - \rho_0/\rho_c)$ yields

$$- k_\perp^2 \hat{E}_1 \exp(k_\perp z) = k_0^2 \frac{\rho_1}{\rho_0} \hat{E}_0^* \exp(ik_\perp z), \quad k_0 = \frac{\omega}{c}. \tag{6.150}$$

$\mathbf{E}_1$ is nearly parallel to $\mathbf{E}_0$ (the sum of both components $k\mathbf{e}_x \pm k_\perp \mathbf{e}_z$ is exactly parallel). Maxima and minima of E_1 and ρ_1 are out of phase by π, in agreement with the prediction from the ray equation. There is an optimum wavelength to be expected for maximum growth of the filamentary instability because the ponderomotive force increases with increasing $k_\perp$; when $k_\perp$ becomes of the order of k or shorter, contrast saturation by diffraction between intensity maxima and minima sets in. Ponderomotive filamentation may be viewed as a special case of Stokes/anti-Stokes stimulated Brillouin forward scattering. The modulations seen in [150] can be interpreted as thermally as well as ponderomotively driven structures. Final clarification is difficult and still missing.

Whole beam ponderomotive self focusing. Self focusing occurs at high laser intensity. This requirement is reached with converging beams. Strong focusing under low *f number* may result in acceptably small foci, however, their extension into axial direction is limited and with it also self focusing action. An optimum in focus quality, self focusing power, and highest possible intensity is reached with paraxial optics, i.e., slender beams with f numbers much less than unity. Under such a constraint the phase in axial direction x may be described by the ansatz $kx + \Phi(x)$, with $\Phi(x = 0) = 0$ and $\Phi(x = \infty) = \pi/2$, as it results for a spherical wave with the focus at $x = 0$. In radial direction $k_\perp(x = -\infty) \ll k$ is set and approximate $\mathbf{E}$ field solutions are determined from the wave equation (5.146). In the specific case a *Gaussian beam* (of lowest order) may be chosen,

$$E(x, r) = E_0 \mathrm{e}^{-r^2/\sigma^2} \mathrm{e}^{ikx + i\Phi - i\omega t}, \quad k = k_0\eta. \tag{6.151}$$

The field amplitude $E(x, r)$ with a Gaussian intensity distribution across the beam resulting from (5.146) is

$$E(x, r) = E_0 \mathrm{e}^{-r^2/\sigma^2}, \quad \sigma^2 = \sigma_0^2(1 + x^2/x_0^2), \quad \theta = \frac{\sigma_0}{R} = \frac{2}{k\sigma_0}, \tag{6.152}$$

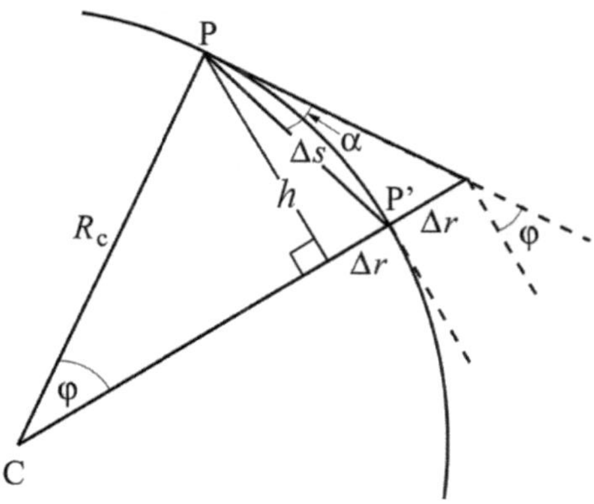

Fig. 6.37 Determination of the curvature radius R_c of a Gaussian ray passing through PP'. Infinitesimal deflection angle $\alpha = \Delta r/\Delta s = \varphi/2$; $\Delta s = R_c\varphi = 2R$, R Rayleigh length. To first order holds $2\Delta r R_c = h^2 = \Delta s^2 \Rightarrow R_c = \Delta s^2/2\Delta r$; $\Delta r = (\sqrt{5}-1)r \simeq r \Rightarrow R_c \simeq 2R^2/r$

$$R = x_0 = \frac{k\sigma_0^2}{2} = \frac{\pi\sigma_0^2}{\lambda}, \quad R_c = \frac{1}{\partial_{xx}\sigma}|_{x=0} = \frac{R^2}{r}.$$

The length of the focus is the *Rayleigh length* R, θ is the diffraction angle in the homogeneous medium, σ_0^2 the beam waist (see any textbook on optics or diffraction), R_c the curvature radius at $(x = 0, r)$. The curvature $\kappa = 1/R_c$ averaged over $2R$ results as

$$\overline{\kappa} = \frac{1}{\overline{R}_c} = \frac{1}{2R}\{\partial_x\sigma(x=0) + \partial_x\sigma(x=2R)\} = \frac{r}{\sqrt{5}R^2} \quad \Rightarrow \quad \overline{R}_c \simeq 2\frac{R^2}{r}.$$

Note, for $r < \sigma_0$ at $x = 0$, σ_0 has to be replaced by r.

It is instructive to observe that the same average curvature $\overline{R}_c$ is alternatively obtained by elementary means, see Fig. 6.37. In the ponderomotively driven inhomogeneity of η a ray through $(x = 0, r)$ undergoes a curvature $1/R_c$ in the direction opposite to the diffraction that is determined from the ray equation (5.156) with the help of the first *Frenet formula* and $\mathbf{k}_0\nabla\eta = 0$, $\partial_r = \partial_z$ as follows,

$$\frac{\mathrm{d}}{\mathrm{d}s}\frac{\eta\mathbf{k}_0}{k_0} = \eta\frac{\mathrm{d}}{\mathrm{d}s}\frac{\mathbf{k}_0}{k_0} = \mathbf{n}\frac{\partial\eta}{\partial r} = -\frac{\mathbf{n}}{2\eta n_c}\frac{\partial n_e}{\partial r}.$$

$$\text{Frenet}: \ \frac{\mathbf{n}}{R_c} = \frac{\mathrm{d}}{\mathrm{d}s}\frac{\mathbf{k}_0}{k_0} \quad \Rightarrow \quad R_c = 2\eta^2\frac{n_c}{\partial n_e/\partial r}; \quad \mathbf{n} \perp \mathbf{k}_0, \ \ \mathbf{n}^2 = 1. \qquad (6.153)$$

From the equilibrium condition $\nabla p_e + \pi = 0$ the spatial variation of n_e and R_c result as

$$\frac{\partial n_e}{\partial r} = \frac{2r}{\sigma_0^2}\frac{n_e}{p_c}\frac{I(r)}{c}, \quad R_c = \eta^2\frac{\sigma_0^2 p_c c}{rI(r)}\frac{n_c}{n_e}. \qquad (6.154)$$

p_c electron pressure at critical density n_c, I intensity. The threshold for self focusing of the limiting ray through $(x = 0, r)$ is reached when the two curvature radii $\overline{R}_c$ and

R_c from (6.154) become equal, i.e., when the local intensity $I(r)$ amounts to

$$I(r) = \frac{2cp_c}{(\sigma_0 k_0)^2}\frac{n_c}{n_e}. \tag{6.155}$$

The higher the intensity $I(\sigma) = \text{const}$ the higher the fraction of the total beam power $P = (\pi/2)\sigma_0^2 I_0$ that is focused; $I_0 = I(r=0)$. From P and (6.154) follows

$$I(r) = \frac{2P}{\pi\sigma_0^2}\mathrm{e}^{-2\frac{r^2}{\sigma_0^2}}, \quad P(r) = \pi\frac{c^3 p_c}{\omega_p^2}\mathrm{e}^{2\frac{r^2}{\sigma_0^2}}. \tag{6.156}$$

Half power ($r = 0.6\sigma_0$) and 86% power focusing ($r = \sigma_0$) require

$$r = 0.6\sigma_0 \ : \ P_{1/2} = 2\pi\frac{c^3 p_c}{\omega_p^2}; \quad r = \sigma_0 \ : \ P_{\sigma_0} = \pi\frac{c^3 p_c}{\omega_p^2}\mathrm{e}^2, \quad \mathrm{e} = 2.718.......$$

For illustration, $n_c = 10n_e = 10^{21}\,\mathrm{cm}^{-3}$, $T_e = 1\,\mathrm{keV}$, $\sigma_0 = 10\lambda$ (6.156) yields $I_0 = 4.3 \times 10^{13}\,\mathrm{Wcm}^{-2}$ on axis for $P_{1/2}$ and $I_0 = 1.6 \times 10^{14}\,\mathrm{Wcm}^{-2}$ for P_{σ_0}. The intensity measured by the experimentalist is $I = P/\pi\sigma_0^2 = I_0/2$.

Relativistic self focusing. Whole laser beam self-focusing is considered in a tenuous underdense plasma under the idealized condition of uniform electron density $n_e = n_0$. Provided it can be assumed that the change of the refractive index is essentially by the relativistic electron mass increase only the refractive index $\eta = (1 - n_0/\gamma n_c)^{1/2}$, $\gamma = (1 + \hat{a}^2/2)^{1/2}$, is higher in regions of high beam intensity, for example along the beam axis or at the center of a filament, and the phase velocity is lower. As a consequence, local self focusing and eventual beam or filament collapse occur. A threshold of self focusing is most easily derived for the Gaussian beam above. Under the condition of $n_0 \ll n_c$, $\hat{a}^2 \ll 1$, from (6.153) and $\hat{a}^2 = 2e^2 I(r)/\varepsilon_0 c^3 m_e^2 \omega^2$

$$\frac{\partial}{\partial r}\frac{n_e}{\gamma n_c} = \frac{n_e}{n_c}\frac{r}{\sigma_0^2}\hat{a}^2, \quad R_c = \eta^2 \frac{n_c}{n_e}\frac{\varepsilon_0 c^3 m_e^2 \omega^2}{e^2 r I(r)}\sigma_0^2.$$

Self-focusing will occur when equality of the mean curvature with the Rayleigh length, $\overline{R}_c = 2R^2/r$, is reached, i.e., at local intensity $I(r)$ and power $P(r)$,

$$I(r) = 2\frac{\varepsilon_0 m_e^2 c^5}{e^2\sigma_0^2}\frac{n_c}{n_e}\hat{a}^2, \ P(r) = \int_0^\infty I(r)\mathrm{e}^{2(r^2-r'^2)/\sigma_0^2} 2\pi r' dr' = \pi\frac{\varepsilon_0 m_e^2 c^5}{e^2}\frac{n_c}{n_e}\mathrm{e}^{2r^2/\sigma_0^2}. \tag{6.157}$$

The total beam power is $P = \pi\sigma_0^2 I(r=0)/2$, the intensity measured by the experimentalist is $I = P/\pi\sigma_0^2 = I(r=0)/2$. The power required for half beam self-focusing $P_{1/2}$ and for 86% beam focusing P_{σ_0} result from (6.157) as

$$P_{1/2} = 2\pi \frac{\varepsilon_0 m_e^2 c^5}{e^2} \frac{n_c}{n_e} = 4.2 \frac{n_c}{n_e}\,\mathrm{GW}; \quad P_{\sigma_0} = e^2 \pi \frac{\varepsilon_0 m_e^2 c^5}{e^2} \frac{n_c}{n_e} = 15.5 \frac{n_c}{n_e}\,\mathrm{GW}. \tag{6.158}$$

Frequently $P_{\sigma_0} = 17 n_c/n_e$ GW for *whole beam focusing* is given in the literature as a result of averaged quantities used for this estimate. This and (6.158) are gross criteria. Geometrical optics is based on the concept of light rays and is of limited applicability. Further limitation is imposed by the use of Gaussian beams. In particular, after self-focusing has occurred once, the beam intensity behind will no longer resume a distribution of this type. However, it is also to be said that Gaussian beams allow for combining diffraction effects with the use of simple ray optics. Pure ray tracing is useful to show the effect of intensity on beam propagation and to get simple criteria for amplification of deviations from ideality. A far more detailed picture is gained by a linearized full wave dynamic treatment of an initially Gaussian beam. Light and density nonuniformities along the axis lead to variations in the focusing strength $1/R_c$ that may give rise to unstable growth of hot spots, transverse and longitudinal coupling of modes, and beam self compression already in the weakly relativistic regime [153].

6.4.8 Modulational Instability

In the oscillation center approximation the ponderomotive force depends on the gradient of the electric field squared and no distinction between transverse or longitudinal polarization with respect to the wave propagation direction results. As a consequence, the steady state equation of motion (3.187) preserves its structure independently of the angle of the irrotational flow relative to the field propagation direction. Therefore unstable nonresonant pulse propagation is expected also for the case that the **K** vector of the pulse perturbation is parallel to the **k** vector of the wave. This instability is named modulational instability and occurs in electromagnetic as well as electron plasma and ion acoustic modes. Consider an electromagnetic pulse with a narrow frequency spectrum centered around (k, ω) propagating in a homogeneous isothermal plasma of density $n = n_0$ along x, $E(x,t) = \hat{E}(x,t)\exp(ikx - i\omega t)$. Evaluating the nonresonant ponderomotively induced plasma density n from (3.187) one is led to $n = n_0 \exp(-\varepsilon_0 \hat{E}^2/4\rho_c s^2)$ for $M \ll 1$ and to $n = n_0/[1 - \varepsilon_0 \hat{E}^2/(2\rho_c v_0^2)]^{1/2}$ for $M \gg 1$. For small differences $n_1 = n - n_0$ the two expressions become

$$M \gg 1: \quad n_1 = n_0 \frac{\varepsilon_0 \hat{E}^2}{4\rho_c v_0^2}; \qquad M \ll 1: \quad n_1 = -n_0 \frac{\varepsilon_0 \hat{E}^2}{4\rho_c s^2}. \tag{6.159}$$

With the current density $j = j_0 + j_1$ from n_0 and n_1 and by observing that (k, ω) obey the dispersion relation (5.154) for a smooth, slowly varying amplitude $\hat{E}$ the wave equation (5.15) reduces straightforwardly to

$$i\frac{\partial}{\partial t}\hat{E} + i\frac{c^2 k}{\omega}\frac{\partial}{\partial x}\hat{E} + \frac{c^2}{2\omega}\frac{\partial^2}{\partial x^2}\hat{E} - \frac{e^2 n_1}{2\varepsilon_0 m}\hat{E} = 0. \tag{6.160}$$

By observing that $kc^2/\omega = c\eta_0$ is the group velocity $v_g = \partial\omega/\partial k$ the second term is recognized as the convective part of the total time derivative of $\hat{E}$. Transforming to the reference system co-moving with the pulse and substituting n_1 from (6.159), (6.160) assumes the structure of the *nonlinear Schrödinger equation*

$$i\frac{\partial\Psi}{\partial t} + P\frac{\partial^2\Psi}{\partial x^2} + Q|\Psi|^2\Psi = 0 \qquad (6.161)$$

with the coefficient $P = c^2\omega'/2\gamma^2\omega_p^2$ in the comoving frame; γ Lorentz factor. It is a simple model equation used to study mild nonlinear phenomena in many branches of physics. In our context $\Psi = \hat{E}$, which without limitation can be assumed as real (for instance, by choosing $t = t_0$ properly). The correctness of (6.161) after transforming to the relativistic co-moving frame follows from substituting $\partial_t + v_g\partial_x = \partial_{t'}/\gamma$ according to (2.231) and by observing that $\partial_{xx} \approx \gamma^2\partial_{x'x'}$ in the context here. From (2.169) follows $\omega = \gamma\omega_p^2/\omega'$. The plasma frequency is a Lorentz scalar and $E' = \gamma E$ for $\mathbf{E} \perp \mathbf{k}$ and $E' = E$ for the electron plasma wave [see (2.165)]. If the modulation n_1 propagates with v_g it transforms like $n_1' = \gamma n_1$ and the coefficient $e^2n_1/2\varepsilon_0 m$ in (6.160) is also a Lorentz scalar. If, however, n_1 and m move at different speed owing to dispersion of n_1, or the quiver motion in the E-field becomes relativistic, the situation is more complex and must be treated properly. The complication arises from the nonlinear velocity addition theorem (2.188).

The general solution of (6.161) is accomplished by the inverse scattering method [154]. Here we consider the steady state pump pulse $E_0 = \hat{E}_0\exp(iQ\hat{E}_0^2t)$ that is a solution of (6.161), perturb it by $E_1 = \hat{E}_1\exp(+iQ\hat{E}_0^2t)$ and linearize (6.161) in $\hat{E} = \hat{E}_0 + \hat{E}_1$,

$$i\frac{\partial}{\partial t}\hat{E}_1 + P\frac{\partial^2}{\partial x^2}\hat{E}_1 + Q\hat{E}_0^2[\hat{E}_1 + \hat{E}_1^*] = 0. \qquad (6.162)$$

Splitting $\hat{E}_1$ into real and imaginary part, $\hat{E}_1 = U + iV$, yields the system of linear equations

$$\frac{U}{t} + P\frac{{}^2V}{x^2} = 0, \quad -\frac{V}{t} + P\frac{{}^2U}{x^2} + 2Q\hat{E}_0^2U = 0, \qquad (6.163)$$

which are solved by the ansatz $\hat{E}_1 = (U_0 + iV_0)\exp[-i(Kx - \Omega t)]$, $U_0, V_0 =$const, provided the determinant is zero, $\Omega^2 + PK^2(2Q\hat{E}_0^2 - PK^2) = 0$. It shows that modulational growth occurs for $\Omega = i[PK^2(2Q\hat{E}_0^2 - PK^2)]^{1/2}$ if $PQ > 0$ and $\hat{E}_0^2 > PK^2/2Q$. The unstable K-interval, growth rate Γ, maximum growth $\Gamma_{\max}$, and related wave number K_{m} are

$$0 \le K \le \left(\frac{2Q\hat{E}_0^2}{P}\right)^{1/2}, \quad \Gamma = [PK^2(2Q\hat{E}_0^2 - PK^2)]^{1/2},$$

$$\Gamma_{\max} = |Q|\hat{E}_0^2, \quad K_{\mathrm{m}} = \left(\frac{Q}{P}\right)^{1/2} \hat{E}_0. \tag{6.164}$$

$PQ < 0$ yields stability. The coefficient $P = c^2\omega'/2\gamma^2\omega_p^2$ is positive and hence n_1 must be negative or $M < 1$ in (6.159).

The electron plasma wave in one dimension exhibits the same structure as the electromagnetic wave, see (5.10), (5.148), (5.145), and the same dispersion with the wave number substituted by $k = k_e = k_0\eta_0 c/s_e$, see (3.95). Hence, in lowest order (6.160) results again for $\mathbf{E}\|\mathbf{k}$; (6.164) follows for the modulationally unstable domain of the Langmuir wave. A relativistic treatment of the modulational instability for longitudinal and transverse polarization with account for nonlinear Landau damping is presented in [155]. There is a whole variety of theoretical papers on the ponderomotively driven modulational instability under various conditions [156–162]. Magnetic field effects have shown a significant effect on the evolution of the instability [163]. By applying the magnetic field in direction of the wave vector the modulational growth Γ is attenuated up to the degree of stabilization [164]. The first direct observation of a modulationally unstable electron plasma wave was accomplished in an electron beam-plasma experiment [165] (authors' claim); the smooth Langmuir pulse evolved into two humps. The initially weak modulation may evolve into a highly nonlinear structure of a finite number of solitons. The existence and stability of such structures in three dimensions is investigated in [166]. The final stage of a pulse after breaking up into solitary humps may be self-contraction with a rate following theoretical predictions and then a collapse after the onset of other nonlinearities, e.g., electron trapping and acceleration. The collapse of the cavity density n proceeds under the field trapped in it also after its decoupling from the outer driver [167].

The model equation (6.161) predicts instability only for $M < 1$. Modulational instability of $\hat{E}$ may occur also at $M > 1$. To see this one simply has to remember that the plasma density perturbation n_1 is in phase with the humps of the electric wave at $M > 1$. For $K \ll k_0$ (5.166) applies. In the maxima of $\hat{E}$ the group velocity is lower and hence, under steady state energy flux, wave amplitude, perturbation n_1, and wave pressure are altogether in phase there; n_1 starts growing.

The ion acoustic wave is also modulationally unstable. The derivation of a nonlinear Schrödinger equation of type (6.161) proceeds in a similar way. When T_e is much higher than T_i a thermoelectric field builds up in the acoustic disturbance, see expression for $\mathbf{E}$ in equation following (5.111). It gives rise to a ponderomotive force acting on the plasma background density and leading to a stabilization or destabilization of it. Detailed analysis shows that at finite angle θ between the acoustic wave and the background modulation its amplitude may become unstable. For θ exceeding $\pi/3$ instability extends to the whole K-domain [168], whereas for $\theta = 0$ stability is predicted [169].

In the plasma nearly all kinds of waves are subject to the modulational instability. It also occurs in many other branches of physics, e.g., nonlinear crystals and fibers.

It is a very versatile phenomenon. An extreme example of electromagnetic pulse modulation by intense laser field ionization may be found in [170].

6.5 Summary

Origin of instabilities. The plasma in general, and the laser generated plasma in particular, is extremely susceptible to a variety of instabilities under the action of external forces. In the laser plasma the external force is the ponderomotive force for which also the terms light or radiation pressure or, more generally, wave pressure are in use. In its nature it is the *multiphoton Compton effect*. In contrast to the high frequency force from the electromagnetic or from the longitudinal electric field this is a secular, i.e., a zero frequency force. For this secular character, although weaker than the Lorentz force, it is the suitable driver of instabilities. The Rayleigh–Taylor (RTI) and the *modulational instability* are the only nonresonant unstable phenomena under the laser action. The RTI arises from the acceleration or gravitation of a heavier fluid superposed on a lighter fluid. By the growth of the slightly perturbed flat interface the overall acceleration of the fluid slab is lowered, the residual work is converted into internal flow energy of the two-component object. As long as the transverse deformation is small compared to its horizontal extension ("wavelength" λ of the perturbation) its growth is exponential. The effective mass per unit area to be accelerated is $\mu = (\rho_1 + \rho_2)\lambda/(2\pi)$, the effective accelerating force per unit area of the interface displacement h is $g(\rho_1 - \rho_2)h$; it equals $\mu\ddot{h}$. The resulting growth rate γ_{RTI} thus is (6.4) from the glossary. The RTI plays an eminent role in dynamic compression of matter by radiation pressure.

In presence of a shear flow two inviscid liquids are Kelvin–Helmholtz (KHI) unstable as soon the interface deviates from a mathematical plane. In the case of an initial harmonic perturbation the single fluid element behaves like a harmonic oscillator driven at resonance by the pressure difference according to Bernoulli's law. The KHI works also in the simplest case of identical fluid densities and no force impressed from outside, see (6.21). The growth rate is modified by the RTI superposed. Surface tension and magnetic fields act as stabilizing effects. In the nonlinear regime RTI and KHI cannot be separated from each other.

Resonance absorption. In the collisionless *resonance absorption* the electromagnetic wave is converted into an electron plasma wave of the same frequency in the ratio one to one, i.e., one photon into one plasmon. Efficient conversion requires resonance: $\mathbf{k}_{em} = \mathbf{k}_{es}, \quad \omega_{em} = \omega_{es}$. By the Bohm–Gross dispersion in a stratified plasma resonance can only occur in the vicinity of the critical plane. In a wide range of intensities the conversion is a linear process owing to strong saturation by energy convection out of the resonance zone. The linear conversion rate formula for A from (6.45) remains valid far beyond the limit (6.56), see Figs. 6.14, 6.15. The resonance width increases with the $1/3$ power of the inhomogeneity length L, see (6.43). Maximum conversion in the linear flat density domain amounts to 50%. Total conversion is also possible at relativistic temperature with $c_{se} \simeq c$ or in a steep transition layer from a low density plasma shelf to a strongly overdense plasma when the two electric fields are parallel to each other.

If an electrostatic wave is strongly driven it *breaks*. It loses periodicity in at least one of its macroscopic variables, density for instance. A superficial criterion for breaking generally accepted is $v_{\rm os} > v_\varphi$. A more rigorous test is obtained by the Coffey criterion (5.96) from the fluid model; it has been derived for warm homogeneous plasma out of resonance. Vlasov simulations on wave breaking at resonance show that the phenomenon is strongly influenced by electron trapping and the coalescence of entire electron bunches. Breaking occurs at the edge of the resonance zone at intensities corresponding to driver strengths E_d indicated in the glossary for cold streaming plasma according to (6.67), generally considerably higher than the Coffey criterion predicts. The relation of hydrodynamic to kinetic breaking is an unresolved question. Frequently wave breaking is invoked for collisionless absorption and efficient generation of fast electrons. The opposite is closer to truth: Collisionless absorption leads to efficient electron trapping and acceleration; wave breaking and breaking of flow appear as implications of them.

Parametric instabilities. All resonant three wave decay processes are ponderomotively driven by radiation and wave pressure. In the reference system in which the plasma mode under inspection is static (*comoving system*) the single fluid element is a harmonic oscillator which is driven at resonance of the Doppler shifted frequency of the electromagnetic *driver*, except the Langmuir decay instability (LDI) in which the driver is provided by an electron plasma wave. From the physical point of view the scattered wave of frequency $\omega'_r = \omega'$ is the reflection of the driver from the static electron density modulation with the wave vector into direction of constructive interference. The ponderomotive force density $\boldsymbol{\pi}$ is in phase with the electron density modulation, see Fig. 6.27. Maximum reflection in the comoving system S' translates into the fundamental *matching*, i.e., *resonance conditions*, in the lab system S where the medium is at rest, see (6.85),

$$S' : \ \mathbf{k}' = \mathbf{k}'_r + \mathbf{k}'_M, \ \omega' = \omega'_r + \omega'_M, \ \omega'_M = 0 \ \Rightarrow \ S : \ \mathbf{k}_1 = \mathbf{k}_2 + \mathbf{k}_3, \ \omega_1 = \omega_2 + \omega_3.$$

With the help of the dispersion relations $\omega(\mathbf{k})$ these conditions on the wave vectors and frequencies allow the decision which combinations of the three plasma modes will grow unstable. The most relevant combinations are listed in the paragraph "Classification". The interaction of the driver $(\mathbf{k}_1, \omega_1)$ with the plasma mode $(\mathbf{k}_3, \omega_3)$ generates the resonantly scattered *Stokes* mode $\omega_1 - \omega_3$ and the nonresonant *anti-Stokes* mode $\omega_1 + \omega_3$. Whenever the dephasing from strong drive becomes larger than the frequency of the slowest mode the anti-Stokes component of the faster mode must be additionally taken into account, see the *strong coupling* regime of *Brillouin scattering* (SBS) and the TPDI and the OTSI. The latter are resonant decays of an electromagnetic wave into a Langmuir wave and an acoustic mode. In long scale length plasmas linear growth of SBS predicts high reflectivity, in special cases up to 100%. In similar gentle gradient plasmas also the opposite behaviour of reflection below 5–10% has been verified experimentally. The theoretical explanation is quite complex. Simulations have shown that the backscattering level may be high for short times, but owing to saturation by ponderomotive self focusing, rapid local time variations of the background plasma and formation of local inhomogeneities may be responsible for the low saturation on the time average. In the *stimulated Raman scattering* (SRS) the incident transverse wave and a strongly red shifted electromagnetic daughter wave drive a fast growing Langmuir wave unstable. Generally it is strongest at $n_c/4$. There it can drive a current of $(3/2)\omega$ frequency in combination with the incident wave that in turn give rise to electromagnetic radiation of the same frequency. Under several aspects the two-plasmon decay (TPD) competes with SRS. From the Bohm–Gross dispersion relation one deduces that the decay into two nearly opposite plasma waves happens with $\omega_{2,3} \simeq \omega_p/2$ close to $n_c/4$, like SRS. Maximum growth occurs under 45° and 135°. SBS, SRS, and TPD are perhaps the most significant instabilities in the laser-plasma interaction.

Instability analysis. The simplest procedure is by normal mode analysis. The linearized equations of motion are Fourier transformed to a linear system of 3 or (if anti-Stokes component included) 4 homogeneous algebraic equations. By elimination of all dynamical variables except one the result is a sixth (eighth) order polynomial equated to zero. In the approximation of constant pump amplitude (no *pump depletion*) at the onset of the instability the degree reduces by 2. Instability manifests itself by the occurrence of a complex frequency $\Omega = \omega + i\gamma, \gamma > 0$ in the ansatz $\exp -i\Omega t$ for the unknown dynamical variable. The procedure, adopted in this chapter, does neither tell about the most unstable mode nor on the type of instability, either *convective* or *absolute*. For the latter distinction the reader is referred to [101]. The *threshold* E_{thres} of the onset of a parametric instability of undamped modes in the homogeneous medium is zero. Linear damping and inhomogeneity make the threshold finite. A corresponding local formula for E_{thres} is given by the *Piliya-Rosenbluth criterion* (6.135).

A more realistic picture of instability is obtained from the study of the temporal and spatial evolution of the moduli of the complex amplitudes squared divided by the frequencies of the three waves involved, see (6.143). They are balance equations of the numbers of photons, plasmons, and phonons. Basically they look very similar to the treatment of the parametric instabilities in the associated quantized fields. The only new elements quantization adds is Planck's constant $\hbar$ and the noise from the zero point energies. The reduction to one degree differential equations (6.143) is accomplished by expanding the second order wave equations to first order in the amplitudes after equating to zero the dispersion relations. The particle numbers $|a_i|^2$, $i = 1, 2, 3$ are adiabatic invariants. Their conservation relations (6.144) in the parametric process are known as the *Manley-Rowe relations*. In the genuine nonlinear regime field quantization and adiabatic invariants are no longer helpful. The remedy is to return to the classical Maxwell equations combined with relativistic dynamics. The criterion for their applicability as mean values is the number of particles per wavelength cube λ^3 much greater than unity.

Relativistic intensities. In this domain of $\hat{a} \simeq 1$ or larger the requirement of high photon and particle numbers is generally well fulfilled. From the extrapolation of the linear analysis to high fields a sensible reduction of the growth rates is to be expected already from the relativistic electron quiver energy change $m_e c^2[\sqrt(1+\hat{a}^2/(2))-1] \rightarrow mc^2\hat{a} \sim \hat{E}$ instead of $\hat{E}^2$. Nonlinear mode-mode interaction, wave pressure, particle trapping, wave breaking, and self focusing act together to destroying coherence. Such tendencies are confirmed already by hydrodynamic simulations. Vlasov simulations of SRS show that (i) damping of instabilities is even much stronger and (ii) that linear polarization of the laser beam leads to lower values of scattering than circular polarization. It can be concluded that parametric instabilities in the relativistic regime play at most a secondary role.

Electron trapping and acceleration. *Nonthermal electrons*, also known under the name *hot electrons*, are a characteristics of laser plasmas from threshold intensities $I \simeq 10^{10}$ up to the highest values of $I \simeq 10^{22}$ Wcm^{-2}. In the intensity range where parametric instabilities and resonance absorption play a role ($I_{\mathrm{Nd,Ti:Sa}} \gtrsim 10^{13}$ Wcm^{-2}) electron plasma waves are excited and electrons are trapped in the potential minima. *Adiabatic trapping* takes place in waves with slowly varying field amplitude at positions where the electron reaches the local phase velocity by ponderomotive acceleration or deceleration, or *nonadiabatic trapping* after acceleration in a discontinuous plasma wave. Once trapped the electron changes its velocity according to the adiabatic invariant $\oint p_e dq_e = \mathrm{const}_a$. The total process of adiabatic trapping, however, is governed by the invariant $\oint (p_e dq_e - E_e dt) = \mathrm{const}_a$. The maximum energy gained by trapping is given by (6.65); it is proportional to the phase velocity at which trapping occurs. It explains why SRS and TPD generate "hotter" electrons than resonance absorption. The primary energy spectrum of the electrons is not Maxwellian, see *Schneider's spoons*. In the experiments and simulations however, as a rule it turns out Maxwellian owing to *post acceleration* by noise in the wave amplitudes and collective electron-electron interaction, for instance due to formation of bunches. In the relativistic regime fast electron generation by *anharmonic resonance* prevails on adiabatic acceleration.

6.6 Problems

▹ (a) Show that $|\mathrm{curl}\,\mathbf{u}| \lesssim kh_0(t)$, except along the interface $h_0(t)$ *Hint:* Apply Stokes' theorem to vanishing loop area. (b) What is the solution of $h_0(t)$ for negative Atwood number? (c) Deduce F in (6.4) from the potential energy change with $h_0(t)$.
▹ Determine γ_{RTI} of a thin shell of thickness $d < k^{-1}$.
▹ Towards what shape will $h_0(t)$ tend when it runs out of the linear validity domain of (6.4)? *Hint:* Try to understand the origin of *bubbles* and *spikes*. Write the equation of motion at a general position x. What can you learn from it?
▹ Execute the passage from (6.11) to (6.12). Sketch ε as a function of $\omega(k)$ for $u_0 =$ const.
▹ What changes in (6.11) and (6.12) if the stability analysis is performed with one

plasma stream at rest?
▹ Repeat the analysis of ε (6.12) with two beams $n_{01}u_{01}$ and $n_{02}u_{02}$. *Hint:* Transform to the center of zero current density and analyze.
▹ An electron beam is injected into a plasma at rest. Analyse the stability behaviour.
▹ (a) What is the change in (6.12) if the ion motion is also taken into account in the electron plasma wave? *Answer:* $m_e \Rightarrow \mu = m_e m_i(m_e + m_i)$. (b) How is (6.12) changed for the electron-positron plasma?
▹ Consider a wave with $k < k_0$ and the corresponding potential energy $e\hat{\Phi} = (\omega_{p1}^2/u_0^2k^2)m_e u_0 \hat{u}_1$. In climbing up the potential hill an electron decelerates only by u_1 from (6.10), not enough to overcome $e\Phi$. Explain the apparent contradiction to the result on two stream growth.
▹ Derive the dispersion equation for a warm plasma of electron temperature $T_e > 0$.
▹ Obtain Bernoulli's law and unstable growth similar to (6.21) with the aid of the centrifugal force for $\rho_1 = \rho_2$.
▹ Why can $\sin kx$ from (6.1) be substituted in (6.13) by $\exp -ikx$?
▹ Justify the mathematical structures of (6.20) and (6.22) by physical arguments, e.g., why stands the RTI term under the square root?
▹ What is the connection of (6.18) with Bernoulli's law (factor $1/2$)?
▹ Show with the aid of Bernoulli's law that the pressures p across the perturbed boundary of the KHI are the same. *Hint:* Transform to the system S' of the center of mass at rest.
▹ Derive (6.42). Give a reason why it is the same formula as (5.166). *Hint:* Think of energy conservation.
▹ Verify criterion (6.56).
▹ Derive the Doppler formulas (6.83) in the homogeneous medium by making use of the known Doppler shift of the plane wave when it enters the vacuum from the medium. *Hint:* Choose a smooth gradient of the medium parallel to the **k** vector in order to avoid refraction and reflection.
▹ Verify the two basic Doppler formulas (6.75) and (6.76).
▹ Calculate the aberration angle of a star due to the earth's revolution around the sun.
▹ If the angle of incidence of a laser beam in the frame comoving with a plasma mode $(\mathbf{k}_M, \omega_M)$ is α, determine the reflection angle β in the lab frame.
▹ An electromagnetic wave propagates collinearly (a) to a moving observer of velocity $\mathbf{v}$, (b) to the emitter moving at $-\mathbf{v}$ and the observer at rest. Calculate the relativistic frequency difference between case (a) and case (b).
▹ A monochromatic electromagnetic wave enters a stationary layered plasma flow $u(x)$ under the angle α and turns back into the vacuum under the angle β. Calculate the red, respectively blue shift and the angle β.
▹ Verify (6.69). *Hint:* Use the steady state fluid model and the linearized standard ponderomotive force.
▹ In the steady state medium in which a local frequency ω can be defined the frequency is not affected by the motion $\mathbf{u}(\mathbf{x})$ of the medium. Give a proof alternative to the three arguments in the text. (*Solution:* The transverse current density $\mathbf{j}$ is insensitive against constant nonrelativistic flow in (5.38) in conjunction with (5.42).)
▹ Derive (6.90) from the velocity addition theorem.

▹ Calculate the ratio $\hat{n}_2/\hat{n}_3$ from (6.97) and (6.96).
▹ Derive (6.97) from the two fluid model.
▹ Show on the basis of the resonant three wave interaction that the Langmuir mode n_2 and the acoustic mode n_3 of the PDI grow at the same rate γ, although their amplitudes $\hat{n}_2$, $\hat{n}_3$ are different and thus violate quasineutrality. What is a sufficient condition for the validity of quasineutrality?
▹ Complex normal mode analysis: Find the general solution of the real equation $\ddot{x} + \gamma\dot{x} + \omega_0^2 x = D\cos\omega t$, including $\omega = \omega_0$, by using real quantities only. What does the shift of the resonance point amount to?

6.7 Self-assessment

- Chose $\rho_1 \simeq \rho_2 = \rho,\ \rho_1 - \rho_2 = \epsilon > 0 \Rightarrow A = \epsilon/2\rho$. Why is $\gamma_{\mathrm{RTI}} \sim \rho^{-1/2}$ while mass in Newton's law is proportional to ρ?
- (a) Derive the Rayleigh–Taylor growth rates γ_{RTI} and γ_d from dimensional analysis. (b) Show $\gamma_d \rightarrow \gamma_{\mathrm{RTI}}$ for $\rho_1 \rightarrow \infty$. *Hint:* For small displacements the fluid flow is self-similar. Take account of the accelerated masses.
- Explain to the undergraduate student without use of mathematics why shear flow is Kelvin–Helmholtz unstable. *Hint:* Think of the centrifugal force.
- Get further feeling for the Kelvin–Helmholtz instability, with particular attention to the gravity-free case of $\rho_1 = \rho_2$, see (6.21): (a) Show that the single fluid element is a harmonic oscillator driven at resonance for arbitrary relative velocity. *Hint:* Pay attention to the phase between displacement and driving force. (b) What is the mechanism behind the restoring force? (c) Why is the RTI term to be added under the square root? *Hint to* (b) *and* (c)*:* Consider the oscillating modes with $\rho_1 = \rho_2$ (b) and $\rho_1 < \rho_2$.
- What is a *Negative energy wave* and what is an indicator for? Find examples.
- Linear resonance absorption. Evaluate the energy transport by the electron plasma wave from (6.29) at its maximum amplitude, chose an absorption coefficient A from Fig. 6.6, determine the magnetic field $B(0)$ at resonance and compare with (6.45).
- In linear optics the tangential electric field component E_t and the normal component $D_\perp = \varepsilon\varepsilon_0 E_\perp$ are continuous at the vacuum-matter interface. In resonance absorption also $E_\perp$ is continuous, think of fs laser pulse interaction in steepened density profiles. How do you resolve the apparent paradox?
- Given an electron plasma wave, what is the cutoff energy of a particle after detrapping? *Hint:* Get inspired by (6.65).
- Try to reproduce formula (6.65) from the corresponding adiabatic invariant.
- Discuss physical scenarios leading to wave breaking. Why does an acoustic wave evolve rather into a shock instead of breaking? Give a meaningful definition of wave breaking.
- Under which physical conditions does a sinusoidal wave evolve into a spiky structure?

- Why is the Lorentz factor γ in the plasma different from the expression in the vacuum? Is Doppler effect possible for $v > c$? *Answer:* yes.
- Write down the Doppler shift of $\mathbf{k}$ and ω in the homogeneous isotropic linear medium.
- What is the effect of constant plasma flow on the frequency Doppler shift of an electromagnetic wave? *Answer:* none. Does $\mathbf{k}$ undergo a change?
- Why is ω_p in (6.90) treated as a constant?
- If the motion of a point charge is studied in the plasma which is the correct Lorentz factor, γ or γ_φ?
- Often it is advantageous (e.g., normal mode analysis, instabilities) to look for complex instead of real solutions. When is the real or imaginary part of the complex result the true physical solution? *Partial answer:* The procedure relies on (i) the linearity of the equation and (ii) on all coefficients being real; otherwise real and imaginary part mix. However, why then complex coefficients, e.g., refractive index η, complex frequencies ω, and wave vectors $\mathbf{k}$, are also admitted in linear equations?
- Show that the Brillouin and Raman effect in the ideal plasma are merely classical light reflection from moving periodic structures.
- (a) Explain in detail: Stimulated three wave parametric effects are resonantly driven by radiation pressure(more general: wave pressure). (b) Convince yourself that the unstable daughter waves of the individual decay process exhibit identical growth rates although their amplitudes $\hat{n}_2, \hat{n}_3$ may be very different from each other
- Justify the use of the standard ponderomotive formula throughout the chapter on three wave interaction.
- In the perturbative derivation of the ponderomotive force $\mathbf{f}_p$ no distinction is made between longitudinal and transverse fields. Why then is $\mathbf{f}_p$ modified in presence of an electron plasma wave from C/ω^2 to $C/(\omega^2 - \omega_{\text{es}}^2)$? See PDI and OTSI.
- Use a simple argument to show that maximum growth of the TPD occurs at (approximately) $\pi/4$ angle between the pump vector $\mathbf{k}_1$ and $\mathbf{k}_2$ of one of the plasmons. Figure out by which effect the electron plasma temperature interferes. *Hint:* Have a look at [122] or [123].
- In normal mode analysis of plasma instabilities the growth of a real field quantity may be obtained from a complex ansatz $\mathbf{x} = \hat{\mathbf{x}} \exp -\Omega t$ with the amplitude $\hat{\mathbf{x}}$ and the frequency Ω both complex. How do the complex quantities relate to the real physical expressions? *Answer:* See (6.101).
- Justify the use of the simpler Fourier instead of the Laplace transform in the normal mode analysis.
- Sketch the difference between a *convective* and an *absolute* instability. Give examples. How could a formal criterion look like?
- The laser beam generated nonuniformity of the plasma refractive index η may be expanded into a series of even powers of the electric field. Why even powers only? *Answer:* Slow, i.e., zero frequency structures depend on intensity, not on the fast varying direction of $\mathbf{E}$.

6.8 Glossary

Rayleigh–Taylor instability (RTI) $\gamma_{\text{RTI}} = (gAk)^{1/2}$; $k = 2\pi/\lambda$,
Atwood number $A = (\rho_1 - \rho_2)/(\rho_1 + \rho_2)$.

(6.4)

Twostream instability occurs if $\varepsilon(k, \omega) = 1 - \frac{\omega_p^2}{(\omega - ku_0)^2} - \frac{\omega_p^2}{(\omega + ku_0)^2} = 0$ with $\varepsilon(k, \omega = 0) = 1 - 2\frac{\omega_p^2}{k^2 u_0^2} < 0$.

(6.12)

Kelvin–Helmholtz instability (KHI)
$\omega = iku_0 \frac{\rho_1}{\rho_1 + \rho_2} \pm \left\{ k^2 u_0^2 \frac{\rho_1 \rho_2}{(\rho_1 + \rho_2)^2} + gkA - \frac{k^{3/2} T}{\rho_1 + \rho_2} \right\}^{1/2}$, T surface tension.

(6.20)

Resonance absorption: Maximum electric field

$$\left| \frac{\hat{E}_{x,\max}}{E_0} \right|^2 = 1.1(1 - \sin^2 \alpha_0)^{1/2} (k_0 L)^{1/3} A \beta^{-4/3}. \tag{6.48}$$

Maximum velocity of trapped electron

$$v_{\max} = v_\varphi \left[1 + 2\frac{\hat{E}_{\max}}{v_\varphi} + \frac{4}{\pi} \left(\frac{2\hat{E}_{\max}}{v_\varphi} \right)^{1/2} \right]^{1/2}. \tag{6.65}$$

Breaking of the electron plasma wave
Cold plasma at resonance:

$$\frac{v_d}{u_0} = \left| \frac{1}{2} + i\frac{\pi}{2} e^{i(\pi/2)^2} \int_{-\infty}^{\pi/2} e^{-i\sigma^2} d\sigma \right|^{-1} = 0.36 \tag{5.110}$$

$$\hat{E}_d > \hat{E}_d^* = 0.72(L/\lambda_D)^{-1/3}. \tag{6.67}$$

Coffey criterion (hydrodynamic, out of resonance):

$$\frac{n_e}{n_0} < \left(\frac{v_\varphi}{c_{\text{se}}} \right)^{2/(\gamma+1)}. \tag{5.96}$$

Doppler effect (a) in vacuum:

$$\mathbf{k}' = \mathbf{k} + \frac{\gamma - 1}{\mathbf{v}^2}(\mathbf{vk})\mathbf{v} - \gamma\omega\frac{\mathbf{v}}{c^2}, \qquad \omega' = \gamma(\omega - \mathbf{kv}). \tag{6.74}$$

$$\omega_r = \omega - (\mathbf{k} - \mathbf{k}_r)\mathbf{v} = \omega\frac{1 - \mathbf{k}^0\mathbf{v}/c}{1 - \mathbf{k}_r{}^0\mathbf{v}/c}; \quad \mathbf{k}^0 = \frac{\mathbf{k}}{|\mathbf{k}|}, \ \mathbf{k}_r{}^0 = \frac{\mathbf{k}_r}{|\mathbf{k}_r|}. \tag{6.76}$$

(b) in medium:

$$\mathbf{k}' = \mathbf{k} + \frac{\gamma_\varphi - 1}{\mathbf{v}^2}(\mathbf{kv})\mathbf{v} - \omega\frac{\mathbf{v}}{v_\varphi^2}, \ \ \omega' = \gamma_\varphi(\omega - \mathbf{kv}); \ \ \gamma_\varphi = \left(1 - \frac{v^2}{v_\varphi^2}\right)^{-1/2}. \tag{6.83}$$

$$\omega_r = \omega - (\mathbf{k} - \mathbf{k}_r)\mathbf{v} = \omega\frac{1 - \mathbf{k}^0\mathbf{v}/v_\varphi}{1 - \mathbf{k}_r{}^0\mathbf{v}/v_\varphi}; \quad \mathbf{k}^0 = \frac{\mathbf{k}}{|\mathbf{k}|}, \ \mathbf{k}_r{}^0 = \frac{\mathbf{k}_r}{|\mathbf{k}_r|}. \tag{6.84}$$

Three wave interactions—Resonance conditions:

$$\mathbf{k}_1 = \mathbf{k}_2 + \mathbf{k}_3, \qquad \omega_1 = \omega_2 + \omega_3. \tag{6.85}$$

Three wave interactions—Growth rates
SBS: Weak coupling

$$\gamma_{\max} = \frac{1}{2^{3/2}\eta_1}\omega_{\mathrm{pi}}\frac{\hat{\mathbf{v}}_{\mathrm{os}}}{(cc_s)^{1/2}}. \tag{6.111}$$

SBS: Strong coupling

$$\Omega = \frac{1}{2^{4/3}\eta_1^{1/3}}\left[\omega_1\omega_{\mathrm{pi}}^2\left(\frac{\hat{\mathbf{v}}_{\mathrm{os}}}{c}\right)^2\right]^{1/3}(1 + i\sqrt{3}). \tag{6.112}$$

SRS:

$$\gamma_{\max} = \frac{k_3}{4}\left[\omega_p^2(\omega_2\omega_3)\frac{e^2}{m_2^2\omega_1^2}\mathbf{E}_1\mathbf{E}_1^*\right]^{1/2} = \frac{k_3}{4}\hat{v}_{\mathrm{os}}\left[\frac{\omega_p^2}{\omega_2\omega_3}\right]^{1/2}. \tag{6.120}$$

SRS: Forward scattering

$$\gamma = \frac{1}{2^{3/2}}\frac{\omega_p^2}{\omega_1}\frac{\hat{v}_{\mathrm{os}}}{c}. \tag{6.121}$$

TPD: $\alpha = \pi/4$

$$\gamma = \frac{\omega_1}{4\sqrt{2}}\frac{\hat{\mathbf{v}}_{\mathrm{os}}}{c}. \tag{6.126}$$

OTSI:

$$\frac{\gamma^2}{k_3^2} = -\frac{\Delta}{\Delta^2 + \nu^2/4}\frac{m_e}{m_i}\frac{\hat{v}_1^2}{4} - c_s^2; \;\; \omega_1^2 - \omega_2^2 = 2\omega\Delta. \tag{6.128}$$

PDI:

$$\gamma = -\left\{\frac{\Delta_+}{\Delta_+^2 + \nu^2/4} + \frac{\Delta_-}{\Delta_-^2 + \nu^2/4}\right\}\frac{m_e\omega}{m_i\omega_3}\frac{\hat{v}_1}{16}; \quad \Delta_\pm = \omega_+ - \omega_\pm, \; \omega_\pm = \omega_2 \pm \omega_3. \tag{6.130}$$

Piliya-Rosenbluth criterion:

$$\frac{C|E_1|^2}{\nu_2\nu_3} = 2\pi C|E_{\text{thres}}|^2\frac{\cos\alpha_2\cos\alpha_3}{|\kappa' v_{g2}v_{g3}|} = 1. \tag{6.135}$$

Maximum (nonrel.) electron velocity by trapping in a discontinuous electron plasma wave:

$$v_{\max} = v_\varphi\left[1 + 2\frac{\hat{E}_{\max}}{v_\varphi} + \frac{4}{\pi}\left(\frac{2\hat{E}_{\max}}{v_\varphi}\right)^{1/2}\right]^{1/2}. \tag{6.65}$$

Rayleigh length

$$R = \frac{\pi\sigma_0^2}{\lambda}; \quad \sigma_0^2 \text{ beam waist.} \tag{6.152}$$

Half power ($r = 0.6\sigma_0$) and 86% power ponderomotive self focusing ($r = \sigma_0$):

$$r = 0.6\sigma_0 \; : \; P_{1/2} = 2\pi\frac{c^3 p_c}{\omega_p^2}; \quad r = \sigma_0 \; : \; P_{\sigma_0} = \pi\frac{c^3 p_c}{\omega_p^2}\mathrm{e}^2, \quad \mathrm{e} = 2.718..... \tag{6.156}$$

Relativistic self focusing

$$P_{1/2} = 2\pi\frac{\varepsilon_0 m_e^2 c^5}{e^2}\frac{n_c}{n_e} = 4.2\frac{n_c}{n_e}\,\text{GW}; \quad P_{\sigma_0} = e^2\pi\frac{\varepsilon_0 m_e^2 c^5}{e^2}\frac{n_c}{n_e} = 15.5\frac{n_c}{n_e}\,\text{GW}. \tag{6.158}$$

Nonlinear Schrödinger equation

$$i\frac{\partial\Psi}{\partial t} + P\frac{\partial^2\Psi}{\partial x^2} + Q|\Psi|^2\Psi = 0. \tag{6.161}$$

Maximum growth rate $\Gamma_{\max}$, K_{m} wave number:

$$\Gamma_{\max} = |Q|\hat{E}_0^2, \quad K_{\text{m}} = \left(\frac{Q}{P}\right)^{1/2}\hat{E}_0. \tag{6.164}$$

6.9 Further Readings

H.-J. Kull, *Theory of the Rayleigh–Taylor instability*. Phys. Rep. **206**, 197–325 (1991).
S. Chandrasekhar, *Hydrodynamic and Hydromagnetic Stability* (Oxford University Press, Oxford, 1961).
W.L. Kruer, *The Physics of Laser Plasma Interactions* (Addison-Wesley Publishing Co., Redwood City, 1988).

References

1. S. Chandrasekhar, *Hydrodynamic and Hydromagnetic Stability* (Oxford University Press, Oxford, 1961)
2. H.-J. Kull, Phys. Rep. **206**, 197–325 (1991)
3. S. Atzeni, A. Schiavi, M. Temporal, Plasma Phys. Contr. Fusion **46**, B111 (2004)
4. A.R. Piriz, J.J Lopez Cela, N.A. Tahir, Nuclear InstMeth. Phys. Res. Sect. **A606**, 139 (2009)
5. A.R. Piriz, Y.B. Sun, N.A. Tahir, Phys. Rev. E **91**, 033007 (2015)
6. H. Takabe, L. Monthierth, R.L. Morse, Phys. Fluids **26**, 2299 (1983)
7. H. Takabe, K. Mima, L. Monthierth, R.L. Morse, Phys. Fluids **28**, 3676 (1985)
8. R. Betti et al., Phys. Plasmas **5**, 1446 (1998)
9. H. Azechi et al., Phys. Rev. Lett. **98**, 045002 (2007)
10. H. Zhang, R. Betti, R. Yan et al., Phys. Rev. Lett. **121**, 185002 (2018)
11. M. Kono, M.M. Skoric, *Nonlinear Physics of Plasmas* (Springer, Heidelberg, 2010)
12. W. Horton, T. Tajima, T. Kamimura, Phys. Fluids **30**, 3485 (1987)
13. A. Bret, M.-C. Firpo, C. Deutsch, Phys. Rev. E **72**, 016403 (2005)
14. W. Brett, *Magnetohydrodynamische Kelvin-Helmholtz-Instabilität (Magnetohydrodynymic Kelvin-Helmholtz instability)*, PhD Thesis (TU Darmstadt, Technische Universität, Darmstadt, 2014), p. 126
15. N.G. Denisov, Sov. Phys. JETP **4**, 544 (1957)
16. A.D. Piliya, Sov. Phys. Tech. Phys. **11**, 609 (1966)
17. J.P. Freidberg et al., Phys. Rev. Lett. **28**, 795 (1972)
18. D.W. Forslund et al., Phys. Rev. A **11**, 679 (1975)
19. H.-J. Kull, Phys. Fluids **26**, 1881 (1983)
20. L. Muschietti, C.T. Dum, Phys. Fluids B **5**, 1383 (1993)
21. E. Sonnenschein et al., Phys. Rev. E **57**, 1005 (1998)
22. B. Nordland, Phys. Rev. E **55**, 3647 (1997)
23. M. Colunga, P. Mora, R. Pellat, Phys. Fluids **28**, 854 (1985)
24. A. Bergmann, H. Schnabl, Phys. Fluids **31**, 3266 (1988)
25. H. Ruhl, *Collective superintense laser-plasma interaction*, Habilitation thesis (TU Darmstadt, 2000)
26. H. Ruhl, Plasma Sour. Sci. Technol. **11**(3A), A154 (2002)
27. M. Abramowitz, I.A. Stegun, *Pocketbook of Mathematical Functions* (Harri Deutsch, Frankfurt/Main, 1984)
28. G.J. Pert, Plasma Phys. **20**, 175 (1978)
29. T. Speziale, P.J. Catto, Phys. Fluids **22**, 681 (1979)
30. W.L. Kruer, *The Physics of Laser Plasma Interactions* (Addison-Wesley, Redwood City, 1988), p. 43
31. D.-E. Hinkel-Lipsker, B.D. Fried, G.J. Morales, Phys. Rev. Lett. **62**, 2680 (1989)
32. F. Brunel, Phys. Rev. Lett. **59**, 52 (1987)

33. F. Brunel, Phys. Fluids **31**, 2714 (1988)
34. O.L. Landen, D.G. Stearns, E.M. Campbell, Phys. Rev. Lett. **63**, 1475 (1989)
35. R. Fedosejews et al., Appl. Phys. B **50**, 79 (1990)
36. R. Fedosejews, R. Ottmann, R. Sigel et al., Phys. Rev. Lett. **64**, 1250 (1990)
37. V.L. Ginzburg, *The Propagation of Electromagnetic Waves in Plasmas* (Pergamon Press, Oxford, 1964)
38. P. Koch, J. Albritton, Phys. Rev. Lett. **32**, 1420 (1974)
39. P. Mulser, H. Takabe, K. Mima, Z. Naturforsch. **37a**, 201 (1982)
40. A. Pukhov, J. Meyer-ter-Vehn, Phys. Plasmas **5**, 1880 (1998)
41. A. Bergmann, P. Mulser, Phys. Rev. E **47**, 3585 (1993)
42. A. Bergmann, *Nonlinear Resonance Absorption and Wavebreaking*, Dissertation (TH Darmstadt, 1990)
43. A. Bergmann, S. Hüller, P. Mulser, H. Schnabl, Europhys. Lett. **14**, 661 (1991)
44. K. Eidmann et al., Europhys. Lett **55**, 334 (2001)
45. F. Sauter, Z. Phys. **203**, 488 (1967)
46. F. Forstmann, R.R. Gerhardts, *Metal Optics near the Plasma Frequency, Chaps. 2, 3* (Springer, Berlin, 1986)
47. J.E. Balmer, T.P. Donaldson, Phys. Rev. Lett. **39**, 1084 (1977)
48. K.R. Manes, V.C. Ruppert, J.M. Auerbach, P. Lee, J.E. Swain, Phys. Rev. Lett. **39**, 281 (1977)
49. R.P. Godwin, M.M. Mueller, Appl. Opt. **18**, 3555 (1979)
50. A.G.M. Maaswinkel, K. Eidmann, R. Sigel, Phys. Rev. Lett. **42**, 1625 (1979)
51. A.G.M. Maaswinkel, Opt. Commun. **33**, 62 (1980)
52. I.V. Igumenshev et al., Phys. Plasmas **14**, 092701 (2007)
53. R. Dinger, K. Rohr, H. Weber, Laser Particle Beams **5**, 691 (1987)
54. W.L. Kruer, K. Estabrook, Phys. Fluids **28**, 430 (1985)
55. J.C. Adam, A. Gourdin Serveniere, G. Laval, Phys. Fluids **2**, 376 (1982)
56. R. Dragila, Phys. Fluids **26**, 1682 (1983)
57. J.C. Adam, A. Heron, Phys. Fluids **31**, 2602 (1988)
58. B. Bezzerides, S.J. Gitomer, Phys. Rev. Lett. **46**, 593 (1981)
59. R.C. Davidson, *Methods in Nonlinear Plasma Physics, Chap. 4* (Academic, New York, 1972)
60. V.E. Zakharov, Sov. Phys. JETP **69**, 334 (1989)
61. M. Borghesi, S. Bulanov, D.H. Campbell et al., Phys. Rev. Lett. **88**, 135001 (2002)
62. K. Büchel et al., *Plasma Production with Nd Laser and Non-thermal Effects*. Laser Interaction, vol. 2, ed. by H.J. Schwarz, H. Hora (Plenum Press, New York, 1972), p. 512
63. S. Weinberg, *Gravitation and Cosmology, Chap. 2* (Wiley, New York, 1972)
64. D.W. Forslund, J.M. Kindel, K. Lee, Phys. Rev. Lett. **39**, 284 (1977)
65. D.W. Forslund, J.M. Kindel, K. Lee, Phys. Fluids **22**, 462 (1979)
66. K.G. Estabrook, W.L. Kruer, Phys. Rev. Lett **40**, 42 (1978)
67. P. Mulser, W. Schneider, *Excitation of Nonlinear Electron Plasma Waves and Particle Acceleration by Laser*. Twenty Years of Plasma Physics, ed. by B. McNamara (World Scientific, Singapore, 1985), p. 264
68. J.S. DeGroot, J.E. Tull, Phys. Fluids **18**, 672 (1975)
69. B. Bezzerides, S.J. Gitomer, D.W. Forslund, Phys. Rev. Lett. **44**, 651 (1980)
70. B. Bezzerides, S.J. Gitomer, Phys. Fluids **26**, 1359 (1983)
71. J.R. Albritton, A.B. Langdon, Phys. Rev. Lett. **45**, 1794 (1980)
72. W.L. Kruer, Phys. Fluids **22**, 1111 (1979)
73. R.D. Brooks, Z.A. Pietrzyk, Phys. Fluids **30**, 3600 (1987)
74. S. Aithal, H. Pepin, Phys. Fluids B **4**, 263 (1992)
75. J.S. DeGroot, K. Mizuno, J.H. Rogers, D. Rasmussen, P.W. Rambo, R.A. Stone, *Microwave Modelling of Moderate Laser-Plasma Interactions*. Laser Interaction, vol. 7, ed. by H. Hora, G.H. Miley (Plenum Press, New York, 1986)
76. M.G. Haines, The physics of the coronal region, in *Laser-Plasma Interactions, Proceedings of the 20th Scottish Universities Summer School in Physics*, ed. by R.A. Cairns, J.J. Sanderson (Edinburgh, 1980), p. 170

77. T.P. Coffey, Phys. Fluids **14**, 1402 (1971)
78. P. Gibbon, A.R. Bell, Phys. Rev. Lett. **68**, 1535 (1992) [see vacuum heating]
79. L. Brillouin, Predicted the effect in 1922; see L. Brillouin: *La theorie des solides er les quanta* (1928), and *Wave propagation in periodic structures* (1940). Experimental verification of the effect by Gross in 1930
80. G. Landsberg, L. Mandelstam, Eine neue Erscheinung bei der Lichtstreuung in Metallen (A new phenomenon in the light scattering from metals). Die Naturwissenschaften **16**, 557 (1928)
81. J. Handke, S.A.H. Rizvi, B. Kronast, Appl. Phys. **25**, 109 (1981)
82. B. Gellert, B. Kronast, Appl. Phys. B **32**, 175 (1983)
83. H.A. Baldis et al., Phys. Fluids B **5**, 3319 (1993)
84. J.L. Kline, J. Phys.: Conf. Ser. **112**, 022042 (2008)
85. S. Hüller et al., Phys. Plasmas **13**, 022703 (2006)
86. D. Pesme, in *La fusion thermonucléaire inertielle par laser*, ed. by R. Dautray, J.P. Watteau (1993)
87. T.W. Johnston et al., Plasma Phys. Contr. Fusion **27**, 473 (1985)
88. B.B. Kadomtsev, *Plasma Turbulence* (Academic, New York, 1965)
89. R.C. Davidson, R. Goldman, Phys. Rev. Lett. **21**, 1671 (1968)
90. S. Depierreux et al., Phys. Rev. Lett. **84**, 2869 (2000)
91. Th Fouquet, D. Pesme, Phys. Rev. Lett. **100**, 055006 (2008)
92. M.V. Goldman, D.F. DuBois, Phys. Fluids **8**, 1404 (1965)
93. D.W. Forslund, J.M. Kindel, E.L. Lindman, Phys. Fluids **18**, 1002 (1975)
94. D.J. Strozzi, *Vlasov simulations of kinetic enhancement of Raman backscatter in laser fusion plasmas* (Massachusetts Institute of Technology, Thesis, 2006)
95. H.X. Vu, D.F. DuBois, B. Bezzerides, Phys. Plasmas **14**, 012702 (2007)
96. L. Yin, W. Daughton, B.J. Albright et al., Phys. Plasmas **13**, 072701 (2006)
97. H.A. Baldis, C.J. Walsh, Phys. Fluids **26**, 1364 (1983)
98. V.P. Silin, Sov. Phys. JETP **21**, 1127 (1965)
99. K. Nishikawa, J. Phys. Soc. Jpn. **24**, 1152 (1968)
100. P. Mulser, A. Giulietti, Europhys. Lett. **12**, 235 (1990)
101. A. Bers, Space-time evolution of plasma instabilities - absolute and convective, in *Handbook of Plasma Physics*, vol.1, Chap. 3.2, ed. by M.N. Rosenbluth, R.Z. Sagdeev (North-Holland Publishing Co., Amsterdam, 1983)
102. W.L. Kruer, *The Physics of Laser Plasma Interactions* (Addison-Wesley, Redwood City, 1988), p. 91
103. S. Hüller, P. Mulser, A.M. Rubenchik, Phys. Fluids B **3**, 3339 (1991)
104. J.F. Drake et al., Phys. Fluids **17**, 778 (1974)
105. R.L. Berger et al., Phys. Fluids B **17**, 2243 (1993)
106. E.S. Cassedy, P. Mulser, Phys. Rev. A **10**, 2349 (1974)
107. Q.S. Feng et al., Plasma Phys. Control. Fusion **59**, 075007 (2017)
108. K. Eidmann, R. Sigel, in *Laser Interaction and Related Plasma Phenomena*, vol. 3B, ed. by H.J. Schwarz, H. Hora (Plenum, New York, 1974), p. 667
109. C.E. Clayton, C. Joshi, F.F. Chen, Phys. Rev. Lett. **51**, 1656 (1983)
110. J.E. Bernard, J. Meyer, Phys. Fluids **29**, 2313 (1986)
111. A.M. Mostovych et al., Phys. Rev. Lett. **59**, 1193 (1987)
112. S. Hüller, P.E. Masson-Laborde, D. Pesme, C. Labaune, H. Bandulet, J. Phys.: Conf. Ser. **112**, 022031 (2008)
113. V.T. Tikhonchuk, S. Hüller, Ph Mounaix, Phys. Plasmas **4**, 4369 (1997)
114. D. Pesme, W. Rozmus, V.T. Tikhonchuk et al., Phys. Rev. Lett. **84**, 278 (2000)
115. P. Michel, C. Labaune, Phys. Fluids **10**, 3545 (2003)
116. P.E. Masson-Laborde, S. Hüller, D. Pesme, M. Casanova, P. Loiseau, J. Phys. IV France **133**, 247 (2006)
117. P.E. Masson-Laborde, S. Hüller, D. Pesme et al., Phys. Plasmas **21**, 032703 (2014)
118. R.P. Drake, R.E. Turner, B.F. Lasinski et al., Phys. Rev. Lett. **40**, 3219 (1989)
119. D.J. Strozzi et al., J. Plasma Phys. **72**, 1299 (2006)

120. F. Ze et al., Comm. Plasma Phys. Contr. Fusion **10**, 33 (1986)
121. D. Mourenas, L. Divol, M. Casanova, C. Rousseaux, Phys. Plasmas **8**, 557 (2001)
122. VYu. Bychenkov, A.A. Zozulya, V.P. Silin, V.T. Tikhonchuk, Beitr. Plasmaphys. **23**, 331 (1983)
123. V.T. Tikhonchuk, Phys. Plasmas **22**, 104701 (2015)
124. C. Stoeckl, H. Baldis et al., Phys. Rev. Lett. **90**, 235002 (2003)
125. Z.-M. Sheng, Y.T. Li, J. Zhang, L. Veisz, IEEE Trans. Plasma Sci. **33**, 486 (2005)
126. W. Seka, J.F. Myatt, R.W. Short et al., Phys. Rev. Lett. **112**, 145001 (2014)
127. D.T. Michel et al., Phys. Plasmas **20**, 055703 (2013)
128. D.T. Michel, A.V. Maximov, R.W. Short et al., Phys. Rev. Lett. **109**, 155007 (2012)
129. W. Seka, D.H. Edgell, J.F. Myatt et al., Phys. Plasmas **16**, 052701 (2009)
130. A.D. Piliya, in *Proceedings of the 11th International Conference on Phenomena in Ionized Gases* (Oxford, 1971), p. 287
131. M.N. Rosenbluth, Phys. Rev. Lett. **29**, 565 (1972)
132. C.S. Liu, in *Advances in Plasma Physics*, vol. 6, ed. by A. Simon, W.B. Thompson (Wiley, New York, 1976), pp. 121–177
133. D. Babonneau, *Instabilites Parametriques en Plasma Inhomogene*, Centre d'Etudes de Limeil B.P.n^0 27, Note C.E.A. n^0 1749 (1974)
134. T.J.M. Boyd, J.J. Sanderson, *The Physics of Plasmas* (Cambridge University Press, Cambridge, 2003)
135. J.M. Manley, H.E. Rowe, Proc. Inst. Radio Engrs. **44**, 904 (1956)
136. J. Brown, Electron. Lett. **1**, 23 (1965)
137. A.J. Brizard, A.N. Kaufman, Phys. Rev. Lett. **74**, 4567 (1995)
138. M. Salimullah, M.H.A. Hassan, Phys. Rev. A **41**, 6963 (1990)
139. S. Guerin, G. Laval, P. Mora, J.C. Adam, A. Heron, Phys. Plasmas **2**, 2807 (1995)
140. S.T. Mahmoud, R.P. Sharma, Phys. Plasmas **8**, 3419 (2001)
141. S.T. Mahmoud, Phys. Scr. **76**, 259 (2007)
142. A. Bers, I.P. Shkarowski, M. Shoucri, Phys. Plasmas **16**, 022104 (2009)
143. G. Shvets, N.J. Fish, J.M. Rax, Phys. Plasmas **3**, 1109 (1996)
144. S. Hain, *Ausbreitung intensiver Laserstrahlung in Materie (Propagation of intense laser radiation in matter)*, PhD thesis, Technische Universität Darmstadt; GCA-Verlag (Herdecke, 1999), pp. 50–62, in German
145. P. Kaw, J. Dawson, Phys. Fluids **13**, 472 (1970)
146. T. Kurki-Suonio, P.J. Morrison, T. Tajima, Phys. Rev. A **40**, 3830 (1989)
147. M.R. Jafari Milani, A.R. Nicknam, A.H. Farahbod, Phys. Plasmas **21**, 063107 (2014)
148. T. Afshar-rad et al., Phys. Rev. Lett. **68**, 942 (1992)
149. H. Hora, J. Opt. Soc. Am. **65**, 882 (1975)
150. O. Willi, P.T. Rumsby, Opt. Commun. **37**, 45 (1981)
151. C. Ren, W.B. Mori, Phys. Plasmas **8**, 3118 (2001)
152. Z. Najmudin et al., Phys. Plasmas **10**, 438 (2003)
153. Ch. Karle, K.H. Spatschek, Phys. Plasmas **15**, 123103 (2008)
154. V.E. Zakharov, A.B. Shabat, Zh Eksp, Teor. Fiz. **61**, 118 (1972)
155. A. Pataraya, G. Melikidze, Astrophys. Space Sci. **68**, 61 (1980)
156. K. Mima, K. Nishikawa, in *Handbook of Plasma Physics*, vol. II, Sec. 6.5.4. and refs. therein, eds. M.N. Rosenbluth, R.Z. Sagdeev (North Holland, Amsterdam, 1984)
157. S.V. Vladimirov, V.N. Tsytovich, S.I. Popel, FKh Kakimov, *Modulational Interactions in Plasmas* (Kluwer Academic Publishers, Dordrecht, 1995)
158. J. Drake, P.K. Kaw, Y.C. Lee, C.S. Lin, G. Schmidt, M. Rosenbluth, Phys. Fluids **17**, 778 (1974)
159. P.K. Shukla, R. Baruthram, L.N. Tsintsadze, Phys. Rev. A **35**, 4889 (1987)
160. V.N. Tsytovich, S.V. Vladimirov, Commun. Plasma Phys. Contr. Fusion **15**, 39 (1992)
161. S.V. Vladimirov, S.I. Popel, V.N. Tsytovich, Phys. Fluids B **5** (1993)
162. P. Jha et al., Phys. Plasmas **16**, 013107 (2009)
163. B.S. Sharma et al., Plasma Phys. Control. Fusion **56**, 075011 (2014)

164. Bhimsen K. Shivamoggi, Phys. Rev. A **31**, 1728 (1985)
165. Yu Ya. Brodskii, A.G. Litvak, S.I. Nechner, Ya Z. Slatsker, Pis'ma. Zh. Eksp. Teor. Fiz. **45**, 176 (1987)
166. E.W. Laedke, K.H. Spatschek, Phys. Rev. A **30**, 3279 (1984)
167. A.Y. Wong, P.Y. Cheung, Phys. Rev. Lett. **52**, 1222 (1984)
168. M.K. Mishra, R.S. Chhabra, S.R. Sharma, Phys. Rev. A **41**, 2176 (1990)
169. P.K. Shukla, Phys. Rev. A **34**, 644 (1986)
170. P. Mulser, F. Cornolti, D. Bauer, Phys. Plasmas **5**, 4466 (1998)

Chapter 7
Transport in Plasma

The Essentials of Transport Theory: Collisions

Interaction of electrons with ions and neutrals and their mutual interactions is at the basis of all kinds of kinetic transport in plasma: collisional absorption, heat conduction, viscosity, friction. It depends on the kind of transport which interactions dominate; in collisional absorption it is the electron-ion collisions. Heat conduction is determined by electron-electron and electron-ion encounters. Viscosity of the plasma is governed by ion-ion interaction, as the derivation of Navier-Stokes equation of Chap. 3 lets suggest. In the dilute neutral gas binary collisions dominate. Thereby the meaning of a collision is that the mutual interaction of two particles is short compared with the time that elapses up to the next close encounter of two partners *and* the laser period $T = 2\pi/\omega$. In the collision integral of the Boltzmann equation (3.83) for instance the collision is modelled as occurring instantaneously. In its most idealized form this is achieved in collisions of hard spheres. The frequency of interaction is described by the binary collisional cross section (2.148), (2.152). The kinetics of binary collisions forms the underlying structure of any microscopic transport theory, including moderately dense fluids also. A systematic treatment of collisional dynamics is offered by the BBWKY expansion. Many body interactions appear as correlations of $n + 1$ order in the distribution function of order n. High power laser interaction generates bunches of fast particles, preferentially electrons. Owing to their shortened time of momentum exchange among themselves and with the ions collisions become inefficient and thermalization is questioned. If Maxwellization of groups surprisingly appears again it is to be attributed to the stochastic interaction with the laser field and with fluctuating fields, either from competing linear modes, or from saturated instabilities. The transport becomes *nonlocal*, it is the domain of numerical modelling and, in first place here, of particle in cell simulations (PIC).

P. Mulser, *Hot Matter from High-Power Lasers*, Graduate Texts in Physics,
https://doi.org/10.1007/978-3-662-61181-4_7

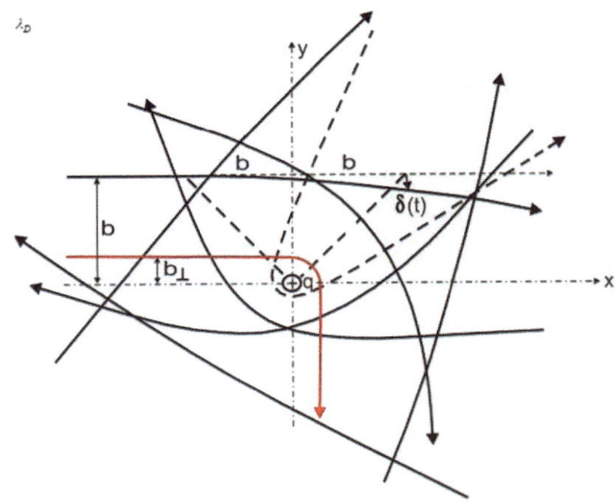

Fig. 7.1 Screening of a positive ion by the attraction of the surrounding free electrons: b impact parameter, $b_\perp$ impact parameter for perpendicular deflection. Fast electrons and large impact parameters lead to nearly straight orbits. Dashed segments under angle $\pi/4$: Jackson's model

7.1 Collision Frequencies

7.1.1 Screening

Transport in the plasma is characterized by the long range Coulomb force and *multiple simultaneous* collisions, 20–100 in the dense laser plasma, 10^6–10^8 in the Tokamak plasma. A typical situation is sketched in Fig. 7.1 between electrons and a positive test ion of charge $q = Ze$. The orbit of the single electron is bent towards the ion and it is disturbed by all electrons within the sphere of the impact parameter b. As a consequence, the Coulomb potential is lowered, the single electron orbit is continuously disturbed by the surrounding *spectator electrons* and also *spectator ions*. Neutral particles also contribute because they are polarized by the interacting electron and ion and exhibit an attractive force in addition to their interaction by the bound electron cloud. If transport theory is developed with the bare Coulomb cross section (2.149) all transport coefficients become infinite.

Let us assume that for the majority of spectators the mean potential energy $\langle E_{\text{pot}} \rangle$ is small compared to the average kinetic energy $\langle E_{\text{kin}} \rangle$. Then the following assumptions will prove to be satisfied: (i) The majority of orbits are nearly straight, (ii) the number of spectators is large enough as to treat them as continuously distributed around the scatterer with a local density $n_{e,i}(\mathbf{x})$, and to apply to them Poisson's law $\varepsilon_0 \nabla \mathbf{E} = q\delta(\mathbf{x}) - e(n_e - Zn_i)$. At small angle deflections by the single test ion there is a one-to-one correspondence between the electron density perturbation $n_1 = n_e - n_0$, induced field $\mathbf{E}$ and the deviation $\boldsymbol{\delta}$ from the straight orbit according to (3.5),

$$n_1(\mathbf{x}, t) = -n_0 \nabla \boldsymbol{\delta}(\mathbf{x}, t), \quad \mathbf{E} = \frac{q}{4\pi\varepsilon_0 r^3}\mathbf{r} + \frac{en_0}{\varepsilon_0}\boldsymbol{\delta}(\mathbf{x}, t) = \mathbf{E}_C + \mathbf{E}_s.$$

The electric field is the sum of the bare Coulomb field $\mathbf{E}_C$ and the induced *screening field* $\mathbf{E}_s$. The total force on an electron of velocity $\mathbf{v}$ is given by $\mathbf{f}_C$ and $-e\mathbf{E}_s = m_e\omega_p^2\boldsymbol{\delta}$. The deflection $\boldsymbol{\delta}$ follows the equation of a driven harmonic oscillator,

$$\ddot{\boldsymbol{\delta}} + \omega_p^2\boldsymbol{\delta} = \frac{\mathbf{f}_C}{m_e}. \tag{7.1}$$

It shows that the interaction with an ion is never Coulomb like; it is modified by the *dynamic* shielding force $-m_e\omega_p^2\boldsymbol{\delta}$. An electron passing by the test charge q with velocity $\mathbf{v}$ is first attracted and then released after passage. At $t = -\infty$ the deflection $\boldsymbol{\delta}$ is zero, at $t = +\infty$ the electron has gained the energy $m_e\dot{\boldsymbol{\delta}}^2/2$. As long as $b \gg b_\perp$ the deflection angle θ is very small for almost all impact parameters b. (iii) Large angle deflections can be treated as true binary events if $b_\perp \lesssim n_0^{-1/3}$.

7.1.2 *Reduction of Simultaneous Interactions to a Sequence of Collisions*

Small angle deflections θ [property (i)] eliminate the problem of multiple simultaneous collisions of an electron with densely packed ions if property (iii) is fulfilled as well. Under these conditions the interaction with N ions in the time interval $\mathrm{d}t$ proves to be equivalent to N successive electron-ion collisions. To this end imagine the ions at the locations $\mathbf{x}_i$. At the time t each of them has produced the shift $\boldsymbol{\delta}_i(t)$. They sum up to a total shift $\boldsymbol{\delta}(t)$ and hence the total kinetic energy of an electron at position $\mathbf{x} = (x = a, r = b)$ is

$$\frac{1}{2}m_e\dot{\boldsymbol{\delta}}^2(a,b) = \frac{1}{2}m_e\left(\sum_{i=1}^{N}\dot{\boldsymbol{\delta}}_i(t)\right)^2 = \frac{1}{2}m_e\sum_{i,j=1}^{N}\dot{\boldsymbol{\delta}}_i\dot{\boldsymbol{\delta}}_j = \frac{1}{2}m_e\sum_{j=i=1}^{N}\dot{\boldsymbol{\delta}}_i^2(t) \tag{7.2}$$

i.e., owing to randomly distributed phases all collisions add up as if they occurred uncorrelated one after another.

7.1.3 *Jackson's Model of Coulomb Interaction*

The introduction of an effective length of the Coulomb interaction is very useful for estimates and approximate calculations. In small deflections, equivalent to linearization of the conservation equations, the orbits are nearly straight. If the true force $\mathbf{f}_C$ exerted by $q(\mathbf{x} = 0)$ is assumed to have the strength $f_C^* = -eq/4\pi\varepsilon_0 b^2$ and the direction perpendicular to $\mathbf{v}$ over the entire distance from $x = -b$ to $x = +b$ and zero outside, the deflection angle ϑ produced by f_C^* coincides with the exact value of

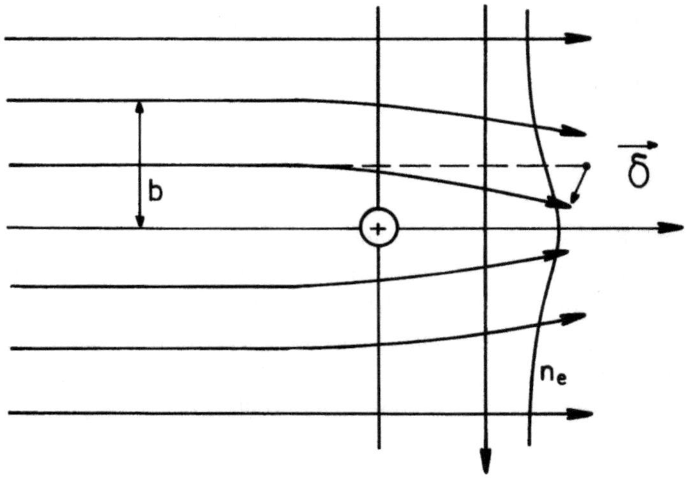

Fig. 7.2 A parallel monoenergetic beam of electrons with flow velocity $\mathbf{v}$ is attracted by an ion at rest. As a consequence, the single orbits deviate by the vector $\boldsymbol{\delta}(b, t)$ from straight lines and the electron density n_e increases towards the symmetry axis, see plot n_e. The nonuniformity of n_e leads to a restoring force acting back on the electrons. Note, the single electron conserves its velocity $|\mathbf{v}|$ in the electron-ion interaction but their sum adds up to an axial flow velocity $\mathbf{u}'$ with $|\mathbf{u}'| < |\mathbf{v}|$ due to bending

(2.149) [1], see dashed lines in Fig. 7.1. Hence, for weakly bent orbits the *effective length* of Coulomb interaction is $2b$.

7.1.4 The Oscillator Model of Uniform Drift

The oscillator equation (7.1) is solved now for a uniform plasma streaming at $\mathbf{v} = \text{const}$ around a charge $q = Ze$, see Fig. 7.2. The induced perturbation can be described in terms of $\mathbf{x}(t) = \mathbf{v}t + \boldsymbol{\delta}(t)$ or, equivalently, by the perturbed electron fluid density $n_e = n_0 + n_1$. In the fluid picture the induced perturbation is the emission of *Cherenkov radiation* of plasmons. Splitting the straight orbit deformation into a component $\delta_\parallel$ parallel to x into direction of $\mathbf{v}$ and a component $\delta_\perp$ perpendicular to $\delta_\parallel$, and doing the same with $\mathbf{E}$, one obtains $\mathbf{E}$ as a function of the collision parameter b,

$$E_\parallel = \frac{en_0}{\varepsilon_0}\delta_\parallel + \frac{\kappa vt}{(b^2 + v^2t^2)^{3/2}}, \quad E_\perp = \frac{en_0}{\varepsilon_0}\delta_\perp + \frac{\kappa b}{(b^2 + v^2t^2)^{3/2}},$$

$$b_\perp = \frac{\kappa e}{m_e v^2}, \quad \kappa = \frac{q}{4\pi\varepsilon_0}. \tag{7.3}$$

The oscillator equation (7.1) translates, under linearization $\ddot{\boldsymbol{\delta}} = \partial_{tt}\boldsymbol{\delta}$ and $x = vt$, into

$$\ddot{\delta}_\| + \omega_p^2 \delta_\| = -\frac{e}{m_e}\kappa \frac{x}{(b^2+x^2)^{3/2}}, \quad \ddot{\delta}_\perp + \omega_p^2 \delta_\perp = -\frac{e}{m_e}\kappa \frac{b}{(b^2+x^2)^{3/2}}. \tag{7.4}$$

The driving Coulomb force $\mathbf{f}_C^*$ on the RHS is the dipole approximation $e\kappa\boldsymbol{\delta}\nabla(1/r) = \mathbf{f}_C(vt, b)\boldsymbol{\delta}$. Ignoring screening corresponds to setting $\omega_p = 0$; it leads to the well known divergence in the Coulomb cross section.

For its usefulness we remember in this place that a particular solution of the inhomogeneous linear differential equation $y'' + a(x)y = f(x)$ is obtained from

$$y(x) = h(x)\int^x \frac{g(\zeta)f(\zeta)}{W(\zeta)}\mathrm{d}\zeta - g(x)\int^x \frac{h(\zeta)f(\zeta)}{W(\zeta)}\mathrm{d}\zeta \tag{7.5}$$

with $W = gh' - g'h$ the Wronskian; g and h are solutions of the homogeneous differential equation.

The solutions of the homogeneous equations are $\boldsymbol{\delta} = \hat{\boldsymbol{\delta}} \exp(\pm i\omega_p t)$, if adiabatically switched on at $t = -\infty$. Hence, $\dot{\boldsymbol{\delta}}(-\infty) = \boldsymbol{\delta}(-\infty) = 0$ and, after introducing the screening length λ_0,

$$\lambda_0 = \frac{v}{\omega_p} \tag{7.6}$$

and the dimensionless variables $\xi = x/\lambda_0,\ \beta = b/\lambda_0$ the solutions are

$$\delta_\|(\xi) = \frac{b_\perp}{\beta}\left\{\sin(\beta\xi)\int_{-\infty}^{\xi} \frac{u\cos(\beta u)\mathrm{d}u}{(1+u^2)^{3/2}} - \cos(\beta\xi)\int_{-\infty}^{\xi} \frac{u\sin(\beta u)\mathrm{d}u}{(1+u^2)^{3/2}}\right\},$$

$$\delta_\perp(\xi) = \frac{b_\perp}{\beta}\left\{\sin(\beta\xi)\int_{-\infty}^{\xi} \frac{\cos(\beta u)\mathrm{d}u}{(1+u^2)^{3/2}} - \cos(\beta\xi)\int_{-\infty}^{\xi} \frac{\sin(\beta u)\mathrm{d}u}{(1+u^2)^{3/2}}\right\}.$$

In this model the cold electron plasma oscillations at $\omega = \omega_p$ persist indefinitely. In reality, owing to finite wave-wave and wave-particle interactions they slowly decay, thereby converting their kinetic and potential energies into thermal energy of the plasma. Their damping is mainly due to profile steepening and cold wavebreaking (see Chap. 6). The relevant amplitudes of $\delta_\perp,\ \delta_\|$ occur at $\xi = \infty$ and, since integrals of the odd functions vanish, they are given by

$$\hat{\delta}_\| = \frac{b_\perp}{\beta}\int_{-\infty}^{+\infty} \frac{u\sin(\beta u)\mathrm{d}u}{(1+u^2)^{3/2}} = b_\perp \int_{-\infty}^{+\infty} \frac{\cos(\beta u)\mathrm{d}u}{(1+u^2)^{1/2}} = 2b_\perp \mathrm{K}_0(\beta), \tag{7.7}$$

$$\hat{\delta}_\perp = \frac{b_\perp}{\beta}\int_{-\infty}^{+\infty} \frac{\cos(\beta u)\mathrm{d}u}{(1+u^2)^{3/2}} = \frac{\pi^{1/2}}{(3/2)} b_\perp \mathrm{K}_1(\beta) = 2b_\perp \mathrm{K}_1(\beta). \tag{7.8}$$

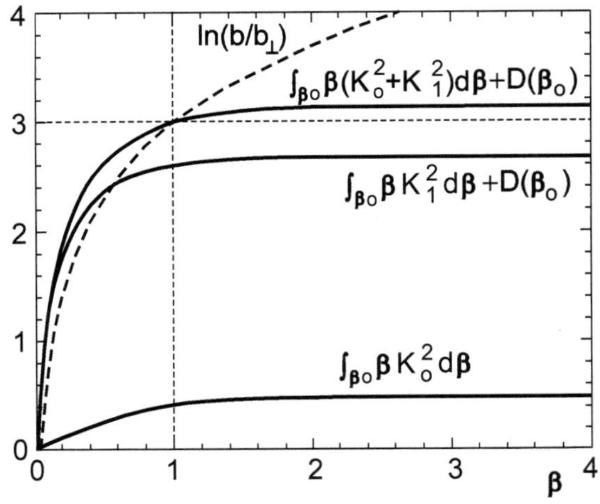

Fig. 7.3 Oscillator model: Transverse and longitudinal oscillation energies $E_\perp \sim \int \beta K_1^2 d\beta$, $E_\parallel \sim \int \beta K_0^2 d\beta$ and their sums as functions of $\beta = b/\lambda_0$, $\lambda_0 = v/\omega_p$. The Coulomb logarithm $\ln \Lambda = \ln(\lambda_0/b_\perp)$ is a good approximation to $E_\perp(\beta = \infty)$ even at the value as low as $\ln \Lambda = 3$. The small deviation derives from regularization $\ln(1 + b_\perp/b_0)$; for b_0 see text. Both, K_0 and K_1 diverge for $\beta \to 0$, and so does $\int \beta K_1^2 d\beta$; $\int \beta K_0^2 d\beta$ remains finite. For β large $\to \beta K_0 K_1 \sim \exp(-2\beta)$

Expression (7.7) is obtained from a partial integration. K_0 and K_1 are modified Bessel functions [2]. For small values of β their expansions are, with $\gamma = 0.57722$,

$$K_0(\beta) = -\left(\ln\frac{\beta}{2} + \gamma\right) + \left(\frac{\beta}{2}\right)^2 \left(1 - \ln\frac{\beta}{2} - \gamma\right) + \cdots$$

$$K_1(\beta) = \frac{1}{\beta} + \left(\frac{\beta}{2}\right)\left(\ln\frac{\beta}{2} + \gamma - \frac{1}{2}\right) + \cdots \tag{7.9}$$

For $\beta = 1$ these expansions for K_0 and K_1 differ by 1% and 7% from the true values. For large values of β the corresponding asymptotic expansions are

$$K_0(\beta) = \left(\frac{\pi}{2\beta}\right)^{1/2} e^{-\beta} \left\{1 - \frac{1}{8\beta} + \frac{9}{128\beta^2} \mp \cdots\right\}, \tag{7.10}$$

$$K_1(\beta) = \left(\frac{\pi}{2\beta}\right)^{1/2} e^{-\beta} \left\{1 + \frac{3}{8\beta} + \frac{15}{128\beta^2} \pm \cdots\right\}. \tag{7.11}$$

Both amplitudes diverge for $\beta \to 0$ (see Fig. 7.3), since

$$K_0(\beta \to 0) \sim -\ln\frac{\beta}{2}, \quad K_1(\beta \to 0) \sim \frac{1}{\beta}.$$

This is a consequence of the straight trajectory, or equivalently, dipole approximation in (7.4). The home made divergence is eliminated by regularization: Choose an integer s of order unity and the reduced *DeBroglie length*,

$$\lambda_B = \frac{\hbar}{m_e v_e}; \qquad \lambda_B[\mathrm{nm}] = \frac{0.185}{(E[\mathrm{eV}])^{1/2}}. \tag{7.12}$$

Integrate the oscillator solution from $b_0 = s\lambda_B \ll \lambda_0$ to infinity and treat the momentum transfer $D(b_0)$ of the close encounters by scattering from the unscreened Coulomb potential,

$$D(b_0) = \frac{m_e v}{\sigma} \int_0^{b_0} \sigma_C(b)(1 - \cos\vartheta)\mathrm{d}\Omega; \quad \sigma = \pi b_0^2. \tag{7.13}$$

The energy irradiated per unit time into plasma waves by a single ion through a plane orthogonal to $\mathbf{v}$ and at a fixed distance $x \to +\infty$ is given by

$$\dot{W} = \frac{1}{2} m_e \omega_p^2 n_0 v \lambda_0^2 \int_{\beta_0}^{\infty} 2\pi\beta(\hat{\boldsymbol{\delta}}_{\|}^2 + \hat{\boldsymbol{\delta}}_{\perp}^2)\mathrm{d}\beta + vD(\beta_0). \tag{7.14}$$

The term $D(\beta_0)$ from (7.13) accounts for the momentum loss by the curved orbits for which $\theta \gtrsim \vartheta_0 = 20$. Consider $1 - \cos\vartheta_0 = 0.060 \ll 1$ and $1 - \cos\vartheta_0 = \theta_0^2/2 = 0.61$. Thus adding simultaneous small deflections $\theta \lesssim \vartheta_0$ agrees with (7.2). $\dot{W}$ can be rewritten as

$$\dot{W} = Zm_e\omega_p^2 vb_\perp \left\{\beta_0 \mathrm{K}_0(\beta_0)\mathrm{K}_1(\beta_0) + \frac{1}{2}\ln\left(1 + \frac{b_0^2}{b_\perp^2}\right)\right\}. \tag{7.15}$$

It gives rise to a frictional force f on the ion into direction of $\mathbf{v}$, $\dot{W} = fv$. The connection with the collision frequency in the cold plasma ν_{Cei} is established through the Drude model,

$$f = \frac{\dot{W}}{v} = \nu_{Cei} m_e v, \qquad \nu_{Cei} = \frac{\dot{W}}{m_e v^2}. \tag{7.16}$$

The discussion of b_0 is postponed to the section on a general consideration of the *Coulomb logarithm* in the ideal plasma.

From the oscillator model the cutoff of the Coulomb potential at the dynamic Debye length λ_0 arises self-consistently. Performing all calculations with a bare Coulomb potential corresponds to ignoring the restoring force in (7.1). It leads to the well-known divergent result. The origin of screening and the reason for convergence become clear from the oscillator model (see Fig. 7.2).

An electron on the distant orbit with b larger than a certain $b_{\max}$ feels a weak attractive Coulomb force over a long time; the oscillators in (7.4) are very slowly shifted from their equilibrium positions and then slowly released. The whole process occurs adiabatically and all energy the oscillators had acquired is given back after the encounter. At an intermediate b, the oscillators associated with $\delta_\parallel$, $\delta_\perp$ are excited since the driving force changes rapidly enough to produce Fourier components not far from the eigenfrequency ω_p. As a result, energy is taken irreversibly from the translational motion and transformed into oscillatory energy. Finally, at very small $b \simeq b_\perp$ the exciting force has frequencies much larger than ω_p so that the oscillator behaves like an attracted particle with zero restoring force.

It is well-known that the deflection angle ϑ of a nearly straight orbit can be calculated exactly if the maximum Coulomb force $qe/4\pi\varepsilon_0 b^2$ is assumed to act over the finite distance of $2b$ ([1], Sect. 1.1.3). Consequently, the effective interaction time in such a collision is $\tau = 2b/v$. A cutoff at $b_{\max} = \lambda_0$ means that interaction times during which the oscillator undergoes more than 1/3 oscillations,

$$\frac{\tau}{2\pi/\omega_p} = \frac{\lambda_0 \omega_p}{\pi v} = \frac{1}{\pi} \simeq \frac{1}{3} \tag{7.17}$$

is ineffective.

The excitation of a plasma wave by an ion moving uniformly through a homogeneous plasma was simulated numerically with a Vlasov code in full generality, i.e., nonlinearities in n_e and bent electron orbits are included [3]. A typical example is shown in Fig. 7.4. For moderate ion charges the nonlinearity in n_e and in the electrostatic potential is small; at high Z-values, however, it is no longer negligible (see Fig. 7.4b) and leads, as a consequence of trapping of free electrons in the ion potential, to a dependence of ν_{ei} on n_i which is smaller than Z^2 of expressions (7.15) and (7.16).

Shielding by a flux of ions $n_0\mathbf{u}$, $n_0 = n_{e0}/Z$ of mass m_i, and charge Z can be treated in the same way. The mass entering in ω_p is the reduced mass $\mu = m_i/2$, the Coulomb repulsion is $\mathbf{f}_{Ci} = -\mathbf{f}_C Z^2$. The corresponding oscillator equation reads

$$\ddot{\boldsymbol{\delta}} + 2\frac{m_e}{m_i} Z\omega_p^2 \boldsymbol{\delta} = -2Z^2 \frac{\mathbf{f}_C}{m_i}. \tag{7.18}$$

It shows ion screening is much weaker compared to (7.1).

The oscillator equation has numerous relevant applications in plasma physics: ion beam stopping in cold ionized matter [4], fast electron transport in laser plasmas [5] and, in general, when the temperature transverse to $\mathbf{v}$ is low. As a consequence, N ions radiate $N\dot{W}$.

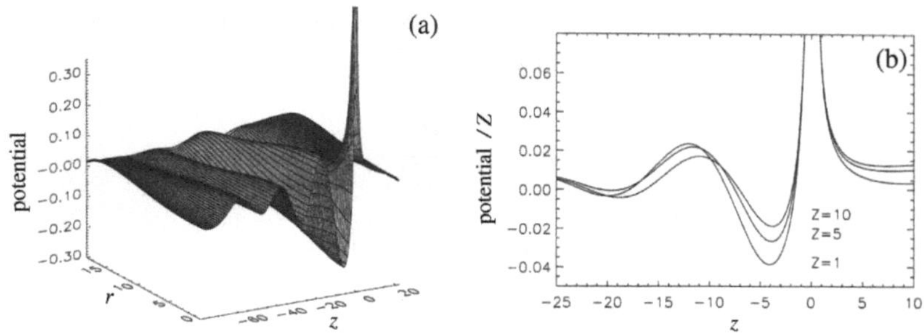

Fig. 7.4 **a** Excitation of an electron plasma wave by an ion at position $z = 0$, $r = 0$ of charge Z (normalized by the number of electrons in the Debye sphere) in a plasma streaming from right to left at speed $v_0/v_{\mathrm{th}} = 4$ (from [3]). **b** Electrostatic potential/Z on the z axis as a function of the normalized charge Z (from [3]). In the linear theory the three curves coincide

In the majority of elementary derivations the electron-ion collision frequency ν_{ei} is based on the momentum change of a single electron in forward direction and only $\delta_\perp$ is taken into consideration. From Fig. 7.3 it is seen that $\delta_\parallel$ also contributes. Imagine two parallel layers of electrons at a suitable distance from each other. When the first layer has reached the position of the ion it has acquired its maximum velocity $v + \varepsilon$ by attraction whereas the layer behind has not yet. This difference in speed leads to a charge nonuniformity oscillating in the direction of $\mathbf{v}$. Its relative importance increases with decreasing Coulomb logarithm $\ln(\lambda_0/b_\perp)$. In fact, for $\mathbf{v} = \mathrm{const}$ the ratio of the two contributions to the absorption coefficient $\alpha_\perp + \alpha_\parallel$ for $\ln \Lambda_0 \gtrsim 2$ is given by

$$\frac{\alpha_\parallel}{\alpha_\perp} = \frac{1}{2 \ln \Lambda_0}. \tag{7.19}$$

In a harmonic laser field the electrons oscillate with the oscillation velocity $\mathbf{v}_{\mathrm{os}}(t)$ and the displacement $\boldsymbol{\delta}(t)$,

$$\mathbf{v}_{\mathrm{os}}(t) = \hat{\mathbf{v}} \mathrm{e}^{-i\omega t}, \quad \hat{\mathbf{v}} = -i \frac{e}{m_e \omega} \hat{\mathbf{E}}, \quad \boldsymbol{\delta}(t) = \hat{\boldsymbol{\delta}} \mathrm{e}^{-i\omega t}, \quad \hat{\boldsymbol{\delta}} = \frac{e}{m_e \omega^2} \hat{\mathbf{E}}. \tag{7.20}$$

As a consequence the collision frequency becomes time-dependent, $\nu_{ei} = \nu_{ei}(t)$. The portion of energy irreversibly transferred to the electrons per unit time under steady state conditions is obtained from the cycle-averaged Poynting term $\overline{\mathbf{jE}}$. From the friction term $\nu m_e \mathbf{u}$ of the Drude model (5.77) a meaningful definition of the cycle-averaged collision frequency $\overline{\nu}_{ei}$ is given by

$$\overline{\mathbf{jE}} = \overline{\nu_{ei} n_e m_e \mathbf{u}^2} \simeq 2\overline{\nu}_{ei} n_{e0} \overline{\mathcal{E}}_{\mathrm{os}}, \quad n_{e0} = \overline{n_e(t)}, \quad \overline{\mathcal{E}}_{\mathrm{os}} = \frac{1}{2} m_e \overline{\mathbf{v}_{\mathrm{os}}(t)^2}. \tag{7.21}$$

With ν_{ei} replaced by $\overline{\nu_{ei}(t)}$ in the foregoing section all absorption formulas remain valid. The cycle averaged collision frequency $\overline{\nu}_{ei}$ defined by (7.21),

$$\overline{\nu}_{ei} = \frac{\overline{\mathbf{jE}}}{2n_{e0}\overline{\mathcal{E}}_{\mathrm{os}}} \tag{7.22}$$

is the collision frequency for *irreversible* energy transfer.

True laser beam absorption is obtained from the cycle averaged collision frequency $\overline{\nu}_{ei}$ only. It regulates the laser intensity attenuation and Beer's law; both are determined by cycle averaged quantities. The work done by the instantaneous quantity $m_e\nu_{ei}(t)\mathbf{v}_{\mathrm{os}}$ goes partially into reversible kinetic energy $m_e\mathbf{v}_{\mathrm{os}}^2/2$, and only the fraction averaged over the laser period $2\pi/\omega$ is *irreversible.*

7.1.5 Debye Shielding

The aim is to find the effective potential $\Phi = q'/4\pi\varepsilon_0 r$ of the screened spherical charge $q' = q - 4\pi\varepsilon_0 \int n_1 r'^2 \mathrm{d}r'$ when the bare ion charge $q = Ze$ is at rest. In the case of thermal equilibrium the answer has been given by (4.74) with λ_D from (4.70). Here a more general derivation is given with the aid of the oscillator model (7.1) in Jackson's approximation for straight orbits. In first approximation the deflection $|\delta(v, r, x = 0)|$ in the plane $x = 0$ perpendicular to the electron velocity $\mathbf{v}$ is the free fall in the bare Coulomb potential over the interaction time b/v,

$$|\delta(v, r, x = 0)| = \frac{1}{2}\frac{f_C^*}{m_e}\frac{b^2}{v^2} = \frac{1}{2}\frac{eq}{4\pi\varepsilon_0 m_e v^2} = \frac{b_\perp}{2} \Rightarrow \nabla\delta = -\frac{b_\perp}{r}. \tag{7.23}$$

For a fixed $\mathbf{v}$ the deflections show cylindrical symmetry. Poisson's equation requires

$$\frac{1}{r}\partial_{rr}(r\Phi) = \frac{e}{\varepsilon_0}n_1 = -\frac{1}{\lambda^2}\Phi; \quad \lambda = \left(\frac{\varepsilon_0 m_e v^2}{n_0 e^2}\right)^{1/2}. \tag{7.24}$$

Relativistic screening. The screening length λ holds also for relativistic v. In fact, the time elements in the lab and comoving frames $\mathrm{d}t, \mathrm{d}t'$ are related by $\mathrm{d}t = \gamma(1 + \mathbf{v}\dot{\delta}/c^2)\mathrm{d}t' = \gamma\mathrm{d}t'$ owing to a good approximation $\mathbf{v} \perp \dot{\delta}$, see (7.19); $\gamma = (1 - v^2/c^2)^{-1/2}$. Therefore the momentum change in the center of mass system ($\sim$ lab frame) reduces to its nonrelativistic expression

$$\frac{\mathrm{d}(\gamma m_e \mathbf{v})}{\mathrm{d}t} = \gamma m_e \frac{\mathrm{d}(\dot{\boldsymbol{\delta}}/\gamma)}{\mathrm{d}t} = m_e \frac{\mathrm{d}\dot{\boldsymbol{\delta}}}{\mathrm{d}t}.$$

For the monoenergetic isotropic electron distribution the relativistic screened potential $\Phi_s = \Phi$ is Debye-like. Φ_s is of considerable relevance since a collimated monoenergetic electron beam first becomes isotropic and only on the next longer time scale it diffuses into energy space. Screening from an arbitrary isotropic distribution function $f(v)$, e.g., a *super-Gaussian* $\exp(-\kappa v^5)$, is obtained from averaging $1/\lambda^2$ in (7.24) (not λ^2, Poisson equation is linear in Φ),

$$\Phi_s = \frac{q}{4\pi\varepsilon_0 r} \exp -(r/\lambda_s); \quad \lambda_s = \left(\frac{\varepsilon_0 m_e}{n_0 e^2 \langle 1/v^2 \rangle}\right)^{1/2}. \tag{7.25}$$

For $f(\mathbf{v})$ Maxwellian, $\langle 1/m_e v^2 \rangle = 1/k_B T_e$. Bare ion screening by the electrons results in the thermal Debye potential Φ_D of range λ_D,

$$\Phi_D = \frac{q}{4\pi\varepsilon_0 r} \exp -(r/\lambda_D); \quad \lambda_D = \left(\frac{\varepsilon_0 k_B T_e}{n_0 e^2}\right)^{1/2}. \tag{7.26}$$

The average in λ_s differs from the kinetic temperature; note $\langle m_e v^2/2 \rangle = 3k_B T_e/2$. Equations (7.1) and (7.23) are valid for $1 - \cos\epsilon = \epsilon^2/2 = (\delta/r)^2/2 \ll 1$. ϵ is the deviation of $\boldsymbol{\delta}$ from the normal to $\mathbf{v}$ in Fig. 7.1. In the weakly coupled plasma $b_\perp \ll \lambda_s$ and ϵ is small for the orbits contributing to screening. For numerical values and two temperature screening see (4.69) and (4.71).

7.2 Collisional Absorption in the Thermal Plasma

The physical mechanism of collisional absorption was analyzed in Chap. 1 under the assumption that electrons and ions are elastic hard spheres. This model explains how irreversibility, i.e., heating, and anisotropy of the electron distribution function come about. It was also shown that all that is required of particle kinetics to determine absorption is the knowledge of the electron-ion (and, perhaps, the electron-electron) collision frequency, provided the model of a fully ionized plasma is valid. As long as the thermal electron speed is nonrelativistic all electrons have a mass close to their rest mass m_e and "see" nearly the same electric field $\mathbf{E}(\mathbf{x}, t)$ with a frequency ω' close to ω so that they all oscillate locally at nearly identical $\mathbf{v}_{\mathrm{os}}$. Hence, the momentum exchange in electron-electron collisions, occurring with frequency ν_{ee}, only leads to isotropization of the distribution function $f(\mathbf{x}, \mathbf{v}_e, t)$, and at later times to thermalization, i.e., Maxwellization. In the reference system comoving at $\mathbf{u} = \mathbf{v}_{\mathrm{os}}(t)$ an electron-electron collision looks the same as in the absence of the laser field. Thus, they do not directly contribute to absorption.

In plasmas produced by high intensity lasers the electron temperatures are generally such that the model of a two-component fully ionized plasma is valid. The frictional force, introduced phenomenologically by (5.77), originates, under the majority of conditions, from long-range, small-angle electron-ion collisions. As a consequence, the collision frequency ν_{ei} becomes highly dependent on the energy of the colliding electrons and, under standard conditions, it is the result of thousands of *simultaneous* collisions rather than of single binary events. Expression (1.30) of ν_{ei} was derived for a Maxwellian velocity distribution, with $|\mathbf{v}_{os}| \ll v_{e,\mathrm{th}}$, and Coulomb logarithm $\ln \Lambda$ appreciably larger than unity. Since in laser plasmas each of these conditions may be violated, a more general derivation of ν_{ei} is needed.

Large radiation field effects, $|\mathbf{v}_{os}| \simeq v_{e,\mathrm{th}}$, were considered first by Silin [6] on the basis of a classical analysis. Early quantum mechanical treatments were given in [7, 8]. A similar treatment, for dense plasmas now, has only later been undertaken [9, 10]. In strong heating by lasers, the relaxation of slow electrons is faster than that of the more energetic electrons, not to mention the *hot electron* component. As a consequence the electron velocity distribution may become non-Maxwellian and nonisotropic.

In the following, the main physical effects determining collisional absorption are described by introducing appropriate models in the test particle approximation. First, the collision frequency ν_{ei} for a monoenergetic unidirectional electron beam has been treated previously in the oscillator model approximation to show the interplay with screening. This model turns out to be particularly useful for explaining dynamical screening of the ion charges and for determining the various cut-offs in a self-consistent way and the validity of the impact approximation in the *ballistic model* used by Pert [11]. The oscillator model is relevant for ion beam stopping and, to a certain degree, for absorption of superintense laser beams in dense targets.

7.2.1 The Ballistic Model of Collisional Absorption

Collisions of individual electrons with a fixed ion of charge number Z are considered. The momentum loss along the original electron trajectory due to deflection is calculated and the result is averaged over all impact parameters b and all velocities $\mathbf{v}(t) = \mathbf{v}_{os}(t) + \mathbf{v}_e$, $\mathbf{v}_e$ taken from $f(\mathbf{x}, \mathbf{v}_e, t)$ in the oscillating frame. As a result the time-dependent collision frequency $\nu_{ei}(t)$ is obtained. Finally, cycle averaging yields $\overline{\nu}_{ei}$. For simplicity the distribution function f is assumed to be homogeneous and locally isotropic; hence $f(\mathbf{x}, \mathbf{v}_e, t) = f(v_e, t)$. The momentum loss $\Delta\mathbf{p}$ in a Coulomb collision along $\mathbf{v}(t)$ follows from the differential Coulomb cross section σ_Ω in (2.149) by integration over b, with $\sigma_t = \pi b_{\max}^2$ the total cross section,

$$\Delta \mathbf{p} = m_e \mathbf{v} \frac{b_\perp^2}{\sigma_t} \int_{\theta=\varepsilon}^{\pi} \frac{(1-\cos\theta)\sin\theta}{4\sin^4\frac{\theta}{2}} \mathrm{d}\theta \mathrm{d}\varphi = 4\pi m_e \mathbf{v} \frac{b_\perp^2}{\sigma_t} \int_{\theta=\varepsilon}^{\pi} \frac{\mathrm{d}\sin\frac{\theta}{2}}{\sin\frac{\theta}{2}}$$
$$= 4\pi m_e \mathbf{v} \frac{b_\perp^2}{\sigma_t} \ln \frac{(b_\perp^2 + b_{\max}^2)^{1/2}}{b_\perp}. \tag{7.27}$$

The momentum loss per unit time is $\dot{\mathbf{p}}(t) = \sigma_t n_i |\mathbf{v}(t)| \Delta \mathbf{p}$,

$$\dot{\mathbf{p}}(t) = -m_e \nu_{ei}(\mathbf{v})\mathbf{v} = -\frac{K}{v^3}\mathbf{v} \ln \Lambda(\mathbf{v}), \quad K = \frac{Z^2 e^4 n_i}{4\pi\varepsilon_0^2 m_e}, \quad \Lambda(\mathbf{v}) = \frac{(b_\perp^2 + b_{\max}^2)^{1/2}}{b_\perp}. \tag{7.28}$$

By the first equality the collision frequency $\nu_{ei}(\mathbf{v})$ for the momentum loss $\mathbf{p} = m\mathbf{v}$ is defined. It is identical with the expression resulting from the Boltzmann collision integral (3.83). Dimensionally $[K] = \mathrm{gcm^3s^{-4}}$. Equation (7.28) implies that simultaneous collisions are reduced to a sequence of successive encounters since small angle deflections obey the relations

$$\langle \Delta \mathbf{v} \rangle = \mathbf{v}\langle [1-\cos\theta(t)] \rangle = \mathbf{v}\langle \theta^2(t) \rangle /2 = \mathbf{v} \sum_{i=1}^{N} \theta_i^2(t)/2 = \mathbf{v}/v^2 \sum \dot{\delta}_i^2(t)/2. \tag{7.29}$$

in perfect analogy to (7.2). To obtain the ensemble-averaged momentum loss $\langle \dot{\mathbf{p}} \rangle$ averaging has to be done on $\mathbf{v}/v^3$ over the thermal velocities $\mathbf{v}_e$. For an isotropic distribution function $f(v_e)$ the velocity $\mathbf{v}$ consists of all vector sums as sketched in Fig. 7.5. The quantity $\dot{\mathbf{p}}$ is parallel to $\mathbf{v}$, whereas $\langle \dot{\mathbf{p}} \rangle$ is directed along $\mathbf{v}_{\mathrm{os}}$. With the angle χ as indicated in the Figure and $f(v_e)$ normalized to unity we find

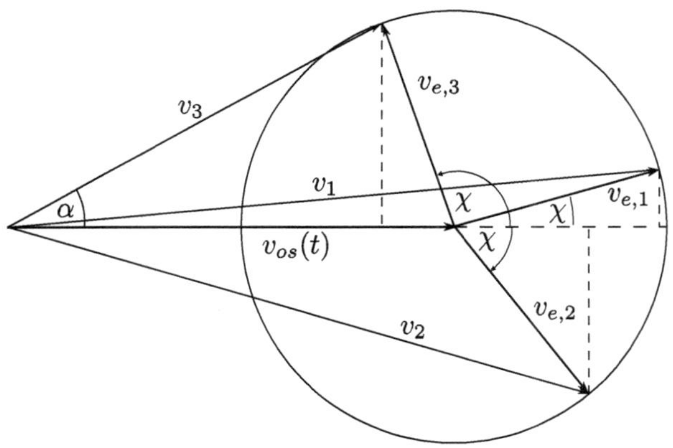

Fig. 7.5 Ballistic model. Individual velocity vector $\mathbf{v}_i$ is the sum of the oscillating laser induced component $\mathbf{v}_{\mathrm{os}}(t)$ and a thermal component $\mathbf{v}_{e,i}$ from an isotropic electron distribution function (indicated by circle): $\mathbf{v}_i = \mathbf{v}_{e,i} + \mathbf{v}_{\mathrm{os}}(t), \;\; i = 1, 2, 3$

$$\langle \dot{\mathbf{p}} \rangle = -m_e \nu_{ei}(t) \mathbf{v}_{os}$$
$$= -K \int_0^\infty \int_0^\pi \frac{\mathbf{v}_{os}(1 + v_e \cos\chi / v_{os})}{v^3} v_e^2 f(v_e) \ln \Lambda(v)\, 2\pi \sin\chi \mathrm{d}\chi \mathrm{d}v_e. \quad (7.30)$$

The combined velocity is $v = |\mathbf{v}| = (\mathbf{v}_{os}^2 + \mathbf{v}_e^2 + 2\mathbf{v}_{os}\mathbf{v}_e)^{1/2}$. The collision frequencies $\nu_{ei}(t)$ and $\overline{\nu}_{ei}$ follow as

$$\nu_{ei}(t) = 2\pi \frac{K}{m_e v_{os}(t)} \int_0^\infty \int_{-1}^1 \frac{v_{os}(t)|1 + v_e \cos\chi / v_{os}(t)|\, v_e^2 f(v_e) \ln \Lambda(v)}{(v_{os}^2 + v_e^2 + 2 v_{os} v_e \cos\chi)^{3/2}} \mathrm{d}\cos\chi \mathrm{d}v_e. \quad (7.31)$$

$$\overline{\nu}_{ei} = 4\pi \frac{K}{m_e \hat{v}_{os}^2} \overline{\int_0^\infty \int_{-1}^1 \frac{|\mathbf{v}_{os}\mathbf{v}|}{v^3} v_e^2 f(v_e) \ln \Lambda(v)\, \mathrm{d}\cos\chi \mathrm{d}v_e}. \quad (7.32)$$

These expressions are simple, the integrals are regular, however they do not handle collective resonances around $\omega = \omega_p$. The integration over b in the Coulomb logarithm runs from $b = 0$ to $b = b_{max}$. No lower cut-off appears, only b_{max} must be determined. Note, (7.32) is the time average of $\nu_{ei}(t)$ according to definition (7.21), with $m_e \hat{v}_{os}^2 = 2\overline{\mathcal{E}}$ for sinusoidal $v_{os}(t)$. It has been tacitly assumed that no dc drift is generated by the laser field and $\mathbf{v}_{os}$ is symmetric around $\mathbf{v}_{os} = 0$. Averaging in time is to be understood as done over the positive half cycle.

Low field limit. The distribution function $f(\mathbf{v})$ is isotropic in the oscillating frame. Under the assumption of $|\hat{\mathbf{v}}_{os}| \ll v_{th}$ expansion in $\mathbf{v}_{os}$ to first order yields

$$f(\mathbf{v}) = f(\mathbf{v}_{os} + \mathbf{v}_e) = f(v_e) + \mathbf{v}_{os} \nabla_v f(\mathbf{v})\Big|_{\mathbf{v}_{os}=0} = f(v_e) + \mathbf{v}_{os} \frac{\mathbf{v}_e}{v_e} \frac{\partial f_e}{\partial v_e}. \quad (7.33)$$

With this dipole expansion, and $|\mathbf{v}_{os} + \mathbf{v}_e|$ replaced by v_e, ν_{ei} is obtained from the average of the projection of (7.28) into the direction of $\mathbf{v}_{os}$,

$$\nu_{ei}(t) = 2\pi \frac{K}{m_e v_{os}(t)} \int_{\cos\chi=-1}^{\cos\chi=+1} \int_0^\infty v_{os}(t) \cos^2\chi \mathrm{d}\cos\chi \times \frac{\partial f_e}{\partial v_e} \ln \Lambda(v_e) \mathrm{d}v_e. \quad (7.34)$$

It is time-independent and thus identical with $\overline{\nu_{ei}(t)}$. No contribution to friction results from $f(v_e)$ in (7.33). With f_e Maxwellian and the slowly varying $\ln \Lambda(v)$ approximated by $\ln \Lambda(v_{th})$ in a wide parameter range the *Spitzer-Braginskii-Silin* low field formula is recovered [6, 12],

$$\nu_{ei} = \overline{\nu}_{ei} = \frac{1}{3}\sqrt{\frac{2}{\pi}} \frac{K}{m_e v_{th}^3} \ln \Lambda(v_{th}); \quad v_{th} = \left(\frac{k_B T_e}{m_e}\right)^{1/2}. \quad (7.35)$$

For an alternative derivation of this expression see Problems.

High field. Straightforward evaluation of $\overline{\nu}_{ei}$ from (7.31) is possible under the simplification of $\ln \Lambda(\mathbf{v}) \simeq \ln \langle \Lambda \rangle$, in analogy to (7.35). Then, the integral in (7.31) assumes the structure of the gravitational attraction of a point mass by a spherical mass distribution centered at distance R. In fact, averaging over all angles χ yields

$$\left\langle \frac{\mathbf{v}}{v^3} \right\rangle = -\left\langle \frac{\partial}{\partial \mathbf{v}_{\text{os}}} \frac{1}{v} \right\rangle = -\frac{\partial}{\partial \mathbf{v}_{\text{os}}} \left\langle \frac{1}{v} \right\rangle$$

for the "gravitational" potential $1/v$. Mass shells of radius $r > R$ do not contribute to attraction. Correspondingly, v_e extends up to $v_{\text{os}}(t)$ and $|\mathbf{v}_{\text{os}}\mathbf{v}| = \mathbf{v}_{\text{os}}\mathbf{v}$ holds. With these observations in mind the instantaneous collision frequency from (7.31) results as

$$\nu_{\text{ei}}(t) = \frac{K}{m_e v_{\text{os}}^3(t)} \ln \langle \Lambda \rangle \int_0^{v_{\text{os}}(t)} 4\pi v_e^2 f(v_e) \mathrm{d}v_e. \tag{7.36}$$

For its time average again $\mathbf{v}_{\text{os}}(t)$ is assumed harmonic (for a short discussion of anharmonic $\mathbf{v}_{\text{os}}(t)$ see [13]). Averaging (7.36) over the positive half period results in

$$\overline{\nu}_{\text{ei}} = \frac{K}{m_e \hat{v}_{\text{os}}^2} \overline{\ln \langle \Lambda \rangle} \overline{\frac{1}{v_{\text{os}}(t)} \int_0^{v_{\text{os}}(t)} 4\pi v_e^2 f(v_e) \mathrm{d}v_e}. \tag{7.37}$$

For a Maxwellian (7.36) can be integrated with the help of the error function $\mathrm{erf}(x) = (2/\sqrt{\pi}) \int_0^x \exp -w^2 \mathrm{d}w$,

$$\begin{aligned} \nu_{\text{ei}}(t) &= 4\pi \left(\frac{\beta}{\pi}\right)^{3/2} \frac{K \ln \langle \Lambda \rangle}{m_e v_{\text{os}}^3(t)} \int_0^{v_{\text{os}}(t)} v_e^2 \exp(-\beta v_e^2) \mathrm{d}v_e \\ &= \frac{K \ln \langle \Lambda \rangle}{m_e v_{\text{os}}^3(t)} \left[\mathrm{erf}\left(\frac{v_{\text{os}}(t)}{\sqrt{2} v_{\text{th}}}\right) - \left(\frac{2}{\pi}\right)^{1/2} \frac{v_{\text{os}}(t)}{v_{\text{th}}} \exp\left(-\frac{v_{\text{os}}^2(t)}{2 v_{\text{th}}^2}\right) \right]. \end{aligned} \tag{7.38}$$

For the evaluation of (7.37) with a Maxwellian it is advisable to expand the exponential in Taylor series and average in time,

$$\overline{\nu}_{\text{ei}} = 2\sqrt{\frac{2}{\pi}} \frac{K \ln \langle \Lambda \rangle}{m_e \hat{v}_{\text{os}}^2 v_{\text{th}}} \sum_{n \geqslant 1} (-1)^{n-1} \frac{\overline{w^{2n}}}{(2n+1)(n-1)!}, \quad w = \frac{v_{\text{os}}(t)}{\sqrt{2} v_{\text{th}}}. \tag{7.39}$$

The averages of even powers of $w \sim \sin \omega t, \cos \omega t$ can be found in [14]. At the drift $\hat{v}_{\text{os}}$ as large as v_{th} from (7.39) a reduction of $\overline{\nu}_{\text{ei}}$ with respect to (7.35) of 21% follows. Thereby additionally $\ln \Lambda(v) \simeq \ln \Lambda(v_{\text{th}})$ is assumed. It shows that the Spitzer collision frequency (7.35) is still an acceptable first approximation.

7.2.2 The Dielectric Model of Collisional Absorption

The random motion of the plasma electrons in the mean field approximation may be described by a classical one particle distribution function $f(\mathbf{x}, \mathbf{v}_e, t)$ when its kinetic temperature T_e is much higher than the Fermi temperature $T_F = E_F/k_B$. In the presence of the laser field a drift velocity is superposed and hence the distribution function $g(\mathbf{x}', \mathbf{v}', t)$ in the lab frame is

$$g(\mathbf{x}', \mathbf{v}', t) = f(\mathbf{x}, \mathbf{v}_e, t), \quad \mathbf{v}' = \mathbf{v}_e + \mathbf{v}_{\text{os}}(t), \quad \mathbf{x}' = \mathbf{x} + \boldsymbol{\delta}_{\text{os}}(t), \quad \boldsymbol{\delta}_{\text{os}}(t) = \int \mathbf{v}_{\text{os}} \mathrm{d}t. \tag{7.40}$$

As a consequence of the attraction by an ion an electric field $\mathbf{E}_{\text{in}}$ is induced by thermal and dynamical shielding. In the case that the screening length λ is much larger than $b_\perp$ and λ_B the majority of the particle orbits is nearly straight and again a collision parameter b_0, $\max(b_\perp, \lambda_B) \ll b_0 \ll \lambda$, can be introduced which separates close encounters from remote ones. The electrons with $b > b_0$ behave fluid-like and are adequately described by the linearized dielectric Vlasov model. The Vlasov equation (3.79) with a test ion of charge $q = Ze$ reads in the oscillatory frame $\mathbf{x} = \mathbf{x}' - \boldsymbol{\delta}_{\text{os}}(t)$

$$\frac{\partial f}{\partial t} + \mathbf{v}_e \frac{\partial f}{\partial \mathbf{x}} + \frac{e}{m_e} \nabla \Phi \frac{\partial f}{\partial \mathbf{v}_e} = 0, \tag{7.41}$$

$$\Phi = \Phi_{\text{C}} + \Phi_{\text{in}}, \quad \Phi_{\text{C}} = \frac{q}{4\pi\varepsilon_0 |\mathbf{x}'|}, \quad \nabla \Phi_{\text{in}} = -\mathbf{E}_{\text{in}}.$$

Φ_{in} is the induced screening potential. The Coulomb potential Φ_{C} of the ion oscillates around its position in the lab frame $\mathbf{x}' = 0$ according to $\mathbf{x} = \mathbf{x}' - \boldsymbol{\delta}_{\text{os}}(t)$. By setting $f = f_0(\mathbf{v}_e) + f_1(\mathbf{x}, \mathbf{v}_e, t)$, with f_1 the disturbance introduced by the test charge, and linearizing around f_0 yields

$$\frac{\partial f_1}{\partial t} + \mathbf{v}_e \frac{\partial f_1}{\partial \mathbf{x}} + \frac{e}{m_e} \nabla (\Phi_{\text{C}} + \Phi_{\text{in}}) \frac{\partial f_0}{\partial \mathbf{v}_e} + \frac{e}{m_e} \nabla \Phi_{\text{C}} \frac{\partial f_1}{\partial \mathbf{v}_e} = 0. \tag{7.42}$$

Poisson's equation requires

$$\nabla^2 \Phi_{\text{in}} = \frac{n_0 e}{\epsilon_0} \int f_1 \, \mathrm{d}\mathbf{v}_e, \qquad \nabla^2 \Phi_{\text{C}} = -\frac{q}{\epsilon_0} \delta(\mathbf{x} + \boldsymbol{\delta}_{\text{os}}(t)). \tag{7.43}$$

It is standard to omit the last term in (7.42) although it is of first order. By its neglect strong ion field effects (close encounters, bent orbits) are excluded. The ∇ operators are removed by a Fourier transform in space, $\tilde{\Phi}(\mathbf{k}, \Omega) = (2\pi)^{-2} \int \Phi(\mathbf{x}, t)$ $\exp(-i\mathbf{k}\mathbf{x} + i\Omega t)\,\mathrm{d}\mathbf{x}\mathrm{d}t$, etc., and (7.42), (7.43) reduce to algebraic relations. Equations (7.43) transform into

$$\tilde{\Phi}_{\text{in}}(\mathbf{k},\Omega) = -\frac{n_0 e}{\epsilon_0 \mathbf{k}^2}\int \tilde{f}_1(\mathbf{k},\Omega,\mathbf{v}_e)\,\mathrm{d}\mathbf{v}_e, \tag{7.44}$$

$$\begin{aligned}\tilde{\Phi}_{\text{C}}(\mathbf{k},\Omega) &= \frac{q}{(2\pi)^2\varepsilon_0\mathbf{k}^2}\int \delta(\mathbf{x}+\boldsymbol{\delta}_{\text{os}}(t))\,\mathrm{e}^{-i\mathbf{k}\mathbf{x}+i\Omega t}\,\mathrm{d}\mathbf{x}\mathrm{d}t \\ &= \frac{q}{(2\pi)^2\varepsilon_0\mathbf{k}^2}\int \mathrm{e}^{i(\Omega t+\mathbf{k}\hat{\boldsymbol{\delta}}_{\text{os}}\cos\omega t)}\,\mathrm{d}t \\ &= \frac{q}{2\pi\varepsilon_0\mathbf{k}^2}\sum_{n=-\infty}^{\infty} i^n\delta(\Omega+n\omega)\,J_n(\mathbf{k}\hat{\boldsymbol{\delta}}_{\text{os}}),\end{aligned} \tag{7.45}$$

where the identities

$$\mathrm{e}^{\pm iz\cos\zeta} = \sum_{n=-\infty}^{\infty}(\pm i)^n J_n(z)\,\mathrm{e}^{\pm in\zeta}, \qquad \delta(z) = (2\pi)^{-1}\int \mathrm{e}^{ikz}\,\mathrm{d}k$$

for the Bessel functions of integer order J_n and for the delta function have been used. The Fourier transform of (7.42) yields

$$(\Omega - \mathbf{k}\mathbf{v}_e)\tilde{f}_1(\mathbf{k},\Omega) = \frac{e}{m_e}(\tilde{\Phi}_{\text{C}} + \tilde{\Phi}_{\text{in}})\mathbf{k}\frac{\partial f_0}{\partial \mathbf{v}_e}. \tag{7.46}$$

In a static linear homogeneous medium the dielectric constant ε connects the external, induced, and total electric fields $\mathbf{E}_{\text{ex}}$, $\mathbf{E}_{\text{in}}$, $\mathbf{E}$ with the polarization $\mathbf{P}$ in the following way,

$$\varepsilon\mathbf{E} = \varepsilon(\mathbf{E}_{\text{ex}} + \mathbf{E}_{\text{in}}) = \mathbf{E} + \frac{1}{\varepsilon_0}\mathbf{P}.$$

By means of Poisson's equation and $\nabla\mathbf{P} = -\rho_{\text{in}}$ this translates into

$$\varepsilon\rho = (\rho - \rho_{\text{in}}) = \rho_{\text{ex}}, \quad \varepsilon\Phi = (\Phi - \Phi_{\text{in}}) = \Phi_{\text{ex}} \tag{7.47}$$

for the total, external, and induced charge densities ρ, ρ_{ex}, ρ_{in} and potentials Φ, Φ_{ex}, Φ_{in}. Thus, $\rho = \rho_{\text{ex}}/\varepsilon$, $\mathbf{E} = \mathbf{E}_{\text{ex}}/\varepsilon$, $\Phi = \Phi_{\text{ex}}/\varepsilon$.

Introducing the *dielectric function* $\varepsilon = \varepsilon(\mathbf{k},\Omega)$, within the validity of linear response these relations hold for a single Fourier component in fields varying in space and time. Substituting $\tilde{f}_1$ from (7.46) in $\tilde{\Phi}_{\text{in}}$ in (7.44) yields

$$\tilde{\Phi}_{\text{C}} = \tilde{\Phi} - \tilde{\Phi}_{\text{in}} = \tilde{\Phi}\left\{1 + \frac{n_0 e^2}{\varepsilon_0\mathbf{k}^2 m}\int \frac{\mathbf{k}\partial f_0/\partial\mathbf{v}_e}{\Omega - \mathbf{k}\mathbf{v}_e}\,\mathrm{d}\mathbf{v}_e\right\}.$$

Hence, owing to $\Phi_{\text{C}} = \Phi_{\text{ex}}$, $\rho_{\text{C}} = \rho_{\text{ex}}$, and $\tilde{\Phi}_{\text{C}} = \tilde{\Phi}_{\text{ex}}$, $\tilde{\rho}_{\text{C}} = \tilde{\rho}_{\text{ex}}$

$$\varepsilon(\mathbf{k},\Omega) = \frac{\tilde{\Phi}_{\text{C}}}{\tilde{\Phi}} = 1 + \frac{\omega_p^2}{\mathbf{k}^2}\int \frac{\mathbf{k}\partial f_0/\partial\mathbf{v}_e}{\Omega - \mathbf{k}\mathbf{v}_e}\,\mathrm{d}\mathbf{v}_e. \tag{7.48}$$

It is analytic in the entire half plane $\Im\Omega > 0$. For a real quantity $h(\mathbf{x}, t)$ the relation $\tilde{h}^*(\mathbf{k}, \Omega) = \tilde{h}(-\mathbf{k}^*, -\Omega^*)$ holds and, consequently, $\varepsilon^*(\mathbf{k}, \Omega) = \varepsilon(-\mathbf{k}^*, -\Omega^*)$. The total potential $\Phi(\mathbf{x}, t) = \Phi_C(\mathbf{x}, t) + \Phi_{in}(\mathbf{x}, t)$ reads now in the frame moving with $\mathbf{v}_{os}(t)$

$$\Phi(\mathbf{x}, t) = \frac{1}{(2\pi)^3}\frac{q}{\varepsilon_0}\sum_{n=-\infty}^{\infty}\int \frac{i^n}{\mathbf{k}^2\varepsilon(\mathbf{k}, -n\omega)} J_n(\mathbf{k}\hat{\boldsymbol{\delta}}_{os})\,\mathrm{e}^{i\mathbf{kx}+in\omega t}\,\mathrm{d}\mathbf{k}. \tag{7.49}$$

The force $\mathbf{f}_q$ acting on the ion at position $\mathbf{x}(t) = -\hat{\boldsymbol{\delta}}_{os}\cos\omega t$ is

$$\begin{aligned}
\mathbf{f}_q &= -q\nabla\Phi(-\hat{\boldsymbol{\delta}}_{os}\cos\omega t, t)\\
&= -\frac{1}{(2\pi)^3}\frac{q^2}{\varepsilon_0}\sum_{n=-\infty}^{\infty}\int \frac{i^{n+1}\mathbf{k}\,J_n(\mathbf{k}\hat{\boldsymbol{\delta}}_{os})}{\mathbf{k}^2\varepsilon(\mathbf{k}, -n\omega)}\,\mathrm{e}^{-i\mathbf{k}\hat{\boldsymbol{\delta}}_{os}\cos\omega t+in\omega t}\,\mathrm{d}\mathbf{k}\\
&= -\frac{1}{(2\pi)^3}\frac{q^2}{\varepsilon_0}\sum_{l,n=-\infty}^{\infty}\int \frac{i^{n+1-l}\mathbf{k}}{\mathbf{k}^2\varepsilon(\mathbf{k}, -n\omega)} J_l(\mathbf{k}\hat{\boldsymbol{\delta}}_{os})\,J_n(\mathbf{k}\hat{\boldsymbol{\delta}}_{os})\,\mathrm{e}^{i(n-l)\omega t}\,\mathrm{d}\mathbf{k}.
\end{aligned}$$

The cycle-averaged energy absorbed per unit volume and unit time is

$$\begin{aligned}
\overline{\mathbf{jE}} &= -\frac{n_0}{Z}\overline{\mathbf{f}_q\mathbf{v}_{os}(t)} = -i\frac{n_0}{Z}\frac{1}{2m_e\omega}\overline{\hat{\mathbf{E}}\mathbf{f}_q(\mathrm{e}^{\{i\omega t} - \mathrm{e}^{-i\omega t})}\\
&= \frac{n_0e^2q}{2m_e\omega\varepsilon_0}\frac{\hat{E}}{(2\pi)^3}\sum_{n=-\infty}^{\infty}\int\frac{i\mathbf{k}\,\mathrm{d}\mathbf{k}}{\mathbf{k}^2\varepsilon(\mathbf{k}, -n\omega)}\\
&\qquad\times\{J_{n-1}(\mathbf{k}\hat{\boldsymbol{\delta}}_{os})J_n(\mathbf{k}\hat{\boldsymbol{\delta}}_{os}) + J_n(\mathbf{k}\hat{\boldsymbol{\delta}}_{os})J_{n+1}(\mathbf{k}\hat{\boldsymbol{\delta}}_{os})\}\\
&= \frac{1}{(2\pi)^3}\frac{\omega_p^2}{\omega}q\hat{\mathbf{E}}\sum_{n=-\infty}^{\infty}\int\frac{i\mathbf{k}\,\mathrm{d}\mathbf{k}}{\mathbf{k}^2\varepsilon(\mathbf{k}, -n\omega)}J_n^2(\mathbf{k}\hat{\boldsymbol{\delta}}_{os})\frac{n}{\mathbf{k}\hat{\boldsymbol{\delta}}_{os}}.
\end{aligned}$$

In the last step the property $J_{n-1}(x) + J_{n+1}(x) = 2nJ_n(x)/x$ was used. $\hat{\mathbf{E}}$ is the real laser field amplitude. Owing to the cylindrical symmetry the integration over $\mathbf{k} \perp \hat{\boldsymbol{\delta}}_{os}$ can be performed easily in polar coordinates,

$$\begin{aligned}
\overline{\mathbf{jE}} &= \frac{1}{(2\pi)^3}\frac{\omega_p^2}{\omega}q\hat{E}\sum_{n=-\infty}^{\infty}\int_0^{\infty}\frac{i\,\mathrm{d}k}{\varepsilon(k, -n\omega)}\frac{2\pi}{\hat{\delta}_{os}}\int_{-1}^{1}nJ_n^2k\hat{\delta}_{os}\cos\chi\,\mathrm{d}\cos\chi\\
&= m_e\hat{v}_{os}Z\frac{\omega_p^2}{\pi^2\hat{\delta}_{os}^2}\sum_{n=1}^{\infty}n\int_0^{\infty}\frac{\mathrm{d}k}{k}\Im\frac{1}{\varepsilon(k, -n\omega)}\int_0^{k\hat{\delta}_{os}}J_n^2(\xi)\,\mathrm{d}\xi.
\end{aligned} \tag{7.50}$$

In the last passage the properties $J_{-n}(\xi) = (-1)^nJ_n(\xi)$, $J_n^2(-\xi) = J_n^2(\xi)$, $\varepsilon(\mathbf{k}, -\Omega^*) = \varepsilon^*(\mathbf{k}, \Omega)$ for $\mathbf{k}$ real and $\Omega = -n\omega$, and $\hat{v}_{os} = e\hat{E}/m\omega$ were used. The electron distribution function $f_0(\mathbf{v}_e)$ is to be used in a system of reference in which no net drift results, i.e., $\int \mathbf{v}_e f_0(\mathbf{v}_e)\,\mathrm{d}\mathbf{v}_e = 0$; all drift is in the ions. The time-averaged collision frequency $\overline{\nu}_{ei}$ follows from (7.21).

For a Maxwellian distribution function f_e (7.48) reads

$$\varepsilon(k,\omega) = 1 + \frac{k_D^2}{k^2}\left(1 - 2\eta \mathrm{e}^{-\eta^2}\int_0^\eta \mathrm{e}^{x^2}\,\mathrm{d}x + i\sqrt{\pi}\eta\mathrm{e}^{-\eta^2}\right), \quad \eta = \frac{\omega}{\sqrt{2}kv_{\mathrm{th}}}. \quad (7.51)$$

The electron thermal velocity is $v_{\mathrm{th}} = (k_B T_e/m_e)^{1/2}$ and $k_D = 1/\lambda_D$, with $\lambda_D = v_{\mathrm{th}}/\omega_p$ the familiar electron Debye length. Simple expressions of $\overline{\mathbf{jE}}$ are attainable for $\hat{v}_{\mathrm{os}} < v_{\mathrm{th}}$ and $v_{\mathrm{os}} \gg v_{\mathrm{th}}$; for intermediate ratios of $v_{\mathrm{os}}/v_{\mathrm{th}}$ (7.50) must be evaluated numerically. The classical Coulomb (or Rutherford) differential cross section σ_Ω, formulated in terms of the impact parameter b, is identical with its quantum mechanical counterpart, formulated in terms of plane waves $\{\mathbf{k}\}$ (e.g. see [15], p. 387). The two variables b and $|\mathbf{k}|$ are related by $k = |\mathbf{k}| = 1/b$ and, correspondingly, $k_\perp = 1/b_\perp$ and $k_B = 1/\lambda_B$. If b is a few times larger than λ_B it is the center of a narrow wave packet around $k = 1/b$. Hence, for $\omega \gtrsim \omega_p$ the relevant k-interval in (7.51 is $k > k_D$ or, equivalently, $\omega_p/kv_{\mathrm{th}} \lesssim \sqrt{2}\eta < 1$. It yields

$$\Im\left(\frac{1}{\varepsilon}\right) \simeq -\left(\frac{\pi}{2}\right)^{1/2}\frac{k_D^2}{k^3}\frac{\omega}{v_{\mathrm{th}}}\frac{\mathrm{e}^{-\omega^2/2k^2v_{\mathrm{th}}^2}}{(1+k_D^2/k^2)^2}. \quad (7.52)$$

Let us take for the moment $\hat{v}_{\mathrm{os}} \ll v_{\mathrm{th}}$ to arrive at $k\hat{\delta}_{\mathrm{os}} \ll 1$ and $J_1(\xi) = \xi/2$ as the leading term owing to $J_n(\xi \ll 1) \simeq (\xi/2)^n/n!$. Hence, in the weak laser field approximation (7.50) contracts to

$$\overline{\mathbf{jE}} = \frac{4}{3}\left(\frac{\pi}{2}\right)^{1/2} Z\left(\frac{n_e e^2}{4\pi\varepsilon_0 m_e}\right)^2 \frac{m_e\hat{v}_{\mathrm{os}}^2}{v_{\mathrm{th}}^3}\int_0^{(\infty)}\frac{k^3}{(k^2+k_D^2)^2}\,\mathrm{d}k. \quad (7.53)$$

The indefinite integral is $[\ln(k^2+k_D^2) + k_D^2/(k^2+k_D^2)]/2$. It diverges for $k \to \infty$. This is a consequence of linearization and omission of the last term in (7.42) or, equivalently, of the straight orbit approximation. In perfect analogy to the treatment of close encounters by $D(\beta_0)$ in (7.13)ff, $k_0 = 1/b_0$ has to be introduced, with b_0 dividing straight orbits from bent trajectories. After folding with a Maxwellian and averaging $v_{\mathrm{os}}(t)\langle D(\beta_0)\rangle$ over one laser cycle the resulting quantity has to be added to (7.53). We conclude by observing that the classical dielectric model yields a lower self-consistent cut-off at $k = k_D$, but it diverges for large k-values associated with close electron-ion encounters. The divergence at large k-values is avoided in a quantum treatment, e.g., following a Kadanoff-Baym [9, 16] or a quantum Vlasov treatment [17]. A complementary derivation in some respect, based on a generalized linear response theory, is given in [18]. The authors arrive at the same expression (7.50) in which now the classical expression of $\varepsilon(k,\omega)$ is replaced by its quantum counterpart $\varepsilon_q(k,\omega)$ [19],

$$\varepsilon_q(k,\omega) = 1 + \frac{k_D^2}{\sqrt{2}k^2}\frac{k_B}{k}\left[D\left(\eta + \frac{k}{\sqrt{2}k_B}\right) - D\left(\eta - \frac{k}{\sqrt{2}k_B}\right)\right]$$
$$+ \{i\left(\frac{\pi}{2}\right)^{1/2}\frac{k_D^2}{\sqrt{2}k^2}\frac{k_B}{k}\mathrm{e}^{-\eta^2}\mathrm{e}^{-\frac{1}{2}(k/k_B)^2}\sinh(\sqrt{2}\eta) \qquad (7.54)$$

with η as before and $D(y) = \mathrm{e}^{-y^2}\int_0^y \mathrm{e}^{x^2}\,\mathrm{d}x$. The evaluation of (7.50) with $\varepsilon = \varepsilon_q$ is complicated. Following again [19] one arrives at

$$\overline{\mathbf{jE}} = \frac{Zm_e\omega_p^4}{\pi^2\hat{v}_{\mathrm{os}}}\int_0^\infty \frac{\mathrm{d}k}{k}F\left(\frac{k}{k_D},\frac{\omega}{\omega_p},\frac{\hat{v}_{\mathrm{os}}}{v_{\mathrm{th}}}\right), \qquad (7.55)$$

$$F = \left(\frac{\omega}{\omega_p}\right)^2\sum_{l=1}^{\infty} l\frac{\Im\varepsilon_q(k,il\omega)}{|\varepsilon_q(k,il\omega)|^2}\int_0^{k\hat{v}_{\mathrm{os}}/\omega}\mathrm{d}\xi\, J_l^2(\xi), \quad \overline{\nu}_{ei} = \frac{\overline{\mathbf{jE}}}{2n_eE_{\mathrm{os}}}$$

for all ratios $\hat{v}_{\mathrm{os}}/v_{\mathrm{th}}$. In many applications holds $\lambda_D \gg b_0 > \lambda_B \gtrsim b_\perp$. Then, for $\hat{v}_{\mathrm{os}} \ll v_{\mathrm{th}}$ with ε_q used in (7.50) the upper cut-off $k = k_{\mathrm{max}} \simeq k_B$ follows. The validity of (7.53) extends, strictly speaking, only up to $k = 1/b_0$. However, when adding the cycle-averaged contribution of the close encounters $v_{\mathrm{os}}(t)\langle D(\beta_0)\rangle$ the sum of both is obtained by extending the interval up to k_{max} and is insensitive to the special choice of the cut b_0 over an adequate range. Hence, the integral yields $\ln(\lambda_D/\lambda_B) + 0.06$ which is nearly identical with the familiar Coulomb logarithm $\ln\Lambda$, $\Lambda = \lambda_D/\lambda_B$, and ν_{ei} becomes identical to the *Spitzer-Braginskii collision frequency* (7.35) [12] for a Maxwellian plasma. It should be stressed that although electron trajectories with $b = \lambda_B$ are generally bent and invalidate any Born approximation or linearization (7.42), under the limitations above the correct result for ν_{ei} is obtained through the cut off at $b_{\mathrm{min}} \simeq \lambda_B$.

7.2.2.1 The Field-Free Coulomb Logarithm

It is standard to interprete Λ in the Coulomb logarithm $\ln\Lambda$ as the ratio of the *cut offs* $b_{\mathrm{max}}/b_{\mathrm{min}}$. In a linearized treatment of the collision frequencies of charged bare Coulomb particles, either plane wave approximation, i.e., Fourier analysis in the dielectric model, or straight orbit approximation in the ballistic model, singularities appear at $b \Rightarrow 0$ and $b \Rightarrow \infty$. In the ballistic model the singularity at $b = 0$ vanishes if bent orbits are included. However, with the bare Coulomb potential and its Rutherford differential cross section the singularity persists at $b = \infty$, see (7.27). The dielectric model can handle large distances selfconsistently (this is its strong point), however the singularity at $b = 0$ reappears as a consequence of linearization. In the ballistic model $b_{\mathrm{max}} = \lambda_D$ is set because any "correlations" are lost beyond. With this recipe and bent orbits included $\ln\Lambda = \{\ln[(b_\perp^2 + \lambda_D^2)/b_\perp^2]\}/2$ follows. Note, $b_\perp$ is the result of integrating b from zero, it is not a lower cut off b_{min}. Nevertheless, a cut off $b_{\mathrm{min}} = b_\perp$ is frequently justified by the *closest approach* $2b_\perp$ although the closest approach for an electron with a positive ion is $b = 0$. Another popular argument for $b_{\min}$ is the uncertainty of localization of a classical orbit better than λ_B. With decreasing impact energy E_r the de Broglie wavelength $\lambda_B \sim E_r^{-1/2}$ exceeds $b_\perp \sim E_r^{-1}$. Therefore the majority of representative textbooks and specialized papers, e.g., [20], adhere to the setting

$$b_{\min} = \max\{\lambda_B, b_\perp\}; \quad \lambda_B[\text{nm}] = \frac{0.185}{(E_r[\text{eV}])^{1/2}}, \quad b_\perp[\text{nm}] = \frac{0.7 \times Z}{E_r[\text{eV}]}. \tag{7.56}$$

No proof exists for such a setting. A special argument for it is by L. Spitzer [12], p. 128. He arrives at the limitation $b \geqslant \lambda_D$ by observing that for impact parameters $b \leqslant \lambda_D$ the Coulomb differential cross section leads to higher diffraction values than an opaque disc of the same radius, which is "unphysical". It seems that for numerous researchers this constitutes an intuitive argument. However, with Spitzer's argument some difficulty arises with Babinet's principle of classical optics. Furthermore, Spitzer's opaque disk generates fringes, the Debye potential does not. Finally, the lengthy discussion in Landau-Lifshitz, Vol.10, Chap. 4 is vague and, in parts it is unsatisfactory. The search for the correct Coulomb logarithm is a complicated problem of mathematical physics. An exact quantum mechanical scattering cross section exists for the Coulomb potential and is identical with the classical Rutherford cross section [15], however, it leads to divergent results for $b \to \infty$. In contrast an analytic quantum cross section for the Debye potential exists merely in first Born approximation. In spite of this deficiency the binary scattering problem for point charges can be solved in two relevant cases under strict adherence to the quantum view, the one is the low field limit (7.34), the other one is the oscillator model of uniform drift. Starting from the two effective Hamiltonians we shall get the correct "lower cut offs" by subjecting them to the standard quantum procedure and to arrive at a coherent interpretation of their meaning that is free of contradictions within the limits of validity of the models under investigation.

On the way to the solution it will appear essential to distinguish between the validity of classical mechanics for the *single* orbit as the limiting quantum case and the correctness of this criterion when properties of orbits are to be determined that are the result of folding on the *totality* of the orbits. There is no general rule for a legitimate transition to the simpler classical model. In one and the same problem it depends on the kind of the variable under consideration whether a given condition for passing to a classical analysis is fulfilled. Examples are the equation of state of an ideal gas and its mixing entropy; the first is classical, the second quantity follows the Sackur-Tetrode formula and not its classical counterpart based on all states accessible in the classical phase space. A particularly simple example is the pressure exerted by a beam of independent particles of momentum $\mathbf{p} = \text{const}$. The classical and the quantum result is $(2)n|\mathbf{p}|$, n particle density, factor 2 for reflection. A further example is the connection between pressure p and energy density ϵ of an ideal gas of $f = 3$ degrees of freedom in thermal equilibrium, $p/\epsilon = f/(f+2)$ for non-relativistic cold Fermi and classical gases, and $p/\epsilon = f/(f+1)$ for them in the superrelativistic regime and for photons. On the other hand, a photon number state

$|n\rangle$ never turns over into "classical light" regardless of how large the photon number is in the mode [21].

Shielding symmetry may determine the lower cut off. The effective Hamiltonian for the spherical Debye-Yukawa potential $\Phi_s(r)$ is

$$H(\mathbf{p}, \mathbf{r}) = \frac{\mathbf{p}^2}{2\mu} + \Phi_s(r), \qquad r = |\mathbf{r}|. \tag{7.57}$$

The standard scattering theory for free-free transitions yields the differential cross section σ_s in first Born approximation [15],

$$\sigma_s(\vartheta) = \frac{b_\perp^2}{4[\sin^2(\vartheta/2) + (\lambda_B/2\lambda_s)^2]^2}; \qquad \tan\frac{\vartheta}{2} = \frac{b_\perp}{b}. \tag{7.58}$$

The term $\rho = \lambda_B/2\lambda_s$ in σ_s is the contribution from the shielding factor $\exp(-r/\lambda_s)$ to the Fourier transform of the Debye potential. For $\lambda_s = \infty$, σ_s shrinks to the well-known Rutherford or bare Coulomb potential σ_C. For large $k \Leftrightarrow$ small $b \sim 1/k$ only the $\{r/\lambda_s \ll 1\}$ region contributes and hence $\sigma_s = \sigma_C$. For small k the outer region $r \simeq \lambda_s$ counts where $\cos\vartheta = 1 - \vartheta^2/2 \lesssim 1 - (\lambda_B/2\lambda_s)^2$. Equation (7.58) inserted in (7.27) and integrated from $b = 0 \Leftrightarrow \vartheta = \pi$ to $b = \infty \Leftrightarrow \vartheta = 0$ yields the Coulomb logarithm

$$L_C = \frac{1}{2}\left[\ln\left(\frac{1}{\rho^2} + 1\right) - \frac{1}{1+\rho^2}\right] \simeq \frac{1}{2}\left(\ln\frac{1}{\rho^2} + 2\rho^2 - 1\right)$$

$$= \ln\frac{\lambda_s}{\lambda_B} + 0.2 + \frac{\lambda_B^2}{4\lambda_s^2}. \tag{7.59}$$

Note, here b is merely an integration variable, related to the scattering angle ϑ by (7.58), and not affected by whether an orbit is classical or not.

Expression (7.59) shows that in the ideal non degenerate plasma with $\lambda_D \gg \lambda_B$ the correct lower cutoff is the de Broglie length and not the parameter for perpendicular deflection. The cut offs $b_{\min}$ and $b_{\max}$ are the result of integration of the impact parameter from 0 to ∞; no additional physical hypotheses are needed. The integral is regular in the whole domain, the neighborhood of $r = 0$ is in no way special. Finally, there is no basis for the simple rule as $b_{\min} = \max\{\lambda_B, b_\perp\}$. The situation with $b_\perp \gg \lambda_B$ is more complicated and not analyzed here. The analysis given here leads to a completely different interpretation of why the de Broglie length comes into play in the Coulomb logarithm. Inspection of (7.58) shows that for small k's, i.e., large b's the outer region $r \simeq \lambda_s$ counts where $\cos\vartheta = 1 - \vartheta^2/2 \lesssim 1 - (\lambda_B/2\lambda_s)^2$. Hence, contrary to the conventional interpretation of $\ln\Lambda$, screening due to the outer regions is responsible for the "lower cut off" and not the singularity of the Coulomb potential at $r = 0$. This interpretation is found also in the representative textbook

Tokamaks by John Wesson [22]. It is the only citation among all papers known to us so far. Apparently its impact has been almost zero.

Validity of the 1st Born approximation. For the first Born approximation to be correct the local, partially scattered state function $\psi(\mathbf{r})$ should be close to the incident wave $\phi(\mathbf{r})$ everywhere. In other words, scattering must be weak for all angles ϑ. We can assume that with the Debye potential this is true for the following reason. The Debye potential Φ_s is smooth and weaker than the bare Coulomb potential in the whole region. The Coulomb-Rutherford cross section σ_C is correct to all orders and it agrees with its first Born approximation (see for instance [15]). Thus, the condition for its use in σ_s is mathematically fulfilled if this argument extends onto real and imaginary part contributing to the modulus of the scattering amplitude separately. Owing to the smooth transition between the two scattering potentials Coulomb $\Leftrightarrow$ Debye this is very likely to be the case. The perfect analogy to classical optics for diffraction from spatial filters may help to convince.

Another qualitative argument is obtained from considering the attenuation of a plane wave of momentum $\mathbf{p}$ incident onto a homogeneously distributed ensemble of ions of density n. The attenuation follows Beer's exponential law $I = I_0 \exp[-\langle n\sigma\rangle x]$ with $\langle n\sigma\rangle = 4\pi b_\perp^2 n \ln \Lambda$ from (7.27), (7.58), and (7.59). The attenuation by a mono-ionic layer of thickness $\Delta x = n^{-1/3}$ and the representative numerical example $n = 10^{21}\text{cm}^{-3}$, $E_r = 100$ eV tells that the first Born approximation is very well fulfilled,

$$I = I_0 \exp[-\langle n\sigma\rangle \Delta x]; \quad \langle n\sigma\rangle \Delta x = 4\pi b_\perp^2 n^{2/3} \ln \Lambda = 3Z^2 \times 10^{-5}; \quad \ln \Lambda = 4.9. \tag{7.60}$$

The qualitative arguments show tendencies in the parameter region.

A rigorous criterion is obtained from wave packet considerations with natural transverse dispersion

$$\phi(\mathbf{r}, t) = (\pi L_t^2)^{-3/4} \exp -\frac{(\mathbf{r} - \mathbf{v}t)^2}{2L_t^2}; \quad L_t^2 = L^2 + \frac{\hbar^2 t^2}{\mu^2 L^2}.$$

Following the mathematical analysis by [23] for the Debye potential ϕ_s we determine the ratio of the two moduli

$$\phi_1 = \lambda_D^{-3} \left| \int \phi_s[r]\mathrm{d}r \right| = 4\pi \frac{\kappa}{\lambda_D}, \; \phi_2 = \left(\lambda_D^{-3} \int |\phi_s(r)|^2 \mathrm{d}r\right)^{1/2} = (8\pi)^{-1/2}\phi_1$$

with $\kappa = Ze^2/4\pi\varepsilon_0$. Let $\alpha = |\psi(0) - \phi(0)|/|\phi(0)|$ be the relative error. Then, with $L = (\lambda_B \lambda_D)^{1/2}$ it is bound by

$$\alpha \leq \left[\frac{\pi^{1/4}}{\sqrt{2}} \left(\frac{\lambda_B}{\lambda_D}\right)^{1/4} + 1\right] \frac{\phi_2}{\phi_1} \frac{b_\perp}{\lambda_B} = 0.2 \left[0.9 \left(\frac{\lambda_B}{\lambda_D}\right)^{1/4} + 1\right] \frac{b_\perp}{\lambda_B}; \quad \alpha \to , 0.2 \frac{b_\perp}{\lambda_B}. \tag{7.61}$$

For our case from above, i.e., $n = 10^{21}\text{cm}^{-3}$, $E_r = 100$ eV, it results

$$\frac{L}{\lambda_B} = 11.3, \quad \left(\frac{\lambda_B}{\lambda_D}\right)^{1/4} = 0.3, \quad \frac{b_\perp}{\lambda_B} = 0.38; \quad \alpha = 0.1.$$

Thus, for this situation close to ideality the use of the first Born approximation is legitimate. For the Tokamak plasma α is of the order of 10^{-2}. Instead of following the demanding wave packet analysis of [23] one could think of proceeding to the second and higher Born approximations. It is not feasible because the second Born approximation already diverges. We conclude that within the validity of screening by the Debye potential $b_{\min} = \lambda_B$. The underlying condition is that bent orbits do not contribute sensibly to the collision frequency.

Coulomb screening in cylindrical symmetry. In contrast to the spherical Debye screening the oscillator model of uniform drift (7.1) exhibits cylindrical screening. The Hamiltonian reads, with the Coulomb interaction in dipole approximation $(Ze^2/4\pi\varepsilon_0)\boldsymbol{\delta}_{\rm os}\nabla(1/r) = \mathbf{f}_C(vt, b)\,\boldsymbol{\delta}_{\rm os}$,

$$H(\mathbf{p}, \boldsymbol{\delta}_{\rm op}, t) = \frac{\mathbf{p}^2}{2\mu} + \frac{\mu}{2}\omega_p^2\,\boldsymbol{\delta}_{\rm op}^2 - \mathbf{f}_C\,\boldsymbol{\delta}_{\rm op} = H_0 + H_C; \quad (\mu = m_e). \tag{7.62}$$

Index "op" stands for operator. The solution is given in terms of coherent or Glauber states (see e.g., [24] or [21]). The ground state $|\psi_i\rangle = |0\rangle$ at $t = -\infty$ is driven by H_C into the coherent Glauber eigenstate $|\psi_f\rangle = |\hat{\boldsymbol{\delta}}\rangle$ at $t = +\infty$. For obvious reasons it is labeled here by the classical amplitude $\hat{\boldsymbol{\delta}}$: the expectation value $\langle\psi_f|\hat{\boldsymbol{\delta}}_{\rm op}|\psi_f\rangle$ of the asymptotic shift at $t = +\infty$ coincides with its classical values $\hat{\boldsymbol{\delta}}$ from (7.7) and (7.8). For b small, K_1 and K diverge both as a consequence of the linearization in polarization $\mathbf{P}$. For vanishing impact parameters b, ω_p reduces smoothly to zero owing to missing screening and interaction goes over into bare Coulomb scattering, as in the former case with Φ_s. Therefore regularization is done by integrating the oscillator solution from $b_0 = sb_\perp \ll \lambda$ to infinity and treating the momentum transfer $D(\beta_0)$ of the close encounters in $0 \le b \le b_0$ by scattering from the unscreened Coulomb potential or, with the same result, from (7.58). The cut $b = b_0$ is to be done in such a way that (i) the single orbits within $b > b_0$ become classical, i.e., the state vectors $|\psi(b)\rangle$ are expressible in the form of an action integral, (ii) these orbits are sufficiently straight owing to $\mathbf{P} = -en_e\boldsymbol{\delta}$ linearized, $n_e = $ const, (iii) the oscillator term in H_0 is much smaller than the driver term H_C in $b \le b_0$, and (iv) b_0 is placed in a region where its individual choice is insensitive within a wide range. Condition (ii) is fulfilled for $1 - \cos\vartheta \simeq \vartheta^2/2 \ll 1$, that is $s = b_0/b_\perp \ge 5 \Rightarrow \cos\vartheta \ge 0.92$. This is also the condition for the fulfillment of (i), see for example [15], p. 103. With the electron-ion interaction time $\tau = 2b/v$ one deduces (iii) that the oscillator term in the Hamiltonian becomes insignificant for the easy condition $b_0^2/\lambda^2 \ll 1/4$. In the ideal plasma requirements (i)–(iii) are simultaneously fulfilled for the setting $b_0 = \sqrt{b_\perp\lambda}$ proposed by [23] for the width L for wave packets. For example, $\ln\Lambda = 5$ yields $s = 12$; from $\ln\Lambda = 10$ follows $s = 150$. Finally, it is a fortuitous circumstance that for $b \to 0$ the driver H_C prevails so strongly on the oscillator term in H_0 that $\omega_p = $ const or $\omega_p \to 0$ makes asymptotically no difference.

From $\beta = b/\lambda_0$ small follows $\beta\mathrm{K}_0\mathrm{K}_1 = -\ln\beta/2 \times [1 + (\beta^2/2)\ln\beta/2] - \gamma$, Euler constant $\gamma = 0.57722$, [2]. Applied to $b = b_0$ the generalized Coulomb logarithm L_C from the curly brackets in (7.15) evolves into

$$L_C = \beta\mathrm{K}_0(\beta_0)\mathrm{K}_1(\beta_0) + D(\beta_0)$$

$$= \ln\frac{\lambda}{b_\perp} + \ln 2 - \gamma - \ln s + \frac{1}{2}\ln\left[s^2\left(1 + \frac{1}{s^2}\right)\right] - \frac{1}{2}\ln\frac{b_0^2 + b_\perp^2}{b_\perp^2}$$

$$- \frac{\beta_0^2}{2}\ln^2\frac{\beta_0^2}{2} = \ln\Lambda + 0.116 + \Delta. \qquad (7.63)$$

The two terms $-\ln b_0 \sim -\ln s$ and $\ln s^2/2$ cancel each other guaranteeing insensitivity with respect to the special choice of b_0. In fact, the difference Δ is

$$\Delta = \frac{1}{2}\ln\left(1 + \frac{1}{s^2}\right) + \frac{s^2}{2\Lambda^2}\left(\ln^2\Lambda - \ln^2 s\right). \qquad (7.64)$$

At $\ln\Lambda = 5$ and $s = 12$ the correction amounts to $\Delta = 6.1 \times 10^{-2}$. At $\ln\Lambda = 10$ and $s = 150$ it results $\Delta = 1.7 \times 10^{-3}$. For $s = 20$ and 200 at $\ln\Lambda = 10$ the deviation is $\Delta = 3.7 \times 10^{-5}$ and 3.0×10^{-3}.

In summary, in the plasma with cylindrical (axial) symmetry of shielding the correct, i.e., quantum Coulomb logarithm is given in leading order by $\ln\Lambda = \ln(\lambda/b_\perp)$ from (7.63) with the lower cut off this time determined by the classical impact parameter of perpendicular deflection $b_{\min} = b_\perp$. In the plasma not far from ideality the 'lower cut off' $b_{\min}$ is not a universal property and not based on the uncertainty principle applied at $b = \lambda_B$. It has its origin in the scattering at large impact parameters; its value depends on the profile of the screening potential and on its geometry. Each screening potential exhibits its individual $b_{\min}$, spherical potential $b_{\min} = \lambda_B$, *linear* axisymmetric screening (7.62) $b_{\min} = b_\perp$ in the absence of external fields [25].

7.2.3 *Ion Beam Stopping*

In a fully ionized plasma an ion projectile p of mass m_p, charge $Z_\mathrm{p}e$, and velocity $\mathbf{v}_0$ is slowed down by collisions with the *free* plasma electrons and its ions of charge Ze. For $\ln\Lambda \gtrsim 1$ small angle deflections prevail and the energy transfer to an electron is by m_i/Zm_e larger than to a plasma ion. At the same time the p-e and p-i collision rates are related to each other by $n_e\sigma_e \simeq n_e\sigma_i/Z_\mathrm{p}$; σ_e, σ_i collision cross sections. This implies that the projectile moves straight, except rare p-i events between nuclei, and the energy loss of the projectile is nearly all to the electrons. In the following p-i collisions, excitation of bound electrons and nucleus-nucleus encounters, are neglected. For the sake of simplicity the energy loss of the projectile ion per unit length

$\mathrm{d}E/\mathrm{d}x$, the so-called *stopping power*, is given in the ballistic model as [compare (7.31)]

$$\frac{\mathrm{d}E}{\mathrm{d}x} = -2\pi Z_\mathrm{p}\frac{K}{v_0}\int_0^\infty\int_{-1}^{1}\frac{\mathbf{v}_0\mathbf{v}}{v^3}\ln\Lambda\, v_e^2 f(v_e)\,\mathrm{d}\cos\theta\,\mathrm{d}v_e, \quad \mathbf{v} = \mathbf{v}_0 + \mathbf{v}_e. \quad (7.65)$$

Again, $f(v_e)$ is normalized to unity. Both cases, moderate drift, $v_0 \lesssim v_\mathrm{th}$, and strong drift, $v_0 > v_\mathrm{th}$, v_th to be understood kinetically, are of interest. From the previous considerations on shielding and the comparison between dielectric and ballistic model it is reasonable to assume spherical Debye shielding at moderate drift and to fix the average Coulomb logarithm as $\langle\ln\Lambda\rangle = \ln\langle\Lambda\rangle = \ln(\lambda_D/\lambda_B)$. The resulting stopping power (7.65) becomes

$$\frac{\mathrm{d}E}{\mathrm{d}x} \simeq -4\pi K Z_\mathrm{p}\frac{\ln\langle\Lambda\rangle}{v_0^2}\int_0^{v_0} v_e^2 f(v_e)\,\mathrm{d}v_e; \quad v_0 \lesssim v_\mathrm{th} : \ \ln\langle\Lambda\rangle = \ln(\lambda_D/\lambda_B). \quad (7.66)$$

It is in agreement with the dielectric result for $k_\mathrm{max} = 1/\lambda_B$. For a Maxwellian $f(v_e) = f_\mathrm{M}(v_e)$ from (7.66), or immediately from (7.38) by setting v_0 for $v_\mathrm{os}(t)$, one obtains

$$\begin{aligned}\frac{\mathrm{d}E}{\mathrm{d}x} &= -4\pi\left(\frac{\beta}{\pi}\right)^{3/2} Z_p\frac{K\ln\langle\Lambda\rangle}{v_0^3}\int_0^{v_0} v_e^2\exp(-\beta v_e^2)\mathrm{d}v_e\\ &= -Z_p\frac{K\ln\langle\Lambda\rangle}{v_0^2}\left[\mathrm{erf}\left(\frac{v_0}{\sqrt{2}v_\mathrm{th}}\right) - \left(\frac{2}{\pi}\right)^{1/2}\frac{v_0}{v_\mathrm{th}}\exp\left(-\frac{v_0^2}{2v_\mathrm{th}^2}\right)\right]. \quad (7.67)\end{aligned}$$

From a comparison of the stopping power here and under vanishing drift, $v_0 \ll v_\mathrm{th}$, (7.35), in the Maxwellian plasma the following ratio results in straight orbit approximation

$$\frac{\mathrm{d}E(v_0)/\mathrm{d}x}{\mathrm{d}E(v_0\to 0)/\mathrm{d}x} = 3\left(\frac{\pi}{2}\right)^{1/2}\frac{v_\mathrm{th}^3}{v_0^3}\left[\mathrm{erf}\left(\frac{v_0}{\sqrt{2}v_\mathrm{th}}\right) - \left(\frac{2}{\pi}\right)^{1/2}\frac{v_0}{v_\mathrm{th}}\exp\left(-\frac{v_0^2}{2v_\mathrm{th}^2}\right)\right]\frac{\ln\langle\Lambda(v_0)\rangle}{\ln\Lambda(v_\mathrm{th})}.$$

Under equality of the two Coulomb logarithms this is a monotonuously decreasing function of v_0. For $v_0 = v_\mathrm{th}$ the reduction of stopping power amounts to 21%.

Although ion beam stopping has been studied for more than one hundred years large uncertainties are reported on the theoretical stopping power modelling in plasmas [3, 26, 27], and a few thorough experimental studies are available. In ion stopping in matter, apart from density effects, one is generally faced with projectiles interacting with free and bound electrons. The degree of ionization of the beam ions by the target electrons is a distance dependent self regulating process which may saturate at a certain depth in the target or may not. In any case it is a complex problem. To get some confidence with the essential aspects as well as with various models of interaction the interested reader may consult P. Sigmund for a general introduction and G.

Zwicknagel et al., for strongly coupled plasmas; see Further Readings, this chapter. As a rule, ion stopping in plasmas is significantly more efficient than in neutral matter. This is easily understood in terms of an oscillator δ of eigenfrequency ω_0 excited to the amplitude $\hat{\delta}$ by the kick $f(|x - v_0 t|) = f(\xi)$ of the projectile passing by,

$$\ddot{\delta} + \omega_0^2 \delta = f(\xi) \;\Rightarrow\; \hat{\delta}(t = \infty) = C \int f(\xi) \cos \frac{\xi}{\omega_0} \mathrm{d}\xi = \frac{C_1}{\omega_0}, \quad \Delta E \sim \delta^2 \sim \frac{1}{\omega_0^2}$$

see (7.5). C, C_1 are appropriate constants. The energy transferred to the bound electron is inversely proportional to the square of its transition (ionization) frequencies. Stopping is highest in the neighbourhood of the mean electron thermal velocity, i.e., at the so called *Bragg peak.* It may exceed typically three times stopping in neutral matter. There a more or less pronounced Bragg peak is found for the projectile velocity v_0 equating $(2\Delta E/m_e)^{1/2}$. At the Bragg peak the Coulomb logarithm $\ln \Lambda(\mathbf{v}_0)$ must be accounted for under the integral in (7.65). If v_0 is expressed by $v_0 = (2E/m_p)^{1/2}$ and the RHS of in (7.65) is denoted by S(E) the position as a function of the projectile energy $x(E)$ results from the numerical integration,

$$x - x_0 = -\int_{E_0}^{E} \frac{\mathrm{d}E'}{S(E')}.$$

Usually the stopping power and its Bragg peak are reported as functions of position. Setting the upper integration limit equal to zero the range of the ion beam $x(E_0)$ is obtained. An oversimplified model of ion beam energy deposition and the Bragg peak may be obtained from $E(x)$ following the simple power law

$$E(x) = C(x_F - x)^\alpha, \quad \alpha < 1 \quad \Rightarrow \quad \frac{\mathrm{d}E}{\mathrm{d}x} = \alpha C(x_F - x)^{\alpha - 1} \to \infty. \qquad (7.68)$$

The range of the beam is indicated by x_F. The infinite Bragg peak there is reduced to finite hight and finite width in reality by the energy dependent ionization Z_p and velocity dependent Coulomb logarithm $\Lambda(v)$. For the study in detail a recent overview on the theory of ion stopping in plasmas with specific numerical results therein is presented by W. Cayzac et al. [27]. Low-Z ion stopping in dense matter is of particular relevance for the calculation of the α particle induced burn front in controlled nuclear fusion. In particular, the ignition threshold of ICF capsules depends on the stopping power modelling [28]. A fully ionized carbon plasma has been probed with nitrogen projectile ions of low energy (typically 0.5 MeV). The data provide a test of stopping power theories, showing that only stopping theories including close collisions, as resulting for example from T matrix theory and to some extent from the ballistic model (however not included here), are consistent with the measurements [29, 30].

Barkas effect. In 1953 F.M. Smith et al. [31] found a minute difference between stopping power in emulsion of positive and negative pions. Subsequently it was verified in countless experiments that the range of a fast particle in matter differs from that of its antiparticle. Generally the difference in stopping of a negative particle from its positive counterpart is called *Barkas effect*. It clearly indicates that stopping is not just due to pure binary encounters because the Coulomb cross section σ_Ω respects charge conjugation. Hence, the difference must have its origin in the presence of spectators involved in the collision. Theoretical investigations have confirmed this interpretation for stopping in neutral matter [32]. In the plasma screening assumes the role of a spectator and the Barkas effect is present there as well [33]. A theoretical treatment of stopping in dense nonideal plasmas going well beyond the standard analysis (Z_b^3-term, see [31]) is presented in [26]. There is an asymmetry between the orbits of a positive and a negative point particle in a screened Coulomb potential as becomes clear from short inspection of Fig. 7.1.

7.2.4 Collision Frequency in the Classical Plasma

The classical dielectric theory is capable of producing reliable electron-ion collision frequencies in the weakly coupled plasma ($\Gamma \ll 1$) under small drift $\hat{v}_{os} \ll v_{th}$ provided the upper cut off k_{max} is identified with $1/\lambda_B$. Such a choice is suggested by our analysis of the Coulomb logarithm, and by the quantum treatment of (7.54). For moderate drift $\hat{v}_{os} \lesssim v_{th}$ and strong drift $\hat{v}_{os} \gtrsim 3v_{th}$ the evaluation of (7.54) and (7.55) has to be done numerically. In order to find simple approximate expressions for $\overline{\nu}_{ei}$ comparison between the ballistic and the dielectric quantum models may be helpful. To this aim in the ballistic model k_{max} must be assigned and the ballistic model needs b_{max} to be fixed. Thereupon their correctness is checked by the quantum formula (7.55).

Cut offs. Let us assume that (i) $b_\perp$ and λ_B are well separated from b_{max} and λ_D. In addition (ii) $b_{max} \gg b_0$, b_0 separating straight from bent orbits, is assumed. For strong drift the time averaged speed is $\overline{v} = v/\sqrt{2}$ owing to $v^2 \sim \sin^2 \omega t$. The upper cutoff $b_{max}(v)$ follows from the oscillator model,

$$\tau_{int} = 2b/v \leqslant \tau_{laser}/3 \quad \Rightarrow \quad \frac{2}{v/\sqrt{2}} = \frac{1}{3}\frac{2\pi}{\omega} \quad \Rightarrow \quad b_{max} = \frac{v}{\sqrt{2}\omega}.$$

We limit ourselves to $\lambda_D > b_\perp$. The reverse is almost of no significance in the high power laser plasma. Guided by the structure of $\sigma_s(\vartheta)$ in (7.58) we redefine $b_{min} = \lambda_B/2 = \hbar/2m_e v$. The formal justification for it may come from the kinetic formulation in terms of Wigner distribution functions and is therefore to be considered of universal character not bound to the special Debye potential of scattering. A physical argument may be that the model of a coherent plane wave is not realistic for lateral extensions larger than $\simeq \lambda_D$. Under the condition $\hat{v}_{os} \ll v_{th}$ the quiver motion induces only small periodic oscillations of the screening Debye sphere. The much faster thermal motion compensates small deformations. This is expected to hold to an acceptable approximation for the whole range of moderate drifts $\hat{v}_{\text{os}} \lesssim v_{\text{th}}$. In the Coulomb logarithm, however, $\lambda_D = v_{\text{th}}/\omega_p$ has to be replaced by v_{th}/ω for the case $\omega > \omega_p$ at arbitrary velocity v, in conformity also with Silin [6, 8]. In case of $\omega < \omega_p$ the plasma frequency as the faster oscillation frequency takes the role of the limiting interaction time for a collision, in agreement with the oscillator model. In summary, the Coulomb logarithm to be used in the ballistic model for the individual velocity $v(t)$, and $k_{\max} = \sqrt{2}/\lambda_B$ in the classical dielectric model are

$$\ln \Lambda(v) = \ln \frac{b_{\max}}{b_{\min}}, \quad b_{\max} = \frac{v}{\sqrt{2}\omega_{\text{m}}}, \quad b_{\min} = \frac{\lambda_B}{2}; \quad \omega_{\text{m}} = \max(\omega, \omega_p). \tag{7.69}$$

The evaluation of $\overline{\nu}$ is done for a Maxwellian as a representative case. For a rough estimate of $\nu_{ei}(t)$ in (7.36) $\ln \Lambda$ may be averaged over the angle θ and treated as a constant

$$\langle \ln \Lambda \rangle \simeq \ln \langle \Lambda \rangle = \ln \frac{\sqrt{2} m \langle v^2 \rangle}{\hbar \omega_{\text{m}}} = \ln \left[\frac{\sqrt{2} m_e}{\hbar \omega_{\text{m}}} (v_{\text{os}}^2 + v_{\text{th}}^2) \right]. \tag{7.70}$$

In cycle averaging of $\nu_{\text{ei}}(t)$ in (7.37) it turns out from numerical tests that $\overline{\ln\langle \Lambda \rangle}$ is best approximated by

$$\overline{\ln\langle \Lambda \rangle} = \ln \overline{\langle \Lambda \rangle} = \ln \left[\frac{\sqrt{2} m_e}{\hbar \omega_{\text{m}}} ((v_{\text{os}}^2/4 + v_{\text{th}}^2) \right]. \tag{7.71}$$

The approximation of an average Coulomb logarithm introduced here, $\overline{\ln\langle \Lambda \rangle} = \ln \overline{\langle \Lambda \rangle}$, is not very accurate. Its main advantage is simplicity of evaluation which can be done analytically with a Maxwellian $f(\mathbf{v}_e)$.

Comparison. With the setting of the cut offs and simplifications introduced here the question arises on their accuracy. We answer the question numerically by calculating the true effective Coulomb logarithm from the quantum formula (7.55) free from cut offs, and evaluate the same quantity from the ballistic model expression (7.32) with the cut offs fixed above. Finally, comparison is made with the averaged Coulomb logarithm (7.71). The result is shown as a function of the drift $\hat{v}_{\text{os}}/v_{\text{th}}$ in Fig. 7.6 with a plasma of $T_e = 100$ eV and density $n_e = n_i = 10^{21}$ cm^{-3} in (a) and $T_e = 1000$ eV and the same densities in (b). The black dots ($r = 0.5$), gray diamonds ($r = 1.5$), and gray triangles ($r = 5.0$) are the result from a thorough evaluation of the ballistic model (7.32). They are to be compared with the individual graphs of the quantum dielectric theory (7.55) for $r = \omega/\omega_p = 0.5$ in black solid, 1.5 gray dotted, and 5.0 gray light dotted. The agreement of the ballistic model (7.32) with the quantum dielectric theory (7.55) is excellent. The dashed lines result from the approximation (7.71). They show an increase above the correct quantities by 20–30% and may be useful for quick estimates in the laboratory. Far from $\omega/\omega_p = 1$ the dielectric values of $\overline{\nu(t)}_{ei}$ are slightly lower than those from the ballistic model. Exception is made

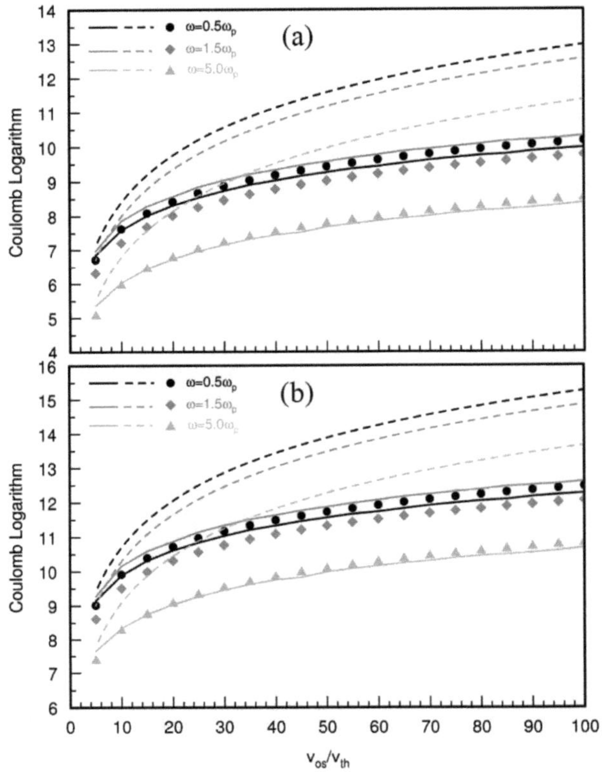

Fig. 7.6 Cycle- and velocity-averaged effective Coulomb logarithm $\ln \Lambda_{\text{eff}} = \overline{\langle \ln \Lambda \rangle}$ "Coulomb Logarithm" as a function of $\hat{v}_{\text{os}}/v_{\text{th}}$ for $r = \omega/\omega_p = 0.5$ (dots), 1.5 (diamonds), and 5.0 (triangles) from *ballistic model*; plasma density is $n_e = n_i = 10^{21}$ cm^{-3} **a** $T_e = 100$ eV, **b** $T_e = 1$ keV. Solid and dotted graphs from (7.55). Dashed lines from (7.71)

at $\omega/\omega_p = 1.5$ where the situation is inverted (see diamonds and gray continuous line). The reason becomes clear from Fig. 7.7 where $\ln \Lambda_{\text{eff}}$ is reported as a function of $r = \omega/\omega_p$. There is a resonant increase around $\omega = \omega_p$ in the dielectric model which the ballistic model is unable to reproduce (at least in the form given here). Such resonances have been reported, in a simplified approximation, for the first time by Dawson and Oberman [34]. In [35] the domain $\omega = \omega_p$ was investigated with more care which led to a qualitative agreement with Fig. 7.7 (maximum at $\omega = 1.2\,\omega_p$ and a plateau for $\omega < \omega_p$). Figure 7.7 shows the validity of the choice $\omega_{\text{m}} = \max(\omega, \omega_p)$ in $b_{\max}$ of (7.69) and (7.70) in a convincing way. At $\omega \simeq \omega_p$ the plasma is parametrically unstable and resonant excitation of high-amplitude Langmuir waves may occur, with the consequence that the *linear* dielectric model (7.55) becomes invalid here also.

Analytic expressions have been derived from (7.54) and (7.55) for $\hat{v}_{\text{os}} \gg v_{\text{th}}$ by making use of the asymptotic forms of $J_n(\xi)$ for large arguments. We list the results of [7, 8, 17], all derived for $\omega^2 \gg \omega_p^2$,

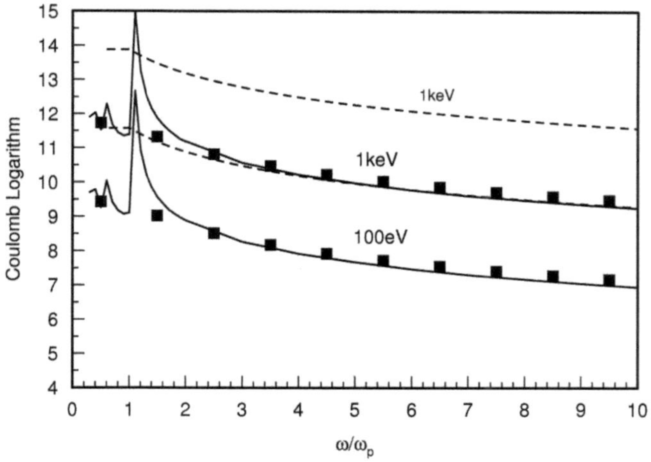

Fig. 7.7 ln Λ_{eff} "Coulomb Logarithm" as a function of ω/ω_p for $n_e = n_i = 10^{21}\,\text{cm}^{-3}$ and $T_e = 100\,\text{eV}$ and 1 keV. Black squares: ballistic model, continuous lines: dielectric model, dashed lines: approximation (7.71)

$$\text{Shima–Yatom} : \overline{\nu}_{\text{ei}} = C \ln \frac{2\hat{v}_{\text{os}}}{v_{\text{th}}} \left(\ln \frac{v_{\text{th}}/\omega}{\lambda_B} + \ln \frac{2\hat{v}_{\text{os}}}{v_{\text{th}}} \right),$$

$$\text{Silin} : \overline{\nu}_{\text{ei}} = 2C \left(1 + \ln \frac{\hat{v}_{\text{os}}}{2v_{\text{th}}} \right) \ln \frac{v_{\text{th}}/\omega}{\lambda_B},$$

$$\text{Kull–Plagne} : \overline{\nu}_{\text{ei}} = 2C \ln \frac{\hat{v}_{\text{os}}}{v_{\text{th}}} \ln \left[\sqrt{2} \frac{v_{\text{th}}/\omega}{\lambda_B} \left(\frac{\hat{v}_{\text{os}}}{v_{\text{th}}} \right)^{1/2} \right]. \tag{7.72}$$

("double Coulomb logarithm") with $C = \omega_p^4/(n_e \hat{v}_{\text{os}}^3)$. The most careful asymptotic analysis is undertaken in the last expression of $\overline{\nu}_{\text{ei}}$. In the domain $\hat{v}_{\text{os}}/v_{\text{th}} > 1$ there is no unique best fit. The special choice depends on the interval to be fitted. For example, the interval $1 < \hat{v}_{\text{os}}/v_{\text{th}} \leq 5$ is not fitted well by any of these formulas. In the ballistic model (7.72) is replaced by (the not very accurate) formula (7.71). Its main advantage is simplicity of evaluation.

The double logarithm allows an interesting interpretation. At $\hat{v}_{\text{os}} \ll v_{\text{th}}$ it is natural to assume that the ion at rest is surrounded by a Debye cloud of electrons so that the single electron interacts with a screened ion of finite interaction length, expressed by the appearance of a single Coulomb logarithm. For encounters with electrons all moving remarkably faster than v_{th} the more natural way to looking at an encounter is to switch to the reference frame of the oscillating electrons onto which the single ion impinges. The single electron is screened by the thermal cloud of the co-oscillating electrons and hence producing $\ln(\hat{v}_{\text{os}}/v_{\text{th}})$. The ion, in turn, is screened by an ion cloud. This ends in the other Coulomb logarithm $\ln v_{\text{th}}/\omega\lambda_B$.

In summary, the various *dielectric* approaches do not need upper cut offs of the collision parameter b, the *ballistic model* does. On the contrary, the ballistic model can handle bent orbits and no lower limit is to be set for $b_{\min}$; integration starts from $b = 0$. The ballistic model is sufficiently exact, it is simpler, its immediate physical interpretation offers easy extension to moderately nonideal plasmas. For example, *overlapping collisions* do not alter $\nu_{\rm ei}$ as long as bending of orbits does not exceed $\theta \simeq 30°$ in (7.29). The ballistic model avoids the decomposition of $\exp(i\mathbf{k}\hat{\delta}\cos\omega t)$ into Bessel functions of the dielectric model and can be used in the *long wavelength approximation* in all cases of interest, including the soft X-ray domain and irradiances at least up to $I\lambda^2 = 10^{18}\,\mathrm{Wcm^{-2}\mu m^2}$. The dielectric procedure is linear, i.e., it is bound to straight orbits, and needs an upper cut off in $k = k_{\max}$ which in a quantum version (e.g. Wigner functions) is given by $k_{\max} \simeq \lambda_B^{-1}$. Under high drift $\hat{v}_{\rm os} \gg v_{\rm th}$ a collision is an instantaneous, $\delta(t)$ function like event consisting of a high number of Fourier components. Correspondingly, the evaluation of (7.55) in the case of drifts $\hat{v}_{\rm os}/v_{\rm th} > 5$ shows convergence only after at least 10^3 terms [19].

So far we have not specified $\mathbf{v}_{\rm os}(t)$. From the kinetic derivation of the first moment conservation equation it becomes clear that in the Drude model $\mathbf{v}_{\rm os}$ and $(\mathbf{v}_0)$ is the velocity of a fluid element and not that of a single electron. As long as $\nu_{ei} \ll \omega$ holds it makes very little difference whether $\mathbf{v}_{\rm os}(t)$ is identified with the oscillatory velocity of an individual electron or with the drift velocity of an oscillating fluid element. However, in the opposite case of $\nu_{ei} \gg \omega$ a consistency problem may arise because the single particle drift from which ν_{ei} is calculated and the fluid drift may differ from each other. Some reasoning shows that at least for Coulomb logarithms somewhat larger than unity they still coincide since most of the orbits contributing to ν_{ei} are straight. At $\hat{v}_{\rm os}/v_{\rm th} > 1$ the electron-ion collision frequency ν_{ei} from (7.31) shows a strong time-dependence. As an immediate consequence higher harmonics appear in $v_{\rm os}$ and, in concomitance, in the current density $\mathbf{j}_e$, in $\mathbf{E}$ and $\mathbf{B}$. The *anharmonicity* is expected to be particularly strong in a dense plasma when T_e is low and $v_{\rm os} \gg v_{\rm th}$.

7.2.5 Supplements

7.2.5.1 Non-Maxwellian Distribution Function

A serious impact of a large $\hat{v}$ on absorption is seen in the distribution function [36]. By comparing the time τ_h of heating the electrons up to the temperature T_e,

$$\tau_h = \frac{3}{2} n_e k_B T_e / \overline{\langle \mathbf{jE} \rangle} = 3\tau_{\rm ei}\left(\frac{v_{\rm th}}{\hat{v}}\right)^2; \quad \tau_{\rm ei} = \nu_{\rm ei}^{-1}$$

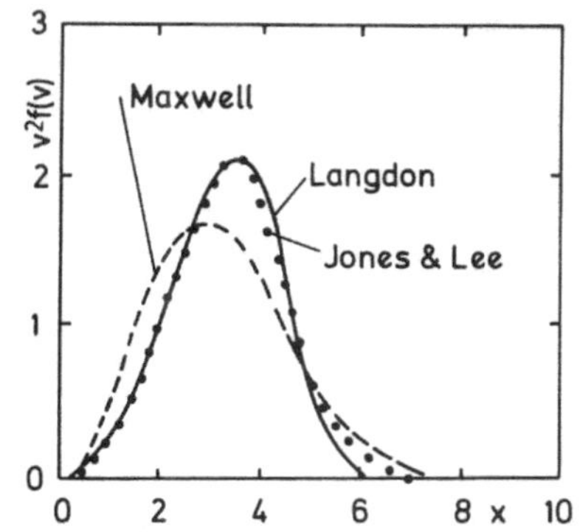

Fig. 7.8 Effect of the laser field on the electron distribution function $v^2 f(v)$ for $\hat{v}_{os}^2/v_{th}^2 = 2.25$ ([39]: dotted, and [36]: solid)

one finds that τ_h becomes shorter than the electron-electron collision time τ_{ee} as soon as $Z\hat{v}^2 \gtrsim 3v_{th}^2$ holds. After one electron-ion or electron-electron collision $f(\mathbf{v})$ becomes nearly isotropic for moderate v_{os}/v_{th} ratios [see also (4.36) and Fig. 4.2]. However, when driven at constant **E** field strength, $f(\mathbf{v})$ evolves into a *self-similar distribution* of the form (Fig. 7.8),

$$f(\mathbf{v}) = \kappa e^{-v^5/5u^5}, \quad u^5 = \frac{10\pi n_i Z^2 e^4 \ln\Lambda}{6m_e^2}\hat{v}^2 t.$$

The tendency towards the evolution of a *Supergaussian* (broadened Maxwellian) has been confirmed in [37] for $\hat{v}_{os}/v_{th} = 2$ and plasma parameter $\Gamma = 0.1$. In a more recent treatment which includes electron-electron collisions self-consistently and allows for an anisotropic part of the distribution function it has been found that $f(\mathbf{v})$ consists of a Supergaussian of slow electrons and a Maxwellian tail of energetic electrons [38]. The systematic study of the evolution of $f(\mathbf{v})$ by Fokker-Planck simulations reveals a departure from a Maxwellian when $\hat{v}_{os}/v_{th} \simeq 1$ and a final returning to a Maxwellian at $\hat{v}_{os}/v_{th} \gg 1$ due to the inefficiency of absorption. This behaviour is easily understood if one bears in mind that with increasing drift motion the electron-ion interaction time decreases whereas the electron-electron collisions are not affected by a drift that is common to all electrons. Finally, significant and surprising enough, at $\hat{v}_{os}/v_{th} = 10$ and $\Gamma = 0.1$ the authors of [37] observed a so far unexplained distribution function narrowing of the respective Maxwellian. They showed that the effect originates from electron-ion collisions.

7.2.5.2 Coulomb Focusing

From the classical collision scheme Fig. 1.5 it follows that, depending on the scattering angle, under the influence of a strong laser field an electron can gain an appreciable amount of energy in a single collision. The gain is highest in the back scatter configuration from an attractive potential and less significant from a

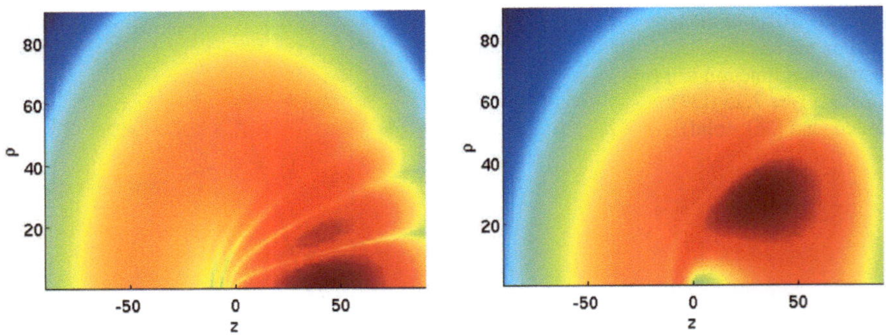

Fig. 7.9 Spatial probability density of a Gaussian wavepacket impinging along z from left with momentum $\hbar k_0 = 0.005\, m_e c$ scattered by an attractive (left graph) and a repulsive (right graph) Coulomb potential. The interference maxima indicate *Coulomb focusing* in the attractive and *defocusing* in the repulsive case [40]

potential that is repulsive. The distinction between positive and negative charge scattering from a Coulomb center clearly contrasts with pure Coulomb scattering for which σ_Ω is indifferent with respect to charge conjugation. Additionally, in presence of the laser field and/or in the effective Debye like potential an electron can be captured temporarily by the ion fulfilling several cycles for a while before being ejected again [41]. In the quantum scattering process interference occurs between the plane incoming and the spherical outgoing wave and, as a consequence, charge accumulation (attractive potential) and rarefaction (repulsive) behind the ion can be observed already in the bare Coulomb field into direction of the laser field (Fig. 7.9). This so called *Coulomb focusing* phenomenon can lead to strongly enhanced scattering [40]. In a refined ballistic model it should be taken into account.

7.2.5.3 Giant Ions

Strong enhancement of collisional absorption can occur due to the coherent superposition of collisions in *clusters*. The absorption coefficient α due to collisions in a plasma is proportional to the ion density n_i and the square of the ion charge, $\alpha \sim Z^2 n_i$. When N ions cluster together the density of the scatterers n_C decreases by N but their charge increases by ZN, resulting in an increase of $\alpha \sim N$. In order to obtain realistic amplification factors of collisional absorption due to clustering $g = \alpha_C/\alpha$, α_C absorption coefficient of the cluster medium, the fraction ξ of ions forming clusters, their *outer ionization* degree η_C, i.e., the net charge of the cluster, and the collision frequency ν_{eC} of an electron with the cluster gas in the presence of the intense laser field need to be known. It holds [42]

$$\nu_{eC} = \xi \eta_C^2 \frac{Z_C}{Z} \frac{L_C}{\ln \Lambda_{\text{ei}}} \overline{\nu}_{\text{ei}}, \qquad \alpha_C = \frac{\overline{\nu}_{eC}}{c} \frac{\omega_p^2}{\omega^2}. \tag{7.73}$$

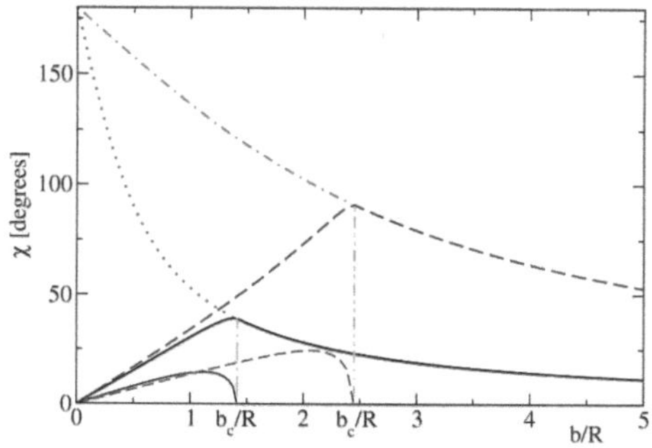

Fig. 7.10 Scattering angle θ as a function of the impact parameter b for a uniform charged spherical cluster. Bold dashed curve for $b_\perp = 2.5R$, and solid curve for $b_\perp = 0.5R$, R cluster radius. The corresponding deflections inside the cluster are indicated by the thin dashed and solid lines ending at their respective $b = b_c$ points. The dotted-dashed and the dotted lines starting from $\theta = 180°$ refer to Coulomb scattering from a point charge q

$\ln \Lambda_{ei}$ is the Coulomb logarithm of the plasma ions and L_C is the generalized Coulomb logarithm of the cluster of radius R, both to be determined. The Coulomb scattering angle θ is a function of the impact parameter b. For $b > b_c = R(1 + 2b_\perp/R)^{1/2}$, $\theta(b = b_\perp) = \pi/2$, it coincides with the Rutherford scattering angle. For orbits crossing the cluster it is determined numerically. For a uniform charge distribution inside the cluster θ is shown in Fig. 7.10 for $b_\perp = 2.5R$ and $b_\perp = 0.5R$. L_C is

$$L_C = \frac{1}{2}\int_0^{b_c} \frac{(1 - \cos\theta) b \, \mathrm{d}b}{R^2} + \frac{1}{2}\ln\left[\frac{b_\perp^2 + b_{\max}^2}{b_\perp^2 + b_c^2}\right]. \tag{7.74}$$

Multiplication factors of order 10^3–10^4 are achievable for realistic cluster parameters. Coherent amplification of collisional absorption is relevant for dusty plasmas, aerosols, sprays, perhaps foams, and small droplets of liquids, all of them with $R \ll \lambda$.

7.2.5.4 Inverse Bremsstrahlung Absorption

To illustrate the physics of absorption of radiation by collisions from a different point of view in this section the absorption coefficient α is deduced from the emission of bremsstrahlung radiation of an optically thin plasma in the weak laser field limit. It may be considered as an application of the photon gas in thermal equilibrium from Chap. 4.

The total power P emitted by an accelerated nonrelativistic electron is given by [1], Sect. 14.2

$$P = \frac{2}{3}\frac{e^2}{4\pi\varepsilon_0 c^3}\dot{\mathbf{v}}^2(t). \tag{7.75}$$

In a nearly thermal plasma velocity changes $\dot{\mathbf{v}}(t)$ are mainly produced by electron-ion and electron-electron collisions. An electron-ion pair constitutes a dipole system to a very good approximation as long as the maximum impact parameter $b_{\max}$ is small compared to the wavelength emitted. For laser plasmas in the intensity range $v_{os} < v_{th}$ and frequencies of $\omega \simeq \omega_{Nd}$ this is always the case because of $\lambda_{Nd} \gg \lambda_D \simeq b_{\max}$. On the other hand, a colliding electron-electron system represents an inert monopole and an oscillating quadrupole which at nonrelativistic speeds and under the above conditions gives negligible contribution to radiation emission. Alternatively, the colliding electron-electron pair can also be viewed as two electric dipoles of relative phase difference π. As a consequence, their dipole emission cancels. Since the principle of detailed balance applies, absorption of radiation does not occur in electron-electron encounters either. The total spectral intensity I_ω emitted by n_e electrons into the unit solid angle is obtained by Fourier-transforming Larmor's formula (7.75), and averaging over all impact parameters b and the corresponding velocity distribution $f(\mathbf{v})$. In this way, for a Maxwellian velocity distribution and the Spitzer-Braginskii-Silin collision frequency ν_{ei} from (7.35) results

$$I_\omega = \frac{\mathrm{d}^2 P}{\mathrm{d}\Omega \mathrm{d}\omega} = \frac{\Re(\eta)\omega_p^2}{4\sqrt{3}\pi^2 c^3}\nu_{ei} k_B T_e \frac{\sigma(\omega, T_e)}{\ln \Lambda} \mathrm{e}^{-\hbar\omega/kT_e}. \tag{7.76}$$

The quantity $\sigma(\omega, T_e)$ is the *Gaunt factor* which provides for the necessary quantum corrections in different (ω, T_e) regimes. For ω_{Nd} and moderate harmonics $\hbar\omega \ll kT_e$ the *Born-Elwert* approximation

$$\sigma(\omega, T_e) = \frac{\sqrt{3}}{\pi} \ln\left(\frac{4}{G}\frac{k_B T_e}{\hbar\omega}\right) = 0.55\left[0.81 + \ln\frac{kT_e}{\hbar\omega}\right]; \; G = 1.781 \tag{7.77}$$

applies if $(\lambda_B/b_\perp)^2 \gg 1$ is fulfilled [43]. Its value is

$$\frac{\lambda_B^2}{b_\perp^2} = \left(\frac{4\pi\varepsilon_0}{Ze^2}\right)^2 \left(\frac{k_B T_e}{m_e}\right)\hbar^2 = \frac{0.13(T_e[\mathrm{eV}])}{Z^2}. \tag{7.78}$$

For instance, with $k_B T_e/Z^2 = 30$ and moderate Z (7.78) applies. The absorption coefficient α_ω can now be determined from *Kirchhoff's law*. Bremsstrahlung radiation from an optically thin plasma generally represents such a small energy loss that (i) the plasma can be in thermal equilibrium kinetically even though thermal radiative equilibrium is not established and (ii) the spectral radiation intensity $I_\omega(T_e)$ is entirely determined, to a high degree of accuracy, by the kinetic electron distribution. Therefore, the relation

$$\frac{I_\omega}{\alpha_\omega} = I_{\mathrm{Planck}} = \frac{\hbar\omega^3}{4\pi^3 c^2}\frac{[\Re(\eta)]^2}{\exp\left(\frac{\hbar\omega}{k_B T_e}\right) - 1} \tag{7.79}$$

holds. Consequently, the absorption coefficient becomes

$$\alpha_\omega = \left(1 - \exp\left(-\frac{\hbar\omega}{k_B T_e}\right)\right) \frac{k_B T_e}{\hbar\omega} \frac{1}{c} \left(\frac{\omega_p}{\omega}\right)^2 \frac{\nu_{ei}}{\Re(\eta)} \frac{\ln\left(\frac{4}{G}\frac{k_b T_e}{\hbar\omega}\right)}{\ln \Lambda}. \tag{7.80}$$

From $\hbar\omega/kT \ll 1$ the factor $\left(1 - \exp(-\frac{\hbar\omega}{kT_e})\right) kT_e/\hbar\omega$ becomes unity. Comparison with (5.189) yields

$$\frac{\alpha_\omega}{\alpha} = \frac{1}{\ln \Lambda} \ln\left(\frac{4}{G}\frac{k_B T_e}{\hbar\omega}\right).$$

To give a numerical example, for $\hbar\omega/k_B T_e = 10^{-3}$ the ratio becomes $\alpha_\omega/\alpha = 5.4/\ln \Lambda$; for $\ln \Lambda = 5.4$ the spectral absorption coefficient α_ω becomes equal to α. In any case these considerations show that

(a) α_ω is, within the uncertainty of $\ln \Lambda$, close enough to α, in this way justifying the identification of *inverse bremsstrahlung absorption* with *collisional absorption*;
(b) the bremsstrahlung intensity $I_\omega(T_e)$ represents the spontaneously emitted amount of radiation, whereas in the absorption coefficient α stimulated re-emission of radiation is also included;
(c) at very low temperatures or higher photon energies (e.g., $\hbar\omega/kT_e \simeq 1$) a purely classical calculation of α begins to fail because then $1 - e^{-\hbar\omega/kT_e} \neq \hbar\omega/kT_e$.

Observation (b) needs two further comments. There are several ways of formulating the principle of detailed balance and Kirchhoff's law. It is convenient to use it in the form of (7.79) where the emitted power density per unit solid angle is indicated by I_ω. This is because experimentally I_ω can be uniquely determined only when the surrounding radiation field is zero; as a consequence I_ω represents the *spontaneously* emitted radiation. On the other hand, in a measurement of α_ω the *true* or effective absorption coefficient is determined which is the difference between absorption and stimulated re-emission of radiation between two energy levels E_1 and E_2 ($E_2 - E_1 = \hbar\omega$). With their populations n_1 and n_2, the Einstein coefficients in the approximation $B_{12} = B_{21}$ for stimulated emission, and ρ_ω the spectral radiant energy density, $\rho_\omega = 4\pi I_\omega/c$, α_ω is determined by

$$\alpha_\omega I_\omega = (n_1 B_{21} - n_2 B_{12})\rho_\omega = \frac{4\pi}{c} B_{12}(n_1 - n_2) I_\omega. \tag{7.81}$$

The refractive index η and degeneracy are assumed to be unity. If κ_ω refers to level excitation by absorption,

$$\kappa_\omega = \frac{4\pi}{c} B_{12} n_1,$$

and the levels are in thermodynamic equilibrium (for instance, because of collisions; Maxwellian distribution in the case of continuous energy spectrum), hence $n_2 = n_1 e^{(-\hbar\omega/kT_e)}$, one obtains

$$\alpha_\omega = \kappa_\omega \left[1 - \exp(-\hbar\omega/kT_e)\right]. \tag{7.82}$$

κ_ω is often used in theoretical work, however, it is not accessible to direct observation. For correctness it has to be mentioned that the rate equation (7.81) holds as soon as the shortest time of interest is much longer than the *transverse relaxation (or phase memory) time* T_2 [44]. In conclusion, when Kirchhoff's law is written in the form $I_\omega/\alpha_\omega = I_\omega^{\text{Planck}}$, I_ω is the *spontaneously* emitted radiation, and α_ω is the *net* absorption coefficient. Both, I_ω and α_ω are directly observable quantities in optically thin samples only.

7.3 Collisionless Absorption from Overdense Plasma Surfaces

7.3.1 Overview and Purpose

Resonance absorption in the critical plasma region is the prototype of *collisionless absorption*. It consists in the one to one transformation of a photon into a plasmon propagating down into the underdense plasma region. In general the electric field of the incident laser wave is not collinear with the longitudinal electric field of the electron plasma wave and therefore a fraction of the incident wave is reflected. However, if the two fields point into the same direction total conversion occurs provided an underdense plasma shelf exists for the plasma wave to propagate. When an intense fs laser pulse impinges onto a solid or liquid surface the skin layer transforms into an almost collisionless, highly overdense plasma in a fraction of a laser cycle. The gradient at the surface remains very steep by inertia and radiation pressure before the ions start to move and to form an underdense plasma shelf. Resonance absorption does not take place because of resonance $\omega = \omega_p$ and underdense plasma both missing. As pointed out in Chap. 1 absorption is finite only if $\overline{\mathbf{jE}}$ differs from zero. Collisions providing for a finite phase shift between current and laser field drop out. On the other hand (i) high absorption ($\geqslant$70–80%), (ii) instantaneous generation of fast electrons, (iii) delay-less absorption at all ratios $n_{e0} > n_c$, and polarization dependent heating, linear versus circular, are experimental facts.

Numerous *collisionless absorption mechanisms* have been proposed, partly not well separated from each other: $\mathbf{j} \times \mathbf{B}$ *heating* [45], already mentioned in Sect. 6.2.3, the *Brunel mechanism* [46], presented in Chap. 1, *vacuum heating* [47], *sheath layer inverse bremsstrahlung*, *anomalous skin layer absorption*, *stochastic heating*, *wave-breaking*, *nonlinear Landau damping*, *excitation of surface plasmons* [48, 49], and others.

The force density $\mathbf{j} \times \mathbf{B}$ heating works best at normal incidence. The majority of researchers seemed to favour it for a while although the model was based on PIC simulations and no physical interpretation was given. In the simulations "electrons are accelerated and then beamed into the plasma by the oscillating ponderomotive force" resulting in up to 17% at $I\lambda = 10^{18}$ Wcm^{-2} under normal incidence. As $\mathbf{j} \times \mathbf{B}$ is a reversible force density for free electrons, the question arises where does "beaming", i.e., irreversibility come from. The value of their work consists in having recognized first the significance of the motion perpendicular to the target surface for absorption and having done in this way the first step towards modelling collisionless absorption. What has to be done is to associate a physical effect which produces "beaming".

This has been accomplished to a certain extent by Brunel's model of "not-so-resonant, resonant absorption". He recognized that the electrostatic forces have to be self consistently combined with the Lorentz force of the laser. All dynamics plays in front of the target; under the combined action of the two fields the single layers do no longer move harmonically and combine with the target with individual delays during a full laser period. Perhaps for this reason Brunels mechanism of the electrons pulled out into the vacuum and then pushed back into the field-free target interior is frequently identified with the term "vacuum heating". To be more specific, the reason why the term vacuum heating played an ominous role in the past and partially still does presently is because, often invoked as the leading collisionless absorption mechanism, it has never been defined properly. To introduce some rating in this respect it seems that two groups of authors can be distinguished. By the concept of vacuum heating the first group addresses the electrons in front of the sharp-edged target that circulate in the vacuum and do not cross the interface during one laser cycle. Identification of vacuum heating with Brunels mechanism is made by the second group to contrast with anomalous skin layer absorption. By the latter all motion is strictly confined to the target inside. Sometimes vacuum heating is interpreted as a consequence of unspecified wave breaking [50, 51]. Meanwhile it has been clarified that Brunel/vacuum heating prevails distinctly on skin layer absorption and that vacuum heating in the restricted sense, i.e., the contribution to absorption of the electrons not entering the target with the periodicity of the laser, is almost insignificant.

A general statement can be given on the origin of collisionless absorption: The dephasing between current density $\mathbf{j}$ and the laser field producing absorption is provided by the electrostatic space charge field induced by the laser beam.

It must be the aim in what follows to characterize the electrostatic field in typical situations and to address the underlying physics. For a more complete understanding of the interaction at high intensities and the form of the fast electron spectra the search for hidden resonances is advisable. So far, little has hitherto been undertaken to evidence them.

7.3.2 Anharmonic Resonance

Collisionless absorption of long ps and fs pulses is, in leading order, by linear resonance at $\omega = \omega_p$. It is therefore quite natural to speculate whether such a resonant conversion is the leading absorption mechanism also in ultrashort pulses on highly overdense targets (solids, liquids, and clusters) before appreciable rarefaction has set in. For more than two decades resonance absorption at $\omega_p \gg \omega$ has been categorically excluded by the scientific community. The insight that this statement is incorrect, because limited to the harmonic oscillator only, is a first step towards a detailed understanding of collisionless absorption [52].

7.3.2.1 Irreversible Energy Gain from a Nonlinear Oscillator

A nonlinear oscillator is governed by an equation of the type

$$\ddot{\boldsymbol{\xi}} + f(\boldsymbol{\xi}) = \mathbf{D}(t), \tag{7.83}$$

where $f(\boldsymbol{\xi})$ is any sufficiently smooth restoring force, and $\mathbf{D}(t)$ is an external driver. The linear oscillator is described by $f(\boldsymbol{\xi}) = \omega_0^2 t \boldsymbol{\xi}$ with constant eigenfrequency ω_0 at all excitation amplitudes. Under the action of a harmonic driver $\mathbf{D}(t) = \hat{\mathbf{D}}(t) \cos \omega t$, $\omega \neq \omega_0$, (7.83) behaves adiabatically as soon as $\hat{\mathbf{D}}$ is an envelope of several cycles $T = 2\pi/\omega$ long. An undamped nonlinear oscillator may exhibit properties differing in many respects from its linear counterpart. For our purpose here its most important difference is the dependence of the eigenperiod T_0 on the excitation level, i.e., on the amplitude. In the case of the very general nonlinear oscillator (7.83) in 1D its eigenperiod T_0 is given by $\int \mathrm{d}t = \int \mathrm{d}\xi/v$, v particle velocity, or

$$T_0 = \oint \frac{\mathrm{d}\xi}{\left\{\left(\frac{2}{m}\right)[V_0 - V(\xi)]\right\}^{1/2}}, \qquad V_0 = \max V(\xi). \tag{7.84}$$

$V(\xi)$ is the potential associated with the restoring force $f(\xi)$ so that $f(\xi) = -\partial_\xi V(\xi)$. If the graph of $V(\xi)$ stays inside the parabola $\omega_0^2 \xi^2/2$ the eigenperiod decreases with the amplitude; if however it widens compared to the parabola its eigenperiod increases with increasing level of excitation. This latter case is of particular interest because Coulomb systems exhibit such a characteristics owing to the $1/r$-dependence of the Coulomb potential at large charge separation and, in concomitance, at a fixed driver period the nonlinear system may enter into resonance when excited to high amplitude. This property opens the new possibility to couple appreciable amounts of energy by adiabatic excitation into such systems originally out of resonance. Since even the shortest laser pulse contains several oscillations absorptive ballistic excitation is not possible. Energy coupling into a system originally out of resonance is illustrated in the following for a cold plasma.

A plane, fully ionized target is assumed to fill the half space $x > 0$. A plane wave $\mathbf{E}(\mathbf{x}, t) = \mathbf{E}_0 \exp[i(\mathbf{kx} - \omega t)]$ in p-polarization in y-direction, wave vector $\mathbf{k}$ and frequency ω, is incident from $-\infty$ under an angle α onto the plasma surface (Fig. 7.11a). After applying a Lorentz boost $v_0 = c \sin\alpha$ the wave impinges normally. The target is thought to be cut into a sufficient number of thin parallel layers, each of which is exposed to a driving force F acting along x of magnitude ev_0B and frequency ω, and an additional component originating from the Lorentz force of the oscillatory motion along y of frequency 2ω (e electron charge, B magnetic field of the laser). With the electron displacement ξ in x-direction and immobile ions of density n_0 (corresponding to ω_{p0}) the motion of a single layer (Fig. 7.11b) in the nonrelativistic limit is determined by

$$\frac{\mathrm{d}^2\xi}{\mathrm{d}t^2} + \omega_{p0}^2\left(1 - \frac{|\xi|}{2d}\right)\xi = D, \qquad |\xi| \le d, \tag{7.85}$$

$$\frac{\mathrm{d}^2\xi}{\mathrm{d}t^2} + \omega_{p0}^2\left(\frac{d}{2|\xi|}\right)\xi = D, \qquad |\xi| \ge d. \tag{7.86}$$

The free plasma oscillator ($D = 0$) oscillates at ω_{p0} if the displacement is small. For a fixed energy $V_0 = \varepsilon$ at high excitation and $\omega_{p0} \gg \omega$, according to (7.84) T_0 is given by $T_0 = 8(\omega_{p0}^2 d)^{-1/2}\xi_0^{1/2}$. The eigenfrequency $\omega_0 = 2\pi/T_0 = (\pi/4)(\omega_{p0}^2 d)^{1/2}\xi_0^{-1/2}$ decreases with increasing oscillation amplitude ξ_0, in contrast to the linear harmonic oscillator, and approaches zero for $\xi \to \infty$. (An exact mechanical analogy is represented by an elastic sphere bouncing on an elastic horizontal glass plate under the influence of gravity, and its mirror image: $\omega_0 \to \infty$ for $\xi_0 \to 0$ and $\omega_0 \to 0$ for

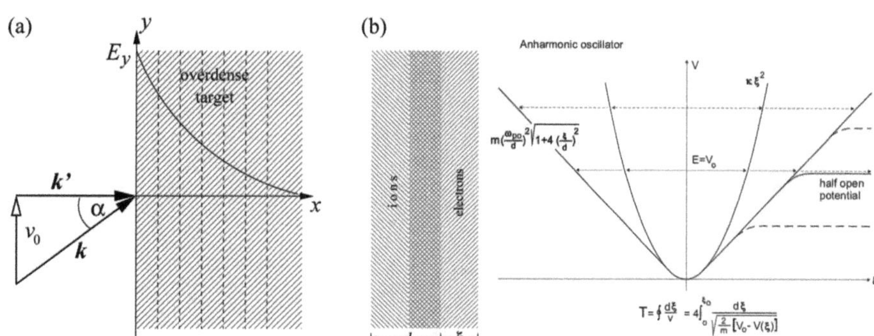

Fig. 7.11 **a** Oblique incidence of parallel laser beam of wave vector $\mathbf{k}$ in the lab frame and $\mathbf{k}' \| x$ in the system boosted by $v_0 = c \sin\alpha$. The overdense target is cut into layers of thickness d. **b** Large electron displacement ξ in plasma layer (LHS) and harmonic (parabola) and anharmonic closed (solid) and half-open potentials (solid, dashed). In wider potentials than harmonic the frequency decreases with increasing ξ

height $\xi_0 \to \infty$). Under a weak driver D and $\omega_{p0} > \omega$, ξ oscillates in phase with D; under a strong driver ξ becomes large and, owing to $\omega_0 \to 0$, it oscillates similarly to a free particle, i.e., dephased by π with respect to D. Principal resonance [53] occurs at $\omega_0 = \omega$ with $\xi_0 = \xi_r = (\pi\omega_{p0}/4\omega)^2 d$. The transition from nonresonant to resonant state at $\omega_0 = \omega$ is irreversible, i.e., resonance breaks adiabaticity. In the neighbourhood of resonance the product $(\mathrm{d}\xi/\mathrm{d}t)D \sim jE$ changes from $-\sin\omega t \times \cos\omega t$ into $\cos^2\omega t$, with nonvanishing cycle average. This is illustrated by a numerical example now.

The two equations of motion above for $|\xi| \le d$ and $|\xi| \ge d$ convert into a single dimensionless equation for the potential $V = m(\omega_{p0}d/2)^2(1+\zeta^2)^{1/2}$,

$$\frac{\mathrm{d}^2\zeta}{\mathrm{d}\tau^2} + \frac{\partial(1+\zeta^2)^{1/2}}{\partial\zeta} = D(\tau), \qquad (7.87)$$

where $\zeta = 2\xi/d$, m electron mass, $\tau = \omega_{p0}t$ and $D \to 2D/(\omega_{p0}^2 d)$. With $D = D_0 \sin^2[\omega t/(2K)]\cos\omega t$, $\sin^2$ for $f(t)$, K number of periods, the results of Fig. 7.12 are obtained numerically. At $D_0 = 0.921$ the layer remains below resonance, the energy gain ε after the pulse is over is negligible (a). Increasing the driver by only $\Delta D_0 = 0.002$, resonance takes place. Much energy, i.e., 43 times more than in (a), is stored now in the oscillator [see the horizontal orbits in (b)]. Under the assumption that resonance lasts half a laser cycle, i.e., when the driver amplitude $E_0 = m\omega_{p0}^2 d/(4e)$, with $d = 0.1 - 0.2$ nm and $\omega_{p0} = 2 \times 10^{16}\,\mathrm{s}^{-1}$, primary resonance

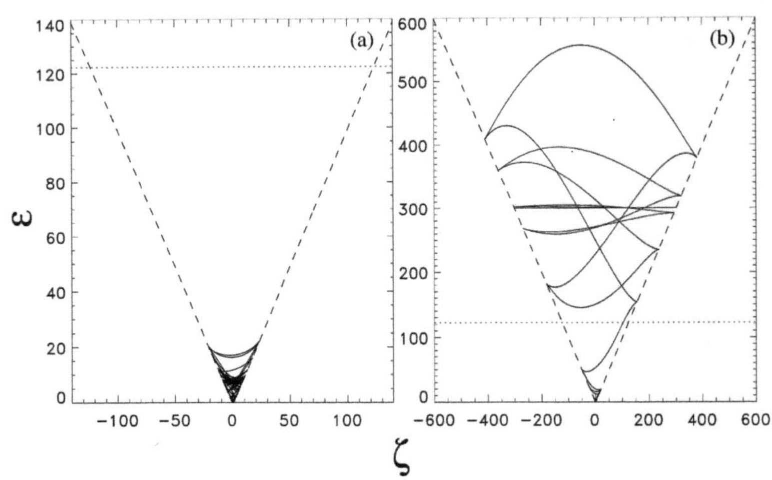

Fig. 7.12 Excitation $\varepsilon(\zeta) = \dot{\zeta}^2/2 + V(\zeta)$ of the oscillator from (7.87) by the driver $D(\tau) = \hat{D}\sin^2[\omega\tau/(2N\omega_{p0})]\cos\omega\tau/\omega_{p0}$, $\omega/\omega_{p0} = 0.1$, $N = 20$, **a** below resonance ($\hat{D} = 0.921$), **b** above resonance ($\hat{D} = 0.923$) with a 43 times higher final energy gain ε_f than in **a** although the driver strength is changed only by 0.2%. The potential and the resonant energy level are indicated dashed and dotted, respectively. Note the different scales in **a**, **b**

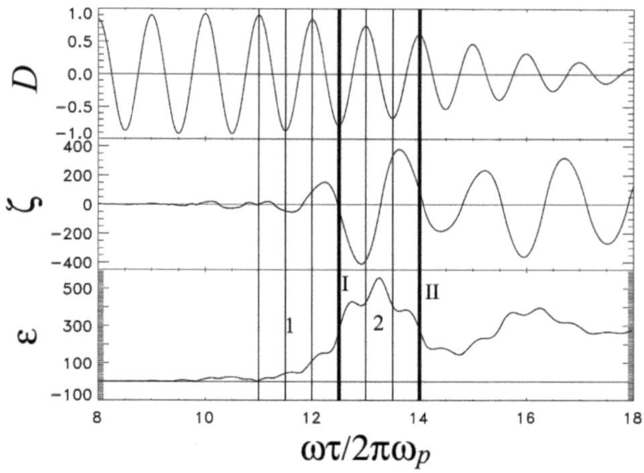

Fig. 7.13 Driver $D(\tau)$, excursion $\zeta(\tau)$ and absorbed energy $\varepsilon(\tau)$ versus time (in driver cycles) for the case in Fig. 7.12. Each time resonance is crossed ζ undergoes a phase shift by π. Bold lines I, II indicate instants of phase lag $\zeta \pm \pi/2$ between D and ζ, i.e., maximum energy gain and loss (points of stationary phase); modulations in ε originate from the $\omega + \omega_0$ spectral component

of a single isolated layer happens at the laser intensity as low as $I = 10^{15}\,\mathrm{Wcm}^{-2}$. Figure 7.13 is of particular relevance. It shows the driving field $D(\tau)$, the displacement $\zeta(\tau)$, and the energy gained $\varepsilon(\tau)$. At position 1 the oscillator is entering resonance (ε starts increasing, $\omega_0 > \omega$), D and ζ are in phase; at 2 it is leaving resonance (ε starts decreasing, $\omega_0 < \omega$), D and ζ are dephased by π. Positions I and II (points of stationary phase) indicate maximum energy gain and maximum energy loss, D and ζ are dephased by $\pm\pi/2$. Thus, the resonance signature is preserved in a rapid transition. The phenomenon repeats in the second maximum of ε, etc. Resonance, i.e., ω_0 becoming equal to ω, is intimately connected with the phase shift by π owing to the different reaction of the oscillator to the driver under a strong restoring force ($\omega_0 \gg \omega$) and a weak one of a nearly free particle. The absorption term $jE \sim \dot{\xi}D$ at resonance transits from $\sim \sin\tau \times \cos\tau$ to $\sim \cos^2\tau$. This guarantees energy transfer from the driver to the oscillator. The time-dependence of ω_0 in (7.87) is accomplished through the amplitude ζ_0, $\omega_0 = \omega_0(\zeta_0[D(t)])$.

7.3.2.2 Anharmonic Resonance Absorption in Cold and Warm Plasma

The extension of the dynamics from one layer of electrons and ions to N layers is accomplished by considering all attractive forces of the fixed ion layers (index l) on all electron layers (index k) and all repulsive forces between the electron layers (indices k, k'). This results into the nonseparable (nonintegrable, chaotic), yet elementary Hamiltonian

$$H = \sum_{k=1}^{N} \left(\frac{p_k^2}{2} + \frac{1}{2} \sum_{k' \neq k}^{N} V_{kk'} + \sum_{l=1}^{N} V_{kl} - D(\tau)\zeta_k \right) \tag{7.88}$$

with $p_k = \mathrm{d}\zeta_k/\mathrm{d}\tau$, and the potentials with the structure of V from Fig. 7.11b (LHS), $V_{kk'} = -[1 + (\zeta_k - \zeta_{k'})^2]^{1/2}$ and $V_{kl} = [1 + (\zeta_k - \zeta_{0l})^2]^{1/2}$, $\zeta_k = 2x_k/d$, $\zeta_{0l} = 2a_l/d$, a_l position of the lth ion layer. When one of the layers is driven into resonance it starts moving oppositely to the coherently moving nonresonant layers, thereby crossing one or several adjacent oscillators. Incidentally, this is a new scenario of very effective breaking of flow (special case: breaking of wave) not described in the literature so far. As a representative case we study the dynamics of a 100 times overdense target, subdivided into layers of $d = 0.125$ nm each, on which $I = 3.5 \times 10^{18}\,\mathrm{Wcm}^{-2}$ at $\lambda = 800$ nm and is impinging with $f(t)$ increasing from zero to unity within 2 laser periods T and then remaining constant. The intensity of D on the kth electron layer is calculated at each time instant by determining the screening due to all layers lying in front of it according to the exponential decay $\exp(-k\eta_i d)$, with η_i the imaginary refractive index. The typical scenario of particles with large amplitude is as follows: After being pulled out into the vacuum and oscillating there for some time ("vacuum heating": no heating!) the layers are pushed back in a *disruption*-like manner into the target (formation of jets; Figs. 7.14, 7.18). Layers leaving from the back of the target are replaced by new layers with zero momentum (cold return current). First indication of resonance: More than half of the layers have gained energies exceeding their quiver energy W. The energy spectrum shows a

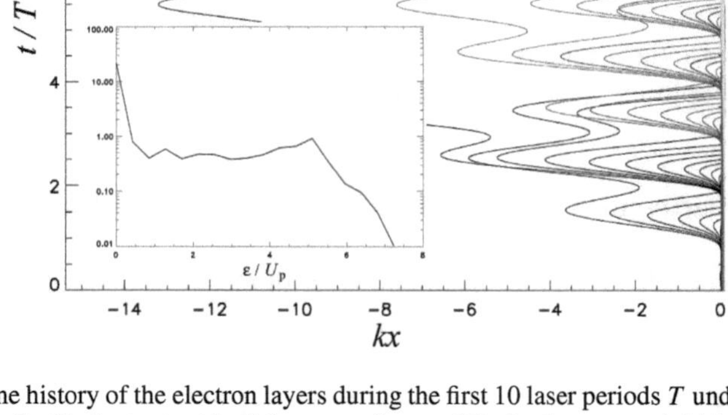

Fig. 7.14 Time history of the electron layers during the first 10 laser periods T under the action of a pulse as described in the text, with $f(t) = \text{const}$ for $t \geq 2T$: displacements $\zeta_k(t)$ in dimensionless units kx, k wave number. No detectable "vacuum heating". Inset: energy spectrum of 2700 layers at $t/T = 100$, $W = U_p$ quiver energy

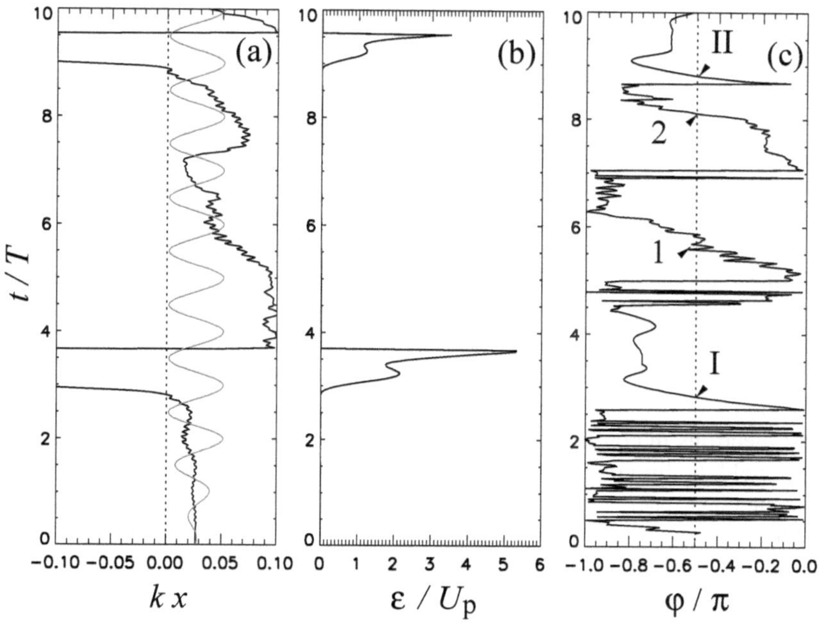

Fig. 7.15 Resonance dynamics of layer #32 as a function of laser periods. Position kx (**a**), k wave number, absorbed energy ε (**b**) and phase φ between velocity v and driving laser field from Fig. 7.14 (mapped into the interval $[-\pi, 0]$) (**c**); ε in units of quiver energy $U_p = W$. Passages through $-\pi/2$ are indicated by I, II, and 1 and 2 (see text). After resonance at 3.5 laser periods the layer is pushed back into the target with high velocity (*disruption*). After crossing the opposite target surface the layer is substituted by a new layer. Note $E \sim -D$

plateau between 1 W and the cut off at 6 W (see inset in Fig. 7.14) and is reminiscent of Brunel's result in Figs. 1.21a and 1.22. In other runs with more realistic driver field and the magnetic field included similar energy spectra were obtained with plateaus extending up to 20 W. To show the occurrence of resonance explicitly the phase of each layer with respect to the driving laser field is investigated. A typical example with $N = 120$ is shown in Fig. 7.15 for layer #32, LHS trajectory and driver field (a), middle absorbed energy (b), RHS phase φ of velocity $v \sim \sin(\omega t + \varphi)$ with respect to the driver $E \sim \cos \omega t$ (c): over $T/2$ there is a continuous and smooth transition of φ through $-\pi/2$ at $t/T = 2.8$ (I) with a simultaneous strong increase in the absorbed energy (b) and the excursion (a), with following disruption of the layer at $t/T = 3.6$. The change of φ is clearly seen also in (a). Another resonance of the same kind is found at $t/T = 8.8$ (II). Other two passages of φ through $-\pi/2$ at $t/T = 5.7$ (1) and 8.1 (2) show rapid fluctuations and hence almost no energy gain [see (b)] and no disruption [see (a)]. Transitions of this latter kind are morphologically clearly distinguishable from the former case, and for none of the 120 layers they are able to accelerate them across the target. In the cold plasma model we find that all layers (no exception) get their energy from resonance and keep it in the underdense region. Only after undergoing resonance each time the layers disrupt, in the present case in chaotic order 6, 5, 4, 3, 2, 15, 17, 16, 9, 11, 10, 8, 12, 24, 31, 34, 28, 32, 14, 30, 1, 33, 45, 26, 36, 27, etc.; layer 113 disrupts before front layer 0. This is one of the fundamental differences in the dynamics in comparison to [46]. The question at which intensity resonance sets in is difficult to answer, if not impossible because dominated by probabilistic behaviour. Simulations with 2, 10, and 20 layers show a decrease and a broadening of its threshold, in agreement with analytical estimates. The advantage of the model lies in its Hamiltonian structure. In combination with simplified and oversimplified drivers it offers much flexibility and considerable help in interpreting PIC simulations. So, for example, the difference in fast electron generation between circular and linear polarization is easily explained: in circular polarization the electron motion is entirely transverse and no resonance is excited in x-direction. Furthermore, anharmonic resonance explains the instantaneous creation of fast electrons in contrast to the cumulative stochastic heating by the fluctuating electric field. In this context Figs. 7.14 and 7.15 show no detectable *vacuum heating*. Resonance, i.e., energy gain during about half an oscillation, implies a phase shift $\varphi_l(t)$ for the individual electron or electron layer. All φ_l at a given position and time $(\mathbf{x}, t)$ sum up to a non vanishing $\phi(t)$ value once at least one electron (charge e, amplitude $\mathbf{v}_{0l}$) is resonant,

$$\begin{aligned}\mathbf{j}(\mathbf{x}, t) &= -e\sum_l \delta(\mathbf{x} - \mathbf{x}_l(t))\mathbf{v}_{0l}\sin(\omega t + \varphi_l) \\ &= -en_e\mathbf{v}_0\sin(\omega t + \phi). \end{aligned} \tag{7.89}$$

Figure 7.15c shows two types of crossing $\phi = -\pi/2$: correlated against cumulative coaction. With respect to resonance variations of D had no qualitative consequences. For quantitative results recurrence must be made to PIC, Vlasov, or molecular dynamics. Anharmonic resonance constitutes a very efficient scenario leading to breaking of regular dynamics of a fluid (special case: wavebreaking). Once a fluid layer undergoes resonance it may move opposite to the adjacent fluid layer and thus destroy the continuous mapping (3.74). We give it the name *resonant (wave) breaking*.

In order to reveal the dominant role of resonance in warm, i.e., realistic plasma, a particle-in-cell (PIC) simulation with 10^6 particles in the collisionless mode under 45° irradiation is performed. In the boosted frame a Gaussian Nd laser beam of $I = 10^{17}\,\mathrm{Wcm}^{-2}$ and halfwidth of 26 fs acts on a plane 80 times overcritical 7.3 μm thick target. In some sense 1D is the most severe case because it limits the number of absorption channels; for example storage of energy in induced ring currents and concomitant magnetic fields decaying later are excluded. On the other hand the absorption values of 70–80% have been achieved in PIC simulations in 1D and in 1D experiments. To reveal resonance single particle orbits must be followed. It is achieved by embedding a sufficient number of test electrons, here 200, at equal distance from each other in the target and following momentum $p_x(t)$ and space coordinate $x(t)$, during 15 Nd laser periods $T_{\mathrm{Nd}} = 3.5$ fs. The outcome of the statistics is convincing: all test electrons interacting with the laser field are resonantly accelerated

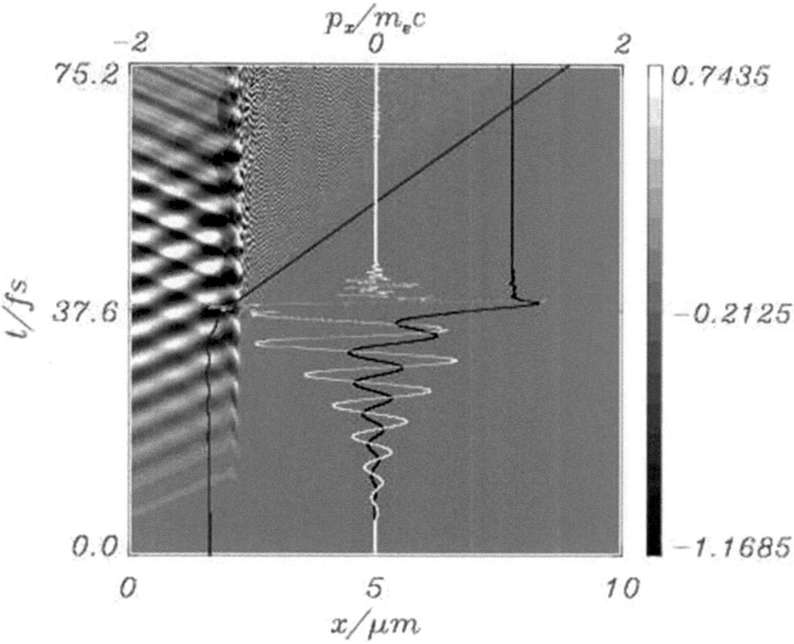

Fig. 7.16 1D3V PIC simulation of a Nd laser beam interacting with a thick overdense target with mobile ions in the boosted frame (parameters see text). Regular shadow structure: laser field, LHS black trajectory: displacement $x(t)$ of test electron #4, RHS black trajectory: corresponding momentum p_x in units of $m_e c$, white line: total electric field at position $x(t)$. Resonance (strong momentum increase and phase shift $\varphi(t)$) and *disruption* are impressive. In all tests: no disruption without preceding resonance. Resonances look all very similar to each other (see following Figure)

and either escape in a disruption-like manner into the target interior or move out into the vacuum from where they are driven back into the target by the space charge. In Fig. 7.16 the time history of a typical test electron is depicted. The electron enters the laser field (shadowed interference pattern), interacts resonantly (see the evolution of momentum normalized to mc) and escapes into the target an instant later with 15.3 times W. Resonance is ensured by the high energy gain, the total E-field deviating only slightly from sinusoidal behavior (white line), and the phase shift in the last half electron oscillation: last maxima of the two lines exactly coincide, the former are dephased by $\pi/2$, as they should at $\omega_p/\omega > 1$. The shadowed fine structure right of the laser field is due to plasmons. All orbits entering the laser field undergo resonance and look very similar to each other, see Fig. 7.17 with other resonant layers. One could object that only the contribution of $\overline{\mathbf{jE}}$ in x direction has been considered. However, owing to the conservation of the canonical momentum in y direction in 1D at normal incidence, $p_y + e\Re(E/i\omega) = \text{const}$, $j_y E_y$ leads only to the change of p_x described above. The simulation analysis tells also important details on the heating mechanism: The fast electrons are generated first by resonance; they excite

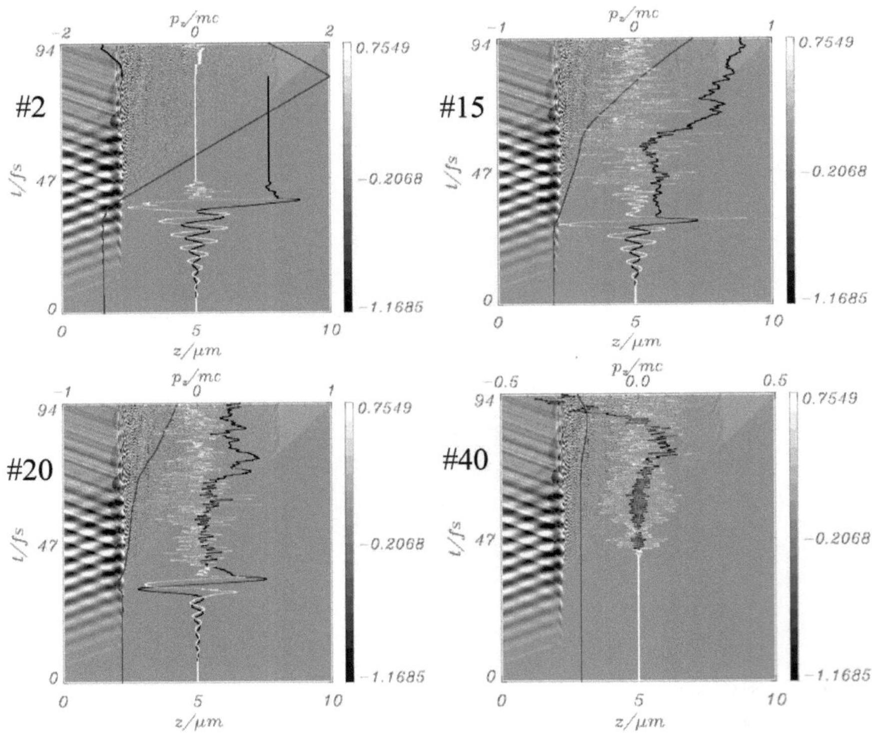

Fig. 7.17 Anharmonic electron resonance and stochastic interaction [52]: a selection of 3 layers (together with layer #2). Regular shadow structure: laser field; black trajectory: orbit $x(t)$ (left) and momentum $p_x(t)$ (right) white traces: electromagnetic/electrostatic field at the particle's position. Primary interaction is by resonance between transverse and longitudinal field during a fraction of laser cycle. Particle in the last picture experiences stochastic acceleration by plasmons only; it never "sees" the laser. Observe the density of plasmons increasing with time

"solitary", i.e., non-Bohm-Gross plasmons in the dense interior which, in turn, heat cold electrons by a mechanism resembling Landau damping. The simulations have also revealed that the electron spectrum is subject to continuous metamorphosis in time, an aspect which may play an important role in *fast ignition*. The mechanism of anharmonic resonance is also active in the collisionless absorption in clusters during the early interaction phase when ion expansion is negligible [42, 54].

Anharmonic resonance has found its first experimental verification in the few laser cycle absorption experiment by Cerchez et al. [55]. After various estimates and cross checks the conclusion is that this represents the most promising candidate for the measured absorption. It should be stressed that the mechanism is active also in long fs or ps pulses when profile steepening is so strong that no linear resonance can take place. The main practical relevance of resonance may be seen in the possibility to tailor the electron spectrum for various applications (electron and ion acceleration, fast ignition, etc.) by designing targets properly, for instance by choosing carefully their thickness, structure, and shape.

7.3.3 *1D PIC Simulations of Relativistic Laser-Overdense Matter Interaction*

Among the various attempts to understand collisionless absorption of intense and superintense ultrashort laser pulses a whole variety of models and hypotheses has been invented to describe the laser beam-target interaction. So far basic modelling and facts have been outlined in Sects. 1.4 and 7.3.2. In terms of fundamental physics collisionless absorption is understood as the interplay of the oscillating laser field with the space charge field produced by it in the plasma skin layer and in front of it. A first approach to this idea is realized in Brunels model the essence of which consists in the formation of an oscillating charge cloud in the vacuum in front of the target, therefore frequently addressed by the vague term *vacuum heating*. The investigation of statistical ensembles of orbits shows that the absorption process is localized at the ion-vacuum interface and in the skin layer: Single electrons enter into resonance with the laser field thereby undergoing a phase shift which causes orbit crossing and braking of Brunels laminar flow. This *anharmonic resonance* acts like an attractor for the electrons and leads to the formation of a Maxwellian tail in the electron energy spectrum. The energetic electrons are generated almost instantaneously during one laser cycle or a fraction of it.

In the following PIC simulations in one dimension (1D) are undertaken in the intensity range from several 10^{16} (corresponding to $a = 0.1$) to $10^{22}\,\mathrm{Wcm}^{-2}$ (corresponding to $a = 100$) [56]. The laser pulse rises from zero to full intensity within 1 fs and then remains constant for 30 cycles. It is incident under 45° and is linearly polarized; the calculations are performed with Nd laser wavelength. The target is well overdense, its thickness is chosen sufficiently thick to avoid multiply crossing the laser beam. The interaction of intense laser beams with dense targets is very complex and rich of peculiar facets. On the other hand, consequences of basic effects, like collisionless interaction under anharmonic resonance, are not clarified as they should. Here we report on phenomena which we believe that most of them will survive in 2D and 3D also.

7.3.3.1 Localization of Heating and the Origin of the Fast Electrons

A compact summary of the interaction outlined so far may be in order. The laser field component perpendicular to the target is assumed to have the structure $E(t) = E_0 \sin \omega t$. It generates electron jets of periodicity $\tau = 2\pi/\omega$. All of them are ejected during the first quarter period and all, except 2.2%, return to the target during the second half period $(\pi, 2\pi)$. During the remaining 3/4 period no further electron

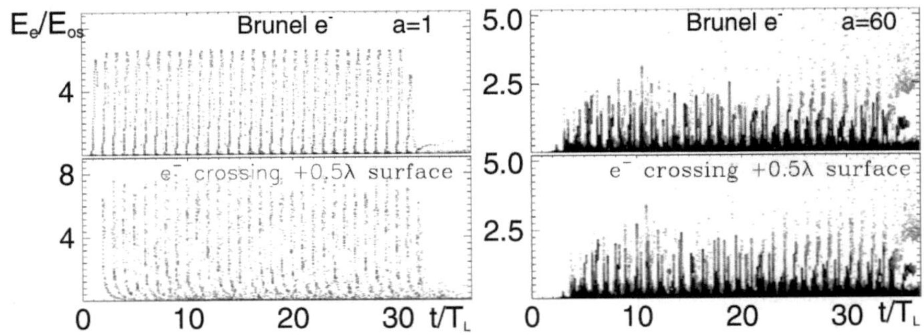

Fig. 7.18 Energy spectra of Brunel jets (upper pictures) and jets at depth $\lambda/2$ as function of time (units in laser cycles) for $a = 1$ and $a = 60$. Double structure is due to $\mathbf{v} \times \mathbf{B}$ acceleration. Strong reduction of E_{max}/E_{os} with increasing a is noticeable

ejection is possible as a consequence of partial screening by the outer layers and driver field inversion. Contrary to a common believe that all electrons in one jet are pushed back by the inverted field, only half of them, lifted in the interval $(0, \pi/4)$, are in phase with the driver, the other half experience a weakened driver due to screening and fall back to the target, attracted by the immobile ions, before the laser field has changed direction. This leads quite naturally to a classification into energetic and less energetic electrons. If not specified differently, throughout the paper we define, somehow arbitrarily, those electrons as hot whose return energy exceeds the quiver energy E_{os} of the free electron. Accordingly, 34% of the Brunel electrons are hot and carry 82% of the energy in the single jet. In contrast to PIC simulations the Brunel spectrum is non-Maxwellian with a pronounced maximum at $E = 9.1E\text{os}$ followed by a sharp cut off. Absorption $A = I_{abs}/I$ is considerably lower than measured at intermediate angles of incidence but reaches unity at 86 of incidence. Crossing of layers among each other during the laser action is excluded, except a few front layers whose contribution to absorption is negligible. The electron flow dynamics is laminar, no wave breaking or, more appropriate in the context, no breaking of flow occurs. Brunels model offered, within limits, the first physical explanation of $\mathbf{j} \times \mathbf{B}$ heating at 2ω.

In Fig. 7.18 the energies of all Brunel particles and of all particles that have crossed the skin layer at a depth of half a vacuum wavelength are plotted at their crossing time for $a = 1$ and $a = 60$. The salient feature is their jet like structure predicted by the Brunel model. At the low intensity ($a = 1$) there is a clear distinction between the Brunel electrons and all electrons having undergone heating. The increase of the energy maxima from $6.5E_{os}$ to $8E_{os}$ and the more diffuse energy profiles of the jets at half wavelength in depth is a clear indication that some heating is localized in the skin layer, in contrast to the Brunel mechanism. At high intensity ($a = 60$) the jets assume a pronounced double structure due to the increased $\mathbf{v} \times \mathbf{B}$ heating operating at 2ω, but the patterns of the two groups appear equally diffuse. The increase in energy of the fastest electrons is almost no longer visible (increase by $0.5E_{os}$). For $a > 1$

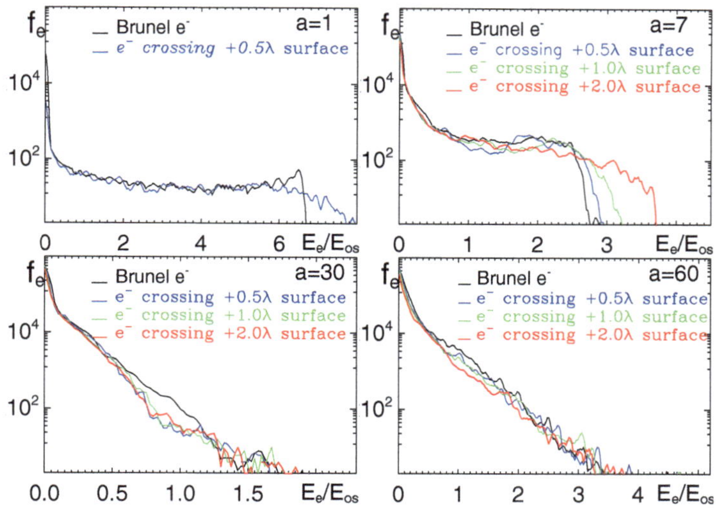

Fig. 7.19 Energy distributions $f(E)$ of Brunel electrons and of electrons crossing positions $x = 0.5\,\lambda$, $1\,\lambda$, $1.5\,\lambda$, and $2\,\lambda$ at intensities $a = 1$, 7, 30, and 60. The distributions are taken after 37 laser cycles when all electrons have returned to position $x = 2\,\lambda$. The non-Maxwellian structure of the Brunel electrons is preserved up to $a = 15$. For higher intensities there is almost no difference between electrons heated in the vacuum and additional heating in the skin layer region

the fraction of Brunel electrons results always higher than the fraction of electrons moving inward and crossing the boundary at $\lambda/2$ in the skin layer. The reason for the difference is to seek in the accumulation of Brunel electrons in the skin layer with increasing laser cycle number, randomized there, and repeatedly crossing the target-vacuum interface before disappearing in the depth of the target.

It is instructive to analyze the spectral distribution function $f(E)$ of the Brunel electrons and all electrons just when crossing positions $x = 0.5\lambda,\ \lambda,\ 1.5\lambda$, and 2λ for the laser intensities corresponding to $a = 1,\ 7,\ 30$, and 60, see Fig. 7.19. Surprising enough, at low intensity ($a = 1$) and, to a minor degree, also at $a = 7$ the Brunel electrons from the PIC simulations resemble much Brunel's analytical spectrum from Fig. 1.21a: The sharp cut offs and the adjacent maxima of $f(E)$ are reproduced, their positions however lie at much lower energies. The maximum of $f(E)$ is still clearly visible for $a = 15$ (not in the Figure), this time at $E = E_{os}$, but the sharp cut off changes into a transition extending over $0.4\,E_{os}$. The formation of a Maxwellian tail in the fast electron spectrum occurs in the skin layer and even deeper inside the target. From $a \simeq 20$ on no difference in the spectra from Brunel and total electrons can be observed, they are all "thermalized". At $a = 30$ the spectra extend up to $1.4\,E_{os}$, at $a = 60$ the maximum energy is shifted to $E = 3.7\,E_{os}$. This is in agreement with the dependence of the fast electron number on laser intensity, see following Section. From Fig. 7.18 we conclude that at moderate intensities ($a \leq 15$) the skin layer contributes sensitively to the production of the most energetic electrons, either by laser-space charge resonance and/or by stochastic Brunel electron-plasmon interactions.

7.3.3.2 The Return Current

Plasma density and velocity distributions as functions of time are the natural outcome in standard PIC simulations. Additional insight in the heating mechanism is obtained from the orbits $\mathbf{x}(t)$ of randomly chosen electrons. We have analyzed numerous such computer runs each with 200 trajectories stochastically selected from (i) all particles heated by the laser and (ii) from the set of the hot electrons only. In Fig. 7.20 their time histories are depicted for the intensities $a = 7$ (left) and $a = 60$ (right). The salient features characteristic of the two groups are the following:

(1) Heating of the energetic electrons is well localized at the vacuum-target interface and takes place during one laser cycle or a fraction of it. This excludes stochastic heating of the hot electrons. Only a low fraction of them gets the high energy in the skin layer without ever emerging into vacuum.
(2) Contrary to the standard assumption the "slow" *return current* is highly irregular as a consequence of the interaction of the jets from Fig. 7.18 with the background. It is clearly recognized in Fig. 7.20 that irregular flow sets in just with the arrival of the first jets and it becomes the more irregular the more jets it is exposed to. The jets are accompanied by strong localized electrostatic fields that force electrons from the return current to reverse their direction towards the back of the target

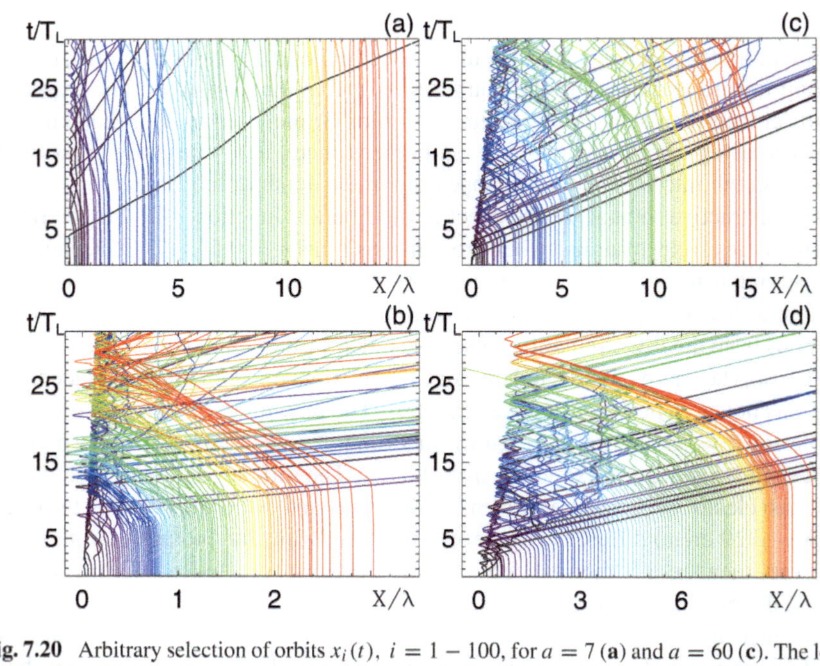

Fig. 7.20 Arbitrary selection of orbits $x_i(t)$, $i = 1 - 100$, for $a = 7$ (**a**) and $a = 60$ (**c**). The lower Figures **b** and **d** show the same number of stochastically chosen orbits from the hot electrons only with $E \geq E_{os}$.Their analysis shows acceleration, i.e., heating almost during one laser cycle or a fraction of it in the laser field at the target front and their strong interaction with the return current

or, if they succeed to cross the charge cloud of an incoming jet they are heavily accelerated towards the target front to interact further with the laser field. In short words, the laminar flow of the return current is heavily perturbed by the *Cherenkov emission* of the *plasmons* excited by the jets. The stochastic interaction, both, return electron acceleration and deceleration, has been observed in test particle models in the past [57].
(3) The plasma flow in the skin layer breaks (like "wave breaking"), i.e., the orbits cross each other, in contrast to Brunel's laminar model of infinite target density.
(4) Excursion into vacuum ("vacuum heating") of the energetic electrons decreases continuously with a increasing to reach a minimum at around $a = 30$ and then to increase again.

Owing to the significance of effects (1)–(3) for localization and understanding hot electron generation, and understanding collisionless laser beam absorption in general, we analyze further the acceleration process.

In Brunel's model the density of the target is assumed infinite. Consequently all electrons start from the same position and no crossing of orbits occurs, the particle flow into vacuum and back to the target is laminar. Despite Brunel's oversimplification his model explains basic properties of the collisionless interaction: formation of steady state jets (Fig. 7.18), two groups of electron energies (energetic electrons co-moving with the laser field, slow free fall electrons), majority of fast electrons stemming from the excursion into vacuum, dominant fraction of laser energy delivered to hot electrons. If therefore *vacuum heating* is identified with Brunel's mechanism it acquires a precise meaning. However there is the missing link to the physics of acceleration in this simple model. Not to forget that in Brunel's model all heated electrons are lifted into vacuum only during the first quarter laser period. The reality with skin layer included is different: The phase for lifting is stochastic, as expected from broken flow; period doubling, tripling, quadrupling, etc., of electron oscillations occurs in the skin layer (see Fig. 7.20); acceleration to high energies is a resonance effect. To see this we must concentrate once more in detail on single orbits selected statistically. In Fig. 7.17 the time history of four particles starting from different depth in the target together with the electric field (white traces) they "see" during their motion is depicted for $a = 1$, i.e., the orbits $x(t)$ and the momenta p_x/m_ec normal to the target. Resonant interaction in the first 3 pictures is clearly recognized by the abrupt changes in $x(t)$ and $p_x(t)$. Out of resonance the phase difference between field and momentum is $\pi/2$, see $p_x(t)$ and laser field (white line) in the first three pictures. The transition to resonance, i.e., field and momentum antiparallel, occurs during half a cycle or less in the kink of $x(t)$, seen best by zooming Fig. 7.17. The essential point of this resonance is its anharmonic character. In contrast to the harmonic oscillator in the oscillator with anharmonic potential resonance is an attractor: Given an excitation by the periodic laser above a certain threshold transition to resonance is unavoidable. The reason for this behavior is as follows. The harmonic potential is the only one in which the degree of excitation does not change its periodicity and therefore it either is driven in or out of resonance. The average stochastically perturbed space charge potential of the plasma is flatter than harmonic and so, depending on the excitation level its eigenfrequency changes continuously from the high level ω_p at low excitation down below the laser frequency ω. At the crossing point resonance occurs. It has two consequences: (i) Driver, when in phase with the electron displacement, transforms it into a runaway particle in general [52]; (ii) the resonant phase switch forces the electron to move against the bulk, the plasma flow breaks. Breaking of flow or wave breaking, respectively, often invoked as acceleration or absorption mechanism [50, 51] is never their origin, it is their consequence.

We conclude that at $I < (5 \times 10^{20} \simeq 1 \times 10^{21})\,\mathrm{Wcm}^{-2}$ the majority of energetic electrons is produced by resonant interaction of the laser field with the longitudinal space charge field over a fraction of one laser cycle in the vacuum as well as in the skin layer. However, there is also indirect acceleration of stochastic nature of the target background, evidenced by the last picture in Fig. 7.17 with an electron heated stochastically by the plasmons emanating from the jets. The rapidly oscillating stochastic field of the plasmons increases with the number of jets produced.

7.3.4 Fast Electrons and Energy Partition

As seen in the previous section there are several thermalizing mechanism: breaking of flow, skin layer noise, Cherenkov plasmons from jets. As a consequence one would expect that such effects dominate the low energy component of the electrons and that this should propagate mainly normally to the target. The more energetic an electron is the more it feels the Lorentz force in $\mathbf{v} \times \mathbf{B}$ direction forcing its motion into laser beam direction, 45° in this paper. For the single free electron starting from rest in a traveling plane wave the maximum energy gain $\Delta\mathcal{E}$ and the lateral angular spread $\tan\alpha$ of the velocity component v_k in propagation direction to the velocity component in the $\mathbf{E}$ field direction v_E are [58]

$$\Delta\mathcal{E} = \frac{1}{2}a^2 m_e c^2, \quad \tan\alpha = \left|\frac{v_E}{v_k}\right| = \frac{2}{a}, \tag{7.90}$$

thus confirming this tendency. In Fig. 7.21 the momenta p_y parallel to the target versus p_x along the target normal of the heated electrons are depicted for $a = 1, 7, 15, 60, 100$. The corresponding distributions of the momenta p_x normal to the target over the space coordinate are shown in Fig. 7.22. From the pictures it is not directly seen that the majority of slow electrons move in the direction of the target normal; however, as expected from (7.90), with increasing energy the electrons follow indeed the direction of the laser beam. In addition, at $a = 60$ and 100 an appreciable percentage is accelerated into specular direction. The reduction of the absolute number of hot electrons with increasing intensity, their almost vanishing at $a = 15$ and their impressive reappearance towards $a = 100$ is particularly striking. This effect will have direct impact on every attempt to formulate scaling laws for the "hot electron" production. We have counted their number as a function of intensity; the result is reported in Table 7.1. Drop and increase with intensity is beyond expectation.

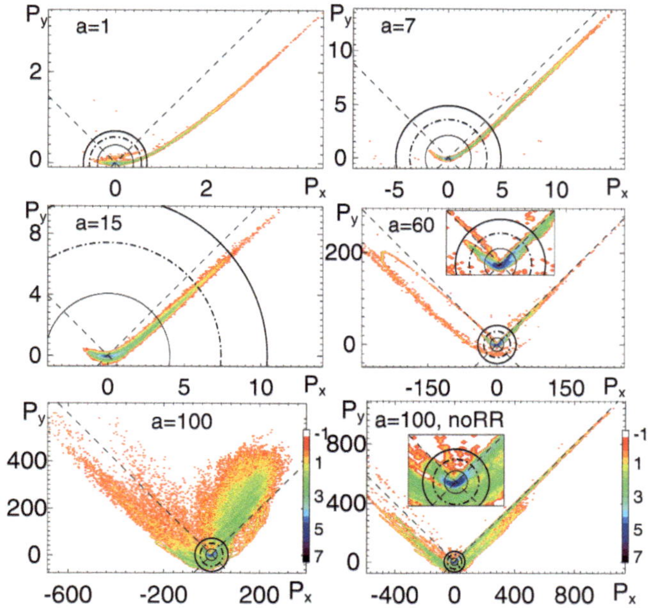

Fig. 7.21 Direction of heated electrons: Momenta p_y/m_ec versus p_x/m_ec for $a = 1, 7, 15, 60$, and 100 (left picture $a = 100$: with radiation damping, right: without) at the end of the standard laser pulse. Electron energies: $E < E_{os}/3$ within inner black circle, $E \in [1/3, 2/3]E_{os}$ within dashed circle, $E \in [2/3, 1]E_{os}$ within bold circle, $E > E_{os}$ ("hot electrons") outside. Circles in the last pictures are very small; therefore see the two insets). The color of the particles in this pictures indicates their number according to the color bar. Low energy electrons within the inner circle (majority in number) penetrate the target normally, energetic electrons follow the laser beam direction (dashed black lines), in $a = 60, 100$ also along the reflected laser beam

The formation of spatial spikes within groups of energies and laser intensities is depicted in Fig. 7.22. It has to be seen as complementary to the spike distribution in time in Fig. 7.18. At low laser intensity only the fastest electrons form jets in space as long as they are "young". As they travel further into the target they become increasingly diffuse as a consequence of their interaction with the *Cherenkov plasmons*. The electrons of varying velocity undergo mixing in phase space, see uniform background in Figs. 7.18 and 7.22, the spikes only are accompanied by strong elecrostatic fields. Their damping by friction is given through the collision frequency $\nu_{coll} = 2(e^4 n_{e0}/\epsilon_0 m_e^2 v_e^3 \gamma) \ln \Lambda$. With the Lorentz factor $\gamma = 1$, $v_e = c$, and $n_{e0} = 10^{23}$ cm^{-3} this results in $\nu_{coll} = 5 \times 10^{10} \ln \Lambda$ s^{-1}, hence, collisional damping of spikes is unimportant. Anomalous interaction of the laser heated electrons with the background has been studied recently [59]. All electrons after having entered the distinctly relativistic regime show a neat spiky structure because they all fly at light speed and behave much stiffer now against their concomitant space charge field. The inclination of groups of spikes with respect to the normal to the abscissa at subrelativistic speeds is self-explaining. We note also that the excursion of the slow electrons into vacuum (p_x negative) reduces with increasing a.

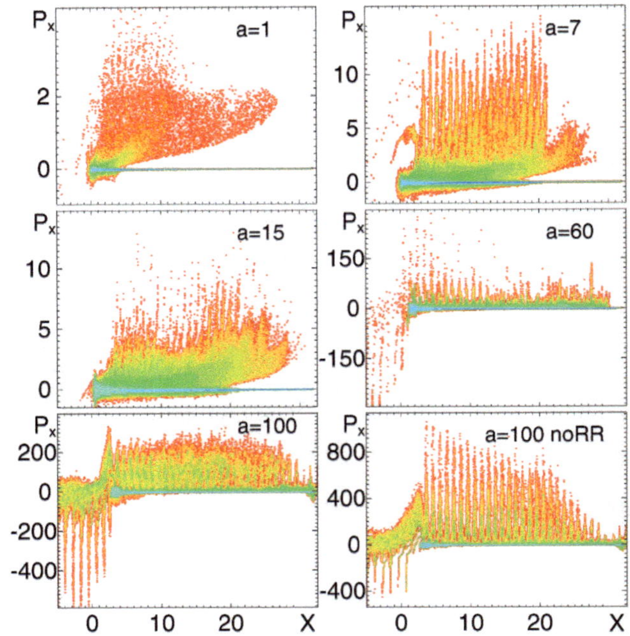

Fig. 7.22 Energy distribution of heated electrons: Momentum p_x/m_ec versus target normal x for $a = 1, 7, 15, 60$, and 100 (left: with radiation damping, right: without) at the end of the standard laser pulse. The color of the particles indicates their number according to the color bar of Fig. 7.21. The spiky structure is a rough indicator of relativistic jets

Table 7.1 Number of hot electrons per unit area (arbitrary units) in dependence of a for $n_{e0} = 100n_c = 100m_e\omega^2\epsilon_0/e^2$

a	1	3	7	15	30	60
N_{hot}	7819	7991	17464	147	265	19273

7.3.4.1 Partition of the Absorbed Energy

Sometimes it is claimed (at least in the past) that all electrons are "hot" in intense laser-solid target interaction. This rises the question on the percentage of the hot electrons with respect to number, to average energy, or to average flux density. Here, we must stress that a percentage in particle number cannot be given, neither in the experiment nor in the simulation for the simple reason that the fraction depends very sensitively on the total number of particles involved: Where to put the lower threshold? Should the shock heated portion of the target be counted also, or is it reasonable to restrict counting on those electrons that have "seen" the laser at least once? However, the situation is totally different with respect to energy fractions because no ambiguity arises on that. In Table 7.2 we present such an absorbed energy partition as a function of laser intensity (parameter $a \sim I^{1/2}$) for two overdense

Table 7.2 Partition of the incident laser energy: fraction of absorbed intensity A = $I_{\rm abs}/I_{\rm inc}$ transmitted to the electrons, the hot and warm electrons, the ions and the plasmons ("fields") at the end of the standard pulse

a_0	n_{e0}/n_c	A	Energy partition				
			All e^-	Hot e^-	Warm e^-	Ions	Fields
0.3	100	0.377	0.25	0.213	0.217	0.009	0.118
	200	0.313	0.24	0.161	0.167	0.007	0.066
0.5	100	0.43	0.28	0.228	0.233	0.01	0.138
1	100	0.358	0.238	0.2	0.211	0.008	0.112
	200	0.354	0.24	0.199	0.206	0.0077	0.106
3	100	0.18	0.122	0.08	0.092	0.003	0.055
	200	0.18	0.123	0.08	0.092	0.003	0.054
5	100	0.2	0.136	0.088	0.102	0.0033	0.061
	200	0.2	0.136	0.089	0.103	0.0026	0.061
7	100	0.19	0.131	0.073	0.089	0.0033	0.056
	200	0.19	0.132	0.076	0.093	0.0025	0.055
15	100	0.067	0.036	0.0002	0.003	0.01	0.021
	200	0.051	0.028	0.0005	0.003	0.007	0.016
	400	0.064	0.039	0.0022	0.008	0.0057	0.019
30	100	0.105	0.0525	0.0002	0.0018	0.0206	0.032
	200	0.045	0.0168	0	0.00012	0.0168	0.011
60	100	0.23	0.126	0.01	0.027	0.033	0.071
	200	0.092	0.031	0.00006	0.0003	0.034	0.027

targets, $n_{e0} = 100n_c$ and $n_{e0} = 200n_c$ (for $a = 15$ also $n_{e0} = 400n_c$): overall fraction of absorption I_{abs}; percentage of energy which is found in the electrons; total fraction of energy absorbed by the hot electrons, $E \geq E_{\rm os}$ and by hot + medium hot electrons of $E > E_{\rm os}/2$ ("warm e^-"); energy fraction transmitted to the ions; energy fraction found in the electrostatic space charge field ("fields"). A first view on the Table tells that the main absorption is accomplished by the energetic electrons (see 5th and 6th column). The absorption by the ions (protons in the Table) remains modest for $a \leq 15$, however, Cherenkov plasmons ("fields") assume a non negligible portion of laser energy, more than predicted.

The increase in ion energy beyond $a = 15$ is due to the deeper penetration of the laser as a consequence of the recession of the electrons by the radiation pressure and hence increased energy coupling to the ions. The overall absorption drops continuously with increasing laser intensity. In contrast to the runaway energy $\mathcal{E}$ in (7.90) the free quiver energy at fixed oscillation center is $E_{\rm os} = mc^2[\sqrt{1 + a^2/2} - 1] \sim I^{1/2}$. However this reduction is counterbalanced by the relativistic increase of the critical density, $n_{cr} \sim \gamma n_c$. As the speed of the moderately hot electrons approaches c the absorption into energetic electrons, which is the major portion, should not change; the drop must have a different, nonrelativistic origin. Although the scaling of $E_{\rm os}$

and n_{cr} may be oversimplified (see [60] for n_{cr} scaling) it is correct in its tendency. Our current explanation attributes the very pronounced reduction of absorption to the limiting effect of the electrostatic field on the oscillation amplitude of the single electron: With increasing intensity I the electrons are pushed more and more inward by the radiation pressure. The electron oscillating in the neighborhood of the vacuum-ion interface oscillates in a narrow anharmonic potential the half width of which towards the target interior is a small fraction of the wavelength ("profile steepening"). A similar reduction of absorption has been reported for normal incidence, with an explanation that agrees qualitatively with ours [61]. Latest beyond $I \simeq 10^{21}\,\mathrm{Wcm}^{-2}$ absorption by the fastest electrons increases again. They are runaway electrons. The electrostatic potential is strong but finite. The phase at which the electrons enter the laser beam is stochastic. Within them there will some of them happen to be in resonance with the field and subject to the Doppler shift

$$\omega' = \gamma(\omega - \mathbf{kv}), \tag{7.91}$$

with γ Lorentz factor, $\mathbf{k}$ wave vector. If such an electron is moving inward from the vacuum it sees the incident wave at a Doppler downshifted low frequency and is accelerated over a longer distance whereas the reflected wave is seen at a highly upshifted frequency and represents merely a high frequency disturbance. For an electron moving outward towards the vacuum the accelerating field is that of the reflected wave. The proof of this acceleration mechanism is based on the study of single particle motions, and is directly confirmed by the appearance of energetic electrons flowing into vacuum in the reflected wave direction in the picture for $a = 60$ and $a = 100$ of Fig. 7.21. To give a numerical example of electron displacement lengthening $\Delta x/\lambda$ in a plane TN:SA laser wave ($\lambda = 800$ nm) during a forth cycle $\Delta\varphi = \pi/2$: $\Delta x/\lambda = 3$ at $I = 10^{21}\,\mathrm{Wcm}^{-2}$ and $\Delta x/\lambda = 30$ at $10^{22}\,\mathrm{Wcm}^{-2}$. For comparison, at $I = 10^{18}\,\mathrm{Wcm}^{-2}$ this shift is 0.03 only. Beyond $I = 10^{22}\,\mathrm{Wcm}^{-2}$ radiation reaction on the electron motion has to be taken into account [56]. A summary of absorption into all plasma channels (electrons, ions, plasmons) its fraction into electrons, the decrease of absorption towards a pronounced minimum close to zero at $a \simeq 15$–20 and its rise beyond is presented in Fig. 7.23.

7.4 On Scaling Laws of the "Hot Electrons"

From intensity scaling the experimentalist and theoretician expect analytical formulas of the shape of the electron spectrum as a function of the laser intensity. As such a goal seems to be beyond reach at present the high power laser community has limited its focus on the energetic electrons. There, the generation of a Maxwellian tail is one of the characteristics of high power interaction. It is also the most interesting part of the spectrum because, as seen from Table 7.2 it contains the main part of the absorbed energy and, last but not least, it is relevant to applications for collective ion acceleration, radiation sources, medical applications, and others. It is aimed at how

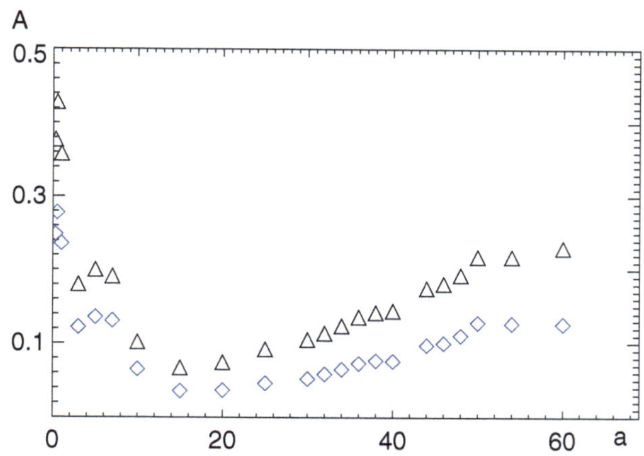

Fig. 7.23 Total absorption (triangles) of a 30 cycles standard laser pulse (see text) and the absorption by electrons (blue diamonds) is given as a function of a. The reduction is due to oscillation inhibition by the induced electrostatic field, its rise beyond is mainly a consequence of entrainment ("runaway electrons")

the number of energetic electrons, the degree of absorption and the mean energy scale with intensity. On the basis of present knowledge scaling of the first two quantities is not feasible. Regarding the mean energy, or the hot temperature $k_B T_{\rm hot}$, respectively, despite the frequent attempts in experiment and theory no convergence has been achieved so far at all. In the light of our foregoing analysis there is not much surprise about.

The frequently invoked ponderomotive scaling ("Wilks' scaling") [62] of $T_{\rm hot} \sim I^{1/2}$ is based on the idea that each laser cycle energetic electrons with energy average in the range of about $E_{\rm os} = m_e c^2(\sqrt{1 + a^2/2} - 1)$ are generated. Subsequently this scaling has been recognized as too strong and, in first place guided by experiments [63], has been replaced by the milder power law [62] $T_{\rm hot} \sim I^{0.34-0.4}$ in the intensity range 10^{18}–10^{21} Wcm^{-2}. It is intended as to be applied to the unidirectional Maxwellian electrons reaching the analyzer at time $t = \infty$. The search for the right mean energy scaling is equivalent to the search for the process of absorption. In the absence of the stochastic element inherent in collisions it is important to understand whether a Maxwellian tail is one of the signatures of the interaction and, if it is, why.

To arrive at a Maxwellian distribution in an ensemble it is sufficient to know that, given a certain amount of particles $n_{\rm hot}$ containing the amount of energy $\mathcal{E}_{\rm hot}$, all possible states in the relevant phase space are equally likely and that the Hamiltonian is given by the sum

$$H = \sum_{i=1}^{n_{\rm hot}} \sqrt{m_e^2 c^4 + c^2 \mathbf{p}_i^2}; \quad |H| = \mathcal{E}_{\rm hot}, \tag{7.92}$$

which expresses the property that the single electrons are uncorrelated. If the relevant phase space is $\{(\mathbf{p}, \mathbf{q})\}$, as for example in statistical thermodynamics, the resulting distribution is the Maxwellian momentum distribution $f(E) \sim \sqrt{E}\exp(-E/k_B T_{\text{hot}})$, in disagreement with Fig. 1.16. However, as outlined in the foregoing chapters, fast electron generation is by anharmonic resonance. This has the important properties: (i) resonance is an attractor for all electrons above a certain oscillation energy, in contrast to harmonic resonance; the always present crossing of trajectories is a clear indicator of it. (ii) All n_{hot} electrons have the same chance to resonate anharmonically regardless of their phase with respect to the laser driver. This makes it very likely that the relevant phase space is the energy acquired at resonance rather than the momentum. Then from (i) and (ii) follows that collisionless absorption is accompanied by a Maxwellian tail of energetic electrons (consequence of anharmonic attractor) and the spectrum scales like $f(E) \sim \exp(-E/k_B T_{\text{hot}})$, without a degeneracy factor $\sqrt{E}$. This is what we also deduce from Fig. 1.16. Let us first examine Fig.1.16 for pulses $I \sim \sin^4$ in the intensity range 10^{18}–$10^{20}\,\text{Wcm}^{-2}$. The uncertainties on the mean energy (slope of log scale) in pictures (a) (30 cycles) and (c) (40 cycles) are considerable, nevertheless we can conclude with certainty that neither Wilks' original ponderomotive scaling [62] ($I^{0.5}$) nor its improved version are met to some extent. They are far too weak. However, the analysis shows that the assumption $k_B T_{\text{hot}} = \kappa \times E_{\text{os}}$ with the constant κ not far from unity works. This means that at these relatively low intensities (from $a \simeq 3$ to $a = 10$) the scaling is

$$k_B T_{\text{hot}} \sim \sqrt{1 + a^2/2} - 1 \tag{7.93}$$

in agreement with [64]. From $a = 12$ on E_{os} is well approximated by $m_e c^2 a \sim I^{1/2}$. In Fig. 7.24 we extended the search for scaling from $a = 10$ up to $a = 60$ the latter is already in the runaway regime of absorption. Satisfactory agreement with [64] is obtained for I from 10^{20} to $10^{21}\,\text{Wcm}^{-2}$. Beyond the change in the absorption mechanism and the stiffer coupling to the ions is noticeable in the increase of slope relative to [64].

Searching for fast electron scaling may have the scope to arrive at a rough characterization of intense laser beam absorption. None of the numerous scaling laws published so far has been successful over a wide range of intensities. Laser-matter coupling depends to a high degree on the structure of the interaction surface (e.g. corrugation, microstructuring) and on the laser pulse duration, its aperture and angle of incidence. From the failure we learn that fast electron scaling is an outmoded concept and is to be substituted by the search for more specific parameters.

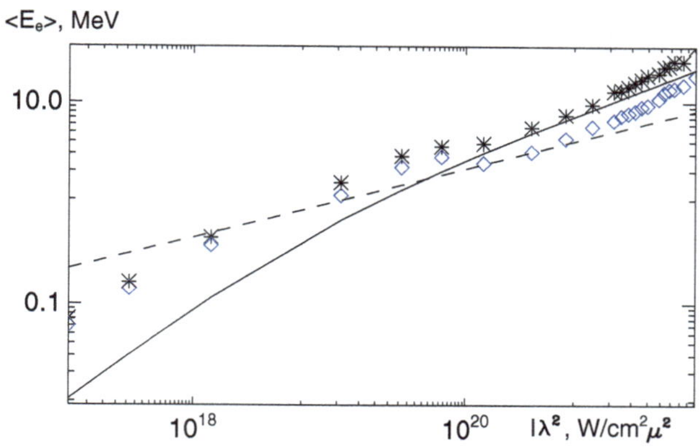

Fig. 7.24 Hot and warm electron energy scaling $\langle E_e \rangle$ with laser intensity $I = 10^{17} - 5 \times 10^{21}$ Wcm^{-2}, standard pulse. Stars: $E_e \geq E_{\rm os}$ ("hot electrons"); diamonds: $E_e \geq 0.5 E_{\rm os}$ ("warm electrons"). In contrast to Fig. 1.16 here $\langle E_e \rangle$ is the average taken over the single energies E_e. Solid line: scaling after [64], $k_B T_{\rm hot} = m_e c^2 [\sqrt{1 + (I\lambda^2)/(2.74 \times 10^{18})} - 1]$. Dashed line: scaling after [63], $k_B T_{\rm hot} = m_e c^2 [I\lambda^2/(1.37 \times 10^{18})]^{0.34}$

7.5 Pressure-Viscosity Tensor and Friction in Plasma

As outlined in Chap. 3 the macroscopic fluid model is the first approach towards a description of the many body problem of the plasma. All memory of the microscopic effects that survive on the macroscopic scale are packed into transport coefficients, an example thereof is the absorption coefficient of radiation in the foregoing sections. The reduction of the plasma to a fluid necessitates the introduction of *surface forces* as a new class of *generalized forces*. Such a force arises on a fluid element ΔV whenever the flux of the associated physical quantity is partially absorbed by ΔV. So, for example, absorption and reflection of radiation generate the radiation pressure onto the surface Σ of ΔV. If the quantity to be transported is a scalar the generalized flow of it is a vector $\mathbf{q}$ and transforms like the position vector $\mathbf{x}$. If $d\Sigma$ is rotated by the angle ϕ from the direction of $\mathbf{q}$ its flux through $d\Sigma'$ is $\mathbf{q}\cos\phi\, d\Sigma$. Examples are the laser intensity $\mathbf{I} = n\hbar\omega\,\mathbf{c}$ and heat flux $\mathbf{q} = -\kappa\nabla T$. If the quantity to be transported through $d\Sigma$ is a vector $\mathbf{q}$ its flux $\mathbf{qw}$ is a second rank tensor and its flux through $d\Sigma'$ is $\mathbf{qw}\cos^2\phi\, d\Sigma$. The pressure-viscosity tensor P_{ij} from Chap. 3 is of this kind. As a particular example radiation pressure $p_L = (1 + R)I/c$ is addressed. Under rotation by Φ it becomes $p'_L = p_L \cos^2\Phi$, in agreement with elementary geometrical reasoning.

7.5.1 Coefficients of Viscosity

The kinetic picture of viscosity uncovers the physics involved in the generation of viscosity in rarefied fluids. It is described in Chap. 3. Recalling the explicit form of the complete isotropic pressure-viscosity tensor $\Pi = \Pi_{ij}$, it has been obtained as

$$\Pi = \Pi_{ij} = (p + \mu'\nabla\frac{\mathbf{u}}{3})\delta_{ij} + P_{ij}; \quad P_{ij} = \mu(\frac{\partial u_i}{\partial x_j} + \frac{\partial u_j}{\partial x_i} - \frac{2}{3}\delta_{ij}\nabla\mathbf{u}); \quad p = m\langle\mathbf{v}^2\rangle. \tag{7.94}$$

In the magnetized plasma the pressure may be anisotropic and is to be replaced by $p = p_i\delta_{ij},\ p_{1,2} = p_\perp,\ p_3 = p_\parallel$, if the magnetic field is oriented along $\mathbf{e}_3$. The coefficients $\mu,\ \mu'$ stand for the *shear* and the *bulk* or *second viscosity*,

$$\mu = mn\langle\mathbf{v}^2/6\rangle\tau; \quad \mu_i' = \mu' = mn\langle\mathbf{v}^2/3\rangle\tau'. \tag{7.95}$$

The single particle velocity averages are to be recovered from individual distribution functions $f(\mathbf{v})$ for electrons, and different species of ions in the plasma locally at rest. $\tau,\ \tau'$ are the collision times $\tau_{ee}, \tau_{ii}, \tau_{ee}', \tau_{ii}'$. In viscous fluids the coefficient of the second viscosity μ' may be greater up orders of magnitude than the shear viscosity coefficient μ. It may become relevant in shock compression. In the fully ionized dilute plasma is $\mu' = 2\mu$ and they differ from each other only by the charge numbers Z. In case of local thermodynamic equilibrium of the plasma constituents holds $\langle m\mathbf{v}^2/3\rangle = k_BT$, with $T = T_e, T_i$ for the electrons and the ions. For the dilute plasma in thermal equilibrium between electrons and ions, $T = T_e = T_i$ follows

$$\mu = \mu_i = \frac{nk_BT}{2\nu_{ii}} = 3\sqrt{2}\pi^{3/2}\varepsilon_0^2 e^{-4}(k_BT)^{5/2}\frac{m_i^{1/2}}{Z^4\ln\Lambda}$$
$$\Rightarrow \mu_i[\mathrm{Ws}^2] = 2.9\times10^7(T[eV])^{5/2}\frac{(m_i[\mathrm{kg}])^{1/2}}{Z^4\ln\Lambda}. \tag{7.96}$$

For moderate Z the viscosity of the electronic component can be discarded because of

$$m_e\langle v_e^2\rangle = m_i\langle v_i^2\rangle \Rightarrow \mu_e = \frac{m_en_e\langle v_e^2\rangle}{\nu_{ee}+\nu_{ei}} \sim Zn_ik_BT\left(\frac{m_e}{m_i}\right)^{1/2}\frac{Z^4}{1+Z^2} \ll \mu_i.$$

The collision frequencies have been taken from the low field limit (7.35).

> The shear (and bulk) viscosity coefficient is proportional to the square root of the particle mass, increases with the 5/2 power of temperature and decreases with the 4th power of its charge; $\mu_{e,i}$ are density independent. For low charge and comparable temperatures T_e, T_i plasma viscosity is dominated by the ions. For the hydrogen plasma the formula for μ is easy to remember, it is the plasma pressure p divided by the collision frequency ν.

7.5.2 Friction

Viscosity is *internal friction* because it is momentum transfer between adjacent elements of the same fluid. In contrast, in a multicomponent fluid *friction* is the momentum transfer from fluid elements of one species to fluid elements of another species. Friction of the plasma electrons with the ions was the subject of the first section of this chapter. It is at the basis of collisional or inverse bremsstrahlung absorption of the laser radiation. Friction between ions of charge Z_1 and mass m_1 with ions of charge Z_2 and mass m_2 becomes relevant when plasma *demixing* under the influence of external forces is investigated. From early studies on laser plasma expansion at low laser intensities the phenomenon has been well known under the name of *ion separation*, see Chap. 1 on multicomponent plasma. It has acquired a re-edition of timeliness in connection with *target normal sheath acceleration* (TNSA) by the fast electrons generated in the superintense laser field.

At all laser intensities a thermoelectric field by the electrons is created which couples to the ions in proportion to their charge number Z. The basic equation governing separation of two species of ions is

$$m_1\frac{\mathrm{d}\mathbf{u}_1}{\mathrm{d}t} = Z_1 e\mathbf{E} - m_1\nu_{12}(\mathbf{u}_1 - \mathbf{u}_2). \tag{7.97}$$

$\mathbf{E}$ is the thermoelectric field after (3.44) and ν_{12} stands for the collision frequency of the two Debye screened ion species. The basic mechanism underlying the collission frequency ν_{12} is identical with that for ν_{ei}. Remember, the collision cross sections are defined in the center of mass system. In the electron-ion system the center of mass is very close to the ion and moves with the low ion speed. Therefore the simplifying *Lorentz model* of infinitely heavy ions at rest applies. In ion-ion collisions the relations (2.147) apply with a non-zero center of mass velocity V. In the Coulomb cross section the ion mass m_1 is to be replaced by the reduced mass $\mu = m_1 m_2/(m_1 + m_2)$.

Let the single ion velocities be $\mathbf{v}_i = \mathbf{u}_i + \mathbf{w}_i,\ i = 1, 2$ and the *isotropic* distribution functions $f_i(\mathbf{w}_i) = f_i(w_i)$ be normalized to the total ion density $n_i = n_1 + n_2$. If before the collision the ion velocities are $\mathbf{v}_1,\ \mathbf{v}_2$, the center of mass velocity $\mathbf{V}$, and the relative velocities and $\mathbf{u} = \mathbf{v}_1 - \mathbf{v}_2$ are fixed. After the collision they are $\mathbf{v}_1',\ \mathbf{v}_2'$ with $\mathbf{V}' = \mathbf{V},\ |\mathbf{u}'| = |\mathbf{u}| = u$. The angle between $|\mathbf{u}|$ and $|\mathbf{u}'|$ is the scattering angle ϑ in the center of mass system. The evaluation of ν_{12} proceeds in close analogy to Sect. 2.1 in this chapter. The momentum loss per unit time of the single particle m_1 to the single particle m_2 is into direction $\mathbf{u}$ and adds up over all scattering angles ϑ to $\dot{\mathbf{p}}_1 = \sigma_t n_2|\mathbf{u}|\Delta\mathbf{p}$, i.e.,

$$\dot{\mathbf{p}} = -m_1\nu_{12}(\mathbf{u})\mathbf{u} = -\frac{K_i}{u^3}\mathbf{u}\ln\Lambda(\langle v_e^2\rangle), \quad K_i = m_1\frac{Z_1^2 Z_2^2 e^4 n_i}{4\pi\varepsilon_0^2\mu^2}. \tag{7.98}$$

In the second step the ensemble average $\langle\dot{\mathbf{p}}_1\rangle$ is calculated. A quick view on Fig. 7.5 tells that $\mathbf{v}_1 - \mathbf{v}_2$ and $\mathbf{u}_1 - \mathbf{u}_2$ assume the role of the former $\mathbf{v}_e$ and the oscillation velocity $\mathbf{v}_{os}$ on which the single quantities $\dot{\mathbf{p}}$ are projected under angle α,

$$\begin{aligned}\langle\dot{\mathbf{p}}\rangle &= -m_1\nu_{12}(\mathbf{u}_1-\mathbf{u}_2) \\ &= -K_i \ln \Lambda(\langle v_e^2\rangle)\int_{\mathbf{v}_1,\mathbf{v}_2}\frac{\mathbf{v}_1-\mathbf{v}_2}{|\mathbf{v}_1-\mathbf{v}_2|^3} f_1(\mathbf{v}_1-\mathbf{u}_1)f_2(\mathbf{v}_2-\mathbf{u}_2)\,\mathrm{d}\mathbf{v}_1\mathrm{d}\mathbf{v}_2.\end{aligned} \tag{7.99}$$

The collision frequency ν_{12} follow as

$$\nu_{12} = 2\pi\frac{K_i\ln\Lambda(\langle v_e^2\rangle)}{m_1|\mathbf{u}_1-\mathbf{u}_2|}\int_0^\infty\int_{-1}^1\frac{|(\mathbf{u}_1-\mathbf{u}_2)(1+\mathbf{u}\cos\chi/|\mathbf{u}_1-\mathbf{u}_2|)|\,w_2^2 f_2(w_2)}{|\mathbf{v}_1-\mathbf{v}_2|^3}\,\mathrm{d}\cos\chi\mathrm{d}w_2 \tag{7.100}$$

$$\mathbf{u}=\mathbf{v}_1-\mathbf{v}_2,\quad \mathbf{w}_{1,2}=\mathbf{v}_{1,2}-\mathbf{u}_{1,2}.$$

The Coulomb logarithm can be taken out of the integral because of $|\mathbf{v}_{1,2}| \ll \langle v_e^2\rangle^{1/2}$. Hence, the ions interact through Debye potentials.

7.6 Particle Diffusion and Thermal Conduction

There exist substantially two classes of transport. In the *convective transport* the transported physical quantity, mass, momentum, energy, is fixed to flow of quasineutral matter. Convection is a macroscopic phenomenon to be described in terms of fluids moving with velocity $\mathbf{u}$. Mixing of different fluids is a prominent example. In contrast, *diffusion* is microscopic transport of particles, momentum, inner energy *without* convection.

Particle diffusion. Consider the distribution of particles of one species n along x for electrons and ions, $n_e \simeq Zn_i$ at constant temperature T. In the time $\mathrm{d}t$ $n(x-\lambda_x)/2$ particles cross the border x from left and $n/x)/2$ cross it from the right at equal average velocity $v_{\mathrm{th,x}}$. It leads to the net current density j,

$$j\mathrm{d}t = \frac{1}{2}[n(x-\lambda_x)-n(x)]v_{\mathrm{th,x}}\mathrm{d}t = \left[n(x)-\lambda_x\frac{\partial n}{\partial x}-n(x)\right]v_{\mathrm{th,x}}\mathrm{d}t = -\frac{\lambda_x}{2}\frac{\partial n}{\partial x}v_{\mathrm{th,x}}\mathrm{d}t.$$

From isotropy of λ and v_{th} follows

$$\lambda_x v_{\mathrm{th,x}} = \lambda v_{\mathrm{th}}\frac{1}{2\pi}\int_0^\pi \cos^2\theta\sin\theta\mathrm{d}\theta\mathrm{d}\varphi = \frac{2}{3}\lambda v_{\mathrm{th}}\mathrm{d}t.$$

The current density $\mathbf{j}$ is parallel to ∇n, hence

$$\mathbf{j} = -\frac{1}{3}\lambda v_{\mathrm{th}}\nabla n = -D\nabla n;\quad D=\frac{1}{3}\lambda v_{\mathrm{th}} = \frac{1}{3}\frac{v_{\mathrm{th}}}{n\sigma_{\mathrm{coll}}} = \frac{1}{3}\frac{v_{\mathrm{th}}^2}{\nu_{\mathrm{coll}}}. \tag{7.101}$$

This is *Fick's first law* of diffusion. Here, v_{th}^2 is $3k_BT/2$. In reality the particles left from x do not start from $x - \lambda_x$ but at random around x with the mean free path λ_x. However this has no consequence for the Taylor expansion as long as the curvature of $n(x)$ over a mean free path is negligible. Particle conservation requires $\partial_t n + \nabla \mathbf{j} = 0$. Combining it with (7.101) leads to *Fick's second law*,

$$\frac{\partial n}{\partial t} = D\nabla^2 n. \tag{7.102}$$

The relations (7.101) and (7.102) apply to the electrons and ions separately in the plasma under the conditions of $\mathbf{u}_{e,i} = 0$. In general the equilibrium electric field $\mathbf{E}$ will be such as to drive also net electron and ion dc currents $\mathbf{j}_{e,i} = \mp n_e \mathbf{u}_{e,i}$ under observation of quasineutrality $n_e \simeq Zn_i$. In *Drude's* approximation the quasistatic dynamics of the two isothermal components is governed by

$$\begin{aligned} \frac{\mathrm{d}\mathbf{u}_e}{\mathrm{d}t} + \nu_{ei}\mathbf{u}_e &= -\frac{e}{m_e}\mathbf{E} - \frac{1}{m_e n_e}\nabla p_e = -\frac{e}{m_e}\mathbf{E} - \frac{k_B T_e}{n_e e}\nabla n_e \\ \frac{\mathrm{d}\mathbf{u}_i}{\mathrm{d}t} - Z\frac{m_e}{m_i}\nu_{ei}\mathbf{u}_i &= +\frac{e}{m_i}\mathbf{E} - \frac{1}{m_i n_i}\nabla p_i = +\frac{e}{m_i}\mathbf{E} - \frac{k_B T_i}{n_i e}\nabla n_i. \end{aligned} \tag{7.103}$$

Diffusion is proportional to ∇n at zero flow $\mathbf{u}$. Therefore, in the stationary case of $d_t\mathbf{u} = 0$, $\mathbf{u}_{e,i}$ is associated with the dc electric currents,

$$\mathbf{u}_{e,i} = \mp\frac{e}{m_{e,i}\nu_{e,i}}\mathbf{E} = \mu_{e,i}\mathbf{E}; \quad \mu_{e,i} = \mp\frac{e}{m_{e,i}\nu_{e,i}} \tag{7.104}$$

μ_e, μ_i are the electron and ion *mobilities*. In the absence of macroscopic flow (7.103) reduces to

$$\frac{en_{e,i}}{m_{e,i}}\mathbf{E} = \mp\frac{k_B T_{e,i}}{e}\nabla n_{e,i}$$

From (7.101) and the mobilities from (7.104) one obtains

$$\frac{D_{e,i}}{\mu_{e,i}} = \frac{1}{3}\frac{m_{e,i}v_{\text{th},e,i}^2\,\nu_{ei}}{e\nu_{\text{coll}}} = \frac{k_B T_{e,i}}{e}\frac{1}{1 + \nu_{ee}/\nu_{ei}}.$$

Generally ν_{ee}/ν_{ei} is significantly smaller than 1 and may be dropped. This leads to the *Einstein relations*

$$\frac{D_{e,i}}{\mu_{e,i}} = \frac{k_B T_{e,i}}{e}. \tag{7.105}$$

Ambipolar diffusion. Electrons diffuse much faster than ions. This leads to a charge imbalance and to a counteracting electric field. As a consequence the electrons are slowed down and the ions are accelerated. Pure diffusion persists under observation of quasineutrality if $\partial_t(Zn_i - n_e) = 0$ or, equivalently, $\mathrm{div}\mathbf{j} = \mathrm{div}(Zn_i\mathbf{u}_i - n_e\mathbf{u}_e) = 0$. According to the definition of diffusion of the combined system electrons plus ions as transport in the absence of convection, locally $\mathbf{j} = 0$ must be satisfied. From (7.101) and (7.103) follows

$$n_e\mathbf{u}_e = -n_e\mu_e\mathbf{E} - D_e\nabla n_e, \quad n_i\mathbf{u}_i = +n_e\mu_i\mathbf{E} - D_i\nabla n_e; \quad n_e \simeq Zn_i = n \quad (7.106)$$

and for the total flux $n_a\mathbf{u}_a$ of the combined system holds

$$n_a\mathbf{u}_a = n_e\mathbf{u}_e + n_i\mathbf{u}_i = -D_a\nabla n \Rightarrow D_a = \frac{\mu_e D_i + \mu_i D_e}{\mu_e + u_i}. \quad (7.107)$$

The *ambipolar diffusion coefficient* D_a fulfils the inequalities $D_e > D_a > D_i$, as expected from the physical picture. The electric field results from subtraction in (7.107),

$$\mathbf{E} = -\frac{D_e - D_i}{\mu_e - \mu_i}\frac{1}{n}\nabla n \simeq -\frac{k_B T_e}{en}\nabla n \quad (7.108)$$

a result well known from (3.44). It holds the two fluids together.

In the derivation of D_a no recurrence has been made to collisions, only quasineutrality has been invoked. As already seen the ratio $D_{e,i}/\mu_{e,i}$ depends only weakly on collisions and coincides with Einstein's relations for $\nu_{\text{coll}} = \nu_{ei}$ and arbitrary ν_{coll}. One may conclude that in its pure structure ambipolar diffusion is collisionless.

7.6.1 Thermal Conduction

In intense beam-matter interaction diffusive energy transport is mainly accomplished by the plasma electrons. In contrast to convective transport which is bound to transport of matter thermal conduction is energy transfer from one position in space to another one by collisions in which more energetic particles give their energy to less energetic particles without involving any mass flow. Energy transport by convection is mediated through mass flow. Convection is *reversible*, heat flow is *irreversible*. Kinetically, heat conduction of one species α in the dilute fluid is given by (see Chap. 4)

$$\mathbf{q}_\alpha = \int \frac{1}{2}m_\alpha(\mathbf{v}_\alpha - \mathbf{u}_\alpha)^2(\mathbf{v}_\alpha - \mathbf{u}_\alpha)f_\alpha d\mathbf{v}_\alpha. \quad (3.92)$$

It depends on the asymmetry of the one particle distribution function f_α. Its shape is the result of collisions and the interaction with macroscopic fields. In the laser plasma thermal conduction is dominated by the quickest particles, i.e., by the electrons. Therefore in the following we limit ourselves to $\mathbf{q}_e$ and to the determination of f_e

from a one particle kinetic equation. For this we use the simplest of all which is the *Vlasov-Boltzmann equation*. However, before proceeding to its solution we use a semikinetic consideration to obtain $\mathbf{q}_e$.

In Chap. 1 *local heat transport* has been obtained from the *Fourier ansatz* $\mathbf{q}_e(\mathbf{x}) = -\kappa \,\mathrm{grad}\, T_e(\mathbf{x})$ with the thermal conduction coefficient to determine from an appropriate model. The reader is already familiar with it in connection with the derivation of the viscosity coefficient μ in Chap. 3 and the simple diffusion model in the preceding paragraph. In complete analogy to particle diffusion and the momentum transfer by viscosity (Fig. 3.2) the heat flow results as

$$\mathbf{q}_e = -n_e v_{\mathrm{th,grad}} \lambda_{\mathrm{grad}} \nabla \frac{1}{2} m_e v_{\mathrm{th}}^2 = \frac{3 p_e}{2(\nu_{ee} + \nu_{ei})} \frac{k_B}{m_e} \nabla T_e = -\kappa_e \nabla T_e. \tag{7.109}$$

Thereby $\nu_{ee} + \nu_{ei} = \lambda^{-1}$ has been used. The index "grad" indicates the projection of $(v\lambda) \sim v_{th}^2$ onto ∇T_e. Local isotropy of $\mathbf{v}^2$ requires $v_{\mathrm{th,grad}} \lambda_{\mathrm{grad}} = \lambda/3$ (see Problems).

7.6.2 κ_e from Boltzmann Equation

A complementary and yet similar picture of $\mathbf{q}_e$ is obtained from the Boltzmann equation with an appropriate collision term. Here we use, for simplicity, the *Bhatnagar-Gross-Krook* (BGK) ansatz in the Vlasov equation with a constant collision frequency ν,

$$\frac{\partial f_e}{\partial t} + \mathbf{v} \frac{\partial f_e}{\partial \mathbf{x}} - \frac{e}{m_e} \mathbf{E} \frac{\partial f_e}{\partial \mathbf{v}} = -\nu f_e. \tag{7.110}$$

In presence of a temperature gradient an electric field inhibits electron-ion demixing and violation of quasineutrality; it provides for $\mathbf{u} = 0$. In the homogeneous plasma of constant temperature $\mathbf{E}$ vanishes in equilibrium and the equation reduces to $\partial_t f_e = -\nu f_e$. Whatever f_e is, in the absence of an electric field it tends to a Maxwellian f_M with decay time $\tau = \nu^{-1}$. The electron thermal heat conduction k_e transport is defined at constant electron density n_{e0} under steady state conditions. Hence, (7.110) reduces to

$$\mathbf{v} \frac{\partial f_e}{\partial \mathbf{x}} - \frac{e}{m_e} \mathbf{E} \frac{\partial f_e}{\partial \mathbf{v}} = -\nu f_e. \tag{7.111}$$

If collisions are frequent enough the situation of LTE is expected to occur with f_e locally deviating little from a Maxwellian. A sufficient condition is a small mean free path λ_e compared to the local temperature gradient length $L_e(\mathbf{x}) = T_e(\mathbf{x})/|\nabla T_e(\mathbf{x})|$ everywhere. Then, the equation can be solved perturbatively by setting $f_e = f_M(\mathbf{x}) + f_1$ with $|f_1| \ll |f_e|$. With such an ansatz (7.111) shrinks in zeroth order to

$$\mathbf{v} \frac{\partial f_M}{\partial \mathbf{x}} - \frac{e}{m_e} \mathbf{E} \frac{\partial f_M}{\partial \mathbf{v}} = 0; \qquad f_M(\mathbf{x}) = \left(\frac{\beta}{\pi}\right)^{3/2} \mathrm{e}^{-\beta v^2}.$$

It is obtained from Boltzmann's collision integral in place of BGK. Integration in $\mathbf{v}$ leads to the well-known thermoeletric field

$$\mathbf{E} = -\frac{\nabla p_e}{n_{e0}e} = -\frac{k_B}{m_e}\nabla T_e, \tag{7.112}$$

see (3.44) with $\mathbf{B}_0 = \Phi_p = \mathbf{u}_{e0} = \mathbf{u}_{i0} = 0$.

Next $\mathbf{q}_e$ is calculated. It holds $\mathbf{q}_e = (m_e/2)\int v^2\mathbf{v} f_e d\mathbf{v} = (m_e/2)\int v^2\mathbf{v} f_1 d\mathbf{v}$. The Maxwellian f_M does not contribute to $\mathbf{q}_e$ because of symmetry in $\mathbf{v}$, but f_1 does. It follows in first approximation of (7.111)

$$-\nu f_1 = \mathbf{v}\frac{\partial f_M}{\partial \mathbf{x}} - \frac{e}{m_e}\mathbf{E}\frac{\partial f_M}{\partial \mathbf{v}} = f_M\mathbf{v}\left\{\beta v^2 - \frac{3}{2} - 1\right\}\frac{\nabla T_e}{T_e}$$

$$\Rightarrow \quad \mathbf{q}_e = \frac{m_e}{\nu}\int\left(\frac{5}{4}v^2 - \frac{\beta}{2}v^4\right)\left(\mathbf{v}\frac{\nabla T_e}{T_e}\right)\mathbf{v} f_M \mathrm{d}\mathbf{v} = -\frac{5n_{e0}k_B^2 T_e}{2m_e\nu}\nabla T_e. \tag{7.113}$$

With Spitzer's $\nu = \nu_S$ the coefficient κ_e becomes

$$\kappa_e = \kappa_{0S}\, T_e^{5/2}. \tag{7.114}$$

Compared to Braginskii's value in (1.61) κ_{0S} here is by the factor $5/2\eta_e = 0.8$ lower. The result is to be expected since, owing to the strong temperature dependence electrons above the thermal mean contribute more to thermal conduction. The next refinement of (7.113) would be to extend the averaging process of $\nu(\mathbf{v})$ under the integral of $\mathbf{q}_e$. On the other hand κ_{0S} is closer to reality than κ_e from (7.109) because it takes the inhibiting thermoelectric field into account; in (7.109) it is missing.

The BGK collision term does not conserve the particle numbers.

Kinetic calculations based on a *Fokker-Planck equation* showed that the Spitzer-Braginskii formula (1.61) becomes invalid as soon as λ_e/L exceeds 2×10^{-3} [65]. Nonlocal effects arise already from the strong velocity dependence of λ_e. The simplest description of the phenomenon is possible in terms of an integral equation with appropriate kernel [66]. Alternative directions have been pursued in [67] with the choice of nearly Maxwellian electron distributions and the development of a collision kinetic model [68]. For improvements and comparison of models see [69–72]. Strong delocalization with concomitant electrostatic field generation is driven by a fast electron component nearly always associated with laser-target interaction from energetic laser pulses. It is a difficult task to select the most relevant and most instructive publications on this subject. To the reader who wishes a deeper insight in what the current state of the art is may start from the review article by T. Bell [65] and then consult additional more recent papers [73–79]. An interesting contribution to heat flux inhibition has been found from Fokker-Planck simulations of fs pulse-solid interactions. The authors found a magnetized energy flow in the evanescent region of the laser field [80].

7.7 *Nonideal Plasma: The BBGKY Hierarchy

So far mainly *ideal plasmas*, with preference for low charge number and limited ionization energies, have been considered. In the interaction of dense matter with energetic long wavelength lasers the energy may be deposited to a high fraction in high density regions far beyond the critical density. It is the domain of *warm* and *hot dense matter* WDM, HDM. Here the *potential energy* of electron-ion and ion-ion interaction can no longer be disregarded. The energy transport to high density is provided by electron thermal conduction, by continuum X radiation and, to a minor extent, by *shock heating*. The prominence of the potential energy fully appears when an intense free electron laser (FEL) interacts with dense material to create the *hot solid* under the shape of *cold degenerate plasma* and WDM. The FEL offers a unique instrument to extend solid state physics into domains far beyond its classical range. The transition from the ideal plasma to the predominance of the potential is best introduced by the so called BBGKY *hierarchy*, named after Born, Bogoliubov, Green, Kirkwood, and Yvon.

7.7.1 The Liouville Equation and Its Reduced Moments

The dynamics of a classical system of N interacting particles is completely described by the Hamiltonian

$$H(\mathbf{p}, \mathbf{q}, t) = H(\mathbf{p}_i, \mathbf{q}_i, t), \quad i = 1, \ldots, N \tag{2.53}$$

and its $6N + 1$ canonical equations of motion (4.79). The variation of an arbitrary *phase function* $f(\mathbf{p}, \mathbf{q}, t)$ is expressed with the help of the *Poisson brackets* from (2.57),

$$\frac{\mathrm{d}f}{\mathrm{d}t} = \frac{\partial f}{\partial t} + \{H, f\}; \quad \{H, f\} = \sum_i \Big(\frac{\partial H}{\partial \mathbf{p}_i}\frac{\partial f}{\partial \mathbf{q}_i} - \frac{\partial H}{\partial \mathbf{q}_i}\frac{\partial f}{\partial \mathbf{p}_i}\Big).$$

Corresponding examples are the canonical equations

$$\dot{\mathbf{p}}_i = \{H, \mathbf{p}_i\}, \quad \dot{\mathbf{q}}_i = \{H, \mathbf{q}_i\}, \quad \frac{\mathrm{d}H}{\mathrm{d}t} = \frac{\partial H}{\partial t} + \{H, H\} = \frac{\partial H}{\partial t}.$$

If H does not depend explicitly on time $f(H)$ is a constant of motion. The phase volume $\Delta\Gamma$ is conserved in time by Liouville's theorem. The number of particles

ΔN contained in it is by definition the N particle distribution function $F_N(\mathbf{p}, \mathbf{q}, t)$ multiplied by $\Delta\Gamma$ and

$$\frac{\mathrm{d}N}{\mathrm{d}t} = \frac{\mathrm{d}F_N}{\mathrm{d}t}\Delta\Gamma + F_N\frac{\mathrm{d}\Delta\Gamma}{\mathrm{d}t} = \frac{\mathrm{d}F_N}{\mathrm{d}t}\Delta\Gamma = 0.$$

The N particle distribution function F_N obeys the *Liouville equation*

$$\frac{\mathrm{d}F_N}{\mathrm{d}t} = \frac{\partial F_N}{\partial t} + \{H, F_N\} = 0. \tag{7.115}$$

F_N is either normalized to the number of particles in $\Delta\Gamma$ or to unity. In the second case it expresses the probability of finding ΔN particles at the instant t in the positions $\mathbf{x}_i$ with momenta $\mathbf{p}_i$. The normalization used in the particular case will result from the context. The N particle distribution function is a *Lorentz scalar* because N and $\Delta\Gamma$ are obviously Lorentz scalars. The distribution function F_N is highly symmetric because its value must not depend on the enumeration of the ΔN particles. A possible representation of F_N is the *Klimontovich distribution*

$$F_N(\mathbf{q}, \mathbf{q}, t) = \frac{1}{N}\sum_i \delta(\mathbf{p}_i - \mathbf{p}_i(t))\delta(\mathbf{q}_i - \mathbf{q}_i(t)). \tag{7.116}$$

It is independent of particle enumeration.

7.7.1.1 Dynamic Variables $b(\mathbf{p}, \mathbf{q}, \mathbf{x}, t)$

The question arises how macroscopic quantities like $\mathbf{E}(\mathbf{x}, t)$, $n(\mathbf{x}, t)$, $T(\mathbf{x}, t)$ can be extracted from F_N. Further, what is the gain in additional insight from Liouville's equation beyond Vlasov and Boltzmann? The answers are given by the dynamic variables b chosen appropriately and illustrated by examples.

(1) Particle density $n(\mathbf{x}, t)$. From setting $b(\mathbf{pq}, \mathbf{x}, t) = \sum_i \delta(\mathbf{q}_i - \mathbf{x})$ follows

$$n(\mathbf{x}, t) = \int \sum_i \delta(\mathbf{q}_i - \mathbf{x}) F_N(\mathbf{p}_i\mathbf{q}_i, t)\mathrm{d}\Gamma \Rightarrow n(\mathbf{x}, t) = \int bF_N(\mathbf{pq}, t)\mathrm{d}\Gamma.$$

(2) Kin. energy from $b_{\mathrm{rel}} = \sum_i(\gamma_i - 1)mc^2\delta(\mathbf{q}_i - \mathbf{x})$, resp. $b_{\mathrm{nrel}} = \sum_i \frac{\mathbf{p}^2}{2m}\delta(\mathbf{q}_i - \mathbf{x})$,

$$\mathcal{E}_{\mathrm{kin}} = \int \sum_i(\gamma_i - 1)mc^2\delta(\mathbf{q}_i - \mathbf{x})F_N\mathrm{d}\Gamma$$

$$\mathcal{E}_{\mathrm{kin}} = \int \sum_i \frac{\mathbf{p}^2}{2m}\delta(\mathbf{q}_i - \mathbf{x})F_N\mathrm{d}\Gamma.$$

(3) In *nonrelativistic* nonideal plasma, for example WDM, the potential energy must be accounted for. Nonrelativistic setting $H_{\text{int}} = \sum_i V_{ij},\ V_{ij} = q^2/4\pi\varepsilon_0|\mathbf{q}_i - \mathbf{q}_j|$, $V_{ij} = V_{ji}$; the forces $\mathbf{f}_i = -\nabla_i V_{ij} = \nabla_j V_{ij} = -\mathbf{f}_i$ respect actio = reactio. $H_{\text{int}} = (1/2)\sum_{i\neq j} V_{ij}$ is symmetric.

$$\mathcal{E}_{\text{pot}} = \frac{1}{2}\int \sum_{i\neq j} V_{ij}\delta(\mathbf{q}_i - \mathbf{x})\delta(\mathbf{q}_j - \mathbf{x})F_N(\mathbf{pq}, t)\mathrm{d}\Gamma.$$

In presence of intense currents or groups of relativistic particles, for example in relativistic HDMH, retardation effects must be taken into account and the electromagnetic interaction is to be expressed in terms of the vector potentials. Finally, the interaction of point particles with external fields is accomplished through

$$H_{\text{ex}} = \sum_i V_{\text{ex}}(\mathbf{q}_i) + mc^2[1 + (\boldsymbol{\pi}_i - \boldsymbol{a}(\mathbf{q}_i, t))^2]^{1/2};\quad \boldsymbol{\pi}_i = \mathbf{p}_i/mc,\ \boldsymbol{a} = q\mathbf{A}/mc. \tag{7.117}$$

The nonrelativistic H_{int} is valid for point particles without electric and magnetic moments. In connection with composed particles (atoms, ions, molecules) terms of the structure $V_{ijk},\ V_{ijkl}$ etc. may appear. Here, we limit ourselves to *binary interactions* and to the Hamiltonian

$$H(\mathbf{pq}, t) = \sum_i H_{\text{kin}} + \frac{1}{2}\sum_{i\neq j} V_{ij}(\mathbf{p}_i\mathbf{q}_i, \mathbf{p}_j\mathbf{q}_j) + \sum_i H_{\text{ex}}(\mathbf{p}_i\mathbf{q}_i, t). \tag{7.118}$$

7.7.1.2 Reduced Distribution Functions

How is the one particle distribution function $f_1(\mathbf{x}, \mathbf{p}, t)$ related to $F_N(\mathbf{pq}, t)$? For this purpose we calculate for instance the kinetic energy density, with $\mathbf{x}_i$ for $(\mathbf{p}_i, \mathbf{q}_i)$,

$$\varepsilon_{\text{kin}}(\mathbf{x}, t) = \int_\Gamma \sum_i (\gamma_i - 1)mc^2\delta(\mathbf{q}_i - \mathbf{x})F_N(\mathbf{x}_1, \ldots, \mathbf{x}_N, t)\mathrm{d}\mathbf{x}_1 \ldots \mathrm{d}\mathbf{x}_N. \tag{7.119}$$

Remember, $\int F_N \mathrm{d}\mathbf{x}_2 \ldots \mathrm{d}\mathbf{x}_N$ expresses the probability to find particle 1 at $\mathbf{x}_1 = (\mathbf{p}_1, \mathbf{q}_1)$ at instant t, and so on for particles from number 2 to number N. Therefore the sum of all integrals in (7.119) expresses the probability for particle 1 to have the kinetic energy $(\gamma_1(\mathbf{p}) - 1)mc^2$, *or any other particle* the same energy at the same position $\mathbf{x}$. Independence of enumeration (standard phrase "indistiguibility"; better "democracy") in F_N and b imposes that all terms are identical. Accordingly,

$$f_1(\mathbf{x}_1, t) = N \int F_N(\mathbf{x}_1, \ldots, \mathbf{x}_N, t) d\mathbf{x}_2 \ldots d\mathbf{x}_N \tag{7.120}$$

is the *one particle distribution function* and

$$\varepsilon_{\text{kin}}(\mathbf{x}, t) = \int (\gamma - 1) mc^2 \delta(\mathbf{q}_1 - \mathbf{x}) f_1(\mathbf{x}_1, t) d\mathbf{x}_1 = \int b_1(\mathbf{x}_1) f_1(\mathbf{x}_1, t) d\mathbf{x}_1 \tag{7.121}$$

is the kinetic energy density. Analoguously, the potential energy density is calculated by extending the integration over all pairs $\{ij\}$. This gives $N(N-1) = N!/(N-2)!$ identical pairs. Setting for the two particle distribution function

$$f_2(\mathbf{x}_1, \mathbf{x}_2, t) = \frac{N!}{(N-2)!} \int F_N d\mathbf{x}_3 \ldots d\mathbf{x}_N \tag{7.122}$$

the potential energy density ε_{pot} results as

$$\varepsilon_{\text{pot}} = \frac{1}{2!} \int b_2(\mathbf{x}_1, \mathbf{x}_2; \mathbf{x}, t) f_2(\mathbf{x}_1, \mathbf{x}_2, t) d\mathbf{x}_1 d\mathbf{x}_2. \tag{7.123}$$

If the symmetric expression for b_2 is used the factor $1/2!$ is to be omitted. Owing to the structure $H = H_1 + H_2 + H_3 + \cdots + H_N$ the dynamic variable b assumes the form

$$b(\mathbf{x}_1, \mathbf{x}_2, \ldots, \mathbf{x}_N) = b_0 + \sum_i b_1(\mathbf{x}_i) + \sum_{i<j} b_2(\mathbf{x}_1, \mathbf{x}_2) + \cdots + b_N(\mathbf{x}_1, \ldots, \mathbf{x}_N).$$

It is to be combined with the reduced s-particle distribution function

$$f_s(\mathbf{x}_1, \ldots, \mathbf{x}_s, t) = \frac{N!}{(N-s)!} \int F_N(\mathbf{x}_i, t) d\mathbf{x}_{s+1} d\mathbf{x}_{s+2} \ldots d\mathbf{x}_N. \tag{7.124}$$

Then, the *mean* or *expectation value* $\langle b \rangle$ is

$$\langle b \rangle = \sum_{s=0}^{N} \frac{1}{s!} \int b_s(\mathbf{x}_1, \mathbf{x}_2, \ldots, \mathbf{x}_s) f_s(\mathbf{x}_1, \mathbf{x}_2, \ldots \mathbf{x}_s, t) d\mathbf{x}_1 d\mathbf{x}_2 \ldots d\mathbf{x}_s. \tag{7.125}$$

In this expression b_s is not symmetrized.

7.7.1.3 The Liouiville Equation

The *Liouville operator*

$$L(\mathbf{p}, \mathbf{q}) = \sum_i \left(\frac{\partial H}{\partial \mathbf{p}_i} \frac{\partial F_N}{\partial \mathbf{q}_i} - \frac{\partial H}{\partial \mathbf{q}_i} \frac{\partial F_N}{\partial \mathbf{p}_i} \right) = \{H, F_N\} \tag{7.126}$$

splits for $H = \sum_i H_i(\mathbf{p}_i) + \sum_i H_{\text{ex},i}(\mathbf{p}_i, \mathbf{q}_i) + \frac{1}{2}\sum_{i\neq j} V_{ij}(\mathbf{q}_i, \mathbf{q}_j, t)$ into

$$L(\mathbf{p}, \mathbf{q}) = \sum_i L_i(\mathbf{q}_i) + \sum_i L_{\text{ex},i}(\mathbf{p}_i, \mathbf{q}_i) + \frac{1}{2}\sum_{i\neq j} L_{ij}(\mathbf{q}_i, \mathbf{q}_j);$$

$$L_i(\mathbf{q}_i) = \dot{\mathbf{q}}_i\nabla_i = \partial_i H_i \nabla_i, \;\; L_{\text{ex},i}(\mathbf{p}_i, \mathbf{q}_i) = \partial_i H_{\text{ex},i}\nabla_i - \nabla_i H_{\text{ex},i}\partial_i, \;\; L_{ij} = \nabla_i V_{ij}\partial_{ji}$$

with the meaning $\partial_i = \partial/\partial\mathbf{p}_i$ and $\partial_{ij} = \partial_j - \partial_i$. Integration of the Liouville equation by $\mathrm{d}\Gamma_{s+1} = \mathrm{d}\mathbf{x}_{s+1}\ldots\mathrm{d}\mathbf{x}_N$ leads to

$$\int\Big[\frac{\partial}{\partial t}\frac{N!}{(N-s)!}F_N + L\frac{N!}{(N-s)!}F_N\Big]\mathrm{d}\Gamma_{s+1} = \frac{\partial}{\partial t}f_s + \frac{N!}{(N-s)!}$$
$$\times\sum_i\int(L_i + L_{\text{ex},i})F_N\mathrm{d}\Gamma_{s+1} + \frac{1}{2}\frac{N!}{(N-s)!}\sum_{i\neq j}\int L_{ij}F_N\mathrm{d}\Gamma_{s+1}. \quad (7.127)$$

If $i \le s$ the terms $L_i + L_{\text{ex},i}$ can be taken out of the integral; if $i \ge s$ partial integration yields zero. Hence, Hence the two terms in (7.127) yield $[\partial_t + \sum\limits_{i\le s}(L_i + L_{\text{ex},i})]f_s$.

Finally, we determine $\sum\limits_{i\neq j}$:

(1) $i, j \le s \;\Rightarrow\; \sum\limits_{i\neq j} L_{ij}f_s$;

(2) $i, j \le s \;\Rightarrow\; \sum\limits_{i\neq j}\int L_{ij}F_N\mathrm{d}\Gamma_{s+1} = 0$ by partial integration;

(3) $i \le s,\; j > s \;\Rightarrow\; \sum\limits_{i,j}\int L_{ij}F_N\mathrm{d}\Gamma_{s+1} = \sum\limits_{j\ge s+1}\sum\limits_{i\le s}\int L_{ij}F_N\mathrm{d}\Gamma_{s+1}$

$$\Rightarrow \quad \frac{N!}{(N-s)!}\sum_{j\ge(s+1)}\sum_{i\le s}\int L_{ij}F_N\mathrm{d}\mathbf{x}_{s+1}\mathrm{d}\Gamma_{s+2}$$
$$= \frac{N!}{(N-s)!}\frac{N!}{(N-s)!}N\sum_{i\le s}\int L_{i\,s+1}F_N\mathrm{d}\mathbf{x}_{s+1}\mathrm{d}\Gamma_{s+2}$$
$$= \sum_{i\le s}\frac{N!}{(N-s-1)!}\int L_{i\,s+1}F_N\mathrm{d}\mathbf{x}_{s+1}\mathrm{d}\Gamma_{s+2} = \sum_{i\le s}\int L_{i\,s+1}f_{s+1}\mathrm{d}\mathbf{x}_{s+1}.$$

From these expressions Liouville equation of motion can be composed for each reduced distribution function f_s from $s = 1$ to $s = N - 1$. The most important reduced distribution functions are the *one particle distribution function* following the conservation equation

$$\partial t f_1 + \dot{\mathbf{x}}_1 + \nabla_1 f_1 + \dot{\mathbf{p}}_1\partial_1 f_1 = -\int L_{12}f_{12}(\mathbf{x}_1, \mathbf{x}_2, t)\mathrm{d}\mathbf{x}_2$$

$$\int L_{12}f_{12}(\mathbf{x}_1, \mathbf{x}_2, t)\mathrm{d}\mathbf{x}_2 = \int\nabla_1 V_{12}(\mathbf{x}_1, \mathbf{x}_2, t)\Big(\frac{\partial}{\partial\mathbf{p}_1} - \frac{\partial}{\partial\mathbf{p}_2}\Big)f_2(\mathbf{p}_1\mathbf{x}_1, \mathbf{p}_2\mathbf{x}_2, t)\mathrm{d}\mathbf{x}_2$$

and the *two particle distribution function* following the conservation equation

$$\frac{\partial}{\partial t} f_2 + (\dot{\mathbf{x}}_1 \nabla_1 + \dot{\mathbf{x}}_2 \nabla_2) f_2 + (\dot{\mathbf{p}}_1 \partial_1 + \dot{\mathbf{p}}_2 \partial_2) f_2 + \nabla_1 V_{12} (\partial_2 - \partial_1) f_2$$
$$= - \int L_{13}(\mathbf{x}_1, \mathbf{x}_2, \mathbf{x}_3) \mathrm{d}\mathbf{x}_3 - \int L_{23}(\mathbf{x}_1, \mathbf{x}_2, \mathbf{x}_3) \mathrm{d}\mathbf{x}_3. \tag{7.128}$$

The source term on the RHS of $(\mathrm{d}/\mathrm{d}t) f_s$ is a linear functional of the next higher distribution function f_{s+1}. To stop the series of equations a suitable functional in terms of the f_σ, $\sigma \le s$ must be found. Boltzmann solved the *closure* for $s = 1$ by the assumption of the *molecular chaos* $f_2(\mathbf{x}_1, \mathbf{x}_2, t) = f_1(\mathbf{x}_1, t) f_1(\mathbf{x}_2, t)$ for point particles interacting in space at $\delta(\mathbf{q}_1 - \mathbf{q}_2)$: Only at the instant of collision two particles are *correlated*. In the general case the two particle distribution function can be written as

$$f_2(\mathbf{x}_1, \mathbf{x}_2, t) = f_1(\mathbf{x}_1, t) f_1(\mathbf{x}_2, t) + g_{12}(\mathbf{x}_1, \mathbf{x}_2, t) \tag{7.129}$$

with g_{12} the *irreducible component* of the two particle correlation function. In a diluted system the single particle functions $f_i(\mathbf{x}_i, t), i = 1, 2$ change in the characteristic collision time τ_{coll}, in contrast the characteristic time of change of g_{12} is the much faster inverse of the plasma frequency. Under this hypothesis Landau's collision term can be deduced from (7.129). The three particle distribution function can be split accordingly, $f_3(\mathbf{x}_1, \mathbf{x}_2, \mathbf{x}_3, t) = f_1 f_2 f_3 + f_1 g_{23} + f_2 g_{13} + f_3 g_{12} + g_{123}$. Here, the new irreducible component reflecting the genuine three-body interactions is represented by the last term of the decomposition.

7.8 Summary

Collision dominated transport is localized. The basic concept is based on the (i) binary collision cross section, (ii) the mean free path between two collisions, and (iii) the folding with the distribution functions of the colliding partners. The Boltzmann collision integral shows exactly such a structure. The three elements provide a physical picture of particle diffusion, thermal conduction, viscous flow, and collisional absorption in rarefied systems of mean free path $\lambda = 1/(n\sigma)$ much larger than $\sqrt{\sigma}$ of the binary collision cross section σ and particle density n. With n increasing three body and multiply collisions become significant.

The plasma is dominated by the long range Coulomb interactions. As a consequence, one particle collides always with many neighbours *simultaneously*, typically of the order of $\Lambda = \exp \ln \Lambda$, with $\ln \Lambda$ the Coulomb logarithm. The concept of linear *screening* and small angle deflections make the transformation of multiple collisions into a *sequence* of binary collisions possible. This is the reason for screening to figure as a basic concept in plasma transport. With increasing density curved orbits become increasingly important, $\ln \Lambda$ decreases and the electron-electron and electron-ion collision frequency $\nu_{ei} = n\sigma\langle v \rangle$ transverse to longitudinal becomes significant, $\nu_{e\perp}/\nu_{e\|} = 1/(2\ln\Lambda)$ (7.19). For a Coulomb collision to become effective for collisional absorption its duration must be shorter than 1/3 of the laser period.

The Debye screening length, generally defined by $\lambda_D = v_{\mathrm{th,e}}/\omega_p$, $v_{\mathrm{th,e}} = \sqrt{k_B T_e/m_e}$, is of more general validity. It is sufficient for the electron distribution function to be isotropic. It is standard to define $\ln\Lambda = \lambda_D/\max\{\lambda_B, b_\perp\}$. No proof for such a choice can be given. In order to decide between the two quantities (1) the difference between them must be large and (2) the contribution of overlapping curved orbits to the collision frequency is not significant in case of $b_\perp \ll \lambda_B$. In the opposite case the Born approximation of the screened Coulomb cross section must be fulfilled. In the laser plasma generally $\lambda_B \gg b_\perp$ is obeyed. The shielding geometry, spherical or cylindrical, plays also a role.

Collision frequency and absorption coefficient of the laser radiation are calculated in the *ballistic model* and compared with the more involved *dielectric model*. The difference is small enough to prefer the ballistic approach, see Fig. 7.5 and (7.32). In particular, strong drifts are easily handled in the latter. They are significant in cold ion beam stopping, see (7.65). Collisional absorption can also be viewed under the emission of photons along the electron orbit bent by the nearest ion. For this reason laser and ion beam absorption is considered as happening by *indirect bremsstrahlung* absorption. Particular enhancement of collisional absorption may happen in clusters and small droplets because the ion charge entering in the absorption coefficient with the high charge number Z_C. Multiplication factors by 10^3–10^4 of absorption in the plasma are realistic.

Latest when T_e exceeds 1 keV$\times Z$ collisional absorption becomes ineffective. On the other hand experiments and simulations show degrees of absorption spanning from 15 to 90%. Meanwhile this *noncollisional absorption* is basically understood. In the absence of any friction the electric field **E** of the laser and the current density **j** are orthogonal to each other in the Poynting theorem (1.74) and (2.16). The strong laser field induces an electrostatic charge separation field; it provides for the compulsory phase shift between **E** and **j**. Numerous models and names have been invented to describe in detail the origin of the electrostatic field. One of the candidates is *anharmonic absorption*. Another source is the excitation of fluctuating longitudinal electric fields fed by plasma instabilities in which the electrons enter statistically and are accelerated to energies above the average electron oscillation energy in the laser field. In medium dense hot matter and plasmas a systematic access to collisional/collisionless absorption is accessible in terms of correlations in the BBGKY hierarchical expansion.

7.9 Problems

- ▹ Prove Jackson's model. *Hint:* $\ddot{\delta}_\perp = Cb/r^3 \;\Rightarrow\; \delta_\perp\big|_{-\infty}^{+\infty} = C \times (2b/b^2)$.
- ▹ Verify (7.7), (7.8).
- ▹ Derive $\dot{W}$ in (7.15).
- ▹ How does the screening length λ_0 change for a relativistic electron beam?

▹ Calculate $\langle 1/v^2 \rangle$ in (7.25). Why is averaging of v^2 incorrect?
▹ Show that the collision frequencies (7.30) and (7.31) from the ballistic model result identical from Boltzmann's collision integral (3.83) in the Lorentz approximation (ions at rest). *Hint:* It is sufficient to show it for (7.28) and to make use of inherent symmetry. For hard spheres you find the proof in Chap. 3.
▹ Derive the Spitzer-Braginskii-Silin formula (7.35) from (7.36).
▹ Show that $b_{\min}$ from (7.69) yields better agreement of (7.32) with (7.55) than $b_{\min} = b_\perp$. *Hint:* Use Fig. 7.6.
$b_{\min} = b_\perp$. Give an estimate of the Barkas effect in the bare Coulomb potential. *Hint:* Calculate a first correction to the straight orbit approximation (and possibly including closer encounters also).
▹ Derive a relation between the net absorption coefficient α_ω and the absorption coefficient κ_ω without induced reemission in thermal equilibrium.
▹ Show that in (7.97) and in (7.98) the momentum transfer is proportional to the ion mass m_1 and not to the reduced mass μ. *Solution:* Use the Boltzmann equation (3.83), multiply by m_1 and integrate over $\mathbf{v}$. The right hand side represents, just by its construction, the net momentum change by all particles of the distribution $f(\mathbf{v})$ from collisions with the particles m_2 of $f(\mathbf{v}_1)$ and results in $-n_1 m_1 \nu_{12}(\mathbf{u}_1 - \mathbf{u}_2)$. Note, $\mathbf{v}, \mathbf{v}_1 = \mathbf{u}_{1,2} + \mathbf{w}_{1,2}$. The left hand side becomes $n_1 m_1 \mathrm{d}\mathbf{u}_1/\mathrm{d}t$ which is the correct fluid equation of motion with m_1 as the mass.
▹ What is the local heat flow density $q(x)$ to keep the 1D rarefaction wave isothermal? *Answer:* $q(x) = p(x)c_{s0}$. Verify.
▹ Show that the potential of the trapped electrons $N_e(x > 0)$ is small compared to $\Phi(x, r = 0)$ from (9.39).

7.10 Self-assessment

• What are the reasons for introducing the concept of shielding? *Answer:* (a) Reduction of multiple simultaneous interactions to a sequence of binary collisions. (b) Elimination of the Coulomb divergences.
• Give a criterion for when the reduction to binary collisions is possible, (a) with small angle deflections, (b) with large angle deflections.
• How does $b_\perp$ depend on Z? At which temperature is $\lambda_B = b_\perp$?
• In an electron-ion collision the electron does not lose energy. Nevertheless, why does Ohmic heating, i.e., energy absorption by the electrons, occur?
• Which conditions must be fulfilled for the Debye shielding to hold? *Hint:* (i) Small deflection angles, (ii) ion at rest. (iii) How many particles have to be in the Debye sphere?
• How are (7.25) and (7.26) to be changed to describe electron-electron screening?
• Does (7.26) hold for the relativistic Maxwellian? *Answer:* Yes. Show why.
• Resolve the following paradox. The Debye length λ_D can be written as the ratio of v_{th} to ω_p. The plasma frequency and the temperature are relativistic invariants, however the electron mass m_e is not. $\Rightarrow$ contradiction to (7.26)?

• Why do the electron-electron collisions not directly contribute to absorption?
• Which physical effect is responsible for saturation of absorption (= oscillator energy) at a finite value as b goes to infinity in Fig. 7.3?
• Is $\nu_{\rm ei}(t)$ from (7.21) correct if (a) $\tau_{ei} = \tau_{\rm Laser}/10$, (b) $\tau_{ei} = 10\tau_{\rm Laser}$? *Answer:* (a) Yes; (b) yes, if $\hat{\mathbf{v}}$ in (7.20) is determined from Drude's formula with ν_{ei} included in $\hat{\mathbf{v}}$ self consistently.
• What is the result of $\nu_{ei}(t)$ in (7.30) if calculated from the Boltzmann integral (3.83)? *Answer:* the same.
• Which qualitative arguments can you give for setting the cut offs according to [7.69), in particular for λ_B instead of $b_{\rm min} = b_\perp$?
• Why does the *Bragg peak* under oscillatory motion reach its maximum at $\hat{v}_{\rm os} = 3\, v_{\rm th}$ in contrast to ion beam stopping? *Hint:* Average drift after (7.71).
• Does the Barkas effect exist in the fully ionized plasma? If so, where is it hidden? *Answer:* Correctly screened Coulomb potential beyond Debye approximation.
• (a) What are the physical conditions for the validity of (7.80)? (b) How does this expression change in the case of $\hat{v}_{\rm os}/v_{\rm th} > 1$?
• The probability for transition from energy level E to $E + \hbar\omega$ is the same as for the inverse transition. What is the reason for net *absorption* $\mathrm{d}I/\mathrm{d}x = -2k_0\Im\eta I > 0$ in thermal equilibrium of the free electrons? *Answer:* The ratio in number of upper to lower states $\sqrt{(E + \hbar\omega)/E} > 1$.
• Consider (7.105). The ratio $D_{e,i}/\mu_{e,i}$ does not depend on collisions. Why is a finite collision frequency needed for its derivation? *Answer:* At $\nu_{\rm coll} = \nu_{ei} = 0$ no steady state evolves.
• At which velocity does $x_F(t)$ from (9.28) propagate? Why is it higher than $u(x_F)$?
• Why does in (7.26) $k_B T_e$ appear for the thermal energy and not $3k_B T_e/2$? If you do not know the answer read again the derivation of λ_D given in this chapter.
• Convince yourself of the transformation properties of fluxes against a rotation by explicitly calculating the Poynting vector $\mathbf{S}$ and the kinetic expression of the pressure-viscosity tensor P_{ij} in the system rotated by the angle ϕ.
• Each of the single ion species is Debye-screened. Why is there the Coulomb logarithm $\ln \Lambda(\langle v_e^2\rangle)$ in (7.100) only present once, i.e., linear and not quadratic? *Hint:* Remember, what is the role of screening?

7.11 Glossary

Cold oscillator model

$$\ddot{\delta} + \omega_p^2 \delta = \frac{\mathbf{f}_C}{m_e}. \tag{7.1}$$

DeBroglie wavelength

$$\lambda_B = \frac{\hbar}{m_e v_e}; \qquad \lambda_B[\mathrm{nm}] = \frac{0.185}{(E[\mathrm{eV}])^{1/2}}. \tag{7.12}$$

Parallel versus orthogonal collisional absorption

$$\frac{\alpha_\parallel}{\alpha_\perp} = \frac{1}{2\ln\Lambda_0}. \tag{7.19}$$

Maximum interaction time

$$\frac{\tau}{2\pi/\omega_p} = \frac{\lambda_0\omega_p}{\pi v} = \frac{1}{\pi} \simeq \frac{1}{3}. \tag{7.17}$$

Single ion emits power $\dot{W}$ into plasmons in cold plasma,

$$\dot{W} = Zm_e\omega_p^2 v b_\perp\left\{\beta_0 \mathrm{K}_0(\beta_0)\mathrm{K}_1(\beta_0) + \frac{1}{2}\ln\left(1 + \frac{b_0^2}{b_\perp^2}\right)\right\}. \tag{7.15}$$

Collision frequency in Drude model

$$\overline{\nu}_{ei} = \frac{\overline{\mathbf{jE}}}{2n_e\overline{\mathcal{E}}_{\mathrm{os}}}. \tag{7.22}$$

Debye screening of isotropic electron distribution

$$\Phi_s = \frac{q}{4\pi\varepsilon_0 r}\exp-(r/\lambda_s); \quad \lambda_s = \left(\frac{\varepsilon_0 m_e}{n_0 e^2\langle 1/v^2\rangle}\right)^{1/2}. \tag{7.25}$$

Collision frequency at arbitrary drift $\mathbf{v}_{\mathrm{os}}(t)$

$$\nu_{ei}(t) = 2\pi\frac{K}{m_e v_{\mathrm{os}}(t)}\int\limits_0^\infty\int\limits_{-1}^1 \frac{v_{\mathrm{os}}(t)|1 + v_e\cos\chi/v_{\mathrm{os}}(t)|\, v_e^2 f(v_e)\ln\Lambda(v)}{(v_{\mathrm{os}}^2 + v_e^2 + 2v_{\mathrm{os}}v_e\cos\chi)^{3/2}}\,\mathrm{d}\cos\chi\mathrm{d}v_e. \tag{7.31}$$

Cycle averaged collision frequency

$$\overline{\nu}_{ei} = 4\pi\frac{K}{m_e\hat{v}_{\mathrm{os}}^2}\overline{\int_0^\infty\int_{-1}^1 \frac{|\mathbf{v}_{\mathrm{os}}\mathbf{v}|}{v^3}v_e^2 f(v_e)\ln\Lambda(v)\,\mathrm{d}\cos\chi\mathrm{d}v_e}. \tag{7.32}$$

$$K = Z^2 n_i K_0, \quad K_0 = \frac{e^4}{4\pi\varepsilon_0^2 m_e} = 7.3\times 10^{-16}\mathrm{gcm}^6\mathrm{s}^{-4}. \tag{7.28}$$

Collision frequency with $\hat{v}_{os}(t) \lesssim v_{th}$, $\hat{w}_{os} = \hat{v}_{os}/\sqrt{2}v_{th}$, $\beta = m_e/2k_BT_e$

$$\overline{\nu}_{ei} = 4\frac{\beta^{3/2}}{\pi^{1/2}}\frac{K}{m_e}\overline{\ln\langle\Lambda\rangle}\sum_{n\geqslant 1}(-1)^{n+1}\frac{2n}{2n+1}\frac{(2n)!}{2^{2n}(n!)^3}\hat{w}_{os}^{2n-2}. \tag{7.39}$$

For $\hat{v}_{os} \sim v_{th}$ the series results

$$\sum_{n\geqslant 1} = \frac{1}{3}\left[1 - \frac{9}{40}\left(\frac{\hat{v}_{os}}{v_{th}}\right)^2 + \frac{15}{448}\left(\frac{\hat{v}_{os}}{v_{th}}\right)^4 - \frac{35}{9216}\left(\frac{\hat{v}_{os}}{v_{th}}\right)^6 \pm \cdots\right].$$

Spitzer-Braginskii-Silin collision frequency (vanishing drift)

$$\nu_{ei} = \overline{\nu}_{ei} = \frac{1}{3}\sqrt{\frac{2}{\pi}}\frac{K}{m_e v_{th}^3}\ln\Lambda(v_{th}); \quad v_{th} = \left(\frac{k_BT_e}{m_e}\right)^{1/2}. \tag{7.35}$$

Dielectric model: True potential induced by Coulomb potential of point charge, Fourier decomposition

$$\tilde{\Phi} = \frac{\tilde{\Phi}_C}{\varepsilon(\mathbf{k},\Omega)}, \quad \varepsilon(\mathbf{k},\Omega) = 1 + \frac{\omega_p^2}{\mathbf{k}^2}\int\frac{\mathbf{k}\partial f_0/\partial\mathbf{v}_e}{\Omega - \mathbf{k}\mathbf{v}_e}\,\mathrm{d}\mathbf{v}_e. \tag{7.48}$$

Impact parameter

$$b_\perp = \frac{Ze^2}{4\pi\varepsilon_0 m_e v_r^2}; \quad b_\perp[\mathrm{nm}] = \frac{0.7\times Z}{E_r[\mathrm{eV}]}. \tag{7.56}$$

Coulomb logarithm L_C in spherical geometry, λ_s screening length, λ_B DeBroglie length

$$L_C = \ln\frac{\lambda_s}{\lambda_B} + 0.2 + \frac{\lambda_B^2}{4\lambda_s^2}. \tag{7.59}$$

Ion (proton) stopping power

$$\frac{\mathrm{d}E}{\mathrm{d}x} = -2\pi Z_p\frac{K}{v_0}\int_0^\infty\int_{-1}^1\frac{\mathbf{v}_0\mathbf{v}}{v^3}\ln\Lambda v_e^2 f(v_e)\,\mathrm{d}\cos\theta\,\mathrm{d}v_e, \quad \mathbf{v} = \mathbf{v}_0 + \mathbf{v}_e. \tag{7.65}$$

Ion stopping with Maxwellian $f(v_e)$

$$\frac{\mathrm{d}E}{\mathrm{d}x} = -Z_p\frac{K\ln\langle\Lambda\rangle}{v_0^2}\left[\mathrm{erf}\left(\frac{v_0}{\sqrt{2}v_{th}}\right) - \left(\frac{2}{\pi}\right)^{1/2}\frac{v_0}{v_{th}}\exp\left(-\frac{v_0^2}{2v_{th}^2}\right)\right]. \tag{7.67}$$

Classical plasma, cut offs

$$\ln \Lambda(v) = \ln \frac{b_{\max}}{b_{\min}}, \quad b_{\max} = \frac{v}{\sqrt{2}\omega_{\mathrm{m}}}, \quad b_{\min} = \frac{\lambda_B}{2}; \quad \omega_{\mathrm{m}} = \max(\omega, \omega_p). \tag{7.69}$$

Cycle averaged Coulomb logarithm

$$\overline{\ln\langle\Lambda\rangle} = \ln\overline{\langle\Lambda\rangle} = \ln\left[\frac{\sqrt{2}m_e}{\hbar\omega_{\mathrm{m}}}((v_{\mathrm{os}}^2/4 + v_{\mathrm{th}}^2)\right]. \tag{7.71}$$

qm collision frequency (Kull)

$$\overline{\nu}_{\mathrm{ei}} = 2\frac{\omega_p^4}{n_e\hat{v}_{\mathrm{os}}^3}\ln\frac{\hat{v}_{\mathrm{os}}}{v_{\mathrm{th}}}\ln\left[\sqrt{2}\frac{v_{\mathrm{th}}/\omega}{\lambda_B}\left(\frac{\hat{v}_{\mathrm{os}}}{v_{\mathrm{th}}}\right)^{1/2}\right]. \tag{7.72}$$

Collision frequency with giant ions

$$\nu_{eC} = \xi\eta_C^2\frac{Z_C}{Z}\frac{L_C}{\ln\Lambda_{\mathrm{ei}}}\overline{\nu}_{\mathrm{ei}}, \quad L_C = \frac{1}{2}\int_0^{b_c}\frac{(1-\cos\theta)b\,\mathrm{d}b}{R^2} + \frac{1}{2}\ln\left[\frac{b_\perp^2 + b_{\max}^2}{b_\perp^2 + b_c^2}\right]. \tag{7.74}$$

Shear viscosity

$$\mu_{e,i} = \frac{nk_BT}{2\nu_{ee,ii}} = 3\sqrt{2}\pi^{3/2}\varepsilon_0^2e^{-4}(k_BT)^{5/2}\frac{m_{e,i}^{1/2}}{Z^3(Z+1)\ln\Lambda}, \quad T_e = T_i,$$

$$\mu_{e,i}[\mathrm{Ws}^2] = 2.9\times10^7(T[\mathrm{eV}])^{5/2}\frac{(m_{e,i}[\mathrm{kg}])^{1/2}}{Z^3(Z+1)\ln\Lambda}. \tag{7.96}$$

Thermoelectric field

$$\mathbf{E} = -\frac{\nabla p_e}{n_{e0}e} = -\frac{k_B}{m_e}\nabla T_e. \tag{7.112}$$

7.12 Further Readings

Peter Sigmund, *Stopping of heavy ions: a theoretical approach*, (Springer, Heidelberg, 2004).
Günter Zwicknagel, Christian Toepffer, and Paul-Gerhard Reinhard, *Stopping of heavy ions in plasmas at strong coupling*, Physics Reports **309**, 117–208 (1999).
Edward Morse, *Nuclear Fusion*, (Springer, Heidelberg, 2019), Chap. 7: Transport.
Radu Balescu, *Transport Processes in Plasmas*, (North-Holland Publ. Co, Amsterdam, 1989).

References

1. J.D. Jackson, *Classical Electrodynamics*, 2nd edn. (Wiley, New York, 1975)
2. M. Abramowitz, I.A. Stegun, *Pocketbook of Mathematical Functions* (Harri Deutsch, Frankfurt a.M., 1994), Sect. 9.6
3. O. Boine-Frankenheim, Ph.D. thesis Technical University of Darmstadt (1996); Phys. Plasmas **3**, 1585 (1996)
4. P. Sigmund, *Stopping of Heavy Ions*, Springer Tracts in Modern Physics, vol. 204 (Springer, Berlin, 2004); *Particle Penetration and Radiation Effects* (Springer, Berlin, 2006); vol. 2 (Springer, Berlin, 2014)
5. L.C. Jarrott et al., Nat. Phys. **12**, 499 (2016)
6. V.P. Silin, Sov. Phys. JETP **20**, 1510 (1965)
7. Y. Shima, H. Yatom, Phys. Rev. A **12**, 2106 (1975)
8. V.P. Silin, S.A. Uryupin, Sov. Phys. JETP **54**, 485 (1981)
9. Th. Bornath et al., Laser Part. Beams **18**, 535 (2000)
10. H. Reinholtz et al., Phys. Rev. E **62**, 5648 (2000)
11. G.J. Pert, J. Phys. B: At. Mol. Phys. **8**, 3069 (1975); At. Mol. Phys. **12**, 2755 (1979); J. Phys. A: Gen. Phys. **9**, 463 (1976)
12. L. Spitzer, *Physics of Fully Ionized Gases* (Interscience, New York, 1967), p. 146
13. P. Mulser, F. Cornolti, E. Bésuelle, R. Schneider, Phys. Rev. E **63**, 016406 (2000)
14. I.S. Gradsshteyn, I.M. Ryzhik, *Tables of Integrals, Series, and Products* (Academic Press, New York, 1965), pp. 131–132, equation (2.513: 1–15)
15. J.J. Sakurai, *Modern Quantum Mechanics* (Benjamin/Cummings, Menlo Park, CA, 1985), Sect. 7.13
16. D. Kremp, Th. Bornath, M. Bonitz, Phys. Rev. E **60**, 4725 (1999)
17. H.-J. Kull, L. Plagne, Phys. Plasmas **8**, 5244 (2001)
18. A. Wierling et al., Phys. Plasmas **8**, 3810 (2001)
19. R. Schneider, Equivalence and unification of the ballistic and the kinetic treatment of collisional absorption, Ph.D. thesis Technical University of Darmstadt, Darmstadt (2002)
20. Wikipedia: Coulomb Collision (2012); I.D. Huba, Coulomb logarithm, NRL Plasma Formulary (Naval Research Laboratory, Washington, 2007), p. 34; wwwlearningace.com/about V Plasma Physics, p. 125; I.P. Skarovski, T.W. Johnston, M.P. Bachynski, *The Partile Kinetics of Plasmas* (Addison-Wesley, Reading, MA, 1966); E.M. Lifshitz, L.P. Pitaevskij, *Physical Kinetics* (Pergamon Press, Oxford, 1981), vol. 10, p. 172f
21. R. Loudon, *The Quantum Theory of Light* (Clarendon Press, Oxford, 1978), Chap. 7
22. J. Wesson, *Tokamaks*, 3rd edn. (Clarendon Press, Oxford), Sec. 14.5, p. 729
23. J.E.G. Farina, J. Phys. B: Atom. Molec. Phys. **10**, 1437 (1977)
24. C. Cohen-Tanoudji et al., *Quantum Mechanics*, vol. 1 (Wiley, New York, 1977), pp. 559–574
25. P. Mulser, G. Alber, M. Murakami, Phys. Plasmas **21**, 042103 (2014)
26. D.O. Gericke, M. Schlanges, Th. Bornath, W.D. Kraeft, Contrib. Plasma Phys. **41**, 147 (2001)
27. W. Cayzac et al., Phys. Rev. E **92**, 053109 (2015)
28. M. Temporal et al., Eur. Phys. J. D **17**, 132 (2017)
29. W. Cayzac et al., Nat. Commun. **8**, 15693 (2017)
30. W. Cayzac et al., Rev. Sci. Instrum. **89**, 053301 (2018)
31. F.M. Smith, W. Birnbaum, W. Barkas, Phys. Rev. **91**, 765 (1953)
32. P. Sigmund, A. Schinner, Nucl. Instr. Meth. B **212**, 110 (2003)
33. Yu.S. Sayasov, J. Plasma Phys. **57**, 473 (1997)
34. J.M. Dawson, C. Oberman, Phys. Fluids **5**, 517 (1962)
35. P.K. Kaw, A. Salat, Phys. Fluids **11**, 2223 (1968); A. Salat, P.K. Kaw, Phys. Fluids **12**, 342 (1969)
36. A.B. Langdon, Phys. Rev. Lett. **44**, 575 (1980)
37. Th. Bornath et al., J. Phys.: Conf. Ser. **11**, 180 (2005)
38. S.M. Weng et al., Phys. Plasmas **13**, 113302 (2006); Phys. Rev. E **80**, 056406 (2009)

39. R.D. Jones, K. Lee, Phys. Fluids **25**, 2307 (1982)
40. G. Rascol et al., Phys. Plasmas **13**, 103108 (2006)
41. S.A. Khrapak, Phys. Plasmas **20**, o54501 (2013)
42. P. Mulser, M. Kanapathipillai, D.H.H. Hoffmann, Phys. Rev. Lett. **95**, 103401 (2005). Correction: Last value in Table II must read 2.3×10^5 (printing error)
43. A.F. Nikiforov, V.G. Novikov, V.B. Ubarov, *Quantum-Statistical Models for Hot Dense Matter: Methods for Computation Opacity and Equation of State* (Birkhäuser, Berlin, 2004), Sect. 5.2.3
44. R.L. Shoemaker, Coherent transient infrared spectroscopy, in *Laser and Coherent Spectroscopy*, ed. by J.I. Steinfeld (Plenum Press, New York, 1978), Sect. 3.3
45. W.L. Kruer, K. Estabrook, Phys. Fluids **28**, 430 (1985)
46. F. Brunel, Phys. Rev. Lett. **59**, 52 (1987)
47. P. Gibbon, A.R. Bell, Phys. Rev. Lett. **68**, 1535 (1992)
48. A. Macchi et al., Phys. Rev. Lett. **87**, 205004 (2001)
49. A. Macchi et al., Phys. Plasmas **9**, 1704 (2002)
50. S. Kato, B. Bhattacharyya, A. Nishiguchi, K. Mima, Phys. Fluids B **5**, 564 (1993)
51. H.-B. Cai, Phys. Plasmas **13**, 063108 (2006)
52. P. Mulser, H. Ruhl, D. Bauer, Phys. Rev. Lett. **101**, 225002 (2008)
53. A.H. Nayfeh, D.T. Mook, *Nonlinear Oscillations* (Wiley, New York, 1979), Chapt. 4
54. M. Kundu, D. Bauer, Phys. Rev. Lett. **96**, 123401 (2006)
55. M. Cherchez et al., Phys. Rev. Lett. **100**, 245001 (2008)
56. T. Liseykina, P. Mulser, M. Murakami, Phys. Plasmas **22**, 033302 (2015)
57. D. Bauer, P. Mulser, Phys. Plasmas **14**, 023301 (2007)
58. P. Mulser, D. Bauer, *High-Power Laser-Matter Interaction* (Springer, Heidelberg, 2010), p. 356
59. M. Sherlock, E.G. Hill, R.G. Evans, S.J. Rose, Phys. Rev. Lett. **113**, 255001 (2014)
60. S.M. Weng, P. Mulser, Z.M. Sheng, Phys. Plasmas **19**, 022705 (2012)
61. J. Sanz, A. Debayle, K. Mima, Phys. Rev. E **85**, 045411 (2012)
62. S.C. Wilks, W.L. Kruer, M. Tabak, A.B. Langdon, Phys. Rev. Lett. **69**, 1383 (1992)
63. F.N. Beg, A.R. Bell, A.E. Dangor et al., Phys. Plasmas **4**, 447 (1997)
64. H. Chen, S.C. Wilks, W. Kruer, P. Patel, R. Shepherd, Hot electron energy distributions from ultra-intense laser solid interactions. LLNL-JRNL-407653
65. A.R. Bell, Transport in laser-produced plasmas, in *Laser Plasma Interactions 5: Inertial Confinement Fusion, Proceedings of the 45th Scottish Universities Summer School in Physics, St. Andrews 1994*, ed. by M.B. Hooper (Institute of Physics, Bristol and Philadelphia, 1995), p. 139
66. J.F. Luciani, P. Mora, J. Virmont, Phys. Rev. Lett. **51**, 1664 (1983)
67. J.R. Albritton, E.A. Williams, I.B. Bernstein, K.P. Swartz, Phys. Rev. Lett. **57**, 1887 (1986)
68. W. Manheimer, D. Colombant, V. Goncharov, Phys. Plasmas **15**, 083103 (2008)
69. A. Bendib, J.F. Luciani, J.P. Matte, Phys. Fluids **31**, 711 (1988)
70. S.I. Krashennikov, Phys. Fluids B: Plasma Phys. **5**, 74 (1993)
71. A. Marocchino, M. Tzoufras, S. Atzeni, Ph. Schiavi, A. Nicolai, J. Mallet, V. Tikhonchuk, J.-L. Feugeas, Phys. Plasmas **20**, 022702 (2013)
72. A.V. Brantov, V.Yu. Bychenkov, Plasma Phys. Rep. **39**, 698 (2013)
73. A. Tahraoui, A. Bendib, Phys. Plasmas **9**, 3089 (2002)
74. O.V. Batishev et al., Phys. Plasmas **9**, 2302 (2002)
75. J.J. Santos et al., Phys. Rev. Lett. **89**, 025001 (2002)
76. E. Martinolli et al., Laser Part. Beams **20**, 171 (2002)
77. N. Jain, A. Das, P. Kaw, S. Sengupta, Phys. Plasmas **10**, 29 (2003)
78. Y. Sentoku, K. Mima, P. Kaw, K. Nishikawa, Phys. Rev. Lett. **90**, 155001 (2003)
79. D. Del Sorbo, J.-L. Feugeas, Ph. Nicolai, M. Olazabal-Loumé, B. Dubroca, S. Guisset, M. Touati, V. Tikhonchuk, Phys. Plasmas **22**, 082706 (2015)
80. M. Sherlock, E.G. Hill, S.J. Rose, High Energy Density Phys. **9**, 38 (2013)

Chapter 8
Radiation from Hot Matter

Hot matter radiates. Low density plasmas also emit radiation in form of spectral lines and inverse continuum bremsstrahlung radiation to a low level. With increasing density and nuclear charge Z production of photons greatly intensifies and is emitted as *black body radiation* or evolves towards such by multiple interaction with cold dense matter. Its transport and evolution towards a Planckian spectrum is governed by the photon mean free path between two "collisions". In contrast to particle-particle encounters here by collision the absorption and emission, or *annihilation*, of a photon and its *creation* at the same position is understood. In its simplest version radiation transport is a diffusion process in close analogy to particle or electron heat diffusion. For a general introduction to the complex field it is advisable to start with the quantized radiation field and to clarify a few fundamental concepts of coherent interaction beyond the validity of the rate equations with spontaneous and stimulated emission and absorption. The *rates* of absorption and emission make up what is generally summarized under *radiation transport*. It describes intensities as cycle averaged quantities. The classical Maxwellian radiation field is more, expressed by field amplitudes and phases. For the characterization in these terms a simple criterion exists. The density n of photons in a given mode must be such as to result in a number N over the volume λ^3 much larger than unity, $N = n\lambda^3 \gg 1$. If the inequality is not fulfilled the fields and their sources must be quantized and the classical $\mathbf{E}$ and $\mathbf{B}$ appear as the *expectation values* of the field operators in Maxwell's equations. In the context of the book classical electrodynamics may be used up to high power laser intensities of some 10^{22} Wcm^{-2} and for the whole secondary frequency spectrum unless the criterion on $N \gg 1$ is not fulfilled for extreme orders of high harmonics.

The chapter starts with the introduction of the coherent Glauber state and some basic coherent phenomena in the *optical Bloch model*. The generation of energetic particles requires a short presentation of radiation emission from the accelerated electron. *Thermal bremsstrahlung* is a relevant application. The last section is dedicated to the fundamentals of black body radiation transport. In standard transport theory

P. Mulser, *Hot Matter from High-Power Lasers*, Graduate Texts in Physics,
https://doi.org/10.1007/978-3-662-61181-4_8

the limitation to intensities pays for the loss of the "kinetic energy", or inertia, of the electromagnetic field. In a highly collisional environment this limitation is justified.

8.1 The Radiating Plasma

8.1.1 The Quantized Maxwell Field

In vacuum all fields are derived from the vector potential $\mathbf{A}(\mathbf{x}, t)$. So far a monochromatic wave component has been described by $\mathbf{A}(\mathbf{x}, t) = \hat{\mathbf{A}} \exp(i\mathbf{kx} - i\omega t)$. Amplitude $\hat{\mathbf{A}}$ and phase $\Phi = i(\mathbf{kx} - \omega t)$ are both sharp, i.e the standard deviations (*variances*) $\Delta\hat{\mathbf{A}} = \sqrt{(\langle\hat{\mathbf{A}}^2\rangle - \langle\hat{\mathbf{A}}\rangle^2)} = 0$ and $\Delta\phi = \sqrt{(\langle\phi^2\rangle - \langle\phi\rangle^2)} = 0$ if evaluated at any given $(\mathbf{x}, t)$. This cannot be the whole truth because we know from quantum mechanics that the uncertainties of the photon number n and the phase have to obey the relations

$$\Delta n \Delta \cos\phi \geq \frac{1}{2}|\langle\sin\phi\rangle|, \quad \Delta n \Delta \sin\phi \geq \frac{1}{2}|\langle\cos\phi\rangle|. \tag{8.1}$$

The number of photons of magnitude $\hbar\omega$ is proportional to $\hat{\mathbf{A}}^2$.

The single Fourier mode $\mathbf{A_k}(t) = \mathbf{A_k} \exp(-i\omega_\mathbf{k} t),\ \mathbf{A_k} = \hat{\mathbf{A}} \exp(i\mathbf{kx})$ obeys the the equation of an isotropic harmonic oscillator

$$\ddot{\mathbf{A}}_\mathbf{k} + \omega_\mathbf{k}^2 \mathbf{A_k} = 0 \quad \Leftrightarrow \quad \ddot{\mathbf{x}} + \omega^2 \mathbf{x} = 0; \qquad \omega = c|\mathbf{k}|. \tag{8.2}$$

It is completely determined by the wave vector $\mathbf{k}$ and the polarization vector $\boldsymbol{\varepsilon}_\mathbf{k}$ perpendicular to $\mathbf{k}$, $\mathbf{A_k} = \boldsymbol{\varepsilon}_\mathbf{k} A_\mathbf{k}$. According to Feynman's principle (identical equations $\Rightarrow$ identical solutions) $A_\mathbf{k}$ exhibits all properties of the onedimensional harmonic oscillator $\ddot{x} + \omega^2 x = 0$. Newton applied this principle to gravity. It let him to the extension of the terrestrial mechanics to the motion of celestial bodies and to the explanation of the tides. Hence it is worthwhile to extend the properties of the quantized harmonic oscillator to Maxwells equations to see what the potential gain may result therefrom. To this aim the characteristics of the quantum oscillator are briefly summarized.

The Hamiltonian of the harmonic oscillator of mass $m = 1$ is $H(\mathbf{p}, \mathbf{q}) = (\mathbf{p}^2 + \omega^2\mathbf{q}^2)/2$, with $\mathbf{p}$ and $\mathbf{q}$ the well known operators of momentum and position. In the single Cartesian components they obey the commutation relations $[p, q] = \hbar/i$. It is convenient to introduce the linear combinations

$$a = (2\hbar\omega)^{-1/2}(ip + \omega q), \quad a^\dagger = (2\hbar\omega)^{-1/2}(-ip + \omega q); \quad (a)^\dagger = a^\dagger. \tag{8.3}$$

In the new Hermitian conjugate quantities a, $a^\dagger$ the Hamiltonian and the commutators read

$$H = \hbar\omega\left(a^\dagger a + \frac{1}{2}\right), \quad [a, a^\dagger] = aa^\dagger - a^\dagger a = 1 \quad \Rightarrow \quad aa^\dagger = a^\dagger a + 1. \quad (8.4)$$

The energy eigenvalues of H are $E_n = \hbar\omega(n + 1/2)$. The operators $a, a^\dagger$ allow the solution of the harmonic oscillator problem entirely by linear algebra if it is assumed that (i) the Hamiltonian is positive definite and (ii) the ground state is not degenerate. If the eigenstates of H associated with the equally spaced energy eigenvalues are indicated by $|n\rangle$, from (8.4) follows that the $|n\rangle$ are eigenstates of the number operator $n = a^\dagger a$,

$$n|n\rangle = a^\dagger a|n\rangle = (H/\hbar\omega - 1/2)|n\rangle = \frac{E_n - \hbar\omega/2}{\hbar\omega}|n\rangle. \quad (8.5)$$

The *number eigenstate* $|n\rangle$ differs from number eigenstate $|n-1\rangle$ by the energy increase of $\Delta E = \hbar\omega$. To the ground state $|0\rangle$ the *zero point energy* $E_0 = \hbar\omega/2$ is associated. It is a consequence of the non-commutation of p with q and *the* difference to the classical oscillator. From the commutator $[a, a^\dagger] = 1$ one deduces straightforwardly that $a, a^\dagger$ are *ladder operators*, for obvious reasons,

$$a|n\rangle = n^{1/2}|n-1\rangle, \quad a^\dagger|n\rangle = (n+1)^{1/2}|n+1\rangle; \quad \langle n|n\rangle = 1; \quad a|1\rangle = |0\rangle. \quad (8.6)$$

$a^\dagger$ is a *creation operator*, a is a *destruction* or *annihilation operator*. The state $|n\rangle$ is generated by n times $a^\dagger$ applied to the ground state,

$$|n\rangle = (n!)^{-1/2}(a^\dagger)^n|0\rangle.$$

8.1.1.1 The Oscillators of the Radiation Field

In order to construct the Hamiltonian of the mode $\mathbf{A_k} = \boldsymbol{\varepsilon} A_\mathbf{k}$ in analogy to the onedimensional oscillator its energy $\mathcal{E}$ is considered in the cube $V = L^3$,

$$\mathcal{E} = \frac{1}{2}\int \varepsilon_0(\mathbf{E}_\mathbf{k}^2 + c^2\mathbf{B}_\mathbf{k}^2)\mathrm{d}V = \int \varepsilon_0(-\dot{A}_\mathbf{k}^2)\mathrm{d}V.$$

As the field quantities enter quadratically they must be taken real,

$$\mathbf{A}_\mathbf{k}(t) = \hat{\mathbf{A}}_\mathbf{k}\mathrm{e}^{i(\mathbf{kx}-\omega_\mathbf{k}t)} + \hat{\mathbf{A}}_\mathbf{k}^*\mathrm{e}^{-i(\mathbf{kx}-\omega_\mathbf{k}t)}$$

$$\mathbf{E}_\mathbf{k}(t) = i\omega\hat{\mathbf{A}}_\mathbf{k}\mathrm{e}^{i(\mathbf{kx}-\omega_\mathbf{k}t)} - i\omega\hat{\mathbf{A}}_\mathbf{k}^*\mathrm{e}^{-i(\mathbf{kx}-\omega_\mathbf{k}t)}, \quad \mathbf{B}_\mathbf{k}(t) = \frac{\mathbf{k}}{\omega} \times \mathbf{E}_\mathbf{k}(t). \quad (8.7)$$

If L is taken large, $L \gg \lambda = 2\pi/|\mathbf{k}|$, periodic boundary conditions (or, alternatively, perfectly reflecting boundaries) can be assumed to obtain simple expressions for $\mathcal{E}$,

$$\mathcal{E} = 2\varepsilon_0 V\omega_\mathbf{k}^2\mathbf{A}_\mathbf{k}\mathbf{A}_\mathbf{k}^* = 2\varepsilon_0 V\omega_\mathbf{k}^2|\mathbf{A}_\mathbf{k}|^2. \quad (8.8)$$

Introduction of the scalar "momentum" P and "position" Q as follows

$$P_{\mathbf{k}} = \frac{1}{i}(\varepsilon_0 V \omega_{\mathbf{k}}^2)^{1/2}(A_{\mathbf{k}} - A_{\mathbf{k}}^*), \quad Q_{\mathbf{k}} = \frac{1}{\omega_{\mathbf{k}}}(\varepsilon_0 V \omega_{\mathbf{k}}^2)^{1/2}(A_{\mathbf{k}} + A_{\mathbf{k}}^*) \tag{8.9}$$

confers $\mathcal{E}$ from (8.8) the structure of the harmonic oscillator Hamiltonian

$$H(P, Q) = \frac{1}{2}(P_{\mathbf{k}}^2 + \omega_{\mathbf{k}}^2 Q_{\mathbf{k}}^2). \tag{8.10}$$

Field quantization is now straightforward:

- Replace $\mathbf{A}_{\mathbf{k}}$ and $\mathbf{A}_{\mathbf{k}}^*$ by the Hermitian conjugate operators

$$\mathbf{A}_{\mathbf{k}} = \frac{1}{(4\varepsilon_0 V \omega_{\mathbf{k}}^2)^{1/2}}(i P_{\mathbf{k}} + \omega_{\mathbf{k}} Q_{\mathbf{k}})\boldsymbol{\varepsilon}_{\mathbf{k}}, \quad \mathbf{A}^{\dagger}{}_{\mathbf{k}} = \frac{1}{(4\varepsilon_0 V \omega_{\mathbf{k}}^2)^{1/2}}(-i P_{\mathbf{k}} + \omega_{\mathbf{k}} Q_{\mathbf{k}})\boldsymbol{\varepsilon}_{\mathbf{k}}. \tag{8.11}$$

- Set the commutator

$$[\mathbf{A}_{\mathbf{k}}, \mathbf{A}^{\dagger}{}_{\mathbf{k}}] = \mathbf{A}_{\mathbf{k}}\mathbf{A}^{\dagger}{}_{\mathbf{k}} - \mathbf{A}^{\dagger}{}_{\mathbf{k}}\mathbf{A}_{\mathbf{k}} = \frac{\hbar^2}{2\varepsilon_0 V \omega_{\mathbf{k}}}. \tag{8.12}$$

In explicit form the field operators $\mathbf{A}_{\mathbf{k}}, \mathbf{E}_{\mathbf{k}}, \mathbf{B}_{\mathbf{k}}$ read, with $[a_{\mathbf{k}}, a^{\dagger}{}_{\mathbf{k}}] = 1$,

$$\begin{aligned}
\mathbf{A}_{\mathbf{k}} &= \left(\frac{\hbar}{2\varepsilon_0 V \omega_{\mathbf{k}}}\right)^{1/2} \boldsymbol{\varepsilon}_{\mathbf{k}} \left\{a_{\mathbf{k}} \mathrm{e}^{i(\mathbf{kx}-\omega t)} + a^{\dagger}{}_{\mathbf{k}} \mathrm{e}^{-i(\mathbf{kx}-\omega t)}\right\}, \\
\mathbf{E}_{\mathbf{k}} &= i\left(\frac{\hbar\omega_{\mathbf{k}}}{2\varepsilon_0 V}\right)^{1/2} \boldsymbol{\varepsilon}_{\mathbf{k}} \left\{a_{\mathbf{k}} \mathrm{e}^{i(\mathbf{kx}-\omega t)} - a^{\dagger}{}_{\mathbf{k}} \mathrm{e}^{-i(\mathbf{kx}-\omega t)}\right\}, \\
\mathbf{B}_{\mathbf{k}} &= i\left(\frac{\hbar}{2\varepsilon_0 V \omega_{\mathbf{k}}}\right)^{1/2} \mathbf{k} \times \boldsymbol{\varepsilon}_{\mathbf{k}} \left\{a_{\mathbf{k}} \mathrm{e}^{i(\mathbf{kx}-\omega t)} - a^{\dagger}{}_{\mathbf{k}} \mathrm{e}^{-i(\mathbf{kx}-\omega t)}\right\},
\end{aligned} \tag{8.13}$$

$a_{\mathbf{k}}, a^{\dagger}{}_{\mathbf{k}}$ are the annihilation and creation operators (8.3) acting on the mode $(\mathbf{k}, \boldsymbol{\varepsilon}_{\mathbf{k}})$. The Hamiltonian acts on the the energy eigenstates $|n_{\mathbf{k}}\rangle$, its expectation value yields the energy eigenvalues

$$H_{\mathbf{k}}|n_{\mathbf{k}}\rangle = \hbar\omega_{\mathbf{k}}(a_{\mathbf{k}}a^{\dagger}{}_{\mathbf{k}} + 1/2)|n_{\mathbf{k}}\rangle; \quad \langle n_{\mathbf{k}}|H_{\mathbf{k}}|n_{\mathbf{k}}\rangle = \hbar\omega_{\mathbf{k}}(n_{\mathbf{k}} + 1/2).$$

The field operator of the Poynting vector is $\mathbf{S}_{\mathbf{k}} = \mathbf{E}_{\mathbf{k}} \times \mathbf{B}_{\mathbf{k}}$. Applied to a photon number state it yields the connection with the momentum and energy density as known from classical electrodynamics

$$\langle n_{\mathbf{k}}|\mathbf{S}_{\mathbf{k}}|n_{\mathbf{k}}\rangle = \frac{c^2}{V}(n_{\mathbf{k}} + 1/2)\hbar\mathbf{k} = c\hbar\omega_{\mathbf{k}}(n_{\mathbf{k}} + 1/2)\mathbf{k}^0.$$

If we postulate $a_{\mathbf{k}}|n'_{\mathbf{k}}\rangle = n_{\mathbf{k}'}^{1/2}|n'\rangle\delta_{\mathbf{k},\mathbf{k}'}$, i.e., $a_{\mathbf{k}}$ acts only on the mode $\mathbf{k}$ of polarization $\boldsymbol{\varepsilon}$, the total Hamiltonian H in a cavity is the sum of all possible $H_{\mathbf{k}}$. Correspondingly, a general photon number state is the direct product of single mode states,

$$\{|n_{\mathbf{k}}\rangle\} = |n_{\mathbf{k}_1}\rangle|n_{\mathbf{k}_2}\rangle|n_{\mathbf{k}_3}\rangle...., \quad H\{|n_{\mathbf{k}}\rangle\} = \sum_{\mathbf{k}} H_{\mathbf{k}}(|n_{\mathbf{k}_1}\rangle|n_{\mathbf{k}_2}\rangle|n_{\mathbf{k}_3}\rangle....).$$

Owing to the equal spacing $\hbar\omega_{\mathbf{k}}$ between the energy levels $|n_{\mathbf{k}}\rangle$ and $|(n+1)_{\mathbf{k}}\rangle$ the $|n_{\mathbf{k}}\rangle$ are the *photon number states* of polarization $\boldsymbol{\varepsilon}_{\mathbf{k}}$; the operator $n_{\mathbf{k}} = a^{\dagger}{}_{\mathbf{k}}a_{\mathbf{k}}$ counts them. The state $\{|0_{\mathbf{k}}\rangle\}$ containing no photons at all is the *vacuum state*.

The photon number states contain a well defined number of photons and are states of well defined amplitude. According to (8.1) the phase ϕ is completely undetermined. In classical physics therefore such states do not exist, $\{|n_{\mathbf{k}}\rangle\}$ represent *nonclassical light*. The first realization of an $|n_{\mathbf{k}}\rangle$ state has been due to H. Walther and his group [2]. Conversely, construction of nonclassical asymptotic states of well defined phase $|\phi\rangle$ and completely undefined amplitude is also possible [3]. Possible phase operators are

$$\cos\phi = \frac{1}{2}(\exp i\phi + \exp -i\phi), \quad \sin\phi = \frac{1}{2i}(\exp i\phi - \exp -i\phi)$$

$$\exp i\phi = (a^{\dagger}{}_{\mathbf{k}}a_{\mathbf{k}})^{-1/2}a_{\mathbf{k}}, \ \exp -i\phi = ...a^{\dagger}{}_{\mathbf{k}}(a^{\dagger}{}_{\mathbf{k}}a_{\mathbf{k}})^{-1/2}. \tag{8.14}$$

It holds

$$\langle n-1|\exp i\phi|n\rangle = 1, \quad \langle n+1|\exp -i\phi|n\rangle = 1. \tag{8.15}$$

All other matrix elements are zero.

Glauber States

So far all theoretical considerations have been based on classical sinusoidal modes of type (8.7). They are coherent to all orders according to the old coherence theory at the price of violating Heisenbergs uncertainty principle. The question arises what kind of state in the quantized field is coherent to all orders and how close it comes to the classical sinusoidal mode. The question is answered by the *coherent* or *Glauber states* $|\alpha_{\mathbf{k}}\rangle$ [4]. They are eigenstates to the destruction operator a (for simplicity the index of the mode is suppressed in the following),

$$|\alpha\rangle = \mathrm{e}^{-|\alpha|^2/2}\sum_{n\geq 0}\frac{\alpha^n}{(n!)^{1/2}}|n\rangle \quad \Rightarrow \quad a|\alpha\rangle = \alpha|\alpha\rangle, \quad \langle\alpha|\alpha\rangle = 1, \ n|\alpha\rangle = |\alpha|^2|\alpha\rangle. \tag{8.16}$$

The eigenvalue α is an arbitrary complex number. The coherent states are not orthogonal,

$$|\langle\beta|\alpha\rangle|^2 = \mathrm{e}^{-|\alpha-\beta|^2}$$

however, with increasing difference $|\alpha - \beta|$ they approach this property. The mean absolute and relative deviations in photon numbers in a cavity mode assume a remarkable value,

$$\langle\alpha|n^2|\alpha\rangle = \mathrm{e}^{-|\alpha|^2}\sum_{n\geq 0}\frac{(\alpha^*\alpha)^n}{n!}n^2 = \mathrm{e}^{-|\alpha|^2}\sum_{n\geq 0}\frac{|\alpha|^{2n}}{n!}[n(n-1)+n] = |\alpha|^4 + |\alpha|^2$$

$$\Rightarrow \qquad \Delta n = |\alpha|, \quad \frac{\Delta n}{n} = \frac{1}{|\alpha|}. \tag{8.17}$$

With real $\alpha = \sqrt{n}$, Δn obeys the Boltzmann statistics of uncorrelated classical particles, in contrast to $\Delta n = 1$ of photons in thermal equilibrium. This paradoxical behavior of noninteracting particles may be paraphrased by: Thermal photons are maximally correlated, coherent photons are maximally uncorrelated.

The expectation values of the vector potential operators A and A^2 in the coherent state $|\alpha\rangle$ are given by

$$\begin{aligned}\langle\alpha|A|\alpha\rangle &= \langle\alpha|\left(\frac{\hbar}{2\varepsilon_0 V\omega_{\mathbf{k}}}\right)^{1/2}\left\{a_{\mathbf{k}}\mathrm{e}^{i(\mathbf{kx}-\omega t)} + a^\dagger{}_{\mathbf{k}}\mathrm{e}^{-i(\mathbf{kx}-\omega t)}\right\}|\alpha\rangle \\ &= 2\left(\frac{\hbar}{2\varepsilon_0 V\omega_{\mathbf{k}}}\right)^{1/2}|\alpha|\cos(\mathbf{kx}-\omega t+\phi)\end{aligned} \tag{8.18}$$

and

$$\langle\alpha|A^2|\alpha\rangle = \left(\frac{\hbar}{2\varepsilon_0 V\omega_{\mathbf{k}}}\right)[4|\alpha|^2\cos^2(\mathbf{kx}-\omega t+\phi)+1] \;\Rightarrow\; \Delta A = \left(\frac{\hbar}{2\varepsilon_0 V\omega_{\mathbf{k}}}\right)^{1/2}. \tag{8.19}$$

It can be shown that for $|\alpha|^2$ large the coherent state assumes its minimum uncertainty in (8.1). We conclude that in a coherent state the expectation values of the Maxwellian field operators approach the classical values for large photon numbers in the mode. The convergence of $\langle A\rangle$ towards the classical sinusoidal field is illustrated by Fig. 8.1 for three different mean photon numbers. A laser well above threshold may come close to a coherent photon state. With a view onto Fig. 8.1 a classical field treatment is justified if the number N_{λ^3} of photons in a cube $V = \lambda^3$ of the wavelength λ is large. For example, at $I_{Ti:Sa} = 10^{10}$ Wcm^{-2} follows $N_{\lambda^3} = 1.1 \times 10^{12}$. This is 10^4 photons along one wavelength, high enough to draw a sin wave with high precision.

8.1.1.2 Line Radiation

Line radiation from atomic transitions is the most familiar picture everybody has in mind. Whenever an electron changes from one energy level to another energy level the energy difference $\Delta E_{if} = E_i - E_f$ is either emitted as a photon $\hbar\omega = \Delta E_{if}$ by upper level deexcitation, or absorbed by excitation of a higher level. The transition

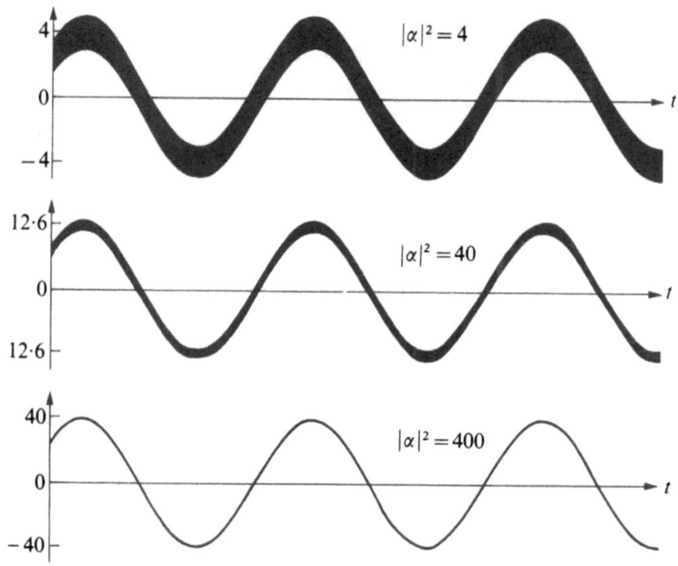

Fig. 8.1 Uncertainty $2\Delta A$ (vertical thickness of graph at fixed point) in the expectation value of the vector potential operator A from (8.13) for number of photons $|\alpha^2| = 4$, 40, and 400 in the cavity **k**. Unit on the ordinate is $(\hbar/2\varepsilon_0 V\omega)^{1/2}$. Courtesy of Loudon [3]

probability w_{if} per unit time is strongest in the electric dipole transition, followed in strength by a magnetic dipole transition and, finally, an electric quadrupole transition. The latter ones are weaker by the dimensionless fine structure constant

$$\alpha = \frac{e^2}{4\pi\varepsilon_0\hbar c} \simeq \frac{1}{137}. \tag{8.20}$$

The electric dipole transition probability is calculated from first order Dirac perturbation theory as the expression [1]

$$w_{if} = \frac{2\pi}{\hbar}|V_{if}|^2\delta(E_i - E_f); \quad V_{if} = \langle i|e(\mathbf{x} - \mathbf{x}_0)\mathbf{E}_{\mathbf{k}\omega}(\mathbf{x}_0)|f\rangle. \tag{8.21}$$

The initial and final states are represented by the state vectors $|i\rangle$ and $|f\rangle$. The electric field mode $\mathbf{E}_{\mathbf{k}\omega}(\mathbf{x}_0)$ is constant over the extension d of the system, i.e. atom or electron in the particular case. The approximation is correct for $d \ll \lambda = 2\pi/|\mathbf{k}|$. The transition is *induced* by the field mode in direction of propagation **k**. For the emission process it means that the photon ejection happens into the direction and the polarization of the mode causing the transition. The probability w_{if} for induced emission is the same as for (induced) absorption, provided the states $|i\rangle, |f\rangle$ are not degenerate. In the case of degeneracy of degree g_i, g_f the principle of detailed balance requires $g_f w_{if} = g_i w_{fi}$, see Fig. 4.4. This is *Fermi's golden rule*. Once the occupation numbers n_1, n_2 of the transition levels are known the Einstein coefficients B_{if}, B_{fi} are related to the transition probabilities $w_{if} = w_{fi}$.

On the basis of thermodynamic arguments Einstein recognized that the upper level E_i decays *spontaneously* into the empty lower level by emitting the photon with equal probability into all modes for which $\mathbf{k}(\omega = \text{const})$ is fulfilled. For the rate A_{fi} of spontaneous emission he found the relation

$$A_{fi} = \frac{\omega^2}{\pi^2 c^3}\frac{g_1}{g_2}(B_{12}\hbar\omega) = \frac{g_1 e^2\omega^3}{3\pi\varepsilon_0 g_2\hbar c^3}|\langle f|(\mathbf{x}-\mathbf{x}_0)|i\rangle|^2. \tag{8.22}$$

The factor $1/3$ is $\langle\cos\theta^2\rangle$ for random orientation of polarization with respect to $\mathbf{E}$. Higher order time dependent perturbation theory describes *multiphoton transitions*. For details and limits of standard perturbation theory see F. H. M. Faisal.

The transition probabilities (8.21) and (8.22) are useful in balance and rate equations. They are not the full truth and are valid under certain limitations only. The general dynamics of atomic level occupations is much richer. To learn about some basic phenomena and conditions for the validity of rate equations the *optical Bloch model* for two level systems, introduced by Feynman, Vernon, and Hellwarth [5] is extremely useful.

8.1.2 The Optical Bloch Model

Dynamics of a Two Level System

The optical Bloch model is the geometric visualization of the dynamics of a non degenerate two level system. The driving electric field is assumed classical and directed along z,

$$E(z,t) = \hat{E}\cos(kz-\omega t) = \frac{1}{2}\hat{E}\left\{e^{i(kz-\omega t)} + e^{-i(kz-\omega t)}\right\}. \tag{8.23}$$

The unperturbed states of matter (solid, atom , ion, molecule) $|1\rangle$ and $|2\rangle$ are assumed to have the energies $E_1 = 0$ and $E_2 = E = \hbar\omega_0$. The Hamiltonian is chosen in electric dipole approximation with H_0 the unperturbed particle operator,

$$H = H_0 + H_1 = H_0 + ezE. \tag{8.24}$$

Thereby ez is the permanent electric dipole moment of the the single particle. The center of mass of the undisturbed material system is assumed to be at $\mathbf{x} = 0$. z is the average electron deformation. One arrives at H_1 from $H = (\mathbf{p} + e\mathbf{A})^2$ under the restriction of the Coulomb gauge $div\mathbf{A} = 0 \Rightarrow \mathbf{pA} = 0$ and dropping the quadratic term $(e\mathbf{A})^2$. The dynamics of the two level system is governed by the von Neumann equation (4.97):

$$\frac{\mathrm{d}\rho_{11}}{\mathrm{d}t}=\frac{1}{i\hbar}(H_{12}\rho_{21}-\rho_{12}H_{21})=-\frac{\mathrm{d}\rho_{22}}{\mathrm{d}t},\quad \frac{\mathrm{d}\rho_{12}}{\mathrm{d}t}=\frac{1}{i\hbar}(H_{12}\rho_{22}-\rho_{11}H_{12})+i\omega_0\rho_{12}. \tag{8.25}$$

Remember $\rho^{\mathrm{op}}=|\psi\rangle\langle\psi|,\ |\psi\rangle=c_1|1\rangle+c_2|2\rangle\Rightarrow\rho_{12}=\langle 1|(c_1|1\rangle)(\langle 2|c_2)|2\rangle=c_1c_2^*,\ \rho_{21}=\rho_{12}^*$. It follows $d(\rho_{11}+\rho_{22})/\mathrm{d}t=0,\ \rho_{11}+\rho_{22}=\mathrm{const}$. In reality there are level deactivations (losses) due to collisions and spontaneous decay. They are included phenomenologically by the damping rates γ_1,γ_2. Collisions and mutual particle interactions lead, in addition, to a weakening of the phase relation between state $|1\rangle$ and state $|2\rangle$. The damping coefficients γ_{12},γ_{21} allow for these losses. Finally, time of flight effects of the individual particles through cavities of finite dimensions (finite laser beam diameter) contribute by additional losses of occupations and phases. In thermal equilibrium at temperature T the two levels are occupied by n_1 and n_2 particles. Hence, the final equations of motion read

$$\begin{aligned}
\frac{\mathrm{d}\rho_{11}}{\mathrm{d}t}&=\frac{1}{i\hbar}(H_{12}\rho_{21}-\rho_{12}H_{21})-\gamma_1(\rho_{11}-n_1),\\
\frac{\mathrm{d}\rho_{22}}{\mathrm{d}t}&=\frac{1}{i\hbar}(H_{21}\rho_{12}-\rho_{21}H_{12})-\gamma_2(\rho_{22}-n_2),\\
\frac{\mathrm{d}\rho_{12}}{\mathrm{d}t}&=\frac{1}{i\hbar}H_{12}(\rho_{22}-\rho_{11})+(i\omega_0-\gamma_{12})\rho_{12},\\
\frac{\mathrm{d}\rho_{21}}{\mathrm{d}t}&=\frac{-1}{i\hbar}H_{12}^*(\rho_{22}-\rho_{11})-(i\omega_0+\gamma_{21})\rho_{21}.
\end{aligned} \tag{8.26}$$

Owing to the cosine dependence in H_1 this system of equations cannot be integrated by standard methods. Simplification is possible by making use of the *rotating wave approximation*. The level spacing ω_0 must lie within the profile of the linewidth of the driver to stimulate appreciable transitions between the two states. On the other hand, for a multilevel system to reduce to two levels the spacing to all other transition frequencies ω_{if} must fulfil the inequality $|\omega_{if}-\omega|\gg|\omega_0-\omega|$. The driver (8.23) is the sum of two circularly polarized waves. By multiplying it with $\exp -i(kz-\omega t)$ the first term reduces to the secular, i.e. slowly evolving amplitude $\hat{E}$, whereas the second term rotates at twice the frequency ω and represents a rapidly oscillating correction only. For this reason it can be dropped. In most of the systems $\gamma_{21}=\gamma_{12}$, or even $\gamma_{12}=(\gamma_1+\gamma_2)$ can be assumed (it depends on the domain of ω_0). Then, setting $\rho_{12}=\hat{\rho}_{12}\exp -i(kz-\omega t)$ transforms the system of (8.26) into a set of secular equations

$$\begin{aligned}
\frac{\mathrm{d}\rho_{11}}{\mathrm{d}t}&=i\kappa\frac{\hat{E}}{2}(\hat{\rho}_{12}-\hat{\rho}_{21})-\gamma_1(\rho_{11}-n_1),\\
\frac{\mathrm{d}\rho_{22}}{\mathrm{d}t}&=i\kappa\frac{\hat{E}}{2}(\hat{\rho}_{21}-\hat{\rho}_{12})-\gamma_2(\rho_{22}-n_2),\\
\frac{\mathrm{d}\hat{\rho}_{12}}{\mathrm{d}t}&=i\kappa\frac{\hat{E}}{2}(\rho_{11}-\rho_{22})+[i(\omega_0-\omega)-\gamma_{12}]\hat{\rho}_{12};\quad \hat{\rho}_{21}=\hat{\rho}_{12}^*.
\end{aligned} \tag{8.27}$$

Thereby $\kappa = ez/\hbar$. The quantity $\mu = e|z|$ is the modulus of the dipole moment of the single particle. μ multiplied with the field amplitude $\hat{E}$ and divided by $\hbar$ is the *Rabi frequency* ω_R,

$$\omega_R = \frac{\mu}{\hbar}\hat{E}. \tag{8.28}$$

System (8.27) is completely analoguos to the slowly evolving system (6.140)–(6.142). It can be solved exactly. However, the solution is so complex that it is of no value. For the electric dipole transition matrix elements to be different from zero the states $|1\rangle$, $|2\rangle$ must differ in parity. In case the electric dipole moment vanishes either the electric quadrupole or the magnetic dipole moments can be used in the equations, however not both simultaneously. For the magnetic dipole transition $\hat{E}$ is to be replaced by $\hat{B} = k\hat{E}/\omega$. High order multipole transitions play an important role in nuclear physics.

The term oscillating with $\omega_0 + \omega$ leads to the *Bloch–Siegert shift* $\Delta\omega_0 = \kappa^2\hat{E}^2/4\omega_0$ of the resonance frequency. It is a ponderomotive energy

The optical Bloch model. The electric dipole moment $P(z, t)$ of the individual particle is

$$P(z, t) = \langle\psi|ez|\psi\rangle = \hbar\kappa(c_{12} + c_{21}) = 2\hbar\kappa\Re\big[\hat{\rho}_{12}\exp i(kz - \omega t)\big] = P_\parallel + P_\perp.$$

The component in phase with E is indicated with $P_\parallel$; $P_\perp$ is the component out of phase. It is convenient to introduce the change of variables

$$2\hat{\rho}_{12} = u + iv,\ \ w = \rho_{22} - \rho_{11},\ \ s = \rho_{11} + \rho_{22}\ \Rightarrow\ P_\parallel = \hbar\kappa u,\ \ P_\perp = \hbar\kappa v. \tag{8.29}$$

Additionally, the frequency mismatch is expressed through $\delta = \omega - \omega_0$ and $\gamma_{12} = \gamma_{12} = 1/T_2, \gamma_1 = \gamma_2 = 1/T_1$. The relaxation times T_1, T_2 are the so called *longitudinal*, and *transverse* or *phase memory* relaxation times, expressions borrowed from nuclear resonance. T_1 is an effective collision time between particles. In these variables system (8.27) reduces to the system

$$\begin{aligned}\dot{u} &= \delta v - \frac{u}{T_2}.\\ \dot{v} &= -\delta u - \kappa\hat{E}w - \frac{v}{T_2},\\ \dot{w} &= \kappa\hat{E}v - \frac{1}{T_1}(n_1 - n_2 + w).\end{aligned} \tag{8.30}$$

Let T_1, T_2 become infinitely long. Then with the vectors

$$\boldsymbol{\Omega} = (\kappa\hat{E},\ 0,\ \omega_0 - \omega),\quad \mathbf{M} = (u,\ v,\ w). \tag{8.31}$$

system (8.30) is cast into the vectorial form of the *optical Bloch model*

$$\frac{\mathrm{d}}{\mathrm{d}t}\mathbf{M} = \boldsymbol{\Omega} \times \mathbf{M}. \tag{8.32}$$

It is isomorphous with the preceding *magnetic Bloch model* introduced to describe the magnetic two level system.

The processes describable by (8.32) all belong to the reversible field of *coherent optical phenomena* and are not accessible to any description in terms of rate equations. In order to be more realistic the model must be generalized to statistical mixtures of states. In a beam for instance the individual particles at a position $\mathbf{x}$ are distributed according to $f(\mathbf{x}, \mathbf{v}, t)$ and are subject to the Doppler shifts $\omega' = \omega - \mathbf{kv}$. All what is to be done is to express the density operator and the polarization by the mixtures

$$\rho = \int f(z, v_z, t)|\psi(v_z)\rangle\langle\psi(v_z)|\mathrm{d}v_z,$$

$$P_{\|} = \hbar\kappa \int_{-\infty}^{+\infty} u(v_z)\mathrm{d}v_z, \quad P_{\perp} = \hbar\kappa \int_{-\infty}^{+\infty} v(v_z)\mathrm{d}v_z; \quad \delta = \langle\omega'\rangle - \omega_0. \tag{8.33}$$

For δ to make sense the individual ω' must not be too widely spread. The study of individual groups of particles from the velocity interval $(\mathbf{v}, \mathbf{v} + \delta\mathbf{v})$ with $\delta = \omega' - \omega_0$ is also possible.

In the electromagnetic wave equations the current density $\mathbf{j}$ appears as the source term. In the unmagnetized electrically neutral medium the (polarization) charge density ρ_P and the (polarization) current density $\mathbf{j}_P$ follow from the well known relations

$$\rho_P = -\nabla\mathbf{P}, \quad \mathbf{j}_P = \frac{\partial}{\partial t}\mathbf{P}, \tag{8.34}$$

A change of the transverse polarization is a generator of light. Standard perturbation analysis of the slowly varying amplitude $\hat{E}_1 = E_{1\|} + iE_{1\perp}$ yields in the rarefied medium ($\eta = 1$)

$$\frac{\mathrm{d}E_{1\|}}{\mathrm{d}t} = \partial_t E_{1\|} + c\frac{\partial E_{1\|}}{\partial x} = -\frac{\omega}{2\varepsilon_0}P_{\perp}(t), \quad \frac{\mathrm{d}E_{1\perp}}{\mathrm{d}t} = \partial_t E_{1\perp} + c\frac{\partial E_{1\perp}}{\partial x} = \frac{\omega}{2\varepsilon_0}P_{\|}(t). \tag{8.35}$$

Consider an electromagnetic wave $E = \hat{E}\cos(kz - \omega t)$ impinging perpendicularly onto a tenuous homogeneous medium of thickness L and producing the polarization $P = \hat{P}\cos(kz - \omega t + \phi)$. It is instructive to determine the total field E_t and the intensity I at position L. Integration of (8.35) results in

$$
\begin{aligned}
E_{1\parallel}(L,t) &= -\frac{\omega}{2\varepsilon_0 c}\int_0^L P_\perp(z,t/z)dz = -\frac{\omega L}{2\varepsilon_0 c}P_\perp(0,t-L/c),\\
E_{1\perp}(L,t) &= \frac{\omega}{2\varepsilon_0 c}\int_0^L P_\parallel(z,t/z)dz = \frac{\omega L}{2\varepsilon_0 c}P_\parallel(0,t-L/c),\\
E_t(L,t) &= \hat{E}\cos[\omega(t-L/c)] + \frac{\omega L}{2\varepsilon_0 c}\left\{P_\parallel \sin[\omega(t-L/c)] - P_\perp\cos[\omega(t-L/c)]\right\}.
\end{aligned}
\tag{8.36}
$$

The ingoing and the outgoing intensities I_0 and I thus are

$$
\begin{aligned}
I &= \varepsilon_0 c\overline{E_t^2} = \varepsilon_0 c\overline{\left\{\left(\hat{E} - \frac{\omega L}{2\varepsilon_0 c}P_\perp\right)\cos\varphi + \frac{\omega L}{2\varepsilon_0 c}P_\parallel \sin\varphi\right\}^2}\\
&= \frac{1}{2}\varepsilon_0 c\hat{E}^2 + \frac{c}{2\epsilon_0}\left(\frac{\omega L}{2c}\right)^2(P_\parallel^2 + P_\perp^2) - \frac{1}{2}\omega L\hat{E}\left(0,t-\frac{z}{c}\right)P_\perp\left(0,t-\frac{z}{c}\right)\\
&\simeq I_0 - \frac{\omega L}{2}\hat{E}\left(0,t-\frac{z}{c}\right)P_\perp\left(0,t-\frac{z}{c}\right) = I_0 - I_{abs};\quad \varphi = \omega\left(t-\frac{z}{c}\right).
\end{aligned}
\tag{8.37}
$$

The induced polarization P generates the wave E_1. After passage the incident wave I_0 appears modulated proportionally to $P_\perp$. The intensity absorbed by the cell is I_{abs}.

8.1.3 Coherent Effects

The system (8.32) can be solved analytically [6]. To obtain an overview on it is not simple and quick. Therefore a pictorial access to some phenomena by (8.33) is chosen here, see Fig. 8.2. A general picture of the optical Bloch model is given in (a): The polarizations u, v and the population inversion w fix the Bloch space. If all particles are in the ground state $|1\rangle$ and no field is switched on $\mathbf{M}$ and $\boldsymbol{\Omega}$ point down along $-|w|$. If the particles are at exact resonance, $\delta = 0$, and the applied electric field amplitude is constant $\mathbf{M}$ rotates with the Rabi frequency ω_R in the (v, w) plane (b). Total inversion is obtained each cycle after $\omega_R t = (2n+1)\pi$. After switching off at $t = (2n+1)\pi/\omega_R$, $\mathbf{M}$ stops in the vertical position and decays exponentially according to $\exp -t/T_1$. In the incoherently pumped system maximum inversion $c_{22} = c_{11}$, i.e. $\mathbf{M}$ in the (uv) plane, is possible for $\hat{E} \to \infty$ only.

Another group of particles from a beam may be Doppler shifted out of resonance. Under the action of the applied constant $\hat{E}$ field to the particles in the ground state $|1\rangle$ the vector $\boldsymbol{\Omega}$ assumes the position of picture (c) and forces the Bloch vector to rotate with constant frequency along the surface of a cone. This is the phenomenon of *optical nutation*. Whether inversion will be reached or not depends on the ratio of ω_r/δ.

Apply a $(\pi/2)$ pulse of strength $\kappa\hat{E} \gg kv_{z,ther}$ to a system in the ground state. The pulse duration is $t_2 - t_1 = \pi/2\omega_R$. After the switching off time t_2 one concludes from (8.30) for the individual particle

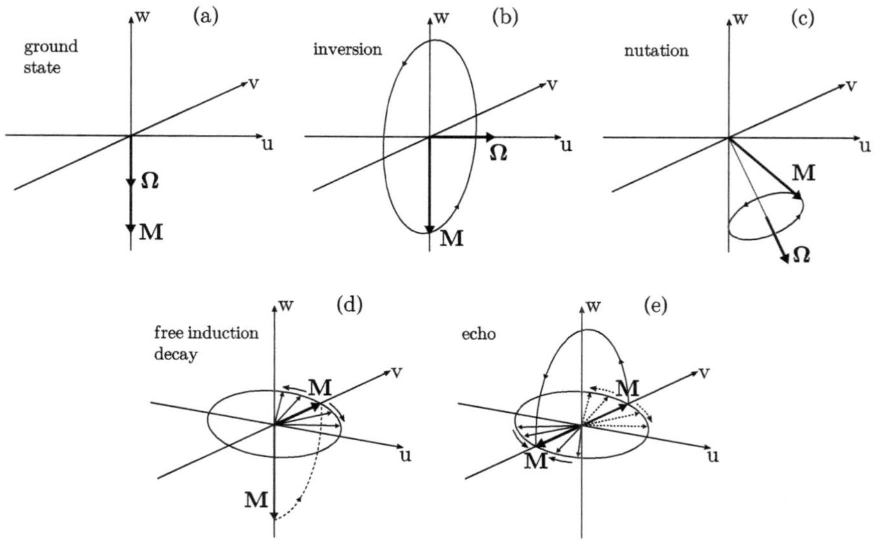

Fig. 8.2 Optical Bloch model: $d\mathbf{M}/dt = \boldsymbol{\Omega} \times \mathbf{M}$. $\mathbf{M} = (u, v, w)$, $\boldsymbol{\Omega} = (\kappa\hat{E}, 0, \omega_0 - \omega)$. **a** All particles in ground state, **b** total inversion with **M** along $w > 0$, **c** optical nutation with $\boldsymbol{\Omega}$ out of resonance, **d** free induction decay due to thermal Doppler effect and collisions, **e** optical echo due to inversion of **M**

$$u(t) = (n_2 - n_1)\exp -\gamma(t - t_2)\sin\delta(t - t_2), \quad v(t) = (n_2 - n_1)\exp -\gamma(t - t_2)\cos\delta(t - t_2)$$

The intensity irradiated at $t > t_2$ follows from (8.37),

$$I = \frac{1}{8}\frac{\omega^2 L^2}{\varepsilon_0 c}|P|^2. \tag{8.38}$$

It decays with time owing to the dephasing of the individual Doppler shifts ω'. This is the phenomenon of the optical *free induction decay*, depicted in (d).

As a final coherent phenomenon the so called *optical echo* is considered, see picture (e). In a first stage a $(\pi/2)$ pulse is applied to a system in the ground state (e1) which by Doppler dephasing starts decaying as in picture (c). At time $t = t_3$ a π pulse of duration $t_4 - t_3 = 2(t_2 - t_1)$ is applied to induce the *umklapp process* of (e2) with the effect that the diverging arrows converge now towards growing polarization $|P|$. This causes a strong signal emission as the echo of the decaying free induction pulse. The calculation of the signal proceeds according to (8.37) and (8.38),

$$I(t) = \frac{\hbar^2\kappa^2\omega^2 L^2}{8\varepsilon_0 c}(n_1 - n_2)^2 \mathrm{e}^{-2\gamma[t - t_4 + (t_3 - t_2) - z/c]}\mathrm{e}^{-(\kappa^2/2\beta)[t - t_4 - (t_3 - t_2) - z/c]}. \tag{8.39}$$

The signal reaches its maximum for $t = t_4 + t_3 - t_2 + z/c$,

$$I_{\max} = \frac{\hbar^2 \kappa^2 \omega^2 L^2}{8\varepsilon_0 c} (n_1 - n_2)^2 \mathrm{e}^{-4\gamma(t_3 - t_2)}.$$

The signal decay in time is by the Doppler dephasing with $\beta = m/2k_B T$ from the Maxwell distribution and γ from particle interactions. Production of echoes is possible also with pulses differering in length from $\pi/2$ and π and with $\kappa\hat{E}$ no longer greater than $k/\beta^{1/2}$. Also in this cases the decay time is $T = T_1 = T_2$. A *stimulated echo* is produced from three $(\pi/2)$ pulses which change the Bloch vector into v direction, from there along w and then again back to v. The *image echo* originates from the first echo and a third pulse. Multiple repetition of a π pulse leads to a chain of echoes, the optical *Carr–Purcell effect* [6].

A salient feature of coherent optics is the quadratic dependence of the intensity of an emitted pulse from the number of emitters. The electric field of the superposition of n identical unidirectional dipoles is proportional to their number n and, consequently, the intensity depends on n^2, in contrast to incoherent light with its linear dependence of radiation on the number of emitters. A stringent example of coherence is the *superradiance*, also called *superfluorescence*, predicted by R. H. Dicke in 1954 [7, 8]. Imagine a medium completely inverted from its ground state $|1\rangle$ to the excited state $|2\rangle$. The dipole moment is zero. In a dilute medium inversion decays exponentially by uncorrelated spontaneous emission of the excited particles. This phenomenon of *fluorescence* emits light isotropically into all directions with $I \sim n$. If the cloud of emitters is dense enough spontaneous phase locking of the dipoles may occur out of noise of the radiation field and result in strong anisotropic emission of superradiance with local tendency of $I \sim n^2$. Owing to the initial absence of a dipole moment and an external radiation field, superradiance is generally labelled as *spontaneous emission.*

Landau plasma echo. In a collisionless plasma spatio-temporal echoes can be produced by exciting an electron plasma wave and waiting until the wave has decayed by Landau damping. By applying a second pulse of suitable strength an electron plasma wave, the *echo*, comes back again. In quantitative terms, impress a weak electrostatic potential

$$\Phi_1(x,t) = \hat{\Phi}_1 \exp ik_1 x\, \delta[\omega(t - T_1)]$$

at time T_1 to the plasma of distribution function $f(x, v)$. After being damped by phase mixing the distribution function is left with a modulation f_1 superposed,

$$f_1 = \hat{f}_1 \mathrm{e}^{i\{k_1[x - v(t - T_1)]\}}$$

but no electric field. After impressing a second pulse

$$\Phi_2(x,t) = \hat{\Phi}_2 \exp ik_2 x\, \delta[\omega(t - T_2)]$$

the modulation f_2,

$$f_2 = \hat{f}_2 \mathrm{e}^{i\{k_2[x - v(t - T_2)]\}}$$

on $f(x, v)$ is is produced. Waiting again until the electric field has cancelled to zero by free particle streaming one is left with a second order modulation on $f(x, v)$ containing terms

$$f_1 f_2 = \hat{f}_1 \hat{f}_2 \mathrm{e}^{\pm i\{(k_2-k_1)x - k_2 v(t-T_2) + k_1 v(t-T_1)\}}.$$

The macroscopic quantities, e.g. n_e, result from the corresponding integrals of $f(x, v)$ over v. The contributions from $\int f_1 f_2 \mathrm{d}v$ are zero unless $-k_2 v(t - T_2) + k_1 v(t - T_1) = 0$. Thus, with $k_2 > k_1$ at time t,

$$-k_2 v(t - T_2) + k_1 v(t - T_1) = 0 \Rightarrow t = \frac{k_1}{k_2 - k_1}(T_2 - T_1) + T_2 \tag{8.40}$$

the echo $n_e = n_1 \mathrm{e}^{\pm i\{(k_2-k_1)x}$ appears.

Einstein coefficients A and B. For the ratio of spontaneous emission A and stimulated emission and absorption b from level $|n_\mathbf{k}\rangle$ to the neighbouring levels $|n_\mathbf{k} - 1\rangle$ and $|n_\mathbf{k} + 1\rangle$, respectively, Einstein has derived the ratio

$$\frac{A}{B\hbar\omega} = \frac{1}{n_\mathbf{k}}; \quad n_\mathbf{k} = \frac{\omega^2 n_\omega}{\pi^2 c^3}.$$

Compare equation (9.7); $n_\mathbf{k}$ is the average photon number of the mode $\mathbf{k}$, the number of modes between $|\mathbf{k}|$ and $|\mathbf{k} + \mathrm{d}\mathbf{k}|$ is $(\mathbf{k}^2/\pi^2)d|\mathbf{k}| = n_\omega \mathrm{d}\omega$. It is instructive to investigate the relation between Einstein's rate equations and the two level dynamics (8.27). Under the assumption of the transverse relaxation time $T_2 \to 0$, i.e. $T_2 = \gamma_{12}^{-1} \ll T_{\text{measurement}}$, follows from (8.27)

$$\frac{\mathrm{d}\hat{\rho}_{21}}{\mathrm{d}t} = \frac{\mathrm{d}\hat{\rho}_{12}}{\mathrm{d}t}$$

and

$$\hat{\rho}_{12} - \hat{\rho}_{21} = \frac{i\kappa\hat{E}(\rho_{22} - \rho_{11})}{2(i\delta - \gamma_{12})} + \frac{i\kappa\hat{E}(\rho_{22} - \rho_{11})}{2(i\delta + \gamma_{12})} = i\delta\kappa\hat{E}(\rho_{22} - \rho_{11})\frac{1}{\delta^2 + \gamma_{12}^2}.$$

Substitution of this expression into the other pair of equations results in the rate equations

$$\begin{aligned}
\frac{\mathrm{d}\hat{\rho}_{11}}{\mathrm{d}t} &= +\frac{1}{2}\frac{\delta\kappa^2\hat{E}^2}{\delta^2 + \gamma_{12}^2}(\rho_{22} - \rho_{11}) - \gamma_1(\rho_{11} - n_1). \\
\frac{\mathrm{d}\hat{\rho}_{22}}{\mathrm{d}t} &= -\frac{1}{2}\frac{\delta\kappa^2\hat{E}^2}{\delta^2 + \gamma_{12}^2}(\rho_{22} - \rho_{11}) - \gamma_2(\rho_{22} - n_2).
\end{aligned} \tag{8.41}$$

In (4.132) the thermal equilibrium of n_1 and n_2 is formulated in terms of the principle of detailed balance. For general populations of a nondegenerate two level system holds according to Einstein

$$\frac{\mathrm{d}\hat{\rho}_{11}}{\mathrm{d}t} = +B_{12}\rho(\rho_{22} - \rho_{11}) + A_{12}\rho_{22}.$$
$$\frac{\mathrm{d}\hat{\rho}_{22}}{\mathrm{d}t} = -B_{21}\rho(\rho_{22} - \rho_{11}) - A_{12}\rho_{22}. \tag{8.42}$$

The radiation energy density ρ is proportional to $\hat{E}^2$. Apart from the depopulation rates with coefficients γ_1, γ_2 the sets of (8.41) and (8.42) coincide with each other.

The general dynamics of the two level system for optical transitions is capable of describing *coherent effects* as well as *incoherent effects* of standard optics. For decay times T_2 of the induced polarization much shorter than the intervals of measurement the optical Bloch equations reduce to the Einstein *rate equations*

8.1.4 Spontaneous Radiation from Single Particles

Accelerated charges radiate. The power emitted by an electron under nonrelativistic motion into the solid angle $\Omega = 4\pi$ is given by Larmor's formula

$$P(t) = \frac{2}{3}\frac{e^2}{4\pi\varepsilon_0 c^3}\dot{\mathbf{v}}^2(t). \tag{7.75}$$

The radiation emitted from an electric dipole $\mathbf{d}$ oscillating at frequency ω can be determined from the Liénard–Wiechert potentials (5.14) in the far field approximation. P follows also from (7.75). With the angle ϑ between amplitude $\hat{\mathbf{d}}$ and the direction of $\mathbf{k}^0 = \mathbf{k}/|\mathbf{k}|$ its intensity $\mathbf{I}$ at distance r and total power P are

$$\mathbf{I} = \overline{\mathbf{S}} = \mathbf{k}^0 \frac{\omega^4 d^2}{32\pi^2\varepsilon_0 c^3 r^2}\sin^2\vartheta, \quad P = \frac{\omega^4 d^2}{12\pi\varepsilon_0 c^3}; \quad d = |\hat{\mathbf{d}}|. \tag{8.43}$$

Applied to the free electron oscillating in the laser field P results with $d = e\hat{\delta}_{\mathrm{os}}$ as

$$P = \frac{e^4}{12\pi\varepsilon_0 m_e^2 c^3}\hat{E}^2 = \frac{e^4}{6\pi\varepsilon_0^2 m_e^2 c^4} I_{\mathrm{Laser}} \Rightarrow P[W] = 6.65 \times 10^{-25} I_{\mathrm{Laser}}[\mathrm{Wcm}^{-2}]. \tag{8.44}$$

The connection of P with I can be written as $P = \sigma_T I$, with σ_T the dimension of an area, and can be interpreted as the power scattered by the electron with cross section σ_T off the incident plane wave. Since σ_T is independent of frequency it has been interpreted as an indicator of the size of the electron. Introducing the *classical electron radius* r_e the relations hold

$$r_e = \frac{e^2}{4\pi\varepsilon_0 m_e c^2} = 2.818\,\mathrm{fm}, \quad \sigma_T = \frac{8\pi}{3} r_e^2 = 0.665\,\mathrm{barn}, \quad P = \sigma_T I_{\mathrm{Laser}}. \quad (8.45)$$

The mean oscillation energy of the single electron is $W = e^2\hat{E}^2/4m_e\omega^2$. The time τ_r elapsed until this energy has shrunk to e^{-1} is

$$\tau_r = \frac{W}{P} = \frac{6\pi\varepsilon_0 m_e c^3}{e^2\omega^2} \quad \Rightarrow \quad \tau_r[s] = 2.25 \times 10^{-8}(\lambda[\mu\mathrm{m}])^2. \quad (8.46)$$

It is equivalent to the number of oscillations $N = \omega\tau_r/2\pi = 6.75 \times 10^6(\lambda[\mu\mathrm{m}])$. In the nonrelativistic regime radiation damping is unimportant; it does not depend on the laser intensity. If the refractive index squared is significantly different from unity (8.43) and (8.44) have to be multiplied by η^{-2}.

From (5.14) one calculates the electric field of arbitrary electron motion in Feynman's representation as

$$\mathbf{E}(\mathbf{x}, t) = -\frac{e}{4\pi\varepsilon_0}\left\{\frac{\mathbf{r}^0}{r^2} + \frac{r}{c}\frac{\mathrm{d}}{\mathrm{d}t}\left(\frac{\mathbf{r}^0}{r^2}\right) + \frac{\mathbf{r}^0 \times [(\mathbf{r}^0 - \boldsymbol{\beta}) \times \dot{\boldsymbol{\beta}}]}{cr(1 - \mathbf{r}^0\boldsymbol{\beta})^3}\right\}_{t'} ; \mathbf{r} = \mathbf{x} - \mathbf{x}', r = |\mathbf{r}|. \quad (8.47)$$

Thereby $\dot{\boldsymbol{\beta}} = c^{-1}\mathrm{d}\mathbf{v}/\mathrm{d}t$, with t in the lab frame S. The bracket has to be calculated at the advanced time $t' = t - r(t')/c$; $\mathbf{r}^0 = \mathbf{r}/r$ is the unit vector at t'. The first two terms in the bracket of (8.47) represent the moving Coulomb field. In the far field they do not contribute to radiation because of their fast decay of $1/r^2$. The Poynting vector results as

$$\mathbf{S}(t) = \varepsilon_0 c^2 \mathbf{E} \times \mathbf{B} = \varepsilon_0 \mathbf{E}^2 \mathbf{r}^0 = \frac{e^2}{16\pi^2\varepsilon_0 c r^2(t')} \frac{\{\mathbf{r}^0 \times [(\mathbf{r}^0 - \boldsymbol{\beta}) \times \dot{\boldsymbol{\beta}}]\}^2}{(1 - \mathbf{r}^0\boldsymbol{\beta})^6}\Big|_{t'} \mathbf{r}^0(t'). \quad (8.48)$$

The magnetic field is $\mathbf{B} = \mathbf{r}^0 \times \mathbf{E}/c$. The intensity radiated into the solid angle $\mathrm{d}\boldsymbol{\Omega} = \mathbf{r}^0(t')\mathrm{d}\Omega$ as a function of t is

$$I(t)\mathrm{d}\boldsymbol{\Omega} = \mathbf{S}(t)r^2\big|_{t'}\mathrm{d}\boldsymbol{\Omega}. \quad (8.49)$$

The radiation is polarized in $\mathbf{E}$ direction with the two components along $\mathbf{r}^0 - \boldsymbol{\beta}$ and $\dot{\boldsymbol{\beta}}$; it is transverse to $\mathbf{r}^0$. The fields and the intensities are "diluted" or "compressed" depending on the motion of the charge relative to the observer. Often one wishes to calculate the amount of radiation emitted during the intrinsic time interval from t_1' to t_2' of acceleration. It differs from t_1, t_2 by the difference in retardation. The two intensities are related by

$$I(t)\mathrm{d}t = I(t')\mathrm{d}t' = I(t)\frac{\mathrm{d}t}{\mathrm{d}t'}\mathrm{d}t' = I(t)(1-\mathbf{r}^0\boldsymbol{\beta})\mathrm{d}t' \;\Rightarrow\; I(t') = (1-\mathbf{r}^0\boldsymbol{\beta})I(t). \tag{8.50}$$

To calculate the total power one may observe that the modulus of the four acceleration squared is the same in any inertial system, $A^2 = a_\alpha a^\alpha = \dot{\mathbf{v}}^2$ and P integrated over any finite time interval is a Lorentz scalar. Hence, the result is the same as for the charge at rest (7.75) if $\dot{\mathbf{v}}^2$ is expressed in covariant form of A^2 from (2.175),

$$P(t) = \frac{e^2}{6\pi\varepsilon_0 c^3}A^2 = \frac{e^2}{6\pi\varepsilon_0 c^3}\left\{\left[\frac{\mathrm{d}(\gamma\mathbf{v})}{\mathrm{d}\tau}\right]^2 - \beta^2\left[\frac{\mathrm{d}|\gamma\mathbf{v}|}{\mathrm{d}\tau}\right]^2\right\} \tag{8.51}$$

$$\Rightarrow \quad P(t) = \frac{e^2}{6\pi\varepsilon_0 c}\gamma^6|\dot{\boldsymbol{\beta}}^2 - (\boldsymbol{\beta}\times\dot{\boldsymbol{\beta}})^2|_{t'}. \tag{8.52}$$

This is the relativistic version of Larmor's formula (7.75), a result obtained already by Liénard in 1898.

In the special case of $\mathbf{v}$ and acceleration $\dot{\boldsymbol{\beta}}$ parallel to the observer in $\mathbf{r}^0$ direction the intensity is rotational symmetric around $\mathbf{v}$ with zero emission along this axis. The total radiation loss per unit time in the lab frame is P,

$$P = \kappa\gamma^6\dot{\beta}^2 = \frac{\kappa}{m_e^2c^2}\left(\frac{\mathrm{d}E}{\mathrm{d}x}\right)^2; \quad \kappa = \frac{e^2}{6\pi\varepsilon_0 c}, \quad E = \gamma mc^2 \tag{8.53}$$

i.e., P depends on the energy loss per unit length squared. P may be compared with the energy gain of the electron by acceleration,

$$\frac{P}{\dot{E}} = \frac{\kappa}{m_e^2c^2v}\frac{\mathrm{d}E}{\mathrm{d}x} \rightarrow \frac{\kappa}{m_e^2c^3}\frac{\mathrm{d}E}{\mathrm{d}x} = \frac{2}{3}\frac{r_e}{m_ec^2}\frac{\mathrm{d}E}{\mathrm{d}x}. \tag{8.54}$$

In *linear acceleration* the fractional radiation loss is $\Delta P = 3\times10^{-13}\Delta\epsilon$ if the change in acceleration energy $\Delta\epsilon$ per cm is expressed in units of the electron energy at rest $m_ec^2 = 0.511\,\mathrm{MeV}$. In *wake field acceleration* with $\Delta\epsilon = 10^4$ [cm^{-1}] it is negligibly small.

Alternatively, *transverse acceleration* leads to a different result. We consider circular motion at constant velocity. From (8.51) and $\mathrm{d}p/\mathrm{d}t = \omega p$ follows

$$P = \kappa\gamma^4\omega^2\beta^2. \tag{8.55}$$

For its application to the circularly polarized laser wave use is made of the familiar relations

$$\omega c\beta = \frac{e\hat{E}}{\gamma m_e}, \quad \gamma = (1+\hat{a}^2)^{1/2}, \quad \hat{a} = \frac{e\hat{A}}{m_e c}, \quad \hat{A} = \frac{\hat{E}}{\omega}, \quad W = m_ec^2\hat{a}$$

$$\Rightarrow \quad P = \frac{2}{3} m_e c r_e \omega^2 \hat{a}^4; \quad \hat{a}^2 = 7.3 \times 10^{-19} I\lambda^2[\text{Wcm}^{-2}\mu\text{m}^2]. \tag{8.56}$$

The radiative loss per cycle in units of the mean rotation energy W thus is

$$\frac{2\pi P/\omega}{W} = \frac{2}{3}(2\pi)^2 \frac{r_e}{\lambda} \hat{a}^3 = 4.7 \times 10^{-35} (I[\text{Wcm}^{-2}])^{3/2} (\lambda[\mu\text{m}^2])^2. \tag{8.57}$$

At an intensity $I_{\text{Nd}} = 10^{22}$ Wcm^{-2} the radiative loss amounts to 5% per cycle. It increases rapidly with intensity. In view of present developments of high power lasers the subject of radiation reaction, radiation losses and back reaction on the electron orbits deserves a more careful analysis in the following. This is true in particular for the most critical linear polarization.

With a look at an accelerated charge moving at velocity $\mathbf{v}$ one could be tempted to calculate the radiation pattern from the velocity addition theorem. Equation (8.48) tells that in contrast to Larmor's formula radiation is intensified by $\dot{\mathbf{v}}$ in collaboration with the acceleration $\mathbf{v}$. In the superrelativistic regime the denominator in (8.48) and γ^6 in (8.52) dominate. Thus, the transform from a charge at rest to its motion is more complex.

8.1.4.1 Directionality of Emission

According to Maxwells stress tensor (5.19) a pressure is associated with the emission of radiation which together with the Lorentz force provides for the momentum balance of the accelerated electron. As a consequence of the momentum conservation the electron moving along a straight line must radiate into forward direction. To investigate this quantitatively the situation of linear acceleration with $\mathbf{v}$, $\dot{\mathbf{v}}$, $\mathbf{r}^0$ all parallel to each other is considered. The intensity distribution $I(t)$, $I(t')$ per unit solid angle is rotationally symmetric relative to $\mathbf{v}$. If θ is the angle of $\mathrm{d}\boldsymbol{\Omega}$ with $\mathbf{v}$, from (8.48) and (8.50) results

$$I(t') = \frac{e^2\dot{\beta}^2}{16\pi^2\varepsilon_0 c} \frac{\sin^2\theta}{(1-\beta\cos\theta)^5}. \tag{8.58}$$

Maximum emission occurs under angle $\theta_{\max}$,

$$\cos\theta_{\max} = \frac{(1+15\beta^2)^{1/2}-1}{3\beta}; \quad \beta \to 1 \Rightarrow \theta_{\max} = \frac{1}{2\gamma}, \quad I(\theta_{\max}) \sim \gamma^8. \tag{8.59}$$

Integration of $I(t)$ reproduces the relativistic Larmor formula (8.52).

The results stipulate a comparison of the radiation losses of a particle at velocity v on an orbit of instantaneous curvature $\kappa = 1/R_c$ and acceleration $\dot{\mathbf{v}}$ normal to the orbit. The tangent vector $\mathbf{t} \parallel \mathbf{v}$ and the normal vector $\mathbf{n} \parallel \dot{\mathbf{v}}$ are the natural choice for the polar angles θ and ϕ of $\boldsymbol{\Omega}$. Equations (8.48) and (8.50) yield

$$I(t') = \frac{e^2\dot{\beta}^2}{16\pi^2\varepsilon_0}\frac{1}{(1-\beta\cos\theta)^3}\left\{1-\frac{\sin^2\theta\cos^2\phi}{\gamma^2(1-\beta\cos\theta)^2}\right\}. \quad (8.60)$$

In the relativistic limit $\gamma \gg 1$ holds $\gamma\dot{v} = (1/\gamma^2)\mathrm{d}\gamma/\mathrm{d}t$ and $I(t')$ is approximated by

$$I(t') \simeq \frac{e^2\dot{\beta}^2}{2\pi^2\varepsilon_0 c}\gamma^6\frac{1}{(1+\gamma^2\theta^2)^3}\left\{1-\frac{4\gamma^2\theta^2\cos^2\phi}{(1+\gamma^2\theta^2)^2}\right\}. \quad (8.61)$$

It indicates strong emission into forward direction. Integration of (8.60) over all angles leads to the emitted power $P(t')$,

$$P(t') = \frac{e^2\dot{v}^2}{6\pi\epsilon_0 c^3}\gamma^4 = \frac{e^2}{6\pi\epsilon_0 m_e^2 c^3}\gamma^2\left(\frac{\mathrm{d}\mathbf{p}}{\mathrm{d}t}\right)^2. \quad (8.62)$$

In collinear acceleration the power irradiated is according to (8.51)

$$P(t') = \frac{e^2}{6\pi\epsilon_0 m_e^2 c^3}\left(\frac{\mathrm{d}p}{\mathrm{d}t}\right)^2.$$

It reveals that at equal applied accelerating force transverse acceleration radiates γ^2 times as much as linear acceleration.

8.1.4.2 Spectral Range

The fundamental relation between the duration of a light pulse Δt and its frequency spread $\Delta\omega$ is derived for a Gaussian intensity distribution of the form $I_0 f(t)$,

$$f(t) = (2\pi\sigma)^{-1/2}\mathrm{e}^{-(\frac{t^2}{2\sigma})}, \quad \int_{-\infty}^{+\infty} f(t)\mathrm{d}t = 1; \quad \Leftarrow \int_{-\infty}^{+\infty} \mathrm{e}^{-y^2}dy = \pi^{1/2}.$$

The quantity $\sigma = \langle(t-\langle t\rangle)^2\rangle$ is the variance of $f(t)$,

$$\langle(t-\langle t\rangle)^2\rangle = \int_{-\infty}^{+\infty} t^2 f(t)\mathrm{d}t = \sigma, \text{ since } \int_{-\infty}^{+\infty} y^2\mathrm{e}^{-y^2}dy = \frac{\pi^{1/2}}{2}. \quad (8.63)$$

The square root of the variance can be taken as a measure of the pulse width $\Delta t = \sqrt{\sigma}$. The Fourier transform $g(\omega)$ of $f(t)$ is also Gaussian,

$$g(\omega) = (2\pi)^{-1/2}\int_{-\infty}^{+\infty} \mathrm{e}^{[-(\frac{t^2}{2\sigma})+i\omega t]}\mathrm{d}t = \sigma^{1/2}\mathrm{e}^{-\frac{\sigma\omega^2}{2}}.$$

$$\langle(\omega-\langle\omega\rangle)^2\rangle = \sigma^{1/2}\int_{-\infty}^{+\infty}\omega^2\mathrm{e}^{-\frac{\sigma\omega^2}{2}}\mathrm{d}\omega = \left(\frac{2\pi}{\sigma}\right)^{1/2} \Rightarrow \Delta\omega = \frac{1}{\sigma}. \quad (8.64)$$

Thus, for a Gaussian pulse the fundamental relation between pulse length and spectral width is

$$\Delta t \times \Delta\omega = 1 \quad \Leftrightarrow \quad \Delta x \times \Delta k = 1. \tag{8.65}$$

The second relation between length and momentum spread follows from $\Delta x = c\Delta t$ and $\Delta x = \Delta\omega/c$, or from the identical derivation (8.63) and (8.64) for Δx and Δk. Note, the variances and widths are calculated from normalized $f(t)$ and $g'(\omega) = (\sigma/2\pi)^{1/2} g(\omega)$. Frequently Δt and $\Delta\omega$ are defined by the width at which the Gaussian pulse assumes half of its maximum value "full half width" δt (*half width at half maximum*, HWHM). It relates to Δt above by

$$\delta t = \sqrt{2\ln 2}\,\Delta t = 1.18\,\Delta t, \quad \delta t \times \delta\omega = 2\ln 2 = 1.4. \tag{8.66}$$

Relation (8.65) is used to estimate the emitted frequency range from transverse acceleration.

From (8.61) the half width of the light cone angle is estimated to be $\Delta\theta'(t') \simeq 1/\gamma(t')$. During the time interval $\Delta t'$ the light beam rotates by the angle

$$\omega\Delta t' = \frac{v}{R_c}\Delta t' = \Delta\theta'(t') \simeq \frac{1}{\gamma(t')}.$$

Keeping in mind $\Delta t = 1 - \beta\cos\theta(t') \simeq 1/\gamma^2(t')$ an observer in the lab frame "sees" the pulse for $\Delta t \simeq \Delta t'/\gamma^2(t') = R_c/c\gamma^3(t')$. Hence, the associated spectral half width follows from (8.65),

$$\Delta\omega = \frac{1}{\Delta t} = \gamma^3(t')\frac{c}{R_c} = \gamma^3(t')\omega_0; \quad \omega_0 = \frac{c}{R_c}. \tag{8.67}$$

The frequency interval spans over $\Delta\omega \simeq \gamma^3$ times the fundamental frequency ω_0 of constant circular motion.

8.1.5 Bremsstrahlung from the Thermal Plasma

Bremsstrahlung in the restricted sense is the radiation from the elastic electron-ion collisions in the transparent plasma. In thermal equilibrium the radiation is isotropic and unpolarized. Hence the total power $P(t)$ from the unit volume is related to the intensity by $P(t) = 4\pi I(t)$. Moreover, the calculation of P is limited here to the nonrelativistic ideal plasma. The restrictions have several beneficial implications on how to do the summation over the single contributions (7.75). The fulfilment of properties (i–iii) in Sect. 7.1.1, and in particular (7.2) and (7.29), imply that the single radiative contributions add together to the *incoherent* intensity $I(t)$. The acceleration $\dot{\mathbf{v}}$ is due to the radial Coulomb force and consists of two contributions, the Coulomb attraction $\ddot{r}$ and the curvature centripetal acceleration $r\dot{\theta}^2$,

$$\dot{\mathbf{v}}^2 = \dot{v}^2 = (\ddot{r} - r\dot{\theta}^2)^2 = \frac{K}{r^4}; \qquad K = \left(\frac{Ze^2}{4\pi\varepsilon_0 m_e}\right)^2. \tag{8.68}$$

The majority of the orbits are nearly straight, therefore the contribution $r\dot{\theta}^2$ from close encounters is dropped and $\ddot{r}$ is calculated from Jackson's model from Sect. 1.1.4,

$$\int_{\infty}^{\infty} \dot{\mathbf{v}}^2 \mathrm{d}t = \frac{2K}{b^3 v}. \tag{8.69}$$

In spite of the time integral from $-\infty$ to $+\infty$ the expression has to be considered as localized at the instant t because the effective acceleration contributing to the integral extends only over $\Delta t \simeq 2b/v$; see also comment on (7.17). Expression (8.69) has to be integrated over all impact parameters b from zero to infinity. The singularity at $b = 0$ is a consequence of the straight line *and* the classical orbit approach. Under the assumption that the radiative contributions for $b < \lambda_B$ is not significant the radiated power from the electron crossing a thin plasma slab of thickness $\mathrm{d}x$ at velocity v is

$$P(v)\mathrm{d}x = \frac{2}{3}\frac{e^2}{4\pi\varepsilon_0 c^3} n_i dx \int_{\lambda_B}^{\infty} \frac{2K}{b^3 v} 2\pi b \mathrm{d}b = \frac{2}{3}\frac{4\pi e^6 Z^2}{(4\pi\varepsilon_0)^3 m_e^2 c^3 \lambda_B}\frac{\mathrm{d}x}{v}; \quad \lambda_B = \frac{\hbar}{m_e v}.$$

Setting $b_{\min} = \lambda_B$ makes $P(v)$ independent of v. The total emitted power from the fully ionized plasma in thermal equilibrium is obtained from folding $n_e v$ with the Maxwellian distribution f_M,

$$P = 12\left(\frac{2}{\pi}\right)^{1/2} \frac{e^6 Z^2 (k_B T_e)^{1/2}}{(4\pi\varepsilon_0)^3 m_e^{3/2} c^3 \hbar} n_e n_i; \quad \langle v \rangle = 2\left(\frac{2}{\pi}\right)^{1/2}\left(\frac{k_B T_e}{m_e}\right)^{1/2}. \tag{8.70}$$

Numerically this is

$$P[\mathrm{Wcm}^{-3}] = 6.2 \times 10^{-49} (T_e[\mathrm{eV}])^{1/2} Z^2 n_e[\mathrm{cm}^{-3}] n_i[\mathrm{cm}^{-3}]. \tag{8.71}$$

It grows with the square of the plasma density, which is characteristic of binary encounters, and with the square root of the electron temperature. In deriving (8.70) several assumptions have been made:

(i) No correction of the electron orbit due to radiative losses is considered. From (8.71) one estimates that at $T_e = 1\,\mathrm{keV}$ the single electron loses $10^{-9} Z^2$ of its average thermal energy through $10^{12} Z^2$ collisions per second in a $n_i = 10^{22}\mathrm{cm}^{-3}$ dense plasma. Under negligible radiation losses only the electrons have to be in thermal equilibrium when using (8.70); it does not matter whether the radiation decouples from matter or not.

(ii) Hard photons are emitted from close electron-ion encounters in which the centripetal acceleration $r\dot{\theta}^2$ prevails. These contributions, neglected in (8.70), are important when the spectrum of the *free - free* radiation is under inspection.

(iii) What is the value of Z if the plasma is not fully ionized? The effective charge

to be used is certainly higher than the ionization degree. Corrections are to be introduced in the Coulomb interaction as soon as the orbit of the colliding electron is deformed by the inner electron cloud.
(iv) No induced emission is considered. In the thermal plasma the overall bremsstrahlung emission may be that of a tenuous plasma. However, for some line emission the plasma may be opaque. Such frequencies must be subject to a special treatment (see Sect. 2.2.1). In applying the smooth bremsstrahlung formulas from above underestimates by a factor of 2 and more may result. Corrections related to points (i)–(iv) plus additional quantum corrections are introduced as adjustments under the name of so-called *Gaunt factors* σ.

8.1.5.1 The Free-Free Radiation Spectrum

The bremsstrahlung intensity emitted per unit frequency interval into the unit solid angle from a thin thermal plasma has been reported in (7.76), Sect. 7.2.5.4, in the context of *inverse bremsstrahlung absorption*,

$$I_\omega = \frac{\mathrm{d}^2 P}{\mathrm{d}\Omega d\omega} = \frac{\Re(\eta)\omega_p^2}{4\sqrt{3}\pi^2 c^3}\nu_{ei} k_B T_e \frac{\sigma(\omega, T_e)}{\ln\Lambda}\mathrm{e}^{-\hbar\omega/kT_e}.$$

The frequency dependent Gaunt factor is indicated by $\sigma(\omega, T_e)$. Most of the limitations for the validity of I_ω have been discussed there and in the foregoing paragraphs of the present chapter. There is one additional quantum correction to be mentioned. In contrast to the classical spectral analysis no photon $\hbar\omega$ exceeding the energy of the colliding electron can be emitted. A. Sommerfeld has derived a general nonrelativistic expression of I_ω respecting this limit for $\omega \gg \omega_p$ in terms of a hypergeometric function [9]. Gaunt factors for a wide interval of frequencies, impact parameters, and temperatures T_e are reported and discussed in [10]. For a comparison of the Born–Elwert approximation with Sommerfeld's result see [11].

8.2 Radiation Transport

The interplay of the radiation field with matter is twofold. Hot plasma emits incoherent unpolarized radiation of frequency ω from each point $\mathbf{x}$ at instant t into all directions and absorbs photons of the same frequency in $(\mathbf{x}, t)$. To avoid solving a 3D wave equation, and to make radiation transport feasible the treatment is limited to geometrical optics. In this model the energy propagates along the rays with group velocity $v_g = c\eta$, η refractive index. In the slowly varying inhomogeneous refractive index field $\eta(\mathbf{x}, t)$ the photon flux widens and narrows along the geometrical rays according to (5.156), (5.165), and Fig. 5.13. This represents the geometrical aspect

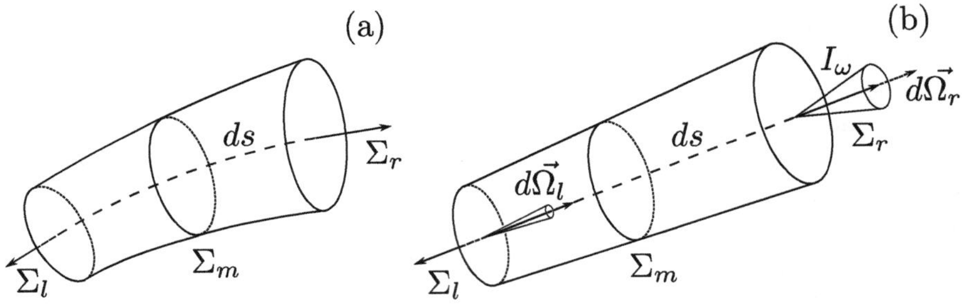

Fig. 8.3 Radiation transport. In the medium of varying refractive index $\eta(\mathbf{x})$ radiation propagates along curved rays. In **a** a ray bundle of intensity $I_{\omega,l}$ and angular spread $\mathrm{d}\boldsymbol{\Omega}_l$ enters the surface $\boldsymbol{\Sigma}_l$, propagates the distance ds under interaction with matter and leaves the surface $\boldsymbol{\Sigma}_r$ of the deformed truncated cone under $I_{\omega,r}\mathrm{d}\boldsymbol{\Omega}_r$. In **b** an equivalent regular truncated cone is sketched which exhibits identical, now parallel surfaces $|\mathrm{d}\boldsymbol{\Sigma}_l|$, $|\mathrm{d}\boldsymbol{\Sigma}_r|$ and identical length ds. The cones (**a**) and (**b**) differ in volume V, mantel, and middle cross section Σ_m only in second order from each other; all considerations on energy fluxes can be done in the rectified Fig. (**b**). The radiation flux is modified by the net absorption $\alpha_\omega V$ and the spontaneous emission $\varepsilon_\omega V$. The energy flow is parallel to the mantel. The refractive index is assumed to decrease from left to right. Correspondingly, the aperture θ of $\mathrm{d}\boldsymbol{\Omega}$ increases along the ray path

of photon transport. With increasing frequency ω the refractive index approaches unity, the rays become straight and the photons propagate with vacuum velocity c.

8.2.1 The Transport Equation

Consider a narrow bundle of rays, sketched in Fig. 8.3a. The energy flux dF measured in the small frequency interval dω depends on the aperture and direction $\mathrm{d}\boldsymbol{\Omega} = (\sin\theta\mathrm{d}\theta\mathrm{d}\varphi)\boldsymbol{\Omega}$ and on the emitting area $\mathrm{d}\boldsymbol{\Sigma}$ parallel to the ray,

$$\mathrm{d}F = I_\omega \mathrm{d}\omega \mathrm{d}\boldsymbol{\Omega}\mathrm{d}\boldsymbol{\Sigma};\quad I_\omega \mathrm{d}\omega \mathrm{d}\Omega = \frac{\mathrm{d}F}{\mathrm{d}\Sigma};\quad \mathrm{d}\Omega = |\mathrm{d}\boldsymbol{\Omega}|,\ \mathrm{d}\Sigma = |\mathrm{d}\boldsymbol{\Sigma}|. \tag{8.72}$$

By the second relation the *spectral intensity* I_ω is defined. In conformity with the units of [Wcm^{-2}] for the Poynting vector of a plane wave the dimensions of I_ω are [Wcm^{-2}s sterad^{-1}] $\leftrightarrow$ [gs^{-2}]. The direction of I_ω is the unit vector $\boldsymbol{\Omega} = \mathbf{k}/|\mathbf{k}|$ of (5.152). The balance of the energy fluxes between the entrance $\mathrm{d}\Sigma_l$ on the left and the exit $\mathrm{d}\Sigma_r$ on the right is

$$\Delta P = I_{\omega r}\boldsymbol{\Omega}_r \mathrm{d}\Omega_r \boldsymbol{\Sigma}_r + I_{\omega l}\boldsymbol{\Omega}_l \mathrm{d}\Omega_l \boldsymbol{\Sigma}_l = I_{\omega r}\mathrm{d}\Omega_r \Sigma_r - I_{\omega l}\mathrm{d}\Omega_l \Sigma_l. \tag{8.73}$$

The vectorial quantity $\mathrm{d}\boldsymbol{\Omega} = (\theta, \phi)$ indicates a unit vector of direction like the often used symbol $\mathbf{n} = (n_1, n_2, n_3) = (\sin\theta\cos\phi, \sin\theta\sin\phi, \cos\theta)$. The scalar quantity $\mathrm{d}\Omega = \sin\theta \mathrm{d}\phi \mathrm{d}\theta$ is the infinitesimal aperture of rays of I_ω emanating from a point $\mathbf{x}$ in space

Dependence on refractive index η. If there is no emission and absorption follows $\Delta P = 0$ and $I_\omega \mathrm{d}\Omega\,\Sigma = \text{const}$ along the propagating energy pulse. The cross section Σ_r is $\Sigma_l + \varepsilon(ds)$. With decreasing cross section all rays in the bundle become nearly parallel, i.e., $\varepsilon(ds)$ is of first order, directly apparent from Fig. 8.3b. In the limit of a slender bundle $\Sigma_l = \Sigma_r = \Sigma_m$ can be set. The aperture $\mathrm{d}\Omega$ behaves differently. It is determined by the aperture of the measuring equipment. In the *layered medium* holds (5.157), $\eta(s)\sin\theta = \text{const}$ along the ray. In a medium with arbitrary isotropic refractive index $\eta(\mathbf{x}, t)$ stratification is fulfilled locally in the neighbourhood of the plane tangent to the ray. For a slender bundle of rays with small aperture the angle θ fulfils

$$\eta\sin\theta = C = \text{const},\;\; \sin\theta \simeq \theta,\;\; \mathrm{d}\Omega = C^2\theta^2,\quad (8.73) \Rightarrow\quad \frac{I_\omega}{\eta^2} = \text{const}. \quad (8.74)$$

Hence, the change of I_ω along $\mathrm{d}s$ caused by the refractive index alone is $\partial I_\omega/\partial s = \text{const}\times\partial\eta^2/\partial s = (I_\omega/\eta^2)\partial\eta^2/\partial s$. The total change of I_ω per unit lengths is given by

$$\frac{\mathrm{d}I_\omega}{\mathrm{d}s} = \frac{\partial I_\omega}{\partial s} + \frac{1}{\eta^2}(\varepsilon_\omega - \alpha_\omega I_\omega) \quad\Leftrightarrow\quad \eta^2\frac{(\mathrm{d}I_\omega/\eta^2)}{\mathrm{d}s} = \varepsilon_\omega - \alpha_\omega I_\omega. \tag{8.75}$$

The partial derivative $\partial_s I_\omega$ accounts for the variable beam spread $\mathrm{d}\Omega$. The source terms ε_ω, $\alpha_\omega I_\omega$ stand for the spectral emission and the net spectral absorption. The equivalent second version in (8.75) tells that the derivative of the intensity I_ω is to be replaced by I_ω/η_ω^2 along curved rays. The conservation of this quantity has to obey (3.3), see Fig. 3.1,

$$\eta^2\frac{\partial}{\partial t}\Big(\frac{I_\omega}{v_g\eta^2}\Big) + \eta^2\nabla\Big(\frac{\boldsymbol{\Omega} I_\omega}{\eta^2}\Big) = \varepsilon_\omega - \alpha_\omega I_\omega \qquad (\nabla = \text{div}).$$

The total change of I_ω per unit time results as (i) change by emission ε_ω into the cylinder and net absorption $\alpha_\omega I_\omega$ there, (ii) change of beam spread by η varying in space, (iii) change in time of η.

If $|\text{grad}\,\eta|/\eta^3 \ll |\text{grad}\,I_\omega|/I_\omega$ the direction $\boldsymbol{\Omega}$ is sufficiently constant and $\nabla(\boldsymbol{\Omega} I_\omega) = \boldsymbol{\Omega}\,\text{grad}\,I_\omega = dI_\omega/ds$ can be set. Then, with $\rho_\omega(\boldsymbol{\Omega}) = I_\omega/v_g$, v_g group velocity, for the spectral radiation energy density the general *spectral radiation transport equation* reads

$$\eta^2\frac{\partial}{\partial t}\Big(\frac{\rho_\omega(\boldsymbol{\Omega})}{\eta^2}\Big) + \eta^2\frac{\mathrm{d}}{\mathrm{d}s}\Big(\frac{I_\omega}{\eta^2}\Big) = \varepsilon_\omega - \alpha_\omega I_\omega. \tag{8.76}$$

The photon density $\rho_\omega(\boldsymbol{\Omega})$ is to be understood as chosen into the ray direction. The emission ε_ω is the *spontaneous spectral emission* and as such it is isotropic. If the *total spectral absorption* is indicated by κ_ω the *net absorption* α_ω is the total spectral absorption diminished by the *induced re-emission* discovered by Einstein, see Chap. 4. For κ_ω and α_ω isotropy can be assumed. If $\rho_\omega(\boldsymbol{\Omega})$ is isotropic it becomes $\rho_\omega/4\pi$, like (4.122) for the Planckian radiator with ρ_ω the spectral radiation density for all directions.

At high frequencies η approaches unity and $v_g \rightarrow c$. The transport law simplifies to

$$\frac{\partial}{\partial t}\rho_\omega(\boldsymbol{\Omega}) + \frac{\mathrm{d}}{\mathrm{d}s}I_\omega = \varepsilon_\omega - \alpha_\omega I_\omega; \quad \eta \longrightarrow 1. \tag{8.77}$$

Spectral photon energy conservation. We define the *spectral radiant energy density* ρ_ω,

$$\rho_\omega = \int_{4\pi} \rho_{\boldsymbol{\Omega}} \mathrm{d}\boldsymbol{\Omega} = \frac{1}{c}\int_{4\pi} I_\omega \mathrm{d}\Omega; \quad \eta = 1 \tag{8.78}$$

and the net *spectral radiant flux density* $\mathbf{S}_\omega$,

$$\mathbf{S}_\omega = \int_{4\pi} \mathbf{I}_\omega \mathrm{d}\Omega; \quad \mathbf{I}_\omega = I_\omega \boldsymbol{\Omega} \tag{8.79}$$

over all directions. The total energy flux density through an oriented surface $\mathrm{d}\boldsymbol{\Sigma}$ is $\mathbf{S}_\omega \mathrm{d}\boldsymbol{\Sigma}$. It is the difference of flux into direction of $\mathrm{d}\boldsymbol{\Sigma}$ and of flux into opposite direction. In the Planckian hohlraum $\mathbf{S}_\omega$ is zero. Upon integrating (8.77) over the solid angle 4π the all direction transport equation becomes

$$\frac{\partial}{\partial t}\rho_\omega + \nabla \mathbf{S}_\omega = 4\pi\varepsilon_\omega - \alpha_\omega \int I_\omega \mathrm{d}\Omega; \tag{8.80}$$

Kirchhoff's law. The source terms can be cast into a form which elucidates the uniqueness of the black body radiator. Regardless of thermalization of the source or what its temperature is Kirchhoff's law (4.135) allows to link spontaneous emission ε_ω with absorption $\alpha_\omega I_{\omega,P}$ by means of the Planck radiator $I_{\omega,P} = I(\omega, T)$,

$$\frac{\varepsilon_\omega}{\alpha_\omega} = I_{\omega,P} = \frac{c}{4\pi}\hbar\rho_{\omega,P}. \tag{8.81}$$

With this (8.77) assumes the standard form

$$\frac{\partial}{\partial t}\rho_\omega(\boldsymbol{\Omega}) + \frac{\mathrm{d}}{\mathrm{d}s}I_\omega = \alpha_\omega(I_{\omega,P} - I_\omega); \qquad \eta = 1. \tag{8.82}$$

Re-emission of radiation, hidden in $\alpha_\omega I_\omega$, is not a temporal process with absorption first and emission with a delay. Stimulated emission happens coherently with the incident radiation. It is therefore physically more appropriate to express the sum of spontaneous emission ε_ω and induced emission altogether by a symbol ε_ω^{tot}. With it Kirchhoff's law reads

$$\frac{\varepsilon_\omega^{tot}}{\kappa_\omega} = I_{\omega,P}; \tag{8.83}$$

The two formulations of Kirchhoff's law (8.81) and (8.83) are equivalent. Spontaneous emission ε_ω is a property of isolated matter and accessible to a measurement. Absorption α_ω depends on the embedding radiation field I_ω. In (8.83) κ_ω can be measured and is a property of matter whereas ε_ω^{tot} is not, except the stimulated emission can be ignored.

It is worth considering once more Kirchhoff's law in terms of the *Einstein coefficients* A_{12} for *spontaneous emission* $\hbar\omega = E_2 - E_1$ from state $|n_2\rangle$ to state $|n_1\rangle$ and B_{12}, B_{21} for the *stimulated processes* between these levels. In these terms (8.81) reads

$$I_{\omega,P} = \frac{c}{4\pi}\frac{A_{12}\,n_2}{B_{21}\,n_1 - B_{12}\,n_2} = \frac{A_{12}\,n_2}{\frac{4\pi}{c}B_{21}\,n_1(1 - n_2/n_1)} = \frac{A_{12}\,n_2}{\kappa_\omega(1 - n_2/n_1)}. \tag{8.84}$$

Kirchhoff's law appears as the ratio of spontaneous emission ε_ω to absorption with re-emission $\alpha_\omega = \kappa(1 - n_2/n_1)$. The number of modes **k** into which a photon $\hbar\omega$ can be spontaneously emitted is $n_\omega = \omega^2/(\pi^2c^3)$. The induced emission per unit volume and unit time is $\varepsilon_\omega^{ind} = B_{12}\rho_\omega\hbar\omega n_2$. We recall from (4.132) and (9.7)

$$\varepsilon_\omega = A_{12}\hbar\omega\, n_2,\ \ \varepsilon_\omega^{ind} = B_{12}\rho_\omega\hbar\omega\, n_2,\ \ \frac{\varepsilon_\omega^{tot}}{\kappa_\omega} = I_{\omega,P}\ \Rightarrow\ \kappa_\omega = \frac{4\pi}{c}B_{12}n_1\hbar\omega. \tag{8.85}$$

In thermal equilibrium the spectral energy density is Planckian, $\rho_\omega = \rho_{\omega,P}$, hence

$$\frac{\varepsilon_\omega + B_{12}\rho_{\omega,P}\hbar\omega\, n_2}{B_{12}n_1\hbar\omega} = \rho_{\omega,P}\ \Rightarrow\ \varepsilon_\omega = \rho_{\omega,P}B_{12}(n_1 - n_2)\hbar\omega\ \Rightarrow\ \frac{\varepsilon_\omega}{\alpha_\omega} = I_{\omega,P}. \tag{8.86}$$

The last equality proves the equivalence. At $n_2 = n_1$ absorption from level $|n_1\rangle$ equals stimulated emission from level $|n_2\rangle$. With the upper level occupied n_2-fold it reduces by the ratio n_2/n_1. The Einstein coefficients of spontaneous and induced emission A_{12}, B_{12} are of general validity for use in rate equations (see Chap. 4 and system of (8.42)), not limited to thermodynamic equilibrium. Under thermal equilibrium the *canonical partition function* applies to the two levels and $\alpha_\omega = \kappa_\omega(1 - \exp\hbar\omega/k_BT_e)$ is recovered.

The proof of equivalence shows that any linear combination $(\lambda_1 \varepsilon_\omega^{\text{tot}} + \lambda_2 \varepsilon_\omega)/(\lambda_1 \kappa_\omega + \lambda_2 \alpha_\omega)$ is a valid formulation of *Kirchhoff's law*. For example

$$\frac{\varepsilon_\omega^{\text{tot}}}{\kappa_\omega} - \frac{\varepsilon_\omega}{\alpha_\omega} = \frac{\varepsilon_\omega^{\text{ind}}}{(4\pi/c)B_{12}n_2} = I_{\omega,P} \tag{8.87}$$

is Kirchhoff's law for stimulated emission

8.2.2 *Thermal Radiation from a Plane Layer*

A plane layer is considered with its surface at $x = 0$ and finite thickness d. The temperature is assumed to vary arbitrarily in the depth but to be uniform in y, z direction, $T = T(x)$. The *emittance* i_ω into direction θ of the surface intensity $I_\omega(\theta)$, respectively of $I_{\omega,P} = I(\omega, T)$ is defined as the quantity

$$i_\omega = I_\omega(\theta)\cos\theta, \quad \text{resp} \quad i_{\omega,P} = I_{\omega,P}\cos\theta. \tag{8.88}$$

It is the energy flux per unit time through the unit surface area under the angle θ into the angle $d\Omega$. The second relation, $i_{\omega,P} = I_{\omega,P}\cos\theta$ expresses *Lambert's emission law* of a radiator or scatterer radiating uniformly into all directions. At depth x the radiation element $\varepsilon_\omega(x)ds = \alpha_\omega(x)I_{\omega,P}(x)ds$ is emitted and attenuated along the segment s from $-x/\cos\theta$ to the surface $x = 0$, see Fig. 8.4. The attenuation is given by Beer's law $\exp[-\int \alpha_\omega(s)ds]$. The quantity $\tau_\omega(s)$,

$$\frac{1}{s}\int_{-s}^{0} \alpha_\omega(s)ds = \tau_\omega(s); \quad l_\omega(s) = \frac{1}{\tau_\omega(s)} \tag{8.89}$$

is the *optical thickness* of the segment s. Its inverse $l_\omega(s)$ is the *mean free path* of the photon $\hbar\omega$. The intensity arriving at the surface from the black body radiator $I(\omega, T)$ is, under steady state conditions and $\eta = 1$, the sum of all contributions along the segment s,

$$\begin{aligned} I_\omega(\theta) &= \int_{-s(-d)}^{s(0)} \alpha_\omega(x)I_{\omega,P}(x)\exp\left[-\int_{-s(x)}^{s(0)} \alpha_\omega(x')ds'\right]ds \\ &= \int_{-d}^{0} \alpha_\omega(x)I_{\omega,P}(x)\exp[-\tau_\omega(x)]d\frac{x}{\cos\theta}. \end{aligned} \tag{8.90}$$

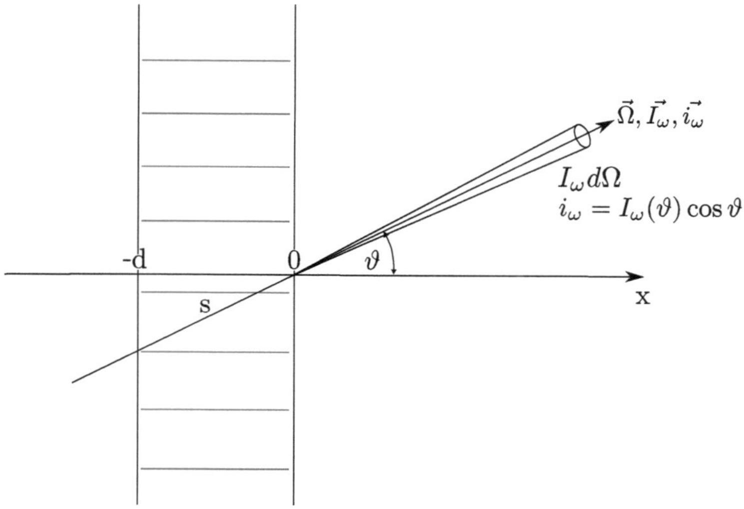

Fig. 8.4 Thermal radiation from a plane layer. The radiation $I_\omega(\theta)$ accumulated along an extended path s into direction $\boldsymbol{\Omega}$ becomes Planckian at the surface ($x = 0$) either under grazing emission or great optical thickness. The emittance into direction $\boldsymbol{\Omega}$ is $i_\omega = I_\omega(\theta)\cos\theta$

Under the conditions of $\alpha_\omega =$ const and $T =$ const in the layer follows $\tau_\omega = \alpha_\omega d/\cos\theta$ and the spectral surface intensity in direction θ becomes

$$I_\omega(\theta) = I_{\omega,P}\left(1 - \mathrm{e}^{-\frac{\tau_\omega}{\cos\theta}}\right). \tag{8.91}$$

I_ω approaches $I_{\omega,P}$ under grazing incidence, and at normal incidence also if its thickness d becomes multiples of the optical thickness, $d \gg l_\omega$. Thus, the surface of the optically thick layer is a black radiator independent of direction.

In the optically very thin layer, $\alpha_\omega d/\cos\theta \ll 1$, the exponent of (8.91) can be expanded, $I_{\omega,P}(1 - \mathrm{e}^{-\frac{\tau_\omega}{\cos\theta}}) = I_{\omega,P}\tau_\omega/\cos\theta \ll I_{\omega,P}$, and the surface emittance $i = I_{\omega,P}\,\tau_\omega$ is angle independent, and much weaker than $i_{\omega,P}$.

The optically thick layer radiates from the surface because photons emitted from depth $d > l_\omega$ are exponentially attenuated. The higher the net absorption α_ω the faster the transition of I_ω to $I_{\omega,P}$ with increasing thickness d

Expression for $I_\omega(x = 0)$ is easily generalized for the unsteady situation with $\partial_t\rho$ included in (8.82). Radiation $\alpha_\omega(-s,t)I_{\omega,P}(-s,t)$ emitted from $-s$ at instant $t = -s/c$ propagates to the surface with attenuation

$$\tau_\omega(-s,-s/c)=\int\limits_{-s,-s/c}^{0,0}\alpha_\omega(-s',-s'/c)\mathrm{d}s'. \tag{8.92}$$

The intensity at the surface is the sum of all attenuated contributions along the distance s_0,

$$I_\omega(s=0,t=0)=\int\limits_{-s_0,-s_0/c}^{0,0}\alpha_\omega(-s,-s/c)I_{\omega,P}(-s,-s/c)\mathrm{e}^{-\tau_\omega(-s,-s/c)}ds. \tag{8.93}$$

Impact of plasma dynamics. Radiation transport is based on geometrical optics. For this to make sense $\omega=-\partial\Phi/\partial t$ and $\eta/\dot{\eta}$ of the refractive index must change much more slowly in time than $1/\omega$ (see the "classical photon" in Chap. 5), and the plasma flow $\mathbf{u}$ should remain well below the light velocity. In addition it can be assumed that changes of the three quantities and the source term α_ω are smooth and limited in space and time. These may be the elements that influence mainly the time derivative in (8.76). The influence of motion on η has been discussed already at the beginning of Sect. 5.5. Here only the impact of motion $\mathbf{u}$ on the partial time derivatives in (8.76) and (8.82) is explicitly shown. For this purpose it is sufficient to estimate it for α_ω in (8.92). We compare its time dependent value for the increment of one period $\Delta t=2\pi/\omega$ with the static expression $\alpha_0=\alpha_\omega(-s',-s/c)$,

$$\int\limits_{-s,-s/c}^{0,0}\alpha_\omega(-s',-s'/c)\mathrm{d}s'=\int\limits_{-s,-s/c}^{0,0}\Big(\alpha_0+\frac{\partial\alpha_\omega(-s',-s/c)}{\partial t}\Delta t\Big)\mathrm{d}s'.$$

It is reasonable to assume $\alpha_\omega\sim n_e$ and to infer $\partial_t n_e\sim n_e u/\lambda$ from electron particle conservation $\partial_t n_e+\nabla n_e\mathbf{u}_e=0$, λ wavelength. Its substitution above yields

$$\int\limits_{-s,-s/c}^{0,0}\Big(\alpha_0+\frac{\partial\alpha_\omega(-s',-s/c)}{\partial t}\Delta t\Big)\mathrm{d}s'\sim\int\limits_{-s,-s/c}^{0,0}\alpha_0\Big(1+\frac{u}{c}\Big)\mathrm{d}s'. \tag{8.94}$$

As long as $u=|\mathbf{u}|$ remains much below c and the constraints of the geometrical optics are fulfilled the term containing the partial time derivative in (8.76) and (8.82) can be dropped; the standard radiation transport equation reduces to

$$\frac{\mathrm{d}}{\mathrm{d}s}I_\omega=\alpha_\omega(I_{\omega,P}-I_\omega);\qquad \eta=1. \tag{8.95}$$

8.2.2.1 The Surface Brightness

Consider a plane layer and the radiation of frequency ω irradiated from it into the vacuum half space per unit time. Its *brightness* B_ω is given by the flux of the surface intensity $I_\omega(\theta)$ from (8.90) integrated over the half space,

$$B_\omega = \int_{2\pi} i_\omega(\theta)\mathrm{d}\Omega = 2\pi \int_0^{\pi/2} I_\omega(\theta)\cos\theta \sin\theta \mathrm{d}\theta. \tag{8.96}$$

In the case of a half space of uniform temperature $B_\omega = c\rho_{\omega,P}/4$. The brightness is frequency dependent. The effective optical thickness $\overline{\tau}_\omega(x)$ and mean free path $\overline{l}_\omega(x)$ result from (8.90) by integration over the emission angle $\theta = \pi/2$. Evidently $\overline{l_\omega}(x)$ is shorter than $l_\omega(x)$ for perpendicular emission. In the case the curvature radius of a nonplanar surface is longer than the characteristic $\overline{l}_\omega$, expression (8.96) applies to curved surfaces also.

A convenient measure of B_ω is the *brightness temperature* $T_{B,\omega}$. It is defined by the equivalent emission of the black body radiator $I_{\omega,P}$ into the half space,

$$B_\omega = I_{\omega,P}(T_{B,\omega}). \tag{8.97}$$

It depends on frequency. An integrated brightness temperature T_B can also be introduced by setting

$$B = \sigma T_B^4, \tag{8.98}$$

where B is the integrated energy flux emitted by the unit surface of the body under consideration. The brightness is of particular relevance for the two cases of a self heated half space and a half space irradiated from outside by heat conduction or soft X ray radiation.

In the first case the standard temperature profile decreases towards the surface to the vacuum, see Fig. 8.5a. For a given frequency the radiation through the surface comes from an optical thickness $\tau_\omega(x) \simeq (2-3)$, radiation from deeper layers undergoes self attenuation and does almost not contribute to B. With respect to the mean brightness temperature T_B from (8.98), dashed Planck curve, in the low frequency domain absorption is stronger and comes from a region $T = T_{B,\omega} < T_B$, see bold curve. Consequently, from some higher frequency on the low frequency deficit is compensated by $T_{B,\omega} > T_B$. By definition, dashed and bold line cover the same area. In the second case, Fig. 8.5b, relevant to the situation of *hohlraum radiation* for indirectly driven inertial fusion the two curves almost coincide.

Starting from an extended optically thin plasma one may ask at which frequencies it become first opaque and Planckian if the plasma density is increased continuously. Comparing free - free absorption, for example (7.80), with the transition probability (8.21) for line radiation from atoms and ions one arrives at the two absorption coefficients α and α_e between non degenerate levels in their simplest form at

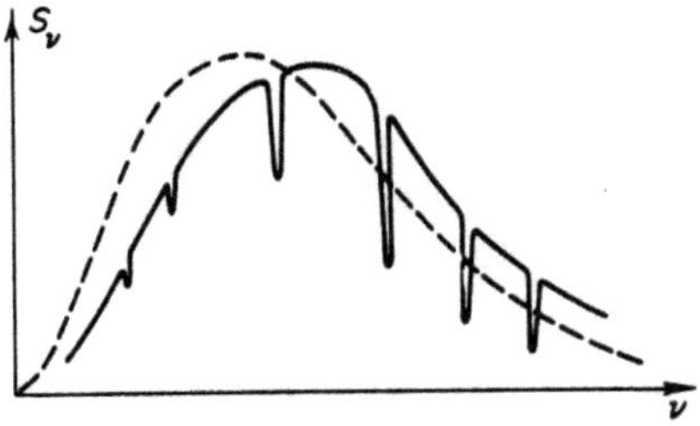

Fig. 8.5 Surface brightness B_ω (**a**) of a temperature distribution falling towards the vacuum boundary (bold curve). Its area equals the Planck distribution reached at some point in the half space interior. The brightness temperature of a half space heated from outside (**b**) almost coincides with the Planck curve of a temperature close to the surface (dashed line). The line radiation exhibits the shortest mean free path with a temperature very close to the surface temperature T_0 in (**a**). The brightness integrated over the line profile equals $I_{\omega,P}(T_0)$. After [20]

Fig. 8.6 With optical thickness increasing spectral lines become black first. After [20]

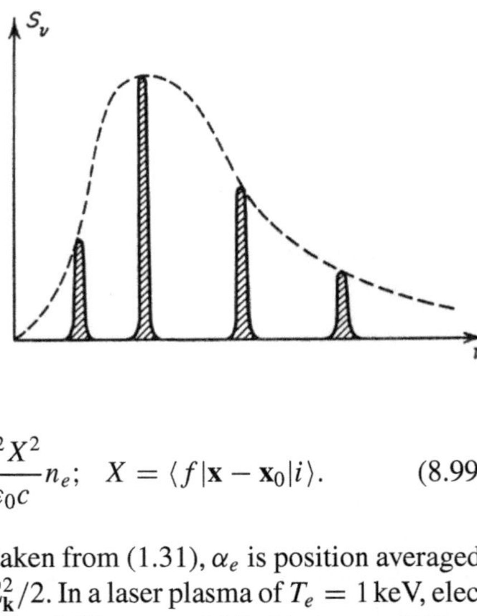

$$\alpha = \frac{\omega_p^2}{\omega^2}\frac{\nu_{ei}}{c}, \qquad \alpha_e = \frac{2\pi}{\hbar}\frac{e^2X^2}{\varepsilon_0 c}n_e; \quad X = \langle f|\mathbf{x}-\mathbf{x}_0|i\rangle. \tag{8.99}$$

The free - free absorption coefficient α is taken from (1.31), α_e is position averaged, both refer to a mode of intensity $I_\mathbf{k} = \varepsilon_0 c\mathbf{E}_\mathbf{k}^2/2$. In a laser plasma of $T_e = 1$ keV, electron density $n_e = 10^{22}$ cm^{-3}, $\ln\Lambda = 5$, $X = 0.01$ nm, and 10 eV photons follows in [cm^{-1}]

$$\alpha = 2.6\times 10^{-43}Zn_e^2 = 26Z; \qquad \alpha_e = 1.2\times 10^{-18}n_e = 1.2\times 10^4. \tag{8.100}$$

As to be expected lines become opaque first, the mean free photon paths l_ω are $400/Z$ µm versus 80 µm. Note $\alpha/\alpha_e \sim n_e$ (Fig. 8.6. In resonance lines X^2 may exceed its value used here easily by a factor of 10. At 10 fold lower electron density l_ω increases to 40/Z mm and 0. 8 mm.

8.2.3 Diffusion Model of Radiation Transport

Heat propagation (1.63) may serve as a prototype of a diffusion equation. For such an equation to be valid the net local energy flux into an arbitrary direction must be small compared to its local modulus and the gradient of the local energy density times the mean free path has to represent only a small deviation from its homogeneous distribution. In order to apply the idea to radiation transport a measure for the local inhomogeneity of the radiation field must found first and then a relation connecting I_ω with the local radiation density ρ_ω must be established.

In vectorial form (8.95) reads

$$\nabla \boldsymbol{I}_\omega = \alpha_\omega (I_{\omega,P} - I_\omega); \qquad \boldsymbol{I}_\omega = I_\omega \boldsymbol{\Omega}, \;\; \nabla(I_\omega \boldsymbol{\Omega}) = \boldsymbol{\Omega}\,\mathrm{grad} I_\omega. \tag{8.101}$$

Multiplication of $\boldsymbol{\Omega}$ with $\nabla \boldsymbol{I}_\omega$ and its source term, and integration of the vector equation over all directions yields

$$\int \boldsymbol{\Omega}(\boldsymbol{\Omega}\,\mathrm{grad} I_\omega)\mathrm{d}\boldsymbol{\Omega} = -\alpha_\omega \int \boldsymbol{\Omega} I_\omega \mathrm{d}\boldsymbol{\Omega} = -\alpha_\omega \mathbf{S}_\omega. \tag{8.102}$$

The term containing $I_{\omega,P}$ vanishes for isotropy. The balance equation is exact within its limits ($\eta = 1$, quasi steady state) and provides a measure of flux anisotropy and spatial inhomogeneity. The integration on $\boldsymbol{\Omega}$ becomes simple under the assumption that I_ω does not depend on direction,

$$\int \boldsymbol{\Omega}(\boldsymbol{\Omega}\,\mathrm{grad} I_\omega)\mathrm{d}\boldsymbol{\Omega} = 2\pi \nabla I_\omega \int_0^\pi \cos^2\theta \sin\theta \mathrm{d}\theta = \frac{4\pi}{3}\nabla I_\omega = \frac{c}{3}\nabla \rho_\omega \tag{8.103}$$

In the real situation isotropy of $\mathbf{I}_\omega$ may not be fulfilled, however quite often the deviations $|\Delta \mathbf{I}_\omega|$ are small in comparison to I_ω so that in this case the integral on the left is still a good approximation and can be equated to the RHS of (8.102) to yield the desired relation

$$\mathbf{S}_\omega = -\frac{l_\omega c}{3}\nabla \rho_\omega. \tag{8.104}$$

It looks similar to Fourier's assumption of heat conduction (1.59) but here it will also depend on density. In optically thick matter l_ω is small and the diffusion approximation is well satisfied. In other words, the smaller the net flux is compared to the spectral radiation density over the gradient length L_ω the smaller the anisotropy in $\mathbf{S}_\omega$ results, $|\mathbf{S}_\omega| = c(l_\omega/3L_\omega)\rho_\omega$. At the matter-vacuum interfaces α_ω drops to zero and the radiation flux is very anisotropic. The difficulties with the diffusion approximation may be bridged by appropriate boundary conditions. For comparison, the quasi steady photon conservation equation integrated over all directions (8.80) reads in the general case

$$\nabla \mathbf{S}_\omega = \alpha_\omega c(\rho_{\omega,P} - \rho_\omega). \tag{8.105}$$

8.2.3.1 The Rosseland Mean Free Path

So far spectral transport equations have been presented for various approximations. In a first study of thermal radiation one may be interested in the distribution averaged over emission angle and the whole spectrum $\mathbf{S} = \int \mathbf{S}_\omega \mathrm{d}\omega$. For this purpose l_ω is to be averaged in a suitable manner. In the radiation field not far from radiative equilibrium (8.104) may be approximated by

$$\mathbf{S}_\omega = -\frac{l_\omega c}{3} \nabla \rho_{\omega,P}. \tag{8.106}$$

Denoting the average mean free path of a photon by l_R one requires

$$\mathbf{S} = -\frac{l_R c}{3} \nabla \rho_P = -\frac{16 l_R}{3} \sigma T^3 \nabla T \tag{8.107}$$

and

$$\mathbf{S} = -\frac{c}{3} \int l_\omega \nabla \rho_{\omega,P} \mathrm{d}\omega. \tag{8.108}$$

Equating the two expressions of $\mathbf{S}$,

$$l_R \frac{\mathrm{d}\rho_P}{\mathrm{d}T} = l_R \int_0^\infty \frac{\partial \rho_{\omega,P}}{\partial T} \mathrm{d}\omega = \int l_\omega \frac{\partial \rho_{\omega,P}}{\partial T} \mathrm{d}\omega$$

leads to the *Rosseland mean free path*

$$l_R = \frac{\int_0^\infty l_\omega \frac{\partial \rho_{\omega,P}}{\partial T} \mathrm{d}\omega}{\int_0^\infty \frac{\partial \rho_{\omega,P}}{\partial T} \mathrm{d}\omega}. \tag{8.109}$$

The partial derivative with respect to the black body temperature results proportional to the difference of two terms

$$\frac{\partial \rho_{\omega,P}}{\partial T} \sim \frac{1}{T^2} \frac{\xi^4 \mathrm{e}^{-\xi} - 4\xi^3 \mathrm{e}^{-\xi}(1 - \mathrm{e}^{-\xi})}{(1 - \mathrm{e}^{-\xi})^2}, \qquad \xi = \frac{\hbar\omega}{k_B T}.$$

Disregarding the smaller negative term one is left with the standard formula for l_R,

$$l_R = \int_0^\infty l_\omega(\xi)\, G(\xi) \mathrm{d}\xi, \qquad G(\xi) = \frac{15}{4\pi^4} \frac{\xi^4 \mathrm{e}^{-\xi}}{(1 - \mathrm{e}^{-\xi})^2}. \tag{8.110}$$

The weighting function G has a maximum at $\hbar\omega \simeq 4 k_B T$, similar to the thermal heat flux density $\mathbf{q}$ which is determined by the electron energy components of the distribution function $f(\mathbf{x}, \mathbf{v})$ beyond $m_e \langle \mathbf{v}^2 \rangle / 2$.

For the Rosseland model to be valid there must be fulfilled (i) the radiation temperature T changes very little over l_R; (ii) α_ω is smooth over the relevant spectrum, no accumulation of sharp spectral lines; (iii) the radiation flux density **S** is nearly isotropic. In contrast to the stellar interior where the temperature of matter almost equals the radiation temperature T, in the laser generated plasma and generally in WDM T is much lower than the free electron temperature T_e.

If in the plasma or in hot dense matter the radiative flux density **S** dominates on **q** this has to be replaced by **S** in the energy equation of motion (3.29) and (3.46). In the absence of motion they reduce to the simple diffusion equation (1.63). Expression (8.107), based on the Rosseland mean free path (8.110), in combination with an appropriate energy conservation equation, is simple enough to study radiation fields generated by laser irradiation of high-Z matter and to admit similarity solutions for relevant boundary conditions [21, 22]. The papers are recommended as further readings for their physical content.

8.3 Radiation Reaction

Under violent acceleration of a moving point charge the back action of the emanating radiation field on the particle trajectory can no longer be ignored. It is assumed that the external force is the Lorentz force on a point charge $F_L = f_{L\alpha} = (q/c^2)F_{\alpha\beta}v^\beta$ from (5.27). In addition the radiation reaction force $F_R = f_{R\alpha}$ determines the trajectory of the point particle,

$$mA = F_L + F_R \iff ma_\alpha = f_{L\alpha} + f_{R\alpha}. \tag{8.111}$$

A fundamental relativistic requirement is $VA = 0$, see (2.175). The Lorentz force fulfils it because of asymmetry,

$$mVA_L = v^\alpha f_{L\alpha} = (q/c^2)F_{\alpha\beta}v^\beta v^\alpha = (q/c^2)F_{\beta\alpha}v^\alpha v^\beta = -(q/c^2)F_{\alpha\beta}v^\alpha v^\beta = 0.$$

Therefore F_R must satisfy also

$$VF_R = v^\alpha f_{R\alpha} = 0. \tag{8.112}$$

One could think of equating the total power irradiated by a point charge according to Larmor's formula $P(t)$ from (8.51) to $VF_R = -P(t)$. Then, with the help of $VA = 0$ follows

$$\frac{\mathrm{d}}{\mathrm{d}\tau}(VA) = A^2 + V\dot{A} = 0 \Rightarrow V\dot{A} = -A^2 \Rightarrow P(t) = \frac{q^2}{6\pi\varepsilon_0 c^3}\dot{A}V = F_R V.$$

Hence

$$\frac{\mathrm{d}}{\mathrm{d}\tau}P = m\frac{\mathrm{d}}{\mathrm{d}\tau}V = F_L + \frac{q^2}{6\pi\varepsilon_0 c^3}\dot{A}.$$

Unfortunately this Larmor based expression of F_R does not satisfy (8.112) owing to $\dot{A} \neq 0$. The reason is that it is obtained from the evaluation of $P(t)$ in the far field which is known to differ from the near field of a moving charge. F_R must contain an additional term. Its correct expression has been derived by Dirac, see Rohrlich [12]. It is demanding.

The simplest heuristic approach to an expression satisfying (8.112) is by assuming a correction term $CV = Cv^\alpha$ [13]. Insertion of $F_R = (q^2/6\pi\varepsilon_0 c^3)\dot{A} + CV$ in (8.112) leads to

$$C = -\frac{q^2}{6\pi\varepsilon_0 c^3}\dot{A}V/V^2 = \frac{q^2}{6\pi\varepsilon_0 c^3}A^2/V^2 = \frac{q^2}{6\pi\varepsilon_0 c^3}A^2/c^2.$$

With this term proportional to the acceleration one arrives at the classical *Lorentz–Abraham–Dirac equation* (LAD)

$$\begin{aligned}\frac{\mathrm{d}P}{\mathrm{d}\tau} &= F_L - \frac{q^2}{6\pi\varepsilon_0 mc^3}\Big[\frac{\mathrm{d}^2P}{\mathrm{d}\tau^2} - \frac{P}{m^2c^2}\Big(\frac{\mathrm{d}P}{\mathrm{d}\tau}\frac{\mathrm{d}P}{\mathrm{d}\tau}\Big)\Big] \Longleftrightarrow \\ \frac{\mathrm{d}^2p^\alpha}{\mathrm{d}^2\tau} &= (q/c^2)F_{\beta\alpha}v^\alpha v^\beta - \frac{q^2}{6\pi\varepsilon_0 mc^3}\Big[\frac{\mathrm{d}^2p^\alpha}{\mathrm{d}\tau^2} - \frac{p^\alpha}{m^2c^2}\Big(\frac{\mathrm{d}p^\beta}{\mathrm{d}\tau}\frac{\mathrm{d}p_\beta}{\mathrm{d}\tau}\Big)\Big]. \end{aligned} \quad (8.113)$$

LAD is an equation containing the acceleration on the RHS. Except special cases it cannot be reduced to the structure of a Newtonian force of type (2.3) which in a general inertial system S depends on position $\mathbf{x}$ and velocity $\mathbf{v}$ but not on acceleration $\dot{\mathbf{v}}$. For small accelerations, generally fulfilled in nonrelativistic dynamics, the term $a^\beta a_\beta$ can be neglected. In the absence of additional prescriptions LAD violates [13]

(1) energy conservation. *Runaway effect*: the particle accelerates indefinitely in the absence of an external field;
(2) causality: the particle motion at proper time τ is partially influenced by forces at $\tau' > \tau$;
(3) locality: owing to the electric field of a point charge its mass is distributed in space according to the field energy $dE = c^2dm$.

In a very recent paper by C. Bild, D.-A. Deckert, and H. Ruhl the problem with the LAD equation is treated from a fundamental point of view (point charge vs. charge distribution, mass renormalization, shortcomings of Dirac's derivation). The authors replace it by a system of second order differential delay equations which are free of most deficits of (8.113), most probably also free of runaway effects [14]. For the purpose of calculating the real single particle orbits in the strong electromagnetic field most researchers favour the runaway-free *Landau–Lifshitz* (LL) equation which represents an approximation for slowly varying external fields [15, 16], see also [17]. In 3-vectorial form the radiation damping term reads

$$\left(\frac{d\mathbf{p}}{dt}\right)_R = \frac{2}{3}r_q^2\Big\{c\big(\mathbf{E}+\mathbf{v}\times\mathbf{B}\big)\times\mathbf{B}+\Big(\frac{\mathbf{v}}{c}\mathbf{E}\Big)\mathbf{E} - \gamma^2\frac{\mathbf{v}}{c}\Big[\big(\mathbf{E}+\mathbf{v}\times\mathbf{B}\big)^2 \\ -\Big(\frac{\mathbf{v}}{c}\mathbf{E}\Big)^2\Big]\Big\} + \frac{2}{3}r_q q\Big[\gamma\frac{1}{c}\Big(\frac{\partial}{\partial t}+(\mathbf{v}\nabla)\Big)\mathbf{E}+\frac{\mathbf{v}}{c}\times\Big(\frac{\partial}{\partial t}+(\mathbf{v}\nabla)\Big)\mathbf{B}\Big], \tag{8.114}$$

r_q is the classical radius of the point charge q. For the electron it is given by (8.45). For many purposes it is further simplified by neglecting the term containing the total time derivatives of **E** and **B** (second row) [16]. The results of Figs. 7.21 and 7.22 are obtained in this way. Implementation of (8.114) in a PIC code and applications to superintense laser-matter interactions are discussed in detail in [16, 18]. A detailed comparison with other variants of radiation reaction, implementation in PIC codes and application to a variety of single particle motions are performed in [19]. Classical radiation damping starts becoming significant for laser intensities $I \gtrsim 10^{22}$ Wcm^{-2}. Thereby the oscillation center motion of the electron plays a decisive role, highest radiation loss in counter-motion, weak in co-motion as a consequence of Doppler frequency up and down conversion.

8.4 Summary

The Lagrangian of the classical radiation field is given by

$$L = \frac{1}{2}\int(\varepsilon_0\mathbf{E}^2 - \mu_0\mathbf{B}^2)\mathrm{d}V; \quad \varepsilon_0\mu_0 = \frac{1}{c^2}. \tag{8.115}$$

It shows that in analogy to mechanics of massive particles only the difference of the fields is the *kinetic energy* density that is free to be varied. By averaging the Poynting vector in time the interaction of the fields with matter is reduced to balance equations with transition rates of the Einstein coefficients of spontaneous transitions B and induced emission and absorption A. No *phases* appear in the balance. This is the standard situation of *radiation transport* where Kirchhoff's principle of detailed balance holds. Depending on the problem to be solved the latter can be formulated in many ways. This is shown in detail in the main text. The difference in formulation has its origin in the spectral coefficients of net absorption α_ω as the difference between total absorption κ_ω and the reemission by the surrounding radiation field. In thermal equilibrium of the electrons the two quantities are connected by

$$\alpha_\omega = \kappa_\omega\big(1-\exp\{-\hbar\omega/k_BT_e\}\big).$$

If the refractive index $\eta(\mathbf{x})$ varies noticeable in space the rays are curved, the spectral intensity I_ω has to account for the variation of of the solid angle $\mathrm{d}\boldsymbol{\Omega}$ and I_ω is to be replaced by I_ω/η^2, see (8.76).

It may happen that a given mode $\mathbf{k}$ does not contain enough photons, for instance if at high energy their production is little, the electromagnetic field has to be quantized. $\mathbf{E}$ and $\mathbf{B}$ become operators acting on the states $|\psi_\mathbf{k}\rangle$ of the radiation fields. Their values result as the expectation values $\langle \mathbf{E_k}\rangle = \langle|\psi_\mathbf{k}|\mathbf{E}_\mathbf{k}^{\mathrm{op}}|\psi_\mathbf{k}\rangle$, and $\langle \mathbf{B_k}\rangle$ and $\langle \mathbf{A_k}\rangle$ accordingly.

The interaction of the laser with matter has been modelled so far by a sinusoidal field in $\mathbf{E}$ which is sharp in amplitude and phase, in contradiction to (8.1). Glauber [4] has shown that the coherent state $|\alpha\rangle$ from (8.16) comes arbitrary close to the classical harmonic wave with increasing photon number α^2. The relative uncertainty in amplitude behalves Boltzmann like $1/\alpha$ and the uncertainty in phase assumes the minimum allowed by (8.16). In addition, with its help a criterion for the use of the classical fields calculated from Maxwell's equations is deduced. To this aim

$$N_\mathbf{k} = n_\mathbf{k}\lambda^3 \gg 1; \quad \lambda = 2\pi/|\mathbf{k}| \tag{8.116}$$

must be satisfied for the photon density $n_\mathbf{k}$. The quantized field and its semiclassical extensions keep the "kinetic" part of the field and make the description of coherent optical phenomena possible; for a selection see Fig. 8.2. Two of most striking phenomena are the realization of *total inversion* in a two level atomic system and of *echoes*. The time scale of coherent optics is given by the *Rabi frequency* $\omega_R = \mu\hat{E}/\hbar$; μ electric dipole moment. In comparison, the time evolution in radiation transport is slow. It can be assumed that for observation times much longer than the *transverse* or *phase memory time* T_2 the dynamics of the optical Bloch model reduces to a system of rate equations with coefficients to be identified with the two Einstein coefficients. As shown by (8.41) and (8.42) this is indeed the case.

Charges if accelerated radiate. Under fulfilment of condition (8.116) in the highly relativistic limit an electron radiates in forward direction with maximum emission under the angle $\theta_{\max} = 1/\gamma^2$. At equal acceleration into longitudinal and into transverse direction the ratio of radiation loss from the circular orbit is γ^2 as high as from the straight orbit. The radiation emission into strictly forward direction is zero. It is forbidden by the transverse character of electromagnetic radiation. When nonrelativistic *bremsstrahlung* is classically calculated from a thin thermal plasma as in (8.70) the result must be subject to *quantum corrections* which are generally expressed in terms of *Gaunt factors* $\sigma(T)$ and $\sigma(\omega, T)$, see (7.76). In contrast to black body radiation the spectrum of thermal bremsstrahlung shows no pronounced maximum at finite ω.

8.5 Problems

▹ Express the Einstein coefficient B_{if} as a function of w_{if} under the assumption that there are n_1, n_2 atoms in levels E_1, E_2.
▹ Verify (8.15).
▹ Verify $a|\alpha\rangle = \alpha|\alpha\rangle$.
▹ Evaluate $\langle\alpha|n^2|\alpha\rangle$. *Solution:* Split n^2 into $n(n-1)+n \Rightarrow \langle\alpha|n^2|\alpha\rangle = \exp -|\alpha|^2 \sum[|\alpha|^{2n}/n!][n(n-1)+n] = |\alpha|^4 + |\alpha|^2$.
▹ Show that the uncertainty ΔE of the electric field is for all states $|n\rangle$ equal to the ΔE of the vacuum state.
▹ An excited atomic state decays exponentially in the absence of any external laser field. How does the decay change if a constant laser field with ω close to resonance ω_0 is on all the time? *Hint:* Consider optical nutation with damping.
▹ Derive the wave equations (8.35) and integrate them under the assumption of no pump depletion: $\hat{E}(z,t) = \hat{E}(0, t - z/c)$. For solutions see (8.36).
▹ A Maxwellian beam of plasma ions exhibits a temperature of $T_i = 10$ eV. In order to observe the free induction decay $\kappa\hat{E} > 3kv_{z,\text{th}}$ should hold. Calculate the laser intensity and the signal at t at the end of the applied pulse.
▹ Show that the Liénard–Wiechert potentials satisfy the Lorentz gauge. Which of **A** and **E** depends on gauge?
▹ Generate the Liénard–Wiechert potentials from a Lorentz transformation.
▹ How do you calculate $\int \exp -y^2 dy$ from (8.63)? *Answer:* By performing it, for example, in R^2 in polar coordinates: $(\int \exp -y^2 dy)^2 = \int \exp -(x^2+y^2) dx dy = 2\pi \int r^2 \exp -r^2 dr = -\pi \exp -r^2|_0^\infty = \pi$.
▹ Verify (8.60) and (8.62).
▹ Estimate the spectral radiation range of a proton in the LHC.
▹ Give an estimate of the angular width of emission in (8.61).
▹ Verify the passage from (8.77) to (8.80) by integration.
▹ Verify the equivalence of (8.81) with (8.83). *Hint:* From $a/b = c/d$ follows $a/b = (a \pm c)/(b \pm d)$.
▹ The optical thickness defined by (8.89) is independent of I_ω. (a) Indicate situations in which it is no longer true. (b) Where in the book is such a dependence explicitly treated? (c) Show that l_ω is indeed the mean free path of the photon $\hbar\omega$
▹ Derive an explicit expression of the θ integrated mean free path l_ω.
▹ Discuss the different structures of (8.104) and (8.105). Under which conditions can be fulfilled both?
▹ A free electron moves at $|\mathbf{v}| = 0.99\,c$. A magnetic field of 100 T orthogonal to $\mathbf{v}$ is switched on adiabatically. Calculate the number of cycles after which the gyroradius has reduced to 1/2 of its initial value.

8.6 Self-assessment

- Does relation (9.7) hold for nonthermal level occupations? *Answer:* yes.
- Find arguments for (8.22) to be of general validity, out of thermal equilibrium.
- Which of the following operators are Hermitian: $a^\dagger$, $\mathbf{A}$, $\mathbf{B}$, $\mathbf{E}$, $(aa^\dagger)^3$, $\cos\phi$?
- What is the variance of $\mathbf{E}$ in the vacuum state $|0\rangle$? *Answer:* $\langle 0|E|0\rangle = i(\hbar\omega/2\varepsilon_0 V)^{1/2}$ $\langle 0|a\exp i\phi - a^\dagger \exp -i\phi|0\rangle = 0$; $\langle 0|E^2|0\rangle = \hbar\omega/2\varepsilon_0 V \Rightarrow \Delta E = (\hbar\omega/2\varepsilon_0 V)^{1/2}$.
- If the number of photons in a volume $\lambda_\mathbf{k}^3$ is much larger than unity, $N_\mathbf{k} = n_\mathbf{k}\lambda_\mathbf{k}^3 \gg 1$, the fields obey the classical Maxwell equations. *Hint:* Make use of the properties of the Glauber state and the field quantization in a finite volume with reflecting or periodic boundary conditions.
- What are the operators of the electric quadrupole and the magnetic dipole transitions? *Answer:* $H_1 = e\sum_j \mathbf{x}_j(\mathbf{x}_j\nabla)\hat{\mathbf{E}}(0)/2$, $H_1 = \sum_j (e/(2m_e)\hat{\mathbf{B}}(0)(\mathbf{x}_j \times \mathbf{p}_j)$.
- Complete (8.34) by the additional terms for a charged magnetized medium.
- In the optical nutation from the ground state inversion depends on one parameter: Which one is this and what is its value?
- Under which condition (a) does Fermi's golden rule hold? (b) What is a sufficient condition for the validity of Einstein's rate equations with A and B? (c) Why does X ray lasing need pump rates by orders of magnitude higher than lasing at infrared frequencies?
- Which time τ after a laser pulse has switched on does the standard refractive index η of linear optics hold? *Hint:* Use the optical Bloch model to discuss relevant situations.
- Explain the corrections by the factor η^{-2} from the plasma refractive index η in (8.43) and (8.44).
- A relativistic charge moves at β close to 1. What is the angular spread of its light cone at the intrinsic time t' and at the lab time t? For definition of t' see (8.108). *Answer:* $\Delta\theta(t') \simeq 1/\gamma$; $\Delta\theta(t) \simeq 1/\gamma^3$. Justify!
- Verify by a geometrical argument or by means of the ray equation (5.156) $\mathrm{d}\Omega = C^2\theta^2$ in (8.74).
- Where is reemission hidden in the source term of (8.82)?
- Discuss the difference of the Poynting vector of frequency ω and $\mathbf{S}_\omega$.
- Write down (8.82) for $\eta \neq 1$.
- Give a sufficient condition for the net absorption α_ω not to depend on direction $\boldsymbol{\Omega}$, for instance in (8.102).
- (a) When does $B_{12}I_\omega = \pi^2c^2/\omega^2 A_{12}$ not hold? (b) Give a sufficient condition for its validity. *Hint:* You find the answers from (8.41) and (8.42).
- Verify (8.87).
- Why is $l_\omega(s)$ in (8.89) the mean free path? *Answer:* Because absorption follows Beer's law. Calculate the mean path between two collisions.

8.7 Glossary

Amplitude and phase

$$\Delta n \Delta \cos\phi \geq \frac{1}{2}|\langle \sin\phi \rangle|, \quad \Delta n \Delta \sin\phi \geq \frac{1}{2}|\langle \cos\phi \rangle|. \tag{8.1}$$

Ladder operators

$$a|n\rangle = n^{1/2}|n-1\rangle, \quad a^{\dagger}|n\rangle = (n+1)^{1/2}|n+1\rangle; \quad \langle n|n\rangle = 1; \quad a|1\rangle = |0\rangle. \tag{8.6}$$

Field operators for **k** mode

$$\mathbf{A}_{\mathbf{k}} = \left(\frac{\hbar}{2\varepsilon_0 V \omega_{\mathbf{k}}}\right)^{1/2} \boldsymbol{\varepsilon}_{\mathbf{k}} \left\{a_{\mathbf{k}} \mathrm{e}^{i(\mathbf{kx}-\omega t)} + a^{\dagger}{}_{\mathbf{k}} \mathrm{e}^{-i(\mathbf{kx}-\omega t)}\right\},$$

$$\mathbf{E}_{\mathbf{k}} = i\left(\frac{\hbar\omega_{\mathbf{k}}}{2\varepsilon_0 V}\right)^{1/2} \boldsymbol{\varepsilon}_{\mathbf{k}} \left\{a_{\mathbf{k}} \mathrm{e}^{i(\mathbf{kx}-\omega t)} - a^{\dagger}{}_{\mathbf{k}} \mathrm{e}^{-i(\mathbf{kx}-\omega t)}\right\},$$

$$\mathbf{B}_{\mathbf{k}} = i\left(\frac{\hbar}{2\varepsilon_0 V \omega_{\mathbf{k}}}\right)^{1/2} \mathbf{k} \times \boldsymbol{\varepsilon}_{\mathbf{k}} \left\{a_{\mathbf{k}} \mathrm{e}^{i(\mathbf{kx}-\omega t)} - a^{\dagger}{}_{\mathbf{k}} \mathrm{e}^{-i(\mathbf{kx}-\omega t)}\right\}.$$

Poynting vector

$$\langle n_{\mathbf{k}}|\mathbf{S}_{\mathbf{k}}|n_{\mathbf{k}}\rangle = \frac{c^2}{V}(n_{\mathbf{k}} + 1/2)\hbar\mathbf{k} = c\hbar\omega_{\mathbf{k}}(n_{\mathbf{k}} + 1/2)\mathbf{k}^0.$$

Glauber state

$$|\alpha\rangle = \mathrm{e}^{-|\alpha|^2/2} \sum_{n\geq 0} \frac{\alpha^n}{(n!)^{1/2}}|n\rangle \quad \Rightarrow \quad a|\alpha\rangle = \alpha|\alpha\rangle, \quad \langle\alpha|\alpha\rangle = 1, \; n|\alpha\rangle = |\alpha|^2|\alpha\rangle. \tag{8.16}$$

Amplitude uncertainty

$$\Delta n = |\alpha|, \quad \frac{\Delta n}{n} = \frac{1}{|\alpha|}. \tag{8.17}$$

Probability of dipole transition

$$w_{if} = \frac{2\pi}{\hbar}|V_{if}|^2 \delta(E_i - E_f); \quad V_{if} = \langle i|e(\mathbf{x}-\mathbf{x}_0)\mathbf{E}_{\mathbf{k}\omega}(\mathbf{x}_0)|f\rangle. \tag{8.21}$$

Ratio of Einstein coefficients A, B

$$A_{fi} = \frac{\omega^2}{\pi^2 c^3}\frac{g_1}{g_2}(B_{if}\hbar\omega) = \frac{g_1 e^2\omega^3}{3\pi\varepsilon_0 g_2 \hbar c^3}|\langle f|(\mathbf{x}-\mathbf{x}_0)|i\rangle|^2 \Leftrightarrow \frac{A}{B\hbar\omega} = \frac{1}{n_{\mathbf{k}}}; \; n_{\mathbf{k}} = \frac{\omega^2 n_\omega}{\pi^2 c^3}. \tag{8.22}$$

Rabi frequency of dipole moment μ

$$\omega_R = \frac{\mu}{\hbar}\hat{E}. \tag{8.28}$$

Optical Bloch model

$$\frac{\mathrm{d}}{\mathrm{d}t}\mathbf{M} = \boldsymbol{\Omega} \times \mathbf{M}, \quad \boldsymbol{\Omega} = (\kappa\hat{E},\, 0,\, \omega_0 - \omega). \tag{8.32}$$

Bremsstrahlung loss in relation to acceleration energy

$$\frac{P}{\dot{E}} = \frac{\kappa}{m_e^2 c^2 v}\frac{\mathrm{d}E}{\mathrm{d}x} \quad \rightarrow \quad \frac{\kappa}{m_e^2 c^3}\frac{\mathrm{d}E}{\mathrm{d}x} = \frac{2}{3}\frac{r_e}{m_e c^2}\frac{\mathrm{d}E}{\mathrm{d}x}. \tag{8.54}$$

Circular bremsstrahlung loss

$$P = \kappa\gamma^4\omega^2\beta^2. \tag{8.55}$$

Loss from circ. pol. laser

$$P = \frac{2}{3}m_e c r_e \omega^2 \hat{a}^4; \quad \hat{a}^2 = 7.3 \times 10^{-19} I\lambda^2[\mathrm{Wcm^{-2}\mu m^2}]. \tag{8.56}$$

Circular vs linear acceleration power ratio

$$\frac{P_{\mathrm{cic}}}{P_{\mathrm{lin}}} = \gamma^2.$$

Spontaneous bremsstrahlung from plasma

$$P(v)\mathrm{d}x = \frac{2}{3}\frac{4\pi e^6 Z^2}{(4\pi\varepsilon_0)^3 m_e^2 c^3 \lambda_B}\frac{\mathrm{d}x}{v}; \quad \lambda_B = \frac{\hbar}{m_e v}.$$

Spont. bremsstrahlung from thermal plasma

$$P = 12\left(\frac{2}{\pi}\right)^{1/2}\frac{e^6 Z^2 (k_B T_e)^{1/2}}{(4\pi\varepsilon_0)^3 m_e^{3/2} c^3 \hbar} n_e n_i; \quad \langle v \rangle = 2\left(\frac{2}{\pi}\right)^{1/2}\left(\frac{k_B T_e}{m_e}\right)^{1/2}$$

$$\Rightarrow P[\mathrm{Wcm^{-3}}] = 6.2 \times 10^{-49}(T_e[\mathrm{eV}])^{1/2} Z^2 n_e[\mathrm{cm^{-3}}] n_i[\mathrm{cm^{-3}}]. \tag{8.71}$$

Emission coefficient from thermal plasma

$$I_\omega = \frac{\mathrm{d}^2 P}{\mathrm{d}\Omega d\omega} = \frac{\Re(\eta)\omega_p^2}{4\sqrt{3}\pi^2 c^3}\nu_{ei} k_B T_e \frac{\sigma(\omega, T_e)}{\ln\Lambda}\mathrm{e}^{-\hbar\omega/kT_e}. \tag{7.76}$$

Spectral radiation transport equation

$$\eta^2 \frac{\partial}{\partial t}\left(\frac{\rho_\omega(\boldsymbol{\Omega})}{\eta^2}\right) + \eta^2 \frac{\mathrm{d}}{\mathrm{d}s}\left(\frac{I_\omega}{\eta^2}\right) = \varepsilon_\omega - \alpha_\omega I_\omega. \tag{8.76}$$

Kirchhoff's law

$$\frac{\partial}{\partial t}\rho_\omega(\boldsymbol{\Omega}) + \frac{\mathrm{d}}{\mathrm{d}s} I_\omega = \alpha_\omega (I_{\omega,P} - I_\omega); \qquad \eta = 1. \tag{8.82}$$

Rosseland mean free path

$$l_R = \frac{\int_0^\infty l_\omega \frac{\partial \rho_{\omega,P}}{\partial T} \mathrm{d}\omega}{\int_0^\infty \frac{\partial \rho_{\omega,P}}{\partial T} \mathrm{d}\omega}. \tag{8.109}$$

LAD radiation reaction, F_L Lorentz force

$$\frac{\mathrm{d}P}{\mathrm{d}\tau} = F_L - \frac{q^2}{6\pi\varepsilon_0 m c^3}\left[\frac{\mathrm{d}^2 P}{\mathrm{d}\tau^2} - \frac{P}{m^2c^2}\left(\frac{\mathrm{d}P}{\mathrm{d}\tau}\frac{\mathrm{d}P}{\mathrm{d}\tau}\right)\right]; \; P = p^\alpha = m\gamma(\mathbf{v}, c). \tag{8.113}$$

8.8 Further Readings

R. Loudon, *The Quantum Theory of Light* (Clarendon Press, Oxford, 1986).
C.C. Tannoudji, J. Dupont-Roc, G. Grynberg, *Atom-Photon Interactions* (Wiley, Hoboken, 1992).
J.J. Sakurai, *Advanced Quantum Mechanics* (Addison-Wesley, Reading,1967).
F.H.M. Faisal, *Theory of Multiphoton Processes* (Plenum Press, New York, 1987).
J.D. Jackson, *Classical Electrodynamics* (Wiley, New York, 1999) (Chaps. 14, 15).
Y.B. Zeldovich, Y.P. Raizer, *Physics of Shock Waves and High-Temperature Hydrodynamic Phenomena* (Academic Press, New York, 1966), Sects. I, II.
R. Pakula, R. Sigel, *Selfsimilar expansion of dense matter due to heat transfer by nonlinear conduction*. Phys. Plasmas **28**, 232 (1985).

References

1. J.J. Sakurai, *Modern Quantum Mechanics* (Addison-Wesley, Reading, 1994)
2. B.T.H. Varcoe, S. Brattke, M. Weidinger, H. Walther, Nature **403**, 743 (2000)
3. R. Loudon, *The Quantum Theory of Light* (Clarendon, Oxford, 1986), p. 146
4. R.J. Glauber, Phys. Rev. **131**, 2766 (1963)
5. R.P. Feynman, F.L. Vernon, R.W. Hellwarth, J. Appl. Phys. **28**, 49 (1957)
6. R.L. Shoemaker, Coherent transient infrared spectroscopy, in *Laser and Coherent Spectroscopy*, ed. by J.I. Steinfeld (Plenum Press, New York, 1978) (Sec. 3.3)
7. R.H. Dicke, Phys. Rev. **93**, 99 (1954)
8. M. Gross, S. Haroche, Phys. Rev. **93**, 301–396 (1982)
9. A. Sommerfeld, *Atombau und Spektrallinien - Atomic Structure and Spectral Lines*, vol. 2 (Friedrich Vieweg und Sohn, Berlin, 1939)

10. I.P. Shkarofsky, T.W. Johnston, M.P. Bachynski, *The Particle Kinetics of Plasmas* (Addison Wesley Publication Co., Reading, 1966) (Chapt. 6)
11. J. Greene, Astrophys. J. **130**, 693 (1959)
12. F. Rohrlich, Ann. Phys. **13**, 93 (1961)
13. E. Rebhan, *Theoretische Physik*, vol. I (Elsevier GmbH, Heidelberg, 2005) (Sect. 21.11)
14. C. Bild, D.-A. Deckert, H. Ruhl, Phys. Rev. D **99**, 096001 (2019)
15. L.D. Landau, E.M. Lifshitz, *The Classical Theory of Fields*, 4th edn. (Elsevier, Heidelberg, 2018)
16. M. Tamburini, F. Pegoraro, A. DiPiazza, C.H. Keitel, A. Macchi, New J. Phys **12**, 123005 (2010)
17. H. Spohn, *Dynamics of Charged Particles and Their Radiation Field* (Cambridge University Press, Cambridge, 2004)
18. M. Tamburini, T.V. Liseykina, F. Pegoraro, A. Macchi, Phys. Rev. E **85**, 016407 (2012)
19. M. Vranic, J.L. Martins, R.A. Fonseca, L.O. Silva, arXiv: 1502.02432v3 (2016)
20. Ya B. Zeldovich, YuP Raizer, *Physics of Shock Waves and High Temperature Hydrodynamic Phennomena* (Academic Press, New York, 1966)
21. R. Pakula, R. Sigel, Phys. Fluids **28**, 232 (1985)
22. R. Sigel et al., Phys. Rev. A **38**, 5779 (1988)

Chapter 9
Applications of High Power Lasers

9.1 The Nonlinear Response of the Plasma to the Laser

At high intensities electromagnetic waves behave strongly nonlinear, at a degree which goes far beyond the familiar nonlinear optics. The origin of the drastically increased nonlinearity has two kinetic roots. The first is the *Lorentz force*. The $\mathbf{v} \times \mathbf{B}$ term generates, at moderate laser strength, electron motion of 2ω and also zero frequency motion. The other root is the convective term $(\mathbf{v}\nabla)\mathbf{v}$ which acts similarly, but with different strength in general. With increasing field intensity higher harmonics of the laser frequency govern more and more the current densities and "contaminate" the $\mathbf{E}$ and $\mathbf{B}$ fields through the Maxwell equations. These, in turn, act back onto the current densities to generate the world of nonlinearities. A third root of nonlinearities stems from the distortion of the fields of atoms and ions by the incident laser field itself. Specifically, from the outset they produce sudden variations of the local electron density due to field ionization (i), or due to ponderomotive or wave pressure effects (ii) and, already at sub-relativistic intensities, they induce the coupling of transverse and longitudinal fields in the homogeneous plasma (iii). Whereas in standard nonlinear optics the number of wave crests of a pulse is invariant after interaction with the plasma an intense electromagnetic pulse can vary its profile as a consequence of (i), e.g. five crests become six or seven crests [1]. Ponderomotive effects (ii) lead to a large number of instabilities, formation of irregular structures, solitons, cavitons, wavebreaking, and chaos. In addition, due to (iii) all effects may even become stronger.

So far, transverse and longitudinal wave propagation at relativistic intensities has been studied extensively only with waves of constant phase velocity in plasmas of constant density [2–4], and recently stationary quasiperiodic solutions have also been presented [5]. Under the influence of (ii) the plasma medium is expected to become heavily deformed and even selfquenching of the electromagnetic modes may occur. Extreme laser intensities are to be expected in the near future ($I > 10^{23}\,\mathrm{Wcm}^{-2}$) and hot dense matter at solid and above solid density may be produced by relativistic overcritical light penetration. The study of intense wave propagation in dense matter is

P. Mulser, *Hot Matter from High-Power Lasers*, Graduate Texts in Physics,
https://doi.org/10.1007/978-3-662-61181-4_9

of vital importance, in itself as well as for applications (equation of state, astrophysics [6]). The beauty in this context is that a great portion of related experiments can already be performed, by analogy, at nonrelativistic and close to relativistic intensities in sub- and super-critical plasma, in foams and in porous media with standard fs laser pulses.

With increasing light intensity laser-based nuclear physics will become a reality [7, 8]. In analogy to atomic systems excitation, stimulated de-excitation and triggering of single decay processes may be controlled by laser. Already with the next generation of photon flux densities ($>10^{23}$, perhaps $\gg 10^{23}\,\mathrm{Wcm}^{-2}$) the direct excitation of nuclear levels and the triggered decay of isomeric nuclei can be studied, provided the photon energy is high enough, as delivered for example by the free electron laser in the keV energy range [9, 10]. Alternatively, the aim may be reached by the coherent superposition of extremely high order harmonics generated from solid targets (order 5000 reached [11]). Such pulses will be used also to study the nuclear Stark effect and fine tuning of nuclear levels for various applications and diagnostics.

All kinds of *vacuum nonlinearities* have their origin in the separation (slang: "creation") of electron-positron or more massive pairs by a corresponding energy supply. The *critical field* or *Schwinger field* for the lightest pair production is $\hat{E} = 2m_e^2c^3/\hbar e = 1.3 \times 10^{18}\,\mathrm{V/m}$, corresponding to the laser intensity $I = 2.3 \times 10^{29}\,\mathrm{Wcm}^{-2}$. There is the claim that by focusing high order harmonics from a fs laser pulse of $I \simeq 10^{22}\,\mathrm{Wcm}^{-2}$ in overdense plasma this limit should be reached [12]. The existence of pair production becomes noticeable already through their virtual appearance at much lower energies (and fortunately, after all "there are no limits" of I to make them real). Nonperturbative vacuum nonlinearities manifest themselves in a variety of processes and effects of the quantum vacuum exposed to superintense photon fields: generation of harmonics, photon splitting, light by light scattering, vacuum polarization [13]. Two favoured candidates, accessible already to the most intense available fs laser installations, are the merging of two photons in the laser field interacting with TeV protons and the dramatic increase by many orders of magnitude of electron bremsstrahlung from a nucleus in the intense radiation field.

9.2 Generation of Radiation

9.2.1 Terahertz Radiation

The generation of intense electromagnetic radiation in the Terahertz (THz) frequency domain has gained growing interest for its multiple applications as a diagnostic tool in medicine for cancer detection, in technology for surface scanning, for remote sensing in terrestrial and extraterrestrial sciences, in security protection for retrieving light element explosives, and for telecommunication. Owing to generation and detection difficulties it was popularly known as *THz gap*. THz radiation operates in the frequency interval from 10^{11} to $10^{13}\,\mathrm{Hz}$, i.e., 0.1 to 10 THz. The corresponding mild photons with energies from 4 meV to 0.4 eV guarantee non invasive, non destructive interaction with probes. Other advantages of THZ radiation are the low elastic scattering level from inhomogeneities which is proportional to the 4th power of the frequency, e.g., in astronomical detection, and the frequent occurrence of resonances in matter at THz level, in particular in polar molecules. In wavelengths this is the interval from 30 μm to 0.3 cm. Thus, THz radiation falls between the infrared and microwave regions of the electromagnetic spectrum. For comparison, black body radiation with its maximum at 1 THz is emitted at a temperature as low as $T = 17\,\mathrm{K}$. Intense THz radiation is delivered from semiconductor lasers, e.g., the THz quantum cascade laser, from linear accelerators with bending magnets, from synchrotons and from free electron lasers. Production of THz radiation from neutral matter suffers from local electric breakdown. It imposes limits on the electric fields and resultant low-to-moderate efficiencies. Characteristic THz output powers lie in the range from μW to mW.

In all isotropic media THz radiation is produced by *(electro-)optical rectification* which, in turn, is driven by the *ponderomotive force*. Therefore optical rectification is *ponderomotive* rectification. Traditionally, optical rectification is understood as the generation of a transverse quasistatic polarization. In our context here quasistatic means in the THz frequency domain. Consider a single monochromatic wave. By its interaction with a charge (or macroscopic medium) it generates the second harmonic 2ω and the secular zero frequency $(\omega - \omega)$ component, this latter giving rise to the static ponderomotive force $\mathbf{f}_p$. If the transverse wave is the sum of two collinear monochromatic waves of frequencies ω_1, ω_2 the static $\mathbf{f}_p$ is generated by selfinteraction of the single waves and the ponderomotive force $\mathbf{f}_{p1}$ at the difference frequency $|\omega_1 - \omega_2|$ by their product. In this sense stimulated Brillouin scattering is the result of optical rectification of the incident laser wave and its Doppler shifted back or side scattered wave. All combinations of wave-wave interactions, transverse-transverse, transverse-longitudinal, longitudinal-transverse, leading to a THz transverse polarization are generators of THz radiation. In plasmas density modulations and static magnetic fields may increase the THZ output through perfect phase matching and resonances.

Strong emission of THz radiation from plasmas was reported for the first time by Hamster et al. [14]. Since then THz radiation generation from plasmas produced by intense short laser pulses has evolved into a broad field of experimental and theoretical research [15]. Intense THz *transition radiation* from energetic laser generated electrons crossing the solid-vacuum boundary has been reported recently [16]; for theory see [17]. The plasma medium has the advantage of sustaining arbitrarily high fields and giving rise to high conversion efficiencies in the percent region [15]. In what follows THz generation from a laser plasma is elucidated for the case of two intense laser beams nonlinearly coupled by their combined ponderomotive force.

The principle is simple. Consider two plane electromagnetic waves propagating in a plasma of constant density n_0 along x and polarized in y direction,

$$E_{yj} = \hat{E}_j e^{ik_jx - i\omega_jt}, \quad j = 1, 2; \quad \omega = \omega_1 - \omega_2, \quad k = k_1 - k_2. \qquad (9.1)$$

The beams are assumed short enough, e.g. sub ps, not to perturb the ion distribution. If the amplitudes $\hat{E}_j$ show spatial variations along y, for instance Gaussian profiles or filamentary structures, they exert a ponderomotive force $f_{py} = -\partial_y \Phi_p$ on the electrons at the beat frequency ω, chosen in the THz interval, according to

$$\partial_t u_j = \frac{-e}{m_e} E_{yj} \Rightarrow u_j = i\frac{-e}{m_e\omega_j} E_{yj} \Rightarrow \Phi_p = \frac{m_e}{2} u_1 u_2^* = \frac{e^2}{2m_e\omega_1\omega_2} E_{y1} E_{y2}^*. \tag{9.2}$$

The perturbed electron distribution n_1 in y direction follows from the familiar linearized equations of motion at the beat frequency ω and negligible electron temperature

$$\partial_t n_1 + n_0 \partial_y u_y = 0, \quad \partial_t u_y = -\frac{1}{m_e}\partial_y \Phi_p - \frac{e}{m_e} E_{\mathrm{es}}, \quad \partial_y E_{\mathrm{es}} = \frac{e}{\varepsilon_0} n_1.$$

E_{es} is the electrostatic field induced by the electron density perturbation n_1 from f_{py}. Elimination of u_y and E_{es} leads to

$$\partial_{tt} n_1 = \frac{n_0}{m_e}\partial_{yy}\Phi_p + \omega_p^2 n_1 \Rightarrow n_1 = -\frac{n_0}{m_e(\omega^2 - \omega_p^2)}\partial_{yy}\Phi_p; \quad u_y = \frac{\omega}{k}\frac{n_1}{n_0}. \tag{9.3}$$

The resulting THz current density $j_T = -en_0 u_y$ drives the THz transverse field E_T from the wave equation (5.15)

$$\nabla \times \nabla \times \mathbf{E}_T + \frac{\partial^2}{\partial t^2}\mathbf{E}_T = -\frac{1}{\varepsilon_0 c^2}\frac{\partial}{\partial t}(\mathbf{j}_\omega + \mathbf{j}_T) \Rightarrow \frac{\partial^2}{\partial x^2} E_T + k_{0T}^2 \eta_\omega^2 E_T = \frac{1}{\varepsilon_0 c^2}\frac{\partial}{\partial t} j_T. \tag{9.4}$$

The linear current $\mathbf{j}_\omega$ is induced by $\mathbf{E}_T$ and gives rise to the plasma refractive index $\eta_\omega = (1 - \omega_p^2/\omega^2)^{1/2} > 0$. The low plasma density $n_0 < n_c(\omega)$ and with it $\mathbf{j}_T$ is responsible for the limited THz output power. In the ideal case k should match with $k_1 - k_2$. This is not the case since

$$\begin{aligned} k_1 - k_2 &= \frac{1}{c}(\omega_1\eta_1 - \omega_2\eta_2) \\ &\simeq \frac{1}{c}\left[\omega_1\left(1 - \frac{\omega_p^2}{2\omega_1^2}\right) - \omega_2\left(1 - \frac{\omega_p^2}{2\omega_2^2}\right)\right] = \frac{\omega}{c}\left[1 + \frac{\omega_p^2}{4\omega_1\omega_2}\right]. \end{aligned} \tag{9.5}$$

For $\omega_j = \omega_{\mathrm{Nd}}$ the dephasing term in the square bracket is smaller than 10^{-5}. Perfect matching can be achieved with $j_T = -en_q u_y$ in a corrugated plasma of density $n = n_0 + n_q \exp -iqx$ and $k_T = k_1 - k_2 - q$. Thereby n_q is assumed much smaller than n_0.

In the current literature numerous studies can be found with the aim to obtain increased THz output from focused coaxial Gaussian and super-Gaussian beams, by optical rectification, and in addition, static magnetic fields superimposed along the laser beam direction and perpendicular to it [18]. A detailed analysis a *single* super-Gaussian laser beam interacting with a preformed density modulation and a static magnetic field perpendicular to the laser field and to the density modulation is presented in [19]. Owing to the difficulties of two beam alignment and their phase matching single beam schemes close to magnetic resonance may offer clear advantages. The calculated efficiency in THz power output is $\lesssim 3 \times 10^{5}$. This is by orders of magnitude lower than the 7.5% from photo conductive emitters at 0.1 THz [20]. However, by orders of magnitude higher THz output can be obtained from fully ionized plasmas because of almost no limits on the laser driver intensity in a plasma environment.

9.2.2 X Ray Lasing

Coherent short wavelength radiation in the extreme ultraviolet (XUV) and soft X ray domain is of eminent significance for its application in various diagnostic fields as well as in fundamental research. In medical science the X ray laser promises fast scanning, higher resolution, and reduction of the dose by at least an order of magnitude owing to coherence. It opens new access to the development of pharmaceutics. In pure research the crystal structure of biomolecules is revealed by the progress in laser micromicroscopy, for example of the picornavirus BEV2 recently [21]. In this context the development of table-top lasers in the *water window* is of special relevance. Here, between the K absorption edges of oxygen at 2.3 nm and carbon at 4.4 nm the large penetration depth of photons ranges between 1 and 10 μm and allows high contrast imaging. In plasma research the X ray laser allows penetration to density regions well beyond the infrared and visible range of the high power lasers. A promising technical application of the X ray laser is the fabrication of minimum size computer chips by radiation microlithography at wavelengths below 13.5 nm of the L edge of silicon. In a nutshell, there is a demand for the X ray laser wherever higher resolution, good collimation, high intensity, coherence, and high repetition rate is asked for.

There are two basic facts to be faced in the development of the X ray laser. Optical transitions between energy levels are of the order of 10 eV or 118 nm. For shorter wavelengths lasing recurrence must be made to highly charged ions. The energy levels E_n^{Z+} of a Z fold hydrogen like ion are proportional to Z^2,

$$E_n^{Z+} = Z^2 \times E_n^H. \tag{9.6}$$

The ground state of fully stripped carbon is $13.6 \times 36 = 490$ eV, the transition corresponding to Lyman α occurs at 365 eV with the emission in the water window at $\lambda = 3.4$ nm. Favourite candidates for lasing are H like, He like, Neon like (10 electrons) and Ni like (28 electrons) owing to the stability of closed shells and the poor background of disturbing neighbouring line transitions, i.e. low opacity. The Ni nucleus exhibits a full K, L, and M shell and two 4s electrons in the N shell. The mean life time of excited neutral atoms is of the order of 10^{-8} s. It is the time an oscillator spends to irradiate spontaneously the excitation energy from the ground state. The frequency and the irradiated power of the excited Z fold ion increase with Z^2 and $\omega^4 \sim Z^8$, the electric dipole moment squared decreases with the forth power of Z. Hence, the mean life time shortens by the factor Z^4; the excited ion life times are measured in ps, the radiative decay is fast. Collisional recombination and deexcitation depend on electron and ion density, more precisely on $n_e^2 n_i$ and $(Z+)^2$.

The same dependence of the mean life time of the excited hydrogen like etc. ion is drawn from (4.137) of the quantized field between the Einstein coefficients A_{12}, B_{12}. Their relation per excited ion,

$$A_{12} = \frac{\omega^2}{\pi^2 c^3}(B_{12}\hbar\omega) \tag{9.7}$$

tells that the possibility to decay spontaneously increases with the square of the frequency, and thus with Z^4, at radiation field fixed. To obtain lasing between states $|2\rangle$ and $|1\rangle$ inversion must exist in the occupation of them, $n_2 > n_1$. Lasing starts with spontaneous decay of the first ions and the emission of photons into the full angle 4π with equal probability. In the cylindrical pumped medium induced emission occurs preferentially in a small coaxial cone because the likelihood to induce an excited ion to emit a photon into the direction of the incident radiation is proportional to the path length. Laterally emitted photons are lost after a small trip through the active medium. This phase of eventual highly coherent X ray pulse forming by *amplified spontaneous emission* (ASE) is characterized by large noise, totally in accordance with its description by the Einstein rate equations which are the result of averaging over all phases. A sufficient condition for their use is that the transverse relaxation time T_2 is short in comparison to the amplification time of the pulse, see (8.41). As a rule, in the visible to XUV domain of a diluted plasma this is the case. The standard technique of laser resonators for multipass beam amplification in the visible and infrared cannot be used in the XUV and soft X ray domain because of poor reflection and refraction of optical components, and because of their low damage threshold. Instead, the X ray laser operates normally in the single pass mode with corresponding trade-off in coherence. Elongated pumping and pumping energy reduction is accomplished by cylindrical focussing [22].

Inversion $n_2 > n_1$ is achieved either by the fast recombination with subsequent decay into the long living upper state $|n_i\rangle = |2\rangle$ and thermal occupation of the lower state $|n_f\rangle = |1\rangle$, or thermal occupation of the upper state and faster depopulation (or decay) of the lower state to create temporary inversion between the two levels.

World's first soft X ray lasing was achieved from Ne like selenium Se^{24+} collisionally pumped by the Novette laser at LLNL in Livermore in 1984 [23, 24]. Pumping occurred Nd doubled from two opposite sides by cylindrical lenses to form a focus of 200 μm in diameter and 2 cm length at intensity $I = 7 \times 10^{13}\,\mathrm{Wcm^{-2}}$. The plasma from the "exploding foil" was $n_e = (3-5) \times 10^{20}\,\mathrm{cm^{-3}}$ dense and $T_e = 900\,\mathrm{eV}$ hot with the ions at half the temperature. Lasing happened at 20.98 nm (59.10 eV) and 20.64 nm (60.07 eV) from the $1s^22s^22p^53p\,(J=2)$ level to the $1s^22s^22p^53s\,(J=1)$ level. The lower level was rapidly depopulated by fast radiative decay (≃0.3 ps) to the ground state $1s^22s^22p^6$. Direct transition from the upper lasing level to the ground state is electric dipole forbidden.

Pumping is accomplished by broad band light sources or lasers, by capillary discharge [25], or by electron beam [26]. Double laser pulse pumping in the ns and sub-ns domain has been used to reduce plasma gradient effects and pumping energy. Saving pumping energy by accurate timing is important when going to ever shorter lasing wavelengths. Ne like lasing from molybdenum at 13.10 and 13.27 nm was accomplished by sequential pumping [27]. Although rapid progress has been made in reducing the pump energy along the scheme utilized to show first lasing desired high repetition rate could not be reached with pump energies ranging from a few to 10 J. By utilizing more sophisticated pump schemes, e.g. simultaneous longitudinal and transverse pumping with elongated stepped mirror pulses, sub-Joule repetition rates of 10 Hz were achieved [28, 29]. By the employment of a novel pump scheme the pump energy could be further reduced to 70 and 80 mJ, respectively, at the same repetition rate [30]. Collisionally pumped plasma based lasers are generally characterized by a very narrow lasing line width of the order of $\Delta\lambda/\lambda = 10^{-5}$. It does not allow to lower the outcome pulse length below 1–2 ps. By applying the chirped pulse amplification technique to a high harmonic fs pulse calculations show pulses as short as 200 fs. In recent experiments along such ideas it has been possible to extract XUV pulses of about 150 fs half width [31], undercutting this way the magic 1 ps limit.

9.2.3 High Harmonic Generation

9.2.3.1 Basics

Fundamentals of ionization and plasma formation in intense near infrared laser fields at fundamental or frequency doubled or tripled frequency are sketched in Chap. 1. In neutral gases and clean solids (or liquids, clusters) the first ionization energy E_I is generally much higher than the energy of the incident photon $\hbar\omega$. The electron is freed either by *multiphoton ionization* or by *tunneling* depending on whether the *Keldysh parameter* is $\gamma_K = (E_I/2W)^{1/2} > 1$ or $\gamma_K < 1$. Multiphoton ionization is described by (at least) nth order time dependent perturbation theory in the unperturbed Coulomb like potential U with the laser field as the perturbing quantity and $n\hbar\omega > E_I$. If there are intermediate resonances $r\hbar\omega \simeq E_l - E_k$ between the levels k, l the ionization rate γ_I is greatly enhanced. As $\gamma_I \gg \omega$ ionization is incoherent.

If matter is exposed to high laser intensity the deformation of the original symmetric potential is deformed in direction of the laser field and the electron can be extracted by *field ionization*. In a fast rising laser pulse, for instance an atom crossing a tiny beam, the Coulomb field is depressed below E_I by the laser field and the electron can escape classically (*barrier suppression ionization* of electron 1 in Fig. 1.3), or it is lowered to form a narrow wall relative to E_I and the electron escapes by tunnelling before the laser field inverts (*tunnelling ionization* of electron 2 and 3, the same Figure). In this strong field limit ionization shows a high degree of coherence. The tunnelling probability is highest in the laser field maxima and minima. As a rule it can be assumed that ionization happens in the neighbourhood of the *critical field* where potential and ionization energy are equal in magnitude. As the bounce frequency of the bound electron is much higher than the laser frequency tunnelling is adiabatic and the velocity there is close to zero. Let us assume that the electron is set free at the instant t_f by a long laser pulse (e.g. ns pulse) acceleration, velocity, position, drift velocity v_d, and energy for $t > t_f$ and $t_f = \infty$, respectively, are determined by

$$\dot{v} = -\frac{e}{m_e}\hat{E}\cos\omega t, \quad v = -\frac{e}{m_e\omega}\hat{E}\sin\omega t + v_f, \quad x = \frac{e}{m_e\omega^2}\hat{E}\cos\omega t + v_f(t - t_f)$$

$$v_d = -v_f = \frac{e}{m_e\omega}\hat{E}\sin\omega t_f; \ \mathcal{E}(\infty) = \frac{1}{2}m_e v^2. \tag{9.8}$$

Except $E(t) = \pm\hat{E}$ the electron is born with a drift velocity v_d. The Coulomb field acts like a phase shift in a collision.

9.2.3.2 Coherent High Harmonics from Gases

Perturbative multiphoton ionization of the single atom/ion shows a sequence of incoherent harmonics rapidly decaying in intensity. With increasing laser intensity and transition towards tunnel ionization the spectrum $I_{n\hbar\omega}$ undergoes a remarkable change: (i) a fast decay in intensity at low harmonics is followed by a plateau of typical intensity of $10^{-5}I_{\text{Laser}}$ of (ii) high order *coherent harmonics*, (iii) ending with a cut off and (iv) an adjacent (super)exponential decay. The shape reminds very much fast electron spectra like Fig. 1.22; see also Fig. 7.19, $a = 1, 7$. In linear polarization the spectrum contains only odd harmonics. This is a consequence of the inversion symmetry; even harmonics would be asymmetric. For circularly polarized laser light the situation is more complicated. A general treatment of *dynamical symmetries* (atomic/molecular + laser field) of harmonic generation is given in [32, 33]. Circular polarization is successfully adopted in detecting chiral structures by absorption spectroscopy. The theory of plateau has been elaborated by Lewenstein et al. for linear polarization and D. B. Milosevic et al. for circular polarization, see Bauer in [34]. "Unexpected" (Eberly) plateau formation has been theoretically observed first in high order multiphoton perturbation theory [35]. The tendency towards such plateau formation at high irradiation intensity can be understood classically. From the Fourier ansatz

$$E = \sum_{0}^{\infty} E_r \exp -ir\omega t + \text{cc}$$

with commonly decreasing coefficients E_r and E_r^* all components contribute to the nth harmonic field E_n if the product of any selection of E_r and $E_{r'}^*$ sum up to $n\omega$ in the associated exponents. The third harmonic for instance is composed of the fields E_3, $E_1 E_2$, $E_1 E_1^* E_3$, ...$E_{n+3} E_n^*$, ... with corresponding coefficients to make the sum dimensionally homogeneous. Due to mixing of high order contributions with small order ones the intensities $I_{n\omega}$ for different indexes n become similar to each other.

Field ionization shows two remarkable properties, detected first by Agostini et al. with a 10 ns Nd laser pulse of $4 \times 10^{13}\,\text{Wcm}^{-2}$ [36]. (i) *Channel closing* in long laser pulses: The first ionization signal appears on the energy scale with $n\hbar\omega$ in excess of the bond energy *plus* the ponderomotive potential $E_I + \Phi_p$. (ii) *Above threshold ionization* (ATI): Additional maxima of the ionization signal appear with decreasing intensity at $(n+s)\hbar\omega,\ s \geq 1$. The number of observable maxima increases with the laser intensity. The appearance of a train of maxima is not surprising because the freed electron still interacts with the ion in the continuum, and hence makes transitions possible, and is periodically modulated at laser frequency ω. The explanation of channel closing in long pulses relies on the adiabatic theorem which starts holding well from 5 laser cycles on. Thus, "long" pulse behaviour for $\omega_{\text{Ti:Sa}}$ may be as short as 10–15 fs.

9.2.3.3 Harmonic Generation Mechanism and Cut Off

A simplified theory of strong field interaction is possible and is, in its essence, accessible even to a classical treatment [37]. Over the years the *rescattering model* has been consolidated as the origin of the ionization based high harmonics (see Fig. 9.1). It consists of the three steps
(i) E_I work is done by the laser field to bring the electron up into the asymptotic free state with zero kinetic energy at the instant t_f.
(ii) In the laser field the electron starts oscillating according to $v(t)$ from (9.8).
(iii) If it returns in the vicinity of its origin it feels the field of the ion and can make a transition into the free state $E_f = E_I$ at time t_c by delivering the energy

$$E_c = E_f + \frac{1}{2} m_e v^2(t_c) = \hbar l \omega \tag{9.9}$$

into the single photon $\hbar l \omega$. If the electron is set free without recombining with the atom its maximum energy is according to (9.8) $\mathcal{E}_{\text{free}} \leq 2\Phi_p$. If however, the electron undergoes rescattering ones the energy just after the event can assume energies up to $10\Phi_p$. For the explicit verification see [34], Fig. 7.12. Finally, the cut off of the plateau of harmonics must be determined. Imposing the condition of return to the

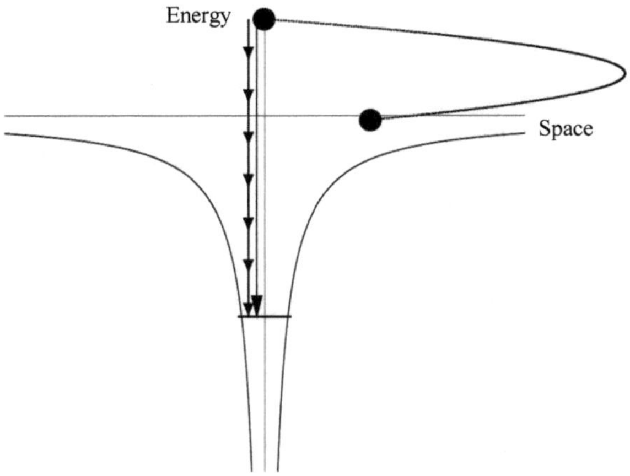

Fig. 9.1 Illustration of the three-step model for high harmonic generation. An electron is (i) released, (ii) accelerated in the laser field, and (iii) driven back to the ion. There it may recombine upon emitting a single photon which corresponds to a multiple of the photon energy of the incident laser light. From Bauer [34], Fig. 7.15

original atom means according to (9.8) $x = \frac{e}{m_e\omega^2}\hat{E}\cos\omega t + v_f(t - t_f) = 0$. Under this constraint the maximum of energies (9.9) results at the time $t_c = 0.61 \times 2\pi/\omega$ as

$$E_c = E_I + 3.17\Phi_p, \tag{9.10}$$

see [37], Fig. 2.5. Actually, closer inspection shows that E_I has to be replaced by $1.32 \times E_I$ [38]. The intensity of harmonics is tightly connected with the rate of ionization. If the incident photon energy resonates with two intermediate Stark shifted atomic levels significant increase of harmonics of one order nj and following $n > n*$ is observed [39]. For a quantum treatment of strong field ionization and harmonics, interferences and limits of validity see [34], Chap. 7. For a critical survey on strong field ionization the review by Popruzhenko [40] is particularly recommended. For multicharged ions with $Z \gtrsim 40 - 60$ at intensities of no less than $10^{22} - 10^{23}\,\mathrm{Wcm}^{-2}$ a relativistic treatment is in order [41].

Coherent HHG into the XUV beyond the water window is possible at high laser intensities. The superposition of a train of harmonics results, thanks to coherence, in intense attosecond pulses. The overwhelming significance of such short pulses lies in the unprecedented possibility to follow inneratomic dynamics by pump-probe experiments [42]. It sets new standards in metrology [43, 44]. Extremely short pulse control, quantum transitions, and atomic motions are accessible now to experiments [45]. For key problems in generating attosecond pulses see [46].

9.2.3.4 Harmonics from the Plasma

Intense high order harmonics find multiple application for diagnostics and spectroscopy in various fields of research. So for instance intense gas harmonics have made it possible to follow electron densities up to $4.3 \times 10^{23}\,\mathrm{cm}^{-3}$ with sub ps time resolution [47]. With the availability of intense ps and sub ps laser pulses the intensity of the harmonics can be increased. In gases there are severe limits to increase the pump; for example, the free electrons from multiphoton ionization may destroy the phase matching. This is the moment of truth for the plasma. A fully ionized carbon plasma is free of such deficiencies. It yields higher harmonics because of increased E_I and Φ_p. In general, alkaline metal ions benefit from the scheme (9.9) for gases owing to their closed shells. At high laser intensities of $I \gtrsim 10^{18}\,\mathrm{Wcm}^{-2}$ generation of high harmonics directly from the plasma nonlinearities is more successful. From plane surfaces harmonics of order beyond 3000 have been produced under the angle of reflection as well as along the incident light beam. At normal incidence the Lorentz force induces the critical surface to oscillate at 2ω and gives rise to the 3rd harmonic in reflection direction. At oblique incidence a Brunel like dynamics of the critical surface at the fundamental ω may prevail (see Chap. 1) to yield the second harmonic at 2ω into direction of reflection. With increasing laser intensity higher order harmonics become significant. Those harmonics with $n\omega > \omega_p$ of the high density target can be observed also into direction of the incident laser. To obtain unambiguous results high contrast of the laser beam is a prerequisite. On the other hand the interaction of the intense laser with an underdense preformed plasma may be advantageous because of the internal coupling with the resonantly excited electron plasma wave at the plasma frequency $\omega_p = \omega$. Coupling of the laser with plasmons excited by energetic bunches of electrons in the interior of the target and modulating the harmonic signal in a peculiar manner also happens [48]. The interaction of the laser with condensed matter offers the advantage of a whole variety of interactions by changing from flat surfaces of solids and liquids to droplets, liquid jets, clusters and fullerenes. In the multi-well potential of an Ar^+ cluster the "universal" constant of 3.17 in (9.10) changed into 3.6 and 4.0 [49].

Odd and even harmonics from overdense flat targets, at least into reflection direction, can be explained by the (relativistic) *moving mirror* introduced by Bulanov in the context of plasma high harmonic generation (PHHG), see [50]. Mutatis mutandis it applies at subrelativistic as well as superrelativistic laser beams. In the highly relativistic regime, e.g. $a = eA/m_ec \sim 20$, Gordienko and Baeva have generalized substantially the moving mirror idea and with its help and with the help of similarity analysis they have arrived at a "*universal spectrum*" of high harmonic generation from plane plasma targets [51].

Universal spectrum. A plane laser wave is perpendicularly incident along $-x$ with polarization of E in y direction. The Lorentz force induces the skin layer to oscillate at 2ω along x with velocity v_x. In this geometry the canonical momentum $p_y = \gamma m_e v_y - eA(x,t) = 0$ is conserved. From $\gamma m_e c^2 = [m_e^2 c^4 + c^2(p_x^2 + p_y^2)]^{1/2}$ follows

$$\frac{v_x}{c} = \frac{p_x}{m_e c[1 + p_x^2/m_e^2c^2 + a^2(x,t)]^{1/2}}; \qquad E(x,t) = -\frac{\partial A(x,t)}{\partial t}. \tag{9.11}$$

At the points where the vector potential A vanishes v_x is highest and approaches c. In the neighbourhood of A passing through zero an observer comoving with an electron against the incoming laser sees ω Doppler upshifted to $\omega' = \gamma_x(\omega + kv_x) \simeq 2\gamma_x\omega \gg \omega,\ \gamma_x = (1 - v_x^2/c^2)^{1/2}$. The skin layer undergoes a periodic distortion during one laser period owing to v_x differing from point to point and perhaps in time. The knowledge of this skin layer dynamics can be circumvented by the correctly formulated *relativistic oscillating mirror*. Its central part is the knowledge of the *apparent reflection point* $X(x, t)$. In region of the validity of the WKB approximation, and only there, a wave can be uniquely splitted into an incident and into a reflected component, in particular in vacuum or homogeneous medium. In the skin layer generally it fails (e.g. see (5.178) and following note). Analogously to $X(x, t)$ in Sect. 5.5.2 x_0 was introduced to show that absorption in the evanescent region is already included if the integration starts from there). From the overdense plasma the incident laser wave is totally reflected and forms a node nearest to (or in) the skin layer. It is the point $X(x, t)$ at which A and E are zero at all times. In the underdense plasma such an apparent reflection point can equally be defined. From the motion of $X(x, t)$ the reflected Doppler shifted wave is uniquely defined. Its dynamic and the generation of coherent harmonics up to wavelengths of the order of interatomic distances in crystals is described step by step in [52]. The author call the central part of harmonics "universal" because they are born in a neighbourhood of the zero vector potential $X(x, t)$, the details of the inner dynamics of the skin layer and the laser pulse shape are of secondary importance. The main production of harmonics occurs around the turning points of E twice per laser period and confers the typical asymmetric shape, with a strong slope into forward direction, to the reflected E field as depicted in Fig. 5.9 of the spiking electron plasma wave. In Fig. 9.2 a PIC simulation of high order harmonics for the pump amplitudes $a_0 = 5$, 10, and 20 is presented. Main results from the relativistic oscillating mirror are the scaling of the intensity of the harmonic order n as $I_n \sim n^{-8/3}$, a cut off with following exponential intensity decay at $n^* \simeq \sqrt{8}\max(\gamma_x)$ and the generation of pulses of zeptosecond length (1 zs = 10^{-21} s). The spatial coherence is such as to allow the harmonics wavelength to distinctly undercut the surface roughness of the plane target. Physically, the latter property is connected with the fact that in the skin layer all points oscillate in phase, corresponding to $c = \infty$, and the refraction index of the high harmonics is almost 1. Experiments supporting a spectral decay $\omega^{-8/3}$ were reported in [53].

The spectrum of high harmonics described in [52] appears in back reflection direction. In forward direction no such Doppler shift $X(x, t)$ exists. Instead in laser beam direction high harmonics from *microbunches* of plasma originating from fast electrons in front of very thin diamond like targets of thickness $d \ll \lambda$ have been registered, see Fig. 9.3. In the particle-in-cell (PIC) simulation a 500 fs Nd laser pulse of amplitude $a_0 = 20$ impinges normally onto a 200 nm thick, 800 times overdense solid target and generates high order harmonic pulses in propagation direction. The fully ionized target is opaque to harmonics of order lower than $28 \simeq \sqrt{800}$. The spectrum recorded during 100 fs in the rising edge of the pulse shows the intensity dependence of $n^{-4/3}$ up to order n = 400, clearly distinguished from $n^{-8/3}$ from the relativistic oscillating mirror (ROM), see inset. According to the authors the generating mechanism of the harmonics is *coherent synchrotron emission* (CSE).

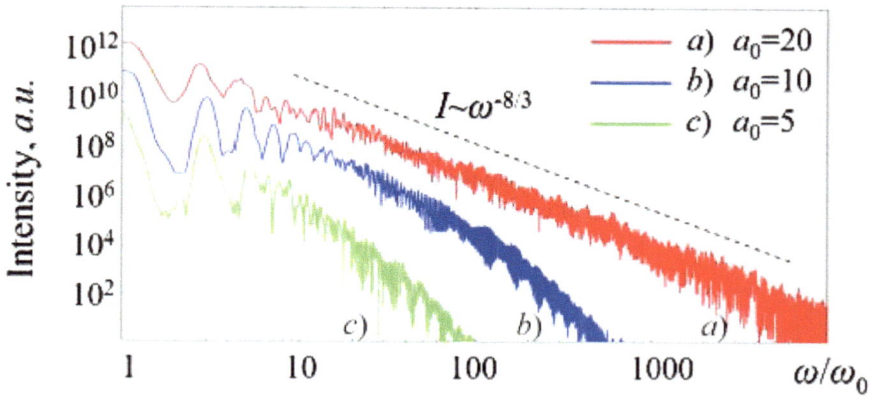

Fig. 9.2 PIC simulation of high harmonics from plasma surface of a plane laser wave of amplitude $a_0 = 5, 10, 20$. At superrelativistic pump strength their intensity I scales like $\omega^{-8/3}$. Courtesy of Baeva [52]

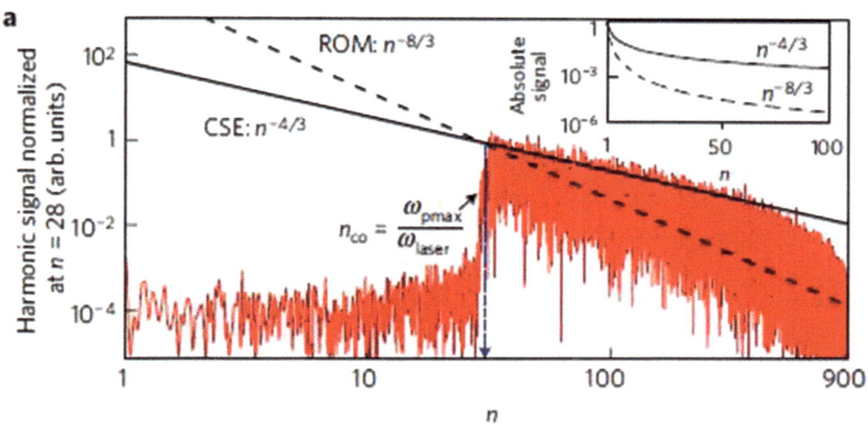

Fig. 9.3 1D PIC simulation of high harmonics from a 200 nm thick 800 times overdense solid carbon target. The incident supergaussian pulse of $a_0 = 20$ is linearly polarized; the harmonic emission into forward direction is due to coherent synchrotron emission (CSE); it scales like $\omega^{-4/3}$. For comparison the scaling from the relativistic oscillating mirror (ROM) is indicated by the dashed black line. Courtesy of Dromey [54]

Universality of ROM spectrum? The ROM model as well as the CSE model are both based on coherent single electron motion; they are purely kinetic. Simultaneously, the plasma is subject to collective response, close to frequency ω_p at low level of excitation and at broadened ω_p and harmonics under strong drive. If the external driver is periodic in ω broad band harmonics $n\omega$ are generated. In p polarization the fast electrons from the target surface act as the periodic driver, see Figs. 1.23 and 7.18. Reentering the plasma target they excite high amplitude plasmons, Figs. 7.17 and 7.20. Examining the literature on harmonics a diversity of scaling laws has been reported. The ROM spectrum of power $-8/3$ was constructed from the only similarity parameter $S = n_e/n_c a_0$. Performing PIC simulations in the interval $1 < S \lesssim 5$ transitions from ROM scaling to 5/3 power were obtained and evidence for plasma oscillations was corroborated "so serving to undermine the notion of universal decay [55]. Plasma waves are excited by bunches of UR electrons, generated by Brunel absorption." In a recent study the authors introduced the model longitudinal field E_l of a plasma soliton of extension D and dipole moment amplitude $\hat{p}$

$$E_l = \frac{2}{3}\pi e n_e D e^{-i\omega_p t}\hat{p} \tag{9.12}$$

to take account of the collective plasma response [56]. The single particle behaviour was then compared with corresponding PIC results, a comparison which reveals a rich palette of scaling. In Fig. 9.4 pictures (a) and (c) refer to PIC simulations with $a_0 = 60, n_e/n_c = 95$ and $a_0 = 85, n_e/n_c = 100$, respectively. In (a) the low order harmonics at $n \simeq 100$ goes over from $p = 5/3$ power to $p = 4$ and ends in the flat decay $p = 2/3$ which is identified as bremsstrahlung emission from the hot electrons. In (c) the bremsstrahlung scaling precedes the fast decay, somehow expected since the energy of fast electrons grows with a_0 increase and reduced S. Pictures (b) and (d) are the results of (9.12). The correspondence is satisfactory. A domain of rapid decay like $p = -4$ has been extracted in one of the early numerical studies of the subject in [57] ($p = -5$ there).

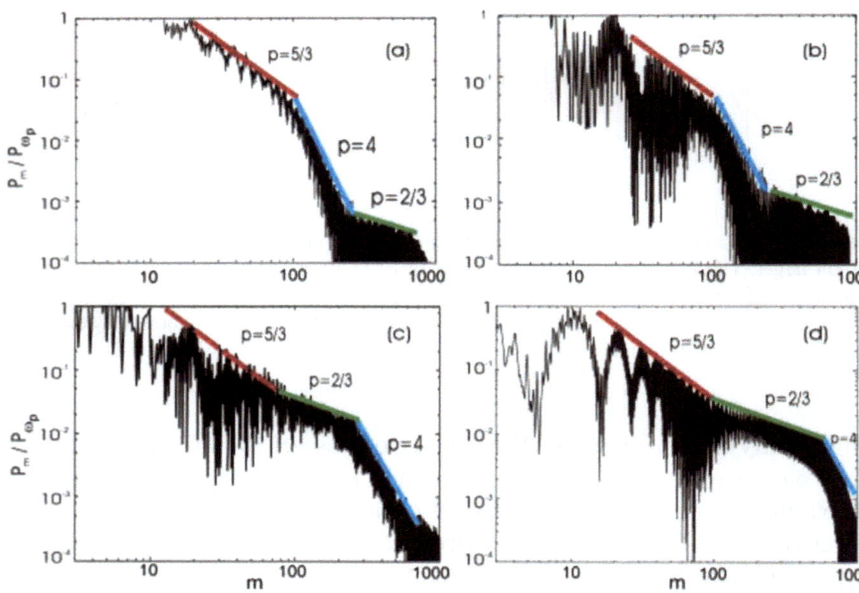

Fig. 9.4 1D PIC simulation of high harmonics from a 200 nm thick 800 times overdense solid carbon target. The incident supergaussian pulse of $a_0 = 20$ is linearly polarized. Pictures (a) and (c) refer to PIC simulations with $a_0 = 60, n_e/n_c = 95$ and $a_0 = 85, n_e/n_c = 100$, respectively; pictures (b) and (d) are the results of (9.12). Courtesy of Boyd [56]

The actual situation with the scaling of high harmonics is not dissimilar to the scaling of fast electron generation by intense laser pulses. Both need more specific investigation. The efficiency of conversion of laser light has been investigated by 1D PIC simulations in plane solid targets in [58]. Wave form controlled HHG has reached a 10% maximum energy conversion into harmonics in the range from 80 to 200 eV. It revealed as largely independent of laser intensity and plasma density. The waveforms most effective at driving harmonics have a broad spectrum with a lower frequency limit set by the width of the incident pulse envelope and an upper limit set by the relativistic plasma frequency.

So far harmonic scaling from flat targets has been considered. One may ask for the influence of the shape of the target on the harmonic spectra and their conversion efficiency. It is known that laser coupling increases in foams and clusters. There are several reasons for the increase, from geometrical optics to the richness of resonances. A thorough study on enhancement of the second and third harmonics in nanoclusters has been presented in a review by Fomichev and Becker [59]. Among a whole variety of phenomena Mie oscillations and multipole and secondary resonances have been treated by the authors.

9.3 Controlled Nuclear Fusion

As early as in 1963, soon after the development of powerful pulsed lasers, R. Kidder and N. G. Basov came out with the first proposals to gain nuclear fusion energy from laser heated and imploded small spheres. With the following decades theoretical and experimental activities started worldwide to realize the ambitious goal. The state of the art in this effort is marked best by the fusion energy output of 54 kJ corresponding to the number of fusion reactions of 1.9×10^{16} in one laser shot from the National Ignition Facility (NIF) in Livermore [60].

From the scientific point of view the inertial fusion project is very ambitious and an excellent training field for young scientists. All disciplines of physics are involved and extrapolated towards the extremes: Thermodynamics and matter far from equilibrium, fast scale unsteady fluid dynamics, equations of state of highly compressed matter and extremely hot diluted plasma, radiation physics at all degrees of opacity from black body radiation strongly coupled with matter to decoupled bremsstrahlung radiation, unstable behaviour of matter, and tasks of stabilization.

9.3.1 *Plan and Requirements*

The working substance of the fusion reactor is a $1:1$ mixture of deuterium and tritium. A deuterium nucleus D combines with a nucleus of tritium T to yield an α particle (He_2^4) of $E_\alpha = 3.5\,\mathrm{MeV}$ and a neutron of $E_n = 14.1\,\mathrm{MeV}$,

$$\mathrm{D}_1^2 + \mathrm{T}_1^3 \Rightarrow \mathrm{He}_2^4 + n + 17.6\,\mathrm{MeV}. \tag{9.13}$$

Thus, 1 kg DT delivers 340 TJ fusion energy; $\rho_{\mathrm{DT}} = 0.2\,\mathrm{gcm}^{-3}$. For comparison, the chemical reaction output of solid hydrogen is $120\,\mathrm{MJkg}^{-1}$ and of benzine is $47\,\mathrm{MJkg}^{-1}$. The comparison reveals the pros and cons of inertial fusion: $10^6 - 10^7$ higher energy output from nuclear forces (+) and miniaturizing to attain non destructive gain (−).

The Lawson Criterion

The reaction rate of (9.13) reaches a flat maximum of $\langle\sigma v\rangle = 8.5 \times 10^{-22}\ \mathrm{m^3s^{-1}}$ at an ion temperature of 60 keV. In the temperature range of interest for us it is described by $\langle\sigma v\rangle\,[\mathrm{m^3s^{-1}\,keV}] = 1.1 \times 10^{-24}T$. From equating the heating $3nk_BT$ to the fusion energy output $(n^2/4)E_\alpha\tau$, τ the confinement time, one arrives at the *Lawson criterion*

$$n\tau \geq \frac{12k_BT}{\langle\sigma v\rangle E_\alpha} = 1.5 \times 10^{14}\,[\mathrm{cm^{-3}s}] \tag{9.14}$$

evaluated at the flat minimum of $n\tau$ around $T = 35$ keV. If a *gain* of energy output G is postulated the limit in (9.14) is to be multiplied by G. Bremsstrahlung radiation from free-free electron encounters with the ions puts a lower limit on the ion temperature $T = T_i = 4.2$ keV. Below this mark radiation losses with proportionality $T_e^{1/2}$ prevail on α particle heating $(n^2/4)\langle\sigma v\rangle E_\alpha$. Normally the neutrons escape. Nuclear ignition and burn start from α particle heating.

For inertial fusion criterion (9.14) is cast into a more appealing form. To this aim let a spherical pellet of solid DT have an initial radius R. The confinement time τ is given by (1.43) as R/c_s with the isothermal ion sound speed

$$c_s = \left(\frac{2k_BT}{m_{\mathrm{DT}}}\right)^{1/2} = 6.6 \times 10^5 (T[\mathrm{eV}])^{1/2}\,\mathrm{cm\,s^{-1}};\quad 1\mathrm{eV} = 10^4\,\mathrm{K}. \tag{9.15}$$

Replacing τ from (9.14) yields for $T = 10$ kev minimum radius $R_{\min}$ and minimum input energy $W_{\min}$ for solid DT compressed κ times,

$$R_{\min} = G\frac{n\tau}{n}c_s = \frac{1.5 \times 10^{14} \times 6.6 \times 10^7}{4.5 \times 10^{22}} = 0.22\,G\,\mathrm{cm}.$$

$$W_{\min} = \frac{4\pi}{3}\left(\frac{R_{\min}}{\kappa}\right)^3 \times 3\kappa nk_BT = 7.0 \times 10^6\frac{G^3}{\kappa^2}\,\mathrm{J};\ \kappa = \frac{n_0}{n}. \tag{9.16}$$

The minimum energy input scales with G^3/κ^2. For solid DT and *breakeven* corresponding to $G = 1$ nearly 7 MJ heating energy of the pellet under idealized conditions are required.

A more appealing criterion for ignition is expressed in terms of the plasma pressure which is also the approximate thermal energy density of the plasma. Multiplication of (9.14) by $2k_BT$ results in

$$nk_BT\tau \geq \frac{24k_B^2T^2}{\langle\sigma v\rangle E_\alpha} = 10\,\tau\,\text{bar s} = 10\,\text{Gbar ns} \tag{9.17}$$

if the product is evaluated at the flat minimum of $T = 14\,\text{keV}$.

The α particles are thermalized by collisions with the electrons and, depending on collective effects approximately above 5 keV, with the plasma ions. For a maximum range of the α particles we determine the mean free pass λ_α from the stopping by the electrons alone in the compressed fuel. The range of the α particles λ_α is calculated as a function of compression κ and temperature in units of 10 keV from

$$\lambda_\alpha = \frac{m_\alpha v_\alpha}{2m_e\nu_{e\alpha}} = \frac{3.5}{\kappa \ln\Lambda}\left(\frac{T[\text{keV}]}{10}\right)^{3/2}\,\text{mm};\quad v_\alpha = 1.2\times 10^6\text{cm s}^{-1}. \tag{9.18}$$

The thermal electron velocity is

$$v_{\text{the}} = \left(\frac{k_BT}{m_e}\right)^{1/2} = 1.3\times 10^9 (T[\text{keV}])^{1/2}\,\text{cm s}^{-1}. \tag{9.19}$$

The simple estimate shows that the energy $W_{\min}$ required to ignite the DT pellet is high and depends crucially on gain $G > 1$ and reduces quadratically on compression κ of solid DT. The realization that only compression can lead to acceptable driver energies (laser, ion beam) for pellet ignition marks the starting point of any realistic route to inertial fusion [61]. The idea has led to a manifold of different approaches: stable *directly* and *indirectly* driven *compressional heating*, *fast ignition*, *shock ignition*, and has finally ended with the decision to build four big installations: *National Ignition Facility* (NIV) in Livermore, *Omega* at the *Institute of Laser Energetics* (ILE) in Rochester, *FIREX/GEKKO12* at the *Institute of Laser Engineering* (ILE) at Osaka University, *Laser Mega Joule* (LMG) in Bordeaux.

9.3.2 Compressional Pellet Heating

The optimum pellet design underlies, besides reaching a minimum ignition temperature, some limits of DT fuel mass to keep the energy release under technical control, of fuel compression κ to limit the laser pdV work by mass ablation, of laser wavelength for good absorption and avoiding fast electron preheat. Efficient compression by long low intensity laser pulses is achieved, in principle, by starting from thin hollow shells of solid DT with large *aspect ratio* $R/\Delta R$ in order to gain a long acceleration path and to keep the laser intensity low. Furthermore the laser pulse must increase in a way that all characteristics intersect simultaneously close to the center. According to (3.217) the acceleration pressure P_a is coupled to the laser intensity I by $P_a \sim I^{2/3}$ of the shell. However, here the Rayleigh–Taylor instability puts severe upper limits. Considering the limits altogether one ends up with a window for design for temperature above 2 keV, outer radius of DT shell $R \lesssim 2\,\text{mm}$, aspect ratio $R\Delta R \lesssim 20$. It is perhaps most instructive to follow the dynamics of pellet compression and heating

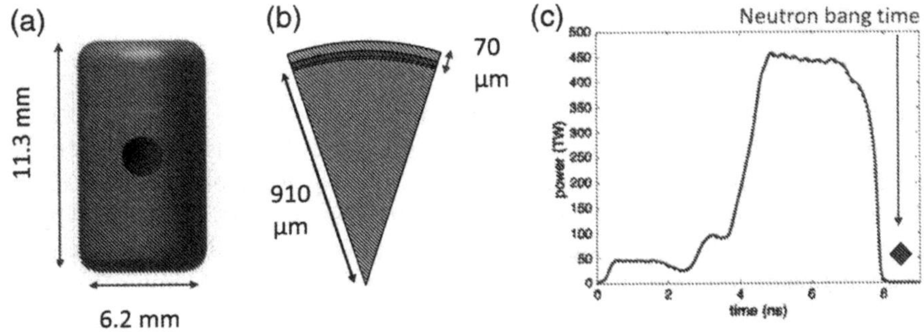

Fig. 9.5 High foot hohlraum driven experiment with the National Ignition Facility (NIF). **a** Cylindrical hohlraum collecting 192 laser beams entering through the holes of the circular end faces. **b** Structure of the spherical shell. **c** The laser pulse. The hydrocarbon capsule of 70 μm shell thickness houses a 20 μm thick tungsten doped layer to shield the fuel from superthermal X rays. Courtesy of LLNL [60]

designed for NIF. The best results so far (shot number N170827) have been achieved with a spherical shell of 70 μm thickness and 910 μm outer diameter irradiated by the 192 beams of NIF in a *hohlraum* sketched in Fig. 9.5. The laser pulse of 7.5 ns length and 450 TW peak power is of *high foot* type, i.e. along a compression adiabate of $\simeq$2.5 Fermi pressure. High foot allows shortening of the pulse length and to bring end of pulse and stagnation closer together compared to the former lower adiabate of 1.5 (*low foot*) experimental series. The pellet is compressed by the Planck radiation of 290 eV temperature. A collection of key achievements realized with the best shot so far are as follows [60]:

Total neutron yield $1.9 \times 10^{16} \pm 3 \times 10^{14}$	Fusion yield 54 kJ
DT ion temperature 4.5 ± 0.15 keV	Flow velocity 3.95×10^7 cms^{-1}
Stagnation pressure 360 ± 45 Gbar	Nuclear burn time 154 ± 30 ps
Hot spot $\rho r = 0.30 \pm 0.034\,\mathrm{gcm}^{-2}$	Alpha particle energy deposition 0.87
Shell max kinetic energy 21 ± 5 kJ	Alpha deposited energy 9.3 ± 1.6 kJ

The α particle range is calculated from

$$\rho\lambda_\alpha[\mathrm{gcm}^{-2}] = \frac{0.025(T_e[\mathrm{keV}])^{5/4}}{1 + 0.0082\, T_e[\mathrm{keV}])^{5/4}}. \tag{9.20}$$

After commissioning of NIF in 2009 the National Ignition Campaign (NIC) started for the next 4 years on the basis of the low foot compression scheme (1.5 adiabate). Unprecedented time and space resolved diagnostics made the comparison of experiments with numerical modelling possible. In several respects reality agreed with expectations, e.g. fuel compression and ρr values, however ignition failed, the hot spot temperature remained far below the requirement. It has been attributed to the complex interplay between hohlraum radiation coupling with the ablation and compression process (energy exchange between laser beams via plasma coupling in the hohlraum cage, back scattering of light, uncontrollable ablation pressure asymmetries), faster Rayleigh–Taylor growth and cooling due to mixing of hot fuel with cold plasma. For a detailed discussion see for instance [62].

Notwithstanding thermonuclear ignition and burn fell short, from the scientific point of view the achievement with NIF are great: twice the gas pressure of the center of the sun and three times higher temperature reached, neutron fluxes that induce to think of experiments with s branch nuclear syntheses and astrophysics in the laboratory. After all the progress from NIC to the recent high foot experiments let vote for optimism.

Two alternative scenarios towards controlled inertial fusion have to be mentioned. *Shock assisted inertial fusion*: At the end of the compression laser beam a powerful shock is launched to the pellet center by the same laser, or a particle beam, under most uniform illumination as possible [63] to form the ignition spark. It should lead to higher gain and better control of the Rayleigh–Taylor instability owing to the higher mass density and lower ablation pressure involved [64–66]. The crucial point of inertial fusion is the igniting spark in connection with nearly perfectly centred gas flow, altogether realized by one laser pulse. For an improved ignition criterion and comparison with large tokamaks, e.g. JET, see [67, 68]. Superintense sub ps lasers should make it possible to circumvent the singularity by decoupling compression from ignition.

9.3.3 Fast Ignition

Assume the fusion capsule moderately precompressed at reduced ablation pressure by the powerful ns laser. A superintense sub ps laser pulse is fired to drill a hole through the compressed shell by light pressure and to deposit its main energy in the high density center just at the moment of stagnation in order to start the thermonuclear burn [69]. The required laser peak intensity was estimated to range between 10^{19} and $10^{20}\,\mathrm{Wcm}^{-2}$. The predictions have proven too optimistic. (i) The required laser intensity has to be chosen one order of magnitude higher, i.e. between 10^{20} and $10^{21}\,\mathrm{Wcm}^{-2}$, the “free ignition energy” necessary to ignite the center [70] ranges from 10–20 kJ, the energy to heat the precompressed pellet varies from optimistic 50 kJ to 70 kJ. (ii) Hole boring at realistic intensities stops latest at 80$\times$ solid DT [71]. In numerical studies the energy is supplied to the relativistic critical surface and then transported to the center by flux limited heat conduction or by energetic electrons. The inward piston motion comes to rest after 20–70 ps and then starts moving outward against the incident beam. At stopping the density of the tip never exceeds $10\,\mathrm{gcm}^{-3}$ DT [70]. Modelling of the laser beam as an impermeable cone of varying aperture angle reproduce quite similar stopping results. Hole boring and fast ignition seem to exclude each other: When there is hole boring, no ignition occurs, and vice versa. The laser beam pressure only causes a more or less deep cone-shaped critical surface that leads to better guidance of the beam and to improved laserplasma coupling. At laser wavelengths of the order of 1 m, successful fast ignition requires strong anomalous laser beam-pellet coupling [72]. Very recently Iwata et al. came to a very similar conclusion [73]. The analysis on hole boring the authors have undertaken by means of analytical modelling and supporting it by extensive particle in cell simulations is convincing. None of the 1D and 2D simulations invade the hole boring region "prohibited by theory". Lateral heat spreading in the low density corona is significant. If the transport from the critical surface to the center is assumed to occur by fast electrons their energy should not exceed 2 MeV. For missing hole boring the original scheme has to be abandoned.

An improved version is *cone guided fast ignition* [74, 75]. The energy deposition zone can be brought as close as possible to the pellet center by mounting the pellet on a hollow cone of high Z material through which the intense pulse is sent. The biggest installation of this kind is FIREX in Osaka/Japan at the Institute of Laser Engineering (ILE). According to theoretical predictions the highest neutron output is obtained from synchronizing the fast ignition pulse with the confluent plasma just a moment before stagnation. Under the condition of supply of 70 kJ the density of the deposition zone has not to be less than 4–5 gcm^{-3} DT; below 1 gcm^{-3} DT no burn wave evolves.

The advantages of fast ignition with powerful short pulse lasers are well known: (i) Decoupling of compression from ignition phase and with it additional freedom in designing inertial fusion scenarios, (ii) less requirements on pulse shaping of the main compression pulse, (iii) lowering of symmetry constraints on peak compression. Asymmetric configurations with center of mass displaced by 0.6 critical radius from geometrical center led to an efficiency reduction from 6.5 % to 5.4 % only [70].

9.4 Ion Acceleration by TNSA

Intense short ion pulses in the MeV energy range are of interest in medical applications, e.g. cancer therapy, and in a whole variety of technical tasks. When a high density target, solid or liquid, is exposed to an intense laser field fast ions escaping from the rear side are observed. The phenomenon has soon found its explanation and has been known under the denomination of *target normal sheath acceleration* or TNSA. It operates in the overdense target-vacuum interface opposite to the incident laser beam. The energetic electrons created by the laser beam penetrate the cold target and quickly ionize it. Some of the most energetic electrons may escape to infinity, however the majority of them are confined in the potential of the ions which at the rear side of the target form a step like distribution. In the sudden transition to vacuum plasma quasineutrality is violated; the electron density falls off over the characteristic length of the hot electron Debye length. The charge imbalance entails an intense localized quasi static electric field which extracts ions from the target and accelerates them to energies of tens of MeV. The hotter the electron cloud the higher the accelerating potential. The aim of the theoretical description of the phenomenon is to arrive at the energy spectrum of the ions for $t = \infty$ and their energy cut off. The problem is too complex as to allow an exact solution. Instead several simplified models have been proposed and compared with accompanying numerical simulations [76]. Three of them will be treated in more detail in the following: Mora's ion dynamics model [77], Passoni's quasi static model [78], and Schreiber's quasistatic hot electron spot model [79]. The models and a variety of alternative ion acceleration schemes, illustrated by a rich selection of experiments, are reviewed in [80] and, in addition, in [81]. According to the author's knowledge the maximum proton energy above 85 MeV so far achieved from the relativistic interaction of laser pulses with micrometer thick CH_2 targets in the TNSA mechanism have been reported in [82]. For extension of the three models, comparison and critical discussion see [83]. For recent experiments in the relativistic regime under oblique incidence see [84].

9.4.1 Dynamic Model of Acceleration

Plane Isothermal Rarefaction Wave

Under the assumption of quasineutrality an isothermal ideal fluid filling the half space $x \leq 0$ expands into vacuum according to

$$\rho = \rho_0 \mathrm{e}^{-u/c_{s0}} = \rho_0 \mathrm{e}^{-(1+x/c_{s0}t)}, \quad u = c_{s0} + \frac{x}{t}; \quad T = T_0. \tag{3.157}$$

In order to keep $T = T_0$ forever at each position x the electron heat flux $q = pc_{s0}$, $p = nk_B T_0$ into the expanding plasma must be provided locally. The quasi static electric field E and its potential follow from (3.44),

$$x \geq -c_{s0}t : \; E = -\frac{\partial_x p_e}{en_e} = \frac{k_B T_0}{ec_{s0}t} \Rightarrow \Phi(x) = \Phi_0(x) - \frac{k_B T_0}{e}(1 + x/c_{s0}t); \;\; n_e = \frac{\rho Z}{m_i}. \tag{9.21}$$

From Φ and ρ one concludes

$$\frac{e(\Phi - \Phi_0)}{k_B T_0} = -\big(1 + \frac{x}{c_{s0}t}\big) = \ln\frac{\rho}{\rho_0} \Rightarrow n_e = n_{e0}\exp\left(\frac{e(\Phi - \Phi_0)}{k_B T_0}\right); \;\; (\Phi - \Phi_0)|_{-\infty} = 0. \tag{9.22}$$

Note, this functional dependence of n_e on Φ is valid as long as $m_e \mathrm{d}u_e/\mathrm{d}t \ll m_i \mathrm{d}u_i/\mathrm{d}t$ holds; quasineutrality is not required for its validity. The electric field is constant in space and decreases $\sim 1/t$; the potential $\Phi(x \to \infty)$ tends to $-\infty$ for all $t \geq 0$. For any finite layer $-d \leq x \leq 0$ at $t = 0$ this is in net conflict with the energy conservation. The isothermal solution must be truncated somewhere. For this in the past a whole variety of proposals has been presented. Looking for the main reason of the failure of expressions (3.157) one finds that quasineutrality must be violated latest at positions where the density scale length $L(x) = \rho/|\partial_x \rho|$ falls below $\lambda_D(x)$. This is indeed the case for $T_i = 0$, $T_e = T_0 > 0$,

$$L = c_{s0}t, \quad \lambda_D = \left(\frac{\varepsilon_0 m_i k_B T_0}{e^2 Z^2 \rho}\right)^{1/2} = \frac{c_{s0}}{\omega_{\mathrm{pi}}}; \quad \omega_{pi}^2 = \frac{n_i (Ze)^2}{\varepsilon_0 m_i}, \;\; Zn_i = n_e.$$

$$L \leq \lambda_D \Leftrightarrow \rho \leq \frac{\varepsilon_0 m_i}{(Ze)^2 t^2} \Leftrightarrow c_{s0} t \leq \frac{c_{s0}}{\omega_{\rm pi}} \Leftrightarrow \frac{1}{\omega_{\rm pi} t} \geq 1. \tag{9.23}$$

For t fixed $L = \lambda_D$ is reached when the density ρ becomes as low as $1/\omega_{\rm pi}^2 t^2$, or when $\omega_{\rm pi} = 1/t$. Equivalently, for all time instants t there exists a position $x = x_F(t)$ from which on quasineutrality is violated. At $x = 0$ it is the case for $t = 0$ owing to $L = 0$ there. As stated above quasineutrality is not required to obtain (3.157) for $x \leq x_F(t)$ (but would be required for (3.157) to be valid in the whole domain $x \leq \infty$). Thus, (9.21) E and Poisson's equation describe correctly the degree of quasineutrality violation at $x = x_F(t)$ and a small neighbourhood around.

We calculate the electric field $E_0 = E(x = 0, t = 0)$. Multiplication of Poisson's equation with $\Phi' = \partial_x \Phi$ yields from (9.22)

$$\frac{\varepsilon_0}{2}(\Phi'^2)' = en_{e0}\left[\exp\frac{e(\Phi - \Phi_0)}{k_B T_0} - 1\theta(0 - x)\right]\Phi'; \qquad x \leq x_F(t), \tag{9.24}$$

θ is the Heaviside function. At $x = \pm\infty$ the electric field $E = -\phi'$ is zero because the total electric charge of electrons and ions is balanced within these limits. We fix $\Phi_0 = 0$ and obtain from (9.22) $\Phi(-\infty) = 0,\ \Phi(+\infty) = -\infty$. Integration of (9.24) from $-\infty$ to $+\infty$ generates

$$\frac{\varepsilon_0}{2}[\Phi'^2(+\infty) - \Phi'^2(-\infty)] = \int_{-\infty}^{+\infty} n_{e0} k_B T_0 \left[\exp\left(\frac{e\Phi}{k_B T_0}\right) - 1\theta(0 - x)\right] \mathrm{d}\frac{e\Phi}{k_B T_0}$$

$$= n_{e0} k_B T_0 \left[\exp\left(\frac{e\Phi(0)}{k_B T_0}\right) - 1\right] - n_{e0}\Phi(0) - n_{e0} k_B T_0 \exp\left(\frac{e\Phi(0)}{k_B T_0}\right) = 0$$

$$\Rightarrow \quad e\Phi(0) = -k_B T_0$$

$$(9.24) \Rightarrow \frac{\varepsilon_0}{2}\Phi'^2(x \geq 0) = -en_{e0}\int_{x\geq 0}^{\infty} \exp\left(\frac{e\Phi}{k_B T_0}\right)\Phi' \mathrm{d}x = n_{e0} k_B T_0 \exp\left(\frac{e\Phi(x)}{k_B T_0}\right). \tag{9.25}$$

Thus, at position x and time $t = 0$ results $\Phi'^2 = (2/\varepsilon_0 \mathrm{e}) n_{e0} k_B T_0$ and

$$E_0 = \left(\frac{2}{\varepsilon_0 \mathrm{e}}\right)^{1/2} (n_{e0} k_B T_0)^{1/2}; \quad \mathrm{e} = \exp(1). \tag{9.26}$$

Breakdown of the Isothermal Rarefaction

The electric field of the quasineutral isothermal rarefaction wave in (3.157) may be expressed through E_0 by

$$E(x, t) = \frac{m_i}{e}\frac{c_{s0}}{t} = \left(\frac{\mathrm{e}}{2}\right)^{1/2} \frac{E_0}{\omega_{\rm pi} t}; \qquad x \leq x_F(t). \tag{9.27}$$

Equality $L = \lambda_D$ is reached after (9.23) with density ρ at time t such as to satisfy $\omega_{\mathrm{pi}} t = 1$. Indicating the ion plasma frequency of the initial density ρ_0 by ω_{p0}, the position of $x_F(t)$ follows from

$$\omega_{\mathrm{pi}} t = \omega_{\mathrm{p0}} \left(\frac{\rho}{\rho_0}\right)^{1/2} t = 1 \Rightarrow \ln(\omega_{\mathrm{p0}} t) - \frac{1}{2}\left(1 + \frac{x_F}{c_{s0} t}\right) = 0$$

$$\Rightarrow \quad 1 + \frac{x_F}{c_{s0} t} = 2\ln(\omega_{\mathrm{p0}} t). \tag{9.28}$$

According to (3.157) the flow velocity at the front is $u(x_F) = c_{s0} + x_F/t = 2c_{s0}$ $\ln(\omega_{\mathrm{p0}} t)$. The point x_F moves with phase velocity $\dot{x}_F = 2c_{s0}\ln(\omega_{\mathrm{p0}} t) + c_{s0} = u(x_F)$ $+ c_{s0}$ as expected since an infinitesimal perturbation has to move with phase velocity relative to the fluid flow. From (9.27) and (9.21) one deduces that in the hydrodynamic picture a positive surface charge $\sigma = \varepsilon_0 E_0(t)/2$ is located at the rarefaction edge of (3.157) and an identical surface charge $\sigma = \varepsilon_0 E_0(t)/2$ is found at $x = 0 - \epsilon$ to compensate -2σ of the electron cloud at $x > x_F(t)$. In contrast to an early paper on the violation of quasineutrality in the isothermal rarefaction wave [85] the author shows that the electron density distribution is monotonous in space at all times, incidentally expected intuitively.

Maximum ion acceleration occurs at the ion front for which the electric field is in excellent approximation given by

$$E_F(t) = E(x_F(t)) = \frac{2E_0}{(2\mathrm{e} + \omega_{\mathrm{p0}}^2 t^2)^{1/2}}. \tag{9.29}$$

Then, an ion at position x_F moves according to

$$\frac{\mathrm{d}v_F}{\mathrm{d}t} = \frac{Ze}{m_i} E_F(t) \Rightarrow v_F(t) \simeq \frac{Ze}{m_i} \int E_F(t)\mathrm{d}t = 2c_{s0}\ln\tau + \sqrt{t^2+1}$$

$$x_F = \int v_F(t)\mathrm{d}t \simeq 2\sqrt{2\mathrm{e}}\lambda_{D0}[\tau\ln[\tau + \sqrt{\tau^2+1}) - \sqrt{\tau^2+1} + 1];$$

$$\tau = \frac{\omega_{\mathrm{p0}} t}{2\mathrm{e}}, \ \lambda_{D0} = \left(\frac{\varepsilon_0 k_B T_e}{n_{e0} e^2}\right)^{1/2}. \tag{9.30}$$

The simplest setting for τ is to insert the laser pulse duration. However, such a setting is certainly too short for pulses of tens of fs and too long for ps pulses [86]. Additional indication may be obtained from an estimate of the thermal energy to be supplied for keeping the acceleration period isothermal (see Problems in this Chapter). Ion acceleration on the basis of an improved rarefaction wave has been presented in detail because the physical insight and the procedure adopted by the author may be a useful guide in other related problems.

9.4.2 *Static Models of Acceleration*

9.4.2.1 Monoenergetic Light Ions

An ion beam of narrow energy spread can be achieved in TNSA by applying a few nm thin layer of light ions, particularly a hydrogen film, on a μm thick heavy ion substrate. The light ions are accelerated in the isothermal hot electron cloud distinctly before a rarefaction wave starts to dilute the thick target. At least this is the given situation when one is primarily interested in the fastest ions. The advantage is that the acceleration process can be described by a static model. In what follows such a model with relativistic electrons is outlined [86].

The intense sub ps laser produces two populations of electrons of density $n_e = n_c + n_h$ with subscripts c for cold and h for hot electrons, as described for instance in Chap. 7. The heavy ions of density n_H (index H) are assumed to be ionized to charge number Z_H. Correspondingly the light ions (index L) are characterized by n_L and Z_L. For their overall numbers N it can be assumed (i) $N_L \ll N_H$ together with $n_h \ll n_c$ and $N_L \ll N_h$ so that Poisson's equation is given by

$$\frac{\partial^2 \Phi}{\partial x^2} = [n_e - Z_H n_{\mathrm{H}} H(-x)]. \tag{9.31}$$

H is the Heaviside function. The contribution of the light ions to the potential is ignored because of assumption (i). In other words, they are considered as test particles. The degree of charge of the target is determined by the 1D relativistic *Maxwell-Jüttner distribution* of the hot electrons of temperature T_e assumed constant in space and time during the acceleration process of interest (ii),

$$f_e(x, p) = \frac{n_{e0}}{2m_e c \mathrm{K}_1} \exp\left\{-\frac{\mathcal{E}}{k_B T_e}\right\}; \ \mathcal{E} = \gamma m_e c^2 - e\Phi, \ \gamma = \left(1 + \frac{p^2}{m_e^2 c^2}\right)^{1/2}. \tag{9.32}$$

The density of the bound electrons is given by

$$n_b(x) = \int_{\mathcal{E} < m_e c^2} f_e(x, p) \mathrm{d}p. \tag{9.33}$$

The degree of charge imbalance of the target is determined by the hot electrons escaping from the target of net positive local charge density $n_{0\mathrm{i}} = (Z_H n_H - n_c)e$. Ionization occurs by the fast electrons and shielding in the target interior is due to the cold electrons of vanishing Debye length [see (4.71)]. Therefore $n_{0\mathrm{i}} \to 0$ well inside can be assumed. The cold electrons are kept back by the heavy ions. The Debye sheath formed in sub ps time by the hot electrons outside the target at $x > 0$ is of the typical spatial dimension of $\lambda_D = (\varepsilon_0 k_B T_e / n_h e^2)^{1/2}$. The relevant electron density n_h in (9.31) is n_b. After introducing dimensionless variables $\xi = x/\lambda_D$, $\zeta = m_e c^2 / k_B T_e$, $\varphi = \Phi / k_B T_e$ Poisson's equation reads

$$\frac{\partial^2\varphi}{\partial\xi^2} = \mathrm{e}^{\varphi}\int_0^{\beta(\varphi)} \mathrm{e}^{-\sqrt{p'^2+\zeta^2}}\mathrm{d}p' - \frac{n_{0\mathrm{i}}}{n_{\mathrm{e}0}}\zeta\mathrm{K}_1(\zeta)\mathrm{H}(-\xi)$$
$$\beta(\varphi) = [(\varphi+\zeta)^2 - \zeta^2]^{1/2},\ p' = \frac{cp}{k_B T_e}. \tag{9.34}$$

The potential $\varphi(\xi)$ is determined by the authors under the assumptions of (iii) $\varphi'' = \varphi' = \varphi = 0$ somewhere in $\xi > 0$, the electric field $\varphi'(\xi = 0)$ to be continuous and $\varphi(\xi \to -\infty) = 0$, $(' = \partial_\xi)$. It reaches its maximum φ^* at a position $\xi^* < 0$ close to the ion edge at $\xi = 0$.

Under conditions (i)–(iii) a first integral $\varphi'(\xi)$ can e obtained from (9.31),

$$\varphi' = -2^{1/2}[\mathrm{e}^{\varphi}I(\varphi) - \mathrm{e}^{\zeta}\beta]^{1/2};\quad I(\varphi) = \int_0^{\beta(\varphi)} \mathrm{e}^{-\sqrt{p'^2+\zeta^2}}\mathrm{d}p'. \tag{9.35}$$

Integration once more yields the potential $\varphi(\xi)$ in implicit form,

$$\int_{\varphi_0}^{\varphi(\xi)} \frac{\mathrm{d}\varphi'}{[\mathrm{e}^{\varphi}I(\varphi) - \mathrm{e}^{\zeta}\beta]^{1/2}} = -\sqrt{2}\xi;\quad \varphi_0 = \varphi(\xi = 0). \tag{9.36}$$

The maximum energy a light ion can get is $\mathcal{E} = k_B T_e Z_L e\varphi^*$. The potentials φ^* and φ_0 are related by

$$\varphi_0 = \frac{\mathrm{e}^{\varphi^*}I(\varphi^*)(\varphi^* - 1) + \mathrm{e}^{-\zeta}\beta^*}{\mathrm{e}^{\varphi^*}I(\varphi^*)}. \tag{9.37}$$

The potential φ is entirely given by (9.36) and (9.37) once φ^* is fixed.

The analytic model presented here can be used to make predictions on maximum TNSA light ion energies attainable from laser pulse energies E_L and corresponding irradiances $I\lambda^2$. Roughly characterizing the acceleration mechanism it can be said that the energy of the fast electrons is, for fixed wavelength, determined by the laser intensity I_L, whereas their number is a function of the energy of the laser pulse. To make predictions on the functional dependence of the hot electron energies the assumption of a reliable scaling law is in order. However, it is precisely this requirement which represents the weakest link of the chain. Based on a "ponderomotive scaling" of the fast electrons and the restrictions (i)–(iii) the authors of the static model arrive at the light ion energies depicted in Fig. 9.6 as a function of irradiances $I_L\lambda^2$ and sub ps laser pulse energies E_L. As an example 100 MeV ions are produced from a 100 J Nd laser pulse focused to $I = 10^{21}\ \mathrm{Wcm}^{-2}$.

9.4.2.2 Maximum Ion Energy

The simpler the model the higher its range of application. We present the model for the determination of the maximum ion energy in the TNSA process introduced by Schreiber [79]. A bunch of hot N_e electrons is assumed to be produced by an intense fs laser beam focused onto an area $A = \pi a^2$ at the front of a flat target,

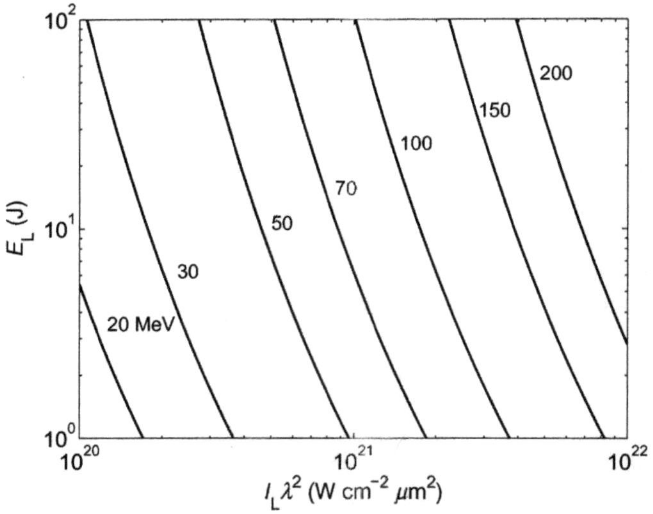

Fig. 9.6 Target normal sheath acceleration of light ions. Relativistic static model of TNSA mechanism for fs laser pulses of irradiance $I_L\lambda^2$ in [Wcm^{-2}µm^2] and energy E_L [J]. Courtesy of Passoni et al. [86]

see Fig. 9.7, laser is incident from left. As a consequence of some divergence in the essentially 1D electron distribution they will map the focus onto an area $B = \pi R^2$, $R = a + d\tan\theta$, if d is the thickness of the target. A small fraction of the N_e electrons will escape through B and leave the target positively charged by Q with a cylindrically symmetric isothermal electron distribution $n_e(x, r, \phi)$. The maximum on the axis $n_e(x, r = 0) = n_e(x)$ is distributed according to (4.73). The spot of area B is located at $x = 0$. The ions are assumed fixed during the acceleration process. The minimum target thickness d is such that $n_e(x = -d) = n_{e0}$ equals the constant ion density Zn_0 (i) and the number Q_{es}/e of the electrons that have escaped through B to infinity is a negligible fraction of $Q > 0$ (ii). It follows from (ii) that the static electron charge $-Q(x > 0) = Q(x \leq 0)$. The ion charge is confined to a disc of the thickness of a Debye length $\lambda_D(< 0)$ in the dense plasma and thus the areal charge density can be set $\sigma_{\mathrm{el}} = Q/B$. A third assumption (iii) is that $-Q(x > 0)$ can be ignored in the determination of the potential $\Phi(x, r = 0)$ owing to large $\lambda_D(> 0)$ in vacuum, $\lambda_D(> 0) \gg \lambda_D(< 0)$. From (5.37) for $\Phi(\mathbf{x})$ and $\sigma = \int \rho_{\mathrm{el}}\mathrm{d}x = Q/B$ the potential on the axis results as

$$\Phi(x, r = 0) = \frac{Q}{2\varepsilon_0 B}\int_0^R \frac{r'\mathrm{d}r'}{(x^2 + r'^2)^{1/2}} = \frac{Q}{2\pi\varepsilon_0 R}\left[(1 + \xi^2)^{1/2} - \xi\right] + \mathrm{const};\ \ \xi = \frac{x}{R}. \tag{9.38}$$

Let $n_e(x = 0) = n_{e0}$ of (4.73). It implies $\Phi(x, r = 0) = 0$ and

$$\Phi(x, r = 0) = \frac{Q}{2\pi\varepsilon R}\left[(1 + \xi^2)^{1/2} - \xi - 1\right] = -E_\infty s(\xi)/e;\ \ s(\xi) = 1 + \xi - (1 + \xi^2)^{1/2}. \tag{9.39}$$

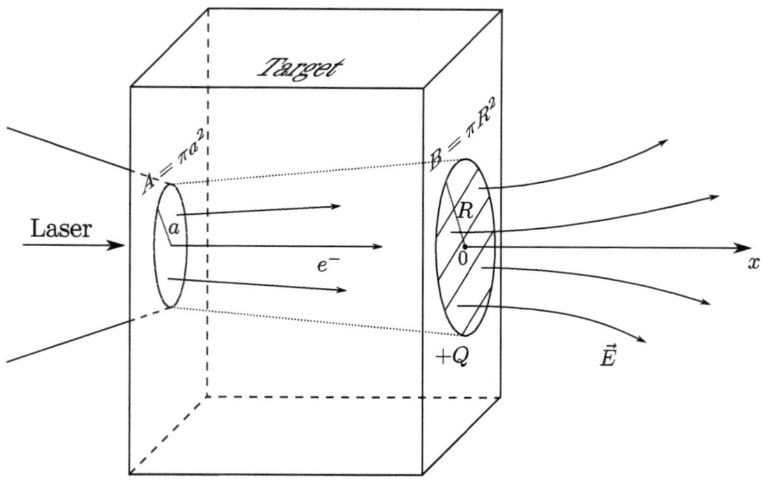

Fig. 9.7 Nonrelativistic model of target normal sheath acceleration (TNSA) of ions according to Schreiber et al. [79]. A laser focused to a radius a onto a flat thin target, left, generates N_e energetic electrons forming a surface charge $Q > 0$; nonrelativistic model after [79]

Electrons with energy $E < E_\infty = Qe/2\pi\varepsilon_0 R$ are trapped. The maximum ion energy is ZE_∞ as the result of the electric field which is highest along the axis. Further, by equating $e\Phi(\xi, r = 0)$ with $k_B T_e$ and $\xi \ll 1$ follows

$$k_B T_e = \frac{\sigma e}{2\pi\varepsilon_0 R^2} x \simeq \frac{n_e e^2}{2\varepsilon_0} x^2 \quad \Rightarrow \quad x = \left(\frac{2\varepsilon_0 k_B T_e}{n_e e^2}\right)^{1/2} \simeq \lambda_D (> 0).$$

Here, $d \simeq x$ has been set because the electron cloud rarefies into the vacuum. The assumption $\xi = x/R \ll 1$ corresponds to the experimental situation.

The acceleration time τ_0 in the potential assumed static during the TNSA process is given in units of the sub ps (ideally a few tens of fs) laser pulse length τ_L by

$$\tau_0 = \int_0^{x=\infty} \frac{dx'}{v(x')} = \tau_L \left\{ X\left(1 + \frac{1}{2}\frac{1}{1-X^2}\right) + \frac{1}{2}\ln\frac{1+X}{1-X} \right\}^{-1},$$
$$X = \left(\frac{E_m(\xi)}{E_\infty}\right)^{1/2}, \; v(x) = \left(\frac{2E(\xi)}{E_\infty}\right)^{1/2}. \tag{9.40}$$

In Fig. 9.8 E_m/E_∞ is plotted as a function of the ratio τ_L/τ_0, see black line. In contrast to (9.30) the ion energy reaches the finite asymptotic energy $E_\infty < \infty$ because of the diverging field of the electrons. The comparison of representative experimental results from various laboratories with X from (9.40) shows acceptable agreement with the static model of TNSA by the mutual repulsion of the ions alone. Under this aspect the essence of TNSA may be seen in the *Coulomb explosion* of bunches of ions.

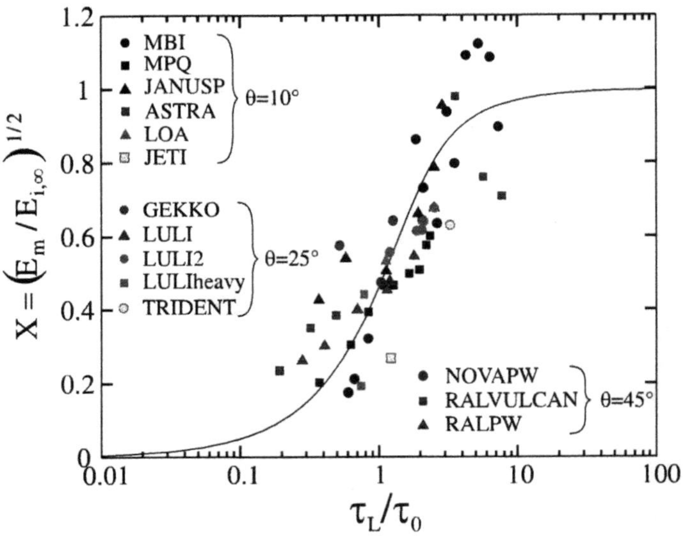

Fig. 9.8 Ion normal sheath acceleration (TNSA) $X = E_m/E_\infty$ of the maximum ion energy E_m as a function of the laser pulse length τ_L in units of the acceleration time τ_0. The black graph is (9.40). The dots, triangles, and squares are representative experimental results from different labs for comparison. For the labs see [79]. Courtesy of Schreiber et al.

How to increase in TNSA the single ion energy? In principle the answer is simple: Increase the potential Φ at the rear side of the target at length of the acceleration track kept constant. Φ increases with increasing T_{hot} of the electrons (i). This may be achieved by calibrating the thickness of the target and giving it a special structure in the zone of laser absorption: hotter electrons by *recirculation*, microstructured target, foams with density close to critical, porous materials, "superponderomotive electrons" generated from nearcritical density plasma [87], and *mass limited targets*, i.e. cross section smaller than laser focus [88]. Φ increases with the number of *hot electrons* (ii). The use of a long pulse to increase the ionization degree and perhaps the absorption in the expanding smooth critical region with subsequently firing the superintense high contrast sub ps main pulse. With all measures employed to keep the acceleration field high care must be taken to prevent the rear edge of the ion profile from flattening as long as possible. Each of the three exemplary models presented has proven that considerable progress is needed to raise the specific ion energy E_i/Z beyond a first magic limit of 100 MeV.

9.5 Radiation Pressure Acceleration (RPA)

Superintense laser pulses are capable of directly accelerating samples of matter by radiation pressure p_L. Under normal incidence of a plane wave of intensity I onto a surface at rest I, Poynting vector $\mathbf{S}$ and p_L in vacuum are related by

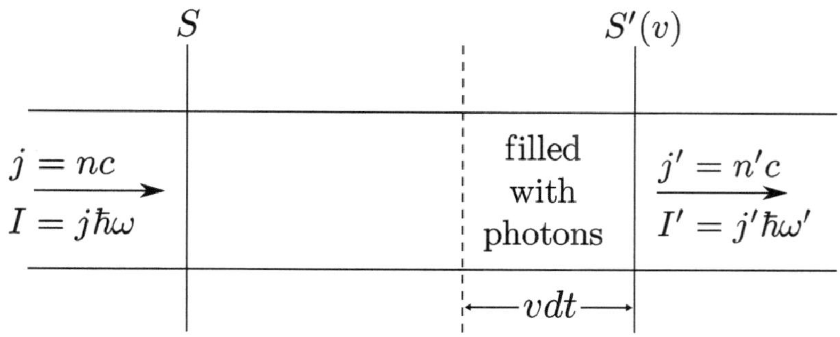

Fig. 9.9 For the observer travelling with S' the number of photons crossing S have to fill the slab $n'vdt' = nvdt$ first; only those left over contribute to the flux through S'

$$I = |\bar{\mathbf{S}}| = \frac{1}{2}\varepsilon_0 c\mathbf{E}\mathbf{E}^*, \quad p_L = (1+R)\frac{I}{c}; \quad \mathbf{S} = \varepsilon c^2 \mathbf{E}\times\mathbf{B} = \varepsilon_0 c\mathbf{E}^2\mathbf{k}^0. \tag{9.41}$$

$\mathbf{k}^0 = \mathbf{k}/|\mathbf{k}|$, R reflection coefficient. In the reference system $S'(\mathbf{v})$, $\mathbf{v} \parallel \mathbf{k}$ the intensity is I',

$$I' = \frac{1}{2}\varepsilon_0 c\mathbf{E}'^2 = \frac{1}{2}\varepsilon_0 c\gamma^2\left|\left[\mathbf{E} + \mathbf{v}\times\left(\frac{\mathbf{k}}{\omega}\times\mathbf{E}\right)\right]^2\right| = \frac{1\mp\beta}{1\pm\beta}I; \; p_L' = \frac{1\mp\beta}{1\pm\beta}p_L; \; \mathbf{v}\parallel\pm\mathbf{k}. \tag{9.42}$$

In the photon picture I transforms according to (3.107) and (6.75)

$$I = nc\hbar\omega; \;\; I' = n'c\hbar\omega' = \gamma(1\mp\beta)nc\hbar\gamma(1\mp\beta)\omega = \gamma^2(1\mp\beta)^2\hbar\omega \tag{9.43}$$

in agreement with (9.42).

Physical interpretation. Geometrical visualization of $I \mapsto I'$ may be instructive. Consider $I = nc\hbar\omega$ and $I' = n'c\hbar\omega'$ crossing per second the planes in the lab system S and $S'(v)$ moving with v in Fig. 9.9. The energy balance is

$$jdt = ncdt = n'vdt' + j'dt' = nvdt + j'\frac{dt}{\gamma} \;\; \Rightarrow \;\; j' = \gamma(j - nv) = \gamma(1-\beta).$$

The observer in $S'(v)$ realizes that the slab $n'vdt'$ has to be filled with photons first and only the difference accounts for j'. It holds $n'vdt' = nvdt$ because as a pure number it is a Lorentz scalar. With $\omega' = \gamma(1 - v/c)$ follows the result of (9.43) for I'.

Steady state momentum balance in $S'(v)$**.** Consider a sufficiently thick slab of matter of density ρ_0 exposed to the radiation pressure $p_L = 2I/c$ of a nonabsorbing laser beam (R = 1) impinging normally on it. Under adiabatic switching on the electrons are pushed forward until the electrostatic field of the ion-electron double layer equals the equilibrium with the inertia of the ions accelerated from zero to velocity v. This is the cold fluid model of a laser piston driving a shock into the target, this time assumed collisionless. The assumption of total reflection is an approximation which a circularly polarized laser beam may come close to because of $\mathbf{v} \times \mathbf{B} = 0$. In the frame $S'(v)$ comoving with the piston the amount $\gamma n_0 \gamma m_i = \gamma^2 \rho_0$ immerses with velocity $-v$ into the electrostatic field, is stopped there and then reaccelerated to $+v$. In the steady state the momentum balance reads as

$$2\frac{I'}{c} = 2\gamma^2 \rho v = 2\frac{1-\beta}{1+\beta}\frac{I}{c}, \; s^2 = \frac{I}{\rho_0 c^3} \Rightarrow \gamma^2\beta^2 = \frac{1-\beta}{1+\beta}s^2, \; v = \frac{s}{s+1}c. \quad (9.44)$$

The compression in the collisionless shock is $(f+1)/f = 2$, $f = 1$ degrees of freedom. The ion velocity v_i and energy $\mathcal{E}_i$ in S are obtained from the velocity addition theorem (2.188) and Einstein's formula, respectively,

$$v_i = \frac{v + v(2\gamma - 1)}{\gamma(1+\beta^2)} = \frac{2v}{1+\beta^2}; \; \mathcal{E}_i = m_i c^2(\gamma - 1) = m_i c^2 \frac{2\beta}{1-\beta^2} = 2m_i c^2 \frac{s^2}{1+2s}. \quad (9.45)$$

The efficiency η of acceleration with N photons is

$$\eta = \frac{N\hbar(\omega - \omega_r)}{N\hbar\omega} = \frac{1-\beta}{1+\beta} = \frac{2s}{1+2s}. \quad (9.46)$$

If the thickness of the target is such that a shock can develop the acceleration process is defined as the *holeboring regime*. An example may illustrate its power. Assume a solid hydrogen target of $\rho = 0.2\,\mathrm{gcm}^{-3}$ and $s = 0.25$. The required laser intensity is $I = 1.35 \times 10^{23}\,\mathrm{Wcm}^{-2}$and the proton energy amounts to $\mathcal{E}_p = 230\,\mathrm{MeV}$. If holeboring is operated with linearly polarized light strong heating of the target takes place accompanied by intense light emission as a cooling mechanism; target heating is observed also in circular polarization [89]. For a recent review of theoretical and numerical studies of holeboring radiation pressure acceleration see [90].

In the opposite situation of a target of thickness d much smaller than the laser wavelength, $d \ll \lambda$, the electrons are pushed ahead of the ion slab. In this electrostatic double layer the mass center of the ions is accelerated according to

$$\frac{\mathrm{d}}{\mathrm{d}t}(\beta\gamma) = \frac{I}{\rho_0 d c^2}\frac{1-\beta}{1+\beta},$$

which gives a rough estimate of ultrathin layer acceleration as a whole. RPA is said to operate in the *light sail regime* [91]. 1D simulations have shown that electrons pile up in a very thin layer at the rear target surface and that only a small fraction of ions is accelerated to high energies. In linear polarization a competition of RPA and strong heating is observed. Again, circular polarization is expected to help at first glance. However, the optimism is reduced over a wide range of laser intensities by observing the onset of longitudinal oscillations in this case in PIC simulations [89, 92]. In [91] a criterion is developed for an intensity limit from which on RPA should prevail on electron heating, even in linear polarization. The author finds $a \gtrsim 19$ corresponding to $I\lambda^2 > 5 \times 10^{20}\,\mathrm{Wcm}^{-2}\mu\mathrm{m}$. Beyond such intensities "unlimited" acceleration is possible with unlimited laser intensities. Highest possible contrast ratio of the superintense laser pulse to prepulse is compulsory. On a restricted number of ions the light sail mechanism is superior to holeboring.

9.6 Wake Field Acceleration

The principle of particle acceleration by a wave, either electromagnetic or electrostatic (or both) is simple and is illustrated by Fig. 9.10 where a point particle is injected into potentials. In (a) the mass point acquires kinetic energy corresponding to its height, $\Delta \mathcal{E}_{\mathrm{kin}} = mgh$. It gains, however, more energy if the previously fixed structure is moving to the right, as indicated in (b). This is because now the force perpendicular to the slope also does work upon accelerating the particle in x direction. In the intense electromagnetic wave an electron is first accelerated by the ponderomotive force in the propagation direction (if the transverse gradient of $\mathbf{E}^2$ is weak). In the relativistic regime, direct acceleration by $\mathbf{E}' = \mathbf{v} \times \mathbf{B}$ sets in preferentially into forward direction over arbitrarily long distances, in principle (c). This is a consequence of the strong frequency Doppler down shift at $v_x \simeq c$. To avoid misinterpretations, in the lab frame all acceleration work is done by the $\mathbf{E}$-field of the laser beam, the Lorentz force, perpendicular to $\mathbf{v}$, merely provides for bending the orbit into forward direction. From the engineering point of view the main problem consists in (i) trapping the particle, (ii) guiding the intense wave over a long distance, and (iii) extracting the particle in the right moment before it starts to be

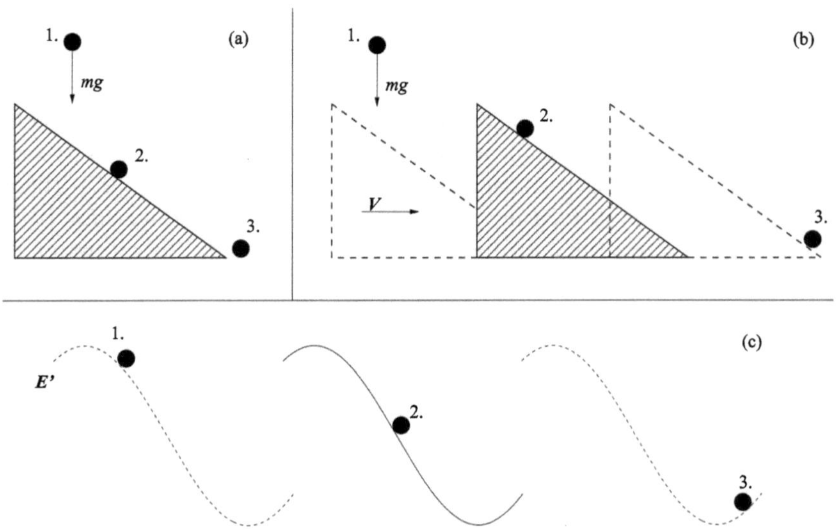

Fig. 9.10 The principle of particle acceleration in a finite potential structure. **a** Fixed slope, **b** moving slope, **c** electron trapped in a traveling wave

decelerated and to lose energy again. As soon as the particle approaches c it does not increase in velocity anymore and it can stay on the slope of the accelerating wave and gain energy forever if the wave also moves with c; it benefits from relativistic time dilation.

9.6.1 The Nonlinear Wake

Electron acceleration in vacuum by a tightly focused petawatt laser beam is possible [93]. The use of the high-intensity longitudinal electric field of a plasma wave, in place of the longitudinal electric field components in the vacuum focus, offers several advantages: (1) intensity increase of the laser pulse by self focusing and its guiding over longer distances than the geometric focal length, (2) tunability of the excited plasma wave length through careful choice of the plasma density, (3) electrostatic field increase by resonance effects, and (4) ponderomotive self-trapping of electron bunches by self-modulation via backward and forward Raman instability. As a result a much higher number of particles is accelerated, typically of the order of 10^9 electrons. On the other hand, efficient extraction of the particles after successful acceleration is in general a more complex enterprise.

In the laser wake field accelerator (LWFA) [94] a highly intense laser pulse is focused into an underdense plasma of n_e ranging typically from some 10^{18} to several 10^{19} electrons per cm^3. Resonant ponderomotive excitation of an electron plasma wave occurs when the (cold) plasma wavelength $\lambda_p = 2\pi c_\varphi/\omega_p, c_\varphi = c/\eta \simeq c$ phase velocity of the plasma wave, is twice the laser pulse length. The laser pulse propagates with group velocity v_g. When the laser pulse power P is close to or above the relativistic self focusing threshold, $P > 17(\omega/\omega_p)^2$ GW, no such matching condition is necessary because the laser beam amplitude undergoes self modulation at the plasma frequency by the Raman forward and side scattering instability [95, 96]. This effect resonantly enhances the creation of wake fields by ponderomotively expelling locally the electrons and subsequent overshooting in their return flow. LWFA in the self modulated regime has been demonstrated in several experiments [97]. Theoretical investigations on self focusing dynamics and stability are presented in [98]. In the wake fields up to several 10^{11} Vm^{-1} could be generated. The spectra of electrons from such fields may vary from purely Maxwellian to genuine spike structures depending in first order on the ratio of laser pulse length to the wavelength λ_p of the excited electron plasma wave. However, next the laser intensity is most important because it determines the regime of operation (e.g. self modulation). The excited nonlinear wake assumes a spiky structure in the electron density, e.g. see Figs. 6.21 and 6.22. In concomitance, the electric field tends to an asymmetric triangular shape, see Fig. 5.9.

9.6.2 Energy Gain from the Electron Plasma Wave

The principle of wake field acceleration can be described in the frame of the relativistic theory of the cold electron fluid. The ions can be assumed to form an immobile background. The electron conservation equations in the lab frame are

$$\frac{\partial}{\partial t}n_e + \nabla(n_e\mathbf{u}_e) = 0; \quad \left(\frac{\partial}{\partial t} + \mathbf{u}\nabla\right)\gamma m_e \mathbf{u}_e = -e\mathbf{E} - \nabla\Phi_p$$

$$\nabla\mathbf{E} = \frac{e}{\varepsilon_0}(Zn_i - n_e), \tag{9.47}$$

Φ_p is the ponderomotive potential of the laser. It propagates at $v_g = \eta c$. In the co-propagating system the wake is quasistationary since ω_p is close to $v_g k_e$. Transforming to $\xi = k_e x - \omega_p t$ yields

$$\partial_x = k_e\partial_\xi, \quad \partial_t = -\omega_p\partial_\xi = -\frac{\omega_p}{k_e} = c\eta\partial_x.$$

We introduce $N_e = n_e/Zn_i$, $\beta_e = u_e/c$, $E_0 = Zn_i e/\varepsilon_0 k_e = (m_e/e)\omega_p^2/k_e$. E_0 corresponds to the field of a capacitor charged by $\sigma = Zn_i e/k_e$. The above system transforms to

$$\frac{\mathrm{d}}{\mathrm{d}\xi}\left(\frac{\beta_e}{\eta} - 1\right)N_e = 0; \quad \left(\frac{\beta_e}{\eta} - 1\right)\frac{\mathrm{d}}{\mathrm{d}\xi}\gamma_e\beta_e = -\eta\frac{E}{E_0} - \frac{1}{c^2\eta}\frac{\mathrm{d}}{\mathrm{d}\xi}\Phi_p,$$

$$\frac{\mathrm{d}}{\mathrm{d}\xi}E = E_0(1 - N_e). \tag{9.48}$$

The solutions are undamped periodic. Hence, the first of these equations is $N_e(1 - \beta_e/\eta) = \text{const} = 1$; N_e is nowhere zero.

9.6.3 Nonlinear Bubble and Monoenergetic Beams

From a good accelerator device low divergence monoenergetic beams are expected. A break-through in this respect has been achieved by Pukhov/Meyer-ter-Vehn with the discovery of the so-called bubble acceleration regime [100]. It consists in a short laser pulse of length $l = N\lambda = 2\pi c/\omega$ satisfying the inequality $l \lesssim \lambda_p/2$ and an intensity such as to generate an electron wake which breaks completely after the first oscillation, see Fig. 9.11. The laser pulse expels the electrons ponderomotively outward from the region of highest intensity $\hat{a}^2$, in forward and lateral direction in much the same way as an electric charge does in the equilibrium plasma (see for comparison Fig. 7.4). In the subsequent restoring motion, owing to weak damping, an

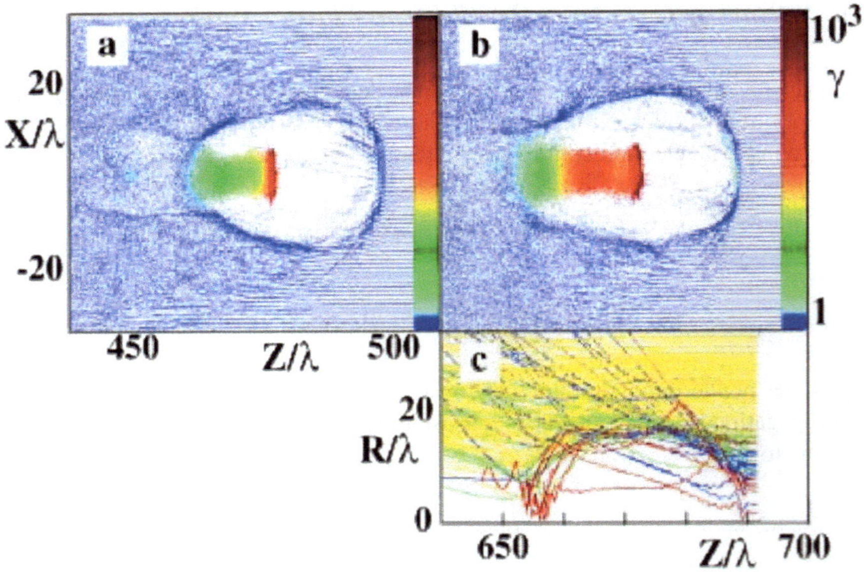

Fig. 9.11 Solitary laser-plasma cavity produced by a 12 J, 33 fs laser pulse in a plasma of density $n_e = 10^{19}\,\mathrm{cm}^{-3}$. **a** $ct/\lambda = 500$, **b** $ct/\lambda = 700$, **c** electron trajectories in the frame moving together with the laser pulse [100]

electron plasma wave develops with wave fronts that move faster on the axis (higher electron density) than on the flanks of the laser pulse. As a consequence, *transverse (geometrical) wave breaking* may set in downstream after a few oscillations. This has been simulated in 3D PIC with a 20 mJ, $\hat{a} = 1.7$ laser pulse of 6.6 fs length propagating through a uniform plasma of $n_e = 3.5 \times 10^{19}\,\mathrm{cm}^{-3}$. Most significant, 10^9 electrons, preferentially along the axis in front of the trailing electron wave crest, with acceleration up to 50 MeV were observed. The conversion of laser energy into fast electrons amounted to surprising 15%. Their energy spectrum had a plateau like shape with an abundance of population in the neighborhood of the cut off. The angular divergence was $\pm 1°$. A second computer run with a 12 J, $\hat{a} = 10$, 33 fs long pulse was performed in the same background plasma as before. At such intensity the ponderomotive expulsion is so violent that a nearly evacuated region of the shape of a bubble is created with no following wave crest owing to destructive self-interaction of the electron wake (Fig. 9.11). From the rear side of this solitary bubble an axial high-density bunch of trapped electrons grows out and gains energy in running down the quasistatic potential hill. As a consequence, successively a unidirectional (2° angular spread) *quasi-monochromatic electron beam* (20% energy spread) of $\mathcal{E}_{\mathrm{free}} \simeq 350\,\mathrm{MeV}$ builds up after $N = 750$ cycles (Fig. 9.12). Starting from the relations $N \simeq \lambda_p/\lambda \sim n_e^{-1/2}$, $\mathcal{E}_{\mathrm{wavebreaking}}/\hat{a} \sim \left(\lambda_p/\lambda\right)/\hat{a} = \mathrm{const}$ [99] the following scaling for laser irradiance $I\lambda^2$, laser power P, pulse energy $\mathcal{E}_{\mathrm{L}}$, and maximum electron energy $\mathcal{E}_{\mathrm{free,max}}$ on the number of laser cycles N is given in [100]:

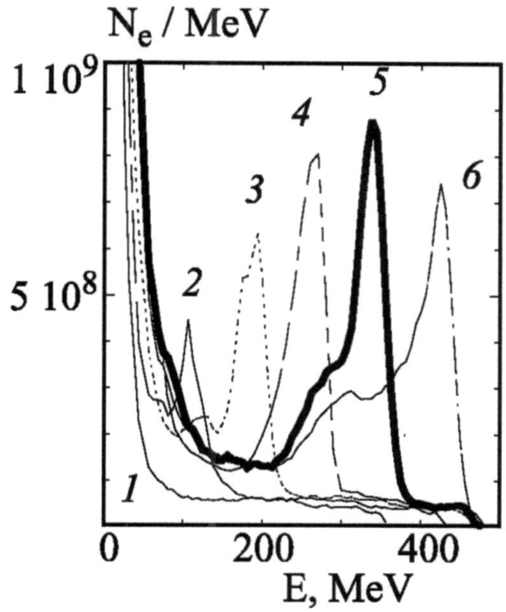

Fig. 9.12 Time evolution of the spectra of accelerated electrons for the same parameters as in Fig. 9.11. (1) $ct/\lambda = 350$, (2) $ct/\lambda = 450$, (3) $ct/\lambda = 550$, (4) $ct/\lambda = 650$, (5) $ct/\lambda = 750$, (6) $ct/\lambda = 850$

$$\begin{aligned} I\lambda^2 \sim \hat{a}^2 \sim N, \qquad P \sim I\lambda^2, \\ \frac{\lambda_p^2}{\lambda^2} \sim N^3, \mathcal{E}_\mathrm{L} \sim PN \sim N^4, \qquad \mathcal{E}_\mathrm{free,max} = e\,E_\mathrm{max}\,L_\mathrm{d} \sim N^{5/2}. \end{aligned} \tag{9.49}$$

L_d is the *dephasing length*, i.e., the distance an electron covers during its acceleration stage in the electrostatic field E_max. L_d plays an important role because, in order not to lose part of its energy gained along L_d, the electron must be extracted after the dephasing time $\tau_\mathrm{d} = L_\mathrm{d}/c$. Finally, it has been found that most efficient acceleration is achieved when the parameters laser pulse length and gas (plasma) density are chosen such that acceleration is entirely by the plasma wave and not by the combined action of the laser and the wake field. For completeness it must be mentioned that wake field acceleration driven by short duration intense electron beams in the bubble regime works in an analogous manner [101, 102].

Scarcely two years after successful generation of relativistic quasi-monochromatic, highly collimated electron beams have been demonstrated in three experiments. Two of them were run in the resonant mode acceleration [103, 104], the third was operated with pulses longer than $2\lambda_p$ in a plasma channel in the self-modulated LWFA mode [105]. The experiments yield surprising results and represent more than one step forward in the development of table-top accelerators for significant applications, as for example generation of fs X-rays and coherent THz and infrared radiation. The main parameters and achievements are summarized in Table 9.1. The technique of LWFA with self-guided laser beams in capillary tubes has made rapid progress since then and electron bunches of 1 GeV and beyond have been produced over lengths of the order of 3 cm [106]. Subsequently considerable electron energy multiplication

Table 9.1 Experiments by Mangles et al. [103], Faure et al. [104], and Geddes et al. [105]. Ti:Sa laser energy $\mathcal{E}_L$, pulse duration τ, intensity I, background electron density n_e, electron energy at peak maximum $\mathcal{E}_{free}$, energy spread $\Delta\mathcal{E}$, angular spread α, number of electrons in the bunch N_e, N_s multiple of acceleration standards $s = 100\,\text{MeVm}^{-1}$, acceleration efficiency $\eta_a = N_e\mathcal{E}_{free}/\mathcal{E}_L$

Exp.	[103]	[104]	[105]
$\mathcal{E}_L$ (J)	0.5	1.0	0.5
τ (fs)	40	30	55
I (W cm^{-2})	2.5×10^{18}	$\cong 10^{19}$[a]	1.1×10^{19}
n_e (cm^{-3})	2×10^{19}	6×10^{18}	1.9×10^{19}
$\mathcal{E}_{free} \pm \Delta\mathcal{E}$ (MeV)	$70 \pm 3\%$	$170 \pm 11\%$	$86 \pm 2\%$
α	$\pm 2.5°$	$\pm 0.3°$	$\pm 1.7°$
N_e	1.4×10^8	3×10^9	2×10^9
N_s	1.3×10^3	570	430
η_a (%)	0.3	8.1	5.5

[a]estimated

up to 4.2 GeV in a plasma electron density of $n_e \simeq 7 \times 10^{17}\,\text{cm}^{-3}$ with a 0.3 PWatt TN:Sa laser could be reached [107]. To compare, in a 1.3 m long electron beam driven plasma wakefield accelerator electron bunches of 28.3 pC have been accelerated to a maximum energy of 9 GeV., corresponding to an acceleration gradient of $4.0\,\text{GeVm}^{-1}$. The root mean square energy spread was 5.0% [108].

LWFA using a single intense laser beam implicates results varying from shot to shot. The high quality nearly monohromatic electron beam depends on the electron injection into the wake; it results unstable and is hard to control. Considerable propress in this respect has been made by the use of two counterpropagating laser beams. They generate a (partially) standing wave in the lab frame from which locally limited electron heating starts. In this way injection relies on a broad energy spectrum of electrons to be trapped; trapping is stabilized and can be controlled [109].

Wake field acceleration with protons. In conventional accelerators the technology sets limits to the specific energy per unit length at about 100 MeV/m. Energies in the desired range can only by obtained by building devices of ever increasing size, and costs. The wake field accelerator is a promising alternative as it promises specific energy increases by factors between 10^3 and 10^4. In order to benefit from the highly developed conventional technology the program AWAKE has been set up at CERN in 2013. It uses a GEV proton beam in place of a powerful laser. Energetic protons are stiff and seem to be particularly appropriate to generate stable and reproducible wakes in a plasma (e.g. Rb) of $(1-10) \times 10^{14}\,\text{cm}^{-3}$ electron density. Successful electron acceleration has been observed recently [110].

9.7 Thomson Scattering as a Plasma Diagnostic Tool

9.7.1 Basics

Scattering of an electromagnetic wave originates from the electrons, bound electrons in neutral matter, free electrons in the plasma. The current induced by the incident wave in the heavy nuclei becomes important at superrelativistic intensities only. *Thomson scattering* in the restricted sense is elastic, i.e. unshifted scattering from free electrons. The elastic scattering from electrons bound in atoms and ions is called *Rayleigh scattering*. The differential Thomson scattering cross section of a monochromatic wave is obtained from the nonrelativistic Larmor formula (8.43) with r_e the *classical electron radius*,

$$\sigma_\Omega = r_e^2 \sin^2\theta, \; r_e = \frac{e^2}{4\pi\varepsilon_0 m_e c^2}; \quad \Rightarrow \mathrm{d}I_{\boldsymbol{\Omega}} = \sigma_\Omega I_{\mathrm{laser}} \mathrm{d}\Omega. \tag{9.50}$$

The cross section is limited to dipole radiation and holds as such for $\hat{\delta}_{\mathrm{os}} \ll \lambda$, with λ the shortest wavelength involved in the scattering process. In the relativistic realm of high intensities, or with hard photons, the recoil on the electrons must be considered. This is the domain of *Compton scattering*. The differential cross section for scattering from the single particle is expressed by the *Klein–Nishina* formula. For small photon energies the total cross section can be expanded,

$$\sigma_{\mathrm{KN}} = \sigma_{\mathrm{T}}\Big(1 - 2\alpha + \frac{56}{5}\alpha^2 + \cdots\Big); \quad \sigma_{\mathrm{T}} = \frac{8\pi}{3} r_e^2, \quad \alpha = \frac{\hbar\omega}{mc^2}. \tag{9.51}$$

The total classical Thomson scattering cross section $\sigma_{\mathrm{T}} = 0.6654$ barn is derived in (8.45). It may be convenient some times to use the expression of Thomson scattering in a more general way by including Compton scattering also.

The scattering cross sections above can be used for an ensemble of particles only if the *intensities* of the single scattering centers can be added. It has been pointed out already in connection with the Doppler effect in Sect. 6.4.2 that in principle the *fields* of the single particles in a scattering volume have to be summed up and only from this result the fluxes can be calculated. In this way account is taken of the degree of coherence between the single contributors to scattering. The simple sum of intensities must result automatically from the correct treatment of fields. This is achieved by decomposing the fluctuations in matter into normal modes $(\mathbf{K}, \boldsymbol{\Omega})$ and to determine the Doppler shifted fields from them. To see how this works let us consider the simple case of a quiet homogeneous plasma and a weak laser probe beam incident onto it. The density $n_e(\mathbf{x}, t)$ is a superposition of all low frequency ion acoustic and high frequency electron plasma modes $n_{\mathbf{K}}$ on n_0. The laser imprints the velocity $\mathbf{v}_{\mathrm{os}} = \hat{\mathbf{v}}_{\mathrm{os}} \exp(i\mathbf{k}_0\mathbf{x} - i\omega_0 t)$ on the electrons. Correspondingly, the current induced by the laser appears modulated,

$$\mathbf{j} = n_e \mathbf{v}_{os} = \hat{\mathbf{v}}_{os} \cos(\mathbf{k}_0 \mathbf{x} - \omega t) \sum_{\mathbf{K}} n_{\mathbf{K}} \tag{9.52}$$

$$= \frac{1}{2} \sum_{\mathbf{K}} n_{\mathbf{K}} \hat{\mathbf{v}}_{os} \Big\{ \cos[(\mathbf{k}_0 + \mathbf{K})\mathbf{x} - (\omega_0 + \Omega)t] + \cos[(\mathbf{k}_0 - \mathbf{K})\mathbf{x} - (\omega_0 - \Omega)t] \Big\}.$$

This is the source term in the electromagnetic wave producing all scattered field terms $E_s(\mathbf{k}_s, \omega_s) \sim \cos[(\mathbf{k}_0 \pm \mathbf{K})\mathbf{x} - (\omega_0 \pm \Omega)t]$. As a consequence of orthogonality *matching conditions* must be fulfilled for the single modes between $(\mathbf{k}_0, \omega_0)$ and $(\mathbf{k}_s, \omega_s)$,

$$\mathbf{k}_0 = \mathbf{K} + \mathbf{k}_s, \qquad \omega_0 = \Omega + \omega_s. \tag{9.53}$$

With the frequencies taken positive (9.53) exhausts the whole spectrum because for the $\mathbf{K}$ mode exists for the same omega also the $-\mathbf{K}$ mode. The down and the up converted frequencies $\omega_s = \omega_0 \mp \Omega$ are the *Stokes* and the *anti Stokes* lines of the spectrum. As we understand from the discussion of stimulated Brillouin and Raman scattering in Chap. 6. Thomson scattering is geometrical reflection from periodic structures. Supposed there are enough electrons in one mode the intensity of $\mathbf{E}_s$ is proportional to the weak $\mathbf{E}_0$ and the individual $n_{\mathbf{K}}$. Additional lines in the spectrum may originate from the polarization of the bound electrons in the neutral plasma component and the partially ionized components. Owing to internal resonances of the shells this contribution to scattering can no longer be interpreted as classical reflection. It must be determined with the help of quantum dynamics. From the considerations on the shortest wavelength of a mode in Chap. 5, see (5.116), a useful criterion on coherent versus incoherent scattering is easily deduced:

$$\alpha = \frac{1}{|\mathbf{K}|\lambda_D} > 1 \Rightarrow \text{coherent}, \quad \alpha = \frac{1}{|\mathbf{K}|\lambda_D} < 1 \Rightarrow \text{incoherent}. \tag{9.54}$$

Thomson scattering into forward direction is coherent.

In the following the elements of coherent and incoherent Thomson scattering are described. Scattering originates entirely from the plasma electrons. Nevertheless one of the major points has to concentrate on when Thomson scattering will reveal the electron and when it will show the ion features. A general theory for plasmas is given in the classical paper by Chihara [111].

9.8 Digression On: Classical or Quantum Treatment?

Traditionally plasma physics and fluid dynamics are domains of classical physics. Particles follow Newton's law, waves exhibit a continuous amplitude and a continuous phase, no quantized steps, Planck's constant $\hbar$ is absent. It is surprising that $\hbar$ appeared in physics late. For three hundred years since Galileo physics evolved entirely without $\hbar$. For one hundred years it found successively its way into the world of science. Looking back from now it is a common believe that the physics of low quantum numbers, e.g. atomic spectra, single photons, is governed by the laws of quantum theory and quantum field theory. At *high quantum numbers* nature follows classical laws whatever that means precisely. On the other hand it is a matter of fact that with progressing research on nonideal plasmas, warm dense matter and the hot solid quantum mechanics has irresistibly invaded classical plasma physics. Some may even wonder now that Planck's constant was not found before 1900.

9.8.1 A Strong Statement

Up to date scientists, in particular theoreticians from the microworld and high energy physics set the contrast:

> There is no experimental fact, not a single one, that contradicts a quantum-theoretical prediction [112]

Quantum theory is the ultimate plan according to which the world evolves. Dynamics follows Heisenberg's equation of motion; $\hbar$ is an essential building block. The fluid and the plasma obey the rules of quantum dynamics. Quantum mechanics is universal. The astronomer has to base his orbit calculations and their perturbations on the Schrödinger equation and yet, the mechanical engineer must use it to build a bridge, unless Newton's law follows from Schrödinger's or Heisenberg's equations. *Tertium non datur*: A third way does not exist. What is the actual reality?

9.8.1.1 Standard Recipes

High quantum numbers. Bohr offers a criterion with his *correspondence principle*, admittedly a very vague criterion, not extending much beyond the personal physical intuition of the single researcher. A better criterion is the DeBroglie wavelength $\lambda_B = \hbar/mv$. If it is short compared to the interparticle distance, to the inverse gradient, or to another characteristic length, the system follows Newton's law of dynamics. However, the system is a single particle only in this case. Let us examine the criteria in more detail. A law is considered classical if it does not contain $\hbar$, otherwise it is quantum mechanical.

The Schrödinger equation of a free particle in the potential $V(\mathbf{x})$ transforms into a wave equation governed by the refractive index $\eta(\mathbf{x}) = [1 - V(\mathbf{x})/E]^{1/2}$,

$$-\frac{\hbar^2}{2m}\nabla^2|\psi\rangle + V|\psi\rangle = E|\psi\rangle \quad \Rightarrow \quad \nabla^2|\psi\rangle + k_B^2\eta^2|\psi\rangle = 0, \;\; \eta^2 = 1 - \frac{V}{E}. \tag{9.55}$$

$k_B = 1/\lambda_B = (2mE)^{1/2}/\hbar$. In the steady state the electric field, transverse or longitudinal, follows the wave equation

$$\nabla^2 \mathbf{E} + k_0^2 \eta^2 \mathbf{E} = 0, \; k_0 = \frac{2\pi}{\lambda}. \tag{9.56}$$

Thereby ∇^2 either reads $divgrad$ for photons, or $graddiv$ for plasmons. Both equations, (9.55) and (9.56), show the same structure. Once the WKB conditions in $\mathbf{x}$ and t are fulfilled the phase $\Psi(\mathbf{x}, t)$ exists, see (5.153). Comparison with $F(t)$ from (2.102) for particles shows that the frequency $\omega = -\partial_t \Psi(\mathbf{x}, t)$ is the Hamiltonian and $\mathbf{k} = \nabla \Psi(\mathbf{x}, t)$ is the momentum. From them the equations of motion (5.159) of the *classical photon* follow. For $\lambda_B \mapsto 0, \; \lambda \mapsto 0$, which in practice is identical with "sufficiently small compared to a characteristic length", both, (9.55) and (9.56), follow the ray equation

$$\frac{\mathrm{d}\mathbf{k}\eta}{\mathrm{d}s} = |\mathbf{k}| \nabla \eta. \tag{5.156}$$

This is a classical relation obeying Newton's second law; see Fig. 5.13.

Macroscopic systems. Going beyond the single particle one may ask for global criteria of complex many particle systems. To this aim let us consider the rarefied classical gas of massive particles or massless photons. Both follow Hamilton's equations of motion, for photons see (5.159). Both follow an adiabatic law $pn^{-\gamma} = \text{const}$. For photons the adiabatic coefficient is $\gamma = 4/3$. In practice macroscopic systems are generally modelled in terms of classical laws with great success. The underlying idea is that when looked at them as eigenstates of their Hamiltonian they are so densely packed that their distribution function appears continuous and the forces acting on them are so strong to induce changes by large quantum numbers. The smallness of $\hbar$ is no longer the adequate unit of measure, it becomes irrelevant. The equations of motion are sufficiently precise in classical terms because with high quantum numbers, often synonymous with high energies or high temperatures, finite differences shrink to differentials. In this sense high energy or high temperature is a criterion for separating classical from quantum systems. There is an additional effect which justify coarse graining. All systems are subject to interactions with the environment that causes perturbations locally and in time on small scales. As a consequence classical laws are spatial and temporal averages. The fluid dynamic equations governing plasmas and fluids are of this type.

The quantum object moon. The argument of coarse graining may be illustrated with the earth—moon system. In crudest approximation it can be described in the center of mass system by a one particle Hamiltonian with reduced mass in perfect analogy to the Schrödinger treatment of the hydrogen atom. The moon is at a distance of $R = 3.81 \times 10^5$ km from the earth and moves at $v_{\text{Moon}} = 1.023$ km/s. Its mass is (1/81) earth mass. From the data one calculates
$\lambda_B = 1.33 \times 10^{-60}$ m
$b_\perp(\pi/2 \text{ deflection}) = 3 \times R_{\text{Moon}} = 1.15 \times 10^6 \text{ km} = 8.6 \times 10^{68} \lambda_B$.

Principal quantum number $n = 2.89 \times 10^{68}$ if it were in an eigenstate $|n, l, m\rangle$. The moon is a ball in the heaven; it must be a superposition of many states for $\langle\psi|\psi\rangle$ to appear as a sphere. With $l = 0$ for example it would be homogeneously distributed over the entire heaven. The orbit is plane. For a given l this implies m is close to l because

$$\lim_{l\mapsto\infty} \frac{L}{L_z} = \frac{\hbar\sqrt{l(l+1)}}{\hbar l} = 1 \quad \Leftrightarrow \quad \frac{\Delta L_z}{L_z} \mapsto 0.$$

If the moon were in a coherent state it would move along its orbit with precision beyond imagination. A 3D Gaussian wave packet of longitudinal spread of $\Delta x =$ 1 km in initial position grows to $\Delta x = 1.5$ km after 5×10^{54} years only.

Assume further that the moon is hit by a meteorite of 1 kg mass and 1 km speed. The consequence is a variation of the principal quantum number by $\Delta n = 2.7 \times 10^{22}$ or by the fraction $\Delta n/n = 9.3 \times 10^{-47}$. The change of n by the impact of 1 µg with the same speed is still $\Delta n = 2.7 \times 10^4$. The moon is exposed to a continuous flow of uncorrelated perturbations of all kinds, e.g. meteorites, dust, cosmic radiation, solar photons. It is therefore quite natural to assume, except low frequency gravitational disturbances by the planets and terrestrial tides, a statistical mixture of states with a sufficiently narrow Gaussian likelihood distribution in the principal quantum number n centered around $n_0 = 2.89 \times 10^{68}$. All criteria for the validity of the ray equation (5.156) are overfulfilled and hence the equivalence of Schrödinger's equation with Newton's second law are shown through (5.159),

$$\frac{\mathrm{d}\mathbf{p}}{\mathrm{d}t} = -\frac{\partial V}{\partial \mathbf{x}}, \quad \frac{\mathrm{d}\mathbf{x}}{\mathrm{d}t} = \frac{\partial H}{\partial \mathbf{p}}, \quad \frac{\mathrm{d}H}{\mathrm{d}t} = \frac{\partial H}{\partial t}.$$

9.8.2 High E and High T Criteria

Going to high quantum numbers does not imply transition to classical behaviour in general. A classical rigid body rotates around a body-fixed stable axis, say with $L_z = \hbar m$. If m is close to l the rotational spread $\Delta L/L_z$ shrinks to zero for $l \mapsto \infty$; for $m = l/2$ it does not. The number of photons of the black body radiator tends to infinity with $T \mapsto \infty$. Photons are the most perfect *ideal gas*, they do not interact in vacuum, they do not know of each other. According to Boltzmann one would expect $\Delta N/N \mapsto 0$. From Planck's law we know $\Delta N/N \to 1$; it has been measured with great precision. Thermodynamics is the science of the entropy S, or the temperature T, or the heat Q. The entropy of the classical ideal gas and of a general classical system, expressed by

$$S(E, V) = \frac{1}{N!(2\pi\hbar)^{3N}} \int_{E\le\mathcal{E}<E+\Delta E} \mathrm{d}^{3N}\mathbf{p}\,\mathrm{d}^{3N}\mathbf{q}$$

never turns into a classical expression, free from the Planck constant. Corresponding formulas not containing it are either ratios of entropies or are differences of entropy not starting from $T = 0$. The same behaviour holds with the free energy of the classical electron gas (4.153) in the Saha equation (4.161) at any temperature and rarefaction. No classical formula was able to describe the spectral energy density of the black body radiation. This difficulty induced Planck to do the first step into quantum mechanics by postulating a new constant $h = 2\pi\hbar$. Also the total black body radiation in the Stefan–Boltzmann law depends on Planck's constant. On the other hand the Rayleigh–Jeans formula is purely classical. What is the rule behind the behaviour of some physical quantities transforming into classical expressions at high energy, high temperature or extreme dilution, like the pressure and specific heat of all gases and plasmas, and other entities never become classical, like entropy and free energy?

An essential criterion can be extracted from the comparison of Planck's formula with the Rayleigh–Jeans law. The latter is valid only in the long wavelength limit in which a *finite* number of states is assumed to describe the physical reality. In contrast, Planck's formula refers to an *infinite* number of states. In this case there exist always rarely occupied levels for any finite energy system to which quantum statistics applies. Stringent evidence of its quantum behaviour is the relative variance of the photon occupations $\Delta n/\langle n\rangle \rightarrow 1$ instead of $1/\langle n\rangle^{1/2}$ for classical particles, see (4.117). The same argument on infinite energy levels holds for the free energy of the classical electron gas.

In modelling phenomena frequently quantum dynamics is combined with classical physics. Most common example is the semiclassical description of radiation emission and absorption in plasmas and neutral matter: the electron dynamics is subject to the quantum laws but the driving fields are described by the classical Maxwell equations, as for instance the monochromatic laser field is represented by plane sinusoidal wave consisting of a slowly varying amplitude and a phase rapidly changing in space and time. Application of the consideration above on the number of states leads to the simple criterion for the use of **E** and **B** as classical averages:

> If the number of photons $N_{\mathbf{k}}$ in the cube λ^3 of a wavelength is much larger than 1 the electromagnetic field may be calculated from Maxwell's equations. In the complementary case the electromagnetic field is to be quantized.

For a quantitative estimate (8.17) may be used; see also Fig. 8.1. The criterion applies to the highest harmonics of the laser light and to electron-positron pair production. Static fields are not quantized.

9.8.3 The Interplay of Quantum Theory and Classical Physics

The quantum theory represents a closed system. There have been attempts to make it more complete [113] and there exist "proofs" of the impossibility to enlarge it without running into contradictions. If it is a complete system the question must be allowed on which objects it applies and how is it to be applied. This leads to the process of measurement which is done by means of classical apparatus. The evolution of a physical system is given by its state vector $|\psi(t)\rangle$. For its calculation generally boundary conditions have to be imposed. In selecting the right ones the scientist follows physical intuition trained in classical physics. Beyond that in quantum mechanics the classical variables have to be substituted by operators; the particle momentum $\mathbf{p}$ becomes $(\hbar/i)\nabla$. Following this rule most of the operators are constructed out of the classical counterparts, as for the example the angular momenta. Their commutators follow the classical algebra of the Poisson brackets. In cases where no classical analogon exists, for example the spin, they follow also the classical commutation rules. The same holds for the symmetries, Noether's theorem applies to quantum systems as to the classical world. However, there exist further symmetries valid only in the quantum world, see e.g. the different symmetries of fermions and bosons. It is the fermionic character that guarantees the stability of matter [114]. In all cases a classical Hamiltonian exists, quantum mechanics starts from there or related analogies. There is no surprise, it is a *must*: all measurements are done by classical apparatus. Furthermore, this implicates that to find an experiment falsifying any of quantum postulates appears, if ever, extremely difficult. The ultimate reason of it may be found in assertions of Bernhard Riemann ("On the axiomas of geometry") and Hermann Helmholtz ("On the axiomas of science"):

Where they are true, they are not applicable; where they are applicable, they are not true.

QM needs classical physics and viceversa.

9.8.4 Conclusion

There exists no general rule for the legitimate passage from the quantum mechanical to a classical description. In the particular case the answer depends on the observables and on the symmetries. The quantum world exhibits symmetries in addition to those known of the classical realm. Classical systems are superpositions of quantum states. Their entanglement evolves into a statistical mixture of states owing to the unavoidable presence of high frequency perturbations from the environment. These are two main reasons why classical observables are different in their nature from quantum systems. Again, the system earth-moon may highlight the difference. The quantum mechanical measurement reduces the state vector $|\psi\rangle$ to an eigenstate. Preparation of an eigenstate and a measurement are identical. The analyzing apparatus contains as many channels as the associated operator produces eigenstates. The astronomer measures the position and the momentum of the moon. However, nobody argues that after this classical measurement the moon is in one of its eigenstates. The astronomer aims at just the contrary as he tries to introduce the minimum possible change by the measurement. The distinction between main system and perturbations from outside is possible because we leave in a rarefied universe.

When and Why Do We Chose Classical Modelling?

- In case the simpler classical treatment can be used without loss of precision.
- If Newton's law is mathematically derivable from Schrödinger's equation, e.g., the orbit of the Moon.
- If the equivalence between the two treatments is intuitively clear.
- If classical modelling leads to agreement with the experiment, however, no theoretical foundation of a classical description exists.
- A classical treatment may be the only feasible procedure, as for example the treatment of the quark-gluon plasma by the classical Navier–Stokes equation.

Whenever possible, comparison of classical modelling with quantum treatment is enlightening and deepens understanding.

9.9 Problems

▹ Estimate the jitter Δt of tunnelling (field) ionization. *Hint:* Calculate the bounce frequency of the bound electron in the (deformed) potential.
▹ Calculate the fraction Φ_p/E_I for the ground state of the first electron in Xe and laser intensity of $I = 4 \times 10^{13}\,\mathrm{Wcm}^{-2}$.
▹ (a) Show that $|\mathrm{curl}\ \mathbf{u}| \lesssim kh_0(t)$, except along the interface $h_0(t)$ *Hint:* Apply Stokes' theorem to vanishing loop area. (b) What is the solution of $h_0(t)$ for negative Attwood number? (c) Deduce F in (6.4) from the potential energy change with $h_0(t)$.
▹ Show that the electric dipole moment $\langle n_f|e\mathbf{x}|n_i\rangle$ of a hydrogen like ion of charge Z scales like Z^2.
▹ Compare the analytic expressions v_F and x_F of (9.30) with their numerical integration. Simplify the analytic expressions for $\tau \gg 1$.
▹ Calculate the energy supply by thermal conduction to keep the electron temperature constant within 10% for various time intervals. Compare with the heat flow limit.
▹ (a) Reproduce $\Phi(x, r = 0)$ from (9.38) without the help of the corresponding text. (b) The electric field of a uniformly charged plate is $E = \sigma/2\varepsilon_0$; in (9.38) it is $E = \sigma/\varepsilon_0$. Why?

▹ Show that in (9.44) the momentum change per unit time in S' is $\rho_0 v^2 \times \gamma^2$. *Hint:* Use the energy-momentum tensor $t^{\alpha\beta} = \rho_0 v^\alpha v^\beta$; $V = v^\alpha = \gamma(\mathbf{v}, c)$.
▹ Thomson scattering. Justify the inequalities of α in (9.54).
▹ Indicate in which angle interval the observed scattering is coherent.

9.10 Self-assessment

- What is the dependence of the THz intensity at the detector on the plasma density n_0 if it is generated from (9.4)? *Hint:* The amplitudes $\hat{E}_T$ in the plasma and $\hat{E}_D$ at the detector are related by $\hat{E}_D^2 = \eta_\omega \hat{E}_T^2$.
- Why is multiphoton ionization and associated harmonics incoherent and field ionization and its harmonics coherent? Can superposition of incoherent harmonics form attosecond pulses? *Answer:* yes, but less efficiently.
- At which laser intensity does the electric field amplitude equal the electric field of H on the Bohr ground orbit? *Answer* for control: $I = 3 \times 10^{16}\,\mathrm{Wcm}^{-2}$.
- Channel closing: Consider a 3–5 fs long Ti:Sa laser pulse. Is it possible to observe ionization below E_I? *yes!* Explain.
- Is (9.9) to be modified for the strong field harmonic generation of an ion owing to the many free energy levels? *Answer:* No, recombination into another energy level is of higher order in general.
- Universal harmonic spectrum: How is $X(x, t)$ to be defined in the case of underdense plasma?
- TNSA: In P. Mora's model (9.26) is obtained under the assumption $e\Phi(+\infty) = -\infty$ which is suggested by (9.21) for E. However (9.21) is invalid at $x \simeq x_F \Rightarrow$ apparent contradiction? *Answer:* no! The model is onedimensional; any negative 1D charge distribution implies $e\Phi(+\infty) = -\infty$.
- Under which conditions do the electrons escaped from the target in Schreiber's model [79] not contribute to the potential Φ of (9.39)? Compare the situation of the mean ion energy (a) in a small Coulomb exploding cluster and (b) in a charged cluster in quasiequilibrium with its electron cloud. Under which conditions does for (b) with T = const a *true* equilibrium exist?
- Why does the Maxwell-Jüttner electron distribution (9.32) contain the potential Φ in the exponent (also in the nonrelativistic case) whereas in LTE it does not? *Answer:* The Maxwellian electron distribution in LTE is a solution of the Boltzmann collision integral in the homogeneous plasma (absence of Φ). Distribution (9.32) is the steady state collisionfree Vlasov evolution of a Maxwellian in the (selfconsistent) potential Φ. The Maxwellian can be thought as prepared in the field free region.
- (a) Justify formula (9.18) for the range of an α particle in a thermal plasma of $T \lesssim 20\,\mathrm{keV}$. *Hint:* in one collision the momentum lost by the α particle is gained by the electron. For the energy transfer use the Drude model or small angle deflections. (b) How large is the contribution from the α particle motion $v_\alpha < v_{\mathrm{the}}$ in general? Give an estimate.

- Ponderomotive acceleration. Under oblique incidence I and p_L change: $I \mapsto I\cos\alpha,\; p_L \mapsto p_L\cos^2\alpha$. What is the projection angle α' in $S'(\mathbf{v})$? *Hint:* Use formula (2.169) or (3.106).
- How do you prove experimentally that the phase Φ of a monochromatic plane wave is a Lorentz scalar? *proposal:* by field ionization showing maxima at max $|\mathbf{E}|$; enables counting the number of oscillations.
- Why the photon density n and the frequency Doppler shift of ω follow both the same transformation law $\{n', \omega'\} = \{n, \omega\}(1 \mp \beta)/(1 \pm \beta)$? *Answer:* $n\mathrm{d}t$ and $\omega\mathrm{d}t$ are both Lorentz scalars expressing the conservation of photon number and number of wavelength contained in $\mathrm{d}t$.
- According to (9.52) the Stokes frequencies are symmetric with respect to the anti Stokes frequencies. Is this also true for their intensities? *Answer:* Not necessarily. Why? Discuss possible differences for the quiet plasma.
- Why is scattering into forward direction coherent?

9.11 Glossary

Critical field

$$\hat{E}_c = \frac{2m_e^2c^3}{e\hbar} = 1.37 \times 10^{18}\,\mathrm{Vm}^{-1} \;\Rightarrow\; I = 2.3 \times 10^{29}\,\mathrm{Wcm}^{-2}.$$

$$\omega_D = \gamma(\omega - \mathbf{kv}) \;\; \text{SLAC}: \gamma_D = 10^5,\; \frac{\lambda_{\mathrm{Nd}}}{2\gamma_D} = 0.05\,\text{A}.\;\; \text{DESY}: \gamma_D = 3.5 \times 10^4,\; \lambda_D = 0.14\,\text{A}. \tag{9.57}$$

$$\frac{\hbar\omega_S}{\hbar\omega_D} = \frac{1}{1 + \frac{\hbar\omega_D}{m_ec^2}(1 - \cos\theta)}. \tag{9.58}$$

$$\sigma_\Omega = \frac{1}{2}r_e^2\left(\frac{\omega_S}{\omega_D}\right)^2\left[\frac{\omega_S}{\omega_D} + \frac{\omega_D}{\omega_S} - \sin^2\theta\right], \quad r_e = 2.88\,\mathrm{fm}. \tag{9.59}$$

Controlled inertial fusion

$$\mathrm{D}_1^2 + \mathrm{T}_1^3 \Rightarrow \mathrm{He}_2^4 + n + 17.6\,\mathrm{MeV}. \tag{9.13}$$

Lawson criterion

$$n\tau \geq \frac{12k_BT}{\langle\sigma v\rangle E_\alpha} = 1.5 \times 10^{14}\,[\mathrm{cm}^{-3}\mathrm{s}]. \tag{9.14}$$

Sound speed

$$c_s = \left(\frac{2k_B T}{m_{\mathrm{DT}}}\right)^{1/2} = 6.6 \times 10^5 (T[\mathrm{eV}])^{1/2}\,\mathrm{cm\,s^{-1}}; \quad 1\,\mathrm{eV} = 10^4\,\mathrm{K}. \tag{9.15}$$

Ignition criterion

$$nk_B T\tau \geq \frac{24k_B^2 T^2}{\langle\sigma v\rangle E_\alpha} = 10\,\tau\,\mathrm{bar\,s} = 10\,\mathrm{Gbar\,ns}. \tag{9.17}$$

α range

$$\lambda_\alpha = \frac{m_\alpha v_\alpha}{2m_e \nu_{e\alpha}} = \frac{3.5}{\kappa \ln \Lambda}\left(\frac{T[\mathrm{keV}]}{10}\right)^{3/2}\,\mathrm{mm}; \quad v_\alpha = 1.2 \times 10^6\,\mathrm{cm\,s^{-1}}. \tag{9.18}$$

Electron speed

$$v_{\mathrm{the}} = \left(\frac{k_B T}{m_e}\right)^{1/2} = 1.3 \times 10^9 (T[\mathrm{keV}])^{1/2}\,\mathrm{cm\,s^{-1}}. \tag{9.19}$$

Radiation pressure acceleration (RPA)
Light pressure

$$I = |\overline{\mathbf{S}}| = \frac{1}{2}\varepsilon_0 c \mathbf{E}\mathbf{E}^*, \quad p_L = (1+R)\frac{I}{c}; \quad \mathbf{S} = \varepsilon c^2 \mathbf{E} \times \mathbf{B} = \varepsilon_0 c \mathbf{E}^2 \mathbf{k}^0. \tag{9.41}$$

Frame $S'(v)$

$$(I', p_L') = \frac{1 \mp \beta}{1 \pm \beta}(I, p_L); \ \mathbf{v} \parallel \pm\mathbf{k}. \tag{9.42}$$

Oblique incidence

$$I' = I\cos\gamma, \quad p_L' = p_L \cos^2\gamma : \quad \angle\,\gamma.$$

Pressure balance

$$\gamma^2\beta^2 = \frac{1-\beta}{1+\beta}s^2; \quad s^2 = \frac{I}{\rho_0 c^3}, \ v = \frac{s}{s+1}c. \tag{9.44}$$

Ion energy

$$\mathcal{E}_i = m_i c^2(\gamma - 1) = m_i c^2 \frac{2\beta}{1-\beta^2} = 2m_i c^2 \frac{s^2}{1+2s}. \tag{9.45}$$

Efficiency η

$$\eta = \frac{N\hbar(\omega - \omega_r)}{N\hbar\omega} = \frac{1-\beta}{1+\beta} = \frac{2s}{1+2s}. \tag{9.46}$$

Thomson scattering
Nonrel. diff. & total cross section

$$\sigma_\Omega = r_e^2 \sin^2\theta,\ \sigma_\mathrm{T} = \frac{8\pi}{3} r_e^2,\ r_e = \frac{e^2}{4\pi\varepsilon_0 m_e c^2};\ \Rightarrow\ \mathrm{d}I_{\boldsymbol{\Omega}} = \sigma_\Omega I_\mathrm{laser}\mathrm{d}\Omega. \tag{9.50}$$

Klein–Nishina expanded

$$\sigma_\mathrm{KN} = \sigma_\mathrm{T}\Big(1 - 2\alpha + \frac{56}{5}\alpha^2 + \cdots\Big);\quad \sigma_\mathrm{T} = \frac{8\pi}{3} r_e^2,\ \alpha = \frac{\hbar\omega}{mc^2}. \tag{9.51}$$

Matching

$$\mathbf{k}_0 = \mathbf{K} + \mathbf{k}_s, \qquad \omega_0 = \Omega + \omega_s. \tag{9.53}$$

Coherent versus incoherent scattering

$$\alpha = \frac{1}{|\mathbf{K}|\lambda_D} > 1 \Rightarrow \text{coherent},\quad \alpha = \frac{1}{|\mathbf{K}|\lambda_D} < 1 \Rightarrow \text{incoherent}. \tag{9.54}$$

9.12 Further Readings

D. Attwood, A. Sakdinawat, *X-Rays and Extreme Ultraviolet Radiation, Principles and Applications* (Cambridge University Press, Cambridge, 2017).
J. Rocca, C. Menoni, M. Marconi (eds.), *X-Ray Lasers 2014* (Springer, Heidelberg, 2016).
P. Mulser, D. Bauer, *High Power Laser-Matter Interaction*. Springer Tracts in Modern Physics, vol. 238 (Springer, Heidelberg, 2010), Chap. 7.
C.J. Joachain, N.J. Kylstra, R.M. Potvliege, *Atoms in Intense Laser Fields* (Cambridge University Press, Cambridge, 2012).
S.V. Popruzhenko, Keldysh theory of strong field ionization: history, applications, difficulties and perspectives. J. Phys. B: At. Mol. Opt. Phys. **47**, 204001 (2014).
R.A. Ganeev, *Nanostructured Nonlinear Optical Materials* (Elsevier, Amsterdam, 2018), Chaps. 6–9.
S. Atzeni, J. Meyer-ter-Vehn, *The Physics of Inertial Fusion* (Oxford University Press, Oxford, 2004).
A. Macchi, *A Superintense Laser-Plasma Interaction Theory Primer* (Springer, Heidelberg, 2013), Chap. 5: Ion Acceleration.
V. Malka, Laser plasma accelerators, in *Laser-Plasma Interactions and Applications*, ed. by P. McKenna et al. (Springer, Heidelberg, 2013), Chap. 11.
R.F. Werner, The classical limit of quantum theory, arXiv:quant-ph/9504016v1. Accessed 24 Apr 1995.

References

1. P. Mulser, F. Cornolti, D. Bauer, Phys. Plasmas **5**, 4466 (1998)
2. T.C. Pesch, H.-J. Kull, Phys. Plasmas **14**, 083103 (2007)
3. P. Kaw, J. Dawson, Phys. Fluids **13**, 472 (1970)
4. C.J. McKinstry, D.W. Forslund, Phys. Fluids **30**, 904 (987)
5. T.C. Pesch, H.-J. Kull in P. Mulser (ed.), Quasiperiodic waves in relativistic plasma. ILIAS Progress Reports No. 3 (GSI Darmstadt, 2008)
6. Y. Kuramitsu et al., High Energy Density Phys. A **17**, 198 (2015)
7. H. Schwörer, L. Magill, B. Beleites (eds.), *Laser and Nuclei* (Springer, Heidelberg, 2006)
8. S. Gales et al., New frontiers in nuclear physics with high-power lasers and brilliant monochromatic gamma beams. Phys. Scr. **91**, 030301 (2016)
9. A. Pálffy, A. Evers, C.H. Keitel, Phys. Rev. Lett. **08**, 172502 (2008)
10. Y. Wu et al., Phys. Rev. Lett. **120**, 052504 (2018)
11. R. Ganeev, Quantum Electron. **45**, 785 (2015)
12. S. Gordienko et al., Phys. Rev. Lett. **94**, 103903 (2005)
13. A. Di Piazza, C. Müller, K.Z. Hatsagortsyan, C.H. Keitel, New frontiers in nuclear physics with high-power lasers and brilliant monochromatic gamma beams. Rev. Mod. Phys. **84**, 1177 (2012)
14. H. Hamster, A. Sullivan, S. Gordon, W. White, R.W. Falcone, Phys. Rev. Lett. **71**, 2725 (1993)
15. F. Bakhtiari et al., Phys. Plasmas **23**, 123105 (2016)
16. G.-Q. Liao et al., Phys. Rev. Lett. **116**, 205003 (2016)
17. S. Casalbuoni et al., Phys. Rev. - Accel. Beams **12**, 030705 (2009)
18. S. Kumar, R.K. Singh, R.P. Sharma, Phys. Plasmas **22**, 103101 (2015); Europhys. Lett. **114**, 55003 (2016)
19. M. Singh, S. Kumar, R.K. Singh, R. Uma, R.P. Sharma, Europhys. Lett. **115**, xxxx (2017)
20. S.-H.-Wang et al., IEEE Trans. Terahertz Sci. Technol. **4**. Accessed 5 Sept 2014
21. P. Roedig et al., Nat. Methods **14**, 805 (2017)
22. D. Bleiner, J.E. Balmer, F. Staub, Appl. Opt. **50**, 6689 (2011)
23. M.D. Rosen, P.L. Hagelstein, D.L. Matthiews et al., Phys. Rev. Lett. **54**, 106 (1985)
24. D.L. Matthiews, P.L. Hagelstein, M.D. Rosen et al., Phys. Rev. Lett. **54**, 110 (1985)
25. J.J. Rocca, Rev. Sci. Instrum. **70**, 3799 (1999)
26. C. Skrobol et al., Eur. Phys. J. D **54**, 103 (2009)
27. B.J. MacGowan et al., J. Appl. Phys. **61**, 5243 (1987)
28. R. Tommasini, J. Nilsen, E. Fill, SPIE **4505**, 85 (2001)
29. E. Fill, J. Bayerl, R. Tommasini, Rev. Sci. Instrum. **73**, 2190 (2002)
30. R. Keenan, J. Dunn, P.K. Patel, D.F. Price, R.F. Smith, V.N. Shlyaptsev, Phys. Rev. Lett. **94**, 103901 (2005)
31. S. Sebban et al., Plasma Phys. Control. Fusion **60**, 014030 (2018)
32. F. Ceccherini, D. Bauer, F. Cornolti, J. Phys. B: At. Mol. Opt. Phys. **34**, 5017 (2001)
33. H.M. Nielsen, L.B. Madsen, J.P. Hansen, J. Phys. B: At. Mol. Opt. Phys. **35**(17), (2002)
34. P. Mulser, D. Bauer, *High Power Laser-Matter Interaction*. Springer Tracts in Modern Physics, vol. 238 (Springer, Heidelberg, 2010)
35. J.H. Eberly, Q. Su, J. Javanainen, Phys. Rev. Lett. **62**, 881 (1989)
36. R. Agostini et al., Phys. Rev. Lett. **42**, 1127 (1979)
37. K. Miller, Classical analysis of high harmonic generation. Thesis, University of Colorado, Boulder, 2015
38. M. Lewenstein et al., Phys. Rev. A **49**, 2117 (1994)
39. P. Ackermann, H. Münch, T. Halfmann, Opt. Express **20**, 13824 (2012)
40. S.V. Popruzhenko, J. Phys. B: At. Mol. Opt. Phys. **47**, 204001 (2014)
41. B. Karnakov, V.D. Mur, V.S. Popov, J. Exp. Theor. Phys. **105**, 292 (2006)
42. T. Brabec, F. Krausz, Rev. Mod. Phys. **72**, 545 (2000)
43. F. Krausz, Phys. Scr. **91**, 063011 (2016)

44. F. Krausz, From quantum transitions to electronic motions. Appl. Phys. B (2017)
45. F. Krausz, M. Ivanov, Rev. Mod. Phys. **81**, 163 (2009)
46. P. Agostini, L. Dimauro, Rep. Prog. Phys. **67**, 813 (2004)
47. W. Theobald, R. Hässner, C. Wülker, R. Sauerbrey, Phys. Rev. Lett. **77**, 298 (1996)
48. T.J.M. Boyd, R. Ondarza-Rovira, Phys. Rev. Lett. **98**, 105001 (2007)
49. L. Na-Na, Z. Zhen, L. Xue-Shen. Chin. Phys. Lett. **25**, 2508 (2008)
50. M. Kando et al., Quantum Beam Sci. **2018**, 2–9 (2018)
51. S. Gordienko, A. Pukhov, O. Shorokhov, T. Baeva, Phys. Rev. Lett. **93**, 115003 (2004)
52. T. Baeva, High harmonic generation from relativistic plasma. Thesis at Heinrich Heine University Düsseldorf, Germany, 2008
53. B. Dromey et al., Phys. Rev. Lett. **99**, 085001 (2007)
54. B. Dromey et al., Nat. Phys. **8**, 804 (2012)
55. T.J.M. Boyd, R. Ondarza-Rovira, Phys. Rev. Lett. **101**, 125004 (2008)
56. T.J.M. Boyd, R. Ondarza-Rovira, Phys. Lett. A **368**, 1368 (2016)
57. R. Gibbon, Phys. Rev. Lett. **76**, 50 (1996)
58. M.R. Edwards, J.M. Mikhailova, Phys. Rev. Lett. **117**, 125001 (2016)
59. S.V. Fomichev, W. Becker, Contrib. Plasma Phys. **53**, 662–696 (2013)
60. S. Le Pape et al., 45 authors. Phys. Rev. Lett. **120**, 245003 (2018)
61. J. Nuckolls, L. Wood, A. Thiessen, G. Zimmerman, Nature **239**, 139 (1972)
62. S. Atzeni, EPL **109**, 45001 (2015)
63. M. Murakami, D. Nishi, Matter Radiat. Extrem. **2**, 55 (2017)
64. R. Betti et al., Phys. Rev. Lett. **98**, 155001 (2007)
65. O. Klimo, V.T. Tikhonchuk et al., EPJ Web Conf. **59**, 05008 (2013)
66. W. Theobald, R. Betti et al., Phys. Plasmas **15**, 056306 (2008)
67. C.D. Zhou, R. Betti, Phys. Plasmas **15**, 102707 (2008)
68. R. Betti et al., Phys. Plasmas **17**, 058102 (2010)
69. M. Tabak et al., Phys. Plasmas **1**, 1626 (1994)
70. S. Hain, P. Mulser, Phys. Rev. Lett. **86**, 1015 (2001). (The calculated 10 - 20 kJ apply to the "free ignition energy", i.e. energy to ignite the spark. Total energy supplied to the corona is 50 (optimistic) - 70 kJ in agreement with other authors [for instance S. Atzeni, Phys. Plasmas **6**, 3316 (1999)]
71. P. Mulser, D. Bauer, Laser Part. Beams **22**, 5 (2004)
72. P. Mulser, R. Schneider, Laser Part. Beams **22**, 157 (2004)
73. N. Iwata et al., Nat. Commun. **9**, 623 (2018)
74. R. Kodama et al., Nature **412**, 798 (2001)
75. R. Kodama et al., Nature **418**, 933 (2002)
76. V.T. Tikhonchuk, Nucl. Instrum. Methods Phys. Res. A **620**, 1 (2010)
77. P. Mora, Phys. Rev. Lett. **90**, 185002 (2003)
78. M. Passoni, M. Lontano, Phys. Rev. Lett. **101**, 115001 (2008)
79. J. Schreiber et al., Phys. Rev. Lett. **97**, 045005 (2006)
80. A. Macchi, M. Borghesi, M. Passoni, Ion acceleration by superintense laser-plasma interaction. Rev. Mod. Phys. **85**, 751–793 (2013), contains 474 references
81. M. Borghesi, Laser-driven ion acceleration: state of the art and emerging mechanisms. Nucl. Instrum. Methods Phys. Res. A **740**, 6 (2014)
82. F. Wagner et al., Phys. Rev. Lett. **116**, 205002 (2016)
83. T. Kiefer, Investigation of the laser-based Target Normal Sheath Acceleration (TNSA) process for high-energy ions: an analytical and numerical study. Ph.D. thesis, Friedrich-Schiller University, Jena/Germany, 2014
84. P.L. Poole et al., New J. Phys. **20**, 013019 (2018)
85. A.V. Gurevich, L.V. Pariiskaya, L.P. Pitaevskii, Sov. Phys. JETP **22**, 449 (1965)
86. M. Passoni, L. Bertagna, A. Zani, New J. Phys. **12**, 045012 (2010)
87. J.H. Bin et al., Phys. Rev. Lett. **120**, 074801 (2018)
88. K.V. Lezhnin et al., arXiv:1603.04232v1 [Phys. Plasmas]. Accessed 14 Mar 2016
89. T. Schlegel et al., Phys. Plasmas **16**, 083103 (2009)

90. S.M. Weng, Z.M. Sheng, M. Murakami et al., Matter and radiation at extremes xx **1–12**, (2017)
91. A. Macchi, arXiv:1403.6273v1 [Phys. Plasmas]. Accessed 25 Mar 2014
92. S.M. Weng et al., New J. Phys. **14**, 063026 (2012)
93. M. Tévenet et al., Nature Phys. **12**, 355 (2016)
94. T. Tajima, J. Dawson, Phys. Rev. Lett. **43**, 267 (1979)
95. P. Sprangle et al., Phys. Rev. Lett. **69**, 2200 (1992)
96. M. Antonsen, P. Mora, Phys. Rev. Lett. **69**, 2204 (1992)
97. A. Modena et al., Nature **377**, 606 (1995); A. Ting et al., Phys. Plasmas **4**, 1889 (1997); M.I.K. Santala et al., Phys. Rev. Lett. **86**, 1227 (2001); V. Malka et al., Phys. Plasmas **8**, 2605 (2001); D. Giulietti et al., Phys. Plasmas **9**, 3655 (2002)
98. A.J.W. Reitsma et al., Phys. Rev. Lett. **88**, 014802 (2002); C.B. Schröder et al., Phys. Rev. Lett. **82**, 1177 (1999)
99. S.V. Bulanov, F. Pegoraro, A. Pukhov, A.S. Sakharov, Phys. Rev. Lett. **78**, 4205 (1997)
100. A. Pukhov, J. Meyer-ter-Vehn, Appl. Phys. B **74**, 355 (2002)
101. J.B. Rosenzweig et al., Phys. Rev. A **44**, R6189 (1991)
102. A. Doche, Particle acceleration with beam driven plasma wakefield. PhD thesis at l'Université Paris-Saclay, No. 572 (2018)
103. S.P.D. Mangles et al., Nature **431**, 535 (2004)
104. J. Faure et al., Nature **431**, 541 (2004)
105. C.G.R. Geddes et al., Nature **431**, 538 (2004)
106. W.P. Leemans et al., Nature Phys. **2**, 696 (2006)
107. W.P. Leemans et al., Phys. Rev. Lett. **113**, 245002 (2014)
108. M. Litos et al., Plasma Phys. Control. Fusion **58**, 034017 (2016)
109. J. Faure, C. Rechatin, A. Norlin, A. Lifschitz, Y. Glinec, V. Malka, Nature **444**, 737 (2006)
110. A. Rao, CERN Search News. Accessed 29 Aug 2018
111. J. Chihara, arXiv:cond-mat/9909372v1
112. B.-G. Englert, Eur. Phys. J. D **67**, 238 (2013)
113. B. Drossel, Stud. Hist. Philos. Mod. Phys. **58**, 12–21 (2017)
114. E. Lieb, R. Seiringer, *Stability of Matter in Quantum Mechanics*, (Cambridge University Press, Cambridge, 2010)

Index

A
absorption, 13
  Brunel's model, 54
  coefficient, 15
  collisional, 13, 15, 19, 59, 429, 561
  collisionless, 23, 48, 54, 461
  inverse bremsstrahlung, 13, 585
  limit of linear resonance, 473
  resonance, 23, 54, 382, 457
  special density profiles, 429
adiabatic
  coefficient, 27
  conservation, 115
  invariant, 98, 102, 105, 114
  theorem, 109
Airy function, 417
anti-Stokes
  component, 525
applications, 677
  Thomson scattering for hot matter diagnostics, 713
Atwood number, 223

B
Beer's law, 16, 424
Bernoulli's law, 186, 454
Bessel functions, 567
black body radiation, 319
blast wave, 222
bounce frequency, 480
breakdown, 3
Buckingham theorem, 35, 220, 221

C
canonical ensemble, 307
Carnot engine, 280
chaos, 526
Chapman–Jouguet condition, 27
characteristics, 196, 233
chemical potential, 309
circular polarization, 524
Clausius, 280
  theorem, 281
collective effects, 189, 260, 431
collision
  binary, 137
  Coulomb collision, 553
  frequency, 60, 139, 552
    ballistic, 562
    dielectric, 566
    giant ions, 584
  frequency with drift, 578
  oscillator model, 554
collision frequency, 80
compressibility, 286
Compton effect
  multiphoton, 536
conduction
  free streeming limit, 39
conductivity
  electric, 16, 372
  thermal, 32
conservation
  canonical momentum, 152
  mass, 180
conservation laws, 180
convective derivative, 202
Cornu spiral, 394
Coulomb logarithm, 20, 570
  cut offs, 570
criterion
  Piliya-Rosenbluth, 538

P. Mulser, *Hot Matter from High-Power Lasers*, Graduate Texts in Physics,
https://doi.org/10.1007/978-3-662-61181-4

critical density, 5
critical field, 11
cross section
  Coulomb, 19
  differential, 14, 138
  hard sphere, 141
  Rutherford, 139
  total, 14
cut off, 18, 378, 381
cyclic variable, 114

D
Debye sphere, 21
density
  critical, 18, 373
  overdense, 20
  underdense, 20
density plateau, 258, 259
dielectric function
  longitudinal, 380
  transverse, 373
dimensional analysis, 39
Doppler effect, 493
  in medium, 496
  in vacuum, 493
Drude model, 18, 59

E
eigenmodes, 403
Einstein
  coefficients, 587
electric field
  amplitude maximum, 466
  transformation, 146
electron distribution function, 525
energy
  densities and fluxes, 455
  inner, 22
  internal, 187, 275
  negative energy wave, 451
  relativistic, 150
entropy, 276
  applications, 282
  heat transfer, 283
  isentropic, 232
equation
  Boltzmann, 206
  canonical, 93
  Lagrange, 85
  of moments, 207
  of state, 240
  Saha, 280
  Vlasov, 204
equation of state
  dense matter, 332
  Grüneisen, 288
equipartition
  degrees of freedom, 301
ergodic theorem, 141
Euler equation, 186

F
Faraday effect, 406
Fermi
  energy, 21
  gas, 21
field
  thermoelectric, 24
flow
  convective, 182
  fluid, 180
  shear, 182
fluid, 179
  collisionless, 180
  dynamics, 179
  ideal, 186
  incompressible, 179, 180
  magnetized, 401
  non-standard, 211
  trajectory, 200
  two fluid model, 21
  unstable, 445
flux
  stationary, 181
force
  Einstein, 149
  external, 179
  internal, 179
  macroscopic, 179
  Minkowski, 149
  Newton, 149
  ponderomotive, 31
  surface, 179
  volume, 179
free electron laser, 3
free energy, 289
Frenet's formulas, 97, 531
friction
  dynamical, 16
fully ionized plasma, 13
function
  generating, 113
  Hamilton, 93, 150

Lagrange, 85, 150
one particle distribution, 204
reduced distribution, 621

G
generalized
coordinates, 88
force, 89
momenta, 90
generation
magnetic field, 200
Gibbs paradox, 311
Gibbs potential, 290
gradient
Fourier ansatz, 33
grand canonical ensemble, 309
group velocity, 131
growth rate, 223
Grüneisen parameter, 288

H
Hall term, 192
Hamilton equations, 150
heat, 261, 277
current density
anomalous, 486
heat flux
inhibition, 246
heating
collisional, 5
collisionless, 599
inverse bremsstrahlung, 5
reheating, 46
return current, 602
high harmonic generation, 683
from gases, 684
from plasma, 686
hohlraum, 29
homeomorphism, 96
hot dense matter, 260
hot electrons, 480, 599
scaling, 608
superthermal, 480
hot matter, 1, 275, 332
Hugoniot, 27

I
ideality prameter, 21
ideal systems
Boltzmann limit, 326
Bose statistics, 317
Fermi statistics, 325
impact parameter, 138
inertial confinement, 24, 219
inertial confinement fusion, 691
fast ignition, 695
pellet compression and heating, 693
requirements, 691
inertial system, 92
instability, 445
absolute, 506
convective, 506
filamentation, 527
Kelvin–Helmholtz, 452
Langmuir decay, 499
mudulational, 533
oscillating two-stream, 499
parametric, 109
parametric decay, 499
physical picture, 502
Rayleigh–Taylor, 223, 445
Richtmyer–Meshkov, 449
self focusing, 527
stimulated Brillouin, 499
stimulated Raman, 499
thermal, 529
two plasmon decay, 499
two-stream, 450
interaction
phonons vs collisions, 63
ultrashort interaction, 599
invariant
adiabatic, 109
Poincaré–Cartan, 109
ionization
above threshold, 6
field ionization, 5
multiphoton ionization, 5
nonsequential, 6
partial, 13
ion separation, 46
demixing, 225
irreversible, 13, 279
isothermal heat flow
plane, 255
spherical, 255

J
Jacobian, 261

K
Keldish prameter, 59

Kelvin, 280
kinetic theory
    BBGKY hierarchy, 619
    reduced moments, 619
Kirchhoff's law, 323, 659

L

Lagrange equation, 150
Landau damping
    electron damping, 383
    nonlinear, 476
Landau echo, 387
laser
    ablation, 27
laser plasma, 4
law
    of mass action, 329
    Ohm's law, 372
    Stefan–Boltzmann, 288
length
    deBroglie, 22
    Debye length, 12
    Rayleigh, 531
libration, 114
light cone, 143
light scattering, 399
    coherent, 399
    incoherent, 399
local thermal equilibrium, 276
local thermodynamic equilibrium, 290
Lorentz
    scalar, 162
Lorentz boost, 147
Lorentz contraction, 144
Lorentz transformation, 142
loss cone, 104

M

Mach number, 30
magnetic field
    frozen, 400
    generation, 200
    transformation, 146
magnetic moment, 97
magnetic pressure, 401
Manley-Rowe relations, 538
Maxwell distribution, 292
Maxwell equations, 362
Maxwellian, 15
    non-Maxwellian, 54, 293
Maxwell relations, 290
mean free path, 139
metric, 143
microcanonical ensemble, 297
mirror
    magnetic, 104
moments, 207
motion
    drift, 99
    guiding center, 103

N

Navier–Stokes equation, 182
Noether's theorem, 91
nonideal gas equation
    van der Waals, 341
nonthermal
    melting, 2

O

Ohm's law, 192, 410
oscillator
    harmonic, 80

P

particle
    monoenergetic beam, 225
    trapping, 136
particle acceleration
    electron bubble regime, 709
    light pressure acceleration, 704
    maximum ion energy, 701
    monoenergetic beams, 700
    monoenergetic electron beams, 709
    TNSA, 696
    TNSA models, 697
    wake field acceleration, 707
particle motion, 73
    nonrelativistic, 74
partition function, 307, 328
phase
    conjugation, 512
    dephasing, 49, 56, 66, 81
    flow, 94
    space, 94, 109
    velocity, 462
photons
    classical, 362
    Hamilton equations, 408
picture
    Eulerian, 200
    Heisenberg, 210

Lagrangian, 200
Schrödinger, 209
plasma
breakdown, 15
classical, 21
degenerate, 21
ideal, 20
magnetized, 399
mirror, 56
multicomponent, 45
nonideal, 179, 619
nonlinear response, 677
preplasma, 50
plasma frequency
electron, 17, 373
ion, 397
plasma theory
classical vs quantum treatment, 714
point explosion, 222
Poisson bracket, 93
ponderomotive force, 117
potential
Coulomb gauge, 370
Lienard–Wiechert potential, 365
Lorentz gauge, 364
ponderomotive, 31
restricted Lorentz gauge, 370
screened, 140
vector potential, 364
pressure, 22
ablation, 27
ponderomotive, 25
radiation, 31, 249
radiation pressure, 244
pressure diffusion term, 192
principle
D'Alembert, 83
extended variational, 94
Hamilton, 84
least action, 84
of action, 105
of detailed balance, 322
profile steepening, 236, 250

Q
quantity
dimensionless, 35
quasineutral, 12

R
radiation
accelerated charge, 648
brightness, 663
coherent phenomena, 644
field quantization, 634
free-free spectrum, 655
Gaunt factor, 655
Glauber states, 638
hot matter, 633
line emission, 638
optical Bloch model, 640
plasma bremsstrahlung, 653
pressure, 249
relativistic Larmor formula, 650
spectral range, 652
radiation generation, 678
radiation reaction, 667
Abraham-Lorentz-Dirac equation, 667
Landau approximation, 668
radiation transport
diffusion model, 665
Rosseland mean free path, 666
Rankine–Hugoniot relations, 237
ray
tracing, 413
ray equation, 412
Rayleigh
equations, 238
Rayleigh–Taylor instability, 223
reduced mass, 137, 192
reflection
coefficient, 408, 419
relativity
Galileian, 75
Lorentzian, 142
relativistic self focusing, 528
resonances, 381
anharmonic, 590
return current
Cherenkov plasmon generation, 604
reversible, 279
Riemann invariants, 232
rotation, 114

S
Saha equation, 330
scale
fast, 192
slow, 192
scattering
induced, 495
scattering angle, 137
screening, 552

secular
force, 101
motion, 101
selfsimilar
solution, 221
separatrix, 136
shielding
Debye shielding, 560
generalized spherical, 560
shock
adiabate, 240
compression, 26
strong, 27, 239
velocity, 26
wave, 26
shock wave, 5
similarity, 220
dimensional matrix, 220
self similar, 37, 235
solution, 221
solution
self similar, 221
similarity, 221
specific heat, 286
state variables, 276
Stirling formula, 297
supergaussian, 18, 293

T
target
corrugated, 56
droplets, 57
foams, 57
microstructured, 56
Target Normal Sheath Acceleration (TNSA), 219
temperature
absolute, 281, 299
cold, 277
hot, 481
kinetic, 208
warm, 277
tensor
contravariant, 162
covariant, 162
Maxwell stress tensor, 366
viscosity, 183
Terahertz radiation, 678
test particle, 140, 306
thermal
local conduction, 614
plasma, 12
radiation, 660
self focusing, 528
thermal equilibrium, 275
criterion, 295
thermal expansion
coefficient, 286
thermal fluctuations, 311
thermodynamic potentials, 289
thermodynamics, 276
first law, 277
second law, 279
third law, 281
zeroth law, 277
thermostatistics, 276
classical, 296
density matrix, 312
fundamental principle, 296
ideal systems, 316
phase space, 296
quantum systems, 312
Thomas–Fermi model, 333
threshold, 11
time
dilation, 143
proper, 143
transformation
Galilei, 75
linear, 142
Lorentz, 142
transport, 551
anomalous, 39
coefficient, 189
collisional, 552
Coulomb focusing, 583
diffusive, 614
friction, 611
ion stopping, 575
local, 552
nonlocal, 39
viscosity, 611
trapping
adiabatic, 135
light, 523
particles, 135

U
uphill acceleration, 135

V
vacuum heating, 56
vector

Poynting vector, 366
space-like, 215
time-like, 215
velocity
addition theorem, 151
Alfvén velocity, 402
electron, 16
electron sound, 378
group velocity, 374
ion, 199
ion sound velocity, 397
phase velocity, 373
Virial theorem, 300
virtual displacement, 84
viscosity
shear, 183
volume, 184
Vlasov equation
relativistic, 204
vortex lines, 109

W
warm dense matter, 2, 288
wave
Alfvén waves, 401
amplitude, 415
blast, 222
breaking, 390, 474
bubbles and spikes, 392, 448
deflagration, 27
eigenmodes, 362, 371, 399
electron plasma wave, 376
geometric breaking, 474
helicity, 79
inhomogeneous plasma, 406
ion acoustic wave, 376
kinetic breaking, 474
linear, 361
magnetoacoustic, 403
plasma waves, 361
rarefaction, 24
reaking, 362
resonant breaking, 490
resonant three wave interaction, 490
self-interaction, 392
unstable, 445
wave equation
Bohm–Gross dispersion, 378
dispersion, 362, 411
eikonal, 411
Stokes, 417
wave frame, 127
WKB approximation, 102, 362, 416

X
X ray lasing, 681

MIX
Papier aus verantwortungsvollen Quellen
Paper from responsible sources
FSC® C105338

If you have any concerns about our products,
you can contact us on
ProductSafety@springernature.com

In case Publisher is established outside the EU,
the EU authorized representative is:
Springer Nature Customer Service Center GmbH
Europaplatz 3, 69115 Heidelberg, Germany

Printed by Libri Plureos GmbH
in Hamburg, Germany